HIGHER LEVEL

Biology

for the IB Diploma Programme

Alan Damon, Randy McGonegal, William Ward

Published by Pearson Education Limited, 80 Strand, London, WC2R 0RL.

www.pearson.com/international-schools

Text © Pearson Education Limited 2023
Development edited by Judith Head and Jess White
Copy edited by Eva Fairnell
Proofread by Eva Fairnell
Indexed by Georgie Bowden
Designed by Pearson Education Limited
Typeset by Tech-Set ltd
Picture research by Integra
Original illustrations © Pearson Education Limited 2023
Cover design © Pearson Education Limited 2023

The rights of Alan Damon, Randy McGonegal, and William Ward to be identified as the authors of this work has been asserted by them in accordance with the Copyright, Designs and Patents Act 1988.

First published 2023

25 24
10 9 8 7 6 5 4 3

British Library Cataloguing in Publication Data
A catalogue record for this book is available from the British Library

ISBN 978 1 29242 774 4

Printed in Slovakia by Neografia

Acknowledgements

The authors and publisher would like to thank the following individuals and organisations for permission to reproduce photographs, illustrations, and text:

Text extracts relating to the IB syllabus and assessment have been reproduced from IBO documents. Our thanks go to the International Baccalaureate for permission to reproduce its copyright.

The "In cooperation with IB" logo signifies the content in this textbook has been reviewed by the IB to ensure it fully aligns with current IB curriculum and offers high-quality guidance and support for IB teaching and learning.

KEY (t – top, c – center, b – bottom, l – left, r – right)

Images:

123RF: Elenamasiutkina 114, Liubomirt 123, Molekuul 212l, Alila 274, 661, Grigory_bruev 350, Arthit Buarapa 356c, Sebastian Kaulitzki 468, James Cumming 541br, Normaals 650b, Nadyginzburg 664, Achim Prill 700, Anna Grigorjeva 822b, Ondrej Prosicky 844, Zhijie Zhuang 856br, Oleg Beloborodov 874, Trish233 905, Veronique Martel 906, sdigitall 907; **Alamy Stock Photo:** Vintage_Space 82, Mint Images Limited 94, Encyclopaedia Britannica/Universal Images Group North America LLC 97, 149, 505, Wayne Lynch/All Canada Photos 142, J & C Sohns/imageBROKER 155, GERAULT Gregory/Hemis.fr 166b, Naeblys 209b, Mathew Taylor 216, Tetiana Zhabska 254l, Kiyoshi Takahase Segundo 264, Monica Schroeder/Science History Images 302, Nucleus Medical Media Inc 310, Blickwinkel/M. Woike 327, Leo Francini 330, Nick Harrison 496, Arindam Ghosh 509, Tom Uhlman 527bl, Peter Groenendijk 550, Photo Researchers/Science History Images 616b, Heather Angel/Natural Visions 686, Pattarawit Chompipat/agefotostock 701, Science History Images 766, Ester van Dam 776, Steven May 805r, Pictorial Press Ltd 810, Rosanne Tackaberry 812, Michael Lidski 813, David Tipling Photo Library 832, Klaus Reitmeier 842t, Enigm 860; **Alan Damon:** Alan Damon 148b, 173, 254r, 324, 362, 402, 470, 581, 680, 774; **Getty image:** David_Slater/iStock Plus 430, Vchal/iStock Plus 574, FREDERIC J. BROWN/Stringer/AFP 624, Micro_photo/Istock Plus 675c, Ryan McVay 856tr, Alain DENANTES 856cl; **Google:** Screenshot from "The Enteric Nervous System", Gastroenterology 487; **International Union for Conservation of Nature and Natural Resources:** International Union for Conservation of Nature and Natural Resources 170; **Ivan Ovcharenko:** Screenshot created byDcode.org: Tools Developed by the Research Group of Ivan Ovcharenko at the National Library of Medicine taken from https://ecrbrowser.dcode.org/xB.php?db=hg19&location=chr2:201491337-201499885# 628b, **M. Turmaine, UCL:** M. Turmaine, UCL 92; **National Human Genome Research Institute (NHGRI):** POINT MUTATION, Chris Gunter, Ph.D. Associate Investigator, Social and Behavioral Research Branch 611; **National Institutes of Health:** https://imagej.nih.gov/ij/download.html 893, **Natur,Kunstformen der:** Natur,Kunstformen der, 'plate 32: Rotatoria'. 861br; **Nextstrain:** Real-time tracking of influenza A/H5N1 virus evolution. Built with nextstrain/avian-flu. 124; **Pearson Education:** Trevor Clifford 8, Oxford Designers & Illustrators Ltd 33, 542, 620t, 638, 650t, 787; **Science Photo Library:** Kateryna Kon 2, 111, 116, 613, Bence Mate/Nature Picture Library 6, Markus Varesvuo/Nature Picture Library 11t, Nasa, Esa, Z. Levy (Stsci) 14, Alfred Pasieka 29, 502, 594, Mikkel Juul Jensen 30, Ron Miller 38, Dr Ken Macdonald 50, Steve Gschmeissner 57, 067r, 449t, 452r, 492, 497, 675t, Niaid/CDC 60, Eye Of Science 499b, 612t, A.B. Dowsett 62, 072r, Biology Media 068c, Dr Kari Lounatmaa 068l, Dr Jeremy Burgess 69, 100, 245, Don W. Fawcett 070l, J.C. Revy, ism 70b, 366, Rogelio Moreno 78, Sheila Terry 96, 717, Roman Uchytel 101, CNRI 103, 67b, 499t, Javier Trueba/MSF 106, 347r, JOHN BIRDSALL SOCIAL ISSUES PHOTO LIBRARY 108, Turtle Rock Scientific/Science Source 109, Peter Gardiner 115, I. Noyan Yilmaz 119, Library Of Congress, Rare Book And Special Collections Division 127, Nicolle Rager Fuller/National Science Foundation 148t, Dr Neil Overy 157, Mark Garlick 166t, animate4.com 180t, Dr Tim Evans 186, Alice J. Belling 190, Gary Carlson 195, Klaus Guldbrandsen 240c, K.R. Porter 244r, Science Photo Library 248, 618, 657, 828, 873, Biomedical Imaging Unit, Southampton General Hospital 249, Jannicke Wiik-Nielsen 271, Biophoto Associates 74t, 289, 305, Dr David Furness, Keele University 295, David Aubrey 341, Mauricio Anton 347l, D. Roberts 349, Clay Coleman 353t, PROFESSORS P. MOTTA & T. NAGURO 386, John Durham 404, Trevor Clifford Photography 410, Omikron 475, Eric Grave 476, Springer Medizin 484, Bob Gibbons 537, Adrian Thomas 539t, Noaa 553, Ted Kinsman 559, DAVID PARKER 580b, Equinox Graphics 616t, Patrick Landmann 617, Kenneth Eward/Biografx 628t, Art for Science 639, National Institute Of General Medical Sciences/National Institutes Of Health 662, Vivid Biology 688, 775, Dr Yorgos Nikas 689, Dr. Keith Wheeler 692, Carolina Biological Supply Company 697, Astrid & Hanns-Frieder Michler 703, Gregory Dimijian 712, Doug Allan 721, Martin Shields 745, Don Fawcett 756, Art Wolfe 778, Paul Zahl 783, Doncaster And Bassetlaw Hospitals 795, British Antarctic Survey 819b, Simon Fraser 823, Katey Walter Anthony/University Of Alaska Fairbanks/Nasa 824t, Nasa/Gsgc 829b, European Space Agency/Atg Media Lab 829t, Adrian T Sumner 839b, Adam Jones 868; **Shutterstock:** T.Thinnapat 4, Aldona Griskeviciene 7, 203, 533b, JohnKwan 9, Polarman 11b, Benny Marty 12, Muratart 13, Corona Borealis Studio 17, 85r, Leonid Andronov 24, 178, Art of Science 25, 73, 3d_man 26, Sirikwan Dokuta 61t, BioMedical 61b, Designua 72l, 83, 231t, 434, 435t, 452l, 535t, 535c, 760t, Schira 74c, Jose Luis Calvo 74r, 235, 247, 259, 474b, Extender_01 84tr, CI Photos 84br, Marina Veder 99, Freya-photographer 125t, Alinabel 125b, Steve Byland 126, Andreas Wolochow 132, M. Schuppich 134, Edward Fielding 138, Chris Howey 145, Photographer DmitryStrong 147, Panaiotidi 151, Sergey Uryadnikov 153cl, Bardocz Peter 153b, Andy333 153cr, Henner Damke 161, jo Crebbin 163, Zern Liew 164, Rich Carey 167, 543,Yusnizam Yusof 168, Barbara Tripp 172, Yandong Yang 174, Christian Colista 175, Yaruna 180b, Sansanorth 181, 189t, BigBearCamera 189b, Biology Education 194, 299l, 299r, Zcxes 196l, 206, Magnetix 196r, Ali DM 201, 210, 504, ibreakstock 209t, 215, Raimundo79 211, Juan Gaertner 212r, Vecton 220, VectorMine 224, 625, 815, Pepermpron 231b, Stanislaw Mikulski 239c, Chakapong 239b, Soleil Nordic 240t, Daniela Barreto 244l, W.Y. Sunshine 258t, Rattiya Thongdumhyu 258b, 343, Steve Cymro 262, Sciencepics 263, Design_Cells 268, Sergio Gutierrez Getino 270, Art of Kosi 272, Pissamai Boonkane 279, Alila Medical Media 281, 298, 620b, 695, Sakurra 282, Songkram Chotik-anuchit 287, Blamb 291t, 297, SciePro 291b, Ilusmedical 292, Michael Hare 293, Hakan.demir 306, Marco Uliana 307, Damsea 309b, ESB Professional 317, Mybox 318, BlueRingMedia 320, Willem van de Kerkhof 326, Chad Zuber 328, Cathy Keifer 334l, leungchopan 334r, Anton Foltin 336t, Vinnie Lauritsen 336b, Ticiana Giehl 337, Klaus Ulrich Mueller 338, Zagorulko Inka 342, Lebendkulturen.de 344, Jolanda Aalbers 345, Achiichiii 348, Wiangya 352, Dirk Ercken 353b, Dr Ajay Kumar Singh 354, Steve Estvanik 355, Picattos 356l, Crystal Eye Studio 435b, Systemoff 449b, Scubaluna 456, Matthew Cole 474t, Terpsychore 479, TimeLineArtist 486, Adastra 488, J. Marini 500, Magic mine 501, C.Lotongkum 511, Kateryna Kon 512, 82841 520, Sergey 402 522, Samib123 523, Raquel Pedrosa 525t, Matt Tilghman 525b, Moolkum 527c, Roman Khomlyak 527br, Yumeee 531, David Osborn 532, TalyaPhoto 533t, Worachat Tokaew 534, Nwdph 536, Wolfen 539c, Critterbiz 541cr, Bildagentur Zoonar GmbH 544, Vadim Sadovski 548, Paul Aniszewski 549, Volkov Alexey 552, Richsouthwales 565, Solarseven 560, Kara Grubis 564, M. Patthawee 580t, Haroon Abbasi 597, Ellepigrafica 612b, Peter Hermes Furian 619, 861tl, Vita khorzhevska 621, Elenabsl 622, Janson George 623, Lightspring 626, Dimarion 634, 642r, OSweetNature 636, Fancy Tapis 637b, 647, 642l, YAKOBCHUK VIACHESLAV 646, Power_J 649, Pikovit 653, UfaBizPhoto 656, Tratong 663r, AkulininaOlga 666, Edgloris Marys 682, Steve Allen 684, Innesslam 710, Udaix 711, Scisetti Alfio 729, Michaeljung 736, Petrroudny43 737, Olivier Le Moal 755, Fabiano Vleira 758, Sebastian Kaulitzki 759, Iv-olga 760b, ParabolStudio 772, Francesco de marco 780t, Zhaoyan 780b, Juefraphoto 781, Szefei 782, IanRedding 788, Marktucan 799, Holly Kuchera 804, Darkydoors 805c, Alexandru V 806, Avigator Fortuner 808, Manishankar Patra 809, Olha1981 811, Sandrina Gomes Saraiva 814, Jeanie333 818, Ziablik 819t, Tony Skerl 822t, Condor36 824b, Paralaxis 825t, Maksimilian Limited 825b, Anticiclo 826, Gentoo Multimedia Limited 827t, Outdoorsman 827b, Evgenii mitroshin 830, Peter Leahy 833l, 833r, Callums Trees 834, ArtMediaFactory 835, Robin Nieuwenkamp 836, Vishnevskiy Vasily 837c, AnutkaT 837b, Heiko Kueverling 838, Incredible Arctic 839t, StockPhotoAstur 840, Protasov AN 842b, JuliusKielaitis 852t, Ververidis Vasilis 852b, Stockagogo, Craig Barhorst 853, Becris 854, IMG Stock Studio 856tl, STILLFX 856tc, Matthew Jacques 856c, Asharkyu 856cr, Ground Picture 856bl, Rawpixel.com 856bc, Julie Phipps 858, Goodluz 859, MJConline 863, MikhailSh 867, Chess Ocampo 870, Wichudapa 888, Natalielme 889, Deyan Georgiev 891, Ket Sang 897, Christian Musat 902; **The Society for Neuroscience:** Provençal, N., Suderman, M. J., Guillemin, C., Massart, R., Ruggiero, A., Wang, D., ... & Szyf, M. (2012). The signature of maternal rearing in the methylome in rhesus macaque prefrontal cortex and T cells. Journal of Neuroscience, 32(44), 15626-15642. 663l.

Text:

American Association for the Advancement of Science (AAAS): Changes in Melatonin Secretion during Day and Night Time (IMAGE), © THE KOREA ADVANCED INSTITUTE OF SCIENCE AND TECHNOLOGY (KAIST) 480; **American Heart Association, Inc:** Marma, A. K., & Lloyd-Jones, D. M. (2009). Systematic examination of the updated Framingham heart study general cardiovascular risk profile. Circulation, 120(5), 384-390. 292; **Centers for Disease Control and Prevention:** Developed by the National Center for Health Statistics in collaboration with the National Center for Chronic Disease Prevention and Health Promotion (2000). 738; **Council on Foreign Relations:** Diana Roy, Deforestation of Brazil's Amazon Has Reached a Record High. What's Being Done?, August 24, 2022. 800; **Frontiers Media S.A:** © Frontiers Media S.A 515; **George E. P. Box:** George E. P. Box (innovator in statistical analysis) Box and Draper, 1987. Copyright © 1987 by John Wiley & Sons, Inc. 864; **Intergovernmental Science-Policy Platform on Biodiversity and Ecosystem Services (IPBES):** SUMMARY FOR POLICYMAKERS OF THE IPBES GLOBAL ASSESSMENT REPORT ON BIODIVERSITY AND ECOSYSTEM SERVICES Copyright © 2019, Intergovernmental Science-Policy Platform on Biodiversity and Ecosystem Services (IPBES) ISBN No: 978-3-947851-13-3 169; **International Baccalaureate Organisation:** © International Baccalaureate Organisation 2015 518; **J. Robert Oppenheimer:** Robert Oppenheimer Barnett 1949 861; **John N. Bahcall:** John Bahcall (commenting on the Hubble space telescope's capabilities) 874; **Karl Popper:** Popper, K. R. (1992). The logic of scientific discovery. United Kingdom: Routledge. 868, 872; **Louis Pasteur:** Louis Pasteur 862; **Marie Curie:** Marie Curie 873; **Mr G's Environmental Systems:** Why are Biomes where they are?, Copyright © 2007-2022 Mr G's Environmental Systems 333; **National Academy of Science:** Freeman, B. G., & Class Freeman, A. M. (2014). Rapid upslope shifts in New Guinean birds illustrate strong distributional responses of tropical montane species to global warming. Proceedings of the National Academy of Sciences, 111(12), 4490-4494. 832; **National Library of Medicine:** Rico, A., Brody, D., Coronado, F., Rondy, M., Fiebig, L., Carcelen, A., ... Dahl, B. A. (2016). Epidemiology of Epidemic Ebola Virus Disease in Conakry and Surrounding Prefectures, Guinea, 2014–2015. Emerging Infectious Diseases, 22(2), 178–183. 519; **National Oceanic and Atmospheric Administration (NOAA):** Trends in Atmospheric Carbon Dioxide, Monthly Average Mauna Loa CO2, National Oceanic and Atmospheric Administration (NOAA) 566, 567; **Niels Bohr:** Niels Bohr (a Danish physicist who helped us understand how atoms work) Ellis 1970. 869; **Our World In Data:** Two centuries of rapid global population growth will come to an end, © Our World In Data 171; **Oxford University Press:** John A. Endler (1980). Natural Selection on Color Patterns in Poecilia reticulata. Evolution, 34(1), 76–91. doi:10.2307/2408316 784; **The Royal Society:** Post, E., & Forchhammer, M. C. (2008). Climate change reduces reproductive success of an Arctic herbivore through trophic mismatch. Philosophical Transactions of the Royal Society B: Biological Sciences, 363(1501), 2367-2373. 840; **The University of Waikato:** © Copyright. 2019. The University of Waikato – Te Whare Wananga o Waikato. All rights reserved. www.sciencelearn.org.nz 329; **Thomas Henry Huxley:** Thomas Henry Huxley 872, 874; **uang Tzu Taois:** Chuang Tzu Taoist text (written more than 2,000 years ago) 865; **University of California Regents:** Ocean acidification, the University of California Museum of Paleontology,© 2022 University of California Regents 332; **World Health Organisation:** Situation by WHO Region, WHO Coronavirus (COVID-19) Dashboard, © WHO 515, World Health Organization 652.

MIX
Paper from responsible sources
FSC www.fsc.org
FSC™ C128612

Contents

Theme B

Theme C

Theme D

D Continuity and change – Ecosystems 772

Syllabus roadmap

The aim of the syllabus is to integrate concepts, topic content and the nature of Science through inquiry. Students and teachers are encouraged to personalize their approach to the syllabus to best fit their interests.

Theme	Level of organization			
	1. Molecules	2. Cells	3. Organisms	4. Ecosystems
A **Unity and diversity**	Common ancestry has given living organisms many shared features while evolution has resulted in the rich biodiversity of life on Earth.			
	A1.1 Water **A1.2** Nucleic acids	**A2.1** Origins of cells *[HL only]* **A2.2** Cell structure **A2.3** Viruses *[HL only]*	**A3.1** Diversity of organisms **A3.2** Classification and cladistics *[HL only]*	**A4.1** Evolution and speciation **A4.2** Conservation of biodiversity
B **Form and function**	Adaptations are forms that correspond to function. These adaptations persist from generation to generation because they increase the chances of survival.			
	B1.1 Carbohydrates and lipids **B1.2** Proteins	**B2.1** Membranes and membrane transport **B2.2** Organelles and compartmentalization **B2.3** Cell specialization	**B3.1** Gas exchange **B3.2** Transport **B3.3** Muscle and motility *[HL only]*	**B4.1** Adaptation to environment **B4.2** Ecological niches
C **Interaction and interdependence**	Systems are based on interactions, interdependence and integration of components. Systems result in emergence of new properties at each level of biological organization.			
	C1.1 Enzymes and metabolism **C1.2** Cell respiration **C1.3** Photosynthesis	**C2.1** Chemical signalling *[HL only]* **C2.2** Neural signalling	**C3.1** Integration of body systems **C3.2** Defence against disease	**C4.1** Populations and communities **C4.2** Transfers of energy and matter
D **Continuity and chance**	Living things have mechanisms for maintaining equilibrium and for bringing about transformation. Environmental change is a driver of evolution by natural selection.			
	D1.1 DNA replication **D1.2** Protein synthesis **D1.3** Mutations and gene editing	**D2.1** Cell and nuclear division **D2.2** Gene expression *[HL only]* **D2.3** Water potential	**D3.1** Reproduction **D3.2** Inheritance **D3.3** Homeostasis	**D4.1** Natural selection **D4.2** Stability and change **D4.3** Climate change

Authors' introduction to the third edition

Welcome to your study of IB Diploma Programme (DP) biology. This is the third edition of Pearson's highly successful Higher Level (HL) biology book, first published in 2007. It has been rewritten to match the specifications of the new IB biology curriculum for first assessments in 2025 and provides comprehensive coverage of the course. It is our intention as authors of this textbook to open a door to biological knowledge that will provide a pathway towards an ever-present curiosity of life, the factors that affect it today, and the factors that may affect it in the future.

While there is much new and updated material in this textbook, we have kept and refined the features that made the previous editions so successful and effective. We hope our knowledge and enthusiasm for biology as well as our understanding of the IB biology requirements will be passed onto you.

Content

This book covers the content that is set out in the IB DP biology subject guide for first assessments in 2025. It utilizes the overarching theme of Nature of Science (NOS) to provide the means for you to accomplish the following aims:

1. to develop conceptual understanding that allows connections to be made between different areas of the subject, and to other DP science subjects
2. to acquire and apply a body of knowledge, methods, tools and techniques that characterize science
3. to develop the ability to analyse, evaluate and synthesize scientific information and claims
4. to develop the ability to approach unfamiliar situations with creativity and resilience
5. to design and model solutions to local and global problems in a scientific context
6. to develop an appreciation of the possibilities and limitations of science
7. to develop technology skills in a scientific context
8. to develop the ability to communicate and collaborate effectively
9. to develop awareness of the ethical, environmental, economic, cultural and social impact of science.

Chapters are presented in the same sequence as provided in the subject guide. There are four main themes:

A. Unity and diversity
B. Form and function
C. Interaction and interdependence
D. Continuity and change

Each theme is then discussed at four different levels of organization. They are:

1. Molecules
2. Cells
3. Organisms
4. Ecosystems

The Understandings are presented in the same sequence as in the subject guide, so that common Standard Level (SL) and HL content is covered first, followed by HL only material. The transition from SL to HL content is shown by icons as follows.

HL

Individual icons to identify HL material such as exercises and practice questions can also be found throughout the book.

Each topic begins with an introductory image and caption supplying a brief entry point into its content. Guiding Questions are then presented for further clarification of chapter content.

Guiding Questions

What plausible hypothesis could account for the origin of life?

What intermediate stages could there have been between non-living matter and the first living cells?

The text covers the course content with all scientific terms explained. We have been careful to apply the same terminology you will see in IB assessments.

Linking Questions that relate topics to one another can be found in each chapter. When encountered, Linking Questions should be considered in order to understand how other concepts from within the course relate to those currently being discussed. When used effectively, Linking Questions can provide an excellent tool for revision.

For what reasons is heredity an essential feature of living things?

Each chapter concludes with Guiding Questions revisited and a summary of the chapter. The summary presents key points from the chapter you should be especially aware of.

Guiding Question revisited

How can viruses exist with so few genes?

Nature of Science

Throughout the course you are encouraged to think about the nature of scientific knowledge and the scientific process as it applies to biology. Examples are given of the evolution of biological theories as new information is gained, the use of models to conceptualize our understandings, and the ways in which experimental work is enhanced by modern technologies. Ethical considerations, environmental impacts, the importance of objectivity, and the responsibilities regarding scientists' code of conduct are also considered here. The emphasis is on appreciating the broader conceptual themes in context. We recommend that you familiarize yourself with these examples to enrich your understanding of biology.

Throughout the book you will find NOS themes and questions emerging across different topics. We hope they help you to develop your own skills in scientific literacy.

Nature of Science

Science has progressed and continues to progress with the development of new study techniques. Not only has the microscope increased our knowledge of the cell, but ultracentrifuges and fractionation of cells have also greatly enhanced our understanding of the cell and its organelles.

Key to feature boxes

A popular feature of our past editions is maintained in this book, that is the different coloured boxes interspersed throughout each chapter. These boxes can be used to enhance your learning.

Global context

The impact of the study of biology is global, and includes environmental, political and socio-economic considerations. Examples of these are given to help you see the importance of biology in an international context. These examples also illustrate some of the innovative and cutting-edge aspects of research in biology.

Thanks to modern communication technologies, it is possible for scientists working all over the world to collaborate and contribute to a scientific endeavour such as sequencing the genome of plants that help feed the world. Rice is one example: biologists from 10 countries contributed to sequencing the first rice genome.

Surface area-to-volume ratio. Full details on how to carry out this activity with a worksheet are available in the eBook.

Skills in the study of biology

These boxes indicate links to the skills section of the course, including ideas for laboratory work and experiments that will support your learning and help you prepare for the Internal Assessment. These link to further resources in the eBook (look out for the grey icon).

When you study the action of sarcomeres, how much is your knowledge limited by two-dimensional models, such as Figure 3?

Theory of Knowledge

These questions, which are mostly from the Theory of Knowledge (TOK) guide, stimulate thought and consideration of knowledge issues as they arise in context. The questions are open-ended and will help trigger critical thinking and discussion.

The sequence of nitrogenous bases in DNA, later transcribed into RNA, forms the basis of the genetic code.

Key fact

Key facts are drawn out of the main text and highlighted in bold. These boxes will help you to identify the core learning points within each section. They also act as a quick summary for review.

You are not required to know all the names of the intermediate molecules of the respiration process. However, you must understand the steps and the overall products.

Hint for success

These boxes give hints on how to approach questions, and suggest approaches that examiners like to see. They also identify common pitfalls in understanding, and omissions made in answering questions.

Challenge yourself

These boxes contain probing questions that encourage you to think about the topic in more depth, and may take you beyond the syllabus content. They are designed to be challenging and to make you think.

Challenge yourself

1. Using Figure 8, showing the DNA profiles from six suspects, can you identify which one matches the DNA profile of the blood stain found at the crime scene?

Interesting fact

These give background information that will add to your wider knowledge of the topic and make links with other topics and subjects. Aspects such as historic notes on the life of scientists and origins of names are included here.

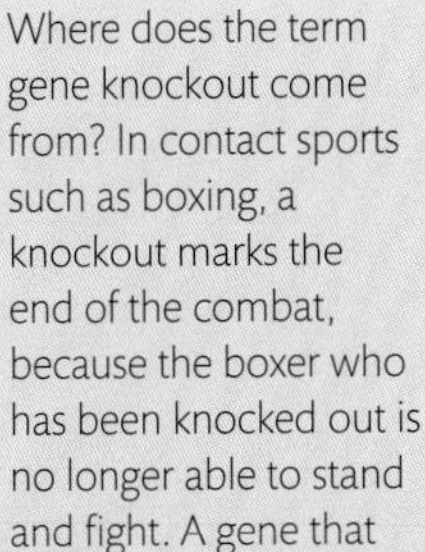

Where does the term gene knockout come from? In contact sports such as boxing, a knockout marks the end of the combat, because the boxer who has been knocked out is no longer able to stand and fight. A gene that has been knocked out will no longer be able to make the protein that produced the original effect or trait

Questions

There are three types of question in this book.

1. Worked examples with solutions

Worked examples appear at intervals in the text and are used to illustrate the concepts covered. They are followed by the solution, which shows the thinking and the steps used in solving the problem.

Worked example

The length of an image you are looking at is 50 mm. If the actual length of the subject of the image is 5 µm, what is the magnification of the image?

Solution

Magnification = 50 mm/5 µm = 50,000 µm/5 µm = 10,000×

Or

Magnification = 50 mm/5 µm = 50×10^{-3} m/1×10^{-6} m = 10,000×

2. Exercises

These questions are found at the end of each chapter. They allow you to apply your knowledge to test your understanding of what you have just been reading. The answers to these are accessed via icons on the first page of each chapter in the eBook. Exercise answers can also be found at the back of the eBook.

Exercises

Q1. Explain why the obligate parasitism shown by viruses may have been a major factor in convergent evolution within the group.

3. Practice questions

These questions are found at the end of each group of chapters displaying a common theme and level of organization. The significance of these questions is that they are IB exam-style questions. The mark schemes used by examiners when marking these questions are accessed via icons in the eBook next to the questions. These questions and mark schemes are essential in providing insight into the depth of comprehension necessary to achieve success in an IB exam.

A2 Practice questions

1. **(a)** An organelle is a discrete structure within a cell with a specific function. In the table below, identify the missing organelles and outline the missing functions.

Name of organelle	Structure of organelle	Function of organelle
Nucleus	Region of the cell containing chromosomes, surrounded by a double membrane, in which there are pores.	Storage and protection of chromosomes.
Ribosome	Small spherical structures, consisting of two subunits.	
	Spherical organelles, surrounded by a single membrane and containing hydrolytic enzymes.	Digestion of structures that are not needed within cells.
	Organelles surrounded by two membranes, the inner of which is folded inwards.	

(4)

(b) The table above shows some of the organelles found in a particular cell. Discuss what type of cell this could be. (2)

(Total 6 marks)

eBook

In your eBook you will find more information on the Skills section of the course, including detailed suggestions for laboratory work, and the answers to the exercises and practice questions found in the text. You will also find links to videos and command term worksheets in the Resources tab of your eBook account. In addition, there are auto-marked quizzes in the Exercises tab of your eBook account (see screenshot below).

We truly hope that this book and the accompanying online resources help you enjoy the fascinating subject of IB biology. We wish you success in your studies.

Alan Damon, Randy McGonegal and William Ward

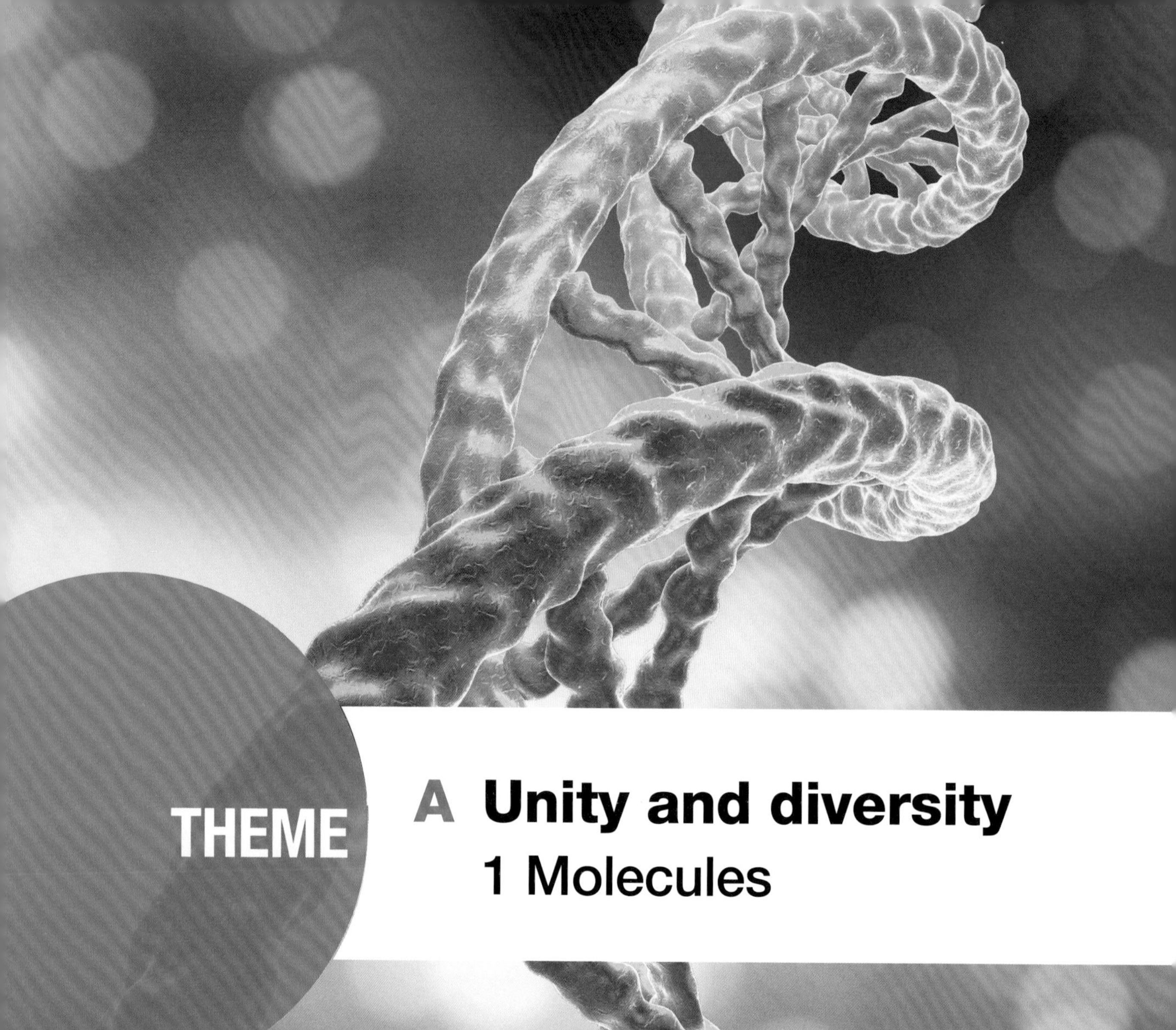

THEME

A Unity and diversity

1 Molecules

This is DNA, one of the molecules classified as a nucleic acid and a molecule that is integral to life on Earth. The molecules that are important to life are diverse and complex. Yet their basic structures are largely consistent from species to species. This allows us to study the fundamental structures and functions of these molecules and apply that knowledge to all living organisms. In this chapter, you will first study the solvent of all biochemically important molecules, water. Later, you will consider the structure of nucleic acids.

A1.1 Water

Guiding Questions

What physical and chemical properties of water make it essential for life?

What are the challenges and opportunities of water as a habitat?

What makes water essential for living organisms? What physical and chemical properties does water have that provide essential benefits to aquatic, marine and terrestrial organisms? What opportunities and challenges does water pose for life? These are not questions designed to be answered in one or more short statements. They are questions that deserve to be explored. A portion of this chapter will attempt to begin that exploration.

Life first evolved in water and all living things are still dependent on this amazing molecule. Fortunately, we live on a planet where water exists in all three states: there is abundant liquid water, water vapour and ice. Water, as a polar molecule, is an excellent solvent for the vast majority of elements and compounds necessary for life. Water molecules are found inside and outside cells, and chemical communication in and out of cells must occur in a water environment.

Water has both advantages and disadvantages for the aquatic and marine organisms that use it as a habitat. Advantages include the fact that water provides buoyancy and stable thermal properties for these organisms. Disadvantages include its relatively high viscosity compared to air. This means that many organisms living in water have adapted their body shape and propulsion mechanisms in order to move easily through an aquatic environment.

A1.1.1 – The medium of life

A1.1.1 – Water as the medium for life

Students should appreciate that the first cells originated in water and that water remains the medium in which most processes of life occur.

Life on Earth has never been possible without water. Imagine a primordial planet slowly cooling from its original molten mass. That primitive Earth would not have had any water because of the extremely high temperatures at its centre *and* on its surface.

The surface of the Earth may have looked like this early in its history, with magma giving off tremendous heat at the surface.

The origin and evolution of the first cells could not begin until temperatures cooled enough for water to form and, later, for the water cycle to begin. We take for granted the changes that water makes as it goes between its solid, liquid and gaseous phases. Earth's varied temperatures allow these changes. That was not the case in our planet's early history.

Approximately 70% of our planet's surface is covered by water. The deepest parts of the Pacific Ocean are deeper than the height of the highest land peaks.

It is thought that the first cells formed and slowly evolved in the oceans. Cells require a complex series of biochemical reactions. This means a **solvent** is needed for reactions to occur. Ocean water provided the source for that solvent. The first cells evolved a membrane to separate the water in the cytoplasm from the "ocean water".

Every solution where water is the solvent is called an aqueous solution. Thus, cytoplasm, rivers, blood and oceans are all aqueous solutions.

When most people think of water, their first thoughts are about the water they drink and bathe or swim in. But water is more widespread than that. Below are a few examples of where the importance of water as a solvent is vital to living organisms.

Water is the solvent that:

- makes up the fluid (cytoplasm) in all cells where all cellular reactions occur
- makes up the fluid inside all organelles in cells
- is found between cells of multicellular organisms (intercellular or tissue fluid)
- permits transport of substances into and out of cells
- is essential to blood and many other body fluids in humans and other organisms
- provides the medium in which all organisms in oceans, lakes and rivers live.

Challenger Deep (the lowest known portion of the Mariana Trench) is 10,984 m below the surface of the Pacific Ocean. Mount Everest (the tallest known land mass) is 8,848 m above sea level. The difference between those points is over 19 km or 12 miles.

Nature of Science

Measurements in science often change over time. If you research the world's deepest and tallest points you may find slightly different numbers (metres below and above sea level). There are various possible reasons including: how recently the data point was taken; what method was used to obtain the data; whether or not the data change over time due to natural causes. Can you think of other reasons for the data to vary?

A1.1.2 – The structure and polarity of water molecules

A1.1.2 – Hydrogen bonds as a consequence of the polar covalent bonds within water molecules

Students should understand that polarity of covalent bonding within water molecules is due to unequal sharing of electrons and that hydrogen bonding due to this polarity occurs between water molecules.

Students should be able to represent two or more water molecules and hydrogen bonds between them with the notation shown below to indicate polarity.

$O^{\delta-}$

$H^{\delta+}$ $H^{\delta+}$

To understand the properties of water and its importance to living organisms, it is necessary to understand the molecular structure of water molecules.

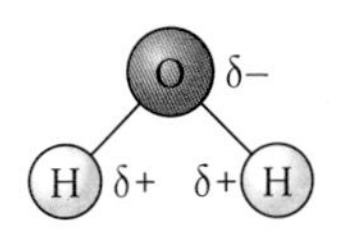

A1.1 Figure 1 This image shows the covalent bonds in a water molecule. Each of two hydrogen atoms is bonded at an angle to a single oxygen atom. Remember that each of the two covalent bonds is a pair of shared electrons.

The covalent bonds between the oxygen atom and the two hydrogen atoms of a water molecule are categorized as **polar covalent bonds.**

You may remember from fundamental chemistry that covalent bonds form when two atoms share electrons. Electrons are negatively charged and the nucleus of an atom is positively charged (because of the protons). So, any equally shared electrons create a **non-polar covalent bond**. This is because neither of the atoms has a higher density of electrons than the other. Good examples of non-polar covalent bonds include the covalent bond between two carbons and the covalent bond between two hydrogens.

Polar covalent bonding results from an unequal sharing of electrons. In water, the single oxygen atom is bonded to two different hydrogen atoms. Each oxygen–hydrogen bond is a polar covalent bond. This results in a slight negative charge at the oxygen end of the molecule and a slight positive charge at the end with the two hydrogens.

Because of the open triangular shape of a water molecule, the two "ends" of each molecule have opposite charges. The oxygen side is slightly negative and the hydrogen side is slightly positive. This is why water is a polar molecule: it has different charges at each end. Because of this, water molecules interact with each other and other molecules in very interesting ways. Many of these interactions are explained by the usually short-lived (ephemeral) attractions between either two water molecules or between a water molecule and another type of charged atom (or ion). These ephemeral attractions are called **hydrogen bonds** and will be explained further in the following sections.

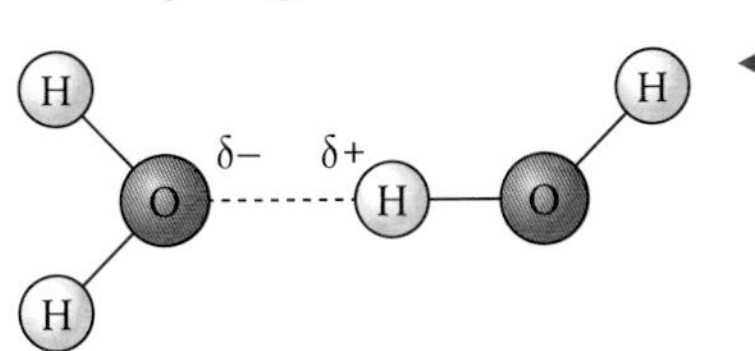

A1.1 Figure 2 Two water molecules showing a single hydrogen bond between them. The bonding force of each hydrogen bond (indicated by the dotted line) is weak. In liquid water, the bond is ephemeral because the water molecules continue to move around.

You may be used to seeing the Greek symbol Δ called delta. Δ is the capital letter symbol and δ is the corresponding small case letter symbol for delta.

The electrons being shared to create the covalent bonds within a water molecule are not being shared equally between the two atoms. In Figure 1, you see the symbols δ^+ and δ^- (delta positive and delta negative). These symbols represent areas of low or high electron density in the sharing of electrons to create a covalent bond. Each hydrogen atom is assigned a δ^+ because that is an area of lesser electron density (thus a small positive charge due to the single proton of the hydrogen atom). The oxygen atom is assigned a δ^- charge due to its high electron density.

Practise sketching from memory a diagram similar to the one shown in Figure 2. Include the hydrogen bond and the delta symbols and charges as shown. Practise adding a third and fourth water molecule with the same symbolism and orientation.

A1.1.3 – Cohesion of water molecules

A1.1.3 – Cohesion of water molecules due to hydrogen bonding and consequences for organisms
Include transport of water under tension in xylem and the use of water surfaces as habitats due to the effect known as surface tension.

Water molecules are highly cohesive. **Cohesion** occurs when *molecules of the same type* are attracted to each other. As you have seen, water molecules have a slightly positive end and a slightly negative end. Whenever two water molecules are near each other, the positive end of one attracts the negative end of another – this is hydrogen bonding. When water cools below its freezing point, the molecular motion of the water molecules slows to the point where the hydrogen bonds become locked into place and an ice crystal forms. Liquid water has molecules with a faster molecular motion, and the water molecules are able to influence each other, but not to the point where molecules stop their motion. This influence is highly important and leads to many of the physical and chemical properties of water. The ephemeral hydrogen bonding between liquid water molecules explains a variety of events, including the following.

You can float a paperclip on water because of the surface tension of the water. Make sure you maximize the surface area of the paperclip on the water if you try this.

- Why water has a **surface tension**. Surface tension is due to the fact that the layer of water molecules at the surface of a body of water does not have molecules of water above it. Because of this, the water molecules show a relatively strong cohesive force to the molecules immediately around and below them (no molecules are pulling upwards). This surface tension must be broken in order for an object to move through the surface from above. It is surface tension that causes you pain when you do a "belly flop" into a body of water. It is also surface tension that creates a habitat for some animals such as water striders and basilisk lizards.

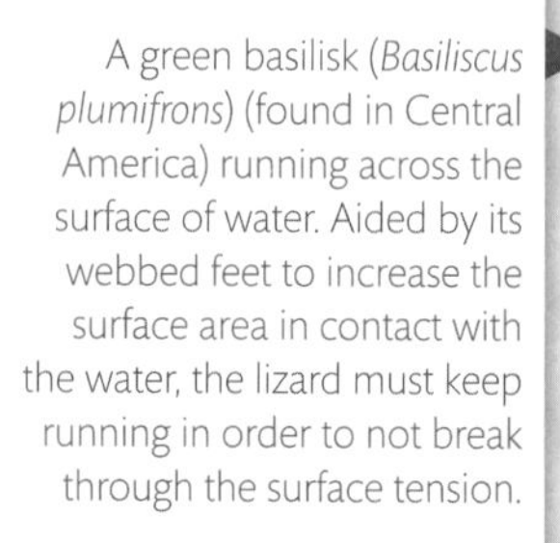

A green basilisk (*Basiliscus plumifrons*) (found in Central America) running across the surface of water. Aided by its webbed feet to increase the surface area in contact with the water, the lizard must keep running in order to not break through the surface tension.

- How water is able to move as a "water column" in the vascular tissues of plants. The majority of water moving upwards in a plant moves within small tubes called **xylem**. Think of xylem as being similar to numerous tiny straws. When water evaporates from a leaf (in a process called transpiration) the water that evaporates in order to exit the leaf has cohesion to the water in a xylem tube that adjoins the exit point. The evaporation with corresponding cohesion creates a low pressure in this area called **tension**. This tension pulls on the other water molecules in the xylem tube so they all move upwards towards the leaf. The molecules are all cohesive to each other and all move up collectively. This evaporation occurs in small, controlled openings called stomata, which are usually found on the underside of leaves. The water that transpires from the leaf is replaced in the xylem in the root system of the plant.

Think of a xylem tube and the upwards movement of water as being similar to what happens when you use a straw in a drink. The suction you provide creates tension (low-pressure pressure area at the top of the straw) and the fluid is moved upwards along the straw. The bottom of the straw in your drink is similar to the bottom of the xylem tubes found in the root system of a plant.

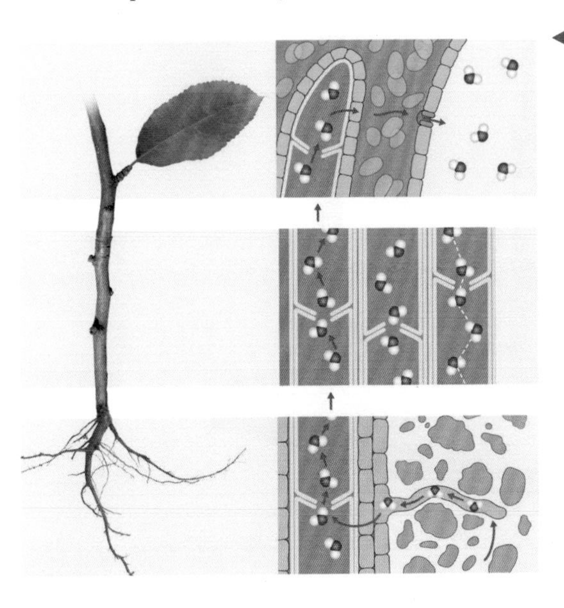

An example of the importance of cohesion. At the top, water is evaporating from a stoma (singular of stomata). Stomata are very small openings that can be opened or closed and are found primarily on the under surface of leaves. The evaporation of water from open stomata is called transpiration. The water is provided to the leaf by many xylem tubes. The transpiration of water creates tension (a low-pressure area in the leaf and xylem tube) and the polarity of water molecules pulls the entire water column to move towards the low-pressure area. The xylem tube within the leaf is continuous with the xylem in the stem and root. The water moving upwards is replaced by ground water moving into the root system.

A1.1.4 – Adhesion between water and other polar substances

A1.1.4 – Adhesion of water to materials that are polar or charged and impacts for organisms
Include capillary action in soil and in plant cell walls.

Water molecules are certainly not the only molecules in nature that exhibit polarity. An attraction between two *unlike* molecules due to hydrogen bonding is called **adhesion**. When water molecules are attracted to cellulose molecules by hydrogen bonding, the attraction is an example of adhesion because the hydrogen bonding is between two different kinds of molecule. Where is this important in nature?

- Water within the xylem. Cohesion and adhesion are both at work in this example. When the column of water is "pulled up", cohesion moves each molecule up; when the column is not being "pulled up", adhesion keeps the entire column from dropping down within the tube. The same phenomenon occurs when water is placed in a capillary tube – you can think of the xylem tissue in plants as being biological capillary tubes.

Cohesion and adhesion are both a result of the polarity of water molecules. Cohesion is an attraction between two water molecules and adhesion is an attraction between a water molecule and another polar molecule that is not water.

A capillary tube is a glass tube (similar to a straw) that has a very narrow inside opening. In this photo, a capillary tube has been inserted into a vessel filled with water with a red dye. The liquid will spontaneously climb upwards into the capillary tube due to adhesion and remain in a fixed position within the tube. The adhesion is the attraction between the inside surface of the glass tube and water molecules.

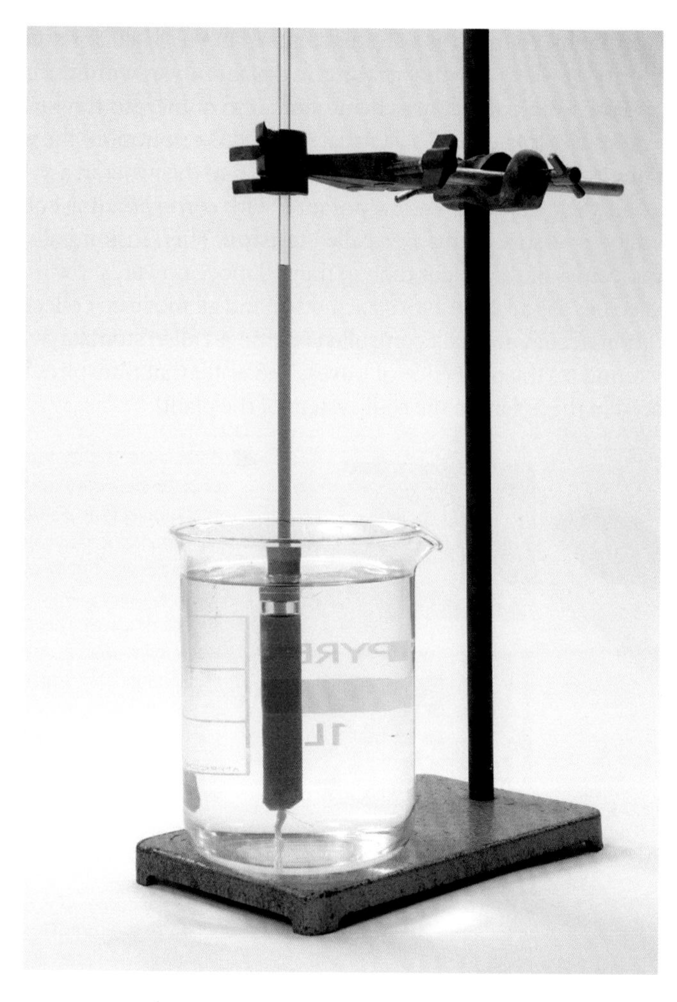

How do the various intermolecular forces of attraction affect biological systems?

- Capillary action in soil. Even soil that appears to be dry contains water in microscopic channels. These small channels act in a similar way to capillary tubes. Water molecules adhere to the polar molecules making up the soil and other water molecules are then sometimes moved by cohesion. The small root hairs of plants intrude into the water-filled spaces and water is taken into the root.

A1.1.5 – The solvent properties of water

A1.1.5 – Solvent properties of water linked to its role as a medium for metabolism and for transport in plants and animals

Emphasize that a wide variety of hydrophilic molecules dissolve in water and that most enzymes catalyse reactions in aqueous solution. Students should also understand that the functions of some molecules in cells depend on them being hydrophobic and insoluble.

As you have seen, water is a polar molecule and thus a polar solvent. In nature, water is almost always found as a solvent carrying one or more of a wide variety of other substances as solutes. Any solution that has water as the solvent is called an **aqueous solution**. Any substance that dissolves readily in water is described as **hydrophilic** (water loving) and any substance that does not dissolve easily is called **hydrophobic** (water fearing).

Hydrophilic molecules

The cytoplasm of a cell is a good example of an aqueous solution and contains a wide variety of water-soluble substances. These hydrophilic solutes include (among others) glucose, ions, amino acids and proteins. Some of the dissolved proteins in cells are the biological catalysts called enzymes. Reactions within the cytoplasm depend on enzymes to proceed at a rate necessary for life and at a temperature tolerated by that type of cell.

The biochemistry of a cell occurs in its cytoplasm and also within membrane-bound organelles such as the nucleus and mitochondria. The fluids of these cellular environments use water as a solvent because most biochemically active molecules are polar and dissolve easily in an aqueous solvent.

Water is an excellent medium for transporting dissolved substances. The water contained in xylem vessels of plants is not pure water. It is an aqueous solution that transports inorganic ions such as sodium, potassium and calcium. These and many other essential substances are hydrophilic; they dissolve easily in water and are transported upwards from the root system to the leaves.

The blood of many animals, including humans, is also an aqueous solution. The red and white blood cells are suspended in plasma. Plasma is an aqueous solution of an incredible array of molecules. Anyone looking at the results of a typical medical blood test can see the variety of solutes in this solution.

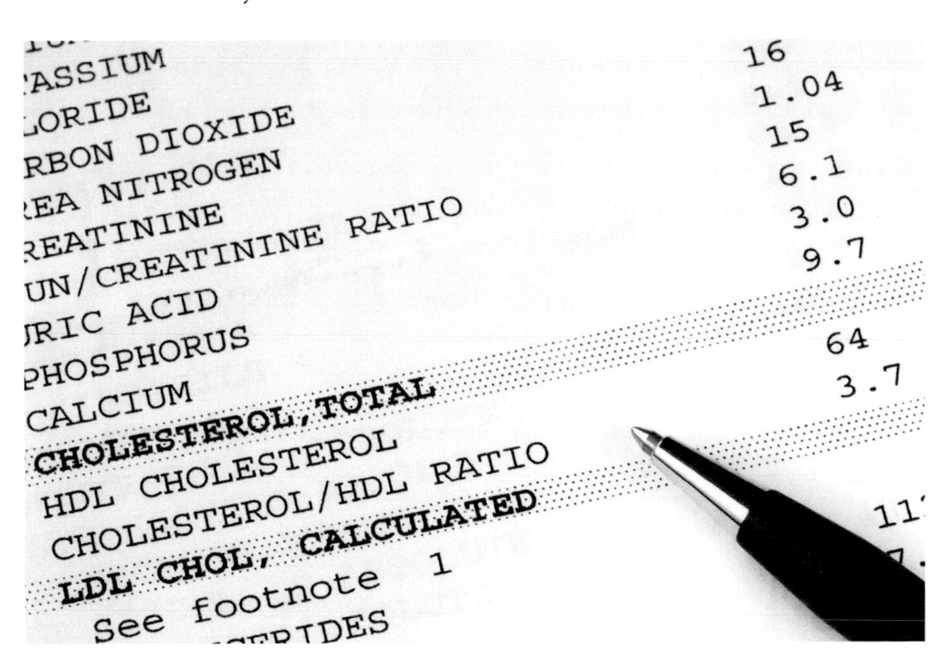

A small section of the results of a human blood test showing some of the dissolved substances in the aqueous portion of blood called plasma.

Hydrophobic molecules

Some non-polar (insoluble) molecules found in nature are important to living organisms. Here are some examples.

- Steroid hormones, such as oestradiol and testosterone, are able to pass directly through the plasma membrane and nuclear membrane of a cell. Steroid hormones can do this because they are hydrophobic and are able to pass directly through the hydrophobic layers of cell membranes.
- Many proteins have some sections that are hydrophilic and other sections that are hydrophobic. Membrane-bound proteins may use one or more hydrophobic areas to embed into the hydrophobic layers of a membrane while their hydrophilic section(s) extends into either the intercellular fluid or cytoplasm. This enables the protein to stay attached to the membrane but still interact with soluble substances in the surrounding cell fluids.

- The epidermal cells of leaves are capable of secreting a wax that is used to coat the leaves and is called the **cuticle**. This wax cuticle is hydrophobic and acts as a barrier to water entering and especially exiting the leaf by evaporation. Without this cuticle, leaves would quickly dehydrate because their function requires a thin, broad surface area exposed to the Sun.

A1.1.6 – The physical properties of water

A1.1.6 – **Physical properties of water and the consequences for animals in aquatic habitats**
Include buoyancy, viscosity, thermal conductivity and specific heat. Contrast the physical properties of water with those of air and illustrate the consequences using examples of animals that live in water and in air or on land, such as the black-throated loon (*Gavia arctica*) and the ringed seal (*Pusa hispida*). *Note: When students are referring to an organism in an examination, either the common name or the scientific name is acceptable.*

Table 1 outlines the important physical properties of water compared with air.

Property	Water	Air
Buoyancy or buoyant force (an upwards force exerted on an object placed in the medium – either water or air)	Buoyant force equals the weight of the water displaced by the object. The buoyant force is upwards because there is more pressure from below (in the water) than above (in the air).	An object placed in air has an almost insignificant buoyant force. This force is equal to the weight of the air displaced by the object.
Viscosity	Water's resistance to an object moving through it.	Air's resistance to an object moving through it. Since air is far less dense than water, air's viscosity is far less.
Thermal conductivity	The ability of a substance to transfer heat. Water has a high thermal conductivity.	The thermal conductivity of air is very low compared to water.
Specific heat capacity	In simplest terms, water can absorb or give off a great deal of heat without changing temperature very much. Think of a body of water on a very cold night: even though the air may be very cold, a nearby body of water is relatively stable in temperature.	Air's ability to absorb or give off heat without changing temperature is very low compared to that of water. The temperature of the air changes easily and rapidly due to weather events.

A1.1 Table 1 Physical properties of water

The physical properties of water have important consequences for animals that live in aquatic habitats, such as the black-throated loon (*Gavia arctica*) and the ringed seal (*Pusa hispida*).

▲ Black-throated loon (*Gavia arctica*).

The black-throated loon is a beautiful bird that lives primarily in very cold regions of the Northern Hemisphere. As with most aquatic birds, the loon transfers regularly between land (for nesting), water (for feeding) and air (for flying). Even though this bird is capable of diving for food, it spends much of its time in water on the surface relying on the buoyant force of the water to float. The bird requires energy to overcome the viscosity of water to move across the water surface and even more when it dives for fish and other food sources below the surface. Webbed feet and efficient, streamlined body shape aid the loon in this movement. When the bird is in water, the high thermal conductivity of the water would cause the loon to lose more body heat than when it is in the air. Like many waterbirds, loons use an adaptation to prevent this. They have an oil gland near their tail and they use their beaks to rub this oil over their feathers to make them waterproof. When the air is very cold (below 1°C) the surrounding water is likely to be warmer than the air because the high specific heat of water allows its temperature to remain relatively stable in comparison to air.

You are not required to memorize the scientific names (genus and species) of example organisms.

The ringed seal is another animal that is common in cold environments of the Northern Hemisphere. This small seal is buoyant, although not as buoyant as a loon – less of its body is above the surface of the water when resting. It is buoyant enough to keep its snout above water easily and thus has an easily available supply of air. Seals spend a great deal of time swimming in and under the water to catch food (fish and invertebrates) and occasionally to escape a predator such as an orca. Their streamlined shape and paddle-like feet are great assets in overcoming the viscosity of water. But water has high thermal conductivity compared to air, so ringed seals need to minimize body heat loss. They do this by having a thick blubber under their skin. The blubber is insulation and reduces heat loss from the seals' internal organs. Like the black-throated loon, ringed seals are protected from very low air temperatures by the relatively high temperature of arctic water (compared to arctic air) which is due to the high specific heat of water.

Melting sea ice due to global warming is threatening many species, including seals, because their habitats are fundamentally changing in a very short period of time. No one country by itself can solve the problem of global warming.

What biological processes only happen at or near surfaces?

◀ Ringed seal (*Pusa hispida*).

HL

A1.1.7 – The origin of water on Earth

A1.1.7 – Extraplanetary origin of water on Earth and reasons for its retention
The abundance of water over billions of years of Earth's history has allowed life to evolve. Limit hypotheses for the origin of water on Earth to asteroids and reasons for retention to gravity and temperatures low enough to condense water.

Earth has excellent conditions for retaining its large volume of water. As stated previously, approximately 70% of the surface of Earth is covered by water and some of that water is very deep. Earth's temperatures allow water to change phases readily, but most remains as liquid water. In addition, our planet is large enough to have a gravitational pull to retain water on or near its surface. Some of the water that helps form our planet is almost permanently trapped deep in the crust. Temperatures deep in the Earth are tremendously high but there are a few opportunities for this water to escape.

Castle Geyser in Yellowstone National Park erupts once every 14–18 hours. The discharge of water is due to underground magma superheating water partially trapped in the crust. There are only a few areas on Earth where water deep in the crust can escape this way.

Water molecules exist in two forms. The difference between the two forms is the number of neutrons in the nucleus of the hydrogen atoms of the molecule. Typically, water contains "ordinary" hydrogen atoms without any neutrons. "Heavy water" contains hydrogen atoms that have a neutron. This hydrogen is called **deuterium**. All bodies of water contain a mix of these two forms, with typical water (no neutrons) being much more common. When researchers calculate the ratio of hydrogen to deuterium in the water of the oceans they get a ratio that is very similar to the ratio found in many asteroids.

Nature of Science

Several countries currently have space craft that are tracking near-Earth asteroids, some of which are known to contain a great deal of water. In addition, several land-based telescopes are also tracking other near-Earth asteroids attempting to pick up signals showing the presence of water. From this and other work it has been estimated that near-Earth asteroids may contain as much as 400 billion to 1,200 billion litres of water.

The water contained within asteroids is not liquid water. The water is in the form of hydrated minerals. These are solid mineral substances molecularly bonded to water molecules or its components.

A common theory is that Earth had an early stage where the surface was nothing other than magma, incredibly hot, molten rock that could not have retained or formed water. Over very long periods of geological time as the Earth cooled, numerous asteroids struck the Earth bringing hydrated minerals that released water, becoming part of the Earth's crust. Our planet's early history had many more asteroid collisions because of the unsettled time period of our early solar system and there were many millions of years in which the collisions could have occurred.

Artist's rendition of a large asteroid striking Earth.

NASA is pairing with some private companies to send an exploratory craft to an asteroid named 16 Psyche. This asteroid is about the size of the state of Massachusetts and is found in an asteroid belt between Mars and Jupiter. 16 Psyche is primarily a metallic asteroid and is projected to contain as much as 10,000 quadrillion US dollars-worth of valuable minerals. The first unmanned mission will be exploratory only, but it is expected that later missions could focus on mining operations for gold and other valuable substances.

Nature of Science

The asteroid theory for the existence of water on Earth is not the only scientific explanation. Some researchers believe that comets were a more important origin. Others believe that hydrogen was trapped in the original cloud of materials that formed the planet. The data is not conclusive for any one theory. Researchers will continue to add data and it is likely that a firm explanation will emerge over time. That is the way science approaches complex questions.

A1.1.8 – The search for extraterrestrial life

A1.1.8 – Relationship between the search for extraterrestrial life and the presence of water
Include the idea of the "Goldilocks zone".

Any planet that could possibly support life must have water. Like Earth, that planet must exist in an area of its solar system that would allow water to exist in its liquid form. This position of Earth in relation to the Sun is called the habitable zone or **Goldilocks zone**.

Earth is in an orbit that is nearly perfect for water retention to occur. There are suitable temperatures for water to exist as a liquid and sufficient gravity for retaining this water. In addition, Earth is of a size that allows a suitable gravitational pull to enable water to remain on and just under the surface. Earth has also developed an atmosphere and magnetic field that protects it from most harmful ionizing radiation being emitted from the Sun.

Collectively, these are rare but necessary conditions for a planetary body to be classified as being in a Goldilocks zone. Our galaxy, the Milky Way, is huge. It has approximately 100,000 million stars possibly acting as the centres of solar systems. The Hubble Space Telescope has shown that our galaxy is only one of 125 billion galaxies in the known portions of the universe. Considering these prodigious numbers of stars, it is reasonable to conclude that Earth is not the only planet in our galaxy or universe that has the conditions necessary to support water and thus life.

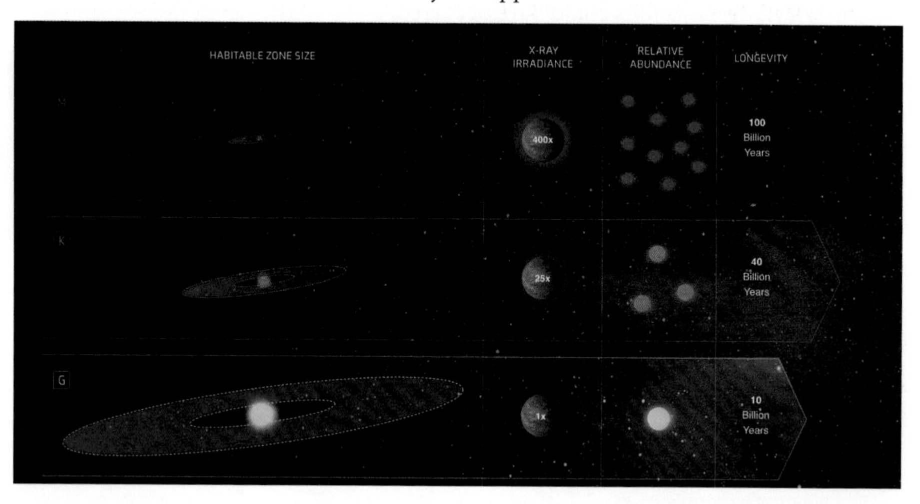

This illustration compares three types of star: yellow G stars (like the Sun), orange K dwarf stars and red M dwarf stars. It shows the relative sizes of habitable zones for each type of star as well as the relative X-ray radiation, their relative abundance and expected star longevity.

The possible existence of water is not the only factor that influences the possibility of life evolving on a planetary body. Our Sun is one of the rarer star types that exist. It is classified as a G type star based on its size and the radiation given off. Two more abundant star types are types K and M. Each of these give off much more radiation than a G star and that radiation would be harmful to life. Notice that K and M stars have a Goldilocks (habitable) zone that is much smaller than a G star like our Sun.

HL end

Guiding Question revisited

What physical and chemical properties of water make it essential for life?

In this chapter we have described how and why water has:

- polar covalent bonds due to an unequal sharing of electrons between oxygen and hydrogen
- cohesive forces attracting one molecule of water to another
- adhesive forces attracting molecules of water to other types of polar molecules
- excellent solvent properties for other polar molecules (solutes)
- properties making water the "solvent of life" as exhibited by cytoplasm, intercellular fluids, blood and many other solutions that are vital to living organisms.

Guiding Question revisited

What are the challenges and opportunities of water as a habitat?

In this chapter we have investigated:

- physical and chemical properties of water that provide both opportunities and challenges for living organisms
  - buoyancy – important to all aquatic and semi-aquatic organisms to keep them at or near the water surface
  - viscosity – the body shape and propulsion mechanisms of animals have become adapted to overcome this resistance that water has for objects moving through it
  - thermal conductivity – organisms living in cold-water environments must have either a physiology adapted for that water temperature or a means of insulation from the cold because water readily conducts heat away from an organism's body.
  - specific heat – water in oceans, lakes and rivers has a very high specific heat that protects many aquatic organisms from much colder surrounding air temperatures

HL

- how water is necessary for life and very few planetary bodies possess conditions necessary to retain water
  - surface temperatures must allow water to exist in liquid form
  - sufficient gravity must exist to prevent water from escaping.

HL end

Exercises

Q1. Describe how a polar covalent bond differs from a non-polar covalent bond.

Q2. Describe the pathway and the forces involved in getting water from the soil surrounding a large tree to a leaf in one of the uppermost branches of that tree (hint: start with the leaf).

Q3. State:

(a) an example of a molecule that is soluble in the cytoplasm of a cell

(b) the function of that same molecule.

Q4. State:

(a) an example of a molecule that is insoluble in the cytoplasm of a cell

(b) the function of that same molecule.

Q5. Describe two adaptations that the black-throated loon (*Gavia arctica*) has evolved for overcoming the viscosity of water.

Q6. HL State three of the conditions necessary for a planetary body to be classified in the Goldilocks zone.

A1.2 Nucleic acids

Guiding Questions

How does the structure of nucleic acids allow hereditary information to be stored?

How does the structure of DNA facilitate accurate replication?

The organisms alive on Earth today have a long history and a very long family tree. Living things do not just appear, rather they are descended from previous generations. This is based on genetics. The information that is being passed from one generation to the next is in the form of DNA. Humans have 46 DNA molecules in each cell in the form of chromosomes. Written in the genetic code of DNA is information that makes a blue whale what it is and makes you what you are.

Along the length of DNA molecules there are chemical messages that code for specific proteins. Most of these protein messages are common to a species, but a few are individual to one single individual of that species. Thus, each living organism is unique. Preceding every cell division, the DNA replicates in an amazingly accurate series of steps that produces two DNA molecules where there was once one. Life has continued in this way for millions of years.

This chapter will introduce you to DNA and other molecules termed nucleic acids. Nucleic acids include DNA and three types of RNA that are all involved in the synthesis of proteins in cells.

A1.2.1 – DNA is the universal genetic material

A1.2.1 – DNA as the genetic material of all living organisms
Some viruses use RNA as their genetic material but viruses are not considered to be living.

Deoxyribonucleic acid (DNA) is the molecule that provides the long-term stored genetic information for all organisms on Earth. When mutations occur that influence evolution, they happen within DNA and are passed on to the next generation. The fact that DNA is universal to all living organisms is evidence of our common ancestry, even back to when the most complex life forms were single cells living in the oceans.

In addition to sugars and phosphate groups acting as a structural framework, DNA has within it four **nitrogenous bases**: adenine, thymine, cytosine and guanine, which are found along the length of the very long molecule. These four bases can be combined in a tremendous variety of orders and lengths. The sequences of nitrogenous bases are the genetic messages or **genes**. The messages are codes for **amino acids**. Amino acids are the "building blocks" of proteins, and a cell's identity and function is determined by the proteins it is able to synthesize. Every cell in a multicellular organism has the same DNA, but each different type of cell only uses the genetic information that is appropriate for that cell.

An artist's rendering of the interior of a cell showing viral particles and a DNA molecule. The spikes on the viral particles are modified proteins that attach to the cells of an organism they infect. Inside each of the viruses is a nucleic acid, either DNA or RNA (ribonucleic acid), that may undergo one or more mutations upon every replication cycle. Some mutations may alter the proteins on the spikes and change how well the protein spikes attach to the host cells.

Viruses are not living organisms. Some viruses contain RNA as their genetic information and some contain DNA. No matter which nucleic acid acts as the genetic code for viral proteins, viruses are not considered to be alive because they cannot survive without a cell of a living organism, and they have no internal biochemistry when they exist as a separate particle. Only when they infect a cell will their nucleic acid (RNA or DNA) become active and use the internal biological components of the cell for their own uses. A virus has absolutely no other function other than to reproduce itself: viruses exist to reproduce. Sometimes that reproduction damages cells to the point of causing great harm to the host organism.

A1.2.2 – The structure of nucleotides

A1.2.2 – Components of a nucleotide

In diagrams of nucleotides use circles, pentagons and rectangles to represent relative positions of phosphates, pentose sugars and bases.

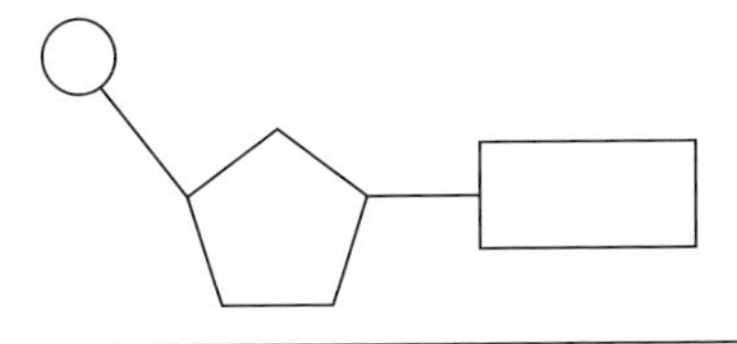

Both DNA and RNA are **polymers** of **nucleotides**. This means that both DNA and RNA have repeating units called nucleotides within the much larger molecule. So, in order to understand the structure of these two molecules important to life, we must first start with the structure of the nucleotides. Individual nucleotides consist of three major parts: one phosphate group, one five-carbon monosaccharide (also called a pentose sugar) and a nitrogenous base. Covalent bonds occur at specific locations in order to produce a functional unit.

It is important to note that in Figure 1 a circle is used to represent a phosphate, a pentagon is used to represent a pentose sugar, and a rectangle is used to represent a nitrogenous base.

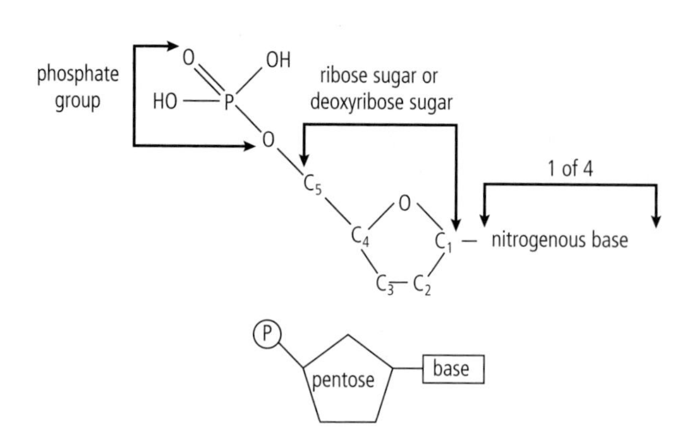

▲ **A1.2 Figure 1** Two representations of a single nucleotide are shown in the diagram. The upper drawing shows more detail, although not every atom and bond are shown of the pentose sugar and only a bonding location is shown for a nitrogenous base. The lower drawing shows the level of detail the IB requires you to draw from memory.

A1.2.3 – Sugar to phosphate "backbone" of DNA and RNA

A1.2.3 – Sugar–phosphate bonding and the sugar–phosphate "backbone" of DNA and RNA
Sugar–phosphate bonding makes a continuous chain of covalently bonded atoms in each strand of DNA or RNA nucleotides, which forms a strong "backbone" in the molecule.

Nucleotides in both DNA and RNA bond together to produce long chains or polymers. In order to form a chain of nucleotides, the pentose sugar of one nucleotide is covalently bonded to the phosphate group of the next nucleotide. This means that there will always be one phosphate group with only one bond to a sugar at one end of the nucleic acid polymer, and a pentose sugar with only one bond to a single phosphate at the other end.

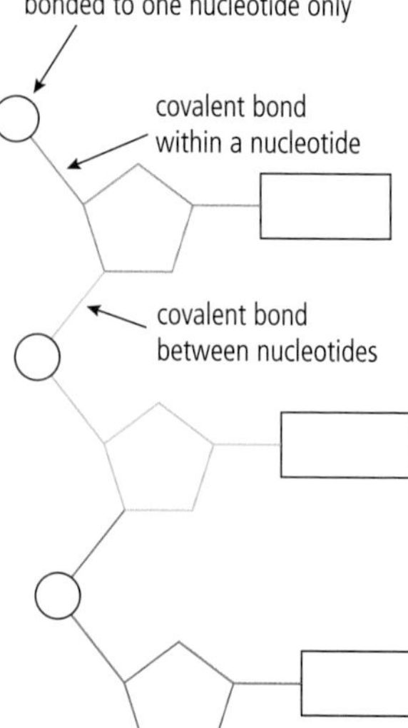

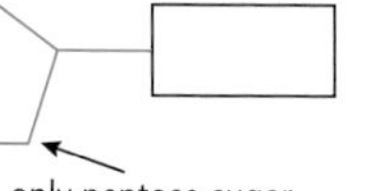

◀ **A1.2 Figure 2** Some nucleic acids are formed from a single chain of nucleotides.

Challenge yourself

Examine Figure 1 on the previous page. Notice that the carbons of the pentose sugar are numbered. Now look at Figure 2, showing six nucleotides bonded together as a single-stranded polymer. Answer the following.

1. Within the polymer of six nucleotides, which *sugar* carbons are bonded to phosphate groups? (Do not consider the first nucleotide.)
2. Within a *single* nucleotide, what number carbon is always attached to the phosphate group?
3. Which carbon number is always attached to the nitrogenous base?

Nucleotides bond to one another to form a chain or polymer as a result of **condensation reactions** forming covalent bonds between the sugar of one nucleotide and the phosphate group of the next nucleotide. The fact that covalent bonds hold the chain together is important as covalent bonds are relatively strong (require a great deal of energy to break) and thus a nucleic acid polymer made of nucleotides is quite stable.

A1.2.4 – Nitrogenous bases within nucleic acids

A1.2.4 – Bases in each nucleic acid that form the basis of a code
Students should know the names of the nitrogenous bases.

In total, there are five possible **nitrogenous bases** in RNA and DNA. Four are found within RNA, and four are found in DNA. Only one of the bases differs in the two types of polymers, as shown in Table 1.

RNA nitrogenous bases	DNA nitrogenous bases
Adenine (A)	Adenine (A)
Uracil (U)	Thymine (T)
Cytosine (C)	Cytosine (C)
Guanine (G)	Guanine (G)

A1.2 Table 1 The five nitrogenous bases found in RNA and DNA

Make sure you know the names of the five nitrogenous bases found in RNA and DNA, and do not just rely on the abbreviated form of a capital letter.

It may look like some of the nucleotides found in RNA and DNA are identical, for example because they both contain the base adenine. However, they are not identical because all the nucleotides found in RNA contain ribose as their pentose sugar, and all the nucleotides in DNA contain deoxyribose. In addition, the base uracil only occurs in RNA, not DNA, and the base thymine only occurs in DNA, not RNA. Thus, there are eight different nucleotides in total. When drawing nucleotides, it is common practice to put the capitalized first letter of the base inside the rectangle, as used by the IB.

The sequence of nitrogenous bases in DNA, later transcribed into RNA, forms the basis of the genetic code.

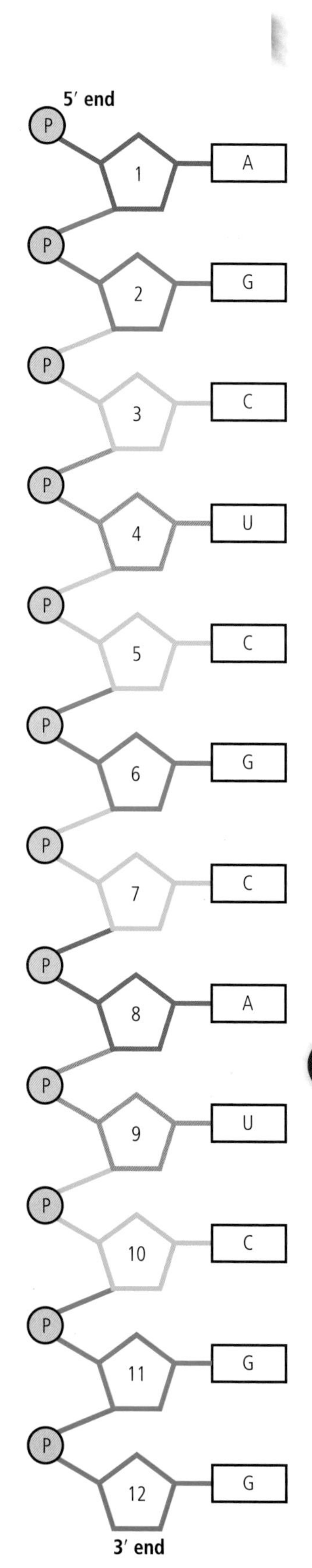

Challenge yourself

4. Use the geometric symbols required by the IB (see Figure 1) to represent all the possible separate nucleotides of DNA. Once you have sketched the four for DNA, do the same for RNA. To remind yourself of the fundamental pentose sugar difference between RNA and DNA nucleotides, you might want to put the letter "R", for ribose, inside the pentose shape of all RNA nucleotides. Then put "DR", for deoxyribose, inside all of the four DNA nucleotides. Make sure you end up with eight different nucleotides in total, one containing uracil and one containing thymine.

A1.2.5 – The structure of RNA

A1.2.5 – RNA as a polymer formed by condensation of nucleotide monomers
Students should be able to draw and recognize diagrams of the structure of single nucleotides and RNA polymers.

RNA is formed when nucleotides become bonded together in very specific sequences. The nucleotides are joined together by a **condensation reaction** between the pentose sugar of one nucleotide and the phosphate group of the next nucleotide. This reaction releases a water molecule (which is why this is called a "condensation" reaction). If an RNA molecule contains 322 nucleotides, 321 molecules of water would have been produced during its **synthesis**, as it would have required 321 condensation reactions to form.

Challenge yourself

5. How many water molecules would have been produced when the condensation reactions occurred that produced the 12 nucleotide RNA sequence shown in Figure 3?

Even though the RNA depiction in Figure 3 has only 12 nucleotides shown, the actual RNA may have as many as a few thousand nucleotides.

A1.2 Figure 3 Twelve nucleotides bonded to form a very small section of a strand of RNA. The molecule is recognized readily as RNA because of the presence of uracil and because it is a single strand. Each adjoining nucleotide has been drawn in a different colour to emphasize the nucleotide structures. Notice that the chain has an alternating pentose–phosphate backbone, with the nitrogenous bases extending outwards from the backbone.

A1.2.6 – The structure of DNA

A1.2.6 – DNA as a double helix made of two antiparallel strands of nucleotides with two strands linked by hydrogen bonding between complementary base pairs

In diagrams of DNA structure, students should draw the two strands antiparallel, but are not required to draw the helical shape. Students should show adenine (A) paired with thymine (T), and guanine (G) paired with cytosine (C). Students are not required to memorize the relative lengths of the purine and pyrimidine bases, or the numbers of hydrogen bonds.

Cytosine
Guanine
Thymine
Adenine

RNA is composed of a single chain or strand of nucleotides, while DNA consists of two chains or strands of nucleotides connected to one another by hydrogen bonds. The strands of both DNA and RNA may involve very large numbers of nucleotides. To visualize DNA, imagine the double-stranded molecule as a ladder (see Figure 4). The two sides of the ladder are made up of the phosphate and deoxyribose sugars. The rungs of the ladder (what you step on) are made up of the nitrogenous bases. Because the ladder has two sides, there are two bases making up each rung. The two bases making up one rung are said to be complementary to each other. Notice that the base pairs are always adenine (A) bonded to thymine (T) and cytosine (C) bonded to guanine (G). There are no exceptions to this in DNA, and these base pairings are known as the **complementary base pairs**. Because the two strands are upside down in comparison to each other, but parallel, they are said to be **antiparallel** to each other.

The nitrogenous bases adenine and thymine are always paired with each other in the double-stranded DNA molecule. Likewise, cytosine and guanine are always paired. These pairings are called the complementary base pairs.

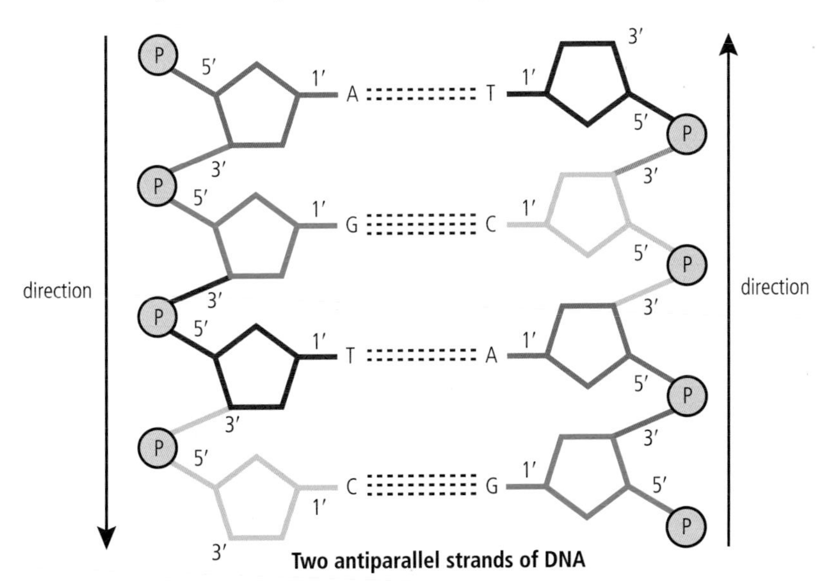

A1.2 Figure 4 A small section of a double-stranded DNA molecule showing hydrogen bonds (dotted lines) between complementary base pairs. This type of representation of DNA is known as a "ladder diagram" and does not attempt to show the helical shape of the molecule.

Always attempt to view DNA and RNA molecules as chains of nucleotides. Identify the first nucleotide with its own phosphate, sugar and nitrogenous base and then visually move to the next, and so on. In Figure 4 you would visually start in the upper left corner for the left strand, and you would start in the lower right corner for the right strand.

Challenge yourself

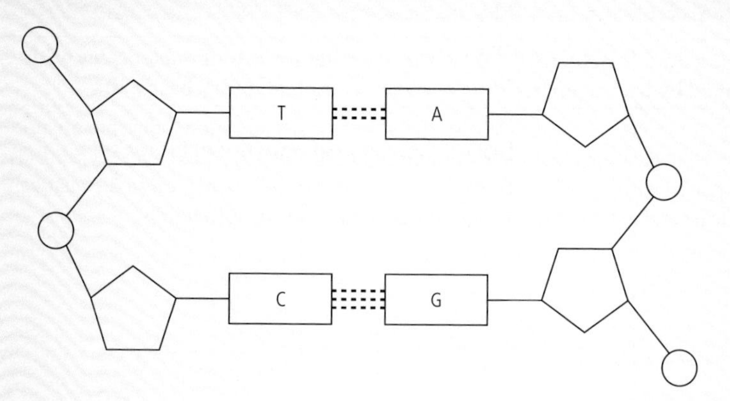

6. On your own paper and using the figure above as a guide, sketch and label the geometric shapes as shown to represent this four-nucleotide section of DNA.
7. Add four more nucleotides to each side by adding to the bottom of your sketch so that you end up with a 12-nucleotide section of antiparallel DNA. Remember to use complementary base pairs, although you can choose the base sequence.
8. Circle two *complete* nucleotides of your *added* nucleotides, one on each side, but do not circle any of the nucleotides in the corners of the figure. Check to make sure that your circles include one phosphate group, one deoxyribose sugar and one nitrogenous base, and that there are no uncircled nucleotides that are incomplete.

A1.2.7 – Distinguishing between DNA and RNA

A1.2.7 – Differences between DNA and RNA

Include the number of strands present, the types of nitrogenous bases and the type of pentose sugar.

Students should be able to sketch the difference between ribose and deoxyribose. Students should be familiar with examples of nucleic acids.

DNA and RNA are both linear polymers, consisting of sugars, phosphates and bases, but there are some important differences between the two molecules. Table 2 summarizes these differences.

DNA	RNA
Double-stranded molecule	Single-stranded molecule
All nucleotides contain deoxyribose sugar	All nucleotides contain ribose sugar
Thymine is one of the four nitrogenous bases	Uracil is one of the four nitrogenous bases
Shaped into a double helix	Variety of shapes depending on type of RNA
Acts as the permanent genetic code of a cell/organism	Does not contain a permanent genetic code, except in RNA viruses

A1.2 Table 2 A comparison of DNA and RNA molecules

Distinguishing between deoxyribose and ribose

Ribose has a molecular formula of $C_5H_{10}O_5$, whereas deoxyribose has a formula of $C_5H_{10}O_4$. Notice that the only difference in the molecular (chemical) formulas is that ribose has one more oxygen compared to deoxyribose. A side-by-side comparison shows where the difference occurs (see Figure 5). In organic chemistry an –OH group bonded to a carbon is called an **alcohol** or **hydroxyl group**. If you remove the oxygen from the hydroxyl group, it simply leaves a hydrogen. This may not look like much, but it is the common difference in all nucleotides of RNA versus DNA.

Practise sketching each of the two molecules shown in Figure 5. Learn the pattern that is common to both molecules and then modify for the single difference between deoxyribose and ribose.

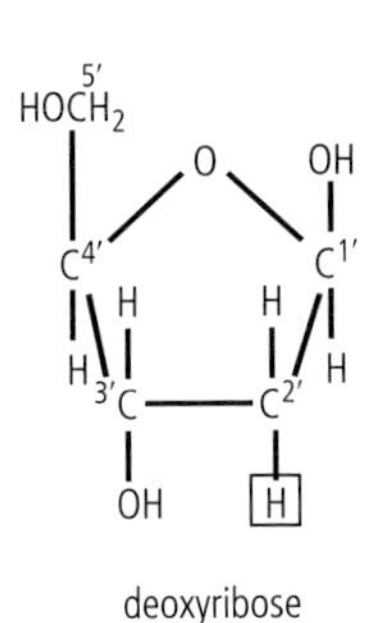

A1.2 Figure 5 A molecular sketch showing the deoxyribose sugar of DNA compared to the ribose sugar found in RNA molecules. Notice the difference in the lower right corners of the two molecules. Ribose has one more oxygen in its structure compared to deoxyribose.

The single "missing" oxygen in the pentose sugar of DNA leads to the name *deoxy*ribose within the full name for DNA (deoxyribose nucleic acid). The full name of RNA is ribonucleic acid.

Specific examples of nucleic acids

All living organisms use DNA as their long-term hereditary storage molecule. DNA stores genetic information as genes, but for that information to become useful to a cell there must be other nucleic acids at work. Here are four of the other nucleic acids as examples.

- **Messenger RNA (mRNA)** – This is an RNA molecule that is synthesized from an area of DNA called a **gene**. In a cell with a nucleus, the mRNA then leaves the nucleus and represents the genetic information necessary to make a protein. This is where it gets its name "messenger" RNA.
- **Transfer RNA (tRNA)** – Special genes of DNA code for tRNA molecules. When a specific protein is synthesized, specific amino acids must be added to the amino acid chain in a specific order. The function of tRNA is to transfer the correct amino acid into a growing chain of amino acids. This is the reason for its name "transfer" RNA.
- **Ribosomal RNA (rRNA)** – Again, special genes of DNA code for rRNA molecules. Along with some previously synthesized proteins, rRNA is used to create an organelle in cells called ribosomes. Cells typically have many thousands of ribosomes, and they are the cellular location where proteins are synthesized.
- **Adenosine triphosphate (ATP)** – This is a single-nucleotide nucleic acid. There are many other single-nucleotide nucleic acids in cells, but we are going to use this one as an example. ATP is used in cells as a type of chemical energy. When a muscle contracts, many ATP molecules are used as an energy source for the movement. The ultimate purpose of cellular respiration is to convert the energy contained within food molecules into the energy of ATP.

Do not concern yourself at this point with the details of these examples of nucleic acid molecules, beyond what is summarized in this section. The function of each of these molecules is explained in much greater detail in other chapters.

Challenge yourself

The figure below shows a molecular diagram of an ATP molecule. You do not need to memorize it, but based on what you have read earlier in this chapter you should be able to look at the diagram and answer the following questions.

9. Why is this molecule called a "triphosphate"?

10. Is the pentose sugar in this molecule ribose or deoxyribose?

11. The "adenosine" portion of the molecule's name comes from the nitrogenous base bonded to the pentose sugar. What is that nitrogenous base?

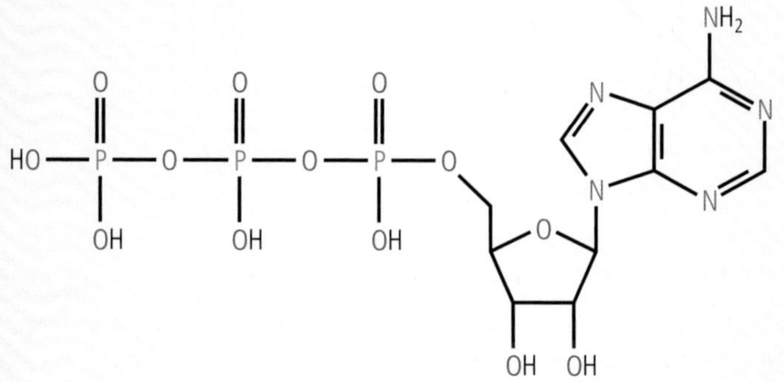

A1.2.8 – The importance of complementary base pairing

A1.2.8 – Role of complementary base pairing in allowing genetic information to be replicated and expressed
Students should understand that complementarity is based on hydrogen bonding.

As you recall, adenine and thymine are complementary to each other in DNA, and cytosine and guanine are complementary as well. This complementarity is based on hydrogen bonding. Adenine and thymine only form hydrogen bonds with each other; adenine does not form hydrogen bonds with any other DNA nucleotide. The same is true for cytosine and guanine.

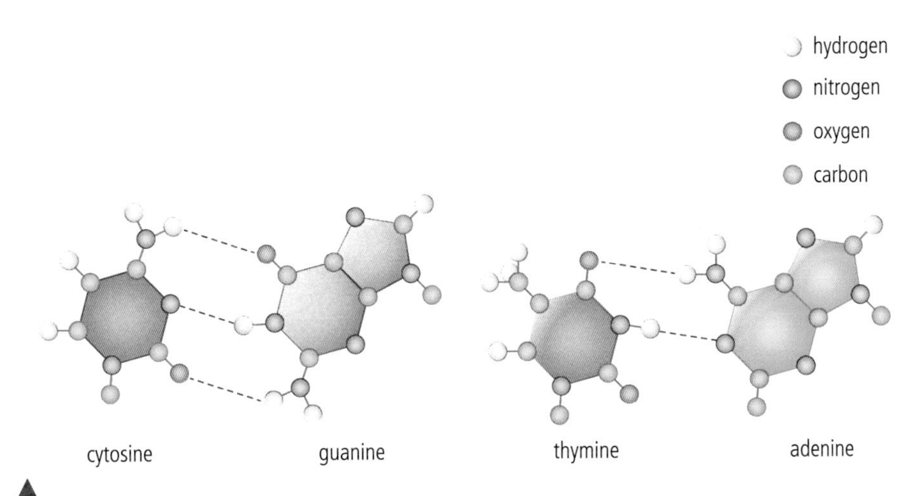

Hydrogen bonding (shown in dotted red lines) between the complementary base pairs within DNA. It is this hydrogen bonding that holds the two antiparallel strands together and ultimately results in the double helix shape.

Complementary base pairing is important in DNA replication. Imagine that an area of DNA has been unzipped (opened up into two single strands). If free-floating individual nucleotides in solution begin pairing with the unmatched nucleotides, an exact copy of the original molecule can be made. In fact, if both sides of the original DNA are used as a template, then two molecules of DNA can be synthesized, each a duplicate of the original. In a simplified form, this is how DNA replication occurs.

A1.2.9 – Storage of genetic information

A1.2.9 – Diversity of possible DNA base sequences and the limitless capacity of DNA for storing information
Explain that diversity by any length of DNA molecule and any base sequence is possible. Emphasize the enormous capacity of DNA for storing data with great economy.

DNA stores genetic information in its sequence of nitrogenous bases. Every three bases represents a meaningful piece of information called a triplet or, more specifically, a **triplet codon.** Many triplets within DNA code for one of the 20 amino acids. There are four different DNA nucleotides that can be arranged as sequenced triplets. So, what are the odds of DNA containing any one triplet in any one gene location? Consider the odds of having G–G–G in one triplet area of DNA. If it was by random chance (although it is not) the odds would be:

$$\frac{1}{4} \times \frac{1}{4} \times \frac{1}{4} = \frac{1}{64}$$

Why? Because there is a one in four chance of the nitrogenous base being guanine, and it occurs in our example three times.

This computation also means that there are 64 combinations of nucleotides within the triplet code system. All of those 64 combinations are used in the genetic code for some purpose, most of them coding for amino acids.

Researchers are working on ways to store data (text files, photos, books, maps) within artificially created DNA molecules. DNA stores information using the very efficient code of four nitrogenous bases, compared to the less efficient 0 and 1 binary code used by computers.

How can polymerization result in emergent properties?

Think about all the ways that the four nitrogenous bases of DNA can be grouped. If DNA was a short molecule (say around 1,000 nucleotides), the number of groupings would be large, but still not unlimited. Now consider that the length of DNA (the number of nucleotides in one strand) is only limited by the amount that will fit efficiently into a cell. The shortest DNA molecule in the human genome is about 50 million base pairs, and the longest about 260 million base pairs.

As you can see, the likelihood that two DNA molecules are identical as a result of random chance approaches zero. DNA can contain a nearly limitless amount of genetic information.

A1.2.10 – Genetic uniqueness

A1.2.10 – Conservation of the genetic code across all life forms as evidence of universal common ancestry
Students are not required to memorize any specific examples.

Identical twins develop when a single fertilized egg or early embryo splits into two portions. Each grows to become a separate person and shares exactly the same DNA sequences.

Imagine a section of DNA that contains the triplet code C–G–A. If that triplet code is used to synthesize a protein, the amino acid that will be produced will be alanine. If the triplet code is A–G–A, the amino acid is serine. A chart listing the triplet codes can provide this information.

It does not matter whether the organism is a species of fungus, an oak tree, or a human being. All living organisms use the same genetic code. The genetic code is therefore said to be universal.

What makes RNA more likely to have been the first genetic material, rather than DNA?

So why are organisms different from each other? The answer to that is the DNA base sequences are different even though the code to read the sequences is the same. Your best friend, although not directly related to you, is related to you by evolution. The two of you share more than 99% of the same gene sequences. If it was 100%, you would not be the unique and different people you are.

A conserved genetic code

Why has the genetic code remained unchanged? The answer to this question lies in the process of evolution. The evolution of living organisms has been occurring for over 3.5 billion years. If you could go back in time far enough you would probably not see any organisms that you recognize today, although some of the organisms you would see will be the ancestors of today's organisms. If you were to keep moving back through time, the organisms would become even less familiar, and eventually they would be nothing more than single-celled organisms living in water.

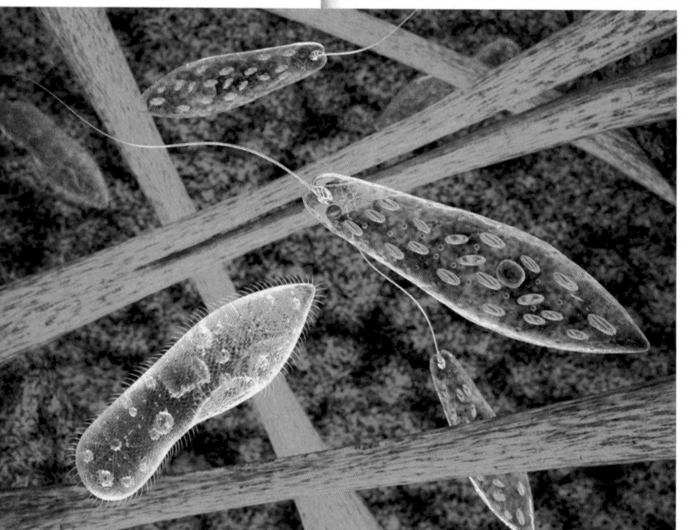

Bacteria and protists were some of the first organisms on Earth to evolve, and thus hold the origin of the genetic code used by all organisms today. Humans and other life forms still have genes in common with these evolutionary pioneers.

These single-celled organisms are the ancestors of all life on Earth today. This is also postulated to be the time period in which the biochemistry of DNA and RNA evolved. All life forms from that point on used DNA to store their genetic information, and RNA to transfer that information to the order of amino acids in their proteins. Evolution changes the DNA sequences slowly, but it always has continued to use the same mechanisms of genetic coding.

Nature of Science

The theory of evolution by natural selection as proposed by Charles Darwin and independently by Alfred Wallace was based primarily on their observations of physical traits. It appeared to them that organisms developed adaptations to fit different ecological niches in the area that they lived in. In 1859, when Darwin published his famous book *On the Origin of Species*, there was absolutely no knowledge of DNA or the molecular basis of heredity and evolution. Scientific ideas that originate in one form can be corroborated by later scientific work if the ideas are sound. Today there is a mountain of evidence supporting evolutionary principles, including a vast amount of information from **molecular genetics**.

HL

A1.2.11 – Directionality of RNA and DNA strands

A1.2.11 – Directionality of RNA and DNA
Include 5′ to 3′ linkages in the sugar–phosphate backbone and their significance for replication, transcription and translation.

The 5′ and 3′ designations refer to the fifth and third carbon atoms in the ribose and deoxyribose sugars (see Figures 5 and 6).

In a DNA molecule there are two strands that run antiparallel to each other. This means that if you compare the sides of DNA, one strand will run 5′ to 3′ and the other will run 3′ to 5′. This does not mean the direction in which they were synthesized is different. Both strands of DNA are synthesized starting with the 5′ nucleotide and working towards the 3′ end.

DNA does not have a single 5′ end or 3′ end. DNA is composed of two strands and each strand has a 5′ and 3′ end.

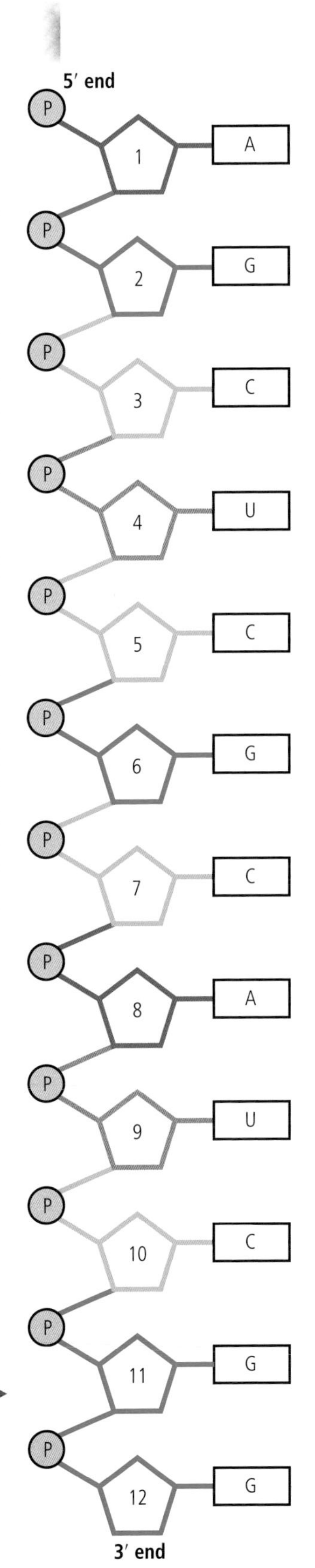

A1.2 Figure 6 This is the same very small section of RNA you studied in Figure 3. Because each nucleotide has one phosphate, one sugar and one nitrogenous base, there are 12 of each of these subunits. Any single strand of DNA or RNA will have an unbound phosphate group at one end (the upper left corner). This end is identified as the 5′ (5 prime) end of the strand. All the other phosphate groups are bonded to two sugars, linking adjacent nucleotides. The opposite end of the strand terminates with deoxyribose sugar with no additional phosphate groups. This end is called the 3′ (3 prime) end of the strand.

Challenge yourself

12. Since the two strands of DNA are antiparallel to each other, you only need to know one end of one strand in order to extrapolate the other three ends.

3′ ______________________________ X

Y ______________________________ Z

Each horizontal line shown above represents a strand of DNA of a double-stranded molecule. Notice that only the 3′ end of one strand has been labelled. Using 5′ and 3′, label each of the ends currently labelled X, Y and Z.

The importance of directionality

When RNA or DNA is formed, one nucleotide at a time is added to the molecule. The nucleotide is not added at a random spot: it is added as the next nucleotide in a growing chain. As all nucleotides have a phosphate group, pentose sugar and nitrogenous base, you can trace this formation by looking for the first nucleotide in the sequence. That first nucleotide will always be the 5′ end of the strand.

This is important when a new nucleic acid strand is formed. When DNA replicates, the two strands separate from each other in a particular area, and each separated strand acts as a template for a new strand to be formed. A *new* strand will always begin with the 5′-end nucleotide first.

Any *new* nucleic acid strand being formed (DNA or RNA) is always formed by starting at its 5′ end and continuing until the final nucleotide is bonded at the 3′ end.

DNA and RNA molecules are both integral to the process of protein synthesis. The first stage of protein synthesis is called **transcription**. This occurs when a gene of DNA (one of the two DNA strands) is opened and an RNA molecule is synthesized using complementary base pairing. The transcription process synthesizes the 5′ end of the RNA first. The resulting RNA is called mRNA, and represents the genetic information of one gene. The mRNA then pairs with an organelle in the cell called a **ribosome**. Another RNA called tRNA, bonded to a specific amino acid, now pairs (by complementary base pairing) to triplets of the mRNA. This process is called **translation**. Translation is accomplished in a sequence starting nearest to the 5′ end of the mRNA molecule. These molecular processes will be difficult to understand completely and visualize until you study Chapter D1.1 on DNA replication and Chapter D1.2 on protein synthesis. For now, visualize nucleic acids as having a directionality (5′ and 3′ ends), and remember that the directionality is important to their structure and function in DNA replication and RNA synthesis.

A1.2.12 – Purine-to-pyrimidine bonding

A1.2.12 – Purine-to-pyrimidine bonding as a component of DNA helix stability
Adenine–thymine (A–T) and cytosine–guanine (C–G) pairs have equal length, so the DNA helix has the same three-dimensional structure, regardless of the base sequence.

The nitrogenous bases of DNA are grouped into pyrimidines and purines, as shown in Table 3. Pyrimidines are smaller as they contain a single-ringed structure, whereas purines are larger as they contain a double-ring structure.

Purines (double ring size)	**Pyrimidines** (single ring size)
Adenine (A)	Thymine (T)
Guanine (G)	Cytosine (C)

A1.2 Table 3 Types of nucleotides found in DNA

When bonding the two strands of DNA together to make the double helix, a purine is always bonded to a pyrimidine. This results in the two strands of DNA being a consistent distance from each other (with three rings in total), leading to an amazing and stable double helix shape as the two strands wind around each other in three dimensions.

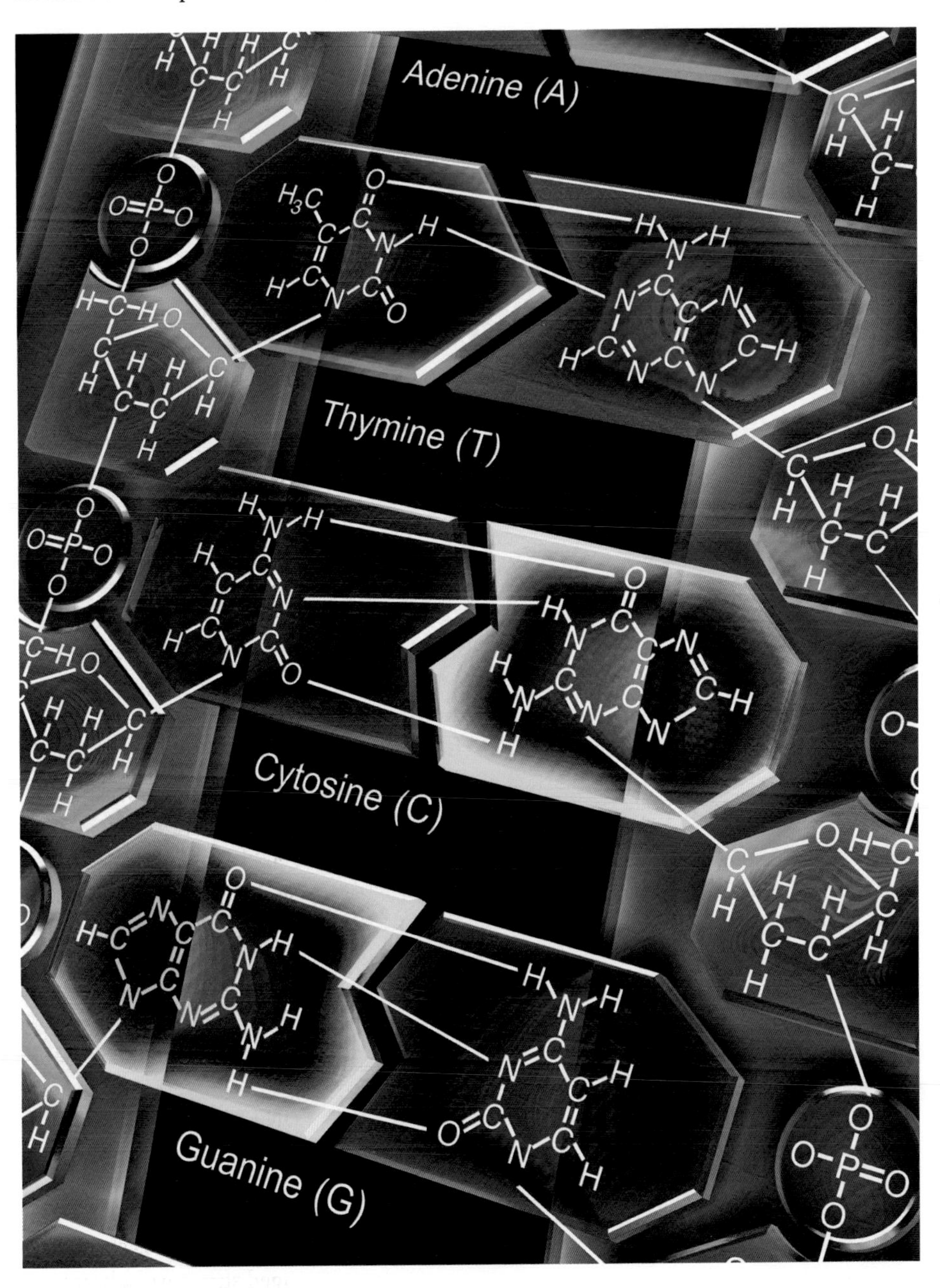

This figure shows complementary base pairing and nitrogenous base structures within a small portion of DNA. You do not have to memorize the nitrogenous base structures for the IB. Notice the double-ring structures of adenine and guanine (purines) and the single-ring structures of thymine and cytosine (pyrimidines).

You learned that complementary base pairing requires A–T and C–G. As you can see, in each case one is a purine and the other a pyrimidine. But, why not G–T and A–C ? Even though that pairing would bond a purine to a pyrimidine, hydrogen bonding between the pairings cannot occur and thus it does not occur in nature.

A1.2.13 – Efficient packaging of DNA molecules

A1.2.13 – Structure of a nucleosome
Limit to a DNA molecule wrapped around a core of eight histone proteins held together by an additional histone protein attached to linker DNA. **Application of skills:** Students are required to use molecular visualization software to study the association between the proteins and DNA within a nucleosome.

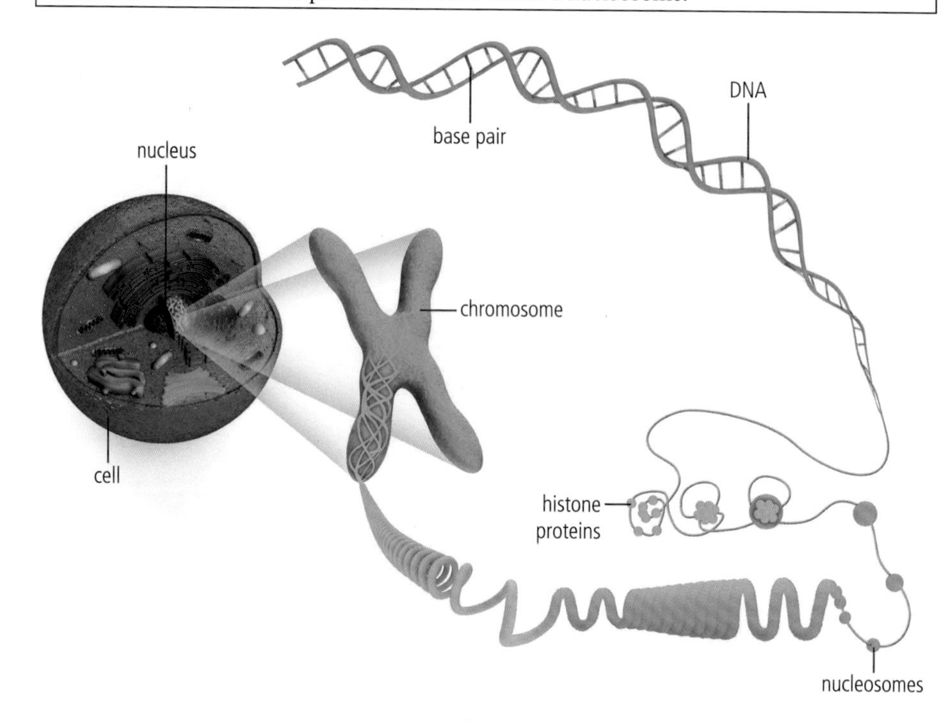

A1.2 Figure 7 One chromosome from a cell's nucleus expanded to show the coiling and structure of nucleosomes.

Biologists often describe DNA organized into nucleosomes as "beads on a string". In this analogy, the DNA (the string) is wrapped around the histones (the beads), rather than running through them as in an actual beads.

DNA is a very, very long molecule. Some DNA molecules in human cells are approximately 2 m in length. In order to fit the very long molecules of DNA inside the nucleus of a cell, a very efficient "packaging" solution has evolved (see Figure 7). Within the nucleus there are proteins called **histones**. First, DNA wraps itself around eight of these histone proteins, and then an additional histone helps hold the structure together. This occurs in many adjoining areas of DNA. Each resulting structure is called a **nucleosome** (see Figure 8). The DNA that extends from one nucleosome to the next is called linker DNA. The multitude of nucleosomes then stack up in an organized pattern and begin coiling around other proteins in a very condensed shape. The overall "packaged" shape is a chromosome. A human cell contains 46 chromosomes.

You can find videos and other molecular visualization aids online to view the use of histone proteins to form nucleosomes and the formation of chromosomes.

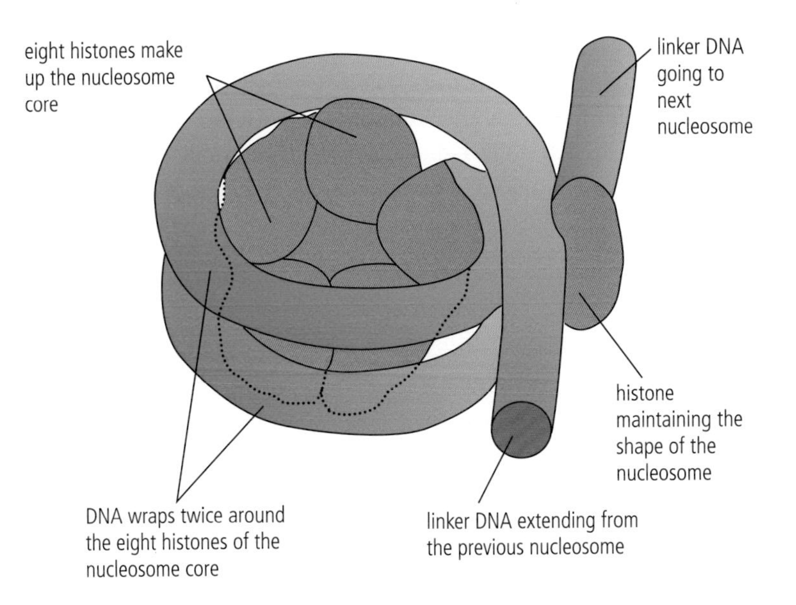

A1.2 Figure 8 A nucleosome showing the DNA wrapped around eight histone proteins with one additional histone helping to keep the structure intact. Linker DNA is found between adjoining nucleosomes. Look back at Figure 7 to see how nucleosomes organize into a chromosome.

A1.2.14 – The Hershey–Chase experiment

A1.2.14 – Evidence from the Hershey–Chase experiment for DNA as the genetic material

Students should understand how the results of the experiment support the conclusion that DNA is the genetic material.

NOS: Students should appreciate that technological developments can open up new possibilities for experiments. When radioisotopes were made available to scientists as research tools, the Hershey–Chase experiment became possible.

Before the middle decades of the 20th century, biologists were unsure of the genetic material of living organisms. Primarily, they debated whether genetics was based on nucleic acids or proteins. Because DNA was composed of only four types of nucleotides, and proteins were composed of 20 types of amino acids, it made sense to many that proteins could better provide the complex genetic variety found in living organisms.

In 1951 and 1952, two researchers, Alfred Hershey and Martha Chase, working at Cold Spring Harbor, NY, carried out experiments that helped confirm DNA as the genetic material. The experiment used a bacteriophage and the bacterium *Escherichia coli*. A bacteriophage is a virus that infects bacteria, and is composed of a protein outer coat and an inner core of DNA. When a bacteriophage infects a cell, the virus takes over the metabolism of the cell, resulting in multiple viruses of that type being formed using molecules such as nucleotides in their synthesis.

Hershey and Chase made use of radioisotopes in their experiment in a process called radioactive isotope labelling. Radioisotopes are radioactive forms of elements that can be detected within molecules. The particles released during their decay allow the specific radioisotope used to be detected. The researchers grew bacteriophage viruses

TOK

Sometimes it is difficult for scientists to let go of long-held beliefs. The beliefs become ingrained as part of their core knowledge, and few even think to challenge that knowledge. How do our expectations and assumptions have an impact on how we perceive things?

in two different types of culture. One culture included radioactive phosphorus 32. The viruses produced in this culture had DNA inside their viral core labelled with the detectable phosphorus 32. Another culture included a radioactive form of sulfur known as sulfur 35. This detectable radioisotope was present in the protein outer coat of the viruses produced. DNA does not include sulfur (because nucleotides do not contain any sulfur atoms). The radioactive sulfur was only detectable in the protein shell of the virus, because two of the 20 possible amino acids that can be present in protein contain sulfur.

The two types of bacteriophages labelled with radioisotopes were each allowed to infect the bacterium *E. coli*. As Figure 9 shows, the *E. coli* infected with the phosphorus 32-labelled bacteriophage had radioactivity detected inside the cells, a location indicating DNA. However, the *E. coli* infected with the sulfur 35-labelled bacteriophage had no radioactivity inside the cell. Because DNA contains phosphorus and not sulfur, this allowed Hershey and Chase to conclude that DNA, not protein, was the genetic material.

It is important for researchers to publish the details and findings of their experimental work. This allows other scientists all over the world to learn and build upon previous work in future research projects.

Once the results of the Hershey–Chase experiment were published, further research involving genetics centred primarily on DNA was possible, and collectively that body of work provides conclusive evidence that DNA is the genetic material.

The Hershey-Chase experiment is sometimes referred to as the "blender experiment" because the experimental process involved using a kitchen-type blender. The blender sheared the viral coatings containing the sulfur radioisotope away from the living bacterial cells containing the phosphorus radioisotope.

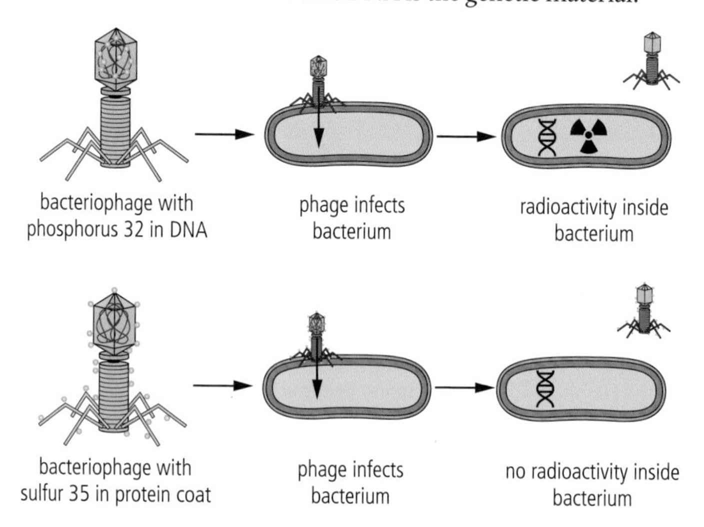

A1.2 Figure 9 A summary of the Hershey-Chase experiment. Radioactivity was detectable inside cells after radioactive isotopes of phosphorus had been used when growing *E. coli* cultures. These radioactive isotopes were used within the nucleotides when *E. coli* cell division occurred and DNA was replicated. Recall that phosphorus is a component of nucleotides within phosphate groups.

Nature of Science

It is not very often that "eureka" moments occur in science. More often scientific advances move along slowly, with one discovery providing information that leads to another. Such was the case with the Hershey–Chase experiments. Their work was made possible by earlier research using radioactive isotopes to trace the pathway of different types of molecules. The Hershey–Chase work showing DNA is the genetic material was not the result of a single experiment but a series of experiments, during which techniques were refined as they worked through the problem one step at a time.

A1.2.15 – Chargaff's rule

A1.2.15 – Chargaff's data on the relative amounts of pyrimidine and purine bases across diverse life forms
NOS: Students should understand how the "problem of induction" is addressed by the "certainty of falsification". In this case, Chargaff's data falsified the tetranucleotide hypothesis that there was a repeating sequence of the four bases in DNA.

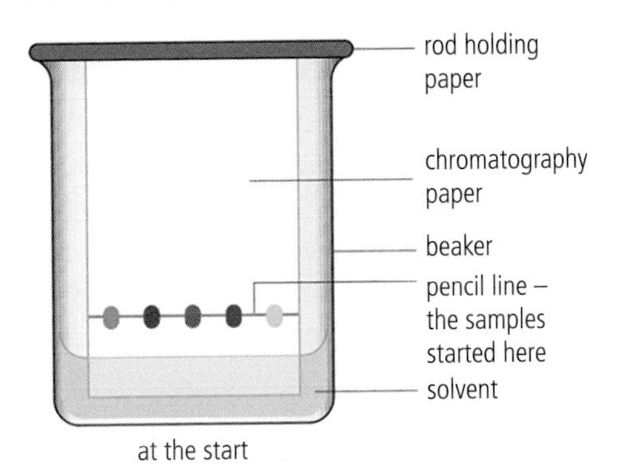

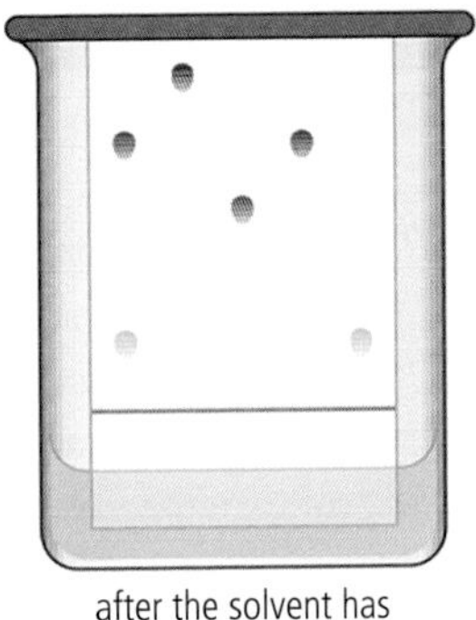

▲ Erwin Chargaff's experimental work determining the nitrogenous base ratios within DNA involved the technique shown in this figure: paper chromatography. A mixture of substances is embedded into chromatography paper. The bottom of the paper, below the embedded substances, is placed into a solvent. The solvent wicks up the paper, carrying components of the mixture with it. Each component travels a different distance based on its molecular size and charge.

As mentioned earlier, scientists in the early part of the 1900s worked under the assumption that protein was responsible for genetic traits. Research showed that there were both DNA and proteins in the nucleus of a cell. The hypothesis was that DNA existed as a tetranucleotide molecule, in other words, in repeating set units of the four nucleotides, and was there to help give structure to chromosomes.

In the late 1940s a researcher named Erwin Chargaff developed a research technique designed to show the proportions of nitrogenous base types found in various sources of DNA. His separation and identification technique involved paper chromatography. Some of the results of his studies are shown in Table 4.

DNA source	Adenine	Thymine	Guanine	Cytosine
Calf thymus	1.7	1.6	1.2	1.0
Beef spleen	1.6	1.5	1.3	1.0
Yeast	1.8	1.9	1.0	1.0
Tubercle bacillus	1.1	1.0	2.6	2.4

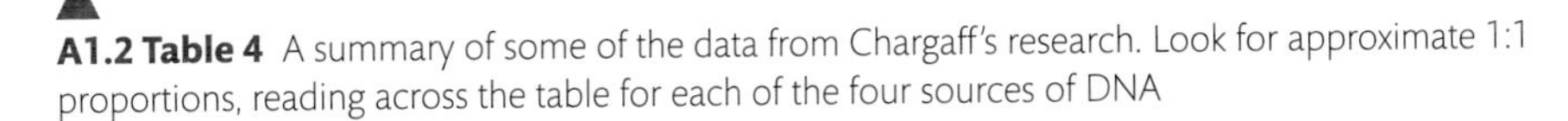

▲ **A1.2 Table 4** A summary of some of the data from Chargaff's research. Look for approximate 1:1 proportions, reading across the table for each of the four sources of DNA

Start by looking at the results for the DNA Chargaff extracted from the thymus gland of a calf. As you can see, there is almost the same ratio of adenine to thymine, and also the same ratio of guanine to cytosine. Very similar ratios exist for the other three biological sources of DNA shown.

Over time, this and similar data became known as Chargaff's rule: DNA contains the same number of adenine as thymine nucleotides, as well as the same number of guanine and cytosine nucleotides.

James Watson and Francis Crick made use of information provided by several other research teams when they proposed and built the first accurate model of DNA in 1953. X-ray crystallography work by Rosalind Franklin showed DNA to be a double helix. Erwin Chargaff's work showed the complementary pairing of the nitrogenous bases.

Nature of Science

Why do you think it is that the ratios of A–T and C–G were not a perfect 1:1 in Chargaff's results? Depending on the experimental methodology used, science often does not give perfect results. This is frequently because of experimental error. Some of that error could be human error, and sometimes it is the best data that the technique can produce. It does not mean that the data is useless. A large data set can minimize some of the error, and statistics can show how significant the errors are.

The importance of Chargaff's rule

The data from Chargaff's work not only showed that the proportion of adenine to thymine, and guanine to cytosine, is equal, it also showed that the tetranucleotide theory is false. The tetranucleotide theory, if correct, would have resulted in the proportion of all of the nitrogenous bases being equal. In other words, the same quantities of A, T, C and G. The data did not show equal proportions of all four bases, and thus the tetranucleotide idea was falsified.

This led to a whole new set of ideas as to what the genetic code could be. Researchers began to consider the sequence of nucleotides in DNA as highly variable and the possibilities of DNA containing the genetic code. We now know the proteins in the nucleus of the cell were primarily histone proteins helping to form nucleosomes.

Nature of Science

The process of science always struggles with the conclusions that can be reached from various data sets. The "problem of induction" is evidenced by assuming something is always true because one or more data sets seems to lead to that conclusion. In actuality, the process of science is often better served by asking questions that lead to falsifications. This is what Erwin Chargaff did when he falsified the idea that DNA was organized into tetranucleotides.

HL end

Guiding Question revisited

How does the structure of nucleic acids allow hereditary information to be stored?

In this chapter we have described how RNA and DNA are structured:

- each is composed of subunits called nucleotides
- nucleotides exist in eight types, four types in RNA and four types in DNA
- each nucleotide contains one of five possible nitrogenous base, adenine, thymine, cytosine, guanine and uracil
- in DNA, the two strands are held together by complementary base pairing between the nitrogenous bases
- the sequence of the nucleotides in sections of DNA called genes allows long-term storage of the genetic code
- RNA molecules are complementary copies of genes of DNA transcribed by using RNA nucleotides.

Guiding Question revisited

How does the structure of DNA facilitate accurate replication?

In this chapter we have described how:

- DNA exists as a double-stranded molecule
- DNA makes copies of itself
- this unwinding allows the nitrogenous bases to make new complementary pairings using the exposed nitrogenous bases as a template
- the pairings are adenine with thymine, and cytosine with guanine
- two DNA molecules are created from one during DNA replication, although neither is completely "new"
- HL nitrogenous base sizes (a purine with a pyrimidine) and hydrogen bonding ensure accuracy when new base pairs are formed.

Exercises

Q1. State how many nucleotide types exist within the structures of DNA and RNA.

Q2. State the structural similarity of the two nitrogenous bases (adenine and guanine) used to classify them as purines.

Q3. Suggest a reason why researchers often give DNA information:

(a) as the sequence of nitrogenous bases without indicating the presence of the phosphate group and sugar component of each nucleotide (for example 5′ATTCCGTGTACGT3′)

(b) from one strand of DNA only.

Q4. You are visualizing a single sequence of nitrogenous bases and you see multiple uracil bases. What does that tell you about the molecule?

Q5. Which of these is not a nucleic acid?

A DNA **B** ATP **C** PCR **D** RNA

Q6. A measurement of a sample of DNA showed that 22% of the nitrogenous bases were cytosine. Calculate the expected percentage of the following nitrogenous bases:

(a) guanine

(b) adenine

(c) thymine.

HL

Q7. Even though cytosine is a pyrimidine and adenine is a purine, they are not complementary to each other. State what prevents these two bases from being complementary to each other within DNA.

Q8. In the experimental work by Hershey and Chase, why was the presence of radioactive phosphorus correlated with DNA and not protein?

Q9. Which statement best describes a single chromosome?

A A length of DNA coiled around eight histone proteins with an additional histone.

B A length of DNA coiled around many groupings of eight histones each having one additional histone.

C Protein structures found within the nucleus of a cell.

D DNA structures found within the nucleus of a cell.

HL end

A1 Practice questions

1. Describe the importance of water to living organisms.

   (Total 5 marks)

2. Draw a labelled diagram showing the structure of three water molecules and how they interact.

   (Total 4 marks)

3. Global warming has changed both the thickness and surface area of sea ice of the Arctic Ocean as well as the Southern Ocean that surrounds Antarctica. Sea ice is highly sensitive to changes in temperature.

   Scientists have calculated a long-term mean for the surface area of sea ice in the Arctic and in the Southern Ocean around Antarctica. This mean value is used as a reference to examine changes in ice extent. The graph shows the variations from this mean (zero line) over a period of time.

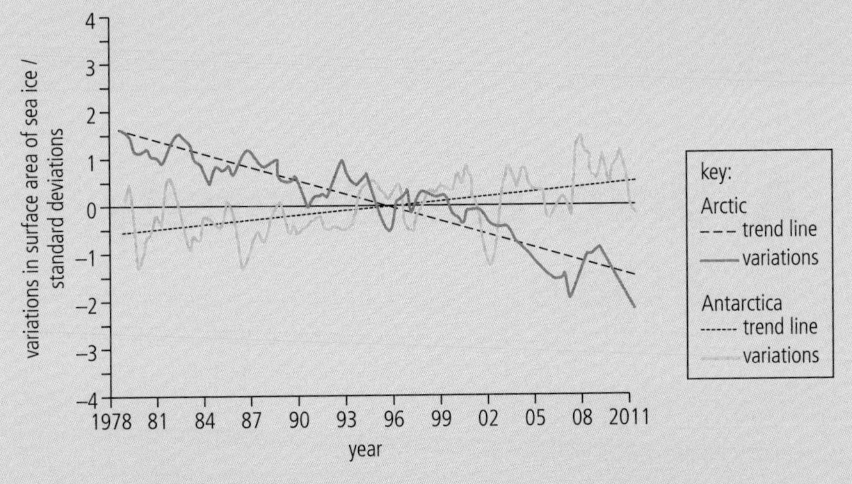

   (a) State the trend in the surface area of sea ice in the Southern Ocean around Antarctica. (1)

   (b) Distinguish between changes in the surface area of sea ice in the Arctic and Antarctica. (2)

   (c) Discuss the data as evidence of global warming. (3)

   (Total 6 marks)

4. Draw a labelled diagram of a section of DNA showing four nucleotides.

   (Total 5 marks)

5. HL Outline the structure and functions of nucleosomes.

   (Total 4 marks)

THEME
A Unity and diversity
2 Cells

Early Earth provided an environment that was extremely inhospitable to life as we know it today. Yet, over long periods of time, conditions changed allowing the building blocks of life to form. Once the building blocks were in place, a slow but steady development occurred. Ultimately, the complexity of life has led to the estimated 8.7 million different species that exist today.

HL

A2.1 Origins of cells

Guiding Questions

What plausible hypothesis could account for the origin of life?

What intermediate stages could there have been between non-living matter and the first living cells?

How exactly did life begin on our planet? How can life come from non-life? How did all the different forms of life present on Earth today come from the first life form? Many, many more questions can be asked. For most of them, we will never have an answer. However, science is about attempting to answer difficult, seemingly impossible questions. We develop hypotheses, we try to imagine and recreate early environmental conditions, we build models, we constantly look for clues in every possible situation, and we gaze at the stars and wonder.

As scientists, we do not stop asking questions and seeking answers. We want to know how we came to be a life form on this third planet from our Sun.

A2.1.1 – The formation of carbon compounds

A2.1.1 – Conditions on early Earth and the pre-biotic formation of carbon compounds
Include the lack of free oxygen and therefore ozone, higher concentrations of carbon dioxide and methane, resulting in higher temperatures and ultraviolet light penetration. The conditions may have caused a variety of carbon compounds to form spontaneously by chemical processes that do not now occur.

Early Earth was quite different from the planet we know now. Our planet is believed to be approximately 4.5 billion years old, and to have been formed from a swirling mass of smaller particles that collided with one another and formed larger masses. Larger masses have more gravity than smaller ones, and so attract other masses; in this way Earth eventually took form. There was no atmosphere on early Earth, allowing various objects from space to impact the surface, causing the planet to grow and the temperature of the surface to rise.

Physicists have provided evidence that the universe was once a single mass. Sometime between 12 and 14 billion years ago, this single mass exploded in a "big bang". Expansion in all directions from this central explosion has been occurring ever since.

Coacervate lab. Full details on how to carry out this activity with a worksheet are available in the eBook.

About 4 billion years ago the number of impacts on the Earth's surface began to decrease. It is thought that the Earth's developing atmosphere at this time was thick with water vapour and other compounds being released by volcanic eruptions. Lightning was most probably a regular occurrence. Early gaseous compounds may have included methane, ammonia, carbon dioxide and hydrogen sulfide, along with many others. These gases, especially carbon dioxide and methane, which were present in higher concentrations than today, allowed ultraviolet light to penetrate the early atmosphere and retained heat, resulting in high surface temperatures. Free oxygen was not present. Had it been, it would have formed a layer of ozone in the upper regions of the atmosphere, thus blocking ultraviolet light penetration and preventing the development of higher temperatures.

Component	Chemical formula
Methane	CH_4
Ammonia	NH_3
Water vapour	H_2O
Carbon dioxide	CO_2
Hydrogen sulfide	H_2S
Hydrogen	H_2
Nitrogen	N_2

Possible components of Earth's early atmosphere

It is thought that the Earth's early atmospheric components coupled with high surface temperatures and lightning, followed by a gradual cooling, resulted in the spontaneous formation of many carbon compounds. This spontaneous formation occurred during the unique conditions early in the geological history of our planet, and is not evident today.

A2.1.2 – Functions of life

A2.1.2 – **Cells as the smallest units of self-sustaining life**
Discuss the differences between something that is living and something that is non-living. Include reasons that viruses are considered to be non-living.

Today, all organisms exist in either a unicellular or a multicellular form. Interestingly, all organisms, whether unicellular or multicellular, carry out all the functions of life. These functions include:

- metabolism
- growth
- reproduction
- response
- homeostasis
- nutrition
- excretion.

These functions act together to produce a viable living unit. **Metabolism** includes all the chemical reactions that occur within the organism. As a result of metabolism, cells can convert energy from one form into another. **Growth** may be limited but is always present. **Reproduction** involves hereditary molecules that can be passed to offspring. **Responding** to stimuli in the environment is essential for an organism to survive. These responses allow an organism to adapt to its environment. **Homeostasis** refers to the maintenance of a constant internal environment. For example, an organism may have to control fluctuating temperature and acid–base levels to create a constant internal environment. Using a source of compounds with many chemical bonds that can be broken down to provide an organism with the energy necessary to maintain life is the basis of **nutrition. Excretion** is essential to life because it enables those chemical compounds that an organism cannot use or that may be toxic or harmful to be released from the organism's system.

The functions of life manifest in different ways in different types of organisms. Because of this variety, you may come across different terms for the same function when you read different sources. However, all organisms maintain the same general functions that allow them to live.

For what reasons is heredity an essential feature of living things?

Viruses are not considered to be living organisms. They cannot carry out the functions of life on their own. However, they can use cells to perpetuate themselves.

Cell theory

Many scientists, over several hundred years, have contributed to the three main principles of today's **cell theory**:

- all organisms are composed of one or more cells
- cells are the smallest units of life
- all cells come from pre-existing cells.

Cell theory has a solid foundation, largely because of the use of the microscope. Robert Hooke first described cells in 1665, after looking at cork through a self-built microscope. A few years later Antonie van Leeuwenhoek observed the first living cells and referred to them as "animalcules", meaning little animals. In 1838, botanist Matthias Schleiden stated that plants are made of "independent, separate beings" called cells. One year later, Matthias Schleiden made a similar statement about animals.

The second principle continues to gain support today, because no one has been able to find any living entity that is not made of at least one cell.

Many famous scientists, such as Louis Pasteur in the 1880s, have performed experiments to support the third principle. After sterilizing chicken broth (soup) by boiling it, Pasteur showed that living organisms would not "spontaneously" reappear. Only after exposure to pre-existing cells was life able to re-establish itself in the sterilized chicken broth.

Nature of Science

Theories are developed after the accumulation of data via observation and/or experimentation. As with most theories, the current cell theory is not without problems. A key characteristic of a scientist is a sceptical attitude towards theoretical claims. To overcome or validate this scepticism, more evidence obtained by observation or experimentation is essential. Whenever possible, controlled experiments are needed to verify or refute theories. Controlled experiments include a control group and a variable group(s); the groups are kept under similar conditions, apart from the factor that is being tested. The factor being tested is referred to as the independent variable. The other factors, the dependent variable, are measured or described using quantitative or qualitative data. Relatively recent findings that have raised questions about the cell theory include observations of striated muscle, giant algae and aseptate fungal hyphae. Sometimes theories will be abandoned completely because of conflicting evidence. When this happens a **paradigm shift** is said to have occurred.

A2.1.3 – Evolution of the cell

A2.1.3 – Challenge of explaining the spontaneous origin of cells
Cells are highly complex structures that can currently only be produced by division of pre-existing cells. Students should be aware that catalysis, self-replication of molecules, self-assembly and the emergence of compartmentalization were necessary requirements for the evolution of the first cells.
NOS: Students should appreciate that claims in science, including hypotheses and theories, must be testable. In some cases, scientists have to struggle with hypotheses that are difficult to test. In this case the exact conditions on pre-biotic Earth cannot be replicated and the first protocells did not fossilize.

Cells are complex structures that carry out the functions of life. The third principle of cell theory states that cells may only come from other pre-existing cells. We have not found an exception to this principle on present-day Earth. Protocells or simple cells had to appear in a pre-biotic (non-life) environment so that the more complex cells we see today could form. Scientists hypothesize that a series of chemical and physical processes have to occur for a cell to evolve. These processes or stages are:

1. The synthesis of small carbon compounds from abiotic (non-living) molecules, such as demonstrated in the Miller–Urey experiment.
2. Small organic molecules joining to form large-chain molecules called **polymers**.
3. Polymers becoming contained by membranes, creating a protective homeostatic environment around the polymers, separate from their surroundings.
4. The development of self-replicating molecules so that inheritance and control can occur.

Nature of Science

Hypotheses about the origin of the first cells are frustrating for scientists, because they are not testable. We do not know the exact conditions of early Earth, so we cannot replicate them for experimentation. Nor have the first protocells been preserved in the fossil record.

The evolution of early cells required the emergence of compartment-alization and self-replication of molecules, and the process would have been helped by the presence of enzyme-like components.

Based on the observation of present-day chemical reactions, small organic molecules can join to form larger chain molecules more rapidly if a compound called an **enzyme** is present. Enzymes are proteins that act as biological catalysts and accelerate chemical reactions. Enzymes are carbon compounds, which means they always contain the elements carbon and hydrogen, usually along with other elements. It is doubtful that enzymes existed when protocells first formed. Scientists have been able to produce polymers from simple compounds by exposing them to hot sand, clay or rock. This vapourizes water from the simple compounds, and molecule chains are formed, suggesting that polymers could have formed on early Earth even if no enzymes were present.

A2.1.4 – Inorganic to carbon compounds

A2.1.4 – Evidence for the origin of carbon compounds
Evaluate the Miller–Urey experiment.

Early biologists and biochemists were puzzled about how simple atmospheric components could be converted or evolve into the more complex compounds necessary for the origin of life. **Inorganic compounds** do not usually contain carbon (an exception is carbon dioxide, which is considered to be an inorganic compound). **Carbon compounds** (often called organic compounds) contain carbon and can be quite complex. It is these more complex compounds or molecules that make life possible, thus the element carbon can be said to be the keystone element for life on Earth.

In 1953, Stanley Miller and Harold Urey conducted an experiment to simulate the conditions thought to be present on early Earth, and to determine if these gases could interact to produce the first stage in the evolution of life. Most scientists believe this first stage will result in simple, non-living organic molecules. The apparatus designed for this experiment is shown in Figure 1.

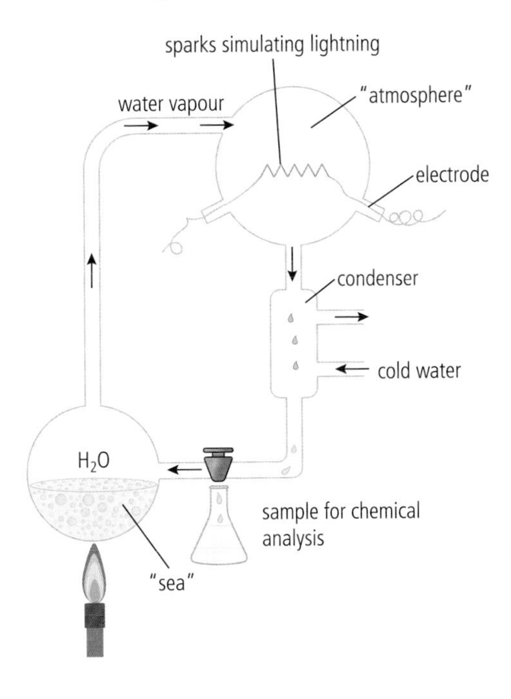

A2.1 Figure 1 The Miller–Urey apparatus used to simulate the conditions on early Earth.

The important features of Figure 1 are:

- the apparatus is initially charged with the simple inorganic compounds CH_4, NH_3 and H_2, representing Earth's early atmosphere
- heat is used to produce water vapour, which rises to the chamber containing the simple inorganic compounds
- two electrodes in this chamber produce 7,500 volts at 30 amps of electricity, representing the lightning that existed on early Earth
- cold water flows into the condenser to allow condensation of gaseous compounds from the chamber
- a sample is collected in the collecting device for chemical analysis.

An organism is an individual living thing consisting of one or more cells. Organisms demonstrate the characteristics of life.

During the experiment, Miller identified several simple organic molecules in the collecting device that are known to exist in **organisms**. These molecules included long chains of hydrogen and carbon, called **hydrocarbons.** He also found some of the building blocks of proteins called **amino acids.** Proteins are major biochemicals essential to all organisms, and are explained in detail in Chapter B1.2. The fluid collected during the experiment suggested to many scientists that life originated in a **primordial soup**, a water-based sea of simple organic molecules.

Nature of Science

The Miller–Urey experiment uses a model to demonstrate how more complex molecules could evolve from simple compounds. Models are much simpler than the complex environments in which actual natural phenomena occur. However, they allow observation of possible reactions that may have occurred on, for example, early Earth. It is important to understand that models have limitations, and these limitations must be considered when making any conclusions or statements.

TOK

In some instances, natural sciences rely on or make assumptions that are not actually provable. The Miller–Urey experiment includes some such assumptions. This would be a wonderful time to discuss with your classmates the value and the limitations of the assumptions made in this experiment. Some concerns to address in this discussion include:

1. What role do models play in the acquisition of knowledge?
2. Do the natural sciences rely on any assumptions that are themselves unprovable by science?
3. What knowledge, if any, is likely to always remain beyond the capabilities of science to investigate or verify?

Many laboratories around the world have reproduced the Miller–Urey experiment with comparable results. Similar experiments with many new variables have also been conducted, resulting in several types and amounts of larger carbon compounds. However, it is important to note that, because we have no definite knowledge of the early Earth's atmosphere, the model/experiment produced by Miller–Urey and other similar experiments do not produce reliable evidence of the first steps in the development of life on our planet. As a result of the unprovable assumptions of this experiment, several other hypotheses have been suggested to explain how more complex carbon compounds came to exist on our planet. One of these alternative hypotheses involves the introduction of carbon compounds by comets and meteorites, as they have struck Earth throughout its history.

Some scientists believe the gases used in the Miller–Urey experiment were not likely to have been present on early Earth. For these scientists, the Miller–Urey results are invalid. Instead, they believe the first atmosphere of our planet formed slowly over extended periods of time as a result of the release of gases from volcanoes. If this is accurate, the gases at this early stage in the Earth's development would have originated from the planet's mantle, an intermediate layer surrounded by the exterior crust and resting on the core. These same scientists believe the chemical properties of the mantle are the same today as they were in the past, and volcanic gases today do not contain methane or ammonia. If this is true, this hypothesis results in an early atmosphere that had the following composition:

- water vapour
- carbon dioxide
- sulfur dioxide
- small amounts of carbon monoxide and hydrogen sulfide.

A hypothesis is an explanation based on reasoning that can be tested by gathering more data. Often, this data is gathered by experimentation.

These atmospheric gases would produce a **non-reducing environment** that, because of the lack of adequate hydrogen for bonding, would not result in simple carbon compounds. The compounds used in the Miller–Urey experiment produce a **reducing environment** favourable for the development of carbon compounds.

Another concern with the Miller–Urey experiment and the primordial soup theory is that, when water is present with proteins, the proteins break down into individual amino acids. In a water environment, amino acids are not observed joining into more complex structures resulting in proteins. This is in conflict with the belief that complex molecules could form in a primordial soup environment, thus allowing the eventual development of life.

Because of these and other concerns, many scientists have questioned the credibility of the Miller–Urey experiment in proposing a means for the first life first to form on Earth. Scientists are now pursuing hydrothermal vents as a possible environment for the origin of cells and life. These hypotheses are discussed Section A2.1.9.

A reducing environment or atmosphere is one in which no free oxygen is present. However, hydrogen is present. If the early Earth's atmosphere was non-reducing, there would have been no hydrogen to allow the production of the first carbon compounds, composed largely of carbon and hydrogen.

A2.1.5 – The formation of vesicles

A2.1.5 – Spontaneous formation of vesicles by coalescence of fatty acids into spherical bilayers
Formation of a membrane-bound compartment is needed to allow internal chemistry to become different from that outside the compartment.

Membrane formation is thought to have been essential to the origin of the first cells on Earth. A membrane provides a barrier or boundary between the inside of the cell and the surrounding environment. This barrier allows the regulation and maintenance of activities within the cell.

Present-day cell membranes are complex in structure and include many different components to allow regulation and maintenance within the cell. This is quite different from the membranes of protocells, which are thought to have been composed of only a single group of compounds.

One of the carbon compounds thought to have been present on early Earth was a simple group known as fatty acids. These fatty acids, when present in water, display a polarity, with one end having water-attracting, **hydrophilic**, properties, and the other end having water-repelling, **hydrophobic**, properties. When large numbers of fatty acids are placed in water, they tend to organize themselves into small, cell-sized double-layer bubbles often referred to as **vesicles**.

These vesicles are often referred to as liposomes, which means "fat bodies". They are highly likely to have been the first protocells.

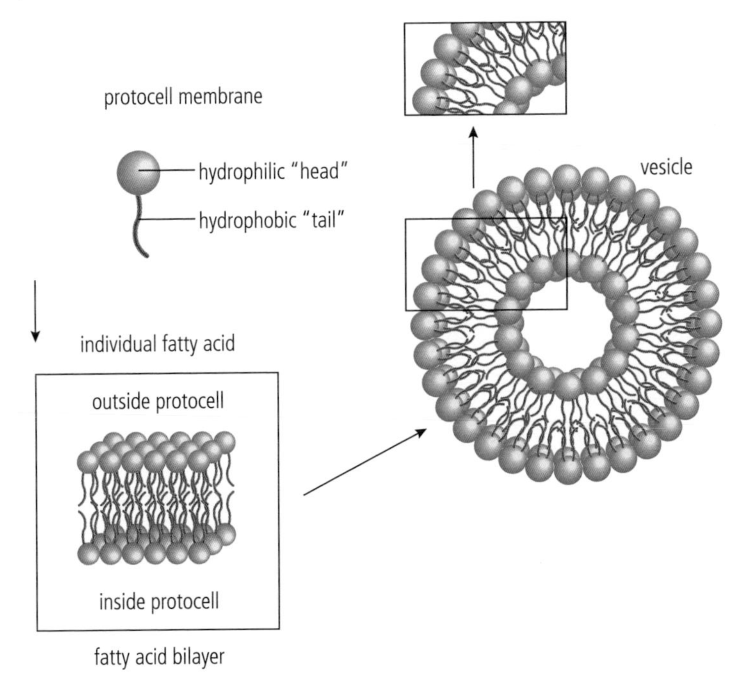

Fatty acids and the development of a protocell bilayer membrane.

Vesicles produced in the laboratory can conduct several key processes, including the ability to engulf other types of organic molecules, grow, and even replicate themselves.

The development of a boundary around a vesicle is called **compartmentalization** and was necessary for the formation of the first cells. As cells continued to progress in complexity, compartmentalization became evident within this outer boundary, allowing specialization of functions in different regions inside the outer protective fatty acid bilayer.

A2.1.6 – RNA as the first genetic material

A2.1.6 – RNA as a presumed first genetic material
RNA can be replicated and has some catalytic activity so it may have acted initially as both the genetic material and the enzymes of the earliest cells. Ribozymes in the ribosome are still used to catalyse peptide bond formation during protein synthesis.

The final process in the formation of the cell is the development of molecules that can allow inheritance and control the cell's functions of life. One hypothesis is that this molecule was **ribonucleic acid**, often referred to as **RNA**.

Short RNA molecules display some interesting properties.

- RNA can assemble spontaneously from simpler organic molecules called nucleotides.
- RNA can form copies of itself, thus acting as a type of genetic material.
- RNA demonstrates the ability to control chemical reactions, thus acting in an enzymatic role.

These properties of RNA would have allowed all the functions necessary for the formation of early cells. A type of RNA called **ribozymes** is highly active today in catalysing activities, such as allowing faster development of peptide bonds. These types of bonds are essential for the formation of proteins. The fact that RNA can spontaneously form when nucleotides are present enables its appearance in early cell formation. The genetic role of RNA is possible because of the varying types of nucleotides present capable of combining to form a larger molecule. This larger molecule of RNA could then carry a code allowing inheritance of specific characteristics.

RNA is a single-chain molecule, while DNA is a double-chain molecule. Both compounds are polymers of nucleotides. However, there are differences in the nucleotides that these two genetic compounds are composed from.

RNA is much simpler than **deoxyribonucleic acid**, also known as **DNA**, which is the predominant genetic material of present-day organisms. Because of its relative simplicity and its ability to form spontaneously, RNA, not DNA, is hypothesized to have been the first genetic and controlling compound of a living cell.

An overview of the major stages in the origin of life is represented by the following:

early Earth → abiotic chemical compounds → small organic molecules → polymers of organic molecules → protocell → cell

A2.1.7 – Evidence for a last universal common ancestor

A2.1.7 – Evidence for a last universal common ancestor
Include the universal genetic code and shared genes across all organisms. Include the likelihood of other forms of life having evolved but becoming extinct due to competition from the last universal common ancestor (LUCA) and descendants of LUCA.

Evidence for the existence of a **last universal common ancestor (LUCA)** for all life on Earth includes:

- a universal genetic code carried by DNA and shared by all cells
- over 300 genes or sections of DNA common to all cells
- the same building blocks for both DNA and RNA in all cells
- common molecular processes within all cells, including the replication of DNA molecules and the production of proteins
- similar transport mechanisms for cellular materials in and out of cells, as well as within cells.

It is hypothesized that other forms of life, demonstrating distinctive characteristics other than these, have evolved over the past 3.5 billion years. However, they are not present today as a result of unsuccessful competition with the LUCA and its descendants.

A2.1.8 – Dating the first living cells and LUCA

A2.1.8 – Approaches used to estimate dates of the first living cells and the last universal common ancestor
Students should develop an appreciation of the immense length of time over which life has been evolving on Earth.

Charles Darwin in his theory of evolution by natural selection utilized the concept of common ancestry. He believed all life on Earth that has existed and does exist can be traced back to a single ancestor, the LUCA.

As mentioned previously, the Earth is thought to be approximately 4.5 billion years old. The earliest evidence of life on Earth comes from **fossils**. Fossils are the remains and/or traces of past life. Most fossils originate from the hard parts of past organisms, such as shells, bones and teeth. However, some fossils can represent the remains of trails and footprints, or even the impressions of soft body parts. Using various dating techniques, it is estimated that the earliest life occurred on Earth about 3.5 billion years ago.

One of the most accurate means of dating fossils involves radiometric techniques, which are based on the **half-life**, the length of time it takes for half of a radioactive isotope to change into another stable element. An isotope is an unstable form of an element. Fossils contain isotopes of elements that accumulated when the organisms were alive. Radioactive isotopes are taken up by organisms at constant rates. Therefore, by measuring the amount of an isotope in a fossil and comparing it with the amount taken up when the organism was alive, we can determine an age for the fossil. This is known as **absolute dating** of a fossil's age, because the half-life of each isotope is not known to vary. Some radioactive isotopes used in this procedure and their half-lives are shown in Table 1.

Palaeontologists study the fossil record to learn about the history of life on our planet as well as the past climate and environment. Other means of dating previous life have also been developed. One recently developed technique involves genomic or genetic analysis. Genomic dating compares the amount of genetic difference between living organisms and computes an age based on rates of genetic mutation over time. However, it is quite difficult to find DNA in fossils, making this technique unfavourable for dating older life forms.

Isotope	Half-life
Uranium-238	4.5 billion years
Uranium-235	700 million years
Carbon-14	5,730 years
Radium-226	1,600 years

A2.1 Table 1 The half-life of various radioactive isotopes

Relative dating of fossils is not as exact as absolute dating. It utilizes the sediment layers of the Earth and **index fossils.** Sediments are particles of eroded and weathered rock and soil that may form layers or strata. These sediment layers occur in sequence, with the oldest layers at the bottom. This layering sequence allows a relative means of working out the age of a fossil. It is important to note, however, when using this method of dating, that geological processes may disturb the original sequence of layers, thus potentially providing inaccurate fossil ages. Index fossils are fossils of the same age that are found in strata in different parts of the world. Therefore, all strata around the world that contain the same fossil(s) must be of the same age.

One way to get an impression of the length of time life has been on our planet is to use a 24-hour time span to represent the 4.5-billion-year existence of Earth, as shown in Figure 2.

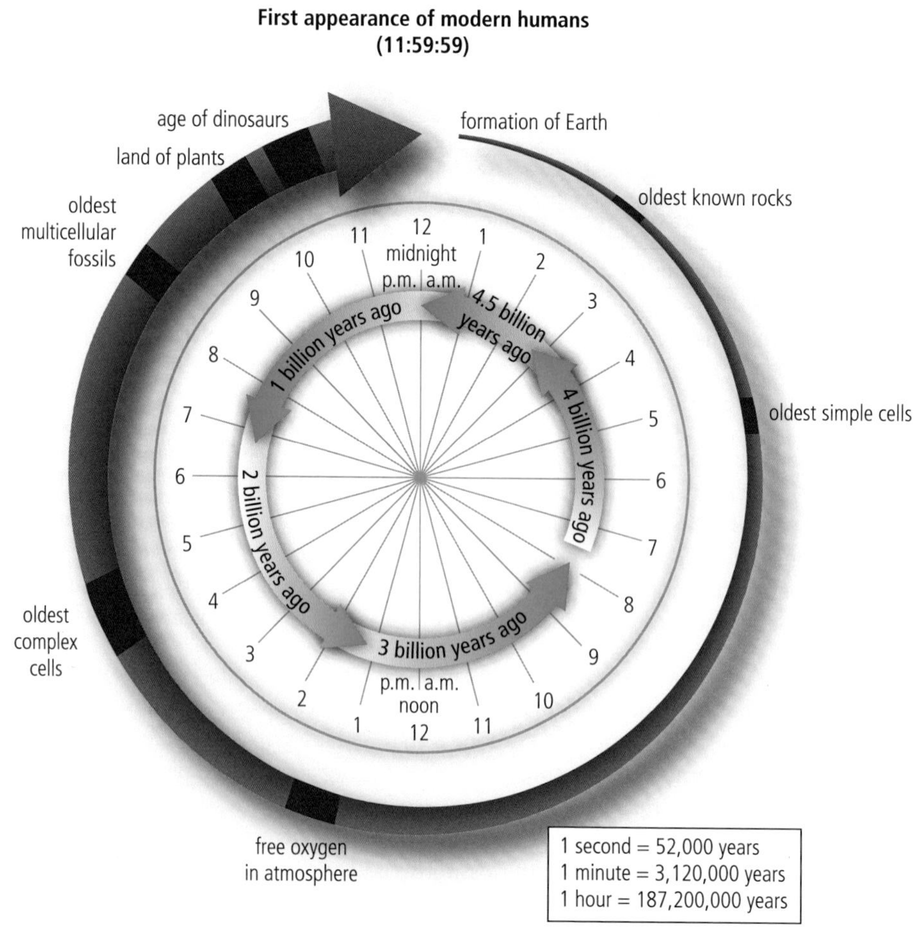

A2.1 Figure 2 Earth's past represented in a 24-hour time span.

Note in Figure 2 the time it took for the first simple cells to evolve or change into the first complex cells. That modern humans do not appear in this 24-hour representation of the Earth's history until 1 second before midnight indicates the immense period of time life has been evolving on our planet, and the relatively short period of time humans have been part of it. Many scientists believe the LUCA existed about 3.5 billion years ago, but it is important to remember that the LUCA is not thought to have been the first life on Earth, but rather the latest that is ancestral to all current existing life (see Figure 3).

What is needed for structures to be able to evolve by natural selection?

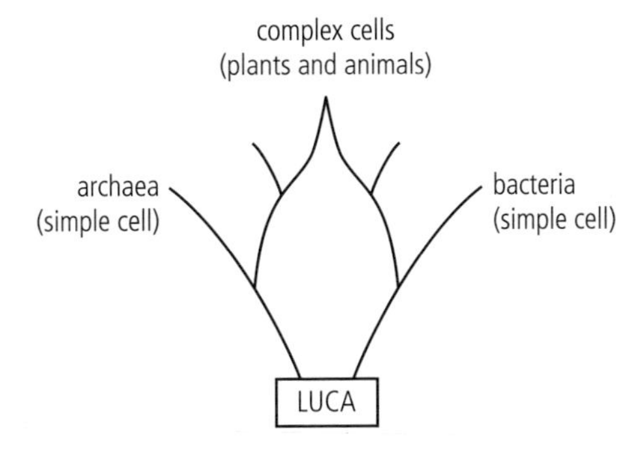

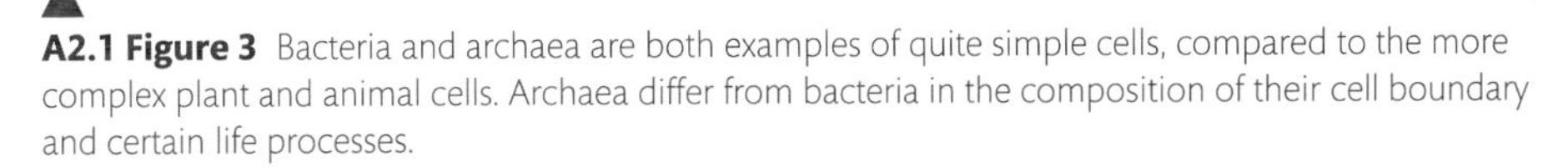

A2.1 Figure 3 Bacteria and archaea are both examples of quite simple cells, compared to the more complex plant and animal cells. Archaea differ from bacteria in the composition of their cell boundary and certain life processes.

A2.1.9 – Hydrothermal vents and the evolution of the LUCA

A2.1.9 – Evidence for the evolution of the last universal common ancestor in the vicinity of hydrothermal vents
Include fossilized evidence of life from ancient seafloor hydrothermal vent precipitates and evidence of conserved sequences from genomic analysis.

One hypothesis suggests that the location for the origin of life on Earth is around **hydrothermal vents**, places where hot water emanates from beneath the ocean floor. The first deep-sea hydrothermal vent was observed in 1977. Such a structure forms when cracks in the crust of the seabed expose seawater to rocks below, which are heated by magma. The hot water rises and picks up countless minerals along the way.

Today, we know about entire communities living around these vents, including creatures never seen before 1977, such as metre-long white-and-red tube worms that absorb the minerals from the water and transfer them to symbiotic bacteria. The bacteria then make food from the minerals, and this food nourishes the tube worms. The discovery of these communities disproves the idea that the bottom of the ocean must be lifeless because there is no sunlight. It also gives credibility to the hypothesis that the earliest forms of life could have formed deep in the ocean around hydrothermal vents.

Hydrothermal vents are sometimes referred to as "black smokers" because the water coming out of them contains so many dark minerals that it looks like smoke.

Hydrothermal vents such as this one are major sites of study. They provide a warm, nutrient-rich environment with the necessary gases, energy and chemical compounds for life to evolve.

Evidence to support the possible appearance of the LUCA at hydrothermal vents include:

- some of the world's oldest fossilized traces or precipitates have originated at these vents
- the commonality of genetic sequences in the organisms near these vents, indicating a likely single ancestor
- the presence of a mineral-rich environment with both the acidic and basic fluids that are necessary for chemical reactions
- the presence of both hydrogen and carbon dioxide at these locations, resulting in the reducing environment essential for carbon compound formation.

A reducing environment has been observed to support an increasing number of bonds between carbon and hydrogen.

Many questions still remain about the how and where of the origin of life on our planet. Various hypotheses offering explanations will continue to be presented in future research. It is important to approach each with an open mind, and rely on experimentation to verify supporting evidence.

Guiding Question revisited

What plausible hypothesis could account for the origin of life?

In this chapter we have presented various hypotheses concerning the origin of life on Earth, some of which may be more plausible than others. Key ideas include:

- it is extremely difficult to explain the spontaneous origin of life because of our lack of knowledge of what the early Earth conditions actually were
- the Miller–Urey experiment resulted in the primordial soup theory for the first protocells
- there is evidence to support a theory that life originated near hydrothermal vents on the ocean floor.

Guiding Question revisited

What intermediate stages could there have been between non-living matter and the first living cells?

In this chapter we have looked at various requirements for the development of life. Key ideas include:

- high concentrations of carbon dioxide and methane were present without free oxygen on early Earth
- the processes that formed the first life on Earth no longer occur today because the present-day environment is so different
- all organisms are composed of a cell or of multiple cells and have common functions
- the formation of a membrane was essential for the formation of the protocell and the more complex cells that followed
- RNA has characteristics that could mean it had the genetic material and control of chemical reactions essential for the first cells
- life has been evolving on Earth for an extremely long period of time
- the last universal common ancestor (LUCA) out-competed other forms of life with different life characteristics, resulting in all the forms of life observed on our planet today.

Exercises

Q1. Define what a radioisotope's half-life is.

Q2. State three conditions of early Earth that Miller and Urey tested in their model, leading to the primordial soup hypothesis.

Q3. Describe the importance of membranes in allowing survival and further development of the first cells.

Q4. Describe how RNA could have been essential to early cells.

Q5. Explain the position of the last universal common ancestor on the modern tree of life.

HL end

A2.2 Cell structure

Guiding Questions

What are the features common to all cells and the features that differ?

How is microscopy used to investigate cell structure?

In the 1660s, Antonie van Leeuwenhoek became interested in the early microscopes being developed by Robert Hooke. The Dutch businessman and scientist used mostly blown-glass lenses to produce his own microscopes, which opened a completely new world to all. His powers of observation led to the first recorded descriptions of bacteria and protozoa. From van Leeuwenhoek's work the science of microbiology took form.

Countless improvements in microscopy since these simple beginnings have led to an understanding of the features common to all cells. We have also learned of the tremendous diversity that exists not only in cells but in all life.

A2.2.1 – Cells and the functions of life

A2.2.1 – Cells as the basic structural unit of all living organisms
NOS: Students should be aware that deductive reason can be used to generate predictions from theories. Based on cell theory, a newly discovered organism can be predicted to consist of one or more cells.

Whether organisms are extremely small or extremely large, understanding their smallest functional units is imperative. These units are known as cells. Organisms range in size from a single cell upwards to trillions of cells. To better understand all the organisms around us we must study their cells.

Cytology is the branch of biology that studies all facets of the cell. As our understanding of the cell has increased, so has our ability to understand all forms of life and diseases that occur on planet Earth. This area of research is extremely active in laboratories all over the world.

HL Chapter A2.1 discusses the cell theory and the functions exhibited by all forms of life.

The cell theory states:

- all organisms are composed of one or more cells
- cells are the smallest units of life
- all cells come from pre-existing cells.

What are the features of a compelling theory?

Nature of Science

Inductive reasoning utilizes specific observations to arrive at broader generalizations. Deductive reasoning works in the opposite direction. It allows you to make an inference using widely accepted facts or premises. Using deductive reasoning, a newly discovered organism can be predicted to carry out the functions of life and demonstrate the principles of cell theory.

A2.2.2 – Cells and the microscope

A2.2.2 – Microscopy skills

Application of skills: Students should have experience of making temporary mounts of cells and tissues, staining, measuring sizes using an eyepiece graticule, focusing with coarse and fine adjustments, calculating actual size and magnification, producing a scale bar and taking photographs.

NOS: Students should appreciate that measurement using instruments is a form of quantitative observation.

Cells are made up of many different subunits. These subunits are often of a particular size, but most are microscopically small.

Unit	Equivalent measurement
1 metre (m)	100 cm = 1,000 mm
1 centimetre (cm)	10^{-2} m (0.01 m)
1 millimetre (mm)	10^{-3} m (0.001 m)
1 micrometre (μm)	10^{-6} m (0.000001 m)
1 nanometre (nm)	10^{-9} m (0.000000001 m)

Commonly used microscope metric equivalents

Microscopes with a high **magnification** and **resolution** are needed to observe cells and especially their subunits. Magnification is the increase in an object's image size compared to its actual size. Pictures or drawings of an image from a microscope include the number of times larger than the actual object they are, for example 500× or 100,000×.

Resolution refers to the minimal distance between two points or objects at which they can still be distinguished as two. As the resolution of a microscope increases, the greater the detail that microscope will reveal. Some like to explain resolution in terms of clarity, with greater resolution providing greater clarity.

Light microscopes use light, passing through living or dead specimens, to form an image. Stains may be used to improve the visibility of structures. **Electron microscopes (EMs)** provide the greatest magnification (over 100,000×) and resolution. These use electrons passing through a specimen to form an image.

A light microscope (above) and an electron microscope (below).

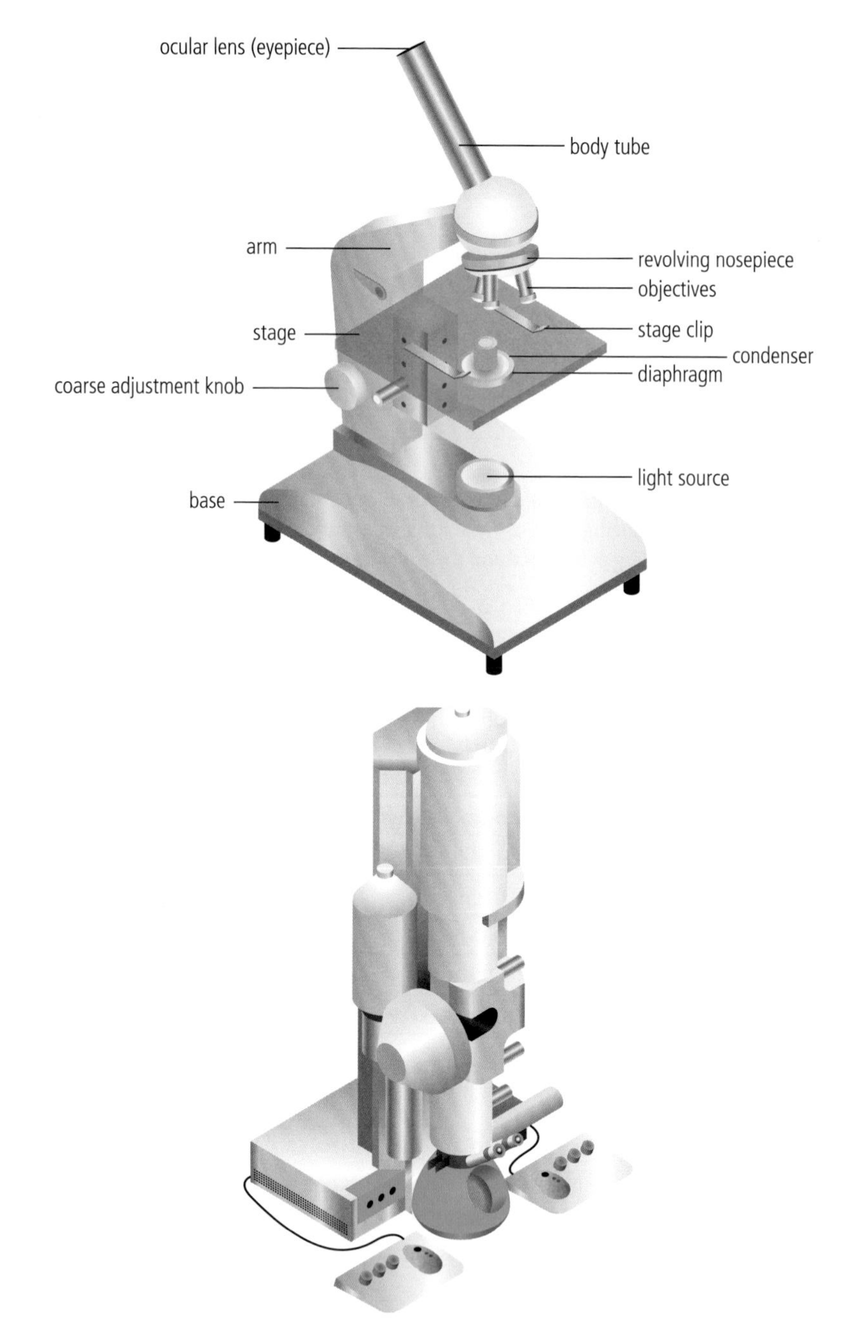

Most cells can be up to 100 micrometres (100 μm) in size. Organelles are up to 10 μm in size. Bacteria are between 1 and 10 μm in size. Viruses are up to 100 nanometres (nm) in size. Cell membranes are 10 nm thick, while molecules are about 1 nm in size. All these structures are three-dimensional.

Light microscope	Electron microscope
Inexpensive to purchase and operate	Expensive to purchase and operate
Simple and easy specimen preparation	Complex and lengthy specimen preparation
Magnifies up to 2,000×	Magnifies over 500,000×
Specimens may be living or dead	Specimens are dead, and must be fixed in a plastic material

A comparison of the light microscope and the electron microscope

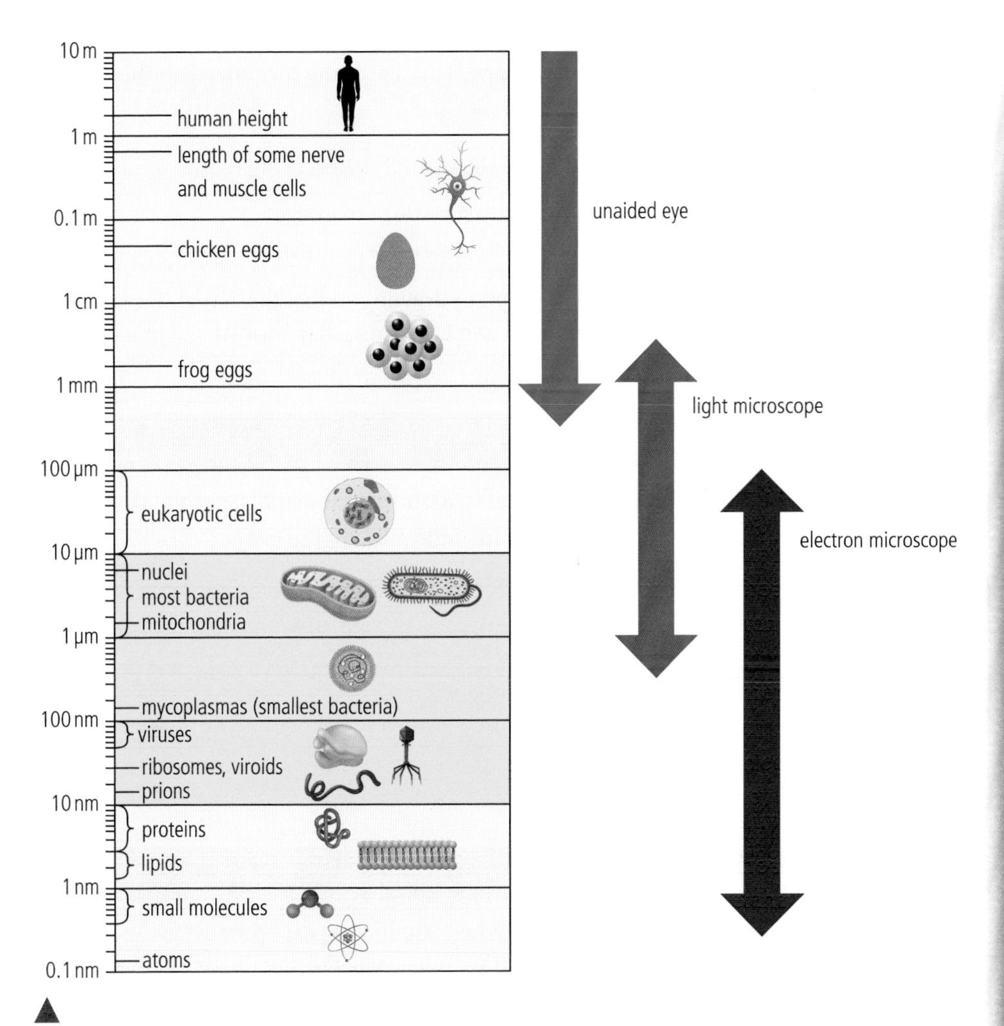

▲ A representation of what can be used to visualize various structures important in biology.

Cells and their subunits are so small they are hard to visualize, so it is important to appreciate relative sizes. Cells are relatively large, and then in decreasing order of size are:

- organelles
- bacteria (some bacteria cells are as large as organelles)
- viruses
- membranes
- molecules.

If you want to calculate the actual size of a specimen seen with a microscope, you need to know the diameter of the microscope's **field of vision**, also called the **field of view**. The field of vision is the total area visible when looking through a microscope's **ocular** or eyepiece, and the diameter can be calculated using special **micrometers**. There are two general types of micrometers: ocular and stage. The **ocular micrometer**, also called a **graticule**, is located in the eyepiece and is engraved with equal units. It is important to note that the units on this micrometer are arbitrary. They are calibrated using a **stage micrometer**. This calibration is often done using a simple ruler or a special slide with defined units, usually millimetres. By comparing the units of the graticule to the known unit size of the stage micrometer, you can determine the size of the image being examined. The ocular micrometer has to be calibrated in this way with each objective power of the microscope. The size of the specimen can then be calculated.

A simple formula can be used to calculate the magnification being used:

$$\text{magnification} = \frac{\text{measured size of image}}{\text{actual size of specimen}}$$

Scale bars are often used with a micrograph or drawing so that the actual size can be determined. Scale bars and magnification will be addressed in more detail in a later practical activity.

Worked example

The length of an image you are looking at is 50 mm. If the actual length of the subject of the image is 5 µm, what is the magnification of the image?

Solution

Magnification = 50 mm/5 µm = 50,000 µm/5 µm = 10,000×

Or

Magnification = 50 mm/5 µm = 50×10^{-3} m/1×10^{-6} m = 10,000×

SKILLS

Use of a light microscope to investigate cells and cell structure sizes. Full details of how to carry out this activity with a worksheet are available in the eBook.

Nature of Science

Scientists need to accumulate data when conducting experiments using scientific methods. Two types of data can be collected. Qualitative data is non-numerical but descriptive. It includes attributes such as colour, presence of a structure or feature (or not) or sex. Quantitative data involves numerical values collected by a specific type of instrument. Examples of quantitative data are mass measured by a laboratory balance or length measured by a ruler. These two types of data, when collected properly, allow meaningful conclusions to be made.

A2.2.3 – Advanced microscopy

A2.2.3 – Developments in microscopy
Include the advantages of electron microscopy, freeze fracture, cryogenic electron microscopy, and the use of fluorescent stains and immunofluorescence in light microscopy.

The microscope has undergone tremendous advancement since the one used by Robert Hooke in 1665. Early microscopes were pivotal in the development of the cell theory, even though they were extremely simple by today's standards. Scientists have also perfected many new techniques in the preparation of materials for study involving the microscope. In this section we will examine some of these developments and techniques.

One significant advancement in microscopy was the development of the electron microscope (EM). The EM utilizes a beam of electrons rather than a beam of light, which the light microscope uses. Electrons have a much shorter wavelength than light. The benefits of the shorter wavelength include a 1,000 times greater resolving power than the light microscope, and the ability to magnify objects over 500,000× compared to the maximum magnification of 2,000× for a light microscope.

There are two general types of EMs – the **scanning electron microscope (SEM)** and **transmission electron microscope (TEM)**. The SEM uses a beam of electrons to scan the surface of a specimen. The TEM aims a beam of electrons through a very thin section of a specimen, allowing its inner structure to be viewed. Both SEM and TEM images provide essential information in cytology investigations.

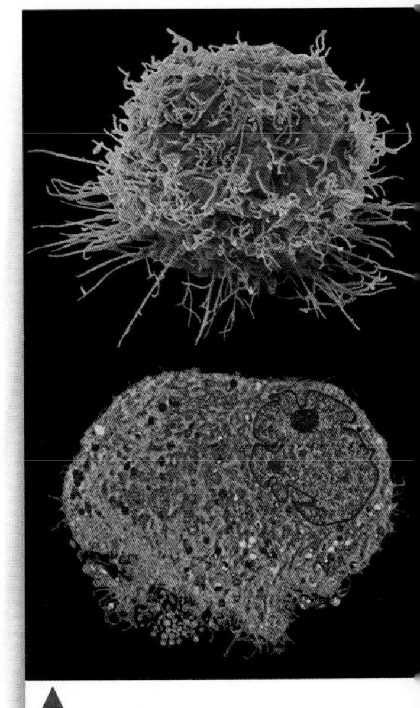

▲ The top image of a leukaemia cell is from a scanning electron microscope (SEM). The bottom image is of the same cell but from a transmission electron microscope (TEM).

Techniques employed when working with an EM include **freeze fracture** and **cryogenic electron microscopy**. Freeze fracture is a process of preparing a sample for observation with an EM. It involves the rapid freezing of a biological specimen followed by physically breaking the specimen apart (fracturing). This technique reveals a plane through the sample that can then be examined. Our understanding of the cell membrane has been greatly enhanced using this technique.

Cryogenic electron microscopy is a recent advancement in EM that has furthered our knowledge of structural biology. It enables an image to be formed using computer enhancement that shows the three-dimensional framework of proteins involved with the function of a cell. It utilizes low temperatures to freeze specimens in ice. Many advances in our understanding of virus composition and structure, cell membrane components and their arrangement, cellular protein synthesis, and even hereditary expression and regulation, are the result of using this technique. New applications of cryogenic electron microscopy are being developed at an amazing pace with enlightening results.

It is obvious that the EM offers tremendous advantages over the light microscope in the study of cells and their structures. However, it is important to note that EMs are expensive, require extensive training to operate, and involve non-living specimens embedded in some sort of matrix such as plastic. Often, structural features called **artefacts** are seen in the pictures produced by an EM. These artefacts do not actually exist in the cell but are produced during the preparation of the samples for an EM.

When living samples are to be studied, the light microscope must be used. Two preparation techniques developed recently for the study of cells using light microscopy involve the use of **fluorescent stains** and **immunofluorescence**. Fluorescent stains are substances or dyes that combine with specific cellular components. When these living samples are then irradiated with ultraviolet or violet-blue light, the parts that accepted the dye will fluoresce. When fluorescence occurs, assorted colours are produced, allowing more detailed visibility. Immunofluorescence also allows greater visibility of living tissue. Immunofluorescence involves antibodies that have dyes already combined with them. Specific antibodies combined with unique coloured dyes recognize and combine with target molecules. This allows the target, usually a protein, to be detected. This technique is often used to detect viral proteins that have infected cells.

Both fluorescent staining and immunofluorescence have been extensively used in the study of severe acute respiratory syndrome coronavirus 2 (SARS-CoV-2) and related viruses. They have provided valuable information about the life cycle of the virus as it attacks living cells.

Fluorescence-based methods have recently been developed to target RNA. We are now able to visualize single RNA molecules within single cells and viruses.

The light microscope has gone through many developments to improve its ability to produce images of living cells and their internal structures. One area of development has involved the part of the microscope called the **condenser**. The condenser is located between the stage and the light source. It possesses a lens that directs light rays from the light source through the specimen. From the specimen, the light rays pass through the objective lens to the ocular lens, where the image is viewed by the researcher. By changing the capabilities of the condenser, we now have some microscope types with unique and valuable features.

Type of microscope	Feature
Brightfield	Visible light is used; the specimen is viewed against a light background; it is the most common and easy to use light microscope
Darkfield	A special opaque lens is used in the condenser, that blocks direct light from entering the specimen; the specimen appears light against a dark background
Phase-contrast	A special condenser with a circular diaphragm and a modified objective lens are used to reveal detailed images of specimens without staining

Types of light microscope

Each advance of the microscope, whether light or electron, leads to a corresponding increase in our understanding of the cell's structures and functions.

A2.2.4 – Structures common to all cells

A2.2.4 – Structures common to cells in all living organisms
Typical cells have DNA as genetic material and a cytoplasm composed mainly of water, which is enclosed by a plasma membrane composed of lipids. Students should understand the reasons for these structures.

As all organisms are composed of one or more cells and demonstrate common functions, all cells possess certain common structures. These include:

- DNA, as their genetic material
- a cytoplasm, composed of mainly water
- a plasma membrane, composed of lipids that surrounds the cytoplasm.

All cells possess three common structures: DNA, cytoplasm and a plasma membrane. Cells usually demonstrate greater complexity than this, with many more structures. However, greater complexity is not required for a cell to carry out the functions of life.

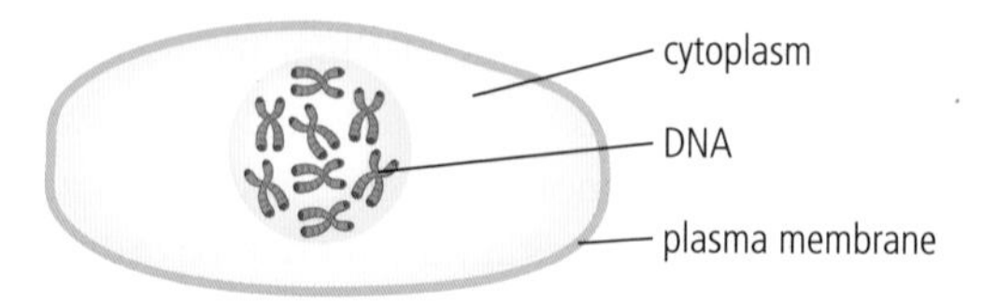

For new cells to be formed from pre-existing cells, there must be a means to store and transfer information. DNA fulfils this role because of its ability to form large molecules from small building blocks called **nucleotides**. Four different nucleotides make up DNA. It is the specific sequence of these unique nucleotides, and their ability to combine to form huge chains, that results in the production of the exact proteins essential for passing on distinctive characteristics from cell to cell and even from organism to organism. DNA also controls the production of enzymes within an organism, which serve a controlling role in chemical reactions.

The cytoplasm is found within the boundary of a cell. This region of a cell consists of a matrix composed mainly of water called **cytosol.** Cytosol contains all the ingredients necessary for a cell to conduct its day-to-day activities. These ingredients include many different carbon compounds, as well as **ions,** which are atoms with a charge, and other inorganic compounds. The cytoplasm of a cell is the location where most chemical reactions take place.

A matrix is an unstructured semi-fluid region within a boundary. The cytosol is a matrix with a gel-like consistency in which other cell structures may be suspended.

The plasma membrane encloses the cell and protects its contents from the surrounding environment. Its major component is two layers of lipids combined as a **bilayer**. Proteins and the element phosphorus are also associated with this bilayer. The membrane controls interactions between a cell's contents and the exterior. Materials needed by the cell are transported into the cell through the membrane, while waste material is transported out of the cell. Membrane proteins provide identity properties to the cell, which is especially important in multicellular organisms. The membrane proteins in multicellular organisms also engage in communication and transport between cells.

What explains the use of certain molecular building blocks in all living cells?

A2.2.5 – The prokaryote cell

A2.2.5 – Prokaryote cell structure
Include these cell components: cell wall, plasma membrane, cytoplasm, naked DNA in a loop and 70S ribosomes. The type of prokaryotic cell structure required is that of Gram-positive eubacteria such as *Bacillus* and *Staphylococcus*. Students should appreciate that prokaryote cell structure varies. However, students are not required to know details of the variations such as the lack of cell walls in phytoplasmas and mycoplasmas.

Two types of organism, bacteria (members of the domain Eubacteria) and archaea (members of the domain Archaea), are made up of prokaryotic cells and are called prokaryotes. Most of these organisms do not cause disease and are not pathogenic (disease-causing). They are an extremely diverse group occupying air, water and soil environments. Prokaryotes are a very successful group of organisms.

What is a prokaryotic cell?

After extensive studies of cells, it has become apparent that all cells use some common molecular mechanisms. There are huge differences between forms of life, but cells are the basic unit and different cells have many characteristics in common. Cells are often divided into groups based on major characteristics. One such division separates cells into **prokaryotic** and **eukaryotic cells.** Prokaryotic cells are much smaller and simpler than eukaryotic cells. In fact, most prokaryotic cells are less than 1 μm in diameter. As bacteria are prokaryotic cells, you can see that such cells play a large role in the world today.

A domain is the highest classification rank of all organisms. Three domains of life are recognized. They are the Eubacteria, the Archaea, and the Eukarya.

Prokaryotic organisms include bacteria and archaea. Bacteria and archaea appear to have followed different branches to eukaryotes (in the domain Eukarya) in the evolution of life. Prokaryotes are mostly small and unicellular. There are thousands of distinct types differentiated by many factors, including nutritional requirements, sources of energy, chemical composition and morphology (shape).

Becoming familiar with common prefixes, suffixes and word roots will help you understand biological terms. For example, the word prokaryotic is from the Greek word "pro", which means "before", and "karyon" which means "kernel", referring to the nucleus.

Features of prokaryotic cells

Study the diagram of a prokaryotic cell (Figure 1) and make sure you can identify:

- the cell wall
- the plasma membrane
- flagella
- pili
- ribosomes
- the nucleoid (a region containing free DNA).

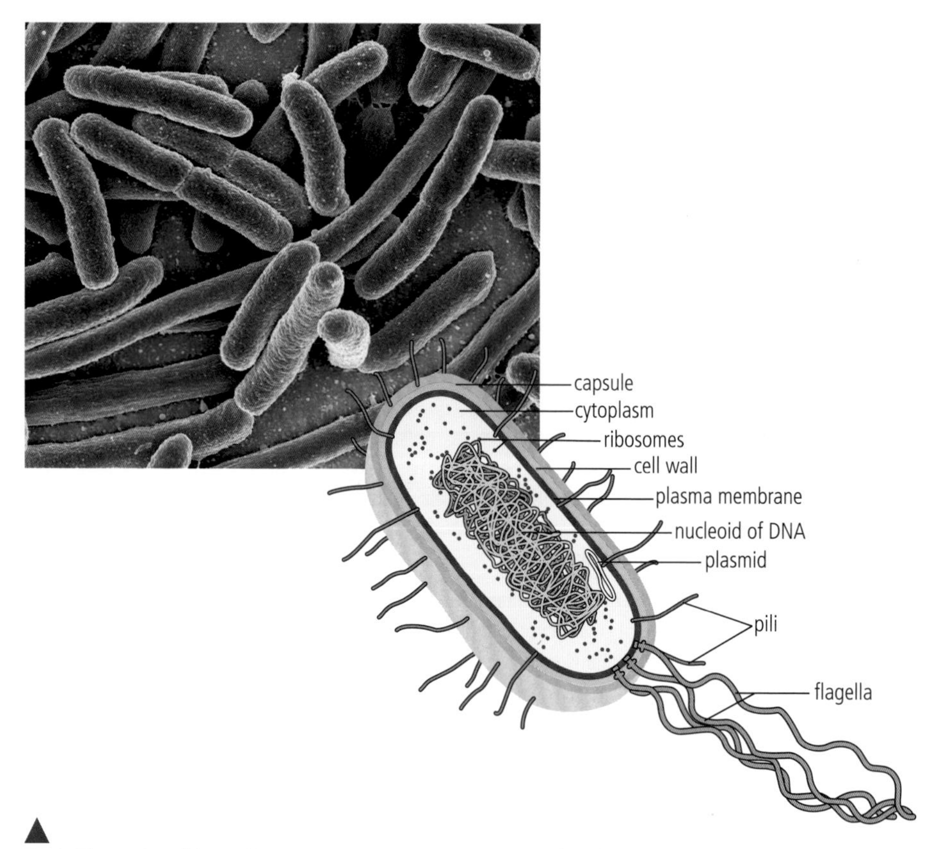

▲

A2.2 Figure 1 A false-colour scanning electron micrograph (SEM) of the bacterium *Escherichia coli*. Below it is a drawing of a prokaryotic cell.

Cell wall and plasma membrane

The prokaryotic cell wall protects and maintains the shape of the cell. It also keeps the bacterial cell from rupturing when water pressure is greater inside the cell than outside. In most prokaryotic cells this wall is composed of a carbohydrate–protein complex called **peptidoglycan**. Some bacteria have an additional layer of a type of polysaccharide outside the cell wall. This layer, called the **capsule**, makes it possible for some bacteria to adhere to structures such as teeth, skin and food.

The plasma membrane is found just inside the cell wall and is similar in composition to the membranes of eukaryotic cells. To a considerable extent, the plasma membrane controls the movement of materials into and out of the cell, and it plays a role in binary fission of the prokaryotic cell.

i

Antibiotics used to treat infections caused by bacteria can attack two areas of the bacterial cell. They may interfere with the proper development of the cell wall, resulting in a weakened outer protective wall. They may also act on ribosomes, to prevent the synthesis of the cell's required proteins. These same antibiotics do not act on eukaryote cell walls or ribosomes, so they can be used to successfully treat bacterially caused infections without harming the cells of the affected eukaryotic organism.

One major way to classify bacteria is by their ability to retain a dye called crystal violet. Bacteria that are "Gram-positive" have cell walls that, when exposed to crystal violet, take on a violet or blue appearance. "Gram-negative" bacteria do not retain this dye and do not appear violet or blue when examined with a microscope. *Bacillus* and *Staphylococcus* are examples of Gram-positive bacteria.

Gram staining is important in medicine as it provides evidence not only of a bacterial infection but also of the type of bacteria causing the infection. This helps in determining a proper treatment plan.

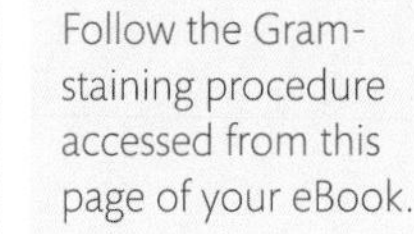

Follow the Gram-staining procedure accessed from this page of your eBook.

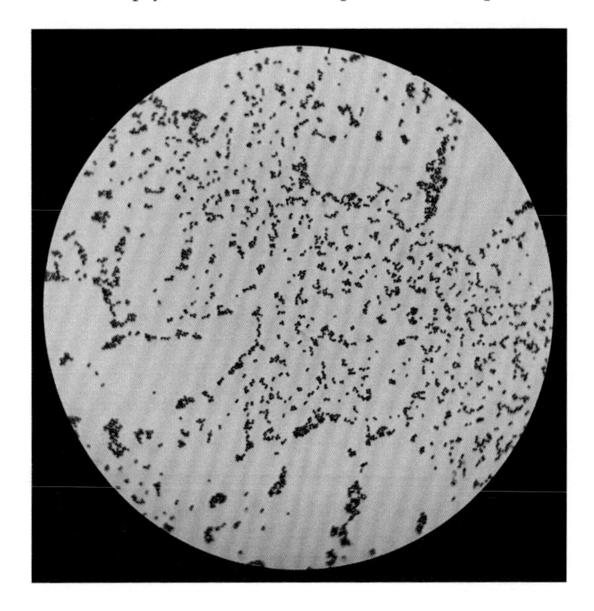

A transmission electron micrograph (TEM) of Bacillus subtilis bacteria. Notice the violet-blue colour indicating that this bacterium is Gram-positive. Had this bacterium been Gram-negative, there would be a pink colour present because of the addition of Gram's safranin, as mentioned in the Gram-staining procedure.

Pili and flagella

Some bacterial cells contain hair-like growths on the outside of the cell wall. These structures are called **pili** and can be used for attachment. However, their main function is in joining bacterial cells in preparation for the transfer of DNA from one cell to another (sexual reproduction).

Some bacteria have flagella (plural) or a flagellum (singular), which are always longer than pili. Flagella allow a cell to move and are anchored to the cell wall and plasma membrane.

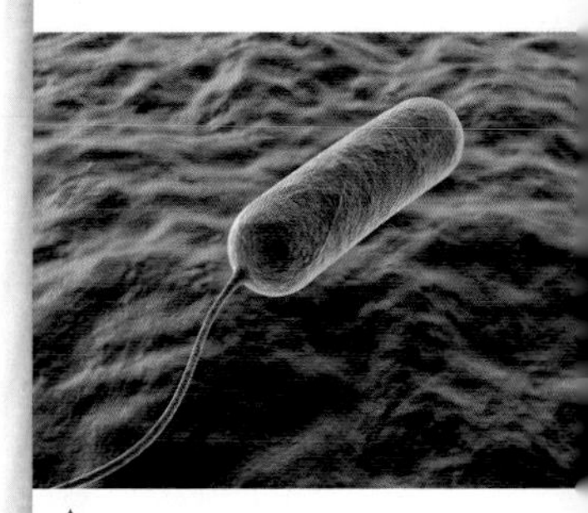

A scanning electron micrograph (SEM) of a bacterial cell with a single flagellum. When flagella are present on a bacterial cell, they are usually involved in movement. Many bacteria have more than one flagellum attached.

Cytoplasm

The cytoplasm occupies the complete interior of the cell. Using a microscope capable of high magnification, the most visible structure of the cytoplasm is the chromosome or a molecule of DNA. There are no specialized areas within the cytoplasm because internal membranes do not exist. All the cellular processes taking place within prokaryotic cells occur within the cytoplasm, without the existence of specialized compartments.

Because there are no specialized areas within prokaryotic cells, chemical reactions are not isolated from one another. This may limit the cell's development and efficiency because of possible interference between the reactions. It is interesting that without this separation of specialized areas, prokaryotes have the most diverse metabolic reactions of all organisms. When areas within a cell take on specific functions and are separated from the surrounding cytoplasm, the cell is said to show compartmentalization. Compartmentalization was a major development as prokaryotic cells gave rise to eukaryotic cells.

Ribosomes

Ribosomes occur in all prokaryotic cells, and they function as sites of protein synthesis. These small structures occur in large numbers in cells that produce a substantial amount of protein, and, when numerous, they give a granular appearance to a TEM of a prokaryotic cell. Ribosomes are composed of two subunits, a protein and a type of RNA called ribosomal RNA. The structure of prokaryotic ribosomes will be explained further in the context of eukaryotic cell structures (Section A2.2.6).

The nucleoid region

The nucleoid region of a bacterial cell is non-compartmentalized and contains a single, long, continuous, circular thread of DNA, the bacterial chromosome. The nucleoid region is not surrounded by a membrane. Prokaryotic cell DNA is not associated with proteins called histones, as the DNA of eukaryotes is; hence bacterial chromosomes can be described as naked loops. This nucleoid region is involved with cell control and reproduction.

In addition to the bacterial chromosome, bacteria may also contain **plasmids**. These small, circular, DNA molecules are not connected to the main bacterial chromosome. The plasmids replicate independently of the chromosomal DNA. Plasmid DNA is not required by the cell under normal conditions, but it can help the cell adapt to unusual circumstances.

Plasmids have especially important roles to play in some techniques involving genetic engineering/modification. Current research into genetic modification is progressing rapidly with the use of a recently discovered biological scalpel called CRISPR. It is hoped that CRISPR will provide a future cure for some genetic diseases.

Binary fission

Prokaryotic cells divide by a very simple process called **binary fission**. During this process, the DNA is copied, resulting in two daughter chromosomes. These daughter chromosomes become attached to different regions on the plasma membrane, and the cell divides into two genetically identical daughter cells. This divisional process includes an elongation of the cell and a partitioning of the newly produced DNA by specialized fibres.

Some types of bacteria go through binary fission every 20 minutes when conditions are ideal. This results in huge populations and greater potential for infections. Refrigeration of foods is often used to reduce ideal conditions for bacteria, resulting in lower bacteria counts in our food and less chance of infection/food poisoning.

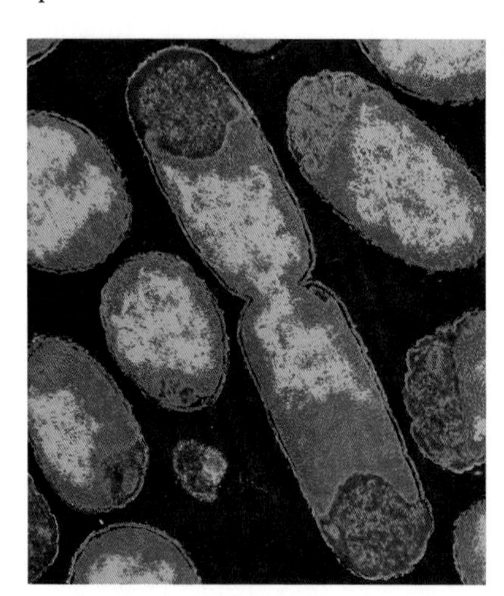

A false-colour transmission electron micrograph (TEM) showing *Escherichia coli* dividing by binary fission.

Challenge yourself

1. Prepare a drawing of the ultrastructure of a prokaryotic cell based on electron micrographs. Remember to use a sharp pencil; use simple, narrow lines, and do not use shading. Label each of the structures, including their function.

A2.2.6 – The eukaryote cell

A2.2.6 – Eukaryote cell structure
Students should be familiar with features common to eukaryote cells: a plasma membrane enclosing a compartmentalized cytoplasm with 80S ribosomes; a nucleus with chromosomes made of DNA bound to histones, contained in a double membrane with pores; membrane-bound cytoplasmic organelles including mitochondria, endoplasmic reticulum, Golgi apparatus and a variety of vesicles or vacuoles including lysosomes; and a cytoskeleton of microtubules and microfilaments.

What is a eukaryotic cell?

Whereas prokaryotic cells occur in bacteria and archaea, eukaryotic cells occur in organisms such as algae, protozoa, fungi, plants and animals. Eukaryotic cells range in diameter from 5 to 100 μm. A "kernel" or nucleus is usually noticeable in the cytoplasm. Other **organelles** may be visible within the cell if you have a microscope with a high enough magnification and resolution. Organelles are non-cellular structures that carry out specific functions (a bit like organs in multicellular organisms); different types of cells may have different organelles. These structures enable compartmentalization in eukaryotic cells, which is not a characteristic of prokaryotic cells. Compartmentalization enables different chemical reactions to be separated, which is especially important when adjacent chemical reactions are incompatible. Compartmentalization also allows chemicals for specific reactions to be isolated; this isolation results in increased efficiency.

The term "eukaryote" comes from the Greek words meaning "true kernel" or nucleus.

Examine Figures 2 and 3, illustrating typical animal and plant eukaryotic cells.

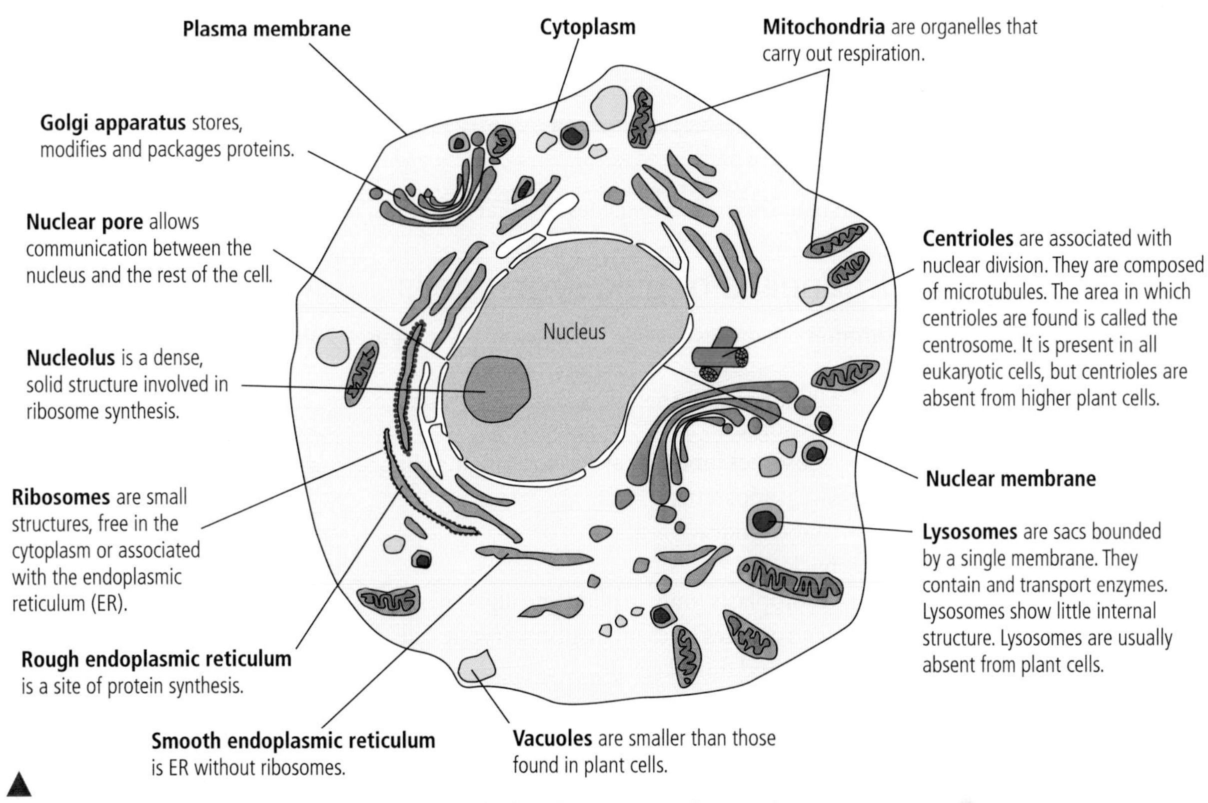

▲ **A2.2 Figure 2** Look at this drawing of a typical animal cell and compare it with Figure 3.

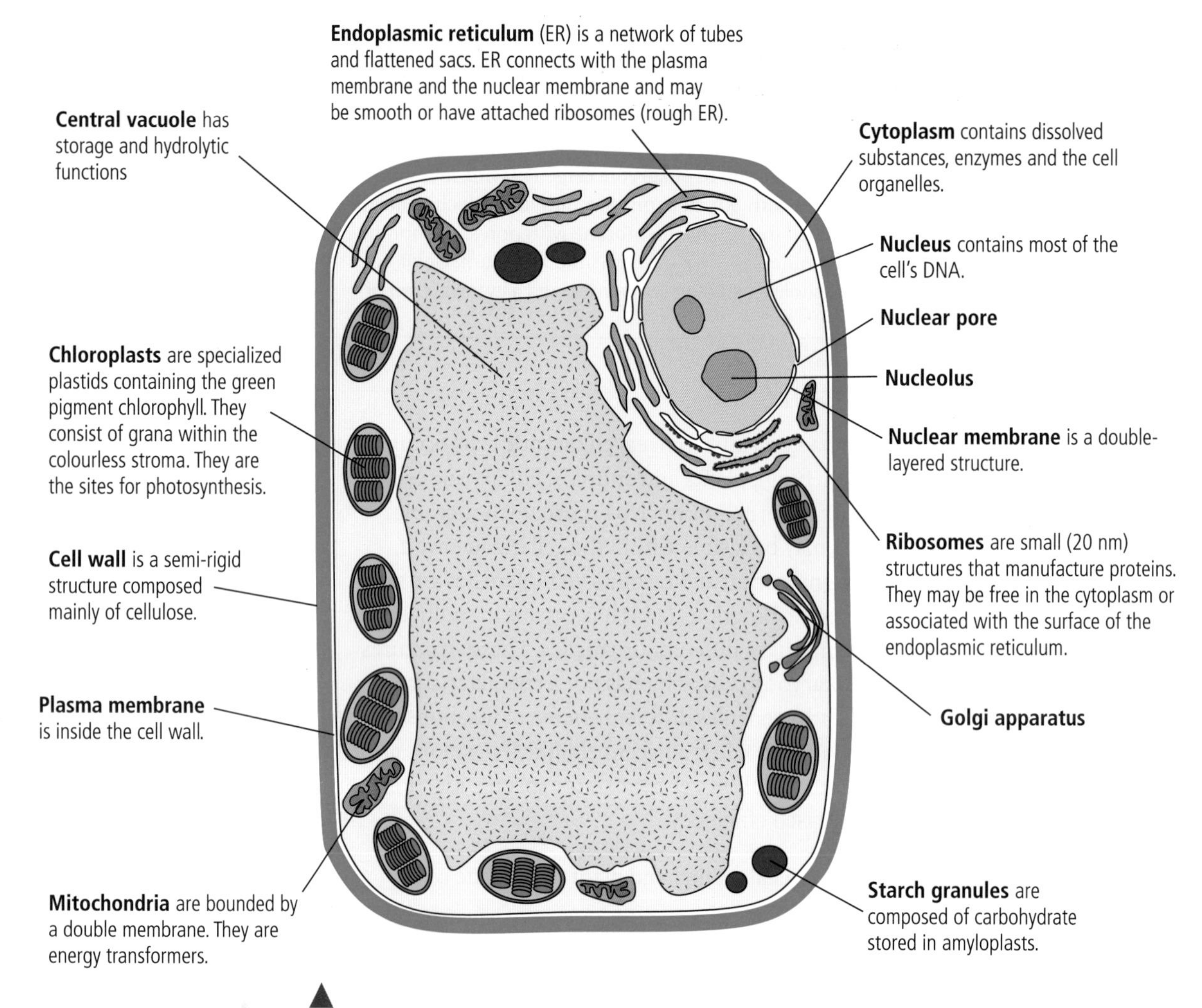

▲
A2.2 Figure 3 What is different and what is similar between this typical plant cell and the animal cell in Figure 2?

As you read about the organelles of eukaryotic cells, refer to Figures 2 and 3.

Organelles of eukaryotic cells

Common organelles include the following (see Figures 2 and 3):

- endoplasmic reticulum
- ribosomes
- lysosomes (not usually found in plant cells)
- Golgi apparatus
- mitochondria
- nucleus
- chloroplasts (only in plant and algal cells)
- centrosomes (present in all eukaryotic cells, but centrioles are not found in most plant and fungal cells)
- vacuoles.

The microscope has given us an insight into the structure and function of eukaryotic cell organelles and characteristics.

Cytoplasm

All eukaryotic cells have a region called the **cytoplasm** that occurs inside the plasma membrane and outside the nucleus. It is in this region that the organelles are found. The fluid portion of the cytoplasm around the organelles is called the **cytosol**. Eukaryotic cytoplasm includes small fibres and rods called a cytoskeleton, which creates a complex internal structure. Prokaryotic cytoplasm lacks a cytoskeleton.

Microscopes have a rich history of international development. Glass lenses were used in the 1st century by the Romans to magnify objects. Savino D'Armate, an Italian, made a magnifying eyeglass in the 13th century to be used with one eye. In the 1590s, two Dutch eyeglass makers, Hans Jansen and his son Zacharias Jansen, produced the first compound microscope by putting two lenses together. Antonie van Leeuwenhoek, also Dutch, improved the Jansen compound microscope in the 1600s. Since this beginning, many individuals in many different countries of the world have contributed to making the present-day microscope extremely effective in the study of the cell and other small structures. Modern technology allowing extensive communication has also been extremely important in the continual improvement of the current microscope.

Cytoskeleton

The eukaryotic cell cytoplasm contains a network of fibres collectively called the **cytoskeleton**. These fibres are composed of protein and provide the following functions within the cell:

- maintaining cell shape
- anchoring some organelles
- aiding cellular movements
- providing a means for some organelles to move within the cell.

The cytoskeleton contains actin filaments, intermediate filaments and microtubules. These fibres can rearrange their protein components so that the cell can respond to changes in both internal and external environments. Actin filaments are also called microfilaments, and function in cell division and cell movement, especially involving contractions, as in muscle cells. Intermediate filaments are found in most animal cells and reinforce cell shape as well as anchoring some organelles. Microtubules shape and support the cell. They also function as movement paths or tracks through the cell for some organelles.

Endoplasmic reticulum

The **endoplasmic reticulum (ER)** is an extensive network of tubules or channels that extends most everywhere in the cell, from the nucleus to the plasma membrane. Its structure enables its function, which is the transportation of materials throughout the internal region of the cell. There are two general types of ER: **smooth ER** and **rough ER**. Smooth ER does not have any of the organelles called ribosomes on its exterior surface. Rough ER has ribosomes on its exterior.

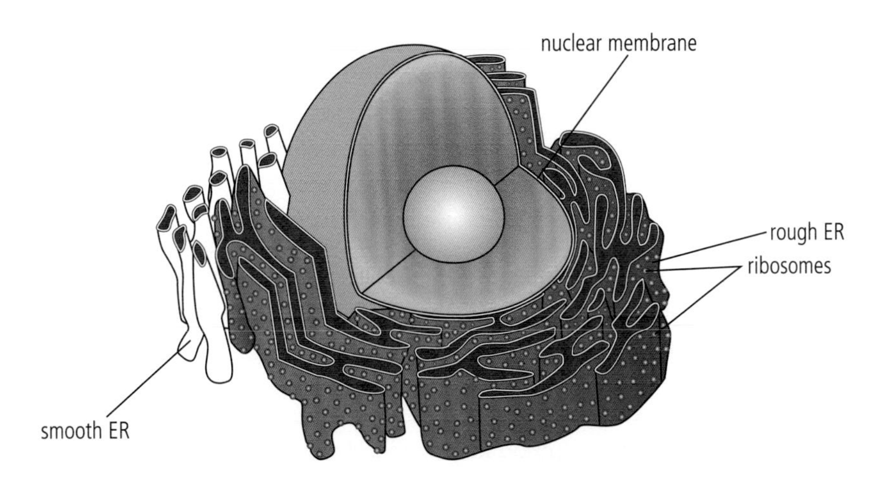

Smooth endoplasmic reticulum (ER) and rough endoplasmic reticulum (ER).

Smooth ER has many unique enzymes embedded on its surface. Its functions include:

- the production of membrane phospholipids and cellular lipids
- the production of sex hormones such as testosterone and oestradiol
- detoxification of drugs in the liver
- the storage of calcium ions in muscle cells, needed for contraction
- transportation of lipid-based compounds
- helping the liver to release glucose into the bloodstream when needed.

Rough ER has ribosomes on the exterior of its channels. The ribosomes participate in protein synthesis, so this type of ER engages in protein development and transport. These proteins may become parts of membranes, enzymes or even messengers between cells. Most cells contain both types of ER, with the rough ER being closer to the nuclear membrane.

The letter S used in the measurement of ribosomes refers to Svedberg units, which indicate the relative rate of sedimentation during high-speed centrifugation. The higher the S value, the quicker the structure will become part of the sediment and the more mass it will have.

Ribosomes

Ribosomes are unique structures that do not have an exterior membrane. They conduct protein synthesis within the cell. These structures may be found free in the cytoplasm, or they may be attached to the surface of ER. They are always composed of a type of RNA and protein. You will recall that prokaryotic cells also contain ribosomes. However, the ribosomes of eukaryotic cells are larger and denser than those of prokaryotic cells. Ribosomes are composed of two subunits. In eukaryotic cells these subunits together equal 80S. The ribosomes in prokaryotic cells are also of two subunits, but they only equal 70S.

Lysosomes

Lysosomes are intracellular digestive centres that arise from the Golgi apparatus. A lysosome does not have any internal structures. Lysosomes are **vesicles** (sacs) bounded by a single membrane that contains as many as 40 different enzymes. The enzymes are all **hydrolytic** and catalyse the breakdown of proteins, nucleic acids, lipids and carbohydrates. Lysosomes fuse with old or damaged organelles within the cell to break them down, so that recycling of the components can occur. Lysosomes are also involved in the breakdown of material that is brought into the cell by **phagocytosis**. Phagocytosis is a type of **endocytosis** and is a means by which materials can enter a cell.

Endocytosis is the uptake of new materials into the cell by invagination of the plasma membrane. If the material entering the cells is solid, the process is known as phagocytosis. When liquid containing dissolved materials enters the cell, it is known as pinocytosis.

The interior environment of a functioning lysosome is acidic; acidic conditions are necessary for the enzymes to hydrolyse large molecules. When **hydrolysis** occurs, large molecules are broken down with the addition of water.

Golgi apparatus

The **Golgi apparatus** consists of flattened sacs called **cisternae**, which are stacked one on top of another. This organelle functions in the collection, packaging, modification and distribution of materials synthesized in the cell. One side of the apparatus is near the rough ER, called the ***cis*** side. It receives products from the ER. These products then move into the cisternae of the Golgi apparatus. They continue to move to the discharging or opposite side, the ***trans*** side. Small sacs called **vesicles** can then be seen coming off the trans side. Lysosomes are an important example of vesicles produced by the Golgi apparatus. The vesicles carry modified materials to wherever they are needed inside or outside the cell. The Golgi apparatus is especially prevalent in glandular cells, such as those in the pancreas, which manufacture and secrete substances.

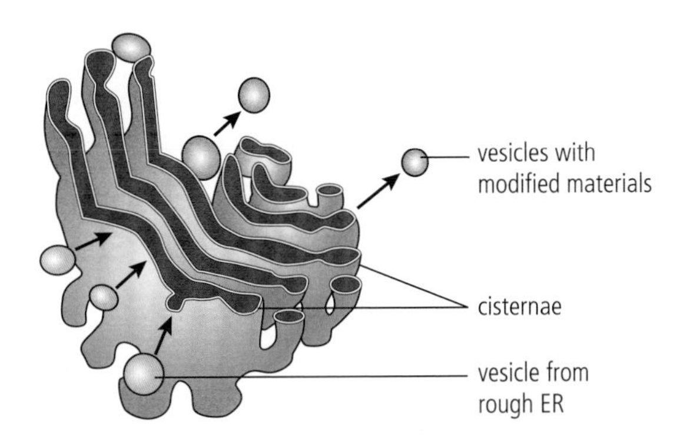

▲ In this drawing of the Golgi apparatus, the movement of the vesicles is shown by arrows. Can you identify which side is the *cis* side and which is the *trans* side?

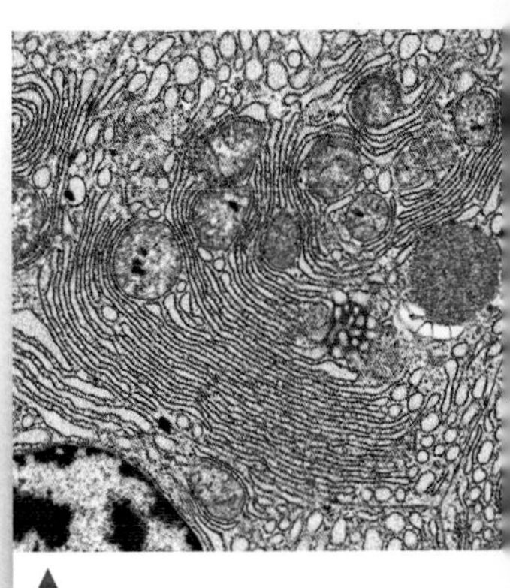

▲ A transmission electron micrograph (TEM) of a pancreatic exocrine cell. Can you tell that this is an animal cell? Locate as many of the structures of an animal cell as you can. How do the structures of this cell reflect the overall functions of the pancreas?

Mitochondria

Mitochondria (singular mitochondrion) are rod-shaped organelles that appear throughout the cytoplasm. They are close in size to a bacterial cell. Mitochondria have their own DNA, a circular chromosome like that in bacterial cells, allowing them some independence within the cell. They have a double membrane: the outer membrane is smooth, but the inner membrane is folded into **cristae** (singular crista). Inside the inner membrane is a semi-fluid substance called the **matrix.** An area called the **inner membrane space** lies between the two membranes.

The cristae provide a huge surface area within which the chemical reactions characteristic of mitochondria occur. Most mitochondrial reactions involve the production of usable cellular energy called **adenosine triphosphate (ATP)**. Because of this, the mitochondria are often called the powerhouse of a cell. This organelle also produces and contains its own ribosomes. These ribosomes are of the 70S type. Cells that have high energy requirements, such as muscle cells, have large numbers of mitochondria.

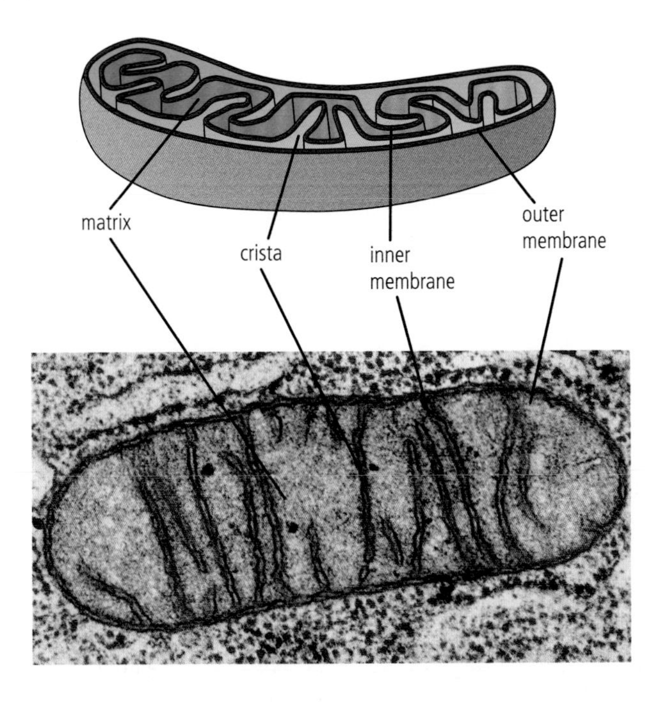

◀ Compare this drawing of a mitochondrion with the corresponding false-colour transmission electron micrograph (TEM) below it.

Nucleus

The **nucleus** in eukaryotic cells is an isolated region where DNA resides. It is bordered by a double membrane referred to as the **nuclear envelope**. This membrane allows compartmentalization of the eukaryotic DNA, thus providing an area where DNA can conduct its functions without being affected by processes occurring in other parts of the cell. The nuclear membrane does not result in complete isolation, because it has numerous pores that allow communication with the cell's cytoplasm.

The nucleus has a double membrane with pores and contains a nucleolus.

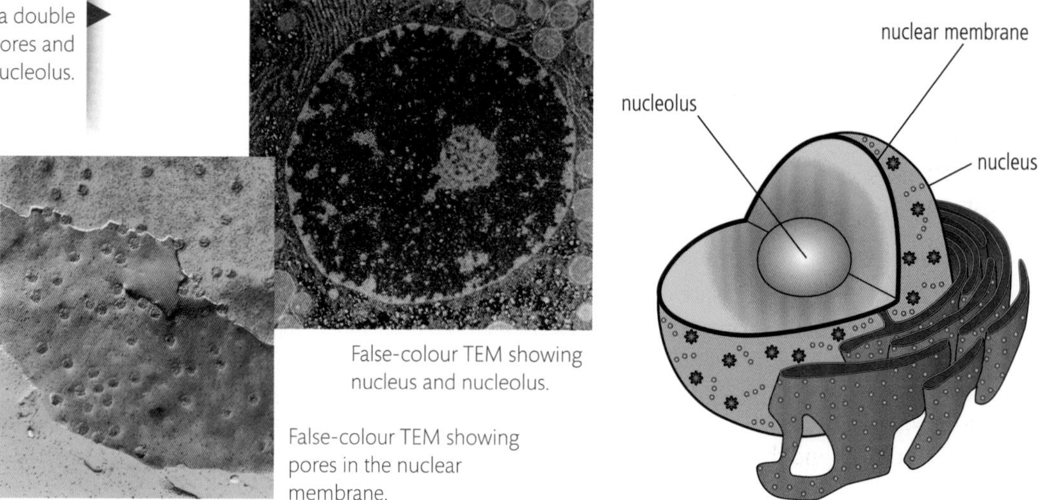

False-colour TEM showing nucleus and nucleolus.

False-colour TEM showing pores in the nuclear membrane.

The DNA of a eukaryotic cell often occurs in the form of chromosomes; chromosomes vary in number depending on the species. Chromosomes carry all the information that is necessary for the cell to exist, thus allowing an organism to survive, whether it is unicellular or multicellular. The DNA is the genetic material of the cell. It enables certain traits to be passed on to the next generation. When the cell is not in the process of dividing, the chromosomes are not present as visible structures. During this phase, the cell's DNA is in the form of **chromatin.** Chromatin is formed of strands of DNA and proteins called **histones**. The DNA and histone combination often results in structures called a **nucleosome.** A nucleosome consists of eight spherical histones with a strand of DNA wrapped around them and secured with a ninth histone. This produces a structure that resembles a string of beads. A chromosome is a highly coiled structure of many nucleosomes.

This drawing shows how DNA is packaged into chromosomes.

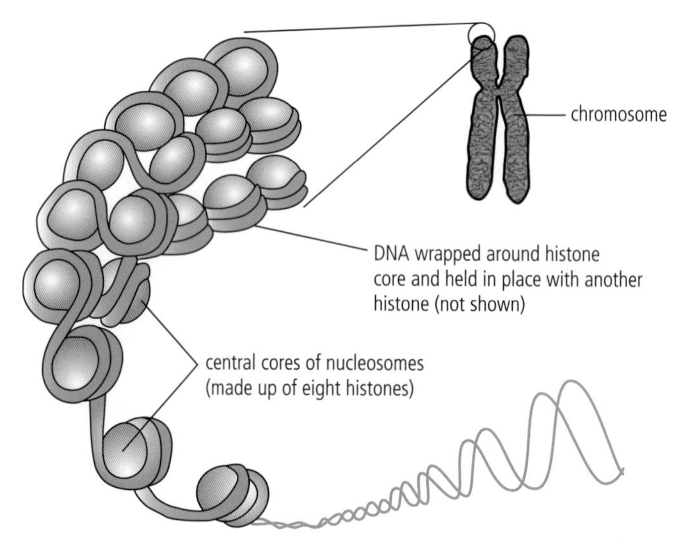

The nucleus is often located centrally within the cell's cytoplasm, although in some cell types it is pushed to one side or the other. The side position is characteristic of plant cells, because these cells often have a large central vacuole. Most eukaryotic cells possess a single nucleus, but some do not have a nucleus at all, and others have multiple nuclei. Without a nucleus, cells cannot reproduce. The loss of reproductive ability is often paired with increased specialization to carry out certain functions. For example, human red blood cells do not have nuclei: they are specialized to transport respiratory gases. Most nuclei also include one or more dark areas called **nucleoli** (singular nucleolus). Ribosome molecules are manufactured in nucleoli. The molecules pass through the nuclear envelope before assembling as ribosomes.

Chloroplasts

Chloroplasts occur only in algae and plant cells. The chloroplast contains a double membrane and is about the same size as a bacterial cell. Like the mitochondrion, a chloroplast contains its own DNA and 70S ribosomes. The DNA of the chloroplast takes the form of a ring.

You should note all the characteristics that chloroplasts and mitochondria have in common with prokaryotic cells.

As well as DNA and ribosomes, the interior of the chloroplast includes **grana** (singular granum), **thylakoids** and the **stroma**, which are labelled in Figure 4. A granum is made up of numerous thylakoids stacked like a pile of coins. The thylakoids are flattened membrane sacs with components necessary for the absorption of light. Absorption of light is the first step in **photosynthesis**. Photosynthesis is a process that converts light energy into chemical energy. The chemical energy is then stored in sugars made from carbon dioxide. The fluid stroma is like the cytoplasm of the cell. It occurs outside the grana but within the double membrane. The stroma contains many enzymes and chemicals necessary to complete the process of photosynthesis. Like mitochondria, chloroplasts can reproduce independently of the cell.

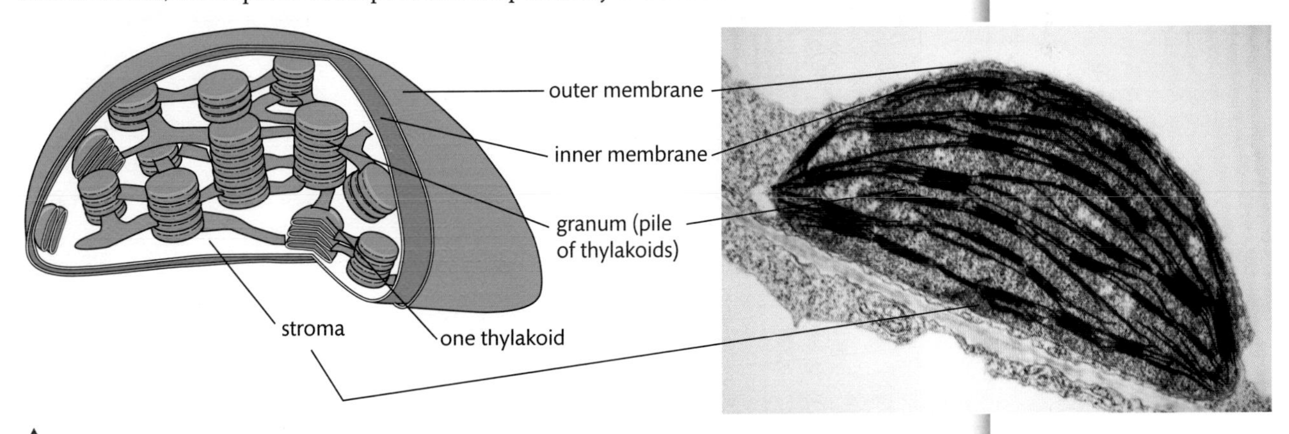

A2.2 Figure 4 Compare the drawing of a chloroplast with the corresponding transmission electron micrograph (TEM) of a chloroplast.

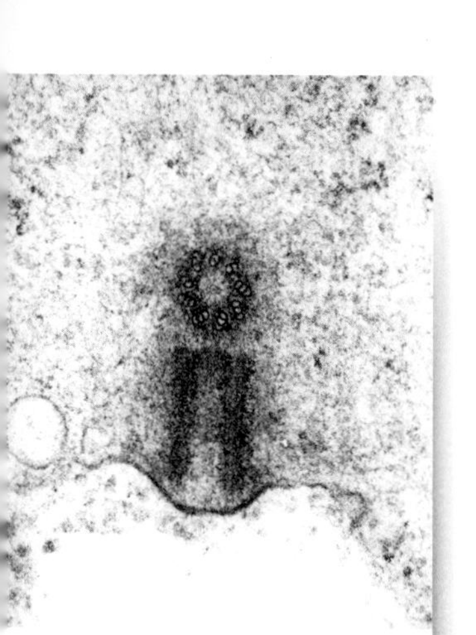

A transmission electron micrograph (TEM) showing the two centrioles of a centrosome. The presence of centrioles indicates that the micrograph is of a eukaryotic cell, but not a plant or fungal cell.

Centrosome

The **centrosome** occurs in all eukaryotic cells. In animal cells it consists of a pair of **centrioles** that are often at right angles to one another. The centrioles are involved with the assembly of **microtubules**, which are important to a cell because they provide structure and allow movement. Microtubules are also important for cell division. Plant and fungal cells do not have centrioles. However, they are able to produce microtubules from their centrosome-like regions, which suggests that centrioles are not necessary for producing microtubules.

The centrosome is located at one end of the cell, close to the nucleus. **Basal bodies** are structures related to the centrosome of eukaryotic cells and are located at the base of cilia and flagella. Not all eukaryotic cells have cilia or flagella, therefore not all eukaryotic cells have basal bodies. The basal bodies are thought to direct the assembly of microtubules within the associated cilia or flagella. When present, centrioles appear to produce basal bodies.

Vacuoles

Vacuoles are storage organelles that are usually formed from the Golgi apparatus. They are membrane-bound and have many possible functions. They occupy a very large space inside the cells of most plants. In animal cells, vacuoles are small and may be numerous. Vacuoles may store several different substances, including potential food (to provide nutrition, as in plant cells), metabolic waste and toxins (to be expelled from the cell) and water. Vacuoles enable cells to have higher surface area-to-volume ratios even at larger sizes. In plants, they allow an uptake of water, which provides rigidity to the organism. When a large vacuole occurs in the central area of a plant cell, it is called a **central vacuole**. Vacuoles are like vesicles except that they are larger.

A coloured transmission electron micrograph (TEM) of a plant cell that has a central vacuole filled with water. Note the central location of the vacuole, with the cytoplasm and all the other organelles pushed to the cell margins.

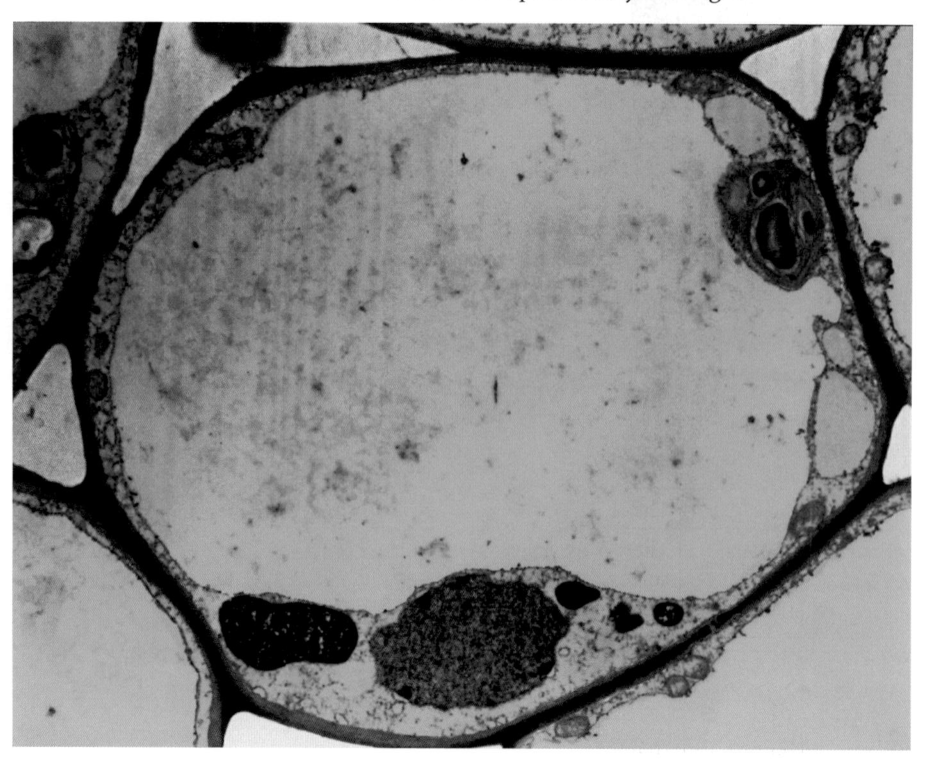

A comparison of prokaryotic and eukaryotic cells

A table is an effective way of summarizing the differences between prokaryotic and eukaryotic cells.

When comparing items, be certain to state the characteristic for each type of item, as shown Table 1 for prokaryotic and eukaryotic cells.

Prokaryotic cells	Eukaryotic cells
DNA in a ring form without protein	DNA with proteins as chromosomes/chromatin
DNA free in the cytoplasm (nucleoid region)	DNA enclosed within a nuclear envelope (nucleus)
No mitochondria	Mitochondria present
70S ribosomes	80S ribosomes
No internal compartmentalization to form organelles	Internal compartmentalization present, forming many types of organelles
Size less than 10 μm	Size more than 10 μm

A2.2 Table 1 Comparing prokaryotic and eukaryotic cells

If asked to state the similarities between the two types of cells, make sure you include the following:

- both types of cells have some outside boundary that always involves a plasma membrane
- both types of cells conduct all the functions of life
- DNA is present in both cell types.

A2.2.7 – Unicellular organisms

A2.2.7 – Processes of life in unicellular organisms
Include these functions: homeostasis, metabolism, nutrition, movement, excretion, growth, response to stimuli and reproduction.

All organisms, whether unicellular or multicellular, carry out all the functions of life. The functions of life are summarized in Table 2.

Metabolism	The sum of all the chemical reactions that occur within an organism
Growth	The development of an organism
Reproduction	The ability to produce offspring
Response to stimuli	As the environment changes, the organism adapts
Homeostasis	Maintenance of a constant internal environment
Nutrition	The ability to acquire the energy and materials needed to maintain life
Excretion	The ability to release materials not needed or harmful into the surrounding environment
Movement	The ability to move or change position

A2.2 Table 2 The functions of life

It is important to note that if the functions of life are evident, then life is said to be present.

Unicellular organisms have unique ways of carrying out the life functions compared to **multicellular** organisms.

- The cell membrane controls the movement of materials in and out of the cell, to help maintain homeostasis.
- Vacuoles isolate and store waste so that it does not harm the organism.
- Cells often possess cilia or flagella that allow movement in response to changes in the environment.
- Vacuoles carry out digestion, to provide nutrition for the organism.
- Mitochondria or areas of enzymes allow energy production to continue for all the functions of life.
- Ribosomes provide the building blocks for growth and repair.

Multicellular organisms often have whole groups of cells called **organs** carrying out these functions.

A2.2.8 – Different types of eukaryotic cells

A2.2.8 – Differences in eukaryotic cell structure between animals, fungi and plants
Include presence and composition of cell walls, differences in size and function of vacuoles, presence of chloroplasts and other plastids, and presence of centrioles, cilia and flagella.

The eukaryotic cells of different types of organisms can vary. Three types of organisms with eukaryotic cells are plant cells, animal cells and fungal cells. There are over 14,000 species of fungi, and it is believed that they were the first eukaryotes to live on land.

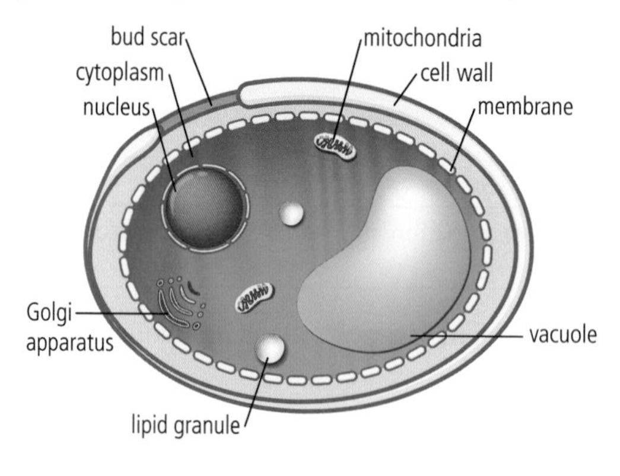

This drawing of a yeast cell illustrates some of the major cell organelles common to fungi.

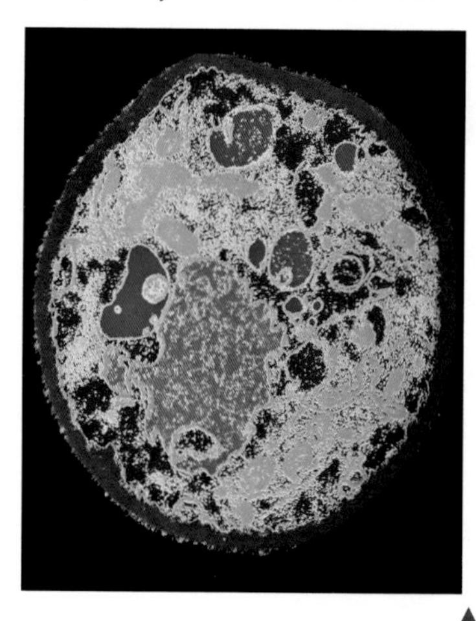

This transmission electron micrograph (TEM) of a yeast cell represents one of the many species of fungi. From our previous work with organelles, identify as many as possible.

Fungi can be unicellular or multicellular. They include yeasts, mushrooms, truffles and bread moulds, plus many more. No fungus can produce its food. Fungi secrete (release into the surrounding environment) digestive enzymes and then absorb the externally digested nutrients as their source of energy. They have major roles in our planet, including decomposing organic debris to enable the recycling of nutrients, being a source of food, being used in medicines, and even controlling many harmful insects.

Most believe fungi are more closely related to animals than to plants. Table 3 summarizes the differences between plant, animal and fungal eukaryotic cells. However, do not forget the similarities between these three cell types as well.

Plant cells	Animal cells	Fungal cells
Exterior of cell includes an outer cell wall composed of cellulose, with a plasma membrane just inside	Exterior of cell includes a plasma membrane. There is no cell wall	Exterior of cell includes an outer cell wall composed of chitin, with a plasma membrane just inside
Chloroplasts are present in the cytoplasm area, enabling the production of carbohydrates	There are no chloroplasts for carbohydrate production	There are no chloroplasts for carbohydrate production
Possess large centrally located vacuoles for the storage of carbohydrates	Vacuoles are generally small and numerous, when present, with many unique functions	Vacuoles are generally small and numerous, with many unique functions
Store carbohydrates as starch	Store carbohydrates as glycogen	Store carbohydrates as glycogen
Usually do not contain cilia, flagella or basal bodies	May have cilia or flagella, with associated basal bodies	May have cilia or flagella, but do not have associated basal bodies
Because a rigid cell wall is present, this cell type has a fixed, often angular, shape	Without a cell wall, this cell is flexible and more likely to be a rounded shape	The cell wall allows a degree of flexibility, along with support for the cell; the cell shape may vary
Possess centrosomes but no centrioles	Possess both centrosomes and centrioles	Possess centrosomes but no centrioles

A2.2 Table 3 Differences between plant, animal and fungal cells

Most of the organelles discussed are present in all eukaryotic cells. When an organelle is present in each of the eukaryotic cell types, it usually has the same structure and function. For example, all three cell types contain mitochondria that possess cristae, a matrix and a double membrane. Also, in all three cell types, the mitochondria function in the production of ATP for use by the cell.

A2.2.9 – Atypical eukaryotes

A2.2.9 – Atypical cell structure in eukaryotes
Use numbers of nuclei to illustrate one type of atypical cell structure in aseptate fungal hyphae, skeletal muscle, red blood cells and phloem sieve tube elements.

The structure of some eukaryotic cells is unique or atypical, which allows them to carry out specialized functions. One example of this atypical structure involves cell nuclei.

Some multicellular fungi produce filaments called **hyphae.** Most of these hyphae consist of chains separated by cross-walls that have pores to allow various organelles and cytoplasm to flow from cell to cell. However, some fungi produce hyphae that lack cross-walls. The result of this is a single mass of cytoplasm (one cell) with more than one nucleus.

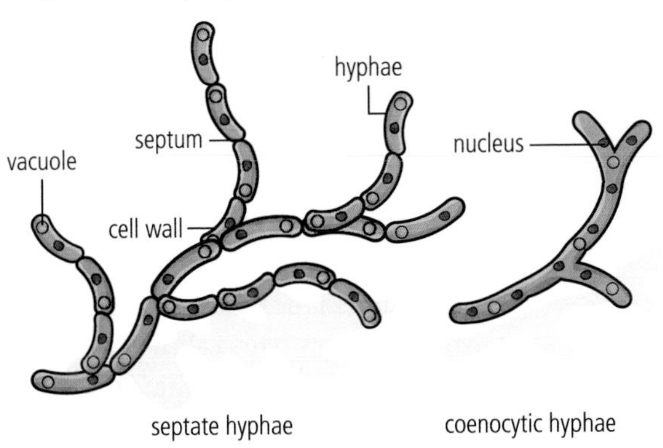

Notice the two types of hyphae shown in this image. The hyphae on the right do not contain cross-walls, while cross-walls are present in the hyphae on the left.

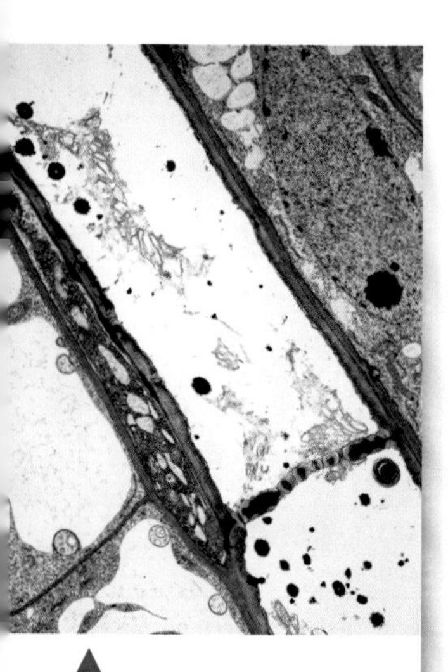

▲
A2.2 Figure 5 A transmission electron micrograph (TEM) of a plant's sieve tube elements and associated companion cells. Notice the lack of substance in the sieve tube elements and the pores in the end wall.

Phloem sieve tube elements, shown in Figure 5, have a specialized function allowing transportation within a multicellular plant. These unique elements/cells have end walls with pores and minimal cellular components such as nuclei, ribosomes, cytoskeleton and cytoplasm. They are connected end to end, forming tube structures. These cells can only remain alive with the help of **companion cells**, which maintain a close connection with the sieve tube elements.

Figure 6 shows a micrograph of human red blood cells. They have the specialized function of carrying oxygen throughout the body. They contain substantial amounts of a molecule called haemoglobin, which easily combines with oxygen. They are shaped to allow a large surface area for the absorption and release of oxygen. They do not have a nucleus, which allows them to carry even more oxygen.

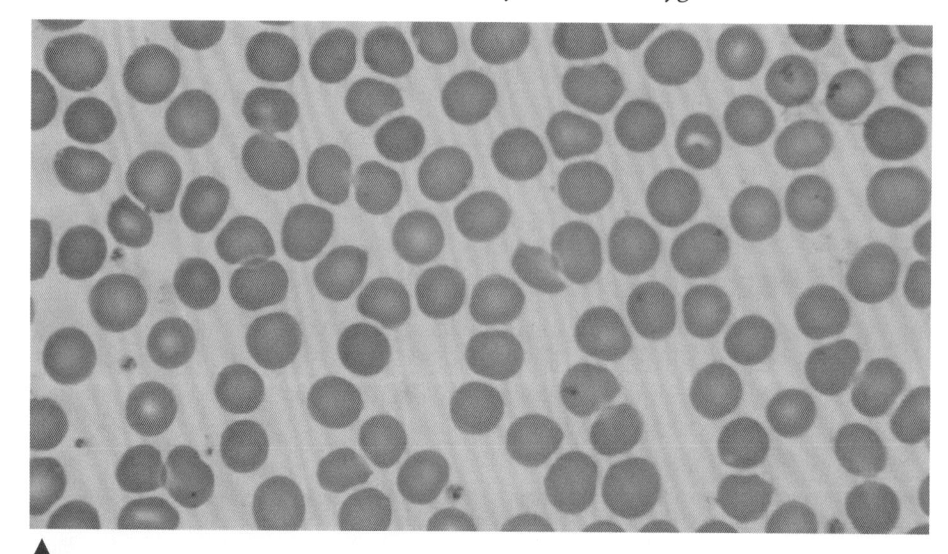

▲
A2.2 Figure 6 A micrograph of a human blood smear.

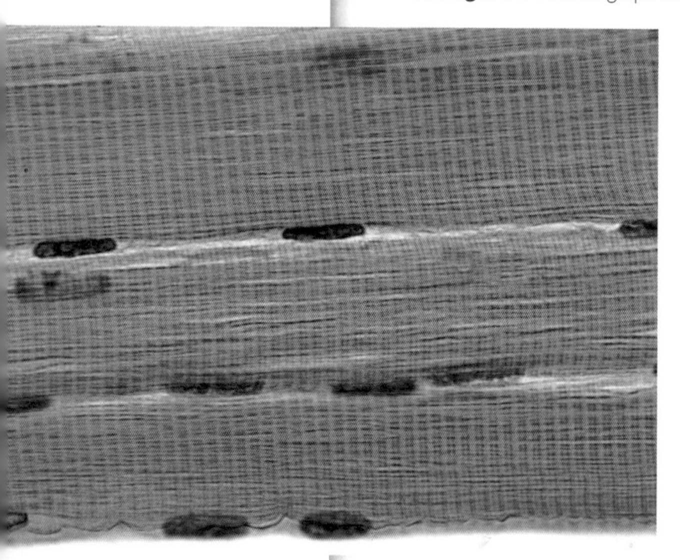

▲
A2.2 Figure 7 An electron micrograph (EM) of human skeletal muscle. Note the large, continuous cells with multiple nuclei.

Figure 7 shows an electron micrograph of human skeletal muscle. This muscle type specializes in allowing body movement. It can carry out this function because of the presence of specialized proteins arranged in bands that contract and relax. The presence of cell membranes is limited, resulting in large, tubular cells with multiple nuclei, allowing more coordinating protein molecules.

Other cells with specialized structures to enable unique functions include:

- nerve cells, which are long and thin with branched connections at each end to transmit electrical impulses
- sperm cells, with many mitochondria and a tail allowing movement and a head with a tip capable of producing an enzyme that facilitates penetration of an egg cell
- cells found in the tubes associated with lungs, which have many tiny hairs called cilia on their exterior that work in unison to move mucus and other particles up and out of the airways.

A2.2.10 and A2.2.11 – Electron micrograph skills

Draw and annotate diagrams of organelles and other cell structures shown in electron micrographs. Full details of how to carry out this activity with a worksheet are available in the eBook.

A2.2.10 – Cell types and cell structures viewed in light and electron micrographs

Application of skills: Students should be able to identify cells in light and electron micrographs as prokaryote, plant or animal. In electron micrographs, students should be able to identify these structures: nucleoid region, prokaryotic cell wall, nucleus, mitochondrion, chloroplast, sap vacuole, Golgi apparatus, rough and smooth endoplasmic reticulum, chromosomes, ribosomes, cell wall, plasma membrane and microvilli.

A2.2.11 – Drawing and annotation based on electron micrographs

Application of skills: Students should be able to draw and annotate diagrams of organelles (nucleus, mitochondria, chloroplasts, sap vacuole, Golgi apparatus, rough and smooth endoplasmic reticulum and chromosomes) as well as other cell structures (cell wall, plasma membrane, secretory vesicles and microvilli) shown in electron micrographs. Students are required to include the functions in their annotations.

It is important that you practise the skills necessary to produce informative drawings throughout the course. Actual laboratory observation of cells using prepared slides and a microscope will help you develop your skills. Draw what you can see in the field of view, and compare your drawings, labels and explanations with those found on appropriate internet sites.

Utilizing the text, diagrams and pictures presented in this chapter, you should be able to differentiate between prokaryotic and eukaryotic cells when presented with light or electron micrographs. You must be able to identify the following cell structures: nucleoid region, prokaryotic cell wall, nucleus, mitochondrion, chloroplast, sap vacuole, Golgi apparatus, rough and smooth endoplasmic reticulum, chromosomes, ribosomes, cell wall, plasma membrane and microvilli. The internet has many sites that show cells of various types, which you can use to develop your skills in this identification process.

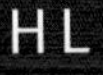

HL

A2.2.12 – The origin of eukaryotic cells

A2.2.12 – Origin of eukaryotic cells by endosymbiosis

Evidence suggests that all eukaryotes evolved from a common unicellular ancestor that had a nucleus and reproduced sexually. Mitochondria then evolved by endosymbiosis. In some eukaryotes, chloroplasts subsequently also had an endosymbiotic origin. Evidence should include the presence in mitochondria and chloroplasts of 70S ribosomes, naked circular DNA and the ability to replicate.

NOS: Students should recognize that the strength of a theory comes from the observations the theory explains and the predictions it supports. A wide range of observations are accounted for by the theory of endosymbiosis.

A common origin for all cells on Earth requires an explanation of how a cell could progress from a simple, non-compartmentalized prokaryote to a complex, highly compartmentalized eukaryote. The **endosymbiotic theory** presents a mechanism by which this progression may have occurred. The key points of the theory are:

- about 2 billion years ago a larger cell that had a nucleus and was capable of sexual reproduction engulfed a smaller prokaryotic cell that could produce energy
- these two cells developed a mutually beneficial (**symbiotic**) relationship, forming a single organism
- the smaller engulfed cell then went through a series of changes to become a mitochondrion.

With this process, the larger cell helped the bacteria prokaryote by providing protection and carbon compounds. The smaller prokaryote, after a series of changes, specialized so that it provided the now more complex larger cell with ATP.

There is much evidence to support this theory, including the characteristics of mitochondria. Mitochondria:

- are about the size of most bacterial cells
- divide by simple binary fission, as do most bacterial cells
- divide independently of the host cell
- have their own ribosomes, which are 70S in size, as prokaryote cell ribosomes are
- produce their own proteins with these ribosomes
- have their own DNA, which occurs in a circular ring, as in prokaryotic cells
- have two membranes on their exterior, which is consistent with an engulfing process
- have an inner membrane with a composition like that of prokaryote membranes, while the outer membrane is more like eukaryotic cell membranes
- have RNA present in mitochondrial ribosomes that closely resembles the RNA present in prokaryotic ribosomes.

Antibiotics used to inhibit protein synthesis in prokaryotes like bacteria also decrease protein production in mitochondria and chloroplasts.

Challenge yourself

2. On a sheet of paper, produce a series of drawings that represent how two membranes could have come to exist on mitochondria and chloroplasts through an engulfing process involving endocytosis.

In addition to the mitochondria, chloroplasts in plant cells also provide evidence for the theory of endosymbiosis. A modern-day protist called *Hatena arenicola* normally fulfils its nutritional needs by ingesting organic matter. However, when it behaves as a predator and ingests a green alga, it switches its method of fulfilling its nutritional needs to one that uses sunlight to convert organic molecules, that is the process known as photosynthesis. The two organisms, the *Hatena* and the green alga, thrive in a symbiotic relationship.

Another organism, *Elysia chlorotic*, demonstrates a similar adaptation. *Elysia* is a slug found in salt and tidal marshes and creeks. The early stage of its life, referred to as its juvenile stage, characteristically involves movement, and it derives its nutrition by ingesting nutrients found in its surroundings. During this juvenile stage it is brown. As *Elysia* develops, if it meets a specific type of green algae, it enters its adult phase, during which chloroplasts from the ingested algae are retained in its digestive tract. The adult stage of *Elysia* is therefore green in colour. This symbiotic relationship between *Elysia* and the green algae allows the adult form of *Elysia* to take on a more sedentary lifestyle, depending on light being available to carry out photosynthesis.

Another source of evidence for the endosymbiotic theory is DNA. DNA provides a code made up of 64 different "words". Interestingly, this code has the same meaning in nearly all organisms on Earth and is said to be "universal". There are only a few variations, which can be explained by changes since the common origin of life on our planet. As already mentioned, the mitochondria of eukaryotic cells have a DNA code that more closely resembles bacteria than eukaryotic cells. Most scientists believe that the more DNA two organisms have in common, the more closely related they are to one another.

Biology is the study of life, yet life is an emergent property. Under what circumstances is a **systems approach** productive in biology, and under what circumstances is a **reductionist approach** more appropriate? How do scientists decide between competing approaches? A systems approach involves study of the larger picture of organisms, while a reductionist approach looks at smaller parts of organisms and then attempts to tie them together to understand the total organism.

Nature of Science

The endosymbiotic theory pulls together a wide range of observations. The strength of this theory results from the many observations as well as the numerous predictions it supports.

A2.2.13 and A2.2.14 – Cell specialization and multicellularity

A2.2.13 – Cell differentiation as the process for developing specialized tissues in multicellular organisms

Students should be aware that the basis for differentiation is different patterns of gene expression often triggered by changes in the environment.

A2.2.14 – Evolution of multicellularity

Students should be aware that multicellularity has evolved repeatedly. Many fungi and eukaryotic algae and all plants and animals are multicellular. Multicellularity has the advantages of allowing larger body size and cell specialization.

With evidence from the fossil record and the absolute dating techniques discussed in Chapter A2.1 a timeline can be constructed for life on Earth (Figure 8).

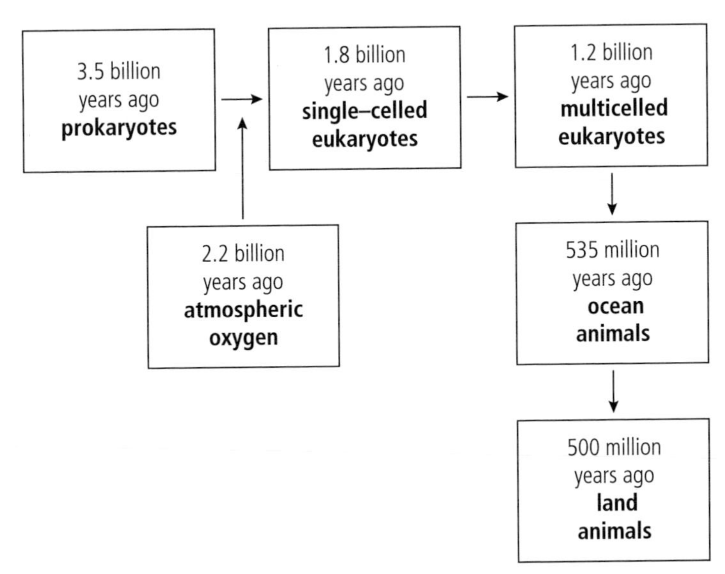

A2.2 Figure 8 A timeline of the major life-related events on Earth.

Endosymbiosis was one factor in the development of the cell. Another very important process that helped the development of the cell was **compartmentalization.** Membranes play a major role in the formation of compartments, enabling efficient reactions and processes to proceed. This compartmentalization resulted in specializations within the cell. One example of this is the nucleus, with a protective membrane, enabling the determiners of heredity, DNA molecules, to function without interference from other reactions taking place in the cytoplasm. Another specialized area of the cell is the mitochondria, where energy is produced so that the cell can perform the functions necessary for life.

Even with specialization, a single-celled organism has its limits, and it has not been successful in all environments. As shown in Figure 8, multicelled eukaryotes appear in the fossil record about 1.2 billion years ago. The presence of one eukaryotic organism possessing many cells eventually led to the differentiation of cells in that organism into highly specialized tissues and organs.

A tissue is made up of a group of similar cells working together to perform a common function. An organ is composed of two or more tissues working together to perform a common function.

The cells, tissues and organs that have developed in eukaryotic organisms coordinate and communicate with each other, resulting in organisms capable of thriving in most environments. This coordinated multicellularity appears to have evolved on more than one occasion, as many fungi and eukaryotic algae, and all plants and all animals, are multicellular.

It is not uncommon in nature to see aggregates of many cells. Often these aggregates are called colonies. Aggregates are not multicellular organisms. They have little differentiation of cells or coordination of function. An example of this is some types of *Volvox*. *Volvox* can exist as a spherical colony in which many green algae band together. The individuals are not organized into tissues or organs. There are also examples of bacteria that band together in a colony. However, each bacteria cell is genetically different and can exist individually.

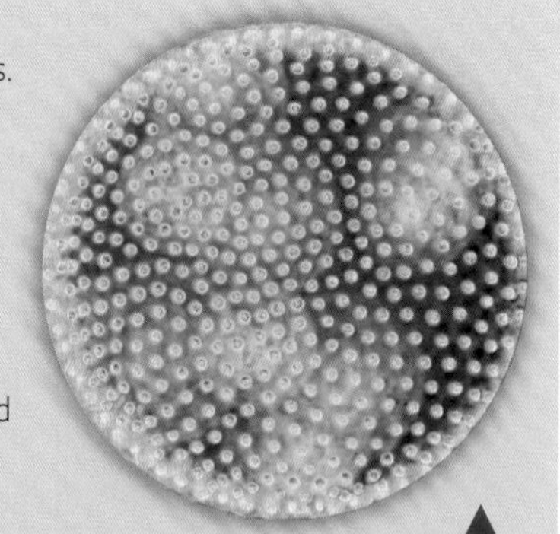

This spherical colony of *Volvox* is composed of hundreds of individuals banded together. However, if the colony is disrupted, each individual organism can exist on its own.

All cells in a multicellular organism have the same genetic information. For specialization and differentiation to occur, mechanisms have developed that control and coordinate gene expression. Differing environments can alter gene expression in the cells of multicellular organisms. All of these developments have allowed most environments on Earth to be inhabited by some form of life.

HL end

Guiding Question revisited

What are the features common to all cells and the features that differ?

In this chapter, we have discovered the following about cells:

- whether unicellular or multicellular, all organisms are composed of cells
- features common to all cells include DNA, cytoplasm and a plasma membrane forming an exterior boundary
- prokaryotic cells display a simple composition, lacking membrane-bounded organelles in their cytoplasm
- eukaryotic cells are compartmentalized, with isolated areas carrying out specialized tasks
- the cytoplasm of eukaryotic cells has many unique organelles working together, exhibiting all the life functions of the cell/organism
- variations of the cell structure result in some unique cellular compositions, such as cells with multiple nuclei and cells with no nuclei

HL

- evidence indicates that all eukaryotes evolved from a common ancestor
- endosymbiosis explains a mechanism for the development of some organelles of eukaryotic cells
- changes in gene expression result in differentiation of cells
- multicellularity appears to have evolved many times in various ways.

HL end

Guiding Question revisited

How is microscopy used to investigate cell structure?

In this chapter, we have discovered the following about microscopes:

- magnification and resolution are two properties of microscopes that are essential for the study of cells
- light microscopes have the advantage that living cells and tissue can be viewed
- EMs have increased the limits of magnification and resolution, allowing views of cells never thought possible even 50 years ago
- freeze fracture and fluorescent stains have furthered the study of cells via microscopy
- immunofluorescence using antibodies and specialized dyes has allowed visualization of the specific tissues viruses attack.

Exercises

Q1. Which pair of organelles is present in plant cells but not in animal cells?

A Chloroplasts and mitochondria.

B Centrioles and central vacuole.

C Chloroplasts and cell wall.

D Lysosomes and plasma membrane.

Q2. What carbon compound is most likely to be transported by rough endoplasmic reticulum?

Q3. Which of the following is not found in eukaryotic cells?

A Microtubules.

B Mitochondria.

C Nucleus.

D Chloroplasts.

Q4. Which cell type is the most likely to possess a capsule?

A Red blood cell.

B Prokaryotic cell.

C Sieve tube element.

D Eukaryotic cell.

Q5. What structure is directly related to prokaryotic cell reproduction?

A Cilia.

B Basal body.

C Centriole.

D Pili.

Q6. Which association is most accurate?

A Red blood cell: nucleus.

B Nucleus: mitochondrion.

C Basal body: ribosome.

D Golgi apparatus: vesicle.

HL

Q7. List three observations that provided a strong basis for the theory of endosymbiosis.

Q8. Compartmentalization was extremely important in the development of cell specialization. What cell organelle structure was instrumental in bringing about compartmentalization?

Q9. If all cells of a multicellular organism possess the same genetic information, how is specialization and differentiation of cells possible in that organism?

HL end

HL

A2.3 Viruses

Guiding Questions

How can viruses exist with so few genes?

In what ways do viruses vary?

Viruses are extremely small and not considered to be living, but their toll on this planet can be huge. The devastating outcomes of the Spanish flu, HIV and COVID-19 on human populations represent some of the staggering effect of these unique entities. Plants are also affected by viruses such as the tobacco mosaic virus and the potato virus Y. Bovine coronaviruses cause respiratory and intestinal infections in cattle and other wild mammals. As we battle the effects of viruses in the present, there is always the fear of what they may cause in the future.

As the world's human population continues to grow, diseases caused by viruses can spread ever more rapidly. Research is needed to understand virus activity and control, to ensure a healthy future.

A2.3.1 – Characteristics of viruses

A2.3.1 – Structural features common to viruses
Relatively few features are shared by all viruses: small, fixed size; nucleic acid (DNA or RNA) as genetic material; a capsid made of protein; no cytoplasm; and few or no enzymes.

Viruses and human history

The years 2019 and 2020 will be forever famous in medicine and history. A virus named SARS-CoV-2, which is the acronym for severe acute respiratory syndrome coronavirus 2, inserted itself into many humans around the world, and everyday life was turned upside down. The disease this virus causes is known as COVID-19. Starting in localized areas the virus soon spread, mostly by infected droplets emanating from sneezing, coughing and even talking, as people travelled from place to place. It rapidly progressed from an **outbreak**, to an **epidemic**, and finally to a **pandemic** designation.

An outbreak is said to occur when there is an unexpected increase in the number of people with a specific condition. An outbreak is given the designation of epidemic when it spreads over a large geographical area. A pandemic is an epidemic that has gone global and displays exponential growth.

Some past pandemics that had a huge impact include:

- the Black Death or Bubonic Plague of 1346–1353, which was caused by a bacterium known as *Yersinia pestis* and killed an estimated 25 million people
- the Flu Pandemic of 1889–1890, which was the result of an influenza virus and killed more than 1 million
- the Spanish Flu of 1918–1920, which was also caused by an influenza virus and killed more than 50 million people
- the Asian Flu of 1957–1958, another disease caused by the influenza virus, which resulted in more than 1.1 million deaths
- AIDS from 1981 to the present, which is caused by the human immunodeficiency virus (HIV) and has claimed nearly 35 million lives so far.

Epidemiology is the study of the occurrence, distribution and control of disease in a population. The World Health Organization (WHO), as well as national health agencies, work to understand disease trends and effective treatments. The goal is the well-being of everyone using the very latest, most advanced science.

The Spanish Flu of 1918–1920 was a devastating pandemic. Records indicate as many as 50 million people died from the disease. Some died within hours of the onset of symptoms. The age group most affected was between 20 and 40 years old. Measures used to slow the spread of the disease included masks, school closures and mandatory quarantines.

After two years, the flu disappeared almost as quickly as it had arrived. Because of this disease, research into the control of viruses using vaccines was given top priority.

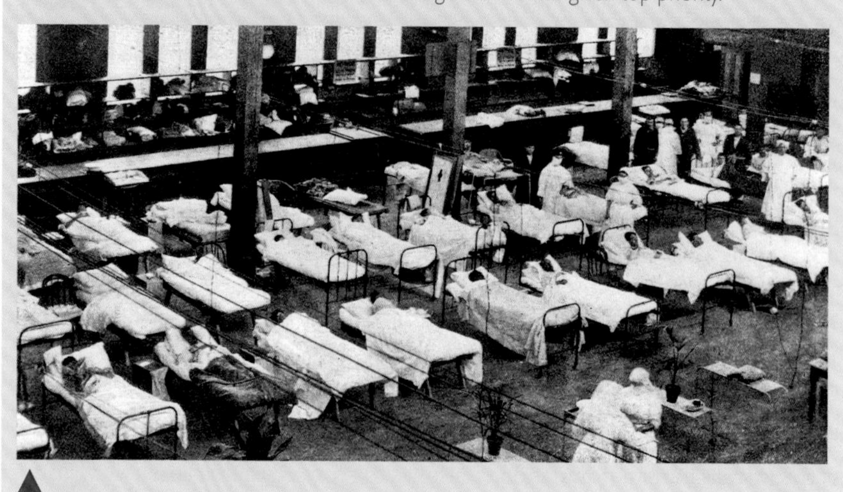

▲ This picture shows the condition of many facilities during the Spanish Flu of 1918. Facilities were over-run with patients. Many of the ill were turned away because of a lack of space. Notice that the health care workers wore masks.

Structural features of viruses

Virologists are medical doctors, scientists and public health professionals who are involved in the many facets of virus study. Their tasks include research, diagnosis, management and prevention of virus infections.

Viruses seem to be everywhere; nearly all living species can be infected by them. Viruses are thought to be the most abundant biological entity on our planet. However, it is important to remember that viruses are not considered to be alive, because they do not carry out all the functions of life (which is discussed in Chapter A2.1). They take over the machinery of infected living cells in order to carry out the activities necessary to ensure their own reproduction and ubiquitous presence. Because viruses have such a huge influence on life on Earth, the study of viruses is extremely important.

Viruses display tremendous diversity. There are certain characteristics, however, that are shown by all viruses:

- they are of a small, fixed size
- they contain a nucleic acid, either RNA or DNA, as genetic material
- they are enclosed by a boundary composed of protein called a **capsid**
- they do not contain cytoplasm inside the capsid
- they possess few, if any, enzymes.

The small, fixed size of each type of virus is a result of its extreme simplicity. The relatively small number of organic molecules present, and the composition of those molecules, prevents the formation of bonds to produce a larger entity.

Even though viruses possess one of the two possible forms of genetic material, they cannot replicate on their own. Regardless of whether they possess single-stranded or double-stranded genetic material, they possess very few genes, the units that control the characteristics of heredity. Most viruses contain fewer than 100 genes within their capsid, and many contain as few as five. The genetic material, whether DNA or RNA, of viruses also demonstrates variation by being linear or circular.

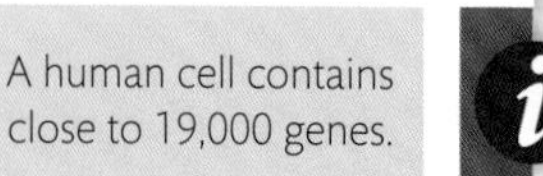

A human cell contains close to 19,000 genes.

The virus protein boundary, the capsid, of each virus is unique, based on the presence of specific amino acids and the structures they produce when bonding. This unique composition and structure determines the ability of a virus to infect a particular **host cell**. There are specialized sites on a virus capsid that allow attachment to specific host cells. A host cell is the cell a virus uses to carry out its metabolic and reproductive functions. In some viruses, the capsid contains specialized proteins that allow the genetic material of the virus to penetrate the host cell membrane. Some viruses also have an envelope outside their protein boundary derived from the host cell's plasma membrane. When present, this envelope is involved in host cell recognition and attachment. Inside the capsid, most of the space is occupied by the one type of nucleic acid present. However, there may also be enzymes, catalysts composed of proteins, present in this inner core area of the virus.

A2.3.2 – Structural diversity in viruses

A2.3.2 – Diversity of structure in viruses
Students should understand that viruses are highly diverse in their shape and structure. Genetic material may be RNA or DNA, which can be either single- or double-stranded. Some viruses are enveloped in host cell membrane and others are not enveloped. Virus examples include bacteriophage lambda, coronaviruses and HIV.

As already mentioned, viruses display variation in their genetic material, which can be either RNA or DNA, and single-stranded or double-stranded. Some viruses possess an envelope outside their capsid, while others are not enveloped. Most viruses are 10–400 nm in size, making them comparable to large protein macromolecules. Because of their small size, the electron microscope is needed to study virus shape and structure. Some viruses are as large as some bacteria, but they are exceptions.

There is also great variation in shape and structure among viruses. They can be threadlike, polyhedral and spherical in shape. Figure 1 illustrates some of the common virus shapes.

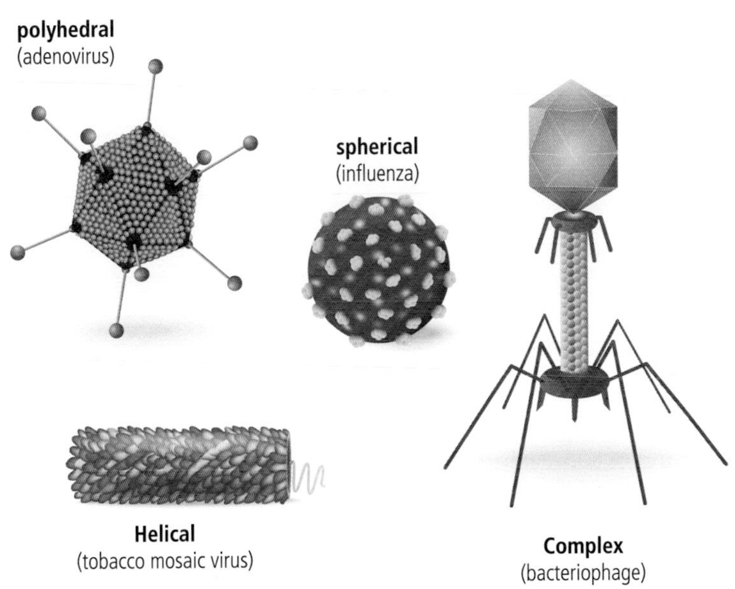

A2.3 Figure 1 The uniqueness and variation in shapes of viruses are illustrated here. Remember that viruses are very small, and these shapes were only revealed after extensive use of electron microscopy.

Three specific viruses and their shapes are the **bacteriophage lambda**, **coronaviruses**, and **human immunodeficiency virus (HIV)**. Figures 2 and 3 illustrate details of the bacteriophage lambda's structure.

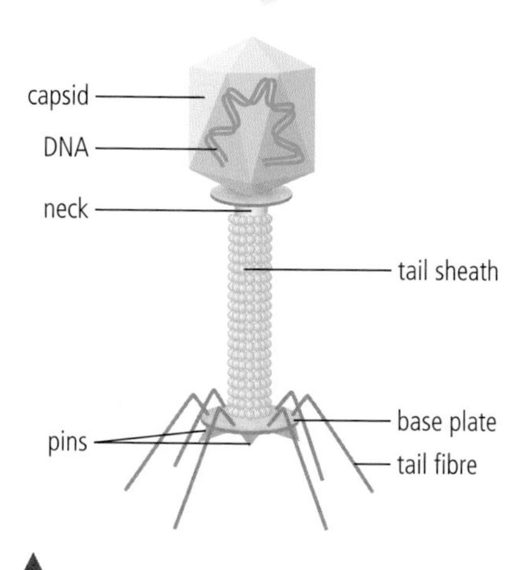

▲ **A2.3 Figure 2** A diagram showing the structural elements of bacteriophage lambda.

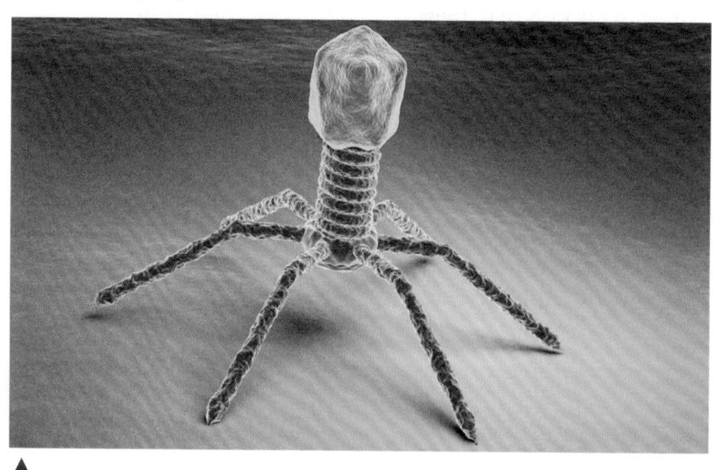

▲ **A2.3 Figure 3** Bacteriophage lambda, shown here, affects the bacterium *Escherichia coli* by attaching to specific regions on the bacterium known as attachment sites. Its DNA is then integrated into the DNA of the bacterium.

Important features of bacteriophage lambda are:

- a capsid head that protects the double-stranded DNA core
- tail fibres that attach the virus to the host cell
- a tail sheath that consists of proteins that contract to drive the tail tube through the host cell's outer membrane
- DNA that is injected through the tail into the host cell.

Figures 4 and 5 are of a coronavirus. Note the structures and the shape.

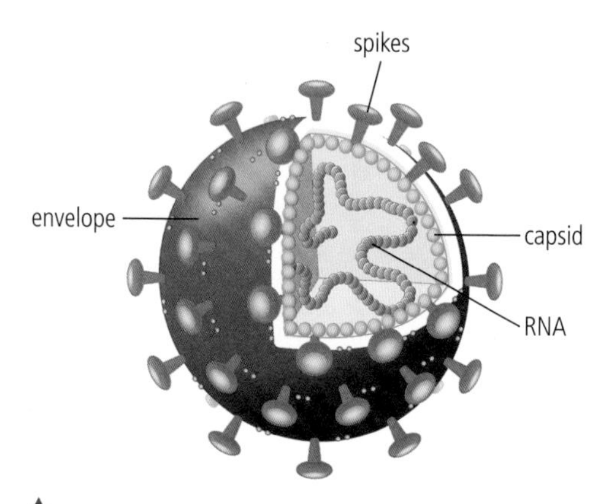

▲ **A2.3 Figure 4** A diagram of a generalized coronavirus.

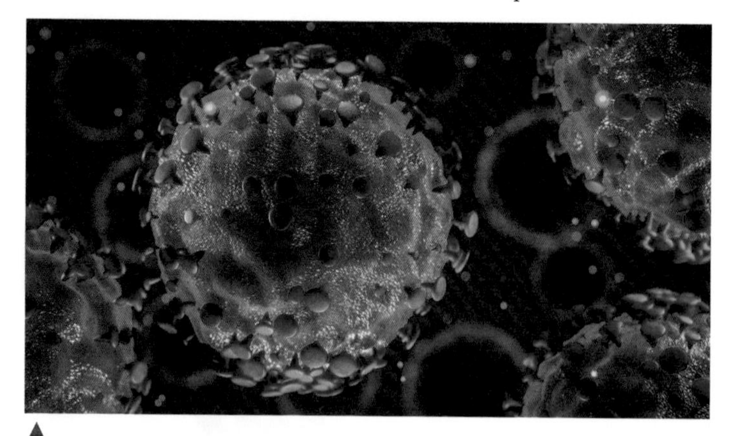

▲ **A2.3 Figure 5** The spikes visible on this image of the coronavirus that causes COVID-19 can bind to a specific receptor on human cells, fuse with the cell's membrane, and incorporate its RNA into the human cell.

Important features of coronaviruses are:

- a spherical shape
- single-stranded RNA as its genetic material
- an envelope outside the capsid
- numerous projections of spike proteins on the envelope, creating a "corona".

The outer spikes of a coronavirus latch onto specific receptors on host cells. Different types of host cells are affected by different viruses. The virus that causes COVID-19 latches onto cells in our respiratory tract because of receptors that exist on the surface of these cells. Once this occurs, the viral envelope merges with the cell membrane, allowing the virus to release its RNA into the host cell.

Figures 6 and 7 show the structures and shape of HIV.

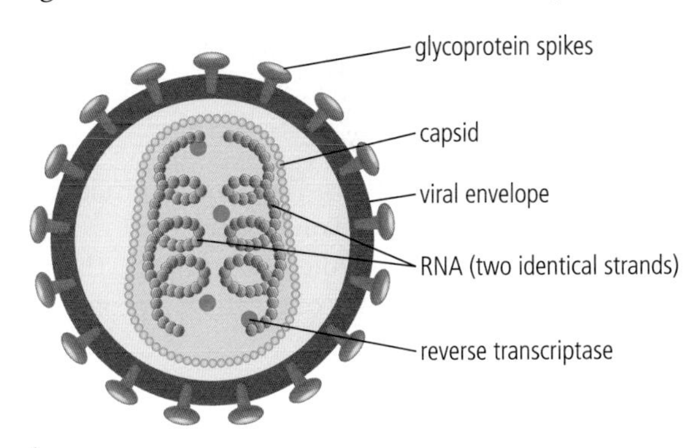

A2.3 Figure 6 A diagram of HIV, the virus that causes acquired immune deficiency syndrome (AIDS).

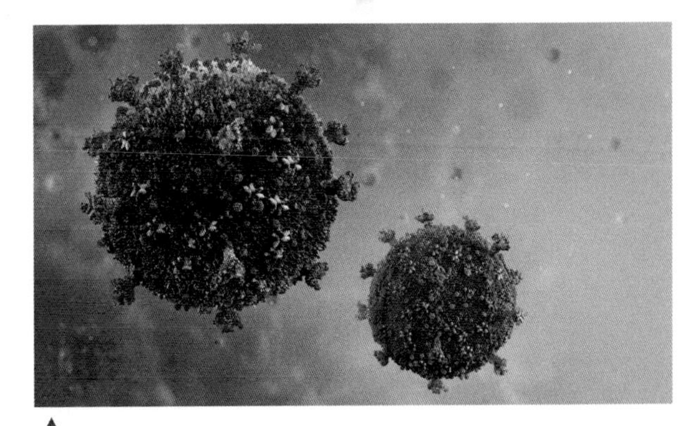

A2.3 Figure 7 A medical illustration showing a close up three-dimensional image of HIV. Notice that it has a similar shape to the coronavirus that causes COVID-19.

Important features of HIV are:

- it has an envelope outside the capsid
- two identical single strands of RNA, protected by the capsid
- within the viral RNA, reverse transcriptase is encoded, which allows the production of DNA using the viral RNA as a model
- it is known as a retrovirus because it makes a DNA copy of its RNA code
- the envelope spikes of HIV are made of protein and carbohydrate.

Retroviruses such as HIV can cause a number of serious diseases, including several types of cancer. The ability of retroviruses to produce DNA from RNA is because of the presence of an enzyme called reverse transcriptase.

Even though there is tremendous diversity in viruses, as shown in Figures 1–7, they are all obligatory intracellular parasites. That is, they rely on living host cells to multiply and to carry out the metabolic functions they require to exist. Each virus has a specific type of cells that it utilizes for its continued existence. This specificity arises from the attachment points of the virus that correspond with the host cell, and the particular properties needed in the host cell for the virus to carry out its necessary functions.

Virus type	Host cell specificity
HIV	White blood cells (CD4 cells)
Polio virus	Spinal nerve cells
Hepatitis virus	Liver cells
Norovirus	Cells of the intestinal system

Types of virus and their corresponding host cells

The challenge of treating a viral infection medically is to control the virus without interfering with the normal functioning of the host cell.

A2.3.3 and A2.3.4 – The life cycle of viruses

A2.3.3 – Lytic cycle of a virus
Students should appreciate that viruses rely on a host cell for energy supply, nutrition, protein synthesis and other life functions. Use bacteriophage lambda as an example of the phases in a lytic cycle.

A2.3.4 – Lysogenic cycle of a virus
Use bacteriophage lambda as an example.

Viruses must have a host cell to perform all their life functions. They cannot produce their own energy because they do not have mitochondria and only rarely have enzymes. They do not have vacuoles or lysosomes to digest potential nutrients. They only possess one nucleic acid, which means they cannot carry out their own protein synthesis. To reproduce, all viruses must:

1. attach to a site on a specific host cell
2. incorporate their genetic material into the cytoplasm of the host cell
3. use the host cell's processes to produce components of themselves
4. assemble the viral components into new functioning virus entities
5. release the new virus entities into the host cell's environment.

Bacteriophages are viruses that infect bacteria. The study of these phage viruses led to the discovery of two unique reproductive cycles.

The lytic cycle

Figure 8 shows the stages of the lytic cycle, following the five steps outlined above.

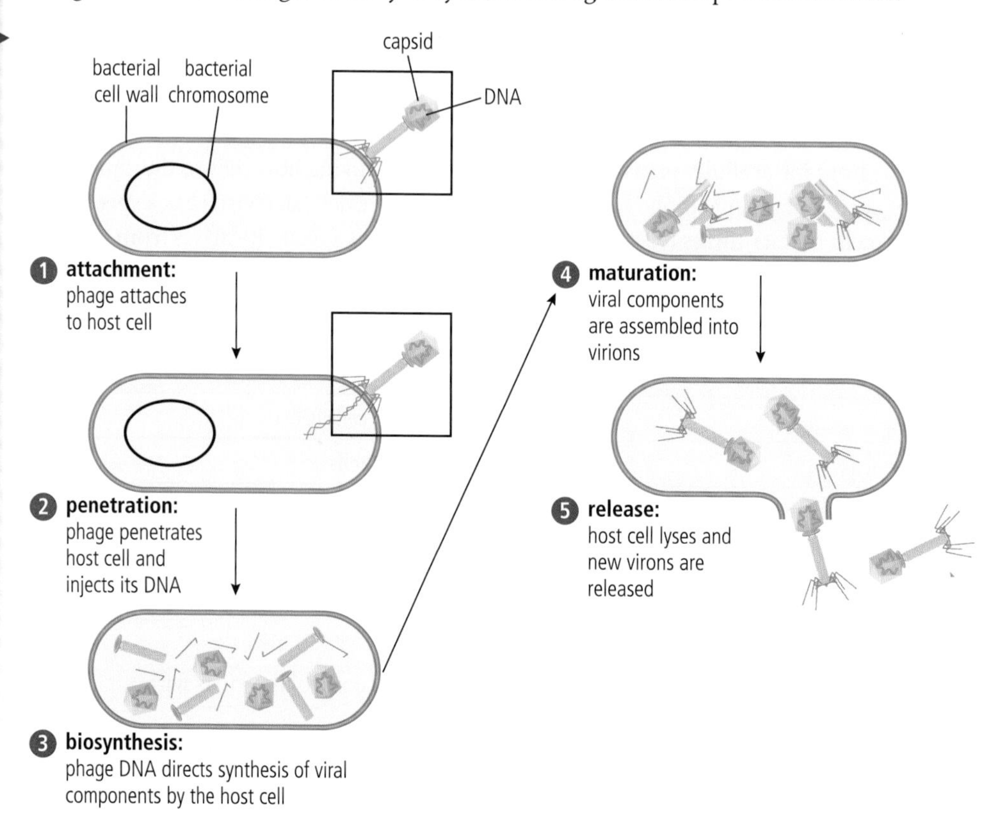

A2.3 Figure 8 The main stages of the lytic cycle when a bacteriophage invades a bacterial cell. The process results in the release of many new virus particles into the surroundings, which are then capable of attacking other bacteria. These new virus particles are also known as virions.

This life cycle is called lytic because the new virus particles are released by the lysis or rupturing of the cell membrane by an enzyme called lysozyme. This enzyme is coded for by the bacteriophage DNA. Lysis only occurs after the production of fully functional virus particles known as **virions**.

It is important to note that in the lytic cycle the bacteriophage DNA is translated or read to produce the proteins necessary for the assembly of new virus particles.

The lysogenic cycle

Unlike the lytic cycle, the lysogenic cycle does not result in the immediate release of new viruses. Instead, the DNA of the bacteriophage combines with the bacterial DNA to form a **prophage**. This prophage, or incorporated bacteriophage DNA, does not alter the bacterial DNA. The bacterium continues its usual life functions, including reproduction. The next generation of bacterial cells will carry the prophage within their **genome**. The genome of a cell is the genetic makeup of that cell.

Figure 9 illustrates the stages of the lysogenic cycle. Note that in the lysogenic cycle the viral DNA is integrated into the bacterial DNA, and as the bacterium replicates its DNA it also replicates the viral DNA. The viral DNA is inactive and does not produce protein. Only when the prophage separates from the bacterial DNA can a lytic cycle begin, with the protein made from the viral DNA. Usually the bacterium involved in this process is unharmed.

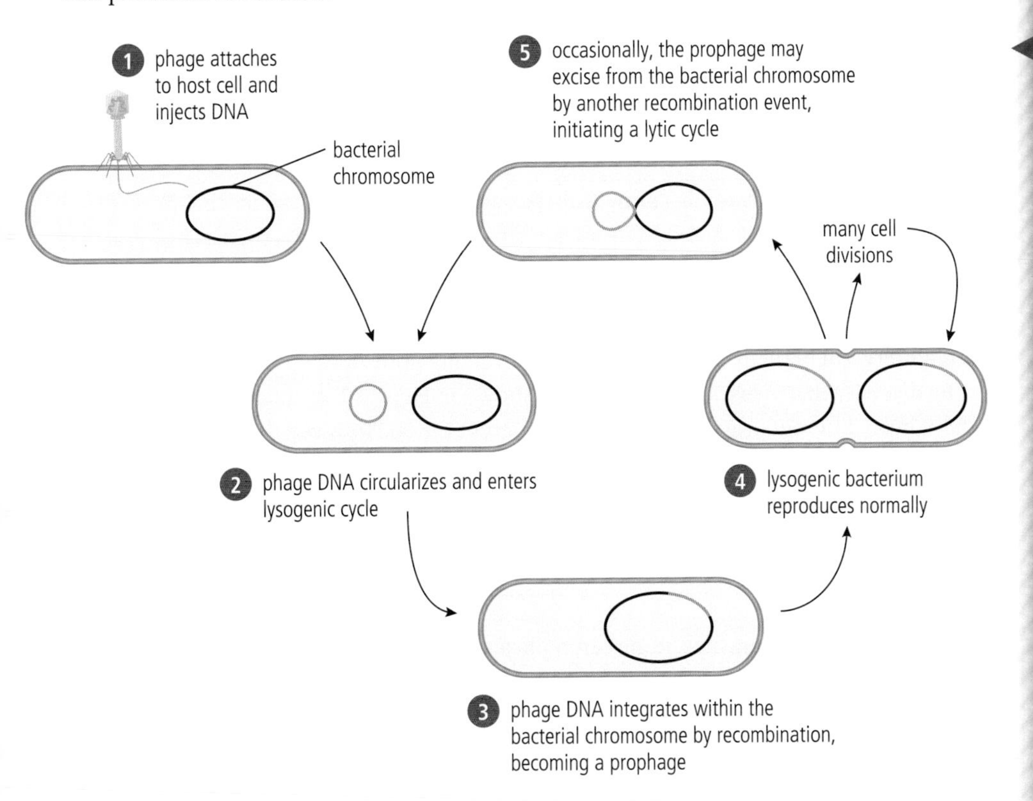

A2.3 Figure 9 The lysogenic cycle results in the bacteriophage DNA being incorporated into the bacterial DNA, producing a prophage. The prophage is an inactive form of the bacteriophage DNA and is replicated along with the bacterial DNA. It is present in all subsequent bacteria.

The integration of the latent prophage with the bacterial DNA can result in the continual production of bacterial cells containing the prophage. The prophage is totally inactive within these cells, unless a spontaneous event occurs that releases the bacteriophage DNA from the bacterial DNA. We do not yet know all the factors that can trigger such as spontaneous event, but ultraviolet light and/or certain chemicals appear to be involved. Once the prophage is released, a lytic cycle is initiated.

A2.3.5 – The origin of viruses

A2.3.5 – Evidence for several origins of viruses from other organisms
The diversity of viruses suggests several possible origins. Viruses share an extreme form of obligate parasitism as a mode of existence, so the structural features that they have in common could be regarded as convergent evolution. The genetic code is shared between viruses and living organisms.

Scientists estimate that 8% of the human genetic makeup comes from viruses. This indicates a coevolution of viruses and our species, especially our defence/immune system. Many researchers believe viruses have played a key role in the evolution of most life forms present today. Even though not able to live on their own, viruses have had, and continue to have, a tremendous impact on our planet. There are three leading hypotheses on the origin of the first viruses.

The **virus first hypothesis** states that viruses originated before cells. This idea is based on the simplicity exhibited by a virus compared to a cell, and that the course of evolution usually proceeds from the more simple to the more complex. This hypothesis also suggests that ancestors of modern viruses could have provided the materials necessary for the development of the first cells.

Convergent evolution is the process by which species of different evolutionary origin come to possess similar structures and functions as a result of sharing habitats and ecological roles that resemble one another. In convergent evolution, it is said natural selection has produced similar adaptations.

The **regressive hypothesis**, also known as the reduction or degeneracy hypothesis, states that viruses were once small cells that became parasites of larger cells. Over time, these small cells shed the structures that were no longer needed via gene reduction.

The **escape hypothesis**, also known as the vagrancy hypothesis, states that portions of genetic material, DNA and RNA, escaped from larger organisms such as bacteria and subsequently became surrounded by an outer boundary.

Each of these hypothesis has to overcome serious challenges to gain adequate credibility. All of them could be correct, and equally none of them may be correct. The diversity that viruses demonstrate, as shown in Figures 1–9, indicates there may have been different origins for viruses at different times. However, there are some interesting common structural, functional and genetic features of viruses that indicate **convergent evolution** may have occurred to a certain extent.

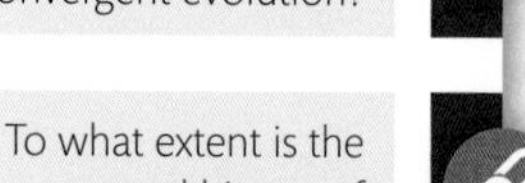

What mechanisms contribute to convergent evolution?

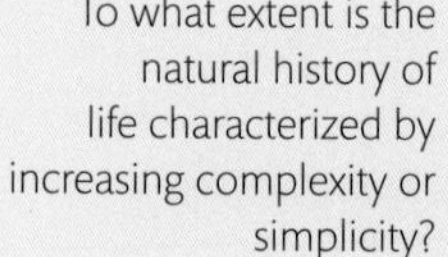

To what extent is the natural history of life characterized by increasing complexity or simplicity?

Features that suggest evidence of convergent evolution in viruses include the fact that they all:

- are obligate parasites, none can replicate or carry out the functions of life individually
- have a protein outer boundary, the capsid, with no cytoplasm
- have genetic material, DNA or RNA, inside the capsid, and the code of this genetic material is shared between viruses and all of Earth's organisms.

There are still many unanswered questions concerning viruses. A major problem with studying the history of viruses is the fact that they do not form fossils. Virology will continue to be a major research area for biologists, not only to better understand the past but also to predict the future of viruses and the diseases they may cause.

A2.3.6 – Rapidly evolving viruses

A2.3.6 – Rapid evolution in viruses
Consider reasons for very rapid rates of evolution in some viruses. Use two examples of rapid evolution: evolution of influenza viruses and of HIV. Consider the consequences for treating diseases caused by rapidly evolving viruses.

The influenza virus and HIV are two types of virus that display very rapid rates of evolution. Like other rapidly evolving viruses, these viruses have large population sizes, short generation times and high **mutation** rates. Mutations are changes in the genetic code, generating variation within a population. Both the influenza virus and HIV have single-stranded RNA as their genetic material, and therefore must produce DNA as part of their gene replication process. This added step increases their chances of mutation, which is a major source of variation within the virus population. DNA-containing viruses do not undergo this step and usually have lower rates of mutation.

It has been estimated that a litre of coastal seawater contains more viruses than there are people on the planet. Another estimation is that, if all the viral genes are unravelled and laid end-to-end, they would extend 250×10^6 million light years from Earth.

Recombination of genetic material in viruses

Although viruses are not alive, they still evolve in similar ways to living organisms. To continue their existence, they must be able to survive the immune system of organisms they infect. One way they can accomplish this is if they undergo genetic change. Genetic changes occur commonly in viruses either by **antigenic drift** or by **antigenic shift**.

Antigenic drift	Antigenic shift
Produces small incremental changes in the viral genetic material over longer periods of time	Two or more different viral strains or even different viruses infect the same cell and recombine genetic material, producing major changes in a relatively short time
This slowly produces variation in the surface proteins of the virus	This causes rapid and major changes in the surface proteins of the virus
The result is accumulated changes that may eventually prevent the immune system from recognizing the virus	The result is a totally new virus capable of creating pandemics because the immune system does not recognize the new viral strain
HIV undergoes very rapid antigenic drift, even within a single individual, creating problems for the immune system to control the virus	The influenza virus has an ability to undergo antigenic shift, creating seemingly "new" viruses that the immune system does not recognize

A summary of antigenic drift and shift

Vaccines can be helpful in the prevention and/or control of many viruses. Because of antigenic drift, the influenza vaccine has to be adjusted each year so that it is effective. This is a successful approach because the changes arising from antigenic drift are small and not particularly rapid. Conversely, antigenic shifts are rapid and bring about major changes. Vaccine adjustments are not as effective in this case, because the changes are not predictable enough to enable successful alterations of the vaccine.

Before vaccines, the virus that causes smallpox could spread rapidly and caused many epidemics. After the development of a vaccine, the virus did not evolve adequate genetic changes to survive, resulting in its extinction in the wild.

HIV is a unique virus because it undergoes antigenic drift at an extremely rapid rate. This is the reason why a successful vaccine has not been developed to prevent its spread.

Influenza viruses and HIV both have high mutation rates, with common recombination and reassortment of genetic material. This results in an ever-changing population of viruses capable of surviving previously effective immune systems and vaccines.

Research into the virus that causes COVID-19 is ongoing. Treatments have been altered to help control of the disease. New treatments for HIV have also emerged, as our knowledge of the disease has progressed.

When treating diseases caused by rapidly evolving viruses such as influenza virus and HIV, it is possible that viruses exist that are resistant to the treatment because of genetic variation. One form of the virus might be controllable, but a variant with resistance to this treatment may survive. This "selected" variant could then thrive.

Ideally, widespread use of a treatment to stop a rapidly evolving virus should completely inhibit virus development and replication. If it does not, new variants of the virus may emerge. The major problem we currently face when treating HIV and influenza viruses is that we are not able to completely inhibit the growth of all the variants of the virus. The result is a continuing occurrence of infection.

Access the eBook for a viruses worksheet.

Because antibiotics are only effective in the treatment of living disease-causing organisms, we cannot use them to stop a virus infection successfully. In some cases, the most effective treatment for a viral infection is to use the organism's own immune or defence system to control the virus. Another means of controlling virus infections in populations is isolation, to effectively stop transmission of the causal agent.

Guiding Question revisited

How can viruses exist with so few genes?

In this chapter we have described how:

- viruses cannot carry out all the functions of life on their own, but depend on a host cell for their continued existence
- the few genes a virus does have are effective in controlling the genes of the host cell so that they can replicate
- the structure of a virus is relatively simple and quite small.

Guiding Question revisited

In what ways do viruses vary?

In this chapter we have seen how:

- viruses show great variation in shape
- viruses can enter lytic or lysogenic cycles
- a virus has only one type of genetic material, either DNA or RNA, whereas all living organisms have both types of genetic material
- some viruses, such as HIV, undergo very rapid antigenic drift, making them difficult to control
- when the influenza virus undergoes antigenic shift, a totally new virus may be produced that is capable of causing extensive epidemics and even a pandemic.

Exercises

Q1. Explain why the obligate parasitism shown by viruses may have been a major factor in convergent evolution within the group.

Q2. Which of the following often play a role in the emergence of novel viruses?

A Mutations of the existing viral genetic material.
B Recombination of genetic material arising from two or more viruses.
C Treating an existing virus population while allowing some variants to survive
D All the above.
E None of the above.

Q3. Which of these functions, characteristics or structures are common to all virus particles and bacteria?

A Self-replication
B Ribosomes
C Independent existence
D Genetic material
E Nucleus

Q4. Compare the lytic and lysogenic cycles of a virus.

Q5. Describe the possible outer boundaries of a virus.

HL end

A2 Practice questions

1. (a) An organelle is a discrete structure within a cell with a specific function. In the table below, identify the missing organelles and outline the missing functions.

Name of organelle	Structure of organelle	Function of organelle
Nucleus	Region of the cell containing chromosomes, surrounded by a double membrane, in which there are pores.	Storage and protection of chromosomes.
Ribosome	Small spherical structures, consisting of two subunits.	
	Spherical organelles, surrounded by a single membrane and containing hydrolytic enzymes.	Digestion of structures that are not needed within cells.
	Organelles surrounded by two membranes, the inner of which is folded inwards.	

(4)

(b) The table above shows some of the organelles found in a particular cell. Discuss what type of cell this could be. (2)

(Total 6 marks)

2. (a) Distinguish between the terms *resolution* and *magnification* when applied to electron microscopy. (2)

The electron micrograph below shows part of a cell.

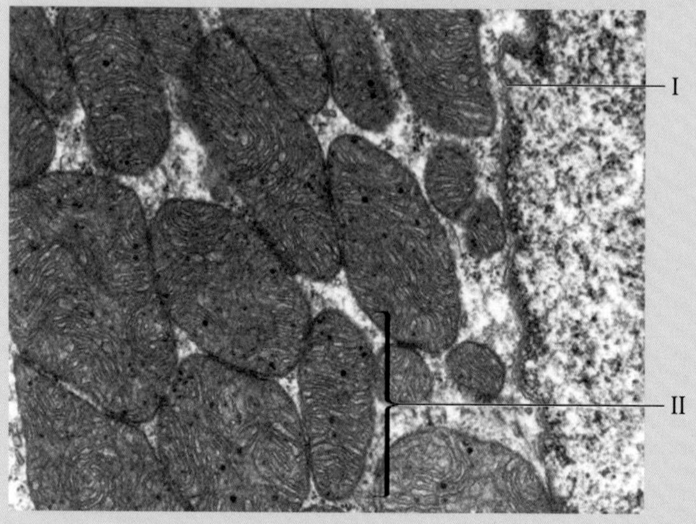

(b) Identify the structures labelled I and II. (2)

(c) State **one** function of the structure labelled II. (1)

(d) Deduce, with a reason, whether this cell is eukaryotic or prokaryotic. (1)

(Total 6 marks)

3. Compare, with the aid of a diagram, the structure of generalized prokaryotic and eukaryotic animal cells.

(Total 8 marks)

4. Explain evidence for the existence of a last universal common ancestor (LUCA).

(Total 4 marks)

HL

5. Draw a bacteriophage lambda and include labels.

(Total 7 marks)

6. The attempt to explain the origin of life on our planet is quite challenging. One of the first tasks faced in this endeavor is to define life itself. Models have often been utilized in possible explanations about how the first living cells appeared.

(a) Describe some key features of early Earth that may have led to the origin of life.

(4)

(b) The Miller–Urey apparatus is one model that has gained favour in presenting a mechanism for the origin of life. Evaluate the effectiveness of this apparatus in presenting a means by which the first life could have spontaneously appeared.

(4)

(c) Recently, there has been scientific support for the origin of life near oceanic hydrothermal vents. List factors providing possible evidence for this hypothesis.

(3)

(Total 11 marks)

7. Important to the formation of the first living cells was a molecule that would allow control and inheritance. Many hypothesize this first molecule was RNA. Outline the properties of RNA that have led to this hypothesis.

(Total 3 marks)

8. Explain the steps in the process by which viruses reproduce.

(Total 4 marks)

9. Compare the general features of the lytic and the lysogenic cycles as first described in the study of bacteriophages that infect bacteria.

(Total 4 marks)

HL end

THEME

A Unity and diversity

3 Organisms

From single-celled organisms to coral reefs to trees, life on Earth shows a remarkable degree of variation. For centuries, physical characteristics have been used to name organisms and to put similar organisms into categories. More recently, thanks to DNA sequencing, we can use the genetic code of an organism (its genome) to help show how closely it is related to other species.

HL Cladistics is a way of classifying life forms based on evolutionary relationships. Cladograms are diagrams similar to a family tree but made with species rather than with parents and offspring, and they allow us to study which species are most closely related by examining recent common ancestors.

A3.1 Diversity of organisms

Guiding Questions

What is a species?

What patterns are seen in the diversity of genomes within and between species?

Although there are at least two dozen definitions for the concept of species in biology, we will examine two: the morphological definition that has been used for hundreds of years, and the biological species concept definition, which has only existed in the past few decades. The first looks at what physical features organisms have, while the second considers whether or not individuals can breed to produce fertile offspring. Each definition has its strengths and weaknesses. No single definition can encompass all living organisms as well as extinct species, because such an astoundingly large diversity exists among the various forms of life on Earth.

When DNA sequences of organisms are compared, it is possible to see that, between individuals of the same species, there are remarkably few differences compared to the differences between individuals belonging to two different species. A single-celled organism with no specialized tissue is likely to have a much smaller quantity of DNA than a multicellular organism with hundreds of different specialized tissues.

A3.1.1 – Variation between organisms

A3.1.1 – Variation between organisms as a defining feature of life
Students should understand that no two individuals are identical in all their traits. The patterns of variation are complex and are the basis for naming and classifying organisms.

If you have pigeons where you live, you might think that they all look the same. But ask pigeon experts and they will tell you that the level of diversity and variation among pigeons is equivalent to the level of diversity and variation in humans. Animal breeders such as pigeon fanciers recognize each individual in the population they are raising, just as you would recognize your dog in a group of similar dogs. No two individuals in a population share all the same traits. Even identical twins have slight differences.

Observing the differences between individuals within one species and observing the differences between one species and another is a daunting task, especially when we consider that there are millions of species on Earth to observe, from invisible microbes to mighty redwood trees over 100 m tall.

How can we classify organisms? There are countless possible ways; a few examples are listed below.

- By feeding habits: it makes its own food/it is a carnivore or herbivore.
- By habitat: land-dwelling/aquatic.
- By movement: sessile (stuck in one place)/free moving.
- By daily activity: nocturnal/diurnal.
- By risk: harmless/venomous.
- By anatomy: plant/animal/vertebrate/invertebrate.

We generally start by categorizing organisms based on **morphology** (the physical appearance of an organism). Is the organism made of a single cell without a nucleus, or does it have a nucleus? If it has a nucleus, is it single-celled or multicellular? Think of these categories as boxes into which the organisms are placed. Each category is called a **taxon** (plural taxa). The biggest taxa are very broad and encompass many species, but as the defining features used become more and more detailed and specific, smaller and smaller boxes are used, containing fewer and fewer species per taxa, until we arrive at a single species. The largest taxon is a "domain" and it contains all the more specific taxa, from "kingdom" down to "species".

Table 1 illustrates the identification of two species from very different kingdoms: one species is an animal, humans, and the other is a plant, garden peas. The science and skill of categorizing life is called **taxonomy** and specialists who do it are called **taxonomists**.

The garden pea (*Pisum sativum*) is the plant Gregor Mendel studied.

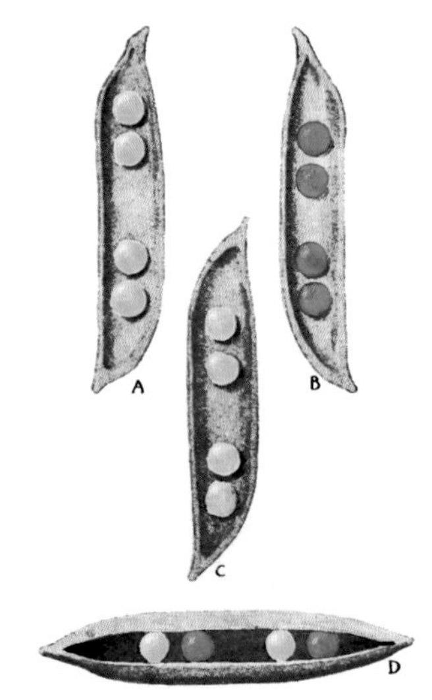

Taxa	Human	Garden pea
Kingdom	Animalia	Plantae
Phylum	Chordata	Angiospermophyta
Class	Mammalia	Dicotyledoneae
Order	Primate	Rosales
Family	Hominidae	Papilionaceae
Genus	*Homo*	*Pisum*
Species	*sapiens*	*sativum*

A3.1 Table 1 The classification of two species

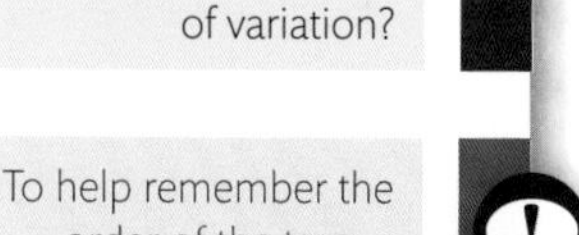

How do species exemplify both continuous and discontinuous patterns of variation?

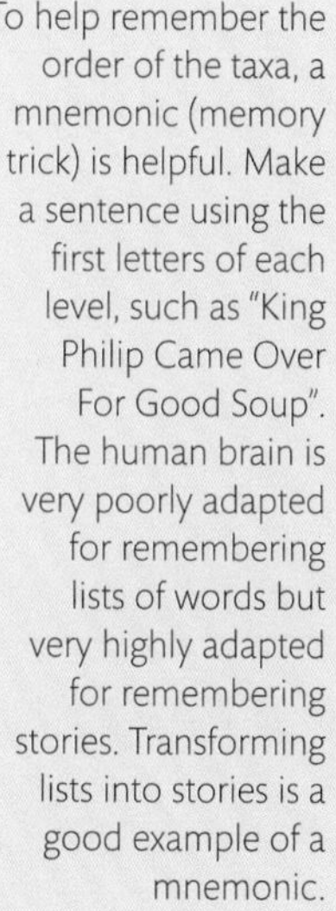

To help remember the order of the taxa, a mnemonic (memory trick) is helpful. Make a sentence using the first letters of each level, such as "King Philip Came Over For Good Soup". The human brain is very poorly adapted for remembering lists of words but very highly adapted for remembering stories. Transforming lists into stories is a good example of a mnemonic.

The variations in characteristics for sorting species into their designated taxon might be obvious (plants have leaves and roots, whereas humans have limbs and a head), but can sometimes be very subtle. Two species of frog might look identical on the outside but can be distinguished by different mating calls. In such a case, the patterns of variation in morphology are not sufficient for classification.

When variation can be placed into distinct categories (type A blood versus type B, for example), we say it is **discontinuous**. When variation has a wide range of possibilities (how tall a tree can grow, for example), we say it is **continuous**. Sometimes we impose categories such as eye colour as if it is an example of discontinuous variation when, in fact, a hundred people who have blue eyes will show a certain amount of continuous variation, from deep blue to very light blue.

A3.1.2 – Species as groups of organisms

A3.1.2 – Species as groups of organisms with shared traits
This is the original morphological concept of the species as used by Linnaeus.

Carolus Linnaeus, an 18th century professor of medicine and botany in Sweden, had difficulty identifying the plants he found on his travels because different botanists used different systems for naming them. This made it difficult to categorize the organisms. Linnaeus then had a remarkable idea: what if we take all the known living organisms, put them into categories, and give them a name using a uniform system? Not just plants, but animals, too. By creating the names using Latin or Greek, no matter what anyone calls the organism in their native language (such as Swedish), it will always have a universally known name.

Linnaeus based the classification system, as well as the names, on the physical features of the organisms. This **morphological classification**, first published in his book *Systema Naturæ* in 1735, was used by generations of botanists and zoologists, and the naming system he created is still used today. Thousands of organisms still carry the scientific name that Linnaeus gave them over two-and-a-half centuries ago, such as the Asian elephant, which he named *Elephas maximus* in 1758.

A3.1.3 – The binomial naming system

A3.1.3 – Binomial system for naming organisms
Students should know that the first part of the name is the genus, the second part of the name is the species. Species in the same genus have similar traits. The genus name is given an initial capital letter but the species name is lowercase.

You have a scientific name based on your species: *Homo sapiens*. This system of naming organisms using two names is called **binomial nomenclature**. "Bi" means two, "nomial" means name and "nomenclature" refers to a system used to name things.

Myrmecophaga tridactyla is a name that literally means "eater of ants" plus "with three fingers". This name refers to the giant anteater of Central and South America. In fact, the animal really has five fingers, but they are hard to see because the animal walks on its front knuckles.

▲ The giant anteater (*Myrmecophaga tridactyla*).

In the early days of classification, all known organisms were classified into only two kingdoms: plants and animals. With the invention of the microscope in the mid-1600s, many new creatures were discovered that were nothing like plants or animals. In effect, the microscope revealed that there is an entire world of invisible organisms living throughout the world's ecosystems.

The first name in the binomial nomenclature system is always capitalized and it refers to the **genus**; the second name always begins with a small letter and refers to the **species**. Both are always written in italics when typed, or underlined when written by hand. Organisms in the same genus will have a higher number of similar characteristics compared to organisms in a different genus.

There are three main objectives and associated rules to using binomial nomenclature:

1. each organism has a unique name that cannot be confused with another organism
2. the names can be universally understood, no matter what nationality or culture is using the name
3. there is some stability in the system, so that people cannot change the names of organisms without valid reasons.

Examples of binomial nomenclature

Sometimes scientific names for organisms are relatively easy to decipher because they contain their common names:

- *Amoeba amazonas*
- *Equus zebra*
- *Gekko gecko* (this lizard gets its name from the sounds it makes)
- *Gorilla gorilla*
- *Paramecium caudatum* (caudate means having a tail).

Homo sapiens

Genus ***species***

The rules about writing binomial nomenclature names are that:

- the genus name is capitalized but the species name is not
- both are written in italics when typed, or underlined when handwritten.

SKILLS

Sometimes, it is more difficult to guess their common name:

- *Apis mellifera* (honeybee, although you might have guessed this if you know that beekeeping is also called apiculture)
- *Aptenodytes patagonicus* (king penguin, although you can probably guess where it lives from its species name)
- *Loxodonta cyclotis* (African forest elephant)
- *Malus domestica* (apple tree).

Scientists naming organisms sometimes have a sense of humour. Here are some examples.

- *Agra schwarzeneggeri* Erwin, 2002. This Costa Rican ground beetle was named after Arnold Schwarzenegger because of the insect's large biceps.
- *Dracula vampira* Luer, 1978. This orchid in Ecuador got its name from the fact that the petals on the flower look like a bat's wings

In taxonomy, there are two opposing philosophies concerning what to do when an organism does not fit easily into existing categories: (1) broaden the definition of an existing category to include the new organism; or (2) invent a new category or subcategory. Specialists who take the first approach are referred to as **lumpers**, while those who take the second approach are referred to as **splitters**.

Challenge yourself

1. Look up the following to find out what their scientific names are:

- your favourite animal
- your favourite fruit or vegetable
- your favourite flower, tree or house plant.

A3.1.4 – Biological species

A3.1.4 – Biological species concept
According to the biological species concept, a species is a group of organisms that can breed and produce fertile offspring. Include possible challenges associated with this definition of a species and that competing species definitions exist.

Another definition of a species that is now often preferred over Linnaeus' morphological definition is the **biological species concept**. This was proposed by Ernst Mayr in 1942. Using this definition, in order to be classified as the same species, individuals must be able to breed together and produce fertile offspring. All modern dogs, *Canis familiaris*, can interbreed to produce fertile offspring, so they are considered to be one species.

▲ All domestic dogs are of the same species.

Not every biologist is happy with this definition, however. How can this definition apply to organisms that reproduce asexually and therefore do not breed? Hybrids produced from parents of closely related but separate species are usually infertile, but not always. Some species are made up of a mosaic of DNA from multiple species. How should they be classified? Should they receive multiple species names if they are composed of more than one? How can we apply the concept to extinct species such as velociraptors when we cannot know from skeletons whether members of a population could interbreed?

Depending on which expert you ask, there are dozens of definitions of the word "species". We have discussed two so far: the morphological definition used in the 18th century, and a more recent definition, the biological species concept, involving the ability to breed and produce fertile offspring. But other characteristics can also be taken into account when deciding on what counts as a species, such as the following.

- The ecological niche of an organism. Because microbes are single-celled, it is challenging to use just morphology to determine what species they belong to. Where they live and what they eat can help classify microbes into different species.
- Genetics. When a sequence of DNA found in a sample of soil from a forest does not match any known sample, it suggests that it is from a species that has not been catalogued yet.
- The types of molecules an organism can produce. This is also useful when classifying microscopic organisms that do not have easily observable features, unlike birds and primates, for example. It is common to find microbes that produce carbon dioxide, but some can make methane or hydrogen gas.
- For extinct species, their lineage. If we find a fossil of an extinct snail that has a shell similar to a modern species, we can use the similarities to assign it a species name based on its position on the same part of the evolutionary tree as the existing species.

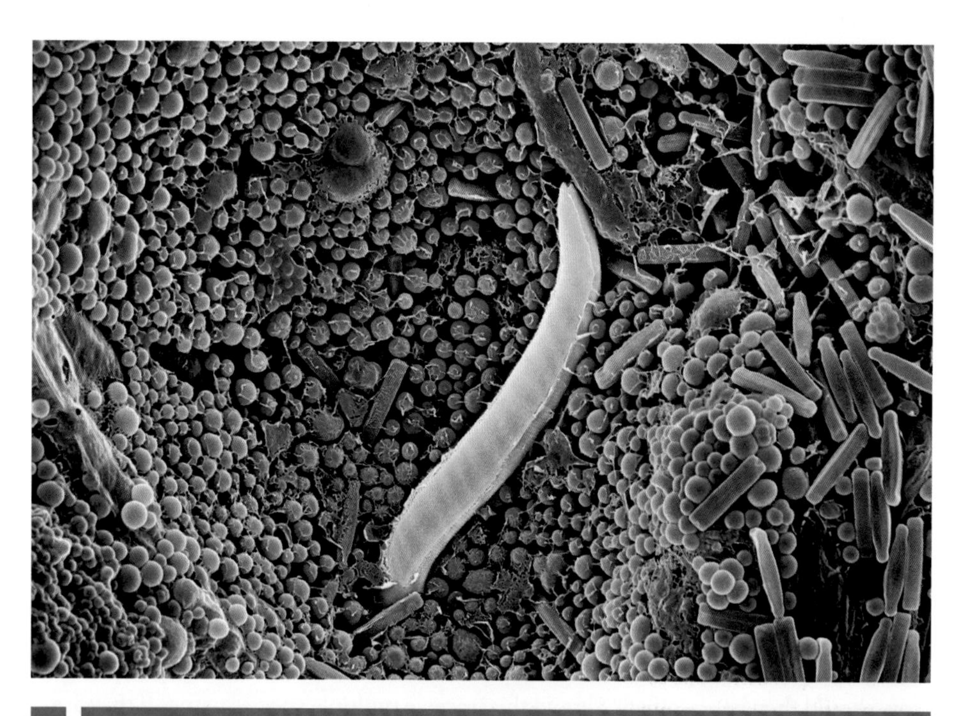

Microscopic soil organisms can be challenging to identify because morphology is insufficient as a criterion to differentiate species.

Nature of Science

To some extent, the debate about what a species really is becomes just as philosophical as biological. "Is all we are doing simply naming things?" "Do the categories we use actually exist in reality or just in our minds?" "Is the difficulty of agreeing on a definition a fault of the limitations of language?" "Is it possible to use the same term (species) for organisms that exist today and to express how their populations evolved over time?" These questions are currently being debated by biologists and, because the variety of life is so diverse, it is difficult to find a consensus.

A3.1.5 – Distinguishing between populations and species

A3.1.5 – Difficulties distinguishing between populations and species due to divergence of non-interbreeding populations during speciation
Students should understand that speciation is the splitting of one species into two or more. It usually happens gradually rather than by a single act, with populations becoming more and more different in their traits. It can therefore be an arbitrary decision whether two populations are regarded as the same or different species.

Speciation, as explored in more detail in Chapter A4.1, is the process by which a population is separated into two groups that can no longer reproduce together. One part of the population evolves one way and the other, living with different selection pressures and producing different sets of mutations, evolves in a different way. The two populations become different enough over time that they can no longer interbreed to produce fertile offspring. As a result, a new species has branched off from the previous one, resulting in two species that have a common ancestor.

Lake Victoria in East Africa is, geologically speaking, a young lake, being only about 400,000 years old. Any fish species that live there have arrived since then. African cichlid fishes, of which there are over 200 species in the lake, all appear to have evolved from a single species introduced about 200,000 years ago. Each one has evolved in its own niche and as a result split off from the others. Some specialize in eating algae, some eat plankton and others eat snails. But each split would have taken many generations and, during those generations, the population that started to explore the new source of food would have continued to interbreed with some success with the original population. As the two populations became more different from each other, the success rates of interbreeding would have diminished until it was no longer possible. It is difficult for specialists to decide when the speciation occurred. When a cut-off point is chosen, it has an arbitrary and subjective aspect to it.

The last woolly mammoth became extinct thousands of years ago. It appeared to share many similar characteristics with today's Asian elephants (*Elephas maximus*), which is why it was originally classified in 1799 in the same genus, as *Elephas primigenius*. Because of the gap in time, it is difficult to apply the biological species concept to decide whether or not the two populations are one and the same species, because there are no living mammoths to test the hypothesis by breeding them with elephants. The mammoth's scientific name has since been changed to *Mammuthus primigenius*, without knowing for sure whether they could breed together or not, so it is a relatively arbitrary decision from the point of view of the biological species concept.

What might cause a species to persist or go extinct?

The woolly mammoth went extinct thousands of years ago. We cannot test whether it was able to breed with modern elephants or not.

Figure 1 shows a common ancestor giving rise to four species. The first speciation event shown happened earlier in time, then the split that generated species B occurred, and, finally, D split from C. Although this type of diagram helps illustrate the sequence of events, it gives the impression that the splits occurred suddenly, which is not always the case.

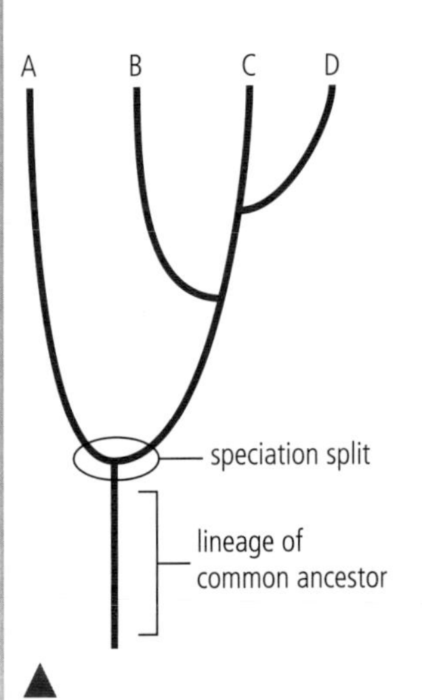

A3.1 Figure 1 Species A, B, C and D evolved from a common ancestor. Three speciation splits led to the generation of these species, the first of which is circled.

A3.1.6 – Diversity in chromosome numbers

A3.1.6 – Diversity in chromosome numbers of plant and animal species
Students should know in general that diversity exists. As an example, students should know that humans have 46 chromosomes and chimpanzees have 48. Students are not required to know other specific chromosome numbers but should appreciate that diploid cells have an even number of chromosomes.

Diploid and haploid cells

The term **diploid** is used to describe a nucleus that has chromosomes organized into homologous pairs. Most cells in the human body are diploid cells, and in such cells the nucleus contains a set of 23 chromosomes from the mother and 23 from the father. There is a category of cells that only contain 23 chromosomes in total: the sex cells, also called **gametes**. Because the chromosomes in sperm and egg cells do not come in pairs, but rather only have a single chromosome from each pair, they are said to be **haploid**. The adult form of animal cells is rarely haploid, but there are exceptions, for example adult male bee, wasp and ant cells are haploid. Generally speaking, the vast majority of cells in sexually reproducing organisms are diploid, and only the gametes are haploid.

Note in Table 2 that diploid cells always have an even number of chromosomes. This is logical because one chromosome in each pair comes from one parent and the other from the other parent.

The variable n represents the **haploid number**, and it refers to the number of sets of chromosomes that a nucleus can have. For a human egg cell, $n = 23$. When an egg cell is fertilized by a sperm cell (a sperm is also haploid and therefore contains 23 chromosomes), a **zygote** is formed and the two haploid nuclei fuse together, matching up their chromosomes into pairs. Hence humans generally have a total of $23 + 23 = 46$ chromosomes. This means that in humans, $2n = 46$, so diploid cells in humans have 23 pairs of chromosomes making a total of 46 chromosomes. Compare this number with some of the other species in Table 2.

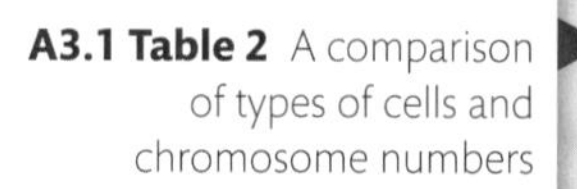
A3.1 Table 2 A comparison of types of cells and chromosome numbers

	Types of cells and chromosome numbers	
Species	**Haploid = n**	**Diploid = $2n$**
Human (*Homo sapiens*)	23	46
Chimpanzee (*Pan troglodytes*)	24	48
Domestic dog (*Canis familiaris*)	39	78
Rice (*Oryza sativa*)	12	24
Roundworm (*Parascaris aquonum*)	1	2

The number of chromosomes is a characteristic of a species

As you can see from Table 2, the number of chromosomes for humans (46) is very different to the number of chromosomes for the roundworm. One of the best-studied worms in genetics laboratories is *Caenorhabditis elegans*, whose genome was first sequenced in 1998. It has six chromosomes, meaning its diploid number, $2n$, is 6, and therefore its haploid number, n, is 3. It would be expected that all the cells in *C. elegans* would have six chromosomes, and, likewise, that all cells in humans would have 46. Although this is true for most cells, we have already seen the exception of haploid cells (n). Note as well that some cells do not contain a nucleus and have no chromosomes, such as red blood cells.

A3.1.7 – Karyotypes

A3.1.7 – Karyotyping and karyograms
Application of skills: Students should be able to classify chromosomes by banding patterns, length and centromere position. Students should evaluate the evidence for the hypothesis that chromosome 2 in humans arose from the fusion of chromosomes 12 and 13 with a shared primate ancestor.
NOS: Students should be able to distinguish between testable hypotheses such as the origin of chromosome 2 and non-testable statements.

A **karyogram** is a representation of the chromosomes found in a cell arranged according to a standard format, as in the example in Figure 2. The chromosomes are placed in order according to their size and shape. The shape depends mainly on the position of the **centromere**. A karyogram is used to show a person's **karyotype**, which is the specific number and appearance of the chromosomes in their cells.

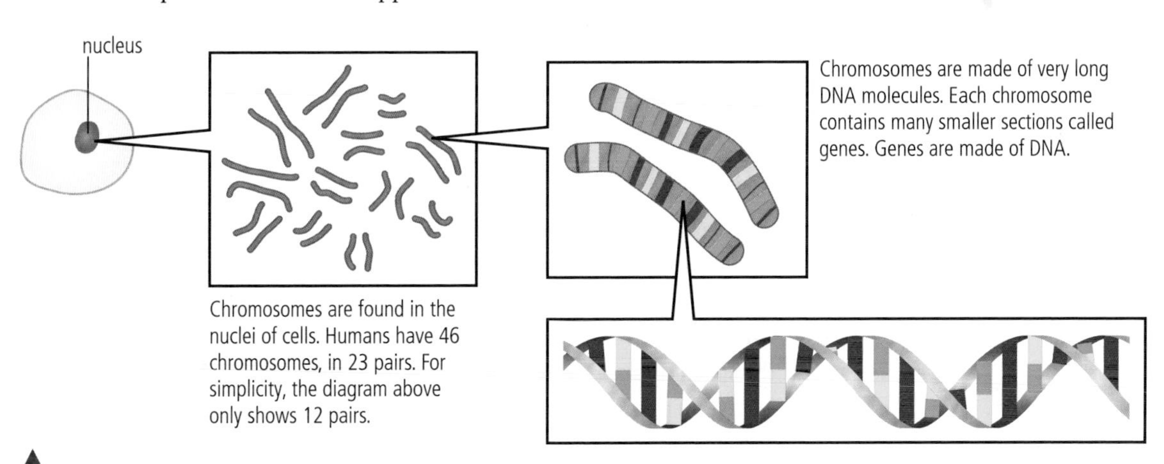

▲ Zooming into a cell reveals where DNA is found.

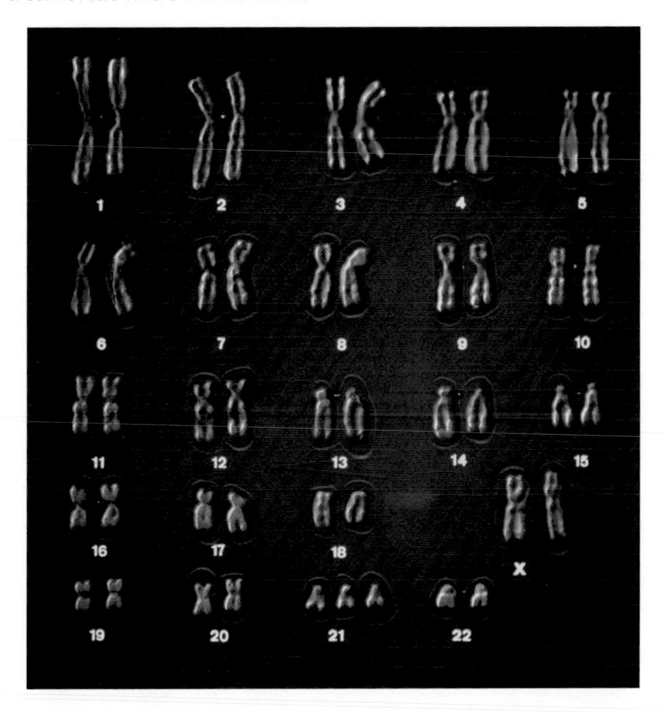

◀ **A3.1 Figure 2** This is a karyogram showing all 23 pairs of chromosomes. What can we learn about the individual's karyotype from this figure? This karyogram was prepared using false-colour imagery.

SKILLS

You can use online tools to prepare your own karyogram by arranging chromosomes by size, banding patterns and the position of the centromere. The website Learn.Genetics from the University of Utah has an activity called "Make a karyotype", for example. Once you have made a karyogram, you can learn certain details about the person. Use the karyogram in Figure 2 to determine whether the individual is a male or a female. How do you know? Does the individual's karyotype include any anomalies? If so, describe what you see. For more about the consequences of extra or missing chromosomes, see Chapter D2.1.

How is a karyogram image obtained? Once the cells of an organism have been collected and grown in culture, a karyogram is made following the steps below.

1. The cells are stained and prepared on a glass slide, to see their chromosomes under a light microscope.
2. Photomicrograph images are obtained of the chromosomes during a specific phase of cell division called the mitotic metaphase (see Chapter D2.1).
3. The images are cut out and separated, a process that can be done using a print out and scissors or on a computer.
4. The images of each pair of chromosomes are placed in order by size and the position of their centromeres. Generally speaking, the chromosomes are arranged in order by decreasing length. The exception is in the 23rd pair of chromosomes, which can contain one or two X chromosomes, which are considerably larger than the chromosomes in the 22nd pair (see the chromosome pair marked X in Figure 2). In addition, the coloured bands that show up in the image can be used to identify which chromosome it is. For example, chromosomes 3 and 4 in the image show very different banding patterns.

The evolution of human chromosome 2

Modern humans have 46 chromosomes. Other human species that no longer exist but whose preserved fossil DNA we can study, such as Neanderthals and Denisovans, also had only 46 chromosomes. Gorillas and chimpanzees are the species most closely related to humans. Our last common ancestor with gorillas existed about 9 million years ago and the speciation split with chimpanzees occurred about 6 million years ago. However, when we prepare a karyogram of the contents of their nuclei, both gorillas and chimpanzees have 48 chromosomes instead of 46. If we shared a common ancestor with them, what happened to our chromosome number?

Two possible hypotheses can be formulated:

1. a complete chromosome disappeared
2. two chromosomes from an earlier common ancestor fused to become a single chromosome.

It is unlikely that an entire chromosome was deleted and disappeared, because removing hundreds of genes in that way would cause a major threat to the viability of the species. To test the second hypothesis, we can look for evidence, and can start by examining the two characteristics that help identify a chromosome: its shape (position of the centromere) and its banding patterns. One shape a chromosome can have is the "X" shape, with the centromere close to the centre. This is called a **metacentric** chromosome. Chromosomes can also have an **acrocentric** shape, meaning the centromere is at one end, making one arm of the chromosome much shorter and the other much longer. All primates have both types.

One hypothesis is that chromosome 2 in humans arose from the fusion of chromosomes 12 and 13 in a shared ancestor. In an article from *Molecular Cytogenetics* by Paweł Stankiewicz in 2016, human chromosome 2 was compared to chimpanzee chromosomes 12 and 13. In terms of shape, these two acrocentric non-human chromosomes, when placed end to end, have a similar length to the human chromosome, although some parts overlap. The position of the centromere in human chromosome 2 lines up with the chimpanzee chromosome 12 but not with chromosome 13. This latter piece of evidence refutes the hypothesis. However, in the zone marked B on the human chromosome in Figure 3, we find the type of DNA we usually encounter in the centromere, known as **satellite DNA**, which consists of short repeating sequences of DNA. This zone corresponds to the position of the centromere in the non-human chromosome 13, giving credibility to the hypothesis. In terms of banding patterns, the long arm of chimpanzee chromosome 12 matches that of the short arm of human chromosome 2, and the long arm of chimpanzee chromosome 13 matches the banding patterns of the long arm of human chromosome 2.

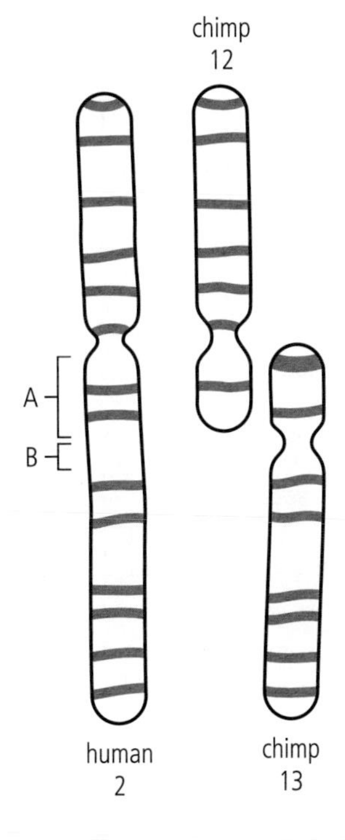

A3.1 Figure 3 A comparison of human chromosome 2 with chimpanzee chromosomes 12 and 13.

Besides shape and banding patterns, other evidence to support the idea of fusion is the presence of telomeric DNA in the centre of human chromosome 2. The **telomeres** are caps at the tips of chromosomes that contain repeating sequences of DNA and provide protection, the same way that bumpers protect cars and aglets protect the ends of shoelaces. Such repeating telomeric DNA is not supposed to be in the centre of chromosomes, only at the tips. And yet, at position A in the human chromosome 2 shown in Figure 3, telomeric DNA is present at the position where the two chromosomes would have fused.

It is very important to understand that this evidence does not say we descended from chimpanzees. The fusion of the chromosomes would have happened after the speciation split of a common ancestor that led to the evolution of chimpanzees on one branch of the tree of life and the evolution of humans on another branch.

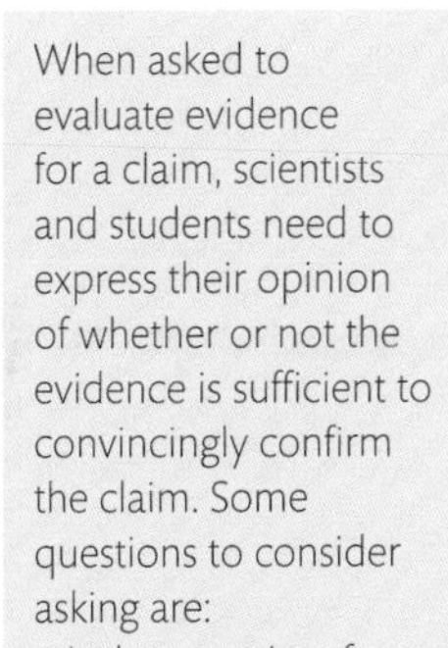

When asked to evaluate evidence for a claim, scientists and students need to express their opinion of whether or not the evidence is sufficient to convincingly confirm the claim. Some questions to consider asking are:

- Is the quantity of evidence sufficient to accept the claim?
- Has the method for collecting evidence been repeated and tested by other scientists, and have they found similar evidence?
- Is the method being used a reliable method?
- Are any counterclaims or refuting evidence enough to doubt the claim?
- Is there a mechanism to explain the cause, or is what we are seeing just a coincidence?

Nature of Science

Some claims are testable and others are not. The hominid fossil nicknamed Lucy, discovered in Ethiopia in 1974, is complete enough to test and confirm claims such as (1) she was a female, (2) she was not a modern human but rather an australopithecine, (3) she could walk on two legs and (4) she lived about 3.2 million years ago. There might be some debate about the details, but the challenges can also be tested. Can you think of any claims about her that would not be testable? For example: "Lucy had a great sense of humour." "Lucy had a recurring dream where she encountered a wildcat." "Lucy spoke three languages." Current tools in science have no way of testing these claims. Statements like these are speculation. What about these: "Lucy had very little meat in her diet." "Australopithecines such as Lucy had strong spiritual beliefs." Are they testable claims?

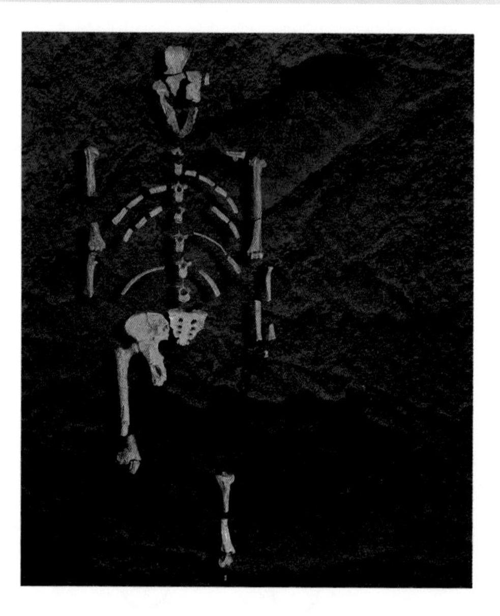

Some claims about the fossil called Lucy are testable and others are not.

A3.1.8 – Unity and diversity of genomes

A3.1.8 – Unity and diversity of genomes within species
Students should understand that the genome is all the genetic information of an organism. Organisms in the same species share most of their genome but variations such as single-nucleotide polymorphisms give some diversity.

It seems counterintuitive, but it is possible to find lots of evidence to support the claim "we are all the same", and it is also possible to find lots of evidence to support the claim "we are all different". From a genetics point of view, humans share many more similarities than differences with each other, especially compared to another species.

If a chimpanzee was walking down your street, you would recognize right away that it was a non-human primate. And yet, the genetic difference between us and chimpanzees is only about 4%. That is a much bigger difference, however, than between you and other humans, which is estimated to be 0.1% to 0.6%. Why does *Homo sapiens* display so many similarities within its global population? Our unity arises largely from the fact that all humans share the same genes. We do not all have the same versions of each of the genes (called **alleles**, see Chapter D3.2); some of us have type B blood and some have type O, for example. But we all possess the genes that determine the ABO blood type.

Where do we find these small but crucial differences between humans? The estimated 3 million to 20 million **base pairs** (e.g. A–T or G–C) of our DNA sequence that can reveal the differences are found scattered all over our chromosomes. Where most people have a T (thymine) nucleotide, for example, a small portion of humans might have a G (guanine) instead at that position. Such variations can start out as mutations (see Chapter D1.3) but are then passed down from generation to generation. Such a variation involving only one base is called a **single nucleotide polymorphism** or SNP (see Figure 4). It is estimated that about every 100 to 300 bases in a human's genetic code contains an SNP. Geneticists interested in the human genome have identified millions of SNPs, and they can be used to help determine ancestry or risk of genetic diseases.

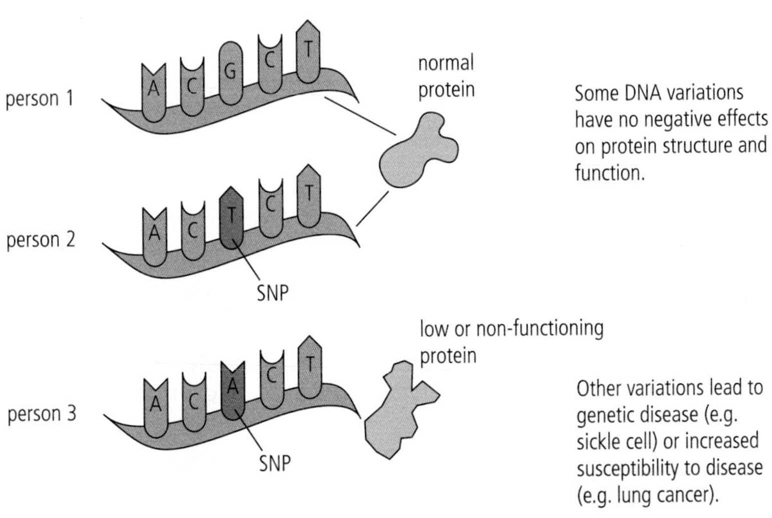

▲ **A3.1 Figure 4** Person 1 has a gene that expresses a normal protein. Person 2 has a T (thymine) nucleotide instead of a G (guanine) in the SNP, but also expresses a normal protein. Person 3, however, has an SNP that causes the protein to not form correctly.

Only about 5% of SNPs are functional, meaning they actually produce a difference in a person's body. Most are neutral, meaning that they will not affect a person's **phenotype** (the physical expression of a gene, such as blood type or colour vision, see Chapter D3.2).

The Human Genome Project

In 1990, an international cooperative venture called the Human Genome Project set out to sequence the complete human **genome**. Because the genome of an organism is a catalogue of all the bases it possesses, the Human Genome Project hoped to determine the order of all the bases A, T, C and G in human DNA. As there were approximately 3,200,000,000 to find, it took over a decade. In 2003, the Project announced that it had succeeded in achieving its goal. Now, scientists are working on deciphering which sequences represent genes and which genes do what. The human genome can be thought of as a map that can be used to show the position of any gene on any one of the 23 pairs of chromosomes.

i In the 1997 science fiction film *GATTACA*, one of the main characters brings a sample of cells to a walk-up window at an establishment that provides anonymous genome services. Within seconds, she gets a full printout and analysis of the genome she is interested in. How far are we from being able to do this today? What ethical implications are there to such a service? Are there laws protecting your genome?

Thanks to modern communication technologies, it is possible for scientists working all over the world to collaborate and contribute to a scientific endeavour such as sequencing the genome of plants that help feed the world. Rice is one example: biologists from 10 countries contributed to sequencing the first rice genome.

The current estimate is that humans have approximately 22,000 genes, and, thanks to advances in technology, the sequencing of a person's genome can be done in hours instead of years.

TOK

Many companies offer genome sequencing for private citizens willing to pay the price. Some of the products reveal ancient family origins and risk factors for some health problems, such as the chances of developing certain types of cancer or heart disease. Would you want to know if there is a chance that your life could be suddenly shortened by the presence or absence of a certain gene? Would you tell your family and friends? Would you want your parents to have such a test? Should people tell their employer or each other about any health-related issues revealed by a genomic analysis? Or, in contrast, is this a private, personal thing that no one else needs to know about? How accurate and reliable are these analyses? Should we believe everything they say? Does all knowledge impose ethical obligations on those who know it?

A3.1.9 – Eukaryote genomes

A3.1.9 – Diversity of eukaryote genomes
Genomes vary in overall size, which is determined by the total amount of DNA. Genomes also vary in base sequence. Variation between species is much larger than variation within a species.

No humans have genes for characteristics such as bioluminescence (glowing in the dark), which many deep-sea organisms do. Although we see some diversity among humans, we do not see such huge ranges of diversity in the human population as wings for flight, gills to breathe underwater, echolocation organs for seeing without light, chloroplasts for photosynthesis, and so on. There is more unity within the human species (comparing any two humans) than diversity compared to other species (comparing humans to non-humans).

Humans are a diverse global population but there are remarkably few differences between any two humans compared to differences between humans and other species.

One major difference between genomes is their size: the quantity of DNA they have in their nuclei. As we will see in Section A3.1.10, some eukaryotic genomes only have a few thousand genes while others can have tens of thousands of genes. This means that one eukaryote will possess genes that another will not have at all. A fish does not need to have genes to produce pollen, and a rose bush does not need genes for making fins to swim. Even with closely related species that have undergone a relatively recent speciation split, they have been evolving separately to the point where the genes are now different enough that they cannot interbreed anymore.

Such differences can be seen in the sequences of base pairs in each genome. Sequencing technology along with databases and computer programs for searching and comparing large data sets have allowed biologists to compare the genomes of organisms from all over the world.

Bioinformatics is a research field that uses both computer science and information technology to help us understand biological processes. Bioinformatics has grown exponentially in recent years. The most data-rich area of bioinformatics is genomics. Genome data is now available in public databases such as The National Center for Biotechnology Information (NCBI). Genetic information can also be explored using the following databases:

- Swiss-Prot, a database of protein sequences
- Ensembl, a database and browser of genomic information about humans and other vertebrates
- GenBank, a National Institutes of Health genetic sequence database that is an annotated collection of all publicly available DNA sequences.

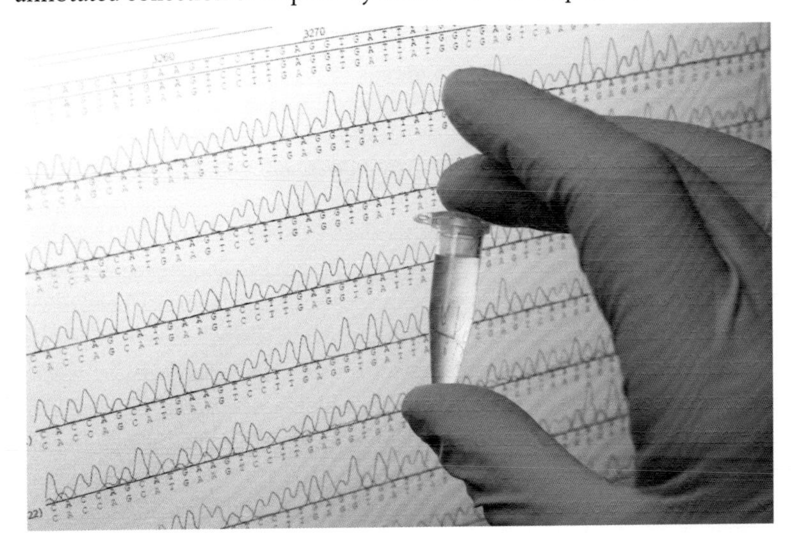

A micropipette containing a DNA sample can be sequenced and added to a database and shared worldwide thanks to web-based information technology.

Instead of sifting through the entire genome of an organism, one way to compare genetic diversity in eukaryotes is to focus on their **mitochondrial DNA**. All eukaryotes have mitochondria, and the way mitochondrial DNA, present only in the egg, not in the sperm cell, is passed down from mother to offspring, means there is not the shuffling and mixing that we see in chromosomal DNA. It is estimated that, within a species, roughly 1 in 1,000 of the genetic code letters is different between individuals' mitochondrial DNA. These genetic differences are expressed in the amino acid sequence that is coded for by the organism's DNA sequence. To see differences between individuals within a species, or to see differences between species, it is possible to look up the amino acid sequences for a particular gene in a database and match them to see if there are amino acids missing, added or modified. Instead of the DNA bases A, T, C and G being displayed, the letters in the databases correspond to the 20 possible amino acids, such as S for serine, G for glycine, A for alanine and V for valine. Some amino acids have a letter that is different from their first letter, such as E for glutamic acid, F for phenylalanine and K for lysine. You will not be asked to memorize the 20 amino acid names and their letters, but you do need to understand that, when comparing genetic differences, it is possible to either use the DNA code or the amino acid sequences.

Table 3 shows part of the sequence for a single gene selected from the online UniProt protein database. The chosen gene is one that all eukaryotes have in their DNA: *cyc1*, the gene for cytochrome c, which is a protein needed by mitochondria to perform their essential task of cellular respiration, to convert sugar into usable energy. Of the hundreds of species available in the database, four species of animal were selected and, rather than looking at all the amino acids that the gene codes for, a short sequence of 60 amino acids was selected for comparison. The differences between the first species and the three other species are highlighted in yellow.

A3.1 Table 3 Comparing a short sequence of 60 amino acids from the mitochondrial gene, *cyc1*, for cytochrome c, in four species

Database codes for specific species	Fragment of the sequence of amino acids coded for in the *cyc1* gene
golden-crowned babbler: TR|A0A7K9SBC6|A0A7K9SBC6_9PASS	SL--ALALSLGGGPLSAGELELHPPNFPWSHGGPLSALDHASVRRGFQVYRQVCSACHSM
brown-headed cowbird: TR|A0A7L3VSC4|A0A7L3VSC4_MOLAT	SLAVALSLSLGGGPVSAGELELHPPGLPWSHGGFLSALDHASVRRGFQVYRQVCSACHSM
green anole: TR|H9GCG1|H9GCG1_ANOCA	GLAVALH-----SAVSAGELELHPPSFPWSHSGPLSSLDHSSVRRGYQVYKQVCSACHSM
big-headed turtle: TR|A0A4D9DRJ9|A0A4D9DRJ9_9SAUR	GLALALH-----TAVSASDLELHPPSYAWSHNGLLASLDHSSIRRGYQVYKQVCAACHSM

The first organism in Table 3 is a bird, the golden-crowned babbler (*Sterrhoptilus dennistouni*), which lives in the Philippines. The next three organisms in Table 3 are a brown-headed cowbird (*Molothrus ater*), a lizard called a green anole (*Anolis carolinensis*), and a big-headed turtle that lives in Southeast Asia (*Platysternon megacephalum*). If we look at the first amino acid in the sequence for the first species, we see S, for serine. Moving down the second column in Table 3, we see that species 2 also has an S but species 3 and 4 have G for glycine instead. Species 1 does not have any amino acids at positions three and four, while the other three do. Of those three, they all have A for alanine in the third position but not all have V for valine in the fourth.

Not surprisingly, compared to the first bird's sequence, there are more differences in the lizard and in the turtle than there are in the other bird species, because the two bird species are more closely related to each other than they are to lizards and turtles. If we looked at the whole amino acid sequence and not just the fragment of 60 amino acids used for Table 3, we would see that the three species in Table 3 have the following percentage of matches with the golden-crowned babbler: 92.9%, 84% and 76.8%, respectively.

Between any two golden-crowned babblers, we would expect more than 99% of the amino acid sequence to be identical, with only one difference every few hundred amino acids. This illustrates that there is much more diversity between organisms in different species compared to organisms within the same species.

The Human Genome Project has shown that there are only a very small number of DNA bases that make one person different from any other person in the world. This creates a feeling of unity. All humans carry inside them a common genetic heritage.
On the other hand, the Human Genome Project has shown that the small differences that do exist make each person unique in terms of skin colour, facial features and resistance to disease, for example. These differences should be appreciated and celebrated as strengths. Unfortunately, they are often the basis of discrimination and misunderstanding.
Can one group of people be considered genetically superior to another? History has shown that many people think so, yet genetics shows that this is not the case. All human populations, whatever slight differences their genomes may have, deserve equal esteem as human beings.

A3.1.10 – Genome sizes

A3.1.10 – Comparison of genome sizes
Application of skills: Students should extract information about genome size for different taxonomic groups from a database to compare genome size to organism complexity.

Using online tools, it is possible to compare the genome of an organism, such as a fruit fly, with other eukaryotes. Table 4 shows data extracted from the NCBI database at the time of writing; because the database is being continually updated, the numbers you find might be different.

A3.1 Table 4 A comparison of genome sizes of various organisms

Species	Genome size in millions of base pairs, Mb
Saccharomyces cerevisiae, baker's yeast	12.1
Drosophila melanogaster, fruit fly	143.7
Mus musculus, house mouse	2,500
Escherichia coli, bacterium	5.12
Homo sapiens, modern human	3,200
Neoceratodus forsteri, Australian lungfish	34,557.6
Plasmodium falciparum, a parasite that causes malaria	22.9
Oryza sativa, rice	420
Caenorhabditis elegans, a nematode worm	100

Escherichia coli, a bacterium that likes to live in your large intestine, has about 5 million letters (base pairs) in its DNA code.

Do you get the impression that the more complex an organism is, the bigger its genome is? For example, we think of humans as being extremely complex and advanced, so when we compare ourselves to the fungus in Table 4, the baker's yeast, we see that our genome size is hundreds of times bigger. But rice has only three times more DNA than the fruit fly. And when we compare our human genome size to the Australian lungfish, it is ten times smaller. Does that mean lungfish are more complex than we are or that we are more complex than yeast? It depends on our definition of complex. Although they may not be capable of doing creative and complex tasks such as sending a spaceship to Mars, both lungfish and yeast can survive in conditions in which humans would die. The examples given and the ones you can find on your own will often give the impression that genome size can indicate complexity, but there are enough exceptions to conclude that it is not a reliable indicator.

A3.1.11 – Whole genome sequencing

A3.1.11 – Current and potential future uses of whole genome sequencing
Include the increasing speed and decreasing costs. For current uses, include research into evolutionary relationships and for potential future uses, include personalized medicine.

Researchers are very excited about genome sequencing because it allows them to identify species and compare them to see evolutionary relationships. They can compare whole genome sequences to see how organisms are related to each other. Such a technique is known as **phylogenetics**. In general, organisms that share similar genomes tend to be more closely related than those that do not.

In Figure 5, the mouse is shown to be much more closely related to the chimpanzee than to the salamander. The DNA sequences (or corresponding amino acid sequences) of the mouse and the chimpanzee would show fewer differences between each other than if one of their DNA sequences was compared to the salamander's genome. In humans, it can tell us about our ancestry, and about possible health risks related to the genes we have inherited.

A3.1 Figure 5 A phylogenic tree of vertebrate chordates.

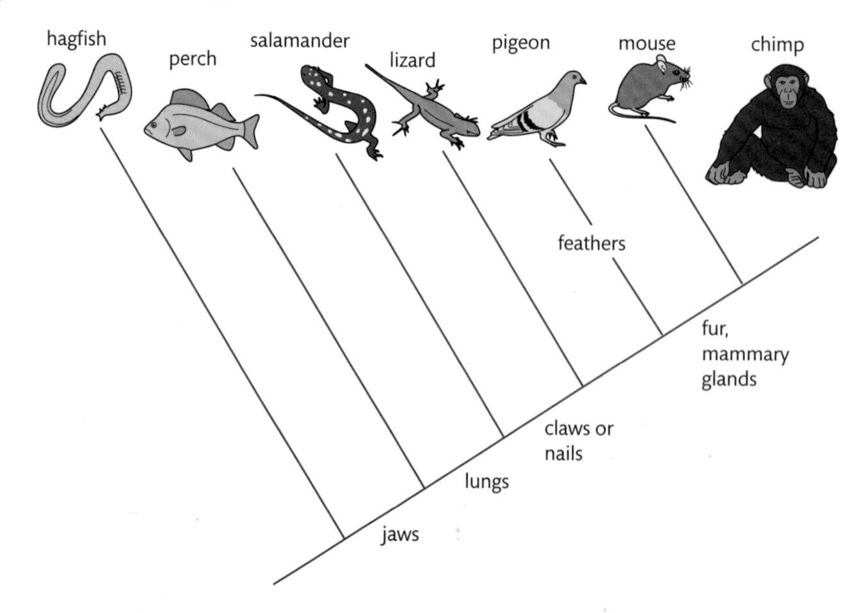

Thanks to **next-generation sequencing techniques**, which use a mix of laboratory hardware, chemical markers and powerful software to increase the speed and decrease the cost of sequencing people's genomes, it is possible for private citizens in some countries to get their genomes sequenced. Other countries have made it illegal to request genome sequencing: laws have been put in place to protect people's privacy. A parent who has put up a child for adoption and does not wish to be identified, for example, might have their identity revealed by this technology even if they do not have their own genome scanned, because a close relative's genome might be sufficient to find the match. In other countries, such services are fully legal and gaining popularity. Several companies in the United States offer genomic testing and provide detailed reports about ancestry and possible health issues related to DNA.

One potential such sequencing holds is the concept of **personalized medicine**, sometimes called precision medicine: information about a person's genetic makeup can be applied to an individual when prescribing treatments. The premise is that, if doctors know a patient's DNA profile, the best adapted treatment can be prescribed. When a doctor prescribes a drug today, the choice of molecule and the dose is based on studies involving people who might not be representative of everyone's genetic makeup. By sequencing the genomes of the participants in drug trials, patterns can be identified that suggest one drug might work better with people who possess a particular genetic sequence, but that for others, another molecule, combination of drugs or different dose would provide better results or perhaps fewer side effects.

A3.1 Figure 6 Knowing that a particular medication produces severe side effects only in people who receive the t version of an identified gene from both parents (*tt*) would allow doctors to know that four people in this group of patients should not be prescribed that medication. All the other patients have received a T from at least one parent (they are either *TT* or *Tt*) and can benefit from the medication without severe side effects.

Personalized medicine is better adapted for diseases that are dynamic, such as cancer, type 2 diabetes or cardiovascular disease, and require different treatments at different stages of the illness. Knowing more about how a patient's genome might cause new proteins to be produced in their cells or trigger certain genes to be turned on or off could lead to breakthroughs in medical treatments. By creating databases of biomarker profiles within a population (such as ***TT***, ***Tt*** or ***tt*** in the example in Figure 6), researchers of personalized medicine hope to provide better diagnoses and more effective treatments with fewer undesirable side effects.

Another advantageous use of the human genome is the production of new medications. This process involves several steps:

- find beneficial molecules that are produced naturally in healthy people
- find out which gene controls the synthesis of a desirable molecule
- copy that gene and use it to instruct synthesis of the molecule in a laboratory
- distribute the beneficial therapeutic protein as a new medical treatment.

This is not science fiction: genetic engineering firms are finding such genes regularly. One current line of research is dealing with genes that control ageing. How much money do you think people would be willing to pay for a molecule that could reverse the effects of ageing and prolong life by several decades?

HL

A3.1.12 – Difficulties with the biological species concept

A3.1.12 – Difficulties applying the biological species concept to asexually reproducing species and to bacteria that have horizontal gene transfer
The biological species concept does not work well with groups of organisms that do not breed sexually or where genes can be transferred from one species to another.

Insects usually reproduce by having males fertilize the eggs of females. The females of certain stick insects in the genus *Phasmatodea*, however, can often produce young without mating with a male. The eggs mature and grow into adult females. This process is called **parthenogenesis**. A similar process happens in plants called **vegetative propagation**, such as when a strawberry plant sends out a runner that takes root near the original plant. Farmers can plant last year's potatoes in their fields to grow new potato plants from them. In such cases, each new plant is an identical copy of the parent plant. This could continue for many generations without the need for sexual reproduction. These generations are continuing to produce offspring but they are not doing it by breeding, so they pose a challenge to the biological definition of what a species is.

▲ Strawberry plants can clone themselves by sending out runners that become new plants.

The examples mentioned can either breed by mixing gametes or produce clones by making copies of themselves, but some organisms can only reproduce asexually. Bacteria reproduce **asexually** using **binary fission** rather than by breeding. There is no such thing as mother/father or male/female, and no gametes are produced. Bacterial cells grow larger, make a copy of their genetic material, and split into two daughter cells that are identical to the original parent cell.

The idea of passing on genes to the next generation is one of the cornerstones of biology. The direction of such a transfer of genes is vertical, from one generation down to the next. But in addition to passing down their genes, bacteria can undergo **horizontal gene transfer**. Whereas other organisms normally only receive genetic material once in their life, when the male and female sex cells that formed them first meet and fuse together, bacteria can exchange genetic material within their lifetime, rather than just at the start. If a bacterium in a population has a mutation that could be useful to another member of the population, the two cells can attach to each other and exchange sequences of DNA. Remarkably, this gene transfer can be done even if the host bacterium is *not* of the same species.

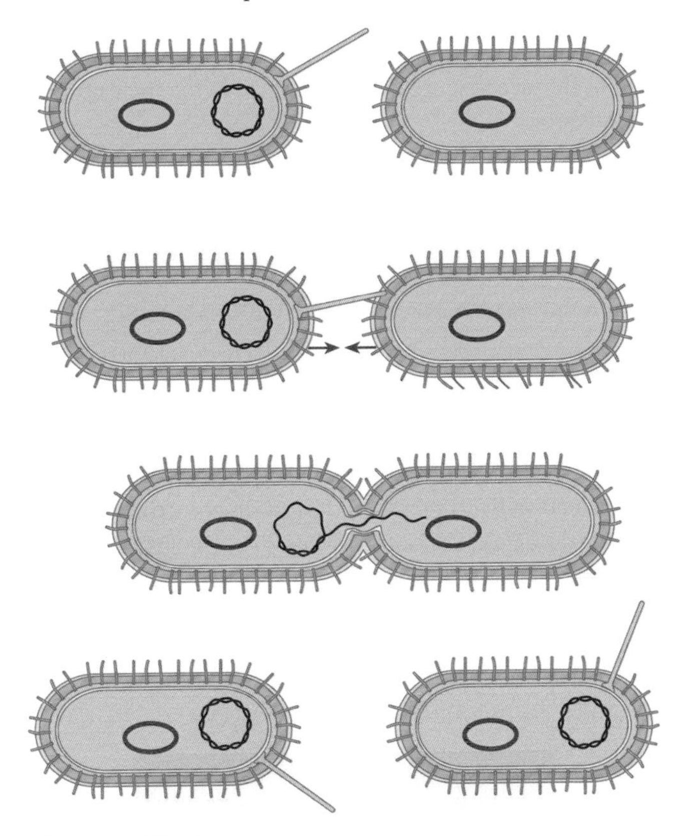

▲ The bacterium on the left, the donor cell, is passing genetic information to the bacterium on the right, the host cell, in a process called **plasmid transfer**.

One of the major assumptions of the concept of a species is that all members of a species have a common lineage and come from a series of common ancestors. This is the basis of the **tree of life** concept. The idea of mixing lineages by trading genes complicates things hugely. When a bacterium has a mix of genes from its own species that it received from previous generations, mixed with genes from a donor species during its lifetime, it poses a challenge to the idea of sharing a common ancestry with the other members of its species. If horizontal gene transfer happened several times in previous generations and again during its lifespan, the genetic material inside the bacterium would be a mosaic of genes from various sources.

Another challenge to the definition of species is the gene transfer that can occur from bacteria to archaea, from viruses to eukaryotes, and bacteria to eukaryotes. We can accept the idea that we all have inside our human cells organelles that were once prokaryotes (e.g. our mitochondria, see Chapter A2.2), and we know that we contain virus DNA, so who are we, really? Can we say that we are pure human, or are we a mosaic of genes from both human and non-human sources?

When sequencing and matching genes, sometimes a sequence of DNA is found that has more in common with another species than the one it is found in. Such genes, known as **xenologs** or **jumping genes**, travel in plasmids from one bacterium to another. Sometimes we find the same identical gene in several very different species

TOK

The tree of life concept is a good example of a well-accepted way of seeing how life evolved over time. From a central point representing the earliest forms of life, biologists believe the tree then branches out in all directions, and each split that creates a new branch is a speciation. Genetic information flows from one generation to the next and, once a split has occurred, there is no going back. The concept of horizontal gene transfer is an example of a paradigm shift, challenging the long-held beliefs of biologists. Evidence accumulated over the last 100 years challenges the original concept of the tree of life, and instead suggests that the tree is, in fact, more like a web. Genetic information can jump from branch to branch, causing an interweaving of the branches into a mesh. Organisms not only share common ancestors but also have unexpected ancestors from completely different branches. How does the way that we organize or classify knowledge affect what we know?

of bacteria. Usually when we see similar DNA sequences, we think that the organism shares a common ancestry but, in this case, the host species are from different branches of the tree of life and do not share the same lineages, such as yeast cells (a fungus) containing bacterial DNA. Such examples challenge the concept of species because it questions the idea that organisms classified in the same genus or species have a common ancestor.

A3.1.13 – Chromosome number as a shared trait

A3.1.13 – Chromosome number as a shared trait within a species
Cross-breeding between closely related species is unlikely to produce fertile offspring if parent chromosome numbers are different.

All domestic dogs belong to the same species, *Canis familiaris*, and dogs have a chromosome number of 78. The pineapple (*Ananas comosus*), on the other hand, has 50 chromosomes. The fruit fly (*Drosophila melanogaster*) has a chromosome number of 8. The red king crab (*Paralithodes camtschaticus*) has 208. You do not need to memorize these numbers, but you do need to know how the number of chromosomes a species possesses is a characteristic that all members of that particular species share. Of course, there are exceptions, which are explored in Chapter D2.1, but when we state the chromosome number of a species, such as 46 for humans, we are referring to the typical diploid number expected in the cells of that species.

Although there are exceptions, generally speaking, the number of chromosomes found in an organism's cells should be identical between all members of the same species.

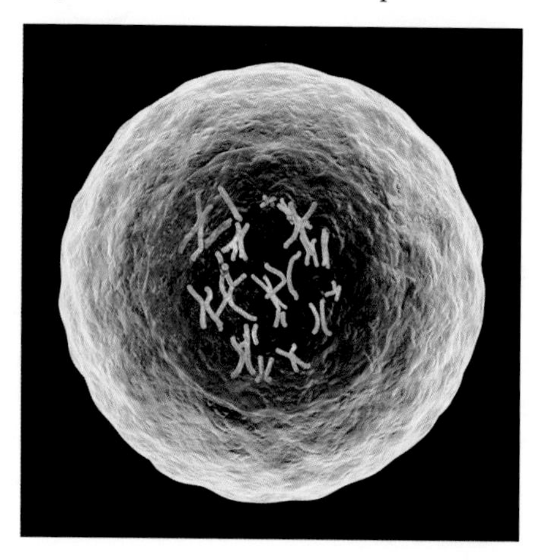

A female horse and a male donkey can mate and produce a mule. However, mules cannot usually mate to make more mules. Because the offspring (the mules) are not fertile, no new species has been created. Instead, a mule is called an **interspecific hybrid**. Hybrids face several challenges to continue as a population. For one thing, the vast majority of animal and plant hybrids are infertile. Even if one generation of hybrids is produced, a second generation is highly unlikely. This presents a genetic barrier between species.

Look at the chromosome number (in parentheses) of a mule and its parents:

female horse (64) + male donkey (62) = mule (63)

Mongolian wild horses (the Przewalski's horse, *Equus przewalskii*) and domesticated horses (*Equus caballus*), which have 66 and 64 chromosomes, respectively, have been known to produce hybrids as well.

What happens when the chromosome number of the parents is different? A mule born with a chromosome number of 63, which is neither that of the mother species nor that of the father species, makes it challenging to successfully mate within the populations of either. The atypical chromosome number makes it difficult for homologous pairs of chromosomes to match up during meiosis (discussed in Section D2.1.9), and thus production of gametes can be difficult. Mules therefore cannot pass on their genes to a subsequent generation, and therefore are not considered a new species. However, in rare cases, interspecific hybrids have been observed to produce fertile offspring.

A3.1.14 – Dichotomous keys

A3.1.14 – Engagement with local plant or animal species to develop a dichotomous key
Application of skills: Students should engage with local plant or animal species to develop a dichotomous key.

When biologists encounter an organism they need to identify, they can use a **dichotomous key** to establish which taxa it belongs to. If you have ever played a guessing game in which the rule is that you can only ask "yes" or "no" questions, then you already know how a dichotomous key works.

Here are the basic principles of how to use a dichotomous key. You can try it out with the example in Figure 7.

1. Look at the first section of the key, which has a pair of sentences, (a) and (b), describing characteristics.
2. Next, look at the organism to see if the particular characteristic described in the first line (a) is present in the organism.
3. If the answer is yes, then go to the end of that line and find the number of the next pair of statements to look at, follow the number given and continue until the end. If the end of the line contains a name, it is the taxon for the organism.
4. If the answer is no, then go to the second statement just below it, (b), and that one should be true; go to the end of that line and find the number of the next pair of statements to look at. Follow the number given and continue until the end.

The idea is to keep going until you get to a name instead of a number: if you have answered each question correctly, that will be the name of the taxon your organism belongs to.

Dichotomous key			
1	a)	No differentiated tissues, no symmetry or identifiable organs	Porifera (sponges)
	b)	Presence of differentiated tissues and organs	go to 2
2	a)	Stinging cells present, can show radial symmetry	Cnidaria (e.g. sea jellies)
	b)	No stinging cells	go to 3
3	a)	Has two-way digestive tract and bilateral symmetry	Platyhelminthes (flatworms)
	b)	Has a one-way digestive tract	go to 4...
etc...			

A3.1 Figure 7 Example of the beginning of a dichotomous key to identify animal taxa.

SKILLS

Develop your own dichotomous key for local plant or animal species. Full details of how to carry out this activity with a worksheet are available in the eBook.

A3.1.15 – DNA barcoding

A3.1.15 – Identification of species from environmental DNA in a habitat using barcodes
Using barcodes and environmental DNA allows the biodiversity of habitats to be investigated rapidly.

When we go to a store to buy a T-shirt or a bottle of orange juice, the barcode on the label is scanned and converted to a number. That number is used to identify the item by matching it against a database of items that includes their prices. Using a similar idea, genetic sequences obtained from organisms can be given a number (a **barcode identification number** or BIN) that is matched against a database of sequences that are known to belong to previously identified and named organisms. A **DNA barcode** is a short sequence of DNA (several hundred base pairs) inside an organism's cells that can be used to quickly identify the species.

To barcode a specimen means to sequence its genetic material and match a specific sequence to a known sequence stored in a genetic library. Thanks to the technology of DNA sequencing, millions of these barcodes have been added to libraries. A category that is commonly used to identify animals is mitochondrial DNA. For prokaryotes, sequences found in ribosomes (ribosomal RNA rather than DNA) are used for the barcodes instead.

In order to be usable by scientists and researchers everywhere, the data about these barcodes needs to be stored in a place that is accessible, such as the Barcode of Life Data System, or BOLD, developed in Canada. If there is a very strong match (99% or more), then we have a high level of confidence that the correct species name has been found for an organism. If the match is less strong, we have less confidence that it is the same species, and other techniques should be used to confirm the true species.

Such databases can be used to rapidly identify the various species present in an ecosystem. Water samples can be taken from lakes, rivers, estuaries and oceans, and soil samples can be taken from fields and forests. The DNA extracted from the water or soil is sequenced and barcodes are isolated for analysis. Such DNA collected from the environment rather than from an organism is called **environmental DNA** or **eDNA**. It is present in the environment because organisms release dead cells, produce faeces or die and start to decay. Think of a wild animal shedding hair, a tree losing a leaf or a fish releasing waste into a river. DNA can be found in any cells shed by an organism. After separating out the different DNA in a sample, it can be amplified using a technique called **polymerase chain reaction (PCR)** (see Chapter D1.3) and then sequenced.

Water or soil samples collected in an ecosystem can be sequenced in a laboratory to identify the organisms that inhabit it.

Specialists studying a zone near a polluted area often want to know if the **biodiversity** of the zone is affected. Biodiversity can be measured by identifying and counting the number of species present. Although dichotomous keys and experienced expert eyes can identify organisms reasonably quickly down to the family or genus level, it can be difficult to identify to the species level with a high degree of accuracy because of very subtle differences in appearance or in differences that might not be noticeable at the time of observation. For animals such as insects, the larval stage often looks very different from the adult stage. For plants, if an unidentified plant is not flowering or a tree has lost its leaves, identification using morphology can be very challenging if not impossible because these features are often crucial to their proper identification. To identify organisms morphologically, they have to be observed directly or captured, which can be time consuming as many animals are elusive. Identifying organisms by physical features is also significantly more difficult for microbes, which cannot be seen unless you use a microscope or can culture them in a laboratory. With DNA barcodes and fast sequencing technology, samples of DNA can be sequenced and matched to identify species with a high level of confidence in a matter of hours.

Certain species can be used as **bioindicators** or **indicator species**: these organisms are so sensitive to certain types of pollution that their presence in an ecosystem indicates a lack of pollution. Conversely, their sudden disappearance from an ecosystem would suggest the appearance of a source of pollution. Caddisfly larva live underwater in streams and can be used as bioindicators for water health. A large caddisfly population is reassuring, but a small or disappearing caddisfly population is a signal to investigators that they should start looking for potential contamination upstream. DNA barcoding of water samples can indicate the presence of one or more species of caddisfly.

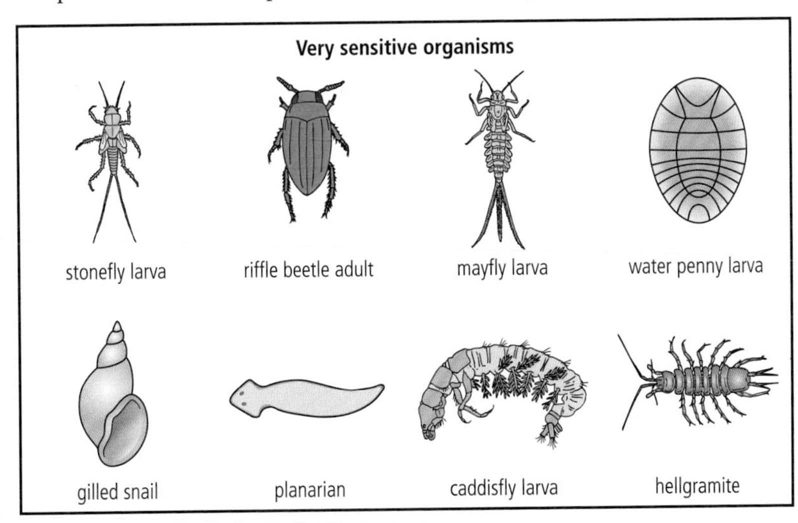

Examples of indicator species that are very sensitive to pollution and whose presence in an ecosystem can reassure us that the habitat is healthy.

TOK

To determine whether biodiversity is increasing or decreasing, a baseline is needed to compare to later: we need to know how many species are present now so that we can see if the number goes up or down later. In most parts of the world, terrestrial, aquatic or marine ecosystems do not have such baselines, so we cannot know how much biodiversity is changing. Barcoding of environmental DNA can help rectify this. Look through the TOK prompts. Do any apply to this example? What are the implications of having, or not having, knowledge?

To perform an ecological survey, one species at a time is identified and counted, and often, for example with insects, birds or fish, this requires capturing them, which can disrupt the population. With eDNA metabarcoding, a single sample can be sequenced for dozens or hundreds of species without having to capture individual organisms. This is more time efficient for the specialists carrying out the ecological surveys. Over months and years, if the number of species in an ecosystem decreases, we know that the biodiversity of that habitat is unhealthy or declining. In contrast, if the number of species is stable or increases over time, we know that biodiversity is stable or improving, suggesting that the ecosystem is healthy and flourishing.

There are some disadvantages to using environmental DNA. Firstly, it only gives an indication of the presence or absence of a species, not the population size. Secondly, the DNA does not indicate if it is from a living organism or a dead one. Thirdly, certain chemical incompatibilities exist with the processing of soil samples because substances in the soil can interfere with the sequencing process, giving rise to inaccurate results.

HL end

Testing for genetic material shed by organisms can be applied in unexpected situations. In the second year of the COVID pandemic, in addition to testing people for viruses, many municipalities were testing wastewater for the presence of SARS-CoV-2. In New York City, some sequences that were identified were from strains of the virus that had never shown up in samples from patients. This suggested that the problem was more complex and widespread than previously understood. Towns and cities that used this technique could target certain neighbourhoods for additional testing, prevention and medical resources.

Guiding Question revisited

What is a species?

In this chapter we have learned that:

- there is no single definition of the term "species" because the sheer variety of currently living species and extinct species is so enormous and complex
- using morphology works up to a point, but this methodology is poorly adapted for microbes or for species that are visually very similar
- the biological species concept works most of the time but is does not work for single-celled organisms that do not breed, or for organisms that are only found in the fossil record.

Guiding Question revisited

What patterns are seen in the diversity of genomes within and between species?

In this chapter we have discussed how:

- there is some diversity in genomes of individuals of the same species
- there is much more diversity when two different species are compared, especially if they were separated in a speciation event that occurred long ago.

Exercises

Q1. The system of giving a scientific or Latin name to organisms such as *Canis familiaris* is used worldwide. State the name of this system and identify the person who perfected and popularized it.

Q2. Distinguish between the morphological definition of species and the biological species concept.

Q3. Explain the features of chromosomes that are taken into consideration when making a karyogram.

Q4. Distinguish between haploid and diploid cells.

Q5. A karyogram can be used to determine if an unborn baby will be a girl or a boy. Explain how a karyogram is analysed to do this.

Q6. Outline the evidence for a fusion of ancestral chromosomes to become human chromosome 2.

Q7. Outline the advantages of personalized medicine using genomes.

Q8. HL Discuss reasons for and against using environmental DNA and barcoding for ecological surveys of biodiversity.

HL

A3.2 Classification and cladistics

Guiding Questions

What tools are used to classify organisms into taxonomic groups?

How do cladistic methods differ from traditional taxonomic methods?

There are many ways to classify organisms. We can use physical features, or we can analyse the DNA or amino acid sequences from organisms and put the most similar ones together, while separating those with more differences into separate categories. Following the taxonomic techniques used by Linnaeus, we can divide organisms into kingdoms, phyla and so forth, down to genus and species. Or, using the more recently developed techniques of cladistics, we can propose testable hypotheses that show the evolutionary relationships between species using cladograms, which attempt to show not only how species are similar or different to each other, but also how they are related to each other.

A3.2.1 – The classification of organisms

A3.2.1 – Need for classification of organisms
Classification is needed because of the immense diversity of species. After classification is completed, a broad range of further study is facilitated.

Life on Earth has an astounding range of diversity. Dolphins live in the oceans and yaks live high in the mountains. Many insects and birds can fly across vast areas, whereas plants and corals spend their adult lives stuck in one place. Emperor penguins can survive in Antarctica at temperatures as low as −50°C, and thermophilic microbes can live deep in the Earth at temperatures as high as 120°C. Photosynthetic organisms are capable of making their own food, whereas heterotrophs have to feed to survive. When a mayfly reaches adulthood and can fly, it only lives a short time, sometimes just one day before dying, whereas some trees have been alive for over four millennia. Putting such diversity into one system of classification is a daunting and time-consuming task, but biologists think it is worth the trouble.

Just as finding a specific book in a library that uses a system of organizing titles by subject and author is faster than trying to find it on a disordered market table, there are great advantages to establishing a classification system for life on Earth. By putting organisms into categories, it is possible to see whether a species has already been discovered, to observe similarities with other species, to deduce common ancestry and to predict features that undiscovered but related species should have. Currently, there are between 1,200,000 and 1,800,000 species that have been catalogued using the binomial nomenclature system. Although there is great debate among specialists as to how a "species" should be defined (see Chapter A3.1) and how many we have found that are still living or extinct, one thing they all agree on is that the estimates are probably far below the actual number of species. The prediction is that there are somewhere between 5 million and 100 million species on Earth.

▲ A large percentage of the species on Earth that have been classified and named are insects.

Studying and classifying organisms helps us to discover ancestries and see which species we are related to and how all species are connected. Current techniques allow us to see which organisms have more recent common ancestors and which ones have more ancient common ancestors.

In 2010, a single finger bone discovered in a cave in Siberia that belonged to a female of an extinct species of humans, the Denisovans, had enough DNA in it to be analysed. Specialists were then able to predict what characteristics she had, such as dark eyes and dark skin and hair. Comparing unknown DNA to libraries of known DNA sequences enables the prediction of such characteristics.

Online databases allow biologists worldwide to identify and compare the organisms they are studying with species that have already been identified. Different databases have different specialties. Some are just for fungi, others just for mammals. Some have libraries of DNA sequences and online comparison tools. The SARS-Cov-2 coronavirus, or COVID, pandemic saw one use of genetic sequencing and shared databases: tracking virus variants that contain new mutations. Tracking and identifying viruses, dangerous bacteria and parasites that cause disease has benefited greatly from the knowledge gained using sequencing techniques. This field is called **genomic epidemiology**. Epidemiology is the study of disease at the population level within a country, a continent or worldwide. Online services, such as those provided by the Nextstrain Project from the University of Basel's Swiss Institute of Bioinformatics, provide genome data and phylogenetic trees that can be used to see how certain pathogens have changed over time.

Real-time tracking of influenza A/H5N1 virus evolution

Built with nextstrain/avian-flu. Maintained by Louise Moncla.

Showing 2265 of 2265 genomes sampled between Dec 1996 and May 2022.

Phylogeny

Region

China
Southeast Asia
South Asia
Japan Korea
West Asia
Africa
Europe
North America

ZOOM TO SELECTED
RESET LAYOUT

75
1980
1985
1990
1995
2000
2005
2010
2015
2020
Date

▲ A screenshot from the Nextstrain website showing how strains of avian influenza (i.e. "bird flu") developed worldwide between 1990 and 2020.

A3.2.2 – Difficulties in classifying organisms

A3.2.2 – Difficulties classifying organisms into the traditional hierarchy of taxa
The traditional hierarchy of kingdom, phylum, class, order, family, genus and species does not always correspond to patterns of divergence generated by evolution. **NOS**: A fixed ranking of taxa (kingdom, phylum and so on) is arbitrary because it does not reflect the gradation of variation. Cladistics offers an alternative approach to classification using unranked clades. This is an example of the paradigm shift that sometimes occurs in scientific theories.

Chapter A3.1 discusses how the system of classification developed by Linnaeus is based on morphology and on the physical characteristics he decided were appropriate for distinguishing one species from another. Such choices, however, are subjective and arbitrary. Although his system is less arbitrary than using an alphabetical order, which would be different in every language, it is a contrived method rather than one that reflects common ancestry or the genetic connections between species.

Another difficulty of Linnaeus' system is the concept of hierarchy whereby a smaller category is placed within a bigger category. One of the principles underpinning the hierarchy is that, if an organism belongs to a certain genus, then, by definition, it must also be in the same family as all the other organisms in that genus. In other words, the system is ranked and each rank must be respected. If a taxonomist decides to move an organism from one genus to another, this could pose a problem for the other organisms that are also in that genus: should any of them be moved as well?

▲ New hybrid plants like this orchid are being developed every year by horticultural specialists, for example to create new ornamental plants or to improve food production.

The more organisms that are discovered, the more often the hierarchy does not work. In plants, for example, hybridization is much more common than it is in animals. New plants can be produced when two parents from different species produce a hybrid. If the hybrids are able to produce offspring, the next generation will be a new species. Or, if plant breeders are able to clone the plants for widespread commercial use, it raises the question of whether those plants should be considered a new species.

Chapter A3.1 considers the assumption that a tree-like mapping of species evolving from common ancestors and branching off into new species works in all cases, but that this assumption is flawed when species exchange material either during plant hybridization, introgression (see below) or bacterial gene transfer. The "tree" becomes a "mesh" or "network", greatly complicating our perception of the evolution of life on Earth.

Another problem with the current system of taxonomy is that we assume that the species should be the basic unit of classification, in the way that an atom is the basic unit of a chemical element. Biological systems are more complicated than this, and there is often a much more nuanced gradient between closely related species than a clear separation.

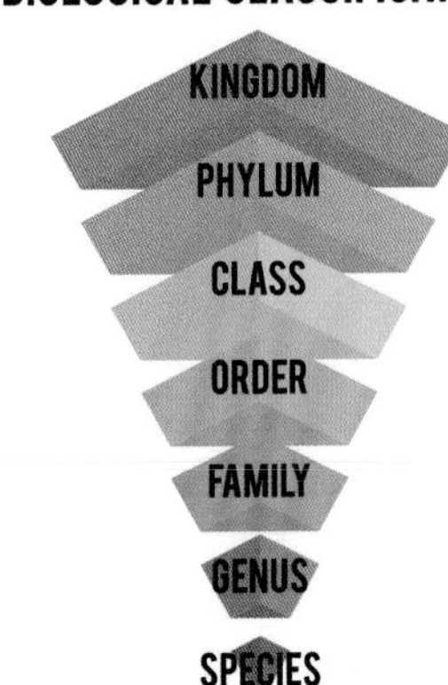

▲ The traditional hierarchical classification system.

Introgression

Introgression is a process by which hybrids form over many generations but, instead of having an equal share of the original two species' genetic information, there is an unequal contribution from each species. This is achieved by the process of backcrossing: the hybrid organism breeds with one of the original parent species to produce offspring. After several generations, an organism resulting from introgression by backcrossing might only possess 5% or 10% of one of the original species' genes. Such zones of genes found in species partly account for the variety we can see in species today.

For example, some of the variety we see in wolf (*Canis lupus*) populations can be explained by introgression with coyotes (*Canis latrans*). In terms of classifying a species, if the species is 90% wolf but 10% coyote, it does not fit particularly well on either the wolf branch or the coyote branch of an evolutionary tree (see Figure 1). This is one of the problems of using a fixed hierarchy with categories that are "either/or" and do not show the full genetic background of the organisms being classified.

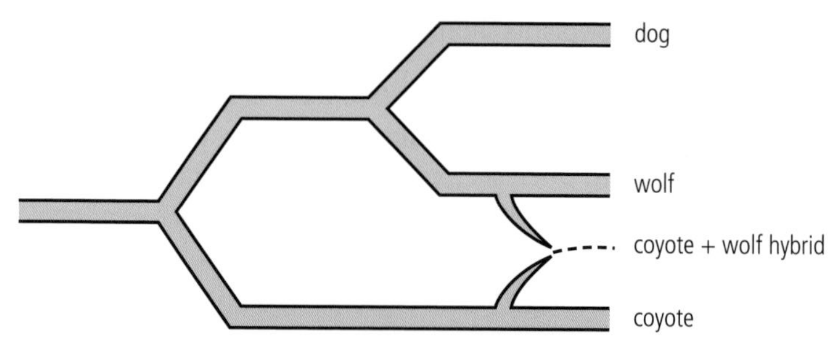

A3.2 Figure 1 Coyotes (*Canis latrans*; pictured) in North America have been known to form hybrids with wolves (*Canis lupus*). If the hybrid then breeds with either a wolf or a coyote, it can generate offspring that will have an unequal share of wolf and coyote DNA, indicating that an introgression has occurred.

Nature of Science

Establishing categories, labels or "buckets" into which we place organisms can be problematic because which characteristics we use are arbitrary, and often an "either/or" approach does not reflect variations that show gradation rather than distinct categories. Cladistics is a different way of classifying organisms. Species are grouped together based on common ancestries. If we learn something new about the ancestry, we can update the groupings. Cladistics is a natural rather than arbitrary system.

A3.2.3 – Classification using evolutionary relationships

A3.2.3 – Advantages of classification corresponding to evolutionary relationships
The ideal classification follows evolutionary relationships, so all the members of a taxonomic group have evolved from a common ancestor. Characteristics of organisms within such a group can be predicted because they are shared within a clade.

What if, instead of looking at the morphology of organisms, we consider other ways of classifying them, such as how they are related to other species? Classifying organisms using molecular differences in protein sequences and DNA is called **molecular systematics**.

Phylogeny is the study of the evolutionary past of a species. Species that are more similar are more likely to be closely related, whereas those that show a higher degree of differences are considered less likely to be closely related.

Comparing the genetic profiles of different species allows us to construct a **phylogenetic tree**, one that shows the evolutionary relationships between species by showing which species developed from a common ancestor. Think of a phylogenetic tree like a family tree but, instead of generations of individuals, it shows relationships between many species.

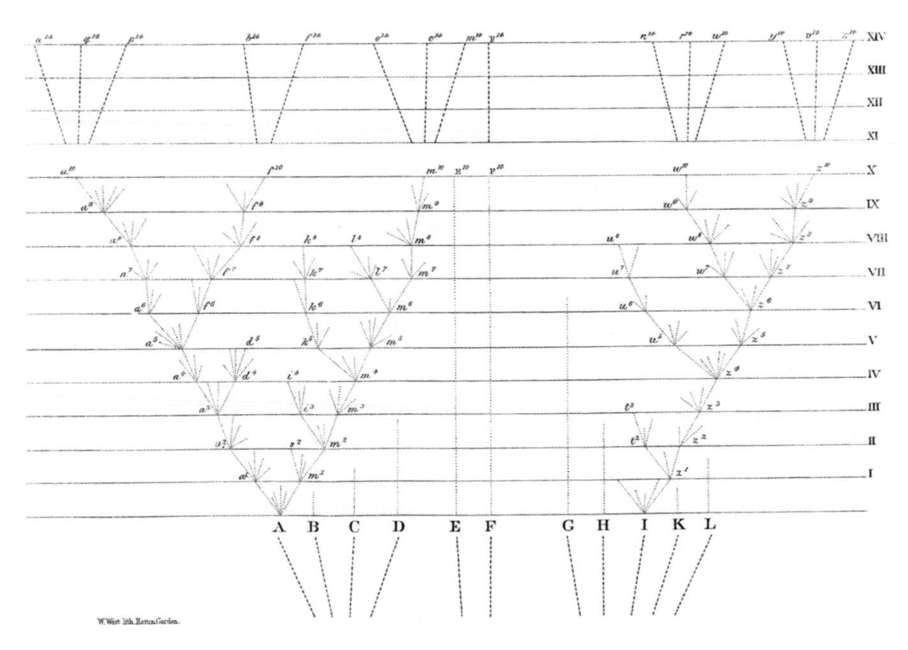

This was the only illustration in Charles Darwin's 1859 book *On the Origin of Species*. It helped lay the groundwork for what we do today using genetics and computer processing (both of which Darwin had no access to) to compare genomes of species and reconstruct their evolutionary past.

The advantages of such a system are that it is not based on arbitrary, subjective or contrived categories. Rather, it reflects species' natural gene sequences. Sometimes both the Linnaean system and a system based on molecular systematics to determine ancestry agree with each other. Other times, as we will see later in this chapter, new evidence prompts taxonomists to change how a species had previously been classified.

When studying the evolutionary relationships between species, we use the concept of **clades**. A clade can also be called a **monophyletic group**. This means it is a group comprising the most recent common ancestor of that group and all its descendants. The characteristics of organisms within such a group can be predicted because they are shared within a clade. A clade can comprise just one species or can be made up of multiple species.

A3.2.4 – Clades display common ancestries and shared characteristics

A3.2.4 – Clades as groups of organisms with common ancestry and shared characteristics
The most objective evidence for placing organisms in the same clade comes from base sequences of genes or amino acid sequences of proteins. Morphological traits can be used to assign organisms to clades.

Cladistics is a natural system of classification for grouping **taxa** (see Chapter A3.1) based on the characteristics that have evolved most recently. In this system, the concept of common descent is crucial to deciding into which groups organisms are classified. To decide how close a common ancestor is, researchers can look at DNA and amino acid sequences, or they can look at which traits are shared between species. Such traits can be divided into primitive and derived traits.

Primitive traits (also called **plesiomorphic traits**) are characteristics that have a similar structure and function (e.g. leaves, with vascular tissue to transport liquids around a plant) and evolved early in the history of the organisms being studied. **Derived traits** (also called **apomorphic traits**) are also characteristics that have similar structure and function but have evolved more recently, in the form of modifications of a previous trait (e.g. flowers, which, according to the fossil record, evolved more recently than leaves with vascular tissue). The resulting classification would show that all the organisms descended from the earliest common ancestor being studied have the primitive trait but, in subgroups of smaller clades, some but not all would have derived traits. These terms are relative, however, and in another clade that considers more species, a primitive trait could be considered a derived trait. In animals, eggs with shells (e.g. in birds) evolved more recently than eggs without shells (e.g. in turtles). By systematically comparing such characteristics, the quantitative results indicate which organisms have undergone a more recent split in the evolutionary past and which have undergone a more distant split. By counting the number of derived traits an organism has, it is possible to determine whether an organism shared a common ancestor more recently or less recently than other organisms.

When a group can be split into two parts, one having certain derived traits that the other does not, the subgroups form two separate clades.

A3.2.5 – The evolutionary clock

A3.2.5 – Gradual accumulation of sequence differences as the basis for estimates of when clades diverged from a common ancestor

This method of estimating times is known as the "molecular clock". The molecular clock can only give estimates because mutation rates are affected by the length of the generation time, the size of a population, the intensity of selective pressure and other factors.

Differences in polypeptide sequences accumulate gradually over time, as mutations occur from generation to generation in a species. Consequently, the changes can be used as a kind of clock to estimate how far back in time two related species split from a common ancestor.

By comparing homologous molecules (e.g. haemoglobin) from two related species, it is possible to count the number of differences between the molecules. The numbers of base pairs along the molecule that do not match are used to quantify similarities and differences.

DNA hybridization

How do we count the differences? One technique is **DNA hybridization.** One strand of DNA is taken from species A, and a homologous strand (a sequence from the same position in the same gene) from species B, which are then fused together using enzymes. Where the base pairs connect, there is a match; where they are repelled and do not connect, there is a difference in the DNA sequence and therefore there is no match (see Figure 2).

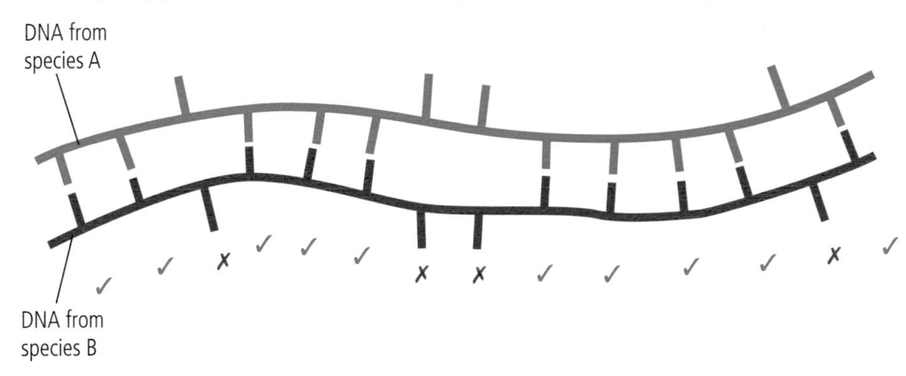

A3.2 Figure 2 DNA hybridization between a strand of DNA from species A (the top strand) and another from species B (the bottom strand). There are four places where a match does not occur.

We can take the idea of counting differences a step further. Looking at Figure 3, as a 36 nucleotide difference is approximately twice a 17 nucleotide difference, we can hypothesize that the split indicated by the number 36 (moth) happened about twice as far back in the past as the more recent split at 17 (duck). This concept uses quantitative biochemical data as a **molecular clock** to estimate the time of the speciation events (see Figure 3).

Experts in various fields of study use the idea of accumulated change over time. For example, linguists look at changes in words and use of vocabulary to trace the evolution of a language. Experts who study chain letters sent by post or by email are interested in the number of modifications made to the original letter over time. By comparing hundreds of versions of the same message, they can analyse what has been added or changed to map its evolution over time. With enough evidence, it is sometimes possible to deduce the origin and approximate date of the original letter in a chain, even if that actual letter is never found.

A molecular clock can only provide estimates because mutation rates are affected by the length of generation time, the size of a population, the intensity of selective pressure and other factors.

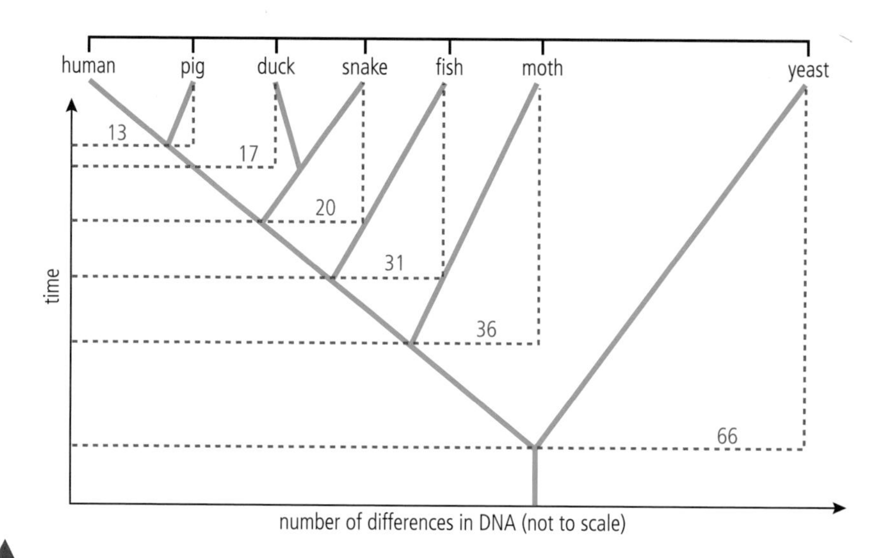

A3.2 Figure 3 Biochemical differences (dotted lines) can be used to see how far apart species are on a phylogenetic tree (solid lines).

However, we need to be careful when using a word such as "clock" in this context. Under no circumstances should we consider the "tick-tock" of the evolutionary clock, which is made up of mutations, to be as constant as the ticking of a clock on the wall. Mutations can happen at varying rates. Consequently, all we have is an average, an estimation or a proportion, rather than an absolute time or date for speciation events. In an attempt to calibrate the timing of the evolutionary clock, biochemical data can be compared to morphological fossil evidence and radioisotope dating (e.g. carbon-14).

A3.2.6 – Constructing cladograms

A3.2.6 – Base sequences of genes or amino acid sequences of proteins as the basis for constructing cladograms
Examples can be simple and based on sample data to illustrate the tool. **NOS**: Students should recognize that different criteria for judgement can lead to different hypotheses. Here, parsimony analysis is used to select the most probable cladogram, in which observed sequence variation between clades is accounted for with the smallest number of sequence changes.

Cladograms

To represent the findings of cladistics in a visual way, a diagram called a **cladogram** is used. A cladogram showing bats, sharks and dolphins, for example, would take into account their skeletal structures and other characteristics, such as the fact that bats and dolphins are mammals. Thus, bats and dolphins are shown as being more similar to each other than sharks are to either.

Figure 4 shows the key characteristics of a cladogram. For example, a **node** is the place where a speciation event happened and the relative position of the common ancestor. The larger clade (shown in yellowy green) is divided into a **sister group**, a group of the closest relatives, and an **outgroup**, which is a group that is less closely related to the others in the cladogram. Sharks are less closely related to bats and dolphins than bats and dolphins are to each other. And yet, if we go back far enough, we will find another node showing that they do have a common ancestor. The base from which other species branch out is called the **root**. The tips of the diagram are called **terminal branches**.

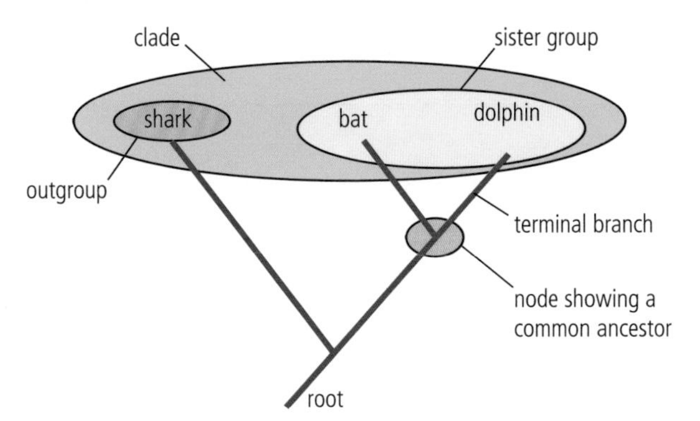

A3.2 Figure 4 A cladogram showing three taxa organized into a clade, of which two form a sister group and one forms an outgroup. Nodes indicate the position of a common ancestor for the descendants. The descendants appear above the nodes in this cladogram.

Chapter A3.1 discusses how a lot of discoveries that used to take months or years to make are today just a few clicks away, because data is available for everyone to consult on the internet. The online genetic database at the National Center for Biotechnology Information (NCBI), for example, has many genes that you yourself can look up. Interested in insulin or haemoglobin? Search for those words or their genes (INS for insulin or HBB for one of the subunits of haemoglobin). Try to see if you can find out at what position and on what chromosome you can find the code for these valuable molecules of life. If you ask for the FASTA data (pronounced "Fast A"), you can see every A, T, C and G that makes up a gene coding for a protein. Also, check out the NCBI 1000 Genome Browser, an online map of human genes chromosome by chromosome. If you get lost, they have video tutorials to help.

The essential idea behind cladograms constructed by studying biochemical differences is that an organism with the fewest modifications of a particular DNA sequence will be the most anciently evolved, and those with the most modifications (mutations) in the same DNA sequence will be the more recently evolved organisms. The former will have nodes at the earliest splits of the cladogram, and the latter will have nodes at the more recent splits.

Cladograms and classification

Every cladogram is a working hypothesis. It is open for testing and for falsification. On the one hand, this makes cladistics scientific, but on the other hand, if it is going to change in the future as new evidence arises, it could be criticized for its lack of integrity.

Each time a derived characteristic is added to the list shared by organisms in a clade, the effect is similar to going up one level in the traditional hierarchy of the Linnaean classification scheme. For example, the presence of hair is part of what defines a mammal, so any species found after the line marked "hair" should be in the class of mammals.

What about feathers? If an organism has feathers, is it automatically a bird? In traditional Linnaean classification, birds occupy a class of their own, but this is where cladistics comes up with a surprise. When preparing a cladogram, it becomes clear that birds share a significant number of derived characteristics with a group of dinosaurs called the theropods. This suggests that birds are an offshoot of dinosaurs rather than a separate class of their own.

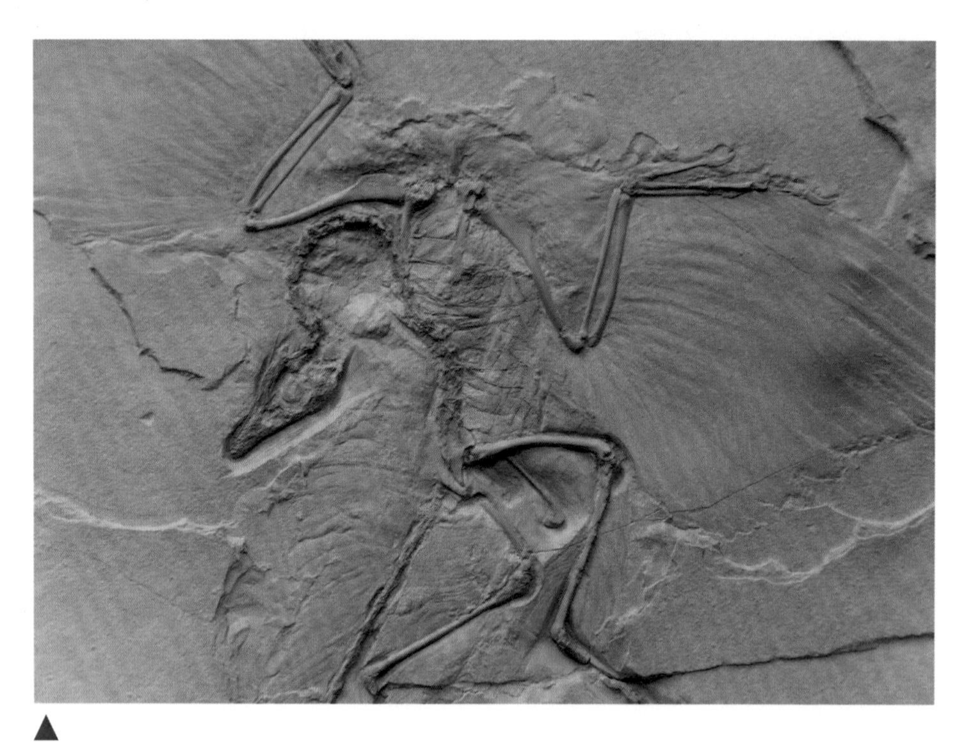

▲
The fossil of a theropod dinosaur called ***Archaeopteryx***. The presence of feathers indicates that it is possibly related to modern birds.

Because birds are one of the most cherished and well-documented classes of organisms on Earth, this idea, when it was first suggested, was controversial to say the least. Some of the derived characteristics used to put birds and dinosaurs in the same clade are:

- a fused clavicle (the "wishbone")
- flexible wrists
- hollow bones
- a characteristic eggshell
- the structure of the hip and leg, notably backward-pointing knees.

Nature of Science

Parsimony is an important logical tool in science: it is the idea that, when given several explanations, it is best to choose simplicity over complexity. If one possible cladogram shows eight splits to explain the relationships between organisms, for example, and another only uses seven for the same organisms, we should choose the one with seven. Or, because it is less likely that something like the development of feathers happened separately in two different parts of a cladogram, it is simpler to build a tree with feathers appearing only once. By applying the idea of parsimony, it is more likely that birds evolved from dinosaurs than trying to explain how feathers could have evolved on two separate occasions. Using parsimony analysis in this way helps researchers find the least convoluted and most likely hypothesis to explain how organisms have evolved.

A3.2.7 – Using cladograms

A3.2.7 – Analysing cladograms

Students should be able to deduce evolutionary relationships, common ancestors and clades from a cladogram. They should understand the terms "root", "node" and "terminal branch" and also that a node represents a hypothetical common ancestor.

Recall the parts of a cladogram:

- node = a split in a cladogram where one branch goes in one direction and the other branch in another direction, representing a hypothetical common ancestor
- root = the base of the cladogram showing the common ancestor to all the clades in the cladogram
- terminal branch = the end of a branch representing the most recently evolved of the organisms in the clade.

A3.2 Figure 5

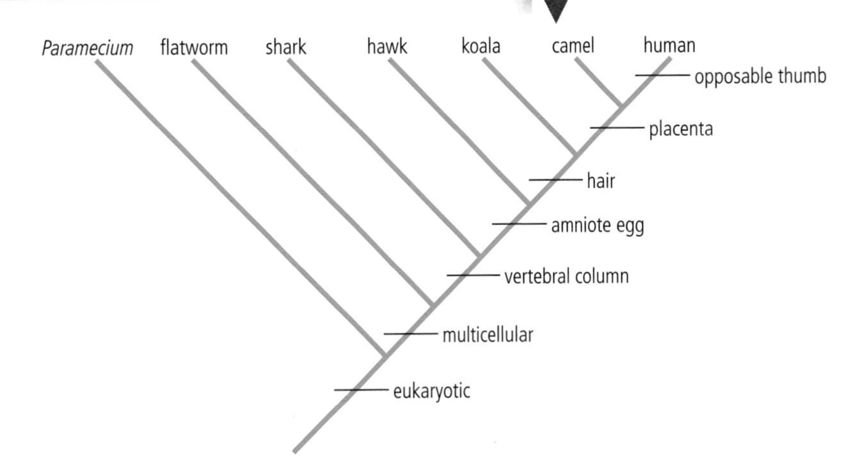

The closer to the root, the further back in the past the cladogram represents, and the fewer derived characteristics the organisms will have. In Figure 5, any organism on a branch above the node with the derived characteristic labelled "'vertebral column" will have a spine, and any organism on a branch below that node and closer to the root will be an invertebrate. Only one species has all the derived characteristics in this cladogram: the human.

A key characteristic of mammals is that they have hair.

What role does authority play in science? If you search the web, you will find lots of sources that claim the following: "Koalas have two thumbs on each hand!" But these tend to be sources that are not written by specialists in animal anatomy. Fact-checking using academic sources reveals that the two digits concerned do not fully qualify as opposable thumbs, implying that in order to declare whether an organism has them, a knower would need to fully understand the definition of "opposable thumbs". This raises the question, does some knowledge belong only to particular communities of knowers?

Worked example

1. What is the primitive characteristic in the cladogram shown in Figure 5?
2. Name the members of the mammal clade shown in this cladogram.
3. What is the outgroup in this cladogram in relation to the clade of multicellular organisms?
4. Do shark eggs have a protective membrane (the amnios) around them?
5. Explain why there are no bacteria shown in Figure 5.

Solution

1. Being eukaryotic is the primitive characteristic shared by all.
2. Koala, camel, human.
3. The *Paramecium*.
4. No. Sharks are not amniotes.
5. Because the most primitive characteristic shown is that the organisms have a nucleus. If bacteria were added to this cladogram, there would be a new primitive characteristic.

A3.2.8 – Cladistics and reclassification

A3.2.8 – Using cladistics to investigate whether the classification of groups corresponds to evolutionary relationships

A case study of transfer of plant species between families could be used to develop understanding, for example the reclassification of the figwort family (Scrophulariaceae). However, students are not required to memorize the details of the case study.

NOS: Students should appreciate that theories and other scientific knowledge claims may eventually be falsified. In this example, similarities in morphology due to convergent evolution rather than common ancestry suggested a classification that by cladistics has been shown to be false.

Note: When students are referring to organisms in an examination, either the common name or the scientific name is acceptable.

From time to time, new evidence emerges about a taxon that requires it to be reclassified. Either the taxon can be moved up or down the hierarchy (family to subfamily, for example), or from one family to another.

Plants commonly known as figworts used to be classified in the family Scrophulariaceae, and many of them have been used in herbal medicine. The name Scrophulariaceae, sometimes affectionately referred to by botanists as "scrophs", comes from a time when plants were frequently named after the diseases they could be used to treat. The medical term "scrofula" refers to an infection of the lymph nodes in the neck. Preparations made with figwort were given to patients who suffered from this infection, which was associated with tuberculosis.

The common foxglove (*Digitalis purpurea*) has been reclassified, so instead of being in the figwort family it is now in the plantain family.

Before the mid-1990s, the family Scrophulariaceae was characterized by morphological features such as how the flower petals are arranged in the bud before the flower opens. This feature is called aestivation, and botanists look for whether the flower petals overlap with each other or whether they are arranged in a spiral. Another characteristic that was used was the morphology of the nectaries, the parts of the flower that make nectar.

Since the mid-1990s, DNA analysis of the plants within this taxon has led botanists to rethink their classification. Analysis of zones of DNA markers, such as the nuclear ribosomal **internal transcribed spacer** (ITS) region, has revealed that the old classification system was not **monophyletic**, meaning the taxa did not share a most recent common ancestor. Rather, the old system was grouping together plants that belonged to separate branches, making it impossible to fit them into a cladogram attempting to show their evolutionary relationships.

The term used to describe species on separate branches is **paraphyletic**, so we now know that the old family Scrophulariaceae was paraphyletic. Plants that were in the Scrophulariaceae family have been given new families to belong to, such as Plantaginaceae, which is where we now find foxgloves. Foxgloves are no longer considered to be figworts.

Moving the branches of the tree of life around and reclassifying a taxon in a new branch in this manner means changing the species' **circumscription**. Circumscription is the process of placing taxa where they clearly show monophyletic groups, indicating that they all share a recent common ancestor.

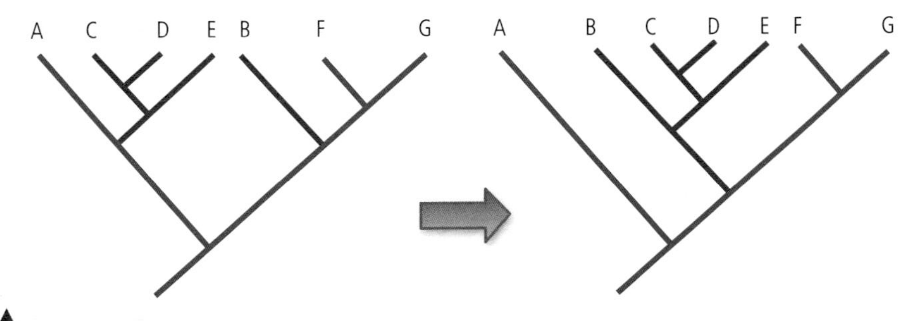

An example of a modification of a species' circumscription. The clade that included species C, D and E on the left was moved from the branch that included species A, and placed on the branch with species B instead, because C, D and E shared a common ancestry with species B. In the old cladogram on the left, B, C, D and E are a paraphyletic clade, whereas in the new cladogram on the right they are shown as a monophyletic clade.

How can similarities between distantly related organisms be explained?

Nature of Science

The reclassification of the foxglove is a good example of how scientists work. Observations were initially based on morphology. The plant was classified into specific categories that included the family Scrophulariaceae, the figwort family. DNA sequencing determined that some plants did not belong with the other figworts but instead belonged in the family Plantaginaceae, along with the plantains. Studies were published in recognized botany journals and now foxgloves have a new family.

In this instance, convergent evolution, similar adaptations in species that are not closely related, was responsible for the similarities used in the original classification, not the fact that the plants were evolutionarily closely related.

Worked example

1. Examine the cladogram in the figure below of four genera of plants.

(a) Name two sister taxa.

(b) Name the outgroup in this cladogram.

(c) Using a clearly marked label, indicate a node.

(d) Which genus possesses characteristics that evolved more recently, *Digitalis* or *Plantago*?

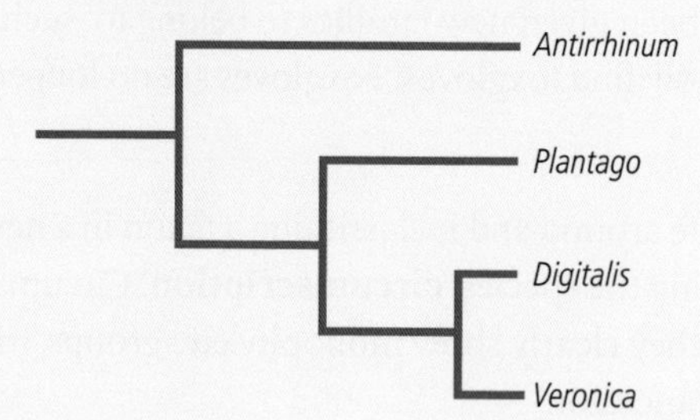

2. Study the phylogenetic tree in the figure below showing some primates and their chromosome numbers. Note that when there is variety between one species and another within a taxon, a range of chromosome numbers is given.

(a) Identify the numbered arrow that indicates a common ancestor for all the primates shown.

(b) Monkeys have tails whereas apes do not. Arrow number 3 shows the point when primates lost their tails. List the apes shown below.

(c) Identify the numbered arrow that indicates when bipedalism completely replaced walking on four legs.

(d) The great apes shown are the four primates that demonstrate the most recently developed derived traits. Identify which taxon represents the lesser apes.

(e) All the great apes shown except one have the same number of chromosomes. Which species has a different number?

(f) Some evidence supports the idea that, in humans, two of our chromosomes fused together at some point in our evolution. What evidence is there in the cladogram to support this?

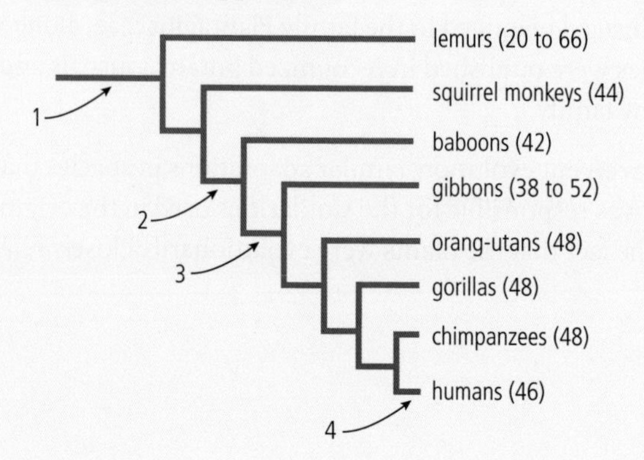

Solution

1. **(a)** *Digitalis* and *Veronica*.

   (b) *Antirrhinum*.

   (c) Answers may vary: anywhere a horizontal line comes to a junction with a vertical line.

   (d) *Digitalis* (it is the product of a more recent speciation).

2. **(a)** 1.

   (b) Gibbons, orang-utans, gorillas, chimpanzees, humans.

   (c) 4.

   (d) Gibbons.

   (e) Humans.

   (f) All of our closest relatives in the great apes clade have 48 chromosomes, whereas we have 46; this would suggest that, if one pair of chromosomes fused with another, we would have gone from 24 pairs to 23 pairs.

A3.2.9 – Three domains of life, not two

A3.2.9 – **Classification of all organisms into three domains using evidence from rRNA base sequences**
This is the revolutionary reclassification with an extra taxonomic level above kingdoms that was proposed in 1977.

At the top of the classification hierarchy are the three largest groupings for organisms, called **domains**. The names of these three domains are **Archaea**, **Eubacteria** and **Eukarya**. All living organisms are classified into one of these three. Note that viruses are not in this list because they are not alive and do not necessarily share a common ancestry with each other, two major conditions required to fit into this classification system.

Archaea are single-celled organisms that are distinct from bacteria; this domain is very ancient. Archean species were first discovered in extreme conditions such as hydrothermal vents and hot springs, but have since been detected in diverse habitats from oceans, to soil, to the guts of mammals. Some of the beautiful colours of hot springs in places such as Yellowstone National Park are because of the presence of archaea. The types of archaea that prefer extreme conditions are called **extremophiles** and include **thermophiles** (heat-loving), **methanophiles** (methane-loving) and **halophiles** (salt-loving).

The Grand Prismatic Thermal Spring in Yellowstone National Park. The bright colours around the edge of the hot water are caused by microbial colonies that include archaea.

Halocins are a type of antibiotic made by halophile (salt-loving) archaea. Just as penicillin was first discovered in a fungus, lots of pharmaceutical drugs come from naturally occurring compounds. Archaea are currently being studied for the types of organic molecules they can produce, and some of them may hold the key to fighting diseases for which we do not yet have a cure.

Eubacteria is the domain in which we find the bacteria you are most familiar with: types of bacteria that make your yoghurt taste good, help your intestines work properly, and also the types that might give you an infection.

Eukarya is the domain in which we find all life other than Archaea and Eubacteria, from the microscopic single-celled yeast that helps bread to rise, to enormous organisms such as sequoia trees and blue whales. A eukaryote is recognizable by its membrane-bound nucleus and membrane-bound organelles.

Classification of Archaea

A major step forward in the study of bacteria was the recognition by Carl Woese in 1977 that Archaea have a separate line of evolutionary descent compared to bacteria. Before his work, it was believed that all life could be classified into Prokaryotes and Eukaryotes. Woese found evidence that required a third domain.

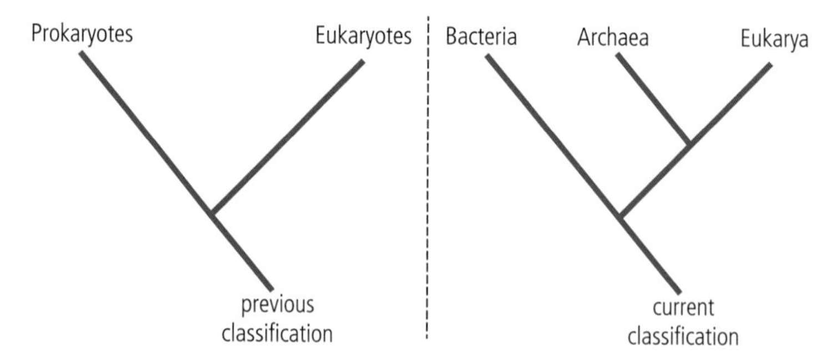

The classification of Archaea.

TOK

In TOK, you are asked to investigate the *assumptions* and *implications* of your arguments. Before Carl Woese, what were the *assumptions* and *implications* of the classification system? The *assumption* was that all life could be divided into either bacteria (prokaryotes) or eukaryotes. This *implied* that no other subdivisions existed and that when any new organism was discovered in the future, it would fit nicely into either of these two categories. That's what Carl Woese thought before he discovered a third category, the Archaea. This paradigm shift shattered the *assumption* and showed that the *implication* was now that any new organism discovered will be a bacterium, or an archaeon or a eukaryote. Despite decades-long efforts to find a fourth or fifth domain, the current understanding is that there are only three. This *implies*, in part, that we should never get too attached to an idea in a science as dynamic as biology.

What justifies separating archaea from the other prokaryotes, bacteria?

- Differences in the subunit of their ribosomes called 16S rRNA make grouping archaea and bacteria together difficult to justify (see Table 1).
- Metabolic reactions carried out by archaea that no bacteria can perform: some archaea can synthesize methane and some can use hydrogen gas as a source of energy.
- The ways in which archaea read DNA to produce RNA (transcription), and the way they produce proteins from RNA (translation), share more characteristics with eukaryotic transcriptions and translation than with bacterial transcription and translation.
- Some of the physical features of archaea are different from bacteria, such as the types of molecules used to build their cell membrane and cell wall.

What are some examples of ideas over which biologists disagree?

	Sequence	Found in methanogens?	Found in typical bacteria?
1	AACCUG	Yes	About a third of the time
	AAUCUG	Yes	Never
	AAG	No	About half the time
2	UAACAAG	Yes	Never
	UAACAAG	Yes	Never
	UAACAAG	No	Almost always
3	AUNCAACG	Yes	Never
	ACNCAACG	Yes	Never
	AXGCAACG	No	Almost always
4	NCCG	Yes	Never
	CCG	Yes	Never
	NCCG	No	Almost always
5	CCCCG	Yes, but less frequently	Almost always

A3.2 Table 1 A summary of Woese's results published in 1977. Woese and his team were surprised that so many sequences that are present in typical bacteria and which should be present in methanogens (if they are considered to be bacteria) were absent, and by how many that were not expected did appear. The unfamiliar letters in the sequences, notably X and N, are variables indicating that more than one nucleotide can be found at that position.

Comparing the last two columns of Table 1, it is possible to see that many of the sequences that Woese found in methanogens, which were expected to match those found in bacteria, did not, in fact, match bacteria. Some that were supposed to be there were absent, and others that were not supposed to be there were present. If more of the last two columns agreed by having "Yes" and "Almost always", for example, or "No" and "Never", then it would be logical to consider methanogens to be a type of bacteria. As this was not the case, it was necessary to reject the hypothesis and concede that Woese and his team was looking at a new, non-bacterial domain. Does this mean we could keep discovering new domains if we continue to analyse DNA? Maybe. But researchers have tried for decades and, for the moment, we only know of three.

Guiding Question revisited

What tools are used to classify organisms into taxonomic groups?

In this chapter we have discussed how there are countless ways to classify organisms:

- traditionally, morphology was used
- today we can also analyse DNA and amino acid sequences, using next generation sequencing techniques at the species level, or by analysing environmental DNA at the community level
- following the taxonomic techniques used by Linnaeus, we can categorize organisms into kingdoms, phyla and so forth, down to genus and species.

Guiding Question revisited

How do cladistic methods differ from traditional taxonomic methods?

In this chapter we have learned that:

- cladistics provides testable hypotheses of evolutionary relationships between species
- cladograms attempt to show not only how species are similar or different to one another, but also show how they are evolutionarily related.

Exercises

Q1. Outline two disadvantages of the traditional taxonomic system using ranked hierarchy of taxa.

Q2. Look at the three sequences below showing amino acids 100 to 116 in one of the polypeptides that makes up haemoglobin. Next to the human's sequence are two other species, A and B.

Amino acid	Human	Species A	Species B
100	PRO	PRO	PRO
101	GLU	GLU	GLU
102	ASN	ASN	ASN
103	PHE	PHE	PHE
104	ARG	LYS	ARG
105	LEU	LEU	LEU
106	LEU	LEU	LEU
107	GLY	GLY	GLY
108	ASN	ASN	ASN
109	VAL	VAL	VAL
110	LEU	LEU	LEU
111	VAL	VAL	ALA
112	CYS	CYS	LEU
113	VAL	VAL	VAL
114	LEU	LEU	VAL
115	ALA	ALA	ALA
116	HIS	HIS	ARG

(a) How many differences are there between the human sequence and the sequence of species A?

(b) How many differences are there between the human sequence and the sequence of species B?

(c) One of the sequences belongs to a horse and the other to a chimpanzee: which is species B more likely to be? Justify your answer.

Q3. Distinguish between analogous and homologous structures.

Q4. Explain why Carl Woese rejected the hypothesis that the methanogens he was studying could be classified as belonging to the domain of bacteria.

HL end

A3 Practice questions

1. The cladogram below shows the classification of species A to D. Deduce how similar species A is to species B, C, and D.

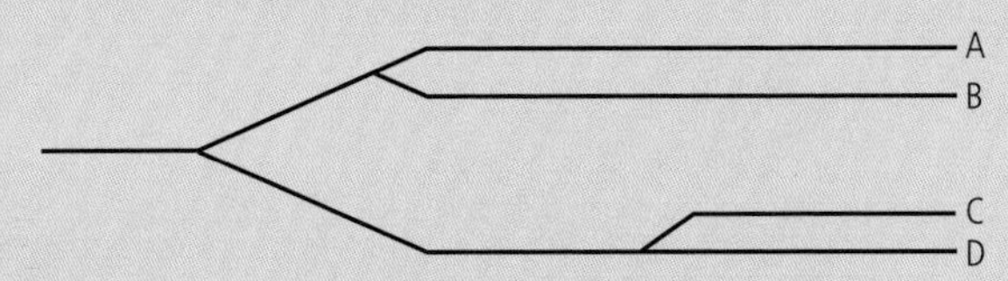

(Total 1 mark)

2. What are *Allium sativa* and *Allium cepa*?
   - **A** Two different species of the same genus.
   - **B** The same species of the same genus.
   - **C** The same species but of a different genus.
   - **D** Two different species of a different genus.

(Total 1 mark)

3. In a pollen grain of a species of flower, there are 20 chromosomes. Which of the following is true of the species?

   A $2n = 10$ **B** $2n = 20$ **C** $n = 10$ **D** $n = 20$

(Total 1 mark)

4. What determines the genomic size of a species?
   - **A** The total amount of DNA.
   - **B** The total number of genes.
   - **C** The total number of alleles.
   - **D** The total number of chromosomes.

(Total 1 mark)

5. The table gives common names and binomial names for some mammals.

Common name	Binomial name
Golden bamboo lemur	*Hapalemur aureus*
Golden jackal	*Canis aureus*
Grey wolf	*Canis lupus*
Red fox	*Vulpes vulpes*

   (a) Identify the **two** species most closely related. (1)

   (b) Identify **two** species from the list that are classified in different genera. (1)

(Total 2 marks)

6. Suggest two reasons for using cladograms for the classification of organisms.

(Total 2 marks)

7. HL Analyse the relationship between the organisms in the following cladogram.

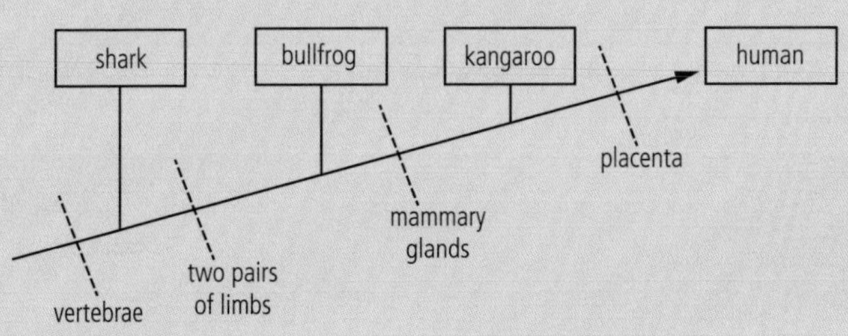

(Total 3 marks)

THEME
A Unity and diversity
4 Ecosystems

◀ The hand on this marine iguana from the Galápagos Islands has five digits. It shares an ancestor with other species that have limbs with five digits. Species adapt to their environment, and when a population finds itself in a unique habitat such as the volcanic beaches of the Galapágos Islands, it can develop adaptations that might transform the genetic makeup of the population enough to make it impossible to breed with other members of the original population. When this occurs, a speciation has happened: where there was once only one species, there are now two.

This process has taken place ever since life first appeared on Earth. As a result the planet is rich in species that fill every available niche. Biodiversity is the variety of life in all its forms. However, humans impact their environment in a variety of ways and many of their actions result in a loss of biodiversity. Scientists fear that we are currently in the middle of the sixth mass extinction. Conservation programmes exist to try to halt the loss of species around the globe. For example, the Galápagos Islands are recognized as an area of particular species richness and the whole area is now a national park. National park status means that the area is carefully managed to preserve the species that live there.

A4.1 Evolution and speciation

Guiding Questions

What is the evidence for evolution?

How do analogous and homologous structures exemplify commonality and diversity?

There is abundant evidence for evolution, and we will examine three types: molecular evidence from genetic data and amino acid sequences; experimental evidence from selective breeding of animals and plants; and morphological evidence from homologous structures, which are features of organisms that reveal they come from a common ancestor. Appendages with five bony digits can be found in animals as diverse as lizards, whales and bats, illustrating the diverse ways in which a limb can be used, such as for walking, swimming and flying. But the uniformity in bone structure and positions within the limbs also reveals that all these organisms had a common ancestor. In addition to homologous structures, there are analogous structures, which evolved on different branches of the tree of life but which serve the same purpose, for example wings in birds and insects. Wings allow flight in both these groups of organisms, but they have not evolved from the same body parts.

A4.1.1 – Evolution

A4.1.1 – Evolution as change in the heritable characteristics of a population
This definition helps to distinguish Darwinian evolution from Lamarckism. Acquired changes that are not genetic in origin are not regarded as evolution. **NOS:** The theory of evolution by natural selection predicts and explains a broad range of observations and is unlikely ever to be falsified. However, the nature of science makes it impossible to formally prove that it is true by correspondence. It is a pragmatic truth and is therefore referred to as a theory, despite all the supporting evidence.

Darwin and Wallace

At the age of 22, Charles Darwin had the opportunity to travel on board the *HMS Beagle* for a scientific exploration mission that started in 1831 and lasted for 5 years. Little did he know that it would allow him to see nature in a new way and come up with what would become one of the most important, controversial and misinterpreted ideas in biology: **the theory of evolution by natural selection.**

Darwin was not the only person to develop a theory to explain evolution. He was surprised to discover in 1858 that Alfred Russel Wallace had independently developed an almost identical theory. The two men presented their ideas jointly to the Linnaean Society in 1858.

TOK

Darwin was very reluctant to publish his ideas, in part because he knew how controversial they were at the time. He knew that other scientists would be highly sceptical of his work and would challenge it strongly. It is only when he read Wallace's ideas outlining a very similar theory that he decided to publish: he was afraid Wallace would get all the credit. Using this example, do you think competition between scientists helps or hinders the production of knowledge?

What is evolution?

Evolution is defined as the process of cumulative change in the heritable characteristics of a population. The word heritable means that the changes must be passed on genetically from one generation to the next, which implies that evolution does not happen overnight. The word cumulative is in the definition to stress the fact that one change is not usually enough to have a major impact on a species. Finally, the word population is in the definition because the changes do not affect just one individual.

Evolution is defined as the process of cumulative change in the heritable characteristics of a population.

Over time, if enough changes occur in a population, a new species can arise in a speciation split (explored in Chapter A3.1). The members of the new population will be different enough from the pre-existing population that they originated from that they will no longer be able to interbreed.

Once evolution by natural selection is understood, many of the mysteries of nature are revealed. When the role of DNA in inheritance (genetics) became known, decades after Darwin's theory had been published, there was a chance that it might have contradicted evolution by natural selection; contradictions often arise with new developments in science, making us rethink and revise our theories. In fact, the opposite happened. DNA evidence provided new support for natural selection beyond anything Darwin could have dreamt of, and led to the **modern synthesis** theory, or neo-Darwinism, a combination of Darwin's ideas with the newer ideas of genetics (based on work by Gregor Mendel, also in the 19th century), which was only confirmed long after Darwin and Wallace had died. One of the fundamental insights of the modern synthesis is the concept of common ancestry (which is explored in Chapter A3.1).

Lamarckism

Darwin and Wallace's theory replaced a previous idea formulated by French naturalist Jean Baptiste Lamarck. His theory was that organisms acquired characteristics through their lifetime and then passed them on to their offspring. For example, Lamarck explained how kangaroos developed more powerful hind limbs and tails during their lifetimes by using them a lot while letting their forelimbs atrophy through underuse. These characteristics were then passed on to their offspring. That sounds plausible, but experiments designed to illustrate the passing on of acquired traits do not produce the results Lamarck expected.

One remarkable feature of kangaroos is the large discrepancy between the size of their forelimbs and hindlimbs.

Nature of Science

The theory of evolution by natural selection predicts and explains a broad range of observations and is unlikely ever to be completely falsified. Some parts of the theory have been falsified, however, such as the pace at which natural selection can work. Darwin thought it was always slow, but we have observed it happening in just a few generations. Darwin also incorrectly predicted that the fossil record would not contribute evidence to support his theory. Scientists do not throw out an entire theory just because there is some evidence against certain aspects of it. When new evidence is presented that contradicts a theory, the theory can be updated rather than being totally invalidated. The role of a theory is to explain the mechanism of how something works in nature, and the theory of natural selection explains evolution very convincingly. No theory has been developed since that has had any success replacing it. Equally, given the nature of science, it is not possible to formally prove that the theory of evolution is true, which means that it is referred to as a theory, in spite of all the evidence supporting it.

A4.1.2 – Biochemical evidence for evolution

A4.1.2 – **Evidence for evolution from base sequences in DNA or RNA and amino acid sequences in proteins**
Sequence data gives powerful evidence of common ancestry.

Your DNA includes genes that go back not just to your parents, grandparents and great-grandparents, but back to when we had a common ancestor with fish (roughly 400 million years ago) and beyond. Some, but not all, of those sequences are still inside you now. This explains how, during the development of human embryos, we have, for a period of time, slits in our neck that are similar to the parts of fish embryos that develop into gills.

Using modern bioinformatic tools, we can compare nucleic acid (DNA or RNA) and protein data from many organisms, including humans, to examine their evolutionary relationships. Computer software can process millions of codes in seconds, and compile the differences and similarities to show how species are related to each other.

Access the link on this page of your eBook for instructions on how to find DNA sequences in the National Center for Biotechnology Information (NCBI) database.

In other words, to show which species are more closely related (e.g. chimpanzees and humans) and which are more distantly related (e.g. humans and fish).

As an example, we can use a protein present in many organisms, haemoglobin, which is the oxygen-carrying molecule found in red blood. It contains four protein chains, two alpha and two beta. Using software that is easily available online, it is possible to compare the protein chains in different organisms. The results of a DNA sequence analysis are shown in Table 1.

Species	DNA nucleotide sequence	Number of base pairs
Rat	ATAATTGGCTTTCAGGCTAAGATGATA-G----GGAAATATATATTTTGCATATAAATTT	960
Mouse	ATAATTGGCTTTTATGCCAGGGTGACAGG----GGAAGAATATATTTTACATATAAATTC	1,003
Dog	ATTGTTGGAGTCA---------------------------------------TATGGATTG	971
Human	AATTTCTGGGTTAAGGCAATAGCAATATCTCTGCATATAAATATTTCTGCATATAAATTG	1,192
Chimp	AATTTCTGGGTTAAGGCAATAGCAATATTTCTGCATATAAATATTTCTGCATATAAATTG * * * * *** ***	1,312

A4.1 Table 1 An excerpt showing 60 DNA nucleotides of compared sequences

Notice that in any position marked with an asterisk (*) under it, all the letters match in that column. In the other positions, there is at least one nucleotide that does not match with the others. Where there is a dash (–), a sequence does not exist there in that species. Only humans and chimpanzees have the middle four letters in this part of the sequence, for example. This implies that humans and chimpanzees are more closely related to each other than they are to the other organisms, and that they have a more recent common ancestor than the others. In fact, in this 60-letter sequence, between the human and chimpanzee there is only one nucleotide that does not match.

Although this line-by-line and base-by-base comparison is helpful, it is not easy to see which organisms are more closely related and which are more distantly related. Instead, we can use **phylogenetic trees**. Figure 1 shows an example.

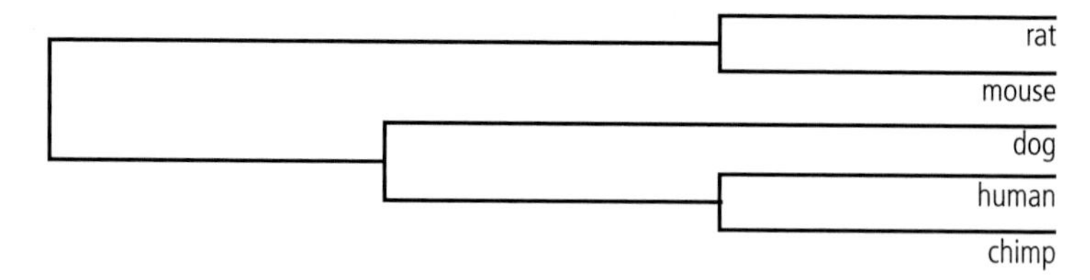

A4.1 Figure 1 An example of a phylogenetic tree.

In addition to comparing A, T, C and G in DNA sequences, it is possible to compare RNA nucleotides or amino acid sequences to reveal evolutionary changes over time.

The phylogenetic tree in Figure 1 indicates the similarities and differences between the sequences of the haemoglobin beta gene, *HBB*, in different species. It suggests that there was a speciation split between rodents and other mammals early on, then a split between dogs and primates. But we must be careful: this is only what *one* gene reveals about our past. Humans have over 20,000 genes, so just looking at evidence from one is unlikely to tell the whole story of how organisms are related.

A4.1.3 – Selective breeding

A4.1.3 – Evidence for evolution from selective breeding of domesticated animals and crop plants
Variation between different domesticated animal breeds and varieties of crop plant, and between them and the original wild species, shows how rapidly evolutionary changes can occur.

Artificial selection and evolution

The breeding of domesticated animals such as cattle, horses, dogs, sheep and pigeons, provides a good opportunity to study changes in heritable characteristics.

By controlling which males mate with which females, animal breeders can make predictions about the characteristics the offspring will have. Over the years, breeders have learned to choose the males and females with the most agriculturally desirable genetic characteristics, and breed them together. This is called **selective breeding**.

This cow has been bred to have a straight back for easier birthing and long legs for easier milking using automated mechanical pumps. She is a product of artificial selection by humans and she never existed in this form before human intervention.

After practising selective breeding for dozens and sometimes hundreds of generations, farmers and breeders realized that certain varieties of animals now had unique combinations of characteristics that did not exist before. Today, the meat or milk available to us is very different from that which was produced thousands of years ago or even only a hundred years ago. This is thanks to the accumulation of small changes in the genetic characteristics of livestock chosen by breeders.

Although selective breeding is evidence that evolution is happening as a result of an accumulation of small changes over time, the driving force is, of course, human choice. The farmers and breeders choose which animals will reproduce together and which will not. This is called **artificial selection** and it should be obvious that it is certainly not the driving force of evolution in natural ecosystems.

TOK

Animal breeding raises ethical questions. From an animal rights activist's point of view, breeding animals involves needless suffering and cruelty, including broiler chickens that grow too quickly for their bones to support their weight, and lifelong respiratory problems in certain dog breeds. From a breeder's point of view, they are providing safe, nutritious and affordable food for billions of people, or providing adorable pets to keep us company. Whose perspective is more convincing? What counts as a good justification for a claim?

Plant breeding

Teosinte is a plant that you may never have heard of, but you probably consume its descendant every day. It is an ancient wild grass, from what is now Mexico, central America and the Andes region, that has small hard edible kernels. About 10,000 years ago, farmers in these regions started saving seeds from the plants that had the most desirable characteristics, and only planted those seeds the following season.

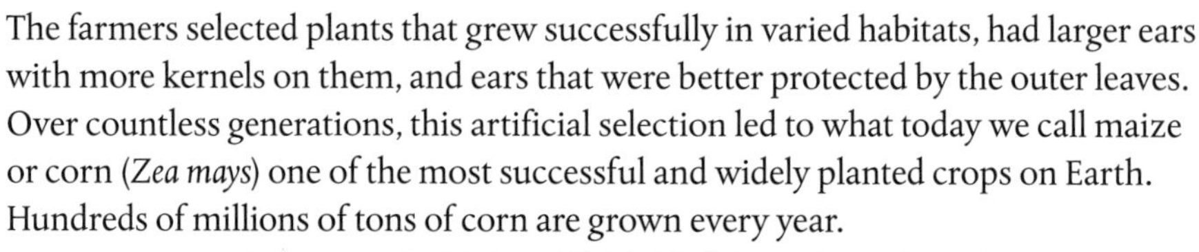

The farmers selected plants that grew successfully in varied habitats, had larger ears with more kernels on them, and ears that were better protected by the outer leaves. Over countless generations, this artificial selection led to what today we call maize or corn (*Zea mays*) one of the most successful and widely planted crops on Earth. Hundreds of millions of tons of corn are grown every year.

Thanks to Neolithic farming techniques of artificial selection, teosinte was transformed into modern corn.

Maize is an ingredient in more foods than you might think. For example, high fructose corn syrup (HFCS), is a food additive found in everything from candy to fast food, to fruit-flavoured drinks and sweet carbonated drinks, all sold worldwide. If you are eating or drinking something that has corn syrup added, you are consuming a product from *Zea mays*.

Selecting seeds with specific desirable traits generation after generation leads to small changes that accumulate over time, and results in a very different plant. The remarkable transformation from teosinte to maize is an example of evolution by artificial selection, and the changes can happen in a geologically short time. Thousands of years or even a hundred years might sound like a long time to you, but compared to the time scale of species (i.e. millions of years), these time scales are extremely short.

A4.1.4 – Homologous and analogous structures

A4.1.4 – Evidence for evolution from homologous structures
Include the example of pentadactyl limbs.

Homologous structures

Homologous structures are structures derived from the same body part of a common ancestor. One of the most striking examples of a homologous structure is the five-fingered limb found in animals as diverse as humans, whales and bats. Such limbs are called **pentadactyl limbs** because "penta" means five and "dactyl" refers to fingers. Although the shape and number of the bones may vary, the general format is the same. However, the specific functions of the limbs may be very different. Darwin explained that homologous structures were not just a coincidence but evidence that the organisms in question have a common ancestor and have therefore evolved from that common ancestor.

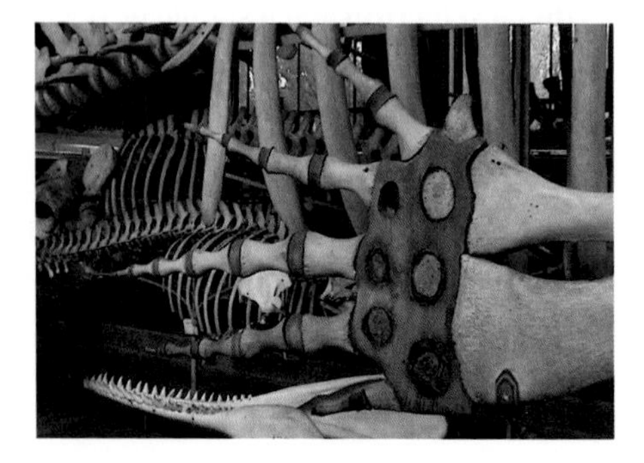

The front right fin of a Southern right whale (*Eubalaena australis*), showing five articulated digits.

Challenge yourself

1. (a) Look at the figure and complete the table.

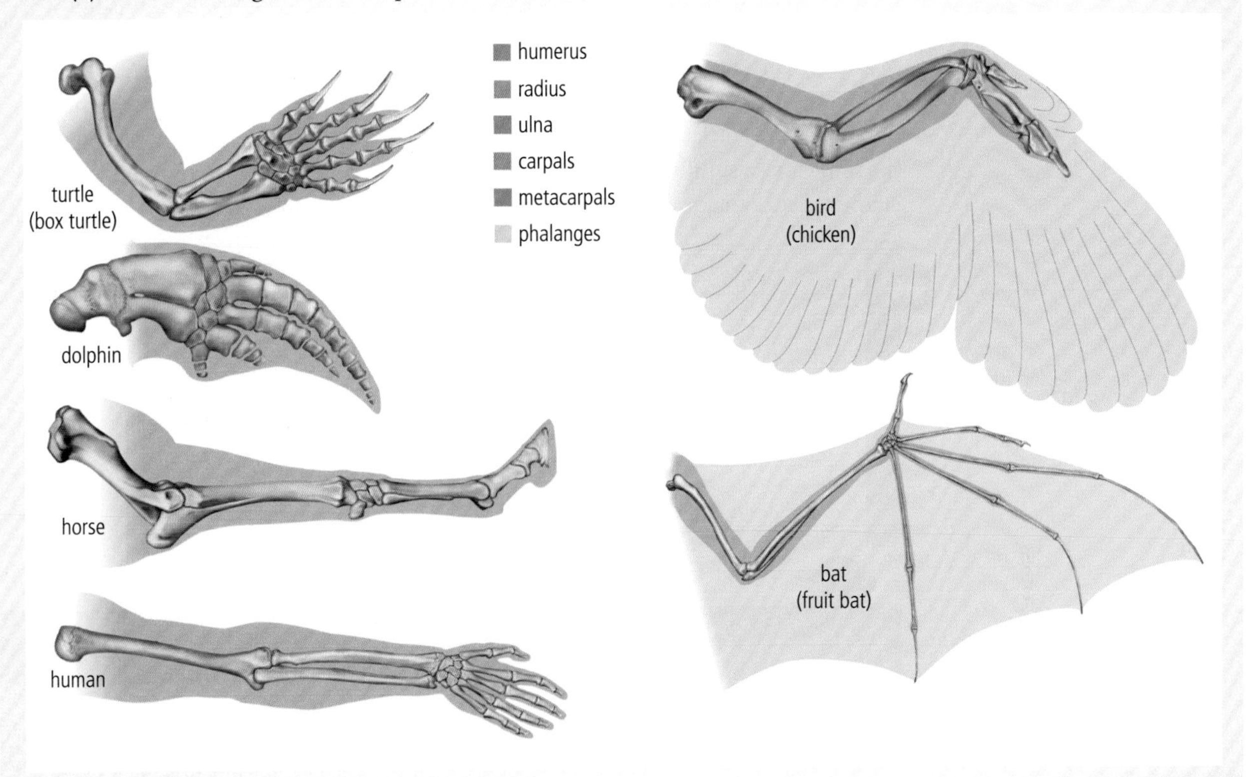

Characteristic	Bat	Bird	Human	Horse	Dolphin	Turtle
Number of digits (fingers)						
Description of phalanges (finger bones) (short/long, wide/narrow)						
Type of locomotion that the limb is best adapted for						

(b) There are two species in the table that have reduced their number of digits over the course of evolution. For these two species, explain why it would have been a disadvantage to keep all 5 digits. Limit your answer to the type of locomotion.

Analogous structures

In contrast, **analogous structures** are those that may have the same function but do not necessarily come from the same body part and do not indicate a common ancestor. Wings, which have developed from different body parts in different groups of organisms, are a good example of analogous structures: eagles, mosquitoes, bats and extinct reptiles such as the pterosaurs, all use (or used) wings to fly. Although these organisms are all classified in the animal kingdom, they are certainly not placed in the same taxon simply because of their ability to fly with wings. There are many other characteristics that must be considered.

A summary of analogous versus homologous characteristics, considering form and function

	Analogous	Homologous
Form: from same body part	No	Yes
Function: used for the same thing	Yes	No, or at least not always
Implies recent common ancestor?	No	Yes

A4.1.5 – Convergent evolution

A4.1.5 – Convergent evolution as the origin of analogous structures
Students should understand that analogous structures have the same function but different evolutionary origins. Students should know at least one example of analogous features.

Analogous structures can also provide evidence for evolution: if a feature such as wings is seen in many different organisms, then it is clearly advantageous and could have evolved in multiple ways over time. It is possible to have two organisms with different **phylogenies** that share similar physical aspects. Phylogeny is the way a species has split from other species.

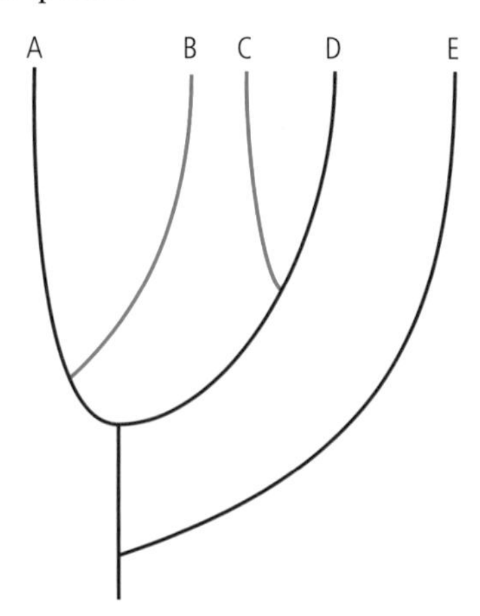

An illustration of divergent evolution (the blue lineages such as A and D, which are becoming less and less similar as time goes on) and convergent evolution (the green lineages B and C, which are becoming more and more similar over time).

Marsupials are mammals that have a pouch instead of a placenta for nourishing their young during early development. The isolated continent of Australia is rich with examples of marsupials, which have developed in similar ways as their distant placental cousins on other continents. For example, the Tasmanian tiger (*Thylacinus cynocephalus*), recently driven to extinction, was a marsupial that looked and behaved similarly to wolves and tigers from other continents.

A thylacine (*Thylacinus cynocephalus*), also known as the Tasmanian tiger, is classified as a marsupial. The species is believed to have been driven to extinction in the first half of the 20th century.

Convergent and divergent evolution can refer not only to entire organisms but also to physical features (such as horns, eyes or wings) and even refer to how organisms use certain molecules. The use of bioluminescent (glowing) chemicals by many deep-sea marine organisms, as well as by some bacteria and fungi, is an example of the convergent evolution of a biochemical.

What counts as strong evidence in biology?

In all these examples, the forces of natural selection used similar pressures on distant phylogenetic lines to favour certain characteristics over others. This process is explained in more detail in Chapter D4.1.

Convergent evolution means that different species start to look or behave more like each other over time. Potentially this allows them to exploit similar niches. Convergent evolution results in organisms developing analogous structures. Divergent evolution results in organisms that look less similar to each other but may have homologous structures.

A4.1.6 – Speciation

A4.1.6 – Speciation by splitting of pre-existing species
Students should appreciate that this is the only way in which new species have appeared. Students should also understand that speciation increases the total number of species on Earth, and extinction decreases it. Students should also understand that gradual evolutionary change in a species is not speciation.

Speciation by divergence of isolated populations

How is a new species formed? Recall that the biological definition of a species is based on the idea that members of the same species can produce fertile offspring together. If a subgroup of a reproducing population gets separated from the main population, it might evolve differently from the main population. For example, if a few iguanas from continental Central or South America were accidentally transported to the Galápagos Islands on a tree branch that broke off in a storm and was transported out to sea, the iguanas would find themselves in a new environment. The conditions on a remote island can be very different to those on

TOK

Extinction is forever. Once the last individual is gone, it is over for that species. There is no going back. It is the end of the line. Or is it? Scientists are working on reviving species by taking the DNA found in fossils and placing it in a similar species alive today. This process is called **de-extinction** or resurrection biology. But if we bring back a species without bringing back the conditions that allowed that species to thrive, it makes it a questionable endeavour. Also, we need to ask ourselves if the research money, time and effort would be better spent on preserving the species that are here now, thousands of which are in danger of going extinct within our lifetimes. In what ways do values affect the production of knowledge?

the continent, for example the types of food available may be different, or there may be no predators. Some of the iguanas on the Galápagos Islands have adapted in ways that allow them to dive for food (algae) on the ocean floor, making them the only marine iguanas in the world.

When we observe today's mainland iguana populations and island iguana populations, we can see that they have evolved differently over millions of years because they have had to adapt to different environments. Little by little, they changed over time and the differences are now large enough such that mainland iguanas can no longer breed with the island iguanas. A speciation split has occurred. Just evolving over time is not what makes a new species. There has to be a split whereby two populations are isolated and exposed to different environments that will select for some traits and against others. Speciation events like this increase the overall number of species, whereas extinctions reduce the total number of species.

Extinction

So far, we have only talked about species that have survived. The vast majority of species that have ever lived on Earth, over 99.99%, are now extinct. Extinction happens when the last individuals of a species die out. Examples include woolly mammoths, the dodo and, of course, *Tyrannosaurus rex*. On a phylogenetic tree, branches of extinct species are cut short and do not reach the extremities of branches the way existing species continue to do.

A4.1.7 – Reproductive isolation and differential selection

A4.1.7 – Roles of reproductive isolation and differential selection in speciation
Include geographical isolation as a means of achieving reproductive isolation. Use the separation of bonobos and common chimpanzees by the Congo River as a specific example of divergence due to differential selection.

Reproductive isolation

In some situations, members of the same species can be prevented from reproducing because there is an insurmountable barrier between them. Such a barrier can be geographical, temporal or behavioural. In each case the effect is the same: over time the two populations will face different selection pressures and will change in different ways. Eventually the two populations will change so much that the individuals from the two separate populations will not be able to reproduce with each other successfully to produce fertile offspring. This is called **reproductive isolation**, and at this point they will have become two separate species, as shown in the previous example with marine iguanas on the Galápagos Islands, which were geographically isolated, and therefore reproductively isolated, from their original population on the mainland.

Geographical isolation

Geographical isolation happens when physical barriers, such as land or water formations, prevent males and females from different parts of a population finding each other, thus making interbreeding impossible. For example, a river, a mountain or a clearing in a forest can separate populations. Tree snails in Hawaii exemplify this geographical isolation: one population lives on one side of a volcano and another population lives on the other side, and they never come into contact with each other.

The barrier can be produced by humans. The Great Wall of China might prevent certain organisms such as salamanders from getting to the opposite side, although it would not stop birds or seeds that are dispersed by wind or birds. Roads and dams can have a similar effect.

Bonobos and common chimpanzees

The Congo River is an example of a physical barrier that prevents the two populations of primates from interacting or interbreeding. The primates to the north and east of the river are chimpanzees (*Pan troglodytes*), and the primates south of the river are bonobos (*Pan paniscus*).

(a) The left image shows bonobos, or pygmy chimpanzees (*Pan paniscus*). **(b)** The right image shows juvenile chimpanzees (*Pan troglodytes*).

Chimpanzees are found north and east of the Congo River (A), whereas bonobos are found south of the river (B). The river is a barrier preventing the two species from encountering each other, and each species has evolved separately.

The differences in habitat, availability of food and the presence of enemies such as snakes have led to differences in traits, notably behavioural traits, between the two separated populations. Chimpanzees are considered to be more aggressive and territorial, while bonobos are more peaceful and nomadic. Chimpanzee social structure is clearly male dominated, whereas bonobo social structure tends to be matriarchal, with older males playing a role in decision making for the group. When one environment favours certain traits, and another environment favours different traits, there is differential selection. Traits such as aggression in defending a territory are selected for in places where resources are scarce, whereas there is no such selective pressure when a group can move around freely to find new food sources.

HL

A4.1.8 – Allopatric and sympatric speciation

A4.1.8 – **Differences and similarities between sympatric and allopatric speciation**
Students should understand that reproductive isolation can be geographic, behavioural or temporal.

Allopatric speciation

When a new species forms from an existing species because the population is separated by a physical barrier, it is called **allopatric speciation**. Geographical isolation leads to allopatric speciation.

Sympatric speciation

Populations do not always have to be separated geographically in order to undergo changes in their genetic makeup that are significant enough to produce a new species. When a new species forms from an existing species living in the same geographical area, it is called **sympatric speciation**. In this case speciation can be caused by temporal or behavioural isolation.

Temporal isolation refers to incompatible time frames that prevent populations or their gametes from encountering each other. For example, if the female parts of the flowers of one population of plants reach maturity earlier or later than the release of pollen of another population, the two will have great difficulty producing offspring together. Or, if one population of mammals is still hibernating or has not returned from a migration when another population of the same species is ready to mate, this would also be a temporal barrier between the two gene pools.

Behavioural isolation occurs when part of the population develops a different behaviour that isolates it from the rest of the population. For example, if part of a population of songbirds gradually develops a different song to attract mates, then over time that part of the population will only mate with those that recognize its song. The two populations will diverge and could eventually develop into different species.

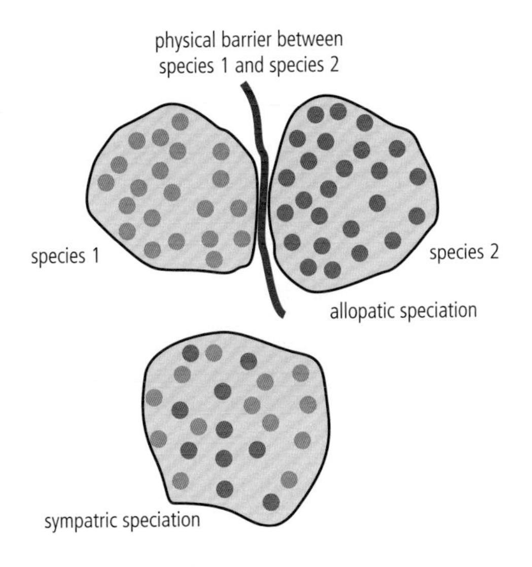

Allopatric speciation and sympatric speciation. Pink (lighter) dots represent individuals from species 1, and red (darker) dots represent individuals from species 2.

A4.1.9 – Adaptive radiation

A4.1.9 – Adaptive radiation as a source of biodiversity
Adaptive radiation allows closely related species to coexist without competing, thereby increasing biodiversity in ecosystems where there are vacant niches.

Adaptive radiation occurs when many similar but distinct species evolve relatively rapidly from a single species or from a small number of species. This happens because variation within a population allows certain members of that population to occupy different niches. A **niche** is a position or role within a community. Through natural selection and the presence of some kind of reproductive isolation, a new species can evolve.

An example of this includes the primates found in Madagascar and the Comoro Islands off the south-east coast of Africa. Millions of years ago, without competition from monkeys or apes, lemurs on these islands were able to proliferate. Large numbers of offspring meant a greater chance for diversity within the population.

The wide range of different lemur species adds to the biodiversity of the islands. Some lemur species are better adapted for living on the ground rather than in trees. Others are better adapted for living in lush rainforests, while some can survive in the desert. Most lemurs are active during the day (diurnal) but some are nocturnal. The reason why there are so many different species of lemur with different specialties is because of adaptive radiation.

Lemurs are primates found in Madagascar. The many species of lemur are a good example of adaptive radiation.

How does the theory of evolution by natural selection predict and explain the unity and diversity of life on Earth?

Other examples of adaptive radiation include the finches on the Galápagos Islands, and birds called Hawaiian honeycreepers. Hawaiian honeycreeper species display a wide variety of beak shapes, some of which are adapted exclusively to sip the nectar of flowers found only on Hawaii. Similar to the Galápagos finches, it is believed that all the Hawaiian honeycreepers alive today are the result of adaptive radiation of a few members of one species that arrived on the islands.

A4.1.10 – Barriers to hybridization, and hybrid sterility

A4.1.10 – Barriers to hybridization and sterility of interspecific hybrids as mechanisms for preventing the mixing of alleles between species
Courtship behaviour often prevents hybridization in animal species. A mule is an example of a sterile hybrid.

A hybrid organism is one that has been generated by fertilization between two different species. On rare occasions, and if the species are similar to each other, offspring may result from two species mating. The animal hybrids we see in zoos are rarely, if ever, produced in the wild because of barriers that prevent hybrids forming. These barriers are similar to those that cause reproductive isolation: geographical, temporal or behavioural.

Chapter A3.1 outlines how hybridization of animals can produce interspecific hybrids such as mules. Here are some examples of animal hybrids:

- female horse + male donkey = mule
- female horse + male zebra = zorse
- female tiger + male lion = liger.

When a sperm cell fertilizes an egg cell, its chromosome 1 needs to be compatible with the egg's chromosome 1. Each gene's position, its **locus**, needs to match in order to allow the genetic information from each parent to contribute to the offspring's features. The same compatibility and matching need to be true for all the chromosomes in the organism for viable offspring. We have also seen that different species usually have different chromosome numbers. When a horse that has 64 chromosomes mates with a donkey that has 62, the mule ends up with an odd number of chromosomes, 63.

It is thought that this mismatch of genes and chromosomes explains the infertility of interspecific hybrids. The mismatch makes the production of egg and sperm cells difficult, if not impossible. Even if some gametes are made, they would not necessarily match the gametes from members of the original two species. All of this means that, although hybrids may occasionally form, they are sterile, which prevents genes mixing between species.

Courtship behaviour

Behavioural isolation can happen when one population's lifestyle and habits are not compatible with those of another population. For example, many species of birds rely on a courtship display for successful mating. Two different species of bird may live in the same area but no reproduction will take place between the populations of the two species because of behavioural differences. The females of one species will not be attracted by courtship rituals performed by males of a different species. Hybridization will therefore not take place between the two species and their gene pools will not mix.

A4.1.11 – Abrupt speciation in plants

You have probably recently eaten a polyploid plant product: examples include peanuts, bread wheat, strawberries, apples and bananas.

A4.1.11 – Abrupt speciation in plants by hybridization and polyploidy
Use knotweed or smartweed (genus *Persicaria*) as an example because it contains many species that have been formed by these processes.
Note: When students are referring to organisms in an examination, either the common name or the scientific name is acceptable.

Polyploidy

You will recall that haploid cells, such as sex cells, contain one set of chromosomes (n). Diploid cells contain two sets of chromosomes ($2n$): one from each parent. **Polyploidy** refers to the situation in which a cell contains three or more sets of chromosomes ($3n$, $4n$, and so on):

- $3n$ = triploid
- $4n$ = tetraploid
- $5n$ = pentaploid.

Polyploidy can arise during the production of sex cells. If the copies of chromosomes are not completely separated into distinct nuclei, they will end up in the same cell. In plants, polyploidy can also happen when two species fuse their genetic makeup to make a polyploid hybrid. Polyploidy is much more common in plants than in animals. In plants, the extra sets of chromosomes can lead to more vigorous plants that produce bigger fruits or food storage organs and are more resistant to disease. In animals, having an entire extra set of chromosomes is usually fatal, although there are exceptions. Animals that demonstrate polyploidy include some fish and African clawed frogs.

▲ The clawed frog of the genus *Xenopus* is one of the rare examples of vertebrates that can demonstrate polyploidy.

However, the change is significant enough that production of offspring with the original population becomes impossible, resulting in speciation. In short, a new species has evolved from the original species, and both can continue. This speciation event can also follow the principle of sympatric speciation (see Section A4.1.8), because the evolving species can coexist without interbreeding.

Allopolyploid speciation

Certain plant species show interesting chromosome combinations that suggest there have been doublings of the chromosome numbers and hybridizations at some point. For example, the genus *Persicaria*, which includes about 120 species, has members that are diploid, tetraploid and even hexaploid. *Persicaria foliosa* is diploid, with two sets of its ten unique chromosomes ($2n = 20$), while *P. japonica* is tetraploid with four sets ($2n = 40$), and *P. puritanorum* is hexaploid ($2n = 60$). When we see multiples of the same number of chromosomes within a genus of plants, it suggests that polyploidy has taken place. This can be confirmed by looking at the chromosomes under a microscope or by sequencing the genes found in them. If there are four copies of each chromosome, this indicates polyploidy.

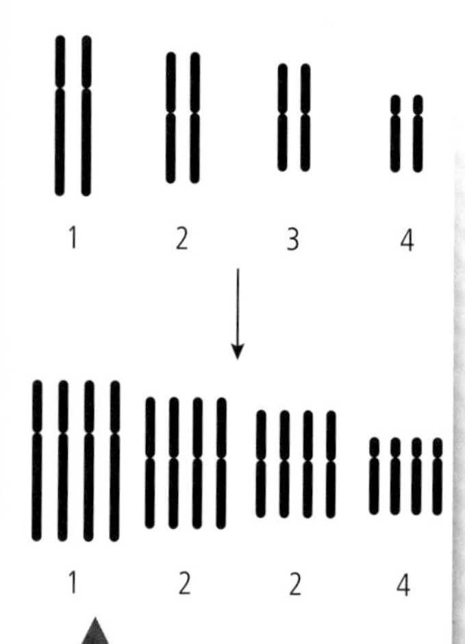

Polyploidy is when one or more complete sets of chromosomes are added to an organism's genetic material. Here a diploid cell becomes a tetraploid cell. This can happen when both sex cells, instead of containing only one gene from each pair, contain two in each pair.

One advantage to polyploidy is that it can allow sterile hybrid plants to become fertile again. With the additional genetic material, it is possible for a plant to produce seeds and pollen with the same number of chromosomes, which can then produce fertile offspring. Two hybrids that have doubled their chromosomes can generate seeds that will grow into a new plant that is different from either of the species that made the hybrid parents. You should know by now what this implies: a new species has been generated.

Another reason for polyploid hybridization in plants is that the extra genetic material can give the plant an advantage over other plants. This is good news for the new species but can be bad news, for example for farmers, whose crops have to compete with more successful weed species. Plants such as *Persicaria maculosa* are considered pests because they are highly invasive and compete with crops for nutrients and sunlight.

One important takeaway message from the processes of hybridization and polyploidy is that they can result in very rapid production of new species. At the time of Darwin and Wallace, this was considered atypical if not impossible: species take a long time to split, and speciation cannot happen in a single generation. The latest research on hybridization and polyploidy suggests that in plants, at least, speciation can be very quick. This is where we get the term **abrupt speciation**, also called instant speciation. The new organism has such a different chromosome number from its parents that it is reproductively isolated. Each time a new species is produced in this way, biodiversity increases.

HL end

Guiding Question revisited

What is the evidence for evolution?

In this chapter we have discovered that evidence for evolution is found in multiple forms, including:

- molecular evidence from genetic data and amino acid sequences
- experimental evidence from selective breeding of animals and plants
- morphological evidence from homologous structures.

Guiding Question revisited

How do analogous and homologous structures exemplify commonality and diversity?

In this chapter we have discussed how:

- the pentadactyl limb can be found in lizards, whales and bats, showing the diverse ways in which such a limb can be used, such as walking, swimming and flying
- uniformity in bone structure and position within the limbs indicates that all these organisms had a common ancestor
- analogous structures have evolved on different branches of the tree of life but serve the same purpose, examples being the wings of birds and insects
- wings allow flight in both groups of organisms, but they have not evolved from the same body parts.

Exercises

Q1. Which of the following is an example of speciation?

A Selective breeding to produce new varieties of the wheat *Triticum aestivum* with higher crop yields.

B Evolution of different courtship behaviours in separate populations of the cricket *Gryllus rubens*.

C Natural selection leading to an increase in the frequency of darker individuals of *Biston betularia*.

D Selective feeding by koalas (*Phascolarctos cinereus*) on eucalyptus species

Q2. Which evolutionary pathway is most likely to result in the evolution of analogous structures in species W and Z?

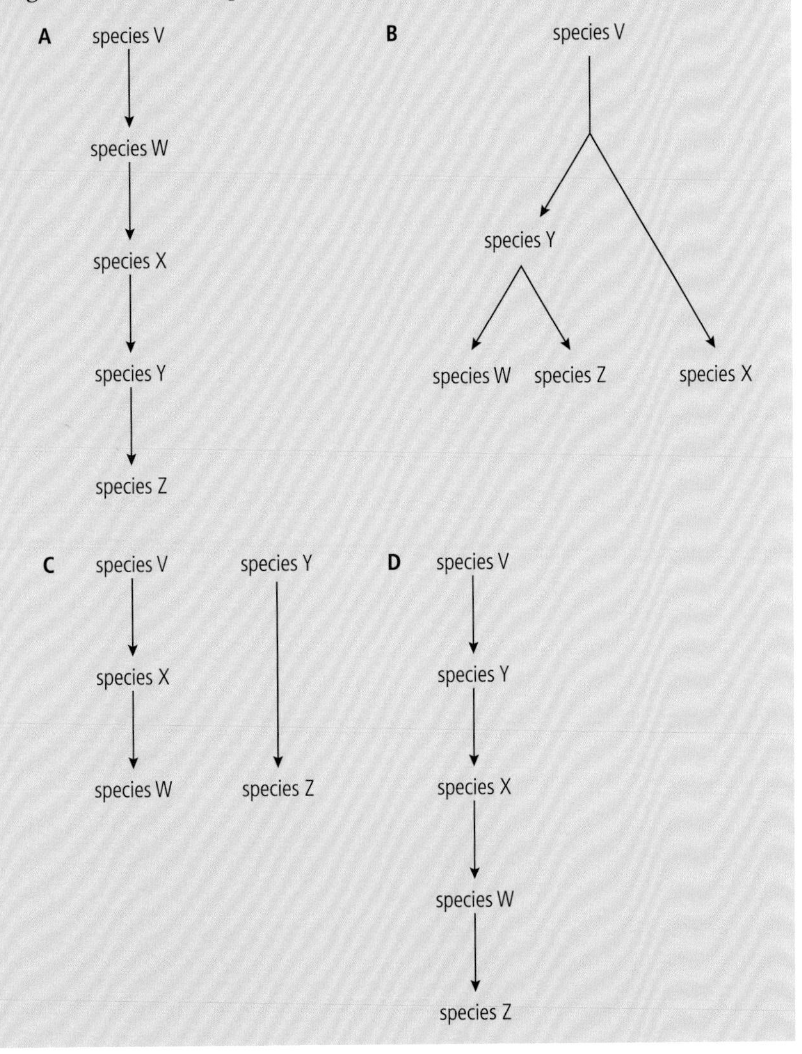

Q3. Which of the following is an example of convergent evolution?

A The pentadactyl limbs of bats and lizards.

B The opposable thumbs of humans and chimpanzees.

C The wings of penguins for swimming and of eagles for flying.

D The front fins of dolphins and sharks for swimming.

Q4. Explain how selective breeding can be a good example of evolution by selection, even though it is not natural selection.

Q5. Outline how modern maize (corn) was developed from teosinte over thousands of years of artificial selection.

Q6. HL Describe the process of adaptive radiation.

A4.2 Conservation of biodiversity

Guiding Questions

What factors are causing the sixth mass extinction of species?

How can conservationists minimize the loss of biodiversity?

The amazing diversity of life on Earth allows it to withstand many environmental pressures. Living things are not designed to be unchanging. Evolution has prepared living things to be adaptable to their environment. If that was not the case, we would find almost all evolutionary genetic lineages only in the fossil record. Instead, we find living species that have changed compared to their ancestral species. Evolution allows life on Earth to adapt and respond to change. Unfortunately, a new powerful factor has been introduced in the last few centuries. That factor is the growth and influence of the human species (*Homo sapiens*).

In this chapter we will look at the influence humans have had on the diversity of life. Our population growth and negative impact on our environment have already led to the beginning of a sixth mass extinction of species, the first to be attributed to human activities. We will discuss a few examples of recent human-caused extinctions and what could be many more to come.

Fortunately, we have begun to document the loss of species richness and have initiated efforts to minimize human impacts on at least some species. These efforts are led by both government and private enterprises, whose efforts are designed to preserve as many species as possible. Humans have treated this planet as if its resources were unlimited for far too long. It is in our best interest, as well as the best interest of all species, if we reconsider the role of humans in the stewardship of our planet.

A4.2.1 – Biodiversity exists in many forms

A4.2.1 – Biodiversity as the variety of life in all its forms, levels and combinations
Include ecosystem diversity, species diversity and genetic diversity.

Biodiversity means the variety of life found in an area. A healthy coral reef has a high level of biodiversity, whereas a recently burned forest does not. Biodiversity is at its best when many types of life forms are present in reasonable numbers. This includes animals, plants, fungi and a variety of microorganisms. It is usually the interactions between these life forms that keep an ecosystem healthy and biodiversity levels high. Biodiversity can be studied at three different levels: ecosystem, species and genetics.

A young emperor angelfish (*Pomacanthus imperator*), a beautiful inhabitant of the Great Barrier Reef.

Ecosystem diversity

Ecosystem diversity considers diversity from the largest overall viewpoint.

The Great Barrier Reef is made up of almost 3,000 individual reefs, along with over 1,000 islands, along the north-east coast of Australia. The entire reef system stretches for over 2,300 km, roughly north to south. This reef system is so long that the climate affecting the northern part of the reef is quite different to the climate affecting the southern part. The coral species and numerous other life forms are somewhat different from one another in the individual reefs, even though they are connected by the waters that they share.

The Great Barrier Reef is an example of one of the most ecologically diverse locations in the world. Each individual reef has its own ecosystem and a high level of biodiversity. This means that the total ecosystem biodiversity in the region is very high (i.e. there are lots of richly diverse ecosystems in one area). This diversity of ecosystems and its inhabitants provides stability in the area, and generates a great deal of species and genetic diversity.

Species diversity

Individual ecosystems have varying degrees of **species diversity**. Species diversity is sometimes known as **species richness** and is simply the number of different species in a community. **Species evenness** is a measure of the relative abundance of each of the species in a community. Some of the healthiest ecosystems have both high species richness and species evenness.

Table 1 presents the species evenness and species diversity in two hypothetical coral reef communities. Both samples were taken from the same sized area.

A4.2 Table 1 Species evenness (number of species) and diversity (types of species)

Number and type of species	
Coral reef community 1	**Coral reef community 2**
11 Hard corals	63 Hard corals
23 Fish	146 Fish
155 Sponges	64 Sponges
118 Echinoderms	21 Echinoderms
307 Total	294 Total

Ecosystem diversity is a measure of how many types of ecosystems there are in a given location. Species diversity is how many types of species exist in single ecosystem. Genetic diversity is concerned with the diversity of the gene pool within a population.

Look at the distribution of the species types in the two communities. Specifically take note that community 1 has a relatively high number of echinoderm and sponge species compared to hard corals and fish species. Even though community 2 has a slightly lower species richness (294 total species compared to 307 species), the species evenness evident in community 2 shows that it may be a healthier ecosystem.

Species evenness (taking into account species proportion) is often more important than species biodiversity (total number of species).

Genetic diversity

Every living organism has its own set of genes, giving that organism a unique set of characteristics within its population. All of the gene types or **alleles** found in the entire population is called the **gene pool** of that population.

Populations with greater **genetic diversity** (or a bigger gene pool) are more stable and can better withstand environmental pressures such as disease and extreme weather events. That does not mean any one randomly selected individual is more likely to survive, but it does mean that at least some of the population is likely to survive.

Generally, larger populations have higher genetic diversity. One of the problems that emerges when the population of an organism falls to low levels is that the gene pool becomes very small. Any genetic diseases contained in that population are then more likely to be expressed.

A4.2 Figure 1 In the late 20th century the Florida cougar (*Puma concolor*) population fell so low that a lack of genetic diversity became a severe problem. In the 1990s, biologists introduced eight female cougars from a population in Texas, and the genetic diversity greatly increased in the Florida population. This was a drastic step to take as biologists are reluctant to change the genetic makeup of a wild population.

The Florida cougar population in the mid-1990s suffered from a variety of genetic weaknesses, including heart failure, susceptibility to a variety of diseases, undescended testes and inability to withstand a variety of parasites.

TOK: Humans have the knowledge to alter the genetic makeup of wild populations. Under what circumstances should we do so? Who should make those types of decisions?

A4.2.2 – Has biodiversity changed over time?

A4.2.2 – Comparisons between current number of species on Earth and past levels of biodiversity

Millions of species have been discovered, named and described but there are many more species to be discovered. Evidence from fossils suggests that there are currently more species alive on Earth today than at any time in the past.

NOS: Classification is an example of pattern recognition but the same observations can be classified in different ways. For example, "splitters" recognize more species than "lumpers" in any taxonomic group.

The rate of extinction is currently very high, and the number of species alive today is lower than it was a few hundred years ago. Most of that loss of diversity can be traced back to human activities that have resulted in extinctions.

However, the fossil record suggests that if you go back further in time there are more species alive today than at any other geological time period. It is worth noting, however, that the number of species alive today and the number of species alive in the past are both estimates. The uncertainty arises because biologists are discovering new species every day and the fossil record is incomplete.

Evolutionary theory helps explain why there are more species alive today compared to any other time period. Speciation is the formation of new species. As part of the process

Some of the most interesting examples of speciation occur when organisms reach one or more islands. Each island may present a different set of resources and environmental challenges. Over a long period of time the ancestral (original) organism undergoes speciation into two or more species in the different island ecosystems. This is a process called **adaptive radiation** and is a type of speciation. Charles Darwin was inspired to understand evolution better when he saw evidence of adaptive radiation while on the Galapagos Islands.

of evolution, **speciation** occurs under certain conditions. Any prolonged period of time where the rate of speciation is greater than the rate of extinction will result in a higher total number of species. Despite the last few hundred years of high extinction rates, there have been many long periods when the speciation rate was higher than the extinction rate.

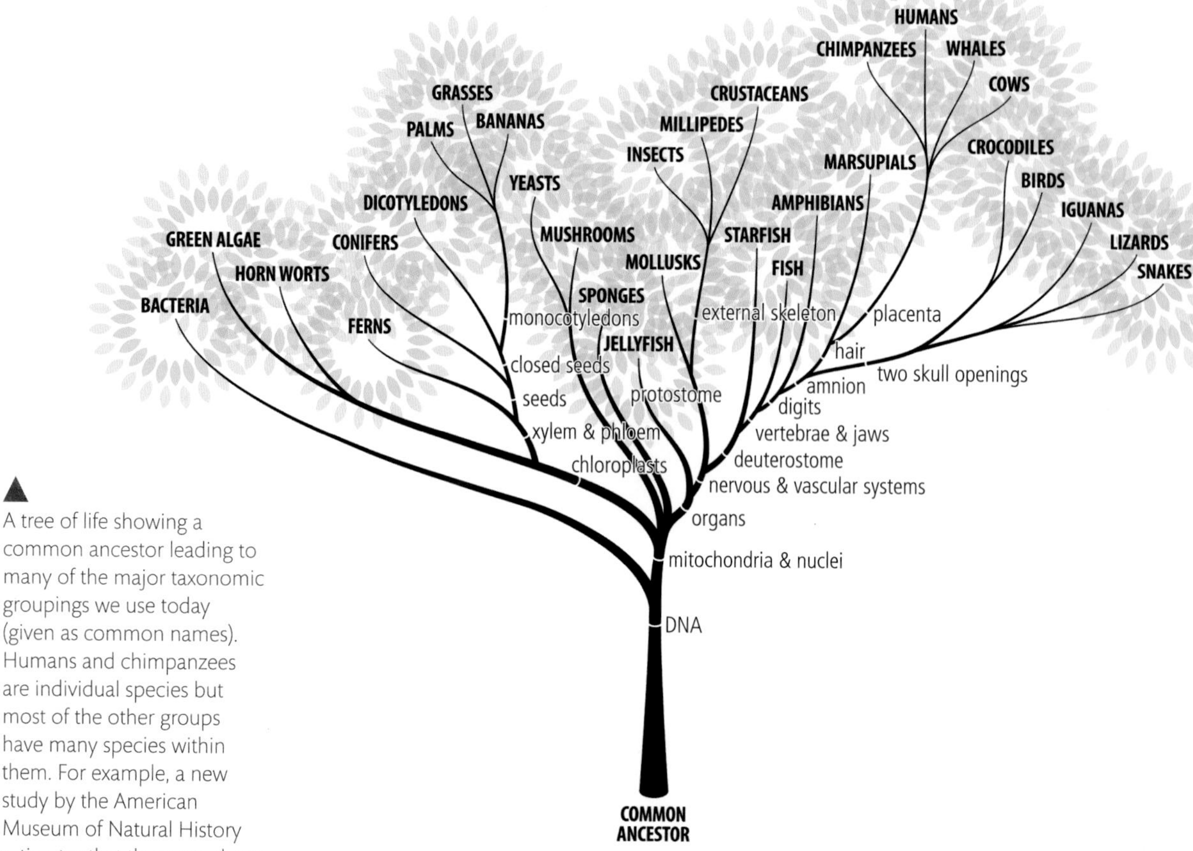

▲ A tree of life showing a common ancestor leading to many of the major taxonomic groupings we use today (given as common names). Humans and chimpanzees are individual species but most of the other groups have many species within them. For example, a new study by the American Museum of Natural History estimates that there are about 18,000 species of birds in the world.

Nature of Science

Early attempts to classify organisms were based on shared physical characteristics and there was much debate concerning the relationships between organisms. New tools for determining genetic descent are primarily based on common DNA sequences and are considered to be far more reliable than physical characteristics. However, that does not mean that there is always agreement on all taxa for every organism.

SKILLS

One of the more common taxonomic grouping systems places each organism into a Kingdom, Phylum, Class, Order, Family, Genus and Species. Memorizing these taxa in sequence allows you to better understand the evolutionary relationships between two or more organisms.

Nature of Science

Classifying organisms into categories called taxa is not always an exact science. There are two general ways of thinking about classification. One approach is represented by the "lumpers". Lumpers generally believe that similarities in organisms are more important criteria for classification than differences. Another approach is represented by the "splitters", who believe the opposite. Splitters end up with more categories of taxa than lumpers. Often a compromise is taken when deciding on taxonomic groupings.

Challenge yourself

1. Figure 1 shows a Florida cougar. Research the classification of the Florida cougar and see how it compares to chimpanzees and humans.

In what ways is diversity a property of life at all levels of biological organization?

A4.2.3 – Human activities and the rate of species extinction

A4.2.3 – Causes of anthropogenic species extinction

This should be a study of the causes of the current sixth mass extinction, rather than of non-anthropogenic causes of previous mass extinctions.

To give a range of causes, carry out three or more brief case studies of species extinction: North Island giant moas (*Dinornis novaezealandiae*) as an example of the loss of terrestrial megafauna, Caribbean monk seals (*Neomonachus tropicalis*) as an example of the loss of a marine species, and one other species that has gone extinct from an area that is familiar to students.

Note: When students are referring to an organism in an examination, either the common name or the scientific name is acceptable.

The extinction of a species caused by human activity is called **anthropogenic** species extinction. There is no doubt that most extinctions in the last few hundred years have had anthropogenic causes.

There have been five previous mass extinction events, and each occurred before humans evolved. The most recent previous extinction event was about 65 million years ago, and there is evidence that it was caused by an asteroid strike. That event killed virtually all of the dinosaurs and a great deal of other life on Earth at the time.

Many scientists propose that we are currently in the midst of a sixth extinction event and that it is the first anthropogenic event.

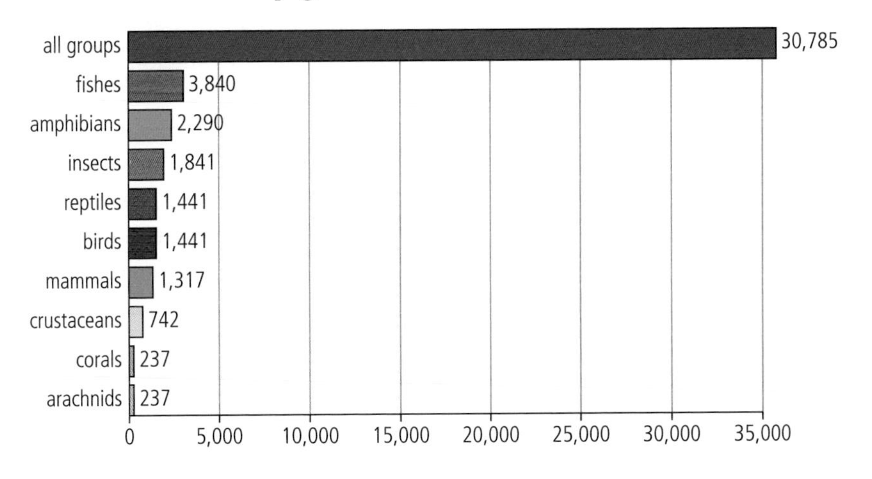

A chart based on data from the International Union for Conservation of Nature's (IUCN) Red List of threatened species, showing some of the types of organisms currently threatened by extinction. This and other data lead biologists to believe that we are now in the midst of a sixth great extinction event.

Case studies of organisms threatened by anthropogenic activities

Case study 1: the North Island giant moa

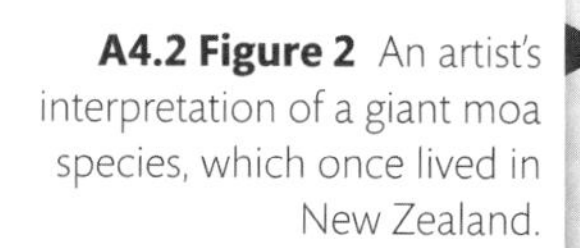

A4.2 Figure 2 An artist's interpretation of a giant moa species, which once lived in New Zealand.

An extinct species called the North Island giant moa (*Dinornis novaezealandiae*) lived in New Zealand up until as recently as 1300 CE. Moas were extremely large herbivorous birds, which swallowed and retained stones in their gizzards in order to grind the plants in their diet to extract more nutrients. As shown in Figure 2, they had no wings. Female moas were much larger than males; skeletal remains indicate that the females reached a height of about 3 m when their necks were stretched upwards. Their bodies were covered by long feathers that were up to 18 cm in length.

New Zealand was first populated by Polynesian people around 1200–1300 CE. It is estimated that the North Island giant moa was hunted to extinction within 100 years of human arrival on the island. This shows that anthropogenic extinction has been occurring for centuries, albeit on a smaller scale than seen today.

Case study 2: the Caribbean monk seal

A photo reconstruction of a Caribbean monk seal.

Caribbean monk seals (*Neomonachus tropicalis*) were declared extinct in 2008 by the US National Marine Fisheries Service, although this species may have actually gone extinct decades earlier. Caribbean monk seals were docile marine mammals living in and around the waters of the Gulf of Mexico and Caribbean islands. Prior to European colonization of the Caribbean area, this seal species was thought to have existed in at least 13 major colonies, with an overall population of approximately quarter of a million.

European colonists killed this seal for its oil, to use in lamps, and for food. The seals often hauled themselves out of water on beaches and rocks, and unfortunately showed little fear of approaching humans. Their behaviour made them easy targets for humans with guns and clubs. Some of the last Caribbean monk seals were killed to provide scientific specimens.

Case study 3: your choice of extinct species

The IB requires you to choose and research a brief case study of a species that has become extinct as a result of anthropogenic factors. You should choose an extinct species from your area of the world, or at least an area that is familiar to you. You can refer to this species by its common or scientific name. Use the two previous case studies as a guide to decide the level of detail required.

A4.2.4 – Human activities and ecosystem loss

A4.2.4 – **Causes of ecosystem loss**
Students should study only causes that are directly or indirectly anthropogenic. Include two case studies of ecosystem loss. One should be the loss of mixed dipterocarp forest in Southeast Asia, and the other should, if possible, be of a lost ecosystem from an area that is familiar to students.

In recent years, human activities have not only resulted in species extinctions, but also the destruction of entire ecosystems. More often than not, ecosystem loss is caused by massive habitat destruction such as deforestation.

Case studies of ecosystem loss by anthropogenic activities

Case study 1: mixed dipterocarp forests in Southeast Asia

The dipterocarps are a family (Dipterocarpaceae) of hardwood, tropical trees comprising about 500 species. Dipterocarp forests once dominated the island nations of Southeast Asia. The ecosystems provided by dipterocarp tree species were rich and varied.

Deforestation of dipterocarp forest in Southeast Asia. The wood is sold to worldwide markets.

Southeast Asia has long been known for its incredible rainforests. Collectively, however, Southeast Asia is losing about 1% of its rainforest every year, and in some individual areas the percentage is much higher. Some regions have already lost over 50% of their dipterocarp forested area. Frequently the forested land is completely stripped of its trees (a practice called clear-cutting) and there is a total loss of the local ecosystem. There are less damaging alternatives for timber removal, but clear-cutting is the least expensive option. By clear-cutting, the land is made available for agricultural use.

In many cases the land is cleared to allow the planting of palm oil trees. The fruit of these trees is used to make an oil that is used in hundreds of products throughout the world. Many of us use products containing palm oil without even knowing it. Sometimes the product ingredient list will use one of the alternative names for palm oil.

▲ A palm oil plantation in Borneo. This area was once most likely a richly biodiverse dipterocarp forest. As much as 50% of areas that were once dipterocarp forest are now being used for agriculture or urban development. Using large areas of land to grow a single crop is called **monoculture**.

Common food products that are likely to use palm oil include margarine, instant noodles, sliced bread and ice cream. Cosmetics such as lipstick, shampoo, toothpaste and moisturizers also contain palm oil, as do cleaning products such as soaps and laundry detergents.

Many of the nations of Southeast Asia have adopted a certification system for palm oil plantations. To receive certification, a plantation must observe certain practices for growing and refining palm oil. Certification requires less forest cover to be removed.

Challenge yourself

2. Suggest specific reasons why monocultures of large land areas are harmful from an ecological perspective.

Case study 2: your choice of ecosystem loss

The IB requires you to choose and research a brief case study of an ecosystem that is under extreme stress because of anthropogenic activity. You should choose an ecosystem from your area of the world or at least an area that is very familiar to you. Use the previous case study on dipterocarp forests as a guide to decide the level of detail required.

A4.2.5 – A biodiversity crisis

A4.2.5 – Evidence for a biodiversity crisis

Evidence can be drawn from Intergovernmental Science-Policy Platform on Biodiversity and Ecosystem Services reports and other sources. Results from reliable surveys of biodiversity in a wide a range of habitats around the world are required. Students should understand that surveys need to be repeated to provide evidence of change in species richness and evenness. Note that there are opportunities for contributions from both expert scientists and citizen scientists.

NOS: To be verifiable, evidence usually has to come from a published source, which has been peer-reviewed and allows methodology to be checked. Data recorded by citizens rather than scientists brings benefits but also unique methodological concerns.

As you know, anyone can create a website or post documents that seem to support a particular position. That does not make the information reliable or correct. All of us, including policymakers who make decisions about ecologically critical issues, need valid and accurate data. Two reliable sources that you and others can use are given on the following page.

IPBES

In 2019, the Intergovernmental Science-Policy Platform on Biodiversity and Ecosystem Services (IPBES, pronounced "Ip – Bes") published a comprehensive report that provides significant and reliable scientific guidance for policymakers. The information was obtained from a wide range of habitats studied by university and governmental research projects, with the input of local sources. The data is sampled regularly in order to identify trends, and findings continue to be updated through IPBES, and are made available via its website and publications. (Search IPBES for further, updated information).

In order to understand the nature of the current biodiversity crisis, we need reliable information. Such evidence can be drawn from IPBES reports and the ICUN's Red List.

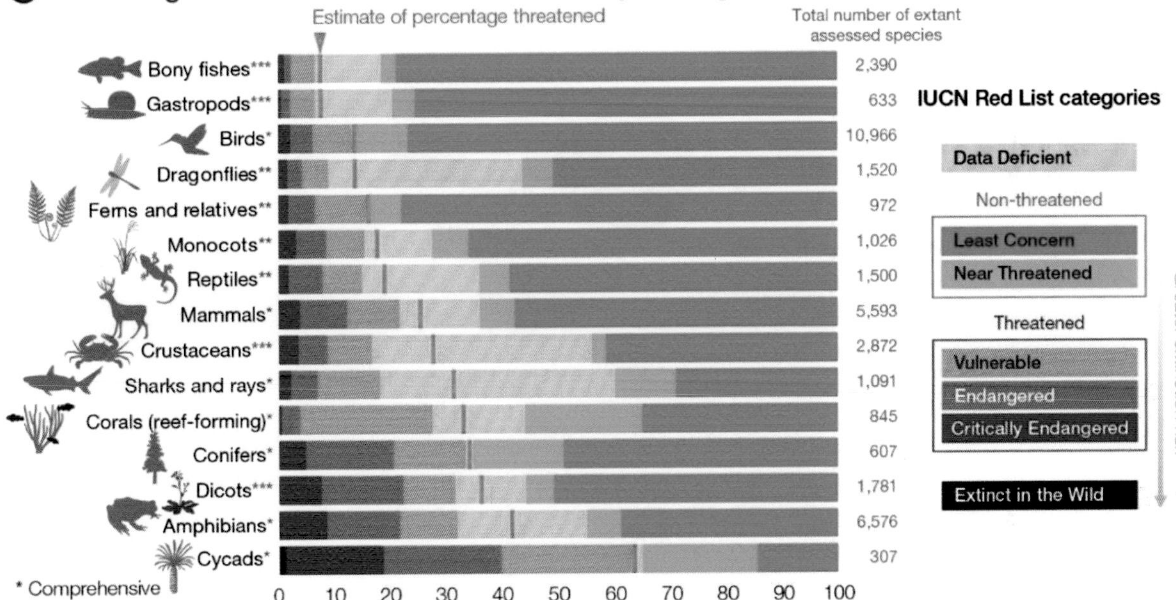

▲ A graphic of data sampled from the 2019 IPBES report.

IUCN Red List

The International Union for Conservation of Nature's (IUCN) Red List was established in 1964. This is a continuously updated list of the world's threatened species. It has become the one of the most comprehensive and trusted sources of information on the extinction status of over 140,000 fungus, plant and animal species. Currently more than 40,000 species are listed as threatened with extinction. Each assessed species is rated on a scale (see Figure 3 on the next page) indicating its ecological health. Each entry to the list contains the details of the research papers used to make the assessment. (Search IUCN Red List for further, updated information.)

Jump to Southwest Bornean Orangutan: In detail

Southwest Bornean Orangutan

Pongo pygmaeus ssp. *wurmbii*

ABSTRACT

Southwest Bornean Orangutan *Pongo pygmaeus ssp. wurmbii* has most recently been assessed for *The IUCN Red List of Threatened Species* in 2016. *Pongo pygmaeus ssp. wurmbii* is listed as Critically Endangered under criteria A4abcd.

THE RED LIST ASSESSMENT

Ancrenaz, M., Gumal, M., Marshall, A.J., Meijaard, E., Wich , S.A. & Husson, S. 2016. *Pongo pygmaeus* ssp. *wurmbii. The ...*

RED LIST

NOT EVALUATED NE | DATA DEFICIENT DD | LEAST CONCERN LC | NEAR THREATENED NT | VULNERABLE VU | ENDANGERED EN | CRITICALLY ENDANGERED CR | EXTINCT IN THE WILD EW | EXTINCT EX

A4.2 Figure 3 Results from a sample search of the IUCN Red List. A rating is given specifying the status of the species. This species is currently rated as critically endangered.

Researchers who participate in creating publications such as IPBES reports and the IUCN Red List often make use of people with local knowledge concerning species in a given area. Indigenous peoples, hunters, people in fishing industries and many other local sources often have information concerning species that no one else can provide.

Nature of Science

Local individuals (citizen scientists) are an important source of information about populations, but they may not be sampling populations in a scientific manner. Data gathered from the local population must be collated by a reliable scientific organization. Such scientific organizations provide established methods of collecting data, and the results are published in peer-reviewed research papers.

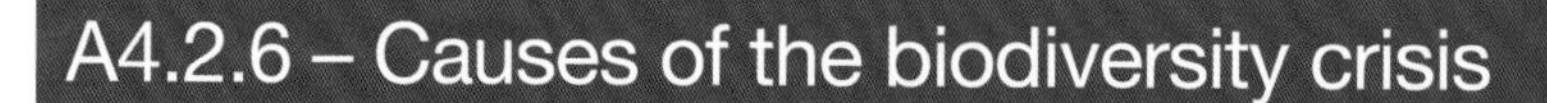

A4.2.6 – Causes of the biodiversity crisis

A4.2.6 – Causes of the current biodiversity crisis
Include human population growth as an overarching cause, together with these specific causes: hunting and other forms of over-exploitation; urbanization; deforestation and clearance of land for agriculture with consequent loss of natural habitat; pollution and spread of pests, diseases and alien invasive species due to global transport.

The current estimate is that there are over 8 billion people on Earth, and our population is continuing to increase. The population of a species at any given point in time is based on the current estimated population size plus the number of births and minus the number of deaths. If the birth rate exceeds the death rate, the population is increasing. That is the situation today for the human population when the data is considered from a global perspective.

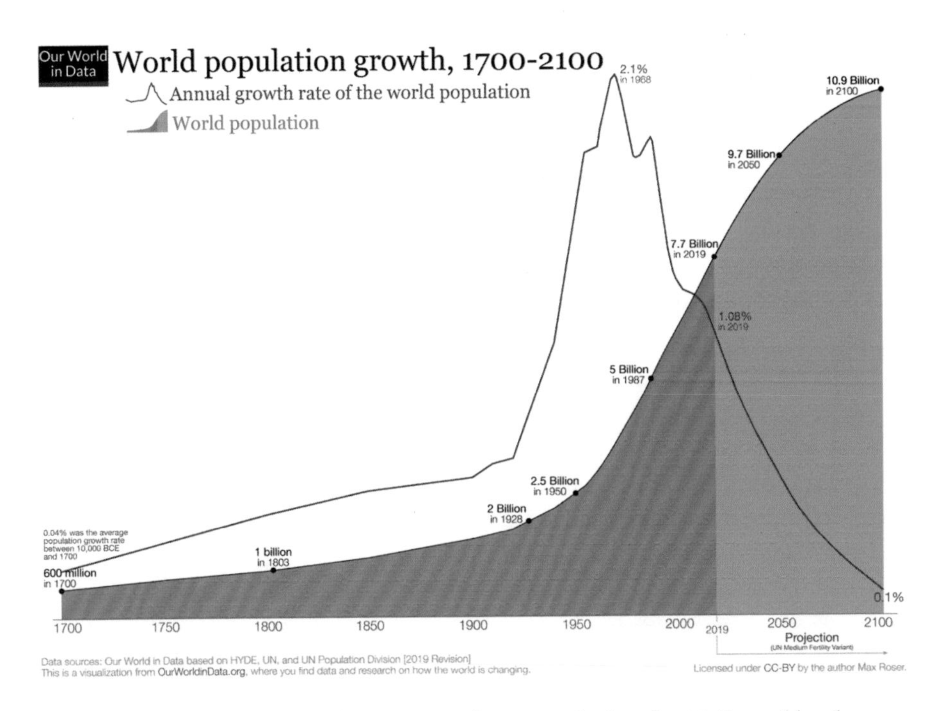

A graph showing two interrelated population factors. The shaded area represents the historic data and future predictions for the human population from 1700 through to a projected 2100. The line shows the past and predicted annual growth rate of the human population.

The annual growth rate of the human population peaked in the 1960s and has been declining since then. Based on this alone, logic would suggest that the total human population is therefore also declining. One of the primary reasons the data does not reflect this pattern is that humans are living for longer, and therefore the death rate is declining as well. This results in a pattern of continued population growth that is projected to continue until the birth rate declines and equals the death rate worldwide. Some scientific projections have this occurring around the year 2100.

The continued growth of the world's population is not evenly distributed around the globe. The continents of Africa and Asia are projected to have much greater population growth rates than other continents. Some European countries, such as Italy, already have declining populations.

The link between an increase in human population and the biodiversity crisis

Humans need resources to survive, and they produce waste and pollution. An increase in population means more resources are necessary and more pollution is produced. An increase in the human population has an impact on ecosystems because resources such as food, minerals and water must be sourced from ecosystems. Each time an ecosystem is damaged, the biodiversity within that ecosystem can be reduced. Some examples of human population growth effects on biodiversity are given below.

- Over-exploitation of resources, e.g. commercial fishing.
- Hunting, e.g. African elephant (*Loxodonta africana*) populations have decreased drastically because the animals have been hunted for their tusks, often illegally.
- Deforestation, e.g. forests have been reduced to extract minerals, hardwoods or to clear the land so that it can be used for agriculture. Often the crops planted are monocultures, which reduces the biodiversity of the area even further.
- Monoculture agriculture practices, e.g. palm oil plantations.
- Pollution, e.g. microparticles of plastics have been found in nearly every corner of the oceans.
- Increased pest species, e.g. the spruce bark beetle (*Ips typographus*) was first found in the UK in 1982. It is thought that the larvae entered the country in some untreated wood from either Europe or Asia.

- Invasive species, e.g. the Burmese python (*Python bivittatus*) was accidently introduced into the Florida Everglades at the end of the last century. Burmese pythons have no natural predators in the Florida Everglades, and they now pose a serious risk to native wildlife, decreasing biodiversity.
- Urbanization, e.g. a growing population means that more houses and services are needed. Towns and cities are growing in size, using land that was previously unused or used for agriculture.
- Spread of disease in both humans and other organisms.

An increasing human population is resulting in increased commercial fishing, placing a strain on many populations of fish throughout the world.

A4.2.7 – Conservation of biodiversity

A4.2.7 – Need for several approaches to conservation of biodiversity
No single approach by itself is sufficient, and different species require different measures. Include in situ conservation of species in natural habitats, management of nature reserves, rewilding and reclamation of degraded ecosystems, ex situ conservation in zoos and botanic gardens and storage of germ plasm in seed or tissue banks.

As you have learned in earlier sections, human activities have resulted in biodiversity declines. There is no single solution for the loss of biodiversity. Many different approaches are required, and each biodiversity problem requires a unique approach. Fortunately, some governments and non-governmental organizations (NGOs) are making concerted and varied efforts to manage and improve biodiversity. These efforts can be classified into two categories:

- in situ conservation efforts, i.e. managing natural areas
- ex situ conservation efforts, i.e. managing one or more species outside their natural area.

In situ efforts to improve biodiversity

Establishment of national parks

A national park is an area of land established by a nation and dedicated to preserving the geology and wildlife of that area. Typically, human visitors are permitted in the park, but development and building within the park are restricted. National parks prevent ecosystems from being lost as a result of extraction of resources, urbanization or many other human activities.

A male gerenuk (*Litocranius walleri*) in Samburu National Park, Kenya.

Establishment of nature reserves

Nature reserves are smaller areas than national parks. Most attempt to provide an area where ecosystems can be protected from urbanization and uncontrolled use. Again, these steps help to maintain biodiversity.

Rewilding of areas damaged by human intervention

The aim of this private and public approach is to let nature take better care of an area than people have. Sometimes the aim is to undo previous damage by removing such things as dams and roads, and reducing active management of wildlife populations. Rewilding promotes forest and aquatic ecosystem regeneration. By leaving an area to regenerate, species that were previously lost from an area may be able to return, thereby increasing biodiversity again.

Reclamation of degraded landscapes

Human activities such as strip mining and clear-cutting of forests leave areas that have little or no possibility of natural regeneration. Reclamation projects aim to rebuild and replant as much of an ecosystem as possible.

Ex situ efforts to improve biodiversity

Breeding programmes by zoos

Some zoological parks have established **animal husbandry** facilities to promote the continuation of species that are threatened and endangered. **Artificial insemination** is a common technique used by zoos, as they typically have very small populations of captive animals. Artificial insemination facilitates the production of offspring from animals in two different zoos (possibly in different countries). This technique promotes genetic diversity within the captive population. Careful pedigrees of animals are kept in order to choose breeding pairs that will increase genetic diversity.

Botanic gardens

Botanic gardens provide a living store of plant material that helps to promote biodiversity and helps conservation efforts. Some plant species now only exist in artificial garden facilities. The plants provide a reservoir of genetic material for restoration efforts, and a source of material for scientific research of a species. Botanic gardens often exchange seeds or pollen in order to help preserve rare, threatened or endangered species.

Seed banks

How does variation contribute to the stability of ecological communities?

There are over 1,000 seed banks scattered around the globe. A seed bank is exactly what it sounds like, a place where you can safely store living seeds. The seeds can then be used to repopulate a species of plant if necessary. Seeds are ideally kept in cool, dark and dry conditions. One of the largest and most famous seed banks is located in Norway. This facility is called the Svalbard International Seed Vault (sometimes called the Doomsday Vault).

Entrance to the Svalbard International Seed Vault, Norway. Most of the facility is underground.

Animal tissue banks

The overall increase in the global population of humans has caused the current biodiversity crisis. Maintaining biodiversity is going to require a wide range of different solutions.

Two types of tissue are stored in animal tissue banks. One type is called **germplasm**, and includes sperm, eggs and embryos. The aim is to collect and store the reproductive cells of various threatened species. One of the challenges is to collect germplasm from wild populations of animals in order to have reproductive cells that can be used in captive breeding programmes. The tissue is stored cryogenically, and can be kept for a nearly indefinite period of time before use. The second type of tissue collected is called **somatic tissue**, and includes non-reproductive tissue samples. This tissue is useful for DNA research and possible cloning.

A4.2.8 – The EDGE of Existence programme

A4.2.8 – Selection of evolutionarily distinct and globally endangered species for conservation prioritization in the EDGE of Existence programme

Students should understand the rationale behind focusing conservation efforts on evolutionarily distinct and globally endangered species (EDGE).

NOS: Issues such as which species should be prioritized for conservation efforts have complex ethical, environmental, political, social, cultural and economic implications and therefore need to be debated.

In 2007 the Zoological Society of London (UK) launched a global programme with the goal of selecting evolutionarily distinct and globally endangered species. A selected species is then promoted for priority status in conservation programmes.

According to the scores generated by EDGE researchers, the aardvark is the world's most evolutionary distinct mammal.

The selection of EDGE of Existence species is a process. First, the IUCN Red List (see Figure 3 on page 170) rating on a species is consulted. A score is then generated from this list indicating how endangered the species is. Next, the species is evaluated for its unique evolutionary history. This is done using DNA sequencing information. Those species that are the most endangered *and* the most evolutionarily distinct are given a high EDGE score, indicating that they should be prioritized.

The aardvark (*Orycteropus afer*). Even though this animal is evolutionarily distinct, it is in the least concern category on the IUCN's Red List and thus does not have an overall high EDGE score.

The EDGE of Existence programme aims to inform governments, conservation organizations and local populations of the ecological peril of different species. It is hoped that this will help ensure that those that are both endangered and evolutionarily distinct will be given the highest priority when deciding which species to protect.

The rapid increase in the global human population is causing the current global biodiversity crisis because humans demand increasing levels of resources. To improve biodiversity multiple approaches are needed, and organisms that are particularly at risk can be identified using the EDGE of Existence programme.

Nature of Science

Species selection for priority status in conservation efforts is a complex issue, with ethical, environmental, political, social, cultural and economic considerations. The selections should have input from many stakeholders.

Guiding Question revisited

What factors are causing the sixth mass extinction of species?

This chapter has discussed how the increasing human global population is leading to:

- deforestation
- monoculture agricultural practices, such as palm oil plantations
- habitat destruction from urban development
- pollution of air, land and water
- excess commercial fishing
- unregulated hunting and poaching of wildlife.

Guiding Question revisited

How can conservationists minimize the loss of biodiversity?

This chapter has considered how the following actions can help to control or minimize the loss of biodiversity:

- establishing national parks and nature reserves
- rewilding and reclamation projects for damaged habitats
- breeding programmes in zoos
- establishing botanical gardens and seed banks
- establishing animal tissue banks
- using the research carried out by a variety of organizations to inform policymakers and the general population of biodiversity problems, including the work of the:
    - International Union for Conservation of Nature's (IUCN) Red List
    - Intergovernmental Science-Policy Platform on Biodiversity and Ecosystem Services (IPBES)
    - EDGE of Existence programme.

Exercises

Q1. Which of these events would most likely lead to adaptive radiation?

A Limitation of a vital resource.

B Immigration of an ancestral species to a diverse group of islands.

C A richly diverse ecosystem.

D Enhanced volcanic activity resulting in numerous sulfur compounds in the atmosphere.

Q2. Distinguish between genetic diversity and species diversity.

Q3. State the primary difference between the sixth (current) mass extinction and the previous five mass extinction events.

Q4. State four ways that coral reefs can be damaged (directly or indirectly) by human activity.

Q5. Briefly describe the two necessary requirements for a species to be given a high rating by the EDGE of Existence programme.

Q6. What is germplasm and why is germplasm being stored by cryogenic techniques?

Q7. Discuss why a zoo captive breeding programme might need reproductive cells extracted from a wild population of animals.

A4 Practice questions

1. Outline how reproductive isolation can occur in an animal population.

   (Total 3 marks)

2. A friend says that the environment on Earth is healthy because there are more species alive today than at any other time in history. Discuss this statement.

   (Total 4 marks)

3. Extensive areas of the rainforest in Cambodia are being cleared for large-scale rubber plantations. Distinguish between the sustainability of natural ecosystems such as rainforests and the sustainability of areas used for agriculture.

   (Total 3 marks)

4. List **two** in situ conservation efforts.

   (Total 2 marks)

5. List **two** ex situ conservation efforts.

   (Total 2 marks)

6. HL Research suggests that many living plant species are polyploid. Explain how polyploidy occurs and, using a named example, how polyploidy can lead to speciation.

   (Total 7 marks)

THEME

B Form and function

1 Molecules

Living things make use of molecules that are similar to each other in that they use carbon as a fundamental component. These carbon-based molecules are called organic molecules. You will study three of these organic molecule groupings in this chapter, specifically carbohydrates, lipids and proteins. Even though these molecules are varied and complex, they have patterned molecular arrangements that you will learn to recognize. At first glance, the computer-generated graphic showing the enzyme invertase (a protein) seems to be beyond comprehension. As you work your way through the study of biochemistry, the structure of this molecule and many others will become clearer to you.

B1.1 Carbohydrates and lipids

Guiding Questions

In what ways do variations in form allow diversity of function in carbohydrates and lipids?

How do carbohydrates and lipids compare as energy storage compounds?

Biochemically important organic molecules are diverse in both structure and function. Carbohydrates and lipids are two groups of vital organic molecules.

The smallest forms of carbohydrates are called monosaccharides. This group of molecules follows the formula, $C_nH_{2n}O_n$, where n equals the number of carbon atoms. Monosaccharides are the monomers of the larger carbohydrates such as disaccharides (two monosaccharides bonded together) and polysaccharides (many monosaccharides bonded together). There are numerous carbohydrates found in nature. They have a variety of different forms and they serve a multitude of functions.

Lipids are molecules that are oils at warmer temperatures and fats at cooler temperatures. Each individual lipid type has its own temperature at which that phase change occurs. The monomers of triglyceride lipids are molecules known as glycerol and fatty acids. The identity of the specific lipid is dependent on the fatty acids as they are highly variable. Lipids also have a variety of functions.

One function that carbohydrates and lipids share is to act as energy storage molecules. Per gram of substance, lipids store approximately twice the chemical energy compared to carbohydrates.

B1.1.1 – The variety of compounds containing carbon

B1.1.1 – Chemical properties of a carbon atom allowing for the formation of diverse compounds upon which life is based
Students should understand the nature of a covalent bond. Students should also understand that a carbon atom can form up to four single bonds or a combination of single and double bonds with other carbon atoms or atoms of other non-metallic elements. Include among the diversity of carbon compounds examples of molecules with branched or unbranched chains and single or multiple rings.
NOS: Students should understand that scientific conventions are based on international agreement (SI metric unit prefixes "kilo", "centi", "milli", "micro" and "nano").

Carbon's name is derived from the Latin word "carbo", meaning charcoal. Only a few compounds that contain carbon are not classified as organic, including carbon dioxide.

The majority of molecules within all living organisms can be categorized into one of four biochemical groups: carbohydrates, lipids, proteins and nucleic acids. These four types of molecules interact with each other in a wide variety of ways in order to carry out the metabolism of each cell. All of these molecules contain carbon and this explains why life on Earth is often described as "carbon-based".

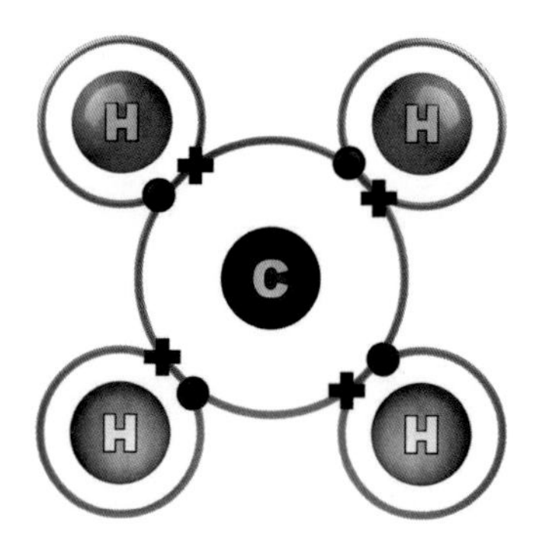

A molecular representation showing all four carbon atoms forming a covalent bond with a hydrogen atom. The resulting compound is methane (CH_4).

Carbon can share its four outer shell electrons in many diverse patterns. For example, two carbon atoms can share electrons with each other forming a carbon–carbon **covalent** bond. Carbon atoms can also share two pairs of electrons forming a double (covalent) bond. A few of the resulting patterns are shown in Figure 1.

methane CH_4

ethene C_2H_4

ethane C_2H_6

propane C_3H_8

butane C_4H_{10}

propene C_3H_6

B1.1 Figure 1 These molecules show you some of the simplest carbon compounds containing carbons and hydrogens only. You do not need to memorize these molecules. The chain of carbons can be much longer than those shown. Also notice that propene and ethene each have a double covalent bond between carbons. This reduces the number of hydrogens in the molecule. In addition, carbons can form branched formations and even ring structures.

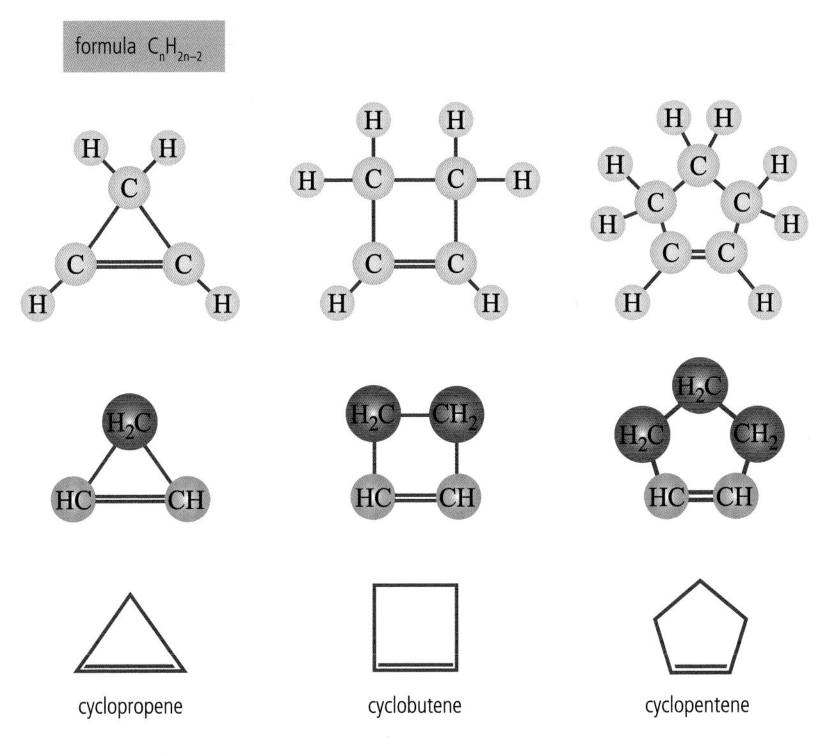

Three examples of organic molecules where the carbon atoms are formed into a ring or cyclic structure. All three happen to have a double bond between two of the carbons. There are three different ways to represent the same molecules. In the bottom row the hydrogens are omitted and are just understood to be present. Carbon atoms always have four bonds and if no other element is shown, it is assumed to be the appropriate number of hydrogens. You will see this in many representations of organic molecules. You are not required to know the names of these compounds.

The following elements are also common within the molecules of living organisms: oxygen, nitrogen and phosphorus. These elements are found in carbohydrates, proteins, lipids and nucleic acids. They often form covalent bonds with carbon, as well as forming covalent bonds with each other. Each of the elements mentioned forms a set number of covalent bonds. Memorizing the number of covalent bonds made by each of these elements will help you understand and draw the molecules that are important to living organisms. Table 1 shows the number of covalent bonds for five elements important in biochemistry.

Element	Number of covalent bonds
Hydrogen	1
Oxygen	2
Nitrogen	3
Carbon	4
Phosphorus	5

B1.1 Table 1 Each of the five most common elements in biochemically important molecules creates a unique number of covalent bonds

In biochemistry, elements are often arranged into **functional groups**. It will help you immensely if you learn to recognize the common functional groups.

Common functional groups (highlighted in blue) shown bonded to a carbon backbone. The bonds to each of the other carbons could be all hydrogens, as shown, or there could be another functional group bonded at that location in the molecule.

Common functional groups

hydroxyl or alcohol (OH)

amino or amine (NH_2)

carboxyl (COOH)

phosphate (H_2PO_4)

As you study biochemistry, you will soon learn to recognize and categorize common biochemical molecules into appropriate categories. Table 2 shows some of the common categories of molecules.

Category	Subcategory	Example molecules
Carbohydrates	Monosaccharides	Glucose, galactose, fructose, ribose
	Disaccharides	Maltose, lactose, sucrose
	Polysaccharides	Starch, glycogen, cellulose, chitin
Proteins		Enzymes, antibodies, peptide hormones
Lipids	Triglycerides	Fat stored in adipose cells
	Phospholipids	Lipids forming a bilayer in cell membranes
	Steroids	Some hormones
Nucleic acids	Nucleotides	Deoxyribonucleic acid (DNA), ribonucleic acid (RNA), adenosine triphosphate (ATP)

B1.1 Table 2 Categories and examples of four biochemically important molecules

Nature of Science

In all scientific disciplines unit conventions are based on international agreement. This includes the use of the following metric prefixes:

kilo = 10^3 centi = 10^{-2} milli = 10^{-3} micro = 10^{-6} nano = 10^{-9}

B1.1.2 and B1.1.3 – Condensation and hydrolysis

B1.1.2 – Production of macromolecules by condensation reactions that link monomers to form a polymer
Students should be familiar with examples of polysaccharides, polypeptides and nucleic acids.

B1.1.3 – Digestion of polymers into monomers by hydrolysis reactions
Water molecules are split to provide the –H and –OH groups that are incorporated to produce monomers, hence the name of this type of reaction.

Most living organisms do not make **macromolecules** one atom at a time. Macromolecules are made up of smaller molecules called monomers. When you ingest food many of the molecules of the food are in the form of macromolecules. Digestion breaks down macromolecules as a result of chemical reactions called **hydrolysis** reactions. Hydrolysis reactions break covalent bonds between monomers.

Macromolecule	Monomer (building blocks)
Carbohydrates	Monosaccharides
Lipids	Glycerol, fatty acids, phosphate groups
Proteins (polypeptides)	Amino acids
Nucleic acids	Nucleotides

Macromolecules and their monomer subcomponents

The resulting monomers are then a suitable size to be absorbed into the bloodstream and circulated to body cells. After entering cells, often the monomers are built up into macromolecules again. This involves forming covalent bonds in reactions called **condensation** reactions.

An example of hydrolysis and condensation is as follows.

- You eat a taco containing beef, a source of protein.

- Hydrolysis reactions occur in the digestive system, resulting in amino acids.

- Amino acids are absorbed into the blood and taken to body cells.

- DNA in a body cell directs specific condensation reactions to produce a specific protein from the amino acids.

Metabolism is best thought about from a molecular perspective. Often, people think only of physiological parameters, such as heart rate and digestion, as their metabolism. But remember that metabolism is all of the reactions within all of the cells of an organism.

In a condensation reaction two products are always formed. One is a larger molecule than either of the two reactants and the other is water. In a hydrolysis reaction, water is always a reactant and two products are formed that are smaller than the initial reactant (other than water).

Condensation and hydrolysis reactions are, in many ways, the reverse of each other. In a condensation reaction, a water molecule is always formed as part of the reaction. In a hydrolysis reaction, a water molecule is split into two components and each component is added in and becomes a part of the two new (smaller) molecules. Both condensation and hydrolysis reactions require specific enzymes.

Linking monomers into polymers

Examples of condensation reactions include the following.

- Condensation reaction of two monosaccharides to form a disaccharide
  glucose + galactose → lactose + water
- Condensation of many glucose molecules to form starch (a polysaccharide)
  (many) glucose → starch + (many) water
- Condensation of amino acids to form a polypeptide
  (many) amino acids → protein + (many) water
- Condensation of nucleotide components to make a DNA or RNA nucleotide
  phosphate group + pentose sugar + nitrogenous base → nucleotide + 2 water

Notice that one water molecule is formed for every condensation reaction that occurs. We will look at the mechanism of a condensation reaction in a little more detail to see where the water molecule comes from and how the two smaller subcomponents are bonded to a larger molecule.

B1.1 Figure 2 The two amino acids shown are undergoing a condensation reaction.

It is possible to predict how many water molecules are going to be produced whenever a polypeptide is formed. You just have to know how many amino acids are being joined together. If there are 443 amino acids being joined, then 442 water molecules are formed and 442 new peptide bonds formed.

The "R" notation in Figure 2 indicates that these two amino acids could be any of the 20 different possibilities. Notice that a portion of the carboxyl group of one amino acid becomes oriented near the amine group of the other amino acid. Stress is placed on the –OH from one amino acid and the H^+ of the other. This results in the covalent bonds breaking. The released –OH (hydroxide ion) and H^+ (hydrogen ion) combine to form a water molecule. The location where the –OH and H^+ were released still contains a pair of electrons that form a new covalent bond. Whenever this occurs between two amino acids, the new covalent bond is called a **peptide bond**. The reaction is catalysed by an enzyme.

Digesting polymers into monomers

Many organisms, including all animals, rely on the foods that they eat to provide energy but also to provide monomers that can be used to make new macromolecules.

Foods are chemically digested in the alimentary canal. The digestive enzymes that accomplish this are **hydrolysing enzymes.** Each individual reaction is called a hydrolysis and requires a molecule of water as a reactant. In a **hydrolysis reaction** water is always "split" as part of the reaction. Examples of hydrolysis include the following.

- Hydrolysis of a disaccharide to two monosaccharides (see Figure 3)
  lactose + water → glucose + galactose

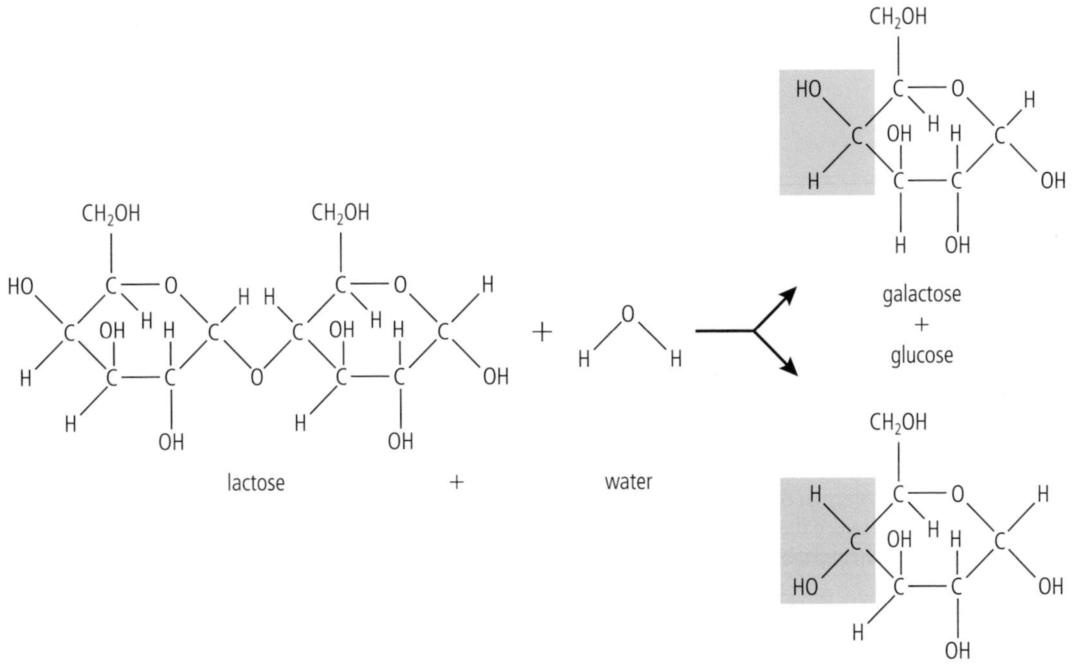

lactose + water → galactose + glucose

B1.1 Figure 3 Hydrolysis of the disaccharide lactose. Notice that water is also a reactant. The products are the two monosaccharides galactose and glucose. The shaded areas of galactose and glucose highlight the only difference between the two monosaccharides. The difference is in the orientation of the hydroxyl group; it is quite subtle, but important enough to distinguish the two monosaccharides from each other.

- Hydrolysis of a polysaccharide to many monosaccharides
  starch + (many) water → (many) glucose

- Hydrolysis of a polypeptide (protein) to amino acids
  protein + (many) water → (many) amino acids

The sum total of all the condensation reactions and all the hydrolysis reactions occurring in your body makes up a large portion of your overall metabolism.

Most enzymes can be recognized as enzymes because their name ends with the suffix "ase". It is now common to incorporate the substrate name and add the suffix - ase when identifying the enzyme of that substrate. However, some enzymes like pepsin were named before that practice was established. The current naming practice makes it much easier to deduce the function of the enzyme. For example: sucrase is the enzyme that catalyses the hydrolysis of the disaccharide sucrose.

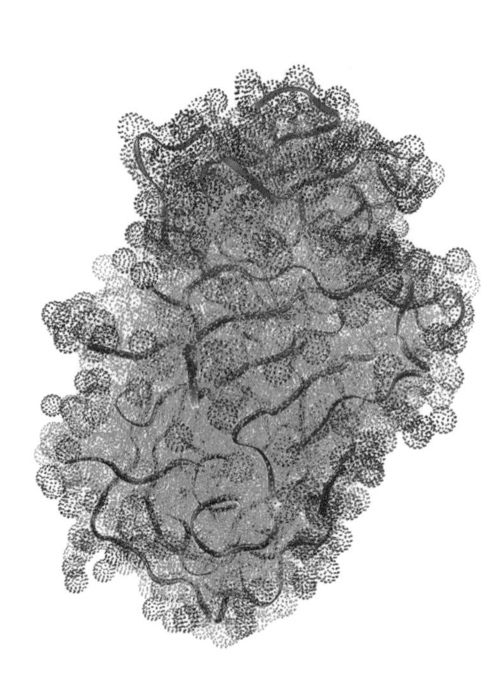

A computer graphic image of the enzyme pepsin. This enzyme is active in the acidic environment of the stomach. It is a hydrolysing enzyme that helps to digest proteins by hydrolysis reactions.

B1.1.4 – Monosaccharides

B1.1.4 – Form and function of monosaccharides
Students should be able to recognize pentoses and hexoses as monosaccharides from molecular diagrams showing them in the ring forms. Use glucose as an example of the link between the properties of a monosaccharide and how it is used, emphasizing solubility, transportability, chemical stability and the yield of energy from oxidation as properties.

You have probably already encountered several monosaccharides. Ribose and deoxyribose monosaccharides are the central components of the nucleotides of RNA and DNA, respectively (see Chapter A1.2). You have read about glucose and galactose within this chapter in reference to condensation and hydrolysis reactions. Figures 4 and 5 show the structure of ribose and glucose, respectively.

Ribose is an example of a **pentose monosaccharide**. This simply means that its carbon backbone is composed of five carbons. The chemical formula for a pentose monosaccharide is $C_5H_{10}O_5$.

B1.1 Figure 4 The structure of ribose.

Glucose is an example of a **hexose monosaccharide** as its carbon backbone is composed of six carbons. Its chemical formula is $C_6H_{12}O_6$.

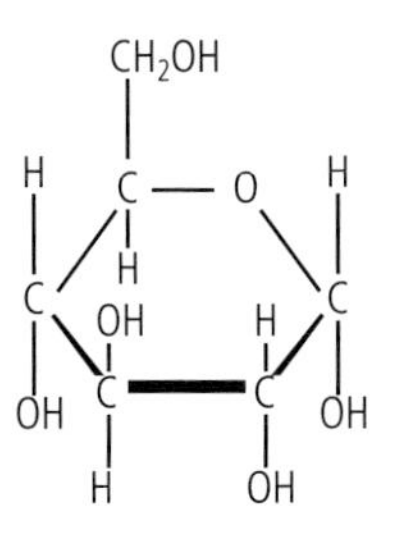

▲ **B1.1 Figure 5** The structure of glucose.

Notice the pattern of the chemical formula for both ribose and glucose. Both of these molecules have the general formula $C_nH_{2n}O_n$ where "*n*" is the number of carbon atoms within the monosaccharide. This means all you need to know is the number of carbon atoms and you can then predict the entire chemical formula. However, this formula does not apply to carbohydrates larger than monosaccharides such as disaccharides and polysaccharides, nor does it apply to modified monosaccharides such as deoxyribose.

Properties and use of glucose

Glucose is one of the most important molecules in nature. It is produced in photosynthesis and used in respiration. Glucose can also be used to make polysaccharides of various types. Some of these polysaccharides are used for structural purposes (for example cellulose) and some are used for energy storage (for example starch).

The monosaccharides within this section are shown as ring structures, also known as cyclic structures. Monosaccharides also have a straight-chain form that you may see in other sources. When in solution, monosaccharides switch between the ring and straight-chain form.

When you look at the structure of glucose (Figure 5), the alcohol (hydroxyl) functional group is found five times within the molecule. Just like in water molecules, the covalent bond between an oxygen atom and a hydrogen atom is a **polar** covalent bond (see Chapter A1.1). This leads to glucose itself being a polar molecule. Glucose molecules have the following properties.

- **Molecular stability**, because the bonds within glucose are stable covalent bonds that do not break easily.
- **High solubility in water**, because glucose is polar and dissolves readily in a polar solvent such as water.
- **Easily transportable**, because of its solubility in water, which means that glucose can easily circulate in blood and in fluids between cells.
- **Yields a great deal of chemical energy** when its covalent bonds are broken. Reactions of this type are called **oxidation reactions** and the high energy yield means that glucose is a good energy store.

What are the roles of oxidation and reduction in biological systems?

B1.1.5 – Polysaccharides and energy storage

B1.1.5 – Polysaccharides as energy storage compounds
Include the compact nature of starch in plants and glycogen in animals due to coiling and branching during polymerization, the relative insolubility of these compounds due to large molecular size and the relative ease of adding or removing alpha-glucose monomers by condensation and hydrolysis to build or mobilize energy stores.

In nature, glucose in a polymer form is often used for energy storage. After glucose is synthesized by photosynthesis, a plant stores much of it as starch molecules. Starch is a polysaccharide made up of hundreds of glucose monomers. In order to make the starch as compact as possible, a plant uses two different kinds of bonds between glucose molecules. One of these bonds is called an alpha 1–4 linkage and the other is called an alpha 1–6 linkage. The numbers refer to the carbon number of the two glucose molecules that are bonded together. For example, in one type of starch called

amylose, carbon #1 is bonded to carbon #4 of the adjoining glucose. When hundreds of glucose molecules are bonded by only 1–4 linkages, the resulting molecule will be linear but in a helix shape. The 1–6 linkages are typical in another type of starch called **amylopectin** and create branches as shown in Figure 6.

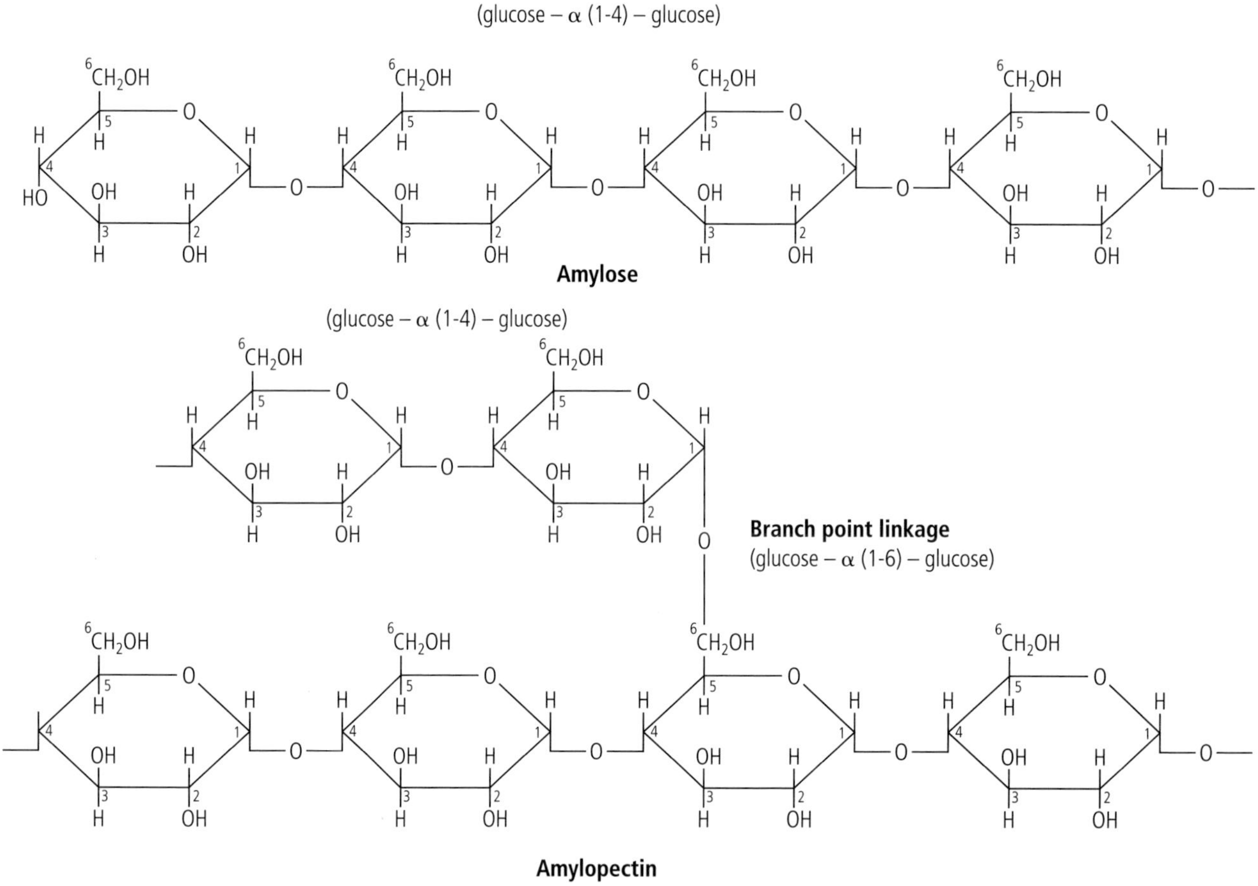

(glucose – α (1-4) – glucose)

Amylose

(glucose – α (1-4) – glucose)

Branch point linkage
(glucose – α (1-6) – glucose)

Amylopectin

▲
B1.1 Figure 6 Molecular bonding within the two carbohydrate molecules making up starch. Notice the numbering of the carbons in each glucose. These numbers are used to describe how the monomers are linked together within each polysaccharide. Amylose uses 1–4 carbon linkages and amylopectin uses 1–6 carbon linkages.

Starch is a polymer of glucose but within starch are two kinds of polysaccharides: amylose and amylopectin. A molecule of starch is a very large molecule and so is not readily soluble in water. This low solubility is important because it means that a plant can easily store the starch. Additionally, although the molecule is large it is also compact. When a plant is photosynthesizing and producing lots of glucose, it can add more glucose molecules to either amylose or amylopectin by condensation reactions. Alternatively, when a plant needs to use its reserves of glucose, hydrolysis reactions are used to break the glucose molecules away from starch.

Glycogen is a polysaccharide made of glucose monomers that are bonded in a very similar pattern as in amylopectin 1–6 linkages. The branching in glycogen is more numerous than the branching in amylopectin. Many animals, including humans, store excess glucose as glycogen. Glycogen reserves are kept within our liver and muscle tissue.

A primary advantage for organisms storing glucose as a polysaccharide is that the macromolecules are not readily soluble in cytoplasm and other fluids. This means that they do not affect the osmotic balance in living tissues, whereas individual (very soluble) glucose molecules would.

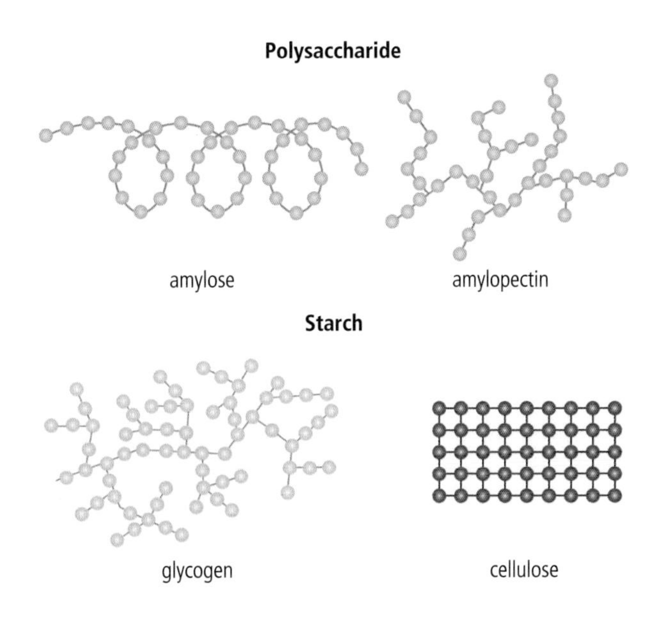

An illustration of four different polysaccharides composed of glucose monomers. Each of the smallest geometric shapes in each of the polysaccharides represents a glucose molecule. The upper two molecules, amylose and amylopectin, are both subcomponents of starch, with amylopectin typically representing a higher percentage of the overall starch molecule. Glycogen is the polysaccharide form used for energy storage in animals. Cellulose is used within the structure of the cell walls of plants and is not considered an energy storage molecule. You will read about cellulose in Section B1.1.6.

B1.1.6 – Cellulose as a structural polysaccharide

B1.1.6 – Structure of cellulose related to its function as a structural polysaccharide in plants
Include the alternating orientation of beta-glucose monomers, giving straight chains that can be grouped in bundles and cross-linked with hydrogen bonds.

alpha - glucose **beta - glucose**

CH_2OH … reversed group

B1.1 Figure 7 When in solution, glucose exists in two forms, alpha and beta. Both are shown here.

Notice in Figure 7 that the two forms of glucose (alpha and beta) are very similar to each other except for the reversal of the atoms shown on the right side of each molecule. This small difference is very important for the polymers that are formed from each of the two molecules. Starch and glycogen both use the alpha form of glucose, while cellulose uses the beta form. Cellulose is a primary component of the cell walls of plants. For that reason, cellulose is estimated to be the most abundant of all organic molecules on Earth.

How can compounds synthesized by living organisms accumulate and become carbon sinks?

In cellulose the 1–4 carbon linkages are between beta-glucose molecules (whereas amylose uses 1–4 linkages between alpha-glucose molecules). The condensation reaction that bonds one beta-glucose molecule to the next requires every second beta-glucose molecule to be oriented "upside down" compared to the glucose it is bonded to. This orients the hydroxyl group of carbon #1 with the hydroxyl group of carbon #4.

When beta-glucose molecules form their 1–4 linkages, the result is a very linear polymer with no branches. You can visualize the resulting molecule as similar to a very long thin fibre. Many of these "fibres" run parallel to each other and form a multitude of hydrogen bonds with the adjoining fibres (see Figure 8). This pattern continues as bundles of fibres made of beta-glucose form even larger bundles held together by the cross-linking attractions of hydrogen bonds. The result is a very stable molecule of cellulose.

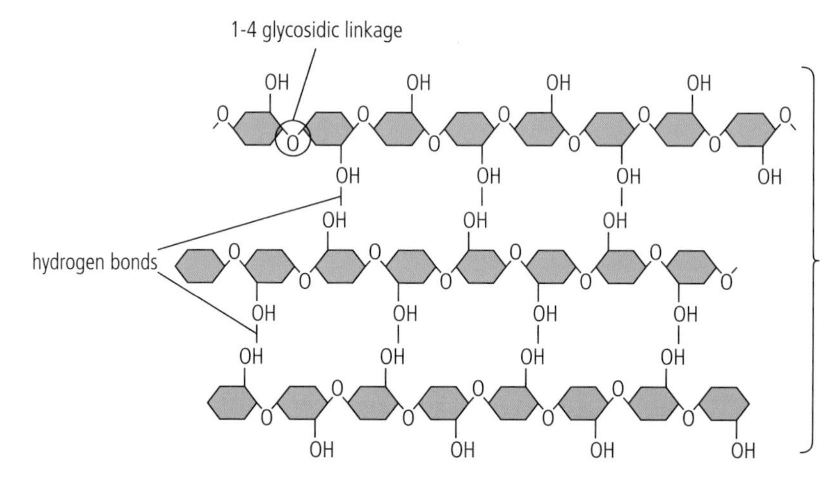

B1.1 Figure 8 Glucose molecules form linear polymers as a result of 1–4 linkages with additional hydrogen bonds, forming a strong and stable macromolecule.

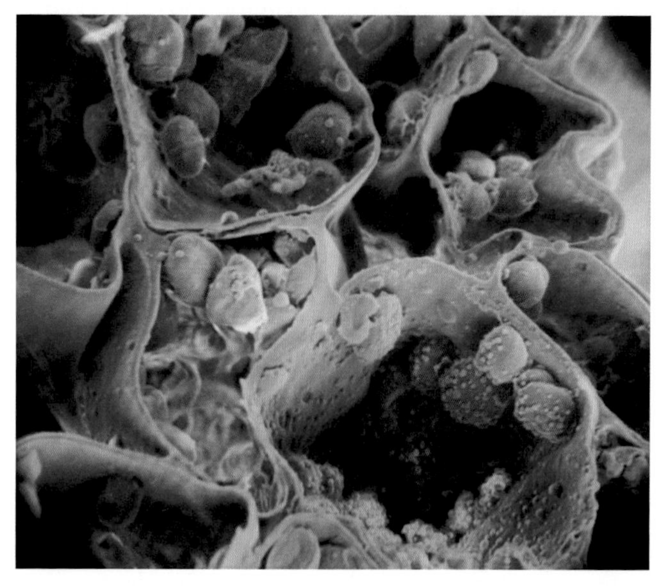

A scanning electron micrograph (SEM) of sliced open plant cells. The plant cell walls are composed mainly of cellulose and are clearly visible. In the interior of the cells are chloroplasts, which produce and store carbohydrates such as starch.

Research is underway to convert cellulose waste (leftover plant materials) into products that would replace items that currently are made of plastics. Plastic pollution is a long-term global problem.

The function of cellulose is to act as a structural molecule in nature, for example cellulose is found in plant cell walls. As well as being strong, cellulose is insoluble in water and the fibres allow water and other substances to pass freely into and out of plant cells. Very few organisms produce the enzyme (cellulase) necessary to digest cellulose, therefore it is not considered to be an energy storage molecule.

B1.1.7 – Conjugated carbon molecules

B1.1.7 – Role of glycoproteins in cell–cell recognition
Include ABO antigens as an example.

So far we have only described the structure and function of a few biochemically important molecules. Each molecule has been a carbohydrate, lipid, protein or nucleic acid. Sometimes two or more of these categories of molecules are bonded together in order to accomplish a specific function. Table 3 shows three general examples.

Carbon components	Type of conjugated molecule
Lipid + protein	Lipoprotein
Carbohydrate + lipid	Glycolipid
Carbohydrate + protein	Glycoprotein

B1.1 Table 3 The composition of some common conjugated molecules

Glycoproteins and cell–cell recognition

Various kinds of proteins are associated with cell membranes (see Chapter B2.1). Membrane proteins may or may not be conjugated. Membrane proteins are responsible for a variety of functions, including:

- cell to cell chemical communication (cell signalling)
- transport of molecules in and out of a cell
- cell to cell adhesion
- catalysis as a result of enzymes adhering to the inside or outside of the cell membrane
- recognition of body cells (self) versus non-body cells (not-self) for immune system functions.

Glycoproteins and ABO blood types

Glycoproteins on the surface of red blood cells determine a person's ABO blood type. Red blood cells can have two possible types of glycoprotein on their plasma membranes. The two proteins, A and B, are called **antigens**, because their presence can trigger the immune system. A major component of the immune system is the ability of some white blood cells to recognize "self" from "not-self".

People that have blood type AB have both the A and B type antigens, and thus their immune system will not be triggered by the presence of either. People with blood type O have neither the A or B protein, and their immune system will be triggered by the presence of either of the A and B antigens. The immune system of people who inherit only A or only B antigen is triggered by the presence of the protein that they do not have.

This is important when a blood transfusion is given. People with blood type O are considered to be universal donors because they can give blood to others with type O and those with types A, B or AB. People with blood type AB are universal recipients because they can receive blood from types AB, A, B and O.

An easy way to understand appropriate blood types for transfusions is to remember that a person cannot receive an A or B antigen unless they have genetically inherited that glycoprotein.

ABO blood type	Glycoprotein found on red blood cell plasma membrane	Can receive a transfusion from	Can give blood to
A	A	A and O	A and AB
B	B	B and O	B and AB
AB	A and B	AB, A, B, O	AB only
O	Neither A or B	O only	AB, A, B, O

The four ABO blood types and their associated glycoproteins and transfusion possibilities

B1.1.8 – Lipid solubility

B1.1.8 – Hydrophobic properties of lipids
Lipids are substances in living organisms that dissolve in non-polar solvents but are only sparingly soluble in aqueous solvents. Lipids include fats, oils, waxes and steroids.

Lipids can be categorized as fats, oils, waxes and steroids. Lipid molecules contain many areas of hydrocarbons, meaning areas containing just hydrogen and carbon. The covalent bond between carbon and hydrogen is a **non-polar covalent bond**. This means that lipids dissolve quite well in non-polar solvents but do not dissolve well in water. Organisms have evolved to take advantage of this very limited solubility, and have come up with some unique solutions to the problem of lipid insolubility when needed. One of those solutions is to conjugate the lipid with another molecule. Examples include glycolipids and lipoproteins.

B1.1.9 – Triglycerides and phospholipids

B1.1.9 – Formation of triglycerides and phospholipids by condensation reactions
One glycerol molecule can link three fatty acid molecules or two fatty acid molecules and one phosphate group.

Lipids are macromolecules composed of subunits. Lipids known as **triglycerides** contain one glycerol molecule and three fatty acid molecules. Lipids are formed from condensation reactions:

$$1 \text{ glycerol} + 3 \text{ fatty acids} \rightarrow 1 \text{ triglyceride} + 3 \text{ water molecules}$$

Molecules known as **phospholipids** are formed if an inorganic phosphate group replaces one of the three fatty acids. The reaction would be:

$$1 \text{ glycerol} + 2 \text{ fatty acids} + 1 \text{ inorganic phosphate} \rightarrow 1 \text{ phospholipid} + 3 \text{ water molecules}$$

Glycerol is a three-carbon molecule, with each carbon bonding to one hydroxyl group initially. The fatty acids vary depending on the number of carbons in each and the possible presence of double bonds between one or more of the carbons. Each fatty acid always contains a terminal carboxyl group that is involved in the condensation reaction.

Formation of a phospholipid or triglyceride

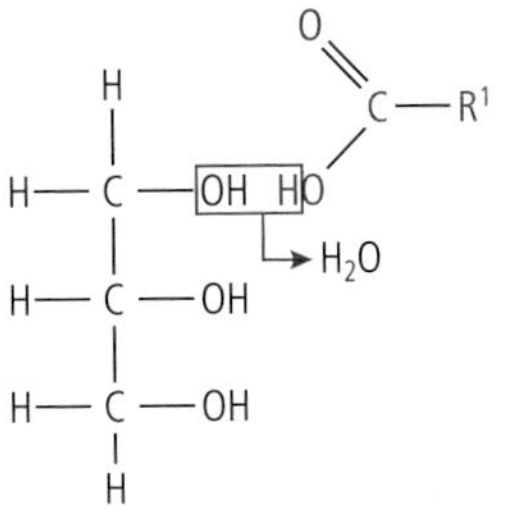

Step 1
The OH group on the glycerol molecule aligns with OH group on the carboxyl group of the general fatty acid number one (R^1). A condensation reaction occurs with the release of one water molecule.

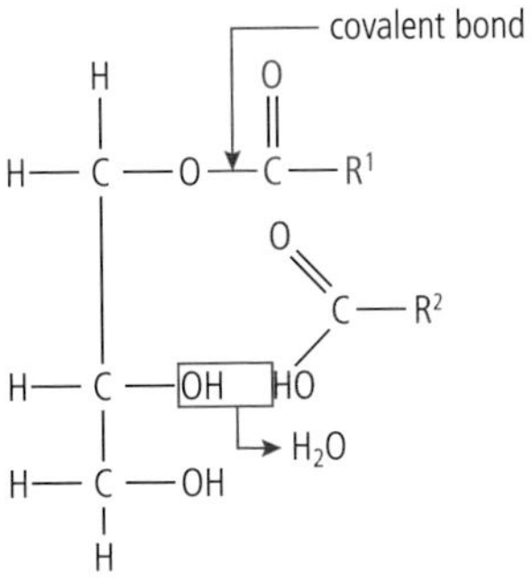

Step 2
A new covalent bond is formed between fatty acid number one and the glycerol molecule. A second fatty acid (R^2) aligns with the second OH group and undergoes the second condensation reaction, with the release of another water molecule.

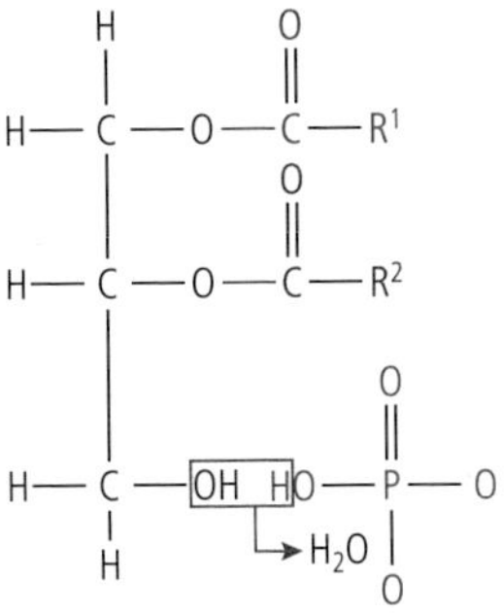

Step 3
A new covalent bond is formed between fatty acid number two and the glycerol molecule. A phosphate group aligns with the third OH group and undergoes another condensation reaction, with the release of one water molecule.

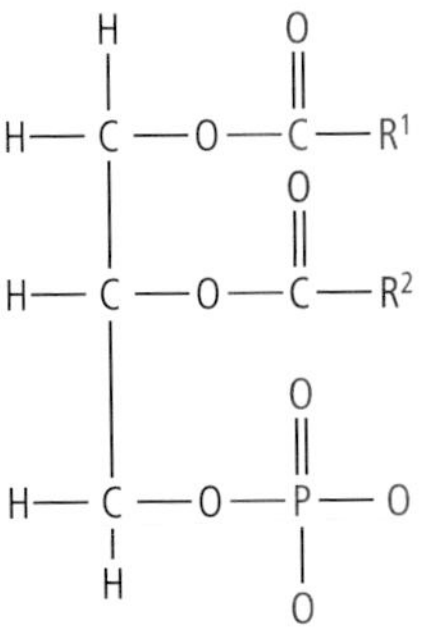

Step 4
A new covalent bond is formed between the phosphate group and the glycerol molecule. A phospholipid has been formed. A triglyceride would have been formed if a fatty acid was used in place of the phosphate group in this step. To fully appreciate the molecular product from these three condensation reactions, you should visualize R^1 and R^2 as long chains of carbons and hydrogens.

B1.1 Figure 9 Three condensation reactions with glycerol. As shown, the reaction could form a triglyceride or a phospholipid depending on the reactants. Each of the fatty acids is shown in an abbreviated form with the letter 'R' representing the long hydrocarbon chain.

B1.1.10 – Properties of fatty acids

B1.1.10 – Difference between saturated, monounsaturated and polyunsaturated fatty acids
Include the number of double carbon (C=C) bonds and how this affects melting point. Relate this to the prevalence of different types of fatty acids in oils and fats used for energy storage in plants and endotherms, respectively.

Fatty acids can be divided into different groups depending on their structure. Three categories of fatty acids are found within lipids.

Saturated fatty acids are fatty acids that contain single bonds between the carbons. This means that all the other carbon bonds are to hydrogens (except for the carboxyl group). In other words, the molecule is "saturated" with hydrogens.

Saturated fatty acids have a relatively high melting point and are solid at typical room temperature. Triglycerides containing only saturated fatty acids are called "fats" and are often used by animals to store excess energy. Fats in animal meat and butter are examples, because they are solids at room temperature.

Monounsaturated fatty acids are fatty acids that have one double bond between two of the carbons in the hydrocarbon chain of the molecule. The location of the double bond can vary.

Triglycerides containing one or more monounsaturated fatty acids have a lower melting point than saturated fatty acids, and are liquid (oil) at typical room temperature. Some animals and many plants store energy in this form.

Polyunsaturated fatty acids are fatty acids that have more than one double bond in the hydrocarbon chain. The number and location of the double bonds can vary.

Triglycerides composed of polyunsaturated fatty acids also have a relatively low melting point and are liquid (oil) at room temperature. Many plants store energy in this form.

Two factors affect the melting point of lipids and the fatty acids they contain. One is the number of carbons in the fatty acids, and the other is the presence and number of double bonds. The highest melting points are found in lipids that contain the most carbons and the fewest number of double bonds.

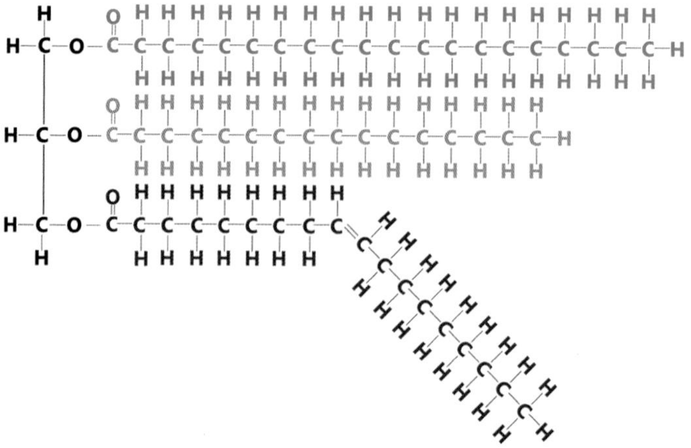

▲ The structure of a triglyceride with two saturated fatty acids and one monounsaturated fatty acid. Saturated fatty acids are more linear in shape, while unsaturated fatty acids have bond angle changes because of the double bond(s). Note the three-carbon glycerol portion of the molecule on the left side.

B1.1.11 – Adipose tissue

B1.1.11 – Triglycerides in adipose tissues for energy storage and thermal insulation
Students should understand that the properties of triglycerides make them suited to long-term energy storage functions. Students should be able to relate the use of triglycerides as thermal insulators to body temperature and habitat.

Adipose tissue is composed of cells that store fat in the form of triglycerides. The quantity of triglycerides that is stored is determined by the organism's caloric intake compared to the calories burned. In Section B1.1.2 you learned about condensation reactions that form triglycerides. The reactions are most common when an organism eats foods that have more calories than the organism is using. Triglycerides can be used to supply energy when sufficient foods are not available for metabolic needs. In that circumstance, the stored triglycerides undergo hydrolysis reactions and the products (glycerol and fatty acids) are made available for energy in the process of cell respiration. Triglycerides are useful for long-term energy storage because they are insoluble in body fluids and thus will not move from their adipose storage sites. Per gram of substance, triglycerides provide approximately twice as much energy as that released by carbohydrates.

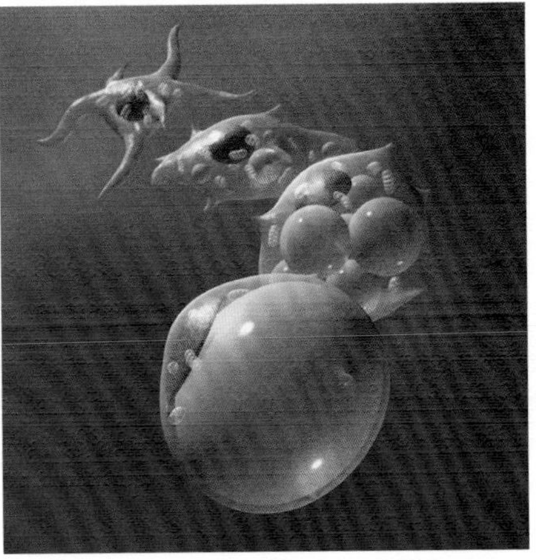

A depiction of an **adipocyte** (fat storage cell) growing larger as it accumulates triglycerides. The triglycerides are stored in one or more large vacuoles.

A thick layer of adipose tissue is typical of many animals that live in cold regions such as the arctic. Birds and mammals are **endotherms**, maintaining a steady internal temperature regardless of their environmental temperature. Seals, walruses and whales are all endotherms. Their thick adipose tissue is called blubber and is found between their skin and muscles. The blubber helps trap the heat generated by the inner metabolic activities of the animal.

The food intake of birds and mammals is high compared to other animals. This is because some of the food energy must be used to maintain a particular internal temperature as part of **homeostasis**.

B1.1.12 – Phospholipid bilayers

B1.1.12 – Formation of phospholipid bilayers as a consequence of the hydrophobic and hydrophilic regions
Students should use and understand the term "amphipathic".

Figure 9 on page 193 shows the stages in the formation of a phospholipid molecule. The presence of a phosphate group has consequences for the polarity of the molecule. Small molecules like water and even glucose are polar because of the polar covalent bonds that they contain. A phospholipid is a much larger molecule in comparison. A

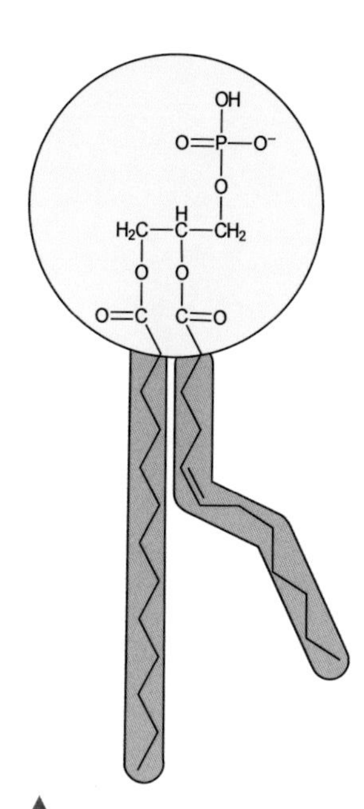

A drawing showing the amphipathic structure of a phospholipid. The highlighted spherical area is polar. The remaining portion of the molecule is non-polar.

phospholipid has a polar end (the end with the phosphate group) and an even longer non-polar end (the two long hydrocarbon tails). Molecules like phospholipids, which have both hydrophilic and hydrophobic regions, are called **amphipathic** molecules.

Phospholipids in an aqueous solution solve the problem of having hydrophobic tails by forming a double layer or **bilayer**. In this bilayer the hydrophobic fatty acid tails extend towards each other in order to keep away from the aqueous solutions inside and outside the cell. The polar phosphate groups are attracted to the aqueous solutions and so arrange themselves on the outside of the bilayer. This is the foundation of the **plasma membrane**. Many organelles within cells have one or more membranes that are used to separate the aqueous fluids within the organelle from the cytoplasm.

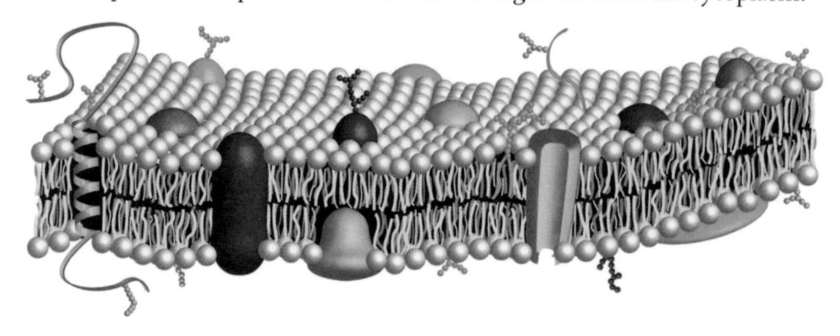

An artist's rendering of a phospholipid bilayer membrane. Notice that the hydrophilic heads of the bilayer are oriented outwards and the two sets of fatty acid tails are oriented inwards. This allows the membrane to interact with water on either side of the membrane and also to have an internal thickness that acts as a barrier to water and other polar molecules. The irregular shapes embedded into and extending through the membrane are a variety of proteins and carbohydrates. Many of these proteins and carbohydrates control the passage of polar molecules through the membrane.

B1.1.13 – Steroid hormones

B1.1.13 – Ability of non-polar steroids to pass through the phospholipid bilayer
Include oestradiol and testosterone as examples. Students should be able to identify compounds as steroids from molecular diagrams.

Hormones are chemical messenger molecules that are produced by a variety of glands in the body. After production, hormones are released into the bloodstream and have access to all body tissues. The body tissues that respond to any one hormone is called a target tissue of that hormone. One group of hormones, called **steroids**, are made from the lipid **cholesterol**. Cholesterol is primarily a hydrocarbon molecule. Steroids retain that hydrocarbon makeup and their fundamental structure is easy to identify.

The IB requires you to recognize the general structure of all steroids but not the atom by atom structure of any one steroid. Look for the four connected ring structures, as shown in Figure 10, to help you identify a molecule as a steroid.

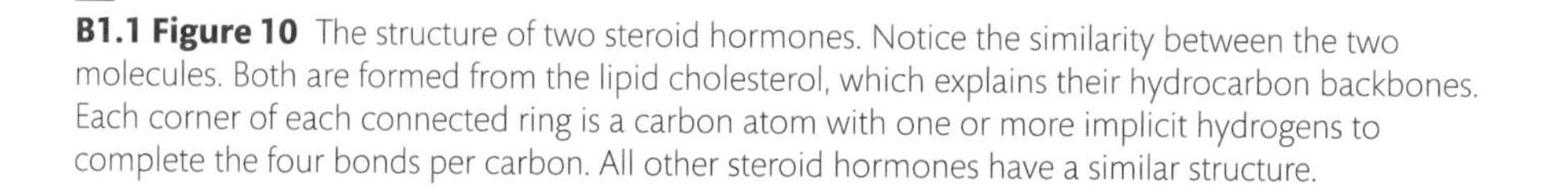

B1.1 Figure 10 The structure of two steroid hormones. Notice the similarity between the two molecules. Both are formed from the lipid cholesterol, which explains their hydrocarbon backbones. Each corner of each connected ring is a carbon atom with one or more implicit hydrogens to complete the four bonds per carbon. All other steroid hormones have a similar structure.

The production and action of **oestradiol** and **testosterone** are similar. Both are produced by **gonadal** tissue and are involved in the development of primary and secondary sex characteristics beginning at puberty. As lipid-based molecules, each of these two hydrophobic hormones is soluble through the lipid bilayer of cells, and directly enters both through the plasma membrane and nuclear membrane of their target tissue cells. Once inside the nucleus, the hormones direct the process of **transcription**, leading to the production of mRNA molecules.

Guiding Question revisited

In what ways do variations in form allow diversity of function in carbohydrates and lipids?

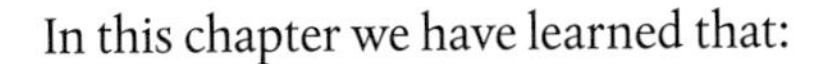

In this chapter we have learned that:

- carbohydrates exist in three forms, monosaccharides, disaccharides and polysaccharides
- monosaccharides are the smallest of the carbohydrate molecules and act as monomers of disaccharides and polysaccharides
- glucose is a monosaccharide and is often used in cell respiration as a direct source of cell energy
- glucose molecules can be joined in different ways to form both energy storage molecules (amylose, amylopectin and glycogen) and structural molecules (cellulose)
- lipids exist in many forms, such as triglycerides, phospholipids, cholesterol and steroid hormones
- lipids have low solubility in aqueous solutions
- amphipathic properties of phospholipids make them ideal for the formation of membrane bilayers
- the hydrophobic properties of steroid hormones enable the molecules to be soluble in the fatty acid layers of cell membranes
- conjugated forms of proteins, carbohydrates and lipids, such as glycolipids and glycoproteins, enable specialized functions in cells.

Guiding Question revisited

How do carbohydrates and lipids compare as energy storage compounds?

In this chapter we have discovered that:

- both carbohydrates and lipids are used in cells for chemical energy storage
- carbohydrate polysaccharide molecules like starch and glycogen are often used for short-term energy storage
- lipids in the form of triglycerides are often used for longer term energy storage
- carbohydrates hydrolyse into glucose, a soluble molecule that can be easily transported within cells and between cells
- one product of triglyceride hydrolysis is fatty acids, which have very low solubility in aqueous solutions and thus are not easily transported
- triglycerides can store approximately twice the chemical energy compared to the same mass of carbohydrates.

Exercises

ATP (adenosine triphosphate)

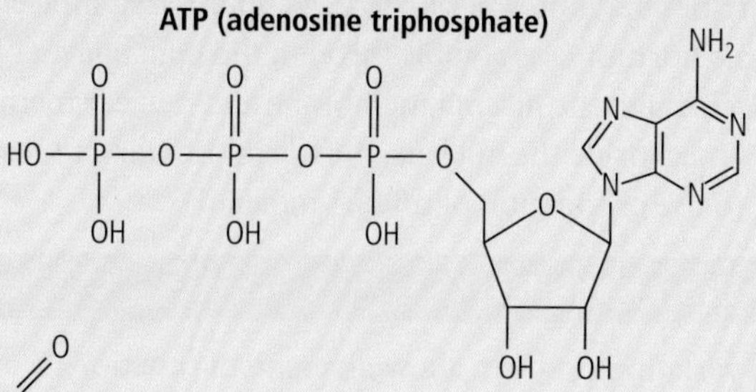

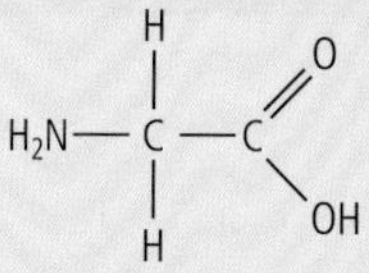

glycine

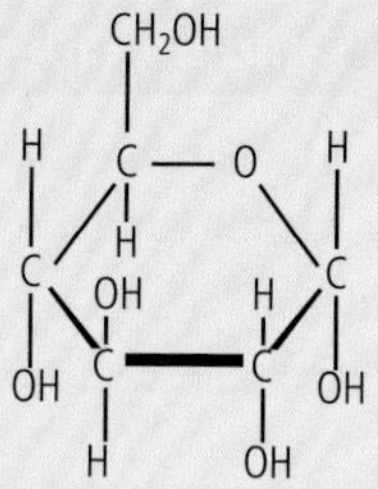

Q1. Study the figure above and answer the following questions.

(a) Which functional group is found repeatedly within glucose?

(b) Which two functional groups are found within the amino acid glycine?

(c) Which three functional groups are found within ATP?

Q2. Predict the complete chemical formula for each of the following monosaccharides using the general formula of $C_nH_{2n}O_n$.

(a) A triose.

(b) A pentose.

(c) A hexose.

Q3. A person has blood type A.

(a) In a transfusion, what type(s) of blood can this person receive?

(b) In a transfusion, to what blood type(s) can this person donate blood?

Q4. Energy storage polysaccharides like amylose, amylopectin and glycogen can add or remove glucose molecules as needed.

(a) What type of reaction would add glucose to an already existing polysaccharide molecule?

(b) Is water a reactant or a product of this reaction?

Q5. **(a)** What type of reaction would remove glucose from an already existing polysaccharide molecule?

(b) Is water a reactant or a product of this reaction?

B1.2 Proteins

Guiding Questions

What is the relationship between amino acid sequence and the diversity in form and function of proteins?

How are protein molecules affected by their chemical and physical environments?

Cells use the naturally occurring 20 amino acids to synthesize polypeptides or proteins. Polypeptides are synthesized under the control of DNA, each polypeptide being coded for by a specific area of DNA molecule called a gene. It is the sequence of amino acids that determines the identity of a polypeptide. Proteins are incredibly diverse in their structure and function as a result of having 20 different building block units for their synthesis. The order of the amino acids in a protein determines its shape and therefore its function.

Protein structure can be changed by the chemical and physical environment. If the structure of the protein changes it may not be able to function correctly. The two most significant environmental factors are pH and temperature. If a protein is heated, or the pH around it changes, chemical bonds within the protein can be affected. The making or breaking of chemical bonds can cause the protein to become an alternative shape, one that may not then function as well.

B1.2.1 – The common structure of amino acids

B1.2.1 – Generalized structure of an amino acid
Students should be able to draw a diagram of a generalized amino acid showing the alpha carbon atom with amine, carboxyl, R-group and hydrogen attached.

In nature there are 20 different amino acids. Each of these is easy to identify as an amino acid because they all have a common structure. The IB requires you to be able to draw the structure that is common to all amino acids. The portion of the amino acid that is unique to each of the 20 is represented by the letter "R". The group of atoms making up the "R" portion of each amino acid can be called the variable group or the side chain (group).

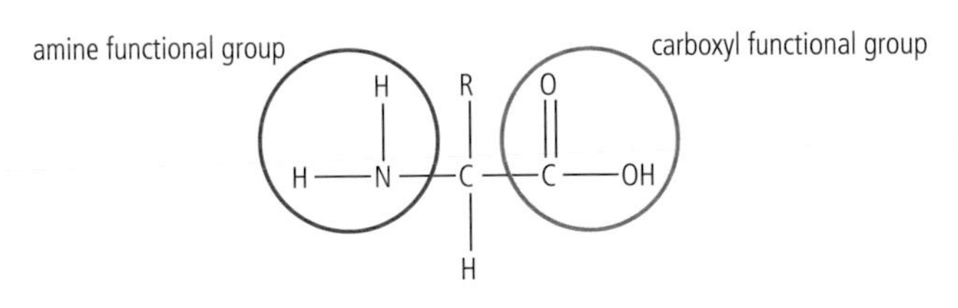

The molecular structure common to all 20 amino acids.

All amino acids are both an acid and a base. In an **aqueous solution** the carboxyl of each amino acid will act as an acid and donate a hydrogen ion, and the amine group will accept a hydrogen ion, therefore acting as a base. By doing this the amino acid becomes ionized.

Drawing the structure of a generalized amino acid is easier if you follow these steps.

1. Draw the single central (alpha) carbon atom and four covalent bonds.
2. Add an amine functional group to this central carbon.
3. Add a carboxyl functional group to the central carbon.
4. Add a hydrogen atom to the central carbon.
5. Finally, add the capital letter "R" to the only remaining covalent bond around the alpha carbon.

Usually the R-group is shown above the central carbon and the H group is shown below the central carbon, with the amine and carboxyl groups on either the left or right. But this is just a convention.

B1.2.2 – Condensation reactions bond amino acids together

B1.2.2 – Condensation reactions forming dipeptides and longer chains of amino acids
Students should be able to write the word equation for this reaction and draw a generalized dipeptide after modelling the reaction with molecular models.

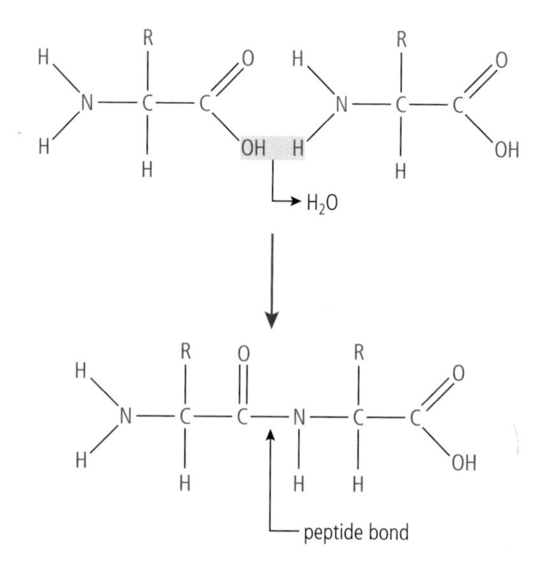

B1.2 Figure 1 The condensation reaction that forms a covalent bond between two amino acids.

Practise drawing the reaction shown in Figure 1 from memory, using the general structure of amino acids. Once you feel confident, practise drawing only the dipeptide from memory without showing the condensation reaction.

SKILLS

Use model kits to practise modelling the reaction in Figure 1. Add as many amino acids to your polypeptide as possible. Each reaction will be identical to every other reaction, but remember the identity of the resulting dipeptide or polypeptide depends on the identity and sequence of amino acids used.

The word equation for the reaction shown in Figure 1 would be:

$$\text{amino acid 1} + \text{amino acid 2} \rightarrow \text{dipeptide} + \text{water}$$

In Figure 1, a molecule called a **dipeptide** is formed. The water molecule that forms comes from a hydroxyl group (–OH) from the carboxyl group of one amino acid, and a hydrogen ion (H^+) from the other amino acid. This frees up electrons to be shared between carbon and nitrogen atoms, so bonding the two amino acids together into a dipeptide.

Notice the carboxyl group on the right side of the dipeptide in Figure 1. This carboxyl group can be used for a condensation reaction with a third amino acid. The third amino acid will now have a carboxyl group that can be used to add a fourth amino acid, and so on. The new covalent bonds linking the amino acids together are called **peptide bonds.** This is where the term "**polypeptide**" comes from.

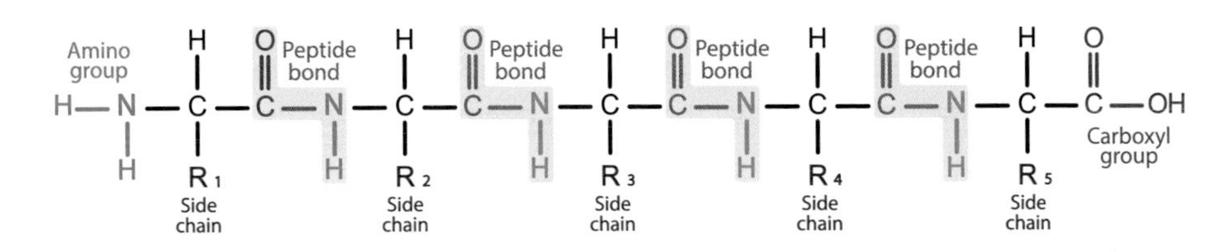

In this figure, five amino acids have been bonded together by four peptide bonds. The identity of the R (side chain) groups determine the identity and purpose of the protein. Notice that one amino group and one carboxyl group are left intact at the two ends of the protein.

TOK Molecular models are just visual representations of structures that we cannot see. Models typically fit the best available data and allow our brain to view an image of what "could" be there. Technology enables us to make computer images of various types, and makes it relatively easy to manipulate the images. Are models misleading in some ways?

B1.2.3 – Essential amino acids

B1.2.3 – Dietary requirements for amino acids
Essential amino acids cannot be synthesized and must be obtained from food. Non-essential amino acids can be made from other amino acids. Students are not required to give examples of essential and non-essential amino acids. Vegan diets require attention to ensure essential amino acids are consumed.

Our cells can synthesize 11 of the 20 amino acids from other amino acids, but nine have to come from our diet. Humans cannot remain healthy unless we eat foods that contain these **essential amino acids.**

A varied diet that includes different sources of protein should provide all the essential amino acids humans need. Meat is a good source of essential amino acids. Vegan diets or diets that rely heavily on just one protein source may have a limited amino acid content. For example, white rice is deficient in the amino acid lysine, so a white rice-rich diet may need to be supplemented with another source of lysine. Legumes, for example beans, contain low levels of the amino acid methionine. However, a diet that includes a daily mix of white rice and beans would offer all of the amino acids. You are not required to memorize the essential and non-essential amino acids.

! The number of peptide bonds within a polypeptide can be easily calculated by knowing the number of amino acids within the polypeptide. A dipeptide contains one peptide bond; a polypeptide with 42 amino acids would have 41 peptide bonds. The number of peptide bonds is always one less than the number of amino acids.

B1.2.4 – The vast variety of polypeptides

B1.2.4 – Infinite variety of possible peptide chains
Include the ideas that 20 amino acids are coded for in the genetic code, that peptide chains can have any number of amino acids, from a few to thousands, and that amino acids can be in any order. Students should be familiar with examples of polypeptides.

The huge variety of polypeptides is possible because:

- DNA codes for the number and order of amino acids within polypeptides
- there are 20 different amino acids
- polypeptides can vary in length, from a few amino acids to thousands of amino acids
- some polypeptides are modified by cells after their initial synthesis
- amino acids can be arranged in any order.

Although each polypeptide synthesized by the same gene is identical, there is an immense number of gene and amino acid combinations. This almost infinite number of possible permutations means that the different polypeptides can also have specific functions.

Some common polypeptides are:

- haemoglobin, an oxygen-carrying protein found in red blood cells
- keratin, found in hair, nails, claws and hooves
- lipase, a digestive enzyme that helps hydrolyse ingested lipids
- collagen, found in connective tissue in the body, including tendons and ligaments
- histones, proteins found in the nucleus of cells that help form chromatin and chromosomes
- insulin, a hormone that helps regulate blood sugar.

There are numerous other examples of polypeptides that you will come across throughout this course. Look for examples as you work your way through the IB biology curriculum.

B1.2.5 – The effect of pH and temperature

B1.2.5 – Effect of pH and temperature on protein structure
Include the term "denaturation".

The function of a protein is very dependent on its structure. Some proteins are shaped like fibres while others are folded to form **globular proteins**. Proteins have a very precise three-dimensional shape resulting from intramolecular bonds between amino acids, for example **hydrogen bonds**.

HL These intramolecular bonds are explored further in Sections B1.2.6–12.

How do abiotic factors influence the form of molecules?

The intramolecular bonds of proteins are susceptible to alterations at above normal temperature. When protein molecules are placed into an environment that is at a higher temperature than their physiological optimum, the increased molecular motion puts a great deal of stress on many of the relatively weak hydrogen bonds. Often the sequence of amino acids connected by peptide bonds remains intact, but the hydrogen bonds that help shape the protein cannot stay in place under the stress. The result is that the protein loses its shape and much or all of its function. This loss of shape and therefore function is called **denaturation**. As long as the bonds between the amino acids remain intact, the protein will return to its normal shape and function if it is returned to its optimal temperature.

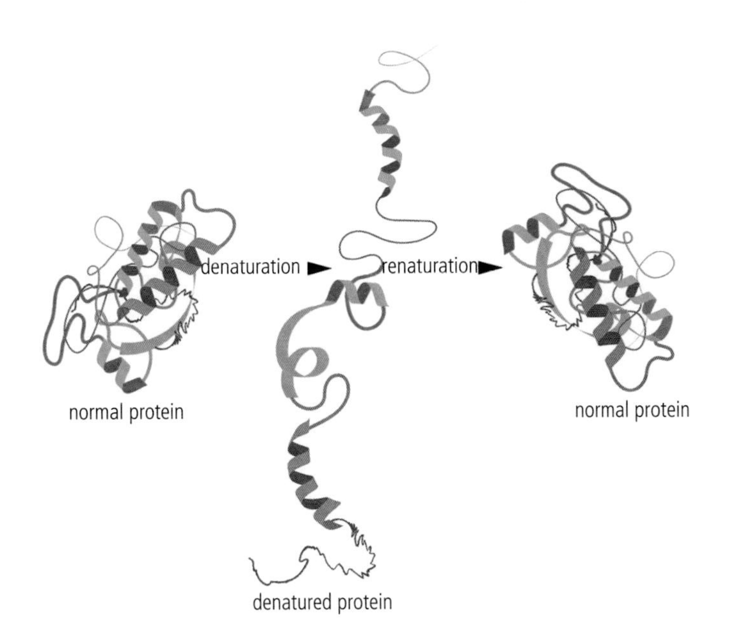

The protein on the left has been denatured by either heat or abnormal pH. The overall shape of the protein is altered and it can no longer function. Upon return to a physiologically normal temperature or pH, the protein re-forms its original shape and its function is restored. However, damage caused by excessive heat or pH conditions may lead to covalent bonds like peptide bonds breaking. If this occurs the protein will not be able to reshape itself.

A similar phenomenon occurs when a protein is placed in a pH environment that is not close to its optimum pH. A protein will lose its normal three-dimensional shape, and thus lose its functionality in these circumstances. When a fluid environment such as cytoplasm or blood plasma is flooded with either H^+ ions (an acid) or –OH ions (a base), the extra charges can prevent normal hydrogen bonding. The protein will not take on its "normal" shape and will not function normally. This denaturation is usually reversible as long as the underlying polypeptide chain is not damaged.

A denatured protein temporarily loses its function because its three-dimensional shape has been altered by unsuitable temperature or pH.

HL

B1.2.6 – R-groups provide diversity

B1.2.6 – Chemical diversity in the R-groups of amino acids as a basis for the immense diversity in protein form and function

Students are not required to give specific examples of R-groups. However, students should understand that R-groups determine the properties of assembled polypeptides. Students should appreciate that R-groups are hydrophobic or hydrophilic and that hydrophilic R-groups are polar or charged, acidic or basic.

In Section B1.2.1 you were asked to practise drawing the common structure of all amino acids, including the letter "R" for the variable group or portion of the molecules that differs between different amino acids. There are 20 of these R-groups in nature, and thus 20 different amino acids. You are not required to memorize the various R-groups but you are expected to understand the varying hydrophobic and hydrophilic properties inherent within them.

As you learned in Section B1.2.1, all amino acids have a carboxyl and an amine group bonded to a central (or alpha) carbon. When in a neutral aqueous solution (as amino acids typically are in most organisms) both the amine and carboxyl groups ionize. This happens because the carboxyl group acts as an acid and donates a hydrogen ion, while the amine group acts as a base and accepts a hydrogen ion. This results in the carboxyl group having a net negative charge and the amine group a net positive charge.

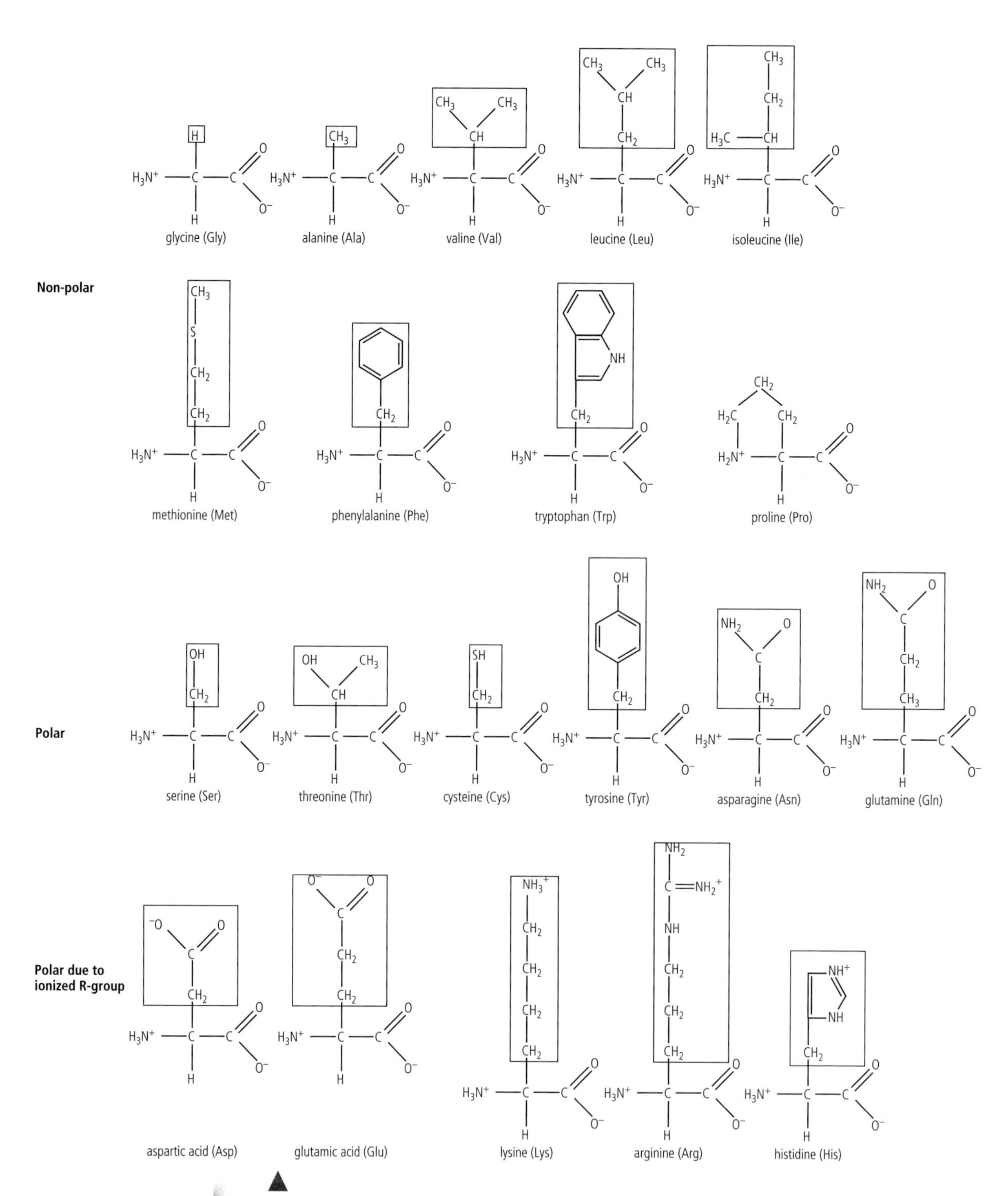

▲ **B1.2 Figure 2** The complete structure of all 20 amino acids. The boxed areas show the R (variable) groups or side chains of each. All the amine and carboxyl groups are shown ionized. Notice that there are two rows of non-polar, one row of polar, and one row of ionized polar R-groups.

The division of these amino acids into polar and non-polar categories is based solely on their R-groups. In other words, it is the R-groups that are being grouped. It is the R-groups that make the amino acids different from each other and create the unique structure characteristic of specific proteins. Those that are non-polar are hydrophobic.

All the others are hydrophilic. Those that are hydrophilic and have charged R-groups can be divided into those that are acidic and those that are basic.

Category of amino acid	Number of amino acids	Explanation for category
Non-polar	9	R-group is a hydrocarbon only
Polar	6	R-group contains element(s) that form a polar covalent bond (oxygen, nitrogen or sulfur)
Polar due to ionization charge (−)	2	R-group acts as an acid
Polar due to ionization charge (+)	3	R-group acts as a base

B1.2 Table 1 Groups of amino acids

B1.2.7 – Primary structure of a protein

B1.2.7 – Impact of primary structure on the conformation of proteins

Students should understand that the sequence of amino acids and the precise position of each amino acid within a structure determines the three-dimensional shape of proteins. Proteins therefore have precise, predictable and repeatable structures, despite their complexity.

DNA contains the genetic code for proteins. The code manifests itself in two ways:

- the number of amino acids in a given protein
- the sequence of amino acids in a given protein.

As you learned earlier in this chapter, amino acids are joined together by peptide bonds. The number and sequence of amino acids held together by peptide bonds is called the **primary structure** of the protein. This sequence is determined by a gene. Each gene can code for the synthesis of the same protein many times and for as long as the protein is needed within the cell. Each time the amino acid sequence will always be the same. Proteins with hundreds and even thousands of amino acids may appear incredibly complex, but each named protein is precise, predictable and repeatable because each time the DNA sequence in the gene will be translated into the same sequence of amino acids: the primary structure. How protein synthesis takes place in the cell is covered in Chapter D1.2.

Nature of Science

In 1941, George Beadle, a researcher working with fruit flies, coined the term "one gene – one enzyme", indicating that a specific portion of DNA codes for a specific enzyme. This assertion has changed over time. We now know that there are many proteins that are not enzymes and the control of protein synthesis is far more complex than George Beadle suggested.

The primary structure of a protein is simply the sequence of the amino acids in the polypeptide chain. However, the sequence of the amino acids then determines which intramolecular bonds will form and therefore the three-dimensional shape that the protein will ultimately take. We will study those shapes in the coming sections.

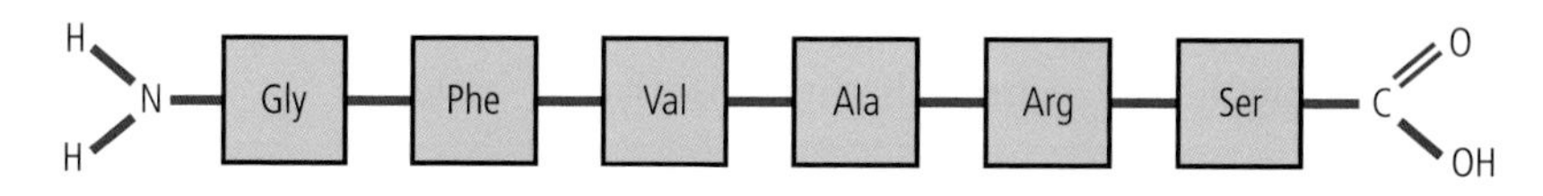

The primary structure of a very small protein containing six amino acids. Three-letter abbreviations are used for the identity of the amino acids. See Figure 2 for these abbreviations. No matter how many amino acids are within the protein there will be one with an unbonded amine group and another with an unbonded carboxyl group.

What is the relationship between the genome and the proteome of an organism?

B1.2.8 – Secondary structure of a protein

B1.2.8 – Pleating and coiling of secondary structure of proteins
Include hydrogen bonding in regular positions to stabilize alpha helices and beta-pleated sheets.

Once peptide bonds are in place, establishing a protein's primary structure, the molecule will begin shaping itself. If all the R-groups are non-polar, the R-groups simply do not influence the shape of the molecule. However, all amino acids have charges, even if their R-groups do not. Those charges come from the amine groups and carboxyl groups (called **residues**) that make up peptide bonds.

Look at the two structures in Figure 3. They show the two possible shapes that can result when the primary structure of a protein contains only non-polar R-groups. One possibility is a **beta-pleated sheet** and the other is an **alpha helix**. These structures are held together in these shapes by hydrogen bonds. Locate the alpha or central carbon atom of each of the amino acids and you will notice that the R-group and a hydrogen have been omitted from each one in the diagram. The reason for this is that the R-groups are not involved in the shaping of the molecule and are not needed to explain the secondary structure. All of the hydrogen bonding is between non-adjacent amine and carboxyl residues, which are portions of peptide bonds.

Beta-pleated sheet

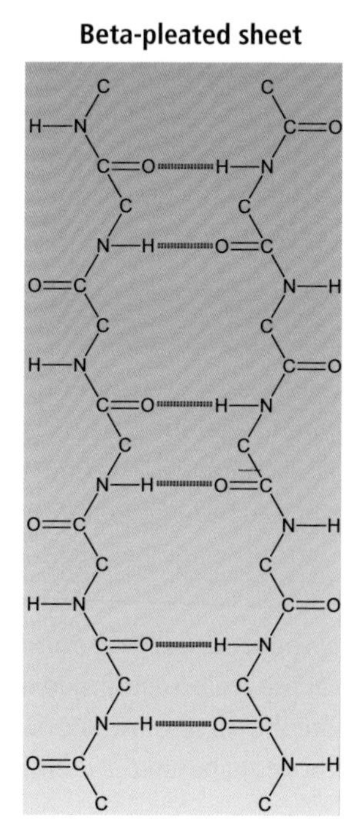

Alpha helix

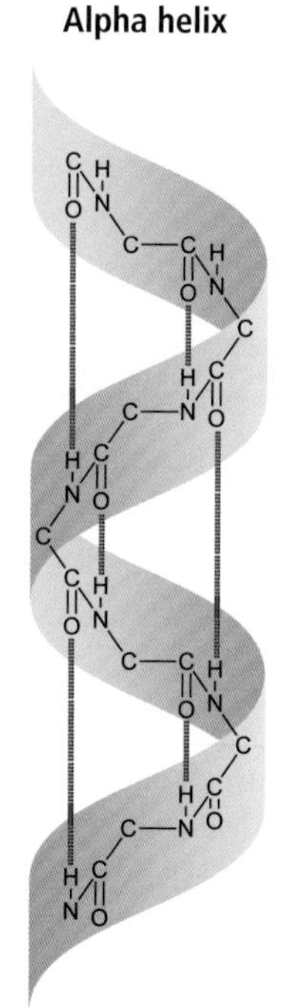

B1.2 Figure 3 The two protein shapes that form when the primary structure of a protein only contains non-polar amino acids (non-polar R-groups). Red lines indicate the hydrogen bonds that maintain the shapes. The central carbon of each amino acid is shown with only two covalent bonds each, to simplify the structures.

You should also notice that hydrogen bonds (red lines) have formed between the oxygen atoms of the carboxyl residues and the hydrogen atoms of the amine residues. Each oxygen atom has a partial negative charge and each hydrogen a partial positive charge.

The protein molecule will take on a shape that maximizes the number of hydrogen bonds. Notice that each of the hydrogen bonds is between amino acids some distance apart. The only possible way the hydrogen bonds can form is for the amino acid chain to shape itself into a beta-pleated sheet or coiled alpha helix.

Proteins that have a purely alpha helix or beta-pleated sheet shape do not form a unique and complex globular shape. Polypeptide chains that are made up of non-polar amino acids create relatively insoluble molecules. Proteins with this structure are often described as being **fibrous** and have a structural function or a function related to movement. Actin and myosin make up muscle fibres and are good examples of this type of protein.

The alpha helix and beta-pleated sheet shapes are often incorporated as portions of larger more complex globular shapes.

B1.2.9 – Tertiary structure of a protein

B1.2.9 – Dependence of tertiary structure on hydrogen bonds, ionic bonds, disulfide covalent bonds and hydrophobic interactions

Students are not required to name examples of amino acids that participate in these types of bonding, apart from pairs of cysteines forming disulfide bonds. Students should understand that amine and carboxyl groups in R-groups can become positively or negatively charged by binding or dissociation of hydrogen ions and that they can then participate in ionic bonding.

Many polypeptides have complex globular shapes. These shapes are the result of a variety of intramolecular bonds and are characteristic of polypeptides that are composed of a variety of different amino acid types. Look back at Figure 2 showing the different categories of amino acids. Below are some of the bonding interactions that can occur between amino acids at the tertiary level.

1. Ionized R-groups (some negative and some positive) will align with each other, to form an ionic bond. Amino acids with a carboxyl in their R-group will dissociate, lose the hydrogen ion and become negatively charged. Amino acids with an amine within their R-group will bind to a hydrogen ion and become positively charged. The two oppositely charged R-groups can then form an ionic bond between them (see 1 in Figure 4).

2. Non-polar amino acids, being hydrophobic, will fold into an area within the interior of the polypeptide in an attempt to avoid the polar water molecules. This is known as a **hydrophobic interaction** (see 2 in Figure 4).

3. Pairs of cysteine amino acids form a covalent bond between themselves and within the polypeptide. Cysteine's R-group contains a sulfur atom bonded to a hydrogen atom. When two non-adjacent cysteine amino acids get close to each other, the two hydrogens can be removed and the two sulfur atoms become covalently bonded to each other. The resulting covalent bond is called a **disulfide bond** and is the strongest of all the bonding forces that influence polypeptide shape (see 3 in Figure 4).

It is not necessary to memorize all 20 amino acids by name and category. However, the IB does require you to remember that pairs of the amino acid cysteine form disulfide covalent bonds within polypeptides.

4. Polar amino acids will form hydrogen bonds with each other and are often found on or near the exterior of the polypeptide because of their hydrophilic properties. Hydrogen bonds are typically the most numerous intramolecular type of bond (see 4 in Figure 4).

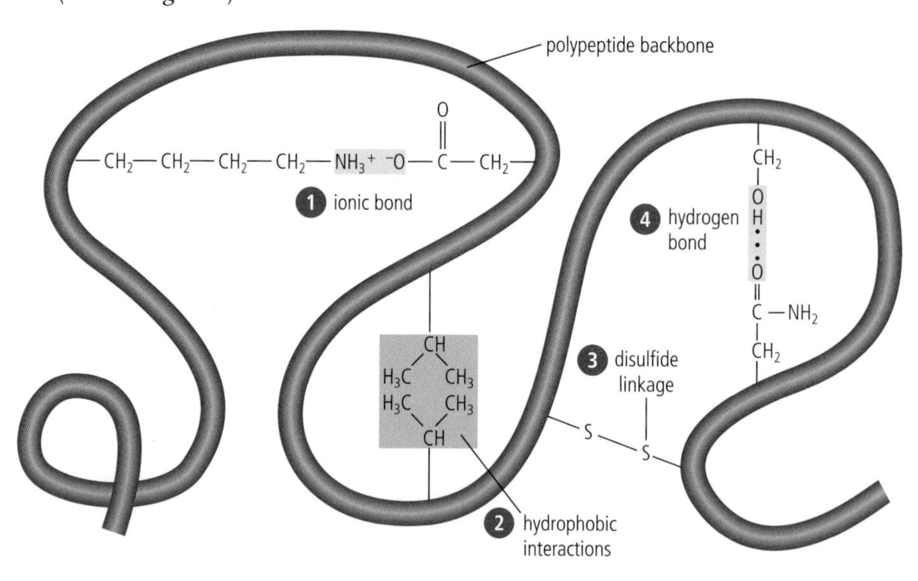

B1.2 Figure 4 Four bonding forces that influence the tertiary shape of a polypeptide. Each polypeptide will fold itself into a shape that maximizes these types of intramolecular forces.

Challenge yourself

1. The folded polypeptide in Figure 4 only shows the eight R-groups of the amino acids involved in each of four interactions influencing the shape of the molecule. Use Figure 2 and Table 1 to identify each of the amino acids and their category. Start on the left of Figure 4 and work towards the right side of the polypeptide.

B1.2.10 – Polar and non-polar amino acids

B1.2.10 – Effect of polar and non-polar amino acids on tertiary structure of proteins
In proteins that are soluble in water, hydrophobic amino acids are clustered in the core of globular proteins. Integral proteins have regions with hydrophobic amino acids, helping them to embed in membranes.

Proteins composed of non-polar amino acids have poor solubility in the cytoplasm and other aqueous solutions. They are primarily structural proteins that are generally stationary and thus do not need to be soluble. Proteins that have polar amino acids have much better solubility and are often found in different locations within the cell or the body. Peptide hormones, for example, are synthesized in a glandular cell and then enter the bloodstream in order to act as a chemical messenger in other areas of the body. We will look at two specific examples of proteins that have interesting solubility properties.

Lipase solubility

Lipase is an enzyme that catalyses the hydrolysis of lipid molecules. You may recall that lipids have long hydrocarbon tail regions that will only interact with an enzyme that is non-polar. If the entire lipase enzyme molecule was hydrophobic it would not be soluble in the solutions of the small intestine. The problem is solved when lipase folds into its three-dimensional shape. The primary structure allows a high concentration of hydrophobic amino acids to fold into the inside or core of the enzyme (to interact with the fat or oil substrate of the enzyme) and a high concentration of hydrophilic amino acids on the outside of the molecule. These outer polar amino acids permit solubility in a water environment.

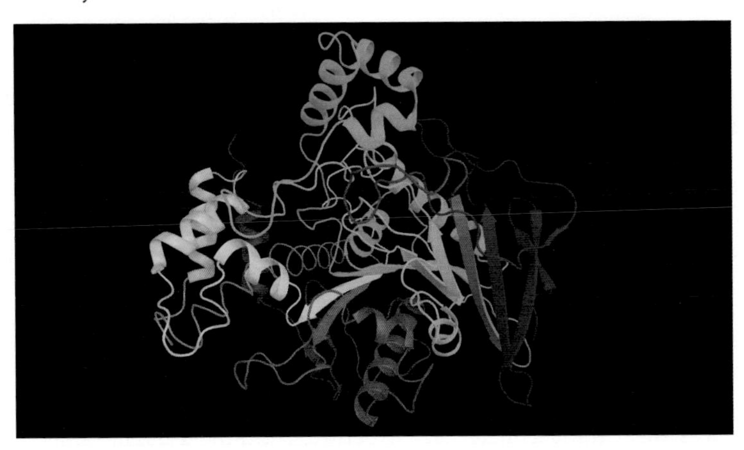

A molecular model of the enzyme lipase. Even though this enzyme has an overall tertiary globular shape it also contains two different kinds of secondary structures within it. The areas that look like springs are areas that have an alpha helix structure. On the far right (in blue) is a representation of a beta-pleated sheet structure shown as "ribbons" running parallel to each other. It is not surprising that lipase has a great deal of secondary structure as its substrate (lipids) is non-polar.

Glycoprotein A and B solubility

Chapter B1.1 discusses two proteins found on the membranes of red blood cells. Specifically, these are glycoproteins called A and B, which determine the ABO blood type. The protein portion of these molecules embeds itself into the phospholipid bilayer, with a portion extending out into the blood plasma. The protein portion of these molecules must fold so that non-polar (hydrophobic) amino acids can interact with the hydrophobic layers of the membrane to remain embedded. The folding also leads to polar (hydrophilic) amino acids bonding to the sugar component of the molecule and interacting with the aqueous blood plasma.

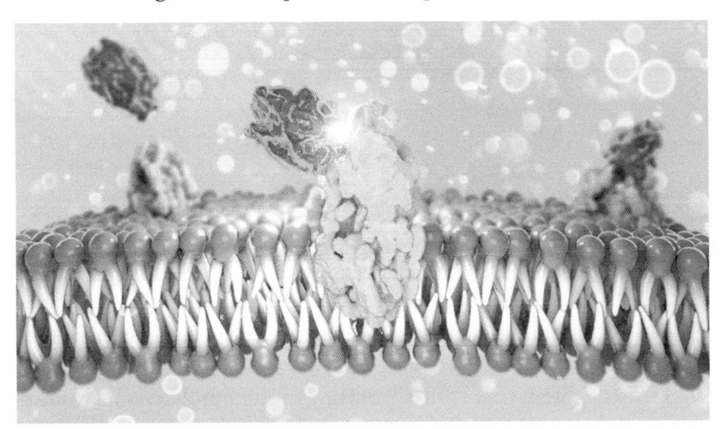

A sketch of phospholipid bilayer with partially embedded proteins. In order to extend into the membrane a protein must fold so that a region of hydrophobic amino acids exists. This leaves hydrophilic amino acids to interact with the aqueous fluid either inside or outside the cell.

B1.2.11 – Quaternary structure of proteins

B1.2.11 – Quaternary structure of non-conjugated and conjugated proteins
Include insulin and collagen as examples of non-conjugated proteins and haemoglobin as an example of a conjugated protein. **NOS:** Technology allows imaging of structures that would be impossible to observe with the unaided senses. For example, cryogenic electron microscopy has allowed imaging of single-protein molecules and their interactions with other molecules.

In the previous sections we have assumed that a protein is a single amino acid chain. That is not always the case. Sometimes two or more amino acid chains bond together into a single molecular structure. If so, that structure is called the **quaternary structure** of the protein.

We will look at three examples of proteins with a quaternary structure.

Insulin structure

There are many websites that allow you to both visualize and even manipulate biochemically important molecules. Do a web search putting in the name of a specific molecule (for example "insulin") or type of molecule (for example "hormone") combined with "3D molecular visualization".

Insulin is a protein hormone secreted by cells of the pancreas. Its function is to promote glucose uptake by body cells. When insulin is first synthesized by pancreatic cells it is a single chain of 110 amino acids. This chain is then modified within the pancreas cell to become two chains of amino acids, one 21 amino acids long and the other 30 amino acids long. These two chains then join together using the same types of bonds that are described in Section B1.2.9 on tertiary structured proteins. These include three disulfide bonds within the 51-amino acid structure.

Insulin is inactive while still within the pancreas cell because it forms **dimers**, two insulin molecules temporarily bonded together. Three of these dimers can also bond together to create a form called a **hexamer**. It is these storage forms of insulin, dimers and hexamers, that exhibit a quaternary structure. The dimers and hexamers will later separate in order for insulin to become a monomer and become active as a hormone.

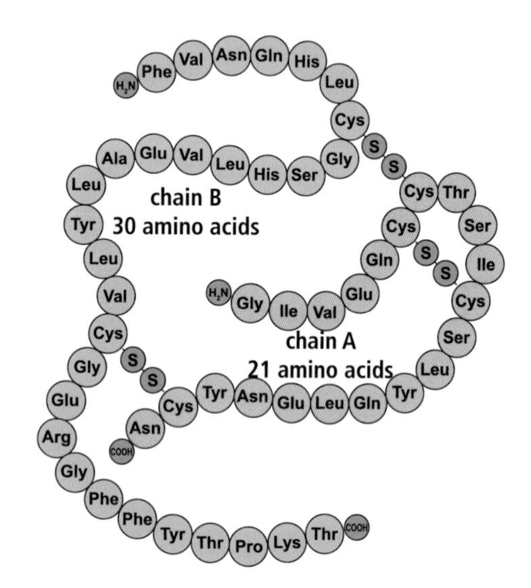

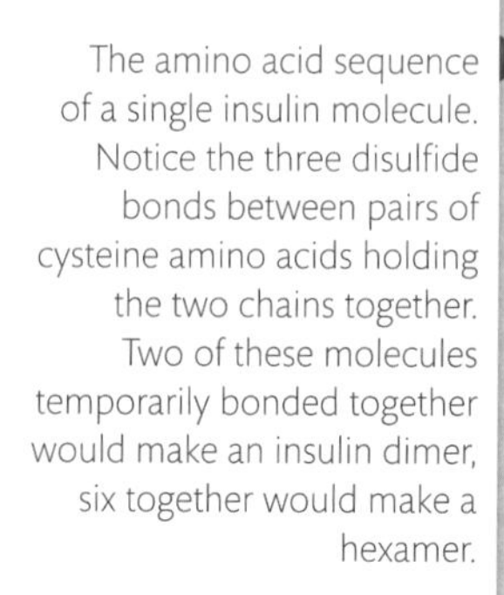
The amino acid sequence of a single insulin molecule. Notice the three disulfide bonds between pairs of cysteine amino acids holding the two chains together. Two of these molecules temporarily bonded together would make an insulin dimer, six together would make a hexamer.

Collagen structure

Collagen is the most abundant protein in the human body. It makes up connective tissue, provides tensile strength to tendons and ligaments and gives elasticity to skin. Collagen is a fibrous protein that consists of three polypeptide chains wound around each other in a helix shape. This arrangement of the three polypeptide chains gives collagen its quaternary structure.

Nature of Science

Technology allows imaging of structures that would be impossible to observe without recent microscopy advances. For example, cryogenic electron microscopy (cryo-EM) has enabled imaging of single-protein molecules and their interactions with other molecules.

Haemoglobin structure

Haemoglobin is a relatively large protein that is found within our red blood cells. Its function is to bind reversibly to oxygen in the lungs and carry the oxygen to body tissues, where it is released for aerobic cell respiration. Haemoglobin can also bind reversibly to carbon dioxide to be taken to our lungs and exhaled.

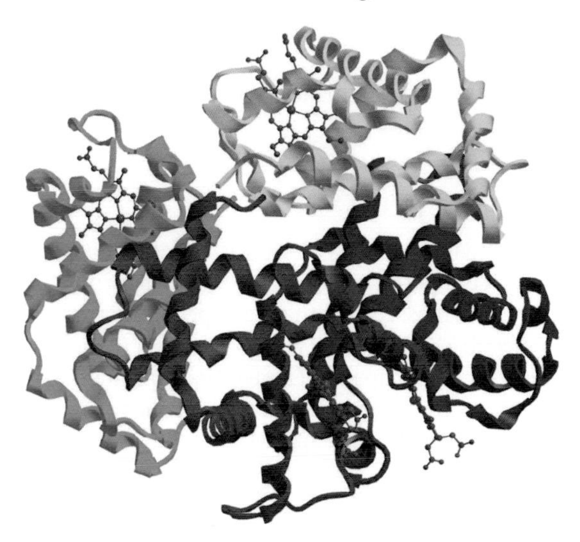

B1.2 Figure 5 A molecular model of a haemoglobin molecule. There are an estimated 270 million haemoglobin molecules in each red blood cell. Each haemoglobin has four polypeptides of two types, with a haem group at the centre of each polypeptide. The haem group is a carbon-based molecule but is not composed of amino acids. Each haem group contains a single iron atom (Fe^{+2}) that reversibly binds an oxygen molecule.

Haemoglobin is an example of a conjugated **quaternary structured protein**. In fact, it is an example of a protein that exhibits each of the different types of protein structures.

- Haemoglobin has two different primary structures, one for each of the two types of polypeptides within it.
- A secondary structure is seen in the alpha helix structures within portions of each polypeptide (see Figure 5).
- A tertiary structure is exhibited by each of the four globular polypeptides that fold and contain intramolecular bonds.
- A quaternary structure is shown by the bonding of the four polypeptides into a single haemoglobin molecule.

In addition, haemoglobin is **conjugated**. This means that the protein has one or more non-protein groups as part of the molecule (in this case the haem group).

Proteins that are composed of just amino acids are called non-conjugated proteins. Insulin and collagen are non-conjugated proteins. Proteins that have one or more non-protein portions are conjugated. Haemoglobin is a conjugated protein because of its haem groups.

B1.2.12 – Protein shape and its function

B1.2.12 – Relationship of form and function in globular and fibrous proteins
Students should know the difference in shape between globular and fibrous proteins and understand that their shapes make them suitable for specific functions. Use insulin and collagen to exemplify how form and function are related.

Proteins are incredibly diverse in both shape and function. As a general rule, fibrous proteins serve a structural role and globular proteins have very specialized roles. The specialized roles include acting as enzymes, antibodies, peptide hormones, cell signalling proteins, and many more. We will now look again at two proteins that illustrate the fundamental difference between fibrous and globular proteins.

▲ A molecular model of the three polypeptide chains that wind around each other to make a collagen fibre. Many of these fibres can run parallel to each other to create even stronger and larger fibres. There are only three types of amino acids within collagen and they are found in repeating sequences.

Collagen: a fibrous protein

You learned earlier that collagen is a fibrous protein with an alpha helix secondary structure that makes up connective tissues. This protein needs to be strong and elastic (able to stretch), but it does not need to be highly variable. In fact, the amino acids within collagen are short repeating sequences of glycine – proline – X. X is a different amino acid depending on the location and specialized function of the collagen molecule. Thi repeating sequence gives collagen a very regular and geometric fibrous shape.

The structural uniformity of collagen makes it an excellent molecule for building tissues that hold the body together.

Insulin: a globular protein

Molecules that have globular shapes have a wide variety of amino acids within their structure. Thus, they can be specialized for a very specific purpose. For example, there are many thousands of enzymes, each with a very specific three-dimensional globular shape that fits the molecule(s) for which it acts as a catalyst (its substrate). Every peptide hormone has a specific shape to fit a membrane protein, so that the hormone can act as a signalling molecule. Every specific antibody has a unique shape that chemically recognizes a molecule called an antigen.

Diabetes is a serious disease that is characterized by disruptions in the signal pathway involving insulin. Type 1 diabetics cannot produce sufficient insulin; type 2 diabetics develop an insulin resistance at some point in their lives. Both conditions require medical treatment.

We have already looked at the molecular structure of insulin, and learned that insulin helps regulate the passage of glucose from the bloodstream into cells. We will now look at how insulin is able to do that. Insulin is released into the bloodstream after blood glucose levels rise, a common occurrence after a meal or snack including carbohydrates. The bloodstream takes insulin to all the body's cells. Each cell has many membrane proteins called insulin receptors. The insulin molecule fits into the insulin receptor proteins, which have a complementary shape to the hormone. The fit is exact and no other molecule can fit properly into the insulin receptors.

When insulin does fit into the membrane receptor, a set of reactions occurs that is called a **signal pathway**. The end result of the signal pathway is that the cell opens channels in the plasma membrane that allow glucose to enter the cell. Glucose is needed in the cell for cellular respiration.

▶ A molecular model image of a plasma membrane with a glucose receptor and glucose channels. The large blue molecule is the glucose receptor. The insulin is the small orange molecule on its upper right side. When insulin is bound to the receptor the portion of the receptor protein that is inside the cell (at the bottom of the image) initiates a series of reactions that ultimately opens glucose channels. These channels are shown in red, with the small yellow molecules representing glucose molecules moving through.

HL end

Guiding Question revisited

What is the relationship between amino acid sequence and the diversity in form and function of proteins?

In this chapter you have learned that:

- DNA codes the amino acid sequence in a protein
- a polypeptide is created when condensation reactions occur between adjoining carboxyl groups and amine groups of multiple adjoining amino acids
- polypeptides vary in length from a few amino acids to thousands of amino acids
- the sequence of amino acids in a protein (primary structure) determines its molecular shape as well as its biological function
- the variety of polypeptides is nearly infinite

HL

- some amino acids have non-polar R-groups and some have polar R-groups
- the secondary structure of a polypeptide can include alpha helixes and beta-pleated sheets, based on hydrogen bonding between non-adjacent amine and carboxyl residues
- secondary-structured proteins are used for structural purposes, such as keratin and collagen
- polypeptides with a tertiary structure have a very precise globular shape
- tertiary-structured proteins have unique functions, such as enzymes and peptide hormones
- quaternary-structured proteins have two or more polypeptide chains joined together to form a single protein.

HL end

Guiding Question revisited

How are protein molecules affected by their chemical and physical environments?

In this chapter you have learned that:

- environmental temperatures higher than an optimum affect the shape and thus the function of many proteins
- many of the bonds that create the shape of a protein are relatively weak hydrogen bonds between polar amino acids
- higher temperatures lead to increased molecular motion and place stress on hydrogen bonds
- breaking hydrogen bonds denatures a protein, leading to decreased activity
- an environment that is too acidic or basic will alter the positive and negative charges that are needed for hydrogen bonding
- protein in a non-optimal pH environment will change shape and its activity will decrease.

Exercises

Q1. What structural features do all amino acids have in common?

Q2. A given polypeptide contains 166 amino acids. How many peptide bonds does this polypeptide have?

Q3. Write the word equation for a condensation reaction between two generalized amino acids.

Q4. **(a)** An enzyme is heated in a laboratory to 8°C above its optimum temperature. When tested at this temperature it had lost its catalytic function. Explain why.

(b) Predict, with a reason, what would happen to the catalytic ability of this enzyme if it is returned to its optimum temperature.

(c) The same enzyme is heated to 50°C above its optimum temperature. When tested, it had lost its catalytic function. Predict, with a reason, what would happen to the catalytic ability of this enzyme if it is returned to its optimum temperature.

HL

Q5. Explain each of the following statements.

(a) When amino acids are grouped by solubility in aqueous solutions, it is the R-group that is being grouped.

(b) All amino acids have some solubility in water when the entire amino acid is considered.

Q6. Which of these statements is true for the R-groups within amino acids found in proteins with a secondary structure?

A They are acidic or basic.

B They are polar.

C They are non-polar.

D They include a wide variety including some non-polar and some non-polar.

Q7. Which of the following statements best describes a disulfide bond?

A A covalent bond between two sulfur atoms of a single cysteine amino acid.

B A covalent bond between two sulfur atoms of two different cysteine amino acids.

C An ionic bond between two sulfur atoms helping to make up a peptide bond.

D An ionic bond between two sulfur atoms in two different polypeptides.

Q8. List four bonding forces that contribute to the complex globular shape of a tertiary-structured protein.

Q9. Explain why proteins that have a secondary structure, but no tertiary structure, are generally unsuited to function as enzymes, peptide hormones or antibodies.

Q10. Identify the molecular characteristic of haemoglobin that qualifies it as:

(a) a quaternary-structured protein

(b) a conjugated protein.

HL end

B1 Practice questions

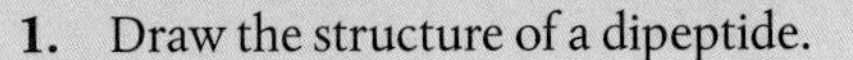

1. Draw the structure of a dipeptide. *(Total 3 marks)*

2. Outline how the structure of cellulose makes it a suitable component of cell walls. *(Total 2 marks)*

HL

3. The figure shows the structure of lactase.

   (a) Identify the protein structures indicated by I and II. (1)

   (b) Describe how structure I is held together. (2)

   (c) This protein is described as a globular protein. Distinguish between globular and fibrous proteins. (2)

   (Total 5 marks)

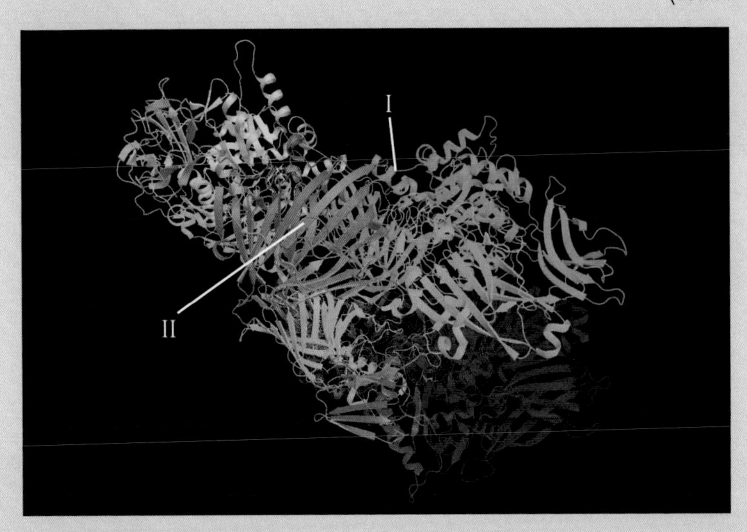

4. Explain how the hydrophobic and hydrophilic properties contribute to the arrangement of molecules in a membrane. *(Total 7 marks)*

HL end

THEME B Form and function

2 Cells

We look at a flower, praise its beauty, but rarely think about all the parts of the plant that played critical roles in its formation. Roots must take in water and nutrients, leaves convert sunlight into usable chemical compounds, and stems conduct essential materials up and down, night and day, before the structure that will become the flower has even formed. Each part of the plant has a structure that allows it to carry out an essential function so that the plant can remain alive. Cells are no different. Cells have many parts that work together to maintain the life of the cell. Each working part has a specific structure that enables it to carry out a function essential to the life of the cell. Some cells have unique tasks that they perform for the organism they are a part of. To accomplish these unique tasks, cells are diverse in size, shape and even organelle composition. We will examine some of these cell forms and functions in this chapter.

B2.1 Membranes and membrane transport

Guiding Questions

How do molecules of lipid and protein assemble into biological membranes?

What determines whether a substance can pass through a biological membrane?

Everywhere we look, we see protective exteriors around almost everything. There certainly are differences in these exteriors, but they all serve similar functions. They keep harmful factors out, they keep beneficial factors in, and they regulate what goes in and out of the structure. Cells are no different. They all have membranes around their exterior protecting them from what are often potentially damaging environments. These membranes are composed of lipids and proteins and are assembled in such a way that protection is provided to the cell interior. As well as providing protection, these lipid–protein complexes also control the movement of substances in and out of the cell.

These amazing membranes have receptors to monitor and respond to the surroundings, channels to allow specific molecules to be transported, carriers to maintain homeostatic conditions, and even structures to allow communication with other cells in the same organism or with other organisms in the same environment.

In this chapter, we will learn about the structure and function of the cell membrane.

B2.1.1 and B2.1.2 – Membrane structure

B2.1.1 – Lipid bilayers as the basis of cell membranes
Phospholipids and other amphipathic lipids naturally form continuous sheet-like bilayers in water.

B2.1.2 – Lipid bilayers as barriers
Students should understand that the hydrophobic hydrocarbon chains that form the core of a membrane have low permeability to large molecules and hydrophilic particles, including ions and polar molecules, so membranes function as effective barriers between aqueous solutions.

A study of beet cell membrane. Full details on how to carry out this activity with a worksheet are available in the eBook.

As early as 1915 scientists were aware that the structure of membranes isolated from cells included proteins and lipids. Further research established that the lipids were phospholipids. Early structural theories suggested that membranes were composed of phospholipids forming a bilayer, and on the inside and outside of this bilayer were thin layers of proteins.

The fact that only slight changes have been made to the Singer–Nicolson model of the cell membrane since 1972 does not mean the model is 100% accurate. Science continually tests theories and models to determine their validity. How can it be that scientific beliefs and knowledge change over time?

We are dependent on properly functioning cell membranes for good health. Cystic fibrosis is an inherited condition in humans. The condition stops cell membranes from functioning correctly. The result is a build-up of thickened mucus in the airways, digestive system and other organs and tissues.

In 1972, Seymour J. Singer and Garth L. Nicolson proposed that proteins are inserted into the phospholipid layer and do not form a layer on the phospholipid bilayer surfaces. They believed that the proteins formed a mosaic floating in a fluid bilayer of phospholipids.

Much of the evidence used to revise the model was obtained using an electron microscope. Another source of evidence was the study of cells and their actions in various environments and solutions. The ability to culture cells in the laboratory allowed many of these studies. Since 1972 further evidence has been gathered about the membranes, and only slight changes to the Singer–Nicolson model have been made.

Nature of Science

Using models is a way in which scientists can explain complex structures such as cellular membranes. Models are based on the knowledge available at the time a theory is suggested. Even though the early models of cell membranes were later proved wrong (because of new data), they helped in the development of the presently accepted model of cell membranes. Discuss why it is important to learn about theories that were later discredited.

The currently accepted model of the cellular membrane is known as the **fluid mosaic model**. The fluid mosaic model is discussed further in Section B2.1.10. All cellular membranes, whether plasma membranes or organelle membranes, have the same general structure. Membranes are flexible, supporting structures. They consist of several different types of molecules that allow them to function correctly.

Phospholipid structure

The "backbone" of the membrane is a bilayer produced from huge numbers of molecules called **phospholipids**. Each phospholipid is composed of a three-carbon compound called glycerol. Two of the glycerol carbons have fatty acids combined with them. The third carbon is attached to a highly polar organic alcohol that includes a bond to a phosphate group. Fatty acids are not water soluble because they are non-polar. However, because the organic alcohol with phosphate is highly polar, it is water soluble. This structure means that phospholipid molecules have two distinct areas when it comes to polarity and water solubility. One part of the molecule is water soluble and polar, and is referred to as **hydrophilic** (water-loving). This is the phosphorylated alcohol side. The other part is not water soluble and is non-polar. It is referred to as **hydrophobic** (water-fearing). Any molecules that have both a hydrophobic and a hydrophilic region are said to be **amphipathic**.

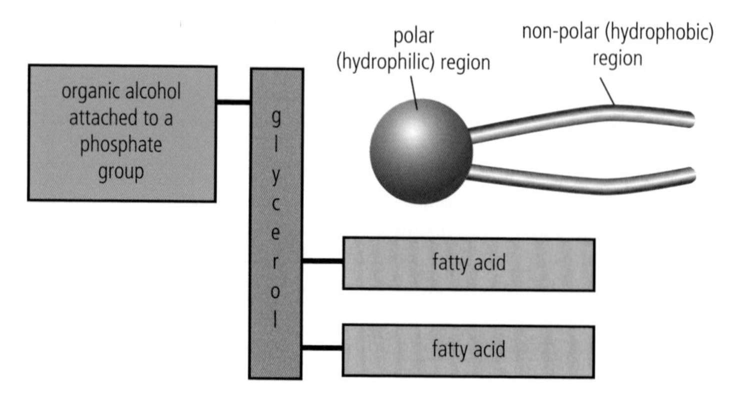

A model of a phospholipid.

Phospholipid bilayer as a barrier

The hydrophobic and hydrophilic regions cause phospholipids to naturally align as a bilayer if there is water present. The hydrophobic regions are attracted to each other and the hydrophilic regions are attracted to the water in the cytoplasm or the extracellular fluid. Because the fatty acid "tails" do not attract each other strongly, the membrane tends to be fluid or flexible. This allows animal cells to have a variable shape and allows the process of **endocytosis** (discussed in Section B2.1.13). What maintains the overall structure of the membrane is the relationship between its chemical makeup and the chemical properties of water.

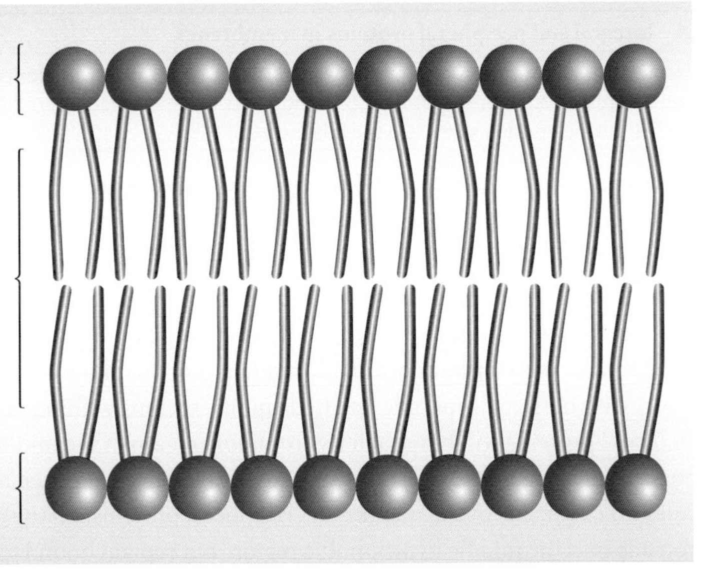

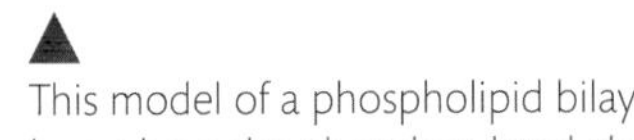

This model of a phospholipid bilayer shows how phospholipid molecules behave in two layers. Both layers have the phosphorylated alcohol end of the molecules towards the outside and the fatty acid tails oriented towards each other in the middle.

Once the bilayer has formed, large molecules cannot pass through it easily because the molecules are tightly packed. Hydrophilic molecules, such as ions, might be smaller but they also find it difficult to move through the membrane because of the hydrophobic region in the middle of the bilayer. The bilayer therefore forms an effective barrier between the inside and outside of the cell. Because the cell does need large and polar molecules to pass into and out of it for certain functions, the structure of the bilayer allows the cell to control what passes through the membrane.

Because phospholipids naturally form continuous sheet-like bilayers in water, they act as a barrier between the inside and outside of the cell. They have low permeability to large molecules and hydrophilic particles. Hydrophilic particles such as ions and polar molecules do not pass easily through the bilayer. This allows the bilayer to function as an effective barrier between aqueous solutions.

B2.1.3 – Diffusion across cellular membranes

B2.1.3 – Simple diffusion across membranes
Use movement of oxygen and carbon dioxide molecules between phospholipids as an example of simple diffusion across membranes.

One type of transport that can take place through the membrane is diffusion. In diffusion, particles move from a region of higher concentration to a region of lower concentration. In a living system, diffusion often involves crossing a membrane. For example, oxygen is used by cells in respiration. There is therefore a lower oxygen concentration inside the cell compared to outside the cell. Oxygen diffuses into the cell as a result. Carbon dioxide diffuses in the opposite direction because carbon dioxide is produced by mitochondrial respiration inside the cell and is present in higher concentrations inside the cell compared to outside the cell. Both CO_2 and O_2 are small, uncharged molecules. They can move between the phospholipid molecules of the membrane, so their diffusion can occur easily.

B2.1.4 – Membrane proteins

B2.1.4 – Integral and peripheral proteins in membranes
Emphasize that membrane proteins have diverse structures, locations and functions. Integral proteins are embedded in one or both of the lipid layers of a membrane. Peripheral proteins are attached to one or other surface of the bilayer.

Another major component of cellular membranes is the proteins. It is these proteins that create the extreme diversity in membrane function. Proteins of various types are embedded in the fluid matrix of the phospholipid bilayer. This creates the mosaic or tile-like effect characteristic of cellular membranes. There are usually two major types of proteins. One type is referred to as **integral** proteins and the other type is referred to as **peripheral** proteins. Integral proteins show an **amphipathic character,** with both hydrophobic and hydrophilic regions within the same protein. These proteins will have their hydrophobic region in the mid-section of the phospholipid backbone. Their hydrophilic region will be exposed to the water molecules on either side of the membrane. Peripheral proteins, on the other hand, do not protrude into the middle hydrophobic region, but remain bound to the surface of the membrane. Peripheral proteins can be found on the surface of both the inner and outer sides of the membrane. Often these peripheral proteins are anchored to an integral protein.

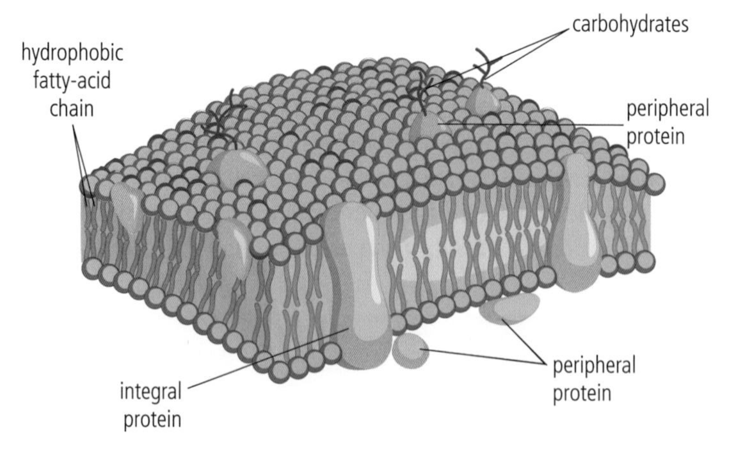

Peripheral and integral proteins of a cell membrane. The peripheral proteins do not extend into the lipid bilayer(s) of the cell membrane like the integral proteins do. Peripheral proteins are attached to one of the two surfaces of the membrane.

Although the protein components of cell membranes differ depending on the cell type and its particular function at a given time, several types of proteins are usually present. Examples are listed in Table 1.

Protein type	Description
Hormone-binding	These proteins have specific shapes exposed to the exterior that fit the shape of specific hormones. The attachment between the protein and the hormone causes a change in the shape of the protein, which results in a message being relayed to the interior of the cell.
Enzymatic	These proteins occur on either the interior or the exterior membrane surface. They are often grouped together so that a sequence of metabolic reactions, called a metabolic pathway, is catalysed.
Cell adhesion	This protein type allows temporary or permanent connections called junctions between cells. There are two types of junctions, gap junctions and tight junctions.
Cell-to-cell communication	Most of these proteins have carbohydrate molecules attached. They provide an identification label so that organisms can distinguish between self and non-self material.
Channel forming	Some proteins span the membrane, providing passageways for substances to be transported through.
Pumps for active transport	In active transport, proteins shuttle a substance from one side of the membrane to another by changing shape. This process requires the expenditure of energy in the form of adenosine triphosphate (ATP).

B2.1 Table 1 Types of protein usually present in a cell membrane

What are the roles of cell membranes in the interaction of a cell with its environment?

B2.1.5 and B2.1.6 – Membrane transport

B2.1.5 – Movement of water molecules across membranes by osmosis and role of aquaporins
Include an explanation in terms of random movement of particles, impermeability of membranes to solutes and differences in solute concentration.

B2.1.6 – Channel proteins for facilitated diffusion
Students should understand how the structure of channel proteins makes membranes selectively permeable by allowing specific ions to diffuse through when channels are open but not when they are closed.

There are two general types of cellular transport:

- passive transport
- active transport.

Passive transport does not require cellular energy (in the form of adenosine triphosphate, ATP) but active transport does. Passive transport takes place when a substance moves from an area of high concentration to an area of lower concentration. Movement is said to occur along a concentration gradient. The source of energy for this movement comes from the kinetic energy of the molecules. If left undisturbed, this directional movement will continue until equal concentrations of the substance are found in both areas and equilibrium is attained.

When active transport occurs, the substance is usually moved against a concentration gradient, so energy expenditure must occur. Equilibrium is not reached with active transport.

Examine Figure 1 illustrating chemical diffusion, an example of passive transport.

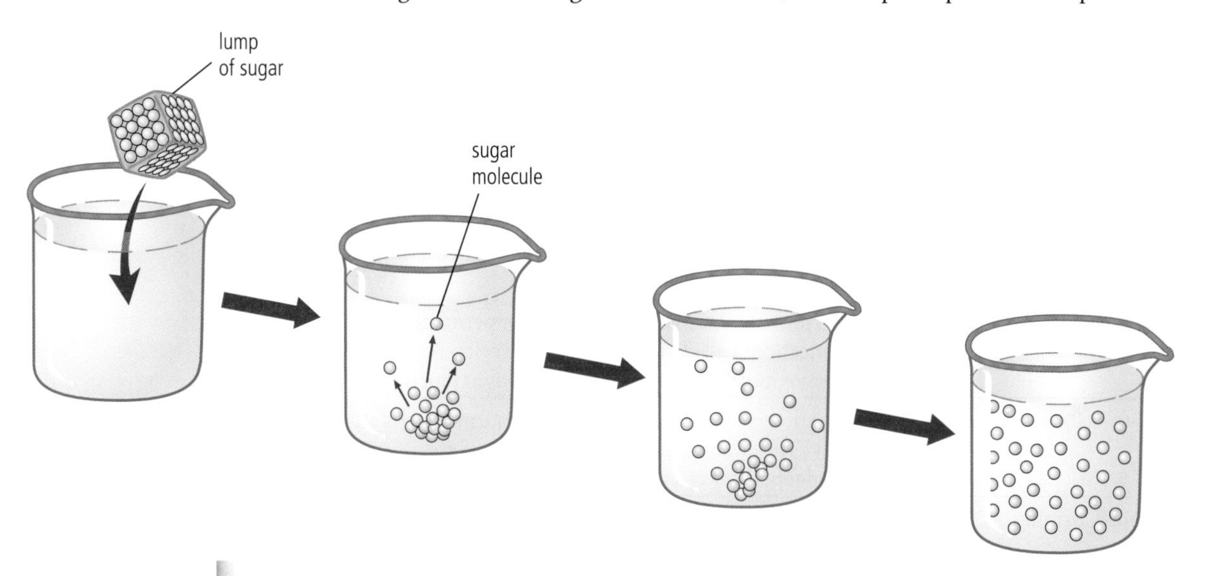

B2.1 Figure 1 Chemical diffusion: note how the sugar molecules move from the area of higher concentration to the area of lower concentration. The particles in the liquid are present in a constant, random motion. When a high concentration of a molecule is present, there are more collisions between the molecules. This creates a net movement of particles into areas that are less concentrated.

Movement of water molecules

Osmosis is another type of passive transport: movement occurs along a concentration gradient. However, osmosis involves only the passive movement of *water* across a **partially permeable membrane**. A partially permeable membrane (also known as a **selectively permeable membrane**) is one that only allows certain substances to pass through (a permeable membrane would allow everything through). The concentration gradient of water that allows the movement to occur is the result of a difference in solute concentrations on either side of the membrane. A **hypertonic** solution has a higher concentration of solutes than a **hypotonic** solution (see Chapter D2.3). Water therefore moves from a hypotonic solution to a hypertonic solution across a partially permeable membrane (see Figure 2). If isotonic solutions occur on either side of a partially permeable membrane, no net movement of water is evident because equilibrium has been achieved.

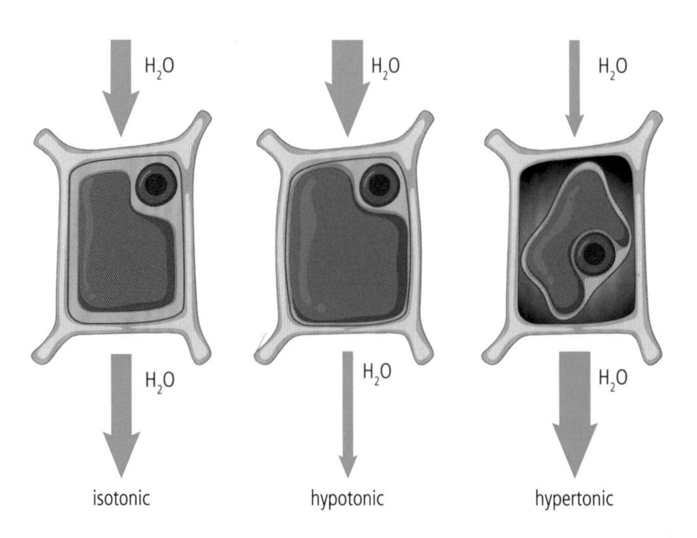

B2.1 Figure 2 The cell wall of a plant makes it difficult to see the many changes that occur inside as a result of water movement. The rigid cell wall resists changes in shape. However, the cell membrane and cell contents are affected by water moving into and out of the cell.

The cell membrane is impermeable to many solute molecules. Therefore, in osmosis, only water moves across the cell membrane. This water moves with more ease than expected of a polar molecule. Usually, polar molecules cannot pass quickly through the cell membrane because of the hydrophobic properties of the middle membrane region. However, most cell membranes have protein channels called **aquaporins**, which allow water molecules to pass through them. Water molecules move randomly (like all other molecules that are diffusing) but if there is a higher concentration of water molecules in one area then there will be more molecules moving randomly towards the area of lower concentration and there will be a net movement towards the area with the lower concentration.

Aquaporins allow water to flow through cell membranes. There are different types of aquaporins depending on the organism and the cell types they are a part of. All aquaporins are embedded in the cell membrane as integral proteins and consist of amino acids producing repeating proteins. These proteins have non-polar areas on their exterior that allow the embedding of the aquaporin in the membrane. They also have polar areas internally associated with the channel to allow water to pass through.

Carrier and channel proteins

Facilitated diffusion is a particular type of diffusion that involves two types of integral proteins: carrier proteins and channel proteins.

Carrier proteins change shape in order to carry a specific substance (usually an ion) from one side of the membrane to the other. If a carrier protein is not working, no transport will occur. Carrier proteins can carry substances along a concentration gradient (in the case of facilitated diffusion) or against a concentration gradient (as in active transport). Carrier proteins can carry both water-soluble and insoluble molecules.

To remember the difference between diffusion and osmosis, think of "H_2Osmosis", linking the water to osmosis.

Channel proteins are different from carrier proteins in that they have pores through which molecules of appropriate size and charge can pass. Most channel proteins have "gates" that open and close in response to chemical or mechanical signals. Channel proteins do not change shape in the way that carrier proteins do: they just open and close a channel through which molecules can diffuse. Channel proteins only carry water-soluble molecules, and are specific for the ion that they carry. The presence of channel proteins and carrier proteins makes cell membranes selectively permeable: specific ions are allowed through the membrane at certain times. The rate of facilitated diffusion depends on several factors, including the concentration difference that exists across the membrane and the number of carrier proteins actively involved in transport and/or the number of channel proteins open.

An example of a disease involving facilitated diffusion is **cystinuria**. This occurs when the protein that carries the amino acid cysteine is absent from kidney cells. The consequence is a build-up of amino acids in the kidney, resulting in very painful kidney stones.

B2.1.7 – Active transport and pump proteins

B2.1.7 – Pump proteins for active transport
Students should appreciate that pumps use energy from adenosine triphosphate (ATP) to transfer specific particles across membranes and therefore that they can move particles against a concentration gradient.

Active transport requires work to be performed. This means energy must be used, so ATP is required. Active transport often involves the movement of substances against a concentration gradient. This process allows the cell to maintain interior concentrations of molecules that are different from exterior concentrations. Active transport can take place because of highly selective proteins in the membrane that bind with the substance to be transported. Different protein carriers involved in active transport differ in the way that they work. Look at Figure 3.

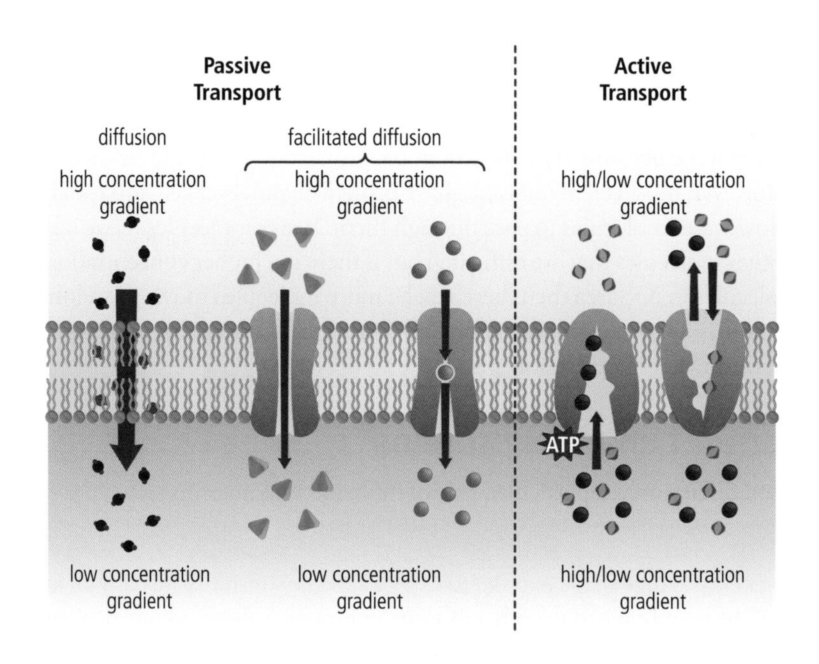

B2.1 Figure 3 A comparison of active and passive transport. Notice the descriptions of the concentration gradients for each type of transport. The active transport example is a protein pump requiring the use of ATP, such as in the sodium–potassium pump.

The sodium–potassium pump

The **sodium–potassium pump** is an extremely important example of active transport. It uses ATP to move ions directly against a concentration gradient. This is especially important in nerve cells, also called **neurons**, so that animals can respond appropriately to environmental stimuli.

B2.1.8 – Membrane permeability

B2.1.8 – Selectivity in membrane permeability
Facilitated diffusion and active transport allow selective permeability in membranes. Permeability by simple diffusion is not selective and depends only on the size and hydrophilic or hydrophobic properties of particles.

Facilitated diffusion and active transport enable selective permeability in membranes through the control of the protein channels involved in the two processes. However, permeability by simple diffusion is not selective and depends on the size and chemical properties of the particles involved.

How easily a substance can move passively across a membrane depends on two major factors: size and charge. Substances that are small and non-polar will move across a membrane with ease. Substances that are polar or large, or both, do not cross membranes easily. Examples of small, non-polar substances are gases such as oxygen, carbon dioxide and nitrogen. Ions such as chloride ions, potassium ions and sodium ions have a great deal of difficulty crossing membranes passively, as do large molecules such as glucose and sucrose. Diffusion of small, simple molecules is therefore not selective. In contrast, the cell can be selectively permeable to large, charged molecules because they must travel through integral proteins.

The size and polarity of molecules determine the ease with which various substances can cross membranes. These characteristics and the ability of molecules to cross membranes are arranged along a continuum:

small and non-polar molecules cross membranes easily  large and polar molecules cross membranes with difficulty

Challenge yourself

This challenge requires knowledge and understanding of cellular transport. Completing it will serve as a review of the cellular transport concepts we have considered so far.

1. A practical example of diffusion and osmosis is kidney dialysis. The kidneys are responsible for removing urea from the blood and also regulate the level of solutes in the body. If the kidneys are not functioning correctly this can be life-threatening. A process called **haemodialysis** can be used to remove urea artificially and restore the correct balance of solutes in the body.

   During haemodialysis, blood is passed through a system of tubes composed of selectively permeable membranes. These tubes are surrounded by a solution that is called the dialysate. The dialysate contains key solutes at levels close to the patient's normal blood levels. Wastes are kept at a low level in the dialysate. As blood moves through the tubes, the dialysate is constantly replaced to maintain ideal levels.

   Using your knowledge of osmosis, diffusion and membrane transport, suggest how haemodialysis works and why the dialysate must be constantly changed.

B2.1.9 – Glycoproteins and glycolipids

B2.1.9 – Structure and function of glycoproteins and glycolipids
Limit to carbohydrate structures linked to proteins or lipids in membranes, location of carbohydrates on the extracellular side of membranes, and roles in cell adhesion and cell recognition.

When cell membrane phospholipids have carbohydrate chains attached to them, they are known as **glycolipids. Glycoproteins** are cell membrane proteins that have chains of carbohydrates attached to them. Carbohydrate chains are only found on the exterior, extracellular side, of the cell membrane. These chains are quite diverse based on their sequences of sugar types and branching structures. Glycoproteins and glycolipids are important for cell identification and cell adhesion (cells sticking to each other).

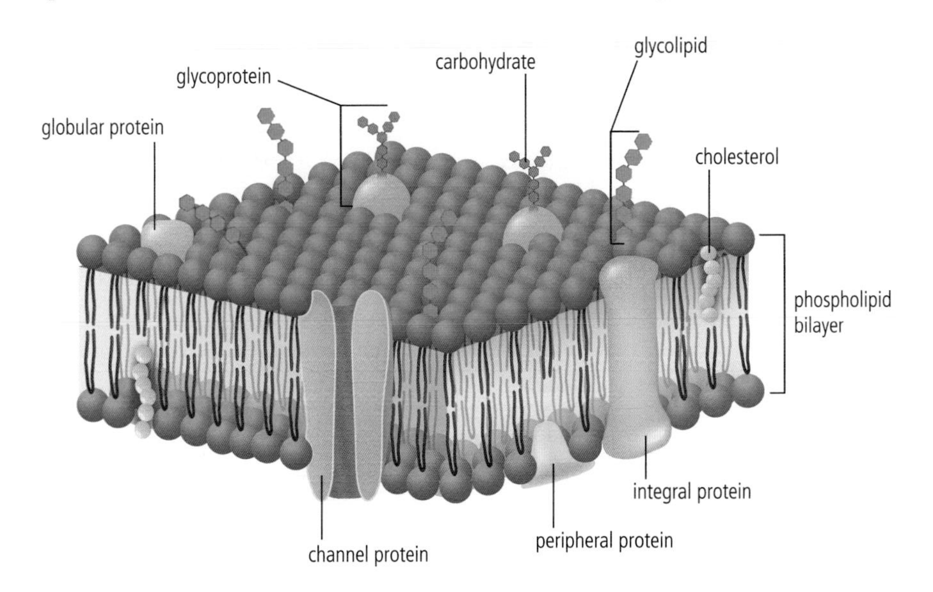

In this diagram of a cell membrane, notice the structures labelled glycolipids and glycoproteins. These structures have carbohydrate chains attached. They occur on the cell membrane outer surface. From our earlier discussion of cell membranes, we know the outermost region and the innermost region of the cell membrane bilayer are hydrophilic. The middle portion of the lipid bilayer is hydrophobic. The relationship of the membrane regions to water allows maintenance of its structure as a rather stable bilayer.

The characteristics of human blood types A, B and O are the result of carbohydrate chains. Carbohydrate chains allow the body to work out which cells belong to it (self) and which cells are from outside the body (non-self). This is especially important in procedures involving transplants. If the carbohydrate chains of the transplanted tissue or organ are not compatible, rejection will occur. Rejection means the receiving patient's immune system attacks the foreign cells, resulting in the possible failure of the transplant.

The **glycocalyx** is a thin sugar layer made up of carbohydrate chains attached to proteins that can cover a cell. It is common in animal cells. This animal cell "sugar coat" has many functions, including cell to cell adhesion, cell to cell recognition and reception of various signalling chemicals. The glycocalyx is also present on the surface of many bacterial and fungal cells, where it may have both adhesion and protective functions. When a glycocalyx occurs in plant cells, it often appears to help anchor the plant cell membrane to the cell wall.

B2.1.10 – The fluid mosaic model

B2.1.10 – Fluid mosaic model of membrane structure
Students should be able to draw a two-dimensional representation of the model and include peripheral and integral proteins, glycoproteins, phospholipids and cholesterol. Indicate hydrophobic and hydrophilic regions.

Figure 4 shows the fluid mosaic model of the cell membrane. You should be familiar with all the parts of a membrane by now, and how they work together to form a selectively permeable barrier.

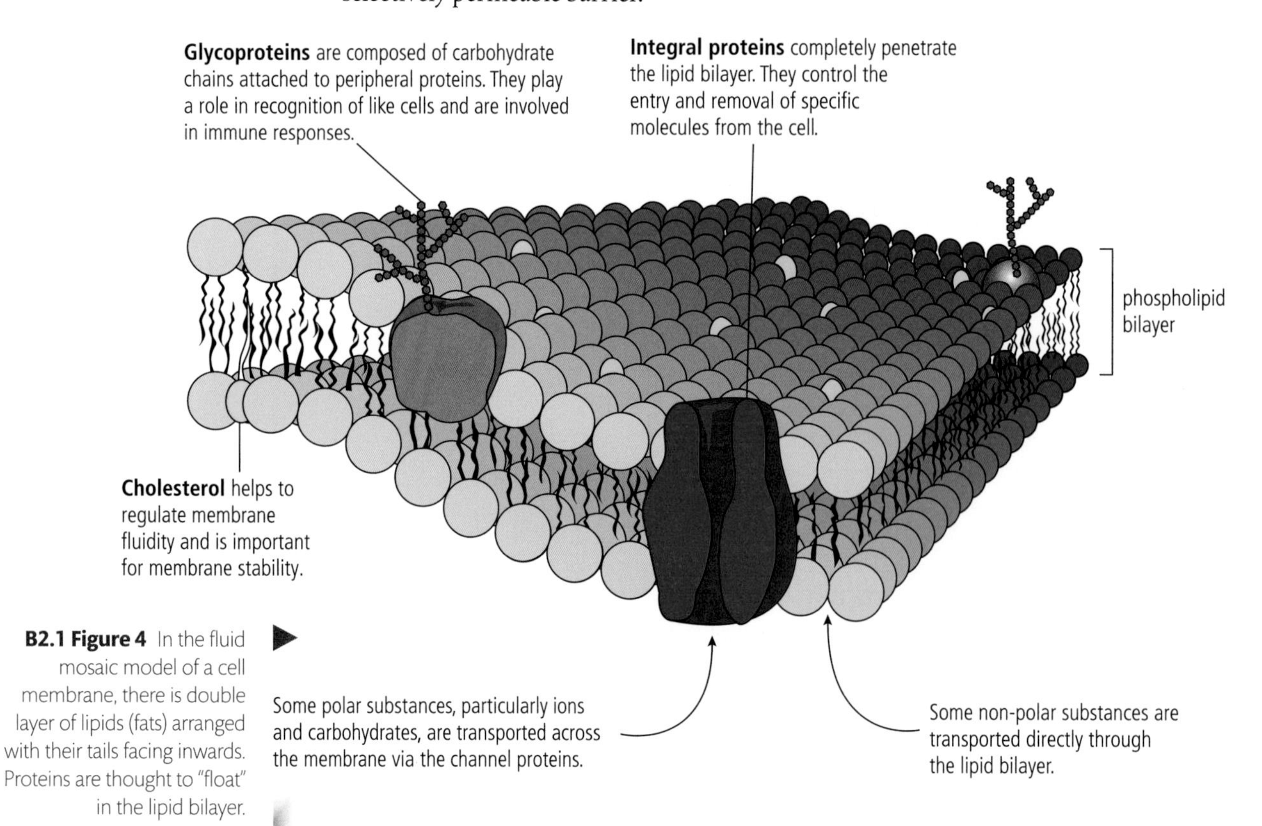

B2.1 Figure 4 In the fluid mosaic model of a cell membrane, there is double layer of lipids (fats) arranged with their tails facing inwards. Proteins are thought to "float" in the lipid bilayer.

Cholesterol

Membranes must be fluid to function properly. They are a bit like olive oil in their consistency. Cholesterol molecules can be found at various locations in the hydrophobic region (fatty acid tails) of animal cells. These molecules have a role in determining membrane fluidity, which changes with temperature. The cholesterol molecules allow membranes to function effectively at a wider range of temperatures than if they were not present. They do this by interacting with the tails of the phospholipid bilayer. Plant cells do not have cholesterol molecules; they depend on saturated or unsaturated fatty acids to maintain proper membrane fluidity.

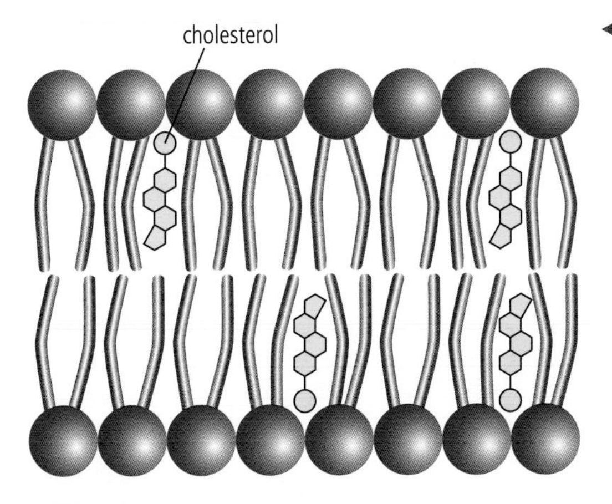

Notice the position of the cholesterol molecules. They are closely associated with the phospholipid tails in animal membranes.

Study the fluid mosaic model in Figure 4. Practise drawing it with the following structures correctly labelled: phospholipids, integral proteins, peripheral proteins, glycolipids, glycoproteins and cholesterol. All the structures must be properly positioned within the drawing to earn marks in an exam. You should show the structure of a phospholipid as a circle with two parallel tails attached. You should also be able to identify the hydrophobic and hydrophilic regions.

HL

B2.1.11 – Fatty acids and membrane fluidity

B2.1.11 – Relationships between fatty acid composition of lipid bilayers and their fluidity
Unsaturated fatty acids in lipid bilayers have lower melting points, so membranes are fluid and therefore flexible at temperatures experienced by a cell. Saturated fatty acids have higher melting points and make membranes stronger at higher temperatures. Students should be familiar with an example of adaptations in membrane composition in relation to habitat.

Phospholipids form bilayers based on their amphipathic structure. The polar heads of the molecules face aqueous environments because of their hydrophilic properties, while the hydrophobic tails form the inner layer away from water. The lipid bilayer is therefore stable because of its relationship with water involving hydrogen bonding. Only the polar heads of the phospholipids are capable of hydrogen bonding with water molecules. Because of the weakness of hydrogen bonds, individual phospholipids and unanchored proteins associated with them are relatively free to move about. This allows fluidity of the cell membrane.

Fatty acids with double bonds and fewer attached hydrogen atoms are said to be **unsaturated**. The double bonds within the fatty acid tails cause the molecules to become less straight and they do not pack together as tightly. They have lower melting points, which allows them to survive cooler temperatures. When the surrounding temperatures are higher, the cell membrane fatty acids possess mostly **saturated** bonds. Saturated bonds mean the fatty acid is straighter in shape, allowing a denser arrangement of the phospholipid layer. This increased density makes the membrane stronger and more able to remain effective at higher temperatures.

The exterior plasma membrane of cells often has a different fatty acid composition to the interior cell membranes, because the interior of the cell is less exposed. Unlike multicellular organisms that maintain constant internal temperatures, individual cells such as bacteria are vulnerable to drastic changes in temperature. For their membranes to always be effective, bacteria have evolved mechanisms to retain the membrane fluidity. Enzymes known as **fatty acid desaturases** exist in bacterial membranes and speed up reactions that result in an increase in double bonds within the fatty acid tails.

Challenge yourself

2. Suggest what the fatty acids in the membranes of the endoplasmic reticulum may be like and how this may affect the membranes they form.

3. Look at the two drawings below. Explain which drawing is of a plasma membrane and which is of an endoplasmic reticulum membrane.

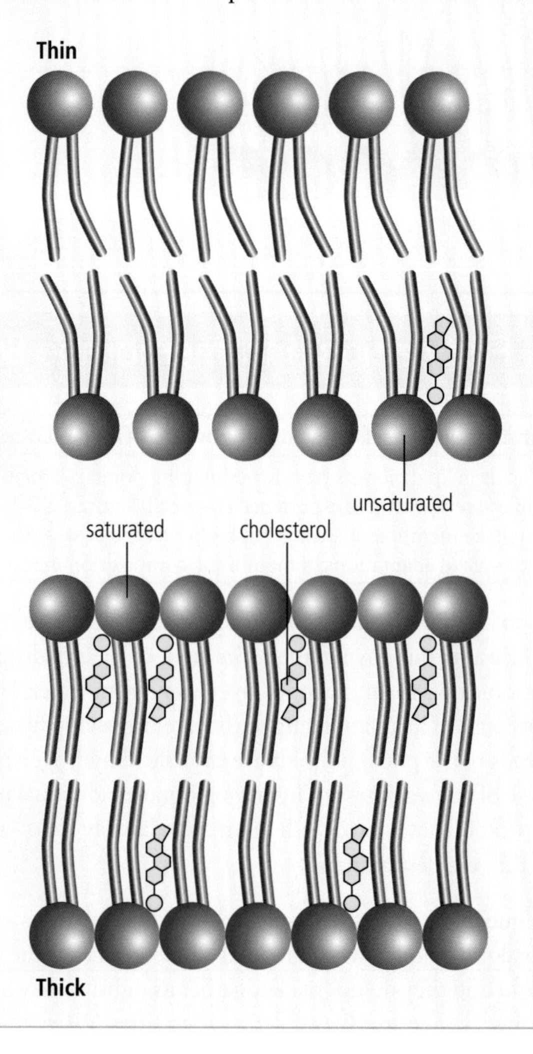

B2.1.12 – Cholesterol and membrane fluidity

B2.1.12 – **Cholesterol and membrane fluidity in animal cells**
Students should understand the position of cholesterol molecules in membranes and also that cholesterol acts as a modulator (adjustor) of membrane fluidity, stabilizing membranes at higher temperatures and preventing stiffening at lower temperatures.

In Figure 4 you may have noticed the cholesterol molecules in the membrane. Cholesterol molecules are closely associated with the fatty acid tails, and they have a large effect on plasma membrane fluidity. They act to stabilize membranes at higher temperatures and maintain flexibility at lower temperatures. There are more cholesterol molecules in the plasma membrane than the endoplasmic reticulum (ER) membrane because the plasma membrane is subjected to more extremes in temperature. Plant cells have very little, if any, cholesterol in their cell membranes. They possess cell walls that help stabilize their plasma membranes, thereby eliminating the need for cholesterol.

B2.1.13 – Bulk transport and membrane fluidity

B2.1.13 – **Membrane fluidity and the fusion and formation of vesicles**
Include the terms "endocytosis" and "exocytosis", and examples of each process.

Endocytosis and **exocytosis** are processes that allow large molecules and large amounts of material to move across the plasma membrane. Endocytosis allows macromolecules to enter the cell, while exocytosis allows molecules to leave. Both processes depend on the fluidity of the plasma membrane. It is important to recall why the cell membranes are fluid in consistency: the phospholipid molecules are not closely packed together, largely because of the rather "loose" connections between the fatty acid tails. It is also important to remember why the membrane is quite stable: the hydrophilic and hydrophobic properties of the different regions of the phospholipid molecules cause them to form a stable bilayer in an aqueous environment.

Examples of endocytosis include:

- **phagocytosis, the intake of large particulate matter**
- **pinocytosis, the intake of extracellular fluids.**

Endocytosis occurs when a portion of the plasma membrane is pinched off to enclose macromolecules or particulates within a vesicle in the cell. This pinching off involves a change in the shape of the membrane. The result is the formation of a vesicle that then enters the cytoplasm of the cell. The ends of the membrane reattach because of the hydrophobic and hydrophilic properties of the phospholipids and the presence of water. This could not occur if the plasma membrane did not have a fluid nature.

Exocytosis is essentially the reverse of endocytosis, so the fluidity of the plasma membrane and the hydrophobic and hydrophilic properties of its molecules are just as important. One example of exocytosis involves proteins that are produced in the cytoplasm of a cell and will eventually be excreted. Protein exocytosis usually begins in the ribosomes of the rough ER and progresses through a series of steps, outlined below.

1. Protein produced by the ribosomes of the rough ER enters the **lumen** (the inner space) of the ER. The protein is packed into a vesicle.
2. The vesicle carrying the protein fuses with the ***cis*** side of the Golgi apparatus.

Examples of exocytosis occur when:
- pancreas cells produce insulin and secrete it into the bloodstream (to help regulate blood glucose levels)
- neurotransmitters are released at synapses in the nervous system.

3. As the protein moves through the Golgi apparatus, it is modified and exits on the ***trans*** face inside another vesicle.
4. The vesicle with the modified protein inside moves towards and fuses with the plasma membrane, resulting in the secretion of the contents from the cell.

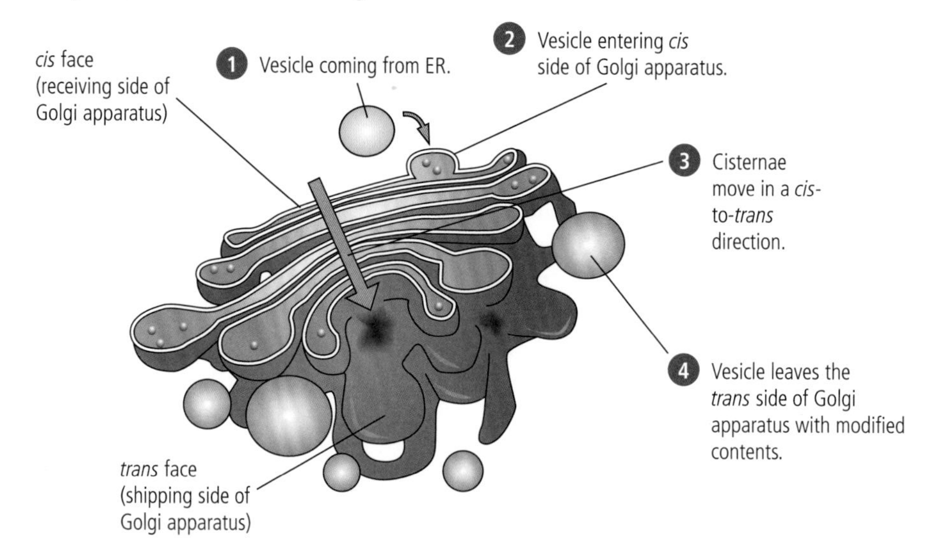

How the Golgi apparatus functions.

The fluidity of the plasma membrane is essential to allow fusion and subsequent secretion of the vesicle contents. At this point the vesicle membrane is part of the plasma membrane.

B2.1.14 – Gated ion channels and cellular transport

B2.1.14 – Gated ion channels in neurons
Include nicotinic acetylcholine receptors as an example of a neurotransmitter-gated ion channel and sodium and potassium channels as examples of voltage-gated channels.

Curare is a muscle relaxant used during surgery. It binds to nicotinic acetylcholine receptors and prevents neurotransmitters from opening the pores to initiate an impulse. Local anaesthetics bind with receptors and prevent conduction through the pores.

There are specialized channels that allow ions to pass quickly through cell membranes. Most of these channels have openings that can be opened or closed as a result of chemical and electrical stimuli, and these channels are said to be **gated**. Movement of ions through these channels controls the electrical potential across membranes, especially in neurons and muscles.

Nicotinic acetylcholine receptors are an example of a **neurotransmitter-gated** or chemically gated ion channel. These gated channels can bind to the neurotransmitter acetylcholine. Neurotransmitters are chemicals that allow signals to pass between two nerves at junctions called synapses (neurotransmitters are discussed in more depth in Chapter C2.1). Neurotransmitters also function at the junctions between nerves and muscles. When acetylcholine attaches to a nicotinic acetylcholine receptor, the channel through the membrane is opened and positive ions such as Na^+, K^+ and Ca^{2+} can pass through. This causes the membrane potential to change so that an impulse can be generated. Nerve impulses can be carried along multiple, connected neurons within the body so that responses are possible. When the neurotransmitter is released at the junction between a nerve and a muscle, the opening of this receptor and the movement of the positive ions can cause muscle movement.

Curare is used by Indigenous peoples in South America. How is it that scientific knowledge is often shared by large, geographically spread and culturally diverse groups?

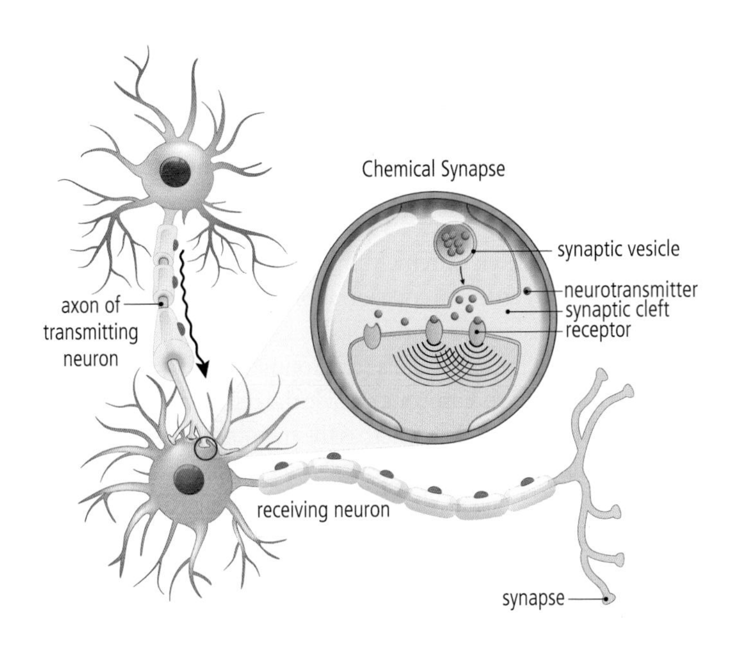

A representation of the transmission of a nerve impulse between two neurons. A neurotransmitter is released by the sending neuron and is bound by receptors on the receiving neuron. A similar process occurs at a junction between a nerve and a muscle cell to cause a response.

The autoimmune disorder known as myasthenia gravis produces antibodies to nicotinic acetylcholine receptors. These antibodies bind to the receptors, reducing the individual's response to the neurotransmitter, acetylcholine. Incomplete muscle movements and responses then become common.

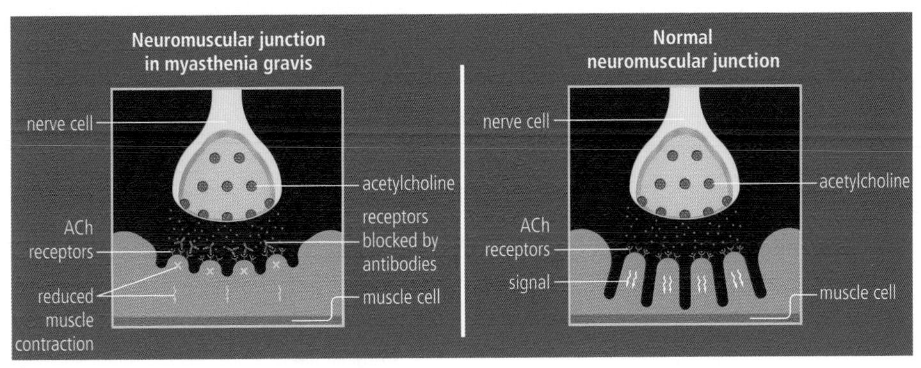

These images illustrate the differences between the normal process of impulse transmission at a nerve–muscle junction and the transmission caused by the condition known as myasthenia gravis. Note the disabling of the receptors by the antibodies produced by myasthenia gravis.

Capsaicin is the chemical in chili peppers that many of us sense as extreme heat. Capsaicin activates the same receptors as high temperatures, resulting in the same sensation.

Channel proteins may also be **voltage-gated**. Voltage-gated channel proteins are opened by changes in membrane polarity. Na^+ and K^+ channels are examples of voltage-gated protein carriers. An electrical stimulus opens and closes the gates on these proteins. Although they only remain open for very short periods of time, they allow the specific ion to move rapidly through them. Sodium channels open first, and sodium ions move from the outside of the neuron to the inside. This depolarizes the membrane. The sodium channels then close quickly. Potassium channels open more slowly. Once the potassium channels are open, potassium ions move from the inside of the cell to the outside and the membrane returns to its normal potential. The process of re-establishing membrane potential after depolarization is called repolarization.

Defective K^+ voltage-gated protein channels can cause a delay in the repolarization of heart muscle cells. When this occurs, the patient is predisposed to abnormal cardiac rhythms and possible sudden death.

B2.1.15 – The sodium–potassium pump

B2.1.15 – Sodium–potassium pumps as an example of exchange transporters
Include the importance of these pumps in generating membrane potentials.

The sodium–potassium pump is an extremely important example of active transport. It uses ATP to move ions directly against a concentration gradient. Animal cells have a much higher concentration of potassium ions inside their cells than outside their cells. Sodium ions are more concentrated in the extracellular environment than within the cells. The higher concentration of positive sodium ions outside the cell than positive potassium ions inside the cell creates a difference in charge across the membrane, called the **membrane potential**. This is especially important in nerve cells (neurons). A nerve impulse takes place as a result of sodium ions diffusing into the cell through specialized channels, creating a change in charge called depolarization. This depolarization spreads down the neuron, generating a nerve impulse. Potassium ions diffuse out of the cell during the second part of the process, to repolarize the cell. The cell maintains the correct concentrations of sodium and potassium using the sodium–potassium pumps. This process needs energy and happens across the cell membrane. The sodium–potassium pumps are exchange transporters (antiporters) and do not directly take part in the formation of a nerve impulse (which relies on voltage-gated channel proteins, covered in Section B2.1.14).

The membrane potential is the electrical charge differential between the interior and exterior of the cell. It is maintained by the sodium–potassium pump. The mechanism for this essential pump has five stages.

1. The pump protein with an attached ATP molecule binds to three intracellular sodium ions. ATP is the chemical that provides usable energy to the cell.

2. The binding of sodium ions causes the pump (protein carrier) to split ATP, which provides usable energy and leaves a phosphate attached to the carrier. The addition of a phosphate is called phosphorylation. ATP has three attached phosphates. When it carries out phosphorylation of the pump, it loses a phosphate and becomes **adenosine diphosphate** (ADP). ADP has only two phosphates. ATP and ADP are discussed more thoroughly in relation to cell respiration in Chapter C1.2.

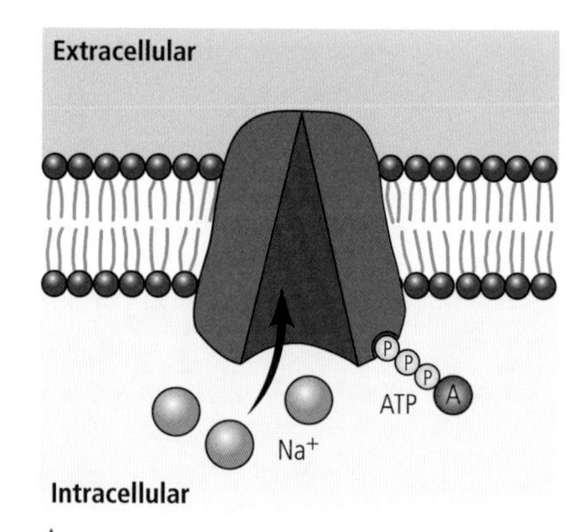

▲ Stage 1: a protein carrier in a phospholipid bilayer opens to the intracellular side and attaches three sodium ions.

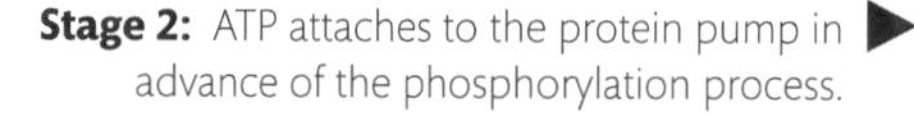

Stage 2: ATP attaches to the protein pump in advance of the phosphorylation process. ▶

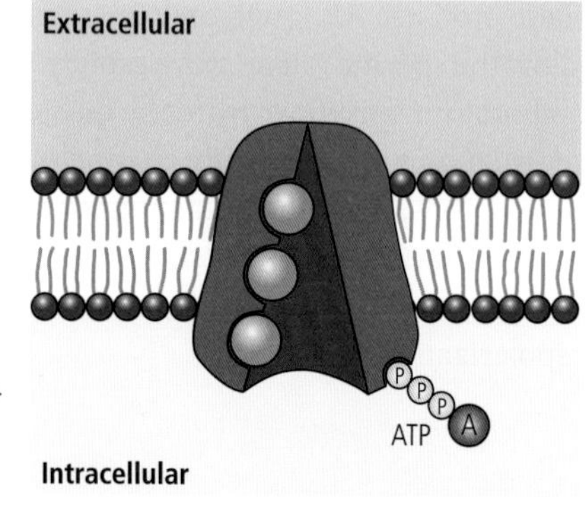

3. The phosphorylation causes the protein to change its shape, thus expelling sodium ions into the extracellular fluid. At this point, the protein pump has a low affinity for sodium ions but the shape change results in a high affinity for potassium ions.

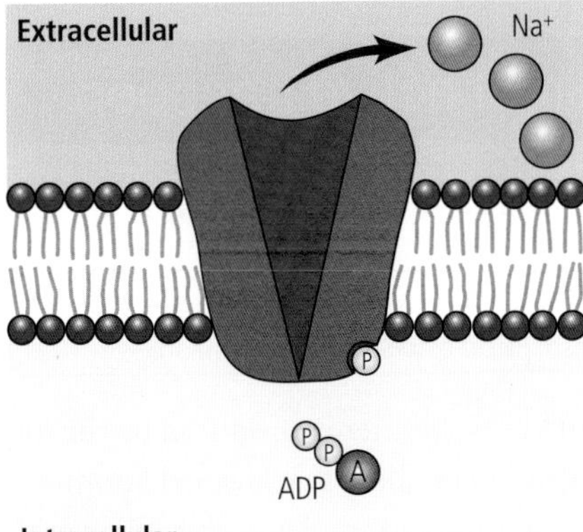

Stage 3: phosphorylation causes the protein carrier to open to the exterior of the cell and the sodium ions are released. The phosphate remains attached to the carrier while the ADP is released to be free in the intracellular space.

4. Two extracellular potassium ions bind to different regions of the protein, causing the release of the phosphate group.

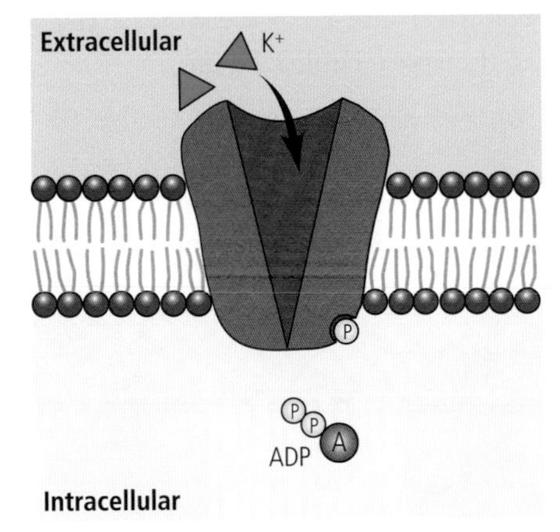

Stage 4: extracellular potassium ions attach to the protein.

5. The loss of the phosphate group restores the protein's original shape, thus causing the release of the potassium ions into the intracellular space. The carrier is now ready to repeat the process.

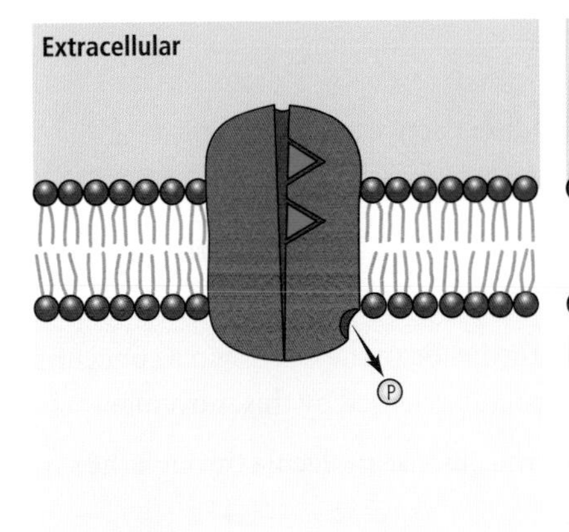

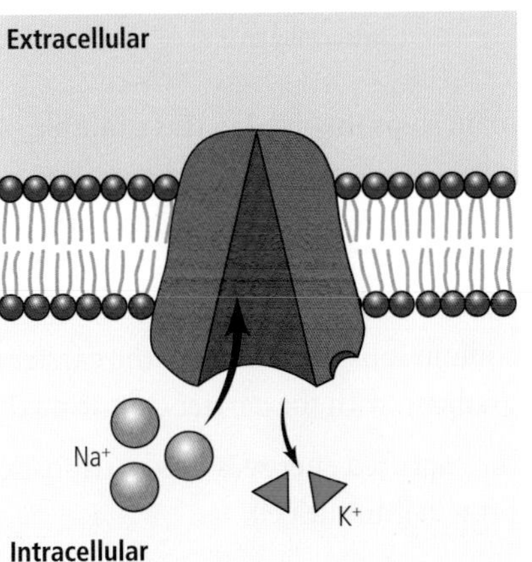

Stage 5: the protein opens towards the cell interior again and releases the potassium ions into the exterior.

In every cycle, three sodium ions leave the cell and two potassium ions enter. Research has shown as many as 300 sodium ions may be transported in one second by this pump. The carrier therefore changes shape rapidly. Different types of animal cells will have varying numbers of these sodium–potassium pump proteins.

B2.1.16 – Indirect active transport

B2.1.16 – Sodium-dependent glucose cotransporters as an example of indirect active transport
Include the importance of these cotransporters in glucose absorption by cells in the small intestine and glucose reabsorption by cells in the nephron.

Indirect active transport uses the energy produced by the movement of one molecule down a concentration gradient to transport another molecule against a gradient. One of the most common examples of this involves the transport of glucose into the cells lining the intestines of animals. Often, there is a higher concentration of glucose inside the cells than outside. Therefore energy in the form of ATP must be provided for this transport to occur. While glucose is being moved, sodium and potassium ions are also being transported by the same carrier proteins. This system is also referred to as coupled transport. Look at Figure 5.

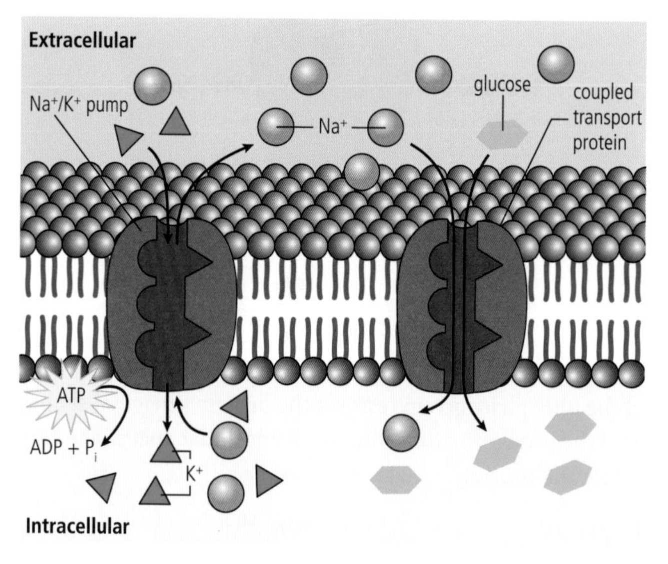

B2.1 Figure 5 This diagram shows how glucose enters the epithelial cells of the intestine against a concentration gradient. At the same time sodium and potassium ions are moved across the cell membrane. This is an example of indirect active transport, also known as **coupled transport**. There are two protein pumps involved here. The ATP produced by the sodium–potassium (Na^+/K^+) pump is needed for the coupled glucose transport protein.

The main steps involved in this example of indirect active transport are:

1. there are more sodium ions outside than inside the intestinal cell
2. sodium ions and glucose molecules bind to a specific transport protein on the extracellular surface
3. sodium ions pass through the carrier to the inside of the cell down a concentration gradient, with the carrier capturing the energy released by this movement
4. the captured energy is used to transport the glucose molecule through the same protein into the cell.

This protein carrier is known as a **sodium-dependent glucose transporter** or sodium-glucose linked transporter (SGLT). The same type of carrier occurs in the functional units of the kidney, nephrons. These two carriers are differentiated by calling the intestinal transporter SGLT1 and the nephron transporter SGLT2. SGLT2 is responsible for most of the glucose reabsorption in the kidney. It works in the same way as SGLT1, needing the sodium ion transport to provide the energy necessary for glucose transport. In the kidney, this carrier decreases glucose in the urine by allowing the uptake of glucose from the kidney filtrate.

In certain types of diabetes, SGLT2 expression is enhanced, resulting in an increase in glucose absorption from the kidney filtrate back into the bloodstream. The result is a condition called **hyperglycaemia**, high blood glucose levels. The aim of research on SGLT2 inhibitors is to decrease the reabsorption of glucose, with the positive result of lowering blood glucose levels.

B2.1.17 – Cell adhesion

B2.1.17 – Adhesion of cells to form tissues
Include the term "cell-adhesion molecules" (CAMs) and the understanding that different forms of CAM are used for different types of cell–cell junction. Students are not required to have detailed knowledge of the different CAMs or junctions.

What processes depend on active transport in biological systems?

Multicellular organisms are dependent on the adhesion of cells to other cells. These adhesive interactions can happen via the plasma membrane and can be stable or temporary. In humans, skin and muscle cells must bind tightly to one another to fulfil their function successfully. A **cell-adhesion molecule** (CAM) is usually involved in cell connections. There are several different types of CAM, and each is used for a slightly different cell to cell junction.

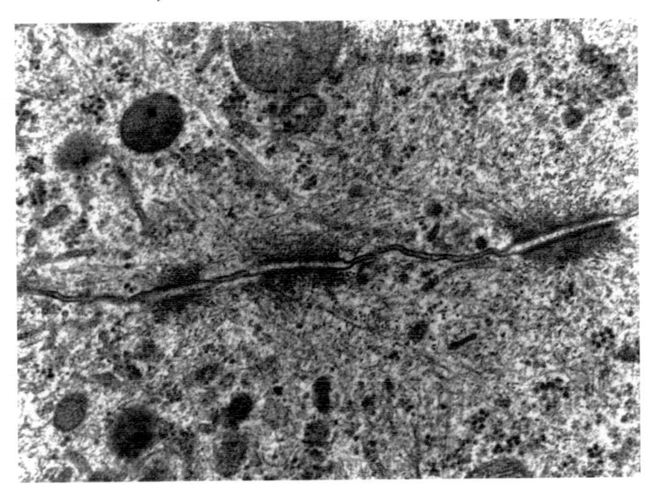

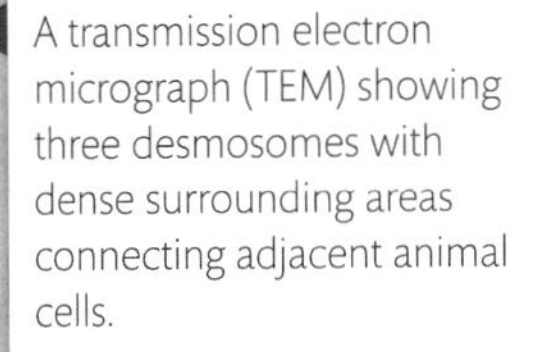

A transmission electron micrograph (TEM) showing three desmosomes with dense surrounding areas connecting adjacent animal cells.

In general, cell connections allow coordinated behaviour and have important structural functions. **Desmosomes** help form sturdy but flexible sheets of cells in organs such as the heart, stomach and bladder. Tissues in these organs get stretched, but the desmosomes hold the cells together. Other types of connections involve channels formed between adjacent cells that allow small molecules and ions to pass between them. This not only joins the cells together but also allows the cells to communicate.

Plant cells often produce **plasmodesmata**, which are tubes connecting the cytoplasm of adjacent cells. These tubes allow the exchange of materials, especially water and small solutes, between connected cells.

HL end

Guiding Question revisited

How do molecules of lipid and protein assemble into biological membranes?

In this chapter we have examined how:

- amphipathic lipids and phospholipids form continuous sheet-like bilayers when in water
- the bilayer forms because the hydrophilic portion of the phospholipid molecules is attracted to water, while the hydrophobic portion of the molecules faces inwards away from contact with water
- integral proteins are embedded in one or both lipid bilayers of a membrane
- peripheral proteins are attached to one or other surface of the lipid bilayer
- cholesterol is often present near the phospholipid tails of the cell membrane and has a role in the control of membrane fluidity
- glycoproteins and glycolipids have carbohydrate structures attached to them and often have roles in cell adhesion and recognition.

Guiding Question revisited

What determines whether a substance can pass through a biological membrane?

In this chapter we have discussed how:

- cell membranes have low permeability to large molecules and hydrophilic molecules
- diffusion and osmosis are examples of passive transport
- protein pumps allow the movement of materials across the cell membrane that otherwise would not be able to pass because of their chemical properties
- aquaporins are important in allowing polar water molecules to pass through cell membranes

HL

- channel proteins have pores, often with controlling gates, that allow ions and other polar materials to pass through cell membranes
- facilitated transport involves carrier proteins or channel proteins that aid in the movement of materials across the cell membrane.

HL end

Exercises

Q1. Which **one** of the following is an example of active transport?

A Facilitated diffusion.

B Osmosis.

C Movement of water through aquaporins.

D Sodium–potassium pump at work.

Q2. Which type of compound occurs only on the surface of the cell membrane?

A Integral proteins.

B Carbohydrates.

C Cholesterol.

D Phospholipids.

Q3. Explain the orientation of the bilayer of phospholipid molecules in the plasma membrane using the terms hydrophobic and hydrophilic.

Q4. Why does a diet high in plants and plant products have relatively low cholesterol levels compared to a diet involving high amounts of animal products?

Q5. Which of the following is *not* a function of the cell membrane?

A Cell adhesion.

B Enzyme synthesis.

C Active transport of specific substances.

D Pump materials against a concentration gradient.

Q6. Name the structures found in the cell membrane that are involved with the transport of water.

Q7. What is the connection between carrier proteins and active transport?

HL

Q8. Describe what happens during the bulk transport of an enzyme out of a cell in the small intestine.

Q9. Explain how the structure of the cell membrane allows exocytosis to take place.

Q10. Explain why the enzymes stay within the vesicle.

HL end

B2.2 Organelles and compartmentalization

Guiding Questions

How are organelles in cells adapted to their functions?

What are the advantages of compartmentalization in cells?

Cells are the building blocks of all life forms. They come in a myriad of sizes and shapes. All are surrounded and protected by a multi-functional membrane, discussed in Chapter B2.1. Within most cells are highly specialized structures carrying out functions essential to the cell and/or organism. These specialized structures are called organelles. Organelles are adapted to their function. For example, mitochondria have infoldings of the inner membrane so that they have a larger internal surface area. This allows the reactions responsible for respiration to take place more efficiently. Some organelles are thought to have originated outside the cells they exist in today. A cell's function is reflected in the types and numbers of organelles present, leading to cell specialization, which is discussed in Chapter B2.3.

Most organelles are membrane bound, allowing compartmentalization within the cell. Compartmentalization allows unique processes to proceed without interference from chemicals or reactions occurring nearby in the cell. We will discuss these discrete cellular compartments in this chapter.

B2.2.1 – Cell compartmentalization

B2.2.1 – Organelles as discrete subunits of cells that are adapted to perform specific functions
Students should understand that the cell wall, cytoskeleton and cytoplasm are not considered organelles, and that nuclei, vesicles, ribosomes and the plasma membrane are. **NOS:** Students should recognize that progress in science often follows development of new techniques. For example, study of the function of individual organelles became possible when ultracentrifuges had been invented and methods of using them for cell fractionation had been developed.

The cell is a small but very busy unit, with different reactions occurring in close proximity within it. Selectively permeable membranes play an important role in allowing these reactions/functions to occur without interfering with one another. This isolation of reactions is referred to as **cell compartmentalization**, the result of which is that cells work much more efficiently than if all the reactions were mixed up together. Much of our research concerning the cell focuses on how cells work at the molecular level. The best way to do this is to reduce the cell to its component parts and study each part individually. This approach is known as **reductionism**. By studying localized parts and reactions, we can develop an understanding of the overall complex reactions of the cell.

Cell compartmentalization refers to the division of a cell into regions or compartments with single or double membranes between them.

Tools for cell research

The development of imaging in cell research is discussed in Chapter A2.2. Our understanding of the cell has advanced tremendously with improvements in light microscopes and electron microscopy and refinements in preparation techniques.

Another tool used in the study of the cell involves a process called **biochemical fractionation**. Fractionation refers to the separation and isolation of specific chemicals and/or structures so that detailed research can be carried out. Several techniques have been developed for cell research, each allowing the separation of different parts of the cell.

Centrifugation or **cell fractionation** allows the extraction of organelles from cells. Ultracentrifuges are often used for this process. Cells are first mixed in a tube with substances that break down the cell membranes. The sample is then spun at high speeds to isolate the different components by size and shape. Larger and heavier cell components can be separated off at lower speeds. Once separated, larger and heavier organelles are found at the bottom of the tube.

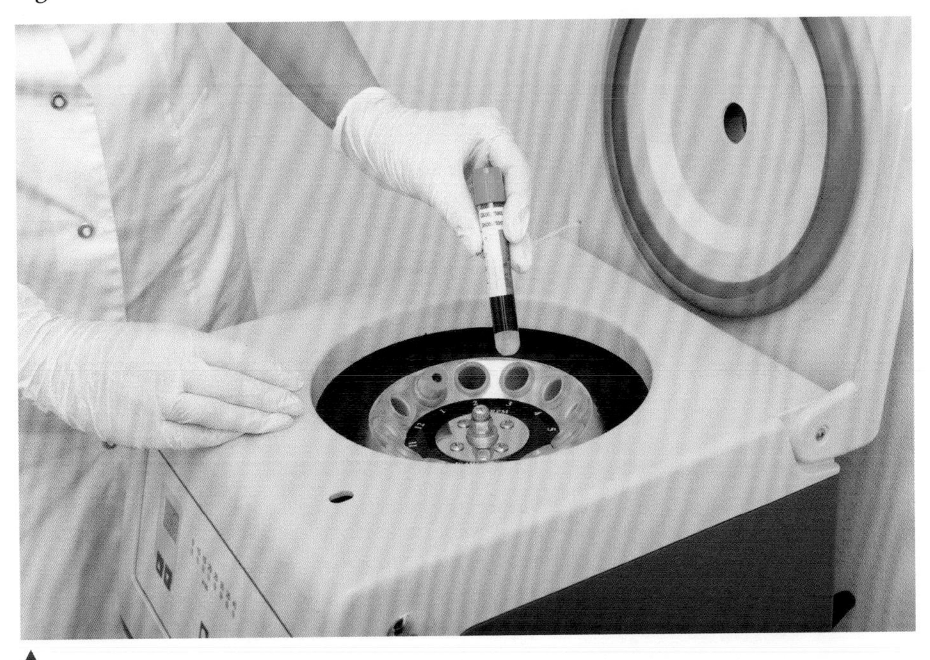

▲ A centrifuge separates components of a sample by spinning it at high speeds. Separation occurs because of the different densities of the component parts.

Chromatography is very effective at isolating pure substances such as amino acids, proteins, carbohydrates and plant pigments. A mixture of molecules is placed in a separating medium. The molecules separate out depending on their size and the speed with which they travel through the medium. There are several different types of chromatography, including gel and ion exchange chromatography.

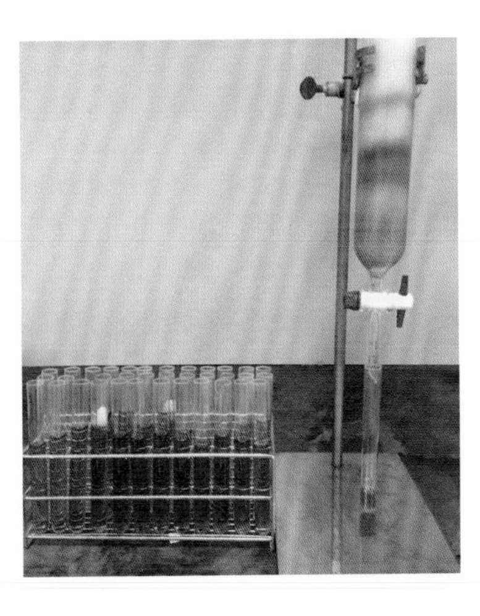

◀ An example of column chromatography. Notice the colours in the column, which indicate different substances isolated from the original compound in the test tubes.

Gel electrophoresis separates molecules of different types by passing them through a gel using an electrical charge. The molecules are separated based on properties such as size and charge. This technique is commonly used in studies involving nucleic acids.

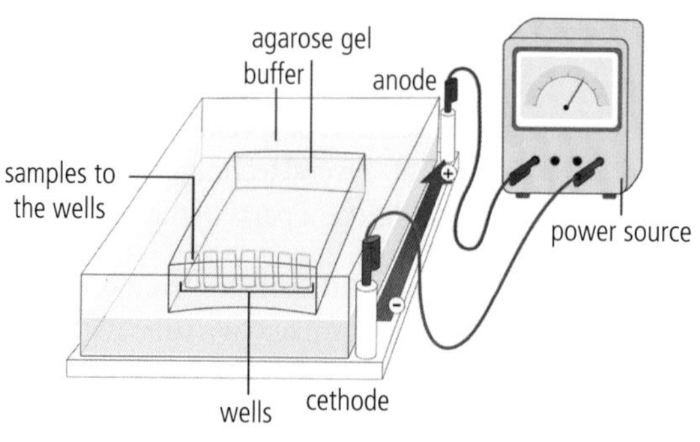

An example of the apparatus used in gel electrophoresis to isolate molecules by size and charge.

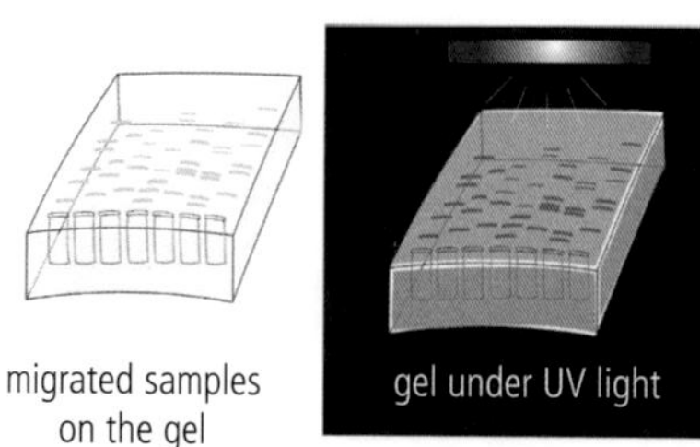

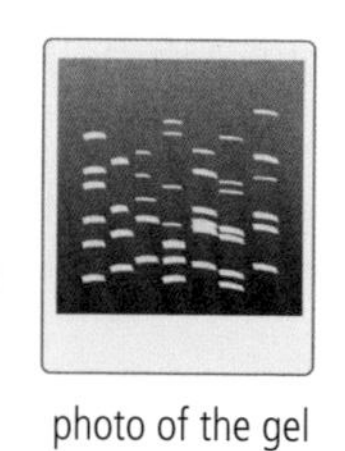

This image of DNA fingerprints shows the high visibility of fragments that can be achieved when using the fluorescent dye known as ethidium bromide in gel electrophoresis. Ultraviolet light is provided to achieve the visibility.

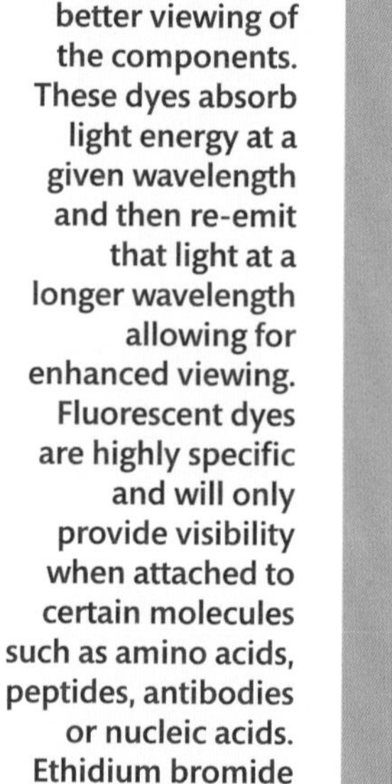

Fluorescent dyes are often used in cell fractionation techniques to allow better viewing of the components. These dyes absorb light energy at a given wavelength and then re-emit that light at a longer wavelength allowing for enhanced viewing. Fluorescent dyes are highly specific and will only provide visibility when attached to certain molecules such as amino acids, peptides, antibodies or nucleic acids. Ethidium bromide is a fluorescent dye often used to observe DNA fragments produced by gel electrophoresis.

DNA fingerprinting using gel electrophoresis techniques is used around the world to convict criminals and identify victims. Specific segments of DNA are examined to determine relationships and identities. It is even possible to determine an individual's ancestral history using DNA analysis.

What separation techniques are used by biologists?

Nature of Science

Science has progressed and continues to progress with the development of new study techniques. Not only has the microscope increased our knowledge of the cell, but ultracentrifuges and fractionation of cells have also greatly enhanced our understanding of the cell and its organelles.

Organelles: the compartments of the cell

Chapter A2.2 discusses the organelles of the cell. They are separate structures within the cell that carry out specialized functions. To carry out these particular functions, each type of organelle has a unique structure. Organelles are separated from the rest of the cell by a protective barrier, sometimes involving two membranes. This barrier is important because it allows the chemical reactions that take place in an organelle to happen without interference from the rest of the cell. However, not all parts of a cell are considered organelles.

What are examples of structure-function correlation at each level of biological organization?

Study Table 1, which summarizes the function of different cell organelles. See Chapter A2.2 for more details about the organelles.

Organelles and compartmentalization. Full details on how to carry out this activity with a worksheet are available in the eBook.

Component	Organelle	General function
Cell wall	No	Encloses and protects plant cells
Cytoskeleton	No	Maintains cell shape, anchors organelles, facilitates cell movement
Cytoplasm	No	The region where most of the metabolic reactions in the cell occur
Nucleus	Yes	Genetic control
Vesicles	Yes	Storage and transport
Ribosomes	Yes	Protein synthesis
Plasma membrane	Yes	Regulates movement in and out of cell, transports materials to maintain the internal cell environment, cell recognition and communication
Cilia/flagella	Yes	Movement
Golgi apparatus	Yes	Modifies and stores endoplasmic reticulum products, forms lysosomes and transport vesicles
Mitochondria	Yes	Cellular energy (ATP) production
Chloroplasts	Yes	Conversion of light energy into chemical energy
Lysosomes	Yes	Digest worn out organelles and debris, digest materials brought into the cell by endocytosis

B2.2 Table 1 Different components of a cell

B2.2.2 – The nucleus and cytoplasm

B2.2.2 – Advantage of the separation of the nucleus and cytoplasm into various compartments
Limit to separation of the activities of gene transcription and translation – post-transcriptional modification of mRNA can happen before the mRNA meets ribosomes in the cytoplasm. In prokaryotes this is not possible – mRNA may immediately meet ribosomes.

The development of the nucleus in eukaryotic cells was a huge advantage compared to prokaryotic cells because it allowed some of the important cell processes to take place more efficiently.

Transcription and **translation** are cell processes responsible for the production of proteins. In transcription, a DNA strand serves as a template or copy strand for the formation of messenger RNA (mRNA). Translation occurs when ribosomes use the code carried by mRNA to produce a polypeptide/protein. Transcription happens in the nucleus of eukaryotic cells, while translation is carried out in the cytoplasm. The separation of these two important cellular processes allows post-transcriptional modification of mRNA to occur in the nucleus before translation happens in the cytoplasm. In prokaryotic cells there is no isolation of these two processes, and mRNA can immediately come into contact with ribosomes and initiate translation without any modification occurring. (Transcription and translation are described in detail in Chapters D1.1 and D1.2. It is important to realize that this additional step of modification of the mRNA in eukaryotic cells decreases the chances of errors happening in the production of polypeptides.) It is the compartmentalization of the cell that allows this greater cell efficiency.

B2.2.3 – Compartmentalization of the cytoplasm

B2.2.3 – Advantages of compartmentalization in the cytoplasm of cells
Include concentration of metabolites and enzymes and the separation of incompatible biochemical processes. Include lysosomes and phagocytic vacuoles as examples.

All eukaryotic cells possess compartments or organelles that are involved with:

- energy production
- metabolism
- biosynthesis
- degradation.

However, it is important to note that the number and size of these compartments and organelles vary depending on the overall function of the cell in which they occur. For example, certain types of pancreatic cells called **acinar cells** specialize in the secretion of digestive enzymes. These cells are essential to life in humans and have a greatly enlarged endoplasmic reticulum (ER), Golgi apparatus and granule storage compartments.

Compartmentalization has allowed a division of labour within the cell, with specific tasks carried out by a single organelle or organelle-like structure. Enzymes can be kept in the areas where they will be most effective. Often, reaction pathways in cells rely on a series of enzyme-controlled reactions. Keeping reactions separate in different parts of the cell means that the metabolites and enzymes for each particular process can be concentrated in a particular area. This ensures that pathways run smoothly, can be easily controlled and do not interfere with each other.

Lysosomes participate in the breakdown of wastes and cellular components that need to be replaced. This breakdown requires some potentially destructive enzymes that could cause severe damage to the cell if they were not isolated by a membrane.

When endocytosis occurs, the result is often a phagocytic vacuole. This vacuole is a means of protecting the cellular contents from potential damage when phagocytosis occurs.

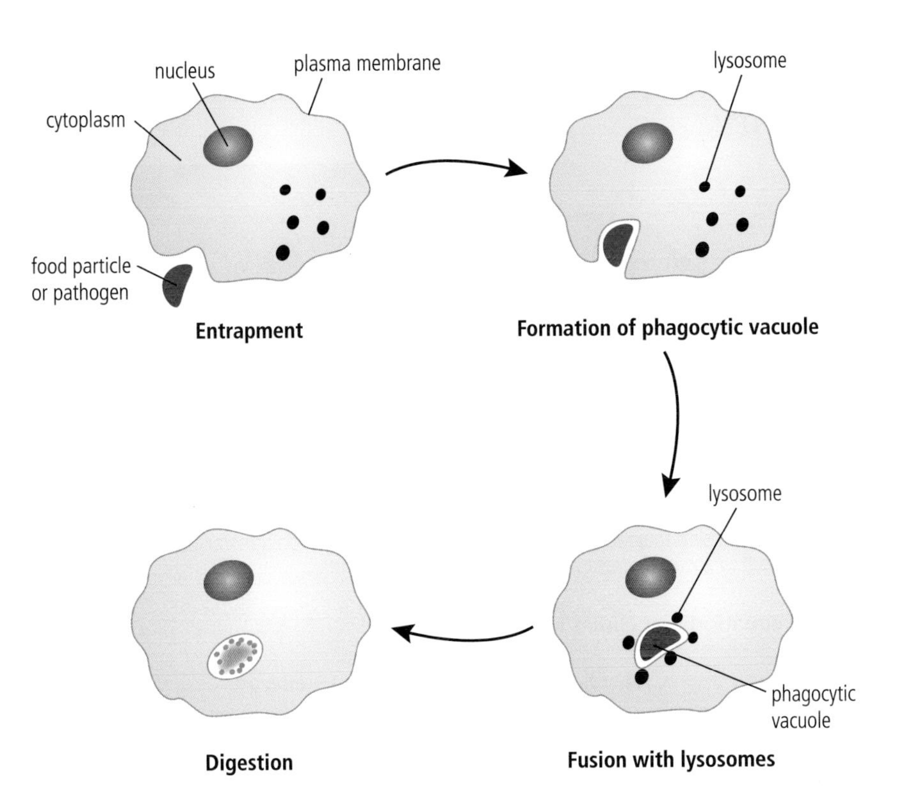

Phagocytosis occurring with a food particle or pathogen being brought into the cell by a form of active transport. Notice the lysosomes fusing with the phagocytic vacuole in order to digest the contents.

A pathogen is a disease-causing organism. Grave harm can be done to a cell if pathogens are not controlled.

Once formed, the phagocytic vacuole will move around in the cell until it contacts a lysosome. The vacuole then fuses with the acidic lysosome, allowing inactivation and digestion of the threat. Phagocytosis plays a key role in defending cells against invading pathogens.

Compartmentalization does present challenges, however. The very fact that each area or organelle carries out one specific function means that the cell must develop a means of integrating all the separate functions. To accomplish this, some organelles are connected in a functional series, allowing the chemical pathways important to the cell to take place. Membrane pumps and carriers have evolved so that the products of one organelle can enter another, and important cell reactions can occur. By overcoming such challenges, compartmentalization has greatly enhanced the successful existence of the cell.

HL

B2.2.4 – The mitochondrion

B2.2.4 – Adaptations of the mitochondrion for production of ATP by aerobic respiration
Include these adaptations: a double membrane with a small volume of intermembrane space, large surface area of cristae and compartmentalization of enzymes and substrates of the Krebs cycle in the matrix.

Cellular respiration is the process by which glucose is used to produce adenosine triphosphate (ATP); this molecule then provides the energy needed for cell reactions. Respiration is a very complex series of chemical reactions, most of which occur in the mitochondrion. You need to know how mitochondria are adapted to produce ATP.

Examine the transmission electron micrograph (TEM) and labelled drawing of a typical mitochondrion in Figure 1. Try to locate all the structures labelled in the diagram on the coloured TEM.

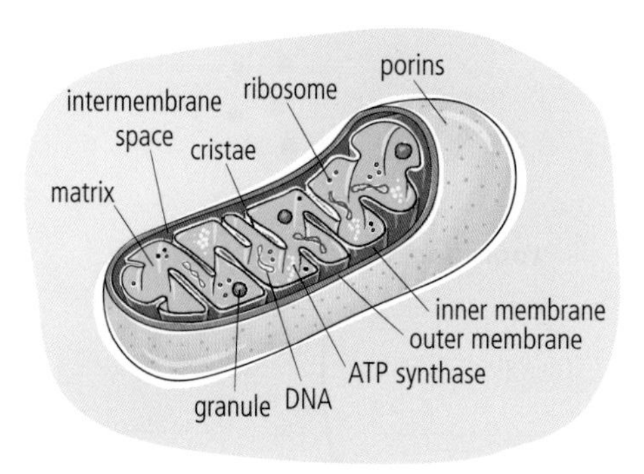

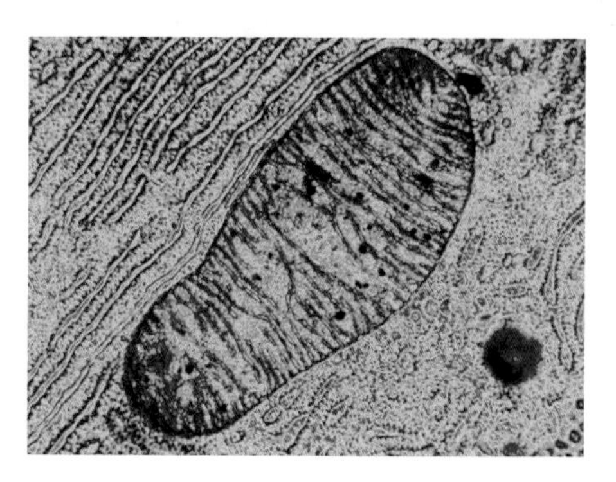

B2.2 Figure 1 A labelled diagram of a mitochondrion, and a coloured transmission electron micrograph (TEM) of a mitochondrion.

The overall equation for cellular respiration is:

$$C_6H_{12}O_6 + 6O_2 \rightarrow 6\ CO_2 + 6H_2O + \text{energy}$$

All organisms must have the ability to produce ATP for energy, and, therefore, all organisms carry out respiration. The equation presented above shows aerobic respiration because it includes oxygen as a reactant. Each part of the mitochondrion has a specific function that allows the process to occur.

The processes involved in respiration are discussed in Chapter C1.2. Table 2 lists the features of a mitochondrion and explains how that feature is linked to a specific function that allows the successful production of ATP by aerobic respiration.

Mitochondrion structure	Description and function
Outer mitochondrial membrane	A membrane that separates the contents of the mitochondrion from the rest of the cell
Matrix	An internal cytoplasm-like substance that contains the enzymes for the first stages of respiration that take place in the mitochondria (the link reaction and the Krebs cycle)
Cristae	Tubular regions surrounded by membranes that increase the surface area for reactions that take place towards the end of respiration (oxidative phosphorylation)
Inner mitochondrial membrane	A membrane that contains the carriers and enzymes for the final stages of respiration (electron transport chain and chemiosmosis)
Space between inner and outer membranes	A reservoir for hydrogen ions (protons), allowing a high concentration of protons

B2.2 Table 2 The functions of different structures in a mitochondrion

If there is a defect in any of the mitochondrial regions or structures, ATP production may be diminished if not eliminated. Mitochondrial defects in children can lead to muscle weakness and affect mental development.

B2.2.5 – The chloroplast

B2.2.5 – Adaptations of the chloroplast for photosynthesis
Include these adaptations: the large surface area of thylakoid membranes with photosystems, small volumes of fluid inside thylakoids, and compartmentalization of enzymes and substrates of the Calvin cycle in the stroma.

Some people refer to the chloroplast as a photosynthetic machine. They are not wrong. Unlike respiration, where some of the steps occur outside the mitochondrion, all of the photosynthetic process occurs within the chloroplast. This includes absorbing light. The chemical reactions involved in photosynthesis are studied in more detail in Chapter C1.3. Chloroplasts, along with mitochondria, represent possible evidence for the theory of endosymbiosis, discussed in Chapter A2.2. Both organelles have an extra outer membrane (indicating a need for protection in a potentially hostile environment), their own DNA, and they are very near in size to a typical prokaryotic cell. Look at the photomicrograph and diagram of a chloroplast in Figure 2.

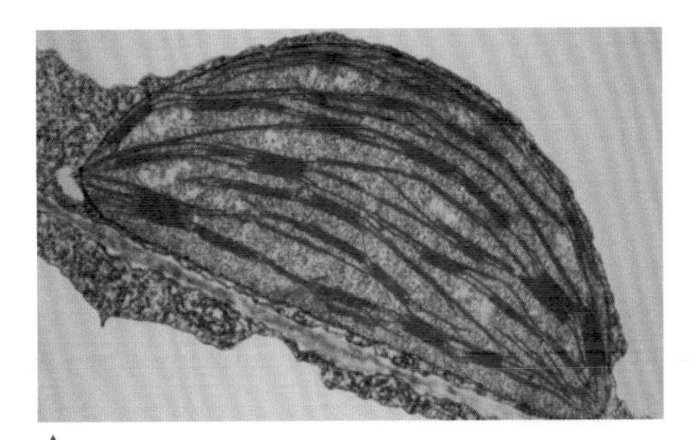

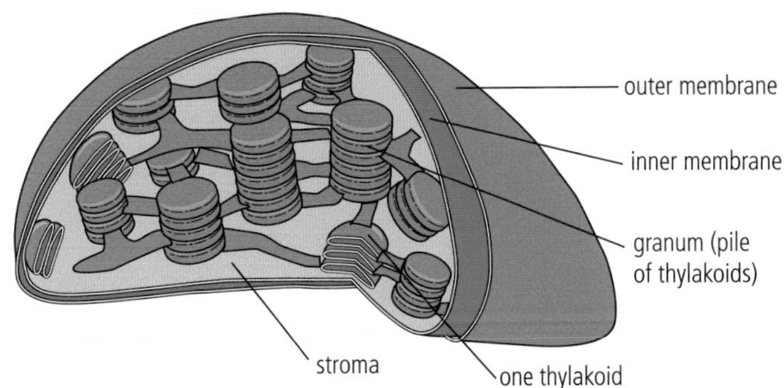

B2.2 Figure 2 A transmission electron micrograph (TEM) and diagram showing the structure of a chloroplast. Can you find all the parts labelled in the diagram in the TEM?

The structure of the chloroplast is discussed in Chapter A2.2. You may want to refer to Chapter A2.2 to remind yourself about this organelle. Chloroplasts occur mostly within the leaves. However, some plants have chloroplasts in the cells of other organs.

In respiration, chemical bonds in glucose are broken and the energy is used to produce ATP. In photosynthesis, chemical bonds are made to produce carbon compounds. The raw materials of photosynthesis are carbon dioxide and water, with light providing energy. Many enzymes are involved to enable the formation of glucose, more water and oxygen. The overall equation is:

$$6\,CO_2 + 12\,H_2O + \text{light} \rightarrow C_6H_{12}O_6 + 6\,H_2O + 6\,O_2$$

Water occurs on both sides of the equation. Twelve molecules are consumed, and six molecules are produced. Clearly, photosynthesis is essentially the reverse of respiration. Whereas respiration is, in general, a **catabolic** process, photosynthesis is, in general, an **anabolic** process. Photosynthesis occurs in organisms referred to as **autotrophs**. These organisms make their own food.

A catabolic process is one that breaks a larger molecule into smaller sub-parts. An anabolic process is the reverse reaction, where sub-parts are combined to form larger molecules.

Scientists in laboratories located around the world are presently working on the development of an artificial leaf. This laboratory-developed leaf would be capable of carrying out photosynthesis. The efforts of these research facilities may one day greatly increase the availability of food resources for the world's hungry human populations.

The chloroplast is vitally important for the overall process of photosynthesis. The structure of the chloroplast allows the anabolic reactions to proceed efficiently. In biology, the relationship of structure to function is a universal theme and is often referred to as structure–function correlation. The chloroplast and photosynthesis are no exception to this, as shown in Table 3.

Description of chloroplast structure	Function
Extensive membrane surface area of the thylakoids	Greater absorption of light by photosystems
Small space (lumen) and low volumes of fluid within the thylakoids	Faster accumulation of protons to create a concentration gradient
Stroma region similar to the cytoplasm of the cell and the matrix of the mitochondrion	Provides a region where the enzymes necessary for the Calvin cycle can work
Double membrane on the outside	Isolates the working parts and enzymes of the chloroplast from the surrounding cytoplasm

B2.2 Table 3 The relationship between structure and function in the chloroplast

Challenge yourself

1. Examine the diagram below of a typical chloroplast. Answer the questions below the diagram with the appropriate letter.

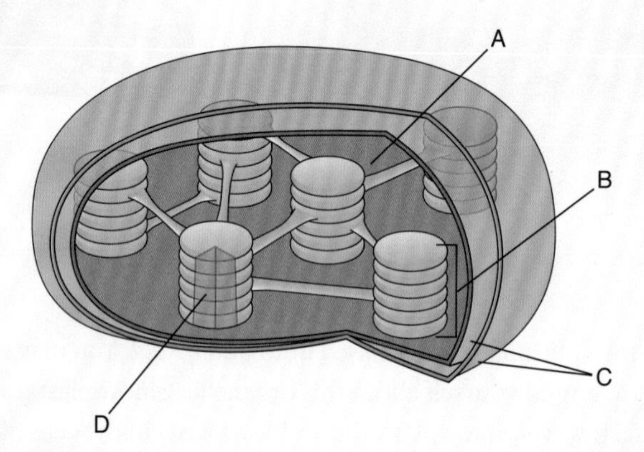

(a) Which letter represents the stroma where all the enzymes necessary for the light-independent reaction occur?

(b) Which letter represents the double membrane that controls the entry and exit of materials for the chloroplast?

(c) What letter indicates the thylakoid that contains the photosystems?

(d) Which letter represents a granum, which is where the light-dependent reaction occurs?

(e) Which two letters represent the areas of the chloroplasts that cause the green colour of chloroplasts? Why do these areas create this colour?

(f) The chloroplasts within some plant cells can often be seen moving in a cyclical pattern near the periphery of the cell. This is called cyclosis or cytoplasmic streaming. What might be the value of such movement to the process of photosynthesis?

B2.2.6 – The double membrane of the nucleus

B2.2.6 – Functional benefits of the double membrane of the nucleus
Include the need for pores in the nuclear membrane and for the nucleus to break into vesicles during mitosis and meiosis.

Chapter A2.2 describes how the nucleus is bordered by a double membrane and is the region where DNA resides. This double membrane, referred to as the **nuclear envelope**, provides an area where DNA can carry out its functions without being affected by processes occurring in other parts of the cell. The nuclear membrane has many pores that extend through both layers of the envelope. These nuclear pores allow ions and small molecules to diffuse between the nuclear material, the **nucleoplasm**, and the cytoplasm. These pores also control the passage of mRNA, proteins and RNA–protein complexes into and out of the nucleus. These RNA–protein complexes often become ribosomes and are produced in a region of the nucleus called the **nucleolus.** mRNA must leave the nucleus in order to be transcribed.

The outer membrane of the nuclear envelope is continuous with the ER of the cytoplasm, and even shares some functions with the ER. Ribosomes are often seen attached to the outer nuclear membrane, and this membrane can also form vesicles just like the ER.

The inner membrane of the nuclear envelope interacts with the inactive form of DNA called **chromatin**, which occurs within the nucleus. It also is important in maintaining the shape of the nucleus.

Chromatin is an elongated form of DNA arranged around proteins. It is invisible during most of the life cycle of the cell. It shortens to become discrete, visible chromosomes prior to cell division. Recent research indicates that the structure of chromatin affects the function of DNA. Changes in chromatin can cause changes in gene expression. These changes are called **epigenetic** changes, and involve differences in the accessibility of the DNA to enzymes, rather than changes in the DNA sequence.

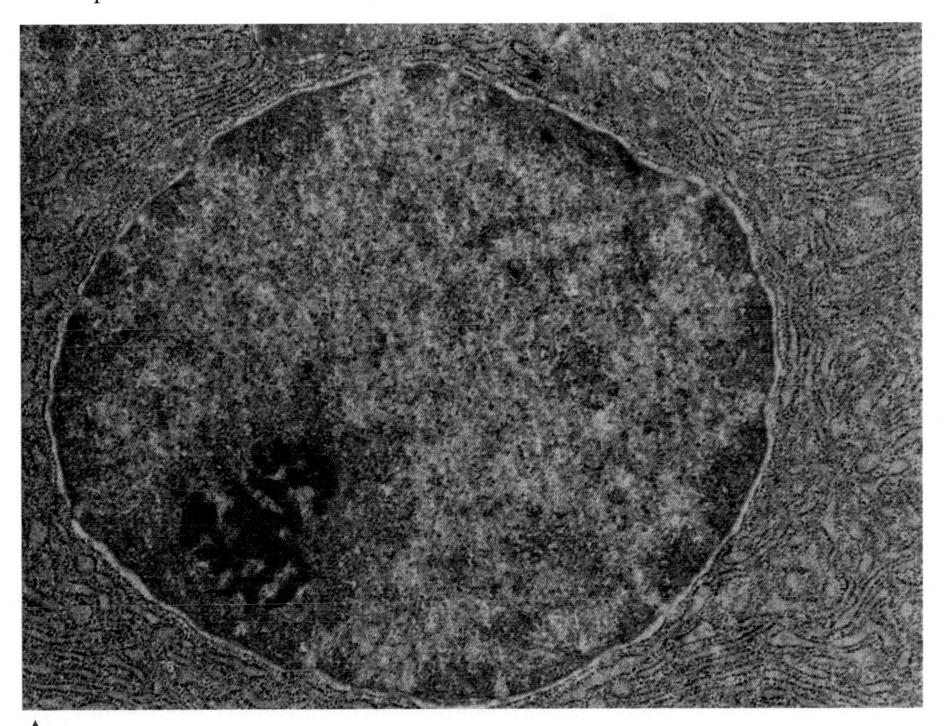

▲ A false-colour transmission electron micrograph (TEM) showing a nucleus. The nuclear envelope is coloured yellow and is an outside barrier; chromatin is green and is located throughout the interior of the nucleus; nucleoplasm is light brown and is throughout the interior; and nucleolus is magenta and is in the lower left region. Rough ER can be seen in the surrounding cytoplasm.

An important feature of the nuclear envelope is seen during the cell division processes, mitosis and meiosis. Early in both processes the nuclear membrane breaks down to allow movement of the DNA structures. In a poorly understood series of steps, the nuclear envelope breaks apart, becoming vesicles freely circulating in the cytoplasm. Once the DNA is correctly positioned at the conclusion of mitosis and meiosis, these vesicles attach to the surface of the highly condensed chromosomes and undergo a series of complex changes to reform the nuclear envelope.

B2.2.7 – The ribosome

B2.2.7 – Structure and function of free ribosomes and of the rough endoplasmic reticulum.
Contrast the synthesis by free ribosomes of proteins for retention in the cell with synthesis by membrane-bound ribosomes on the rough endoplasmic reticulum of proteins for transport within the cell and secretion.

Ribosomes are cytoplasmic organelles found in both prokaryotic and eukaryotic cells. The ribosomes of prokaryotic cells are slightly smaller than those of eukaryotic cells. Ribosomes are composed of proteins and a specific type of RNA known as **ribosomal RNA** (rRNA). Two subunits make up each ribosome, as shown in Figure 3. There are specialized attachment sites located on the two subunits that allow multiple amino acids to be bonded into highly specific proteins.

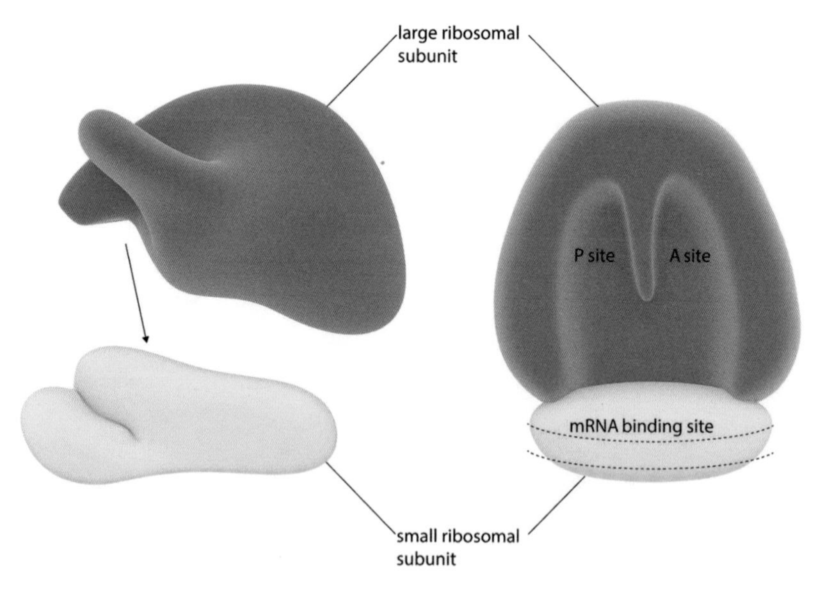

B2.2 Figure 3 This diagram of a ribosome shows the two component subunits. The P site and A site are locations for attachment to molecules carrying amino acids. Notice the mRNA binding sites. The ribosome is the site of protein production.

All ribosomes produce proteins. Here we will discuss eukaryotic ribosomes. Ribosomes are responsible for protein synthesis and are found in two locations within cells. They may be attached to the ER, or they may be free in the cytoplasm of the cell. ER that has ribosomes attached to it is called rough ER (rER).

Free and membrane-bound ribosomes produce different proteins. Free ribosomes produce proteins that are used within the cell, such as in the supporting cytoskeleton of the cell. Proteins produced by free ribosomes are also used in the nucleus, by mitochondria, and in other organelles not derived from the cell's inner membrane system.

Ribosomes that are bound to membranes produce proteins that are transported through the ER and often are exported from the cell. Proteins called secretory proteins are produced by membrane-bound ribosomes and are sent to the Golgi apparatus, where they are properly packaged for cellular exit. Hormones and enzymes are examples of secreted proteins.

Protein synthesis and the ribosome are discussed in much more detail in Chapter D1.2.

B2.2.8 – The Golgi apparatus

B2.2.8 – Structure and function of the Golgi apparatus
Limit to the roles of the Golgi apparatus in processing and secretion of protein.

The Golgi apparatus usually consists of flattened sacs. There can be as few as one in single-celled organisms, and more than 20 in some animal cells. These sacs are especially numerous in cells engaged in producing and secreting substances. The stacks of flattened sacs are called cisternae. Actively secreting cells will also have more Golgi apparatus than non-secreting cells.

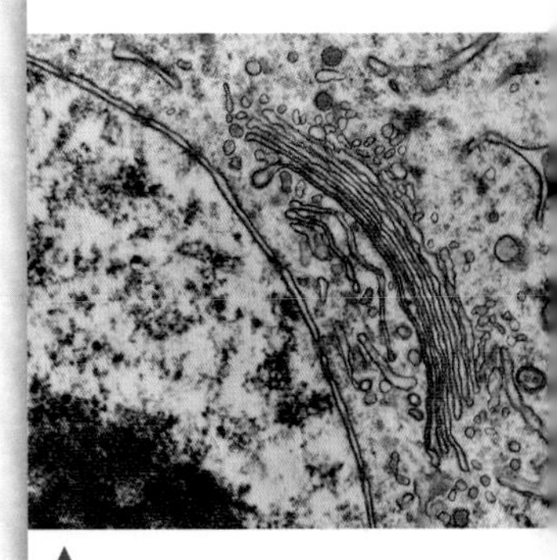

A false-coloured transmission electron micrograph (TEM) of a Golgi apparatus showing the characteristic cisternae, with one side oriented towards the ER and the other side oriented towards the plasma membrane. The ER is between the nuclear membrane and the Golgi apparatus, although it is difficult to see in this image. Notice the small round sacs, vesicles, leaving the side closest to the plasma membrane.

The position of the Golgi apparatus in the cell provides evidence for its cellular function. One side of the stack of flattened sacs is located near the ER (this is called the ***cis*** side). The other side is directed towards the plasma membrane (the ***trans*** side).

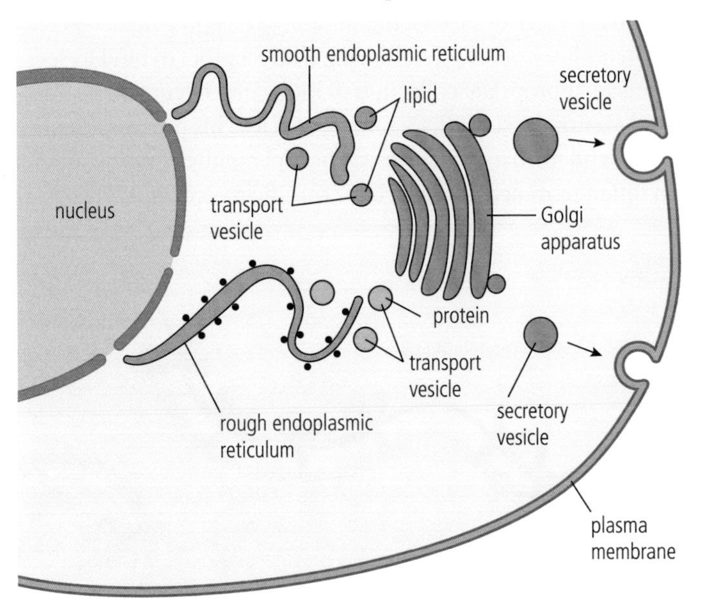

Note the position of the Golgi apparatus in this diagram. The *trans* side of the apparatus has secretory vesicles coming off it and moving towards the plasma membrane. The vesicles will fuse with the plasma membrane, allowing the emptying of their contents to the cell's exterior (exocytosis).

Protein- or lipid-filled transport vesicles are received on the *cis* side of the Golgi apparatus from either the rough ER or smooth ER. The difference between rough and smooth ER is discussed in Chapter A2.2. As the protein or lipid moves through the cisternae, they are modified so that they can carry out the specific function needed at that time. Once modified the final product is then packaged into vesicles that depart on the *trans* side. Often the modification of the substance includes the attachment of a signal chemical that directs the destination of the product. Some vesicles may become lysosomes within the cell, and others may return to the ER. Many will combine with the plasma membrane and go through the process of exocytosis, resulting in the secretion of the substance out of the cell.

B2.2.9 – Cellular vesicles

B2.2.9 – Structure and function of vesicles in cells
Include the role of clathrin in the formation of vesicles.

Vesicles are small membrane-bound sacs in which various substances are transported or stored in the cell. Do not confuse these with vacuoles, which are larger than vesicles. Some common examples of vesicles are:

- peroxisomes, which contain enzymes used to break down fatty acids
- lysosomes, which contain enzymes necessary for cellular digestion and also for destroying defective or damaged organelles
- transport vesicles, which move molecules within the cell
- secretory vesicles, which contain materials that are to be excreted from the cell, such as neurotransmitters (involved with nerve and muscle action) and hormones (control many general functions in plants and animals).

Clathrins are proteins in the cell membrane that anchor certain proteins to specific sites, especially on the exterior plasma membrane in receptor-mediated endocytosis. The clathrin proteins line coated pits, allowing the receptors to bind to specific molecules. When an appropriate collection of molecules occurs in the lined pit, the pit deepens and will eventually seal off, forming a vesicle. This process is highly specific and the sealing off and formation of a vesicle occurs rapidly. Because different cells need to take in different molecules, they have specific receptors on their surface.

The specific and efficient nature of receptor-mediated endocytosis is especially important in the exchanges that take place between maternal and foetal blood at the placenta. This allows essential nutrients to enter the baby's bloodstream.

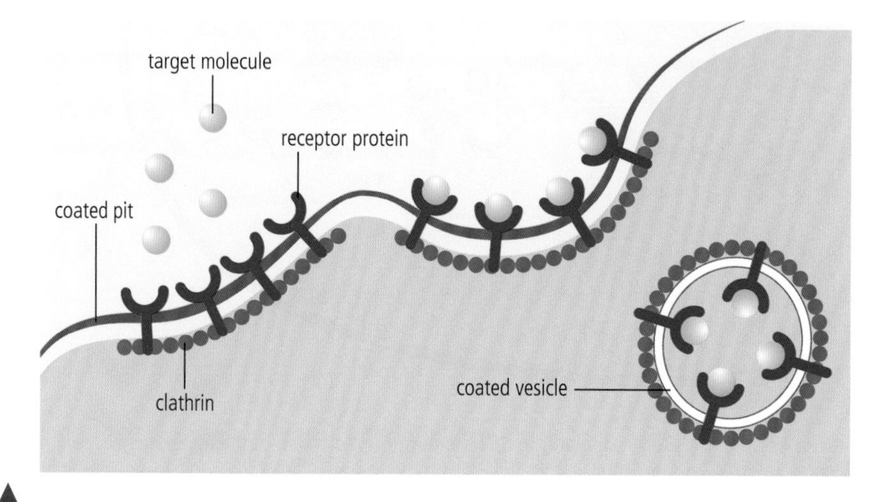

▲ A diagram of receptor-mediated endocytosis. Note the anchoring clathrin proteins allowing the receptors to bind to specific molecules.

The advantage of receptor-mediated endocytosis is that it is quite selective and efficient, especially compared to ordinary endocytosis.

HL end

Guiding Question revisited

How are organelles in cells adapted to their functions?

In this chapter we have discovered that:

- there is a structure–function correlation with all cell organelles
- each organelle of the cell has a unique structure that allows that organelle to perform its function, for example lysosomes are packages of digestive enzymes surrounded by a double membrane
- the cell membrane around the lysosomes prevents the digestive enzymes from damaging healthy structures in the cell
- the lysosomes can move anywhere in the cell and have a flexible membrane that can fuse with vesicles when necessary

HL

- mitochondria have a small intermembrane space so that a hydrogen ion concentration gradient can be quickly achieved
- the Krebs cycle of the mitochondria occurs in the matrix, and substrates are present so that it can continue
- the Calvin cycle of the chloroplast occurs in the stroma, where all the necessary substrates for the cycle are present
- the membranes of chloroplast thylakoids have a large surface area with photosystems present to convert light energy efficiently into the chemical energy of ATP
- rough ER is most often seen transporting the proteins produced by its attached ribosomes.

HL end

Guiding Question revisited

What are the advantages of compartmentalization in cells?

In this chapter we have discovered that:

- compartments are areas of the cell that have been isolated from other parts of the cell so that specialized functions can be carried out within them
- splitting the cell up into compartments allows the concentration of metabolites and enzymes in a particular area to be controlled
- this makes cells processes more efficient and stops the enzymes from one cell process interfering with another process, for example lysosomes contain enzymes used to break down substances ingested into the cell and worn out cell components, and by packaging them in an organelle the digestive enzymes are contained and controlled
- separating the nucleus from the rest of the cell means that mRNA can be processed before it reaches the ribosomes which translate it

HL

- organelles are cellular compartments and include the nucleus, vesicles, ribosomes, mitochondria, chloroplasts and cell membrane
- the cell wall, cytoskeleton and cytoplasm are not considered to be organelles
- clathrins are anchor proteins that function in the process of endocytosis
- the nuclear membrane is a double membrane that has pores that are essential for the overall function of the cell.

HL end

Exercises

Q1. Why is post-transcriptional modification possible in eukaryotic cells but not in prokaryotic cells?

Q2. Which cell fractionation process uses an electrical current to separate molecules of different size and electrical properties? What groups of cell molecules are often separated using this technique?

Q3. What are the advantages of compartmentalization to the cell?

HL

Q4. What type of substance is most likely to be found in vesicles leaving the smooth ER?

A Proteins. **B** Carbohydrates.
C Nucleic acids. **D** Lipids.

Q5. Which cell organelle contains structures called cisternae that are involved in the transport and modification of substances?

A Golgi apparatus. **B** Nucleus.
C Chloroplast. **D** Ribosome.

Q6. A mitochondrion within a eukaryotic cell has been damaged. What cell structure will be seen moving towards it and, eventually, connecting with it?

A Vacuole. **B** Lysosome.
C Ribosome. **D** Centrosome.

Q7. Which cellular organelle will most likely have pores in its surrounding membrane when examined under an electron microscope?

A Ribosome. **B** Vesicle.
C Chloroplast. **D** Nucleus.

Q8. Which of the following matches an organelle to its correct structure?

A Chloroplast – matrix. **B** Chloroplast – cristae.
C Ribosome – membrane. **D** Nucleus – thylakoid.
E Mitochondrion – double membrane.

Q9. Which of the following processes would the molecule known as clathrin be most associated with?

A Ultracentrifugation. **B** Exocytosis.
C Receptor-mediated endocytosis. **D** Phagocytosis.

Q10. What features allow the efficient absorption of sunlight in the chloroplast?

Q11. Explain the value of the low volume intermembrane space in the mitochondrion.

HL end

B2.3 Cell specialization

Guiding Questions

What are the roles of stem cells in multicellular organisms?

How are differentiated cells adapted to their specialized functions?

Cells have an amazing capability to specialize. They carry a genetic code that allows the development of specific traits and functions that contribute to the survival of the organism. In humans we see specialization in cells such as those in muscles, the lungs, eggs and sperm. Plants also show specialization, with the development of unique characteristics portrayed by roots, stem, leaves and flowers. Each cell develops in a specific way so that it is best suited to its function. For example, root cells have a large surface area to help absorb water and nutrients from the soil. They also have many mitochondria so that they can carry out the active transport of ions into cells.

Stem cells have captured the interest of many people because they maintain at least some degree of versatility. Some types of stem cell formed early in an organism's development can differentiate into every type of cell that can exist in the adult form of the organism. Some cells remain unspecialized within the organism in order to be able to develop into new cells. For example, in plants meristematic tissue is found in buds and stems. This tissue can differentiate into any of the many types of tissue that plants need to grow. In this chapter, we will examine stem cells and some of the factors involved in cell specialization within a multicellular organism.

B2.3.1 – Cell reproduction and organism development

B2.3.1 – Production of unspecialized cells following fertilization and their development into specialized cells by differentiation
Students should understand the impact of gradients on gene expression within an early-stage embryo.

From the formation of the single-celled zygote until the body structures begin to appear in approximately the ninth week of gestation, a developing human is called an embryo. Once the body structures appear, the embryo is then called a foetus, from the ninth week until birth.

Many cells have the ability to reproduce themselves. In multicellular organisms this allows growth to happen. It also means damaged or dead cells can be replaced.

Multicellular organisms usually start their existence as a single cell called a **zygote**. The zygote is formed as a result of fertilization, which is part of sexual reproduction. The two cells that fuse in sexual reproduction are called **gametes**. Each gamete has one-half the genetic material of a zygote. The following shows the development progression for humans:

gametes → zygote → embryo → foetus → infant

The single-celled zygote can divide at a very rapid rate. Initially the cells produced are unspecialized. However, the cells of the zygote rapidly start to differentiate, a process that results in the formation of specialized cells. The number of different cell types that arise from the one original cell can be staggering. This differentiation process is the result of the expression of some genes but not others. Each body cell contains all the genetic information needed to produce the complete organism. However, each cell will develop in a very specific manner depending on which genes become active. What causes some genes to become active depends on the signals that the cell receives.

Human embryonic development

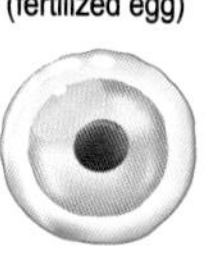

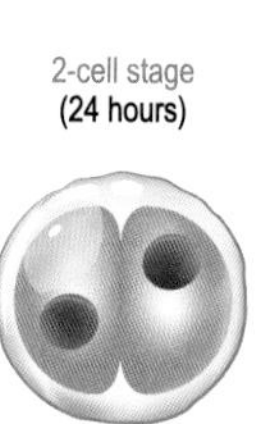

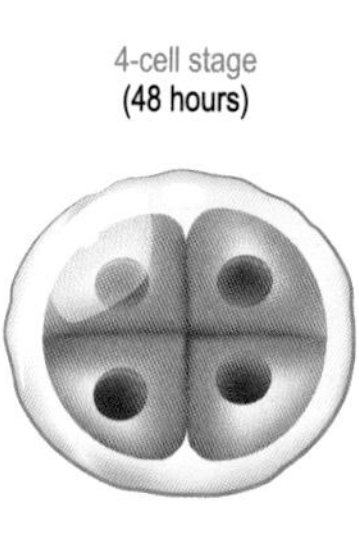

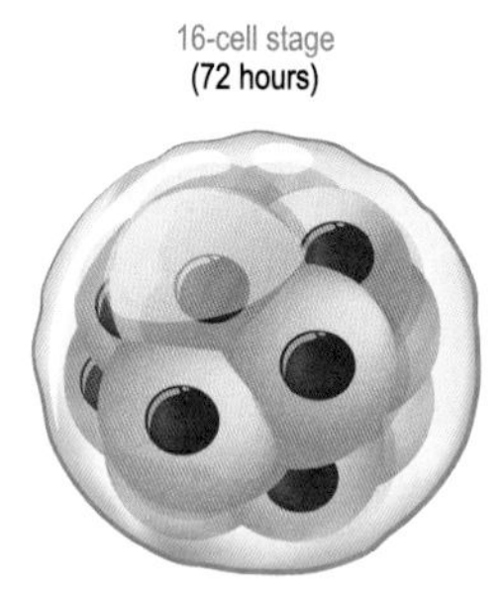

The protein known as bicoid is a morphogen that determines the anterior, head, end of a fruit fly. It is produced through specific directions from the *bicoid* gene of the fruit fly. A fruit fly embryo with defective *bicoid* genes will have posterior structures at both ends.

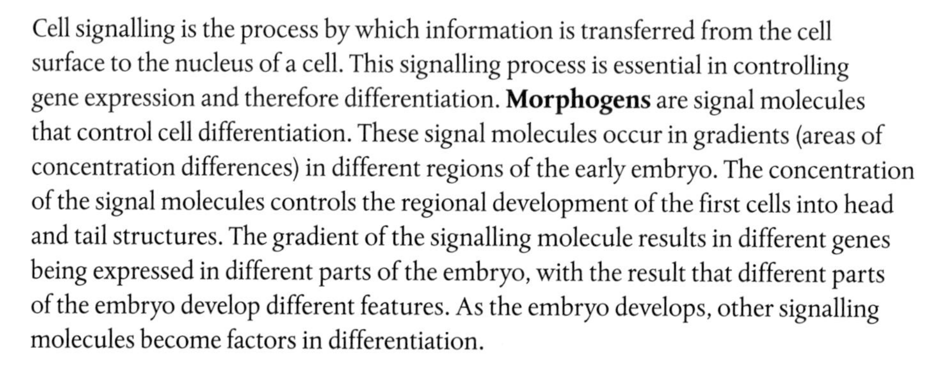

Cell signalling is the process by which information is transferred from the cell surface to the nucleus of a cell. This signalling process is essential in controlling gene expression and therefore differentiation. **Morphogens** are signal molecules that control cell differentiation. These signal molecules occur in gradients (areas of concentration differences) in different regions of the early embryo. The concentration of the signal molecules controls the regional development of the first cells into head and tail structures. The gradient of the signalling molecule results in different genes being expressed in different parts of the embryo, with the result that different parts of the embryo develop different features. As the embryo develops, other signalling molecules become factors in differentiation.

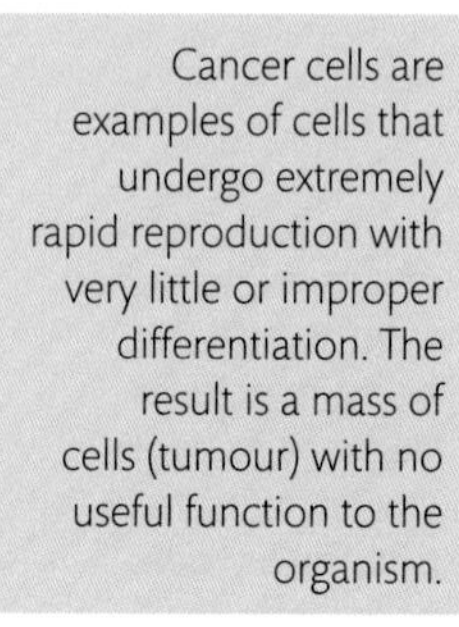

Cancer cells are examples of cells that undergo extremely rapid reproduction with very little or improper differentiation. The result is a mass of cells (tumour) with no useful function to the organism.

Some cells have a greatly diminished ability to reproduce once they become specialized, or lose the ability altogether. Nerve and muscle cells are good examples of this type of cell. Other cells, including epithelial cells such as skin, retain the ability to reproduce rapidly throughout their life. The cells formed from these rapidly reproducing cells will be the same cell type as the original cell.

How do cells become differentiated?

B2.3.2 – Stem cells

B2.3.2 – Properties of stem cells
Limit to the capacity of cells to divide endlessly and differentiate along different pathways.

There are populations of cells within organisms that retain their ability to divide and differentiate into various cell types. These cells are called **stem cells**. Stem cells retain the ability to divide indefinitely and can differentiate along different pathways, resulting in all the cell types an organism possesses.

Plants contain such cells in regions of **meristematic tissue**. Meristematic tissues are found near root and stem tips. The tissues are composed of rapidly reproducing cells that can become various types of tissue within the root or stem. Gardeners take advantage of these cells when they take cuttings from stems or roots and use them to grow new plants.

B2.3.3 – Stem cell niches

B2.3.3 – Location and function of stem cell niches in adult humans
Limit to two example locations and the understanding that the stem cell niche can maintain the cells or promote their proliferation and differentiation. Bone marrow and hair follicles are suitable examples.

When stem cells divide to form a specific type of tissue, they also produce some daughter cells that remain as stem cells. Figure 1 illustrates a common method stem cells employ when they divide. Note that this method allows continual production of a particular type of tissue while also providing for the continuation of stem cells, a process called **self-renewal**.

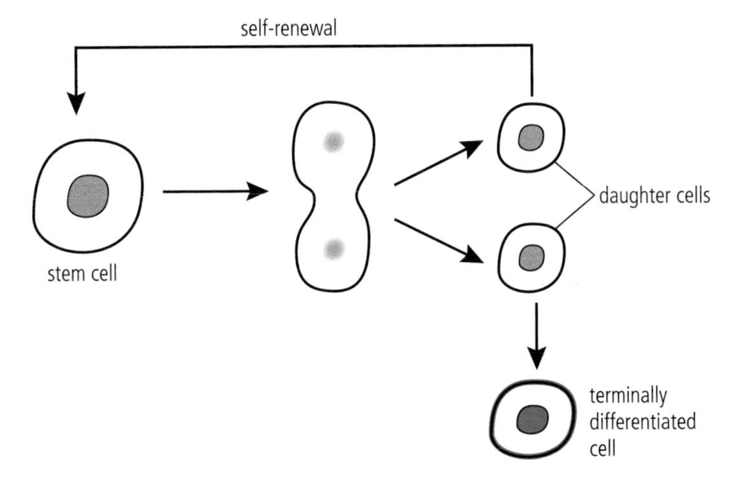

B2.3 Figure 1 Stem cell division.

Stem cells have two unique properties.

1. They can self-renew. This is shown in Figure 1. When a stem cell divides, there are several possible outcomes: both daughter cells may remain as stem cells, or a stem cell and a differentiated cell may be formed, or both cells may be differentiated. Whatever the outcome, stem cells are maintained.
2. They can recreate functional tissues. Cells become differentiated when cell signalling ensures that specific genes are expressed as the cell develops.

In the same way that niche environments provide all the support needed for normal stem cell function, malignancies or cancer may also occur as a result of changes in a particular niche. Tumour growth is often associated with blood vessel development. The added vasculature provides nutrients and transport that may increase the severity of the cancer. One approach to anticancer therapy includes attempting to limit this blood vessel support to tumours and cancer regions.

For stem cell research, scientists examine certain locations or **stem cell niches** in humans. In a stem cell niche the stem cells are present in high numbers as a result of regular proliferation, but they also demonstrate differentiation. Bone marrow and hair follicles are both stem cell niches in humans.

In the bone marrow, the stem cells that produce blood cells are found alongside self-renewing stem cells. As the blood cells are produced, the differentiated cells are transported away via a large array of supporting blood vessels. The renewal process ensures a constant supply of stem cells to continue differentiation.

Hair follicles exist in the skin, and large numbers of epithelial stem cells are found in the bottom, rounded area of a hair follicle. These stem cells are **multipotent**. They are involved with hair growth, skin and hair follicle regeneration, and the production of sebaceous (oil-producing) glands associated with hair follicles.

Stem cell niches in humans have also been studied in the central nervous system, the intestinal system and in muscle fibre bundles. One feature that all these niche areas have in common is the presence of signalling factors that bring about both self-renewal and cell differentiation.

B2.3.4 – Types of stem cell

B2.3.4 – Differences between totipotent, pluripotent and multipotent stem cells
Students should appreciate that cells in early-stage animal embryos are totipotent but soon become pluripotent, whereas stem cells in adult tissue such as bone marrow are multipotent.

In the early 1980s scientists found and described embryonic stem cells in mice. Similar stem cells have since been discovered in the embryo stage of many organisms. There are different types of stem cells.

Name of stem cell	Major characteristics
Totipotent	Capable of continued division and possesses the ability to produce any tissue in the organism. Very low numbers of cells are totipotent. Only exist in the very early stages of embryo development. They may form a complete organism.
Pluripotent	Arise from totipotent cells and only exist in the early embryonic stage. They can mature into almost all the different cell types that exist in an organism. Unlike totipotent cells, they cannot produce a complete organism.
Multipotent	Only forms a limited number of cell types. Bone marrow tissue that produces different types of blood cell is multipotent. They occur later in the development of the embryo and are present during the remainder of an organism's life.
Unipotent	Only forms a single cell type, such as sperm cells in mammals. They usually form late in the embryonic stage and exist in the functioning organism.

Different types of stem cell

At about 4-6 days into development, totipotent cells specialize and become pluripotent stem cells. In adult tissue, specialization has progressed to the point that multipotent stem cells such as in bone marrow are now present rather than pluripotent cells. Even though unipotent stem cells may form only a single cell type, they retain the property of self-renewal that distinguishes them from non-stem cells.

Because of their unique properties, medical experts saw the possibilities of using stem cells to treat certain human diseases. However, one problem discovered early on in stem cell research was that stem cells cannot be distinguished by their appearance. They can only be isolated from other cells based on their behaviour.

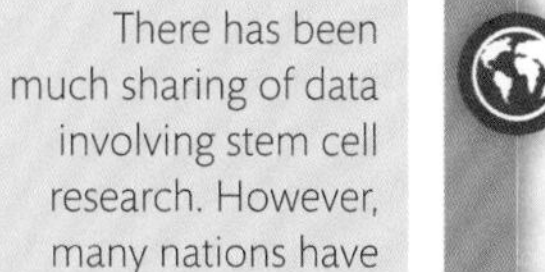

Recently some very promising research has been directed towards growing large numbers of embryonic stem cells in culture so that they can be used to replace differentiated cells lost because of injury and disease. This involves **therapeutic cloning**. Parkinson's and Alzheimer's diseases are caused by the loss of proper functioning brain cells, and it is hoped that implanted stem cells could replace many of these lost or defective brain cells, thus relieving the symptoms of the disease. With some forms of diabetes, the pancreas is depleted of essential cells, and it is hoped that a stem cell implant in this organ could have positive effects. However, currently most of the research on stem cells is being carried out in mice: it will probably be some time before this approach to treatment becomes widespread in humans.

Stem cells are being utilized in several ways by scientists around the world. One area of research involves using human embryonic stem cells to understand human development better. This research involves studying cell division and differentiation. Other scientists are using stem cells to test the safety and effects of experimental therapeutic drugs. Information in this area is essential to our understanding of how these drugs might affect differentiating cells in existing organisms. Another interesting area of study involves cell-based therapies, especially as they may have a positive influence on the treatment of diseases and traumas such as Alzheimer's disease, spinal cord injuries, heart disease, diabetes, burns and strokes.

TOK

There are ethical issues involved in stem cell research. The use of pluripotent stem cells is particularly controversial. These cells are obtained from embryos, largely obtained from laboratories carrying out **in vitro fertilization (IVF)**. Harvesting these cells involves the death of an embryo, and some people argue that this is the taking of a human life. Others argue that this research could result in a significant reduction in human suffering and is, therefore, acceptable. Where do you stand in the debate about the nature of stem cell research? How do you feel about the sources of pluripotent stem cells? Should scientific research be subject to ethical constraints or is the pursuit of all scientific knowledge intrinsically worthwhile?

B2.3.5 – Cell size and specialization

B2.3.5 – Cell size as an aspect of specialization
Consider the range of cell size in humans including male and female gametes, red and white blood cells, neurons and striated muscle fibres.

The size of cells and organelles is discussed in Chapter A2.2. The microscope is an essential tool for the study of cells because of their small size. The function of a cell determines how large the cell must be.

Cell type	Size
Sperm cell	3 μm in diameter, 50 μm in length
Egg cell	120 μm
Fat cell	50–150 μm
Red blood cell	7.5 μm
White blood cell	12–15 μm
Skeletal muscle cell	10–50 μm in width, 40 mm in length
Neuron (nerve cell)	350 μm in length

Sizes of various human cell types

The male and female gametes in humans, the egg and sperm cells, form a zygote during fertilization. This zygote may then go through development, as discussed earlier, to become an embryo, then a foetus, and finally an infant. This development involves a tremendous increase in the number of cells present as well as the differentiation of cells in various regions to become specialized tissues. The sperm cells are relatively small because they only carry out the function of transporting genetic material so that a viable zygote can be formed.

Red blood cells carry oxygen through the organism and have several important adaptations:

- they contain haemoglobin that can combine with and release oxygen
- they have a biconcave disc shape that allows more surface area for oxygen absorption
- they lack mitochondria as well as a nucleus
- they are flexible and size limited because they need to move through narrow blood capillaries.

White blood cells are larger than red blood cells. Their main function is defence against infections. They retain their nucleus throughout their lifetime. There are several distinct types of white blood cells and each type has a specific function. Blood cells are discussed fully in Chapter B3.2. Many possess vesicles with enzymes that can kill microorganisms. The enzymes present are also used in the breakdown of harmful cellular debris brought into the cell by phagocytosis. Their increased size is because of the necessary presence of the nucleus, granules and organelles such as mitochondria.

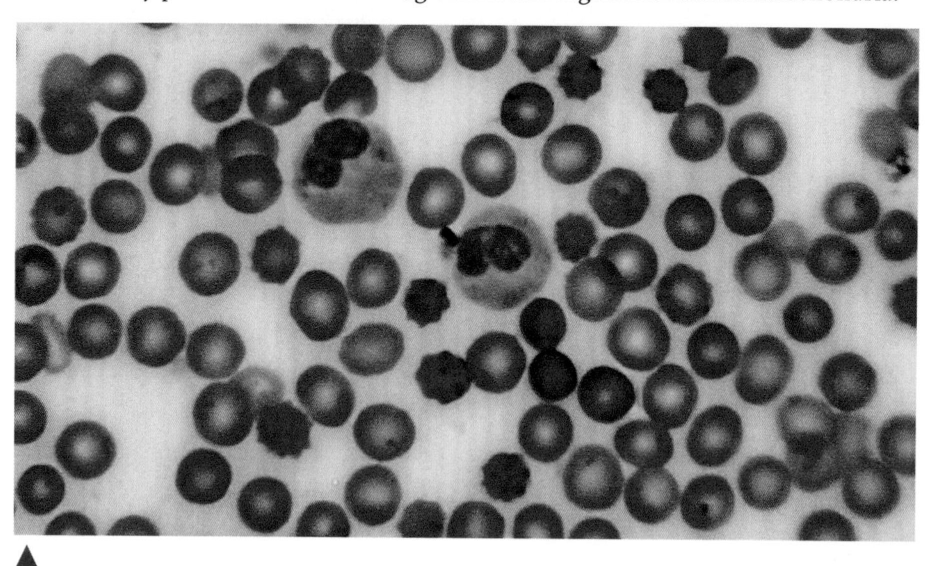

▲ A human blood smear under a light microscope. The larger cells are the white blood cells. The smaller cells are red blood cells. There are normally 4.2–6.2 million red blood cells per cubic millimetre of blood. In the same cubic millimetre, there are normally 5,000–10,000 white blood cells.

There are several types of neurons (nerve) cells. The distinct types of neuron exhibit adaptations for specialized roles. Motor neurons carry impulses from the brain or spinal cord that allow muscles to respond appropriately to stimuli. This nerve cell type has long fibres called axons that can carry impulses up and down the body over long distances.

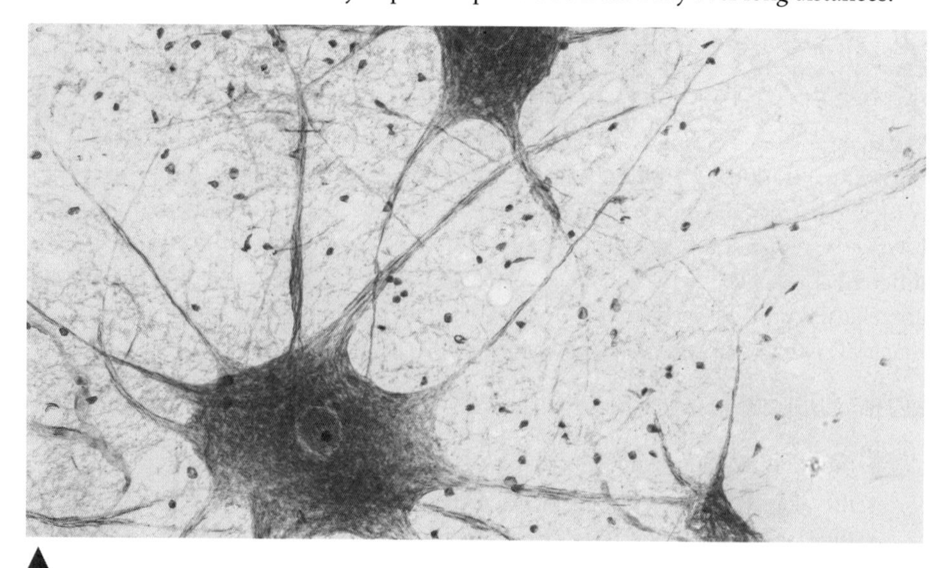

▲ A false-coloured micrograph of a motor neuron. The extensions leaving the main cell body allow it to carry impulses in multiple directions. The nucleus is obvious in the cell bodies shown.

The axons of the motor neuron can extend up to 1 m in the human body. This allows efficient and rapid transmission of nerve impulses from the brain and spinal cord to muscles in the limbs to produce movement.

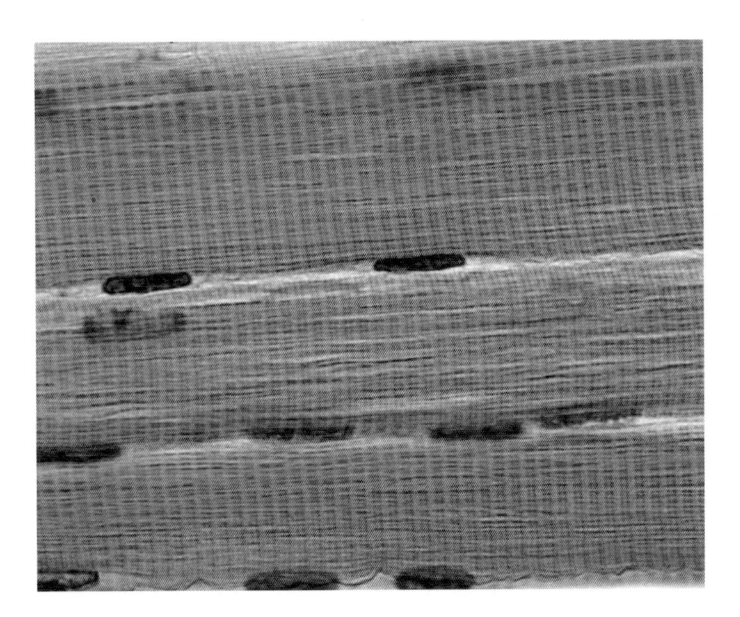

B2.3 Figure 2 Bands of striated skeletal muscle fibres are shown in this light microscope micrograph. Notice the bands, position of nuclei and lack of plasma membranes in the picture.

Striated muscle fibres are also specialized cells found in skeletal muscle. Each muscle fibre is a single muscle cell. The fibres are cylindrical and surrounded by membranes capable of impulse propagation. Striated muscle fibres can be up to 12 cm long and are longer than the muscle fibres found in smooth or cardiac muscle. The bands visible in the micrograph (Figure 2) represent units of contraction within the muscle fibre. This muscle type can only produce movement by contraction or shortening. Because the fibre is relatively long, all the units contracting together produce a significant movement.

As seen with these examples, different cell types not only show unique adaptations involving shape and structure but also in their size. Cell size is largely dictated by two major factors.

1. Basic processes of cell physiology, such as the need for materials to move in and out of the cell. This usually involves the surface area-to-volume ratio, which is discussed in Section B2.3.6 (and Section B2.3.7).
2. Cell division apparatus. If cells are too large or too small the mitotic spindle will not function properly.

Cell size is set as the cell goes through its differentiation process to become a particular type of cell within an organism. All the adaptations come together to produce the most efficient cell possible for the specific function it has.

B2.3.6 – Constraints on cell size

B2.3.6 – Surface area-to-volume ratios and constraints on cell size
Students should understand the mathematical ratio between volume and surface area and that exchange of materials across a cell surface depends on its area whereas the need for exchange depends on cell volume. **NOS:** Students should recognize that models are simplified versions of complex systems. In this case, surface area-to-volume relationship can be modelled using cubes of different side lengths. Although the cubes have a simpler shape than real organisms, scale factors operate in the same way.

Surface area-to-volume ratio. Full details on how to carry out this activity with a worksheet are available in the eBook.

The cell is obviously a small object. You may wonder why cells do not grow to larger sizes. The **surface area-to-volume ratio** of a cell limits the size a cell can reach. In a cell, the rate of heat and waste production, and rate of resource consumption, are functions of (depend on) the volume. Most of the chemical reactions of life occur inside a cell, and the size of the cell affects the rate of those reactions. The surface of the cell, the membrane, controls what materials move in and out of the cell. A cell with more surface area per unit volume can move more materials in and out of the cell, for each unit volume of the cell.

As the width of an object such as a cell increases, the surface area also increases, but at a much slower rate than the volume. This is shown in Table 1: the volume increases by a factor calculated by cubing the radius; at the same time, the surface area increases by a factor calculated by squaring the radius.

B2.3 Table 1 Surface area-to-volume ratio

Factor	Measurement		
Cell radius (r)	0.25 units	0.50 units	1.25 units
Surface area ($4\pi r^2$)	0.79 units	3.14 units	19.63 units
Volume ($4/3\pi r^3$)	0.06 units	0.52 units	8.18 units
Surface area : volume	13.17	6.04 : 1	2.40 : 1

Sphere formulae:

surface area = $4\pi r^2$

volume = $4/3\pi r^3$

This means that a large cell, compared to a small cell, has less surface area to bring in materials that are needed and to get rid of waste. Because of this, cells are limited in the size they can reach and still be able to carry out the functions of life. This means that large animals do not have larger cells; instead, they have more cells.

Nature of Science

Models are useful to help us understand complex systems. For example, you may have looked at surface area-to-volume ratios using cubes that have different lengths of side. Clearly most cells are not regular cubes but this is still a good model because the scale factors work in the same was as they do in cells. The surface area-to-volume ratio relationship can also be modelled using spheres of different radii.

HL

B2.3.7 – Surface area-to-volume adaptations

B2.3.7 – Adaptations to increase surface area-to-volume ratios of cells
Include flattening of cells, microvilli and invagination. Use erythrocytes and proximal convoluted tubule cells in the nephron as examples.

Cells that are larger in size have modifications that allow them to function efficiently when taking in gases and nutrients and getting rid of metabolic wastes.

Modifications that give more favourable surface area-to-volume ratios in cells include:

- changes in cell shape
- cellular projections, both inwards and outwards
- location relative to sources of nutrients and means of transporting away wastes
- how the cells fit together at a specific location.

The biconcave disc shape of red blood cells (**erythrocytes**) allows a greater surface area-to-volume ratio. Their size coupled with their flexibility also allows them to squeeze through small capillaries to deliver oxygen to all cells of the body.

The cells that line the part of the human kidney called the **proximal convoluted tubule** have several unique adaptations that increase their ability to reabsorb fluids and secrete ions.

- The cube-shaped cells are closely packed together in order to use space efficiently.
- The cells have tiny projections called microvilli pointing outwards into the lumen of the tubule in which fluid flows, and this **brush border** increases the surface area of the cell.
- Large numbers of mitochondria are found in the cells, allowing active transport of ions and other substances.
- Channels on the opposite side of the cell to the lumen increase the surface area of the cell to help transport.

The human kidney is composed of basic functional units called **nephrons**. Each nephron has three parts: the renal corpuscle, which absorbs fluids from the circulating blood; the filtering component, which carries out reabsorption of fluids and ion secretion; and the collecting duct, which carries out the final reabsorption of water with the formation of urine.

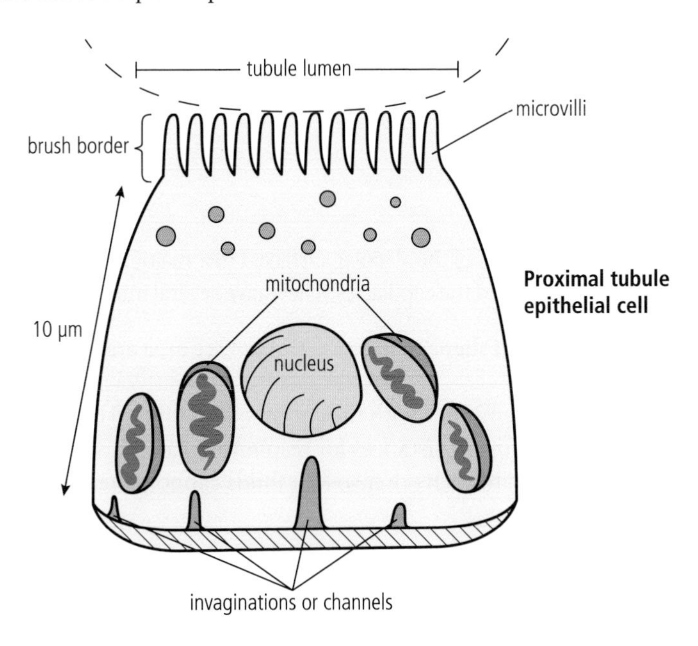

A drawing of a proximal tubule epithelial cell. Notice the brush border and the channels or invaginations on the opposite side of the cell.

B2.3.8 – Lung alveoli

B2.3.8 – Adaptations of type I and type II pneumocytes in alveoli
Limit to extreme thinness to reduce distances for diffusion in type I pneumocytes and the presence of many secretory vesicles (lamellar bodies) in the cytoplasm that discharge surfactant to the alveolar lumen in type II pneumocytes. Alveolar epithelium is an example of a tissue where more than one cell type is present, because different adaptations are required for the overall function of the tissue.

The functional unit of the lungs is the **alveolus** (plural alveoli). The alveoli increase the surface area of the lung to maximize gas exchange. The alveoli must also be able to function as the lungs expand and contract during the breathing process. They are at the very end of the respiratory tract and can be thought of as empty sacs lined by a wall made up of a layer of single cells. Figure 3 on the next page illustrates the structure of an alveolus.

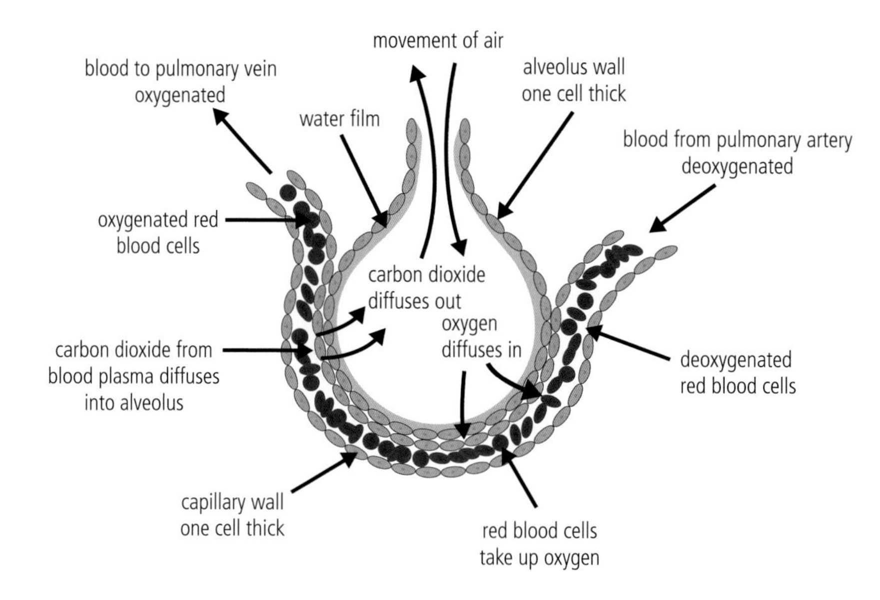

B2.3 Figure 3 A diagram of a single alveolus in the lungs, showing the process of gas exchange.

The alveoli contain **macrophages**, which are white blood cells. You do not need to know about these for the IB, but they perform endocytosis to remove harmful substances and organisms from the inhaled air.

The cell surface of each alveolus is covered by three very active cell types:

- type I pneumocytes
- type II pneumocytes
- alveolar macrophages.

Type I pneumocytes cover 95% of the alveolar surface. Their major function is to allow gas exchange between the alveoli and the capillaries. They have several interesting adaptations:

- they are thin and flat in shape to increase the surface area and minimize diffusion distance
- a shared basement membrane with the endothelium (lining) of lung capillaries, which minimizes diffusion distances for respiratory gases
- they are tightly joined to each other so that fluids cannot enter the alveoli from the capillaries.

Type II pneumocytes are not as prevalent as the type I pneumocytes, making up less than 5% of the surface. They are found between the type I pneumocytes. The role of type II pneumocytes is to produce **pulmonary surfactant**, which reduces surface tension and prevents the alveoli from collapsing and sticking to each other during the breathing process. Adaptations of type II pneumocytes include:

- a cube shape, providing a larger cytoplasmic area for the organelles producing the surfactant
- microvilli oriented towards the alveolar sac, increasing the surface area and allowing more surfactant secretion
- a cytoplasm that contains many organelles involved with surfactant production and its secretion, including secretory vesicles (called lamellar bodies)
- the ability to transform into type I pneumocytes when needed.

It is important to recognize that lung tissue has more than one cell type present. The different cell types present in the lung allow its many functions to be carried out efficiently. Most tissues of an organism have multiple cell types so that they can carry out all their overall functions successfully.

B2.3.9 – Muscle fibres

B2.3.9 – Adaptations of cardiac muscle cells and striated muscle fibres
Include the presence of contractile myofibrils in both muscle types and hypotheses for these differences: branching (branched or unbranched), and length and numbers of nuclei. Also include a discussion of whether a striated muscle fibre is a cell.

Muscle fibres are cells. Some unique adaptations of striated skeletal muscle cells have already been mentioned. These include:

- a long, cylindrical shape
- a membrane capable of impulse propagation
- multiple nuclei (multinucleated)
- visible bands capable of shortening to produce voluntary movement.

Cardiac muscle cells. Note the branching, striated cells.

branching cell
cardiac muscle fibre
striations
nucleus

Cardiac muscle fibres occur in the heart, and they have banded cells like the striated skeletal muscle cell. However, they also have further adaptations to perform their unique functions. They:

- are composed of branching, striated cells
- have a single nucleus per fibre/cell
- are connected at the ends by intercalated discs.

Cardiac muscle fibres are shorter in length than striated skeletal muscle fibres. Because of their branching cells, and the connections between them, cardiac muscle fibres coordinate a contractile process involving the whole heart in order to pump blood successfully throughout the body.

It has been debated whether striated skeletal muscle fibres should be considered cells. These cells are larger than most animal cells. Once they are formed, skeletal muscle fibres do not follow the usual pattern of cell division. They do not expand by producing more cells in adult organisms, but grow when their supporting satellite cells fuse with the existing fibres. Also, when muscle fibre cells are severely damaged, they do not go through the usual process of cell death (apoptosis) followed by cell replacement.

Apoptosis refers to programmed cell death, which involves a series of cellular events leading to the death and destruction of a cell. This is part of the normal cycle of cell replacement as a result of cell ageing, but can also occur so that body parts can develop and change as an organism moves from one life phase to another, for example when a tadpole changes into an adult frog.

However, even with all these unique characteristics, striated muscle fibres still have many of the organelles and features discussed when describing cells.

B2.3.10 – Sperm and egg cells

B2.3.10 – Adaptations of sperm and egg cells
Limit to gametes in humans.

You will recall that egg and sperm cells are referred to as gametes, and when they fuse a zygote is formed. The zygote then proceeds with multiple cell divisions, leading to the formation of an embryo, then a foetus, and ultimately an infant.

B2.3 Table 2 Adaptations of sperm and egg cells

Sperm cell adaptations	Egg cell adaptations
One of the smallest human cells	One of the largest human cells
3 µm in width, 50 µm in length	120 µm
Flagellum (mid-piece) present that allows motility. Mitochondria located near the flagellum to supply energy for movement	No flagellum present, the cell is non-motile
Shape includes a head and a tail region and is streamlined for speed and efficiency	Shape is non-streamlined. It is spherical
Very few cytoplasmic organelles such as endoplasmic reticulum, Golgi apparatus or ribosomes	Most cytoplasmic organelles are present plus specialized storage structures for initial embryo development
Continually produced in vast numbers throughout the life of a male	All the early gamete-forming cells are present before birth. No new egg-forming cells are produced after birth
Head has a specialized secretory vesicle called the acrosomal vesicle that helps the sperm penetrate the egg's outer coat	Has special secretory vesicles just under the plasma membrane that release their contents after one sperm penetrates the egg to prevent other sperm from entering
Contains a haploid nucleus	Contains a haploid nucleus

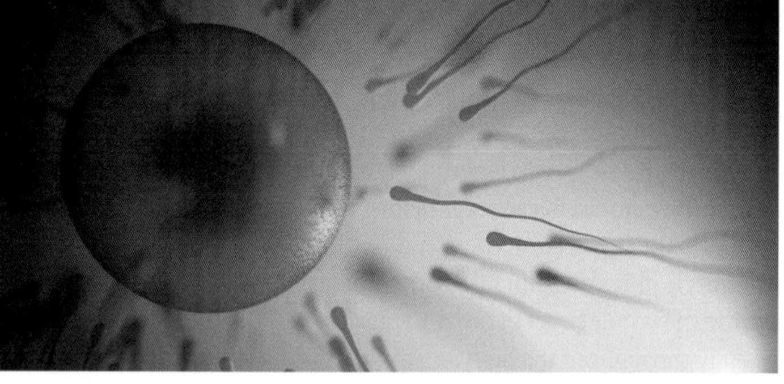

A conceptual image of multiple sperm trying to fertilize an egg.

Note the specific cellular structures mentioned in Table 2, describing the adaptations of both sperm and egg cells.

Human egg (ovum)

nucleus
cytoplasm
first polar body

Human sperm (spermatozoa)

tail/flagellum
mid-piece
head
mitochondria
centriole
acrosome
nucleus
axoneme

SKILLS

Cells have adaptations to carry out their designated functions. Many of these adaptations can be observed by looking at cells under a microscope or by studying photomicrographs of cells available on the internet. While carrying out your observations and studies, you should make drawings of the cells emphasizing the adaptations visible. Use a pencil and centre your drawings on a page. Give the drawing a title and note the magnification of the view. Make the drawing large enough that the necessary detail and labelling can be shown. Use simple, narrow lines when labelling, making sure the lines to different structures do not cross.

What are the advantages of small size and large size in biological systems

HL end

Guiding Question revisited

What are the roles of stem cells in multicellular organisms?

In this chapter we have described how:

- stem cells can reproduce throughout their life
- stem cells can differentiate along different pathways
- totipotent stem cells are the rarest in number and they can differentiate into any type of cell in the organism
- stem cells divide to form a specific type of cell, and they also produce some daughter cells that remain as stem cells
- stem cell niches exist in areas of an organism, and these areas have signalling factors that bring about self-renewal and differentiation of the stem cells present.

Guiding Question revisited

How are differentiated cells adapted to their specialized functions?

In this chapter we have learned how:

- cell signalling is the process that controls gene expression at various stages of embryo and organism development
- morphogens function in the embryo to control regional development of specific cell types
- cell function has a great influence on the size of a cell
- cells involved in secretion are small so that their surface area-to-volume ratio is larger
- many different cells exist in an organism, and all have a size and shape that means they can carry out their essential function efficiently

HL

- the unique structures of erythrocytes and kidney cells allow specialized functions
- lung cells possess properties that bring about efficient oxygen/carbon dioxide exchange
- striated muscle fibres have unique features that allow them to bring about movement
- specialized features of egg and sperm are essential to the production of a zygote.

HL end

Exercises

Q1. The very first embryonic cells would be which of the following kind of stem cell?

A Multipotent. **B** Totipotent.
C Differentiated. **D** Pluripotent.

Q2. Which of the following cells lacks a nucleus in its differentiated stage?

A Neuron. **B** Gamete.
C Muscle cell. **D** Red blood cell.

Q3. What normally happens to a spherically shaped cell as it grows larger?

A Surface area-to-volume ratio stays the same.
B Surface area-to-volume ratio decreases.
C Surface area-to-volume ratio increases.
D Its ability to bring in adequate nutrients increases.

HL

Q4. What type of cell would include an acrosome as a necessity to perform its main function?

A Sperm. **B** Egg.
C Cardiac muscle. **D** Red blood cell.

Q5. Which type of gradient is most important in initiating gene expression in early embryo development?

A Nutrient. **B** Oxygen.
C Morphogen. **D** Niche.

Q6. Which of the following would not be considered a specialized feature of a differentiated cell?

A Single nucleus. **B** Axon.
C Many nuclei. **D** Presence of haemoglobin.

HL end

B2 Practice questions

1. Describe how the properties of the molecules that make up a cell membrane help to maintain the cell membrane structure. *(Total 3 marks)*
2. Describe the process of active transport. *(Total 3 marks)*
3. Explain how a channel protein allows a cell membrane to be selectively permeable. *(Total 4 marks)*
4. Describe the advantages of compartmentalization in the cytoplasm of cells. *(Total 4 marks)*
5. Explain how the surface area-to-volume ratio influences cell sizes. *(Total 3 marks)*
6. Compare and contrast totipotent, pluripotent and multipotent stem cells. *(Total 3 marks)*

HL

7. List the adaptations of the chloroplasts for carrying out photosynthesis. *(Total 4 marks)*
8. Explain the process by which sodium–potassium pumps generate and maintain membrane potentials within nerve cells. *(Total 3 marks)*
9. For each of the following human cell types, list an adaptation it possesses to carry out its major function:
   (a) type I pneumocytes
   (b) muscle fibres
   (c) red blood cells. *(Total 3 marks)*

HL end

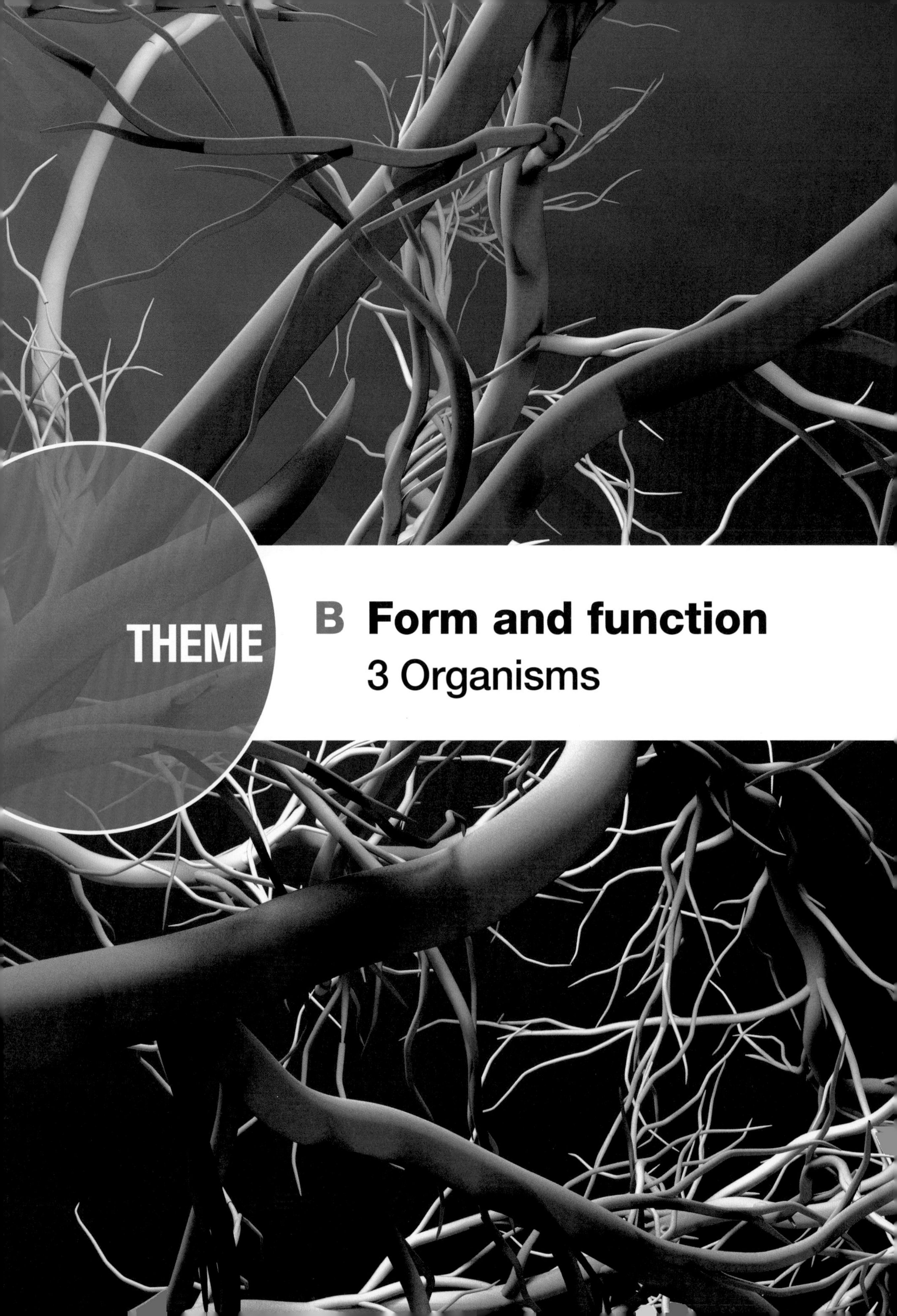

THEME

B Form and function

3 Organisms

Blood vessels form a characteristic "tree" pattern as they branch out into a network whilst feeding tissues with oxygen and nutrients. One of the fundamental challenges that organisms faced in their evolutionary development from single celled to multicellular creatures was the distribution of nutrients and subsequent removal of waste products. In large multicellular organisms, such as humans, each cell must have access to molecules that are only available in their environment. Many of those cells are in the interior of the organism and thus cannot rely on direct molecular transfer from and to the outside environment. Both animals and plants have evolved adaptations that allow them to take in molecules from the environment, circulate the molecules to the interior cells within branching vessels, and then use the branching vessels to take waste products to locations that facilitate removal from the organism.

B3.1 Gas exchange

Guiding Questions

How are multicellular organisms adapted to carry out gas exchange?

What are the similarities and differences in gas exchange between a flowering plant and a mammal?

Multicellular organisms have the problem of getting the air molecules that they need from their environment to cells that may be deep within them. Diffusion alone cannot solve this bioengineering problem. Thus, many multicellular life forms have adaptions that combine a respiratory gas exchange system with a fluid transport system. Lungs for gas exchange coupled with a circulatory system for transport is a common example.

Plants, like animals and other life forms, require an exchange of gases with the atmosphere. Photosynthesis typically comes to mind when people think about the life processes associated with plants. Plants are also aerobic organisms and require oxygen for their cells just like animals do. Plants have adaptations that allow the atmospheric gases of both cell respiration and photosynthesis to be exchanged with their environment.

B3.1.1 – The exchange of gases between organisms and their environment

B3.1.1 – Gas exchange as a vital function in all organisms
Students should appreciate that the challenges become greater as organisms increase in size because surface area-to-volume ratio decreases with increasing size, and the distance from the centre of an organism to its exterior increases.

Most organisms are aerobic, meaning that they require oxygen to metabolize energy from organic substances such as glucose. In addition, organisms need to remove metabolic waste products such as carbon dioxide. A few organisms, including many single-celled life forms, can exchange oxygen and carbon dioxide directly with the atmosphere through their plasma membranes. However, that is not an option for larger multicellular organisms where metabolically active tissues may lie deep within the organism and far away from their environment. These organisms have evolved complex adaptations to exchange respiratory gases between the atmosphere or water habitat and their tissues.

How do multicellular organisms solve the problem of access to materials for all their cells?

Surface area alone cannot be considered an important factor for solving biological problems like providing respiratory gas exchanges. It is the ratio of surface area to volume of the organism that is the important factor. A microscopic protist has a lower surface area than an elephant but the surface area-to-volume ratio of a protist is much higher than that of an elephant.

The problem of getting gases directly to and from an organism's interior cells is compounded by the **surface area-to-volume ratio**, which changes as an organism gets larger. Surface area is a squared function of its dimensions and that is why we give surface area a square unit (such as cm^2 or m^2). Volume is a cubed function (such as cm^3 or m^3). Another way of expressing this idea is that the surface area-to-volume ratio decreases with increasing size.

The volume of an organism is a reflection of its metabolic need to exchange respiratory gases. An organism's ability to take in and release substances is limited by its outer layer surface area. Only the smallest organisms can rely on direct exchange of respiratory gases with their environment, all others must have anatomical and physiological adaptations to get oxygen to internal tissues and take carbon dioxide away.

B3.1.2 – Gas exchange surfaces

B3.1.2 – Properties of gas-exchange surfaces
Include permeability, thin tissue layer, moisture and large surface area.

Organisms that have evolved adaptations for gas exchange must have specialized tissues designed for the molecular exchanges. The specialized tissues are found in the skin of some small organisms, gills of many aquatic organisms, and the lungs of some larger terrestrial organisms. The exchange of gases sometimes occurs between the air and the living tissue (lungs) or between water and the living tissue (gills). In many organisms the gases are immediately exchanged to blood vessels to be circulated to body tissues.

Gas exchange surfaces are characterized by:

- being thin (often only one cell layer), to keep diffusion distances short
- being moist, to encourage gas diffusion
- having a large surface area, for maximum diffusion
- being permeable to respiratory gases (oxygen and carbon dioxide).

These properties allow the maximum volume of gases to be exchanged across the surface in the smallest amount of time.

A freshwater salamander known as an axolotl (*Ambystoma mexicanum*). The axolotl only lives in one small area in Mexico and is critically endangered. This amphibian, even as an adult, has six external gills for respiratory gas exchange with water.

B3.1.3 – Concentration gradients at exchange surfaces in animals

B3.1.3 – Maintenance of concentration gradients at exchange surfaces in animals
Include dense networks of blood vessels, continuous blood flow, and ventilation with air for lungs and with water for gills.

Oxygen and carbon dioxide are exchanged by diffusion. This means that **concentration gradients** must be maintained for oxygen to diffuse into the blood and carbon dioxide out of the blood.

B3.1 Figure 1 A scanning electron micrograph (SEM) of fish gills. Inside the thin gill tissue are numerous blood capillaries. A higher concentration of oxygen must be maintained in the water passing over these gills compared to the oxygen in the blood of the internal blood capillaries. This allows diffusion of oxygen into the blood to continue.

In Figure 1 there are two fluids to take into account for respiratory gas concentrations. One fluid is the environmental water passing over the gill tissue. The other is the blood within the capillaries of the gills. The concentration of respiratory gases in the environmental water does not change as long as the body of water maintains good ecological health and the water is not stagnant around the gills. The concentrations of oxygen and carbon dioxide do change within the blood of the organism, however.

When the blood is first circulated to the gills, it has recently been within capillaries of the muscles and other body tissues. The body cells are continuously respiring, which utilizes oxygen and produces carbon dioxide. The blood that leaves body tissues contains a higher concentration of carbon dioxide and a lower concentration of oxygen compared to levels before the blood reached the active body tissues. The blood will then be transported to the gill tissue and the exchanges occur once again.

All animals that use gills are **exothermic** (cold blooded). One of the reasons for this is the relatively high metabolic rates necessary to be **endothermic** (warm blooded). The low oxygen levels available in bodies of water would not support the metabolic rate needed for a constant internal body temperature.

The only blood vessels that permit the exchange of substances are capillaries. Capillaries are only one cell thick.

Diffusion gradients also explain the gas diffusion that takes place in animals with lungs. Within the lungs are numerous dense capillaries that contain blood that has recently come from respiring body tissues. The concentration of oxygen in the lung capillaries is lower than that of air inspired into the lungs. In addition, the concentration of carbon dioxide in the lung capillaries is higher than that in the air inspired.

Two events must occur to keep concentration gradients in place:

- water must be continuously passed over the gills/air must be continuously refreshed (ventilated) in the lungs
- there must be a continuous blood flow to the dense network of blood vessels in both the body tissues and the tissues of the gills or lungs.

B3.1.4 – Gas exchange in mammalian lungs

B3.1.4 – Adaptations of mammalian lungs for gas exchange
Limit to the alveolar lungs of a mammal. Adaptations should include the presence of surfactant, a branched network of bronchioles, extensive capillary beds and a high surface area.

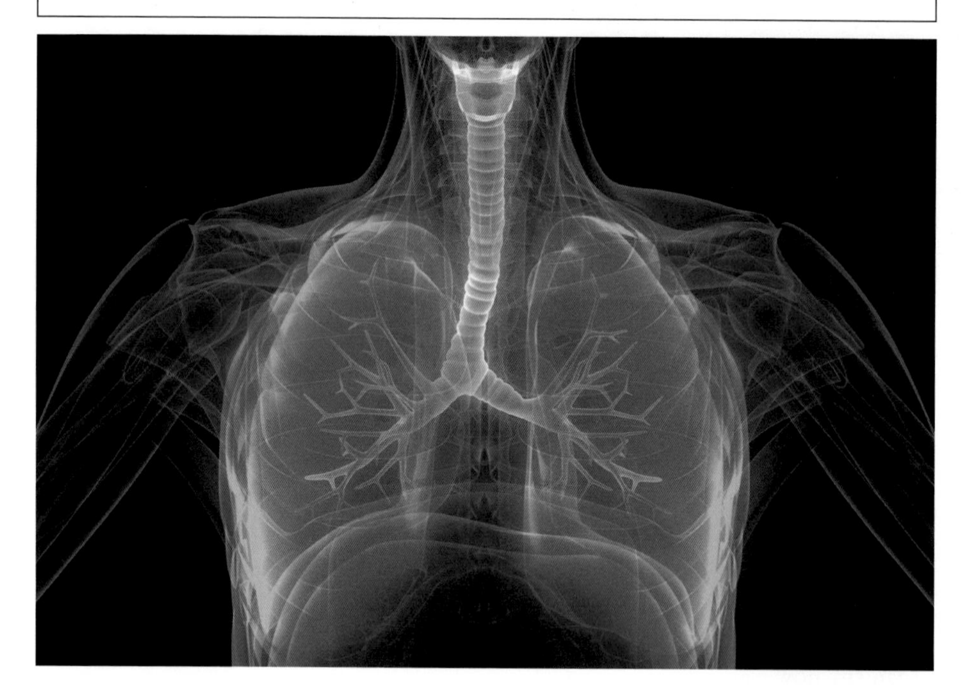

A colorized X-ray showing the anatomy of the respiratory system. The large central air tube is the trachea, which branches into the right and left bronchi leading to each lung. Within the lungs further branching occurs repeatedly, into small bronchioles. At the end of each tiny branch is an air sac called an alveolus. Each lung contains about 300 million alveoli.

Our lungs have an amazing capacity to expose life-giving air to an incredibly large surface area of gas exchange tissue. The lungs do this by subdividing their volume into microscopic spheres called alveoli. Each **alveolus** is at a terminal end of one of the branches of tubes that started as the trachea. Every time you breathe in (**inspire**) and breathe out (**expire**) you replace most of the air in millions of alveoli.

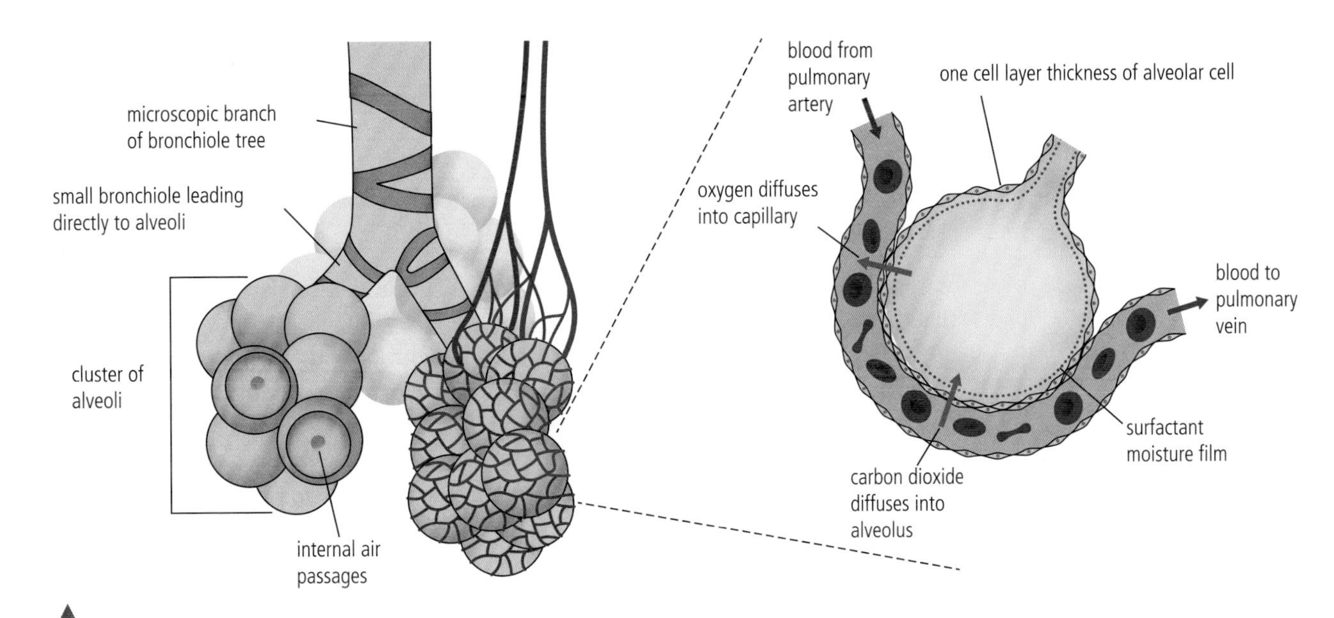

▲ **B3.1 Figure 2** On the left is a cluster of alveoli connected to the smallest bronchi of a bronchiole tree. Much of the air in each alveolus is refreshed every time you inspire and expire. One alveoli group is shown with a dense surrounding capillary bed. On the right side is a sectioned view of a single alveolus and capillary.

The inner surface of each alveolus is lined by a thin phospholipid and protein film called a **surfactant**. Specific alveolar cells secrete the surfactant, which coats the inside of each alveolus. The surfactant acts to reduce the surface tension of the moist inner surface and helps prevent each alveolus from collapsing each time air is expired.

The inside of each lung is subdivided into several lobes, which are in turn subdivided into the millions of spherical alveoli all connected by small tubes called **bronchioles**. All of the bronchioles are ultimately connected into the trachea for access to inspired and expired air. The spherical shape of the alveoli provides a vast surface area for the diffusion of oxygen and carbon dioxide.

The diffusion of respiratory gases is also helped by the dense network of capillaries surrounding the alveoli. Each alveolus has close access to a capillary (see Figure 2).

Air inspired into the alveoli has a higher concentration of oxygen and lower concentration of carbon dioxide compared to the blood in a nearby capillary. The two types of gas diffuse according to their concentration gradient (see Figure 2). Because capillaries are just one cell thick and each alveolus is just one cell thick, the respiratory gases only need to diffuse through two cells to enter or exit the blood stream. The oxygen-rich blood is now ready to return to the heart to be pumped out into actively respiring tissues. The entire process is ongoing as long as the heart continues to send blood to the capillaries within the lungs, and air continues to be refreshed within the alveoli.

Visible Human Project worksheet. Full details of how to carry out this activity are available in the eBook.

B3.1.5 – Lung ventilation

B3.1.5 – Ventilation of the lungs
Students should understand the role of the diaphragm, intercostal muscles, abdominal muscles and ribs.

We breathe in and out continuously all our lives. The mechanism is a series of events that thankfully usually happens without our conscious thought. The tissue that makes up our lungs is passive and not muscular, therefore the lungs themselves are incapable of purposeful movement. However, there are muscles surrounding the lungs, including the diaphragm, muscles of the abdomen, and the external and internal intercostal muscles (surrounding your ribs). All of these muscles work collectively to either increase or decrease the volume of the thoracic cavity, leading to pressure changes in the lungs.

The mechanism of breathing is based on the inverse relationship between pressure and volume. **Boyle's law** states that an increase in volume will lead to a decrease in pressure, and vice versa. Your lungs are located within your thoracic cavity (or **thorax**). The thoracic cavity is closed to the outside air. Inside the thoracic cavity are the lungs, which only have one opening, through your trachea (via your mouth and nasal passages). The diaphragm is a large, dome-shaped muscle that forms the "floor" of the thoracic cavity. When it contracts it flattens the dome shape and increases the volume of the thoracic cavity. See Figure 3.

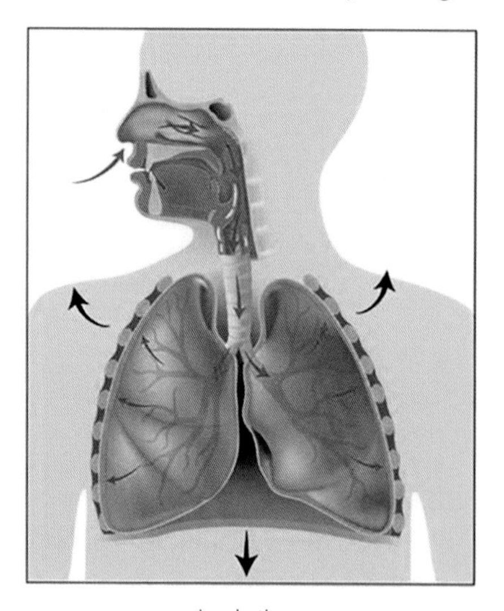

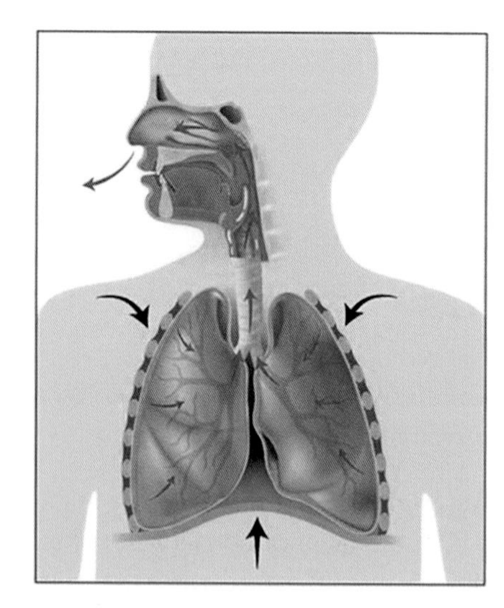

Expiration

B3.1 Figure 3 On the left, contraction of the diaphragm and external intercostal muscles raising the rib cage results in an increased volume of the thoracic cavity. This allows the lungs to expand, lowering their internal pressure. Air enters the nose or mouth and continues down to each of the alveoli. On the right, relaxation of diaphragm and contraction of the internal intercostal results in an increased pressure on the lung. This leads to an expiration of air.

Inspiration (breathing in)

Inspiration occurs in a series of steps.

1. The **diaphragm** contracts, increasing the volume of the thoracic cavity.

2. At the same time, the **external intercostal muscles** and one set of abdominal muscles both contract to help raise the rib cage. These actions also help increase the volume of the thoracic cavity.
3. Because the thoracic cavity has increased in volume, the pressure inside the thoracic cavity decreases. This leads to less pressure "pushing on" the passive lung tissue.
4. The lung tissue responds to the lower pressure by increasing its volume.
5. This leads to a decrease in pressure inside the lungs, also known as a **partial vacuum**. Air comes in through the open mouth or nasal passages to counter the partial vacuum within the lungs, and fills the alveoli.

Contraction of the diaphragm increases the volume of the thoracic cavity by flattening the curvature of the muscle.

All the steps become more frequent and exaggerated when you are exercising and thus breathing deeply. For example, the abdominal muscles and intercostal muscles achieve a greater initial thoracic volume. This leads to deeper breathing and more air moving into the lungs.

Following very heavy exercise, the pain you may feel in your midsection may be because you have overused the various muscles involved in frequent and deep breathing. These muscles include the diaphragm, abdominal muscles and intercostal muscles.

Challenge yourself

1. The process leading to an expiration (breathing out) is the result of the opposite action of muscles compared to inspiration. Using the five steps above, list the steps of an expiration. The layer of muscles that contract and move the rib cage are the **internal intercostal muscles**. The first step is given to you:

   Step 1 The diaphragm relaxes, decreasing the volume of the thoracic cavity.

B3.1.6 – Lung volume

B3.1.6 – Measurement of lung volumes
Application of skills: Students should make measurements to determine tidal volume, vital capacity, and inspiratory and expiratory reserves.

A device known as a **spirometer** is used to measure lung volume. A range of air volumes can be measured, including the following.

- Tidal volume – the volume of air that is breathed in or out during a typical cycle when a person is at rest. The term "tidal" volume comes from the idea of an ocean tide coming in and out.
- Inspiratory reserve volume – the maximum volume of air that a person can breathe in (measured from the maximum point of the tidal volume).
- Expiratory reserve volume – the maximum volume of air that a person can breathe out (measured from the minimum point of the tidal volume).
- Vital capacity – the sum of the inspiratory reserve volume, the tidal volume and the expiratory reserve volume.

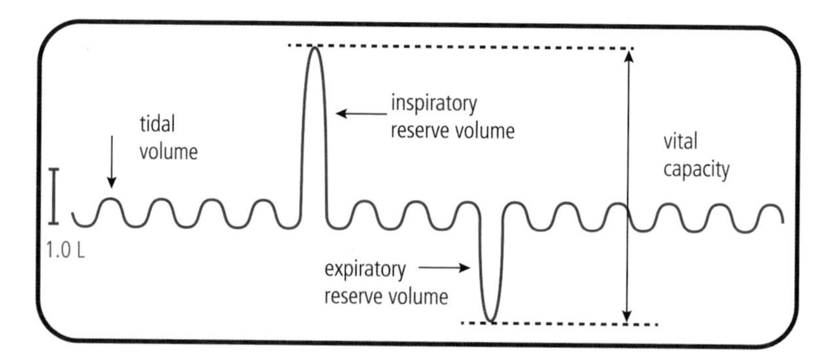

Figure 4 A graphic representation of the various lung volumes that can be measured or calculated with the help of a spirometer.

SKILLS

The IB asks you to take the measurements shown in Figure 4 using an inexpensive spirometer.

Study your spirometer and take note of the units and graduations that you will be using. Most spirometers do not hold a position on these markings, so you need to be able to read numbers from the device very quickly.

1. First, use the spirometer to measure a person's tidal volume. Ask your subject to stand and breathe in and out using the device for about 15 seconds. When you think their breathing pattern is stable, measure the low and high volume readings for each breath. The difference between low and high readings is their tidal volume.
2. Give your subject a rest period of at least one minute. Ask your subject to breathe in and out normally again (recording the tidal volume), then at some point, when they seem comfortable, ask them to take in the maximum volume of air they can. Record this maximum value. Calculate their inspiratory reserve volume by subtracting the maximum reading of their tidal volume from this maximum inspiration volume.
3. Once again give your subject a rest period, so that they can relax. This time ask them to breathe in and out normally (tidal volume) and at some point, when they are comfortable, ask them to release the maximum volume of air out they can. Calculate their expiratory reserve volume by subtracting the minimum reading of their tidal volume from this maximum expiration volume. (Note that not all spirometers can measure the volume of air taken in.)
4. Calculate your subject's vital capacity by summing the inspiratory reserve volume, tidal volume and expiratory reserve volume.

i According to the Institute for Health Metrics and Evaluation (IHME), 13% of deaths worldwide are the direct result of smoking tobacco products, with an additional 2% of deaths the result of second-hand smoking. Lung cancer and chronic obstructive pulmonary disease (COPD) are two of the principal diseases leading to these fatalities.

B3.1.7 – Gas exchange in leaves

B3.1.7 – Adaptations for gas exchange in leaves
Leaf structure adaptations should include the waxy cuticle, epidermis, air spaces, spongy mesophyll, stomatal guard cells and veins.

Plants are adapted to exchange respiratory gases with the atmosphere. A typical leaf is thin, comprising only a few cell layers, so that the diffusion of gases can be quick and efficient. This also permits a relatively large surface area-to-volume ratio for efficient diffusion. There are two primary energy-related processes within plants, cell respiration and photosynthesis. The cells of a plant are always using aerobic cell respiration to synthesize adenosine triphosphate (ATP) molecules for energy-requiring reactions. In addition, when light is available, plants are using photosynthesis to make sugars as fuel for cell respiration. The summary reactions for cell respiration and photosynthesis are the opposite of each other.

Cell respiration: glucose + oxygen → carbon dioxide + water

Photosynthesis: carbon dioxide + water → glucose + oxygen

The rates of these two sets of reactions are not equal, however. In a plant the rate of cell respiration is fairly constant, while the rate of photosynthesis is heavily dependent on light availability. When conditions are optimal for photosynthesis its rate is far greater compared to cell respiration.

Look at the summary reactions again. During the day, when photosynthesis can occur at a high rate, a plant requires carbon dioxide and releases oxygen. A leaf has various structural adaptions to help facilitate this exchange of gases. Refer to Figure 5 in the next section as you read about these adaptations.

- A **waxy cuticle**: a wax lipid layer that covers the surface of leaves and prevents uncontrolled and excessive leaf water loss by evaporation.
- An upper epidermis: small cells on the upper surface of leaves that secrete a waxy cuticle.
- **Palisade mesophyll**: a densely packed region of cylindrical cells in the upper portion of the leaf. These cells contain numerous chloroplasts and are located to receive maximum sunlight for photosynthesis.
- **Spongy mesophyll**: these loosely packed cells are located below the palisade layer and just above the stomata. They have few chloroplasts and many air spaces, providing a large surface area for gas exchange.
- Veins: these structures enclose the fluid transport tubes called **xylem** and **phloem**. Water moves up from the root system to the leaves in the xylem. Water and dissolved sugars are distributed to other parts of the plant in the phloem. Veins are located centrally within a leaf, to provide access to all the cell layers.
- A lower epidermis: small cells on the lower surface of leaves that secrete a waxy cuticle. Guard cells forming stomata are embedded in this layer.
- **Stomata** (singular stoma): numerous microscopic openings on the lower surface of leaves. Each stoma is composed of two **guard cells**. A pair of guard cells can create an opening or close it, as needed. When open, stomata permit carbon dioxide to enter the leaf and at the same time water vapour and oxygen to exit the leaf. These three gases move by diffusion as a result of their concentration gradients. At night many plants close their stomata. Their location on the lower surface of leaves limits water loss as a result of transpiration, because the lower surface of leaves experiences lower temperatures compared to the upper surface.

B3.1.8 – Leaf tissue distribution

B3.1.8 – Distribution of tissues in a leaf
Students should be able to draw and label a plan diagram to show the distribution of tissues in a transverse section of a dicotyledonous leaf.

Figure 5 on the next page shows how the different tissues are distributed within a leaf.

The IB requires you to be able to draw and label a sectioned plan diagram of a dicotyledonous leaf, similar to Figure 5. However, the internal detail of the cell types shown here is not required for a plan diagram, and a two-dimensional diagram is acceptable.

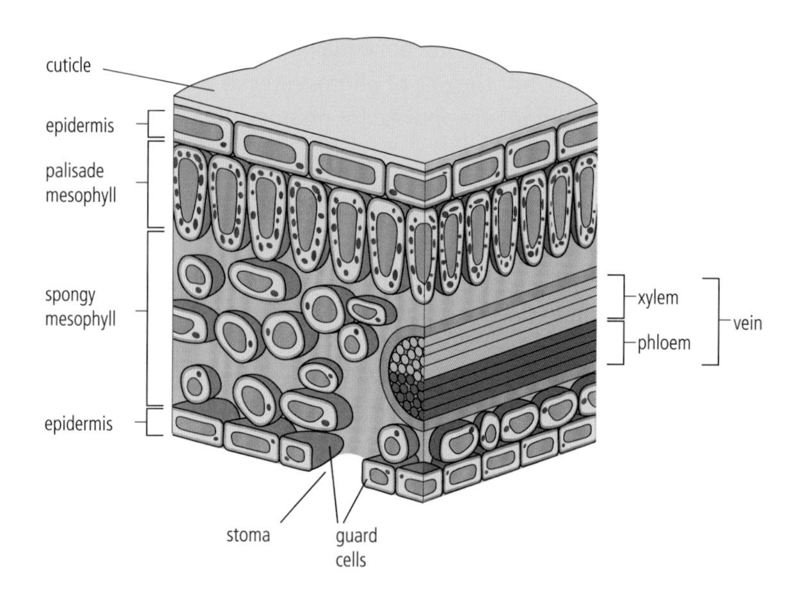

B3.1 Figure 5 A transverse section of a dicotyledonous leaf.

B3.1.9 – Transpiration

B3.1.9 – Transpiration as a consequence of gas exchange in a leaf
Students should be aware of the factors affecting the rate of transpiration.

The evaporation of water through open stomata is called **transpiration**. Transpiration is a natural consequence of a leaf's function to accomplish photosynthesis. A leaf must open its stomata to allow carbon dioxide to enter as a reactant of photosynthesis. Excess oxygen is then diffused out while the stomata are opened. The mesophyll area of the leaf is very humid and water will also evaporate through any open stomata. The leaf can open or close its stomata but it cannot filter which gases pass through the openings. The water evaporated can be traced back to the water that entered the roots and has now reached the upper sections of the plant. Transpiration can amount to a significant volume of water when conditions are optimal. The factors that influence the rate of transpiration are described in Table 1.

B3.1 Table 1 Factors affecting the rate of transpiration

Environmental factor	Effect on rate	Explanation
Increased light (see note below)	Increases	Light stimulates guard cells to open stomata. Increased light also stimulates the rate of photosynthesis to increase. Open stomata permit diffusion of carbon dioxide in and oxygen out
Increased temperature	Increases	Increased molecular movement, including increased evaporation of water
Increased wind speed	Increases	Wind removes water vapour at the entrance of stomata, thereby increasing the water concentration gradient between the inside and outside of the leaf
Increased humidity	Decreases	Increased humidity lessens the water concentration gradient between the inside and outside of the leaf

Note that if a lack of light results in stomata being closed, the rate of transpiration will be zero. In that situation changing the other three factors will have no effect.

B3.1.10 – Stomata

B3.1.10 – Stomatal density
Application of skills: Students should use micrographs or perform leaf casts to determine stomatal density. **NOS:** Reliability of quantitative data is increased by repeating measurements. In this case, repeated counts of the number of stomata visible in the field of view at high power illustrate the variability of biological material and the need for replicate trials.

Studies have shown that the density of stomata varies between species of plants and even varies within a single species based on long-term environmental factors. In order to study any factor that may be correlated with stomata density, you need to be able to view the stomata and measure the area you are viewing. Stomata density can be expressed as number of stomata mm^{-2} or number of stomata μm^{-2}.

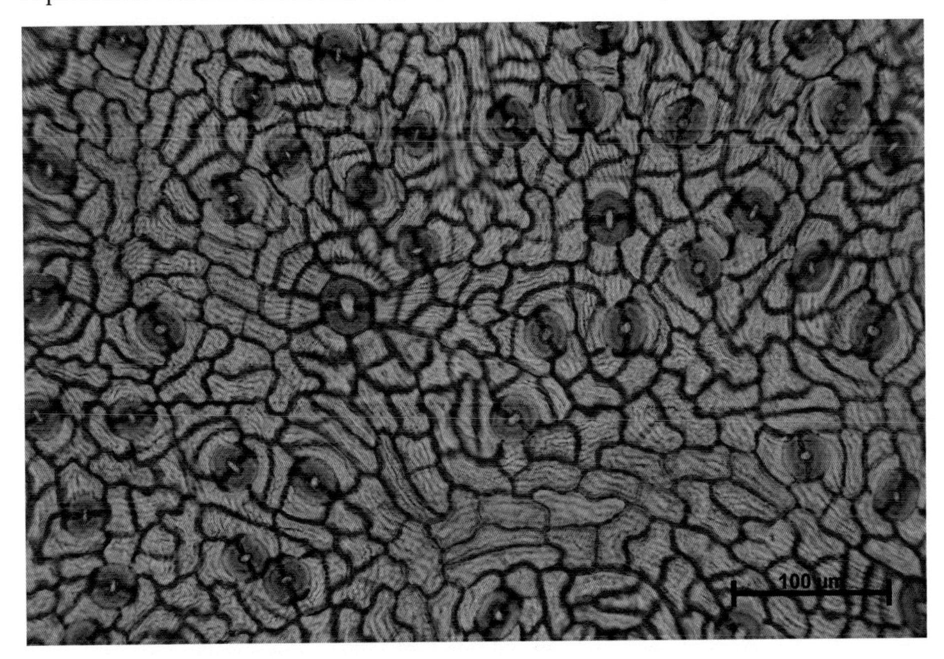

B3.1 Figure 6 A light micrograph of an unknown species of leaf showing paired guard cells forming stomata. The scale bar in the lower right corner allows you to calculate the area of the photograph.

Challenge yourself

2. Calculate the density of stomata (stomata mm^{-2}) in Figure 6. You will need a metric ruler. Measure the scale bar showing 100 µm using the metric ruler, and then measure the length and width of the photograph. Use the proportions of the measured photo to the measured scale bar to calculate the actual length and width. Express your stomata count to the nearest whole number of stomata; count any stoma where at least one half of the stoma can be seen. Express your stomata density to the nearest whole number of stomata mm^{-2}.

Nature of Science

When working with living material, it is normal for a great deal of variation to exist in factors such as stomatal density. Replicated counts can minimize any variation arising from human error.

The IB requires you to use microscopic leaf sections to determine stomata density. This can be accomplished in a variety of ways, and some suggestions for viewing stomata and taking the measurements necessary for calculating the density are given here. These suggestions are not intended to provide step by step directions, but instead are hints that will help you in an investigation of your own design.

- If available, learn to use a micrometre slide to help calculate areas. This is a microscope slide that has very precise length measurements etched into it.
- Alternatively, access online sources for the diameter of field of view at various powers of magnification. Apply the formula for the area of a circle once you know the diameter.
- A cast of a stomata can be made using clear fingernail polish and clear sticky tape. Paint the underside of a leaf with clear fingernail polish and let it dry for 8–10 minutes. Stick clear tape over the dried polish area, gently smooth it to get a good adhesion, and then peel back the tape. Transfer the tape to a clean microscope slide and view it under a light microscope (use clear tape not "magic" tape).
- If available, use a set of micrographs of stomatal tissue that have scale bars included. The scale bar will allow you to calculate the total area of the photograph, as you did in the Challenge yourself.

HL

B3.1.11 – Haemoglobin and oxygen transport

B3.1.11 – Adaptations of foetal and adult haemoglobin for the transport of oxygen
Include cooperative binding of oxygen to haem groups and allosteric binding of carbon dioxide.

Haemoglobin is the protein molecule found within red blood cells (**erythrocytes**) that is responsible for carrying most of the oxygen within the bloodstream. Each erythrocyte is basically a plasma membrane surrounding cytoplasm filled with haemoglobin molecules. Erythrocytes have no nuclei and few organelles. Each haemoglobin molecule is capable of reversibly binding to both oxygen and carbon dioxide molecules.

Chapter B1.2 discusses how each haemoglobin molecule is composed of four polypeptides, and has a quaternary structure. Each polypeptide has a haem group near its centre, and each haem group has an iron atom within it. When haemoglobin reversibly binds to an oxygen molecule, it is the iron atom within the haem group that is bonding with the oxygen. Because haemoglobin has a total of four iron atoms within four haem groups, it has the capacity to transport four oxygen molecules ($4O_2$). When in that form, haemoglobin is said to be **saturated**.

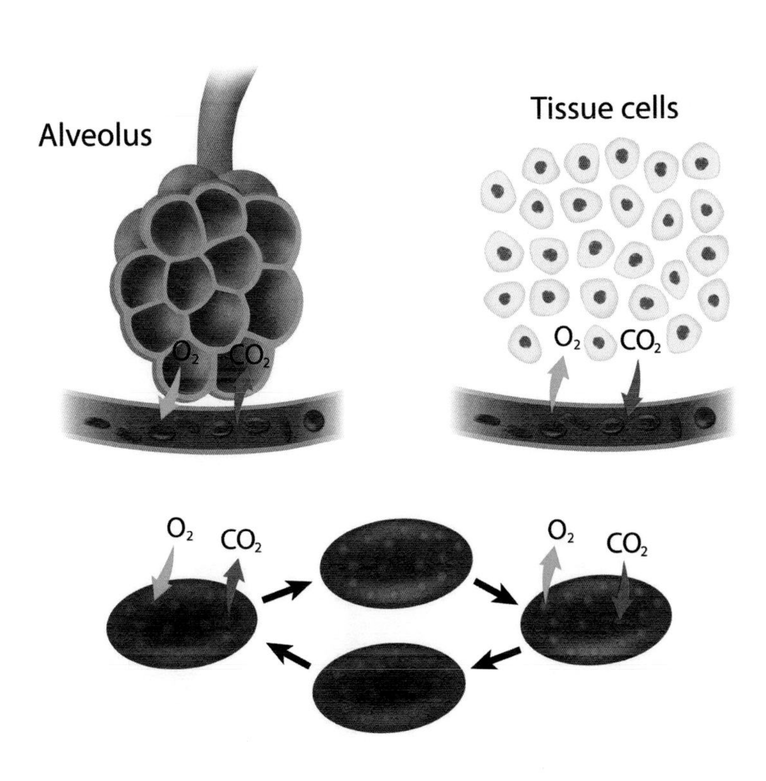

There are only two locations where the blood gases oxygen (O_2) and carbon dioxide (CO_2) are exchanged. Diffusion in the lungs leads to more oxygen in the blood stream and less carbon dioxide. Diffusion in body tissues leads to a decrease of oxygen in the blood and an increase in carbon dioxide. The only blood vessels capable of molecular exchanges are capillaries.

Cooperative binding of oxygen to haemoglobin

The molecular shape of haemoglobin is influenced by its bonding with oxygen molecules. Any oxygen molecule bonded to haemoglobin increases its affinity (attraction for) more oxygen. This phenomenon is called **cooperative binding**, because the oxygen molecules are acting in concert with each other to increase the haemoglobin's affinity for oxygen.

Haemoglobin molecules that are already carrying three oxygen molecules have the greatest affinity for oxygen. Conversely, haemoglobin molecules that are carrying no oxygen molecules have the least affinity for oxygen. You might think that this does not make sense, but it does when you remember that each oxygen molecule that binds to haemoglobin changes the haemoglobin's shape in a way that increases its affinity for another oxygen molecule. Haemoglobin can carry a maximum of four oxygen molecules, so one that is already carrying four oxygens has no affinity for oxygen.

Oxygen molecules bind to identical areas within haemoglobin (iron atoms) and their synergistic effect of increasing oxygen affinity is called cooperative binding. Carbon dioxide binds to a different area of haemoglobin called an allosteric site and the resulting decrease in oxygen affinity is an example of allostery.

Allosteric binding of carbon dioxide to haemoglobin

As already mentioned, haemoglobin can bind to carbon dioxide as well as oxygen.

The binding of carbon dioxide to the polypeptide chain(s) of haemoglobin and the resulting change in haemoglobin's affinity for oxygen is called **allostery**. Unlike oxygen, which binds to the iron of the haem group, carbon dioxide binds to the polypeptide regions of the molecule. The area of each polypeptide where carbon dioxide binds is known as the **allosteric site** of the polypeptide. The binding of carbon dioxide to haemoglobin results in an increase in the release of oxygen molecules and is known as the **Bohr shift**. This is covered in more detail in Section B3.1.12.

Some enzymes have **allosteric sites**. If an allosteric molecule decreases or stops the activity of the enzyme it is called an **allosteric inhibitor**. Cyanide is an allosteric inhibitor of an enzyme involved in aerobic cell respiration and can be deadly.

Foetal haemoglobin

The molecular structure of haemoglobin in a foetus is slightly different compared to haemoglobin in an adult. That structural difference enables foetal haemoglobin to have a higher affinity for oxygen compared to the haemoglobin of the mother.

In the **placenta** of a pregnant female, her capillaries come very close to the capillaries of the foetus. This allows molecular exchanges between the mother and foetus, including oxygen and carbon dioxide. Remember that the mother is breathing but the foetus is not. The foetus is actively carrying out cell respiration, and the blood sent to the foetal side of the placenta is relatively low in oxygen and high in carbon dioxide. The concentration gradient between the blood of mother and foetus, aided by the foetal haemoglobin's greater affinity for oxygen, encourages diffusion of the mother's oxygen to the foetus.

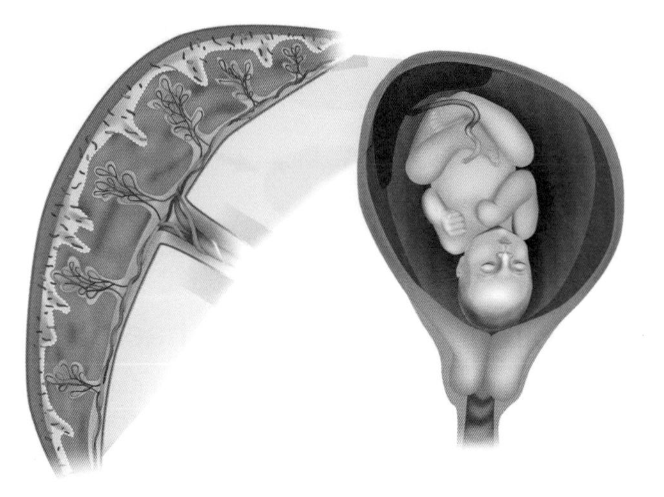

▲ On the right side is a human foetus inside a uterus. The placenta is shown in the upper left of the uterus. The left diagram is a sectioned view of the placenta with a small portion of the umbilical cord extending outwards. Within the placenta there are both foetal capillaries and maternal capillaries. These sets of capillaries do not share blood but they are the site of vital chemical exchanges, including the respiratory gases, oxygen and carbon dioxide, between the mother and the foetus.

B.3.1.12 – The Bohr shift

B3.1.12 – Bohr shift
Students should understand how an increase in carbon dioxide causes increased dissociation of oxygen and the benefits of this for actively respiring tissues.

When haemoglobin bonds to a carbon dioxide molecule, its affinity for oxygen decreases. Another way of saying this is that haemoglobin has a greater tendency to give up oxygen molecules in the presence of carbon dioxide. This change in affinity of haemoglobin in the presence of carbon dioxide is called the **Bohr shift**.

Physiologically this makes perfect sense, as binding to carbon dioxide is most likely to occur where carbon dioxide is at greater concentrations, in muscles and other body tissues, as a product of cell respiration. This is also where oxygen is most needed. The opposite is true in lung tissue. The alveoli of the lungs have relatively low concentrations of carbon dioxide and high concentrations of oxygen, permitting haemoglobin to lose carbon dioxide and thereby giving the haemoglobin a renewed affinity for bonding to oxygen.

Many animals, including humans, make short-term journeys or migrations through high-altitude locations. The short-term physiological response to high altitude includes an increased ventilation rate. Over a longer period of time increases in both haemoglobin and red blood cell production will occur.

Increasing altitude does not alter the percentage of any one gas in the mixture that makes up air. At any altitude oxygen will represent approximately 20% of the gas molecules. However, there are fewer molecules of all the gases at higher altitudes, making oxygen availability challenging.

B3.1.13 – Oxygen dissociation curves

B3.1.13 – Oxygen dissociation curves as a means of representing the affinity of haemoglobin for oxygen at different oxygen concentrations
Explain the S-shaped form of the curve in terms of cooperative binding.

The characteristics of haemoglobin described in Sections B3.1.11 and B3.1.12 can be explained using a graph known as an **oxygen dissociation curve**. The effects of the cooperative binding of oxygen molecules is shown in Figure 7. Notice that the *y*-axis shows the percentage of haemoglobin saturation. This is the percentage of haemoglobin that is transporting the maximum of four oxygen molecules. The *x*-axis shows the **partial pressure** of oxygen. Oxygen partial pressure varies depending on where the blood is within the body. The partial pressure of a gas is the pressure exerted by a single gas within a mixture of gases. The partial pressure of oxygen decreases in the body and blood as it is used for aerobic cellular respiration.

When working out the meaning of an oxygen dissociation curve, the most useful approach is to start with the values for the partial pressure of oxygen on the **x**-axis, then look up to where it intersects the plotted line(s) and then over to the intersect on the **y**-axis for the percentage of haemoglobin saturation. Study the range of oxygen partial pressures shown in Figure 7.

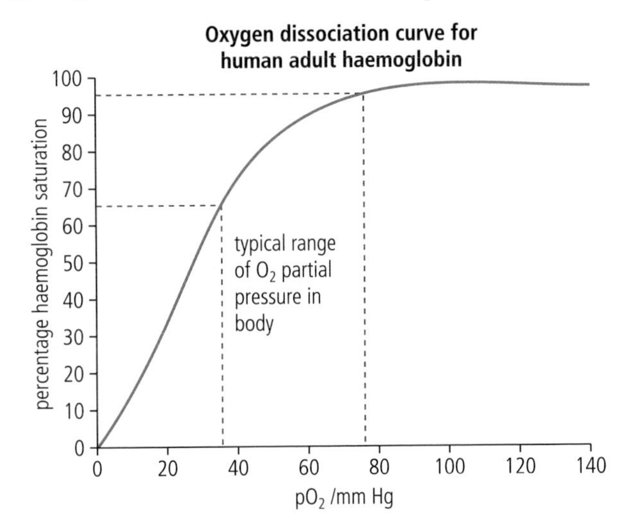

B3.1 Figure 7 This oxygen dissociation curve shows the range of oxygen partial pressure found in the body. The shape of the graph is the result of the cooperativity of oxygen molecules as haemoglobin is loaded to become saturated. Without cooperativity the graph would be linear, based on oxygen partial pressures only. Visualize connecting the first plotted point of the graph to the last plotted point with a straight line. That would be the shape of the graph without cooperativity. Partial pressures are sometimes given in kPa (kilopascals) rather than mm Hg (millimetres of mercury).

Look at the two vertical dashed red lines in Figure 7. The dashed line to the right is the partial pressure that corresponds to the oxygen partial pressure in the lungs. The dashed line to the left corresponds to the oxygen partial pressure in body tissues where cell respiration is utilizing oxygen. If you subtract the left *y*-axis intersect point on the graph (~65%) from the right *y*-axis intersect point (~95%) you can calculate the percentage of oxygen released to the tissues (~30%). These values vary from person to person and are only meant to show the pattern demonstrated by haemoglobin's release of oxygen to respiring tissues.

In Section B3.1.11, you learned about the enhanced affinity of foetal haemoglobin over maternal haemoglobin. We can see this effect by comparing their oxygen dissociation curves (Figure 8). Notice that the foetal haemoglobin's curve is shifted to the left of the mother's, indicating a greater affinity for oxygen at almost every partial pressure of oxygen.

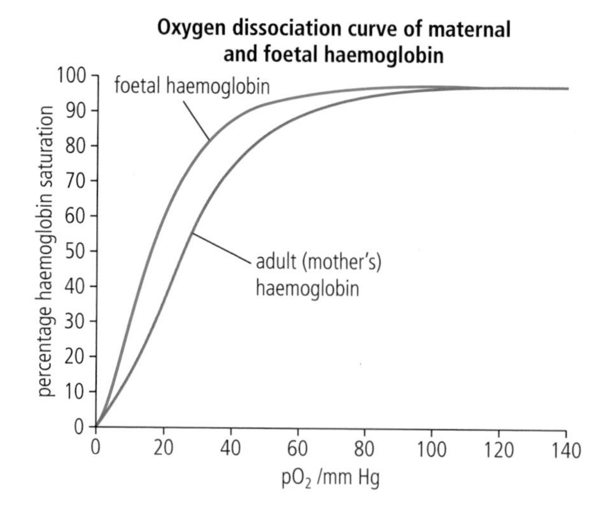

B3.1 Figure 8 Foetal haemoglobin has a greater affinity for oxygen than adult haemoglobin. At all meaningful levels of oxygen partial pressure, foetal haemoglobin is more highly saturated with oxygen compared to the mother's haemoglobin.

You also learned about a phenomenon known as the Bohr shift (Section B3.1.12). When haemoglobin binds to one or more carbon dioxide molecules, its affinity for oxygen is reduced. Figure 9 shows what this looks like using oxygen dissociation curves. Notice that increased carbon dioxide leads to a shift to the right.

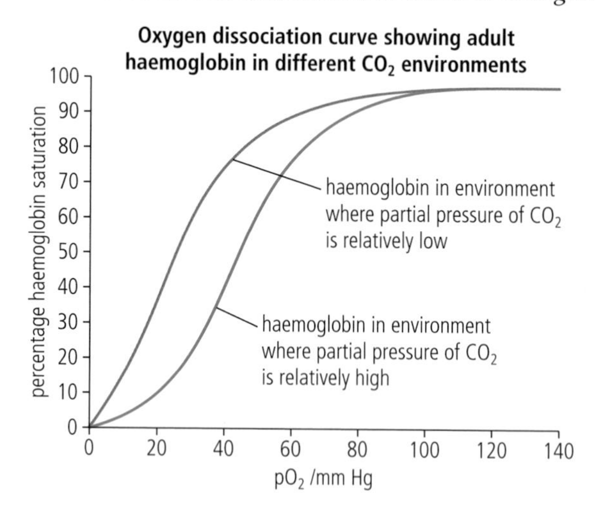

B3.1 Figure 9 This graph illustrates the Bohr shift. Haemoglobin is more likely to release oxygen in an environment where the partial pressure of carbon dioxide is high. This shift to the right represents haemoglobin's ability to release more oxygen once it has bonded to carbon dioxide in the tissues.

Consider the physiological advantage of the Bohr shift when actively exercising. During strenuous activity, your muscles need more oxygen for cell respiration. The increased rate of cell respiration leads to increased carbon dioxide production, which will bind to haemoglobin in the nearby capillaries. The Bohr effect will then lead to an increased release of oxygen to the muscles, where it is needed.

What is the relationship between gas exchange and metabolic processes in cells?

HL end

Guiding Question revisited

How are multicellular organisms adapted to carrying out gas exchange?

In this chapter we have described how:

- multicellular organisms must have adaptations that solve the problem of a decreased surface area-to-volume ratio as organisms gain size
- animals use lungs and gills to exchange respiratory gases between the exterior air and interior tissues
- transport vessels within animals move respiratory gases from the lungs or gills to tissues deep within the organism
- gas exchange surfaces are adapted for efficient diffusion, with continuous blood flow and thin tissue layers
- air within lungs is constantly renewed, with changing volume and pressures within the thoracic cavity

HL

- haemoglobin is a quaternary-structured protein that uses cooperative binding of oxygen molecules to ensure sufficient oxygen is brought to respiring tissues
- carbon dioxide binds to haemoglobin at allosteric sites and results in more oxygen dissociating from haemoglobin (the Bohr effect).

HL end

Guiding Question revisited

What are the similarities and differences in gas exchange between a flowering plant and a mammal?

In this chapter we have discussed how:

- gas exchange of oxygen and carbon dioxide occurs in both flowering plants and mammals
- the exchanges occur in both as a result of diffusion along concentration gradients
- the exchanges occur between thin, moist tissues with minimal cell layers
- mammalian gas exchange occurs between alveoli within the lungs and adjacent blood capillaries
- exchanges in flowering plants occur through multiple small leaf openings called stomata
- flowering plants can stop gas exchanges for periods of time by closing stomata, whereas mammals must have a constant exchange
- during daylight hours, the diffusion of gas exchange in flowering plants shows a net diffusion of oxygen out and carbon dioxide in as a result of photosynthesis
- the diffusion of gas exchange in mammals shows a net diffusion of oxygen in and carbon dioxide out as a result of aerobic cell respiration.

Exercises

Q1. When living organisms are relatively large, the distance from their exterior to the organism's tissues at its centre increases, creating a problem for respiratory gases to be exchanged. A second problem is the small tissue surface area exposed to the air (or water) in relation to the organism's large volume. In your own words, explain this second problem using a comparison of a single-celled amoeba to a multicellular rabbit.

Q2. In terms of diffusion, what happens in a mammal's lungs if its heart stops beating and the blood is no longer transported to and from the capillaries surrounding the alveoli?

Q3. Specific cells inside alveoli secrete a substance called a surfactant. What is the function of this secretion and what could happen if the surfactant was not coating the inside of the alveoli?

Q4. You are looking for a good visual representation of the interior of a small portion of a mammal's lungs. Which of these would be the best physical representation?

A A balloon blown up with air.

B Two balloons blown up.

C A large cluster of grapes.

D A collection of small boxes arranged to look like a lung.

Q5. What is a spirometer used for?

Q6. Leaves are usually broad, flat and thin. What prevents a leaf from desiccating in the hot sun?

Q7. Assume a leaf is in bright sun and its stomata are open. Identify three other environmental factors that would lead to a relatively high rate of transpiration.

HL

Q8. **(a)** In your own words, what is the Bohr shift?

(b) Suggest a reason why the Bohr shift is particularly important when you are exercising strenuously.

Q9. Cooperative binding of oxygen molecules to haemoglobin explains the steep slope in the middle of an oxygen dissociation curve. What does the term cooperative binding mean in this context?

Q10. A plot of foetal haemoglobin shows a curve that is to the left of the mother's curve when both are plotted together on an oxygen dissociation curve. What does this mean and why is it important?

HL end

B3.2 Transport

Guiding Questions

What adaptations facilitate transport of fluids in animals and plants?

What are the differences and similarities between transport in animals and plants?

Chapter B3.1 discusses the adaptations of organisms for efficient gas transport within multicellular organisms. In this chapter the focus will be the transport of fluids in both animals and plants.

In most animals there are blood vessels called arteries and veins that circulate fluids. These two types of blood vessels are connected by capillaries, thin-walled microscopic vessels that permit chemical exchanges with the cellular tissues. The heart pumps blood within these vessels.

In plants transport of fluids is accomplished within vessels known as xylem and phloem. The fluid is water with dissolved substances unique to each type of vessel. The two types of vessels use different mechanisms to ensure water movement.

B3.2.1 – Capillaries and chemical exchange

B3.2.1 – Adaptations of capillaries for exchange of materials between blood and the internal or external environment
Adaptations should include a large surface area due to branching and narrow diameters, thin walls, and fenestrations in some capillaries where exchange needs to be particularly rapid.

Capillaries receive their blood from the smallest of arteries called **arterioles**. Within body tissues an arteriole branches into what is called a **capillary bed**. This is a network of capillaries that all receive blood from the same arteriole. There are millions of arterioles and capillary beds in your body. A single capillary bed will drain its blood into the smallest of veins called a **venule**.

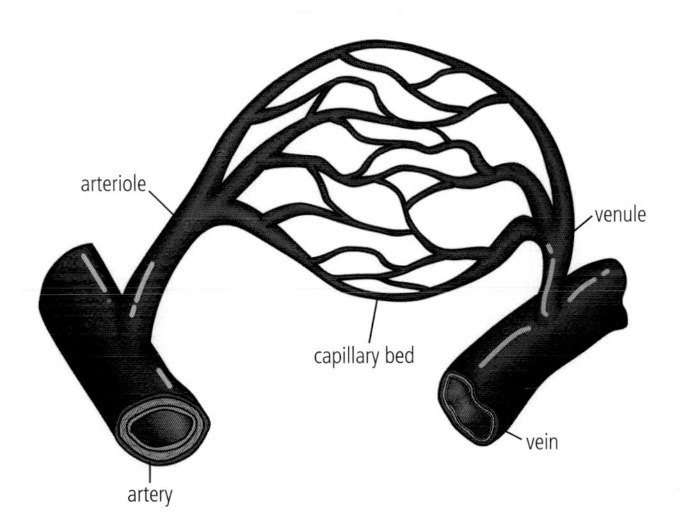

An illustration of an arteriole feeding blood into a capillary bed that drains blood into a venule. All chemical exchanges in the lungs and body tissues happen within these capillary beds.

Many diagrams of the circulatory system use the colour blue for vessels that contain deoxygenated blood, or blue for the deoxygenated blood itself. This is misleading because mammalian blood is never blue and blood vessels themselves are neither blue nor red. It is true, however, that deoxygenated blood is a darker red than oxygenated blood.

When blood enters a capillary bed much of the pressure and velocity of the fluid is lost. Blood cells line up in single file because the **lumen** (inside diameter) of each capillary is only large enough to accommodate one cell at a time. Each capillary is a small tube composed of a single-cell thickness of inner tissue and a single-cell thickness of outer tissue. Both of these cell layers are very permeable to many substances, either through the membranes or between the membranes forming the tube. The total surface area and extensive branching of capillary beds is very high, so no cell in the body is far from a capillary. Some metabolically active tissues in the body are especially enriched with capillary beds. This is referred to as **highly vascular tissue**. Most capillaries exchange molecules within the tissues of an organism, although the capillaries within lungs and gills exchange molecules between the blood and the external environment.

Some tissues have capillary beds that are designed to be even more permeable to substances than a typical capillary. These capillaries are said to be **fenestrated**. The fenestrations are small slits or openings that allow relatively large molecules to exit or enter the blood and allow increased movement of all molecules in a given period of time. Examples of fenestrated capillaries include the numerous small capillaries of the kidneys and areas of the intestine where movement of molecules needs to be rapid.

Capillaries are adapted to their function by:

- having a small inside diameter
- being thin walled
- being permeable
- having a large surface area
- having fenestrations (in some).

B3.2.2 – Arteries and veins

B3.2.2 – **Structure of arteries and veins**
Application of skills: Students should be able to distinguish arteries and veins in micrographs from the structure of a vessel wall and its thickness relative to the diameter of the lumen.

Arteries take blood away from the heart and veins take blood back to the heart. The identification is not based on the level of blood oxygenation.

Arteries and veins are identified according to whether the vessel receives blood from the heart and takes that blood to a capillary bed (artery) or receives blood from a capillary bed and takes that blood back to the heart (vein).

Because arteries receive blood directly from the heart and the blood is under relatively high pressure, they are lined with a thick layer of smooth muscle and elastic fibres. The lumen of arteries is relatively small compared to veins. Veins receive low pressure blood from capillary beds. They are relatively thin walled with a large lumen to carry the slow-moving blood.

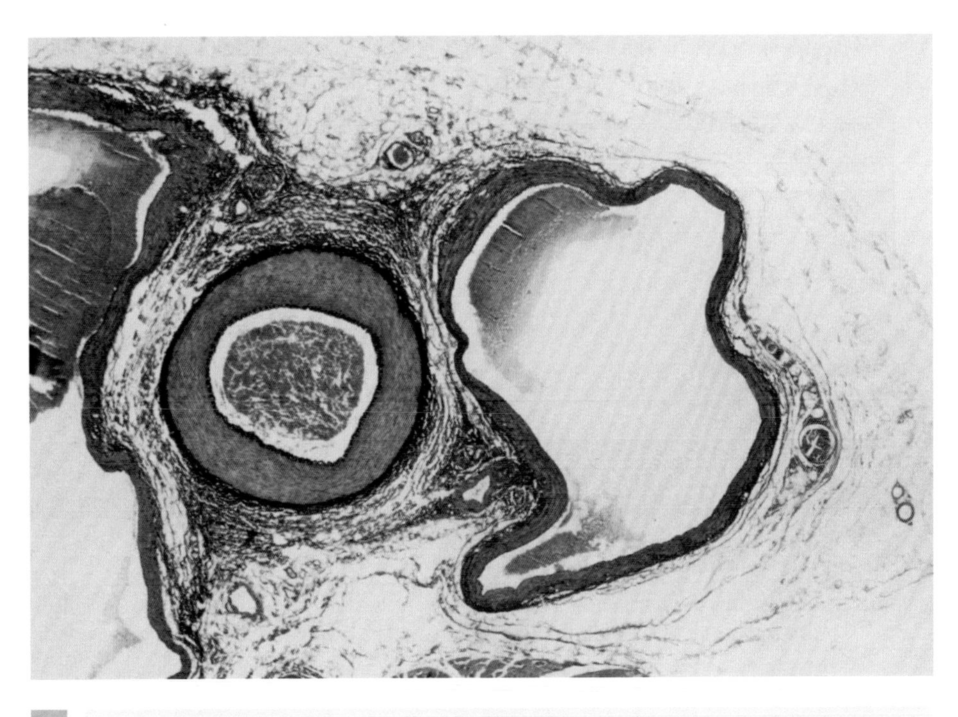

B3.2 Figure 1 Light micrograph of an artery (on the left) and a vein (on the right). There is residual blood filling most of the lumen of the artery. Notice that the artery has a relatively thick wall and small lumen. In comparison, the vein is thin walled and has a large lumen.

You need to be able to identify an artery and a vein from a micrograph similar to Figure 1. The thicknesses of the walls and the lumen size will help you identify the two types of vessels.

B3.2.3 – Adaptations of arteries

B3.2.3 – Adaptations of arteries for the transport of blood away from the heart
Students should understand how the layers of muscle and elastic tissue in the walls of arteries help them to withstand and maintain high blood pressures.

Arteries are adapted to transport high pressure blood away from the heart. The heart contracts and relaxes on a rhythmic schedule. When the heart contracts a surge of blood enters an artery and its branches. Each artery has a relatively thick layer of **smooth muscle** controlled by the **autonomic nervous system** (ANS). The ANS controls those functions in your body that are necessary but not controlled consciously. The smooth muscle changes the lumen diameter of arteries to help regulate blood pressure.

In addition to smooth muscle, the wall of each artery contains the proteins elastin and collagen. The muscular and elastic tissues permit arteries to withstand the high pressure of each blood surge and keep blood continuously moving. When blood is pumped into an artery, the elastin and collagen fibres are stretched and allow the blood vessel to accommodate the increased pressure. Once the blood surge has passed the elastic fibres recoil and provide further pressure, propelling the blood forwards within the artery. In this way the blood in arteries maintains a high pressure between pump cycles of the heart.

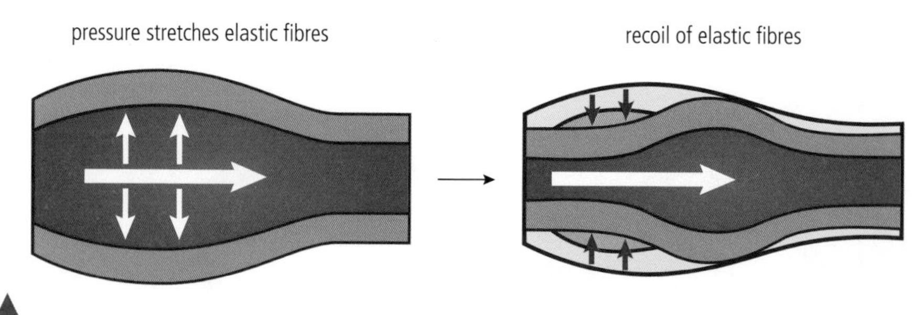

High pressure caused by each contraction of the heart pushes outwards on the elastic wall of each artery. In between contractions the stretched area recoils and helps maintain the high pressure characteristic of arteries.

B3.2.4 – Measuring the pulse rate

B3.2.4 – Measurement of pulse rates
Application of skills: Students should be able to determine heart rate by feeling the carotid or radial pulse with fingertips. Traditional methods could be compared with digital ones.

The pulse rate or heart rate is a measurement of the number of times your heart beats in a minute. Each time the heart contracts and sends blood directly into arteries, the "pulse" of pressure described in Section B3.2.3 can be felt in an artery.

You can take your own pulse rate by feeling for the pulse using your index and middle fingers at two possible locations.

- The carotid artery – feel for this artery on either side of your trachea (windpipe) in your neck.
- The radial artery – feel for this artery on your wrist with the palm of your hand facing upwards. You should feel the pulse 2 cm from the base of your thumb.

It may take a little practice, but you can learn to identify the feeling of the arteries as they "pulse" with each heartbeat. Once you are confident in identifying the pulse, use a clock or timer to determine the pulse rate. You can count for the full 60 seconds or alternatively count for 30 seconds and then multiply your count by two. Remember that a data set requires repetition to minimize uncertainty in measurement. Once you have a good data set using the traditional method of feeling for the pulse, compare that data set with one or more generated using digital methods.

B3.2.5 – Adaptations of veins

B3.2.5 – Adaptations of veins for the return of blood to the heart
Include valves to prevent backflow and the flexibility of the wall to allow it to be compressed by muscle action.

Veins are blood vessels that return blood back to the heart after the blood has passed through a capillary bed. Blood loses a great deal of pressure and velocity in capillary beds. To account for this, veins have thin walls and a larger internal diameter. The unidirectional flow of the relatively slow-moving blood in veins is aided by internal valves that help prevent backflow of the blood. In addition, the thin walls of veins are easily compressed by surrounding muscles. One of the many reasons to stay active!

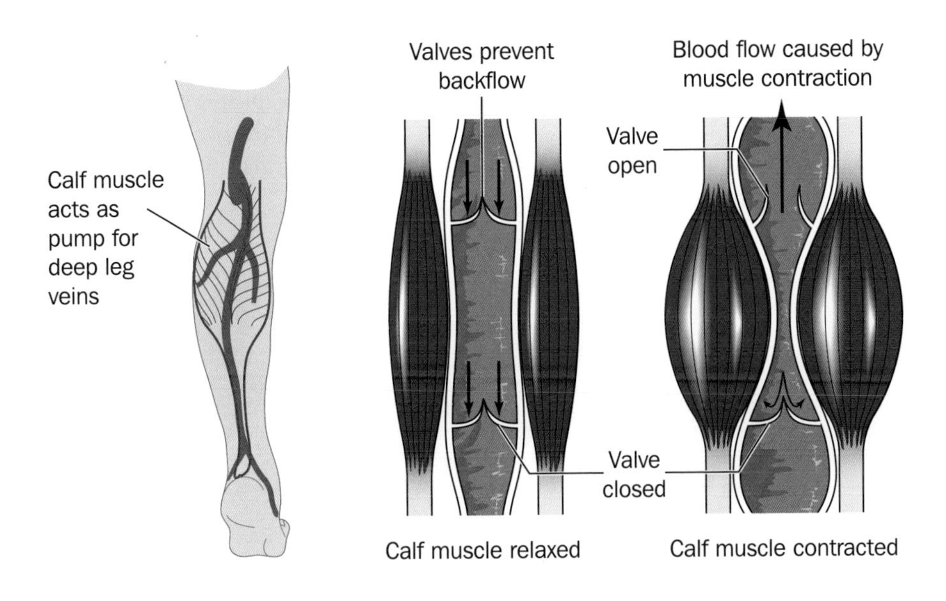

Contraction of skeletal muscle stimulates blood flow by squeezing veins. Internal valves ensure a one-way flow.

B3.2.6 – Occlusion of coronary arteries

B3.2.6 – Causes and consequences of occlusion of the coronary arteries
Application of skills: Students should be able to evaluate epidemiological data relating to the incidence of coronary heart disease. **NOS:** Students should understand that correlation coefficients quantify correlations between variables and allow the strength of the relationship to be assessed. Low correlation coefficients or lack of any correlation could provide evidence against a hypothesis, but even strong correlations such as that between saturated fat intake and coronary heart disease do not prove a causal link.

The heart is a very active and thick muscle. Like any muscle it requires oxygen and nutrients to stay active and healthy. The arteries that supply blood to cardiac muscle are called **coronary arteries**.

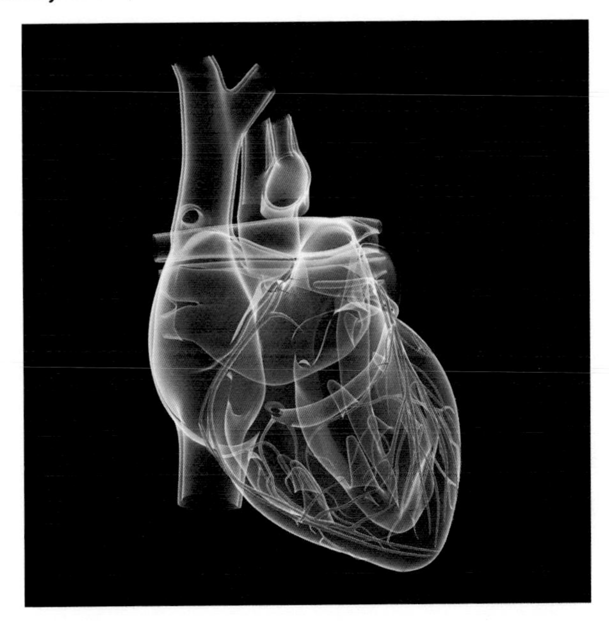

A depiction of the coronary arteries (in red) that feed oxygen and nutrients directly into the muscle tissue of the heart.

Coronary heart disease is a term often used to describe the narrowing of coronary arteries by plaque.

Over time a person may develop a build-up of cholesterol and other substances in the lumen of arteries. This build-up is called **plaque** and the restriction in blood flow it causes is called an **occlusion**. As you can see in Figure 2, plaque build-up is progressive and can severely decrease the artery's blood flow. If the occluded artery is a coronary artery, it may result in a heart attack because the cardiac muscle in one or more areas of the heart will be deprived of oxygen.

B3.2 Figure 2 Progressive stages in the build-up of plaque, consisting of cholesterol and other lipids. If blood flow is restricted within one or more of the coronary arteries, an occlusion can lead to a heart attack.

You need to be able to evaluate **epidemiological** data relating to the incidence of coronary heart disease (CHD). Epidemiological studies deal with the incidence, distribution and control of conditions such as heart disease. Many factors have been correlated with CHD, including sex, age, family history, diet, diabetes, hypertension, high cholesterol, weight and smoking. It is very difficult to measure the effects of any one factor and its impact on the incidence of CHD. Almost all factors have an impact on one or more other factors. One of the most important considerations in evaluating data with any of these risk factors is to remember that you are considering a correlation and not a "cause and effect" relationship.

Challenge yourself

Answer the following questions relating to the figure.

1. When comparing sex and smoking only, which factor appears to have a higher correlation coefficient with the predicted 10-year risk of developing coronary heart disease?
2. At what age do the correlations appear to increase most significantly?

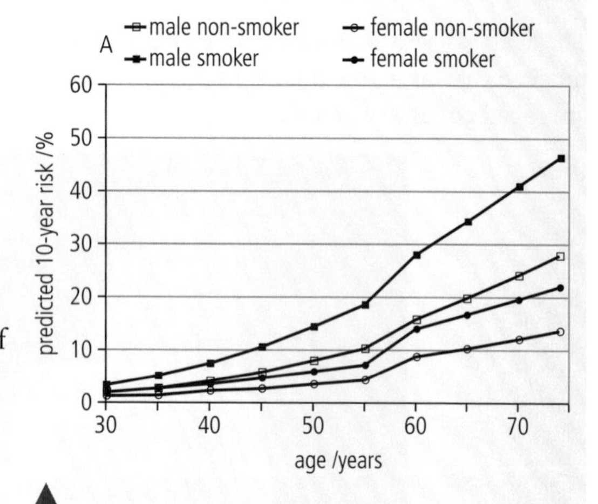

Predicted correlation of age, sex and smoking status with a future 10-year risk of developing coronary heart disease.

3. At what age is the predicted 10-year risk of coronary heart disease 10% for (a) a male non-smoker and (b) a male smoker?

4. One of the methods used for collecting the data was a questionnaire. Evaluate the use of this data collection method.

5. Does the data show that smoking causes an increase in coronary heart disease?

Nature of Science

Correlation coefficients quantify correlations between variables and allow the strength of the relationship to be assessed. Low correlation coefficients or a lack of any correlation can provide evidence against a hypothesis, but even strong correlations such as that between saturated fat intake and coronary heart disease do not prove a causal link.

B3.2.7 – Water transport from roots to leaves

B3.2.7 – Transport of water from roots to leaves during transpiration
Students should understand that loss of water by transpiration from cell walls in leaf cells causes water to be drawn out of xylem vessels and through walls by capillary action, generating tension (negative pressure potentials). It is this tension that draws water up in the xylem. Cohesion ensures a continuous column of water.

Unlike many animals, plants do not have a heart to pump fluids for distribution to various tissues. In order to bring water and dissolved minerals up from the roots, a plant relies on a tension force generated by transpiration. Chapter B3.1 describes how transpiration is the evaporation of water from leaves through open stomata. The water is located in the air spaces created by the spongy mesophyll layer of the leaf. The loss of water by transpiration causes water to be pulled through the cell walls of nearby xylem tissue by capillary action. This creates tension (a negative pressure) at the upper end of each xylem tube. The tension results in the movement of water up the xylem, and the entire column of water moves up because of cohesion. This upwards movement of water with dissolved minerals is called the **cohesion-tension theory**. Chapter A1.1 provides the background to this topic.

In a typical plant more than 90% of the water taken in by the roots is lost by transpiration. A mountain range in the USA is called the Great Smokey Mountains because of the continual haze above it created by transpiration from the abundant trees present.

In southwest Utah in the United States there is a quaking aspen tree with so many "tree shoots" that it covers more than 43 ha (108 acres). Each shoot appears to be a separate tree, but DNA analysis has shown that all of the "trees" are in fact a single organism sharing a common root system.

The moisture haze above a portion of the Great Smoky Mountains, USA.

B3.2.8 – Adaptations of xylem vessels

B3.2.8 – Adaptations of xylem vessels for transport of water
Include the lack of cell contents and incomplete or absent end walls for unimpeded flow, lignified walls to withstand tensions, and pits for entry and exit of water.

Plants that form wood have many concentric rings of once active xylem tissue. As the plant grows by increasing its girth, the xylem near the outside (under the bark) still conducts water but the interior xylem does not. The interior xylem provides excellent support, allowing the growth of massive trees.

Imagine many cylinder-shaped plant cells stacked up on each other to make a long tube. When alive these cells would have had complete cell walls, plasma membranes and typical plant cell organelles. Now imagine that all of these cells die leaving behind only their thick cylinder-shaped cell walls. Even the end walls where the cells were joined to each other in the tube completely or partially degenerate. This describes the formation of xylem tubes.

The dead xylem tubes have cell walls fortified with **lignin** for strength. The lignin provides resistance to collapse of the tubes because of the tension created by transpiration. The partial or total lack of cell walls between adjoining cells of the xylem tube allows unobstructed water flow upwards. Xylem also has small pits (microscopic holes) in its sidewalls that allow the easy flow of water in and out as needed.

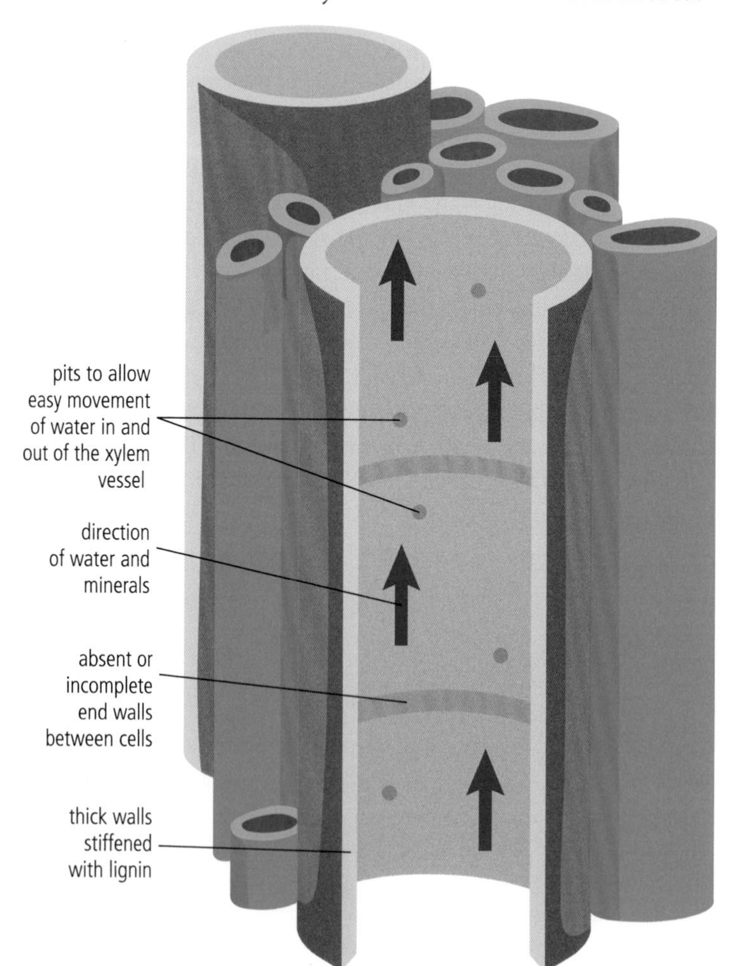

A xylem vessel shown in section. Each vessel extends from a root to the upper parts of the plant where the leaves are located.

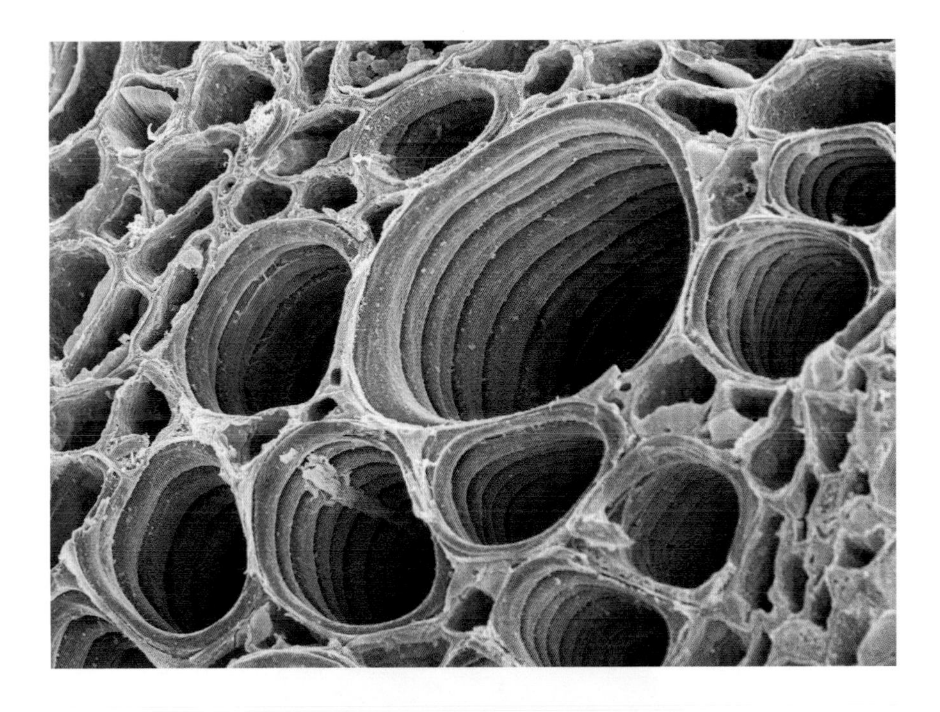

A scanning electron micrograph (SEM) of xylem tissue. You are looking at the dead remains of once living cells that now form tubes. The cell walls are composed of cellulose with a high content of lignin for strength.

B3.2.9 – Tissues in a dicotyledonous stem

B3.2.9 – Distribution of tissues in a transverse section of the stem of a dicotyledonous plant
Application of skills: Students should be able to draw plan diagrams from micrographs to identify the relative positions of vascular bundles, xylem, phloem, cortex and epidermis. Students should annotate the diagram with the main functions of these structures.

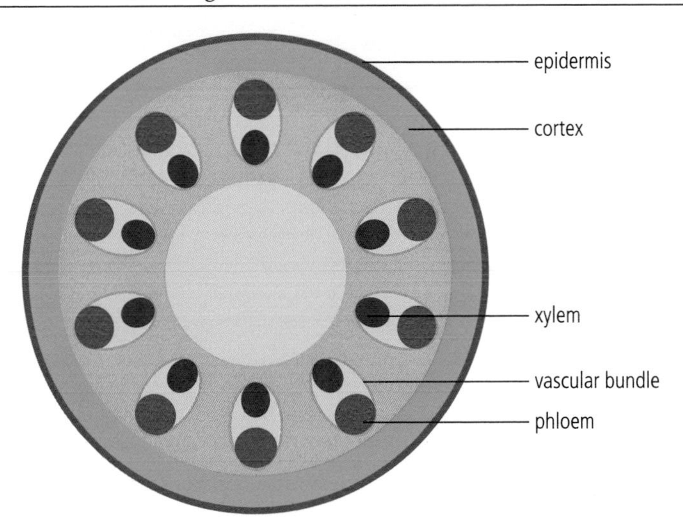

B3.2 Figure 3 A transverse section of a typical dicotyledonous stem. Dicotyledons are one of two categories of flowering plants; monocotyledons are the other category. Only dicotyledons have the arrangement of tissues shown here.

You will be required to draw, label and annotate a transverse section plan diagram of a dicotyledonous stem by looking at microscopic sections of prepared slides. A plan diagram does not show individual cells. Your plan diagram should look similar to the one shown in Figure 3, but keep in mind that actual tissue is likely to be more complex. The annotations should provide brief functions for all the labelled structures. Table 1 outlines the functions of the different tissues.

B3.2 Table 1 A summary of the functions of five dicotyledonous stem tissues

Tissue	Function
Epidermis	Prevents water loss and provides protection from microorganisms
Cortex	An unspecialized cell layer that sometimes stores food reserves
Xylem	Transport tubes that bring water up from the roots
Phloem	Transports carbohydrates, usually from leaves to other parts of the plant
Vascular bundle	Contains multiple vessels of both xylem and phloem

Xylem is the dead leftover walls of cells, while phloem is a living tissue.

B3.2.10 – Tissues in a dicotyledonous root

B3.2.10 – Distribution of tissues in a transverse section of a dicotyledonous root
Application of skills: Students should be able to construct diagrams from micrographs to identify vascular bundles, xylem and phloem, cortex and epidermis.

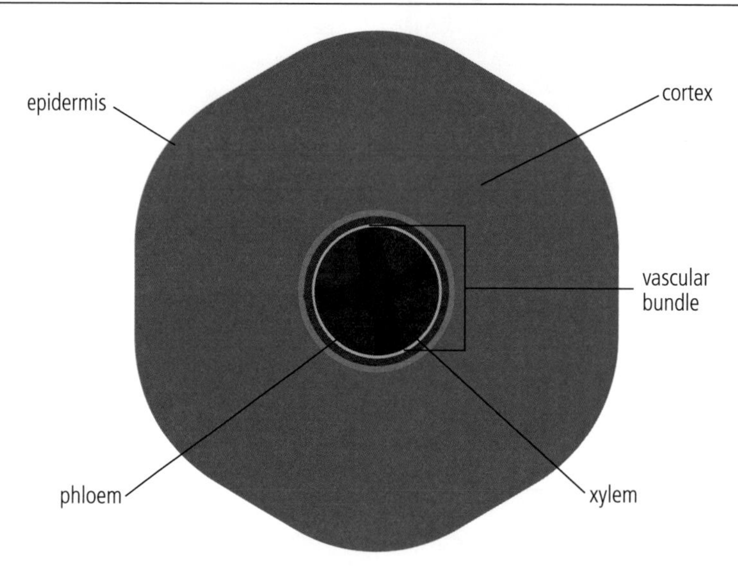

B3.2 Figure 4 A transverse section of a typical dicotyledonous root.

You will be required to draw, label and annotate a transverse section plan diagram of a dicotyledonous root by looking at microscopic sections of prepared slides. A plan diagram does not show individual cells. Your plan diagram should look similar to the one shown in Figure 4, but keep in mind that actual tissue is likely to be more complex. The annotations should provide brief functions for all the labelled structures. Table 2 outlines the functions of the different tissues.

B3.2 Table 2 Tissues and their functions within a dicotyledonous root

Tissue	Function
Epidermis	Grows root hairs that increase the surface area for water uptake
Cortex	An unspecialized cell layer that stores food reserves
Xylem	Transport tubes for water and minerals, starting in the roots
Phloem	Transport tubes that receive sugars from leaves
Vascular bundle	The area in the centre of the root containing xylem and phloem

You may be presented with plan diagrams of a dicotyledonous stem and/or root and asked to provide labels and annotations. Practise for this possibility.

HL

B3.2.11 – Capillaries and tissue fluid

B3.2.11 – Release and reuptake of tissue fluid in capillaries
Tissue fluid is formed by pressure filtration of plasma in capillaries. This is promoted by the higher pressure of blood from arterioles. Lower pressure in venules allows tissue fluid to drain back into capillaries.

You have learned that blood is a fluid (plasma) and that cells contain a fluid (cytoplasm). In order for cells to chemically exchange substances with blood, there has to be a fluid between the cells and blood. That fluid is called **tissue fluid**. Think of tissue fluid as the solution that bathes all cells.

Tissue fluid is constantly renewed by being released from the side of a capillary bed closest to the arteriole. An arteriole is the smallest of all arteries and directly branches out into a capillary bed. Within the capillary bed, blood pressure is highest at this end and the release of tissue fluid is called **pressure filtration**. The pressure at the arteriole end of the capillary bed is high enough to open gaps between the cells that make up the wall of the capillary.

At the other end of the capillary bed is the smallest of all veins, a venule. Pressure at this end of the capillary bed is relatively low because it is further away from the direct pulse of the heart. This lower pressure of the capillary bed nearer the venule allows much of the tissue fluid to drain back into the capillaries. See Figure 5.

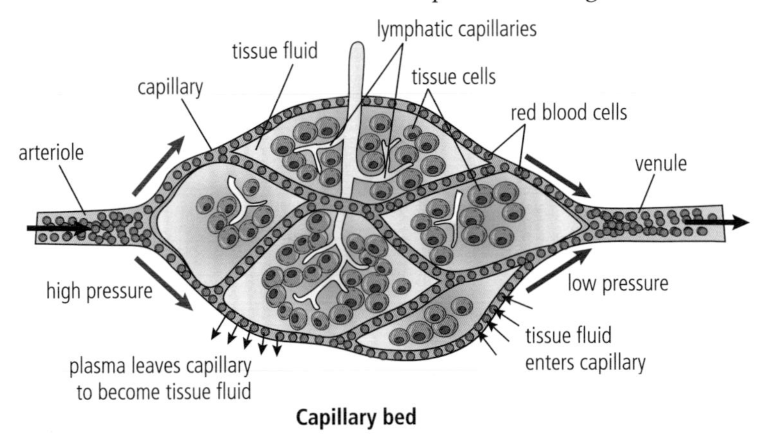

B3.2 Figure 5 The formation and reuptake of tissue fluid by capillaries.

B3.2.12 – Exchange between cells and tissue fluid

B3.2.12 – Exchange of substances between tissue fluid and cells in tissues
Discuss the composition of plasma and tissue fluid.

The chemical makeup of blood plasma and tissue fluid is very similar, because of the largely unregulated passage of substances through very porous capillary membranes and gaps under arteriole pressure. Red blood cells and large proteins do not exit the capillaries, and thus remain in the blood stream, because they are too large to exit through the capillary walls. Some white blood cells are able to squeeze through capillaries into tissue fluid.

Body cells are in constant need of oxygen and a variety of nutrients. In turn, body cells produce waste products such as carbon dioxide and a waste product of amino acid metabolism called urea. Many of the molecules that have natural concentration gradients, such as oxygen, carbon dioxide and glucose, diffuse directly through the cell's plasma membrane or diffuse through protein channels in a process called **facilitated diffusion**.

Unlike the very porous membranes of capillaries, plasma membranes of cells often use active transport mechanisms to regulate the presence of various ions. The concentration of potassium ions (K^+) is typically many times higher in cytoplasm compared to tissue fluid. Conversely, the concentration of sodium ions (Na^+) is many times higher in the tissue fluid compared to cytoplasm. The cell must use adenosine triphosphate (ATP) in active transport mechanisms to keep the high concentrations of these and other ions more concentrated on one side of the plasma membrane. The variety of reasons why there are ion concentrations is beyond the scope of this chapter.

Lymph fluid enters microscopic lymph capillaries within body tissues and the fluid is later returned to blood via larger lymph ducts.

B3.2.13 – Lymph ducts

B3.2.13 – Drainage of excess tissue fluid into lymph ducts
Limit to the presence of valves and thin walls with gaps in lymph ducts and return of lymph to the blood circulation.

Lymphoedema is severe swelling of body tissues, usually in the arms or legs, as a result of the accumulation of tissue fluids that normally drain into the lymphatic system.

Look again at Figure 5 showing a capillary bed producing and reabsorbing tissue fluid. Some of the tissue fluid does not re-enter the venous side of the capillary bed but does enter into small tubes called **lymphatic capillaries**. The small lymphatic capillaries are very thin walled and contain gaps between adjoining cells to facilitate easy movement of water and solutes. Fluid that enters lymphatic capillaries is called **lymph**. The collection of tissue fluid in lymph vessels prevents fluid build-up around body cells. You can see one of the lymph vessels and its small lymphatic capillaries in Figure 5.

Lymph vessels are similar to veins in that they have internal valves to keep fluid moving in one direction. Like veins they rely on skeletal muscle contractions to squeeze the vessels and one-way valves to keep the lymph fluid moving. Also like veins, lymph vessels join together into larger and larger lymph ducts, eventually taking lymph fluid back to veins so that it can become part of blood plasma once again.

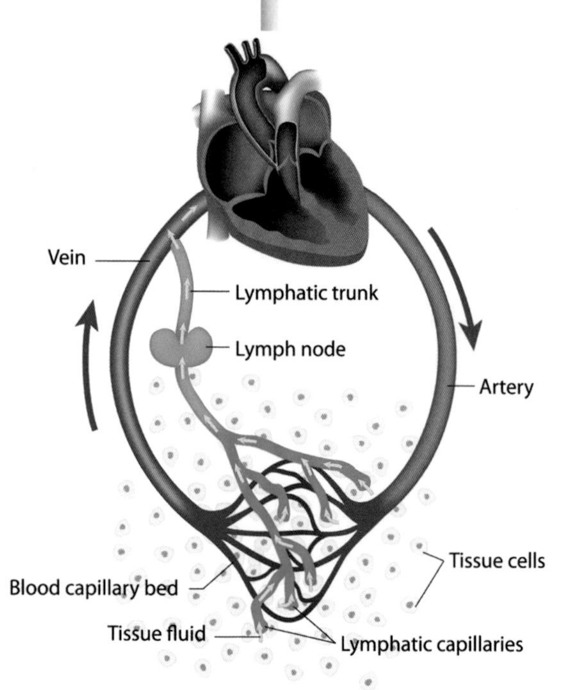

A schematic diagram showing the interaction between the circulatory system and the lymphatic system. Fluid entering small lymphatic vessels is often routed through structures called **lymph nodes** before returning to a vein. Lymph nodes filter bacteria, viruses and sometimes even cancer cells out of the lymph fluid and are considered to be part of the immune system.

B3.2.14 – Single and double circulation

B3.2.14 – Differences between the single circulation of bony fish and the double circulation of mammals
Simple circuit diagrams are sufficient to show the sequence of organs through which blood passes.

The circulation pattern of fish is simple but limiting. Fish have a two-chambered heart, one chamber to receive blood and another chamber to pump the blood out. When blood is pumped out it is sent to the gills for oxygen and carbon dioxide exchange. The reoxygenated blood is collected from the gill capillaries and sent to capillary beds in body tissues. The deoxygenated blood is then returned to the heart to be pumped to the gills once again. The limitation of this circulatory pattern is the loss of blood pressure when the blood is within the capillaries of the gills.

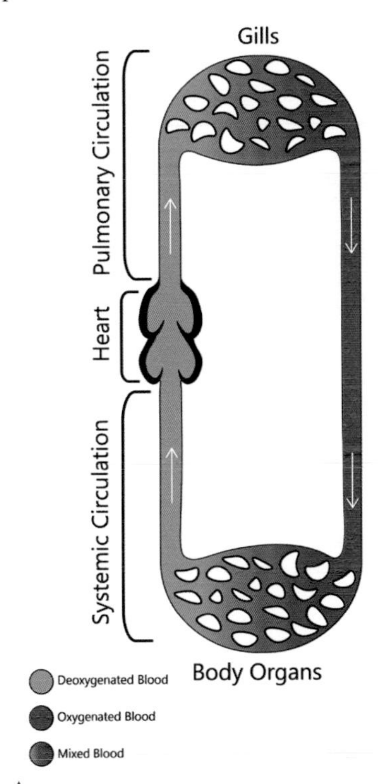

A schematic diagram showing the single circulation pattern of fish. Two capillary beds receive blood from only one contraction of the heart.

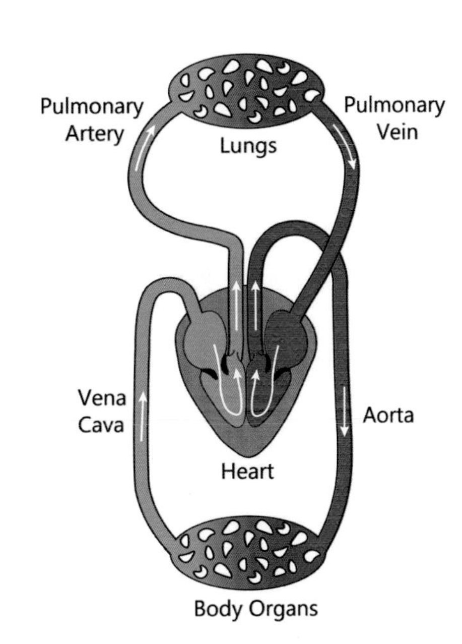

A schematic diagram showing the double circulation pattern of mammals. Each capillary bed in both the lungs and body tissues is supplied with high pressure blood.

Mammals use a double circulation pattern via a heart that has four chambers. One side of the heart is used to pump the blood to capillaries in the lungs for reoxygenation. This is called the **pulmonary circulation**. The blood is returned to the other side of the heart to be pumped out to capillaries in body tissues, to supply the oxygen to where it is needed. This is called the **systemic circulation**. The additional trip to the heart allows the blood pressure to be restored.

When looking at sections of the heart and other body organs, the left and right appear to be the opposite of what you might expect. The sides are relative to the organism and not the observer. Touch your right arm with your left hand and think about what that would look like on a labelled sketch from the front.

B3.2.15 – The mammalian heart

B3.2.15 – Adaptations of the mammalian heart for delivering pressurized blood to the arteries
Include form–function adaptations of these structures: cardiac muscle, pacemaker, atria, ventricles, atrioventricular and semilunar valves, septum and coronary vessels. Students should be able to identify these features on a diagram of the heart in the frontal plane and trace the unidirectional flow of blood from named veins to arteries.

The two sides of the mammalian heart form two major routes for blood to flow along. The right side of the heart sends blood to and from the lung capillaries in a route that is called the pulmonary circulation. The left side of the heart sends blood to and from body tissues via a route called the systemic circulation. The advantage of this double-sided pattern is that both lung and body capillaries can receive blood from arteries and arterioles. This allows pressure filtration to occur in all capillaries, as described in Section B3.2.11.

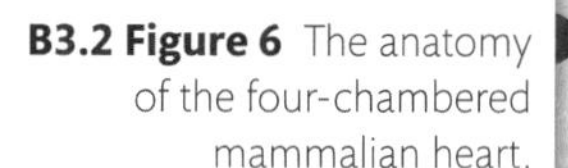

B3.2 Figure 6 The anatomy of the four-chambered mammalian heart.

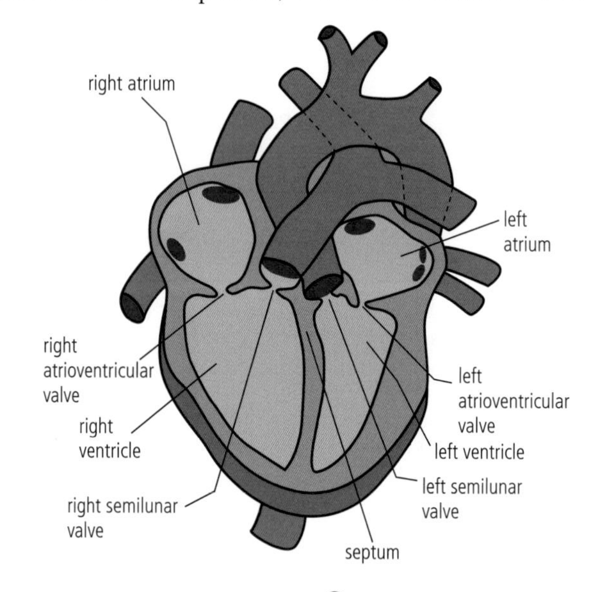

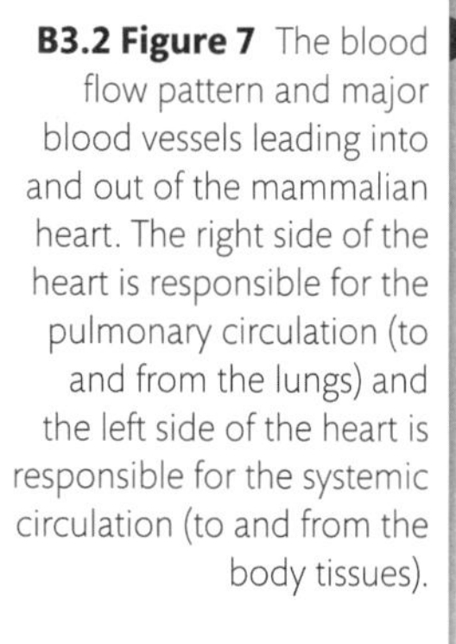

B3.2 Figure 7 The blood flow pattern and major blood vessels leading into and out of the mammalian heart. The right side of the heart is responsible for the pulmonary circulation (to and from the lungs) and the left side of the heart is responsible for the systemic circulation (to and from the body tissues).

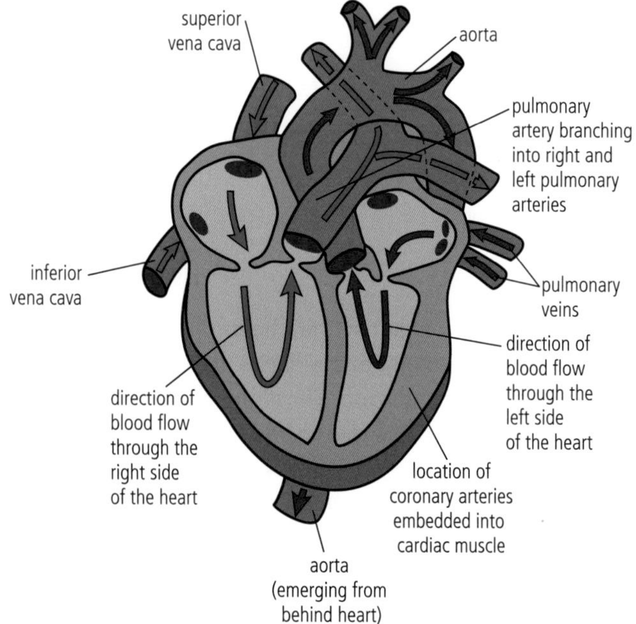

You should learn the names of all the structures labelled in Figures 6 and 7. In addition, you should be able to trace the blood flow as shown in Figure 7, including the names of all the chambers, valves and blood vessels immediately entering and leaving the heart.

Challenge yourself

6. Practise learning blood flow pattern and anatomy by making a sequenced list of all the chambers, valves and blood vessels of one complete circulation, starting with the deoxygenated blood entering into the right atrium. Create your sequence as if you are following one possible circuit of a single red blood cell.

Adaptations for efficient blood flow

The human heart is a double-sided pump, with the right side involved with pulmonary circulation (to and from the lungs) and the left side the systemic circulation (to and from the body tissues). The heart has a variety of adaptations to ensure that both atria contract simultaneously, followed by both ventricles contracting simultaneously. In addition there are four heart valves to make sure that there is only a one-way flow of the blood. Below are some of the more notable adaptations for efficient blood flow.

- Cardiac muscle – a highly vascular tissue making up the heart muscle. This muscle is especially thick in the ventricles of the heart. The muscle making up the wall of the left ventricle is the thickest, as it pumps blood out to locations in the entire body.
- A pacemaker – known as the **sinoatrial node** or SA node. An area of specialized cells in the right atrium generate a spontaneous electrical impulse to start each heartbeat.
- Atria – thin muscular chambers of the heart designed to receive low pressure blood from the capillaries of the lungs or body tissues by way of large veins entering the heart. The atria send blood to the ventricles.
- Ventricles – thick muscular chambers that pump blood out under pressure to the lungs or body tissues.
- Atrioventricular valves – valves located between the atria and ventricles that close each heart cycle to prevent any backflow of blood into the atria.
- Semilunar valves – valves that close after the surge of blood into the pulmonary artery or aorta, to prevent backflow of blood into the ventricles.
- Septum – a wall of muscular and fibrous tissue that separates the right side of the heart from the left side.
- Coronary vessels – blood vessels that provide oxygenated blood to the heart muscle.

Heart valves open and close as a result of differences in blood pressure on either side of the valves. For example, when the left ventricle contracts the initial pressure build-up forces the left atrioventricular valve to close, and then further pressure build-up forces the left semilunar valve to open. This allows blood to be pumped through the aorta without backing up into the left atrium.

B3.2.16 – The cardiac cycle

B3.2.16 – Stages in the cardiac cycle
Application of skills: Students should understand the sequence of events in the left side of the heart that follow the initiation of the heartbeat by the sinoatrial node (the "pacemaker"). Students should be able to interpret systolic and diastolic blood pressure measurements from data and graphs.

If you are presented with blood pressure data, the lowest blood pressure would be associated with the rest period of the ventricles (diastole) and the increase leading to the highest blood pressure would be the muscular contraction of the ventricles (systole). Atrial systole and diastole help keep blood moving through the heart but do not directly affect blood pressure. For example, a blood pressure measurement of 115 / 75 would be a **systolic pressure** of 115 and a **diastolic pressure** of 75. Most pressure measurements are typically given in millimetres of mercury (mm Hg)

The cardiac cycle is a series of events that is commonly referred to as one heartbeat. The frequency of the cardiac cycle is your heart rate, and is typically measured in beats per minute. If you have a resting heart rate of 72 beats min^{-1}, you are performing 72 cardiac cycles each minute.

When a chamber of the heart contracts there is an increase in pressure on the blood within the chamber, and the blood leaves the chamber through any available opening. This is called **systole** (pronounced sis-tol-ee). When a chamber is not undergoing systole, the cardiac muscle of the chamber is relaxed. This is called **diastole** (di-astol-ee). Both atria contract at the same time, therefore you can say that both undergo systole at the same time. Both ventricles also undergo systole simultaneously, just a fraction of a second after atrial systole.

Control of the cardiac cycle by the SA node

SKILLS

How long does it take to return to resting heart rate after mild exercise? Full details of how to carry out this experiment with a worksheet are available in the eBook.

The SA node (sinoatrial node) is a group of modified cardiac muscle cells located in the thin muscle wall of the right atrium. Cardiac muscle is capable of spontaneous contractions without stimulation from the nervous system, but it is not able to control the timing of the contractions. The SA node or "pacemaker" provides an electrical stimulation to regulate the contractions, and is capable of generating electrical impulses at a regular frequency. If your **myogenic** or resting heart rate is 72 beats min^{-1}, your SA node is generating an electrical impulse every 0.8 second. The **action potentials** from the SA node spread out almost instantaneously and result in the thin-walled atria undergoing systole. The SA node action potential also reaches a group of cells known as the **atrioventricular (AV) node**. This node is also located in the right atrium, in the septum between the right and left atria.

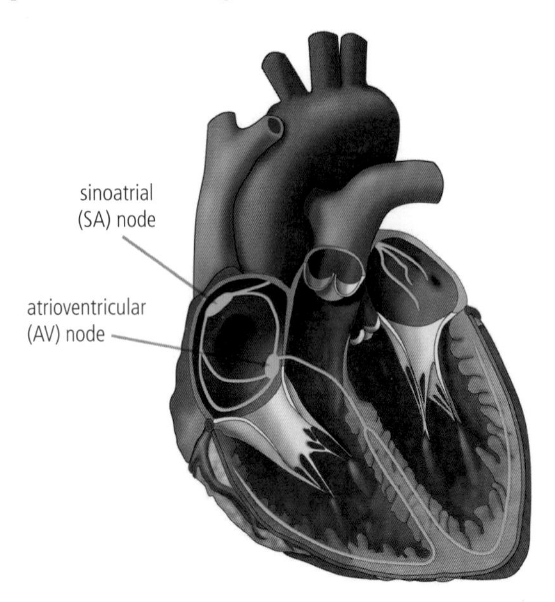

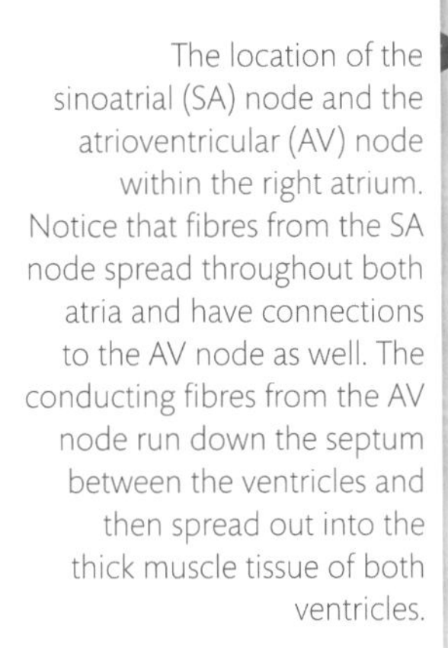

The location of the sinoatrial (SA) node and the atrioventricular (AV) node within the right atrium. Notice that fibres from the SA node spread throughout both atria and have connections to the AV node as well. The conducting fibres from the AV node run down the septum between the ventricles and then spread out into the thick muscle tissue of both ventricles.

The AV node receives the impulse coming from the SA node and delays for approximately 0.1 second. The AV node then sends out its own action potentials that spread out to both ventricles. As you learned earlier, the walls of the ventricles have much thicker muscle tissue than the walls of the atria. In order to get the action potentials to reach all of the muscle cells in the ventricles efficiently, there is a system of conducting fibres that begin at the AV node and then travel down the septum between the two ventricles. At various points these conducting fibres have branches that spread out into the thick cardiac muscle tissue of the ventricles. Reception of this impulse results in both ventricles undergoing systole simultaneously.

An electrocardiogram (ECG)

An **electrocardiogram** (ECG) is a graph plotted in real time, with electrical activity (from the SA and AV nodes) plotted on the *y*-axis and time on the *x*-axis. Electrical leads are placed in a variety of places on the skin in order to measure the small voltage given off by these two nodes of the heart. Every repeating pattern on an ECG is a representation of one cardiac cycle.

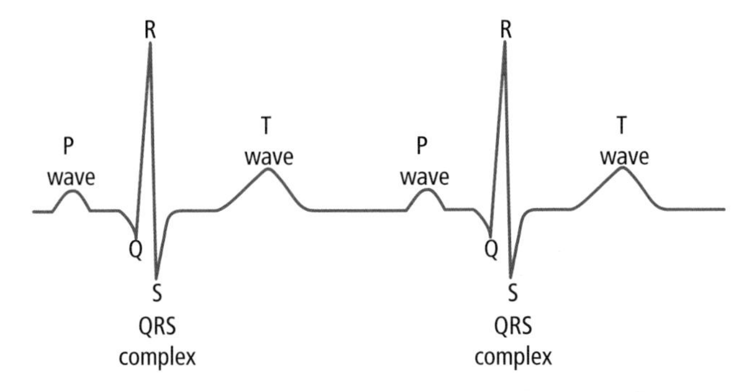

B3.2 Figure 8 An electrical trace of two cardiac cycles (note the repetition from left to right). Think of this as a graph with electrical activity (measured in millivolts) plotted on the *y*-axis and time plotted on the *x*-axis.

What processes happen in cycles at each level of biological organization?

SKILLS

How to "read" a "normal" ECG trace (Figure 8).

- P wave: this part shows the voltage given off by the SA node, thus it marks atrial systole.
- Point Q: this is the point at which the AV node sends its impulse.
- QRS complex: this is where the impulse from the AV node spreads down the conducting fibres in the septum between the ventricles and out to the cardiac muscle of the ventricles. This is known as ventricular systole.
- T wave: the AV node is repolarizing in preparation to send the next set of electrical signals.

The characteristic sounds of the heart are "lub dub" followed by a silence. The sounds represent one cardiac cycle. Heart sounds are made by the heart valves closing.

B3.2.17 – Xylem root pressure

B3.2.17 – Generation of root pressure in xylem vessels by active transport of mineral ions
Root pressure is positive pressure potential, generated to cause water movement in roots and stems when transport in xylem due to transpiration is insufficient, for example when high humidity prevents transpiration or in spring, before leaves on deciduous plants have opened.

If you recall, transpiration through stomata of leaves creates a negative tension that acts to pull water columns up xylem tissues. Does a plant cease water movement in xylem when transpiration is not possible, for example when stomata are closed or leaves are not present?

It has been shown that many plants are able to move water into their roots and up the xylem tissue even in spring, when leaves on **deciduous** plants have not yet grown. The same phenomenon can occur when leaves are present but transpiration is not possible, such as in highly humid conditions. This means that plants must have some ability to move water into and up xylem without the use of negative tension provided by transpiration.

Osmosis is the movement of water through cell membranes from an area of higher water potential to lower water potential. See Chapter B2.1 for information on osmosis and other membrane transport mechanisms. Root cells can create a low water potential by the active transport or diffusion of mineral ions into cells. Water will then follow by osmosis. Root hairs growing from even the smallest branches of a root system will take in water by osmosis in this manner. The mineral ions can then diffuse or be actively transported across the epidermis and cortex of the root until the minerals reach the xylem tubes in the centre of the root. Figure 4 on page 296 shows the tissues that water and dissolved minerals would pass through in a root to get from the soil to the xylem tissue. Specifically, the epidermis, cortex and then into the xylem in the central part of the root. Some mineral ions can be selectively moved cell by cell by active transport. Remember that water will always follow these solutes by osmosis, as the presence of mineral ions creates an area of low water potential. This will not only allow the water to enter the xylem but also create a positive fluid pressure pushing the column of water upwards.

B3.2.18 – Phloem translocation of sap

B3.2.18 – Adaptations of phloem sieve tubes and companion cells for translocation of sap
Include sieve plates, reduced cytoplasm and organelles, no nucleus for sieve tube elements and presence of many mitochondria for companion cells and plasmodesmata between them. Students should appreciate how these adaptations ease the flow of sap and enhance loading of carbon compounds into phloem sieve tubes at sources and unloading of them at sinks.

Phloem is the vascular tissue in plants that transports required organic molecules from one location to another. The fluid is often rich in sugars and is called **sap**. The direction of movement is based on a single principle: the movement from a **source** to a **sink**. A source is a plant organ that is a net producer of sugar, either by photosynthesis or by the hydrolysis of stored starch. Leaves are the primary sugar sources as they are responsible for photosynthesis. A sink is a plant organ that uses or stores sugar. Roots, buds, stems, seeds and fruits are all sugar sinks. It is possible for some structures to be both a source and a sink. For example, a root structure can store sugar or break down starch to provide sugar, depending on the season: root storage structures such as potatoes act as sinks in the summer and as sources in the early spring.

Nature of Science

Researchers continue to use novel approaches to study biological concepts. Traceable radioactive isotopes have been used to better understand the pathways organic molecules follow in phloem under varying environmental conditions.

The cellular structure of phloem is based on two types of cells that pair together into a functional unit and ultimately create a tube-like vascular network throughout the plant. The two cell types of phloem are called phloem **sieve tubes** and **companion cells**. Individual phloem sieve tube cells are connected to one another by porous sieve plates to form sieve tube elements. Both sieve tube elements and companion cells are living cells. However, the sieve tube elements cannot remain alive without the numerous metabolic activities of the companion cells. Sieve tube elements do not contain a nucleus and many other important cell organelles, because they are designed to be nearly empty in order to serve their function as vessels carrying a fluid. Companion cells and sieve tube elements have multiple connections called **plasmodesmata**. These allow the cytoplasm of the tube cells to be shared and are the origin of the proteins and ATP needed by the highly specialized sieve tube elements. Sap does not travel through the cytoplasm of either of these two types of cells, but through the tube-like area of the sieve tube elements, where the plasma membrane and cytoplasm are greatly reduced.

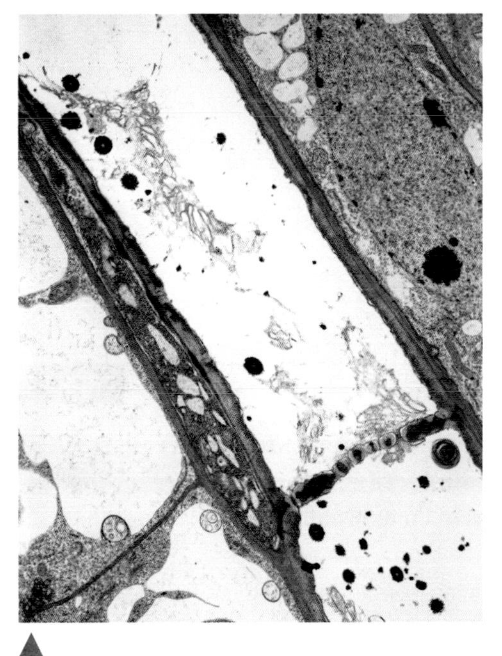

A transmission electron micrograph (TEM) of a phloem sieve tube and accompanying companion cell.

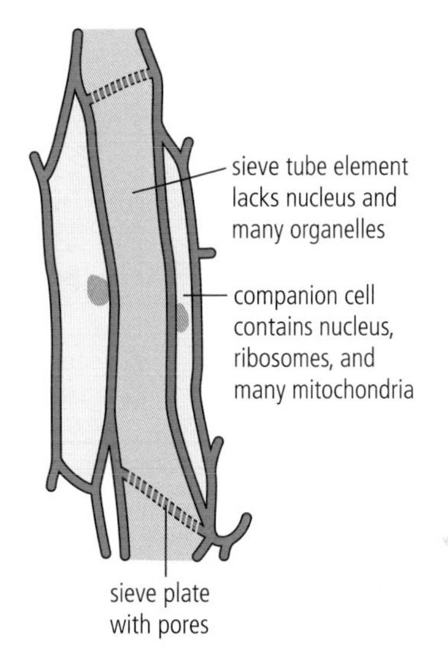

The structure of phloem, including the sieve tube element and accompanying companion cell.

The movement of sap within the sieve tube elements is called **translocation** and occurs because a water pressure is created at the source. The pressure begins at any portion of the plant that has sugars (such as sucrose) that need to be transported elsewhere. Companion cells in the source actively transport sugar molecules in, and the sugars pass through the plasmodesmata into the sieve tube elements in that area (see Figure 9).

The movement of sugars into an area of sieve tube elements creates an area of low water potential because of the high number of solutes. A nearby xylem vessel will release water into the sieve tube element, as water will move from an area of high water potential (xylem) to low water potential (phloem) by osmosis. The influx of water into the sieve tube elements results in the cells expanding outwards because of the increased pressure. Because of this pressure the water, now rich in sugars, will begin moving through the

How do pressure differences contribute to the movement of materials in an organism?

sieve plates within the tube created by the sieve tube elements. The water will go to wherever along the tube there is the lowest pressure. That area will be wherever sugars are being downloaded out of the sieve tube elements, into companion cells and then into an area where the sugar is needed for energy or storage. In the area of the sink, because solutes are being removed, water will return to a xylem vessel by osmosis.

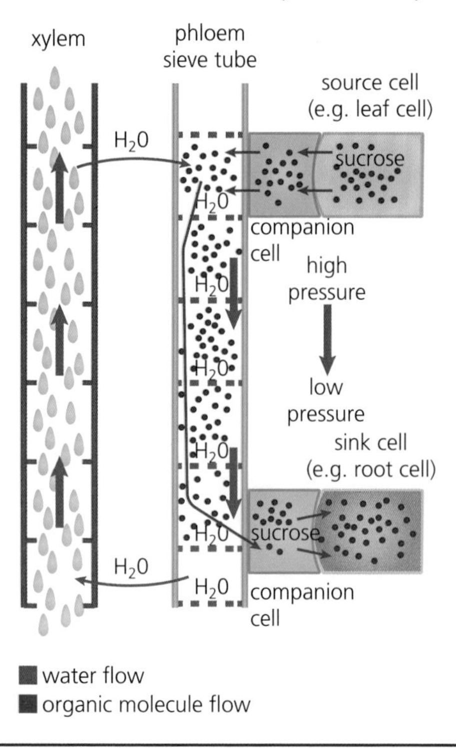

B3.2 Figure 9 A schematic diagram showing the flow of sap in phloem tissue from source to sink.

HL end

Guiding Question revisited

What adaptations facilitate transport of fluids in animals and plants?

In this chapter we have described the following in relation to both animals and plants:

- the use of water as a solvent to carry dissolved substances to tissues
- the use of tube-like structures (arteries, veins, xylem and phloem) for fluid transport
- multiple branches of tube structures leading to an increase in surface area in tissues
- the use of fluid pressure to provide the force necessary to move liquids.

Guiding Question revisited

What are the differences and similarities between transport in animals and plants?

In this chapter we have described how:

- both have fluid moving in tube-like structures
- both have mechanisms for creating a pressure differential to move fluid
- both allow the chemical exchanges necessary for life processes

- both increase the surface area by having multiple vessel branches
- animals use blood as a transport medium, while plants use water with dissolved substances.

HL

- animals use a muscular heart for positive pressure flow, while plants use cellular transport processes to create positive and negative tension
- plants have some vessels (phloem) capable of a two-way directional flow when needed.

HL end

Exercises

Q1. State whether each of these descriptions applies to an artery or a vein.

(a) Thin muscular wall.

(b) Capable of changing internal diameter (lumen).

(c) Carries blood away from the heart.

(d) Has internal valves to ensure one-way flow.

(e) Carries high pressure blood.

Q2. Data has been collected correlating a high fat diet with eventual occlusions of coronary arteries. Why is it incorrect to say that high fat diets cause heart attacks?

Q3. Why is it important for a plant to have a **continuous** water column all the way from the roots to the leaves?

Q4. Why is more tissue fluid released from the arteriole end of a capillary bed compared to the venule end of that same capillary bed?

HL

Q5. State the primary difference in flow pattern between lymph fluid and blood.

Q6. State three adaptations of capillaries to facilitate rapid molecular exchanges.

Q7. Why is it important that phloem is living tissue?

Q8. What is the primary benefit of a mammal's double blood circulation compared to the single circulation of a fish?

Q9. In this chapter you learned that the heart sounds (lub dub) are caused by the valves closing. Suggest a reason why there only two heart sounds when there are four valves?

HL end

HL

B3.3 Muscle and motility

Guiding Questions

How do muscles contract and cause movement?

What are the benefits to animals of having muscle tissue?

Animals are usually easily distinguished from other living things by their ability to freely move. This is accomplished by muscles, in combination with a skeletal system. In addition, the nervous system controls when muscles contract. These three systems of the body work together so that you can run, swim or play a piano.

Vertebrates, including humans, have an internal skeleton known as an endoskeleton, made of bones. Some animals, such as insects, have an exoskeleton. The animals with exoskeletons use their outer covering as an anchor for muscles, while vertebrates use their endoskeleton.

Animals use skeletal muscles to move bones or exoskeleton segments. Skeletal muscle is under conscious control by an animal's nervous system. Evading predators, pursuing prey, finding grazing areas and locating mates are just a few examples of the benefits muscles provide animals. In addition, smooth muscle is controlled by those areas of the brain that control involuntary movements, such as changing the inside diameter of blood vessels, shivering, focusing the lens of the eye, and many other functions.

B3.3.1 – Adaptations for movement

B3.3.1 – Adaptations for movement as a universal feature of living organisms
Students should explore the concept of movement by considering a range of organisms including one motile and one sessile species.

Movement is a universal feature of living organisms. You are familiar with the way many organisms move about their environment because you have the same ability. Some organisms, however, only use movement to capture prey while their body remains in one location. Organisms that have adaptations allowing movement within their habitat are called **motile**. **Sessile** organisms cannot move from place to place, but are still able to alter their body form in response to environmental stimuli. We are going to consider an example of both a sessile organism and a motile organism.

Venus flytrap: a sessile organism

A Venus flytrap (*Dionaea muscipula*) is a carnivorous plant native to the subtropical wetlands found in North and South Carolina in the United States. This plant species lives in soils that are deficient in minerals, especially nitrogen. The plant's "trap" is a pair of leaves with short but sturdy trigger hairs. A Venus flytrap waits for an insect to crawl or fly inside its paired leaves and trigger the hairs. Within about a second of the hairs being triggered, the leaves close around the prey animal, to prevent it from escaping. The internal portion of these leaves then secretes enzymes to digest the trapped insect.

One misconception about the Venus flytrap is that it uses the prey animals as its primary nutrition. The plant gets most of its nutrition from photosynthesis, just like any other plant. It obtains nitrogen and other minerals from the prey animals it traps.

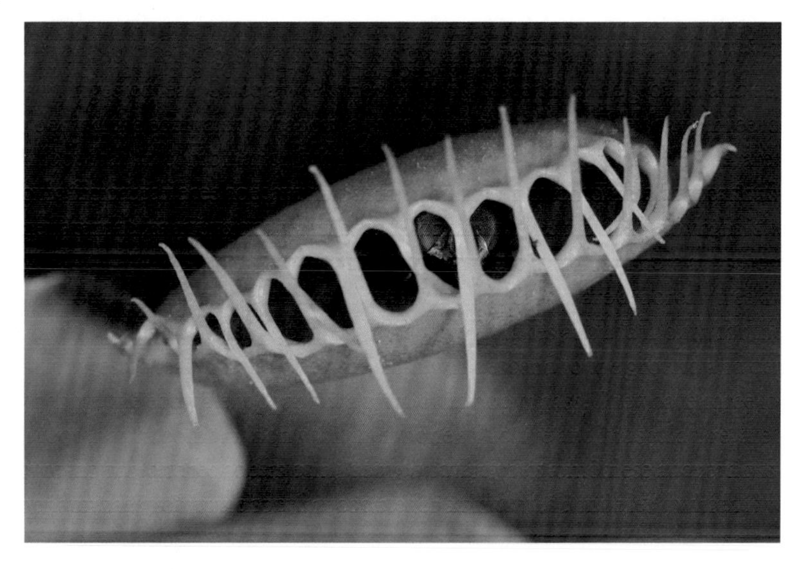

A Venus flytrap (*Dionaea muscipula*) with an ensnared fly. The leaves will secrete hydrolytic enzymes and after approximately one week the fly will have been digested.

Brown-throated three-toed sloth: a motile organism

The brown-throated three-toed sloth (*Bradypus variegatus*) is a motile but very slow-moving mammal. Its average speed is about 0.25 km h^{-1} (0.15 mile h^{-1}), which is why the word "sloth" has become synonymous with slow movements. There are several species of sloth, all native to Central or South America. Sloths are arboreal (tree dwelling) and herbivorous. Their digestive process is slow and it takes about a month for them to process an ingested leaf. Once a week the sloth will descend to the ground and defaecate, leaving behind the equivalent of about a third of its body mass. They have three long toes on each foot that, in combination with their bone structure and musculature, are adapted to hanging from branches and moving using a pulling motion. These adaptations make movement on the ground almost impossible for a sloth.

Even though their movement on the ground is very cumbersome, sloths are surprisingly good swimmers.

What are the relative advantages of versatility and specialization in biological mechanisms?

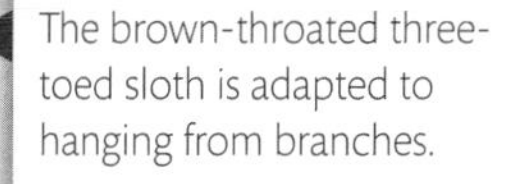

The brown-throated three-toed sloth is adapted to hanging from branches.

B3.3.2 – The sliding filament theory

B3.3.2 – Sliding filament model of muscle contraction
Students should understand how a sarcomere contracts by the sliding of actin and myosin filaments.

Like any other tissue in the body, muscle tissue is made up of cells. These cells are highly modified for contraction, and thus their cellular structure is not as apparent as in other types of cells. Each muscle is composed of thousands of cells, called **muscle fibres** because of their elongated shape. Muscle fibres are multinucleate because each fibre actually represents several cells that have merged together.

Each muscle fibre is composed of many protein filaments called **myofibrils** that run parallel to each other. Sequentially placed along each myofibril are contracting units called **sarcomeres**.

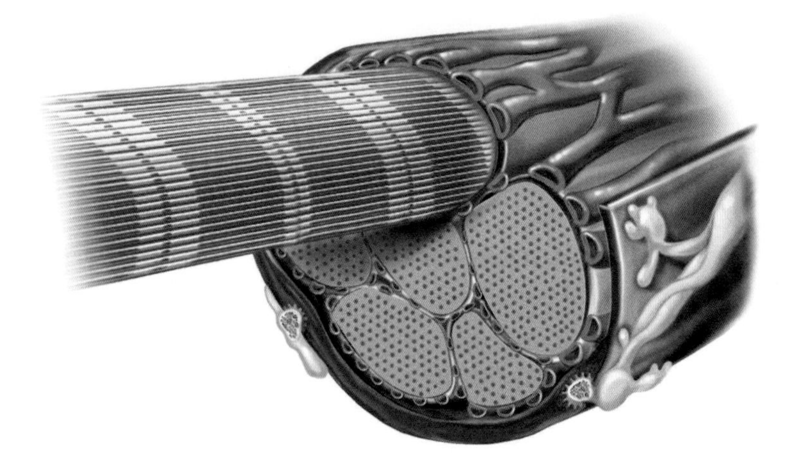

B3.3 Figure 1 Muscle tissue showing an extended myofibril running parallel to several others. Each myofibril shows the striations typical of skeletal muscle. Skeletal muscle is sometimes referred to as **striated muscle**.

All sarcomeres are attached to each other end to end. When one sarcomere contracts, all the sarcomeres in that same muscle contract. The resulting action makes the muscle fibre and entire muscle shorter. The key to understanding the contraction of a muscle is to understand how an individual sarcomere contracts. The striations of skeletal muscle are the result of alternating fibres of two proteins called **myosin** and **actin**. Figure 2 shows a diagram of one sarcomere in a relaxed state.

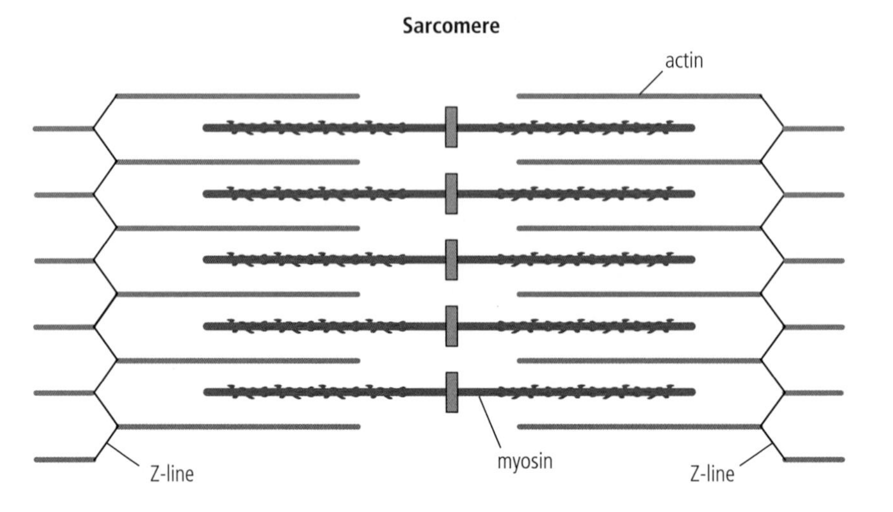

B3.3 Figure 2 A diagram of a single sarcomere. You can identify one sarcomere by locating the "Z lines" in a relaxed sarcomere. The Z lines are located in the centre of the longer, lighter shaded areas of a sarcomere; this area is lighter because it only has one protein, actin, present. Centrally located between Z lines will be a shorter, light area, where only myosin is located. Look back at Figure 1 and identify the Z lines indicating one sarcomere.

When a sarcomere contracts, visualize the myosin remaining stationary and the two sides of the actin moving towards the centre of the sarcomere. The myosin has movable heads that interact with the actin to accomplish this, using the energy of adenosine triphosphate (ATP) in very specific pathways.

Each Z line is shared by two sarcomeres, one to the right of the Z line and one to the left. It is these connections that allow muscle fibres to contract as a unit and shorten the entire muscle.

When sarcomeres contract, the actin and myosin fibres do not shorten. Instead the actin filaments slide over the myosin fibres. This results in each sarcomere shortening. This model of muscle contraction is often called the sliding filament theory.

The dark areas of sarcomeres are a result of the presence of both actin and myosin in those areas. The light areas are a result of the presence of either actin or myosin, but not both.

Understanding the sliding filament theory requires you to understand why the light areas of a sarcomere are present in relaxed muscle tissue but cannot be seen in contracted muscle tissue.

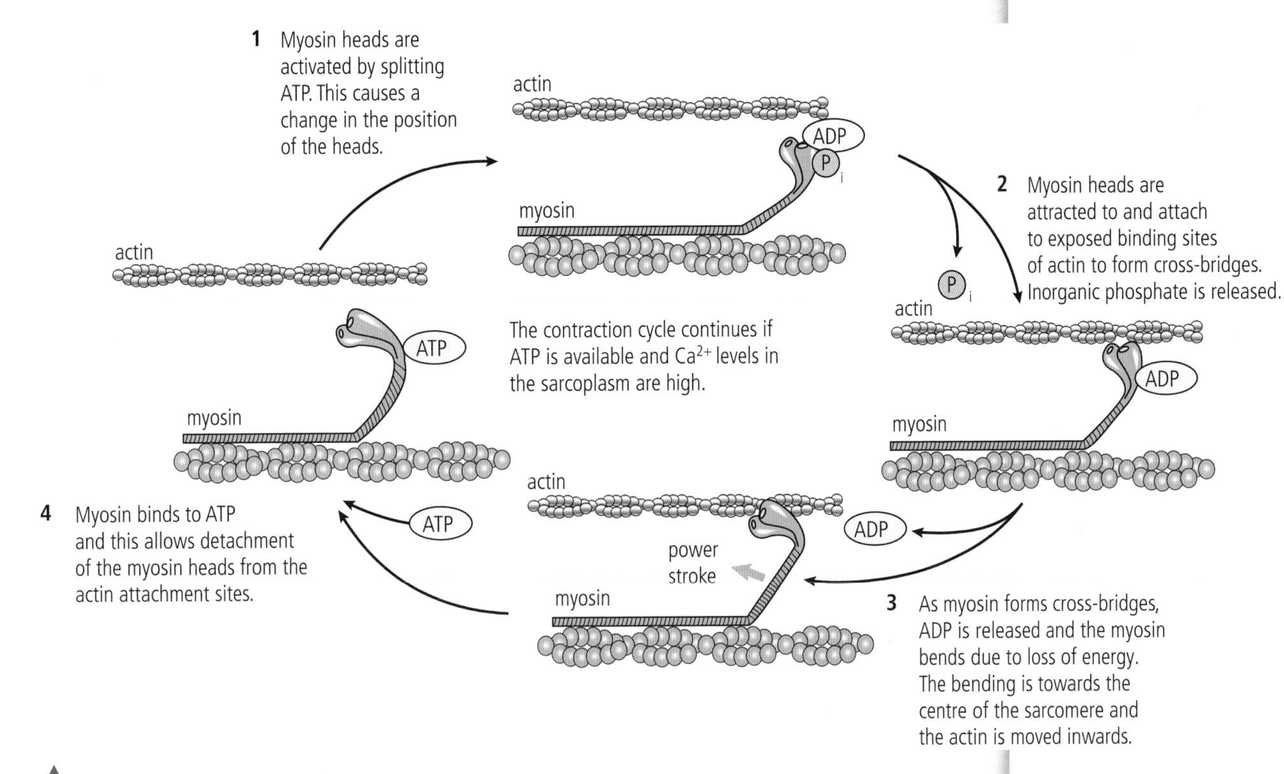

▲ The interaction of one myosin head with one actin filament. Note that when ATP binds to the myosin head it releases the myosin from the actin of the previous contraction. When ATP is hydrolysed, it prepares the myosin head by changing its position. The movement of the myosin head is initiated by releasing the adenosine diphosphate (ADP) leftover from the hydrolysis of the ATP. In order to release the myosin from the actin, another ATP must bind to the myosin.

Many animals' bodies, including humans, become very rigid a few hours after death, a condition called **rigor mortis**. This is the result of no new ATP being generated after death, and thus the myosin heads cannot detach from the actin-binding sites, resulting in the rigid muscle condition. Rigor mortis starts to decrease about 36 hours after death as the proteins degenerate.

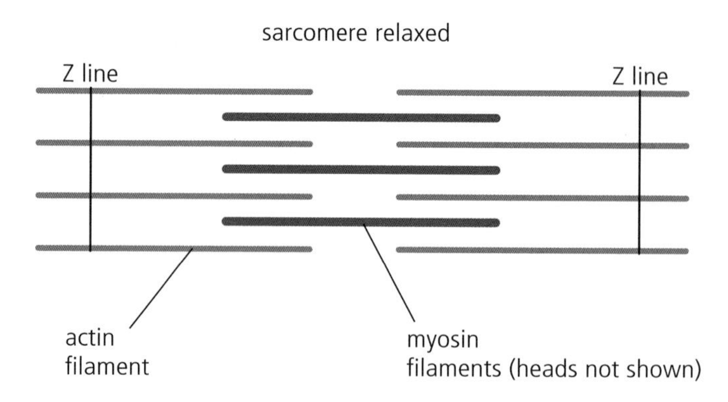

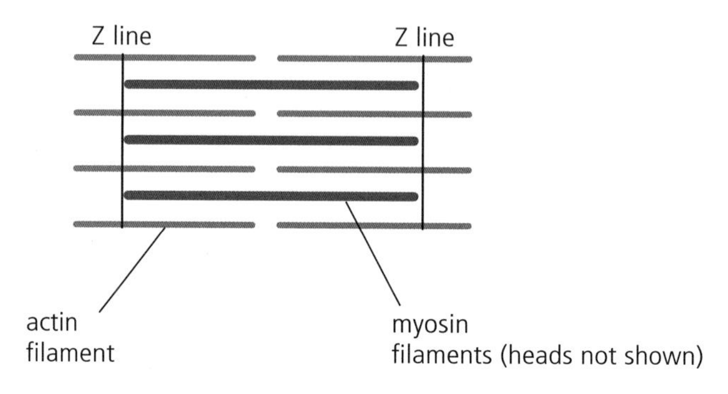

B3.3 Figure 3 A sarcomere shown at rest and during a contraction. Notice that the Z lines in the contracted sarcomere are closer together after the contraction. There are many sarcomeres in a series, and when all become shorter, the entire muscle fibre becomes shorter. This will result in the movement of a bone.

TOK: When you study the action of sarcomeres, how much is your knowledge limited by two-dimensional models, such as Figure 3?

B3.3.3 – Antagonistic muscle pairs and titin

B3.3.3 – Role of the protein titin and antagonistic muscles in muscle relaxation
The immense protein titin helps sarcomeres to recoil after stretching and also prevents overstretching. Antagonistic muscles are needed because muscle tissue can only exert force when it contracts.

Muscles use connective tissues called **tendons** to attach to two bones, one at each end of the muscle. One bone acts as an immovable anchor (called the **origin**), while the other bone (called the **insertion**) moves as a result of the muscle contraction. A muscle can only exert a force when it contracts, so once a bone has been moved the opposite movement requires a different muscle. The two muscles that accomplish opposite movements are said to be **antagonistic** to each other.

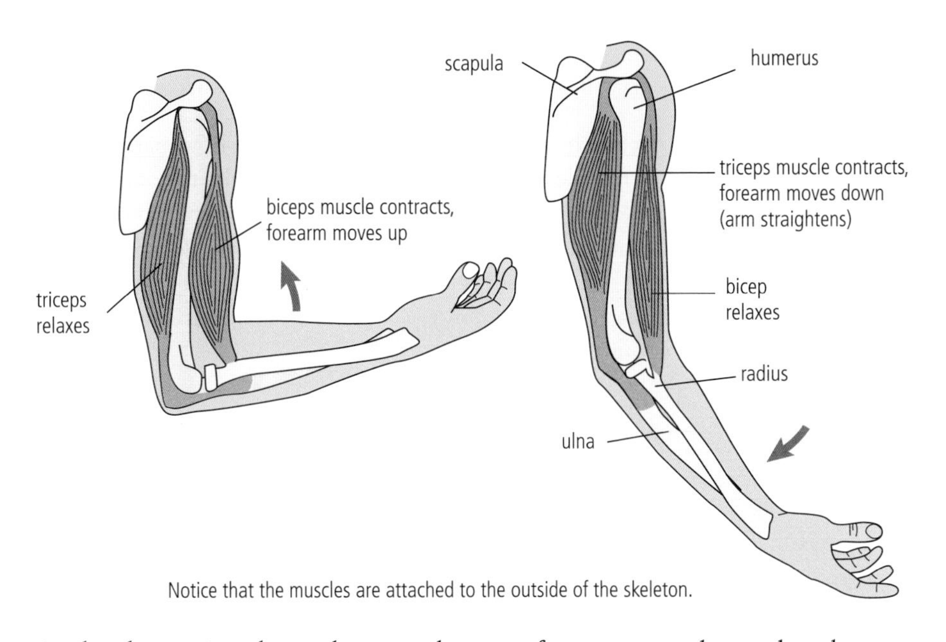

B3.3 Figure 4 The contraction of the biceps muscle results in the curling of the forearm in an upwards direction. The contraction of the triceps results in the opposite movement. These two muscles are antagonistic to each other.

As already mentioned, muscles can only exert a force to move a bone when they contract. Muscles also use a force to help with relaxation, as a result of the spring-like action of a protein called **titin**. Titin is an immense protein that has multiple folds that allow it to act as a spring.

Titin is the largest known protein in the human body.

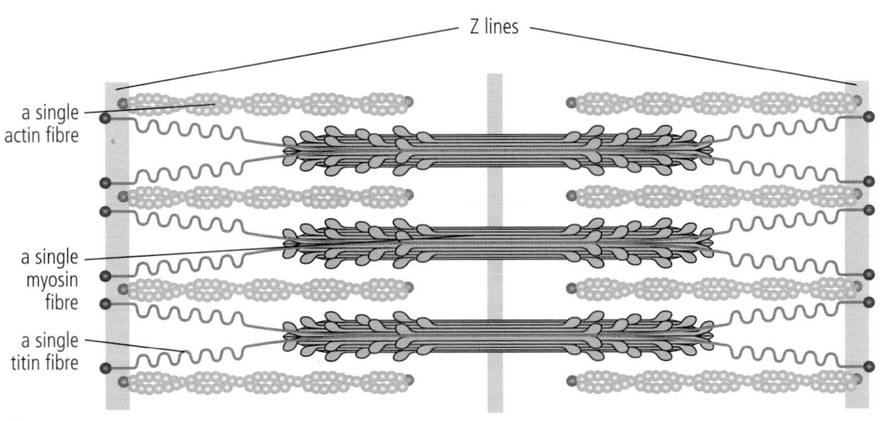

A small portion of one relaxed sarcomere showing the position of titin molecules.

When sarcomeres shorten during a contraction, the two sides of each sarcomere move towards the centre. This creates a spring-like tension in titin that is released when the muscle relaxes. This allows each sarcomere of the muscle to undergo a contraction once again. Titin also holds myosin fibres in place in the sarcomere and prevents muscle fibres overstretching.

B3.3.4 – Motor units

B3.3.4 – Structure and function of motor units in skeletal muscle
Include the motor neuron, muscle fibres and the neuromuscular junctions that connect them.

Skeletal muscle contractions are under the control of the nervous system. Every movement made requires many electrical impulses originating in your brain and terminating at synapses called **neuromuscular junctions**. These junctions are a

Synapses are locations where the electrical activity of a neuron is converted into a chemical release at the **terminal end** of the neuron. The chemical that is released is called a **neurotransmitter**. The nervous system uses many neurotransmitters, but the neurotransmitter released at a neuromuscular junction is always **acetylcholine**.

Occasionally your brain predicts a greater force of contraction is needed for a particular motion than is necessary. For example, if you lift a nearly empty carton of milk that you thought was full, the intensity of your muscle contraction may be too high because more motor units were activated than you really needed. Until you actually feel the mass of the object you are moving, your lifting motion may feel awkward.

type of synapse where a chemical message is sent into the muscle tissue to stimulate a contraction. Neurons that carry these "messages" are called **motor neurons.**

Each muscle is able to contract with varying intensity, depending on how many of the total muscle fibres within the muscle receive a nervous system impulse to contract. Each single motor neuron has a set number of muscle fibres that it controls and is called a **motor unit.** If a low intensity contraction is needed, a relatively low number of motor units is activated by the brain. If your brain predicts a high intensity contraction will be needed, more motor units receive impulses. The ratio of motor neurons to muscle fibres varies from about 1:10 to 1:200.

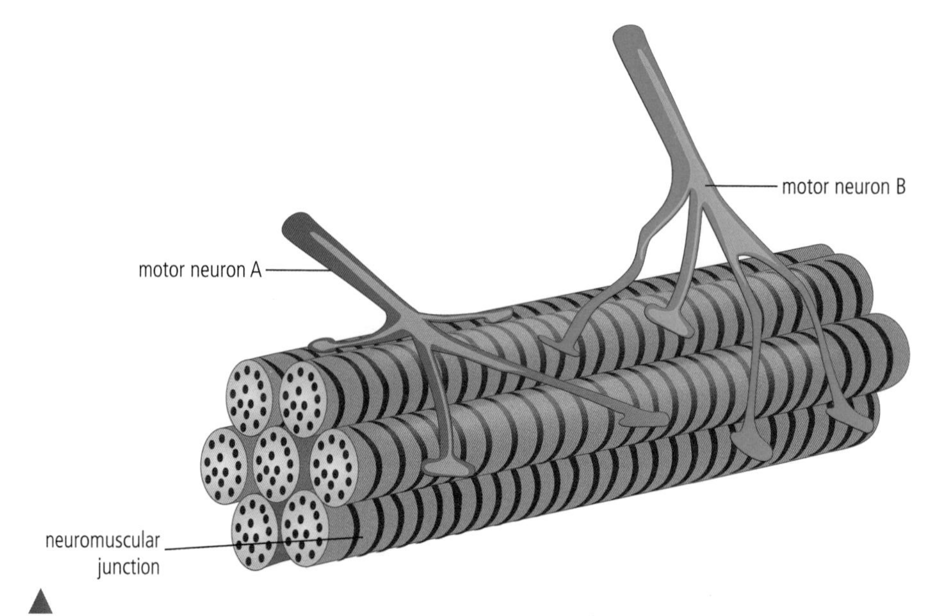

▲ In this diagram, the axons of two motor neurons carrying electric impulses to skeletal muscle tissue are shown. Two motor units are shown: a single motor unit is one motor neuron and the muscle fibres that neuron controls. In this simplified example, if a low intensity contraction is required, only motor neuron A would send an impulse and only those fibres would contract. A higher intensity contraction would require impulses from motor neurons A and B.

B3.3.5 – Skeletons as levers and anchor points

B3.3.5 – Roles of skeletons as anchorage for muscles and as levers
Students should appreciate that arthropods have exoskeletons and vertebrates have endoskeletons.

Vertebrates attach muscles to the outside of the bones of their endoskeleton. Arthropods attach muscle to the inside of their chitinous exoskeleton.

Vertebrate animals have an internal skeleton or endoskeleton made of bones. Muscles are attached to the bones at various points to allow the movements characteristic of that animal. Relative differences in bone length and muscle attachments result in different movements, such as the hopping of a kangaroo compared to the walking of a human.

Arthropods, such as insects, have an exoskeleton made of a substance called **chitin**. Because the skeleton is on the outside of the animal's body the muscle attachment points are on the inside of the hollow skeleton.

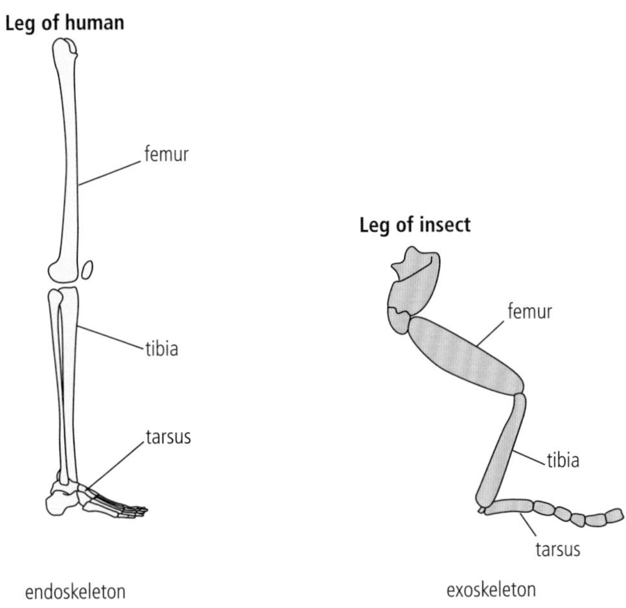

The bones of the endoskeleton of a human and the segments of the exoskeleton of an insect are similar in function and even share some anatomical names.

Many individual bones and segments of skeletons act as **levers**, to maximize efficiency for a variety of movements. A lever is a rod (a bone) able to rotate about a fixed point known as a **fulcrum** (a joint). Levers lower the force necessary to accomplish work, for example the work to move an arm, leg, or perhaps fingers. The lever action of bones and joints allows muscles to exert a lower force to accomplish any one movement.

Your head is able to move up and down in a "nodding" movement because the attachment point of the cranium to the vertebrae acts as a fulcrum. One set of muscles contracts to bring the head down from the fulcrum point and another set brings it back up. This is another example of how muscles work as antagonistic pairs. Look back at Figure 4 to visualize how the muscles and bones of your arm use your elbow as a fulcrum when you use your forearm to lift objects.

Arthropods, with their exoskeletons, often take full advantage of leverage. Arthropods not only have jointed legs but also jointed body parts. It is as if they are medieval knights in body armour made of chitin. The muscles that attach to the inside of this "armour" are in antagonistic pairs, just as in animals with endoskeletons. Arthropods are capable of an amazing range of motion by maximizing leverage. Fleas have been measured jumping 200 times their body length using the muscles and leverage provided from their jointed "toes". A mantis shrimp can throw out its front appendage to stun a prey at a speed of about 80 km h^{-1} (50 miles h^{-1}).

B3.3.6 – Synovial joints

B3.3.6 – Movement at a synovial joint
Include the roles of bones, cartilage, synovial fluid, ligaments, muscles and tendons. Use the human hip joint as an example. Students are not required to name muscles and ligaments, but they should be able to name the femur and pelvis.

Synovial joints occur in the body where two bones need to move against each other, and are notable for the wide range of motions that they allow. Common examples include the joints at your elbow, knee, shoulder and hips.

The hip joint is a ball-and-socket synovial joint.

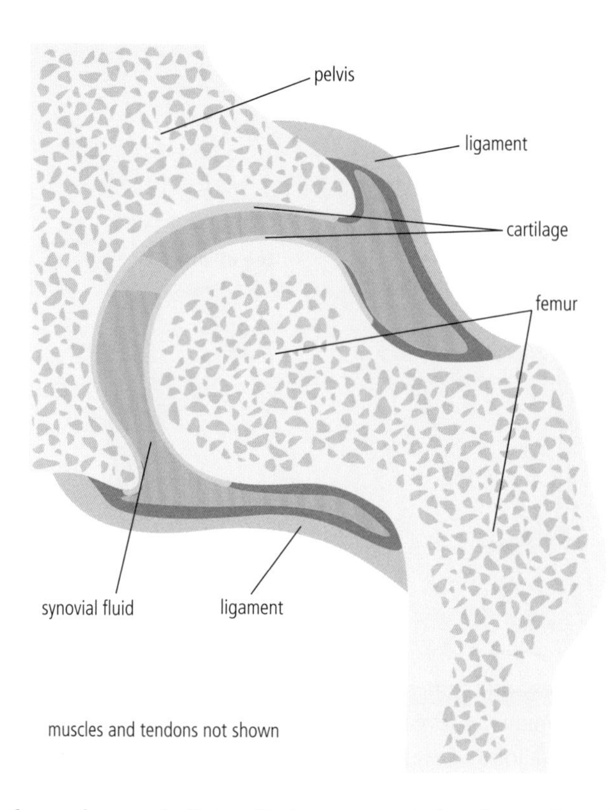

The IB does not require you to know the names of the specific muscles and ligaments of the hip. You are required to know the pelvis and femur, and the functions of cartilage, synovial fluid, ligaments, muscles and tendons.

The head of a femur forms a ball that fits into a rounded socket in the pelvis bone, thus it is called a **ball-and-socket** joint. Cartilage covers both bones to avoid bone on bone contact. The entire joint is encased by a membrane that contains a lubricant known as **synovial fluid**. The hip joint is encircled by tough, fibrous **ligaments** that hold the bones in place but also allow a range of motion. There are numerous muscles that control the movements of the hip joint, each with **tendons** that connect the bones.

Pelvis structure	Summary of function
Pelvis and femur	Bones forming the ball-and-socket joint of the hip
Cartilage	A smooth protective connective tissue that lines both the pelvis and femur within the hip joint
Synovial fluid	Lubricating fluid within the hip joint that reduces friction
Ligaments	Tough connective tissue that holds the bones of the hip joint in place
Tendons	Connective tissue that connects each of the muscles of the hip joint to its appropriate bones
Muscles	Muscle tissues that contract and relax to enable movement of the femur within the socket of the pelvis

The function of different pelvis structures

B3.3.7 – Range of motion

B3.3.7 – Range of motion of a joint
Application of skills: Students should compare the range of motion of a joint in a number of dimensions. Students should measure joint angles using computer analysis of images or a goniometer.

Physical therapists often use an inexpensive device known as a **goniometer** to measure the range of motion of a joint. Range of motion is the distance and direction that a joint can move, and is usually measured in degrees. The measurements can be used to document improvements in joint movement after an injury or surgery. Some joints such as the hip are capable of movement in a number of dimensions.

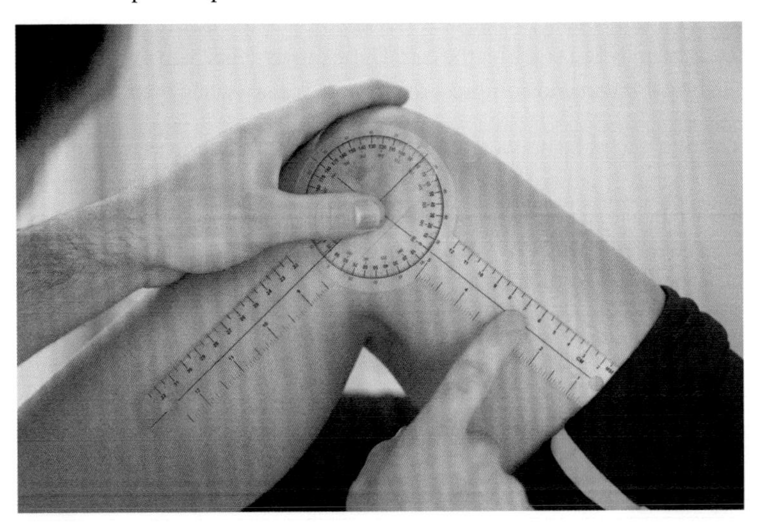

A goniometer being used to measure the angle of a knee joint.

SKILLS

Compare the range of motion of a joint in a number of dimensions. This can be done using computer analysis of images or with the use of inexpensive goniometers.

To use a goniometer, choose an arm or leg joint, where it is easy to locate the two bones on either side of the joint. The elbow or knee is ideal.

- Start in a position where the joint has no bend. If you are measuring the knee joint, your subject should be laying down on their back, with their legs extended straight.
- Align the extended arms of the goniometer to each of the two bones on either side of the knee joint of one leg. The circular body of the device will be located at the centre of the joint. Take a reading, which should be at or close to 0 degrees. You read the range of motion from the circular body of the device.
- To measure the range of motion of the joint, ask your test subject to move their joint as far as possible. If you are measuring the knee, the subject should slide their foot towards their buttocks.
- Move one arm of the goniometer, to realign both arms as they were originally on the two bones either side of the joint, and take the new reading from the circular "protractor" portion of the device.
- When measuring other joints, remember to first align the two arms of the goniometer with the two bones of the joint when there is no flexion of the joint.

B3.3.8 – Antagonistic muscle action

B3.3.8 – Internal and external intercostal muscles as an example of antagonistic muscle action to facilitate internal body movements
Students should appreciate that the different orientations of muscle fibres in the internal and external layers of intercostal muscles mean that they move the ribcage in opposite directions and that, when one of these layers contracts, it stretches the other, storing potential energy in the sarcomere protein titin.

You learned earlier that muscles act in antagonistic pairs to accomplish opposite movements of the skeleton. An excellent example of this are the paired muscles known as the intercostals located between the ribs. To better understand these muscles, wrap of your hands around the front of your ribcage making sure that you can feel your ribs on both sides. Now breathe deeply and notice the movement of your ribcage on each breath in and out. You should observe that your ribcage moves up and out when you breathe in, and down and in when you breathe out. These movements help expand the thoracic cavity when breathing in and compress the thoracic cavity when breathing out.

The intercostal muscles lie between each pair of ribs and use the ribs as their origin and insertion points. Each set works collectively to change the shape of the entire ribcage. If looking at the ribcage from the outside, the first intercostal muscles that you would see would be the **external intercostal muscles**. The origin and insertion points on each pair of ribs lie at an angle. Beneath this set would be the **internal intercostal muscles**, also using the ribs as their origin and insertion points, but they lie at an angle almost opposite that of the external set. Both sets of muscles use the attachment of the ribs to the vertebrae as the fulcrum point.

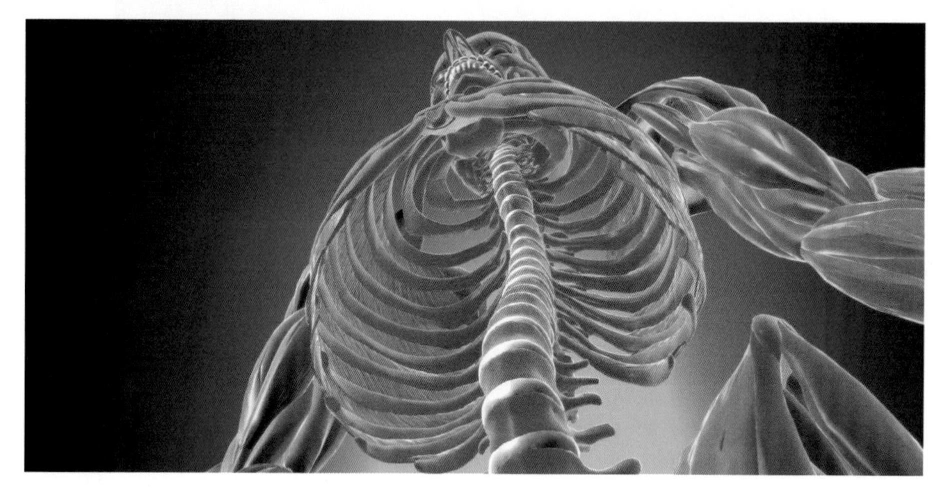

An artist's drawing of the ribcage, with the intercostal muscles shown in orange. Because you are looking up through the inside of the ribcage, the muscles you can see are the internal intercostal muscles, which help lower the ribcage as part of an expiration.

When you are enjoying a "rack of ribs" at a restaurant you are eating intercostal muscles.

When the external intercostal muscles contract, the rib cage is pulled upwards and out. If you recall, when you felt your ribcage, this movement occurs when you breath in (an **inspiration**). The antagonistic internal intercostal muscles move the ribcage down and inwards. This is typical of a breath out (an **expiration**).

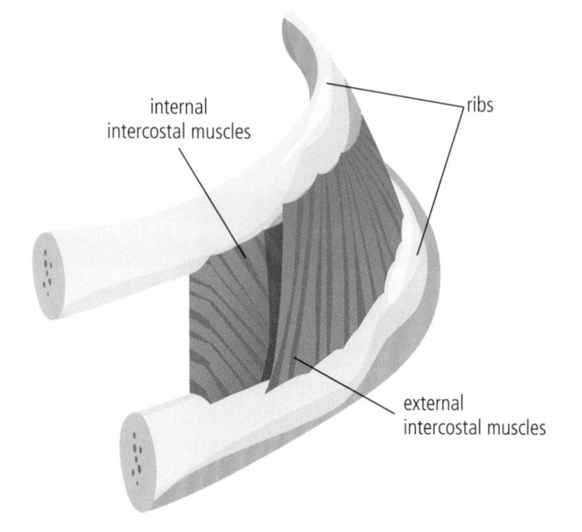

The orientation of the muscle fibres of the external and internal intercostal muscles. The different orientations encourage stretching of the titin fibres in the layer not currently contracting.

The movement of the ribcage and different orientations of the muscle fibres permit a stretching of the muscle layer that is not being contracted. In other words, when the external intercostal muscles contract the expansion of the ribcage results in stretching of the internal intercostal muscles. This stretches the titin fibres in each sarcomere of this muscle layer, creating potential energy that will be used for the next contraction of the internal intercostal muscles. The contraction of the internal intercostal muscles stores potential energy by stretching the titin in the external intercostal muscles.

B3.3.9 – The need for locomotion

B3.3.9 – Reasons for locomotion
Include foraging for food, escaping from danger, searching for a mate and migration, with at least one example of each.

Most animals rely on locomotion, the ability to move from place to place. Animals move for a variety of reasons, such as to find food, find a mate, escape predators or migrate (Table 1).

Reason	Animal	Brief description
Foraging for food	Honey bees	Flying from flower to flower to collect nectar and pollen
Escape danger	Flying fish	Escaping predators by swimming fast and extending their very long pectoral fins to glide over the water
Searching for mate	Loggerhead sea turtle	Both males and females swim back to the beach where they were hatched to mate and lay eggs
Migration	Arctic tern	Migrating from their Arctic breeding grounds to the Antarctic region and back each year, to take advantage of available food
Dispersal	Hoary bat	North American populations have established permanent colonies on the Hawaiian Islands

B3.3 Table 1 Reasons for animal locomotion

Starting with the four reasons for movement given in Table 1, research your own examples of animals using movement.

What are the advantages and disadvantages of dispersal of offspring from their parents?

B3.3.10 – Swimming adaptations

B3.3.10 – Adaptations for swimming in marine mammals
Include streamlining, adaptation of limbs to form flippers and of the tail to form a fluke with up-and-down movement, and changes to the airways to allow periodic breathing between dives.

Marine mammals such as dolphins, whales and seals are all descended from ancestral species that once lived on land. Their internal anatomy is adapted to a marine environment but still has many similarities with their land ancestors.

B3.3 Figure 5 The skeleton of a dolphin shown within an outline of the body.

Consider the adaptations that have allowed dolphins to be successful inhabitants of ocean waters. They:

- have a streamlined body, allowing the animal to move through the viscous water with relative ease and at great speeds
- have lost almost all body hair, to reduce drag through the water
- have a tail adapted to form a fluke, which allows an up-and-down motion for propulsion
- have lost their rear legs, because movement is provided by the fluke
- have front limbs adapted to become flippers primarily used for steering
- have an airway called a "blowhole" (not shown in Figure 5) located on the dorsal (top) surface of the head, to exchange air at periodic intervals with a minimum of the body leaving the water
- can seal the blowhole tightly between breaths so that water does not enter the airway
- can stay underwater for several minutes without breathing, so they can make deep dives
- have retained mammalian characteristics, such as being endothermic, producing milk for their young, having an advanced two-sided circulatory system, and long-term parental care of their young.

Guiding Question revisited

How do muscles contract and cause movement?

In this chapter we have described the following:

- muscles can only contract, resulting in a single movement
- muscles are found in antagonistic pairs, so that an opposite movement can be made
- the endoskeleton of vertebrates and the exoskeleton of arthropods provide anchor points for muscles
- the nervous system controls muscle activity by sending electrical signals to neuromuscular junctions
- muscle contraction units called sarcomeres are arranged in long fibres

- sarcomeres shorten during a contraction by the sliding action of actin filaments over myosin filaments using ATP
- during a contraction, a protein called titin is stretched and acts as a spring to help restore a sarcomere to its relaxed position.

Guiding Question revisited

What are the benefits to animals of having muscle tissue?

In this chapter we have described the following uses for muscles:

- internal body movements associated with breathing
- moving objects, as illustrated by the biceps and triceps moving the forearm using the elbow as a fulcrum
- movement to find food
- movement to evade a predator
- movement to find a mate
- movement to be able to migrate.

Exercises

Q1. Specify whether the following statements apply to actin, myosin and/or titin. The answers may include one or more of the molecules.

(a) Made up of protein.

(b) Has movable heads.

(c) Moves when a sarcomere shortens.

(d) Stretches like a spring.

(e) Stabilizes myosin fibres.

Q2. What is the immediate effect of ATP binding to a myosin head?

Q3. What are the antagonistic muscles to:

(a) the triceps?

(b) the external intercostal muscles?

Q4. In an insect, what are the muscles attached to?

Q5. The hip joint is an example of a ball-and-socket synovial joint.

(a) Which bone of the hip forms the socket?

(b) What is the function of synovial fluid?

Q6. Identify the role of each of these types of connective tissue.

(a) Ligament.

(b) Tendon.

HL end

B3 Practice questions

1. Outline the process of inspiration in humans.

(Total 4 marks)

2. HL Describe the functions of valves in the mammalian heart.

(Total 4 marks)

3. Coronary heart disease (CHD) is common in some families, with men being more susceptible to the disease than women. Researchers in Finland carried out an investigation to determine whether the pattern within families was the same for women as for men. The graph shows how the risk of developing CHD in men and women of certain ages depends on whether they had a brother or sister with the disease.

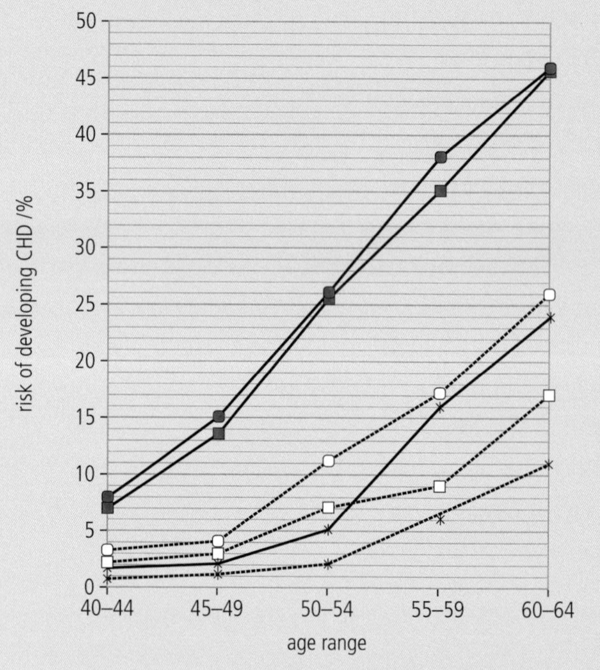

(a) State the risk of a man developing CHD between the ages of 55–59 if his brother had CHD. (1)

(b) Calculate the increase in risk over the control group for a woman of 60–64 of developing CHD if her sister had the disease. (1)

(c) Compare the results for the men and the women. (3)

(d) Suggest **two** reasons why a man is more likely to develop CHD if his brother had the disease. (2)

(Total 7 marks)

4. **(a)** Xylem and phloem contain structures that are adapted for transport. Outline the differences between these structures in xylem and phloem. (2)

   (b) Explain how the properties of water allow it to move through xylem vessels. (2)

   (Total 4 marks)

5. HL Explain how skeletal muscle contracts.

   (Total 6 marks)

6. Explain the relationship between the structure and functions of arteries, capillaries and veins.

   (Total 9 marks)

7. HL Which vessel carries deoxygenated blood?

   A The pulmonary artery

   B The coronary artery

   C The aorta

   D The pulmonary vein

   (Total 1 mark)

8. Explain the need for, and the mechanism of, ventilation of the lungs in humans.

   (Total 8 marks)

9. The leaves of plants are adapted to absorb light and use it in photosynthesis. Draw a labelled diagram to show the arrangement of tissues in a leaf.

   (Total 6 marks)

THEME

B Form and function

4 Ecosystems

A reticulated giraffe (*Giraffa reticulata*) in its native habitat in Kenya. This species of giraffe and all of the other organisms that live in this ecosystem are well adapted for their environment. With the exception of a few invasive species, organisms have lived in specific ecosystems for many thousands of generations. Each generation is a genetic "package" that provides efficient adaptations for that environment. Evolution provides very small changes to these genetic packages in keeping with the changes that occur to the environment. The living organisms of an ecosystem are highly dependent on each other for survival. Sometimes this can be as simple as one type of organism providing nutrition for another. Other times it involves much more complex interactions that we must study carefully to truly understand the nuances of the interspecies dependencies.

B4.1 Adaptation to environment

Guiding Questions

How are the adaptations and habitats of species related?

What causes the similarities between ecosystems within a terrestrial biome?

Organisms have complex lives. Few people think about the conditions necessary for earthworms, gopher tortoises, giraffes and other species to stay alive, but the environmental conditions they need are numerous and quite specific. A habitat must provide an organism with the basic requirements to stay alive. Organisms develop adaptations over time that allow them to be successful in their environment. The mechanisms for developing those adaptations explain evolution. Any genetic variation that permits a greater survivability in a given habitat will be passed on by the process of reproduction. Habitats change over time, and the adaptations of the organisms that live within those habitats must also undergo change.

A combination of mean annual precipitation and air temperature creates predictable terrestrial land areas called biomes. These biomes are found in various locations on Earth, but the ecosystems each one supports are also predictable, with similar characteristics. Convergent evolution leads to organisms living in each type of biome solving physiological challenges in similar ways.

B4.1.1 – What is a habitat?

B4.1.1 – Habitat as the place in which a community, species, population or organism lives
A description of the habitat of a species can include both geographical and physical locations, and the type of ecosystem.

A **habitat** is a place where organisms live. If more than one species have similar requirements, then a habitat can be a place where a **community** of multiple species lives. Habitats provide the organisms that live there with the basic requirements they need to stay alive long-term. Organisms need shelter, food, water, oxygen and often light.

Habitats can be described by their geographical or physical location *and* by the type of ecosystem they exemplify. Imagine you are visiting the Everglades National Park in Florida, USA, and use your phone or other global positioning system (GPS) device to find your location. You would be working out your geographical location. Even though that GPS location might be accurate, it would give a very incomplete description of your surroundings. More useful information can be provided by describing your surroundings, for example as containing shallow water, large areas of sawgrass plants, alligators, and numerous bird species. Such a description would give others a much better idea of the type of ecosystem you are visiting rather than just your location.

Living organisms do not live in isolation, instead they share habitats with each other. Each living organism has an impact on the other living organisms with which it shares a home.

B4.1.2 – Adaptation to the abiotic environment

B4.1.2 – Adaptations of organisms to the abiotic environment of their habitat
Include a grass species adapted to sand dunes and a tree species adapted to mangrove swamps.

We are going to consider two examples of how organisms have adapted to a relatively harsh abiotic environment.

Sand dune grass species

The sea oat (*Uniola paniculata*) is a species of grass that lives on and creates sand dunes along the eastern seaboard of the United States.

Sea oats (*Uniola paniculata*) helping to form a sand dune. Blowing sand accumulates around the base of the plants. As the sand gets higher, the plants grow higher, to make sure that the seeds stay above the level of the sand. Look for the seed heads on the upper portions of the plants.

Sea oats are drought resistant and, like other dune grasses, have a large shallow root system. They also have narrow leaves, to help reduce transpiration. Sea oats will close their stomata if soil/sand conditions around the roots remain dry for an extended period of time. The sandy "soil" they live in does not hold water for very long, so dense interwoven roots are needed to maximize the take-up of water during the short period of time it is available after rain. This intricate root system is also important because the massive intertwined roots help to hold the sand in place and prevent beach erosion. The reason that sand dunes grow taller is that blowing sand accumulates and is held by the root system of sea oats.

Sea oats thrive in full sun and easily tolerate salt spray; they can even survive complete immersion in saltwater for a short period of time. Sea oat plants produce **nodes** and **rhizomes** near their base, above the sand line. When covered by blowing sand, these asexual growth shoots are stimulated and produce shoots above the newly accumulated sand. Sexual reproduction is accomplished with the production of seed heads that resemble those of a true oat plant.

Mangrove tree species

Red mangrove (*Rhizophora mangle*) is a tropical and subtropical tree that grows along the saltwater tidal zone in Bermuda, Florida, the West Indies and other areas of tropical America. The prop roots of this tree extend above the water line, forming a "spider-like" support system. The roots above the water line also absorb air. The air is used to oxygenate the root tissues, which are below the water line and buried in mud. The roots below the water line filter salt out of the water, so that the tree has access to fresh water. Red mangroves are adapted to the changing water levels characteristic of saltwater tides. The tangled root growth under the trees provides a protective habitat for many fish and other marine animals. Marine animals often use this habitat as a nursery for their young.

Their adaptations for growing in a salty environment allow sea oats and mangroves to live in an environment that is inhospitable to most other plant species.

Red mangroves (*Rhizophora mangle*) are adapted for their saltwater tidal habitat

Red mangroves produce an unusual fruit, containing a seed that germinates and begins to grow before falling from the parent plant. The young plant is called a **propagule**. The propagule eventually falls from the tree and floats in the water below.

After absorbing water, the propagule orientates itself in shallow water, with its roots downwards (the same orientation as shown in Figure 1), and begins its early root growth. A shoot with early leaves grows from the opposite end. This is an adaptation for plant dispersal in a marine environment.

Mangrove species are now legally protected in most of the areas where they are found. The roots of mangroves prevent erosion and encourage a build-up of sediments. The thickets of roots absorb and dissipate the energy of major storms such as hurricanes, helping to minimize coastal erosion.

B4.1 Figure 1 A propagule of a red mangrove tree. What appears to be a seed pod is actually a young growing tree waiting to drop and begin its life away from the parent tree.

B4.1.3 – Abiotic variables

B4.1.3 – Abiotic variables affecting species distribution

Include examples of abiotic variables for both plants and animals. Students should understand that the adaptations of a species give it a range of tolerance.

Abiotic factors are the non-living components of an ecosystem.

Common abiotic factors include:

- water availability
- temperature range
- light intensity and duration
- soil composition
- pH range
- salinity.

Because of the complexity of habitats, the distribution of living organisms is dependent on many abiotic factors. Any one of those abiotic factors can act as a **limiting factor** if that factor is outside the tolerance zone of an organism. Organisms do not need an abiotic factor to be held at a constant level, but instead adapt to tolerate an acceptable range of values. Figure 2 shows the predicted population size of a species along a gradient for a particular environmental factor.

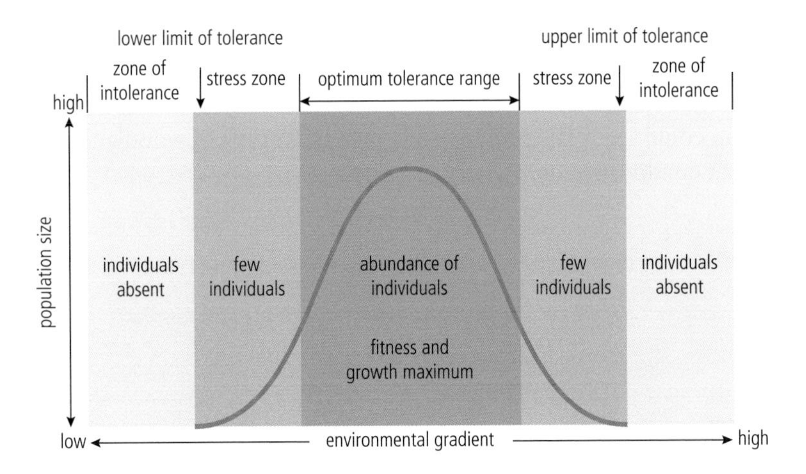

B4.1 Figure 2 The environmental gradient on the *x*-axis could be one of any number of abiotic factors, such as water availability or temperature.

As you can see from Figure 2, abiotic factors can and do affect population sizes, but often the abiotic factor can be far from the optimum before an organism is excluded from an area. Some organisms have developed special adaptations that extend their tolerance range within their habitat. Many species of catfish can take in oxygen through their skin, which permits them to live in oxygen-poor habitats. There are even species of catfish that can burrow into wet mud in order to survive a drought.

Being able to tolerate high or low values for certain abiotic factors can provide habitat opportunities for some organisms. Here are some examples of organisms that tolerate unusual habitats:

- red mangroves – high salinity shorelines
- sea oats – sandy soil along beaches
- polar bear – low air temperatures in arctic regions
- thermophilic bacteria – natural water sources at temperatures of 60–80°C.

Each of these organisms not only has a wide range of tolerance for the abiotic factor listed, but its optimum value is also unusually high or low compared to many other similar organisms. This allows the organism to experience less competition within a given habitat. For example, red mangrove trees have little competition for the saltwater shorelines that they inhabit, as few other plants can live in a high saline water habitat.

What are the properties of the components of biological systems?

B4.1.4 – Limiting factors

B4.1.4 – **Range of tolerance of a limiting factor**
Application of skills: Students should use transect data to correlate the distribution of plant or animal species with an abiotic variable. Students should collect this data themselves from a natural or semi-natural habitat. Semi-natural habitats have been influenced by humans but are dominated by wild rather than cultivated species. Sensors could be used to measure abiotic variables such as temperature, light intensity and soil pH.

A limiting factor is an abiotic (or biotic) factor that limits the population size or even presence of a particular species in a habitat. It is possible to work out the point at which a factor starts to limit the abundance of a species by carrying out practical experiments. The data collected needs to include a measure of the abundance of the species being studied and the level of the abiotic factor. For example, you could study the presence of a particular type of woodland plant in different light conditions.

You should be able to design and carry out a study where an abiotic limiting factor is correlated to the **distribution** of an animal or plant species. This is best accomplished as a small group or class project.

You can collect this data from a natural or semi-natural habitat. A semi-natural habitat is one that may have been influenced by humans but is still dominated by wild, rather than cultivated, species. The design should be based on counting population numbers along a **transect**. A transect is a scaled line (such as a long tape measure) that is laid along the entire length of the area you plan to investigate. The organism of interest is counted at specific intervals along the transect.

There are several types of transects. Two that you may be interested in using are the **line transect** and **belt transect**. A line transect is usually used to simply determine whether an organism is present or not at set intervals. When using a belt transect, a quadrat is placed at regular intervals along the transect and the number of individuals within each area counted.

Before you start you will need to decide:

- which abiotic factor you will measure (ideally you will choose one that is variable along the transect line)
- which organism will be counted
- where exactly you will set the transect
- how long the transect will be
- what intervals you will use
- what type of transect you will use (if you use a belt transect, how wide the belt will be).

Many of these preparatory steps are interrelated, such as the choice of area and the organism to be counted. For example, a transect running from the edge of a lake to higher ground might be used to measure soil water content and the presence of a native plant species. In an area where there is light and shade, you might want to measure light levels and the abundance of an invasive plant species.

If you carry out this study as a small group or entire class, a discussion of preparatory steps will be helpful. Remember to measure the abiotic factor at each interval, as well as the abundance of the organism. This should allow you to see whether there is a correlation between them, and at which point the abiotic factor becomes a limiting factor for the organism.

A quadrat counting grid placed along a transect line in a coral reef environment. A photograph is being taken of the randomly selected grids, which will be used to count organisms at a later date.

Nature of Science

Measurements can be taken using sensors for data logging. The measurements can be taken rapidly and/or automatically over longer periods of time. Measurements taken this way are accurate and reliable. Sensors are available for light, temperature, pH, carbon dioxide levels, and many other abiotic factors.

B4.1.5 – Coral reef formation

B4.1.5 – Conditions required for coral reef formation
Coral reefs are used here as an example of a marine ecosystem. Factors should include water depth, pH, salinity, clarity and temperature.

Coral reefs are found in less than 1% of the ocean's surface area yet, amazingly, an estimated 25% of all marine species live in and around coral reefs. Corals are the result of a **symbiotic** relationship between coral polyps and a microscopic algae called zooxanthellae. Both organisms in this mutualistic relationship require suitable growing conditions. The small size of the ocean surface area populated by coral reefs is an indication that the combination of all the right abiotic factors for these symbiotic species is rare.

Abiotic factor	Limiting effect
Water depth	Light only penetrates to relatively shallow depths. Zooxanthellae are photosynthetic and require adequate light levels. Most of the ocean floor is too deep to allow enough light to support coral reef growth.
Water temperature	Corals only survive in a narrow range of water temperatures (between 20°C and 28°C). Global warming is resulting in temperatures that are too warm for corals to tolerate. When the water becomes too warm, corals become stressed, and they expel the symbiotic zooxanthellae living in their tissues. Bleached coral is the result.
Salinity	Corals need the correct amount of salt in the water around them. Areas with freshwater run-off may not be of the correct salinity.
Water clarity	Water needs to be clear for light to pass through it. If there is a lot of sediment or pollution in the water, the clarity decreases and the zooxanthellae may not receive enough light.
Water pH	Increased carbon dioxide from fossil fuel emissions is being absorbed into ocean water, resulting in a lowered pH that is detrimental to coral growth. A lower pH (**acidification**) results in less calcium carbonate compounds being available in the water for corals to use when building reefs.

Abiotic factors affecting the growth and health of coral reefs. Reefs have a range of tolerance for each of these factors, but human activities are resulting in the upper or lower tolerance limits for many of the factors being exceeded

Challenge yourself

Use the figure below to answer the following questions.

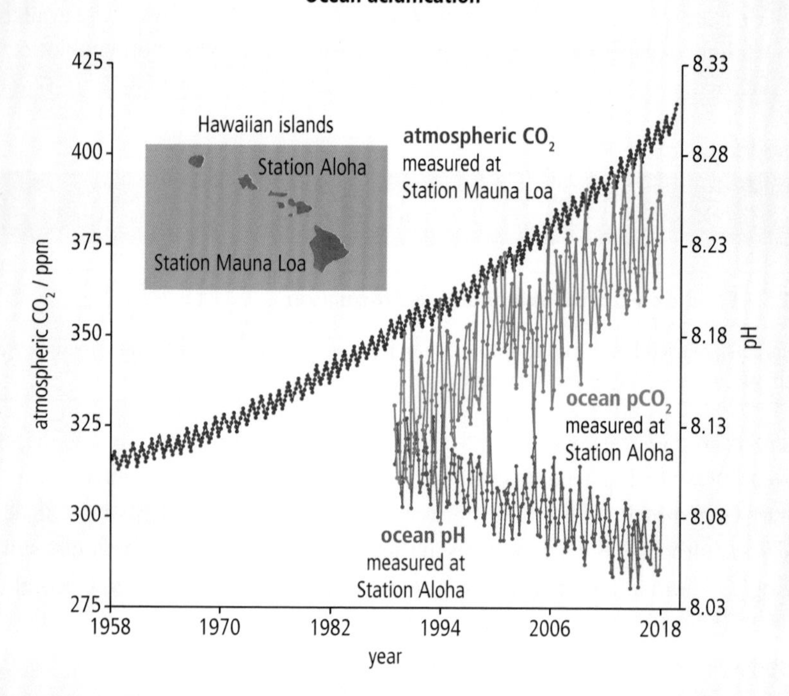

Atmospheric and ocean carbon dioxide content plotted with ocean pH over a selected period of time. Carbon dioxide is given in parts per million (ppm). As carbon dioxide is in solution in the ocean, the value given is the partial pressure of carbon dioxide (partial pressure is the pressure of one gas within a mixture of gases).

1. What is the apparent correlation between carbon dioxide in the atmosphere and the partial pressure of carbon dioxide in the Pacific Ocean around the Hawaiian islands?
2. Why do you think a mountain top in the Hawaiian islands was chosen as a long-term monitoring station for carbon dioxide?
3. Atmospheric measurements show regular cycles, with a minimum and maximum carbon dioxide level each year. Suggest a reason why these cycles are so consistent.
4. Does this data show a causal link between atmospheric carbon dioxide levels and ocean pH?
5. Some people suggest that oceans are a good "sink" or reservoir for excess carbon dioxide in the atmosphere. Describe one potential issue with increased carbon dioxide levels in oceans.

B4.1.6 – Terrestrial biomes

B4.1.6 – Abiotic factors as the determinants of terrestrial biome distribution
Students should understand that, for any given temperature and rainfall pattern, one natural ecosystem type is likely to develop. Illustrate this using a graph showing the distribution of biomes with these two climatic variables on the horizontal and vertical axes.

A **biome** is a large geographical area that contains communities of plants and animals that are adapted to living in that environment. Biomes are often named after the dominant vegetation type that is found within the biome. For example, a grassland biome contains many different grasses. The desert and tundra biomes are exceptions, but biologists can still predict what plant and animal species will be present in those biomes. Biomes of any one type can be found in various locations on Earth, because they are characterized by specific temperatures and rainfall levels, which are not restricted to one geographic location. Deserts, for example, are found in Africa, Asia, America and Australia. Biomes can be subdivided based on other environmental conditions. For example, there are hot deserts and cold deserts. All tropical forest biomes have plentiful rain, but some have more than three times the rainfall that others receive. For any given temperature and rainfall pattern, one natural ecosystem type called a biome is likely to develop.

Because biomes are created by varying conditions of precipitation and temperature, they can be plotted on a graph using the two environmental conditions as the horizontal and vertical axes.

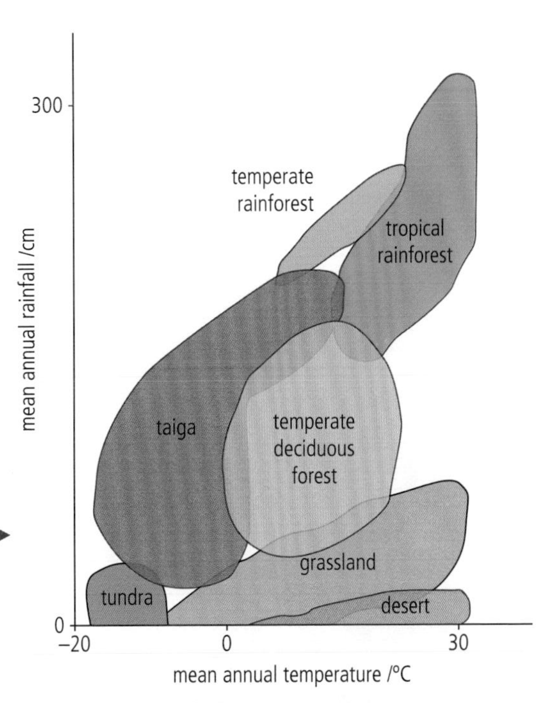

B4.1 Figure 3 Biomes identified by mean annual precipitation and mean annual temperature. Taiga is also known as boreal or coniferous forest. Notice that some environmental conditions lead to an overlap of biomes.

Spend some time making sure that you understand Figure 3, including what is shown on each axis. Note that the average annual temperature starts at −20°C.

Average annual temperatures and average annual rainfall determine global biomes.

B4.1.7 – Biomes, ecosystems and communities

B4.1.7 – Biomes as groups of ecosystems with similar communities due to similar abiotic conditions and convergent evolution

Students should be familiar with the climate conditions that characterize the tropical forest, temperate forest, taiga, grassland, tundra and hot desert biomes.

Biomes have identifiable abiotic characteristics but any one type of biome can be scattered across many different locations on Earth. These locations do not usually share geographic borders. Biomes can contain many ecosystems. An ecosystem is made up of the physical environment and the plants and animals that live there and interact with each other. Biomes are often much too large for each set of shared interactions.

The plants and animals found in similar biomes that are geographically separated will have different genetic backgrounds. If you visit a desert community anywhere in the world you will probably find similar organisms in each. However, while their **morphology** and **physiology** will be quite similar these organisms will usually have little genetic similarity.

The reason for this similarity is a type of evolution called **convergent evolution**. Convergent evolution occurs when two or more organisms solve an environmental problem by independent (unshared) genetic adaptations. Similar species that live within the same ecosystem are often genetically related to each other as often they are a result of **adaptive radiation.** Thus, they have a fairly recent common ancestor and a very similar set of genes. In contrast, species that live in the same type of biome but in different parts of the world may solve challenges in a similar way but have distinct genetic differences. Species evolve in order to adapt to challenges, and can do so by different sets of "trial and error" adaptations that lead to a very similar solution. The adaptations that work best will be similar in different locations because the abiotic conditions they are responding to are similar.

Carnivorous plants of different species have independently solved the challenge of living in poor soils by developing adaptations to capture and digest insects as a source of nitrogen. These different adaptations for the same purpose are examples of convergent evolution.

Left: A sundew plant (*Drosera capensis*) secretes a sticky digestive juice on the ends of filaments to entrap and digest a fly. The fly will be used as a source of nitrogen.

Right: Pitcher plants (*Nepenthes spp.*) have evolved a jug-like shaped container that holds digestive secretions in its base. The plants produce a slippery substance around the lip of their container. Insects slide from the lip into the container, where they are digested. This provides a source of nitrogen for the plants.

Convergent evolution will never result in two or more species becoming one.

How much has evolutionary theory advanced since molecular biology has been used to determine the relatedness of organisms? Common structural features of organisms were originally the only evidence that indicated organisms were related. Convergent evolution often gives a false representation of relatedness between organisms. DNA similarities can not only show relatedness, but also provide evidence about how long ago two organisms diverged in their evolutionary path.

Biome	Climatic conditions	Communities include
Hot desert	Very low annual rainfall (less than 300 mm per year), hot temperatures during the daytime but cold at night.	Sparse vegetation, often with spines for leaves, burrowing animals only active during the cooler night time.
Grassland	Semi-arid climate with somewhere between 500 mm and 950 mm rainfall per year. Temperatures vary depending on latitude. The annual range can be between –20°C and 30°C. Grasslands can have seasons (i.e. a wet season and a dry season).	Vegetation dominated by grass species. Little significant tree growth because of the lack of water. Animal species dominated by grazers and few predators.
Tundra	Cold temperatures (between –40°C and 18°C). Low precipitation (150 mm to 250 mm per year). In the winter it is dark for long periods.	No trees because of the lack of water and short growing season. The soil is frozen for most of the year. Animals have adapted to hibernate for long periods of time or to migrate when the conditions on the tundra become too difficult.
Taiga or conifer forest	Very cold winters and relatively high precipitation in the form of snow. Temperatures can range from –40°C to 20°C. Usually between 300 and 900 mm of rain per year.	Evergreen forests dominated by conifer trees. Animals must have adaptations for a very cold climate. The largest terrestrial biome on Earth by landmass.
Temperate forest	Four seasons with no extremes of temperatures, abundant year-round precipitation (somewhere between 750 and 1500 mm of rain). Soil enriched by leaf drop each year.	Area dominated by **deciduous** broad-leafed trees. Rich variety of animal species.
Tropical forest	High annual rainfall (from 2,000 mm to 10,000 mm per year). Warm temperatures (around 20°C to 25°C). Nutrient-poor soil as plants are rarely deciduous.	Very high plant and animal biodiversity.

Climatic conditions and community types typical of biomes

B4.1.8 – Hot deserts and tropical rainforests

B4.1.8 – Adaptations to life in hot deserts and tropical rainforest
Include examples of adaptations in named species of plants and animals.

We are going to consider examples of how organisms have adapted to hot deserts and tropical rainforests.

Hot desert biome adaptations

The saguaro cactus

The saguaro cactus (*Carnegiea gigantea*) is native to the Sonoran desert in southwestern United States. As a desert plant, most of its adaptations are related to water gathering and retention. The thick waxy skin is completely waterproof and is covered by bristles as a defence against grazers. The saguaro cactus has a single long taproot that it sends down to retrieve deep water when it is available, but it also has a massive shallow root system to absorb occasional rainwater. After a rare rain shower, the water taken up is stored in sponge-like tissue. This stored water maintains the low water needs of the plant until the next rainfall.

The growth of the saguaro cactus is very slow. At ten years of age, its height is only about 2 cm. The cactus will not reach its full height of about 14 m until it is about 200 years old.

The saguaro cactus (*Carnegiea gigantea*), sometimes called the organ pipe cactus. The larger cacti shown in this photograph will be more than 100 years old.

The fennec fox

The fennec fox (*Vulpes zerda*) is a small fox native to the desert areas of the Sahara in North Africa. It has many adaptations for desert life, the most notable being its very large ears. The ears are highly **vascular** and help dissipate heat. In addition, the large ears help the animal locate small prey animals moving underground. Like most desert animals, fennec foxes are nocturnal hunters. They spend their daylight hours in large underground dens shaded from the Sun. They obtain their water primarily from their food, although they will drink from a water source, if available. Fennec foxes have kidneys adapted to reabsorb most of the water that passes through them, and they only rarely urinate.

A fennec fox (*Vulpes zerda*) showing its characteristic large, highly vascular ears..

Sometimes a species is introduced into a new habitat that is very well suited to that habitat, but is also very damaging to that habitat. Europe alone is estimated to have over 100 terrestrial vertebrates, 600 terrestrial invertebrates, and 300 aquatic species that are invasive.

Tropical rainforest biome adaptations

The kapok tree

A tropical rainforest is characterized by high temperature and abundant rainfall. In tropical latitudes there is also plentiful sunlight. Collectively, those abiotic factors lead to abundant and varied plant growth. The abundant growth creates competition for available sunlight, and those species that can grow the tallest will have access to more sunlight. One of the species that can grow very tall is the kapok tree (*Ceiba pentandra*). This tree forms part of the upper canopy layer of rainforests in Costa Rica and the Amazon. In order to support rapid growth and a very tall trunk in a relatively shallow soil, a kapok tree makes a strong foundation from **buttress roots** that extend above ground.

Is light essential for life?

Indigenous people in the Amazon use kapok fibre from the seed coats to wrap around their blowgun darts. The fibres create a seal that allows pressure to build-up before the dart is forcefully expelled through the blowgun tube.

▲ The buttress roots of a kapok tree (*Ceiba pentandra*) are an adaptation that creates a strong foundation for the trees to grow very large in shallow rainforest soil.

Poison-dart frogs

Poison-dart frogs are well adapted to their tropical rainforest environments. As amphibians, they must reproduce by laying their eggs in water. The rainforest provides small pools of water inside the many bromeliad plants that are found in the canopy. Predators such as snakes and lizards are numerous, but these small frogs have developed highly toxic chemicals in their skin as a result of their diet of poisonous insects. They have evolved to have very bright colours and body patterns as a warning to predators. This is known as warning coloration because the predators have coevolved an instinct to avoid brightly coloured frogs.

A poison-dart frog species (*Dendrobates tinctorius azureus*) showing bright coloration and distinctive body pattern. Predators evolve instincts that help them avoid prey animals with this type of bright coloration.

The toxic chemicals produced by poison-dart frogs are used by Indigenous peoples to coat the darts they use in blowguns for hunting.

Guiding Question revisited

How are the adaptations and habitats of species related?

In this chapter you have learned that:

- different habitats offer various ranges of abiotic factors
- organisms often have a range of tolerance for any one abiotic factor
- any one abiotic factor can exclude a species from a habitat if the factor is outside the range of tolerance of those organisms
- adaptations help to extend the range of tolerance for organisms, for example
  - a kapok tree can grow to great heights because of its buttress roots
  - a fennec fox can tolerate a hot desert environment because of its large and highly vascular ears
  - a saguaro cactus can survive drought periods in the desert because of its adaptations to store water
  - carnivorous plants can survive nitrogen-deficient soils because they can capture and digest insects.

Guiding Question revisited

What causes the similarities between ecosystems within a terrestrial biome?

In this chapter you have learned that:

- a biome is the largest geographic biotic unit
- biomes are often named after their dominant vegetation type
- biomes are created by predictable rainfall levels and temperature ranges
- the organisms making up the ecosystems within any type of biome (e.g. tropical rainforest) are likely to have many similar adaptations even if they have limited genetic similarity

- this is because the organisms are all adapting to the same environmental conditions
- convergent evolution is the driving force for these similar adaptations
- convergent evolution leads to similar solutions to challenges within similar ecosystems.

Exercises

Q1. Identify three adaptations of sea oats that make this species well suited to growing along beach shorelines.

Q2. Describe what is meant by the optimum tolerance range as related to an organism and an abiotic factor?

Q3. Why are there desert biomes in many different locations around the globe? Choose one of the following answers.

A Deserts are often surrounded by mountainous regions.

B Deserts are found in any area of land with very low annual rainfall.

C Deserts are inhabited by organisms with water conservation adaptations.

D Deserts are found in any area of land that has very high temperatures.

Q4. Explain the relationship between increased atmospheric carbon dioxide levels and the pH of the oceans.

Q5. Outline why water clarity is important to coral reefs.

Q6. Some species of tropical rainforest frogs have bright colours and obvious body markings even though they are not poisonous to predators. Suggest a reason why these adaptations are still an effective defence against predators?

B4.2 Ecological niches

Guiding Questions

What are the advantages of specialized modes of nutrition to living organisms?

How are the adaptations of a species related to its niche in an ecosystem?

If all organisms ate the same types of food, there would be considerable competition for the food. Specializing in one category of food or method of feeding ensures that the number of competitors for a species is reduced. Making your own food, as photosynthetic organisms do, allows organisms to be somewhat more independent. However, we will see that even these organisms need help getting certain nutrients such as nitrogen. If an organism can become well adapted to what and how it feeds, it can occupy a niche that no other organism occupies, and hopefully flourish as a consequence.

In order to occupy a specific niche in an ecosystem, organisms have to adapt. Physical adaptations, and sometimes behavioural adaptations, are crucial. The morphology of an organism's body and teeth, or stems, roots and leaves, allows it to obtain certain resources from its environment. Physical and behavioural adaptations help predators find, pursue and kill prey, but also help the prey to hide, deter predators or escape. In this way, each species plays a particular role within an ecosystem.

B4.2.1 – Species and ecosystems

B4.2.1 – **Ecological niche as the role of a species in an ecosystem**
Include the biotic and abiotic interactions that influence growth, survival and reproduction, including how a species obtains food.

Each species plays a unique role within a community

The unique role that a species plays in the community is called its **niche**. The concept of niche includes where the organism lives (its spatial habitat), and what its role is in nature: what and how it eats (its feeding activities), and its interactions with other species. The ecologist Eugene Odum once said "If an organism's habitat is its address, the niche is the habitat plus its occupation".

Spatial habitat

Every type of organism has a unique space in an ecosystem. The physical area inhabited by any particular organism is its **spatial habitat**. For example, leopard frogs (*Rana pipiens*) live in the ponds in dunes in Indiana, USA. They burrow in their spatial habitat, which consists of mud in between the grasses on the edge of the ponds. Some of the aspects of an organism's habitat are **abiotic factors**, meaning they are made up of non-living things. Sand and water are two important components of the mud where leopard frogs live, and are both abiotic. Abiotic factors also include sunlight, soil type, pH and temperature.

The leopard frog (*Rana pipiens*).

There are also **biotic factors** in an ecosystem, meaning other living organisms. Biotic interactions can include feeding relationships, the provision of shelter such as nest sites, or the presence of parasites within the environment. Feeding relationships can be complex and involve many other species in the ecosystem. For example, an organism may be in competition with another organism for the food supply. It may itself be the prey for a larger predator. It may harbour parasites in its intestines. These interactions are complex and difficult to piece together, but they indicate the important role of the organism in the ecosystem. Interactions between the abiotic factors and biotic factors greatly influence the health, growth, survival rate and reproduction of an organism. A change in acidity of rainwater, for example, could greatly affect the ecosystem where the leopard frog lives, and if one organism is affected by this change it can have an effect on others, because the biotic and abiotic factors are all connected.

B4.2.2 – Obligate anaerobes, facultative anaerobes and obligate aerobes

B4.2.2 – Differences between organisms that are obligate anaerobes, facultative anaerobes and obligate aerobes
Limit to the tolerance of these groups of organisms to the presence or absence of oxygen gas in their environment.

Many species on Earth not only survive perfectly well without oxygen gas, some are poisoned by it. How well a species reacts to the presence of something in its environment is called **tolerance**. An organism with low tolerance to oxygen gas means it does not survive well when the gas is present. No tolerance at all would lead to death in the presence of oxygen.

The chemical transformation of food nutrients into energy that requires oxygen is referred to as **aerobic respiration**. The chemical transformation of food into energy that does not require oxygen is called **anaerobic respiration**. Some organisms are adapted to do one or the other, and some can do both.

Obligate anaerobes

Obligate anaerobes are single-celled organisms that have no tolerance to the presence of oxygen and are poisoned by it. The prokaryotes present on Earth for the first billion years of life were intolerant of oxygen. Initially that was not a problem, because it was not until photosynthesis evolved that oxygen started to collect in the atmosphere and in the water. Today, to escape from Earth's atmosphere, obligate anaerobes live in places where the air cannot reach them, such as in soil, deep water or the intestines of animals, including humans.

Facultative anaerobes

Facultative anaerobes are capable of carrying out both anaerobic and aerobic respiration. Baker's yeast (*Saccharomyces cerevisiae*) is a single-celled fungus that can use oxygen to convert sugar to energy when oxygen is available. When oxygen is not present, the yeast cells can switch to anaerobic respiration. They are neither hurt by nor killed by the presence of oxygen.

Sourdough bread is made using a starter, visible in the glass container, that contains yeast as well as food for the yeast, such as wheat flour. The bubbles in the starter as well as the bread are filled with carbon dioxide gas that the yeast cells have produced.

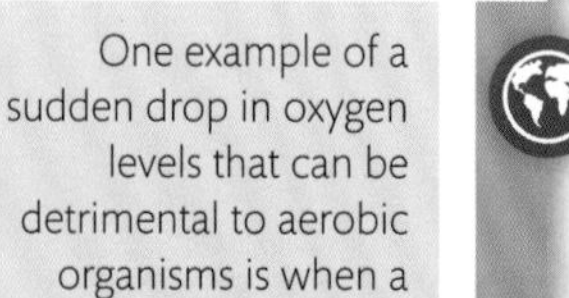

One example of a sudden drop in oxygen levels that can be detrimental to aerobic organisms is when a power station releases warm water into a river. Gasses dissolve better in cold water, so if the water is suddenly warmed, the oxygen level will drop.

Obligate aerobes

Obligate aerobes require oxygen and cannot convert food nutrients into energy without it. If oxygen in their environment is greatly reduced (**hypoxia**) or absent (**anoxia**), these organisms die. Fish, such as trout, that live in freshwater streams, survive well when dissolved oxygen levels in the water are between 7 and 12 mg L^{-1}. But if the dissolved oxygen levels drop below 3mg L^{-1}, the fish could die. Unlike yeast cells, fish cells in hypoxic environments cannot switch to anaerobic respiration.

B4.2.3 – Photosynthesis

B4.2.3 – Photosynthesis as the mode of nutrition in plants, algae and several groups of photosynthetic prokaryotes
Details of different types of photosynthesis in prokaryotes are not required.

Roughly a thousand million years after life first evolved on Earth, photosynthesis began. Bacteria living 3,500,000,000 years ago developed the ability to convert carbon dioxide and water into sugar using energy from sunlight. The waste product of photosynthesis is oxygen gas, O_2, and it is thanks to photosynthetic organisms that we have oxygen in the atmosphere today. The process of photosynthesis provides a remarkable bridge between non-living matter and organic matter as it transforms air and water into food.

For about 3 thousand million years, the only organisms photosynthesizing were single-celled organisms. We can still find various forms of photosynthetic bacteria today. Examples include not only cyanobacteria but green sulfur bacteria such as those in the genus *Chlorobium*, or purple bacteria such as those in the genus *Rhodospirillum*.

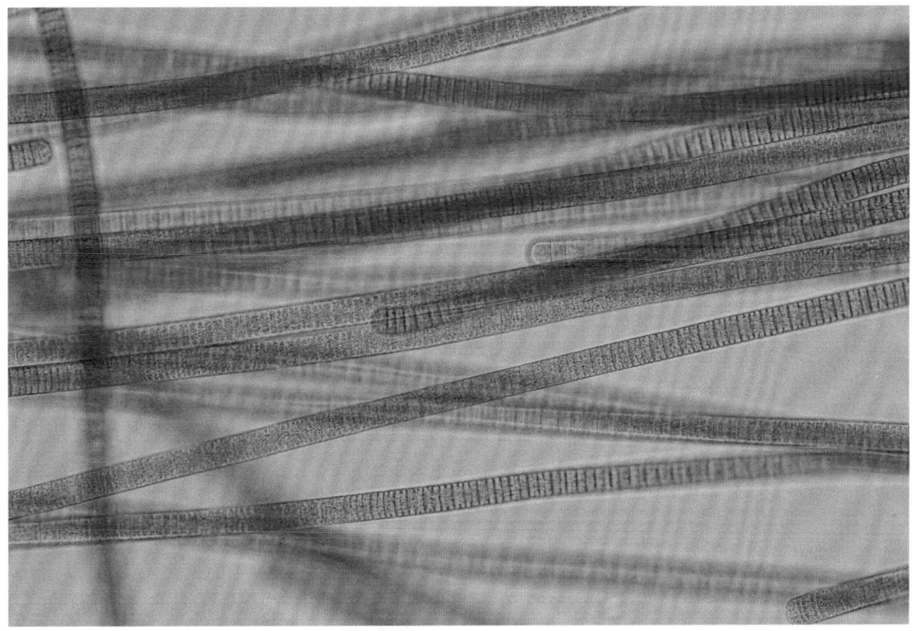

Microscopic view of filaments of cyanobacteria that can photosynthesize.

In the last 400 million years or so, other organisms, such as algae and, of course, aquatic and terrestrial plants, have adopted this type of metabolism. Algae are eukaryotes and can be single-celled, such as *Chlorella vulgaris*, or they can be multicellular such as kelp (e.g. *Ecklonia maxima*), a type of seaweed. In these organisms, the green pigment **chlorophyll** is used for the process of photosynthesis (see Chapter C1.3).

Chlorophyll is also found in aquatic plants and terrestrial plants and it is what gives leaves their green colour. The vast majority of ecosystems in the world get their initial energy from sunlight, which is used to convert inorganic matter into food using photosynthesis. If you greatly reduced or turned off sunlight from our planet, most of life as we know it would die. Almost all living things rely either directly or indirectly

In 1815 a volcano in Indonesia sent up so much volcanic dust into the atmosphere that a haze persisted in the sky for over a year, causing temperatures to drop worldwide. 1816 was declared the "year without a summer", and a massive reduction in solar energy reaching the surface meant reduced photosynthesis. This caused crop failures and led to famine in many countries.

on energy from the Sun. Organisms that can make their own food from inorganic substances using techniques such as photosynthesis are called **autotrophs**. Because they not only produce food for their own growth but can also be eaten by other organisms, they are often referred to as **producers**.

B4.2.4 – Holozoic nutrition

B4.2.4 – Holozoic nutrition in animals
Students should understand that all animals are heterotrophic. In holozoic nutrition food is ingested, digested internally, absorbed and assimilated.

Organisms need to either make their own food or get it from other organisms. In Section B4.2.3 we saw that photosynthesis means that certain organisms are able to make their own food. Organisms that cannot make their own food but rely on eating other organisms are called **heterotrophs**.

Examples of heterotrophs include:

- zooplankton
- sheep
- fish
- birds.

There are different types of heterotrophs, depending on how they get their nutrients. **Holozoic nutrition** refers to a way of getting nutrients by ingesting all or part of an organism. The eaten organism's parts are ingested and broken down into nutrients (digested) that can then be absorbed into the bloodstream (absorption) and used within the body (assimilation). Humans use holozoic nutrition, as do our pet cats and dogs. Organisms that obtain their food in this way are called **consumers**.

B4.2.5 – Mixotrophic nutrition

B4.2.5 – Mixotrophic nutrition in some protists
Euglena is a well-known freshwater example of a protist that is both autotrophic and heterotrophic, but many other mixotrophic species are part of oceanic plankton. Students should understand that some mixotrophs are obligate and others are facultative.

Organisms that are both autotrophic and heterotrophic are capable of making their own food *and* ingesting nutrients from other organisms. Such a mode of nutrition is called **mixotrophic nutrition**. This is useful if levels of sunlight are too low at times to support the organism through photosynthesis alone, and there is not enough food in the environment for a heterotrophic existence. The genus *Euglena* is made up of species that are single-celled protists that have photosynthetic pigments but also can ingest food from the water around it. This is an example of mixotrophic nutrition.

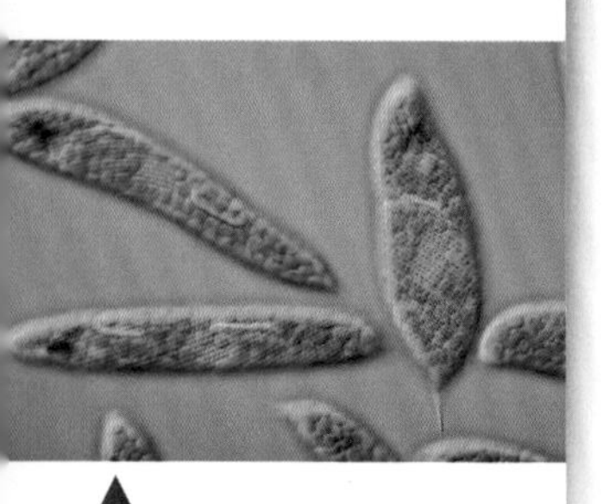

▲ *Euglena spirogyra* is a eukaryote and has membrane-bound chloroplasts filled with green chlorophyll.

Obligate mixotrophs need both systems to grow and thrive. **Facultative mixotrophs** can survive on one system but use the other as a supplement.

A lot of different terminology is used in this chapter. Make sure you know what unfamiliar terms mean. As we saw with both anaerobes and mixotrophs, when the term obligate is used, it means that the organism must use that mode and no other. When the term facultative is used, it means that the organism can sometimes use that mode but is capable of using another.

What are the relative advantages of specificity and versatility?

Humans can get all the vitamin D they need from a healthy, well-balanced diet. But, facultatively, we can obtain vitamin D by exposing our skin to ultraviolet rays.

B4.2.6 – Saprotrophic nutrition

B4.2.6 – Saprotrophic nutrition in some fungi and bacteria
Fungi and bacteria with this mode of heterotrophic nutrition can be referred to as decomposers.

Organisms called **saprotrophs** live on or in non-living organic matter, secreting digestive enzymes and then absorbing the products of digestion. Saprotrophs play an important role in the decay of dead organic materials. The fungi and bacteria that are saprotrophs are also called **decomposers**, because their role is to break down waste material. A mushroom that is growing on a fallen tree is secreting enzymes into the dead tissue of the tree trunk, in order to break down the complex molecules within the tree tissue. The mushroom then absorbs the simpler energy-rich carbon compounds that are released by the action of the enzymes. Slowly, over time, the tree trunk decomposes as the molecules inside the wood are digested and reused.

Fungi such as mushrooms on the forest floor are saprotrophs, helping to decompose material such as this fallen tree.

Holozoic consumers ingest part or all of an organism by swallowing and extracting nutrients using their gut. Saprotrophic decomposers release enzymes onto their food and the digestion happens outside their bodies. They then absorb the digested nutrients.

B4.2.7 – Diversity of nutrition in archaea

B4.2.7 – **Diversity of nutrition in archaea**
Students should understand that archaea are one of the three domains of life and appreciate that they are metabolically very diverse. Archaea species use either light, oxidation of inorganic chemicals or oxidation of carbon compounds to provide energy for ATP production. Students are not required to name examples.

Living things can be classified into three domains: Bacteria, Archaea and Eukarya. Organisms in the domain Archaea show remarkable diversity in the methods they use to obtain nutrients and energy, including:

- photosynthesis – generating cellular energy with the help of sunlight
- chemosynthesis – generating cellular energy from reactions involving inorganic molecules (without the help of sunlight)
- heterotrophic nutrition – obtaining nutrition by eating other organisms.

The archaea in the genus *Halobacterium* are able to perform a type of photosynthesis that is very different from that of organisms that use chlorophyll. Another pigment, bacteriorhodopsin, is used in these archaea to generate cellular energy, i.e. adenosine triphosphate (ATP). The "halo" part of their name refers to salt, because these microbes like to live in very salty environments such as the Great Salt Lake or the Dead Sea. These archaea are not considered autotrophs in the way plants are because, although they can generate ATP from sunlight, they get the carbon they need from other organisms.

Bioremediation is the concept of using microbes to clean up a toxic environment, such as those around mines and certain industrial sites. Because some archaea can survive in extreme conditions and have such diverse ways of using substances in their environment for food, they can be used to convert toxic compounds into safer ones.

When an organism is capable of producing its own food using chemical reactions, without the need for sunlight, it is called a **chemoautotroph**. This way of generating energy is called **chemosynthesis**. The archaeon *Ferroplasma acidiphilum*, for example, lives in very acidic environments (e.g. lower than pH 2) and gets its energy by oxidizing ferrous iron. It can be found living in wastewater runoff from iron mines. Archaea in the oceans and in soils use ammonia, NH_3, as a source of energy, and allow bacteria to convert the nitrogen compounds they generate into forms that can be used by plants.

Some archaea rely on organic food sources for their energy needs. These heterotrophs include members of the genus *Pyrococcus*, which can use amino acids, starch or maltose for food.

B4.2.8 – The relationship between dentition and diet

B4.2.8 – **Relationship between dentition and the diet of omnivorous and herbivorous representative members of the family Hominidae**
Application of skills: Students should examine models or digital collections of skulls to infer diet from the anatomical features. Examples may include *Homo sapiens* (humans), *Homo floresiensis* and *Paranthropus robustus*.
NOS: Deductions can be made from theories. In this example, observation of living mammals led to theories relating dentition to herbivorous or carnivorous diets. These theories allowed the diet of extinct organisms to be deduced.

We are primates. We belong to the family **Hominidae**, the great apes, which include the following genera (singular genus):

- *Pongo*, orangutans, of which there are three extant (living) species
- *Gorilla*, of which there are two extant species
- *Pan*, chimpanzees, of which there are two extant species
- *Homo*, of which there is one extant species, modern humans (*Homo sapiens*).

Dozens of species in the family Hominidae are now extinct, so the only evidence we have of their presence on Earth is their fossil remains, such as bones, skulls and teeth, and even, in some cases, fossil DNA. Occasionally we may also find evidence of the tools that they used. Examples of extinct species of Hominidae include:

- *Australopithecus africanus*, which was present on Earth about 3 million years ago
- *Homo erectus*, which was present on Earth about 2 million years ago
- *Paranthropus robustus*, which was present on Earth about 1 million years ago
- *Homo floresiensis*, which was present on Earth about 100,000 years ago and, because of the first specimen's small stature (with a height just over 1 m), has been nicknamed "the Hobbit"
- *Homo neanderthalensis*, which also was present on Earth about 100,000 years ago, and encountered (and interbred with) modern humans.

What an organism eats tells us a lot about its place in an ecosystem. Knowing what eats what can help us determine which species occupy which niches, for example. Working out what extinct species ate can therefore help us work out what niches they may have occupied. Palaeontologists often look at teeth and jawbones, which are much better preserved in the fossil record than, for example, soft digestive organs.

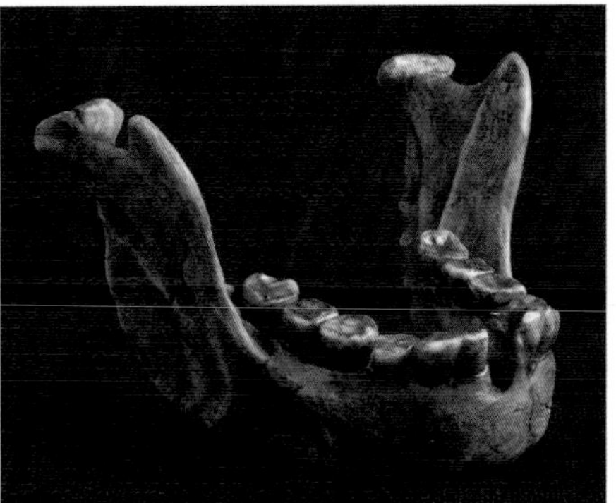

A fossil jawbone of *Homo floresiensis*. The small canines and flat, broad molars suggest a plant-based diet, although tools and other archaeological evidence indicates that they were hunters, so they probably consumed some meat as well.

Paranthropus robustus also has the dentition of a herbivore.

Tooth anatomy

Incisors, your front teeth, are found at the front of your mouth; your **canines** are to the side of the incisors; then your **premolars** are in the middle of your mouth, followed finally by the **molars** at the back. Incisors are used for cutting off bite-sized pieces of food. Next time you eat an apple or a sandwich, notice how the incisors act as scissors. Primates that eat mostly plant material such as leaves (leaf eaters are called **folivores**) and fruit (fruit eaters are called **frugivores**) tend to have large incisors. Canines are sharper and are used for ripping and tearing tougher materials such as meat. Premolars are for crushing or slicing up food, and molars are for grinding food and reducing it to a paste before swallowing. Generally speaking, the narrower and more serrated (pointed) the crowns (tops) of the premolars and molars are, the better adapted they are for eating meat, whereas the more rounded or blunt (not pointed) they are, the better adapted they are for eating plant material. Herbivores tend to have bigger incisors than carnivores.

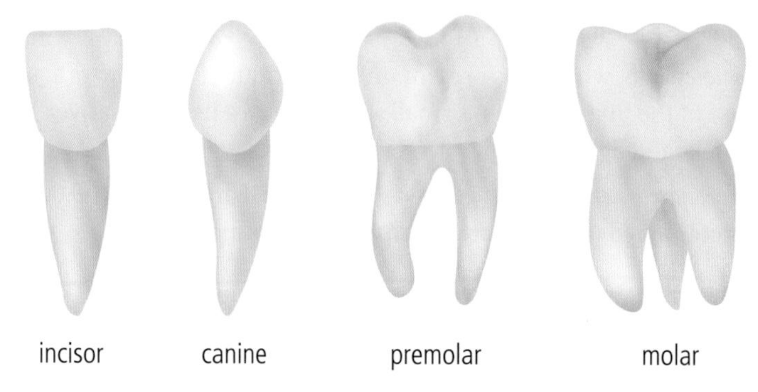

Each type of tooth has a name and a specific function, such as chopping, tearing off or grinding up food. Human teeth are shown here, but the same names are used in other mammals.

Diets of the great apes

If we look at the diets of extant species (those that are not extinct, at least not at the time of writing), we see that orangutans eat mostly fruit, which explains why they live in trees and occasionally also eat leaves. Some orangutans supplement their diet with insects, eggs or honey, so they can be considered omnivores, but they are essentially frugivores. Gorillas eat almost exclusively plant material: leaves, roots and stems. Some occasionally eat ants or termites, but gorillas are herbivores and more specifically folivores. Chimpanzees are omnivores, preferring fruit but also eating leaves and stems from plants, as well as meat. Chimpanzees consume invertebrates such as ants, termites and bees (and their honey), but also vertebrates such as monkeys, birds (and their eggs), antelope and warthogs, especially the young, which are easier to catch. Humans are also omnivores, eating fruit and grains, but consume vertebrates such as birds, fish, pigs and cattle as well.

Connecting dentition with diet

If we look at herbivores in general, they tend to have large incisors and wide premolars and molars that have rounded peaks and valleys for shearing and crushing plant material. Carnivores tend to have sharp, pointy teeth, not just their incisors and canines, but even their premolars and molars can be serrated and narrow rather than wide and rounded. Omnivores' teeth are somewhere in between. Their canines are not as long and pointed as carnivores. Their molars are of an intermediate width, not as wide as herbivores but not as narrow as carnivores. Their premolars and molars are usually rounded rather than serrated.

Looking at the teeth of chimpanzees, it appears that their dentition matches their diet. They have small incisors and long pointed canines for eating meat, for example. However, if we look at human teeth, even though meat plays a big role in many people's diet, our canines, premolars and molars are not shaped like carnivore teeth. Orangutans have long pointed canines and yet they do not eat meat. The complication is that teeth are not only used for eating. Some animals use sharp teeth to intimidate rivals or fend off intruders. Gorillas are herbivores but have very intimidating canines. Male chimpanzees tend to have much more prominent canines than females. Such evidence should be kept in mind when trying to determine diet solely from dentition. While there are general theories about the morphology of teeth and the diets of the animals, there are also exceptions.

One other aspect that experts look at in both extant and extinct species is **microwear**, small abrasions or removal of a tooth's surface, made as organisms chew, which can reveal the type of food they were eating. Softer foods will leave different marks compared to harder foods, and foods that have grit in them from soil will scratch teeth in a particular way that can be seen and analysed under a microscope.

TOK

When it is claimed that the dentition of an extinct hominid skull indicates that it was a herbivore, how much evidence is necessary before we decide that the claim can be considered valid? Do three or more experts have to agree? Or is it enough that one eminent, highly respected expert declares it? Is it important that their results are published in peer-reviewed journals, or is it sufficient that they made their observations and declarations without publishing? Is it necessary to compare their evidence to evidence from other fields, such as archaeology, to corroborate the findings, or are the results of a single specialty enough? How can we judge when the evidence provided is adequate?

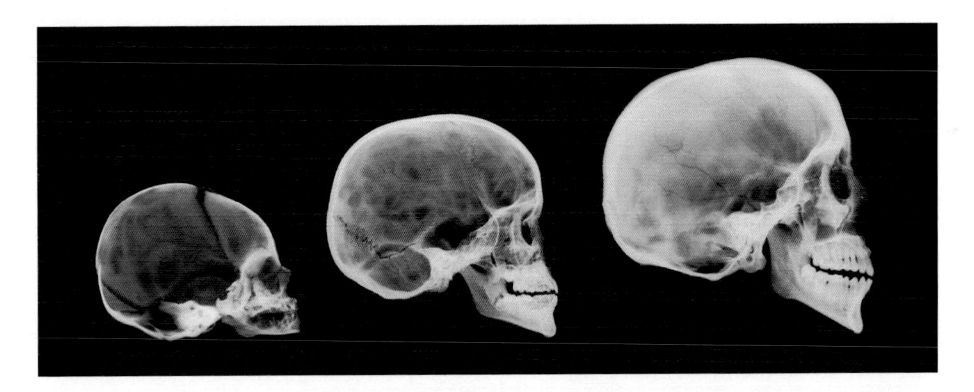

X-ray images of the skulls of a gorilla (*Gorilla gorilla*), a male chimpanzee (*Pan troglodytes*) and a human (*Homo sapiens*).

Nature of Science

Deductions can be made from theories. For example, observation of living mammals has led to theories relating dentition to herbivorous and carnivorous diets. These theories allowed the diet of extinct organisms to be deduced. But nature is full of diversity, and theories have to be modified as new evidence arises. As we have seen, teeth can be used for self-defence, not just for eating, and certain hominids that lack classic carnivore teeth can still be meat eaters, by using tools instead of teeth to cut and tear flesh. The ability to use fire to cook meat also reduces the need for specialized carnivore teeth, as cooked meat is much more tender and easier to chew.

B4.2.9 – Adaptations of herbivores and plants

B4.2.9 – Adaptations of herbivores for feeding on plants and of plants for resisting herbivory
For herbivore adaptations, include piercing and chewing mouthparts of leaf-eating insects. Plants resist herbivory using thorns and other physical structures. Plants also produce toxic secondary compounds in seeds and leaves. Some animals have metabolic adaptations for detoxifying these toxins.

How herbivores are adapted to eating plant material

Plants are not always easy to eat. Their leaves tend to be protected by thick layers of cells with semi-rigid cell walls, and not many organisms possess the enzymes necessary to break down cellulose, the chains of carbohydrates that make up plant fibre. Some insects solve this problem by piercing the plant and drinking the sugar dissolved in the sap inside. This is the case for aphids, small insects that you might find on a rosebush, which use modified mouthparts called **stylets**. Other insects, such as grasshoppers and caterpillars, use their sharp pinching **mandibles** to cut into grass blades and leaves, to help ingest them. Such insects are considered pests by farmers who do not want to see their crops damaged.

Herbivorous vertebrates such as cows and sheep have specialized back teeth that are broad and flat for grinding plant matter, and their digestive systems are adapted for digesting it. Cows are **ruminants**, which means that they swallow grass or hay before fully chewing it, then regurgitate it later when they are resting in order to chew it some more, a process called **chewing the cud**. Cows and many other herbivores harbour bacteria and archaea in their digestive systems that help them break down the cellulose. Giraffes' long legs and necks allow them to access leaves from their favourite tree, the acacia, and their tough tongues can resist the pointy thorns.

How plants protect themselves from herbivores

▲ The stinging nettle (*Urtica dioica*): notice the pointy hairs on the underside of the leaf.

Herbivory means to feed on plants. Because plants cannot run away from the animals that want to eat them, they have other adaptations for defending themselves. Thick bark is difficult for many insects and some animals to penetrate, and therefore protects the plant against animals like aphids. Thorns and spikes are useful for deterring herbivores.

The common nettle (*Urtica dioica*) has tiny hairs of silica on its stem and on the underside of leaves, which are filled with chemical irritants. When animals approach the plant and rub against them, the silica breaks, scratching the skin, and the irritant inside is released into the damaged skin. This generates an unpleasant stinging sensation like a burn, as well as swelling of the skin, which the animal is likely to remember next time it considers eating the plant.

Plant poisons, called **phytotoxins,** are made from secondary compounds (see the Global context box) and can cause nausea, cardiac problems or hallucinations when ingested. Foxgloves, in the genus *Digitalis*, for example, produce toxins that will make many types of mammals very sick, including humans. The castor bean (*Ricinus communis*) produces seeds that are rich in nutritious oil, making them tempting to eat. Humans make castor oil from these seeds and use it for food as well as industrial purposes. To protect the seeds from animals that want to eat them, the plant produces a phytotoxin called ricin, which is highly toxic. Fortunately, in the process of manufacturing castor oil, this toxin is removed.

Secondary compounds, also called secondary metabolites, are molecules that are not necessary for the normal growth or reproduction of the organism, but which can be used by the organism as a toxin for defence. Humans sometimes use secondary compounds as medicines or stimulants. Examples include quinine, penicillin and caffeine. Foxgloves have long been used as medicinal plants, and a heart medication has been developed from them.

As plants evolve chemical deterrents such as alkaloids and tannins, animals evolve ways of neutralizing the toxins so that they are not poisoned. In ruminants and insects that rely on microbes for digestion, the microbes that live in the gut can detoxify many plant poisons. If a ruminant eats a small quantity of a toxic plant, colonies of microbes that can degrade the poison will start to grow in its gut. As it eats more of the plant, it can cope with more toxins as the colonies proliferate. However, it would eventually be poisoned if it continued to eat the toxic plant. Animals seeking out new food sources use a technique called cautious sampling, in which they do not eat too much of a plant the first time it is encountered. In browsing herbivores such as the moose (*Alces alces*), proteins in their saliva have evolved to neutralize tannins. If a mammal is not killed by a toxin, the toxin will travel through the blood to the liver, where it will be neutralized.

B4.2.10 – Adaptations of predators and prey

B4.2.10 – Adaptations of predators for finding, catching and killing prey and of prey animals for resisting predation
Students should be aware of chemical, physical and behavioural adaptations in predators and prey.

How predators find, catch and kill prey

Chemical adaptations

When we think of predators, we often picture a cheetah running after a gazelle, but not all predators rely on speed and physical strength to catch their prey. Some use surprise tactics, or inject chemicals into their prey, while others can use chemical compounds to lure their prey by trickery.

The black mamba (*Dendroaspis polylepis*) is a venomous snake that lives in southern and eastern parts of Africa. The venom in its bite contains neurotoxins that paralyse its prey. After biting and injecting the venom, the black mamba waits until its victim, such as a small bird or rodent, is no longer moving, then it will ingest it whole.

Pheromones are organic molecules used to send messages through the air, and some of them are intended to attract mates. Certain species of orb-weaver spiders are capable of producing chemicals that mimic the sex pheromones of moths. They release the pheromones and wait for their prey to arrive. After following the scent, the moth finds that it has been invited to dinner not by a mate but by a predator.

Physical adaptations

To find and chase down prey, predators need to be able to detect their prey using senses such as sight, smell or even electrolocation. Birds of prey such as hawks and eagles have excellent eyesight for detecting prey. Owls have eyes that are well adapted for seeing in low light at night. Bats and dolphins use echolocation, which involves sending out ultrasonic vibrations, and their brains process how the waves bounce off objects (including prey) in the environment, in order to perceive their environment with sound. Sharks have specially adapted organs in their heads called ampullae of Lorenzini. These sensing organs detect changes in electromagnetic fields, allowing them to detect prey. As a fish or seal swims, its nervous system releases small discharges of electricity that can be detected by sharks. An acute sense of smell helps birds like vultures find rotting flesh, or fish find prey in low-light conditions.

The Malaysian orchid mantis is well adapted for attracting prey by mimicking a flower that has delicious nectar, but it is also well adapted for catching insects (using appendages that grasp) as well as eating insects (with sharp mouthparts to chew through tough exoskeletons).

But finding prey is only part of the story. Catching the prey and then eating it also present challenges. The ability to fly, run or swim not only rapidly but also with precision is key in chasing down prey. Then claws, beaks, teeth and a well-adapted digestive system are needed to kill the prey and extract nutrients. Predators also need a brain that can quickly assess rapidly changing circumstances and make complex decisions involving the time, energy and risk involved in pursuing prey. If the risk of exhaustion or injury is too great, the predator must know when to give up and try again another time.

Behavioural adaptations

Some predators are **ambush predators**. They hide and wait for prey to come near and then pounce on them. This is true for many spiders, notably those that build a web, when they hide at one end of the web and wait. Anglerfish such as frogfish hide on the ocean floor and use a lure called an **illicium** (a long thin appendage protruding from their head) to attract prey. They then open their mouths in a fraction of a second to engulf the prey, before the prey is even aware that they are there.

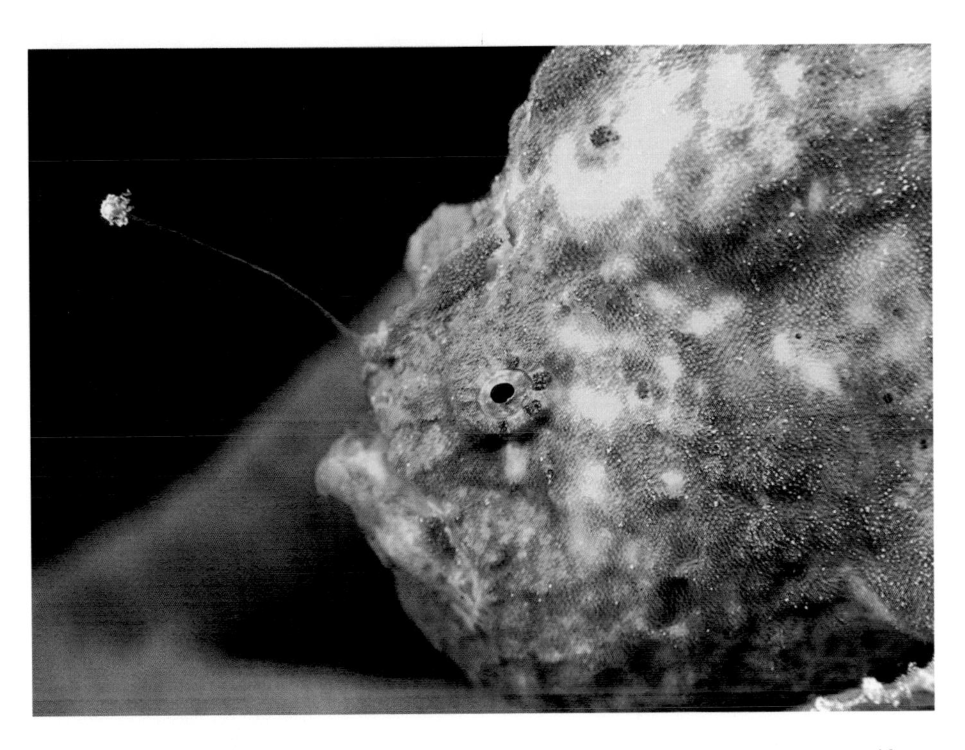

This frogfish uses a lure on the end of an appendage that it wiggles about to attract the attention of unsuspecting prey.

Teamwork is sometimes a successful adaptation. **Pack hunting** is common in wolf species, and the chances of bringing down a large animal are greater if several wolves work together than if one wolf hunts on its own. For this to work, there must be an established relationship of trust between individuals. The group knows which animal is the leader of the pack, and which animals are subservient. Other social animals that engage in group hunting, or even what could be considered warfare, are the hymenoptera: ants, termites, bees and wasps. Soldier ants, for example, participate in raiding parties and use their large mandibles to kill and dismember prey, then bring the dead prey back to the colony for a feast.

A type of predator you have probably seen on nature documentaries is the **pursuit predator**, which relies on speed to outrun its prey. Cheetahs are the fastest land mammal, at least over short distances, and they are well-adapted to chasing down gazelles, especially those that are very young, very old or unhealthy. But speed is not the only strategy: endurance can work sometimes, too. The idea then is to keep pursuing the prey for many hours until it drops from fatigue. This is known as **persistence hunting**. Humans living as hunter gathers use persistence hunting.

How animals that are preyed upon resist predation

Chemical adaptations

Organisms that are preyed upon can try to run away, but they can also produce chemicals to dissuade or fool predators. One adaptation is to produce chemicals that taste bad or that poison the predator. This is the strategy used by poison dart frogs. One highly poisonous dart frog, *Phyllobates terribilis*, produces an alkaloid on its skin that can interfere with muscle function, including heart muscles, causing death. First Nations people of Colombia have successfully used this poison for hunting by applying it to the tips of the darts they use in their blowguns.

Poison dart frogs like this one secrete toxins on their skin that can paralyse or kill animals that try to catch it. The bright pigments in their skin act as a warning sign

The harlequin filefish (*Oxymonacanthus longirostris*) has an interesting adaptation: it picks up the characteristic smell of the coral reefs it feeds on, so that it smells like coral. Predators such as cod fish are not attracted by the smell of coral, so will search elsewhere for prey.

Physical adaptations

One of the best ways to not get eaten is to not be detected by a predator. **Camouflage**, the ability of an organism to take on the appearance of its surroundings, can work well against predators that rely on vision. Some organisms only have one fixed adaptation, such as the coloration patterns on the wings of moths, whereas others can adapt to their environments. Certain species of octopus can not only modify cells in their skin to adapt to the colours of their background, but they can generate bumpy or smooth textures on their skin to mimic the surfaces they hide on.

Poison dart frogs use dramatically bright and unusual colours, such as yellow, blue and red, to inform potential predators that they are poisonous. This technique is called **aposematism**. Some other prey animals mimic these warnings. Non-venomous kingsnakes look like coral snakes, which are poisonous. Warning vocalizations can deter predators and warn fellow prey that there is danger. Many birds such as jays and blackbirds have specific warning calls, and they are not alone. Non-human primates such as monkeys can use different types of calls, depending on the type of threat.

Growing a protective shell can be another way of deterring some predators. Invertebrates such as grasshoppers and lobsters have exoskeletons, and clams and mussels have hard shells for protection. Among vertebrates, turtles and tortoises have shells to reduce the chances that a predator might consider them as prey. An alternative way of dissuading a predator is to have sharp spines, like the porcupine.

Behavioural adaptations

Many animal behaviours are instinctive and encoded in their DNA. These can include fleeing at the sight of a predator, hiding, forming groups or using certain types of dissuasive behaviour to ward off predators.

The expression "there is safety in numbers" is not just a figure of speech. It has been tested using both field observations and controlled experiments. A large group attacked by a predator will suffer fewer kills than solitary individuals or small groups that are attacked. For example, when threatened, a herd of elephants will group together, with the largest adults placing themselves facing outwards and the young positioned in the centre. Predators will often be dissuaded from attacking a group because the risk of getting injured is greater than the chance of successfully taking down a vulnerable, solitary, juvenile, for example.

Wildebeests know instinctively that there can be safety in numbers. When bison or wildebeest stick together in a group on the prairie, there are more eyes watching out for predators in all directions, so the chances of being surprised by an ambush are lessened.

B4.2.11 – Harvesting light

B4.2.11 – Adaptations of plant form for harvesting light

Include examples from forest ecosystems to illustrate how plants in forests use different strategies to reach light sources, including trees that reach the canopy, lianas, epiphytes growing on branches of trees, strangler epiphytes, shade-tolerant shrubs and herbs growing on the forest floor.

In order for leaves to photosynthesize at their optimal rate, they need to catch as much sunlight as possible, using the best angle possible for the longest possible period of the day. Leaves tend to be flat and angled towards the Sun. Chloroplasts are concentrated on the top surface to catch as much light as possible and convert it to food.

Trees

One adaptation to maximize access to sunlight is to position leaves far above the ground, so that they are above their competitors. This is the case for the tallest tree species, whose sturdy woody trunks allow them to dominate the **canopy**. The canopy is the upper layer of a forest where the crowns (tops) of trees are found; zones below the canopy are called the **understorey**, where shorter trees can be found. The **shrub layer** contains the shortest trees and shrubs, while the **forest floor** is home to smaller, non-woody plants. In a dense forest, every square centimetre of the forest floor is almost always in shade, because leaves from plants at every level above it absorb the sunlight and shield the zones below.

Even with some trees cleared to make a path through this forest, most of the sunlight is blocked by leaves, and only a few small patches of light reach the forest floor.

But there is a price to pay for reaching high above the ground: building a sturdy trunk and strong supporting branches requires a sizeable investment in energy and nutrients.

Lianas

Plants that cannot build trunks big enough to reach the canopy can use another adaptation: borrowing support from a nearby tree. Lianas are vines that take root on the forest floor and use trees as a scaffold, allowing them to grow into the canopy to obtain more light. When seeds germinate, most seedlings seek out light and bend towards it. Liana seedlings do the opposite: they grow towards shade, which means they grow towards tree trunks, and can then start to climb.

It should be obvious that liana vines are direct competitors for trees, not only for sunlight but also for minerals in the soil and space on the forest floor for germinating seeds. The bigger and more entangled the lianas get, the more harm they do to the trees, causing the trees to grow less well. Lianas can eventually kill a tree.

Epiphytes

Similar to lianas, epiphytes take advantage of the height and strength of trees to get up into the understorey or canopy to access sunlight. The difference is that their roots are not in the soil on the forest floor. Have you ever seen moss growing on a tree branch? It is getting all the moisture it needs from water trickling along the branch when it rains, or from humidity in the air. The orchids we can buy at a florist's are epiphytes. In nature, they attach their roots to tree trunks and are well adapted to survive on very little water. One of the best ways to kill an orchid that you have adopted as a houseplant is to overwater it.

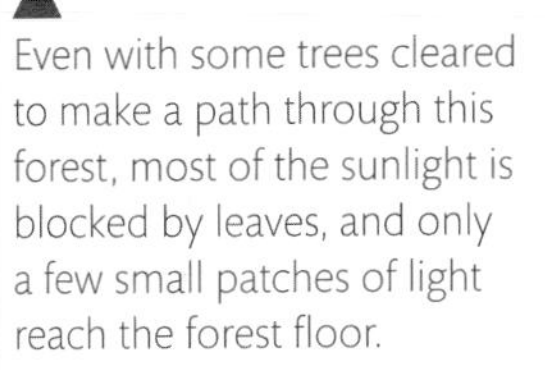

For each form of nutrition, what are the unique inputs, processes and outputs?

Orchids are epiphytes for their entire lives: their roots do not need to be in soil down on the forest floor, they can survive on very little water while living on tree trunks or branches.

The strangler fig growing around this tree is an example of a hemi-epiphyte, because it can take root in the soil at the base of the tree.

Strangler figs are examples of **hemi-epiphytes**. They spend the early part of their life in a tree without any roots in the soil. In addition to pushing their stems upwards to get more sunlight, they then push some of their stems downwards to reach the ground and start growing roots. The intertwining of the strangler fig's stems and branches can completely encircle the host tree's trunk.

Growing in the shade

Shade-tolerant shrubs grow on the forest floor between trees and are well adapted to absorbing the wavelengths of diffuse sunlight that remain after passing through other leaves, notably the longer wavelengths in the red part of the spectrum.

Some well-known and popular foods such as bananas and ginger are from herbaceous plants that grow in the understorey of forests in the tropics, and are well adapted to growing in the shade. **Herbaceous plants**, otherwise called **herbs**, are those that do not produce a woody stem with bark the way trees do. Banana plants are herbs because the part that looks like a woody stem is, in fact, not made of wood but of rigid layers of the bases of the leaves. Other examples of herbaceous plants that can grow on the forest floor are wildflowers and berries such as strawberries.

TOK

When someone declares, "I don't need to know what relationships there are between organisms in a distant tropical rainforest or at the bottom of the sea", how would an expert in ecosystems respond? Probably by saying that ignorance of these systems means that if the conditions in those places are modified over time by processes such as deforestation and human-induced climate change, we won't know what existential threats there are for other parts of the world or other species, including our own. What are the implications of not having knowledge?

B4.2.12 – Ecological niches

B4.2.12 – Fundamental and realized niches
Students should appreciate that fundamental niche is the potential of a species based on adaptations and tolerance limits and that realized niche is the actual extent of a species when in competition with other species.

The **fundamental niche** of a species is the potential niche that it could inhabit, given the adaptations of the species and its tolerance limits. The **realized niche** of a species is the actual niche that it inhabits. The realized niche can be different to the fundamental niche because of competition with other species.

The habitat of the red fox (*Vulpes vulpes*) in the USA is the forest edge. Its food consists of small mammals, amphibians and insects. It interacts with other species, such as the mosquitoes that suck its blood and scavengers that eat its leftovers. Its physical characteristics and behaviour allow it to survive in all seasons, including cold, snowy winters. The forest edge is the fundamental niche of the red fox.

What has happened to the red fox's fundamental niche in recent decades? The forest edge has been turned into farmland in many places. Some of the species eaten by the red fox no longer live there. For example, amphibians are particularly sensitive to changes in their environment and to the pesticides that farmers use. The red fox has less physical space and there is less food availability. In addition, there is direct competition from the coyote (*Canis latrans*), whose own niche has also been modified by changes in the environment cause by human activity. This new and narrower niche is called the fox's realized niche.

An activity on fundamental and realized niches in two species of *Paramecium* can be found on this page of your eBook.

B4.2.13 – Competitive exclusion

B4.2.13 – Competitive exclusion and the uniqueness of ecological niches
Include elimination of one of the competing species or the restriction of both to a part of their fundamental niche as possible outcomes of competition between two species.

Do not confuse the Russian ecologist Gause with the German mathematician Gauss, who lived a hundred years earlier and is notable for many reasons, among them the idea of a normal distribution around a mean represented by a bell-shaped curve, also called a Gaussian curve or Gaussian distribution.

The **principle of competitive exclusion** states that no two species in a community can occupy the same niche. If they do coexist for a certain period of time, as is currently happening with the fox and the coyote mentioned in Section B4.2.12, the numbers of both populations will tend to decrease. In the long run, it is often the case that one species will replace the other. This is easier to see in microbial populations, which reproduce at a very fast rate.

The competitive exclusion principle was demonstrated in 1934 by a Russian ecologist, G. F. Gause. He performed a laboratory experiment with two different species of *Paramecium*: *P. aurelia* and *P. caudatum* (see Figure 1). His experiments showed the effects of **interspecific competition** between two closely related organisms. Interspecific means between two or more different species. When each species was grown in a separate culture, with the addition of bacteria for food, they did equally well. When the two were cultured together, with a constant food supply, *P. caudatum* died out while *P. aurelia* survived. *P. aurelia* out-competed *P. caudatum*. The experiment supported Gause's hypothesis of competitive exclusion. When two species have a similar need for the same resources in the same space at the same time, one will be excluded. One species will die out in that ecosystem and the other will survive. *Paramecium aurelia* must have had a slight advantage that allowed it to out-compete *P. caudatum*.

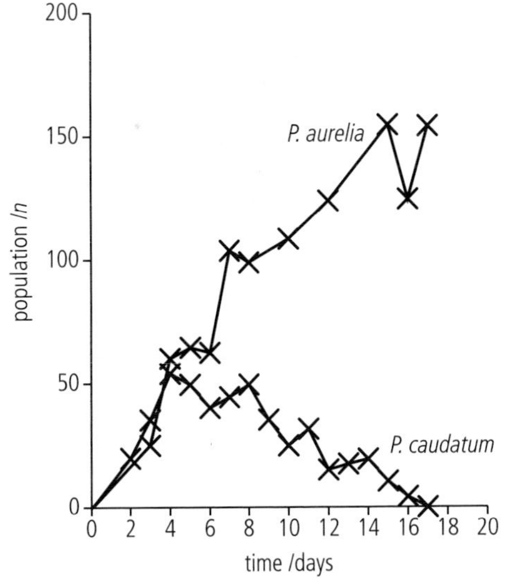

B4.2 Figure 1 The results of Gause's experiment to demonstrate the competitive exclusion principle.

It is not easy to observe this phenomenon happening in ecosystems, because it has usually already had its effect: when we look at a niche, we find only one species occupying it within an ecosystem. If we base our observations on what happens when an introduced or invasive species takes over, it is hard to know if it is only competitive exclusion that is happening or if there are other factors too. Currently in Britain the population of eastern grey squirrels (*Sciurus carolinensis*) is growing (the species was introduced from North America), while the population of red squirrels (*Sciurus vulgaris*; the native species) is falling to such low levels that there is a worry this species will be permanently driven out of some areas. It is difficult to know if this change is caused by the inability to share a niche, or if it is the impact of, for example, human activity or disease.

Guiding Question revisited

What are the advantages of specialized modes of nutrition to living organisms?

In this chapter you have learned that:

- some organisms use a particular form of nutrition so that they do not compete with other organisms for the same food
- some organisms have adaptations for eating a diverse range of foods
- autotrophs make their own food, whereas consumers such as heterotrophs and saprotrophs need to eat other organisms to get their nutrition
- some organisms are mixotrophic, and can use autotrophic nutrition at certain times and heterotrophic nutrition at others
- each organism is adapted to obtain nutrition for itself but avoid being eaten, and those adaptations can be seen in morphological features such as dentition, production of secondary metabolites and position of leaves.

Guiding Question revisited

How are the adaptations of a species related to its niche in an ecosystem?

In this chapter we have discussed how:

- an organism's niche is where it is best adapted to survive
- normally a particular niche can only be occupied by one species
- in order to continue to occupy a specific niche in an ecosystem, organisms have to adapt.

Exercises

Q1. Which of the following terms describes the place where an organism lives and the role it plays in nature?

A Niche. **B** Habitat.
C Ecosystem. **D** Community.

Q2. Which of the following primates does not belong to the great apes?

A Orangutan. **B** Gorilla.
C Lemur. **D** Human.

Q3. The Venus fly trap plant can photosynthesize but it needs to take in nitrogen by trapping and digesting insects. Which of the following terms describe it?

I. Obligate mixotroph. II. Obligate aerobe. III. Decomposer.

A I and II only. **B** I and III only.
C II and III only. **D** I, II and III.

Q4. Outline two behavioural adaptations found in predators to help them catch their prey.

Q5. Using a named example, outline the advantages of mixotrophic nutrition.

Q6. Compare and contrast the concept of an organism's fundamental niche with its realized niche.

B4 Practice questions

1. In South Korea, flocks of birds of the tit family (Paridae) forage together on trees for food. Researchers observed four species of Paridae to determine whether they shared the same habitat in the trees and whether their position on the tree depended on their size. The leafy part of the tree (crown) was divided into nine sections, three according to height from the ground and three according to the distance from the tree trunk. Observations were also made of birds foraging in the bushes surrounding the trunk and on the ground below the tree. The chart shows the relative use of each section of the habitat by the birds.

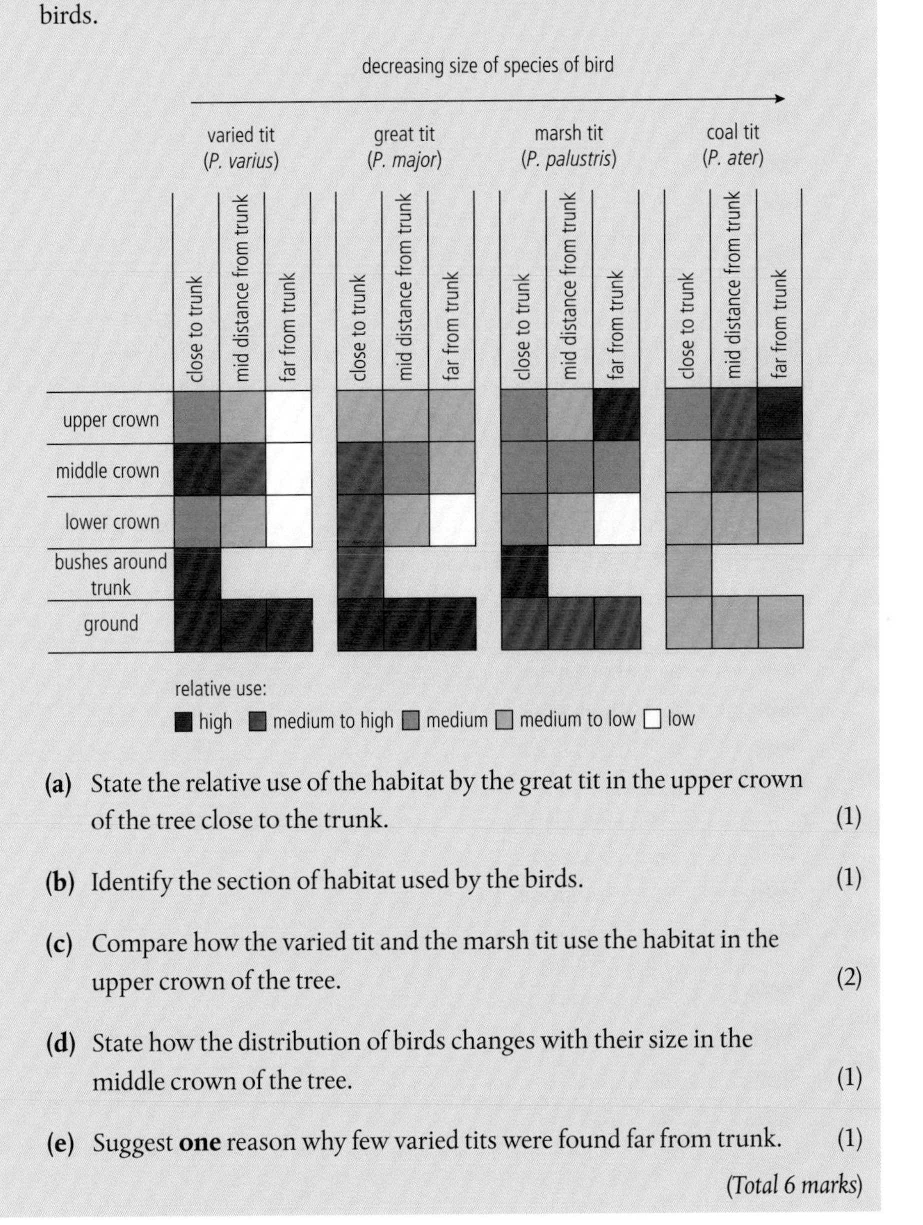

(a) State the relative use of the habitat by the great tit in the upper crown of the tree close to the trunk. (1)

(b) Identify the section of habitat used by the birds. (1)

(c) Compare how the varied tit and the marsh tit use the habitat in the upper crown of the tree. (2)

(d) State how the distribution of birds changes with their size in the middle crown of the tree. (1)

(e) Suggest **one** reason why few varied tits were found far from trunk. (1)

(Total 6 marks)

2. During the 1980s, a tiny invasive crustacean *Bythotrephes cederstroemii* entered the eastern Great Lakes from Europe (probably via freshwater or mud in the ballast water of merchant ships) and eventually colonized Lake Michigan. *Bythotrephes* reproduces very quickly and eats common zooplankton, disrupting the food web by directly competing with small juvenile resident fish. *Bythotrephes* avoids predation by larger fish through the timing of its activities, which have been investigated in offshore waters of Lake Michigan at various depths during the day and night.

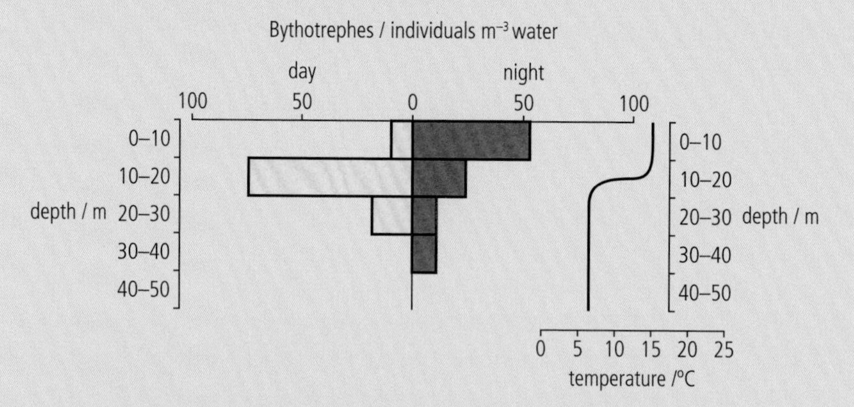

(a) State the depth range showing the most *Bythotrephes* during the night. (1)

(b) Describe the distribution of *Bythotrephes* during the day. (2)

(c) Deduce the responses of *Bythotrephes* to temperature and light. (2)

(d) Explain the change in distribution of *Bythotrephes* between day and night in terms of its position in the lake food chain. (2)

(Total 7 marks)

3. An experiment was set up to investigate how populations change over time. Rotifers and water fleas are small protists that are found in freshwater plankton. Two flasks, X and Y, were set up where each could grow alone and they each thrived when fed a constant supply of food. A third flask, Z, was set up in the same conditions into which both species were introduced. The graph shows the results from flasks X and Z.

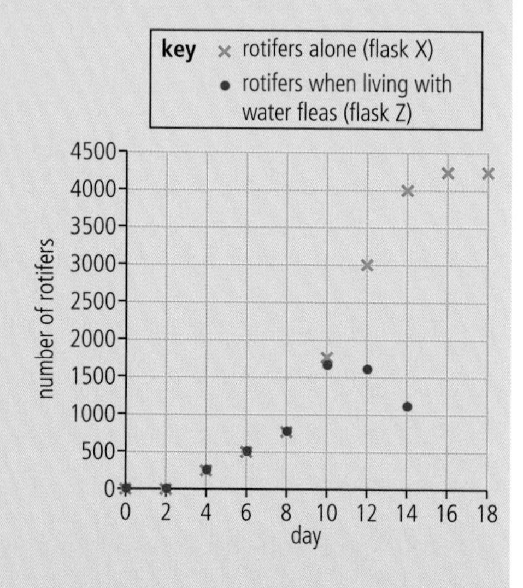

(a) Describe the changes in the experiments in flasks X and Z over time. (2)

(b) Explain the differences in population changes in flasks X and Z. (3)

(Total 5 marks)

4. Ecologists studied the distribution of five species of insectivorous wood warblers of the genus *Dendorica* living on different parts of coniferous trees in mature forests.

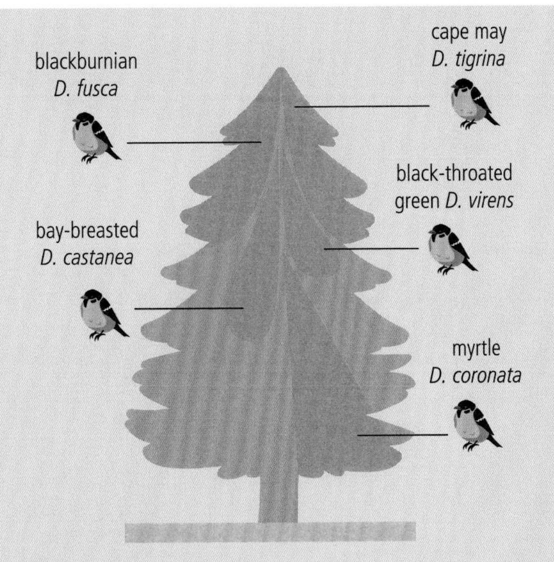

(a) Distinguish between the distribution of *D. tigrina* and that of *D. coronata*. (1)

(b) Outline the principle of competitive exclusion. (2)

(c) Other than position in the tree, suggest two ways in which the niches of the warblers in the ecosystem may differ. (2)

(d) The diagram shows the realized niches of the five species of warbler. Suggest how the fundamental niche of *D. castanea* might differ from its realized niche. (2)

(Total 7 marks)

5. Explain the importance of saprotrophic nutrition in a woodland.

(Total 3 marks)

6. Using a named example, outline how a plant can protect itself from herbivory by producing certain chemicals. *(Total 3 marks)*

THEME

C Interaction and interdependence

1 Molecules

The butterfly *Anartia jatrophae* utilizing the leaves of a plant for stability and rest. The interdependence of plants and animals is evident all around us every day.

Molecules are a connecting thread for all life on Earth. Waste products from one organism become essential building blocks for other life forms. For example, respiration requires oxygen, which plants release into the atmosphere as a waste product from photosynthesis. In turn, photosynthesis needs carbon dioxide, which is one of the products formed during respiration. Chemical reactions occur continuously in all living organisms.

Chemical reactions are essential to life and can occur on their own. However, it is essential that these reactions take place at a rate that is favourable to maintaining life. As you will see, enzymes control the rate of all chemical reactions in organisms, including cellular respiration and photosynthesis.

C1.1 Enzymes and metabolism

Guiding Questions

In what ways do enzymes interact with other molecules?

What are the interdependent components of metabolism?

Enzymes are catalysts. They do not work in isolation, however; they interact with substrates and other molecules. These interactions can control the rate of reactions.

Living organisms are dependent on a relatively small number of different atoms, mainly carbon (C), hydrogen (H) and oxygen (O) (with others in smaller amounts). These molecules are taken in from the surrounding environment and then rearranged to make all the molecules that the organism needs. All components of metabolism are interdependent. Living organisms depend on complex sequences of reactions.

Adenosine triphosphate (ATP) provides energy for many of the cellular activities essential for life. The production of ATP is complex and involves many pathways of sequential chemical reactions. The control of these reactions is the focus of this chapter.

C1.1.1 – Enzymes as catalysts

C1.1.1 – Enzymes as catalysts
Students should understand the benefit of increasing rates of reaction in cells.

Most reactions within a cell proceed too slowly on their own to sustain the life processes. However, in the presence of **catalysts** these reactions occur much faster, so that essential life functions can be maintained. In organisms, organic catalysts are known as **enzymes**.

C1.1.2 – Metabolism

C1.1.2 – Role of enzymes in metabolism
Students should understand that metabolism is the complex network of interdependent and interacting chemical reactions occurring in living organisms. Because of enzyme specificity, many different enzymes are required by living organisms, and control over metabolism can be exerted through these enzymes.

Metabolism includes all the chemical reactions that occur in an organism. These chemical reactions may be independent of one another, or they may interact with other reactions. Each chemical reaction is controlled by a specific enzyme. Because of this specificity, there are many, many enzymes in each organism. All chemical reactions involve **reactants** and **products**. Reactants are the substances that participate in a reaction, while products are the substances that are formed.

$$\underset{\text{reactants}}{A + B} \rightarrow \underset{\text{products}}{C + D}$$

C1.1.3 – Anabolism and catabolism

C1.1.3 – Anabolic and catabolic reactions
Examples of anabolism should include the formation of macromolecules from monomers by condensation reactions including protein synthesis, glycogen formation and photosynthesis. Examples of catabolism should include hydrolysis of macromolecules into monomers in digestion and oxidation of substrates in respiration.

Some metabolic reactions use energy to build complex organic molecules from simpler organic molecules. These reactions are said to be **anabolic**, and the process is called **anabolism**. The metabolic reactions that break down complex organic molecules, with the release of energy, are called **catabolic** reactions and the process is called **catabolism**. Table 1 summarizes anabolic and catabolic reactions.

C1.1 Table 1 Anabolic and catabolic reactions

Anabolic reactions	Catabolic reactions
Build macromolecules (and release water) from monomers by condensation reactions	Break down macromolecules into monomers by hydrolysis (the splitting of molecules by adding water)
Require energy input to occur	Release energy as they occur
Examples include photosynthesis, protein synthesis and glycogen formation	Examples include digestion and the oxidation of substrates in respiration

Chapter B1.1 discusses how condensation and hydrolysis reactions are common in living systems, allowing all the functions of life to be maintained. As you work through this chapter, you should keep in mind the characteristics of these two general metabolic reactions.

Energy and life

All organisms maintain their structure and function through chemical energy. In general, energy is the capacity to cause change, to do work.

Table 2 lists several forms of energy particularly important to organisms

Form of energy	Description
Kinetic energy	Energy of motion, including movement of molecules within objects
Potential energy	Stored energy or energy in a form that is not being used at a point of time
Chemical energy	A form of potential energy that is available for release when a chemical reaction occurs
Thermal energy	A form of kinetic energy stored within objects. Capable of being transferred from one object to another as heat

C1.1 Table 2 Different forms of energy

A kilocalorie is 1000 calories (cal). A calorie is the amount of heat necessary to raise the temperature of 1 gram of water by 1 degree Celsius (°C). Food labels use the symbol C for calories, which is the same as a kilocalorie. The joule is another unit of energy. One joule equals 0.239 cal.

There are also other forms of energy, including mechanical, sound, radioactivity and electric current. These different forms of energy can require different methods of measurement, but one that is often used is heat. The unit used to measure heat in biology is the **kilocalorie** (kcal).

Adenosine triphosphate (ATP) is the energy currency of a cell. Study the diagram of ATP and the similar molecule adenosine diphosphate (ADP) (Figure 1).

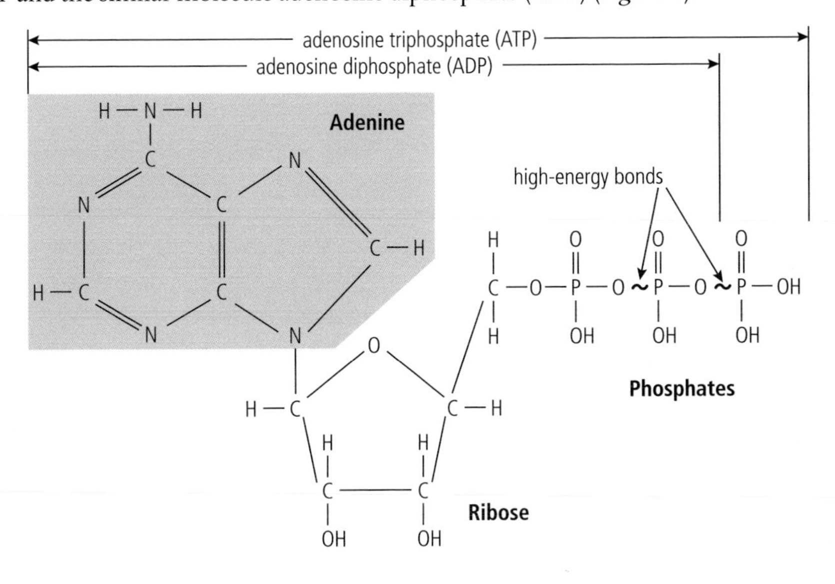

C1.1 Figure 1 This diagram shows the structure of both ADP and ATP. Adenine and ribose combine to form the molecule adenosine. ATP has three phosphate groups attached to adenosine, while ADP has only two. Note the locations of high-energy bonds represented by wavy lines. The high-energy bonds, especially the one located between the second and third phosphate in ATP, are the source of energy for chemical reactions within a cell.

ATP has many functions in an organism, including:

- supplying the energy needed to synthesize large molecules called macromolecules
- supplying the energy necessary for mechanical work, such as muscle action, chromosome movement and cilia or flagellum motion
- providing energy to move substances across the cell membrane, such as the sodium–potassium pump.

What are examples of structure-function relationships in biological macromolecules?

C1.1.4 – Globular proteins and active sites

C1.1.4 – Enzymes as globular proteins with an active site for catalysis
Include that the active site is composed of a few amino acids only, but interactions between amino acids within the overall three-dimensional structure of the enzyme ensure that the active site has the necessary properties for catalysis.

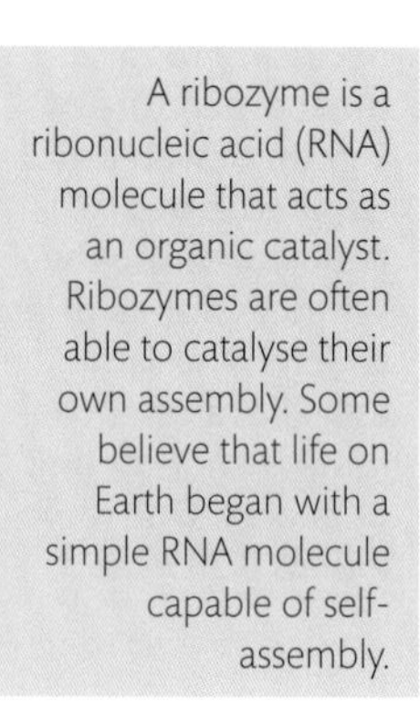

A ribozyme is a ribonucleic acid (RNA) molecule that acts as an organic catalyst. Ribozymes are often able to catalyse their own assembly. Some believe that life on Earth began with a simple RNA molecule capable of self-assembly.

Almost all enzymes are proteins. Protein enzymes are long chains of amino acids that have taken on a very specific three-dimensional shape. Think of a flexible metal wire that can be bent many times into what is called a **globular** shape. This shape is complex and at first glance appears to be random, but in enzymes (and other **globular proteins**) the complex shape is not random: it is very specific. Somewhere in the three-dimensional shape of the enzyme is an area that matches the shape of that enzyme's substrate. This area of the enzyme is called the **active site**. The shape of the active site closely matches the shape of one particular substrate. It is important to note that the active site is composed of only a few amino acids. It is the interaction between the amino acids in the overall three-dimensional enzyme shape that provides the active site with the properties necessary to carry out catalysis. This three-dimensional shape is essential to the action of the enzyme. If it is changed in any way, the enzyme is said to be **denatured**, and it will no longer function as a catalyst. Several factors can cause denaturation by affecting the chemical bonds amongst the amino acids present. These factors will be discussed later in this chapter.

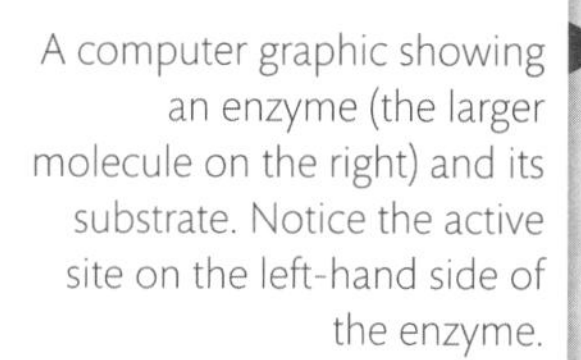

A computer graphic showing an enzyme (the larger molecule on the right) and its substrate. Notice the active site on the left-hand side of the enzyme.

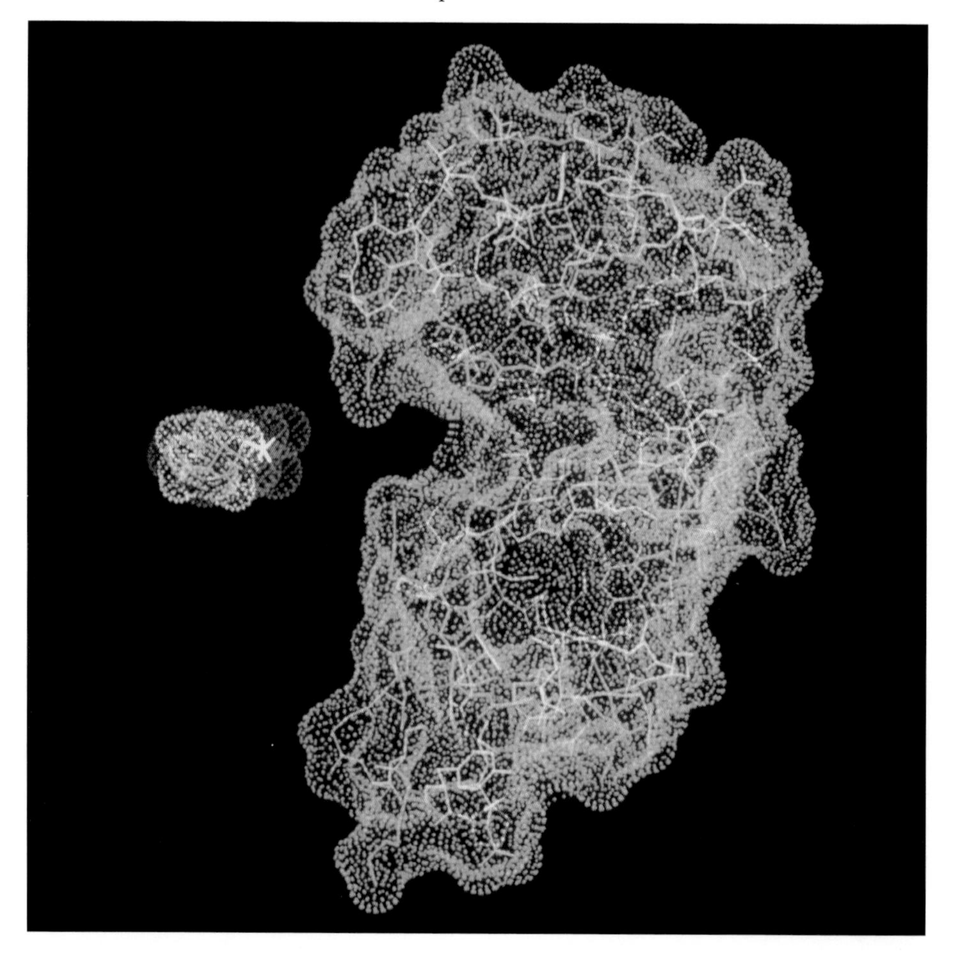

C1.1.5 and C1.1.10 – Enzyme activation

C1.1.5 – Interactions between substrate and active site to allow induced-fit binding
Students should recognize that both substrate and enzymes change shape when binding occurs.

C1.1.10 – Effect of enzymes on activation energy
Application of skills: Students should appreciate that energy is required to break bonds within the substrate and that there is an energy yield when bonds are made to form the products of an enzyme-catalysed reaction. Students should be able to interpret graphs showing this effect.

The induced-fit model

In the 1890s Emil Fischer proposed the **lock-and-key** model for enzyme action. In this model, the lock represents the enzyme's active site, and the key represents the substrate. Because the three-dimensional shape of the internal portion of the lock is complex and specific, only one key will fit it. At the time this model provided a good explanation of the specificity of enzyme action. However, as knowledge about enzyme action has increased, Fischer's model has been modified into what is now known as the **induced-fit model** of enzyme action.

TOK

The work by Emil Fischer was carried out before details of enzyme structure were known. What is the role of imagination and intuition in the creation of hypotheses in the natural sciences?

Nature of Science

Models often change over time as more evidence is gathered, resulting in changing hypotheses, theories and predictions. Peer review amongst researchers is essential to develop and verify the most accurate models possible.

Research has shown that many enzymes undergo significant changes in their conformation (shape) when substrates combine with their active site. A good way to visualize this model of enzyme action is to think of a hand and glove, the hand being the substrate and the glove being the enzyme. The glove looks a bit like the hand. However, when the hand is placed in the glove, there is an interaction that results in shape changes of both the hand and the glove, thus providing an induced fit. The changes in the shape of the substrate (the hand in this analogy) causes stresses upon its chemical bonds. The bonds become destabilized, which favours reactions and increases reaction rates.

Activation energy

It is not enough for an enzyme's substrate(s) to enter an active site. The substrate(s) must enter with a minimum rate of motion, kinetic energy, that will provide the energy necessary for the reaction to occur. Enzymes do not provide this energy; they simply lower the energy minimum that is required. The energy being referred to is called the **activation energy** of the reaction. Thus, enzymes lower the activation energy of reactions. Enzymes are not considered to be reactants and are not used up in the reaction. An enzyme can function as a catalyst many, many times.

Activation energy is the energy necessary to destabilize the existing bonds in a substrate so that a reaction can proceed. Reactions that require larger amounts of activation energy tend to proceed more slowly than those requiring smaller amounts.

Catalysts lower the activation energy needed for a reaction to proceed. Chemical reactions are reversible, which means they can occur in both directions. By reducing activation energy, catalysts increase the rate of a chemical reaction in both the forward and reverse directions. You can think of a catalyst as lowering the energy barrier that is preventing or hindering a reaction from occurring. In living systems enzymes act as catalysts.

Exergonic and exothermic reactions release energy when they occur. The products of an exergonic reaction have less energy than the reactants had because of this released energy. Endergonic and endothermic reactions result in products that have a higher energy level than the reactants.

This is because there are fewer molecules colliding with sufficient energy to overcome the initial energy requirement. There are two ways of overcoming the energy barrier and increasing the rate of these chemical reactions.

1. Increasing the energy of the reacting molecules and thus increasing the rate of collisions, usually by the addition of heat.
2. Lowering the activation energy that is required to stress particular chemical bonds in the reactants so that the bonds can be broken more easily.

Because living systems are vulnerable to higher temperatures, most chemical reaction rates in organisms are increased by the action of enzymes.

In addition, an enzyme cannot force a reaction to occur that would not otherwise happen without the enzyme. However, the reaction will be much more likely to occur with an enzyme present because the input of energy (activation energy) required will be lower.

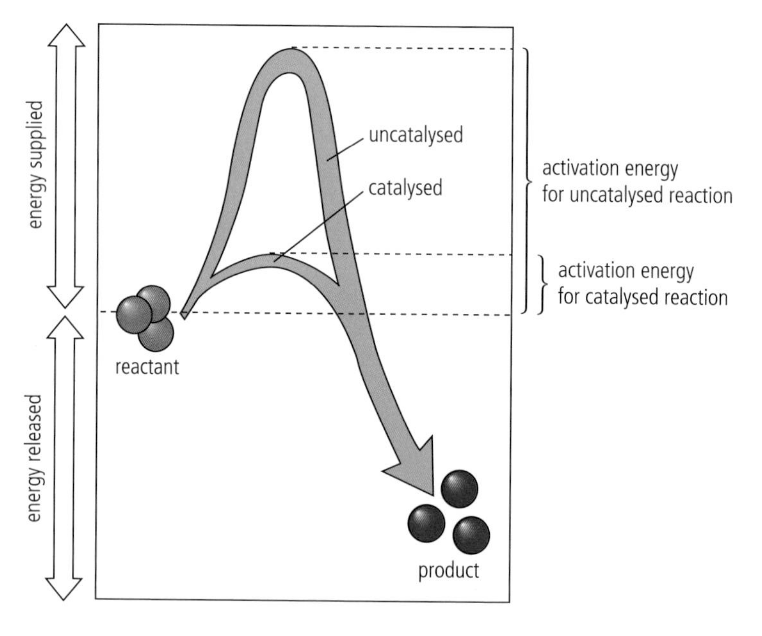

Enzymes accelerate exothermic reactions by lowering the activation energy required. The activation energy is needed to destabilize the chemical bonds in the reactant. The upper curve shows the activation energy when no enzyme is involved. The lower curve shows the activation energy required when an enzyme is present to catalyse the reaction.

Reactions proceed until there is equilibrium between the relative amounts of reactants and products. It is important to note that, even though enzymes lower the activation energy of a particular reaction, they do not alter the proportion of reactants to products at equilibrium. Some reversible chemical reactions require a different enzyme to lower the activation energy in the reverse direction.

Exergonic and endergonic reactions are **bioenergetic reactions**. Molecules are rearranged and energy can be used to do work. Exothermic and endothermic reactions are primarily **thermodynamic reactions.** The energy is given out or taken in in the form of heat. In chemistry it is more usual to have exothermic and endothermic reactions. In biology, exergonic and endergonic are more common.

SKILLS

Breaking chemical bonds requires energy, and when chemical bonds form they release energy. Look at the figure below. In A, more energy is released when the chemical bonds form in the products than is needed to break the bonds in the reactants. Overall, the reaction releases energy (it is exergonic). In B, more energy is needed to break the bonds in the reactants than is released when the products form. Overall, the reaction takes in energy (it is endergonic).

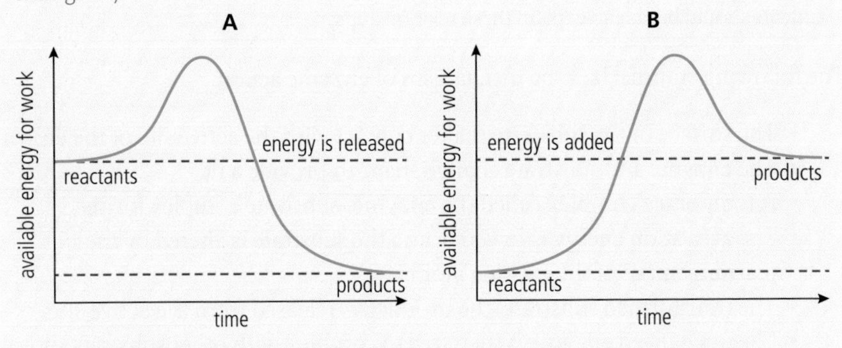

1. What is the initial energy called that starts the reaction in both graphs?
2. What type of reaction results in products that have a higher energy level than the substrates?
3. Photosynthesis is an endergonic reaction. Explain why an endergonic reaction is useful to cells.
4. What would be the effect of adding an appropriate enzyme to each reaction type?
5. Some reactions will occur spontaneously after a level of activation energy is provided. This means no additional energy will be required for the reaction to proceed. Which of the above reactions represents a spontaneous reaction?
6. An additional supply of energy beyond activation energy is required for the reaction shown in B. Where is that energy stored in the product produced?

C1.1.6 – Molecular motion

C1.1.6 – Role of molecular motion and substrate-active site collisions in enzyme catalysis
Movement is needed for a substrate molecule and an active site to come together. Sometimes large substrate molecules are immobilized while sometimes enzymes can be immobilized by being embedded in membranes.

Active sites and movement are both important to enzyme action and control of chemical reactions. The active site must join with the substrate based on shape. However, movement is also important. In order for substrates to react, they need to collide and they need to find the enzyme's active site. Molecules therefore need enough energy to move and collide. Often, the substrate or enzyme is anchored or immobilized in a membrane, allowing a more efficient joining of substrate and active site. The processes of cellular respiration and photosynthesis utilize enzymes embedded in membranes to carry out essential reactions efficiently. This is discussed in Chapters C1.2 and C1.3. The evolution of life has progressed largely as the result of the development of more efficient chemical reactions.

Immobilized enzymes are utilized in many present-day industrial practices. Attached to stationary surfaces, they are used in the following processes:

- conversion of carbohydrates to ethanol-based fuels (biofuels)
- production of dairy products and several beverages
- diagnosis of various diseases
- manufacture of penicillin and other antibiotics.

C1.1.7 – Mechanism of enzyme action

C1.1.7 – **Relationships between the structure of the active site, enzyme–substrate specificity and denaturation**
Students should be able to explain these relationships.

The following summarizes the mechanism of enzyme action.

- The surface of the substrate makes contact with the active site of the enzyme.
- The enzyme and substrate change shape to provide a fit.
- A temporary complex called the enzyme–substrate complex forms.
- The activation energy is lowered, and the substrate is altered by the rearrangement of the existing atoms.
- The transformed substrate, the product, is released from the active site.
- The unchanged enzyme is then free to combine with other substrate molecules.

Enzyme action can also be summarized by the following equation:

$$\mathbf{E + S \leftrightarrow ES \leftrightarrow E + P}$$

where E is the enzyme, S is the substrate, ES is the enzyme–substrate complex, and P is the product.

As you have seen, the structure of the enzyme is key to its function. The active site must be a specific shape to allow the formation the enzyme–substrate complex. Each enzyme is specific for its substrate and anything that changes the shape of the enzyme (including denaturization) will affect the rate at which the enzyme works.

C1.1.8 – Factors affecting enzyme-catalysed reactions

Collision theory states that reactants of a chemical reaction must collide with one another with sufficient energy to react. They must also collide in the correct orientation so that chemical bonds are affected, allowing the chemical reaction to proceed. Many factors play a role in these collisions, including the concentration of reactants, temperature, nature of the reactants and catalysts present.

C1.1.8 – **Effects of temperature, pH and substrate concentration on the rate of enzyme activity**
The effects should be explained with reference to collision theory and denaturation. **Application of skills:** Students should be able to interpret graphs showing the effects. **NOS:** Students should be able to describe the relationship between variables as shown in graphs. They should recognize that generalized sketches of relationships are examples of models in biology. Models in the form of sketch graphs can be evaluated using results from enzyme experiments.

When you are considering the various environmental factors that affect enzyme-catalysed reactions, you must first remember that all chemical reactions are fundamentally caused by molecules colliding. If the molecules that are colliding do so at a high enough speed, and the molecules have the capability of reacting with each other, then there is a chance that a reaction will occur. Enzymes cannot change those fundamentals.

Effect of temperature

Imagine an enzyme and its substrate floating freely in a fluid environment. Both the enzyme and substrate are in motion and the rate of that motion is dependent on the temperature of the fluid. Fluids with higher temperatures will have faster moving molecules (more kinetic energy). Reactions are dependent on molecular collisions and, as a rule, the faster molecules are moving, the more often they collide, and with greater energy. Reactions with or without enzymes will increase their reaction rate as the temperature (and thus molecular motion) increases. However, reactions that use enzymes do have an upper limit (see Figure 2). That limit is based on the temperature at which the enzyme (as a protein) begins to lose its three-dimensional shape because the intramolecular bonds are being stressed and broken. When an enzyme loses its shape, including the shape of the active site, it is said to be denatured. **Denaturation** can be temporary, as in many instances the intramolecular bonds will re-establish when the temperature returns to a suitable level. Denaturation will be permanent if the increasing temperature is such that it prevents a return to the native, biologically shaped molecule.

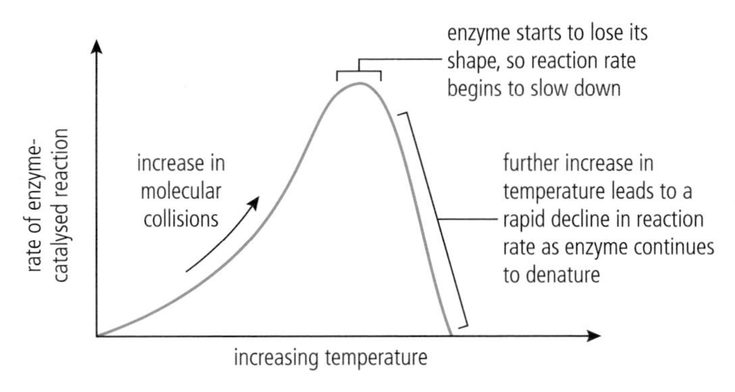

C1.1 Figure 2 The effect of increasing temperature on the rate of an enzyme-catalysed reaction.

Effect of pH

The active site of an enzyme typically includes a few amino acids. Some amino acids have areas that are charged either positively or negatively. The charged areas of the substrate must match the charged areas of the amino acids in the active site. When a solution has become acidic, the concentration of the hydrogen ions (H^+) rises. The hydrogen ions can bond with the negative charges of an enzyme or a substrate, and prevent matching of the charges between the two. A similar scenario occurs when a solution has become too basic: hydroxide ions (OH^-) can bond with the positive charges of a substrate or an enzyme, and once again prevent proper charge matching between the two. Either of these scenarios will result in an enzyme becoming less efficient. If the pH changes a great deal, bonds within the enzyme may start to break, causing the enzyme to lose its shape and thus become denatured.

There is no one pH that is best for all enzymes (see Figure 3). Many of the enzymes operating in the human body are most active when in an environment that is near neutral. There are exceptions to this, however; for example, pepsin is an enzyme that is active in the stomach. The environment of the stomach is highly acidic, and pepsin is more active in an acidic pH.

> *i* Whether or not an enzyme is permanently destroyed by denaturation is largely dependent on whether covalent bonds (such as peptide bonds) have broken. Deoxyribonucleic acid (DNA) determines the order of amino acids, and they have no way of reassembling properly if they become detached from each other.

	pH
strongly acidic	1
	2
	3
	4
weakly acidic	5
	6
neutral	7
weakly alkaline	8
	9
strongly alkaline	10
	11
	12
	13
	14

▲ The pH scale. Most fluids within the human body are close to neutral. The pH of blood plasma is typically 7.4, making it very slightly alkaline.

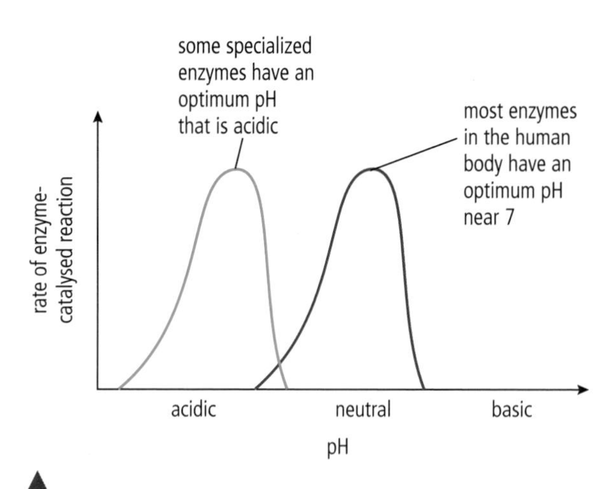

▲ **C1.1 Figure 3** The effect of pH on the rate of an enzyme-catalysed reaction. This illustrates that there is no single pH that is best for all enzymes.

The pH scale is a logarithmic scale. This means that each whole number on the pH scale represents an increase or decrease by a power of 10. Thus, a solution with a pH of 4 has a 10 times greater concentration of hydrogen ions compared to a solution with a pH of 5. That same solution with a pH of 4 has a 100 times greater concentration of hydrogen ions compared to a solution with a pH of 6.

Effect of substrate concentration

If there is a constant amount of enzyme, as the concentration of a substrate increases, the rate of reaction will increase as well (see Figure 4). This is explained by the **collision theory**. If the concentration of reactant molecules increases, there are more molecules to react and collide with each other and the enzymes. There is a limit to this, however, because enzymes have a maximum rate at which they can work and only one active site. If every enzyme molecule has an active site that is occupied, adding more substrate to the solution will not increase the reaction rate further (see Figure 4).

What biological processes depend on differences or changes in concentration?

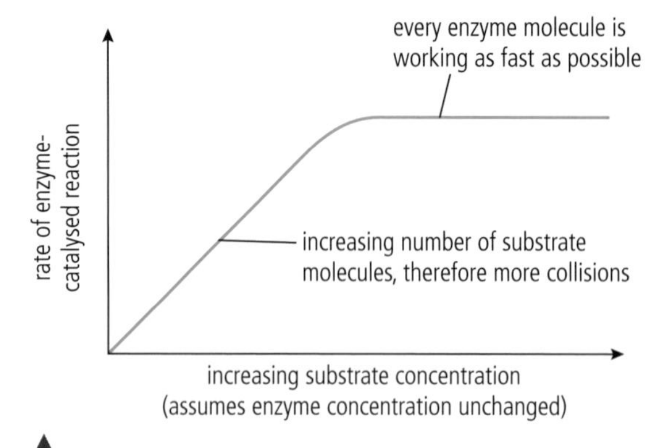

▲ **C1.1 Figure 4** The effect of increasing the substrate concentration on the rate of an enzyme-catalysed reaction.

Nature of Science

Graphs are often used to show the relationship that exists between variables. Graphs such as the ones we have just examined concerning temperature, pH and substrate concentration on enzyme activity are examples of models in biology. These models in the form of sketch graphs can be evaluated using data from enzyme experiments.

An important skill to develop in science is the ability to interpret graphs. In this exercise you will examine and interpret two graphs involving enzyme activity. As you now know, temperature and pH all affect enzyme activity. The temperature at which a particular enzyme works most efficiently is known as the enzyme's **optimum temperature.** Similarly, the pH at which an enzyme works most efficiently is known as its **optimum pH**. Study the two graphs and answer the following questions.

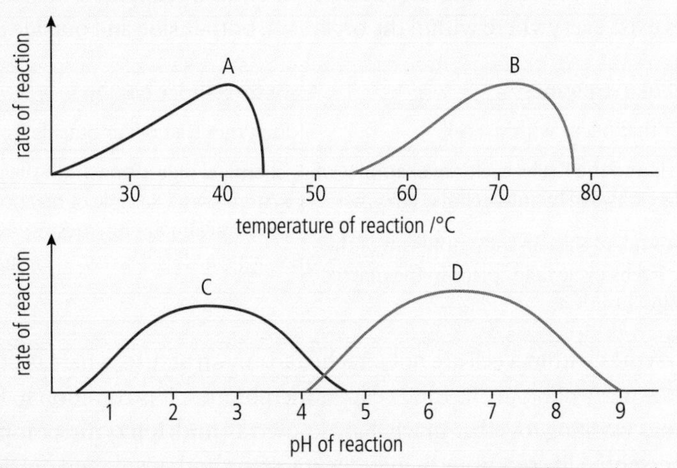

1. Which of the two lines in the top graph represents an enzyme working within the human body?
2. The human stomach is strongly acidic while the human intestine is much less acidic. Which letter represents an enzyme functioning in the human intestine?
3. Pepsin is one of the enzymes that functions in the human stomach. What is likely to be the cause of its drastic decrease in rate of reaction as it moves into the small intestine?
4. In the top graph, the enzyme represented by letter B is placed in an environment with the temperature held at 40°C. It is then placed back into its optimum temperature environment. Will the enzyme work with a higher rate of activity at the higher temperature?

C1.1.9 – Measuring enzyme-catalysed reactions

C1.1.9 – Measurements in enzyme-catalysed reactions

Application of skills: Students should determine reaction rates through experimentation and using secondary data.

Measuring the rate of an enzyme-controlled reaction is similar to measuring the rate of any other reaction. It is possible to measure either the rate at which the substrate is used or the rate at which the product is produced. For example, in the investigation provided (see the link to eBook) the enzyme used is lactase, which is an enzyme that digests lactose sugar into glucose and galactose. The rate of the enzyme activity is measured by monitoring the rate at which glucose is produced.

Investigation of factors affecting enzyme activity. Full details of how to carry out this experiment with a worksheet are available in the eBook.

See Section C1.1.5 for Section C1.1.10: they have been presented together.

HL

C1.1.11 – Intracellular and extracellular reactions

C1.1.11 – **Intracellular and extracellular enzyme-catalysed reactions**
Include glycolysis and the Krebs cycle as intracellular examples and chemical digestion in the gut as an extracellular example.

As mentioned previously, it is mostly highly specific proteins called enzymes that catalyse the reactions allowing organisms to exhibit the functions of life. These enzymes exist everywhere within the organism, both inside and outside cells.

A comparison of intracellular and extracellular enzymes

Intracellular enzymes	Extracellular enzymes
Enzymes that occur within a cell	Enzymes that occur outside a cell
Glycolysis and the Krebs cycle are examples of reactions catalysed by intracellular enzymes. Glycolysis takes place in the cytoplasm of the cell. The Krebs cycle takes place in the matrix of the mitochondria.	Chemical digestion within the gut/digestive system is an example of reactions catalysed by extracellular enzymes.

Multienzyme complexes add to the overall efficiency of catalysed chemical reactions by keeping the products of sequential reactions within the complex. By doing so, side reactions involving intermediate products are prevented, allowing a higher output of the end product. Also, products of each intermediate reaction are not allowed to diffuse away from the complex, allowing the generation of more end product.

Many enzymes within a cell are not attached to any structure, whereas others function as essential parts of organelles and cellular membranes. It is common to find groups of enzymes working together in what are known as **multienzyme complexes**, to catalyse essential life reactions in incremental steps both inside and outside the cell.

C1.1.12 – Metabolic efficiency

C1.1.12 – **Generation of heat energy by the reactions of metabolism**
Include the idea that heat generation is inevitable because metabolic reactions are not 100% efficient in energy transfer. Mammals, birds and some other animals depend on this heat production for maintenance of constant body temperature.

The energy transfer in metabolic reactions is not 100% efficient. There is also a great deal of variation in the efficiency of the different metabolic reactions. Many biologists believe that roughly 35% of the energy available to an organism is used for all cellular activities. ATP provides usable energy to the cell. The remaining energy is transferred as heat.

Several species of a plant genus called *Caladium* can generate heat and/or direct heat transferred out of chemical reactions to their reproductive structures to successfully carry out their reproductive cycles. However, there are no true "warm-blooded" plants.

In animals known as **endotherms** this heat is essential to maintain their constant internal body temperature. Birds and mammals are examples of true endotherms, often called warm-blooded animals. Without the release of heat from inefficient chemical reactions, these organisms could not survive the low temperature extremes they can currently tolerate.

C1.1.13 – Metabolic pathways

C1.1.13 – Cyclical and linear pathways in metabolism
Use glycolysis, the Krebs cycle and the Calvin cycle as examples.

Almost all metabolic reactions in organisms are catalysed by enzymes. Many of these reactions occur in specific sequences and are called **metabolic** or **biochemical pathways**. Figures 5 and 6 shows two different types of biochemical pathway.

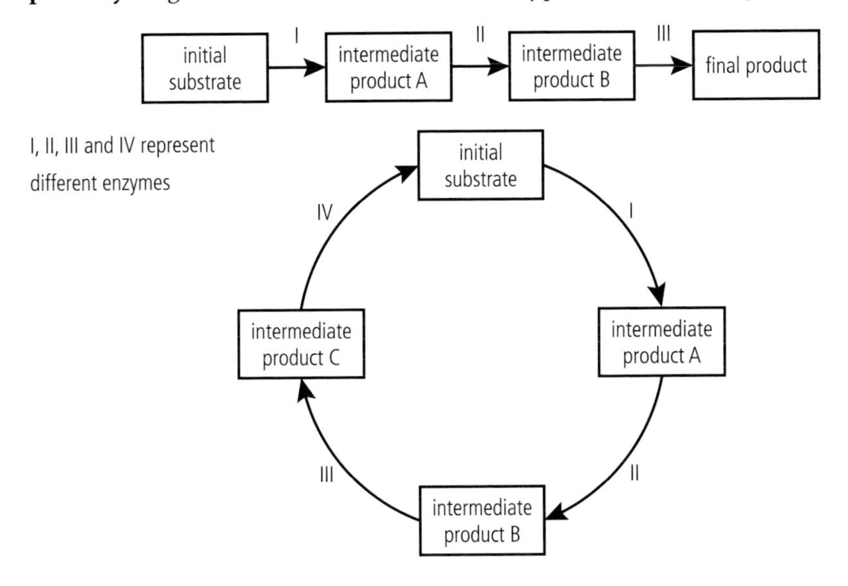

C1.1 Figure 5 A general representation of a linear metabolic pathway.

C1.1 Figure 6 A general representation of a cyclic metabolic pathway.

In both Figures 5 and 6, each arrow represents a specific reaction and its enzyme that causes one substrate to be changed to another, until the final product of the pathway is formed. The product of one reaction becomes the substrate for the next. Figure 5 shows a chain or **linear metabolic pathway**, while Figure 6 shows a cyclical or **cyclic metabolic pathway**. In reality the pathways are rarely this simple. They are often highly branched and it is common to find interactions between different pathways. Each step in a pathway is regulated. This allows fine control of the overall process and means that only small amounts of usable molecular energy are released or used at any one time (rather than releasing a large amount all at once, which could damage the cell/organism). It is important to note that the cyclical pathway begins and ends with the same substance. The linear pathway ends with a product that is different from the initial reactants/substrates.

The colourless and odourless gas carbon monoxide interferes with the metabolic pathway of cellular respiration, resulting in inadequate usable energy (ATP) production. It also binds with the oxygen-carrying molecules of red blood cells called haemoglobin. The result of both effects is loss of consciousness and possibly death.

Cellular respiration and photosynthesis include metabolic pathways. These pathways are discussed in Chapters C1.2 and C1.3. **Glycolysis** is an example of a complex linear metabolic pathway. It is part of cellular respiration, and starts with one 6-carbon compound and ends with the final products of two 3-carbon compounds. The **Krebs cycle** is cyclical, as its name infers. It is also part of a series of complex metabolic pathways that occurs in cellular respiration. It begins and ends with the same 4-carbon compound. In photosynthesis, a cyclic metabolic pathway called the **Calvin cycle** occurs. It begins and ends with the same 5-carbon compound.

C1.1.14 – Non-competitive inhibition

C1.1.14 – Allosteric sites and non-competitive inhibition
Students should appreciate that only specific substances can bind to an allosteric site. Binding causes interactions within an enzyme that lead to conformational changes, which alter the active site enough to prevent catalysis. Binding is reversible.

The effects of pH, temperature and substrate concentration on the action of enzymes were discussed in Section C1.1.8. We will now discuss the effects that certain types of molecules have on enzyme active sites. Enzyme inhibition occurs when a molecule, an **inhibitor**, binds to an enzyme and decreases the enzyme's activity by altering the active site.

Many enzymes need an inorganic ion or a non-protein organic molecule at the active site to work properly. These necessary ions or molecules are called **cofactors**. Vitamins often function as enzyme cofactors.

Non-competitive inhibition involves an inhibitor that does not compete for the enzyme's active site. In this case, the inhibitor interacts with another site on the enzyme. Non-competitive inhibition is also referred to as **allosteric inhibition**, and the site the inhibitor binds to is called the **allosteric site**. Binding at the allosteric site causes a change in the shape of the enzyme's active site, making it non-functional. Metallic ions, such as mercury, are examples of non-competitive inhibitors. These ions bind to the sulfur groups of the amino acids that make up the enzymes. This binding results in shape changes in the enzyme and therefore its active site. This shape change causes inhibition of the enzyme.

Normally, the effects of non-competitive inhibitors are reversible and the enzyme is not damaged. These inhibitors are not influenced by the concentration of the substrate.

C1.1.15 – Competitive inhibition

C1.1.15 – Competitive inhibition as a consequence of an inhibitor binding reversibly to an active site
Use statins as an example of competitive inhibitors. Include the difference between competitive and non-competitive inhibition in the interactions between substrate and inhibitor and therefore in the effect of substrate concentration.

In **competitive inhibition**, a molecule called a **competitive inhibitor** competes directly with the usual substrate for the active site. If the competitive inhibitor occupies the active site instead of the substrate, the rate of the chemical reaction will be decreased. The competitive inhibitor must have a structure similar to the substrate in order to function in this way.

Competitive inhibition is affected by substrate concentration and can be overcome by increasing the substrate concentration. Increasing the substrate concentration allows more substrate molecules to bind with the active sites as they become available, and the chemical reaction may proceed more rapidly.

Study Figures 7 and 8 demonstrating the difference between non-competitive and competitive inhibition.

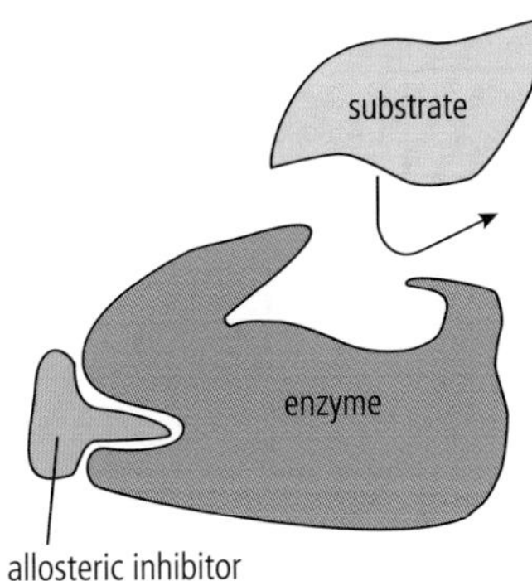

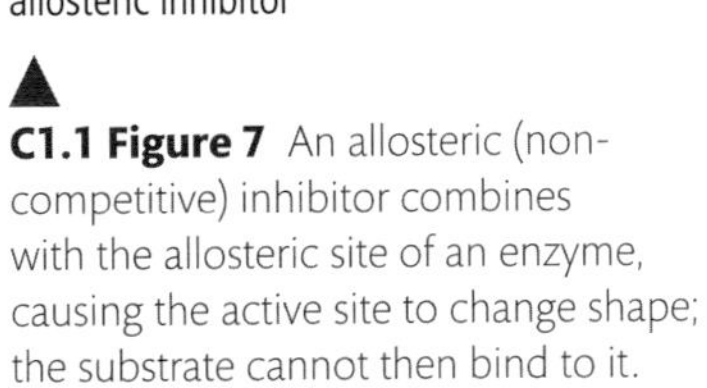

C1.1 Figure 7 An allosteric (non-competitive) inhibitor combines with the allosteric site of an enzyme, causing the active site to change shape; the substrate cannot then bind to it.

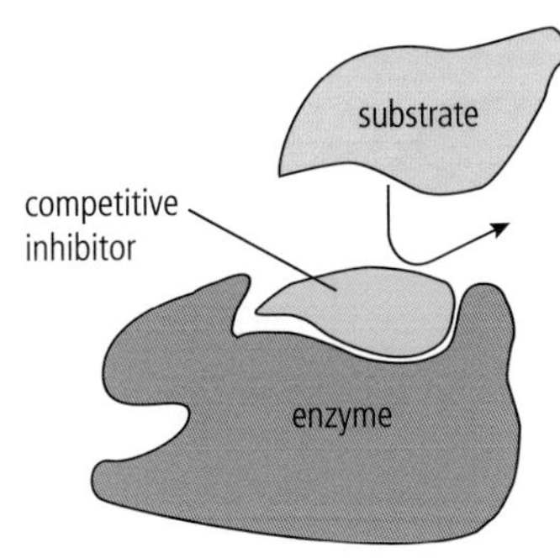

C1.1 Figure 8 A competitive inhibitor blocks the active site of an enzyme so the substrate cannot bind to it.

Many people have high blood cholesterol levels, which can result in blockages in blood vessels in the body, especially in the heart. Cardiovascular disease is a major cause of death in the world today. When people have high levels of cholesterol, doctors may recommend a group of drugs known as **statins**. Statins act as competitive inhibitors because they can combine with the active site of an enzyme essential in catalysing the biosynthesis of cholesterol within the liver. By competing for the active site of this enzyme, they cause a reduction in the production of cholesterol, therefore lowering the risk of cardiovascular disease.

Statins can have severe side effects in some people. It is therefore important to monitor their use carefully. It may be necessary for a person to stop taking a statin because of side effects. There are other medications to help decrease cholesterol, and dietary changes coupled with exercise have been found to be effective in many cases.

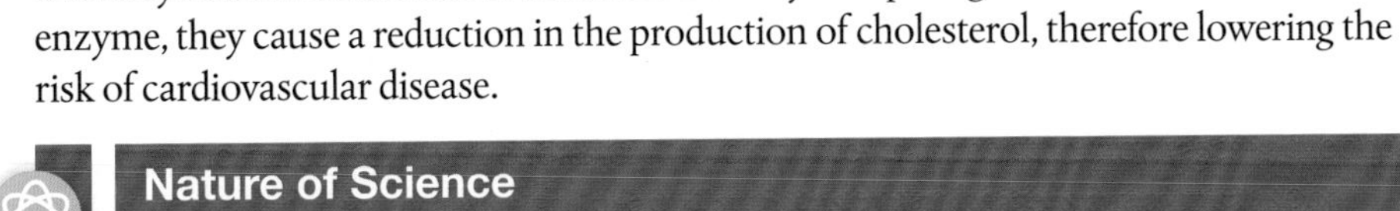

Nature of Science

Much testing goes into the development of a medicine before it is released as a treatment. Scientists have an obligation to access the risks associated with a potential medicine. Hopefully, there will be no harmful side effects when using a medicine. Statins are controversial because they do have positive effects by lowering cholesterol, but they have also been found to cause rather serious side effects in some people.

C1.1.16 – Feedback inhibition

C1.1.16 – Regulation of metabolic pathways by feedback inhibition
Use the pathway that produces isoleucine as an example of an end product acting as an inhibitor.

End product inhibition, or **feedback inhibition**, prevents the cell from wasting chemical resources and energy by making more of a substance than it needs. Many metabolic reactions occur in a linear pathway and result in a specific end product. Each step is catalysed by a specific enzyme. When the end product is present in a sufficient quantity, the pathway shuts down. This is usually because the end product inhibits the action of the enzyme in the first step of the pathway. When present in high concentrations, the end product binds with the allosteric site of the first enzyme, thus bringing about inhibition. As the existing end product is used up by the cell, the first enzyme is reactivated. The enzyme that is inhibited and reactivated is an allosteric enzyme.

A short pathway of metabolic reactions with a specific end product that, when in sufficient quantity, causes end product inhibition. This is also a form of negative feedback. The intermediates are essential molecules produced in the step-by-step pathway to achieve the end product. A represents a normal pathway with several enzymes producing intermediate compounds along the way. B represents feedback inhibition. In this condition, a large amount of end product is present. The end product inhibits enzyme 1 in the pathway. The result is that the pathway is halted.

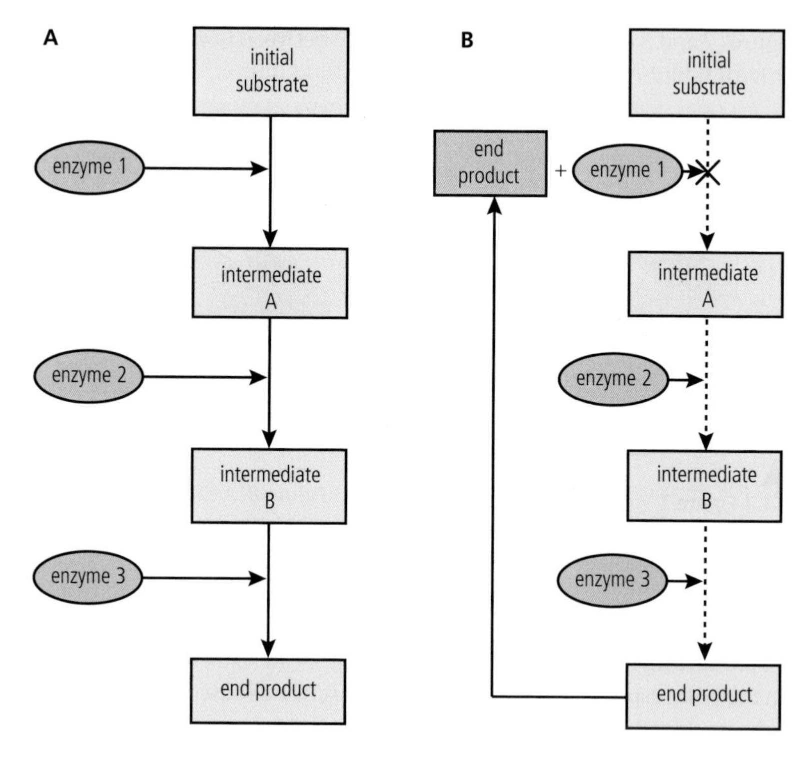

An essential amino acid is one that cannot be made by the body. As a result, they must come from our diet. We need nine essential amino acids.

Humans do not have the ability to synthesize the amino acid known as isoleucine. Isoleucine is critical to many functions of the body, including growth, glucose transportation and protein metabolism. It is also essential to our immune system. Because we cannot synthesize it, we must ingest it in our diet. Plants and bacteria can synthesize this essential amino acid and do so using a metabolic pathway controlled by **feedback inhibition**.

This is the metabolic pathway controlled by feedback inhibition that allows the production of the essential amino acid isoleucine. As the concentration of isoleucine increases, it binds to the allosteric site of the enzyme threonine deaminase. The result of this binding is an alteration of the enzyme's active site conformation, and the pathway is rendered inactive.

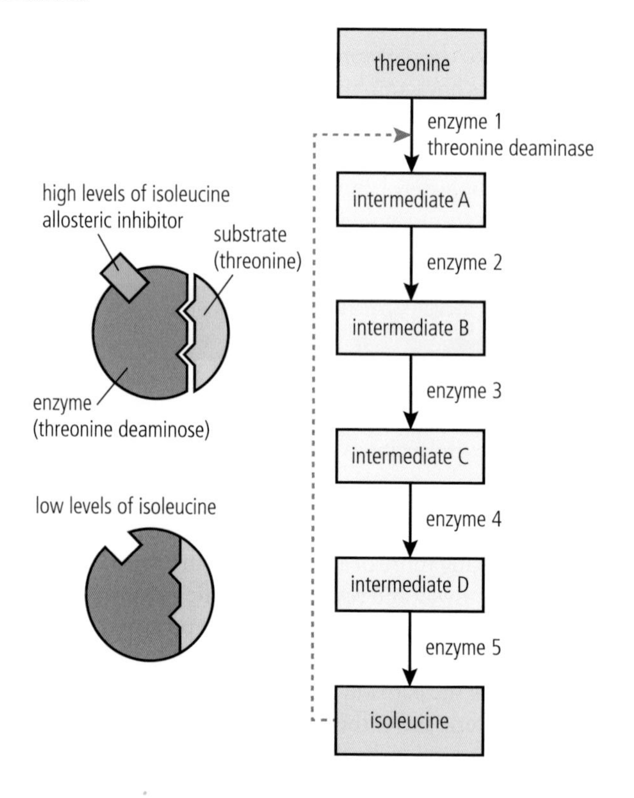

The pathway is inactive when large amounts of isoleucine are present. This is because the isoleucine combines with the allosteric site of threonine deaminase, the first enzyme in the metabolic pathway. When this happens, the active site of the enzyme is altered and it can no longer combine with the initial substrate, threonine. If the concentration of isoleucine is low, threonine deaminase can combine with threonine, allowing the pathway to proceed.

Worked example

Competitive and non-competitive inhibitors are examples of reversible enzyme inhibitors. When graphs of their effects are produced, certain characteristics can be seen. When a chemical is a competitive inhibitor, it competes for the active site of an enzyme, and its concentration must be kept high to keep the chemical reaction occurring at a slower rate. Non-competitive inhibitors do not compete for the active site of the enzyme. The result of this is that the rate of reaction will only increase if the enzyme concentration is increased.

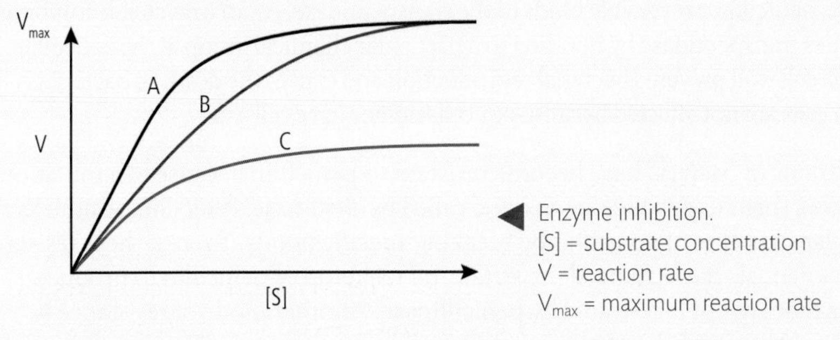

Enzyme inhibition.
[S] = substrate concentration
V = reaction rate
V_{max} = maximum reaction rate

Curve A represents a chemical reaction catalysed by an enzyme without the effect of an inhibitor. Curves B and C represent chemical reactions catalysed by enzymes affected by inhibitors.

1. Which curve represents the reaction in which a competitive inhibitor is active?
2. Which curve represents the effects of a non-competitive inhibitor?
3. Explain your answers.

Solution

1. Curve B.
2. Curve C.
3. Curve C shows the action of a non-competitive inhibitor because it results in a lower maximum reaction rate. This occurs because the inhibitor binds to the enzyme present and is not released. The reaction rate will not increase as the substrate increases because there is a limited amount of enzyme still active. Curve B represents competitive inhibition because, as the substrate increases, the rate of reaction also increases. This is because of the larger concentration of substrate out-competing the inhibitor for the active site of the enzyme. Curve B will eventually equal the maximum rate when enough substrate is added.

When asked to differentiate between competitive and non-competitive inhibition curves on a graph, look to see if the maximum reaction rate is achieved as the substrate is increased. If it is, then competitive inhibition is being represented. If it is not achieved and is significantly less, non-competitive inhibition is being represented.

C1.1.17 – Mechanism-based inhibition

C1.1.17 – Mechanism-based inhibition as a consequence of chemical changes to the active site caused by the irreversible binding of an inhibitor
Use penicillin as an example. Include the change to transpeptidases that confers resistance to penicillin.

Penicillin was discovered by Alexander Fleming in 1928, when he noticed a mould growing on a Petri dish that had antibacterial properties. However, it was not until 1944 that penicillin was produced in high enough quantities to effectively treat bacterial infections. Research conducted by Dorothy Hodgkin provided the structure of penicillin, allowing greater production levels for the infection-treating medicine. Hodgkin used X-ray studies of crystallized penicillin to determine its structure. For this and other similar work, she was elected to the Royal Society (an ultra-prestigious group of select UK scientists) in 1947 and received the Nobel Prize in 1964. Penicillin and related antibiotics inhibit an enzyme called transpeptidase that catalyses the last step in the formation of bacterial cell walls. In this case, penicillin irreversibly binds to the transpeptidase. As an irreversible inhibitor, it inactivates transpeptidase by bonding to a particular chemical group at the active site. The defective cell wall prevents bacterial reproduction and causes the death of bacterial cells. Human cells are not affected because our cells do not have cell walls.

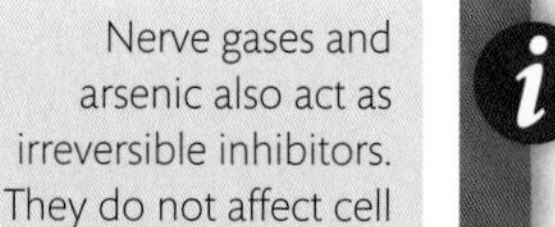

Nerve gases and arsenic also act as irreversible inhibitors. They do not affect cell walls. However, they affect critical metabolic pathways within our bodies, often with fatal results.

Some strains of bacteria have become resistant to penicillin because of a mutation that allows them to produce an enzyme called penicillinase. Penicillinase attacks the molecular structure of penicillin by breaking specific bonds. To treat these resistant strains, scientists have changed the structural make-up of penicillin to produce variants that are not inactivated by penicillinase. Another mode of resistance is occurring because of change in the transpeptidases of bacteria. Transpeptidases are enzymes that function in the formation of bacteria cell walls. Penicillin inhibits these enzymes, thus leading to inhibition of bacterial cell wall synthesis. If mutations occur within the transpeptidase active site, penicillin will no longer bind to its active site, resulting in bacterial resistance. New research is working to alter the shape of specific areas of penicillin so affinity for the transpeptidase active site can be restored.

HL end

Guiding Question revisited

In what ways do enzymes interact with other molecules?

In this chapter we have looked at how:

- enzymes increase the rates of reactions in cells
- enzymes are globular proteins with an active site, and it is the active site that combines with a substrate to catalyse a reaction
- both enzymes and substrates change shape when binding occurs
- if the active site of an enzyme is changed, it will not be able to bind to the substrate and increase the rate of the reaction
- extremes of pH and temperature can change the shape of the enzyme and therefore the active site

HL

- an active site of an enzyme may be changed if a specific molecule binds to an allosteric site on the enzyme, thus preventing the enzyme from functioning

- enzymes are essential to metabolic pathways involving both catabolic and anabolic reactions.
- enzymes may catalyse intracellular and/or extracellular reactions.

HL end

Guiding Question revisited

What are the interdependent components of metabolism?

In this chapter we have looked at how:

- metabolism often involves both cyclical and linear pathways
- the enzymes of metabolism and, therefore, the rates of metabolism are greatly affected by changes in pH, temperature and substrate concentration
- enzymes lower activation energy, allowing chemical reactions to occur faster
- feedback inhibition prevents the cell from wasting resources and energy by producing more products than are needed

HL

- feedback inhibition usually involves an effect on the first enzyme in a metabolic chain of reactions
- several types of metabolic pathway inhibition can occur, including feedback inhibition, competitive and non-competitive inhibition, and mechanism-based inhibition.

HL end

Exercises

Q1. Explain why enzymes only work with specific substrates.

Q2. What similar response in enzymes is caused by changes in both temperature and pH?

A Lowering of activation energy.
B Rate of movement of the enzyme.
C Rate of movement of the substrate.
D The three-dimensional shape of enzyme is altered.

Q3. What is activation energy?

A Energy released from the hydrolysis of a molecule.
B Energy required to initiate a chemical reaction.
C Energy of motion.
D Energy produced by a condensation reaction.

HL

Q4. What determines whether an enzyme is competitively or non-competitively inhibited?

Q5. Feedback inhibition is a common mechanism for controlling metabolic pathways. Explain why inhibition usually occurs in one of the very first steps in the reaction.

Q6. Why is inhibition brought about by allosteric site binding usually reversible?

HL end

C1.2 Cell respiration

Guiding Questions

What are the roles of hydrogen and oxygen in the release of energy in cells?

How is energy distributed and used inside cells?

All cells can convert organic molecules into the usable chemical energy known as adenosine triphosphate (ATP) through the process of cell respiration. This conversion involves a cascade of chemical reactions all with controlling mechanisms, so that efficiency and lack of damage to the cell are achieved. These chemical reactions occur in specialized cellular regions and involve molecules known as enzymes. Enzyme actions (discussed in Chapter C1.1) are essential to the process of cell respiration.

Hydrogen and oxygen play important roles in the production of ATP, the key energy-providing molecule of the cell. Energized hydrogen is an essential part of the ATP-generation process, while oxygen acts as the final electron acceptor in respiration to form water.

Oxidation is a general type of chemical reaction resulting in products with lower potential energy than the reactant(s). When it occurs in cells, there is an increase in compounds with carbon to oxygen (C-O) bonds. The opposite reaction to oxidation is **reduction**. When reduction occurs, the products contain more potential energy than the reactants. Reduction in cells creates a higher number of compounds with carbon to hydrogen (C-H) bonds. Because oxidation and reduction reactions always occur together, these chemical reactions are known as **redox reactions**. Oxidation results in loss of electrons, while reduction involves a gain of electrons.

C1.2.1 – ATP structure and function

C1.2.1 – ATP as the molecule that distributes energy within cells
Include the full name of ATP (adenosine triphosphate) and that it is a nucleotide. Students should appreciate the properties of ATP that make it suitable for use as the energy currency within cells.

Organic molecules contain energy stored in their molecular structures. Each covalent bond in a molecule of glucose, an amino acid, or a fatty acid represents stored chemical energy.

Cells break down (or metabolize) their organic nutrients by slow oxidation. A molecule, such as glucose, is acted on by a series of enzymes. The function of these enzymes is to catalyse a series of reactions in which the covalent bonds are broken (oxidized) one at a time and new products are formed that have a lower energy. The goal of releasing energy in a controlled way is to store the released energy in the form of ATP molecules. If a cell does not have glucose available, other organic molecules may be substituted, such as fatty acids or amino acids.

The structure of ATP is discussed in Chapter C1.1. Nucleotides are compounds that consist of a 5-carbon sugar bonded to a nitrogenous base and a phosphate group. ATP is a nucleotide because it contains the 5-carbon sugar ribose, the nitrogenous base adenine, and three phosphate groups.

ATP has a specific chemical structure that allows it to function as the energy currency of the cell. The last two phosphate groups of ATP are attached to the main molecule by high-energy bonds.

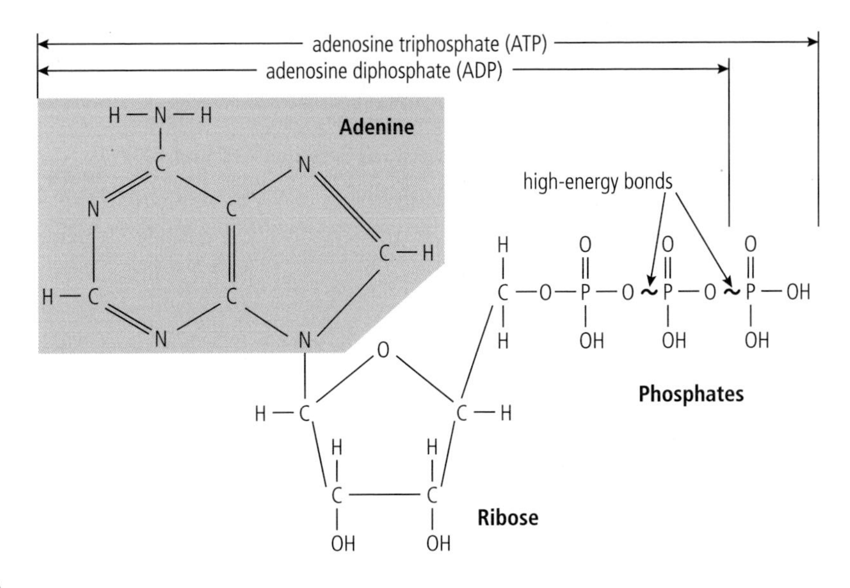

▲ The two high-energy bonds (shown as wavy lines) attaching the final two phosphates to the molecule are important to its function. If the last phosphate is removed, the molecule becomes adenosine diphosphate (ADP). When two phosphates are missing, the molecule is known as adenosine monophosphate (AMP).

Because the phosphate groups are negatively charged, they repel one another, resulting in an unstable covalent bond between the two, referred to as a high-energy bond. These unstable bonds have a low activation energy and are easily broken by hydrolysis. This hydrolysis reaction is exergonic or energy releasing. The released energy is then free to perform cellular work.

In what forms is energy stored in living organisms?

C1.2.2 – Life processes within cells require ATP

C1.2.2 – Life processes within cells that ATP supplies with energy
Include active transport across membranes, synthesis of macromolecules (anabolism), movement of the whole cell or cell components such as chromosomes.

Cellular work carried out using the energy released from the high-energy bonds of ATP includes:

- active transport across cell membranes (discussed in Chapter B2.1)
- synthesis of macromolecules by anabolism (discussed in Chapter C1.1)
- movement of the whole cell by cilia or flagellum action
- movement within the cell of cell components, such as chromosome movement in mitosis or meiosis.

C1.2.3 – ATP and ADP

C1.2.3 – Energy transfers during interconversions between ATP and ADP
Students should know that energy is released by hydrolysis of ATP (adenosine triphosphate) to ADP (adenosine diphosphate) and phosphate, but energy is required to synthesize ATP from ADP and phosphate. Students are not required to know the quantity of energy in kilojoules, but students should appreciate that it is sufficient for many tasks in the cell.

Because ATP is needed for all cell activities, it must be continually produced. This production involves a cycle, as shown in Figure 1.

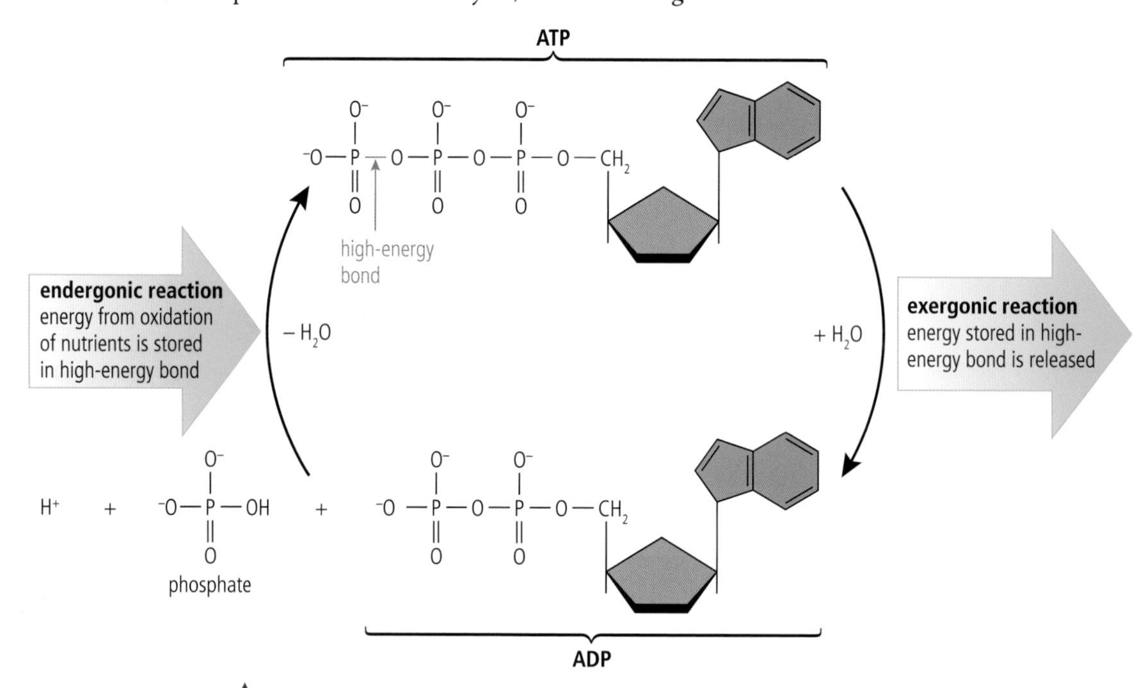

C1.2 Figure 1 The ATP cycle. A cyclic pathway allows the formation of ATP from ADP and inorganic phosphate. This reaction requires energy. When hydrolysis of ATP occurs, the third phosphate group and energy are both released. This energy is then available for cellular work. A molecule of ADP is also formed.

Most cells do not store large amounts of ATP. Cells can fulfil their continuous energy needs because the ATP cycle is constantly turning, as long as ADP, phosphate groups and energy are available. Because of this cycle, ATP is known as a renewable resource.

In the ATP cycle it is important to note that energy is required to synthesize ATP from ADP and phosphate. This energy is then stored in the high-energy bond that exists between the second and third phosphate groups. Because of the input of energy and the higher potential energy of ATP than ADP, this reaction is an **endergonic** reaction. When ATP undergoes hydrolysis to form ADP plus a separated phosphate group, energy is released that makes this an **exergonic** reaction.

Nature of Science

Quantitative measurements are more objective in scientific research than qualitative measurements. Scientists use sensors to measure the amount of energy released when ATP undergoes hydrolysis. The sensors used are limited in precision and accuracy. However, with repeated procedures, measurements and peer reviews, more reliability in the measurements is possible.

C1.2.4 and C1.2.5 – ATP and cell respiration

C1.2.4 – Cell respiration as a system for producing ATP within the cell using energy released from carbon compounds
Students should appreciate that glucose and fatty acids are the principal substrates for cell respiration but that a wide range of carbon/organic compounds can be used. Students should be able to distinguish between the processes of cell respiration and gas exchange.

C1.2.5 – Differences between anaerobic and aerobic cell respiration in humans
Include which respiratory substrates can be used, whether oxygen is required, relative yields of ATP, types of waste product and where the reactions occur in a cell. Students should be able to write simple word equations for both types of respiration, with glucose as the substrate. Students should appreciate that mitochondria are required for aerobic, but not anaerobic, respiration.

Cell (cellular) respiration is the process by which most organisms on Earth synthesize ATP for cellular functions. It involves the release of energy from carbon compounds, especially glucose ($C_6H_{12}O_6$) and fatty acids. Carbohydrates (other than glucose), proteins and many other carbon-containing compounds can also be used in respiration.

We will focus on glucose as the carbon-containing compound because it is commonly catabolized in cellular respiration. It is important to keep the following equation in mind as we discuss this life-sustaining process for the cell:

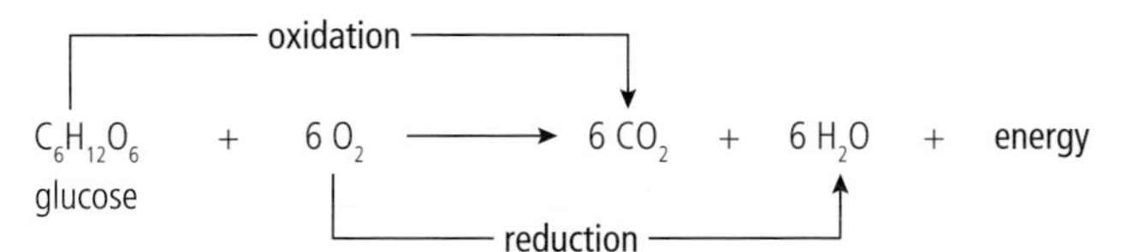

The complete breakdown of glucose in cell respiration results in carbon dioxide, water and energy release. This catabolic reaction involves the removal of electrons from glucose (oxidation) and the acceptance of those electrons by oxygen (reduction).

It is important to recognize that cellular respiration is different from gas exchange. Gas exchange is the process by which an organism obtains sufficient oxygen (O_2) and disposes of the waste gas carbon dioxide (CO_2). Cell respiration is the enzyme-controlled metabolic pathway that produces ATP.

Glucose is a high-energy molecule compared to carbon dioxide and water. Therefore, as this reaction proceeds, energy is released. The pathways of cellular respiration allow the slow release of energy from the glucose molecules so that ATP can be produced more efficiently.

Anaerobic cellular respiration

All cells begin the process of cell respiration in the same way. This initial stage is called **glycolysis**. Glucose enters a cell through the cell membrane and is found in the cytoplasm. Enzymes then catalyse reactions to ultimately cleave the 6-carbon glucose molecule into two 3-carbon molecules. Each of these 3-carbon molecules is called **pyruvate**. Some, but not all, of the covalent bonds in the glucose are broken during this series of reactions. Some of the energy that

C1.2 Figure 2 A simplified version of the events of glycolysis.

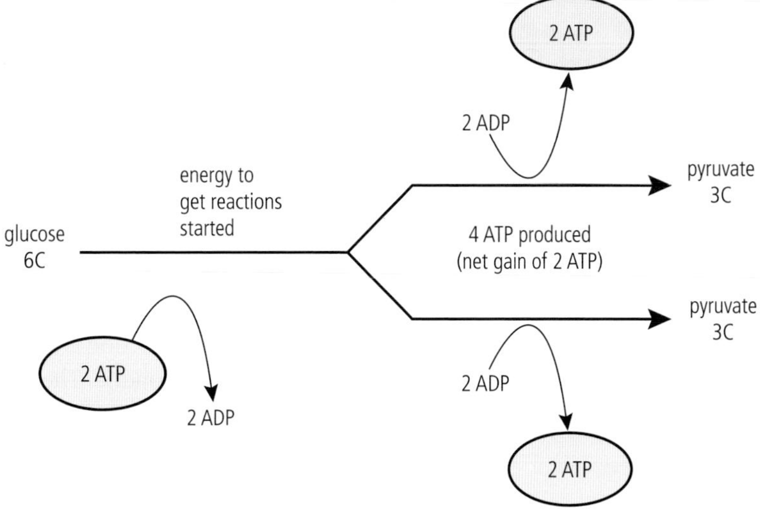

is released from the breaking of these bonds is used to form a small number of ATP molecules. Study Figure 2 on the previous page.

Notice in Figure 2 that two ATP molecules are needed to begin the process of glycolysis, and a total of four ATP molecules is formed. This is a net gain of two ATPs.

Alcoholic fermentation is a type of anaerobic respiration that occurs in yeast. Yeast cells take in glucose, and glycolysis occurs as shown in Figure 2. As there is no oxygen present, the pyruvate molecules are then converted into two molecules of ethanol. Ethanol is a 2-carbon compound. The third carbon of pyruvate is lost from the system, ultimately combining with oxygen to form carbon dioxide. The carbon dioxide is released into the cellular environment.

As we have said, glycolysis is the metabolic pathway that is common to most organisms on Earth. As can be seen in Figure 2, oxygen is not needed for glycolysis to proceed. Some organisms derive all their ATP without the use of oxygen. These organisms are said to carry out **anaerobic** cell respiration. The breakdown of organic molecules for anaerobic ATP production is called fermentation. There are two types of fermentation: **alcoholic fermentation** and **lactic acid fermentation**.

Our focus here is on the cell respiration that occurs in humans. If oxygen is not present after the initial stage of cell respiration, then in humans lactic acid fermentation commences. If your exercise rate exceeds your body's capacity to supply adequate oxygen, at least some of the glucose entering cell respiration will follow the anaerobic pathway called lactic acid fermentation. Study Figure 3.

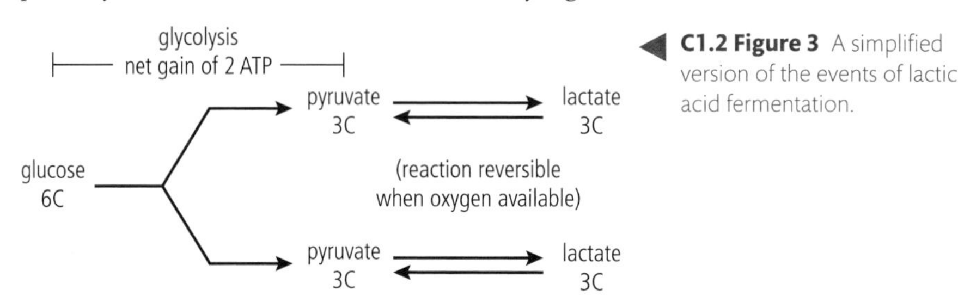

C1.2 Figure 3 A simplified version of the events of lactic acid fermentation.

The lack of adequate oxygen results in the conversion of each pyruvate produced by glycolysis into lactic acid molecules. Like pyruvate, lactic acid molecules are 3-carbon molecules. Lactic acid fermentation allows glycolysis to continue because there is not a build-up of pyruvate. However, only two ATP molecules are generated from anaerobic respiration.

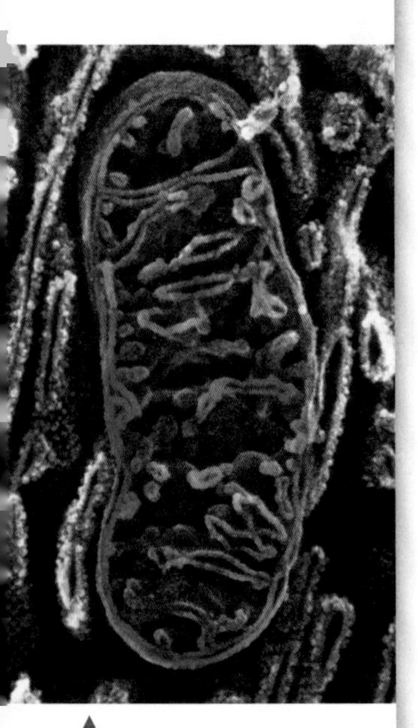

This high-resolution, false-colour transmission electron micrograph (TEM) shows a single mitochondrion. It is in this organelle that the preparatory (link) reaction and the Krebs cycle occur.

You may have experienced the muscle burn that occurs as a result of lactic acid accumulation during intense exercise. The burn goes away when adequate supplies of oxygen are provided to the muscle so that aerobic cell respiration can occur. The lactic acid is carried to the liver via the bloodstream, where it is converted to glucose, then glycogen.

Aerobic cellular respiration

Aerobic cell respiration produces many more ATP molecules than anaerobic cell respiration, and it occurs in the presence of oxygen. Like anaerobic respiration, aerobic respiration begins with glycolysis producing two molecules of pyruvate in the cytoplasm of the cell. The two pyruvates then enter the mitochondrion.

Once inside the mitochondrion, the pyruvate molecules are turned into a 2-carbon compound that enters the next stage of respiration, which is called the **Krebs cycle**. The preparatory reaction is known as the **link reaction** and takes place in the matrix of the mitochondria. The Krebs cycle also takes place in the matrix of the mitochondrion, and is a series of reactions that begins and ends with the same molecule. A net gain of two ATPs occurs in the Krebs cycle.

The final stage of aerobic respiration is the **electron transport chain**, which occurs in the cristae of the mitochondrion. Most ATP molecules produced from the breakdown of glucose are made in the electron transport chain: 30–34 ATPs are produced in this stage. More details about the stages of aerobic respiration are given in later sections of this chapter.

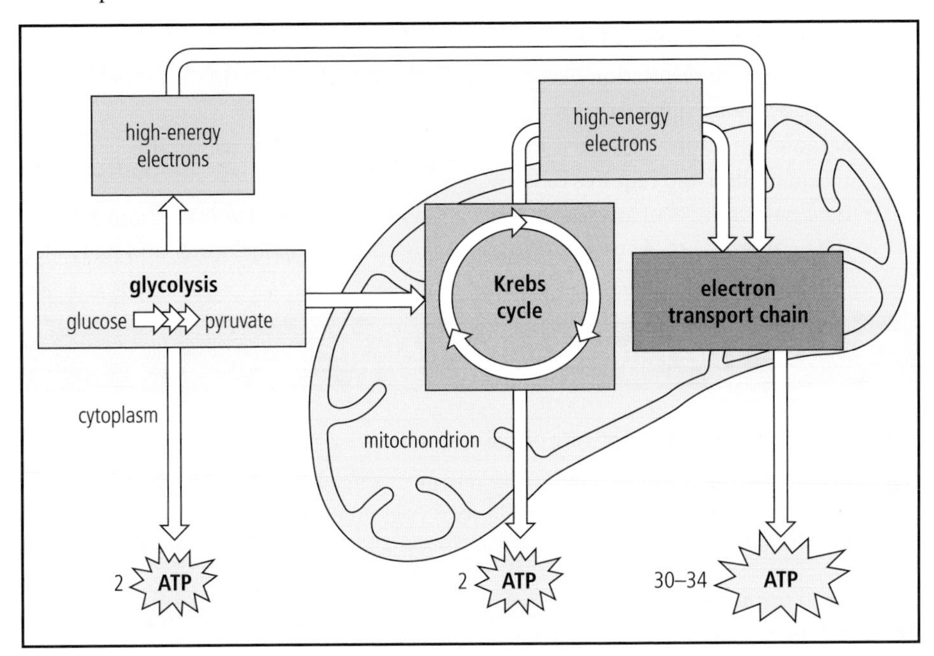

An overview of aerobic cell respiration. There is a link or preparatory reaction that changes the pyruvate produced by glycolysis into a 2-carbon compound that enters the Krebs cycle. The link reaction and Krebs cycle occur in the matrix of the mitochondrion, while the electron transport chain occurs in the cristae.

Research is continuing to determine the exact number of ATP molecules produced by aerobic cellular respiration. Roughly 30–40% of the available energy in a glucose molecule is transferred to ATP. The rest is transferred out of the cell in the form of heat.

A summary of anaerobic and aerobic respiration

Examine the overview given in Table 1 of anaerobic and aerobic respiration in humans.

Anaerobic cell respiration	Aerobic cell respiration
Does not require oxygen but does require glucose	Requires oxygen and glucose
Takes place in the cytoplasm of the cell	Begins in the cytoplasm
Glucose is split into two molecules of pyruvate	The product of the first part of respiration is two molecules of pyruvate made from glucose
If the oxygen supply is inadequate, in humans, pyruvate is made into lactic acid. This fermentation will occur in the cytoplasm	If oxygen supply is adequate, pyruvate will move into the mitochondria
No mitochondria are needed	Pyruvate is converted into a 2-carbon compound in the matrix of mitochondria
Net gain of two ATPs	The 2-carbon compound enters the Krebs cycle, also in the mitochondrial matrix
	Carbon dioxide is produced as a waste product of the Krebs cycle
	30–34 ATPs are produced in the cristae of the mitochondria

C1.2 Table 1 Anaerobic and aerobic cell respiration

Key points to remember are:

- both types of cellular respiration initially take place in the cytoplasm
- in both cases glucose (a 6-carbon molecule) is broken down into two molecules of pyruvate (a 3-carbon molecule)
- the production of ATP is very low in anaerobic cellular respiration compared to aerobic cellular respiration
- anaerobic cell respiration occurs outside the mitochondria (in the cytoplasm) and does not require oxygen
- aerobic cell respiration starts in the cytoplasm but finishes within the mitochondria and requires oxygen
- the final products of anaerobic respiration in humans are lactic acid and ATP
- the final products of aerobic respiration are carbon dioxide, water and ATP.

C1.2.6 – The rate of cell respiration

C1.2.6 – Variables affecting the rate of cell respiration
Application of skills: Students should make measurements allowing for the determination of the rate of cell respiration. Students should also be able to calculate the rate of cellular respiration from raw data that they have generated experimentally or from secondary data.

Because cell respiration involves a series of chemical reactions, there are many factors that affect its overall rate. Some of these factors are listed below.

- Temperature: the optimum temperature for the rate of cell respiration is 20–30°C. Significantly higher and lower temperatures greatly decrease the rate.
- Carbon dioxide concentration: an increase in carbon dioxide concentration adversely affects the rate of cell respiration.
- Oxygen concentration: lower concentrations of oxygen lower the rate of cell respiration. The absence of oxygen results in anaerobic respiration.
- Glucose concentration: low levels of glucose in the cell will decrease the rate of cell respiration.
- Type of cell: some types of cells require more energy than others. Those that require more energy have higher cell respiration rates.

Factors that affect cell respiration can be determined experimentally by calculating the rate of cell respiration using raw or secondary data. **Respirometers** are often used to calculate the rate of cell respiration. The following Worked example demonstrates how the rate of cell respiration can be calculated using a respirometer.

Worked example

Respirometers are devices used to measure an organism's rate of respiration by measuring the oxygen rate of exchange. They are sealed units in which any carbon dioxide produced is absorbed by an alkali such as soda lime or potassium hydroxide. Absorbing the carbon dioxide allows an accurate measurement of oxygen exchange. These devices may work at a cellular level or at a whole-organism level. Look at the graph and answer the questions. The *y*-axis of the graph represents the relative amount of oxygen used.

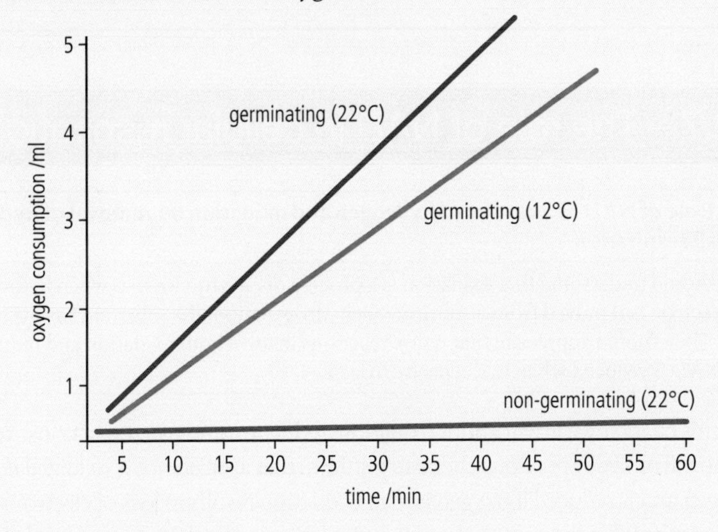

Oxygen consumption by germinating and non-germinating pea seeds at 12°C and 22°C.

1. In the germinating pea seeds, what type of respiration is occurring? What is the evidence for this answer?
2. Why is the oxygen consumption of non-germinating pea seeds very low?
3. Why would the germinating seeds show a greater oxygen consumption at 22°C than at 12°C?
4. Predict how the graph would look for non-germinating seeds at 12°C.

Solution

1. Aerobic. There is a significant amount of oxygen consumption occurring.
2. They are not carrying out respiration and have a low metabolic rate.
3. At 22°C the rate of respiration is faster than at 12°C. Therefore, there is a greater oxygen consumption at the higher temperature.
4. The line of the graph would be almost right on the non-germinating (22°C) line that exists now. A prediction that it would be just slightly lower is best.

Experiment to determine the effect of environmental factors on anaerobic respiration rates. Full details of how to carry out this experiment with a worksheet are available in the eBook.

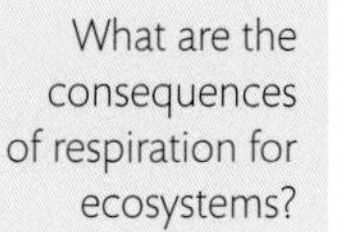

Nature of Science

The use of animals in such experiments has ethical implications. It is essential to refer to the IB animal experimental policy before carrying out any procedures on animals. Scientists have obligations to assess the risks associated with their work. Most aim to do no harm. Ethical and environmental consequences must be constantly considered when carrying out research.

What are the consequences of respiration for ecosystems?

HL

C1.2.7 – The role of NAD in cellular respiration

C1.2.7 – Role of NAD as a carrier of hydrogen and oxidation by removal of hydrogen during cell respiration

Students should understand that oxidation is a process of electron loss, so when hydrogen with an electron is removed from a substrate (dehydrogenation) the substrate has been oxidized. They should appreciate that redox reactions involve both oxidation and reduction, and that NAD is reduced when it accepts hydrogen.

Earlier in this chapter we learnt about oxidation, reduction and redox reactions. You should know from your previous chemistry studies that a substance is oxidized if it gains oxygen and is reduced if oxygen is lost. Oxidation results in loss of electrons, while reduction results from the gain of electrons. Oxidation and reduction reactions always occur together in what are called redox reactions. Both oxidation and reduction happen in respiration. **Nicotinamide adenine dinucleotide** (NAD) is a **coenzyme** utilized by the enzymes of cell respiration to carry out oxidation and reduction. NAD is also known as a hydrogen carrier. When hydrogen is added to NAD, the molecule is said to be reduced. When hydrogen is removed the molecule is said to be oxidized or **dehydrogenated**.

Summary of oxidation and reduction reactions

Oxidation	Reduction
Loss of electrons	Gain of electrons
Gain of oxygen	Loss of oxygen
Loss of hydrogen	Gain of hydrogen
Results in many C–O bonds	Results in many C–H bonds
Results in a compound with lower potential energy	Results in a compound with higher potential energy

A coenzyme is an organic, non-protein, molecule that helps an enzyme in its catalysis functions. The coenzyme is loosely bound to the enzyme. Coenzymes may be used repeatedly. Vitamins often function as coenzymes, or they provide the raw materials from which coenzymes are made.

A useful way to remember the general meaning of oxidation and reduction is to think of the words "oil rig".

- Oil = oxidation is loss (of electrons or hydrogen)
- Rig = reduction is gain (of electrons or hydrogen)

It is important to remember that hydrogen atoms consist of a proton and an electron. Therefore, when NAD receives hydrogen it is actually receiving one proton and one electron.

Earlier in this chapter, an equation representing the breakdown of glucose in cellular respiration was presented. The reactants of the equation were glucose and oxygen, while the products were carbon dioxide, water and energy being released. The energy released can be tracked through changes in the location of hydrogen atoms. As glucose ($C_6H_{12}O_6$) is catabolized/oxidized, it loses hydrogen atoms to become carbon dioxide (CO_2). As this happens, the reactant oxygen gains hydrogen atoms as it is reduced to water (H_2O). The movement of electrons in the conversion of glucose to carbon dioxide is an energy-releasing process. This energy, created by the movement of electrons, is carried by the coenzyme NAD in its reduced form (NADH). The reduced NAD produced may then carry the energy-rich electrons to further cell respiration processes to ultimately produce ATP, the usable energy form most preferred by the cell.

C1.2.8 – Glycolysis, ATP and NAD

C.1.2.8 – Conversion of glucose to pyruvate by stepwise reactions in glycolysis with a net yield of ATP and reduced NAD
Include phosphorylation, lysis, oxidation and ATP formation. Students are not required to know the names of the intermediates, but students should know that each step in the pathway is catalysed by a different enzyme.

Earlier, we discussed the overall process and products of glycolysis. We will now look in more detail at this important metabolic pathway. The word glycolysis means "sugar splitting" and this pathway is thought to be one of the first biochemical pathways to evolve. It uses no oxygen and occurs in the cytoplasm of the cell. No organelles are required. The **lysis** (splitting) of the sugar proceeds efficiently in both aerobic and anaerobic environments, and glycolysis occurs in both prokaryotic and eukaryotic cells. A hexose sugar, usually glucose, is split in the process. This splitting involves many steps and each one of those steps is controlled by a different enzyme. Three of these steps or stages are outlined below.

1. Two molecules of ATP are used to begin glycolysis. In the first reaction, the phosphates from the ATP molecules are added to glucose to form fructose-1,6-bisphosphate, a process called **phosphorylation**. This step is important because it creates a less stable molecule.

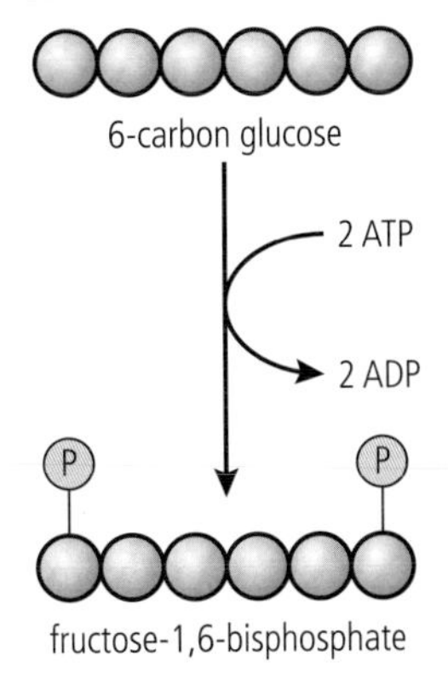

The first stage of glycolysis; the grey circles represent carbon atoms. The circles labelled P represent phosphate atoms.

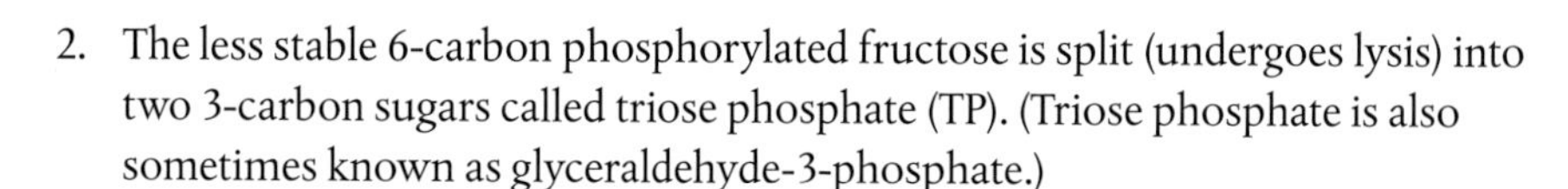

2. The less stable 6-carbon phosphorylated fructose is split (undergoes lysis) into two 3-carbon sugars called triose phosphate (TP). (Triose phosphate is also sometimes known as glyceraldehyde-3-phosphate.)

The second stage of glycolysis.

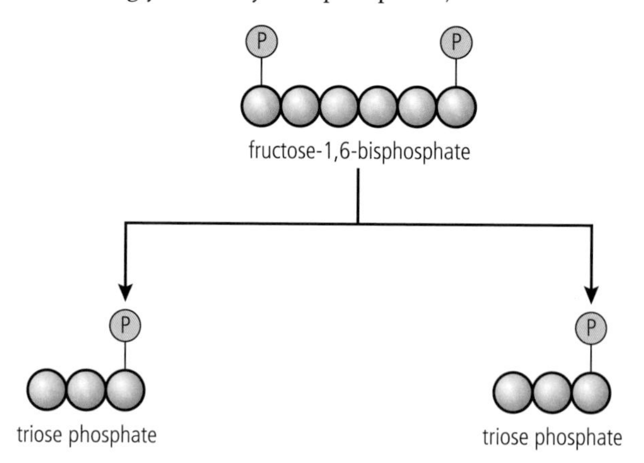

TOK

Many molecules have multiple names, such as triose phosphate also being known as glyceraldehyde-3-phosphate. Why is it essential for scientists to have a common language within the profession? Does scientific language have a primarily descriptive, explanatory or interpretative function?

3. Each TP molecule undergoes oxidation to form a reduced molecule of NAD (NADH). As reduced NAD is being formed, released energy is used to add an inorganic phosphate to the remaining 3-carbon compound. This results in a compound with two phosphate groups. Enzymes then remove the phosphate groups so that they can be added to ADP to produce ATP. The result is the formation of four molecules of ATP, two molecules of reduced NAD (NADH) and two molecules of pyruvate. Pyruvate is the ionized (electrically charged) form of pyruvic acid.

The third stage of glycolysis.

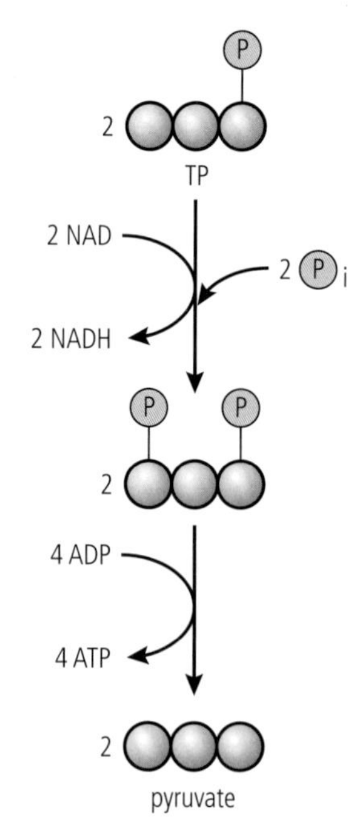

You are not required to know all the names of the intermediate molecules of the respiration process. However, you must understand the steps and the overall products.

To summarize, in the process of glycolysis we see phosphorylation, lysis, oxidation and reduction, and ATP formation. The final products are two pyruvate molecules, four ATP molecules, and two molecules of reduced NAD (NADH).

C1.2.9 – The fate of pyruvate

C1.2.9 – Conversion of pyruvate to lactate as a means of regenerating NAD in anaerobic respiration
Regeneration of NAD allows glycolysis to continue, with a net yield of two ATP molecules per molecule of glucose.

Once pyruvate is obtained, the next pathway is determined by the presence or absence of oxygen. If oxygen is present, pyruvate enters the mitochondria and aerobic respiration occurs. If oxygen is not present, anaerobic respiration occurs in the cytoplasm. In the latter case, pyruvate is converted to lactate in animals, and ethanol and carbon dioxide in plants.

When oxygen is not present, anaerobic cell respiration in humans allows muscles to work vigorously for a short period of time. However, as mentioned earlier in this chapter, the lactic acid that builds up in the muscle during this fermentation process creates a sensation of burning if intense exercise continues, and can be toxic in higher concentrations. Anaerobic respiration provides energy when it is needed but is hard to sustain for long periods of time. Anaerobic cell respiration also allows the regeneration of NAD. This is important because without NAD being available to accept hydrogen, glycolysis would have to stop. The regeneration of NAD occurs when reduced NAD (NADH) donates its hydrogen and electrons to the pyruvate molecules being formed. In humans, this reduction of pyruvate results in the formation of lactate. In anaerobic respiration the only ATP being produced is that formed by glycolysis. The net yield of ATP from glycolysis is two molecules of ATP for every glucose molecule. Glycolysis must therefore continue if life functions are to continue.

C1.2.10 – Anaerobic cell respiration in yeast

C1.2.10 – Anaerobic cell respiration in yeast and its use in brewing and baking
Students should understand that the pathways of anaerobic respiration are the same in humans and yeasts apart from the regeneration of NAD using pyruvate and therefore the final products.

Yeast is a common, single-cell fungus that uses alcoholic fermentation for ATP generation when oxygen is not present. You will recall that all organisms use glycolysis to begin cell respiration. Yeast cells take in glucose from their environment and generate a net gain of two ATPs through glycolysis. The organic products of glycolysis are always two pyruvate molecules. Yeast then converts both 3-carbon pyruvate molecules to molecules of 2-carbon ethanol. The "lost" carbon atom is given off in a carbon dioxide molecule. Both the ethanol and carbon dioxide that are produced are waste products and are released into the environment. Bakers' yeast is added to bread products because the generation of carbon dioxide helps the dough to rise. Yeast is also commonly used in the production of alcoholic drinks.

All alcohol that is sold to be drunk is ethanol. Beer, wine and spirits contain different proportions of ethanol, plus other ingredients for flavouring.

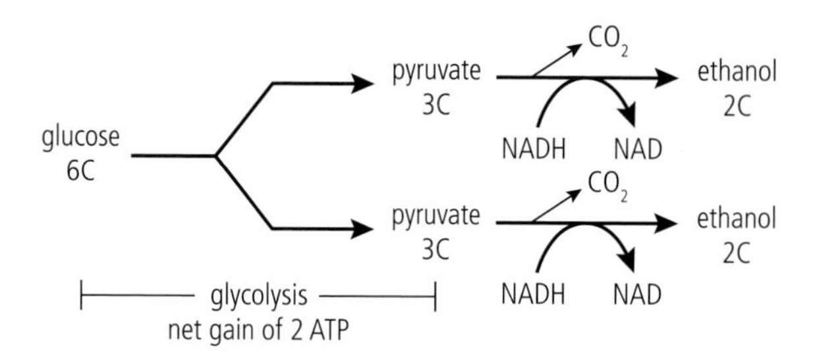

C1.2 Figure 4 A simplified version of the events of alcoholic fermentation.

Figure 4 shows that ethanol and carbon dioxide are produced from ethanol fermentation. NAD is also regenerated in this process, when the reduced NAD (NADH), from glycolysis donates its electrons and hydrogens to two acetaldehyde molecules to form the two molecules of ethanol.

Bioethanol is ethanol produced by living organisms as a renewable energy resource. Corn is most often used for this production process, which occurs in large structures called fermenters. These fermenters provide an anaerobic environment and the enzymes and environmental conditions needed for maximum ethanol production. This is one approach being applied around the world to reduce fossil fuel usage in an attempt to reduce harmful emissions into our atmosphere.

C1.2.11 – The link reaction

C1.2.11 – Oxidation and decarboxylation of pyruvate as a link reaction in aerobic cell respiration
Students should understand that lipids and carbohydrates are metabolized to form acetyl groups (2C), which are transferred by coenzyme A to the Krebs cycle.

Once glycolysis has occurred, and if there is oxygen present, pyruvate enters the matrix of the mitochondria via active transport. Inside the matrix, pyruvate is decarboxylated to form the 2-carbon acetyl group. The removed carbon is released as carbon dioxide, a waste gas. The acetyl group is then oxidized and reduced NAD (NADH) is formed. Finally, the acetyl group combines with **coenzyme A** (CoA) to form **acetyl-CoA**. Acetyl-CoA can then enter the **Krebs cycle**. This **link reaction** is controlled by a system of enzymes.

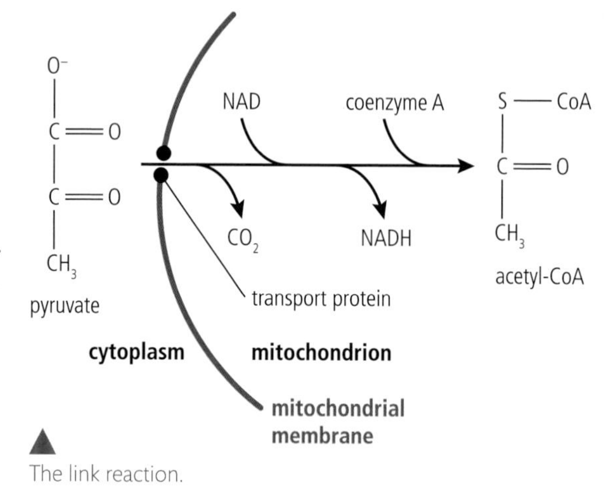

The link reaction.

So far we have focused on glucose as the respiratory substrate. However, acetyl groups can be produced from most carbohydrates and fats. Once acetyl groups are formed, they can be transferred by CoA into the Krebs cycle.

C1.2.12 – Oxidation and decarboxylation in the Krebs cycle

C1.2.12 – Oxidation and decarboxylation of acetyl groups in the Krebs cycle with a yield of ATP and reduced NAD

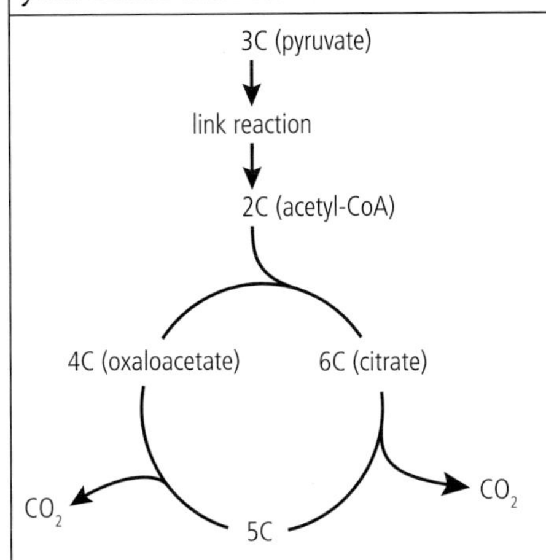

Students are required to name only the intermediates citrate (6C) and oxaloacetate (4C). Students should appreciate that citrate is produced by transfer of an acetyl group to oxaloacetate and that oxaloacetate is regenerated by the reactions of the Krebs cycle, including four oxidations and two decarboxylations. They should also appreciate that the oxidations are dehydrogenation reactions.

The Krebs cycle occurs in the matrix of mitochondria. You do not need to remember the names of all the compounds formed in the cycle. However, you should remember the roles of **citrate** and **oxaloacetate**.

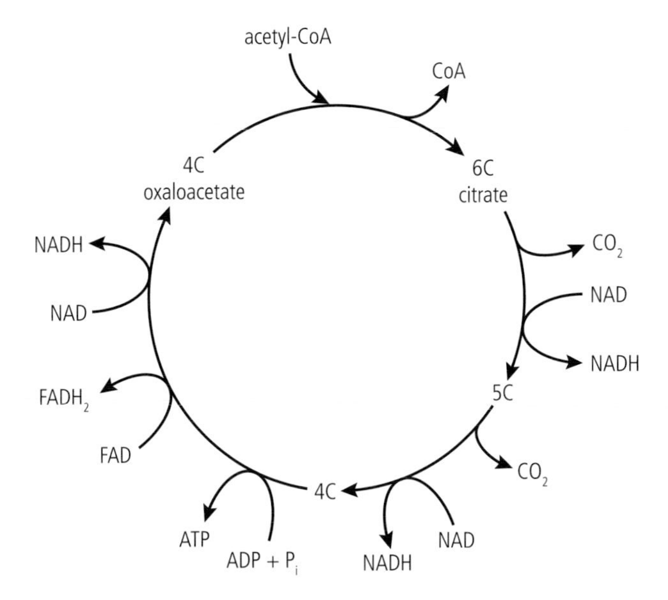

The Krebs cycle. Note the many oxidation reactions and the decarboxylations.

The Krebs cycle includes two **decarboxylation** reactions and four points at which the carbon compound is oxidized for each acetyl group brought into the cycle by CoA. Notice that citrate is a 6-carbon compound produced by the combination of the acetyl group from the link reaction with the 4-carbon oxaloacetate from the cycle itself. Also, notice that there is an additional coenzyme involved in the Krebs cycle. The additional coenzyme is flavin adenine dinucleotide (FAD), and it acts as a hydrogen/electron acceptor just as NAD does.

Two molecules of carbon dioxide are released for each glucose molecule during the link reaction. Four carbon dioxides are released during the Krebs cycle. This accounts for all six of the carbon atoms that were present in the initial glucose molecule. Glucose is completely catabolized, and its original energy is now carried by reduced molecules of NAD or FAD, or is in ATP.

What are the consequences of respiration for ecosystems?

It is important to remember that the Krebs cycle will run twice for each glucose molecule entering aerobic cell respiration. This is because a glucose molecule forms two pyruvate molecules. Each pyruvate produces one acetyl-CoA that enters the cycle. Here is an overview of the products produced from the breakdown of one glucose molecule in the Krebs cycle:

- two molecules of ATP
- six molecules of reduced NAD
- two molecules of reduced FAD
- four molecules of carbon dioxide.

Challenge yourself

1. Examine the figure below.

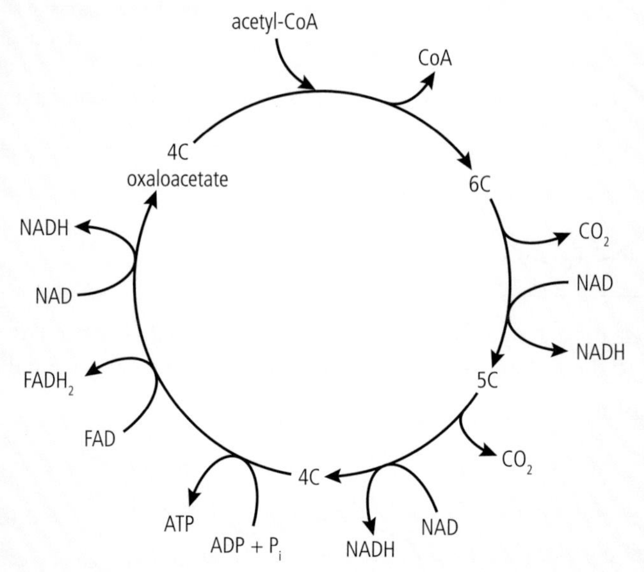

(a) Redraw the figure and place an arrow and the letter D at the two locations where decarboxylation occurs. On the same figure place an arrow and the letter R at the five locations where reduction of a cyclic intermediated compound occurs.

(b) How did you determine the two locations where decarboxylation occurred?

(c) How did you determine the five locations where reduction occurred?

C1.2.13 – Reduced NAD

C1.2.13 – Transfer of energy by reduced NAD to the electron transport chain in the mitochondrion
Energy is transferred when a pair of electrons is passed to the first carrier in the chain, converting reduced NAD back to NAD. Students should understand that reduced NAD comes from glycolysis, the link reaction and the Krebs cycle.

The reduced NAD (NADH) coenzymes from glycolysis, the link reaction and the Krebs cycle each carry a pair of hydrogen atoms. As mentioned earlier in this chapter, a hydrogen atom is made of a proton and an electron. The hydrogen atoms are split in the matrix of the mitochondria. The electrons have a high energy level and pass into the electron transport chain. The electron transport chain occurs on the inner mitochondrial membrane and on the membranes of the cristae. Embedded in these membranes are electron carrier molecules that are easily reduced and oxidized. The reduced NAD is converted back into NAD, which can then be used again in earlier parts of the process.

C1.2.14 and C1.2.15 – The electron transport chain and chemiosmosis

C1.2.14 – Generation of a proton gradient by flow of electrons along the electron transport chain
Students are not required to know the names of protein complexes.
C1.2.15 – Chemiosmosis and the synthesis of ATP in the mitochondrion
Students should understand how ATP synthase couples release of energy from the proton gradient with phosphorylation of ADP.

Reduced NAD molecules transfer high-energy electrons to the first carrier in the **electron transport chain**. These electrons are then passed from one carrier to another, with small amounts of energy released at each exchange. Study Figure 5.

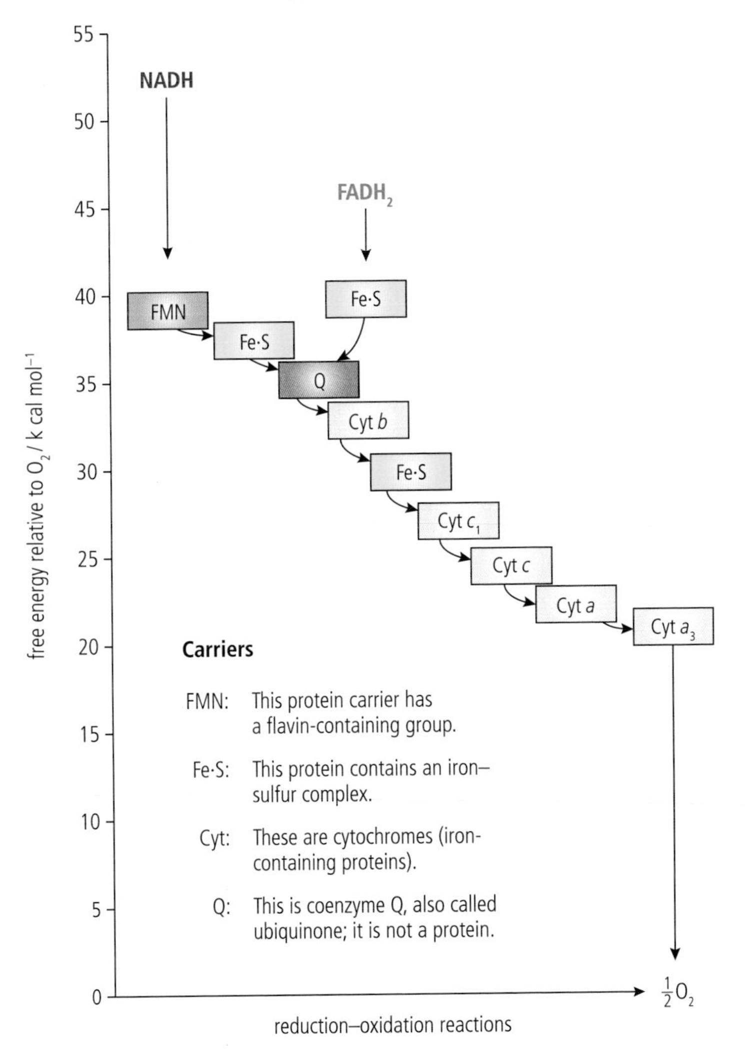

C1.2 Figure 5 The oxidation–reduction reactions of the electron transport chain. It is not necessary for you to remember all the names of the carriers.

The *y*-axis in Figure 5 represents the amount of energy available, while the *x*-axis represents the occurrence of redox reactions. Notice that as the high-energy electrons are passed from carrier to carrier there is a decrease in available energy. Also note that $FADH_2$ brings its electrons into the chain at a slightly lower energy level than NADH. As the electrons move down the chain, the energy released in small increments is used to transfer protons across the inner membranes from the matrix into the space between the membranes.

The electron transport chain does not directly produce any molecules of ATP. However, the electron transport chain does produce a **proton gradient**, which provides the energy for the formation of ATP from ADP and inorganic phosphate (P_i). The formation of ATP driven by the proton gradient is called **chemiosmosis.** Examine Figure 6 of the electron transport chain and chemiosmosis working together to produce ATP.

C1.2 Figure 6 Oxidative phosphorylation occurs at the inner membranes of the mitochondria of a cell. The pumping actions of the carriers result in a high concentration of hydrogen ions in the intermembrane space. This accumulation allows movement of the hydrogen ions through the enzyme ATP synthase. The enzyme uses the energy from the hydrogen flow to couple phosphate with ADP to produce ATP.

intermembrane space
inner mitochondrial membrane
mitochondrial matrix
protein complex of electron carriers
electron carrier
Q
Cyt *c*
H^+
NADH + H^+
(carrying electrons from food)
NAD
$FADH_2$
FAD
2 H^+ + $\frac{1}{2}O_2$
H_2O
ADP + P_i
ATP
ATP synthase
electron transport chain
chemiosmosis

The electron transport chain and chemiosmosis work together to result in ATP production. Refer to Figure 6 as you read through the different steps involved.

- While electrons move down the electron transport chain, energy is released in small amounts and used to pump protons (hydrogen ions) out of the matrix and into the intermembrane space.
- This creates a proton gradient, with a high concentration of protons in the intermembrane space compared to the mitochondrial matrix.
- Protons move down the concentration gradient by diffusion through channels in an enzyme called ATP synthase.
- As the protons move passively from the intermembrane space to the matrix, ATP synthase harnesses this energy, allowing the phosphorylation of ADP to form ATP.

Because of the hydrophobic region of the membrane, the protons (hydrogen ions) can only pass through the ATP synthase channels. Some poisons that affect metabolism act by establishing alternative pathways through the membrane, thus preventing ATP production.

Remember that there are protons in the matrix as a result of NAD transporting hydrogen atoms to the electron transport chain. These hydrogen atoms are split, and the electrons enter the electron transport chain. The protons (hydrogen ions) are free in the matrix.

$$H \rightarrow H^+ + 1e$$

C1.2.16 – The role of oxygen

C1.2.16 – Role of oxygen as terminal electron acceptor in aerobic cell respiration

Oxygen accepts electrons from the electron transport chain and protons from the matrix of the mitochondrion, producing metabolic water and allowing continued flow of electrons along the chain.

Oxygen is the final electron acceptor. When the electrons combine with the oxygen, so do two protons (hydrogen ions) from the matrix. The result is water. Because of the way it is formed, this water is known as **water of metabolism**. This acceptance of electrons by oxygen is what allows the continued flow of electrons along the electron transport chain.

The kangaroo rat from a desert region of the USA gets 90% of its daily water intake from water of metabolism. In contrast, a typical human only gets 12% of their daily water intake from metabolism.

C1.2.17 – Respiratory substrates

C1.2.17 – Differences between lipids and carbohydrates as respiratory substrates

Include the higher yield of energy per gram of lipids, due to less oxygen and more oxidizable hydrogen and carbon. Also include glycolysis and anaerobic respiration occurring only if carbohydrate is the substrate, with 2C acetyl groups from the breakdown of fatty acids entering the pathway via acetyl-CoA (acetyl coenzyme A).

The focus up until now has been on glucose, a carbohydrate, in aerobic respiration. However, other organic molecules, particularly proteins and lipids, are also important sources of energy. We will focus on lipids in this section. Lipids are carbon compounds that are insoluble in water. They include fats, waxes, oils, hormones and certain components of membranes. An important function of lipids is energy storage. Lipids have less oxygen and more oxidizable C–H bonds in their structure than carbohydrates.

In what forms is energy stored in living organisms?

Triglycerides are simple lipids composed of glycerol and three molecules of fatty acids. Triglycerides are the storage forms of lipid and can be used for energy production. Long-chain fatty acids usually have an even number of carbon atoms with many C–H bonds.

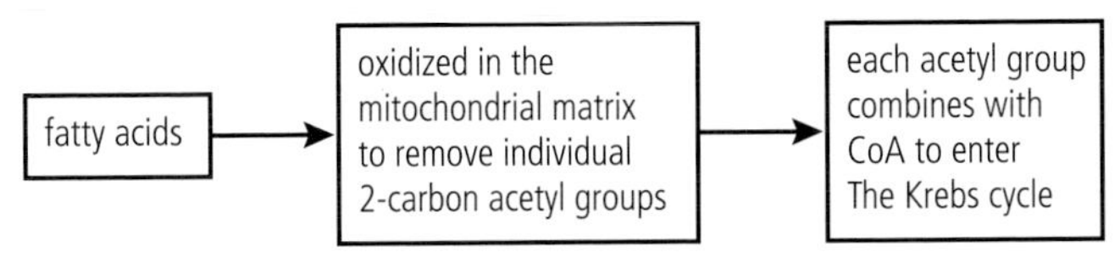

An example of lipids as a respiratory substrate.

When measuring energy output from a 6-carbon fatty acid and a 6-carbon carbohydrate such as glucose, we find that the fatty acid yields 20% more ATP than the carbohydrate.

It is important to note that the full pathway of glycolysis and anaerobic respiration can only occur if a carbohydrate is the substrate. When acetyl-CoA enters the pathway, aerobic cell respiration is occurring and the Krebs cycle is activated.

Other organic molecules may enter the cell respiration pathway as various intermediates. Amino acids can enter the pathway either as pyruvate or as an acetyl group.

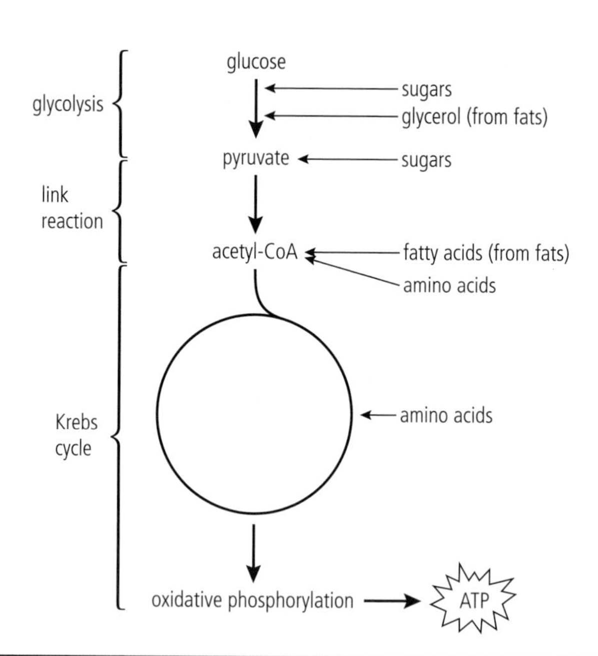

This diagram shows the locations where different carbon compounds can enter the cell respiration pathway.

HL end

Guiding Question revisited

What are the roles of hydrogen and oxygen in the release of energy in cells?

In this chapter we have described how:

- oxidation of carbon compounds occurs when hydrogen is removed from a molecule, each hydrogen atom is made of one electron and one proton
- oxygen is required in aerobic cell respiration, but not for anaerobic cell respiration
- glycolysis does not require oxygen to proceed
- the energy to produce ATP from ADP comes from the energy present in the electrons of hydrogen removed during oxidation of carbon compounds

HL

- hydrogen ions (protons) are pumped across the inner membrane of mitochondria to produce a proton gradient, this proton gradient is then used to provide the energy for chemiosmosis (the formation of ATP)
- aerobic respiration uses oxygen as the final electron and hydrogen ion acceptor
- electron carriers such as NAD can be reversibly oxidized and reduced, allowing a continuous flow of high-energy electrons to the electron transport chain.

HL end

Guiding Question revisited

How is energy distributed and used inside cells?

In this chapter we have discussed how:

- ATP is known as the energy currency of the cell and is used to distribute energy within cells
- ATP is a nucleotide with a high-energy phosphate bond

- ATP is essential for active transport across cell membranes, the synthesis of macromolecules needed by the cell, and all types of movement involving the cell
- cell respiration is the metabolic pathway that produces ATP in the cell

HL

- lactic acid fermentation is a form of anaerobic cell respiration, it results in fewer ATP molecules than aerobic respiration but the cell is producing some ATP and can therefore sustain life processes for a period of time
- most ATP is produced in the final stage of aerobic cell respiration called chemiosmosis
- a hydrogen ion (proton) gradient and the enzyme ATP synthase are essential for the production of ATP in aerobic cell respiration.

HL end

Exercises

Q1. State the final products of anaerobic respiration.

Q2. State the final products of aerobic respiration in humans.

Q3. Explain reasons for referring to ATP as the energy currency of the cell.

Q4. Explain the products and major locations within the cell where the anaerobic phases of cellular respiration occur.

HL

Q5. Describe the location of the electron transport chain.

Q6. Explain why NAD is needed during aerobic cell respiration.

Q7. State where each of the following stages of respiration takes place.

(a) Glycolysis. **(b)** Link reaction.

(c) Krebs cycle. **(d)** Oxidative phosphorylation.

Q8. Which of the following stages of respiration produce the greatest number of ATP molecules?

A Glycolysis. **B** Link reaction.

C Krebs cycle. **D** Oxidative phosphorylation.

Q9. Complete the following sentence using the correct option below.

In glycolysis, ________ is oxidized and ________ is reduced.

A NAD ____ glucose. **B** ATP _____ ADP.

C Glucose ______ NAD. **D** Glucose ______ oxygen.

Q10. Complete the following sentence using one of the four options given below.

The link between the electron transport chain and ATP synthesis ________

A depends on the absence of oxygen.

B is the transfer of electrons to ATP synthase.

C is a proton gradient.

D is a high-energy intermediate like pyruvate.

HL end

C1.3 Photosynthesis

Guiding Questions

How is energy from sunlight absorbed and used in photosynthesis?

How do abiotic factors interact with photosynthesis?

Plants are an amazing group of organisms. Their ability to absorb light energy and convert it into chemically stored energy opened the world to the extraordinary development of all other organisms on our planet. Photosynthesis is the cornerstone of almost all life forms that exist today. We are continuously studying plants to better understand their ability to absorb and use sunlight. Artificial leaves and solar panels are just a couple of the spinoffs from this research.

Of course, plants need more than just sunlight to carry out photosynthesis. They need water, the correct temperature, carbon dioxide and a host of chemicals to convert sunlight into chemical energy efficiently. Many of the factors required are abiotic or non-living factors. Today we are able to manipulate some of these abiotic factors in greenhouses and other agricultural settings, so that crop yields are higher than ever before.

C1.3.1 – Light energy and life processes

C1.3.1 – Transformation of light energy to chemical energy when carbon compounds are produced in photosynthesis
This energy transformation supplies most of the chemical energy needed for life processes in ecosystems.

Life on the eastern Colorado plains (USA) and everywhere else in the world ultimately depends on the process of photosynthesis. Nearly every oxygen molecule we breathe in was once part of a water molecule that was turned into oxygen through photosynthesis. All the energy present on Earth has a connection to this process. An understanding of photosynthesis is essential to appreciate the intertwined ecosystems of our planet.

Plants and other photosynthetic organisms produce foods that start food chains. Most living things rely on the Sun to provide the energy needed for both warmth and food production. However, the sunlight that strikes Earth must be converted into a form of chemical energy if it is to be useful. It is this transformation of light energy into chemical energy that fuels the life processes in ecosystems. The chemical energy produced by photosynthesis can be in the form of glucose. If you recall, glucose is also the molecule that organisms use for the process of cell respiration.

Chapter C1.2 discusses respiration and considers how the cell breaks down chemical bonds in glucose to produce adenosine triphosphate (ATP). In this chapter we are going to find out how chemical bonds are made to produce carbon compounds. The raw materials of photosynthesis are carbon dioxide and water but the process also requires light. The overall word equation is:

$$\text{carbon dioxide + water} \xrightarrow{\text{light}} \text{glucose + oxygen}$$

HL Chapter A2.1, discusses how early Earth had an atmosphere of carbon dioxide, methane and ammonia, amongst many other gases. There was also plenty of water vapour in the early atmosphere. About 2.7 billion years ago, oxygen began to appear in our atmosphere as a result of the process of photosynthesis.

Organisms that use light to produce their own food are called **photoautotrophs**. The United States Department of Agriculture estimates that up to 200 billion metric tons of sugar are produced each year by these organisms. **Heterotrophs** are organisms that cannot produce their own food. **Autotrophs** are organisms that can produce their own food. Not all autotrophs are photoautotrophs, some are **chemoautotrophs** (chemotrophs), which use chemical reactions to obtain their energy.

C1.3.2 and C1.3.3 – The equation for photosynthesis

C1.3.2 – Conversion of carbon dioxide to glucose in photosynthesis using hydrogen obtained by splitting water
Students should be able to write a simple word equation for photosynthesis, with glucose as the product. $\text{carbon dioxide + water} \xrightarrow{\text{light}} \text{glucose + oxygen}$

C1.3.3 – Oxygen as a by-product of photosynthesis in plants, algae and cyanobacteria
Students should know the simple word equation for photosynthesis. Students should know that the oxygen produced by photosynthesis comes from the splitting of water.

What gas is produced during photosynthesis? Full details of how to carry out this experiment with a worksheet are available in the eBook.

During photosynthesis water is split to give hydrogen and oxygen. Oxygen is a by-product of photosynthesis: it is not needed by the plant and is released into the atmosphere. The hydrogens released during the splitting of water reduce carbon dioxide to form glucose. The chemical equation for photosynthesis showing the accompanying redox reactions is shown in Figure 1.

$$6\,CO_2 + 6\,H_2O \xrightarrow{\text{energy}} C_6H_{12}O_6 + 6\,O_2$$

(becomes reduced: CO_2 → $C_6H_{12}O_6$; becomes oxidized: H_2O → O_2)

C1.3 Figure 1 The redox reactions of photosynthesis.

As we proceed in our discussion of photosynthesis, we will refer to this equation often. Plants, algae and cyanobacteria all carry out photosynthesis, and as a result all release oxygen into the atmosphere as a by-product.

What are the consequences of photosynthesis for ecosystems?

C1.3.4 – Photosynthetic pigments and light absorption

C1.3.4 – Separation and identification of photosynthetic pigments by chromatography
Application of skills: Students should be able to calculate R_f values from the results of chromatographic separation of photosynthetic pigments and identify them by colour and by values. Thin-layer chromatography or paper chromatography can be used.

Most plant leaves appear green to our eyes. If you were able to zoom into leaf cells and look around, you would see that the only structures in a leaf that are actually green are the **chloroplasts.** Plants contain a variety of pigments in the chloroplasts. The photosynthetic pigment that dominates in most plant species is the molecule **chlorophyll**. There are actually several different types of chlorophyll, and these produce the characteristic green colour of most plants. Of the different types of chlorophyll, chlorophyll *a* and chlorophyll *b* are the most common. We can also find a group of pigments called carotenoids. The carotenoids usually include the specific pigments known as carotene and xanthophyll. Each of the pigments and their concentrations are unique to a plant species. The pigments that a particular plant contains can be separated by a process known as chromatography.

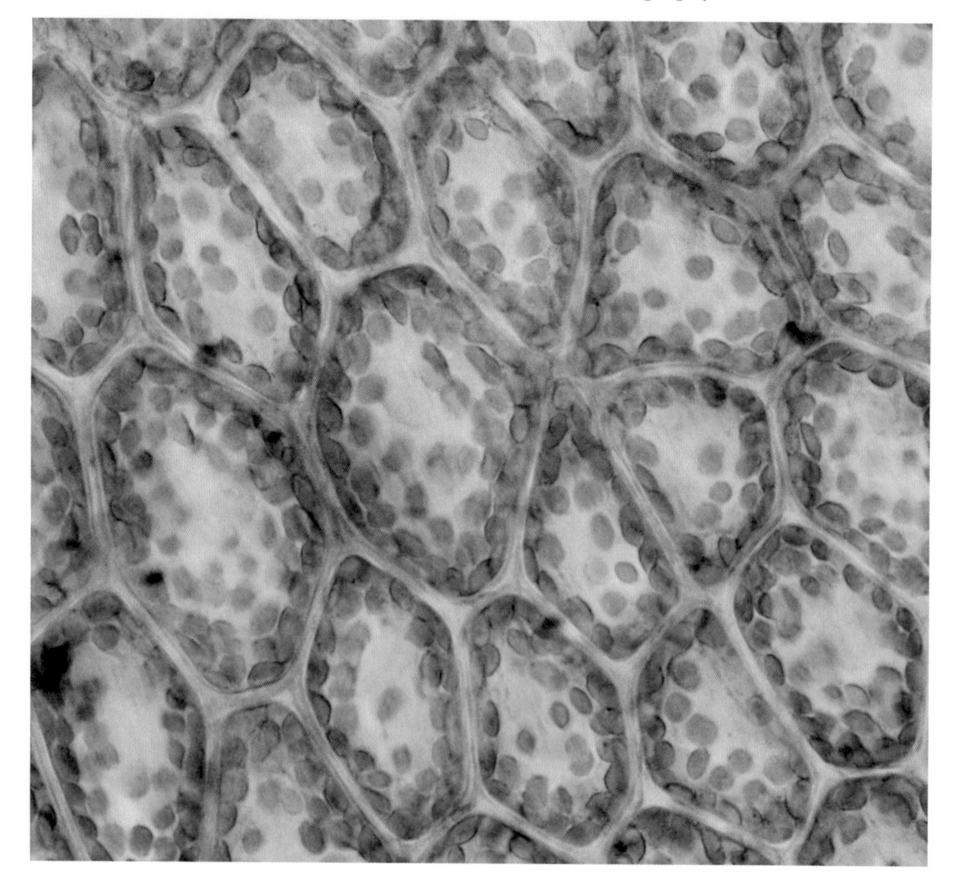

Inside each of these plant leaf cells are many green chloroplasts. Each chloroplast is loaded with light-absorbing pigments.

In Chapter B2.2, **chromatography** is presented as one of the techniques that allows the isolation or separation of pure substances from a complex material. Gel and ion-exchange chromatography are two types that are often used. Another very common type of chromatography is paper chromatography. This technique uses a paper sheet or strip through which a solution can pass to separate pure substances. The paper is referred to as the stationary phase because it does not move. The mobile phase is a solvent that moves up the stationary phase (paper) by capillary action. As the mobile phase moves through the stationary phase, it carries the components of the mixture being analysed with it. During this movement, pure substances separate from each other because they have different migration rates across the stationary phase. It is an inexpensive yet very powerful means of analysing the components of a solution. This technique is used often to separate the different pigments found in plants. By using this technique, a variety of pigments have been found in different plants. Examine the chromatogram on the right.

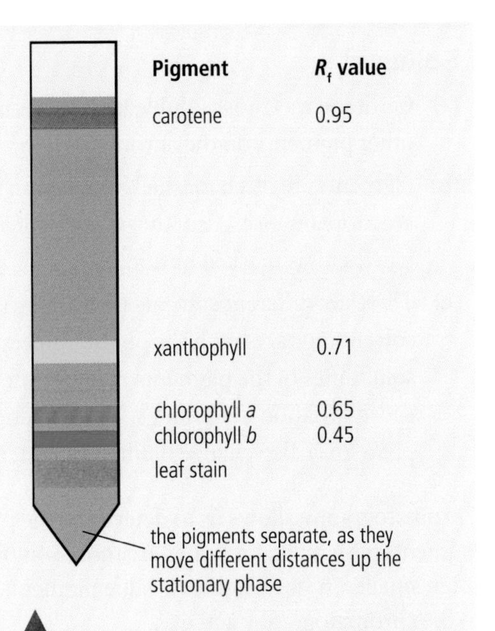

▲ A paper chromatograph obtained from the leaf of a plant showing the various pigments that are necessary for photosynthesis. Each of the pigments migrated at a different rate to produce the separate bands, which are labelled with the name of the pigment. The R_f value is the ratio of the distance travelled by the pigment to the distance travelled by the solvent (mobile phase).

Worked example

The chromatogram on the right was obtained in the laboratory. Calculate the R_f value for each of the pigments by using the following formula:

$$R_f = \frac{\text{distance moved by substance}}{\text{distance moved by solvent}}$$

Once you have calculated the R_f value for each pigment, answer the questions below.

(a) Explain which of the pigments was most soluble in the solvent.

(b) These colours were not visible in the leaf of the plant. Explain why they are visible on the chromatograph.

(c) Suggest why the calculated R_f values are different to those in the chromatogram above.

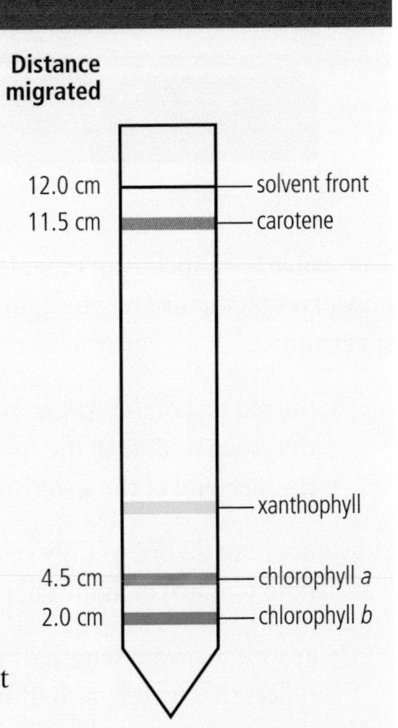

Solution

(a) Carotene was most soluble in the solvent because it moved farther than the other pigments on the chromatogram.

(b) Pigments are often masked by other pigments in a plant because some colours are more intense. Also, the quantity of a pigment present in a plant may affect whether it is masked or not.

(c) There are different solvents that can be used to separate plant pigments. Each solvent interacts with the pigments present in different ways. For example, the solubilities of the pigments may be different in alternative solvents. This would cause variation in R_f values. The attraction of the pigment to the stationary phase may also change if different materials are used.

Separation of photosynthetic pigments by chromatography. Full details of how to carry out this experiment with a worksheet are available in the eBook.

Chromatography allows us to determine the pigments present in chloroplasts. Pigments with higher R_f values are more soluble in the solvent being used, and they are often smaller in size. Higher R_f value pigments have a lower affinity for the paper used in the chromatography activity.

Plants make use of the same part of the electromagnetic spectrum that our eyes can see. We call this the visible portion of the spectrum. Sunlight is a mixture of different colours of light. You can see these colours when you let sunlight pass through a prism.

increasing energy
increasing wavelength
0.001 nm 1 nm 10 nm 1000 nm 0.01 cm 1 cm 1 m 100 m
gamma rays X-rays UV light infrared radio waves
visible light
400 nm 430 nm 500 nm 560 nm 600 nm 650 nm 740 nm

The electromagnetic spectrum. Notice that the visible light portion of this spectrum has colours with wavelengths between 400 nm and 740 nm.

The visible light spectrum includes many colours but, for the purpose of considering how plant pigments absorbs light energy, we are going to consider three regions of the spectrum:

- the red end of the spectrum
- the green middle of the spectrum
- the blue end of the spectrum.

Substances can do one of only two things when they are struck by a particular wavelength (colour) of light. They can:

- absorb that wavelength (if so, energy is being absorbed and may be used)
- reflect that wavelength (if so, the energy is not being absorbed and you will see that colour).

Worked example

You are walking outside with a friend who is wearing a red and white shirt. Explain why the shirt appears to be red and white.

Solution

Sunlight is a mixture of all the wavelengths (colours) of visible light. When sunlight strikes the red pigments in the shirt, the blue and the green wavelengths of light are absorbed, but the red wavelengths are reflected. Thus, our eyes see the colour as red. When sunlight strikes the white areas of the shirt, all the wavelengths of light are reflected, and our brain interprets the mixture as white.

C1.3.5 and C1.3.6 – Absorption and action spectra

C1.3.5 – Absorption of specific wavelengths of light by photosynthetic pigments
Include excitation of electrons within a pigment molecule, transformation of light energy to chemical energy and the reason that only some wavelengths are absorbed. Students should be familiar with absorption spectra. Include both wavelengths and colours of light in the horizontal axis of absorption spectra.

C1.3.6 – Similarities and differences of absorption and action spectra
Application of skills: Students should be able to determine rates of photosynthesis from data for oxygen production and carbon dioxide consumption for varying wavelengths. They should also be able to plot this data to make an action spectrum.

The ability of photoautotrophs to absorb light energy is determined by the pigments present on the membranes of the chloroplasts. It is the pigments that absorb the light. The amount of light absorbed plotted against the wavelength of light produces the **absorption spectrum** for that pigment. The **action spectrum** indicates the rate of photosynthesis at different wavelengths of light. The rate of photosynthesis can be calculated from either the rate of oxygen produced, or the rate of carbon dioxide used up (see Section C1.3.7). Study the figure on the next page showing the absorption and action spectra of some common photosynthetic pigments.

When reading any graph it is essential you note what each axis represents. By looking at the two *y*-axes in the figure, it is clear which one represents the absorption spectrum and which represents the action spectrum. In the case of the *x*-axes, note that the blue light has a shorter wavelength than red light.

Determining the photosynthetic rate and production of an action potential

The rate of photosynthesis can be determined in a number of ways. Two of the most commonly used methods involve measuring oxygen production over time and carbon dioxide consumption over time. Once the photosynthesis rate has been calculated using either of these two methods, an action spectrum can be constructed, such as shown in the second graph in the figure below. The first graph is an example of an absorption spectrum, showing the absorption of light by various photosynthetic pigments of a plant.

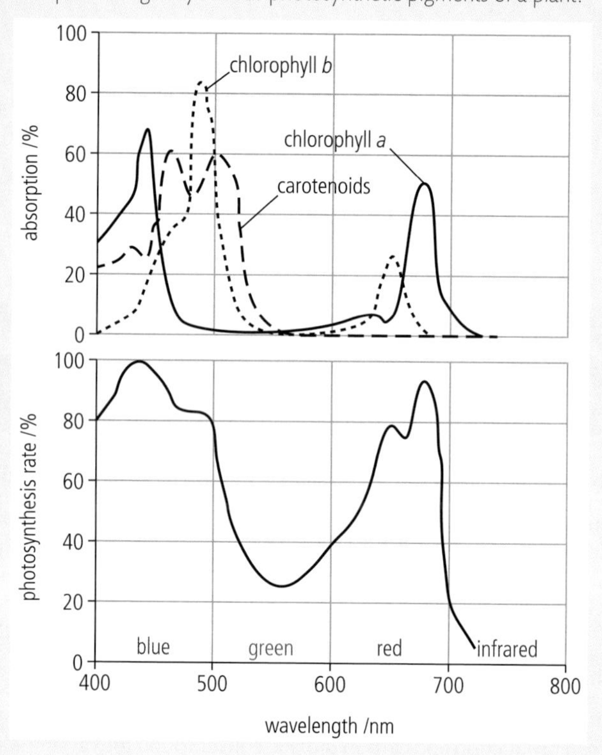

The top graph represents the absorption spectrum of common photosynthetic pigments. Carotenoid pigments are yellow, orange and red pigments, and the absorption spectrum for them as a group is presented here. The bottom graph represents the action spectrum of photosynthesis, plotting the overall photosynthetic rate at different wavelengths (colours).

1. At wavelengths where there is a higher percentage of light absorption, the rate of photosynthesis is also high. Absorption is greatest for the blue (400–500 nm) and red (600–700 nm) wavelengths.
2. The least light absorption occurs at with the green (500–600 nm) wavelength. Plants reflect green light rather than absorb it: this is why they look green to us.
3. When absorption of light by a pigment is low, the contribution of that pigment to the rate of photosynthesis will be low. Green light is therefore the least effective type of light for photosynthesis.

A summary of the differences and similarities between absorption and action spectra is presented in Table 1.

Absorption spectrum	Action spectrum
Varies depending on type of photosynthetic pigment present	Varies depending on type of photosynthetic pigment present
Represents the amount of light energy being absorbed by the photosynthetic pigment	Represents the rate of the photosynthetic process being carried out by the pigment
For the plant, this spectrum represents the light absorbed by all the pigments present	For the plant, this spectrum represents the rate of photosynthesis as a result of all the pigments present
Chlorophylls *a* and *b* have a high absorption of light energy in the violet-blue and red light wavelengths	Chlorophylls *a* and *b* create a relatively high efficiency rate of photosynthesis
Pigments like carotenoids absorb light energy at different wavelengths compared to chlorophyll *a* and *b*	Pigments like carotenoids allow photosynthesis at different wavelengths
Other pigments are not as efficient at absorbing light energy as chlorophylls *a* and *b*	Other pigments are not as effective at achieving high rates of photosynthesis as chlorophylls *a* and *b*

C1.3 Table 1 A comparison of absorption and action spectra

Light energy behaves as if it exists in discrete packets called photons. Shorter wavelengths of light have greater energy within their photons than longer wavelengths. Photons can transfer their energy upon interaction with other particles. This transfer of energy occurs many times during photosynthesis.

When a pigment absorbs light, the energy is used to raise an electron in the pigment to a higher energy level. This is known as **excitation of electrons**. Raising an electron to a higher energy level requires a specific amount of energy (or specific photons of light). This explains why different pigments absorb different wavelengths of light: they each need a different wavelength to excite electrons. Once the electrons are excited to a higher energy level, this energy can then be used to make chemical bonds. The net result is that light energy is transformed into chemical energy.

What are the functions of pigments in living organisms?

C1.3.7 – Measuring the rate of photosynthesis

C1.3.7 – Techniques for varying concentrations of carbon dioxide, light intensity or temperature experimentally to investigate the effects of limiting factors on the rate of photosynthesis
Application of skills: Students should be able to suggest hypotheses for the effects of these limiting factors and to test these by experimentation.
NOS: Hypotheses are provisional explanations that require repeated testing. During scientific research, hypotheses can either be based on theories and then tested in an experiment or be based on evidence from an experiment already carried out. Students can decide in this case whether to suggest hypotheses for the effects of limiting factors on photosynthesis before or after performing their experiments. Students should be able to identify the dependent and independent variable in an experiment.

Look again at the summary reaction for photosynthesis:

$$6CO_2 + 6H_2O \rightarrow C_6H_{12}O_6 + 6O_2$$

This balanced equation shows us that carbon dioxide molecules are reactants and oxygen molecules are products of photosynthesis. If you recall some of the information you have learned about cell respiration, you will see that the reverse is true for that process. In other words, in cell respiration oxygen is a reactant and carbon dioxide is a product.

At any given time of year, any one plant has a fairly consistent rate of cell respiration. Not only is this rate consistent throughout the day and night, but it is also at a relatively low level. Plants need ATP for various biochemical processes, but the levels they need are typically far lower than animals need.

The same consistency is not true regarding the rate of photosynthesis. The photosynthetic rate is highly dependent on many environmental factors, including the intensity of light and air temperature. During the daytime, especially on a warm, sunny day, the rate of photosynthesis may be very high for a particular plant. If so, the rate of carbon dioxide taken in by the plant and the rate of oxygen released will also be very high. Because the plant is also carrying out cell respiration, a correction needs to be made to the carbon dioxide and oxygen levels. At night, the rate of photosynthesis may drop to zero. At that time, a particular plant may be releasing carbon dioxide and taking in oxygen to maintain its relatively low but consistent rate of cell respiration.

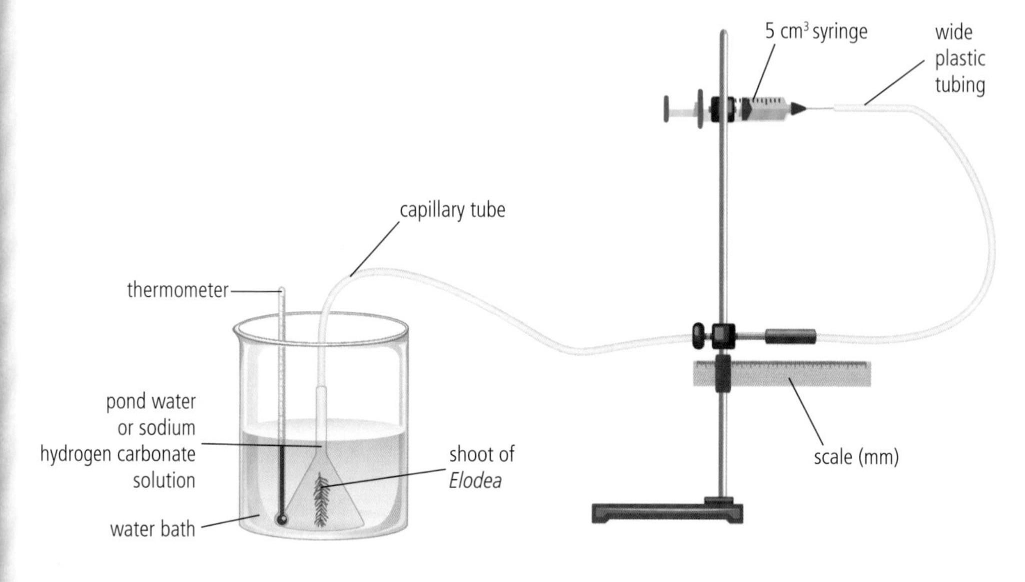

▲ A diagram showing the elements of a photosynthometer, used to measure the rate of photosynthesis with various independent variables.

Measuring the rate of oxygen production or carbon dioxide intake allows direct measurement of photosynthetic rate, if a correction is made for cell respiration.

This student is measuring oxygen production by an aquatic plant. Oxygen is collected and the volume measured using a **photosynthometer** centre. With this apparatus, you can determine the effect of various environmental conditions, such as varying light intensities or temperature, on the rate of photosynthesis. ▶

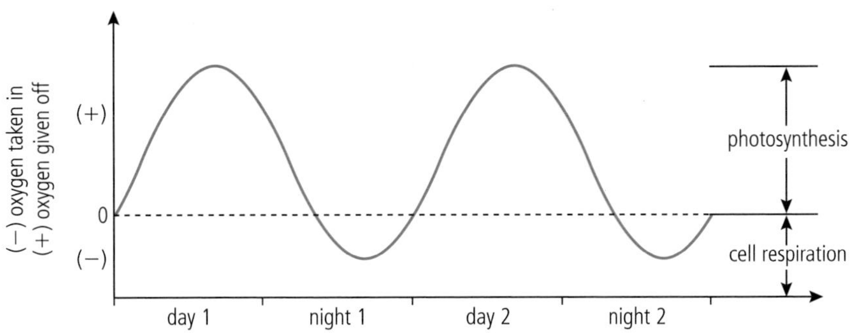

A graph showing the oxygen given off and taken in by a hypothetical plant over a 48-hour period. When the line intersects 0, the oxygen generated by photosynthesis is equal to the oxygen needed for cell respiration.

Another common method for measuring photosynthesis is to track changes in the biomass of experimental plants. However, the mass of plants is an indirect reflection of photosynthetic rate, as an increase or decrease in biomass can be caused by a variety of factors as well as the photosynthetic rate.

SKILLS

Experimental methods for measuring the rate of photosynthesis

Many techniques can be used to determine the rate of photosynthesis, including the following.

- Counting bubbles given off by aquatic plants. The bubbles are of oxygen and their number per unit of time can be recorded. The volume of the oxygen produced can also be measured.
- Solutions such as hydrogen carbonate indicators or universal indicators can be used with aquatic plants to measure pH changes. Carbon dioxide in water produces carbonic acid. As carbon dioxide is consumed, the pH of the solution becomes less acidic, with a corresponding colour change in the indicator. Colour changes can be recorded over time.
- Electronic sensors can also be used to measure the amount of carbon dioxide and oxygen, in both aquatic and carefully controlled atmosphere settings.

The effects of temperature, light wavelength, light intensity and other independent variables can all be determined using the above techniques. When looking at wavelengths of light, the data can then be used to produce an action spectrum.

Limiting factors

In 1905, British plant physiologist Frederick Frost Blackman proposed the **law of limiting factors**. It stated that a process that depends on multiple factors will have a rate limited by the factor at its least favourable (lowest) value. In many cases, this is the factor that is in "shortest" supply. Photosynthesis as a metabolic process has many potential limiting factors, including the amount of water, sunlight, temperature, carbon dioxide, chloroplasts and chlorophyll. Scarcity of any one of these factors influences the rate of photosynthesis. Study Figures 2–4, illustrating how light intensity, temperature and carbon dioxide concentration can affect the rate of photosynthesis. After becoming familiar with the graphs, complete the Challenge yourself activity.

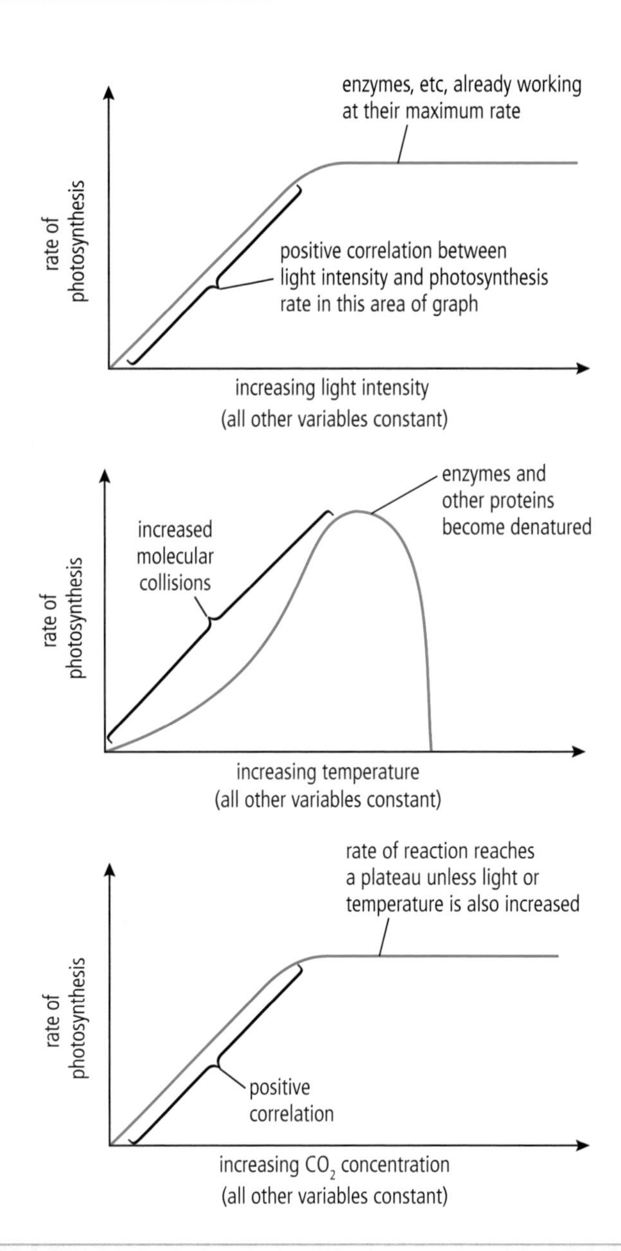

C1.3 Figure 2 The effect of increasing light intensity on the rate of photosynthesis.

C1.3 Figure 3 The effect of increasing temperature on the rate of photosynthesis.

CC1.3 Figure 4 The effect of increasing carbon dioxide concentration on the rate of photosynthesis.

Challenge yourself

Use Figures 2–4 above to answer the following questions about photosynthesis and limiting factors.

1. Look at Figure 2. Explain why the early part of the graph is labelled as a positive correlation.
2. Look at Figure 3. Why does the denaturing of enzymes and other proteins at higher temperatures dramatically lower the rate of photosynthesis?
3. For Figure 4, suggest what could cause a change from the plateau to an increasing rate, besides increasing CO_2 concentration.
4. Design a procedure to investigate the effect of one of the limiting factors mentioned above on the rate of photosynthesis. Note that water for photosynthesis experiments can be made to be free of dissolved carbon dioxide by boiling and then cooling it.

5. Suggest a hypothesis that can be tested using your procedure. Remember that your hypothesis for one of these experiments should be based on your knowledge and understanding of photosynthesis.

Nature of Science

When designing your procedure to investigate the effect of a limiting factor on the rate of photosynthesis, you may have had an idea about what would happen. A hypothesis is a provisional explanation that requires repeated testing to verify it, and is based on knowledge. During scientific research, hypotheses can be based on theories and then tested in an experiment, or they may be based on evidence from an experiment that has already been carried out. It is acceptable to suggest hypotheses for the effects of limiting factors on photosynthesis before or after performing experiments. You should be able to identify the dependent and independent variables in an experiment. **Dependent** variables are the outcome of the procedure or the value that you are measuring, in this case the rate at which carbon dioxide is used or oxygen is produced. The **independent** variable is the variable that is being changed in the experiment, for example the temperature, or the wavelength of the light.

The ability to ask a meaningful question or create a hypothesis is often quite challenging to the scientist. What is the role of imagination and intuition in the creation of hypotheses in the natural sciences?

C1.3.8 – Carbon dioxide levels and future rates of photosynthesis

C1.3.8 – Carbon dioxide enrichment experiments as a means of predicting future rates of photosynthesis and plant growth
Include enclosed greenhouse experiments and free-air carbon dioxide enrichment experiments (FACE).
NOS: Finding methods for careful control of variables is part of experimental design. This may be easier in the laboratory but some experiments can only be done in the field. Field experiments include those performed in natural ecosystems. Students should be able to identify a controlled variable in an experiment.

Many experiments have been carried out or are ongoing to determine the effects of increasing carbon dioxide levels in our atmosphere on plants and photosynthetic rates. From our discussion of carbon dioxide as a limiting factor, you might predict that increasing carbon dioxide levels would increase plant productivity. This has been verified. However, it is interesting to note that the growth rate of weeds increases more than the growth rate of crop plants and trees.

A study recently conducted by Peter Wayne and colleagues in controlled greenhouses involved ragweed (*Ambrosia artemisiifolia*) plants, which produce pollen that causes allergy symptoms in many people. The results of the study showed that a doubling of carbon dioxide levels stimulated ragweed pollen production by 61%. To a person with ragweed pollen allergy, this is not good news.

As well as many greenhouse-based experiments, studies are being conducted in natural settings. Natural settings often provide conditions that cannot be controlled in the same way as laboratory-based procedures can. These studies are known as free-air carbon dioxide enrichment experiments (FACE). FACE allows us to examine the effects of increasing carbon dioxide levels on plants in natural and agricultural ecosystems. By artificially increasing the carbon dioxide levels in these natural systems, we gain a more reliable picture of what may happen in the real-world future of steadily increasing carbon dioxide levels.

Nature of Science

Control of variables is an essential part of experimental design. Controlling as many variables as possible increases our ability to arrive at valid and reliable conclusions after an experimental procedure. **Reliability** is the degree to which an assessment tool produces stable and consistent results. **Validity** refers to how well a test measures what it is intended to measure. Controlling variables is usually much easier in the laboratory, but some experiments can only be done in the field. Field experiments include those performed in natural ecosystems, such as described in FACE studies.

Studies are being carried out all over the world to better understand the present and future effects of an ever-increasing amount of carbon dioxide in our atmosphere. Early hypotheses included the possible beneficial effects of higher carbon dioxide levels. However, at present relatively few plants are able to increase photosynthetic rates in a beneficial way to our planet. Most studies have found serious problems with increased levels of carbon dioxide. As carbon dioxide has increased, so has temperature, resulting in reduced water and mineral availability and lowered photosynthesis rates. Also, with increased carbon dioxide levels, studies have shown that many food crops (such as wheat) have lower protein levels. Increased carbon dioxide levels also seem to lead to greater plant ingestion by animals. This may be because the plants are producing less defensive chemicals (toxins). Overall, the rate of photosynthesis by plants appears to be dropping.

HL

C1.3.9 – Light-dependent reactions and photosystems

C1.3.9 – Photosystems as arrays of pigment molecules that can generate and emit excited electrons
Students should know that photosystems are always located in membranes and that they occur in cyanobacteria and in the chloroplasts of photosynthetic eukaryotes. Photosystems should be described as molecular arrays of chlorophyll and accessory pigments with a special chlorophyll as the reaction centre from which an excited electron is emitted.

Cyanobacteria are found in most places on our planet. They are even present in harsh environments such as deserts and hot springs. They are one of the few groups of organisms that can convert atmospheric nitrogen into an organic form, such as nitrate or ammonia, that plants can use. They are often referred to as blue-green algae and have been found in fossils dated to more 1.9 billion years ago.

Photosynthesis can be broken down into two separate stages, the **light-dependent** reactions, and the **light-independent** reactions. The light-independent reactions are also known as the **Calvin cycle**. The light-dependent reactions capture light energy, which is then needed for the synthesis of carbohydrates in the Calvin cycle. The Calvin cycle does not use light directly. We will look at the light-dependent reactions first.

Photosystems are essential to the light-dependent reactions of photosynthesis. A photosystem is always located in a membrane. In plants, the photosystems are found in the thylakoids of chloroplasts. Photosystems also exist in cyanobacteria, a type of bacteria capable of photosynthesis.

The centre of a photosystem is known as the **reaction centre**. Around the reaction centre are various pigment molecules bound to proteins that are known as light-receiving complexes. This array of molecules in the membrane is highly structured and allows energy to be captured and passed to the central molecule. Different species of plants possess different pigments, both in number and in type.

The light-receiving complexes are composed of chlorophyll and other accessory pigments. When a light-receiving pigment absorbs a photon, the energy is transferred from pigment molecule to pigment molecule until it reaches the reaction centre complex. The reaction centre complex includes a pair of special chlorophyll molecules whose electrons can move away from the molecules when energy reaches them. These energized electrons are then received by a primary electron acceptor. See Figure 5.

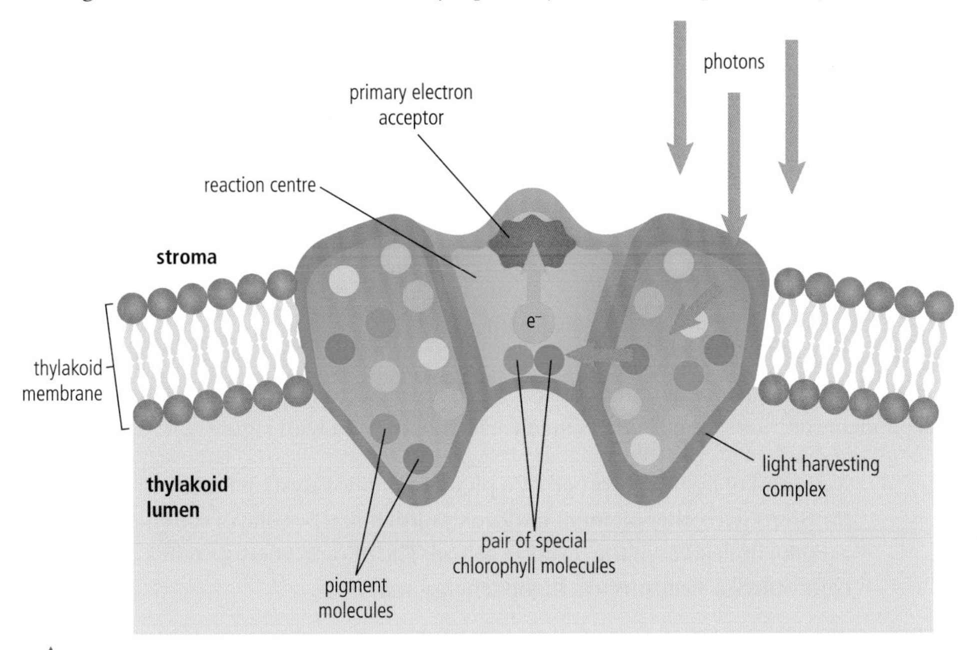

C1.3 Figure 5 A diagram of a photosystem. When light of the appropriate wavelength is received by a light-receiving pigment, the energy of the photon is passed from pigment to pigment until it reaches the reaction centre. Once the energy reaches the reaction centre, energy is transferred to electrons of a chlorophyll *a* molecule. These energized electrons leave the reaction centre and are received by the primary electron acceptor.

Plants and cyanobacteria utilize two photosystems in the photosynthetic process. Each photosystem absorbs light most efficiently at a different wavelength. Photosystem I is most efficient at 700 nanometres (nm) and is labelled as P700. Photosystem II is most efficient at 680 nm and is labelled as P680. These two photosystems work together. Figure 6 shows the overall light-dependent reactions of photosynthesis.

Photosynthetic bacteria utilize the plasma membrane as the photosynthetic membrane. Often, the plasma membrane in these bacteria possesses numerous folds to increase the surface area. Many of these bacteria use a single photosystem in their photosynthetic process.

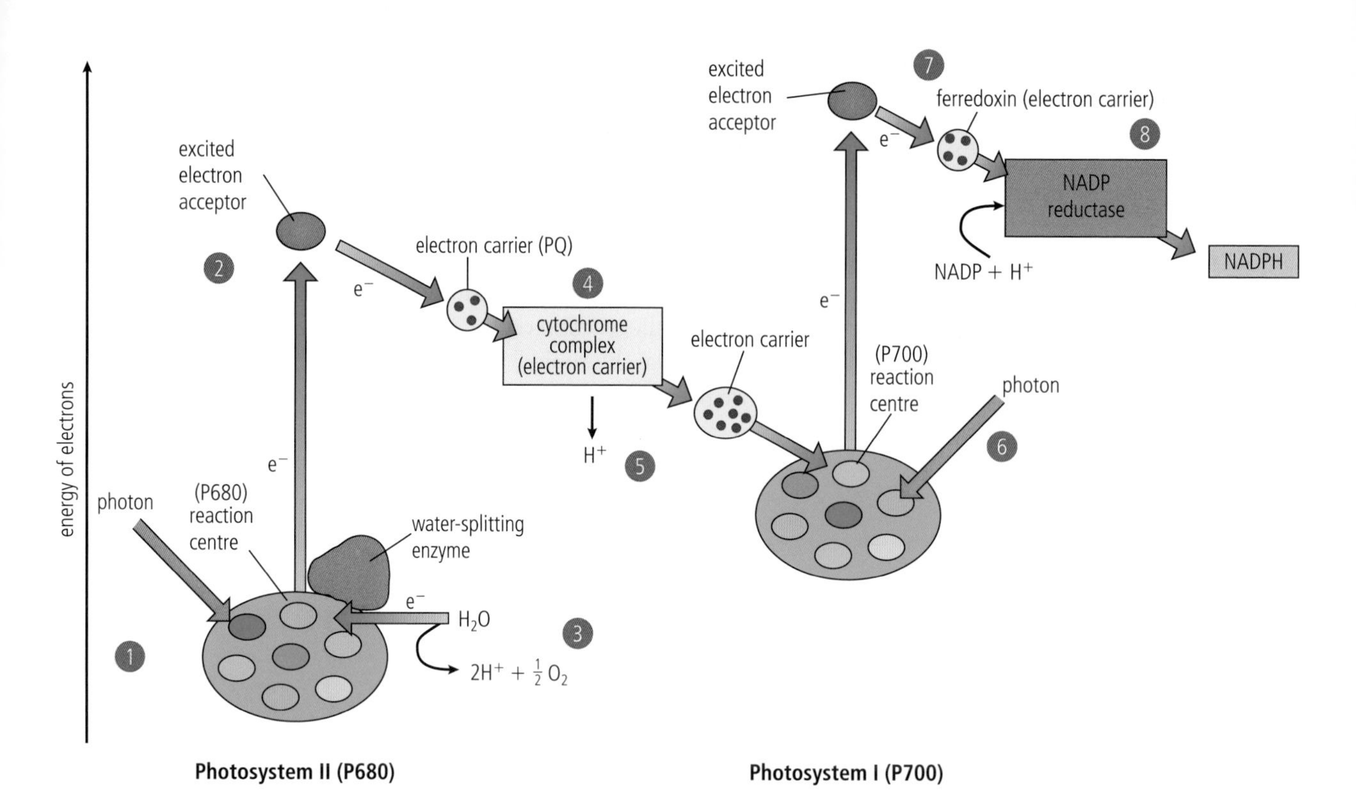

C1.3 Figure 6 Photosystems I and II in the light-dependent reactions of photosynthesis.

As you go through these steps, keep in mind the final products. It is not essential that you know the name of each intermediate substance, but you should recognize how each step in the overall process results in the products at the end.

The following sequence describes the role of the two photosystems in the light-dependent reactions. Look at this sequence in conjunction with Figure 6.

1. A photon of light is absorbed by a pigment in photosystem II and is transferred to other pigment molecules until it reaches one of the chlorophyll *a* (P680) molecules in the reaction centre. The photon energy excites one of the chlorophyll *a* electrons to a higher energy state.
2. This electron is captured by the electron acceptor of the reaction centre.
3. Water is split by an enzyme to produce electrons, hydrogen ions and an oxygen atom. This process is driven by the energy from light and is called **photolysis**. The electrons are supplied one by one to the chlorophyll *a* molecules of the reaction centre. These electrons replace the electrons lost from chlorophyll *a* to the electron acceptor.
4. The excited electrons pass from the primary acceptor down an electron transport chain, losing energy at each exchange. The first of the three carriers is plastoquinone (PQ). The middle carrier is a **cytochrome complex**.
5. The energy lost from the electrons moving down the electron transport chain drives **chemiosmosis** (like that in respiration) to bring about phosphorylation of ADP to produce ATP (see Section C1.3.12).
6. A photon of light is absorbed by a pigment in photosystem I. This energy is transferred through several accessory pigments until received by a chlorophyll *a* (P700) molecule. This results in an electron with a higher energy state being transferred to the electron acceptor. The de-energized electron from photosystem II fills the void left by the newly energized electron.

7. The electron with the higher energy state is then passed down a second electron transport chain that involves the carrier ferredoxin.

8. The enzyme NADP reductase catalyses the transfer of the electron from ferredoxin to the energy carrier NADP. Two electrons and one hydrogen ion from the stroma are required to reduce NADP fully to NADPH (see Section C1.2.13).

Chloroplast electron transport. Full details of how to carry out this experiment with a worksheet are available in the eBook.

C1.3.10 – Advantages of the structured array

C1.3.10 – Advantages of the structured array of different types of pigment molecules in a photosystem
Students should appreciate that a single molecule of chlorophyll or any other pigment would not be able to perform any part of photosynthesis.

Accessory pigments exist as different types and in large numbers in a photosystem so that all the parts of the light reactions of photosynthesis can occur. If only one pigment existed in a photosystem, photosynthesis would not proceed beyond the light-dependent reactions. All the accessory pigments work together to increase the rate of photosynthesis (the action spectrum). Each accessory pigment absorbs light of a different wavelength, allowing the array of all the pigments in a photosystem to work towards the goal of the highest rate of photosynthesis possible. The greater the number of accessory pigments in the structured array, the greater the potential rate of photosynthesis as long as chlorophyll *a* is present. Chlorophyll *a* is referred to as the primary pigment.

C1.3.11 – Oxygen as a waste product

C1.3.11 – Generation of oxygen by the photolysis of water in photosystem II
Emphasize that the protons and electrons generated by photolysis are used in photosynthesis but oxygen is a waste product. The advent of oxygen generation by photolysis had immense consequences for living organisms and geological processes on Earth.

Reduced NADP and ATP are the final products of the light-dependent reactions. They supply chemical energy for the Calvin cycle to occur (see Sections C1.3.15–17). However, step 3 outlined in Figure 6 also shows the origin of the oxygen released by photosynthesizing plants. When photolysis occurs during the light-dependent reactions, the protons and electrons are used in photosynthesis. The oxygen is given off as a waste product. It is this waste oxygen that altered the atmosphere of planet Earth. Geological processes were also affected by the release of oxygen as a waste gas.

The appearance of oxygen in our atmosphere as a result of photosynthesis in early organisms is known as the **Great Oxygenation Event**. It had many effects on our planet, including the mass extinction of many anaerobic bacteria. It also coincides with a greater number of minerals being formed. A significant effect was the thinning action of oxygen in our atmosphere. This allowed more sunlight to reach the Earth's surface. We also see evidence of more moisture in the atmosphere, probably as a result of greater water evaporation from the surface.

C1.3.12 and C1.3.13 – Photophosphorylation

C1.3.12 – ATP production by chemiosmosis in thylakoids
Include the proton gradient, ATP synthase, and proton pumping by the chain of electron carriers. Students should know that electrons are sourced, either from photosystem I in cyclic photophosphorylation or from photosystem II in non-cyclic photophosphorylation, and then used in ATP production.

C1.3.13 – Reduction of NADP by photosystem I
Students should appreciate that NADP is reduced by accepting two electrons that have come from photosystem I. It also accepts a hydrogen ion that has come from the stroma. The paired terms "NADP and reduced NADP" or "$NADP^+$ and NADPH" should be paired consistently.

ATP production in photosynthesis is very similar to ATP production in respiration. **Chemiosmosis** allows the process of phosphorylation of adenosine diphosphate (ADP). In this case, the energy to drive chemiosmosis comes from light. As a result, we refer to the production of ATP in photosynthesis as **photophosphorylation.**

A comparison of chemiosmosis in respiration and photosynthesis is shown in Table 2.

C1.3 Table 2 A comparison of chemiosmosis

Respiration chemiosmosis	Photosynthesis chemiosmosis
Involves an electron transport chain embedded in the membranes of the cristae	Involves an electron transport chain embedded in the membranes of the thylakoids
Energy is released when electrons are exchanged from one carrier to another	Energy is released when electrons are exchanged from one carrier to another
Released energy is used to pump hydrogen ions actively into the intermembrane space, creating a proton gradient	Released energy is used to pump hydrogen ions actively into the thylakoid space, creating a proton gradient
Hydrogen ions come from the matrix	Hydrogen ions come from the stroma
Hydrogen ions diffuse back into the matrix through the channels of ATP synthase	Hydrogen ions diffuse back into the stroma through the channels of ATP synthase
ATP synthase catalyses the phosphorylation of ADP to form ATP	ATP synthase catalyses the photophosphorylation of ADP to form ATP

In both cases the ATP synthase is embedded along with the carriers of the electron transport chain in the membranes involved.

In photosynthesis, the production of ATP occurs between photosystem II and photosystem I. Look at Figure 7, which illustrates the production of ATP at the electron transport chain between photosystems II and I.

Notice that the b_6–f complex, which is a **cytochrome complex**, pumps the hydrogen ions into the thylakoid space. This increases the concentration of the hydrogen ions, which then move passively through the ATP synthase channel, providing the energy to phosphorylate ADP.

The process is known as **non-cyclic photophosphorylation**. The electrons move through the electron transport chain and eventually are used to reduce NADP with hydrogen ions from the stroma. Another way light-dependent reactions of photosynthesis can produce ATP is **cyclic photophosphorylation**. This proceeds only when light is not a limiting factor and when there is an accumulation of reduced NADP in the chloroplast. In this process, light-energized electrons from photosystem I flow back to the cytochrome complex of the electron transport chain between photosystem II and photosystem I (look back at Figure 6). From the cytochrome complex, the electrons move down the remaining electron transport chain allowing ATP production via chemiosmosis. These ATPs are then shuttled to the Calvin cycle so that it can proceed more rapidly. In cyclic photophosphorylation, oxygen and reduced NADP are not produced: only ATP is produced.

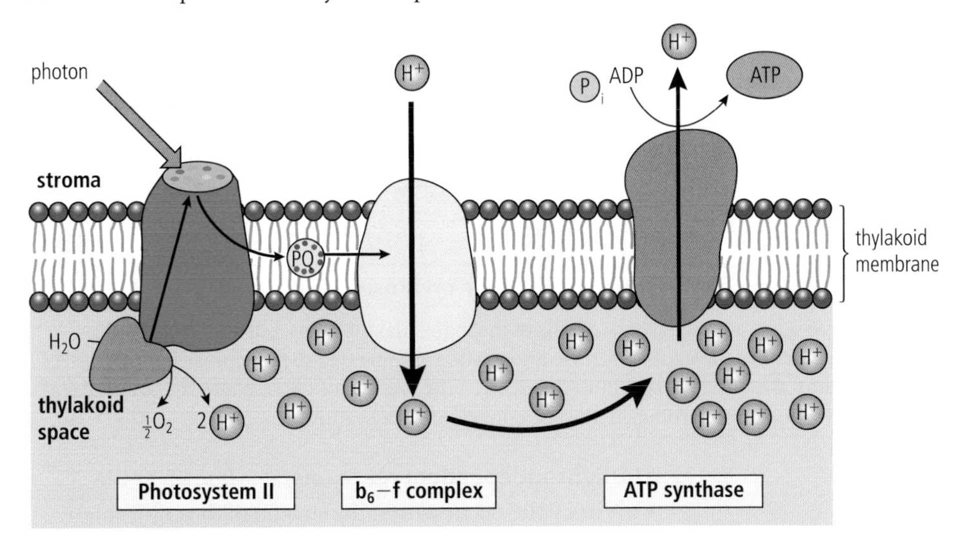

C1.3 Figure 7 Chemiosmosis in a plant cell chloroplast.

C1.3.14 – The role of thylakoids

C1.3.14 – Thylakoids as systems for performing the light-dependent reactions of photosynthesis

Students should appreciate where photolysis of water, synthesis of ATP by chemiosmosis and reduction of NADP occur in a thylakoid.

As discussed, the light-dependent reactions of photosynthesis take place in the thylakoid membranes of plants and algae. The structure of the chloroplast is discussed in Chapters A2.2 and B2.2 and it might be a good idea to review the relevant sections.

The light-dependent reactions capture light energy, which is then needed for the synthesis of carbohydrates in the Calvin cycle. The light-dependent reactions take place in the thylakoids of the chloroplasts, while the Calvin cycle occurs in the stroma.

The two stages of photosynthesis.

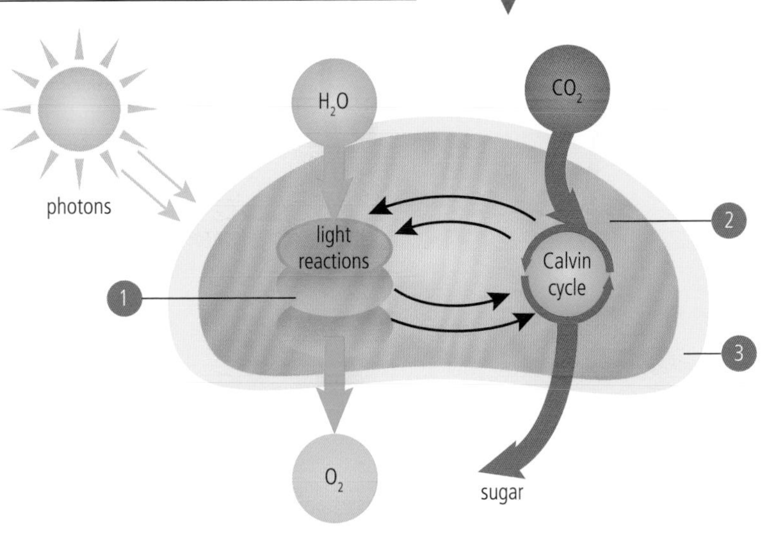

C1.3.15, C1.3.16 and C1.3.17 – Light-independent reactions and the Calvin cycle

C1.3.15 – Carbon fixation by Rubisco
Students should know the names of the substrates RuBP and CO_2 and the product glycerate 3-phosphate. They should also know that Rubisco is the most abundant enzyme on Earth and that high concentrations of it are needed in the stroma of chloroplasts because it works relatively slowly and is not effective in low carbon dioxide concentrations.

C1.3.16 – Synthesis of triose phosphate using reduced NADP and ATP
Students should know that glycerate-3-phosphate (GP) is converted into triose phosphate (TP) using NADPH and ATP.

C1.3.17 – Regeneration of RuBP in the Calvin cycle using ATP
Students are not required to know details of the individual reactions, but students should understand that five molecules of triose phosphate are converted to three molecules of RuBP, allowing the Calvin cycle to continue. If glucose is the product of photosynthesis, five-sixths of all the triose phosphate produced must be converted back to RuBP.

C1.3 Figure 8 The Calvin cycle.

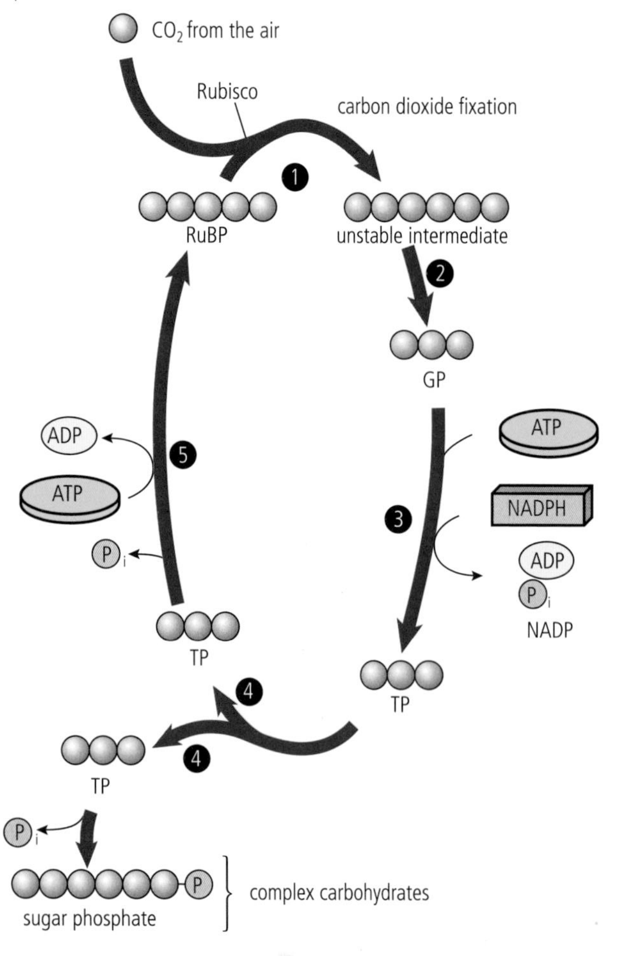

The **light-independent reactions**, also known as the **Calvin cycle**, occur within the stroma of the chloroplast. Although the Calvin cycle is not directly dependent on light, it is indirectly dependent on light because it requires the energy carriers ATP and reduced NADP (NADPH), which are produced by the light-dependent reactions. These two carriers provide the energy and reducing power for the light-independent reactions to occur.

Because it is a cycle, the Calvin cycle begins and ends with the same substance. You should recall that a similar cyclic metabolic pathway occurs in cell respiration: the Krebs cycle. Study Figure 8, and refer to it as you read about the steps of the Calvin cycle.

1. **Ribulose bisphosphate** (RuBP), a 5-carbon compound, binds to an incoming carbon dioxide molecule in a process called carbon fixation. This fixation is catalysed by an enzyme called RuBP carboxylase (**Rubisco**). The result is an unstable 6-carbon compound.
2. The unstable 6-carbon compound breaks down into two 3-carbon compounds called **glycerate 3-phosphate** (GP).
3. The 3-carbon molecules of GP are acted upon by ATP and reduced NADP (NADPH) from the light-dependent reaction to form two other 3-carbon molecules called **triose phosphate** (TP). This is a reduction reaction.

4. The molecules of TP may then go in either of two directions. Some leave the cycle to become sugar phosphates, which may then become more complex carbohydrates. Most, however, continue in the cycle to reproduce the original molecule in the cycle, RuBP.
5. In order to reform RuBP molecules from TP, the cycle uses ATP.

Rubisco (RuBP carboxylase) is the most abundant enzyme on Earth. It is needed in high concentrations because it works relatively slowly, and it is not effective in low carbon dioxide concentrations. It is extremely important to life on Earth because it is responsible for catalysing the reaction that converts inorganic carbon into organic carbon.

In Figure 8, spheres are used to represent the carbon atoms so that they can be tracked through the cycle. For every 12 TP molecules, the cycle produces one 6-carbon sugar and six molecules of the 5-carbon compound RuBP. It is important to note that 18 ATPs and 12 reduced NADP are necessary to produce six RuBP molecules and one molecule of a 6-carbon sugar. Now study Figure 9, which is the same as Figure 8 but with the addition of the numbers of molecules involved in the cycle.

TP is the pivotal compound in the Calvin cycle. It can be used to produce sugars such as glucose, disaccharides such as sucrose, and polysaccharides such as cellulose and starch. However, most of it is used to regain the starting compound of the Calvin cycle, RuBP. It is essential that five-sixths of all the TP produced is converted back to RuBP. If less RuBP is produced, the cycle will not be able to continue to its full potential. Also, there will not be sufficient resources for the light-dependent reactions to proceed.

As in other complex biological reaction pathways, you should keep in mind the end products while studying the intermediate products. Key points to focus on are the ATP molecules produced, the reformation of the starting RuBP for the cycle, the role of reduced NADP and the sugar phosphate product.

Some diagrams use one-half the coefficients for each product shown in Figure 9. That is just as correct. The benefit of using the coefficients shown in Figure 9 is that it shows how one molecule of the 6-carbon sugar phosphate is achieved.

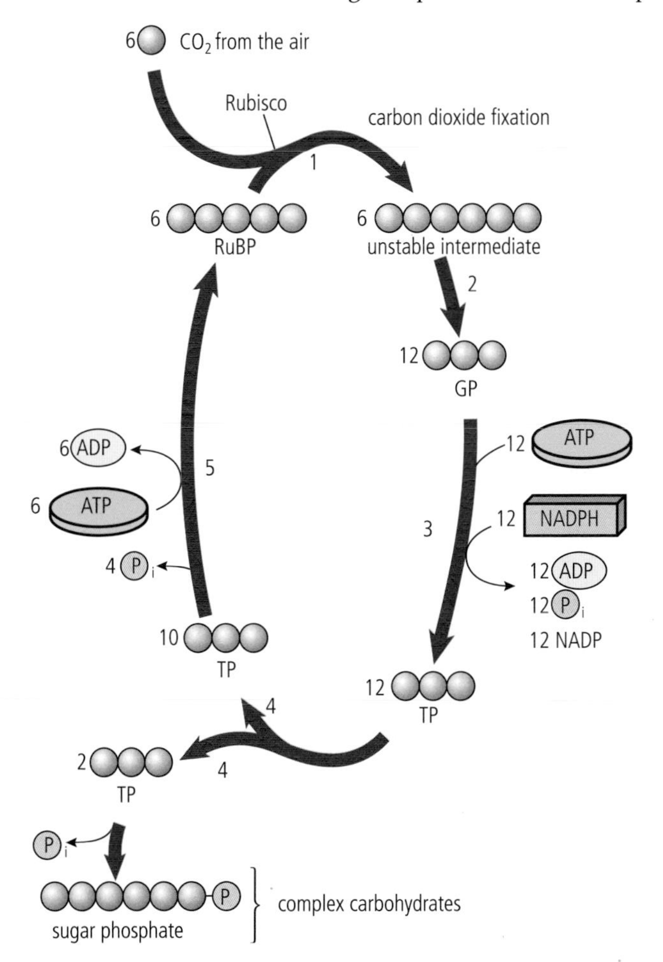

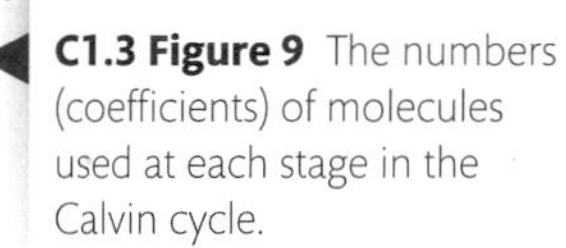
C1.3 Figure 9 The numbers (coefficients) of molecules used at each stage in the Calvin cycle.

C1.3.18 – Synthesis of other carbohydrates and amino acids

C1.3.18 – Synthesis of carbohydrates, amino acids, and other carbon compounds using the products of the Calvin cycle and mineral nutrients
Students are not required to know details of metabolic pathways, but students should understand that all of the carbon in compounds in photosynthesizing organisms is fixed in the Calvin cycle and that carbon compounds other than glucose are made by metabolic pathways that can be traced back to an intermediate in the cycle.

Note that all the carbon in photosynthesizing organisms is fixed in the Calvin cycle. Carbon compounds other than glucose are made by metabolic pathways that are linked to an intermediate product in the Calvin cycle.

Other carbon compounds can be produced by metabolic pathways originating from intermediates of the Calvin cycle. Glucose can be combined with fructose to form sucrose. Glucose phosphate is the starting point in the production of cellulose and starch. Fatty acids and glycerol are produced from the intermediate glycerate 3-phosphate (GP). When nitrogen is added to the hydrocarbon backbone of GP, amino acids can be formed, leading to protein production. All the carbon found in plant tissues (which is subsequently passed through the food chain) is fixed in the Calvin cycle.

Nature of Science

A team led by Melvin Calvin in the late 1940s and early 1950s worked on experiments to find the early products of photosynthesis. They were successful in working out the details of carbon fixation (the Calvin cycle). Calvin made use of improvements in apparatus design and developments in radioactive tracers and autoradiography to devise the "lollipop" apparatus. This is a thin, almost bulb-shaped, glass vessel with a supporting stem. The vessel was designed to mimic the shape of a leaf: thin and broad. He then carried out the following procedures.

- *Chlorella pyrenoidosa* (a type of green algae) was placed inside the lollipop.
- The algae cells were then exposed to ^{14}C (radioactive carbon) and light.
- Samples of *Chlorella* were released from the apparatus at short intervals.
- Each removed sample was immediately placed into a boiling methanol solution to denature the enzymes and stop the photosynthetic process.
- The compounds within the algae were then separated. Two-way paper chromatography was used for this separation. One solvent was used to separate the first set of components, then the paper was turned and placed in a different solvent to obtain further separation of components.
- The final radioactive products were identified using autoradiography.

Because Calvin carried out this procedure with algae released at different time intervals during the process, he obtained different products at different times. This allowed him to sequence the steps of the overall process and elucidate the pathways of carbon fixation (the Calvin cycle).

C1.3.19 – Overview of photosynthesis

C1.3.19 – Interdependence of the light-dependent and light-independent reactions
Students should understand how a lack of light stops light-independent reactions and how a lack of CO_2 prevents photosystem II from functioning.

The process of photosynthesis includes both light-dependent and light-independent reactions. The products of the light-dependent reactions are ATP and reduced NADP (NADPH), which are then needed to allow the light-independent reactions to proceed. Thus, light is needed for the light-independent reaction to occur, but not directly, and a lack of light will stop the light-independent reactions. If carbon dioxide is absent, the photosystems of the light-dependent reaction cannot occur. A summary of the two reactions is shown in Figure 10 and Table 3.

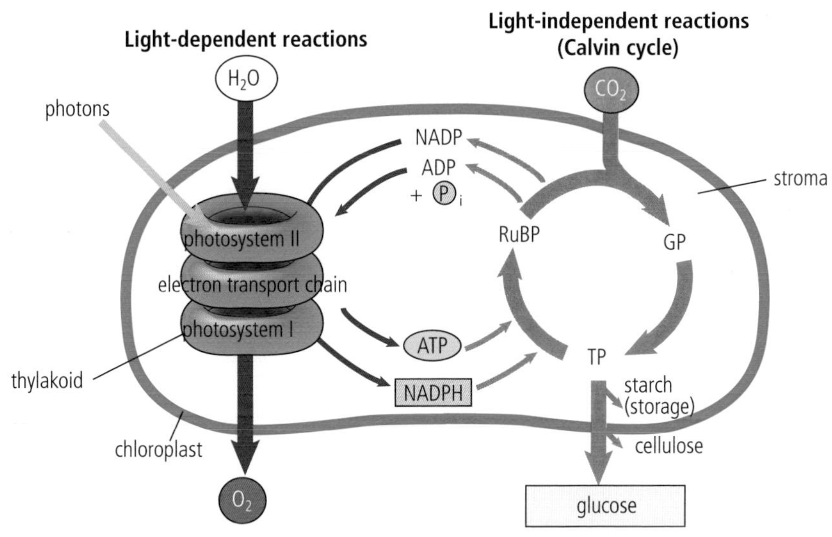

C1.3 Figure 10 A summary of the complete process of photosynthesis.

Note that NADP and ATP move back and forth in the chloroplast from the thylakoids to the stroma in their reduced and oxidized forms.

Light-dependent reactions	Light-independent reactions
Occur in the thylakoids	Occur in the stroma
Use light energy to form ATP and reduced NADP	Use ATP and reduced NADP to form triose phosphate
Split water in photolysis to provide replacement electrons and H^+, releasing oxygen to the atmosphere as waste	Return ADP, inorganic phosphate and NADP to the light-dependent reaction
Include two electron transport chains and photosystems I and II	Involve the Calvin cycle

C1.3 Table 3 A summary of the complete process of photosynthesis

Challenge yourself

Examine the figure, then answer the questions with the appropriate letter.

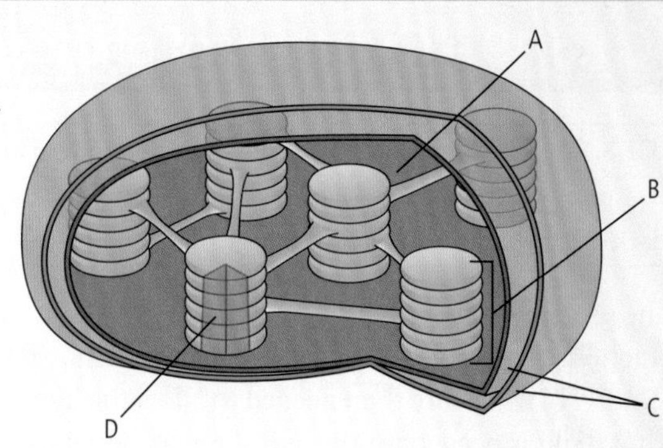

6. Which letter represents the stroma?
7. What is the function of the stroma?
8. Which letter represents the double membrane?
9. What is the function of the double membrane?
10. Which letter is pointing to a thylakoid?
11. What is the function of the thylakoid?
12. Which letter points to a granum?
13. What is the function of the granum?
14. Which two letters point to areas that are green in colour?
15. Explain why chloroplasts are green.
16. The chloroplasts within some plant cells can often be seen moving in a cyclical pattern near the periphery of the cell. This is called cyclosis or cytoplasmic streaming. What might be the value of such movement to the process of photosynthesis?

To become more confident in your understanding of the chloroplast, look online to find some electron micrographs of chloroplasts from several different plants. On these micrographs, annotate names of structures and their functions.

The first international conference dedicated to the creation of an artificial "leaf" was held in 2011. This conference addressed the goals of the Global Artificial Photosynthesis (GAP) project. Research centres are active throughout the world, and the objective of this conference was to allow researchers to share their discoveries. Energy capture, energy conversion and storage, and carbon fixation using modified and synthetic biological processes, were all discussed. It is hoped that the GAP project will lead to enhanced crop production, reduced atmospheric carbon dioxide levels, and increased availability of fuels for heating and cooking.

HL end

Guiding Question revisited

How is energy from sunlight absorbed and used in photosynthesis?

In this chapter we have discussed how:

- absorption of light for photosynthesis is essential to produce carbon compounds (chemical energy) and atmospheric oxygen
- oxygen is formed from the splitting of water during photolysis and is a waste product of photosynthesis
- photosynthetic pigments such as the carotenoids and chlorophylls absorb specific wavelengths of light
- the light-dependent reactions require the pigments of photosystems I and II to proceed efficiently

HL

- accessory pigments increase the absorption of light so that the primary pigment of chlorophyll *a* can be energized more, resulting in an increase in reduced NADP for use in the light-independent reactions

- reduced NADP and ATP are produced by the light-dependent reactions and are essential for the Calvin cycle (light-independent reactions)
- the electron transport chain is driven by energized electrons from photosystem II and produces ATP while delivering the resulting de-energized electrons to photosystem I
- photosystem I results in the production of reduced NADP, which can be used in the light-independent reactions.

HL end

Guiding Question revisited

How do abiotic factors interact with photosynthesis?

In this chapter we have discussed how:

- carbon dioxide and water are two essential reactants of photosynthesis
- carbon dioxide is converted to glucose after reduction during carbon fixation
- temperature, light intensity and carbon dioxide concentration are all potential abiotic limiting factors of photosynthesis
- HL the Calvin cycle uses the hydrogen ions and electrons produced by the photolysis of water in the light-dependent reaction.

Exercises

Q1. Compare and contrast action and absorption spectra.

Q2. List the reactants and the products of photosynthesis.

Q3. Explain two reasons why the splitting of water in the photosynthetic process is so important to ecosystems.

HL

Q4. In which part of the chloroplast does chemiosmosis occur?

A Stroma. **B** Matrix.

C Thylakoids. **D** Cytoplasm.

Q5. Order the following events in photosynthesis.

(a) ______ photolysis

(b) ______ carbon dioxide fixation

(c) ______ chemiosmosis

(d) ______ formation of triose phosphate

Q6. Which photosystem generates reduced NADP (NADPH)?

Q7. What specific process is the enzyme ATP synthase involved in?

Q8. Give two reasons why photolysis is so important to the photosynthetic process.

HL end

C1 Practice questions

1. Which of the following statements is **true** about enzymes?

   A They are used up in the reactions they catalyse.

   B Allosteric inhibitors bind to the active site.

   C They lower the energy of activation for a reaction.

   D They supply the energy of activation for a reaction.

   (Total 1 mark)

2. At the start of glycolysis, glucose is phosphorylated to produce glucose 6-phosphate, which is converted into fructose 6-phosphate. A second phosphorylation reaction is then carried out, in which fructose 6-phosphate is converted into fructose 1,6-bisphosphate. This reaction is catalysed by the enzyme phosphofructokinase. Biochemists measured the enzyme activity of phosphofructokinase (the rate at which it catalysed the reaction) at different concentrations of fructose 6-phosphate. The enzyme activity was measured with a low concentration of ATP and a high concentration of ATP in the reaction mixture. The graph below shows the results.

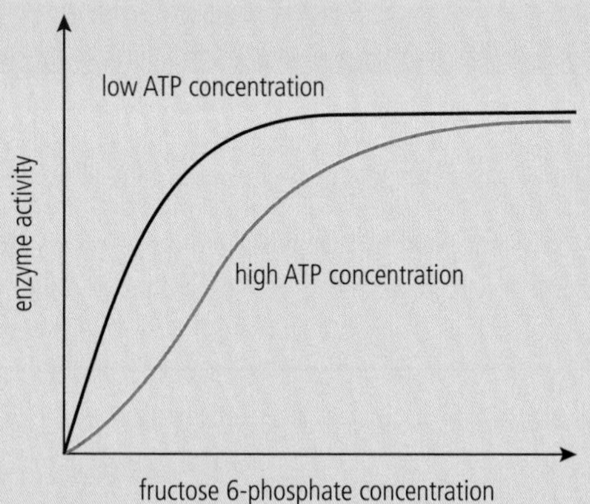

(a) **(i)** Using **only** the data in the above graph, outline the effect of increasing fructose 6-phosphate concentration on the activity of phosphofructokinase, at a low ATP concentration. (2)

(ii) Explain how increases in fructose 6-phosphate concentration affect the activity of the enzyme. (2)

(b) **(i)** Outline the effect of increasing the ATP concentration on the activity of phosphofructokinase. (2)

(ii) Suggest an advantage to living organisms of the effect of ATP on phosphofructokinase. (1)

(Total 7 marks)

3. Biosphere 2, an enormous greenhouse built in the Arizona desert in the USA, has been used to study five different ecosystems. It is a closed system so measurements can be made under controlled conditions. The effects of different factors, including changes in carbon dioxide concentration in the greenhouse, were studied. The data shown below were collected over the course of one day in January 1996.

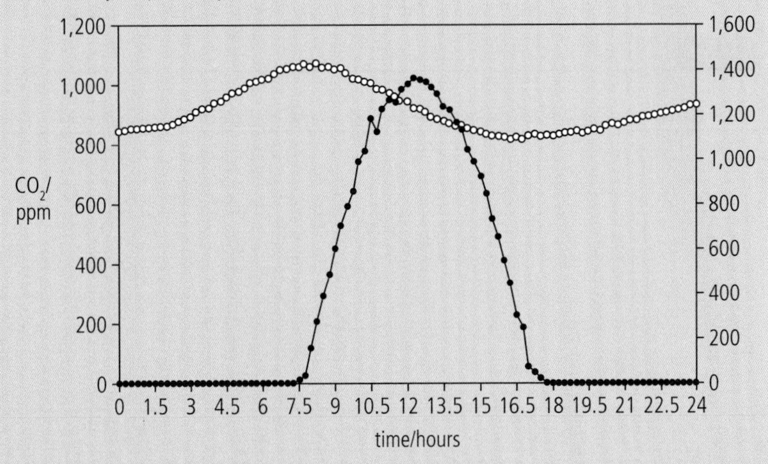

(a) (i) Identify the time of day when the Sun rose. (1)

(ii) Identify the time of minimal CO_2 concentration. (1)

(b) Determine the maximum difference in the concentration of CO_2 over the 24-hour period. (1)

(c) Suggest reasons for changes in CO_2 concentration during the 24-hour period. (2)

(Total 5 marks)

4. Inflammation of human tissues often causes pain. Cyclooxygenases (COX) are a group of enzymes that play a role in causing inflammation. Analgesics are drugs that can reduce pain. The graph below shows how increasing concentrations of the analgesic drug dipyrone affects the activity of three different cyclooxygenases, COX-1, COX-2 and COX-3.

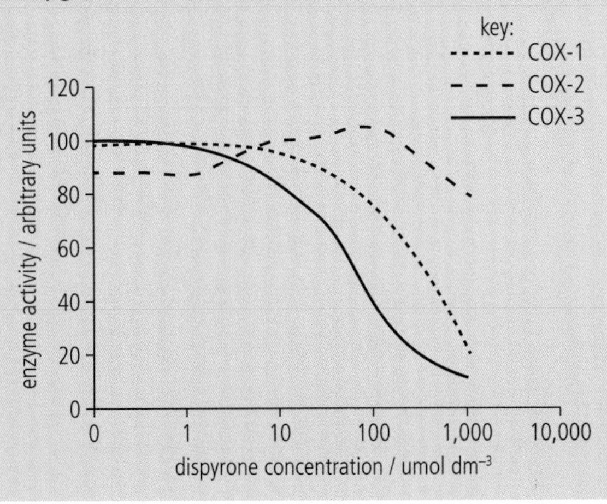

(a) Outline the relationship between dipyrone concentration and COX-3 activity. (2)

(b) Deduce whether dipyrone is an inhibitor of COX-2. (2)

(c) Evaluate the potential of dipyrone as an analgesic using the data in the graph. (2)

(Total 6 marks)

HL

5. Where is carbon dioxide produced in the mitochondrion?

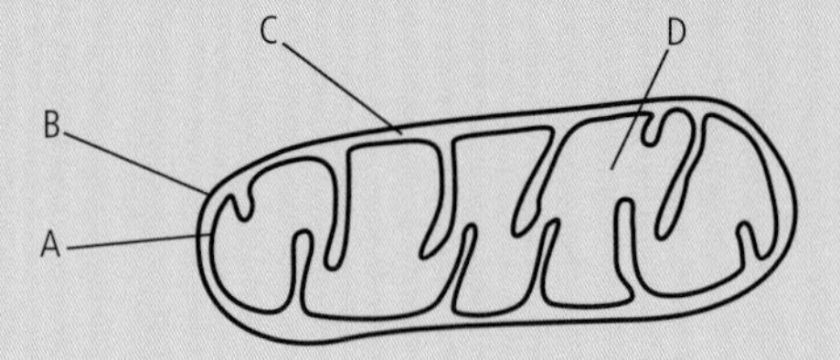

(Total 1 mark)

6. In the mitochondrial electron transport chain, what is the last electron acceptor?

A CO_2 **B** H_2O **C** O_2 **D** NAD

(Total 1 mark)

7. What is the role of NADH + H^+ in aerobic cell respiration?

A To transfer hydrogen to the electron transport chain.

B To reduce intermediates in the Krebs cycle.

C To accept electrons from the electron transport chain.

D To combine with oxygen to produce water.

(Total 1 mark)

8. What reaction, involving glycerate 3-phosphate, is part of the light-independent reactions of photosynthesis?

A Glycerate 3-phosphate is carboxylated using carbon dioxide.

B Two glycerate 3-phosphates are linked together to form one hexose phosphate.

C Glycerate 3-phosphate is reduced to triose phosphate.

D Five glycerate 3-phosphates are converted to three ribulose 5-phosphates.

(Total 1 mark)

9. What is the advantage of having a small volume inside the thylakoids of the chloroplast?

A High proton concentrations are rapidly developed.

B High electron concentrations are rapidly developed.

C Photosynthetic pigments are highly concentrated.

D Enzymes of the Calvin cycle are highly concentrated.

(Total 1 mark)

10. During glycolysis a hexose sugar is broken down to two pyruvate molecules. What is the correct sequence of stages?

A Phosphorylation → oxidation → lysis

B Oxidation → phosphorylation → lysis

C Phosphorylation → lysis → oxidation

D Lysis → oxidation → phosphorylation

(Total 1 mark)

11. Which is correct for the non-competitive inhibition of enzymes?

	Inhibitor resembles substrate	Inhibitor binds to active site
A	yes	yes
B	yes	no
C	no	yes
D	no	no

(Total 1 mark)

12. Where are the light-dependent and light-independent reactions taking place in the diagram below?

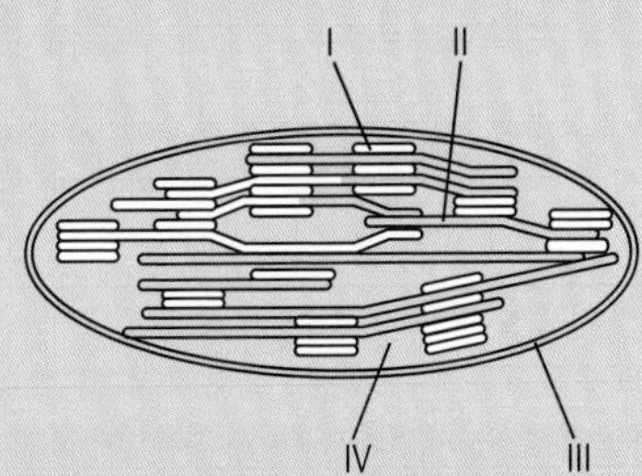

	Light-dependent	Light-independent
A	I	IV
B	II	III
C	III	II
D	IV	I

(Total 1 mark)

13. What is the link reaction in aerobic respiration?

A Pyruvate is carboxylated, acetyl reacts with coenzyme A, reducing NADH + H^+.

B Pyruvate is decarboxylated, acetyl reacts with coenzyme A, forming NADH + H^+.

C Pyruvate reacts with coenzyme A, forming NADH + H^+.

D Pyruvate is decarboxylated, reacting with coenzyme A, reducing NADH + H^+.

(Total 1 mark)

HL end

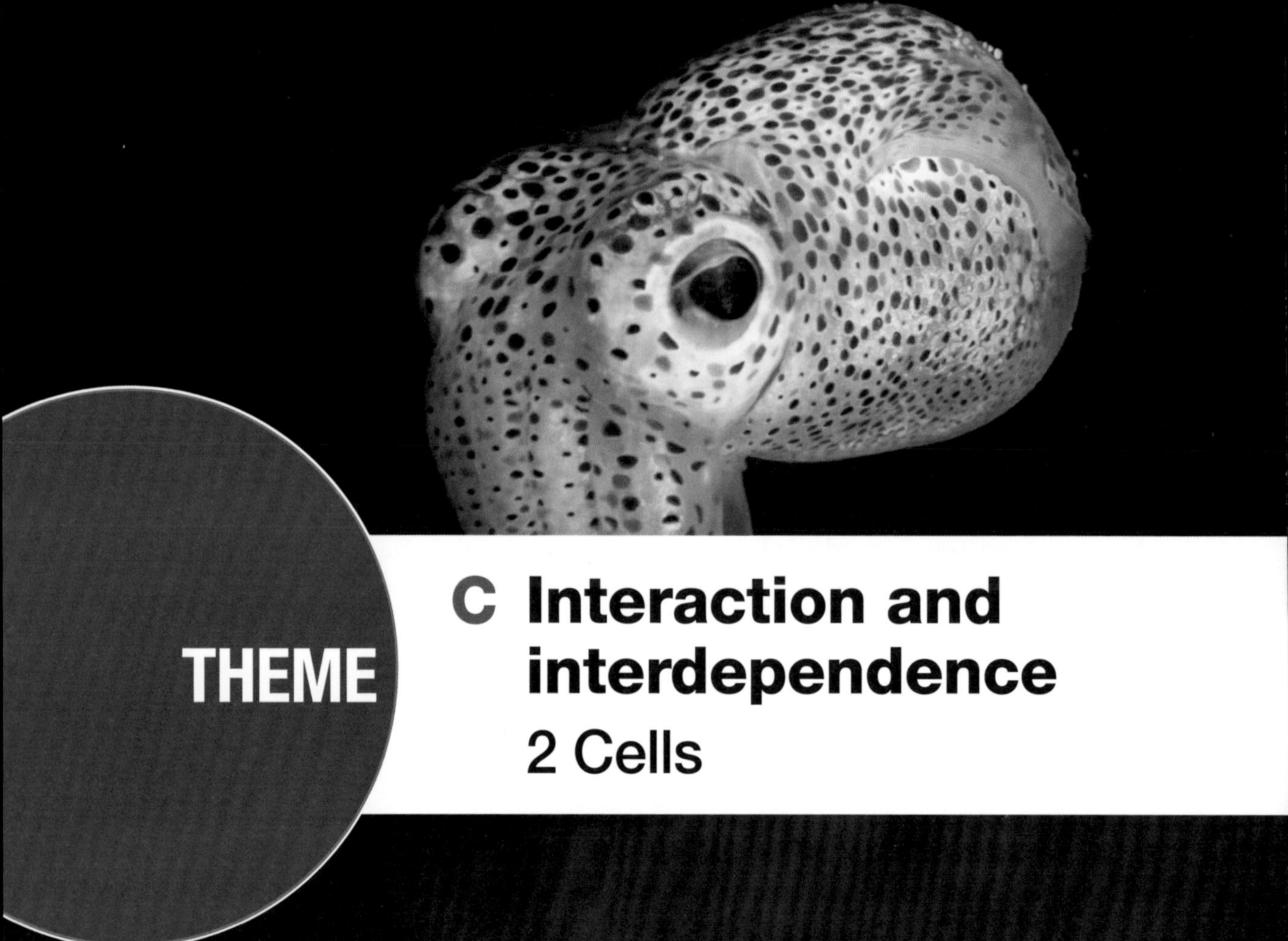
THEME
C Interaction and interdependence
2 Cells

The Hawaiian bobtail squid (*Euprymna scolopes*) has a symbiotic relationship with the bacterium *Vibrio fischeri*. The bacteria reside in the light organ of the squid's mantle and produce light by bioluminescence. The light produced by the bacteria provides protection against predators for the squid. The *Vibrio* bacteria control the production of this light using a process of cell signalling called quorum sensing.

All living things receive and deliver signals of several different types. Two major types of signalling are chemical and neural. Chemical signals are produced by structures in specialized regions of the cell and/or organism. Once produced, they can be transported by many different means. Receptors are then capable of receiving these chemical signals and initiating a response within the same or different cell and/or organism. The *Vibrio* bacteria mentioned above produce such a chemical that initiates the result of bioluminescence within the squid.

Neural signalling involves electrical impulses and activity. Nerve and muscle cells within animals demonstrate neural signalling. This neural signalling is essential to maintaining proper and coordinated functioning within the cell and/or organism. In medicine, the electrical impulses of neural signalling are often utilized to study activity within various organs. Some common examples of medical applications of neural signalling within animals and humans include studies of the heart known as electrocardiograms (ECG) and studies of the brain known as electroencephalograms (EEG).

HL

C2.1 Chemical signalling

Guiding Questions

How do cells distinguish between the many different signals that they receive?

What interactions occur inside animal cells in response to chemical signals?

We hear about the importance of communication every day. Too often stories are told where a misunderstanding led to a path of mistrust and/or conflict between individuals. In contrast, when a group of people are working together effectively, there is constant efficient communication between them concerning plans, processes and results. Barriers may be encountered during the communication process that result in a decrease in effectiveness. When this happens, further communication is needed to get back on the right path. Cells in organisms are no different.

Cells participate in a complex system of communication. This communication governs and coordinates the basic cellular activities in multicellular organisms. This communication is referred to as cell signalling.

Just like a person in present-day society, our cells are constantly transmitting and being bombarded by many types of signals. Some of the signals control localized, specific actions, while others elicit general and far-ranging actions. These signals come in many forms, including a myriad of chemicals. Hormones, neurotransmitters and cytokines are just a few of the more common types of chemicals. For these chemicals to have an effect, the cell must possess receptors to receive them and translate them into the correct actions.

Chemical signals are called ligands, and enable a diversity of responses and their control, including bioluminescence, nervous system coordination and response, transport of various materials into and out of cells, proper and effective muscle action, and even transcription within the nucleus of the cell. The regulation of these chemical signals is dependent on feedback systems, which can be positive or negative.

C2.1.1 – The requirements for cell signalling

C2.1.1 – Receptors as proteins with binding sites for specific signalling chemicals
Students should use the term "ligand" for the signalling chemical.

Cell signalling involves **ligands**, **receptors** and **signal transduction pathways**. Ligands are molecules that bind reversibly to specific proteins. Usually, the ligand is smaller than the protein. When a ligand binds to a protein, this can deliver a signal. The protein to which a ligand binds is the receptor. When a ligand binds to its receptor, there is a specific cellular response. The signal transduction pathway is the series of steps that causes the cell to respond to the binding of the ligand.

Cells with receptors that bind a particular ligand are known as target cells for that ligand. If a cell does not have a receptor for a specific ligand, then it is not affected by the signal. Ligands can act over varying distances within the organism. A simple overview of cell signalling is given in Figure 1.

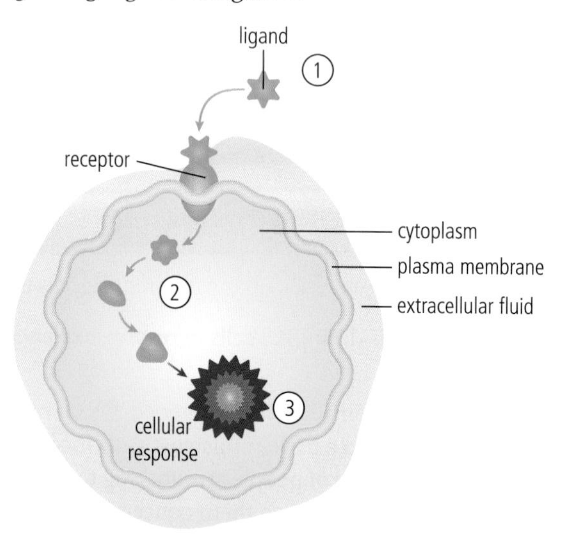

C2.1 Figure 1 An overview of cell signalling. First, the ligand binds with the receptor. Second, this binding begins the signal transduction pathway. Third, there is a cellular response to the ligand binding. The cell in the diagram is the target cell for the specific ligand involved.

C2.1.2 – Quorum sensing in bacteria

C2.1.2 – Cell signalling by bacteria in quorum sensing
Include the example of bioluminescence in the marine bacterium *Vibrio fischeri*.

Quorum sensing occurs in pathogenic (disease-causing) bacteria. In some cases quorum sensing means that bacteria do not produce harmful substances until there are sufficient bacteria to overwhelm the host's immune system and cause serious illness.

Quorum sensing is a mechanism by which bacteria can alter group behaviour depending on population density. It allows individual bacteria within colonies to coordinate their actions. This coordination is the result of the production of ligands known as **autoinducers**. As the number of bacterial cells increases, so does the concentration of the autoinducer. Typically, autoinducers are released into the environment by individual bacteria. When the concentration of the autoinducer reaches a threshold level, the gene expression by the entire population of bacteria is altered. Quorum sensing controls a variety of functions in bacteria, including the production of bioluminescence, sporulation, virulence and biofilm formation.

Quorum sensing was first discovered in the marine bacterium *Vibrio fischeri*. These bacteria can produce light through bioluminescence when autoinducer concentrations are high. The bacteria are found free-floating in oceans, and as a symbiont of marine animals, especially a type of squid known as the Hawaiian bobtail squid (*Euprymna scolopes*). Hatching squid capture the free *Vibrio* bacteria from the environment and house them in what is known as a light organ. Once inside the light organ, the bacteria progress through the following steps to produce light.

1. During their reproductive cycle, individual *V. fischeri* bacteria produce a quantity of autoinducer molecules.
2. The autoinducer molecules pass through their cell membrane and cell wall into the exterior environment.
3. Because the bacteria are reproducing, there is an increasing number of bacterial *Vibrio* cells.
4. With a higher population of *Vibrio* cells, there is a corresponding increase in the concentration of autoinducer molecules in the surroundings.
5. When the number of autoinducer molecules reaches a threshold level, they move into the bacterial cells and bind to a protein known as LuxR.
6. LuxR binds to a DNA binding site called a lux box.
7. The lux box, a section of DNA, is activated by this binding, and begins the production of a luminescent protein known as luciferase.

The light produced by the production of luciferase allows the squid to illuminate its underside to match the surrounding light from the Sun. This disguises the squid from predators.

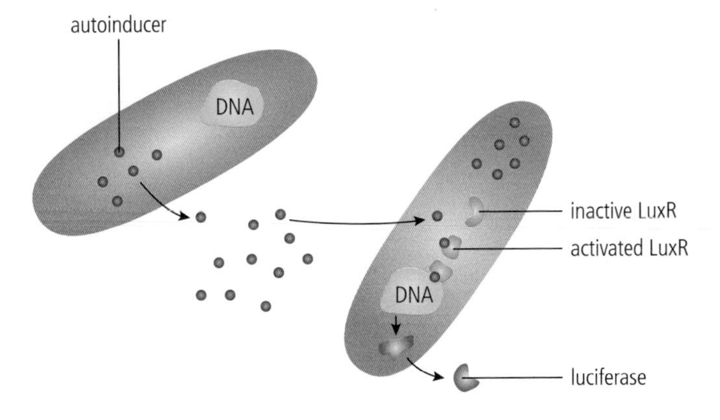

Quorum sensing in the bacterium *Vibrio fischeri*. When this bacterial type is present in the Hawaiian bobtail squid (*Euprymna scolopes*), the light produced by the protein luciferase provides protective camouflage for the squid.

It is a good idea to understand some of the key terms used in this section. Bioluminescence refers to the production of light by an organism. Sporulation is the production of a nearly dormant, inactive, form of a bacterium. Virulence refers to the severity of a disease, in this case caused by a bacterium. Biofilm formation occurs when organisms adhere to one another and to a surface of some type, such as bacteria forming dental plaque on teeth.

Symbiosis is when two different species interact in a unique way. The two species can interact with one another in a beneficial, neutral or detrimental way. The interaction can also affect just one or both species. In the case of the bobtail squid and the bacterium *Vibrio*, the symbiotic relationship is known as **mutualism** because both benefit from the relationship.

Challenge yourself

1. *Vibrio fischeri* only produces luciferase and glows when inside the light organ of the Hawaiian bobtail squid (*Euprymna scolopes*). When free-living in the ocean, they do not glow. Explain the reasons for this.

C2.1.3 – Functional categories of animal signalling chemicals

C2.1.3 – Hormones, neurotransmitters, cytokines and calcium ions as examples of functional categories of signalling chemicals in animals
Students should appreciate the differences between these categories.

Earlier we discussed cell signalling receptors. We mentioned that they are almost always proteins and can bind with specific signalling chemicals called ligands to initiate a signal transduction pathway. We are going to examine four functional categories of animal ligands:

- hormones
- neurotransmitters
- cytokines
- calcium ions.

Hormones are signalling chemicals secreted by specialized endocrine cells; they are carried through the circulation system to act on target cells at distant body sites. A certain hormone can only bind with a specific receptor on a target cell. Therefore, not all cells will respond to a hormone ligand. A classic example is the hormone oestradiol, which is produced by the ovary in animals. This hormone stimulates the development and maintenance of the female reproductive system and secondary sex characteristics.

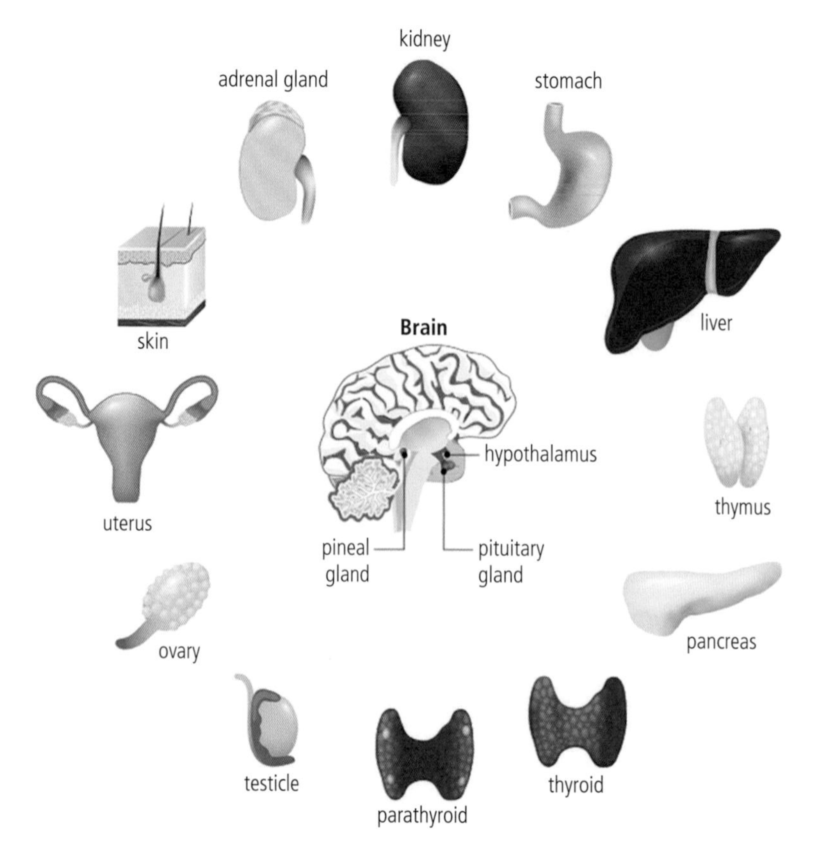

Parts of the human endocrine system. Each of these parts produces ligands known as hormones.

Neurotransmitters are signalling chemicals released from one cell that act on neighbouring target cells. A good example of this occurs at the synapse of nerve cells. A synapse is part of the nervous system in animals. It is a junction between two nerve cells.

When an electrical impulse travels along a nerve cell and reaches a synapse, it triggers the release of a neurotransmitter that moves into the gap that exists between the two cells at the junction. The neurotransmitter then binds to the receptors on the receiving cell, causing an electrical impulse in this nerve cell. These neurotransmitters act locally in their signalling function.

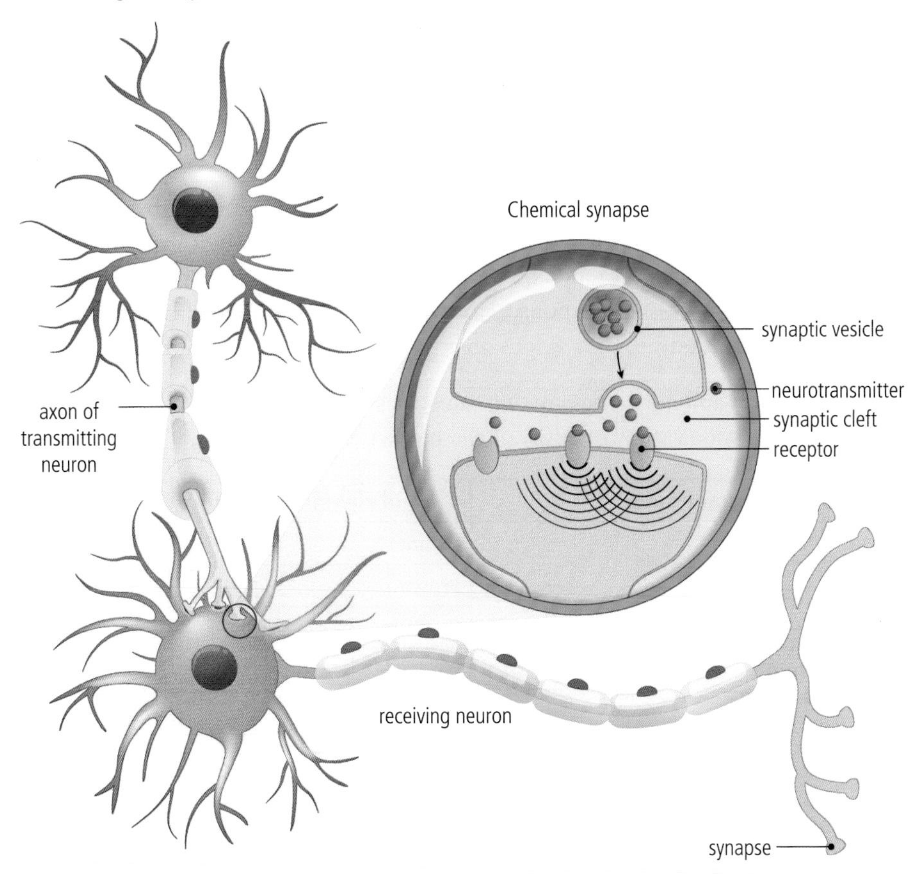

Neuron communication involving neurotransmitters. A neuron is a nerve cell with a unique structure. Notice the gap, called the synaptic cleft, between the two neurons in the enlarged section. The neurotransmitter moves from one neuron to another across this cleft.

Cytokines are glycoproteins that act as messengers between cells. There are more than 50 different cytokines, but most increase cell reproduction rate and have inflammatory actions. When a cytokine binds to a receptor on the cell membrane a signal is transmitted to the inside of the cell and many proteins within the cytoplasm are activated. This is a cascade reaction, where one protein causes the activation of another protein.

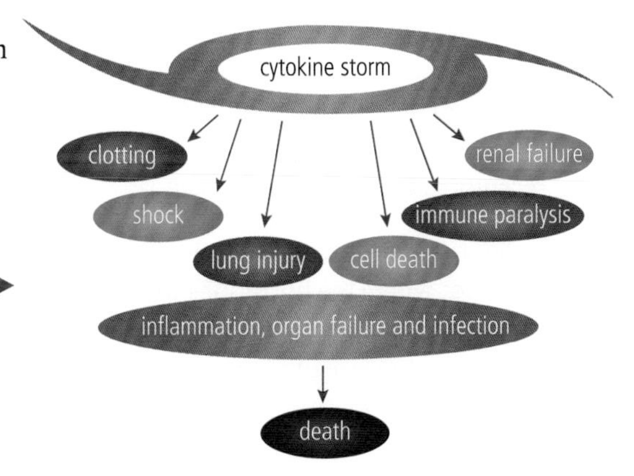

Cytokine storm. As shown in this diagram, none of the results of a cytokine storm is good. It is often difficult to stop the cascade of reactions that leads to such a storm within an organism.

Many complications attributed to COVID-19 involved what is known as a cytokine storm. The viral infection caused the immune system to flood the bloodstream with cytokines, which in this case acted in an inflammatory role. The resulting effects of this inflammation included tissue death and organ damage.

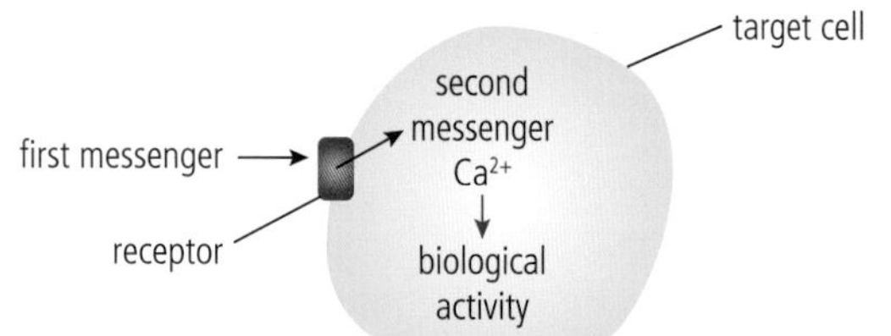

▲ Calcium ions acting as a second messenger. The calcium ions act within the cell to support biological activities.

Calcium ions are another category of animal ligand. Calcium ions are usually involved in cell signalling as one of the steps in a signal transduction pathway. They are often called **second messengers**. However, in the muscles of animals, calcium ions are involved in the initial stages of muscle contraction. Additionally, in the endocrine system, calcium ions are involved in the control of hormone secretion.

Plant growth and development are regulated by a group of signalling chemicals called plant hormones. The common groups of plant hormones are auxins, gibberellins, cytokinins, abscisic acid and ethylene. It is essential we learn more about these plant ligands so that we can try and increase food production to feed an ever-increasing world population. Research has found that seed germination is enhanced by hormones produced by certain types of soil bacteria. These hormones shorten the dormancy period for the seeds of many species, resulting in a shorter germination process. This enables greater plant growth and more world food production.

C2.1.4 – The diversity of signalling chemicals

C2.1.4 – Chemical diversity of hormones and neurotransmitters
Consider reasons for a wide range of chemical substances being used as signalling chemicals. Include amines, proteins and steroids as chemical groups of hormones. A range of substances can serve as neurotransmitters including amino acids, peptides, amines and nitrous oxide.

Neurotransmitters are signalling molecules that bind to receptors on nearby cells to cause a localized response. Neurotransmitters diffuse across a synaptic gap, a distance of approximately 20 nm. Hormones, on the other hand, are ligands carried in the bloodstream to cause a distant response. Hormones produced in the brain can cause an effect all over the body.

Neurotransmitters and hormones are two categories of signalling molecules. Within both categories a wide variety of different chemicals is used.

Type of signalling molecule	Class of chemical substance
Hormone	Amines
	Proteins
	Steroids
Neurotransmitter	Amino acids
	Peptides
	Amines
	Nitrous oxide

▲ Chemical diversity in signalling molecules

Both categories of ligand are essential to the life of an organism. Each group is defined by a *common function* rather than a *common chemical structure*.

Hormones have the general function of regulation. Many different life functions, including homeostasis, development and reproduction, are controlled by hormones. Hormones can have wide-reaching effects triggered by just small amounts of the signalling molecule. We stated earlier that hormones only cause a response in cells that have receptors able to bind with them.

So why do so many different chemical substances act as hormones? One major reason has to do with where the hormone binds to the receptor. Amines and proteins are hydrophilic, which makes it difficult for them to cross a membrane. They bind to receptors on the surface of the cell membrane. Steroid hormones are non-polar, which means that they are hydrophobic. Steroid hormones easily cross the cell membrane and bind with receptors located within the cell.

Neurotransmitters carry signals between neurons, and from neurons to other types of target cells. Most neurotransmitters are hydrophilic, which means they bind to receptors on the surface of the cell membranes of target cells. Some neurotransmitters, such as epinephrine, act as hormones as well. Epinephrine produced by adrenal glands helps regulate blood glucose by breaking down glycogen in muscle cells.

Nitrous oxide (NO) is responsible for signalling the dilation of blood vessels. Some heart conditions are treated with a medicine called nitro-glycerine. Nitro-glycerine is converted to nitrous oxide in the body, which dilates the coronary blood vessels, increasing blood flow to the heart muscle.

C2.1.5 – The range of effects of signalling molecules

C2.1.5 – Localized and distant effects of signalling molecules
Contrasts can be drawn between hormones transported by the blood system and neurotransmitters that diffuse across a synaptic gap.

Some of the reasons for chemical diversity in hormones and neurotransmitters are:

- the ligands in these two groups must act on many different target cells, so a variety of signalling chemicals makes it easier for them to act on different cells
- having different signalling chemicals means that there are many different receptors, which allows specificity
- some ligands must travel long distances in the body, while others act only on nearby cells
- different ligands work in different ways on target cells
- some signalling chemicals create an effect that persists over a long period of time, while others have very short-term effects.

It is important to remember that the chemical signalling processes that exist in organisms today have existed in some form since the appearance of the first cells on our planet. There are many differences in the chemical substances that form ligands. This is largely because remnants from the past are still present in organisms now. Many changes have accumulated in organisms over a very long period of time, resulting in their current communication and regulation systems. These systems work in all their complexity to maintain the existence of life today and in the future.

Neurotransmitters are signalling molecules that bind to receptors on nearby cells to cause a localized response. Hormones, on the other hand, are ligands carried in the bloodstream to cause a distant response.

C2.1.6 and C2.1.7 – Differences between transmembrane and intracellular receptors

C2.1.6 – Differences between transmembrane receptors in a plasma membrane and intracellular receptors in the cytoplasm or nucleus
Include distribution of hydrophilic or hydrophobic amino acids in the receptor and whether the signalling chemical penetrates the cell or remains outside.

C2.1.7 – Initiation of signal transduction pathways by receptors
Students should understand that the binding of a signalling chemical to a receptor sets off a sequence of responses within the cell.

Receptors are protein molecules that respond to a specific ligand. A ligand will bind to a receptor with a complementary shape, much like a substrate binds with an enzyme's active site. This binding of the ligand to the receptor produces a change in the shape of the receptor, which initiates a signal transduction pathway. Receptors are found in two locations:

- **transmembrane receptors** are embedded in the cell membrane
- **intracellular receptors** occur within the cell cytoplasm.

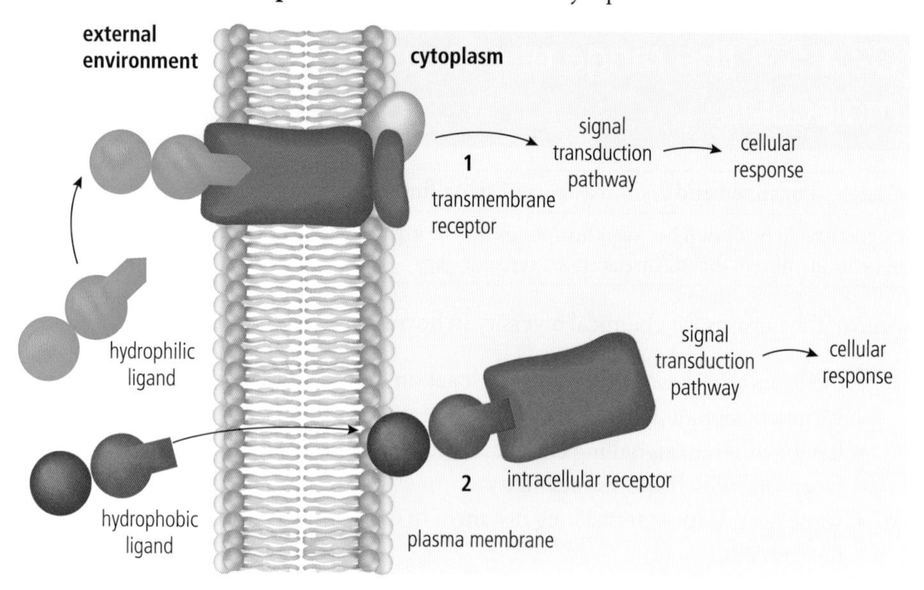

C2.1 Figure 2 The two receptor types. (1) A transmembrane receptor, which occurs in the cell membrane. (2) An intracellular receptor, which occurs inside the cell.

From Figure 2 you can see that hydrophilic ligands attach to transmembrane receptors because they cannot cross the cell membrane. Hydrophobic ligands can cross the cell membrane and they bind with intracellular receptors.

A protein domain is a portion of a protein's polypeptide chain that is self-stabilizing and that folds independently from the rest. Domains are the structural and functional units of proteins. Receptors possess both hydrophobic and hydrophilic domains.

Transmembrane receptors have cytoplasmic and extracellular domains, as shown in Figure 2. This structure allows them to interact with molecules both outside and inside the cell. Hydrophobic amino acids are found in the interior of the cell, while hydrophilic amino acids are found outside the membrane, in contact with the aqueous environment.

Intracellular receptors are located within the cytoplasm or nucleus. Lipid-soluble, hydrophobic, signalling chemicals bind with these receptors because only hydrophobic or very small molecules can pass through the cell membrane.

As also shown in Figure 2, both transmembrane and intracellular receptors can initiate a signal transduction pathway, with a resulting cellular response after binding with the specific ligand.

Types of transmembrane receptors

With transmembrane receptors it is important to remember that it is the receptor and *not* the ligand that is responsible for information crossing the membrane. There are three general transmembrane receptor types based on their structure and function.

Type	Structure	Function
Chemically gated ion channel receptors	Multi-pass protein with a central pore	Possess gates that open and close
Enzymatic receptors	Single-pass protein	Bind ligands extracellularly, activate enzymes intracellularly
G protein-coupled receptors (GPCRs)	Multi-pass protein with intracellular binding site for G protein	Bind with ligands that activate the G protein

C2.1 Table 1 Types of transmembrane receptor

C2.1.8 – Acetylcholine and changes to membrane potential

C2.1.8 – Transmembrane receptors for neurotransmitters and changes to membrane potential

Use the acetylcholine receptor as an example. Binding to a receptor causes the opening of an ion channel in the receptor that allows positively charged ions to diffuse into the cell. This changes the voltage across the plasma membrane, which may cause other changes

The acetylcholine receptor is an example of a chemically gated ion channel receptor. This receptor is said to be a **multi-pass protein** because it is composed of many domains that thread back and forth across the cell membrane several times. In the centre of this protein is a pore large enough for ions to pass through. The channel is said to be chemically gated because it is open only when the neurotransmitter acetylcholine binds to it. When the acetylcholine is attached, the ion channel opens and positively charged ions diffuse into the cell. This influx of positive ions changes the voltage inside the cell membrane, which causes a cellular response. The acetylcholine receptors in muscle cell membranes allow sodium ions (Na^+) to pass into the cell, resulting in contraction. Study Figure 3.

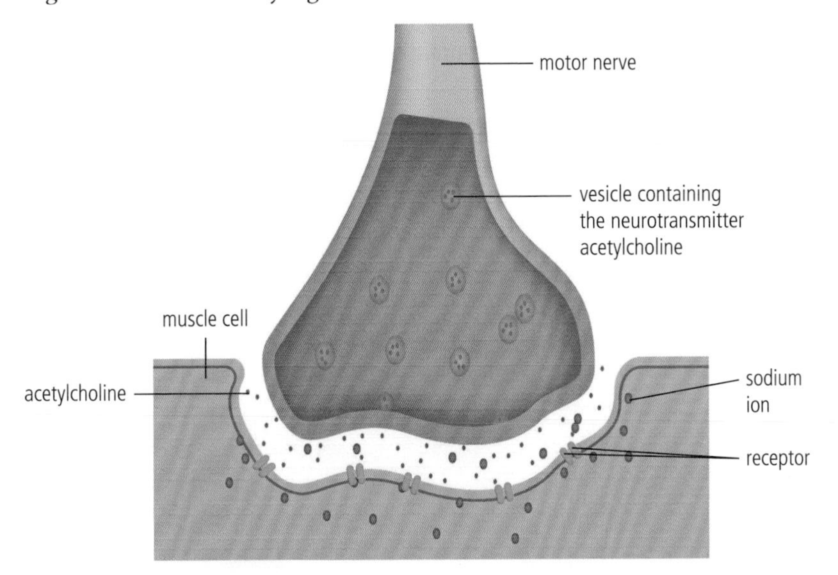

C2.1 Figure 3 A neuromuscular junction showing chemically gated ion channel receptors on the muscle cell membrane. When the molecules of acetylcholine (the ligand) bind with the receptors on the cell membrane, the gate of the receptors open, allowing sodium ions to enter the muscle cell. This changes the membrane potential of the muscle, causing a contraction.

C2.1.9 and C2.1.10 – G protein-coupled receptors

C2.1.9 – Transmembrane receptors that activate G protein
Students should understand how G protein-coupled receptors convey a signal into cells. They should appreciate that there are many such receptors in humans.

C2.1.10 – Mechanism of action of epinephrine (adrenalin) receptors
Include the roles of a G protein and cyclic AMP (cAMP) as the second messenger.
NOS: Students should be aware that naming conventions are an example of international cooperation in science for mutual benefit. Both "adrenaline" and "epinephrine" were coined by researchers and are based on production of the hormone by the adrenal gland; "adrenaline" comes from Latin *ad* = at and *ren* = kidney and "epinephrine" comes from old Greek *epi* = above and *nephros* = kidney, respectively. Unusually, these two terms persist in common use in different parts of the world

A G protein-coupled receptor is a type of transmembrane receptor that acts indirectly on enzymes or ion channels with the aid of a protein called G protein. This protein is called G protein because it binds to the nucleotide known as guanosine triphosphate (GTP). GTP attaches to G protein and results in its activation when it loses its third phosphate by hydrolysis. When GTP loses a phosphate it is then called GDP or guanosine diphosphate.

The sequence of actions with the receptor is as follows:

ligand binds to receptor → receptor activated → G protein activated → effector protein activated.

This sequence of actions allows the signal to be conveyed into the cell. The receptors are known as G protein-coupled receptors (GPCR) because of their association (coupling) with the G protein. The signal would not be carried into the cell if this intermediary protein was not involved.

GPCRs are the largest type of signal receptor in animals. They can bind many different ligands, such as peptides, proteins, ions and lipids. These receptors are a unique group because of their characteristic structure that anchors them to the cell membrane. In Figure 4, notice the transmembrane domain of the receptor showing multiple (seven) passes through the cell membrane.

C2.1 Figure 4 Activity at a G protein-coupled receptor. Note that the G protein is intracellular. When a ligand binds to the receptor outside the cell, the G protein is activated by phosphorylation. The activated G protein then activates an enzyme or an ion channel, bringing about the specific intracellular reaction. ▼

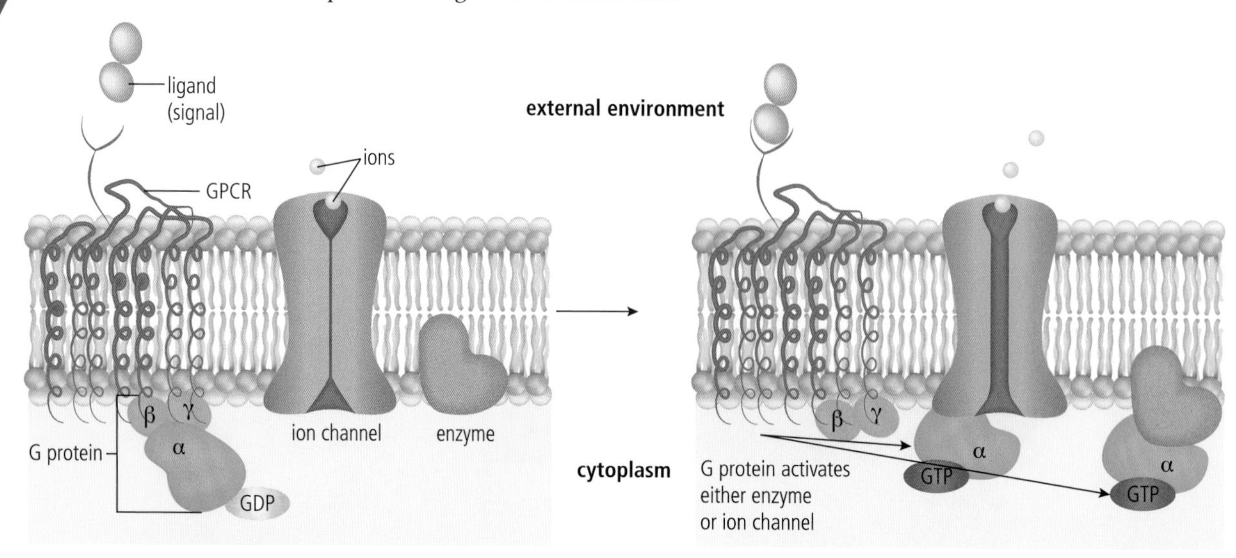

Many GPCRs utilize a second messenger in the pathway to produce a cellular response. An example of this involves the action of epinephrine (adrenaline) receptors. Study Figure 5.

The **cyclic AMP (cAMP)** that acts as a second messenger in the pathway shown in Figure 5 is produced when the enzyme adenylyl cyclase acts on ATP. This continues the cascade involving multiple phosphorylation reactions until the final cell effect is attained.

The human liver has many epinephrine receptors. Epinephrine is released in the "fight-or-flight" response by the adrenal glands, which sit on the top of each kidney. When epinephrine binds to the appropriate receptors of the liver, large amounts of glucose are released into the bloodstream so that a potential life-saving response can occur.

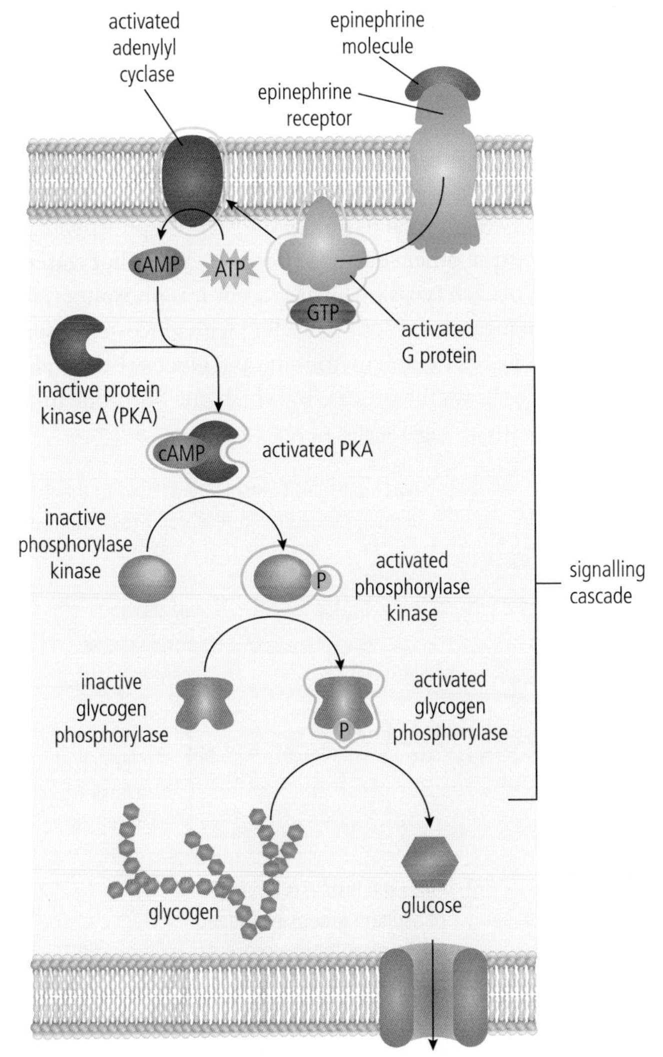

C2.1 Figure 5 The signalling cascade initiated by the ligand epinephrine (adrenaline). When epinephrine binds to the appropriate receptor on the cell membrane, a G protein is activated. This activated G protein then begins a signalling cascade that involves cyclic AMP (cAMP) as the second messenger. This signalling cascade depends on phosphorylation to activate each step. The final cellular response shown is the breakdown of glycogen (the storage form of glucose) so that individual glucose molecules are available to the organism for energy.

Nature of Science

Naming conventions in science are an example of international cooperation, and are for mutual benefit. These conventions allow accurate and meaningful communication between scientists from different countries. Both the terms "adrenaline" and "epinephrine" were coined by researchers and are based on where the hormone is produced, by the adrenal gland which is above (epi-) the kidney (nephrons). Unusually, both terms are commonly used by scientists in different parts of the world.

C2.1.11 – Tyrosine kinase activity

C2.1.11 – Transmembrane receptors with tyrosine kinase activity
Use the protein hormone insulin as an example. Limit this to binding of insulin to a receptor in the plasma membrane, causing phosphorylation of tyrosine inside a cell. This leads to a sequence of reactions ending with movement of vesicles containing glucose transporters to the plasma membrane.

Many transmembrane receptors work as enzymes or are directly linked to enzymes. A good example of this type of receptor is tyrosine kinase. Like other enzymatic receptors, it is a single-pass protein (see Table 1) and consists of three domains, the extracellular domain, the transmembrane domain and the intracellular domain. This receptor occurs in pairs. The transmembrane domain passes through a cell membrane only once. The extracellular domain binds with the hormone called insulin (the ligand). When insulin binds with tyrosine kinase, the intracellular domain acts as a **kinase.** A kinase is an enzyme that catalyses the transfer of a phosphate group from ATP to another substance (the other substance becomes **phosphorylated**). In this case the intracellular domain of each tyrosine kinase receptor is phosphorylated inside the cell. This leads to a sequence of reactions ending with glucose transporter vesicles such as Glut-4 moving towards the cell membrane and glucose being released into the bloodstream. Figure 6 shows the process by which one hormone molecule can produce a large response involving a series of reactions.

C2.1 Figure 6 Insulin and the tyrosine kinase receptor. Note that there are two receptors side by side. When the ligand binds to the extracellular domain of each tyrosine kinase receptor, phosphorylation occurs on the intracellular portion/domain of each receptor. This creates a series of many sequential reactions ending with the response of glucose transporter vesicles moving towards the cell membrane. The phosphorylated pair of receptors work together to produce the necessarily amplified reactions to produce the movement of the transporter vesicles.

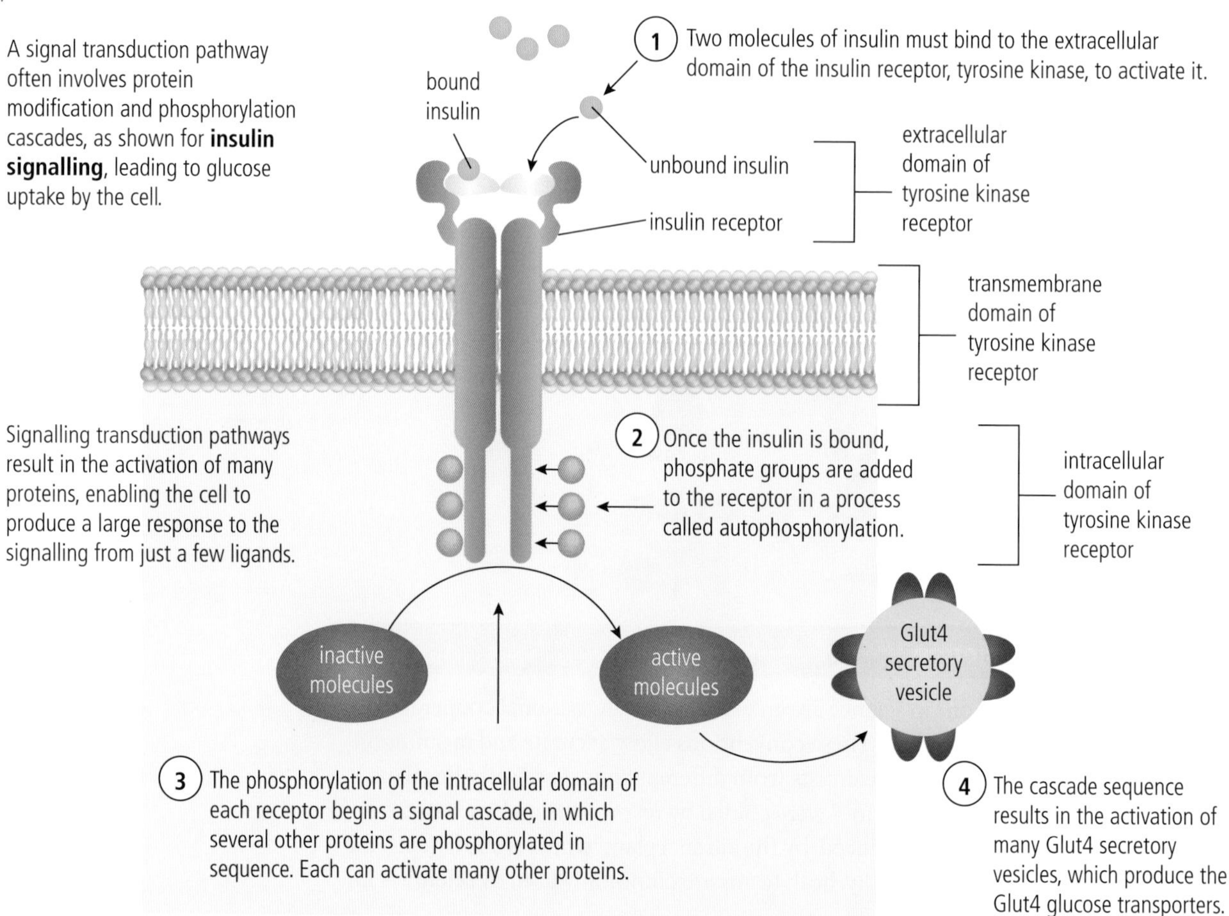

C2.1.12 – Intracellular receptors and gene expression

C2.1.12 – Intracellular receptors that affect gene expression

Use the steroid hormones oestradiol, progesterone and testosterone as examples. Students should understand that the signalling chemical binds to a site on a receptor, activating it. The activated receptor binds to specific DNA sequences to promote gene transcription.

Most cell receptors are transmembrane receptors, and they bind with hydrophilic ligands. Intracellular receptors bind hydrophobic ligands and are located inside the cell membrane in the cytoplasm or nucleus. Steroids are examples of hydrophobic ligands, and they can cross the cell membrane and enter the cytoplasm of target cells. When a steroid molecule binds with a receptor inside the cell membrane it forms a **receptor–signal complex**. This complex then moves into the nucleus, where it binds to the DNA. Binding with DNA results in a complex that affects **transcription** within the nucleus. Transcription is the first step in protein synthesis, and is discussed in detail in Chapter D1.2. Common steroid ligands include the sex hormones oestradiol, progesterone and testosterone.

Figure 7 illustrates the general action of a steroid binding with an intracellular receptor.

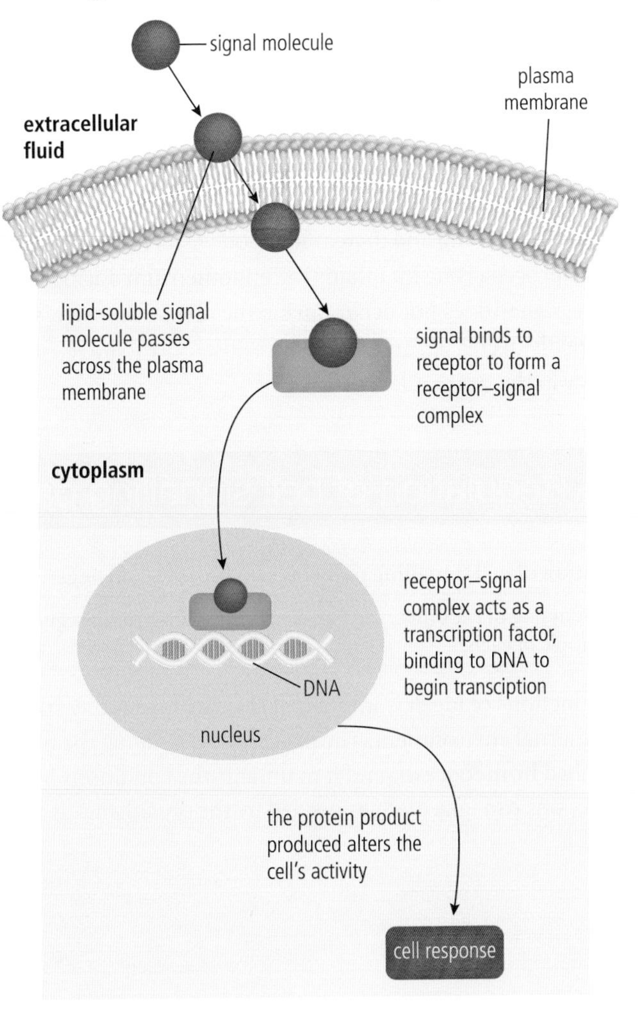

C2.1 Figure 7 A steroid ligand and intracellular receptor action pathway. Notice that the ligand passes freely into the cytoplasm. Once there it binds to an intracellular receptor, forming an activated complex. This activated receptor–signal complex then moves through the nuclear membrane and binds to DNA, where it affects transcription.

C2.1.13 – Effects of oestradiol and progesterone on target cells

C2.1.13 – Effects of the hormones oestradiol and progesterone on target cells
For oestradiol, limit to cells in the hypothalamus that secrete gonadotropin-releasing hormone. For progesterone, limit to cells in the endometrium.

Because they affect transcription, the steroid hormones **oestradiol** and **progesterone** directly affect gene expression. Oestradiol and progesterone are produced in the ovaries of human females and are responsible for the development of secondary sexual characteristics at puberty, such as breasts and body hair patterns. These same hormones also maintain the menstrual cycle in females and are important during pregnancy.

The **hypothalamus** of the brain influences heartbeat, blood pressure, appetite, body temperature and water balance. It also controls secretions of the **pituitary gland**. Oestradiol influences the hypothalamus of the brain, especially the secretion of gonadotropin-releasing hormone. The gonadotropin-releasing hormone stimulates the release of gonadotropic hormones from the pituitary gland, which in turn stimulate the ovaries to produce eggs, the female sex cells. These hormones also stimulate the release of more oestradiol from the ovaries.

What patterns exist in communication in biological systems?

Progesterone plays a significant role in preparing the wall of the uterus for pregnancy. The interior layer of the uterus is called the endometrium and it is essential for implantation of a fertilized egg and proper development of that fertilized egg. Progesterone causes thickening and increased blood vessel formation in the endometrium. The progesterone maintains the endometrium during the whole pregnancy. If a pregnancy does not occur during the menstrual cycle, secretion of progesterone is greatly decreased and the lining is lost, only to be re-established for a possible pregnancy in the next cycle.

C2.1.14 – Regulation of cell signalling pathways

C2.1.14 – Regulation of cell signalling pathways by positive and negative feedback
Limit to an understanding of the difference between these two forms of regulation and a brief outline of one example of each.

To maintain the functions of life, it is important that organisms have the means to regulate their internal environment. The process of regulating the internal environment is called **homeostasis**, and it maintains the conditions inside the body so that chemical reactions can take place as needed by the organism.

To maintain the proper internal environment, there must be constant monitoring of the environment so that any necessary adjustments can take place. These changes must be coordinated throughout the organism. To accomplish this, organisms possess feedback systems. There are two general types of feedback systems:

- negative feedback systems
- positive feedback systems.

Negative feedback systems prevent fluctuations outside a set range. Positive feedback systems serve to reinforce or amplify a response, often to achieve a particular response. Positive feedback is not stable, so it is rarer in biological systems than negative feedback.

Negative feedback: regulation of blood glucose

The regulation of blood glucose in humans involves two hormones: insulin and glucagon. These hormones are produced in the pancreas by specialized cells known as islet cells. Beta islet cells produce and secrete insulin. Alpha islet cells produce and secrete glucagon. These two hormones act in opposite ways to control blood glucose levels. Figure 8 illustrates the regulation of blood glucose.

Type I diabetes is a disease affecting more than 9 million people in the world today, according to the World Health Organization. Diabetes is caused by problems with the control of blood sugar levels. Symptoms of the disease include excessive excretion of urine, thirst, constant hunger, weight loss, vision changes and fatigue. Diabetes treatments often involve administering insulin. However, this treatment is expensive and often unobtainable in some parts of the world.

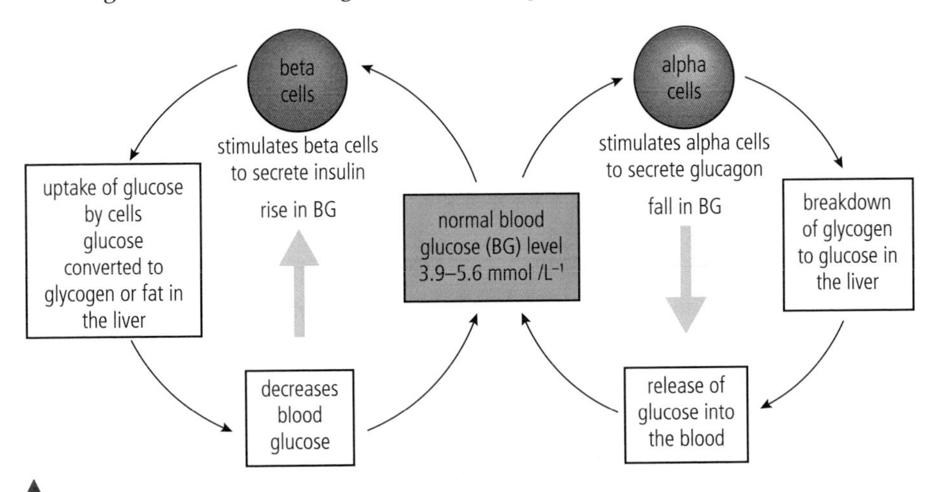

C2.1 Figure 8 The control of human blood glucose levels. The alpha cells of the pancreas produce glucagon when blood glucose levels are low, so that glycogen in the liver can be hydrolysed to form glucose units, which, when released into the blood, elevates glucose levels. Insulin is released from the beta cells of the pancreas when blood glucose levels are high, resulting in an uptake of glucose into the liver to form fat or glycogen. This decreases blood glucose levels.

In what ways is negative feedback evident at all levels of biological organization?

Positive feedback: fever

Positive feedback *increases* a response to achieve a particular result. Examples of this include childbirth, blood clotting and fruit ripening. A positive feedback mechanism will stop when the right effect has been achieved. One example of positive feedback is during a fever, as shown in Figure 9.

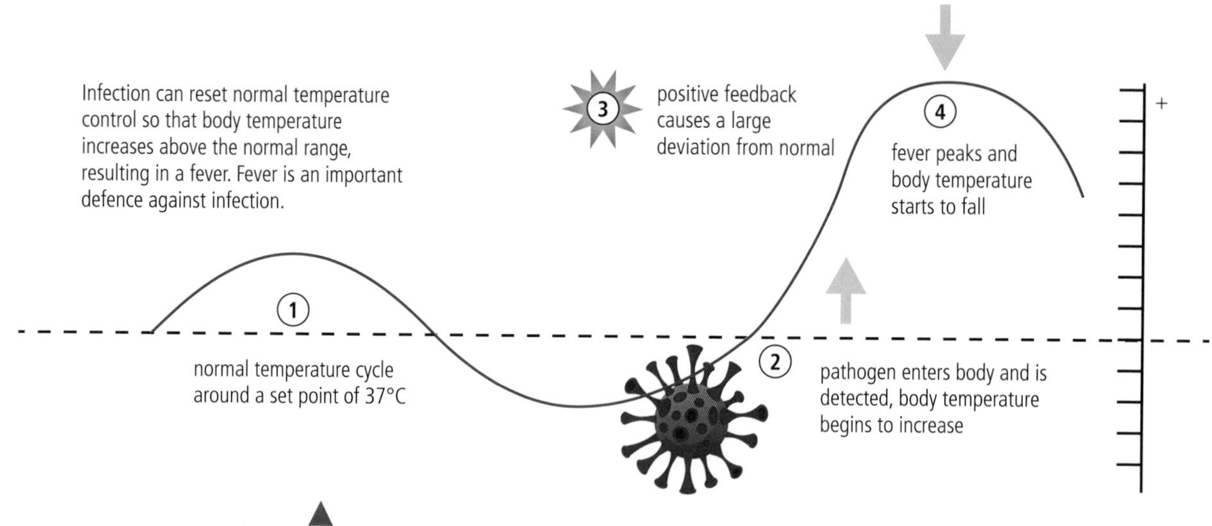

C2.1 Figure 9 Positive feedback and an individual's response to a fever. The aim of this positive feedback system is the destruction of the pathogen, the disease-causing organism. The feedback causes an increase in the individual's body temperature, which can kill the pathogen. However, a problem with this mechanism is that it can result in a body temperature that is also too high for the individual, causing damage to the host as well as the pathogen.

Guiding Question revisited

How do cells distinguish between the many different signals that they receive?

In this chapter we have discussed how:

- receptors are proteins that have binding sites for specific signalling chemicals
- transmembrane receptors are bound to the cell membrane, while intracellular receptors are found in the cytoplasm or the nucleus of the cell
- hormones, neurotransmitters, cytokines and calcium ions are examples of functional categories of cell chemical signals
- receptors only bind with specific signalling chemicals, to initiate a series of reactions within the cell that will bring about a cell effect
- hydrophilic cell signals bind with transmembrane receptors because they cannot pass through the cell membrane
- hydrophobic cell signals such as hormones pass through cell membranes and interact with receptors that can then control transcription
- quorum sensing involves autoinducers that allow communication among bacterial cells in a region, for example bioluminescence in the bacteria *Vibrio fischeri* is a biological effect controlled by quorum sensing.

SKILLS

Chemical signaling. Full details of how to carry out this activity are available in the eBook.

Guiding Question revisited

What interactions occur inside animal cells in response to chemical signals?

In this chapter we have learned that:

- signal conduction pathways are the result of a ligand binding to a receptor, and a cascade of responses is then initiated within the cell
- feedback systems exist in the cell so that homeostasis can be maintained

- feedback systems can be negative or positive, with negative feedback systems being far more common
- neurotransmitters are cell signals that create a change in cell voltage to bring about a cell effect
- cell signals can utilize second messengers in their signal conduction pathways, for example one common second messenger is cAMP
- transmembrane receptors can change the membrane potential because they allow ion channels to open within the membrane.
- transmembrane receptors can activate G proteins, which then interact with enzymes within the cell to produce second messengers
- insulin can bind with specific transmembrane receptors to move glucose transporter vesicles to the plasma membrane
- specific steroid hormones can bind to transmembrane receptors to ultimately promote gene expression.

Exercises

Q1. Describe where hydrophobic and hydrophilic ligands bind with their specific receptors.

Q2. What general type of ligands operate at neuromuscular junctions?

Q3. Describe the effects of hormone ligands.

Q4. What must happen to the kinase domain of tyrosine kinase before a signal transduction pathway can be initiated?

Q5. Describe how a hormone causes DNA transcription in a cell.

Q6. What does a protein kinase do?

Q7. Which of the following is not a transmembrane receptor?

A Enzymatic receptor.

B Ion channel receptor.

C Steroid hormone receptor.

D G protein-coupled receptor.

HL end

C2.2 – Neural signalling

Guiding Questions

How are electrical signals generated and moved within neurons?

How can neurons interact with other cells?

Very few people think of the human body in terms of electrical pathways. However, life would not occur without them. Chemicals, especially charged atoms known as ions, are constantly on the move within our systems. The movement of charged particles into and out of neurons results in electrical impulses, and these impulses can then be propagated through the nerve cells.

To maintain the functions of life, our body must constantly monitor and evaluate changes in our environment, both internally and externally, so that chemical reactions and electrical pathways can persist. Once changes are detected, messages (impulses) must be conducted to the spinal cord and/or brain so that the appropriate responses can occur. A complex system of nerve cells known as neurons interacts with all areas of the body to allow this control and communication. Neurons connect with one another at junctions known as synapses so that electrical pathways can connect all cells. Neurotransmitters are signalling chemicals that allow impulse transmission across synapses. Neurotransmitters are also used at the junctions between nerves and muscles.

C2.2.1 – The role of neurons

C2.2.1 – Neurons as cells within the nervous system that carry electrical impulses
Students should understand that cytoplasm and a nucleus form the cell body of a neuron, with elongated nerve fibres of varying length projecting from it. An axon is a long single fibre. Dendrites are multiple shorter fibres. Electrical impulses are conducted along these fibres.

The brain and spinal cord make up the **central nervous system** (**CNS**). These two structures receive sensory information from various receptors around the body, and then interpret and process that sensory information. If a response is needed, some portion of the brain or spinal cord initiates that response.

The cells that carry this information are called **neurons**. **Sensory neurons** carry information to the CNS, and **motor neurons** carry response information to muscles.

Together, the sensory neurons and motor neurons make up the **peripheral nerves**. A neuron is an individual cell that carries electrical impulses from one point in the body to another, and does so very quickly. When many individual neurons group together into a single structure, that structure is called a **nerve**. You can think of a nerve as being like a large cable made up of many individual, smaller cables. Each smaller cable within that large cable represents a single neuron. Study Figure 1.

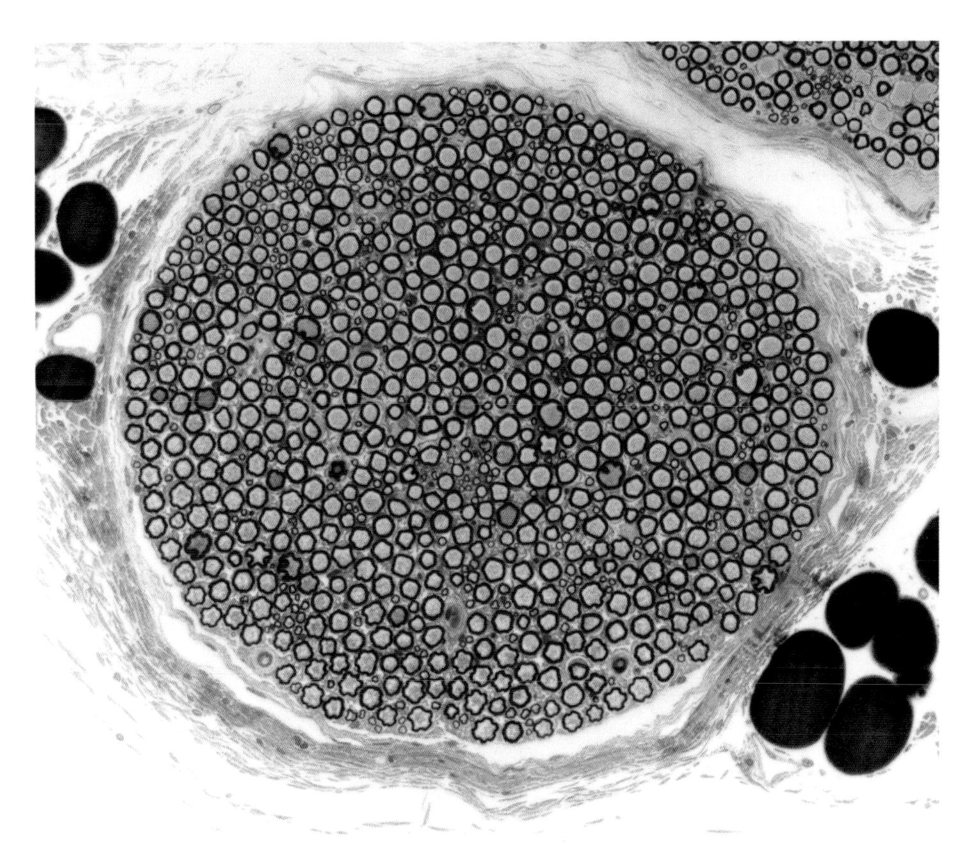

C2.2 Figure 1 A light microscope photograph of a section of a nerve. The very large circle is the entire nerve, and each small circle within it is one of the axons of a neuron contained within that nerve.

Neurons can be very long. In the human body, there are neurons that extend from the lower portion of the spinal cord all the way to the big toe: single cells that extend about 1 m! Of course, not all neurons are that long. In fact, some neurons are quite short.

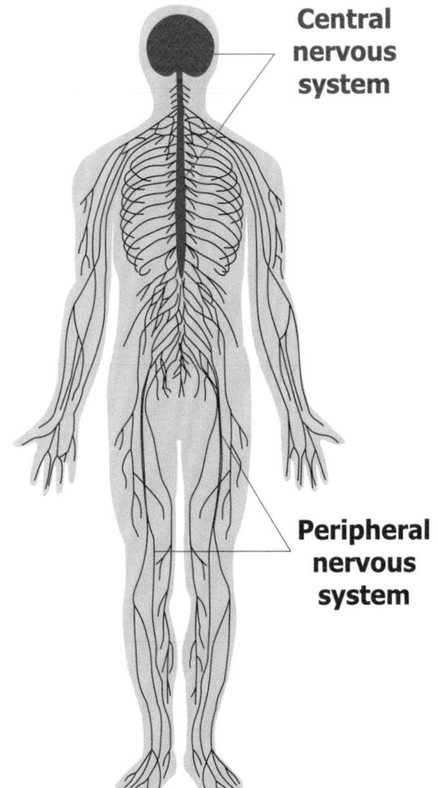

The central nervous system (CNS) consists of the brain and spinal cord. The peripheral nervous system (PNS) is made up of the nerves and branches that enter and leave the spinal cord and brainstem.

Neuron structure

A single neuron is made up of **dendrites**, a **cell body** and an **axon**. Axons are long, single fibres; dendrites are multiple, shorter fibres. At the end of the axon are **synaptic terminal buttons**, which release chemicals called **neurotransmitters** that continue the impulse chemically to the next neuron(s) or a muscle. An impulse is always carried from the dendrite end of a neuron along the membrane of the cell body down the axon, and results in a release of a neurotransmitter.

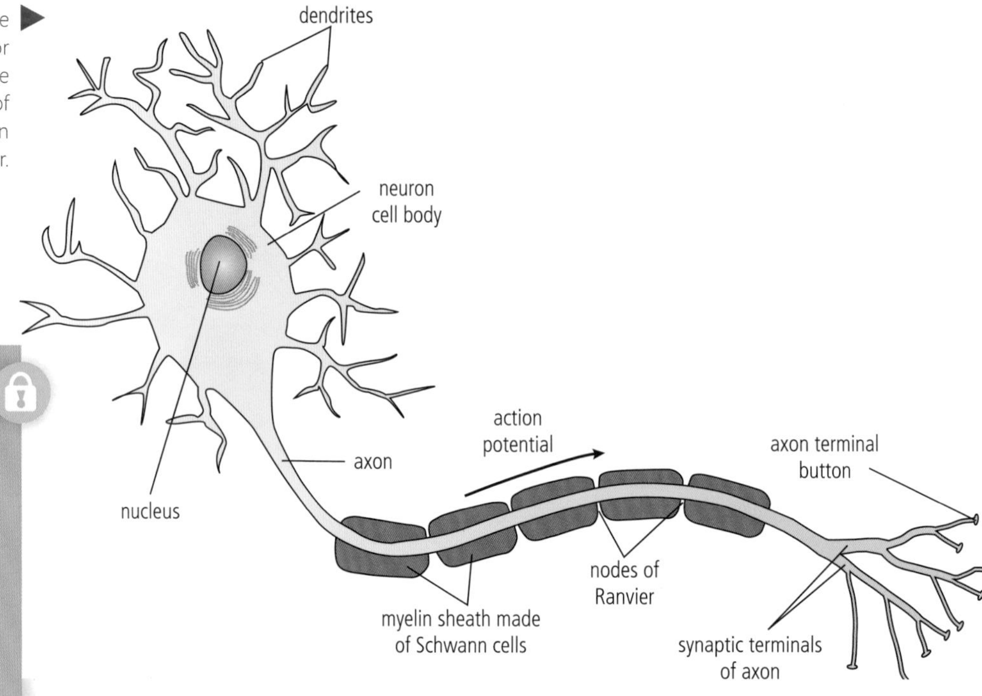

C2.2 Figure 2 The structure of an individual motor neuron. The function of the myelin sheath and nodes of Ranvier are discussed later in this chapter.

An axon is a relatively long single fibre. Dendrites are multiple shorter fibres. The cell body of the neuron connects the dendrites with the axon and contains the organelles to keep the neuron alive and functioning. Electrical impulses are conducted along these fibres.

Figure 2 shows an **action potential** moving from the cell body towards the axon terminal buttons. Action potential is another name for a nerve impulse, and it is electrical in nature because it involves the movement of positively charged ions.

The term "nerve impulse" is misleading because it is not the nerve that carries the impulse: the individual neurons within the nerve are each capable of carrying the impulse. Because axons of neurons are often relatively long, it is convenient to think of the axon as the conductor of a neuron impulse. The axons of neurons in some organisms, including humans, have surrounding membranes that are collectively called the **myelin sheath**. We will talk more about the myelin sheath in the following sections. However, to learn how a neuron conducts an impulse, we will first discuss an axon without a myelin sheath.

How is the structure of specialized cells related to function?

C2.2.2 and C2.2.3 – Generation and transmission of an impulse along a neuron

C2.2.2 – Generation of the resting potential by pumping to establish and maintain concentration gradients of sodium and potassium ions
Students should understand how energy from ATP drives the pumping of sodium and potassium ions in opposite directions across the plasma membrane of neurons. They should understand the concept of a membrane polarization and a membrane potential and also reasons that the resting potential is negative.

C2.2.3 – Nerve impulses as action potentials that are propagated along nerve fibres
Students should appreciate that a nerve impulse is electrical because it involves movement of positively charged ions.

An **action potential** is a sequence of events that allows an impulse to travel through a neuron. When a neuron is ready to send an impulse, it is **polarized** and has a **resting potential** across the membrane. The resting potential is created by the **active transport** of sodium and potassium ions in two different directions across the cell membrane of the neuron. Most of the sodium ions are actively transported out of the axon's cytoplasm and into the intercellular fluid, and most of the potassium ions are transported into the cytoplasm from outside the cell. This transport of sodium and potassium in opposite directions is the result of an active transport mechanism called the **sodium–potassium pump** (described in Chapter B2.1). The sodium–potassium pump works by transporting three sodium ions out of the cell for every two potassium ions transported in. It is an example of active transport, which means that adenosine triphosphate (ATP) is required to provide the energy for this pumping action to occur. In addition, there are negatively charged organic ions permanently located in the cytoplasm of the axon. The net result is that the outside of the cell becomes positively charged in relation to the inside. This potential difference in charge across the cell membrane is called the **membrane potential**.

The nerve impulse is the action potential that is propagated through the neurons, immediately followed by the sodium–potassium pump to restore the resting potential, in a wave-like action. This allows the neuron to propagate another action potential straight away.

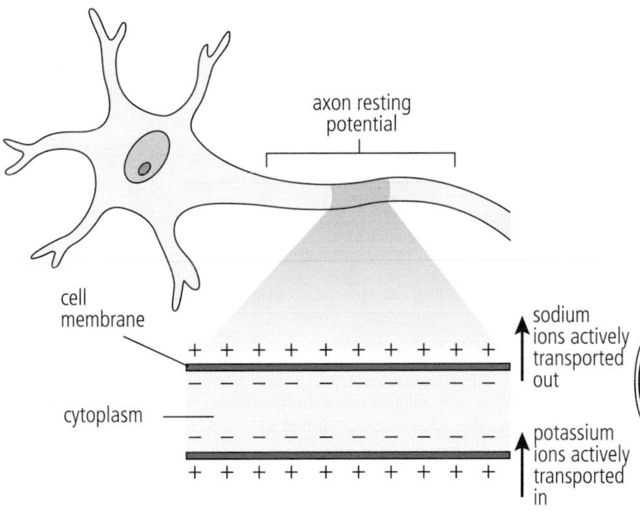

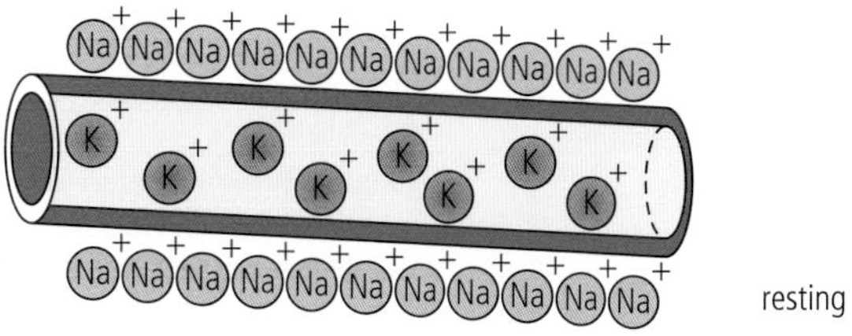

A neuron axon at resting potential. Think of the axon as a three-dimensional tube, and thus the ion movements shown are occurring all around the tube. Notice that there is a negative net charge inside the fibre relative to the outside because there are more positive sodium ions (Na^+) outside than potassium ions (K^+) inside. The diagram on the right is a model showing the actual number of K^+ and Na^+ ions inside and outside the neuron relative to one another.

C2.2.4 – The speed of nerve impulses

C2.2.4 – Variation in the speed of nerve impulses
Compare the speed of transmission in giant axons of squid and smaller non-myelinated nerve fibres. Also compare the speed in myelinated and non-myelinated fibres.
Application of skills: Students should be able to describe negative and positive correlations and apply correlation coefficients as a mathematical tool to determine the strength of these correlations. Students should also be able to apply the coefficient of determination (R^2) to evaluate the degree to which variation in the independent variable explains the variation in the dependent variable. For example, conduction speed of nerve impulses is negatively correlated with animal size, but positively correlated with axon diameter.

Several factors contribute to the speed at which an action potential moves along a nerve fibre. One factor is whether there is a myelin sheath surrounding the fibre or not. Figure 2 shows a neuron with a myelin sheath made of **Schwann cells**. Schwann cells wrap themselves around a nerve fibre (axon) and act as insulators: there is no ion movement in an axon covered by Schwann cells. Schwann cells are spaced evenly along an axon, with small gaps between them; these gaps are called **nodes of Ranvier**.

When myelin sheaths are present, action potentials skip from one node of Ranvier to the next as the impulse progresses along the axon towards the synaptic terminals. In this case, the action potential does not have to undergo the time-consuming and energy-expensive ion movements in the membrane underneath the myelin material. The myelin sheath prevents charge leakage through the membrane, thus acting as an insulator. The advantage of this type of impulse transmission is two-fold:

- the impulse travels much faster compared to an impulse in non-myelinated fibres, because the in/out movement of ions only occurs at the nodes of Ranvier
- less energy (in the form of ATP) is expended for the transmission of impulses, because the sodium–potassium pump is only working at the nodes.

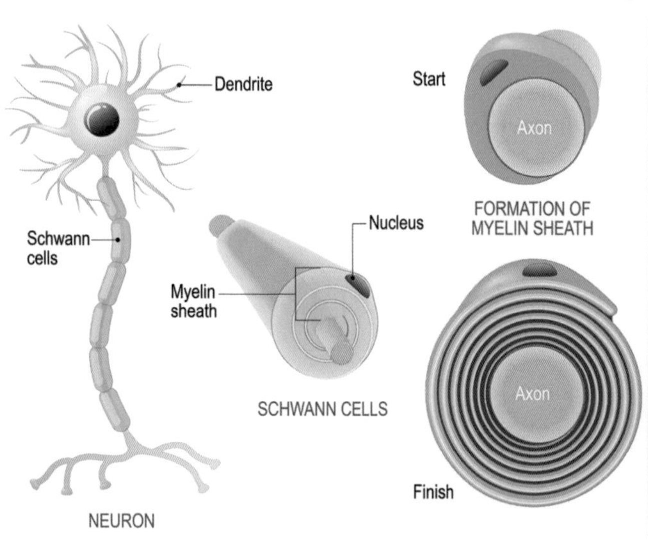

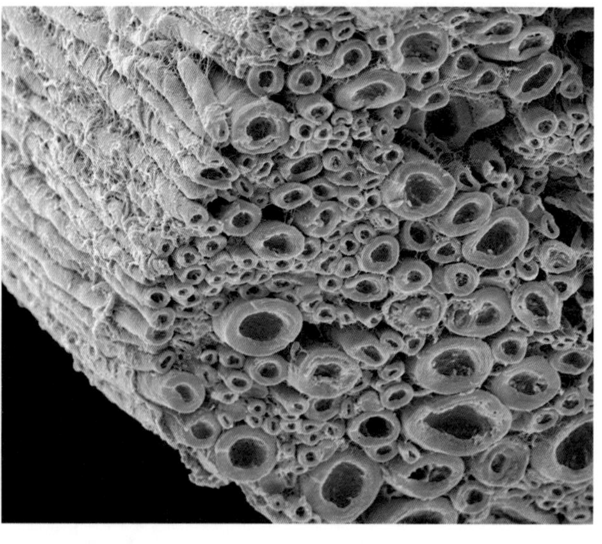

Neurons with myelinated axons and nodes of Ranvier. Action potentials only occur at these nodes. The photomicrograph is a false-colour scanning electron micrograph (SEM) of a nerve (bundle of neurons) with myelin sheaths. The darker (blue) colour indicates the axons, and the surrounding paler (yellow) colour indicates the myelin sheath of each axon.

Table 1 compares the conduction velocities of neurons in different species. Axons found in squid are particularly large and are unmyelinated.

C2.2 Table 1 Axon conduction velocities ▼

Axon source	Axon diameter / μm	Myelin present?	Conduction velocity / m s^{-1}
Giant squid axon	500	No	25
Human leg axon	20	Yes	120
Human skin temperature receptor axon	5	Yes	20
Human internal organ axon	1	No	2

Two observations can be made from Table 1:

- myelinated axons conduct action potentials faster than non-myelinated axons
- axons with a greater diameter have a faster transmission velocity than those with a smaller diameter.

Observations are being made all the time in science. From Table 1 we can see that axon diameter is positively correlated with impulse conduction velocity. Correlation means there is a statistical association between two variables. Causation means that a change in one variable actually causes a change in another variable. Remember, just because there appears to be a correlation does not mean there is causation.

A scatter graph plots two continuous variables on opposite axes and can be used to look for a correlation. If one variable increases as the second increases, this is a positive correlation. If one variable increases while the second decreases, this is a negative correlation.

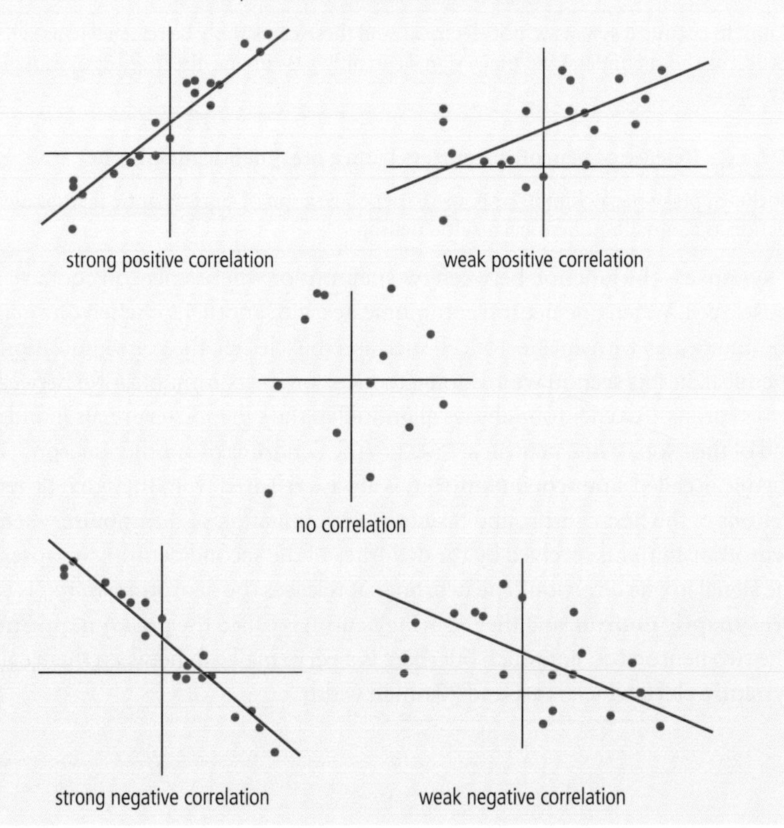

Plotting a line of best fit helps visualize the correlation. Statistical tests can also be used to analyse correlation. A correlation coefficient quantifies the strength of the linear relationship between two variables, and can be denoted using the symbol R. For a perfect positive correlation, $R = 1$. If there is a weak correlation, R is closer to 0.

Look at the figure on the previous page. If a data point is a long way from the line of best fit, then there is a lot of variance in that data point. In a data set with lots of variance, the correlation will be low. Squaring the correlation coefficient (to get R^2) provides the coefficient of determination: this value gives you some idea of the shared variance in the data set as a whole. If there is a smaller amount of variance, all the data points are closer to the line of best fit. This means that changes in the independent variable can be used to predict changes in the dependent variable. For example, if we studied the relationship between the number of hours of biology revision a student did and their exam score, we might find a correlation of 0.9. This very high value suggests that spending time on revision helps you gain a good exam score. In this case the coefficient of determination is $R^2 = 0.81$. To look at it another way, 81% of the variance in exam scores can be explained by the amount of revision done.

In the case of impulses in axons, scientists have found that the conduction velocity of a nerve impulse is negatively correlated with animal size. They also have shown a strong positive correlation between conduction velocity of a nerve impulse and axon diameter.

C2.2.5 and C2.2.6 – Synapses, neurotransmitters, and their actions

C2.2.5 – Synapses as junctions between neurons and between neurons and effector cells
Limit to chemical synapses, not electrical, and these can simply be referred to as synapses. Students should understand that a signal can only pass in one direction across a typical synapse.

C2.2.6 – Release of neurotransmitters from a presynaptic membrane
Include uptake of calcium in response to depolarization of a presynaptic membrane and its action as a signalling chemical inside a neuron.

Approximately 50 different neurotransmitters have been identified as active in the human brain. An imbalance of just one can result in conditions such as schizophrenia or severe depression. Many pharmaceuticals have been developed to treat these conditions, based on knowledge of how synapses and neurotransmitters work.

A **synapse** is the junction between two neurons or where a neuron contacts a muscle cell. When one neuron communicates with another, or when one neuron communicates with an effector cell such as a muscle cell, the communication is chemical. In this section we are going to focus on the communication between two neurons. Two neurons always align so that the synaptic terminals found at the end of the axon of one neuron are next to the dendrites of another neuron. The chemical, called a **neurotransmitter**, is always released from the synaptic terminal buttons of the first neuron, and results in a continuation of the impulse when the neurotransmitter is received by the dendrites of the second neuron. Synapses only pass the signal in one direction. The neuron that releases the neurotransmitter is called the **presynaptic neuron**, and the receiving neuron is called the **postsynaptic neuron**. The two neurons do not touch but there is a very small gap between them called the **synaptic cleft**, which is around 20 nm in width.

When an action potential reaches the area of the terminal buttons, it initiates the following sequence of events.

1. The action potential arrives at the terminal button, which results in depolarization of the presynaptic membrane and the uptake of calcium ions (Ca^{2+}) into the terminal buttons.
2. The calcium ions act as a signalling chemical, activating a pathway that moves vesicles containing the neurotransmitter through the cell. The vesicles then fuse with the presynaptic membrane.
3. The neurotransmitter is released from the fused vesicles into the synaptic cleft.
4. The neurotransmitter binds with a protein receptor on the postsynaptic neuron membrane.
5. This binding results in an ion channel opening, and sodium ions diffusing in through this channel.
6. This initiates the action potential to begin moving down the postsynaptic neuron, because it is now depolarized.
7. Any neurotransmitter that is bound to the protein receptor is released back into the synaptic cleft. The neurotransmitter in the synaptic cleft is degraded (broken into two or more fragments) by enzymes.
8. The ion channel in the postsynaptic membrane closes to sodium ions.
9. Neurotransmitter fragments diffuse back across the synaptic cleft, to be reassembled in the terminal buttons of the presynaptic neuron.

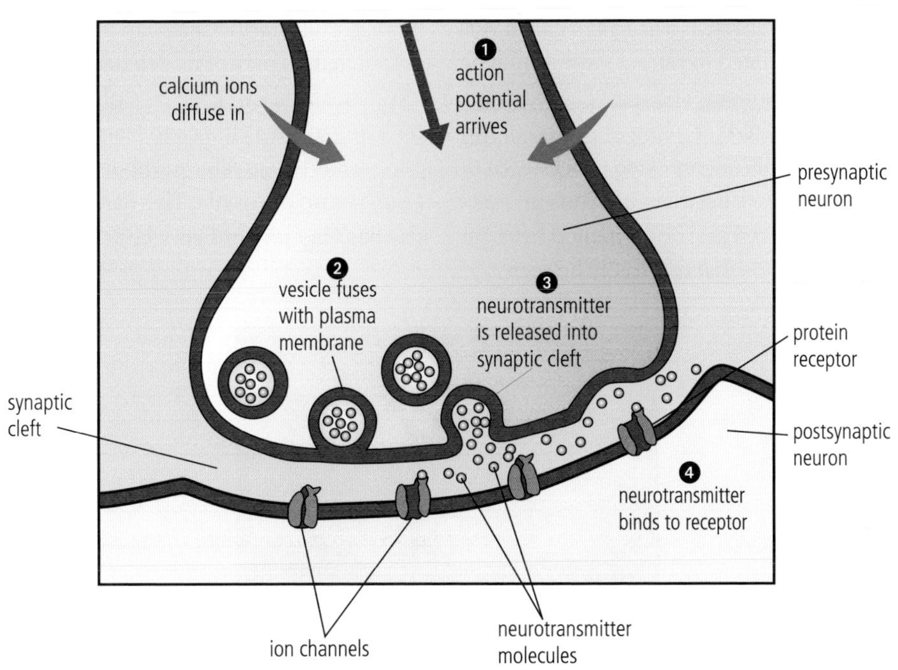

The mechanism of synaptic transmission.

C2.2.7 – Acetylcholine and the generation of a postsynaptic potential

C2.2.7 – Generation of an excitatory postsynaptic potential
Include diffusion of neurotransmitters across the synaptic cleft and binding to transmembrane receptors. Use acetylcholine as an example. Students should appreciate that this neurotransmitter exists in many types of synapse including neuromuscular junctions.

Acetylcholine is a very common neurotransmitter and is found at synapses between two neurons, and at synapses between a neuron and a muscle cell (a **neuromuscular junction**). When an impulse passes across a neuromuscular junction, the stimulus causes the muscle to contract. **Acetylcholinesterase** is an enzyme found in many synaptic clefts of muscles and nerves. It immediately breaks down the neurotransmitter acetylcholine into fragments, so the transmission of the action potential from the presynaptic membrane to the postsynaptic membrane occurs only once.

Nature of Science

The fields of psychology, chemistry, biology and medicine all combine to contribute to our knowledge of memory and learning. One of the many complications for research on memory and learning is the sheer complexity of the human brain. Often, complex biological systems are best studied by using simpler "models" that represent the more complex activity.

Biologists often use invertebrates, which have a simpler nervous system compared to humans and other vertebrates, as a model. One interesting invertebrate is a sea snail called *Aplysia parvula*. This marine snail can be stimulated to retract its siphon when it is touched, as part of its defence mechanism. The snail can learn from experience, and can keep its siphon protected for a longer period of time after being given a chance to learn. In addition, repeated touching of the siphon leads to a greater number of synapses between neurons in the very simple brain of *Aplysia*. This can be observed and documented because *Aplysia* has very few, but very large, neurons that can easily be seen

The freckled sea snail (*Aplysia parvula*).

HL

C2.2.8, C2.2.9 and C2.2.10 – Neuron depolarization and repolarization

C2.2.8 – Depolarization and repolarization during action potentials
Include the action of voltage-gated sodium and potassium channels and the need for a threshold potential to be reached for sodium channels to start to open.

C2.2.9 – Propagation of an action potential along a nerve fibre/axon as a result of local currents
Students should understand how diffusion of sodium ions both inside and outside an axon can cause the threshold potential to be reached.

C2.2.10 – Oscilloscope traces showing resting potentials and action potentials
Application of skills: Students should interpret the oscilloscope trace in relation to cellular events. The number of impulses per second can be measured.

Neuron depolarization

An action potential is often described as a self-propagating wave of ion movements in and out of the neuron membrane. The ions do not move along the length of the axon, but instead diffuse from outside the axon to the inside, and from inside the axon to the outside. The resting potential requires active transport (the sodium–potassium pump) to set up a concentration gradient of both sodium and potassium ions. As sodium ions are actively transported to the outside of the membrane, a concentration gradient is created. When the voltage changes across the membrane, voltage-gated sodium channels open and sodium diffuses in, down the concentration gradient. Refer to Chapter B2.1 for more detail concerning sodium channels. This diffusion of sodium ions results in the inside of the axon becoming temporarily positive in relation to the outside. It is a nearly instantaneous event that occurs in one area of an axon, and is known as **depolarization**. The depolarized area of the axon then initiates the next region of the axon to open the voltage-gated channels for sodium, and thus the action potential continues down the axon. This is the self-propagating part of an action potential: once you start an impulse at the dendrite end of a neuron, that action potential will self-propagate to the axon end of the cell, where the synaptic terminals are located. The action potential will then cause neurotransmitters to be released at the synapse.

Each action potential must reach a minimum **threshold potential** in order to be self-propagated. Usually this starts with the first receptor neuron that began the chain of events. A receptor neuron is a neuron that converts a physical stimulus of some kind into the first action potential. For example, some of the cells that make up the retina of your eyes are receptor cells. Each type of retinal cell needs a minimum physical stimulus to begin an impulse. For some retinal cells this is a minimum intensity of light. If that minimum intensity is not reached, no action potential begins. If the minimum is reached, an action potential is initiated and begins to self-propagate.

There is no such thing as a strong impulse or a weak impulse: if the minimum threshold for that type of receptor is reached, an impulse begins. This is because of the **all-or-nothing** action of depolarization. When a nerve impulse is being self-propagated along a neuron, each successive area of the neuron membrane reaches

Typically we are not aware of single impulses that reach our brain. If we sense a small amount of pressure on an area of our skin, it is because a few pressure receptors in that area have reached their threshold potential. If we feel a greater pressure, it is because the pressure has caused even more receptors in that area to reach their threshold potential. Often, after a period of time, you may no longer notice this pressure because of the ability of the brain to "filter" out information that it does not need, to prevent us from being overwhelmed. However, if many stimuli are detected, and impulses propagated, our brain continues to inform us of the impulse because of the seriousness of the originating stimulus.

its threshold and causes the next area of the membrane to reach its threshold. The intensity of a signal travelling down an axon is the result of the number of impulses generated per given period of time.

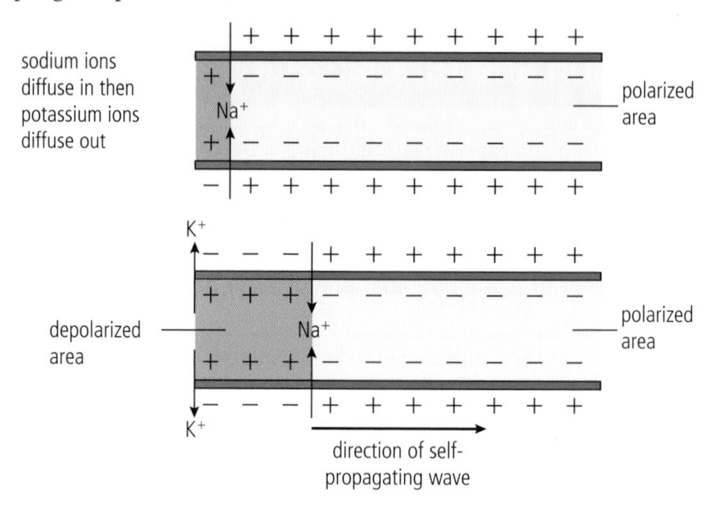

A neuron axon during and shortly after a depolarization.

Neuron repolarization

A neuron may be required to send dozens of action potentials during a very short period of time. When one area of an axon has opened a channel to allow sodium ions to diffuse in, that area cannot send another action potential until the resting potential has been reached again.

Depolarization takes place when sodium ions diffuse through the axon membrane from the outside to inside. This means that, for a very short period, both sodium ions and potassium ions are inside the axon. The inside of the membrane becomes positive relative to the outside. Immediately following a depolarization, voltage-gated potassium ions open to potassium ions and they diffuse out of the axon. This is the first step of **repolarization**. The inside of the axon becomes negatively charged relative to the outside. The sodium–potassium pump begins to actively transport sodium and potassium ions across the membrane. This entire series of events, beginning with potassium ions diffusing out of the localized area of the membrane, is called repolarization, and the whole process is necessary for the neuron to be ready to send another impulse.

The **refractory period** is the period of recovery after depolarization during which the neuron is unable to respond to additional stimulation. Ongoing research is being conducted involving electrical stimulation as an alternative to current interventions for diabetic peripheral neuropathy. Transcutaneous electrical nerve stimulators (TENS) send electrical pulses through the skin in the hope of relieving pain. This type of therapy is thought to stimulate the body's own natural pain killers, and possibly interferes with impulse transmission by producing an extended refractory period.

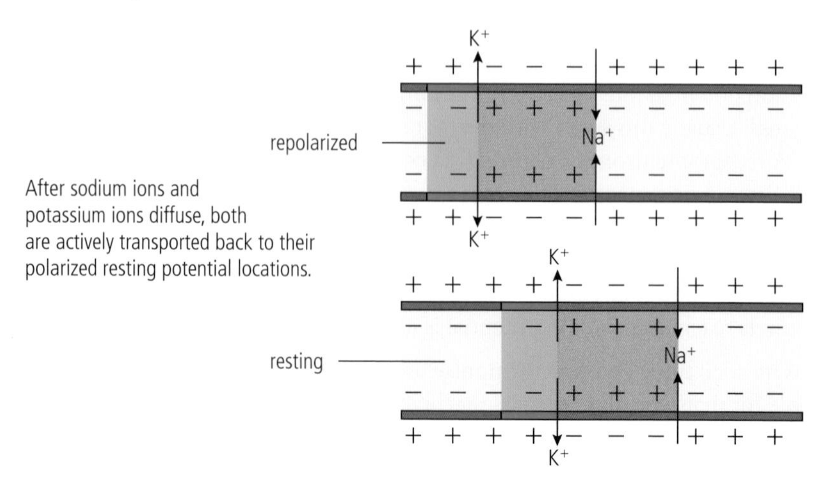

Repolarization of a neuron axon.

An oscilloscope is an electronic test instrument that graphically displays changing voltages. It is used in the study of axons to demonstrate the voltage changes that occur as part of an action potential. Study the figure, which shows the cellular events and the voltage versus time relationship of an action potential progression at a particular axon location. The middle diagram represents the oscilloscope tracing, while the peripheral diagrams illustrate what is actually happening in the neuron to obtain the tracing.

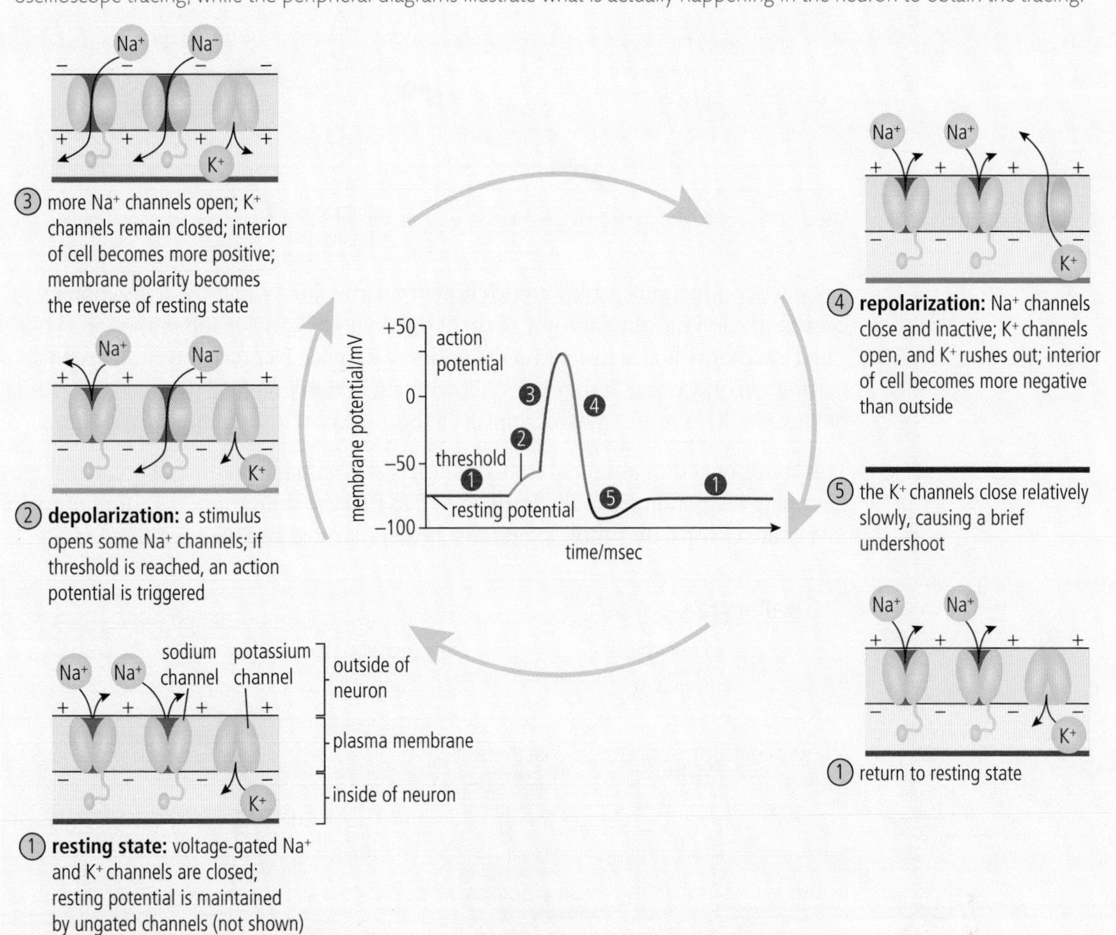

Key to understanding neural signalling is being able to explain precisely what is happening at a neuron to produce each major section of the oscilloscope tracing.

Neural signaling. Full details of how to carry out this activity are available in the eBook.

C2.2.11 – Saltatory conduction in myelinated axons

C2.2.11 – Saltatory conduction in myelinated fibres to achieve faster impulses
Students should understand that ion pumps and channels are clustered at nodes of Ranvier and that an action potential is propagated from node to node.

Saltatory conduction is the name given to the phenomenon whereby an action potential jumps from one node of Ranvier to the next as an impulse progresses along a myelinated axon. Specialized cells called **Schwann cells** produce myelin by wrapping themselves around an axon multiple times, creating layers of the same cell membrane. The Schwann cells are spaced evenly along any one axon, with small gaps between them called **nodes of Ranvier**.

Figure 3 illustrates the process of saltatory conduction along an axon.

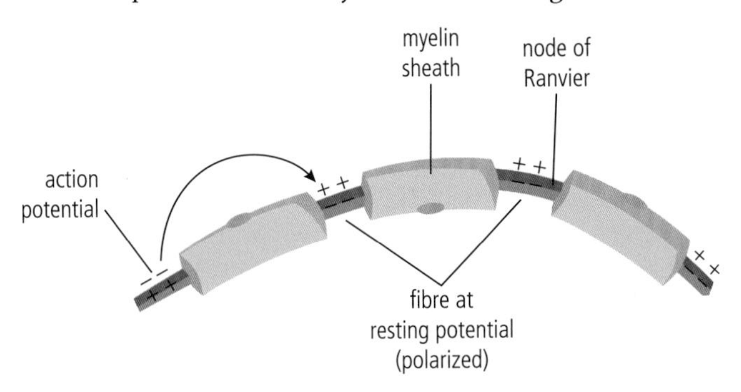

C2.2 Figure 3 Saltatory conduction at nodes of Ranvier along an axon.

Saltatory conduction enables a much faster transmission velocity of action potentials because the ion movements only occur at the nodes of Ranvier. Ion pumps and voltage-gated ion channels are clustered at the nodes of Ranvier. Depolarization and repolarization of the axon only occur at the nodes, allowing the action potential to jump along the length of the axon. This results in what appears to be an almost instantaneous response.

This mode of action potential transmission is also valuable to the cell because it requires less energy expenditure in the form of ATP to fuel the sodium–potassium pump. In non-myelinated axons, the whole axon must be depolarized and then repolarized.

Challenge yourself

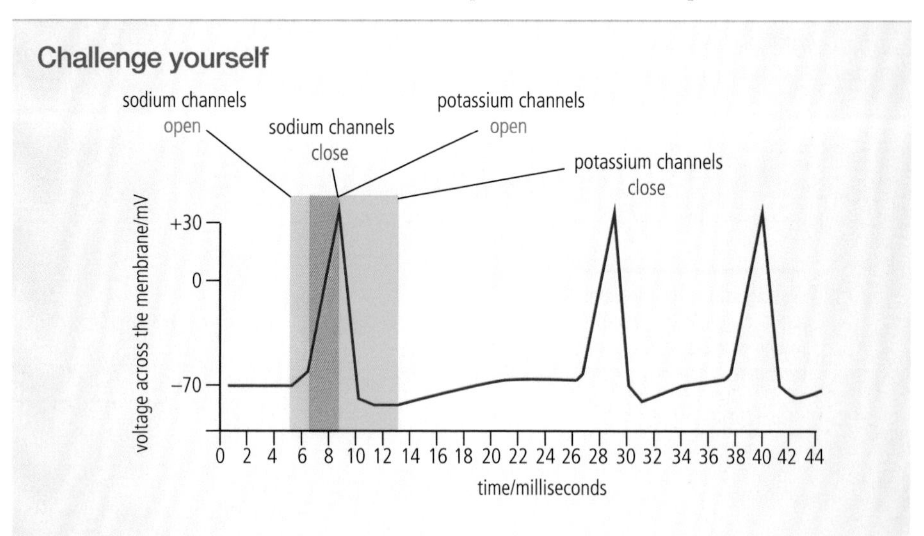

Use the figure above, showing the change in voltage for a neuron sending impulses down its axon, to answer the following questions.

1. If each spike on this graph shows an impulse somewhere in the middle of an axon, what event must have occurred in the area of the axon just preceding this one?
2. If the axon shown is myelinated, where along the axon did these voltage changes occur?
3. If this graph shows an impulse somewhere in the middle of an axon, and this is a myelinated fibre, what area of the axon will next undergo an action potential?
4. Where along the *x*-axis of the graph would the sodium–potassium pump be beginning to work to re-establish a resting potential?
5. What do you think would happen if discrete sensory information from a receptor was being received repeatedly at a faster rate than about 5 milliseconds apart?

C2.2.12 – The effect of exogenous chemicals

C2.2.12 – Effects of exogenous chemicals on synaptic transmission
Use neonicotinoids as an example of a pesticide that blocks synaptic transmission, and cocaine as an example of a drug that blocks reuptake of the neurotransmitter.

Exogenous chemicals are chemicals produced outside the body, while **endogenous chemicals** are produced within the body. Many exogenous chemicals can have an effect on the transmission of an action potential at a synapse. We are going to look at two examples of exogenous chemicals: neonicotinoids and cocaine.

Neonicotinoid insecticides are a relatively new class of insecticide that are chemically similar to nicotine. This type of insecticide works by binding to the postsynaptic receptors that normally accept the neurotransmitter acetylcholine. When acetylcholine binds to the receptor protein, the result is the opening of sodium channels and the propagation of the action potential along the postsynaptic neuron. When neonicotinoid molecules bind to the same receptor proteins, the action potential is not propagated. In addition, the neonicotinoid molecules are not released by the receptor and are not broken down in the synaptic cleft. The receptor becomes permanently blocked. This leads to paralysis of the affected insect, and eventually death.

Cocaine affects the action of a neurotransmitter called dopamine, which is associated with feelings of reward, pleasure, motivation and being productive. Cocaine prevents the removal of dopamine from the synapse and stimulates dopamine-releasing neurons to release dopamine that is usually held in reserve. Normally, dopamine is removed from the synaptic cleft by a specialized protein called the dopamine transporter. When cocaine is present, it attaches to the dopamine transporter and blocks the removal process. This causes a build-up of dopamine in the synaptic cleft, flooding the brain with an elevated response. This is why cocaine is highly addictive.

Repeated use of cocaine often results in the brain adapting to an unnatural reward pathway and becoming less sensitive to natural reinforcers. This increases the likelihood of the user seeking the drug instead of relationships, food or other natural rewards. Also, as cocaine use continues, a tolerance may develop so that higher doses and more frequent use are sought, to produce the same level of pleasure and relief from the withdrawal that may be experienced. The use of cocaine as well as many other addictive drugs can also seriously damage essential organs of the body.

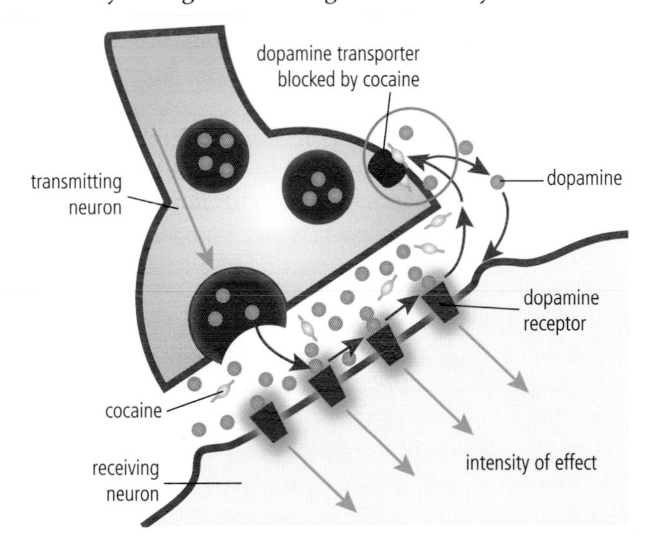

The effect of cocaine on the neurotransmitter dopamine at the synaptic cleft. It is important to notice that the dopamine transporter is blocked by cocaine. This allows dopamine to remain in the cleft, elevating the effects of dopamine.

Early studies of neonicotinoid pesticides suggested that they were relatively safe from an ecological viewpoint. More recent studies have revealed some possible links to the "colony collapse syndrome" being experienced by honeybee colonies. Each country must consider the mounting evidence itself, but chemicals in our environment have ways of crossing international borders, through water, air and many other means. If neonicotinoids are shown to be damaging to honeybee colonies, an international effort to curtail or stop their use will be necessary.

Cocaine illustrates how a drug can affect a neurotransmitter (dopamine) at a synapse in ways that promote intensified drug use, dependence and even addiction. Recent studies have shown certain DNA sequences in humans confer addiction resistance. It appears that individuals with a specific and rare variation of the gene for the mu opioid receptor have a genetic resistance to cocaine and opioid addiction.

Excitatory neurotransmitters:

- increase the permeability of the postsynaptic neuron to Na^+
- causing Na^+ to diffuse in
- resulting in the neuron being depolarized
- allowing the impulse to be carried forward.

Inhibitory neurotransmitters:

- make the inside of the neuron more negative
- because of Cl^- moving in or K^+ moving out
- resulting in the neuron being hyperpolarized
- allowing inhibition of the impulse.

C2.2.13 – Neurotransmitters and postsynaptic potentials

C2.2.13 – Inhibitory neurotransmitters and generation of inhibitory postsynaptic potentials
Students should know that the post-synaptic membrane becomes hyperpolarized.

Acetylcholine is an example of a neurotransmitter that is excitatory. **Excitatory neurotransmitters** generate an action potential by increasing the permeability of the postsynaptic membrane to positive ions. This causes the positive sodium ions (Na^+) that are in the synaptic cleft to diffuse into the postsynaptic neuron. The postsynaptic neuron is depolarized locally (just in that area) by the influx of positive sodium ions. During depolarization, the inside of the neuron develops a net positive charge compared to the outside. An action potential is generated that then moves along the axon.

GABA (gamma-aminobutyric acid) is an example of an **inhibitory neurotransmitter**. Inhibitory neurotransmitters cause **hyperpolarization** of the neuron, which inhibits action potentials. Hyperpolarization refers to the inside of the neuron becoming more negative than normal, making it even more difficult for an action potential to be generated.

An inhibitory neurotransmitter binds to a specific receptor. This causes negatively charged chloride ions (Cl^-) to move across the postsynaptic membrane into the postsynaptic neuron, or it can cause positively charged potassium ions (K^+) to move out of the postsynaptic neuron. This movement of chloride ions into the neuron or potassium ions out of the neuron causes hyperpolarization.

C2.2.14 – Summation of inhibitory and excitatory effects

C2.2.14 – Summation of the effect of excitatory and inhibitory neurotransmitters in a postsynaptic neuron
Multiple presynaptic neurons interact with all-or-nothing consequences in terms of postsynaptic depolarization.

In what ways are biological systems regulated?

A postsynaptic neuron can receive many excitatory and inhibitory stimuli at the same time. If the sum of the signals is inhibitory, then the impulse is not carried forward. If the sum of the signals is excitatory, then the impulse is carried forward. The summation of impulses in this way enables processing in the CNS.

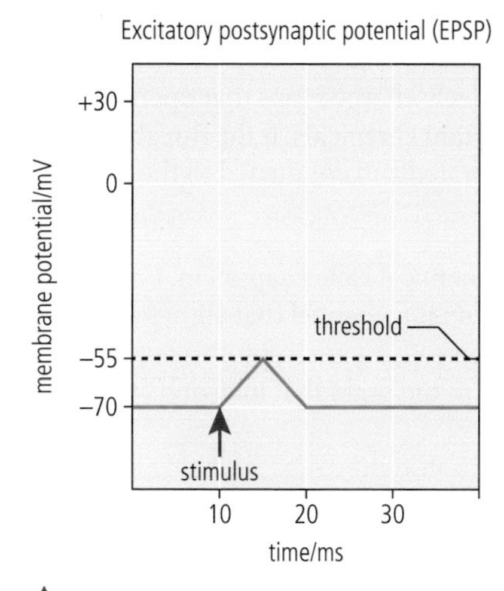

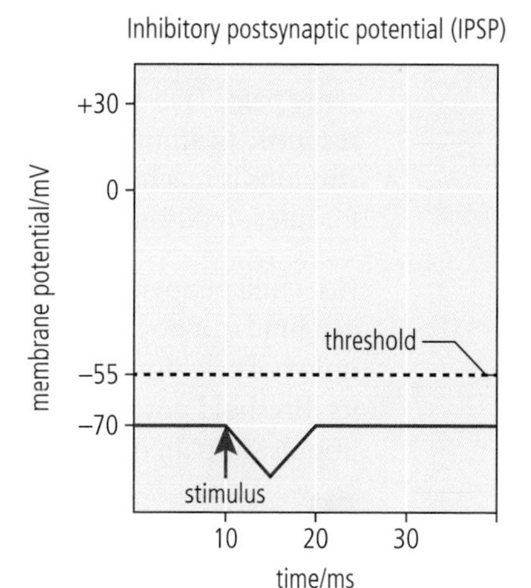

The effects of excitatory and inhibitory postsynaptic potentials.

C2.2.15 – Perception of pain

Human nociceptors communicate with other pain pathway sensory neurons. Unique neurotransmitters called substance P and glutamate are involved in this communication process. When tissue is damaged, prostaglandins are released, which increase the sensitivity of the nociceptors. With increased sensitivity comes greater pain. Analgesics such as aspirin can inhibit the production of prostaglandins and decrease pain.

C2.2.15 – Perception of pain by neurons with free nerve endings in the skin

Students should know that these nerve endings have channels for positively charged ions, which open in response to a stimulus such as high temperature, acid, or certain chemicals such as capsaicin in chilli peppers. Entry of positively charged ions causes the threshold potential to be reached and nerve impulses then pass through the neurons to the brain, where pain is perceived.

Sensory receptors are crucial for the survival of all animals. They are essential for an organism to maintain life processes, avoid danger, find food, and even locate mates. Sensory receptors allow organisms to respond to information received from both inside their bodies and the external environment. These sensory receptors can simply be free nerve endings, or they can be specialized neurons found in sense organs. The membrane of sensory receptors includes protein receptors that react to a stimulus. When a specific stimulus is received by a sensory receptor, ion channels open and ions flow across the cell membrane. If the stimulus is sufficient and the threshold potential is reached, a nerve impulse if propagated. This nerve impulse (action potential) will then be carried by sensory neurons to the CNS. The brain allows perception of the action potentials that reach it.

Class	Function
Thermoreceptors	Detect changes in temperature
Nociceptors	Detect actual or potential tissue damage (pain)
Electromagnetic receptors	Detect electric currents and magnetic fields
Photoreceptors	Detect light waves
Baroreceptors	Detect blood pressure
Chemoreceptors	Detect chemicals, allowing the senses of taste and smell

Classes of sensory receptors

Our focus in this section is on the nociceptors, which are sensory receptors that sense pain. These receptors have channels for positively charged ions that open in response to stimuli such as temperature, acid or certain chemicals. If the threshold stimulus is reached, an action potential will be generated and conducted to the CNS for interpretation.

Hot chilli peppers, such as the jalapeno, contain a chemical called capsaicin. Capsaicin can bind to a nociceptor and trigger the opening of an ion channel that allows an influx of calcium ions into the neuron. Capsaicin can cause a nociceptor to reach its threshold potential and send an action potential to the brain that interprets the impulse as pain or heat.

Congenital insensitivity to pain is a rare condition that results in an individual not being able to perceive pain. Anyone born with this condition never feels pain in any part of their body when injured. As a result, they often end up with bruises or even broken bones. The condition appears to affect the sodium channels of the neurons that generate and conduct the electrical signals of pain. This condition is inherited as an autosomal recessive disease.

TOK

There are many explanations of what consciousness actually is. Some say consciousness is the state of being awake and alert to what is occurring around an individual, or an awareness of feelings. Some say higher consciousness is a state of elevated awareness in which an individual has a deeper understanding of reality. What knowledge, if any, is likely to always remain beyond the capabilities of science to investigate or verify? Is human knowledge confined to what the natural sciences discover, or are there other important inquiries that are not covered by the natural sciences?

C2.2.16 – Consciousness as an emergent property

C2.2.16 – Consciousness as a property that emerges from the interaction of individual neurons in the brain
Emergent properties such as consciousness are another example of the consequences of interaction.

Biology is concerned not only with life, but also anything that affects life. When studying organisms, it is possible to take two approaches. One approach, often referred to as **reductionism**, reduces the complex phenomena of organisms to the interactions of their parts. Essentially, this viewpoint considers the sum of the parts make up the complex system recognized as an organism. Another approach is that of **emergence**, or of looking at all the systems together. This approach believes that the whole is greater than the sum of the parts.

Chapters A2.2 and B2.2 look at how all the different parts of a cell, each carrying out specific functions, work together to allow life to not only occur, but to flourish. To simply add all the cell functions together would never produce the sum of life. However, to understand the complexity of life, we reduce it to a study of the component parts.

Consciousness is very similar. How do the millions of chemical reactions constantly occurring in the human body add up to produce an individual consciousness? Perception of the world around us involves the reception of external and internal environment stimuli by specific receptors, conversion of these received stimuli from a chemical nature to an electrical nature, conduction of these electrical impulses, and interpretation of these impulses. We then need to work out what the proper response is, and ensure that that response takes place. This seems impossible, but somehow all these interactions working together produce a consciousness in a functioning human organism. Emergence is very much demonstrated by the presence of consciousness.

HL end

Guiding Question revisited

How are electrical signals generated and moved within neurons?

In this chapter we have looked at how:

- neurons are specialized cells that carry electrical impulses
- as part of their specialization they have a long axon
- for an action potential to occur, depolarization must first occur within the neuron
- depolarization involves the opening of voltage-gated sodium and then potassium channels, with a more positive electrical charge resulting within the neuron
- after depolarization of a neuron region occurs, it immediately undergoes repolarization
- the sodium–potassium pump is required for repolarization to occur
- electrical impulses are propagated through neurons and only move in one direction, i.e. from the dendrites to the other end of the axon
- myelin is an insulating material that speeds up the movement of nerve impulses through neurons
- larger axons carry impulses at a faster rate than smaller axons

HL

- a neuron only depolarizes when a threshold potential is reached
- there is no variation in electrical impulse intensity, based on the all-or-nothing principle.

HL end

Guiding Question revisited

How can neurons interact with other cells?

In this chapter we have learned how:

- electrical impulses are transmitted from one neuron to another at a synapse
- an impulse in a neuron can be passed to another neuron
- an impulse can also be passed via a synapse to a muscle, where the impulse causes the contraction of the muscle, or to another effector cell, causing a specific reaction
- neurotransmitters are chemicals that diffuse across the synaptic cleft
- impulses or signals can only pass in one direction across a synapse
- depolarization at the presynaptic membrane causes an influx of calcium ions, which in turn causes vesicles carrying neurotransmitters to release their contents into the synapse
- the neurotransmitter diffuses across the synapse and binds to the postsynaptic membrane, generating an excitatory postsynaptic potential
- neurotransmitters such as acetylcholine exist at many different types of synapse, including neuromuscular junctions

HL

- inhibitory neurotransmitters cause hyperpolarization at postsynaptic membranes of synapses, preventing the transmission of an action potential
- action potentials travel to the central nervous system, where appropriate responses are determined to maintain the functions of life in the organism
- there are many sensory receptors, for example light receptors in the eye
- sensory receptors that detect pain stimuli are called nociceptors
- pain stimuli result in the opening of sodium channels, causing depolarization of the nociceptor.

HL end

Exercises

Q1. Explain how a resting potential is produced in a neuron.

Q2. Name two variables that affect the speed of action potentials in neurons.

Q3. Explain why saltatory conduction is a faster means of electrical impulse transmission in a neuron.

Q4. Name the neuron parts labelled on the diagram.

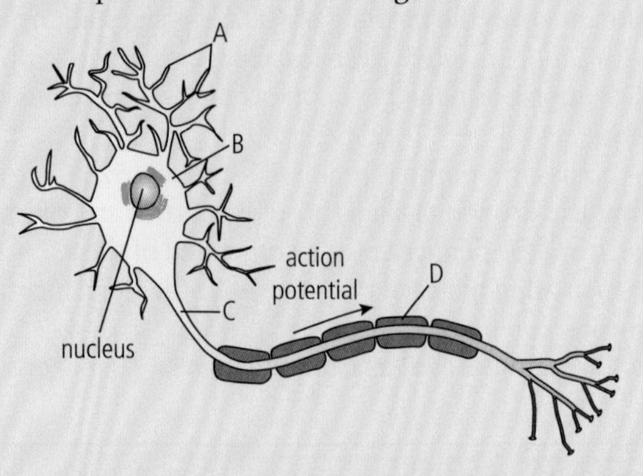

Q5. What is the role of calcium ions in the transmission of an action potential at a synapse?

Q6. Which letter of the following represents an axon that is demonstrating depolarization?

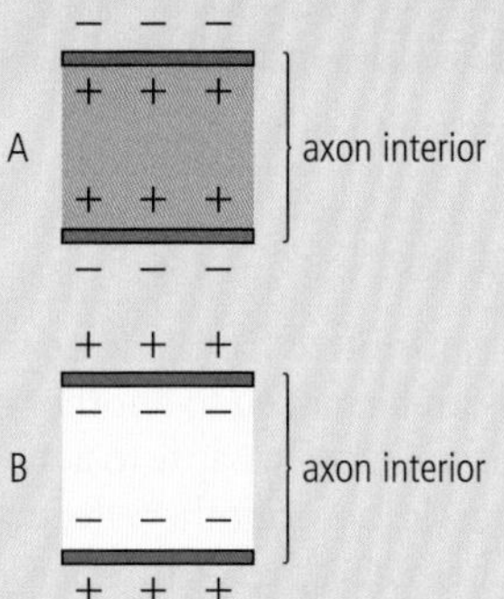

HL

Q7. Complete the sentence below using one of the four options.

Inhibitory neurotransmitters ____________________

A hyperpolarize presynaptic membranes.

B hyperpolarize postsynaptic membranes.

C depolarize presynaptic membranes faster.

D depolarize postsynaptic membranes more slowly.

Q8. Explain the following terms:

(a) All or nothing.

(b) Threshold potential.

(c) Refractory period.

HL end

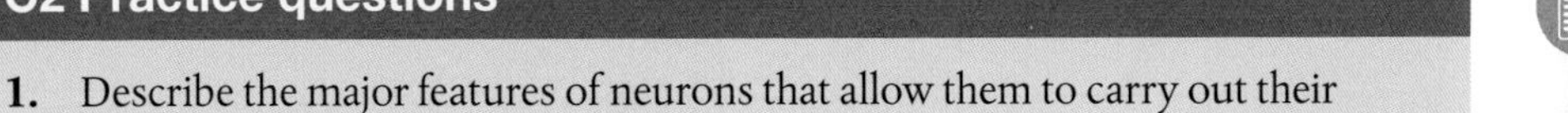

C2 Practice questions

1. Describe the major features of neurons that allow them to carry out their function.

(Total 3 marks)

2. Explain the steps required to produce an excitatory postsynaptic potential.

(Total 4 marks)

3. Which of the following is not a factor in the transmission of an action potential from a presynaptic neuron to a postsynaptic neuron?

A Neurotransmitter is reassembled in the presynaptic neuron.

B Calcium ions act as a signalling chemical.

C The neurotransmitter is degraded in the synapse.

D The sodium–potassium pump is active, to cause depolarization in the postsynaptic membrane.

E The length of the single axon.

(Total 1 mark)

HL

4. Which of the following is not a membrane receptor for ligands carrying out chemical signalling?

A Steroid hormone receptors.

B Enzymatic receptor.

C Channel-linked receptor.

D G protein-coupled receptor.

(Total 1 mark)

5. Explain how different receptors can have the same effect on a cell.

(Total 3 marks)

6. Describe some common factors found in all examples of cellular signalling.

(Total 3 marks)

7. Explain quorum sensing. Give an example.

(Total 3 marks)

8. Explain saltatory conduction.

(Total 3 marks)

HL end

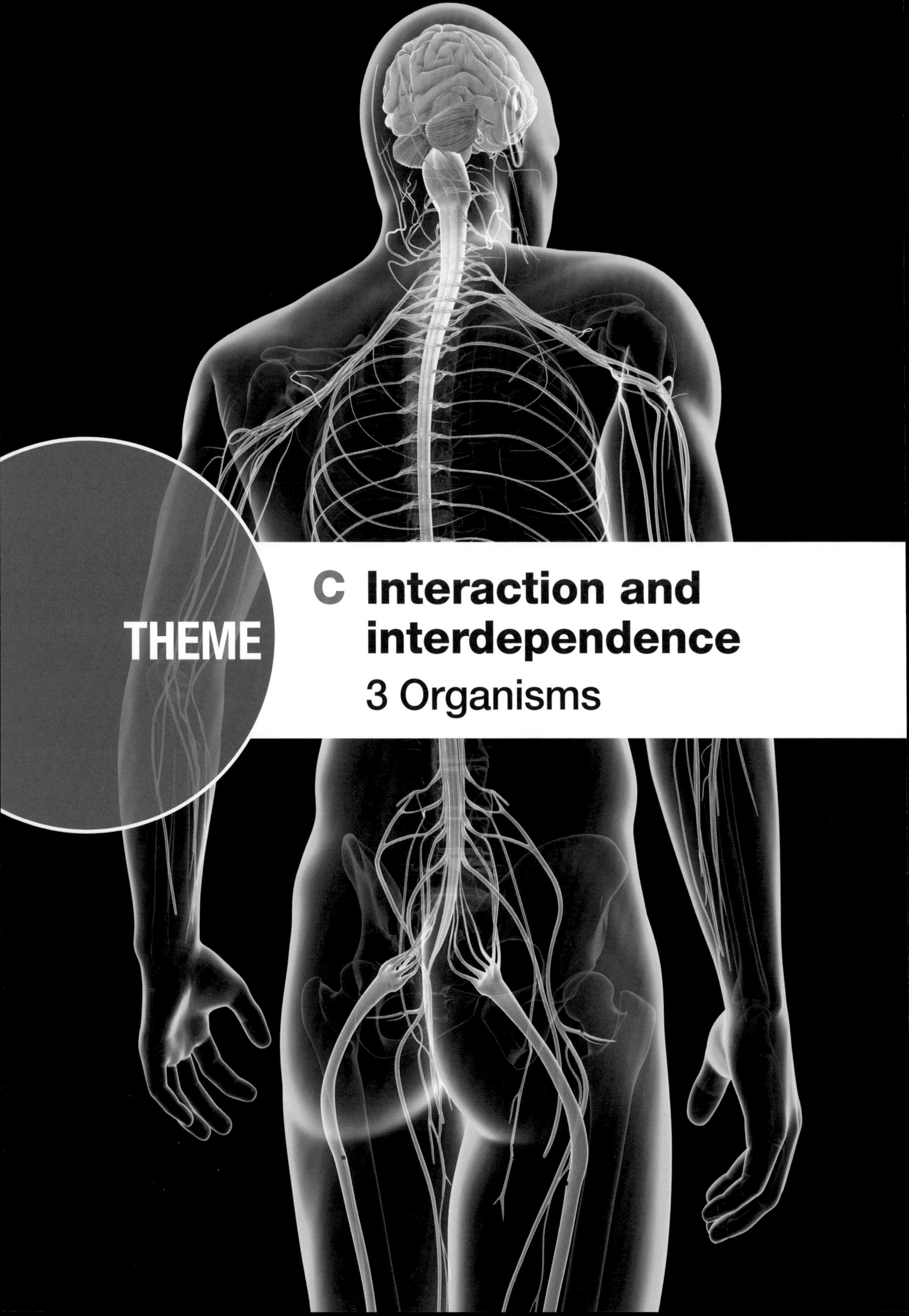

THEME

C Interaction and interdependence

3 Organisms

The human nervous system is just one of the many systems in the body that enable communication and thus integration of body systems. We are aware of only some actions of the nervous system, as others are accomplished by the autonomic nervous system. In addition, chemical communication between tissues by hormones and the endocrine system, and cell-to-cell communication by the immune system, provide different means of transmission for important signals around the body.

C3.1 Integration of body systems

Guiding Questions

What are the roles of nerves and hormones in integration of body systems?

What are the roles of feedback mechanisms in regulation of body systems?

Multicellular organisms have specialized cells that form tissues. Tissues are responsible for specific functions within an organism and, as a consequence, make up organs. In order for the tissues within organs to work in concert with each other, there must be communication between the tissues. Without a means of communication, the emergent property represented by an entire organism would not be possible. A human being or a magnolia tree can only exist because its cells, tissues and organs have evolved ways of communicating with each other.

Two systems have evolved within animals specifically to enable communication between, and thus integration of, body systems.

- A nervous system, designed to receive sensory information through structures called receptors and send motor responses to muscles, resulting in movement.
- An endocrine system, consisting of glands that respond to chemical signals in the body with the production and release of a variety of hormones. Hormones affect the activity of specific cells known as the target tissue of a particular hormone.

HL Plants also use hormones for chemical communication, resulting in a variety of responses.

Humans and many other organisms have evolved feedback mechanisms that inform the body of the need for action by one or more body tissues. Usually these feedback mechanisms work to keep a factor within a normal, homeostatic, range. Examples of factors kept within a normal range are body temperature and levels of glucose in the bloodstream. If a factor rises above a particular level, or set point, the body will initiate one set of actions to bring levels down. If a factor goes below the set point, another set of actions will be initiated to bring that factor back to the set point. This is called negative feedback control.

C3.1.1 – Coordinating systems

C3.1.1 – System integration
This is a necessary process in living systems. Coordination is needed for component parts of a system to collectively perform an overall function.

Living organisms are complex. That complexity is the result of millions of years of evolutionary adaptations that represent only the relatively few mutations that gave an organism a greater chance to survive and reproduce. All life started as single-celled organisms and has expanded into the rich variety of single-cell and multicellular life that exists today.

The Atlantic puffin (*Fratercula arctica*) and many other animals use the same specialized systems as humans. All of its cells, tissues, organs and body systems work collectively as a puffin.

Coordination is needed for the component parts of organisms to collectively perform complex functions.

In order for complex organisms to evolve to survive in their environments, it was necessary for cells to become specialized for certain functions. Groups of specialized cells became specialized **tissues**, and groups of specialized tissues became **organs**. Some organs have evolved to work collectively to accomplish certain functions, and have become **body systems**. Body systems are specialized for functions such as obtaining nutrients, discarding waste and reproduction. All of the body systems working in unison represent the entire organism.

Even though cells in a multicellular organism are specialized, they are all using the same DNA as their genetic code. A specialized cell, such as a muscle cell, uses some genes that other cell types do not.

Some organisms have evolved into organized collections of billions of cells. The cells in one part of the organism often need to communicate with other cells where cell-to-cell communication is impossible. Two systems of communication have evolved to enable communication within organisms and facilitate efficient processes. One is found in both plants and animals, and involves chemicals called **hormones** that are produced in one location and then carried within fluids to other locations in the body. The other communication system is specific to animals and involves **electrical signals** sent from one location to another by a nervous system.

C3.1.2 – Hierarchy of body subsystems

C3.1.2 – Cells, tissues, organs and body systems as a hierarchy of subsystems that are integrated in a multicellular living organism
Students should appreciate that this integration is responsible for emergent properties. For example, a cheetah becomes an effective predator by integration of its body systems.

Multicellular animals and plants have evolved a common hierarchy of organization that permits effective communication and functioning within their environments. It is because of this hierarchy of subsystems that living organisms are able to survive and interact with their surroundings (Table 1). Each level of organization allows greater efficiency and complexity. All living organisms continue to evolve to improve adaptations to the environment in which they exist at each level of cellular organization.

Subsystem	Animal example	Plant example
Cell	Smooth muscle cell	Guard cell
Tissue	Muscular wall	Stoma
Organ	Bladder	Leaf
Organ system	Urinary	Vascular
Organism	White-tailed deer (*Odocoileus virginianus*)	Magnolia tree (*Magnolia grandiflora*)

C3.1 Table 1 Levels of organization within an animal and a plant

Emergent properties are those that exist when the sum of all the parts creates features that do not exist within the individual components. This is the advantage of an organism level of complexity. A puffin does not exist as a puffin until it is a complete being with all its component parts working collectively. The organism level of organization results in a combination that is said to be greater than the sum of its parts.

C3.1.3 – Integration of organs in animals

C3.1.3 – Integration of organs in animal bodies by hormonal and nervous signalling and by transport of materials and energy
Distinguish between the roles of the nervous system and endocrine system in sending messages. Using examples, emphasize the role of the blood system in transporting materials between organs.

Organs in the body must work together in order to maintain body processes. Body processes include digestion, maintaining the heart rate, blood glucose levels, blood pressure, and many others. Sometimes the communication is by the **nervous system**, and often we have no idea that the communication is actually occurring. The part of your nervous system that communicates with your body tissues without your conscious knowledge is called your **autonomic nervous system** (ANS).

In addition, humans and many other animals use hormone production and secretion in an **endocrine system** to help communication between organs and to respond to special situations. The body tissue where a hormone exerts an effect is called the **target tissue** of the hormone. Because hormones are produced in endocrine glands, they must be transported to their target tissues by the bloodstream.

The nervous system and endocrine system often work together to integrate body processes. An example is the release of epinephrine (adrenaline) from the adrenal glands. Sensory organs transmit information to the nervous system that indicates epinephrine is needed as part of the fight-or-flight response. The autonomic nervous system then sends impulses to the adrenal glands to release epinephrine. This hormone leads to a variety of body responses, including increased heart rate and increased flow of blood to muscles, to prepare the body for immediate increased activity. This integration between body systems is thought to have evolved as a survival mechanism.

Even though the nervous and endocrine systems often work together to integrate the systems in the body, each has its own characteristics.

In the nervous system:

- electrical impulses are used to send messages
- cells called neurons are used to transmit and receive impulses
- portions of the system control voluntary actions while other portions control involuntary actions
- responses occur quickly but are short lived.

In the endocrine system:

- hormones are used to send messages
- hormones travel through the bloodstream
- only involuntary functions are controlled
- responses are typically slow but are long lasting.

Most multicellular organisms have become so large that it is impossible for nutrients and waste products to be efficiently and directly moved from cell to cell. Transport vessels and aqueous fluids have evolved to serve that purpose. Humans and many other animals use blood circulating in arteries and veins to transport a variety of substances throughout the body tissues. The oxygen needed by leg muscles will be supplied by blood that has received that oxygen from lung tissues a short time before it is used. Urea produced as a by-product of protein metabolism in the liver will be transported by blood to the kidneys to be filtered out and become part of urine.

C3.1.4 – The brain and information processing

C3.1.4 – The brain as a central information integration organ
Limit to the role of the brain in processing information combined from several inputs and in learning and memory. Students are not required to know details such as the role of slow-acting neurotransmitters.

The brain is the most complex organ in the body. The brain regulates and monitors unconscious body processes such as blood pressure, heart rate and breathing. It receives a flood of messages from the senses, and responds by controlling balance, muscle coordination and most voluntary movements. Some parts of the brain deal with speech, emotions and problem solving. The brain relies on a variety of receptors to receive information. Some of these receptors relay information that we process at the conscious level, such as:

- **photoreceptors**, located within the retina of the eyes for visual information
- **chemoreceptors**, many located within our tongue for tasting
- **thermoreceptors**, located in the skin to provide information on changes in temperature
- **mechanoreceptors**, located in inner ear and sensitive to sound vibrations.

Other receptors send information to the brain at the subconscious level and are important components of our autonomic nervous system. Examples include:

- **osmoreceptors**, located in carotid arteries and the hypothalamus of the brain, which sense solutes and the water content of blood
- **baroreceptors**, located in carotid arteries and the aorta, which sense blood pressure based on how much a blood vessel is being stretched by internal pressure
- **proprioceptors**, located in muscles and joints, which provide the brain with a sense of balance and coordination, especially when the body is moving.

While receiving sensory information from a variety of receptors, the brain continuously processes the information and generates responses as needed. The process involves filtering the sensory information that is important from the information that is unimportant. This is necessary because the electrical signals travelling from sense organs are continuous and vast.

The brain communicates with the body in two ways. One involves impulses sent in and out of the spinal cord by 31 paired nerves called **spinal nerves**, which emerge directly from the spinal cord. The other involves the brain's own nerves, called **cranial nerves**. This set of 12 paired nerves connects various body parts to the brain through the **brainstem**. An example of cranial nerves is the pair of optic nerves that carry impulses from the retinas of the eyes to the brain.

All parts of the brain must interact with each other for proper neural functioning, but the frontal lobe of the cerebrum is the dominant structure for learning and memory.

The brain is divided into three main areas, each of which is subdivided further based on location and function.

- **Cerebrum**: the cerebrum is divided into right and left sides called **cerebral hemispheres**. Each hemisphere consists of four lobes: frontal, temporal, parietal and occipital, as shown in Figure 1 on the next page. The neural processing carried out by these four lobes dominates our conscious activities. Although the four lobes interact with each other continuously, learning and memory activities are largely coordinated by the frontal lobe.
- **Cerebellum**: the cerebellum coordinates voluntary movements, and controls balancc and equilibrium.
- **Brainstem** – in addition to relaying impulses between the cerebrum, cerebellum and spinal cord, the brainstem is responsible for most functions associated with the autonomic nervous system. The functions controlled by the brainstem are necessary for life, yet occur at a subconscious level. The **medulla** (or medulla oblongata) is an important part of the brainstem because it regulates both breathing and heart rate.

Very specific neural functions have been mapped by location on and within the brain. Specific brain activities are also associated with specific neurotransmitters that carry signals from one brain neuron to another. Some examples of neurotransmitters used within the brain are serotonin, dopamine and endorphins. There are many others.

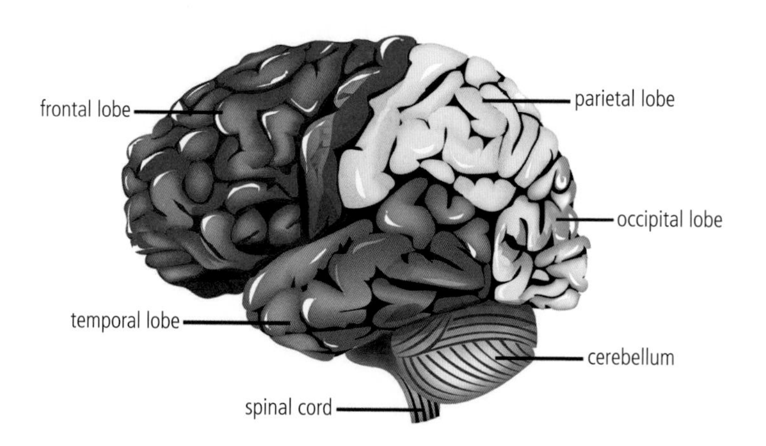

C3.1 Figure 1 The left cerebral hemisphere, cerebellum and spinal cord are shown in this illustration. The brainstem extends up from the spinal cord and is hidden in this view because it is covered by the temporal lobe of the cerebrum.

C3.1.5 – The spinal cord and unconscious processes

C3.1.5 – The spinal cord as an integrating centre for unconscious processes
Students should understand the difference between conscious and unconscious processes.

Together, the brain and spinal cord make up the **central nervous system** (**CNS**). The spinal cord is a neural pathway between the body and the brain, but is capable of information processing on its own. The spinal cord controls some unconscious reflexes associated with balance and other skeletal muscle functions independently of the brain.

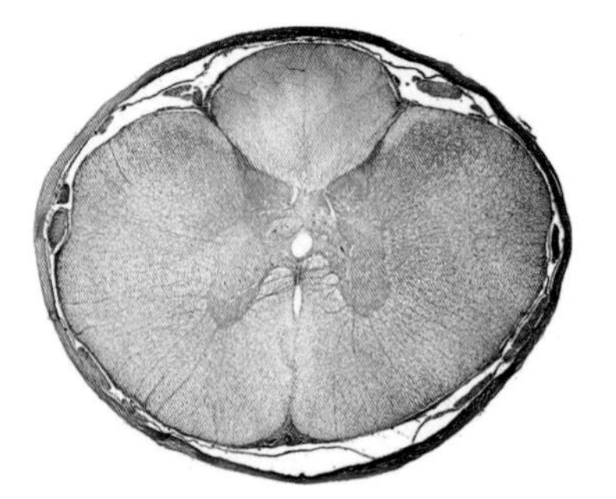

C3.1 Figure 2 A photograph of a section of the human spinal cord. In the centre, the darker "butterfly" shape is the grey matter of the spinal cord, capable of integration processing. Outside the grey matter is white matter, which carries impulses to and from the brain.

The 31 pairs of spinal nerves bring sensory information into the CNS from the body and allow motor (muscular) information to be sent out. Some of the information travelling in and out of the CNS is processed at the conscious level, meaning that the cerebrum of the brain is involved in the pathway.

The spinal cord has two types of neural tissue, **white matter** and **grey matter**. The white matter is composed primarily of axons of neurons and carries neural impulses to and from the brain. The grey matter of the spinal cord contains neurons and synapses involved in spinal cord integration processes. When sensory information enters the grey matter of the spinal cord and motor information is immediately sent back out, the pathway of the impulse is called a **reflex arc**. You will study a specific spinal reflex arc in Section C3.1.9.

C3.1.6 – Sensory neurons and conveying information

C3.1.6 – Input to the spinal cord and cerebral hemispheres through sensory neurons
Students should understand that sensory neurons convey messages from receptor cells to the central nervous system.

The beginning of a sensory neural pathway begins with a **receptor**. A receptor is a modified neuron that is capable of **transduction**. Transduction is the conversion of a physical stimulus into an electrical signal called an **action potential** that is carried along a neuron. Each type of receptor is specialized to transduce one specific type of physical stimulus. The retinal cells of the eyes transduce light, thermoreceptors in your skin transduce heat or lack of heat, baroreceptors transduce pressure; there are many other receptor types. The neurons that carry impulses from receptors to the brain or spinal cord are called **sensory neurons**.

The quantity and variety of sensory information being sent to your brain at any given moment is vast. In fact, most sensory information has to be filtered out as unimportant by a process called **sensory gating**. At this moment try to make yourself aware of some of the sensory information that previously you were subconsciously filtering by sensory gating. This may include sights, sounds, things you are touching, temperature sensory information and a great deal more.

Example of a specific sensory path

The skin of your fingertips contains mechanoreceptors that sense pressure. When you touch your phone screen to type a message, the mechanoreceptors at the tips of your fingers and thumbs send action potentials up the small nerves in your hand and arm, to eventually join one of the 31 spinal nerves. Specifically, the action potentials travel to a spinal nerve that enters your spinal cord in the region of your chest. The neurons that carry this information enter the grey matter of the spinal cord but are directed out into the white matter (see Figure 2). The pathway of this spinal cord neuron carries the impulses up through your spinal cord, through the brainstem and into the parietal lobe of your cerebrum. This is where the sense of touch is interpreted or integrated. You are aware of this sensation, thus it is a conscious process, although it is only one sensation of many that you would be experiencing at that moment.

Each receptor that we perceive at the conscious level has a specific pathway to the CNS. Many receptors send their action potentials to a particular lobe of the cerebrum, for example:

- information from sound mechanoreceptors travels to the temporal lobe
- information from chemoreceptors for taste travels to the parietal lobe
- information from photoreceptors for visual information travels to the occipital lobe.

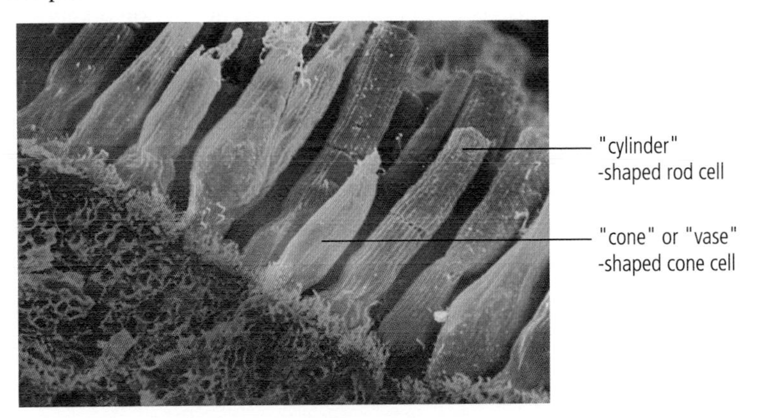

A colorized scanning electron micrograph of a small section of the retina of an eye. The photoreceptors of the retina are shown. Rod cells that can sense black and white vision are shown in blue (they are rod-shaped). Cone cells responsible for colour vision are shown in green (they are cone-shaped).

C3.1.7 – Motor neurons and muscle stimulation

C3.1.7 – Output from the cerebral hemispheres of the brain to muscles through motor neurons
Students should understand that muscles are stimulated to contract.

The cerebrum uses the sensory information it receives to make decisions concerning movements. The action potentials are carried to muscle tissue by **motor neurons**. The portion of the cerebrum that sends the action potentials is called the **motor cortex** and is located in the most posterior portion of the frontal lobe of the cerebrum.

Motor neurons terminate in muscle tissue. Each axon of a motor neuron branches repeatedly and forms synapses with multiple fibres of a muscle.

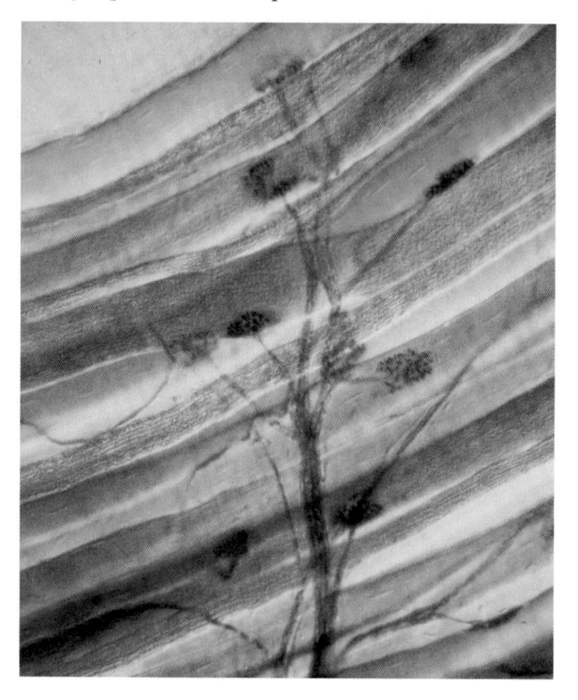

▲ A light micrograph showing a single motor neuron axon branching to form multiple synapses with fibres of skeletal muscle. Each synapse is called a motor end plate or neuromuscular junction.

Motor neurons form synapses with muscle fibres called **motor end plates** or **neuromuscular junctions** (discussed in Chapter B3.3). When action potentials reach a motor end plate, a neurotransmitter called **acetylcholine** is released. This is the chemical signal that initiates contraction of the **sarcomeres** in that area of the muscle, leading to the shortening of the muscle and movement of a bone.

C3.1.8 – Nerve fibres

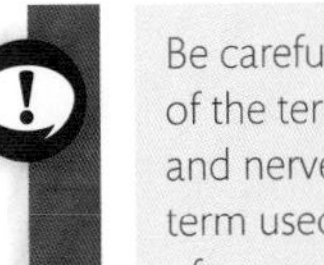

Be careful with the use of the terms neuron and nerve. Nerve is the term used for a bundle of neurons surrounded by a protective sheath.

C3.1.8 – Nerves as bundles of nerve fibres of both sensory and motor neurons
Use a transverse section of a nerve to show the protective sheath, and myelinated and unmyelinated nerve fibres.

An individual neuron is either sensory or motor, depending on whether it is carrying action potentials towards or away from the CNS. An individual neuron cannot be mixed.

The terms **nerve** and **neuron** are not interchangeable. A neuron is an individual cell of the nervous system, whereas a nerve is a collection of neurons surrounded by a protective sheath. Neurons may be sensory, motor or interneurons.

- **Sensory neurons** carry action potentials from receptors to the CNS.
- **Motor neurons** carry action potentials from the CNS to a muscle.
- **Interneurons** are located between sensory and motor neurons and are only found within the CNS.

Some of the 12 pairs of cranial nerves only contain sensory neurons, while others only contain motor neurons and others contain some of each type. The 31 pairs of spinal nerves contain both sensory and motor neurons, carrying action potentials in opposite directions. Nerves that contain both sensory and motor neurons are called **mixed nerves**.

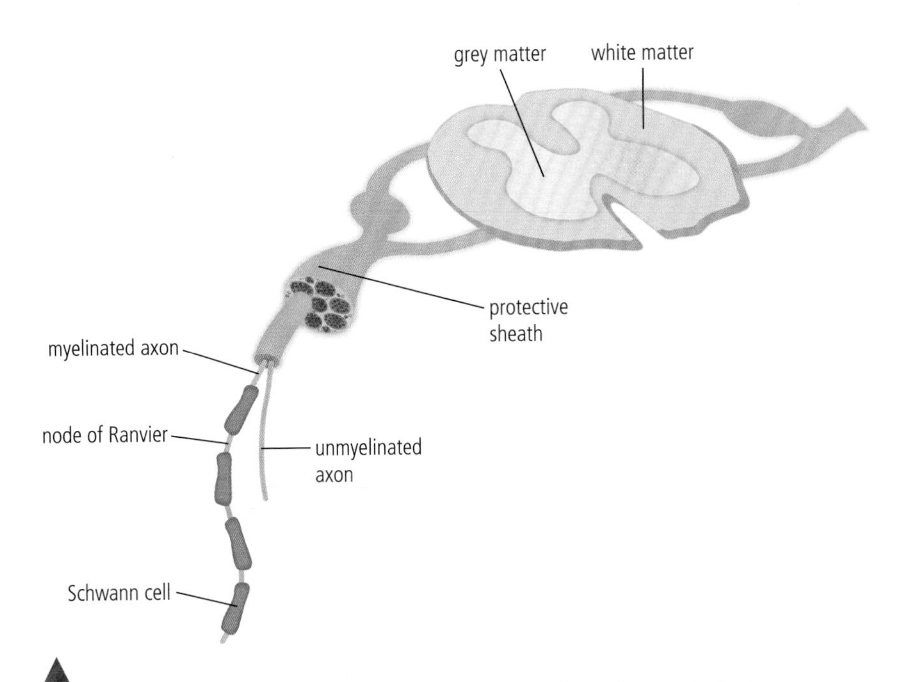

A section view of an area of the spinal cord showing one of the 31 pairs of spinal nerves.

Neurons can be either **myelinated** or **unmyelinated**. Those that are myelinated have cells called **Schwann cells** wrapped around their axon, and intervening areas where there are no Schwann cells (also discussed in Chapter C2.2). The areas between Schwann cells are called **nodes of Ranvier**. The action potentials of myelinated axons are able to skip from one node of Ranvier to the next, making transmission of the action potential much faster compared to unmyelinated axons. Groupings of myelinated and unmyelinated axons are surrounding by protective sheaths.

C3.1.9 – Pain reflex arcs

C3.1.9 – Pain reflex arcs as an example of involuntary responses with skeletal muscle as the effector
Use the example of a reflex arc with a single interneuron in the grey matter of the spinal cord and a free sensory nerve ending in a sensory neuron as a pain receptor in the hand.

A **pain reflex arc** is an example of an involuntary response and involves only three neurons. The first of the neurons is a receptor neuron known as a **nociceptor** or pain receptor. Imagine that you accidently hold a finger too close to an open flame. This results in nociceptors located in the skin of your finger initiating **afferent** (sensory) action potentials. These action potentials travel through your hand and eventually join one of the spinal nerves. After entering the spinal cord, the afferent neuron synapses with a short **interneuron** (also called a relay neuron) located entirely within the grey matter of the spinal cord. The interneuron synapses with a motor neuron and the resulting action potentials go directly to arm muscles (the **effector**), which moves quickly to pull your finger away from the flame.

A schematic showing the three-neuron pathway of a pain reflex arc.

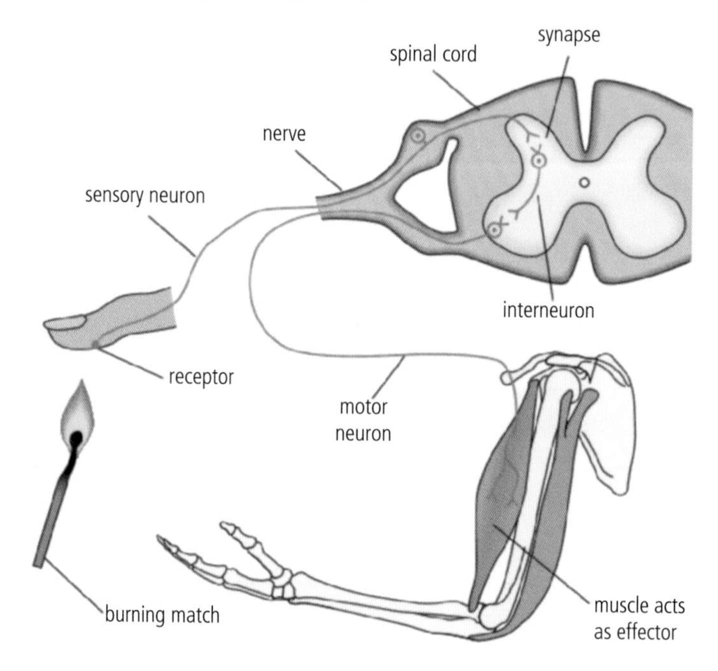

The action of pulling your finger away from the source of pain occurs much faster than truly sensing the pain. The reason for this is that the sensation of pain must travel to your cerebrum to be integrated by many neural synapses before a sensation is felt and a motor response formulated. The pain reflex arc has evolved to limit damage to body tissue by generating a quick reaction involving only three neurons. The unusual aspect of this is that the reflex arc uses skeletal muscle as the effector, tissue that is normally innervated by the frontal lobe of the cerebrum.

C3.1.10 – The cerebellum and skeletal muscle coordination

C3.1.10 – Role of the cerebellum in coordinating skeletal muscle contraction and balance
Limit to a general understanding of the role of the cerebellum in the overall control of movements of the body.

Although the **cerebellum** is a very important part of the brain associated with body movements, it does not initiate those movements. The initiation of muscle contractions and thus body movements is accomplished by the **motor cortex** of the cerebrum. As soon as a movement begins, the cerebellum receives feedback impulses from the area of the body that is moving and many sense organs. The cerebellum then sends out impulses to coordinate the movement. This results in smooth and balanced muscular activity, leading to coordinated movements. The cerebellum coordinates posture, balance, walking, hand and finger movements, eye movements, speech and much more. The term "muscle memory" often used by athletes is more to do with training coordinated movements by the cerebellum than actually training muscle.

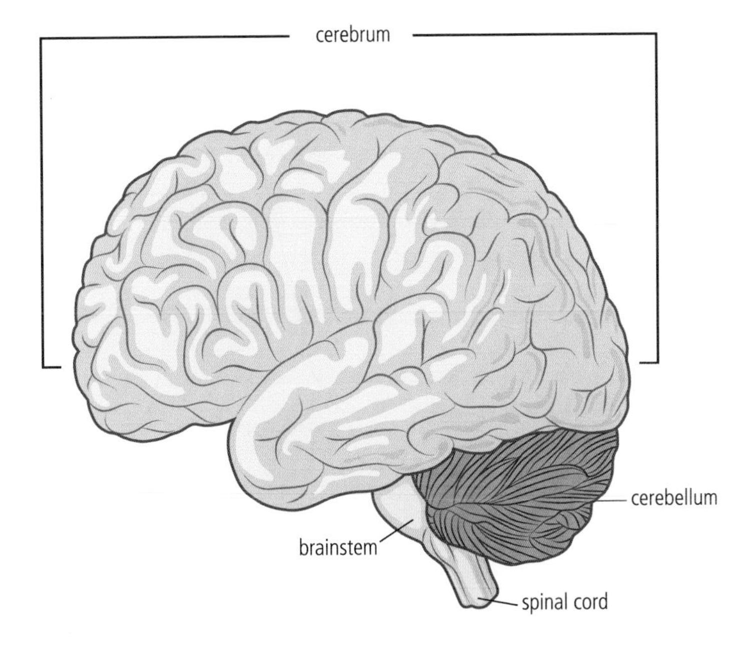

▲ The three main portions of the human brain are shown, with the tissue of the cerebellum highlighted in a darker colour. Many body functions, including maintaining balance and coordinating movements, require interactions between all three of these parts plus the spinal cord, which extends down from the brainstem.

Nocturnal animals have specialized adaptations for sensory input during the night. Many nocturnal animals have a structure called the **tapetum lucidum** behind their retina. This is a layer that sends light back through the retina sensory receptors a second time by reflection. This doubles the intensity of light striking the sensory cells of the retina. You can tell when an animal has a tapetum lucidum because its eyes appear to glow when you shine a light at them.

C3.1.11 – Melatonin secretion and sleep patterns

C3.1.11 – Modulation of sleep patterns by melatonin secretion as a part of circadian rhythms
Students should understand the diurnal pattern of melatonin secretion by the pineal gland and how it helps to establish a cycle of sleeping and waking.

A **circadian rhythm** is any pattern of behaviour or physiology that is based on a 24-hour cycle. The most obvious pattern of a circadian rhythm is our wake and sleep cycle. Many other organisms follow a circadian rhythm for sleep. Some animals, like ourselves, are **diurnal**, meaning that we are more active in daylight hours. Other organisms are **nocturnal** and are more active at night.

Evidence suggests that the circadian rhythm is largely controlled or modulated by a small endocrine gland called the **pineal gland**. This small gland is located near the centre of the brain between the cerebrum and brainstem. Its function is to produce a hormone called **melatonin**. This hormone regulates the sleep schedule. Studies have shown that melatonin levels are high during the night for diurnal animals and high during the day for nocturnal animals. Other studies have shown that light striking the retina of the eye inhibits melatonin production.

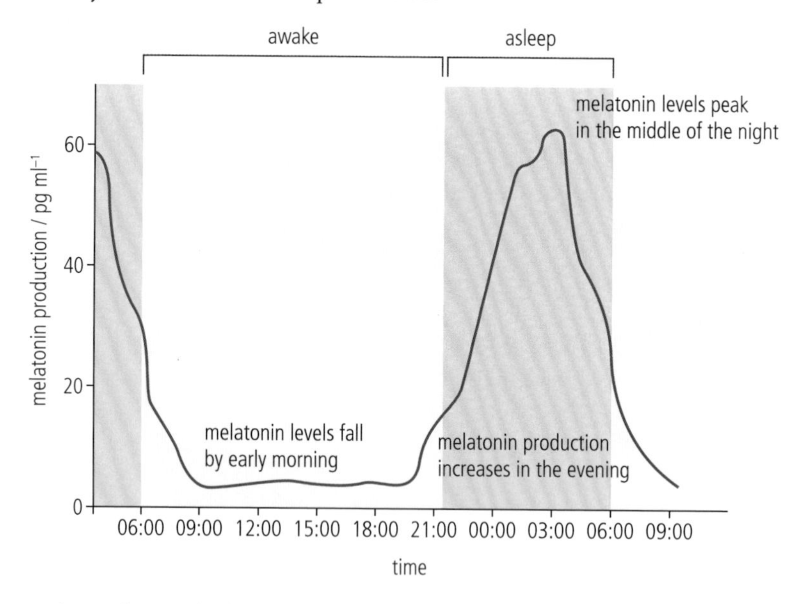

A graph showing melatonin production by the human pineal gland over a time period of about 30 hours. Humans are diurnal and show increasing melatonin production soon after sunset.

Over a prolonged period of time our bodies become naturally regulated to a circadian rhythm that is only interrupted by atypical events. One of those events is travelling through several time zones in a short period of time, sometimes known as "jet lag". Extended viewing of television, mobile phone and computer screens in the evening has also been shown to alter the natural circadian rhythm.

C3.1.12 – Epinephrine and vigorous activity

C3.1.12 – Epinephrine (adrenaline) secretion by the adrenal glands to prepare the body for vigorous activity
Consider the widespread effects of epinephrine in the body and how these effects facilitate intense muscle contraction.

When humans encounter a stressful situation, a hormone called **epinephrine** (also called adrenaline) is released from glands located on the upper or superior side of each kidney. These glands are called **adrenal glands**. Epinephrine release occurs as a result of a potentially harmful event, such as an animal attack, but also because some people choose recreational activities that stimulate release of the hormone. Like all hormones, epinephrine is released into the bloodstream and results in numerous responses by the body.

Epinephrine has widespread effects in the body, including:

- increasing the heart rate and blood pressure
- increasing the diameter (dilation) of air passages so more air can be received by the lungs
- dilation (increased size) of the pupils of the eyes
- increasing blood sugar levels by stimulating glycogen conversion to glucose in the liver
- increasing the blood supply to muscles.

Epinephrine prepares us for the fight-or-flight response, so called because the body's resources are called upon for immediate action in response to a threat or other stimuli that require a vigorous and immediate response. Intense muscle contractions are associated with epinephrine release.

C3.1.13 – The hypothalamus, pituitary gland and endocrine system

C3.1.13 – Control of the endocrine system by the hypothalamus and pituitary gland
Students should have a general understanding, but are not required to know differences between mechanisms used in the anterior and posterior pituitary.

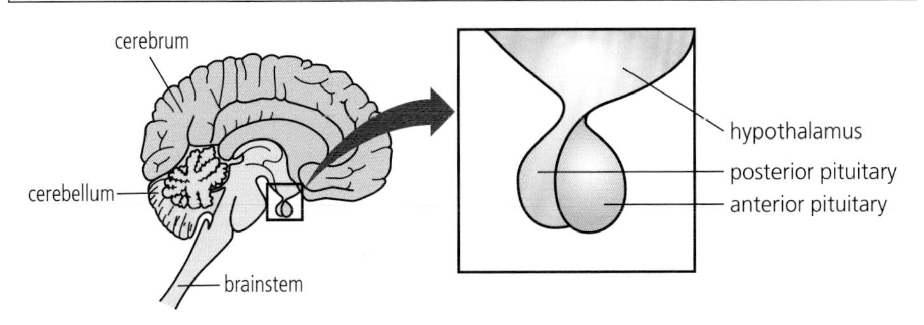

A section of the human brain showing the location of the hypothalamus and the two portions of the pituitary gland.

The **hypothalamus** is an area of the brain that acts as a link between the nervous system and the endocrine system. The hypothalamus contains some receptors that are associated with autonomic nervous system functions, and receives action

Each hormone produced by an endocrine gland has one or more tissue types in the body that is the "target tissue" of that hormone. In many instances the target tissue is located far away from the endocrine gland. Thus endocrine glands secrete hormones into the blood for dispersal to all cells of the body, even though only the target tissue cells are affected by the hormone.

In Figure 3, note that one of the hormones secreted by the posterior pituitary is oxytocin. Oxytocin is produced during childbirth and is controlled by a positive feedback mechanism. Oxytocin induces uterine contractions, and uterine contractions stimulate further secretion of oxytocin.

potentials from other areas of the body that also contain this type of receptor. The hypothalamus is composed of both neurons and **glandular cells**. The glandular cells of the hypothalamus produce hormones that either stimulate hormone release by the **pituitary glands**, or inhibit their release.

Often the pituitary is referred to as a singular gland, but it is actually two glands that exist as different "lobes". The anterior and posterior lobes of the pituitary communicate with the hypothalamus in different ways, and each secretes its own hormones. The majority of these hormones are chemical signals released into the bloodstream that regulate the homeostasis of various physiological factors, such as metabolic rate, reproductive cell formation and water balance.

A good example is **antidiuretic hormone** (**ADH**) produced by the hypothalamus, which is sent to the posterior pituitary and when needed is secreted by the posterior pituitary. ADH helps control homeostatic levels of water in the body. The hypothalamus has specialized receptors called **osmoreceptors** that are capable of sensing the water content of blood as it passes through the hypothalamus. If the water content is relatively low, the hypothalamus will send action potentials to the cells in the posterior pituitary which then secrete ADH into the bloodstream. The target tissue of this hormone is the collecting tubules of nephrons in the kidneys. When the collecting tubules detect ADH, they reabsorb water that would have been released as part of urine. The target tissues of some other pituitary hormones controlled by the hypothalamus are shown in Figure 3.

Many hormones, such as ADH, work using a mechanism called **negative feedback**. The goal of negative feedback is to maintain homeostasis. ADH and kidney function maintain a homeostatic level of water in the body. If water in the body rises above the homeostatic level, more urine is produced. If the water level becomes too low, ADH is produced and water is reabsorbed before becoming part of urine.

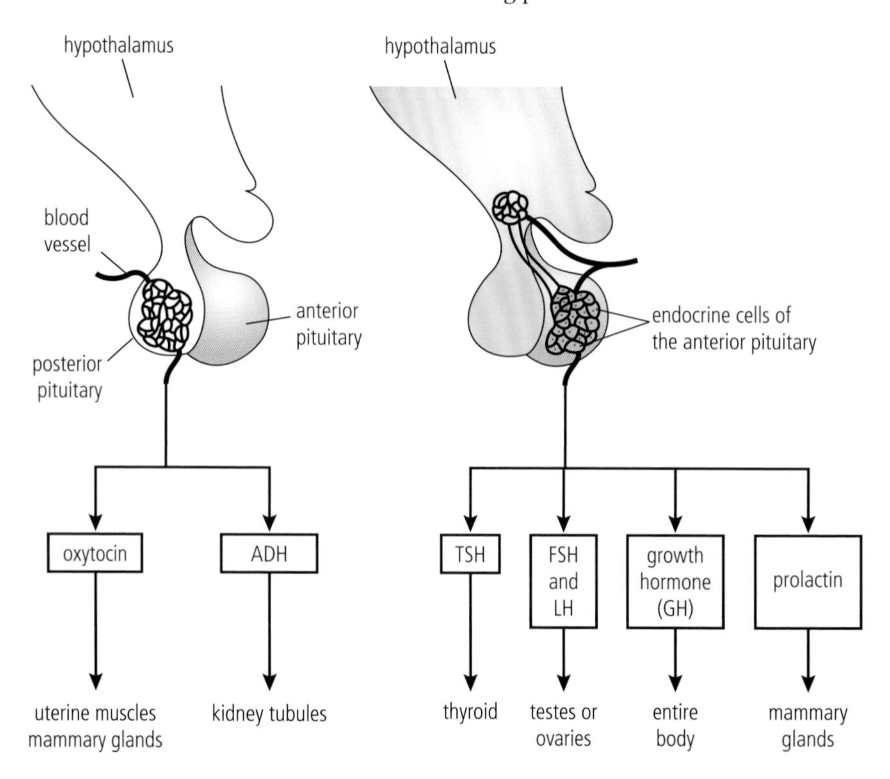

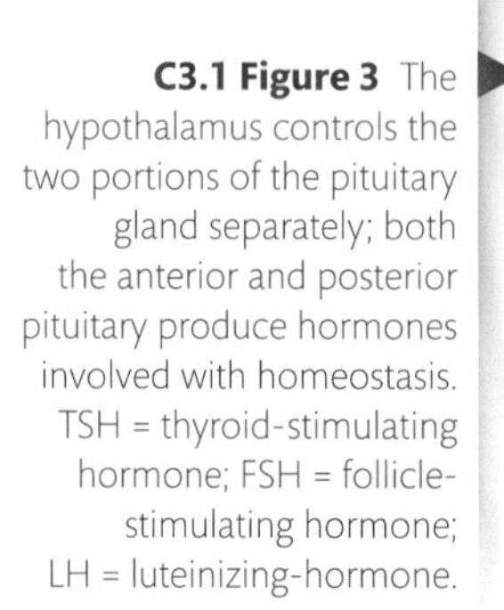

C3.1 Figure 3 The hypothalamus controls the two portions of the pituitary gland separately; both the anterior and posterior pituitary produce hormones involved with homeostasis. TSH = thyroid-stimulating hormone; FSH = follicle-stimulating hormone; LH = luteinizing-hormone.

C3.1.14 – Feedback control of heart rate

C3.1.14 – Feedback control of heart rate following sensory input from baroreceptors and chemoreceptors
Include the location of baroreceptors and chemoreceptors. Baroreceptors monitor blood pressure. Chemoreceptors monitor blood pH and concentrations of oxygen and carbon dioxide. Students should understand the role of the medulla in coordinating responses and sending nerve impulses to the heart to change the heart's stroke volume and heart rate.

Check your understanding of a topic by explaining diagrams to yourself or a friend. If you can verbalize a topic, you will better understand it.

Many physiological factors in the body change depending on environmental conditions and body activity. Changes to ventilation rate, body temperature and heart rate are brought back to set points by **feedback control** mechanisms. As an example, when the body is "at rest", the heart rate is under the control of the natural pacemaker within the heart known as the **sinoatrial** (**SA**) node. When you become active, muscle tissue requires additional oxygen and releases additional carbon dioxide as a result of the increased rate of cell respiration. An increase in heart rate and **stroke volume** is required in order to carry the additional respiratory gases to and from the lungs. Stroke volume is the volume of blood pumped out of the heart with each ventricular contraction.

Receptors known as **baroreceptors** and **chemoreceptors** are able to detect changes in the blood vessels and contents of the blood associated with an increase in the rate of cell respiration. Both baroreceptors and chemoreceptors are located in similar but not identical locations, as shown in Figure 4. The largest artery in the body is the **aorta**, and it forms an arch shape as it exits the left ventricle of the heart. One location for baroreceptors is on the arch of the aorta. Almost immediately, other major arteries begin to branch from the aortic arch. Two of those major branches are the **carotid arteries**, which carry oxygenated blood to your head and brain. Just before the two carotid arteries branch, they form an enlargement called a **sinus**. Both of the carotid sinuses also have baroreceptors on the walls of the blood vessels.

C3.1 Figure 4 The location of baroreceptors and chemoreceptors associated with changes in blood pressure, heart rate and stroke volume of the heart.

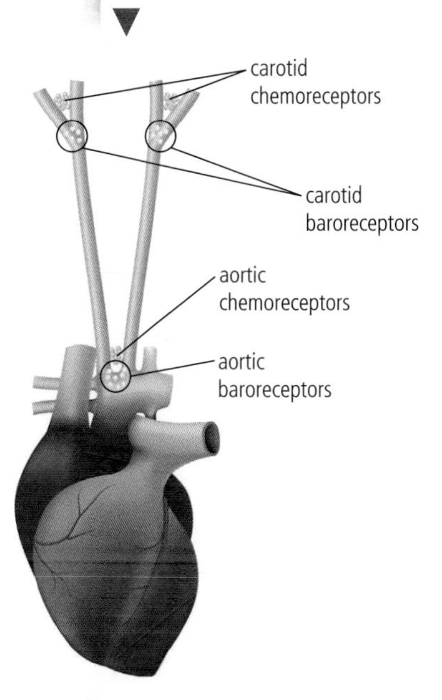

Chemoreceptors are located in tissue near where the baroreceptors are located but outside the blood vessels. Each of the major arteries has small arteries that extend from the vessel and immediately branch into capillaries. The chemoreceptors monitor oxygen, carbon dioxide and pH levels in these capillaries. The chemoreceptor cells are capable of releasing a neurotransmitter that initiates action potentials that are carried to the **medulla**.

What are examples of branching (dendritic) and net-like (reticulate) patterns of organization?

Baroreceptors

Baroreceptors are sensitive to pressure changes in arterial blood vessels. When blood pressure increases, the wall of an artery is distended or stretched outwards. This distention results in an increase in the rate of action potentials sent to the medulla. The medulla responds by sending impulses to the SA node to decrease the heart rate and decrease the force of contraction, leading to a lower stroke volume. When the blood pressure falls below normal, a decrease in action potentials sent to the medulla will lead to an increase in heart rate and stroke volume.

Baroreceptors and chemoreceptors are neurons that are part of the **autonomic nervous system (ANS)**. Any responses arising from these receptors that decrease heart rate and stroke volume are associated with the **parasympathetic division** of the ANS, whereas responses arising from the same receptors that lead to an increase in heart rate and stroke volume are part of the **sympathetic division** of the ANS.

Chemoreceptors

Chemoreceptors are sensitive to levels of three different factors in the bloodstream that change as a result of an increase in cell respiration rate:

- oxygen levels, which decrease because oxygen is a reactant of cell respiration
- carbon dioxide levels, which increase because carbon dioxide is a product of cell respiration
- pH, which lowers as most carbon dioxide entering the blood combines with water and forms carbonic acid.

Each of these chemical changes has its own chemoreceptors that send an increased rate of action potentials to the medulla when there is an increase in rate of cell respiration. An increase in heart rate and stroke volume will result from the action potentials sent from the medulla to the SA node. The same chemoreceptors can sense the opposite physiological effects when exercise ceases, and action potentials will be sent to the heart to slow the heart rate and lower the stroke volume.

C3.1.15 – Feedback control of ventilation rate

C3.1.15 – Feedback control of ventilation rate following sensory input from chemoreceptors
Students should understand the causes of pH changes in the blood. These changes are monitored by chemoreceptors in the brainstem and lead to the control of ventilation rate using signals to the diaphragm and intercostal muscles.

When resting, the ventilation rate of the lungs is controlled by groups of cells called **respiratory centres** located in your medulla. When at rest, spontaneous action potentials are released by these cells, which travel to your diaphragm and intercostal muscles to maintain breathing at a relatively slow and controlled pace.

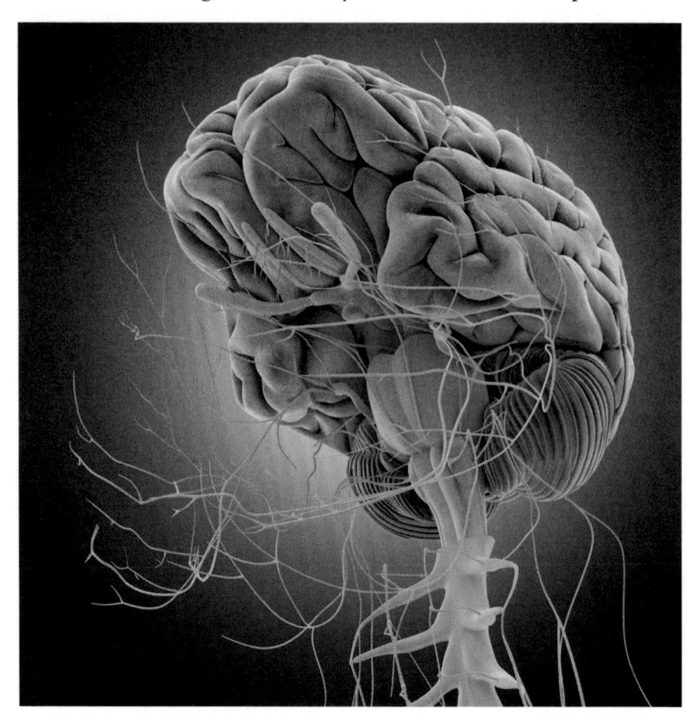

An illustration showing a portion of the brain and the 12 pairs of cranial nerves. The cranial nerves all connect into the brainstem and, among many other functions, carry both sensory and motor impulses related to heart and ventilation rates. If you look carefully, you can see branches of the cranial nerves forming the shape of the upper and lower jaws.

Chemoreceptors located in the medulla allow feedback control of the ventilation rate during and after exercise. Many chemoreceptors monitor the levels of carbon dioxide and pH in the blood passing through the medulla. The pH of blood typically falls within the small range of 7.35 to 7.45. In other words, blood is normally slightly alkaline. The response to exercise serves to keep the blood pH in this slightly alkaline range.

Body activity increases the rate of cell respiration and thus leads to an increase in carbon dioxide production. When carbon dioxide enters a red blood cell the following reaction, catalysed by carbonic anhydrase, occurs:

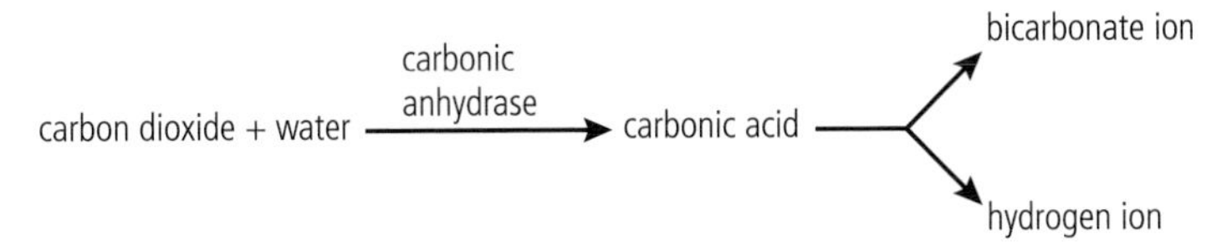

The production of carbonic acid and the resulting bicarbonate ion and hydrogen ion that occurs when carbon dioxide enters a red blood cell.

Strenuous exercise will lead to high levels of carbon dioxide being produced and a large number of hydrogen ions being produced. Chemoreceptors in the medulla sense this increase in hydrogen ions and send action potentials at a higher rate proportional to the number of hydrogen ions. All of the muscles associated with lung ventilation, especially the diaphragm and intercostal muscles, respond to these action potentials. Not only is the rate of ventilation increased but also the volume of air moving in and out is increased. When exercise decreases, hydrogen ion concentrations will also decrease and action potentials sent from the respiratory centres will decrease.

C3.1.16 – Control of peristalsis in the alimentary canal

C3.1.16 – Control of peristalsis in the digestive system by the central nervous system and enteric nervous system
Limit to initiation of swallowing of food and egestion of faeces being under voluntary control by the central nervous system (CNS) but peristalsis between these points in the digestive system being under involuntary control by the enteric nervous system (ENS). The action of the ENS ensures passage of material through the gut is coordinated.

Swallowing food is a voluntary action and is controlled by the **CNS**. Many hours later, after nutrients have been removed from the food, solid waste called **faeces** are egested, and that process is also under control of the CNS. A separate nervous system called the **enteric nervous system (ENS)** keeps the food (at various stages of digestion) moving along the **alimentary canal**. This movement of food is called **peristalsis** and is under **involuntary control**. The ENS is a web of sensory neurons, motor neurons and relay neurons embedded in the tissues of the alimentary canal, stretching from the lower portion of the oesophagus all the way to the rectum.

The alimentary canal is a long tube that begins at the mouth and finishes at the anus. Each part of the tube is innervated by neurons of the enteric nervous system (ENS). In addition, the tube is composed of smooth muscle (for peristalsis) and receives a rich blood supply for the absorption of nutrients from digested foods.

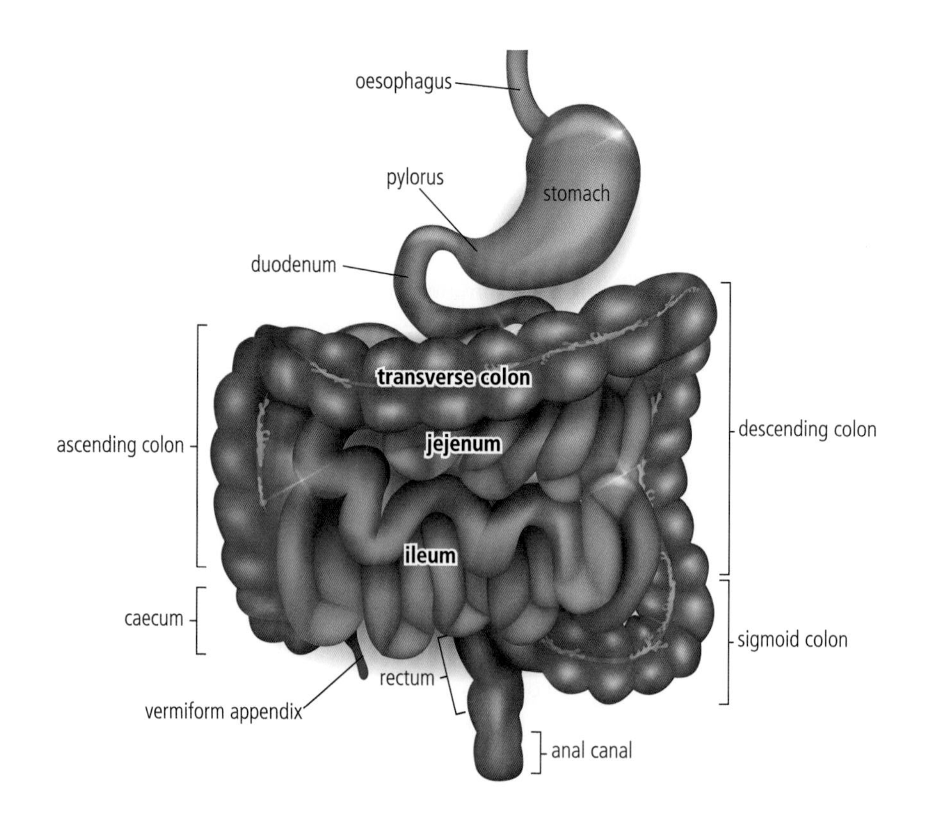

Peristaltic reflex

The **peristaltic reflex** is a series of smooth muscle contractions that occurs along the entire length of the alimentary canal to keep foods moving. Most smooth muscle in the body is controlled by areas of the brainstem. Peristalsis is controlled by the ENS because it has its own sensory, relay and motor neurons spread across the entire area of the alimentary canal.

When food moves through the alimentary canal it forms into rounded masses, each called a **bolus**. It is these food masses that initiate the peristaltic reflex. Wherever a bolus is located in the tube of the alimentary canal, that area of the tube becomes distended, stimulating stretch receptors in the ENS. These (sensory) stretch receptors then synapse with nearby relay neurons. The relay neurons in turn synapse with two different types of motor neurons. One type of motor neuron releases an **excitatory neurotransmitter** to an area of smooth muscle "behind" the bolus of food. This stimulates that area of smooth muscle to contract, pushing the bolus along. At the same time, another type of motor neuron releases an **inhibitory neurotransmitter** "ahead" of the bolus. The smooth muscle ahead of the bolus relaxes in response, and opens the lumen (central space of the tube) for the food bolus to slide through. This reflex occurs many times along the gut, ensuring that the movement of the food material moves forwards in a coordinated manner.

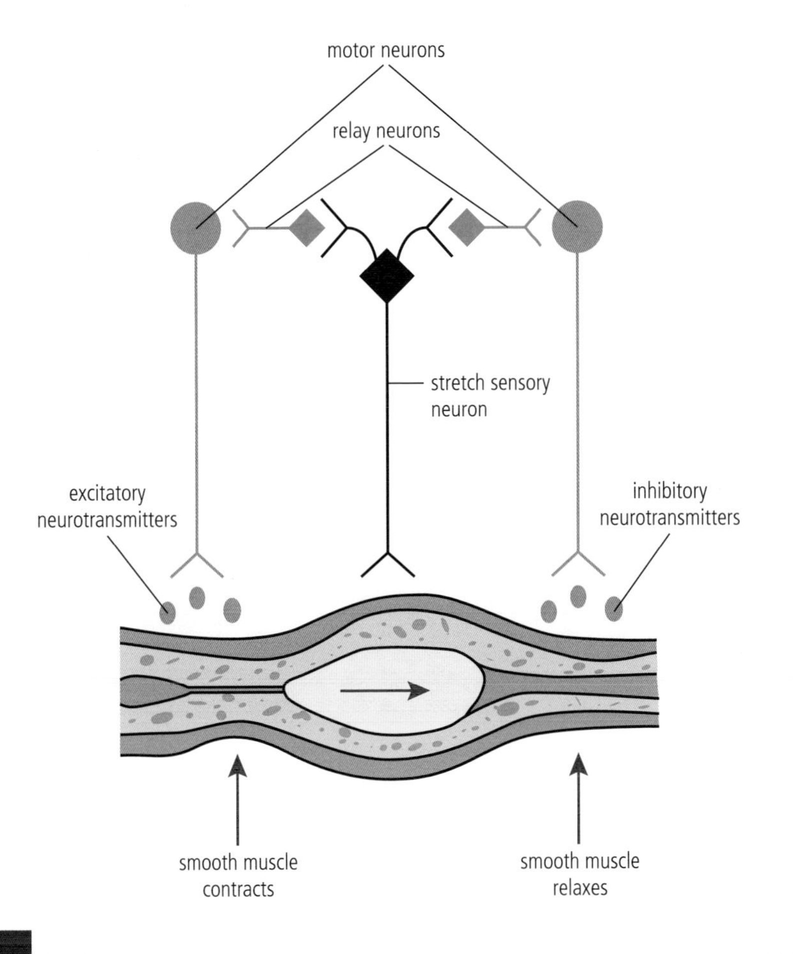

A bolus of food keeps moving in a single direction within the alimentary canal because of the peristaltic reflex. Peristaltic movements are quite rapid in the oesophagus, creating a churning motion while food is in the stomach, and then slow down in the small and large intestine.

There are many terms that end with "nervous system". Think about the meaning of each, as some are anatomical and some are functional in origin. Here are few used within this text:

- central nervous system (CNS), the term for the brain and spinal cord together
- peripheral nervous system (PNS), which comprises neurons and nerves outside the brain and spinal cord
- autonomic nervous system (ANS), which is any part of the nervous system that controls unconscious activities within the body
- enteric nervous system (ENS), which comprises those neurons that are located in, and control the unconscious peristaltic movement of food in, the alimentary canal.

HL

C3.1.17 and C3.1.18 – Tropic responses

C3.1.17 – Observations of tropic responses in seedlings

Application of skills: Students should gather qualitative data, using diagrams to record their observations of seedlings illustrating tropic responses. They could also collect quantitative data by measuring the angle of curvature of seedlings.

NOS: Students should be able to distinguish between qualitative and quantitative observations and understand factors that limit the precision of measurements and their accuracy. Strategies for increasing the precision, accuracy and reliability of measurements in tropism experiments could be considered.

C3.1.18 – Positive phototropism as a directional growth response to lateral light in plant shoots

Students are not required to know specific examples of other tropisms.

Plants often compete for light. As immobile (sessile) organisms, plants tend to live out their lives wherever a seed takes root. As the seedling shoot begins growth, it is often in an area where the greatest light availability is in one direction only. The root system may not be able to move location, but plants have evolved a mechanism that enables directional growth towards light. This phenomenon of growing towards a source of light is known as **phototropism**. It is categorized as a **positive tropic** response because the plant growth is towards the light stimulus. A tropism is any directional growth response to an external stimulus.

Tomato plant seedlings grown near a window demonstrating positive phototropism.

SKILLS

The IB requires you to grow seedlings of a plant species in order to observe and measure a tropic response. It should be possible to gather both **qualitative** and **quantitative** data during this investigation. Qualitative (descriptive) data can be in the form of daily descriptions of the plant growth, taking note of factors such as early shoot growth, appearance of **cotyledons**, appearance of first true leaves, and direction of growth towards or away from a stimulus. Diagrams can be useful for recording qualitative data.

Quantitative data is based on measurements, and in this case should focus on the curvature of growth of the stems of the seedlings. Curvature can be measured as an angle of growth, with greater angles of curvature indicating a greater tropic response. If the curved growth is in the direction of a stimulus, you will be measuring a positive tropic response; if the curved growth is away from a stimulus, you will be measuring a negative tropic response.

You should use your own experimental design or a design planned and discussed by a group, but here some design factors to consider.

- Phototropism is the only plant tropic response detailed within the IB curriculum. The IB suggests you use directional light as your stimulus, but if you want to research other plant tropic responses you may.
- What planting conditions will you establish? Consider the type of seeds and consistent soil and watering conditions.
- Consider using varying intervals or intensities of the stimulus (e.g. different plant groups with varying light intervals, varying light intensities, or varying light angles). This will be your independent variable.
- How many seedlings will make up any one group?
- What are the controlled variables that will be held constant across all the plant groups? What design can incorporate all those controlled variables?
- Plant one or more seedling groups that will act as a control group(s).
- What qualitative data do you plan on measuring? One or more labelled diagrams of plants showing the tropic response may be useful.
- How will you measure the curvature of the stems (quantitative data)? How often will you take those measurements? Stem curvature will be your dependent variable.
- What strategies will you use to increase the precision, accuracy and reliability of your measurements?

Nature of Science

All measuring tools have a degree of uncertainty that should be reported with the data set. That uncertainty is based on the smallest increments for which the tool is marked. If you use a protractor where the smallest markings are one degree, then you should record any measurements as a number of degrees followed by +/– 1°. This can be given in the heading of those measurements within a data table.

TOK

Why is it important for researchers to note the degree of uncertainty within their published works?

Nature of Science

In science, good experimental design and practice use strategies that attempt to maximize:

- precision – precise data is data where all measurements within a single grouping are close to each other
- accuracy – accurate data comes close to an accepted value
- reliability – reliable data achieves the same or very similar results consistently by using the same methods under the same circumstances.

In biological experimentation, it is not always possible to determine the accuracy because an accepted value may not be available. If your data sets show poor precision, you may have to reconsider your procedure and/or measuring tools and begin again. Most importantly, always report the precision of your measurement tools and discuss the precision, accuracy and reliability of your data sets as part of your laboratory report, including their possible importance to any conclusions you may draw.

C3.1.19 – Plant hormones

C3.1.19 – Phytohormones as signalling chemicals controlling growth, development and response to stimuli in plants
Students should appreciate that a variety of chemicals are used as phytohormones in plants.

Plants do not have a nervous system, but they can respond to environmental cues by producing hormones. Hormones produced by plants are called **phytohormones**. Phytohormones are used as signalling molecules to control growth and the development of flowers, fruit and seeds, and to help the plant respond to environmental stimuli. The specifics of some phytohormones are discussed in Sections C3.1.20–23.

Common phytohormones

Phytohormone	Brief function
Auxin	Results in plant cell elongation
Cytokinin	Increases rate of cell division
Ethylene	Promotes fruit ripening
Gibberellin	Controls stem elongation, seed germination, flowering and dormancy

C3.1.20 – Auxin efflux carriers

C3.1.20 – Auxin efflux carriers as an example of maintaining concentration gradients of phytohormones
Auxin can diffuse freely into plant cells but not out of them. Auxin efflux carriers can be positioned in a cell membrane on one side of the cell. If all cells coordinate to concentrate these carriers on the same side, auxin is actively transported from cell to cell through the plant tissue and becomes concentrated in part of the plant.

Auxin is a phytohormone that is produced in the growing regions of plants, specifically in the tips of shoots and roots and growth buds. Under certain environmental influences, auxin is evenly concentrated in an area of growth and cell elongation is uniform in that area. Auxin can enter phloem tissue and be moved throughout the plant within phloem sap.

Plant tissues can change the position of the auxin efflux carriers to change auxin concentrations when light availability changes.

Auxin can also be concentrated in certain areas of a plant or even on one side of growing tissue, leading to differential growth. Auxin easily enters into cells by diffusion but requires membrane proteins known as **auxin efflux carriers** to exit a cell. If needed, a plant can distribute efflux carriers predominantly on one side of a series of adjoining cells to encourage a one-way movement of auxin through that series of cells. This cell-to-cell movement is a type of active transport because adenosine triphosphate (ATP) is required.

An auxin source is any area of growing tissue in a plant, such as the tip of a plant shoot. Plants only grow in length at their tips. An auxin sink is the area of plant tissue that the auxin is moving towards. Auxin efflux carriers are distributed within each cell so that they are closer to the auxin sink. Auxin cannot move back towards the auxin source, because the carrier proteins are not located on that side of the cell.

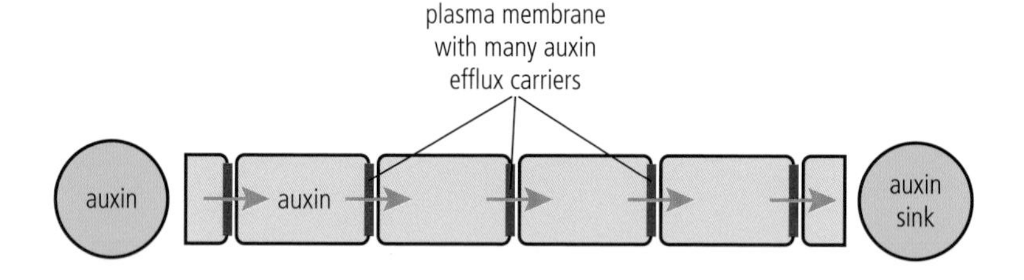

C3.1.21 – Plant cell elongation

C3.1.21 – Promotion of cell growth by auxin
Include auxin's promotion of hydrogen ion secretion into the apoplast, acidifying the cell wall and thus loosening cross links between cellulose molecules and facilitating cell elongation. Concentration gradients of auxin cause the differences in growth rate needed for phototropism.

Cell elongation in plants necessitates an increase in length of plant cell walls. This can only occur if the cross-linking fibres of cellulose are first loosened. Auxin is the phytohormone that promotes this loosening. When auxin enters a cell, it promotes the synthesis of hydrogen ion (proton) pumps. Auxin also binds to the proteins making up the hydrogen pumps and stimulates their insertion into the plasma membrane of the cell.

Using ATP as an energy source, the pumps move hydrogen ions from the interior of the cell through the plasma membrane. This concentrates the hydrogen ions (or acidifies the fluids) in an area called the **apoplast**. The apoplast is any area of adjoining plant cells outside the plasma membranes. This would include the cell wall and any intercellular spaces between cell walls. The hydrogen ions activate a protein called **expansin** already found within the cell wall. Expansin loosens the hydrogen bonds that cross-link cellulose fibres, allowing the fibres to move past each other into new positions. Absorption of water will create high internal **turgor pressure** creating the force necessary for the fibres to slide past each other. The fibres of cellulose make new hydrogen bonds in their new positions. The result is longer cell walls and elongated cells.

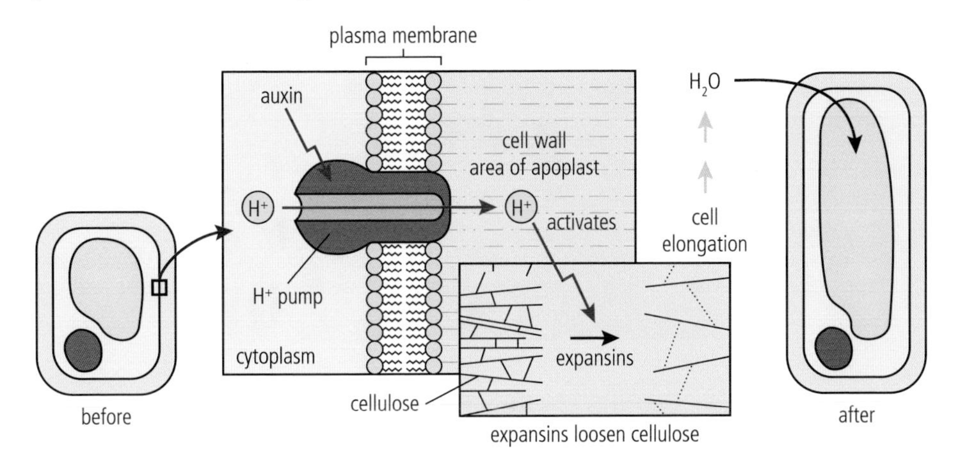

Auxin's effect on the elongation of a plant cell. Auxin promotes hydrogen ion pumps to release protons into the apoplast surrounding the plasma membrane. Protons activate expansin, and as a result cellulose fibres are loosened. Note the absorption of water creating a high turgor pressure that helps move the cellulose fibres into new positions.

In Sections C3.1.17–18 you studied phototropism, a positive directional growth response to light. Plants position auxin efflux pumps on the side of a growing shoot away from the direction of light (the shaded side). The presence of auxin results in elongated cells on the shaded side and results in a curved growth pattern. If the availability of light changes as the plant grows taller, auxin efflux pumps can be repositioned on the new shaded side and new growth will once again turn towards the availability of light.

C3.1.22 – Integration of root and shoot growth

C3.1.22 – Interactions between auxin and cytokinin as a means of regulating root and shoot growth
Students should understand that root tips produce cytokinin, which is transported to shoots, and shoot tips produce auxin, which is transported to roots. Interactions between these phytohormones help to ensure that root and shoot growth are integrated.

The growth phytohormones **auxin** and **cytokinin** are produced in the growing regions of plants. Specifically, auxin is produced in the shoot growing region and cytokinin is produced in the root growing region. That means that both hormones must be transported to affect a region of growth where they are not produced. Cytokinin is transported in xylem fluid, because the direction of that fluid is from the root towards the shoot. Auxin is transported from the shoot to the root in the sap within phloem tissue.

The growing regions at the tips of shoots, buds and roots is called **meristem tissue**.

At certain concentrations, these two hormones work synergistically with each other. In other words, the best plant growth is when the growing tips of shoots and roots have access to both cytokinin (for cell division) and auxin (for cell elongation).

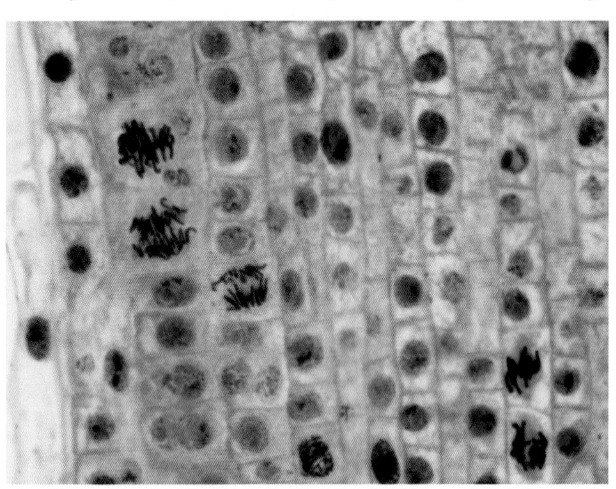

A light micrograph section of an onion (*Allium cepa*) root tip showing several cells undergoing mitosis. Rapid cell division in this area of the plant is stimulated by the synergistic effect of cytokinin in the presence of auxin. Notice that many of these cells are shaped like small boxes; these will soon elongate, primarily as a result of the effects of auxin.

C3.1.23 – Feedback control of fruit ripening

C3.1.23 – Positive feedback in fruit ripening and ethylene production
Ethylene (IUPAC name: ethene) stimulates the changes in fruits that occur during ripening, and ripening also stimulates increased production of ethylene. Students should understand the benefit of this positive feedback mechanism in ensuring that fruit ripening is rapid and synchronized.

Fruits go through a series of developmental stages. One of the later stages is **ripening**. During ripening, fruits produce a gas called **ethylene** (called ethene by the International Union of Pure and Applied Chemistry, IUPAC). Because ethylene is a gas, when one fruit begins to ripen and produce ethylene, any adjacent fruit is exposed to the gas, and all the fruit then ripen quickly and at the same time. This is an example of a **positive feedback mechanism** because the release of ethylene is a self-perpetuating event. In other words, the release of ethylene leads to a higher release of ethylene.

Why have plant species evolved to have all of their fruit ripen at the same time? The answer lies in the purpose of the fruit from the plant's perspective. A fruit is an ovary with one or more seeds contained within it. Most fruits are green when they start their development, and then become a bright and less camouflaged colour as they ripen. Most fruits also become sweeter as they ripen. This strategy is to attract herbivores that eat the fruit and disperse the seeds when they deposit their faeces. A display of bright fruit is intended to be a colourful and tasty advertisement to attract helpful animals.

What are the consequences of positive feedback in biological systems?

HL end

Guiding Question revisited

What are the roles of nerves and hormones in integration of body systems?

Within this chapter you have learned:

- coordination and communication is necessary between cells, tissues and organs for them to work as an entire organism
- the nervous system and endocrine system are both important for information integration within many animals
- the brain is a central information integration organ in animals
- sensory information is carried to the brain and spinal cord through sensory neurons
- motor information is carried away from the brain and spinal cord to muscles through motor neurons
- the cerebellum coordinates impulses going to muscles to make movements smooth and efficient
- the hypothalamus and pituitary glands control many hormones secreted from endocrine glands
- HL plants use chemicals called phytohormones to control many processes, including directional growth and fruit ripening.

Guiding Question revisited

What are the roles of feedback mechanisms in regulation of body systems?

Within this chapter you have learned:

- melatonin secretion by the pineal gland helps regulate the circadian rhythm of the sleep/wake cycle
- epinephrine secretion by the adrenal glands prepares the body for immediate rigorous activity
- baroreceptors and chemoreceptors send information to the medulla in order to regulate contraction rate and stroke volume of the heart
- the brainstem uses chemoreceptors to send information to regulate ventilation rate
- peristalsis is controlled by neurons within the alimentary canal that form the enteric nervous system
- most hormones help maintain homeostasis within the body and use negative feedback control to maintain a physiological variable within a narrow range
- HL fruit ripening is controlled by a positive feedback mechanism where ethylene production is enhanced by previous ethylene production.

Exercises

Q1. Identify the portion of the human brain most closely associated with the following functions.

(a) Recalling memories from earlier in the day.

(b) Coordinating motor action potentials for running.

(c) Increasing heart rate during running.

(d) Interpreting action potentials from the retinal cells of the eyes.

(e) Learning new information.

Q2. What is the advantage of having one or more muscle movements occur through a pain reflex arc rather than initiated by the cerebrum?

Q3. Which of these is not a part of the enteric nervous system?

A Stretch sensory neuron

B Relay neuron

C Interneuron within brainstem

D Inhibitory neurotransmitter

Q4. Outline what happens to your blood pH and why when you exercise.

Q5. Outline the difference between a "nerve" and a "neuron".

Q6. Nervous system receptors are neurons that are specialized to convert a specific physical stimulus to action potentials. What is the physical stimulus for each of these named receptor types?

(a) Photoreceptors

(b) Baroreceptors

(c) Chemoreceptors

(d) Nocireceptors

Q7. Briefly describe the mechanisms for movement of auxin for each of the following.

(a) Movement of auxin from the growing tip of a plant shoot to the roots.

(b) Movement of auxin within the growing tip of a plant.

Q8. A common adage is "one rotten apple spoils the barrel". Explain how and why there is truth in that saying.

Q9. HL How do auxins cause plant shoots to grow towards light?

A Increase cell division on the side of the stem near the light source.

B Increase cell division on the side of the stem away from the light source.

C Increase cell elongation on the side of the stem near the light source.

D Increase cell elongation on the side of the stem away from the light source.

C3.2 Defence against disease

Guiding Questions

How do body systems recognize pathogens and fight infections?

What factors influence the incidence of disease in populations?

Pathogens are viruses, bacteria and other small organisms that can cause disease. Our first defence against pathogens is to prevent them from entering the body, our skin and mucous membranes acting as barriers. When a pathogen is able to enter the body, we have a two-layered immune system that responds to fight the infection. Our innate immune system consists of white blood cells called phagocytes that recognize pathogens as foreign and engulf them by endocytosis resulting in their digestion.

True immunity is built up over time by our adaptive immune system. This component of our immune system chemically recognizes the specific molecules that make up pathogens. These molecules are called antigens. A type of white blood cell called lymphocytes cooperates in the presence of specific antigens. This leads to cell cloning of specific lymphocytes to fight off the pathogens that carry the identified antigen. One important component of this response is the production of proteins called antibodies. Long-lived memory lymphocytes remain after an infection, providing long-term immunity.

There are many factors that influence the incidence of disease in populations. We only began to understand the causes of diseases about two hundred years ago. Since then, great advances have been made in disease prevention, especially in the way that human wastes are treated and sanitary conditions have been improved for water sources and preparing foods. Identification and treatments for viral diseases have been the most difficult to understand, with many viral diseases remaining and spreading in the population. Vaccines are our best protection against viruses, but there are challenges to their acceptance in some human populations. Bacterial diseases have been successfully treated by antibiotics, but their overuse is leading to the emergence of antibiotic-resistant strains of bacteria.

C3.2.1 – Infectious diseases are caused by pathogens

C3.2.1 – Pathogens as the cause of infectious diseases

Students should understand that a broad range of disease-causing organisms can infect humans. A disease-causing organism is known as a pathogen, although typically the term is reserved for viruses, bacteria, fungi and protists. Archaea are not known to cause any diseases in humans.

NOS: Students should be aware that careful observation can lead to important progress. For example, careful observations during 19th-century epidemics of childbed fever (due to an infection after childbirth) in Vienna and cholera in London led to breakthroughs in the control of infectious disease.

How fast does a disease spread? Full details of how to carry out this activity with a worksheet are available in the eBook.

Pathogens are disease-causing organisms. Pathogens are any viruses, bacteria, fungi and protists that result in disease upon entry into the body. The vast majority of these types of organisms are (thankfully) not pathogenic to humans. There are some that are only pathogenic to other animals, which explains why cats, dogs and farm animals are susceptible to their own diseases. Most of these diseases will not transmit to humans, although there are exceptions.

▲ This community water pump was responsible for 616 deaths in the 1854 cholera outbreak in London, UK.

Nature of Science

Meticulous observations can lead to breakthroughs. In 1854, there was a major cholera outbreak in a suburb of London, UK. The *Vibrio cholerae* bacteria that causes cholera is found in the faeces of infected individuals. It was common practice in the 1800s for residents to empty human waste into areas in front of their homes. The bacteria moved down through the soil and infected a drinking well used by many in the community. A physician by the name of John Snow suspected cholera was being transmitted in water supplies. He created a map of all known infections and found one particular well had been used by all the infected people. The well was closed, and the number of cholera infections fell. The information and map created by Snow formed the basis of modern epidemiology studies that trace outbreaks of disease.

Pathogenic organisms are not inherently "evil". They just happen to use human tissues as food and shelter. Their means of growth and secretions, however, can cause us harm. We have begun to learn a great deal about pathogenic organisms and have devised treatments for both before and after infection. Improvements in public health policies also help prevent the spread of pathogenic organisms. Throughout most of recorded history, humans had no real knowledge of the presence of pathogens and blamed infectious diseases on factors that now seem almost nonsensical to us. We must always remember that we live in an age where science is providing information not available to us even one or two generations ago.

Nature of Science

In the mid-1800s, a physician called Ignaz Semmelweis, in Vienna, Austria, began to study a lethal disease commonly known as childbed fever. The disease had alarmingly high rates of infection and death in Vienna's maternity wards. Semmelweis began to note a difference in infection rates between two maternity wards. One ward was staffed by midwives and the other by physicians and medical students. The rate of infection and death on the physicians' ward was much higher than the ward staffed by midwives. Semmelweis began to narrow down and eliminate specific differences between the two wards, besides who staffed them. The one difference that proved to be consequential was that the physicians had often carried out autopsies before helping women in childbirth. Semmelweis postulated that the physicians were carrying small "particles" that caused disease from the autopsies to the pregnant women. He ordered the physicians to wash their hands with a chlorine solution before treating patients, and there was an immediate improvement in infection rate in the maternity ward.

Modern classification systems place all living organisms in one of three domains. One of these three domains is Archaea, comprising small single-celled prokaryotic cells that are fundamentally different from bacteria. No member of the Archaea domain is known to cause a human disease.

C3.2.2 – Skin and mucous membranes as the first line of defence

C3.2.2 – Skin and mucous membranes as a primary defence
The skin acts as both a physical and chemical barrier to pathogens. Students are not required to draw or label diagrams of skin.

The best way to stay healthy is to prevent pathogens from having the opportunity to cause disease. One way to do this is to try to stay away from sources of infection. This is why it is still common to isolate (or quarantine) people who have highly transmittable diseases. Obviously, it is not possible to isolate yourself from every potential source of infection. Therefore, the human body has evolved some ingenious ways of making it difficult for pathogens to enter and start an infection. One of those ingenious ways is your skin.

Think of your skin as having two primary layers. The underneath layer is called the **dermis** and is very much alive. It contains sweat glands, capillaries, sensory receptors and dermal cells, which give structure and strength to the skin. The layer on top of this is called the **epidermis**. This epidermal layer is constantly being replaced as the underlying dermal cells die and are moved upwards. This layer of mainly dead cells forms a physical barrier against most pathogens because it is not truly alive. As long as our skin remains intact, we are protected from most pathogens that can enter living tissues. This is why it is important to clean and cover cuts and abrasions of the skin when they do occur.

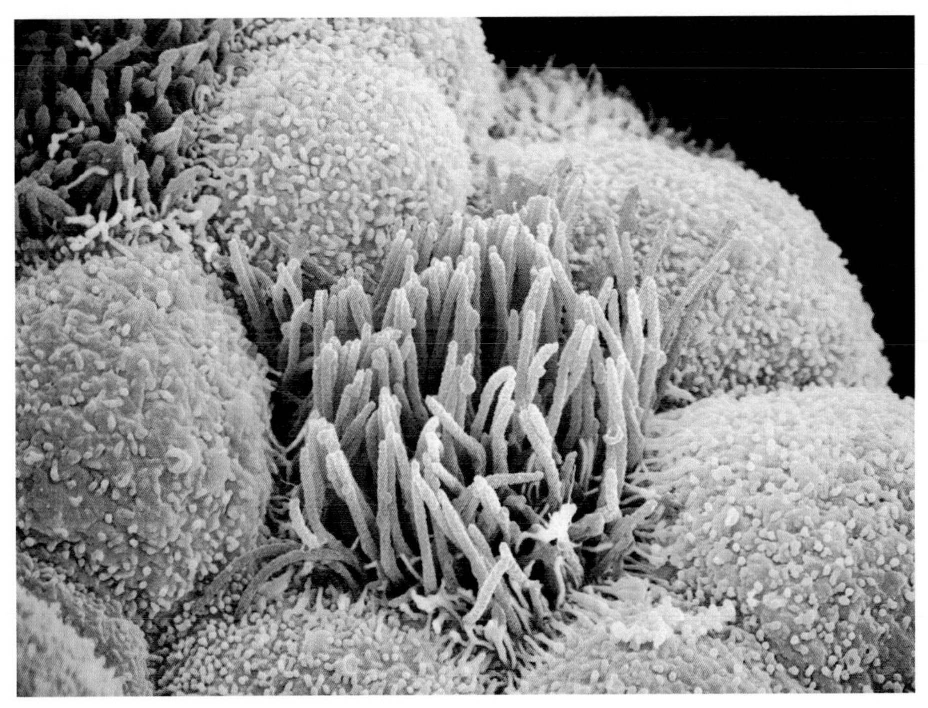

A false-colour scanning electron micrograph (SEM) of the mucous membrane lining of the trachea. The large white cells are called goblet cells and they secrete mucus. Hair-like cilia (in pink) are also visible.

Pathogens can enter the body at the few locations that are not covered by skin. These entry points are lined with tissue cells that form a **mucous membrane**. The cells of mucous membranes produce and secrete a lining of sticky mucus. This mucus can trap incoming pathogens and so prevent them from reaching cells that they could

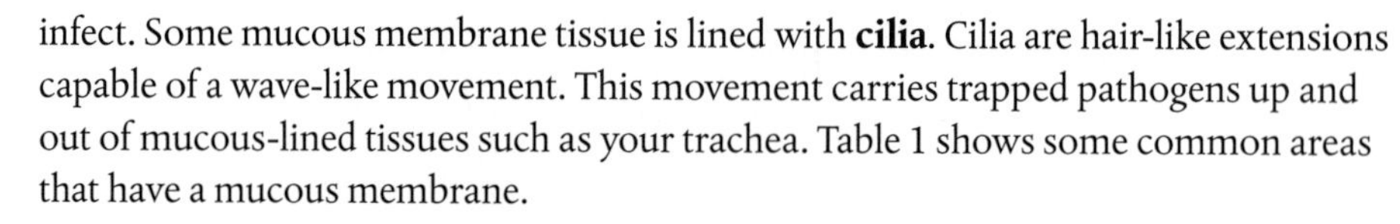

infect. Some mucous membrane tissue is lined with **cilia**. Cilia are hair-like extensions capable of a wave-like movement. This movement carries trapped pathogens up and out of mucous-lined tissues such as your trachea. Table 1 shows some common areas that have a mucous membrane.

C3.2 Table 1 Areas of the body that have a mucous membrane

Area with a mucous membrane	What it is and does
Trachea	The tube that carries air to and from the lungs
Nasal passages	Tubes that allow air to enter the nose and then the trachea
Urethra	A tube that carries urine from the bladder to the outside
Vagina	The reproductive tract leading from the uterus to the outside

According to an article published by the National Institutes of Health (NIH), bacteria outnumber their human hosts by about 10 to 1 cells. In a typical human adult, bacteria therefore account for about 2% of the human's body mass.

C3.2.3 – Blood clotting minimizes blood loss and infection

C3.2.3 – Sealing of cuts in skin by blood clotting
Include release of clotting factors from platelets and the subsequent cascade pathway that results in rapid conversion of fibrinogen to fibrin by thrombin and trapping of erythrocytes to form a clot. No further details are expected.

When small blood vessels such as capillaries, arterioles and venules are damaged, blood escapes from the closed circulatory system. Often the damaged blood vessels are in the skin, and so pathogens are then able to enter the body. Our bodies have evolved a set of responses to create a clot that "seals" the damaged blood vessels, so preventing excessive blood loss and helping prevent pathogens from entering the body.

Circulating in the blood plasma are a variety of molecules called **plasma proteins**. These proteins serve many purposes, including some that are involved in clotting. Two of the clotting proteins are **prothrombin** and **fibrinogen**. These two molecules are always present in blood plasma, but remain inactive until "called to action" by events associated with bleeding. Also circulating in the bloodstream are cell fragments known as **platelets**. Platelets form in the bone marrow, along with red blood cells (**erythrocytes**) and white blood cells (**leucocytes**), but do not remain as entire cells. Instead, one very large cell breaks down into many fragments, and each of the fragments becomes a platelet. Platelets do not have a nucleus and they have a relatively short cellular life span of about 8–10 days.

Consider what happens when a small blood vessel is damaged. The damaged cells of the blood vessel release chemicals that stimulate platelets to adhere to the damaged area, forming a "plug". The damaged tissue and platelets release chemicals called **clotting factors** that convert prothrombin to **thrombin**. Thrombin is an active enzyme that catalyses the conversion of soluble fibrinogen into the relatively insoluble **fibrin**. The appropriately named fibrin is a fibrous protein that forms a mesh-like network that helps to stabilize the platelet plug. More and more cellular debris becomes trapped in the fibrin mesh, and soon a stable clot has formed, preventing both further blood loss and the entry of pathogens.

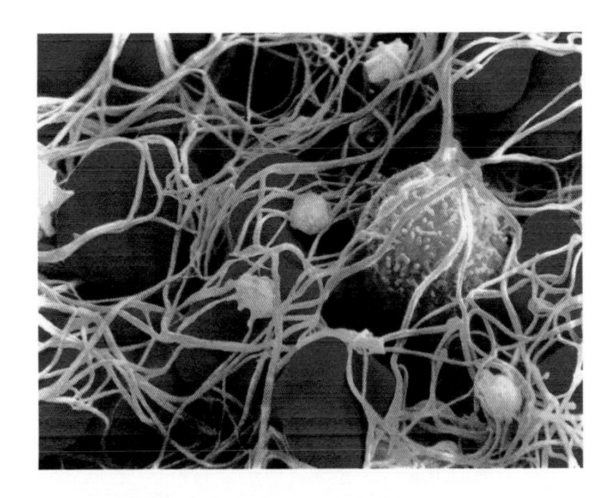

This false-colour SEM shows the formation of a blood clot. Small platelets (roughly spherical in shape and shown in pale green) have triggered the formation of insoluble fibrin protein fibres. Trapped in the fibrin are several red blood cells, platelets and one white blood cell (a larger sphere shape shown in yellow).

C3.2.4 – A two-layered immune system: innate and adaptive

C3.2.4 – Differences between the innate immune system and the adaptive immune system
Include the idea that the innate system responds to broad categories of pathogen and does not change during an organism's life whereas the adaptive system responds in a specific way to particular pathogens and builds up a memory of pathogens encountered, so the immune response becomes more effective. Students are not required to know any components of the innate immune system other than phagocytes.

Humans are born with an immune system called the **innate immune system**. This first layer of the immune system responds to broad categories of pathogens and does not change during a person's lifetime. For example, the innate immune system would recognize any bacterium as a bacterium, rather than a specific species of bacterium.

The basis of the innate immune response is the ability to recognize those things that belong in the human body versus those that do not belong. In other words, it can recognize and respond to things that are "not-self". This includes bacteria, viruses, protists and fungi, and even things such as pollen and dust. The molecules of these foreign or not-self entities that can trigger an immune response are called **antigens**. The innate immune response involves activation of a group of leucocytes called **phagocytes**, which area capable of engulfing invading material by **endocytosis**.

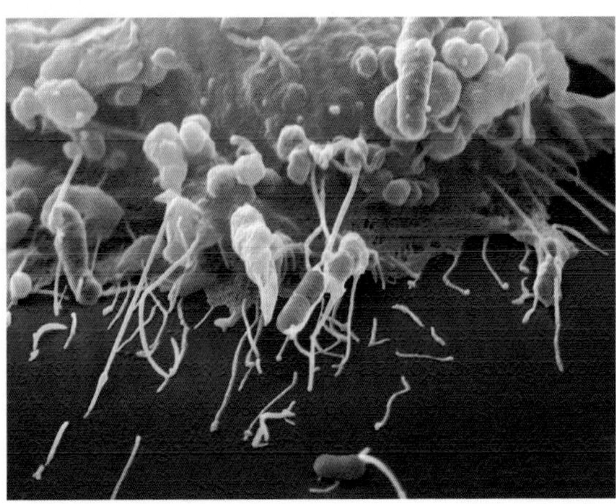

A false-colour SEM of a large phagocyte (yellow) that has recognized a group of bacteria (pink rod shapes) as "not-self" and is in the initial stages of endocytosis.

The second layer of human immunity is called the **adaptive immune response**. This portion of our immune response develops over time and only after exposure to specific antigens of specific pathogens. The first exposure to a specific antigen leads to a series of cellular events culminating in molecules and cells that are long-lived and have the ability to defend the body against a specific pathogen. The specific long-lived white blood cells that are formed during the first exposure are called **memory cells**. Upon a second exposure to the same pathogen, these specific memory cells can be activated quickly. They can be so effective in fighting a pathogen that a person may not even realize that they were exposed a second time. The adaptive immune response becomes more effective with age, as a person becomes exposed to more pathogens.

C3.2.5 – The role of phagocytes

C3.2.5 – Infection control by phagocytes
Include amoeboid movement from blood to sites of infection, where phagocytes recognize pathogens, engulf them by endocytosis and digest them using enzymes from lysosomes.

Phagocytes are leucocytes (white blood cells) that are capable of an action called **amoeboid movement**. Cells capable of amoeboid movement can purposefully extend sections of their plasma membrane, followed by their cytoplasm and organelles. Phagocytes use this type of motion to squeeze their way through capillaries so that they can leave and enter the bloodstream in order to move through body tissues. When a phagocyte encounters something in body tissues that contains antigens and thus is not-self, it sends out plasma membrane extensions to engulf the foreign body in a process called **endocytosis**. The foreign body is brought inside the phagocyte, where the hydrolytic enzymes of **lysosomes** digest the potential invader. This response by phagocytes is non-specific and is part of the innate immune response.

A portion of the plasma membrane of a phagocyte engulfs a bacterium by endocytosis. Two stages are shown. The bacterium ends up being encased in a vesicle that is later digested by enzymes from one or more lysosomes.

C3.2.6 – The role of lymphocytes

C3.2.6 – Lymphocytes as cells in the adaptive immune system that cooperate to produce antibodies

Students should understand that lymphocytes circulate in the blood and are contained in lymph nodes. They should appreciate that an individual has a very large number of B-lymphocytes that each make a specific type of antibody.

There are many types of leucocyte that contribute to the human immune system. Two major types are called **B-lymphocytes** and **T-lymphocytes.** Sometimes their names are shortened to just B-cells and T-cells. Lymphocytes continuously circulate in the blood stream and are also contained within our lymphatic system, especially within lymph nodes. We are going to look at the function of B-lymphocytes first, and then explore the functions of T-lymphocytes.

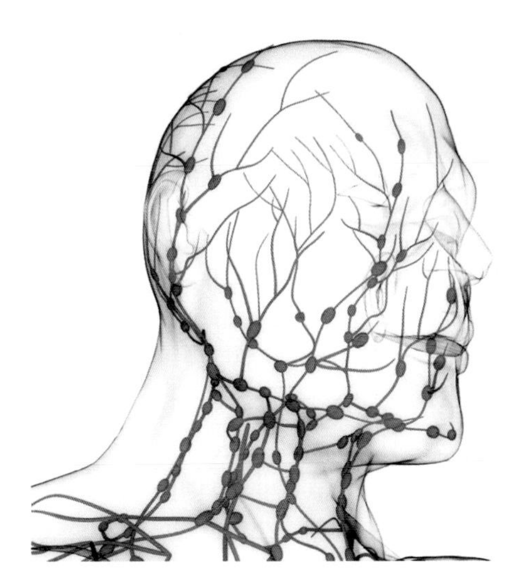

A diagram showing the location of lymph nodes in the neck and head area. B-lymphocytes accumulate in lymph nodes. Each lymph node has lymph vessels bringing lymph fluid in, and other lymph vessels taking lymph fluid away.

The specific leucocytes called B-lymphocytes produce protein molecules called **antibodies** as part of the adaptive immune response. There are many types of B-lymphocytes, and each type is able to synthesize a specific antibody. Each specific antibody is able to recognize and bind to a specific antigen. If you had a measles infection, you would produce one type of antibody, and if you contract a virus that gives you influenza (flu), you would produce another type of antibody. Each type of antibody is different because each type has been produced in response to a different pathogen.

Each type of B-lymphocyte is a biological factory for synthesizing only one type of antibody.

Antibodies are a Y-shaped proteins. At the end of each of the branches of the Y is a **binding site**. The binding sites are where an antibody attaches itself to an antigen. Because the antigen is a protein on the surface of a pathogen (such as a bacterium), the antibody thus becomes attached to the pathogen. Each of us has many different types of antibody-producing B-lymphocyte cells and each can produce only one type of antibody.

The specificity that an antibody has for a certain antigen is not unlike an enzyme's specificity for a certain substrate or a hormone's specificity for a target protein found on only certain cells.

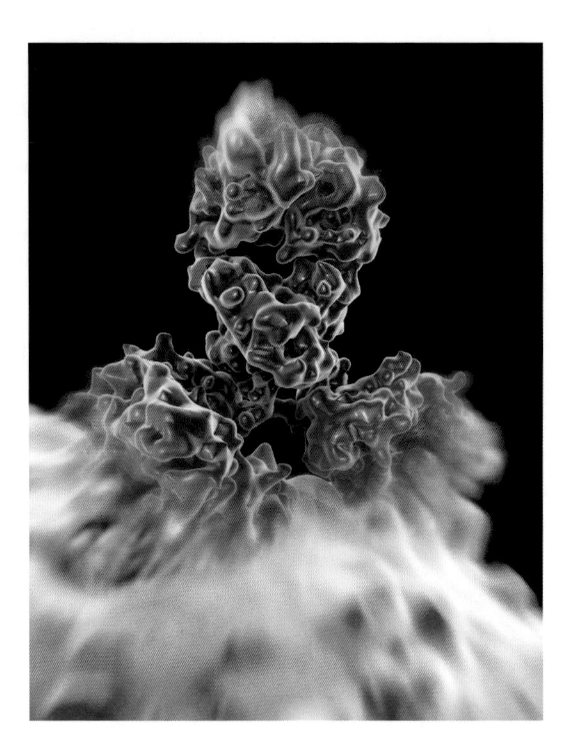

Computer artwork showing a single antibody (a blue, upside down Y shape in this picture) attached to an antigen on the surface of a pathogen.

Antibodies have specific mechanisms for fighting off infection by a pathogen. Any one antibody has two binding sites. If multiple antibodies bind to a cluster of pathogens, a clump is created because each antibody can potentially bind to two different pathogens. The antibody-bound cluster makes it easier for phagocytes to find and engulf the entire clump.

Many viruses attach to the plasma membrane of a body cell. The DNA or RNA of the virus is then injected into the cell. This cell then becomes a cellular factory to make more viruses. The protein coat of the virus, called a **capsid**, often remains on the outer plasma membrane of the cell and can be recognized by antibodies. Multiple antibodies of the same type use their binding sites to attach to proteins of the capsids. This is a way of marking the infected cell to be engulfed later by phagocytes.

C3.2.7 – Antigens trigger antibody production

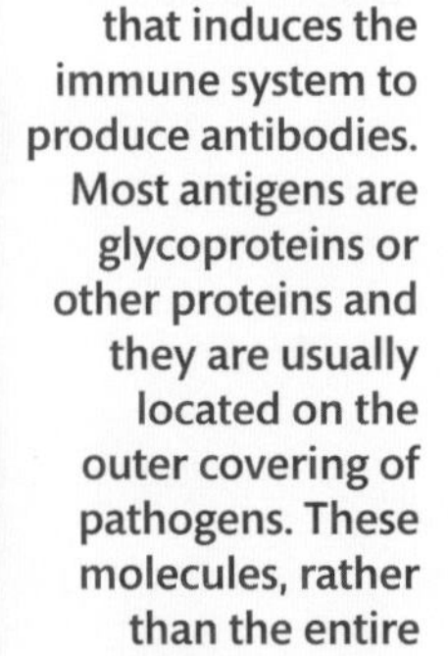

An antigen is any substance that induces the immune system to produce antibodies. Most antigens are glycoproteins or other proteins and they are usually located on the outer covering of pathogens. These molecules, rather than the entire pathogen, are the molecular antigens that result in an immune response.

C3.2.7 – Antigens as recognition molecules that trigger antibody production
Students should appreciate that most antigens are glycoproteins or other proteins and that they are usually located on the outer surfaces of pathogens. Antigens on the surface of erythrocytes may stimulate antibody production if transfused into a person with a different blood group.

The adaptive immune response is based on many specifics:

- each type of B-lymphocyte makes a specific type of antibody
- each antibody is specific for one antigen
- each antigen is part of a specific group of molecules of a specific pathogen.

Antigens are usually proteins, and many are **glycoproteins** found embedded in the outer membrane of a pathogenic organism. This could be the plasma membrane of a bacterium or the outer cells of a protist or fungus. Viruses do not have a plasma membrane, but they do have a protein coat called a capsid and the capsid proteins act as antigens.

Other molecules, besides those in pathogens, can be recognized as antigens by our immune system. When organs are transplanted surgically, the organ or tissue transplanted must be "matched" very carefully by comparing the proteins of the donor and recipient. Transplanted hearts, kidneys and skin are examples of organs that can be transplanted. Many proteins must be taken into consideration, and rarely is there a perfect match. The exception is when identical twins are used as both donor and recipient.

Because pathogens have many antigens, a strong immune response may be a response to more than one of those antigens. In other words, more than one type of antibody can be produced in response to a single pathogen.

Blood transfusions should only occur after the blood types of both donor and recipient have been tested and are known to be compatible. When a blood type is indicated by the notation AB^+, for example, this notation is actually providing information about two blood types. One is called the ABO blood type, and the other is the Rh blood type. In this example, AB represents the ABO type, and + represents the Rh type. Blood typing is based on the presence or absence of three different antigens that are genetically inherited and found on the surface of erythrocytes. The three antigens are the A protein, B protein and Rh protein. The presence or absence of the three antigen proteins indicates a person's blood type.

	Antigen found on the plasma membrane of erythrocytes
ABO blood type	
A	A protein
B	B protein
AB	A and B proteins
O	Neither A nor B protein
Rh blood type	
Positive	Rh protein
Negative	No Rh protein

The antigens associated with different blood types

For example:

- a person with blood type B^- has the B protein on their erythrocytes but does not have either the A protein or the Rh protein
- a person with blood type O^+ has neither the A nor the B protein but does have the Rh protein.
- For blood transfusions to be successful a person must not receive a protein that they do not already have, as determined by their own genetics. A person with blood type AB^+ can receive blood from anyone because they already have all three antigens. A person with blood type O^- can only receive blood from someone who also has blood type O^- because they have none of the three antigens on their erythrocytes.

In a blood transfusion, a person cannot receive any of the three possible erythrocyte antigen proteins that they do not already have. This includes the A, B and Rh antigen proteins.

Challenge yourself

1. For each of the potential recipients shown, state all of the blood types that they could safely receive in a blood transfusion.

(a) Recipient with blood type A^+.
(b) Recipient with blood type O^+.
(c) Recipient with blood type AB^-.

The first blood transfusions were carried out before people had any knowledge of blood types. Transfusions were even carried out using farm animals as donors. As you can imagine, sometimes this accomplished more harm than good.

If someone receives a blood transfusion of an incompatible blood type, a transfusion reaction will occur. The reaction is an immune response to what the body identifies as an antigen. If someone receives type B blood and they do not genetically produce the B antigen protein, antibodies will be produced that bind to the donated cells and **agglutination** (clumping) can occur. The resulting transfusion reaction may lead to minor effects but has been known to be fatal.

C3.2.8 – The role of helper T-lymphocytes

C3.2.8 – Activation of B-lymphocytes by helper T-lymphocytes
Students should understand that there are antigen-specific B-cells and helper T-cells. B-cells produce antibodies and become memory cells only when they have been activated. Activation requires both direct interaction with the specific antigen and contact with a helper T-cell that has also become activated by the same type of antigen.

Two important types of leucocytes that respond in the adaptive immune response are **helper T-lymphocytes** (T-cells) and **B-lymphocytes** (B-cells).

Helper T-lymphocytes chemically communicate with other leucocytes, including B-lymphocytes, to signal the presence of a specific antigen.

- There are many types of helper T-lymphocytes.
- Each type can only activate a specific B-lymphocyte.
- The same antigen that activates a specific B-lymphocyte will also activate a helper T-lymphocyte.
- Helper T-lymphocytes display antigens on their own plasma membrane.
- Helper T-lymphocytes release molecules called **cytokines** after finding a specific antigen to help activate a specific B-lymphocyte.
- Some helper T-lymphocytes are long-lived and are called **memory cells.**

B-lymphocytes produce a specific antibody that binds to a specific antigen.

- There are many types of B-lymphocytes.
- Each type produces an antibody specific to one antigen.
- Each type that produces a specific antibody must be activated before it can make antibodies.
- Activation requires exposure to an antigen of the pathogen and also exposure to an activated T-lymphocyte that is displaying the antigen and releasing chemicals called cytokines.
- Some B-lymphocytes are long-lived and are called memory cells.

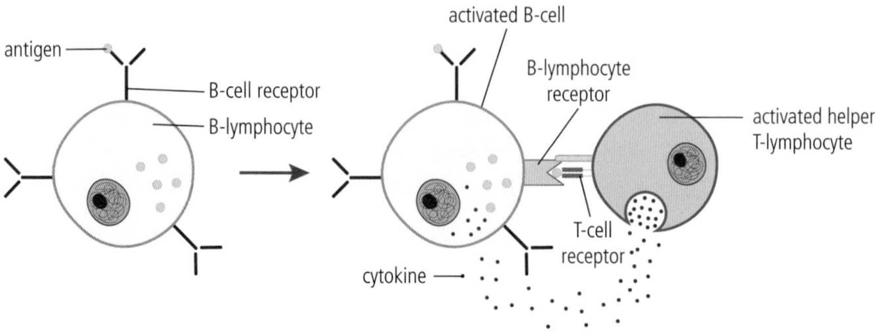

C3.2 Figure 1 Activation of a B-lymphocyte by a helper T-lymphocyte. Both the B-lymphocyte and the helper T-lymphocyte have already encountered the pathogen. An antigen from the pathogen is displayed on the plasma membrane of the B-lymphocyte and on a receptor of the helper T-lymphocyte. A protein receptor on the B-lymphocyte must match a receptor on the helper T-lymphocyte. Cytokines from the helper T-lymphocyte are released and taken in by the B-lymphocyte. When all of these events have occurred, the B-lymphocyte is activated.

C3.2.9 – Activation of a B-lymphocyte results in cloning

C3.2.9 – Multiplication of activated B-lymphocytes to form clones of antibody-secreting plasma cells
There are relatively small numbers of B-cells that respond to a specific antigen. To produce sufficient quantities of antibody, activated B-cells first divide by mitosis to produce large numbers of plasma B-cells that are capable of producing the same type of antibody.

The helper T-lymphocytes and B-lymphocytes described in Section C3.2.8 are antigen specific. The problem is that there is an incredible number of different antigens that may require a response. The immune system can only maintain a relatively low number of each type of cell that can respond to any one antigen. When specific B-lymphocytes are needed in an immune response, they first become activated and then undergo numerous mitotic cell divisions. In effect, they create **clones** of cells that have the genetic instruction to synthesize mass quantities of the antibodies that can bind to the antigens of a pathogen.

The basis of the adaptive immune response is that a few specific lymphocytes of each type are present in the body at all times. When a specific cell type is needed for an immune response, activation of that cell type leads to cloning to make many copies of that type of cell.

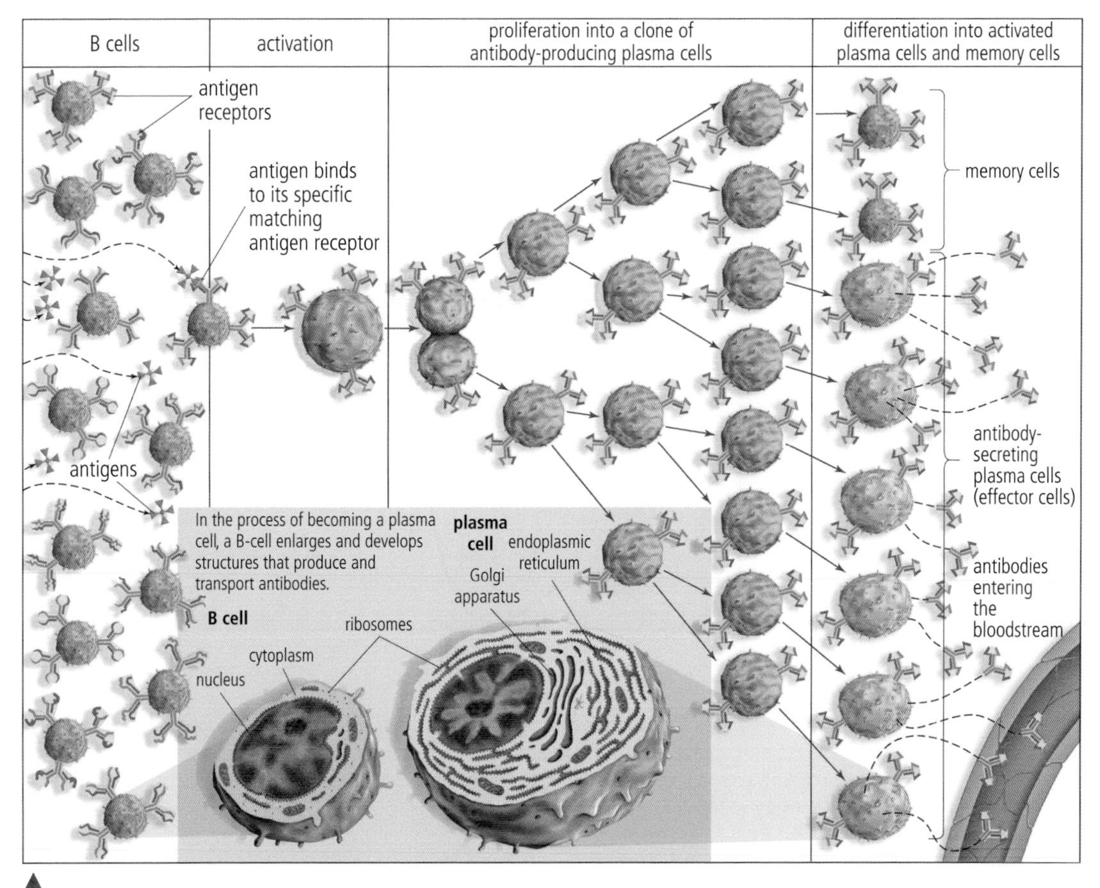

▲ On the left is a representation of many types of B-lymphocytes. Only one type is activated by an antigen (and helper T-lymphocyte, not shown). The activated B-lymphocyte undergoes repeated mitosis to create a small "army" of the same type of B-lymphocyte. A few of these, shown on the upper right, are memory cells and will not produce antibodies during the current infection. The rest form antibody-secreting plasma cells that produce antibodies that can be useful in body tissues or be circulated in the bloodstream. Activated B-lymphocytes become larger, and develop many ribosomes, endoplasmic reticulum and Golgi bodies, all used for antibody production and secretion.

C3.2.10 – The role of memory cells

C3.2.10 – Immunity as a consequence of retaining memory cells
Students should understand that immunity is the ability to eliminate an infectious disease from the body. It is due to the long-term survival of lymphocytes that are capable of making the specific antibodies needed to fight the infection. These are memory cells.

The adaptive immune response requires an initial exposure to the antigen(s) of a particular pathogen. That first response is relatively long and is called the **primary immune response**. During this first exposure there are no memory cells, as the few lymphocytes that can respond to this pathogen have not yet been activated or cloned. The length of time it takes for a primary immune response to occur varies depending on the pathogen, but there is almost always sufficient time for symptoms of disease to develop. For example, if it is the first time a person is exposed to a specific cold virus, they will have symptoms of the cold. The primary immune response is taking place while the person is experiencing the symptoms, and will eventually result in the symptoms disappearing as the pathogen is eliminated from the body.

The second or any subsequent exposure to that same cold virus will trigger a **secondary immune response**. The memory cells that were produced during the primary infection continue to circulate in the bloodstream. These very long-lived cells, now in relatively large numbers, are capable of responding to the same pathogen very quickly. It is usually so quick that symptoms of the disease do not present, or are quite minor. The secondary immune response not only occurs faster it also produces many more antibodies than the first exposure.

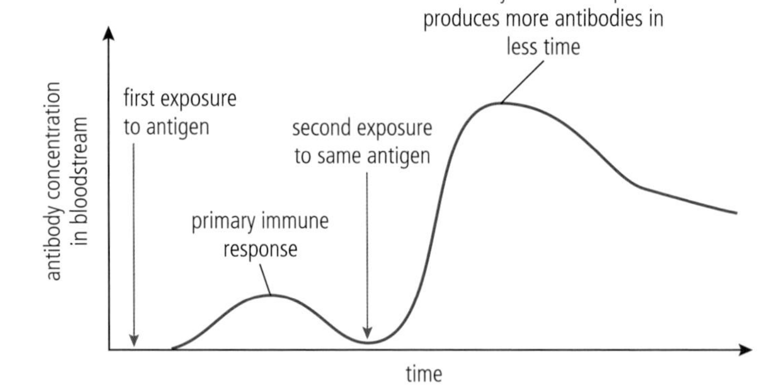

Antibody production by the primary and secondary immune responses. The second exposure to the same antigen may be months or years after the first exposure. The production of antibodies is quicker after a second infection, and the number of antibodies is greater.

We have no true immunity to a pathogen during the first infection, but we do have an immune system that is usually able to eliminate a new pathogen. True immunity begins with a subsequent infection as a result of the activity of memory cells.

As shown by the recent **pandemic** caused by a coronavirus (SARS-CoV-2), pathogens are easily spread across the globe. SARS-CoV-2 is not the first pathogen to have moved from country to country despite interventions to stop its spread. Some previous pandemics have been referred to as plagues.

Sometimes the term immunity refers to the body's ability to eliminate an infectious disease during a primary immune response, when symptoms are likely to occur. This is because our immune system has both primary and secondary immune responses.

How do animals protect themselves from threats?

C3.2.11 – HIV transmission

C3.2.11 – Transmission of HIV in body fluids
Include examples of the means and implications of HIV (human immunodeficiency virus) transmission.

HIV is the abbreviation for a virus called **human immunodeficiency virus.** Just like any virus, HIV is very specific about which organisms and which cell types in an organism it infects. Unfortunately, the (host) cells it infects in humans is one of the key lymphocyte cell types involved in the human immune response.

HIV does not survive outside the body and is not transmitted by saliva, tears or sweat. It is also not transmitted by insects such as via mosquito bites. The fluids that can transmit HIV are blood, semen, rectal fluids, vaginal fluids and breastmilk.

The two most common ways that HIV is spread from person to person is by having unprotected sex with an infected person, and by using a hypodermic needle that has previously been used by someone who is infected. In addition, it is possible for an HIV-positive mother to infect her child during pregnancy, labour, delivery or breastfeeding. In some countries, receiving a blood transfusion can spread HIV, but this is no longer a risk in countries where blood and blood products are routinely tested for contamination. Some medical treatments, such as injections for treating haemophilia, have been known to spread HIV when the injected material was purified from human blood. In many areas of the world, these products are now produced by genetically engineered bacteria and there is no risk of transmitting HIV.

C3.2.12 – The result of HIV infection

C3.2.12 – Infection of lymphocytes by HIV with AIDS as a consequence
Students should understand that only certain types of lymphocyte are infected and killed, but that a reduction in these lymphocytes limits the ability to produce antibodies and fight opportunistic infections.

HIV is very specific about "choosing" which cell to infect. The host cells of HIV are known as helper T-lymphocytes or **CD4 T-lymphocytes.** CD4 is the name of the glycoproteins that are found on the plasma membrane of helper T-lymphocytes and are used by HIV in its mechanism for entering a cell.

Unfortunately for anyone infected with HIV, the helper T-lymphocytes will eventually be killed by the virus, but they are the same cells used to activate B-lymphocytes and some cells involved in an immune response (see Figure 1 on page 504). A person who has a very low helper T-lymphocyte count in their blood stream will not have a strong immune response to pathogens. This is the disease called **AIDS** or **acquired immune deficiency syndrome**. People that are HIV-positive and have progressed to AIDS are susceptible to **opportunistic infections**. These are infections that occur more often or with more severity in people with weakened immune systems. Opportunistic infections include tuberculosis, salmonella, pneumonia and several others.

> *i* Deaths from HIV have greatly decreased in countries with advanced medical care where HIV-positive patients take daily doses of medicines that reduce the potential for damage to the immune system.

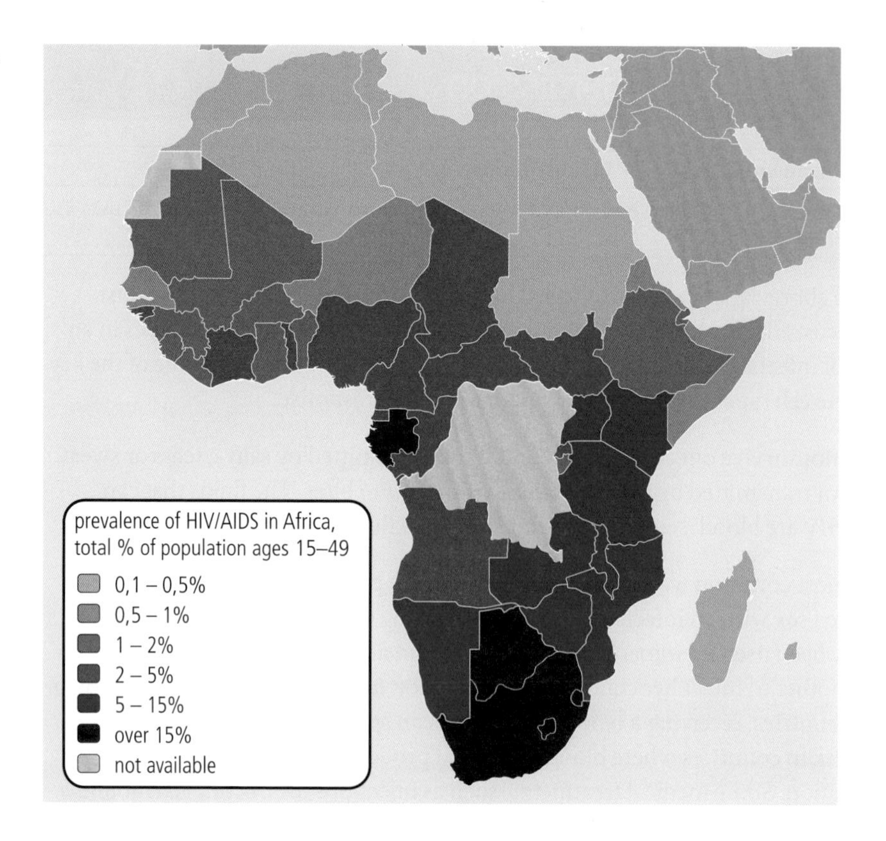

The prevalence of people infected with HIV is very unevenly distributed around the world. Many sub-Saharan African countries continue to show a high percentage of people infected. In this graphic, published in the peer-reviewed journal *Nature*, the data points are shown as much smaller, more accurate clusters than ever shown before. This can be used to help distribute resources to fight AIDS.

C3.2.13 – Antibiotics against bacterial infections

C3.2.13 – Antibiotics as chemicals that block processes occurring in bacteria but not in eukaryotic cells
Include reasons that antibiotics fail to control infection with viruses.

Bacteria are **prokaryotic** cells. Humans and other animals are composed of **eukaryotic** cells (see Chapter A2.2). There are many structural and biochemical differences between prokaryotic and eukaryotic cells. For example, while protein synthesis occurs in both types of cells, the processes are different. Also, bacteria have a cell wall, a structure that is not characteristic of eukaryotic animal cells.

There are chemicals that have been developed as **antiviral medications**. These chemicals suppress a virus's ability to infect and multiply in the host's cells. Antiviral medications have become the standard of care for people with HIV and hepatitis C infections.

Antibiotics are chemicals that take advantage of the differences between prokaryotic and eukaryotic cells: they selectively block some of the biochemical pathways needed by bacteria while having no effect on human or other animal cells. There are many categories of antibiotics, depending on the biochemical pathway that is being targeted. One type of antibiotic selectively blocks protein synthesis in bacteria, but has no effect on eukaryotic cells' ability to manufacture proteins. Another type of antibiotic inhibits the production of a new cell wall by bacteria, thus blocking their ability to grow and divide.

Viruses have no metabolism, which explains why antibiotics have no effect on them. Viruses make use of our own body cells' metabolism to create new viruses. Any chemical that could inhibit viral metabolic activity would also be damaging to our own body cells. Antibiotics should not be routinely prescribed for viral diseases.

C3.2.14 – Pathogenic resistance to antibiotics

C3.2.14 – Evolution of resistance to several antibiotics in strains of pathogenic bacteria
Students should understand that careful use of antibiotics is necessary to slow the emergence of multiresistant bacteria. **NOS:** Students should recognize that the development of new techniques can lead to new avenues of research; for example, the recent technique of searching chemical libraries is yielding new antibiotics.

Bacterial resistance to antibiotics is a serious problem around the world. For too long, antibiotics have been used improperly and too frequently. The fundamental principles of evolution explain how pathogenic bacteria become resistant to any one antibiotic. When bacteria find their way into living tissues, frequently the environment is nearly perfect for their growth and cell division. Tissues are moist, full of nutrients, and relatively warm. In this type of growing environment some pathogenic bacteria can grow exponentially and double their numbers in as small a time period as 20 minutes.

Every cell division requires the DNA of a bacterial cell to replicate. Mutations occur spontaneously when DNA replicates. Most of the mutations are of no consequence, but when the DNA replication rate is very high, one or more mutations is likely to occur that is consequential. One of those random, but consequential, mutations may give a bacterial cell protection from the biochemical action of a particular antibiotic. One mutation is all that is required, as that bacterial cell will undergo binary fission repeatedly to grow into many cells. All bacterial cells that arise from the mutated cell will have the resistance to the antibiotic. In addition, the mutated strain may now cause infections in other people.

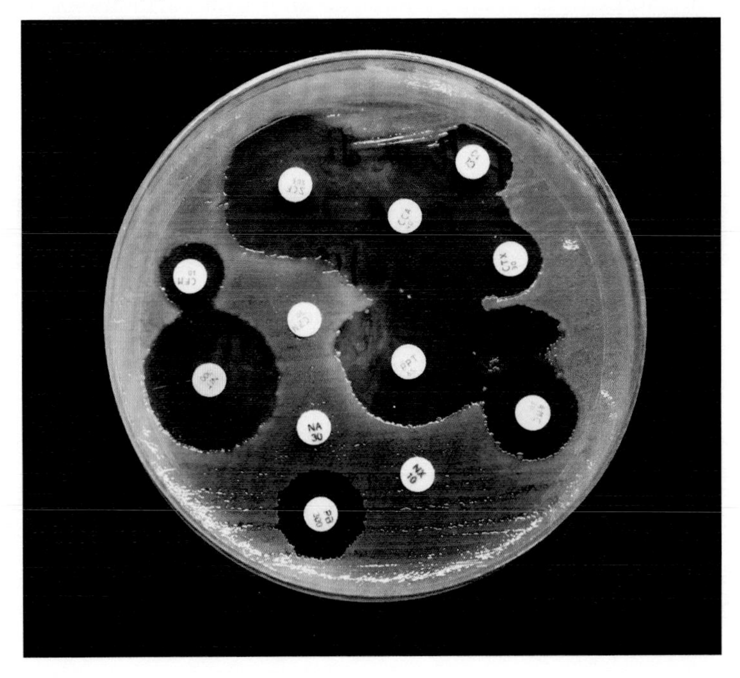

Bacterial species grown in a Petri dish showing resistance to antibiotics. The dull cloudy area is where bacteria are growing. The small white discs are impregnated with antibiotics. The clear area around many of the discs indicates that the antibiotics are preventing bacterial growth. The clear area is called a **zone of inhibition**. The presence of bacterial growth around three of the discs indicates those antibiotics that have little to no ability to prevent growth.

Nature of Science

A great deal of scientific research precedes the release of a new antibiotic before it can be prescribed. Antibiotics always come with directions for use that are based on this research. These directions include how long you should take the medicine for. If someone stops taking an antibiotic early, for example because their symptoms have improved, only the bacteria that are most sensitive to the antibiotic have been killed. A few resistant cells may not have been killed and can grow into a new resistant strain. You should always take an antibiotic for the full prescribed duration.

A few strains of pathogenic bacteria have emerged over the years that are resistant to several antibiotics. One example is methicillin-resistant *Staphylococcus aureus* (MRSA). This bacterial strain causes **staphylococcal infections** that are very difficult to treat. Responsible use of existing antibiotics is needed to prevent the emergence of more multiresistant bacteria. Responsible use includes:

- only prescribing an antibiotic when necessary
- taking the full course of an antibiotic, and not stopping when symptoms first subside
- reducing the spread of bacterial diseases by vaccination, hand-washing and proper food hygiene
- reducing or stopping the practice of adding antibiotics to farm animal feed.

Methods using artificial intelligence (AI) are being used to screen for new antibiotics. This involves predicting the interaction of chemicals using AI and the known chemicals in chemical libraries. Halicin, a chemical screened by AI, is a new and promising antibiotic that is being prepared for clinical trials.

Nature of Science

The development of new techniques can lead to new lines of research. The use of chemical libraries is a new approach to antibiotic development. Chemical libraries store chemicals along with all the known information about those chemicals. Searching chemical libraries is yielding new antibiotic treatments because synergistic effects of antibiotic combinations can be explored at the molecular level.

C3.2.15 – Zoonotic diseases

C3.2.15 – **Zoonoses as infectious diseases that can transfer from other species to humans**
Illustrate the prevalence of zoonoses as infectious diseases in humans and their varied modes of infection with several examples including tuberculosis, rabies and Japanese encephalitis. Include COVID-19 infection as an infectious disease that has recently transferred from another species, with profound consequences for humans.

Many infectious diseases are species specific. Those that can cross species, specifically animal to human, are called **zoonotic diseases**. The pathogen may be a virus, bacterium, protist or fungus. Some examples are described below.

Rabies

Rabies is a disease caused by a virus. Most human cases of rabies are the result of dog bites, although the dog may have received the virus from a wild animal. Cases of rabies occur throughout the world, but are more common in Africa and Asia. The rabies virus causes a progressive and fatal inflammation of the human brain and spinal cord. By the time symptoms begin to show in an infected person, it is too late for treatment. The best defence against rabies in humans is preventative vaccination of dogs. Seeking medical treatment shortly after a bite from a rabid animal can prevent death if the treatment is received quickly.

Tuberculosis

Zoonotic tuberculosis is a bacterial disease caused by *Mycobacterium bovis*. Humans are exposed to this bacterium through cattle. Ingestion of unpasteurized milk and milk products and infected meat are the primary means of transmission. Airborne transmissions are also possible, especially for those that work with cattle. The main symptom is damaged lung tissue, but other human tissues are also affected. The name tuberculosis comes from growths called tubercles that occur in the lymph nodes of an infected person.

Japanese encephalitis

Japanese encephalitis is caused by a virus that is transmitted through the bite of a species of *Culex* mosquito. The mosquito receives the virus from either a pig or wading bird. Most cases occur in southeast Asia and are quite mild, although a few cases have progressed to more serious symptoms, including coma and eventually death. There is a vaccine that prevents symptoms but it is not widely used in the rural areas where the disease is typically transmitted.

COVID-19

COVID-19 is a disease caused by a coronavirus known as **SARS-CoV-2**. It is almost certainly zoonotic, although no specific species has yet been identified as the first to infect humans. All other known coronaviruses can be zoonotic. SARS-CoV-2 has been shown to transfer easily from humans to other animals, such as dogs, cats and deer, and many other animals have tested positive for the virus. Most researchers classify COVID-19 as an "emerging infectious disease of probable animal origin", until more is learned about its origins. The virus quickly caused a global pandemic by spreading from person to person in 2019–2020. COVID-19 symptoms vary from **asymptomatic** (no symptoms) to severe and fatal respiratory damage. Variants of the virus are continuing to emerge, resulting in increased transmission rates. Vaccines have been developed but are not readily available in all areas of the world.

Current practices of keeping very dense populations of domesticated animals can lead to both increased animal-to-animal and animal-to-human zoonotic disease transmission.

C3.2.16 – Vaccines and immunity

C3.2.16 – Vaccines and immunization
Students should understand that vaccines contain antigens, or nucleic acids (DNA or RNA) with sequences that code for antigens, and that they stimulate the development of immunity to a specific pathogen without causing the disease.

In Section C3.2.4 you learned about the primary and secondary immune responses that result in the production and use of memory leucocytes. For many diseases, vaccines have been developed that act as the first exposure to a pathogen. A vaccine is typically composed of the chemical components of a pathogen after eliminating the disease-causing abilities of the pathogen. In traditional vaccine production, the pathogenic virus or bacterium is inactivated so that it cannot cause the disease. The pathogen or selected antigens from the pathogen are then injected into a person, resulting in the same immune response as if the pathogen had entered the host's body. This injection results in a primary immune response that then leaves behind memory cells that can be quickly triggered into action upon reinfection by the pathogen.

Recent advances in vaccine research and technology have led to a new approach. Instead of injecting an inactivated pathogen or antigen, the DNA or RNA molecules that code for the synthesis of specific protein antigens are injected. Body cells take in the nucleic acid and use their normal cell protein synthesis organelles and enzymes to produce antigens. These antigens, although produced by body cells, are recognized as foreign and stimulate a primary immune response without exposure to the pathogen. As with traditional vaccines, memory cells are produced to provide immunity.

A depiction of SARS-CoV-2 sitting in a protein receptor of a cell. Proteins that make the "spike" structures of the virus are synthesized by human cells after injection with one of the RNA vaccines.

TOK

Social media has become a major news source for many people. What is the responsibility of social media platforms to ensure that shared information is fact based?

As part of the COVID-19 response, RNA vaccines were rapidly developed and manufactured. The protection provided by these vaccines has been excellent, especially for reducing the more serious symptoms that can result in hospitalizations. However, acceptance of the vaccines as safe and effective tools against COVID-19 has not been universal in some countries. Widespread misinformation concerning the safety and value of the vaccines has contributed to limited acceptance.

C3.2.17 – The role of herd immunity

C3.2.17 – **Herd immunity and the prevention of epidemics**
Students should understand how members of a population are interdependent in building herd immunity. If a sufficient percentage of a population is immune to a disease, transmission is greatly impeded. **NOS:** Scientists publish their research so that other scientists can evaluate it. The media often report on the research while evaluation is still happening, and consumers need to be aware of this. Vaccines are tested rigorously and the risks of side effects are minimal but not nil. The distinction between pragmatic truths and certainty is poorly understood.

Many pathogenic diseases spread as a result of person-to-person contact. When a large percentage of people in a given area (a herd) achieve immunity to a disease, there is a far reduced chance of the disease spreading. Even someone with no immunity is far less likely to get the disease when **herd immunity** has been achieved. The percentage of immune people that is needed to achieve herd immunity differs depending on the disease. Generally, the more contagious a disease, the higher the percentage needs to be. Measles, a highly contagious disease, requires 92–94% of the population to be immune. It is not yet certain what percentage of people need to be immune to COVID-19 to achieve herd immunity, or even if herd immunity is achievable. COVID-19 continues to produce new variants that may or may not be recognized by the immune response from a previous variant.

Nature of Science

Scientists publish their research so that other scientists can evaluate it. The media often report on the research while evaluation is still happening, and consumers need to be aware of this. Vaccines are tested rigorously and the risks of side effects are minimal but not nil. The distinction between pragmatic truths and certainty is poorly understood. A pragmatist can consider something to be true without needing to confirm that it is universally true.

C3.2.18 – Evaluating COVID-19 data

C3.2.18 – **Evaluation of data related to the COVID-19 pandemic**
Application of skills: Students should have the opportunity to calculate both percentage difference and percentage change.

There are many tools available to scientists for evaluating data. Two of the more common tools are calculating the **percentage difference** and **percentage change.** These very different calculations are often confused.

Calculating the percentage difference is useful when you are comparing two values that mean the same thing at the same time, for example if you are comparing the height of two people on the same day.

The percentage difference is the difference between two values divided by the average of the two values expressed as a percentage.

Worked example

Noah has a height of 176 cm and Mithun has a height of 184 cm. What is the percentage difference in their height?

Solution

Percentage difference = (184 cm − 176 cm) / ((184 cm + 176 cm) / 2) × 100

= 8 / 180 × 100 = 4.4%

There is a 4.4% difference between their heights.

Calculating the percentage change is useful when you are comparing two values that are separated by time.

Percentage change is the difference between the new and old values divided by the old value expressed as a percentage.

Worked example

Two years ago, Noah had a height of 160 cm. He now has a height of 176 cm. What is Noah's percentage change in height over the two years?

Solution

Percentage change = (176 cm − 160 cm) / 160 cm × 100

= 16 cm / 160 cm × 100 = 10%

There has been a 10% increase in Noah's height.

Challenge yourself

Herd immunity is difficult to achieve when a disease is highly contagious. How contagious a disease is represented by a value denoted as R_0, pronounced "R nought" or "R zero". The R_0 value is the estimated number of people that will be infected by a single infected person if everyone they contact is susceptible to the disease. The higher the R_0 number, the more contagious the disease is.

Use the data in the following table to answer the following questions.

2. What is the correlation between the R_0 value and the threshold percentage necessary for herd immunity?

3. Young people no longer receive a vaccine for smallpox. Suggest a reason why they do not need one.

4. In 1955, one company that produced a polio vaccine released some batches that contained active polio virus. Over 250 people contracted polio, with some resulting in paralysis. Discuss why information like this should be publicly available.

5. Calculate the percentage difference between the R_0 values of H1N1 and SARS-CoV-2 given in the table.

6. The virus causing COVID-19 has become more contagious over time as a result of mutations. The R_0 value of the original SARS-CoV-2 that emerged from China in 2019 was calculated to be 2.8 by the National Institutes of Health. Use the more recent R_0 value for the SARS-CoV-2 variant given in the table to calculate the percentage change in R_0 value.

Infectious diseases	R_0 value	Herd immunity threshold
Smallpox	5–7	80–85%
Mumps	4–7	75–86%
Measles	12–18	92–94%
Diphtheria	6–7	85%
Pertussis	12–17	92–94%
Polio	4–13	75–92%
Rubella	6–7	83–85%
H1N1 (2009 Pandemic)	1.6	40%
SARS	2–4	50–75%
SARS-CoV-2 (COVID-19)	5.7	82.5%

The concept of herd immunity based on how contagious a disease is. The R_0 and herd immunity values for SARS-CoV-2 are based on 2022 estimations

The World Health Organization (WHO) is the United Nations agency that promotes good health practices and care throughout the world. Figure 2 shows the data collected by WHO, as of June 2022, on the number of SARS-CoV-2 cases in different areas of the world. Note, however, that testing protocols and reliability differed between the regions.

Total number of SARS-CoV-2 cases in June 2022

Europe	222,715,473 confirmed
Americas	158,984,066 confirmed
Western Pacific	62,063,771 confirmed
South-East Asia	58,251,326 confirmed
Eastern Mediterranean	21,828,710 confirmed
Africa	9,043,241 confirmed

C3.2 Figure 2 The total number of SARS-CoV-2 cases in June 2022 since virus transmission began in six WHO regions of the world.

How can false-positive and false-negative results be avoided in diagnostic tests?

Guiding Question revisited

How do body systems recognize pathogens and fight infections?

In this chapter you have learned:

- humans and many other animals have both innate and adaptive immune systems
- the innate immune response attempts to remove anything recognized as foreign or "not-self" without identifying it
- the adaptive immune response recognizes specific foreign entities and the response leaves behind immunity to that entity in the form of memory cells
- adaptive immunity involves recognition of molecules, called antigens, that make up pathogens
- specific helper T-lymphocytes are needed to recognize an antigen, and clone themselves to create higher numbers of that cell type
- specific B-lymphocytes are activated by helper T-lymphocytes, and produce antibodies that bind to antigens making up the pathogen
- long-lived cells of both types of lymphocyte act as memory cells to provide long-term immunity to the antigen
- vaccines are created by using inactive pathogens or nucleic acids injected into the body, which leads to the production of memory cells.

Guiding Question revisited

What factors influence the incidence of disease in populations?

In this chapter you have learned:

- skin and mucus membranes can often prevent entry of a pathogen into the body
- HIV/AIDS is a viral disease that can greatly lower the ability of a person's body to mount an effective immune response
- antibiotics are chemicals that selectively target the growth processes of bacteria and have been instrumental in treating bacterial infections
- antibiotic misuse and overuse have led to some pathogenic bacteria becoming resistant to one or more antibiotics
- vaccines are effective in creating long-term immunity in a population
- some pathogens are known to pass from species to species and are known as zoonoses
- herd immunity is achieved for a specific pathogen when a high percentage of the population has achieved immunity, making it unlikely that a pathogen would infect an unprotected person.

Exercises

Q1. Briefly state the function of each of the following during the process of blood clotting.

(a) Platelets.

(b) Fibrinogen.

(c) Thrombin.

Q2. Which one of these is an unsafe transfusion of blood?

A O^+ to A^+

B A^+ to AB^+

C O^+ to O^-

D AB^- to AB^+

Q3. B-lymphocytes must be exposed to two things before becoming activated. What are those two things?

Q4. What is the cellular process that produces clones of selected lymphocytes?

Q5. Antibiotics will not help control a viral infection. Why?

Q6. Why do symptoms develop when the body undergoes a primary immune response?

Q7. A single pathogen can result in multiple adaptive immune responses in a person. Which answer best explains this.

A A pathogen may be related to a previous pathogen.

B A pathogen may contain many antigens recognizable by the immune system.

C A pathogen will always mutate inside the body.

D A pathogen is altered by the adaptive immune response.

C3 Practice questions

1. Annotate the diagram of the reflex arc to show the name and function of the structures labelled I and II.

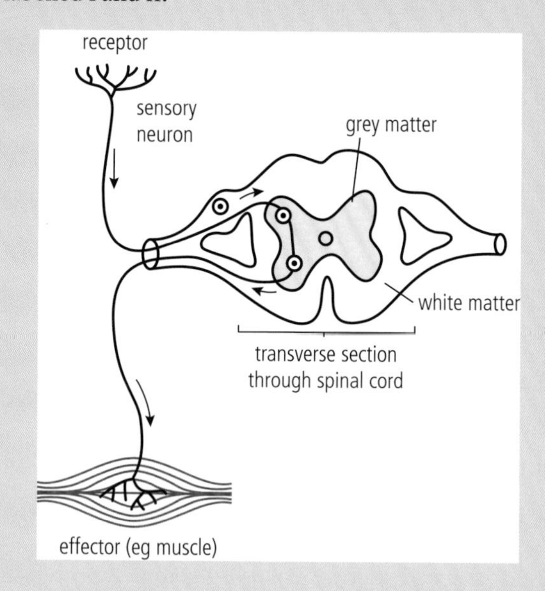

(Total 2 marks)

2. The heart responds quickly to physical activity. Describe how heart rate is controlled to meet increased circulatory demands.

(Total 2 marks)

3. **HL** How do auxins cause plant shoots to grow towards light?

A Increase cell division on the side of the stem near the light source.

B Increase cell division on the side of the stem away from the light source.

C Increase cell elongation on the side of the stem near the light source.

D Increase cell elongation on the side of the stem away from the light source.

(Total 1 mark)

4. Bacteria are prokaryotes that sometimes act as pathogens. Describe how the body can defend itself against pathogens.

(Total 7 marks)

5. Explain the evolution of antibiotic resistance in bacteria.

(Total 6 marks)

6. Ebola virus disease (EVD) is the disease in humans and other primates that is caused by the Ebola virus. Fruit bats are the reservoir for the virus and are able to spread the disease without being affected. Humans can become infected by contact with fruit bats or with people infected by the virus, their body fluids or equipment used to treat them.

The stacked bar graph shows the epidemiological data for EVD cases in Conakry, the capital city of Guinea, the surrounding suburbs, and rural areas in Guinea, from the beginning of January 2014 to the end of March 2015.

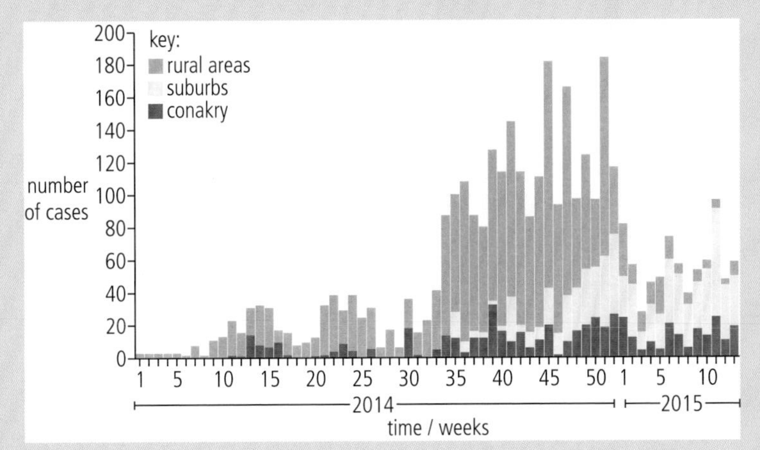

(a) Identify the week and year in which the first cases were recorded in the suburbs. (1)

(b) Based on the graph, compare and contrast the progress of the epidemic in the suburbs and rural areas. (3)

(c) Suggest **two** reasons for the overall decline in the epidemic after week 51. (2)

(Total 6 marks)

THEME

C Interaction and interdependence

4 Ecosystems

◀ Ecosystems are made up of complex interactions between organisms and their environment. In the Sonoran Desert, USA, plants and animals compete and cooperate in different ways.

Notice the word "system" in ecosystem: systems are made up of components that interact and are dependent on each other. Put many individuals together and you have a population; put many populations together and you have a community. New properties arise as the number of individuals and the number of species increase. These can include cooperation and competition, or herbivory and predation. Communities interact with their environment and usually depend on sunlight, temperature and water supplies to thrive.

A population can change over time. Researchers can use various sampling techniques to find out whether a population is growing or shrinking. Because organisms are connected in food chains and food webs, if one species can no longer survive in an ecosystem because of climate change or human activity, other organisms are also affected.

C4.1 Populations and communities

Guiding Questions

How do interactions between organisms regulate sizes of populations in a community?

What interactions within a community make its populations interdependent?

In this chapter we will see that nature has multiple ways of ensuring that no single species takes over an ecosystem and that each population has the potential to reach its maximum, but that many factors can limit populations or decrease their numbers. Such factors include the availability of food and water, presence of predators and the introduction of diseases. Some organisms engage in chemical competition by secreting molecules into their environment to reduce the chances of their competitors being successful.

No species lives in an isolated bubble. Plant eaters need plant material to survive, carnivores need other animals to eat, and some plants rely on microorganisms to provide nutrients such as nitrogen compounds to help them grow. This last example shows that sometimes two very different species can cooperate for mutual benefit.

C4.1.1 – Populations

C4.1.1 – Populations as interacting groups of organisms of the same species living in an area
Students should understand that members of a population normally breed and that reproductive isolation is used to distinguish one population of a species from another.

In biology, a **population** is defined as a group of individuals of the same species living in the same geographical area at the same time and able to interbreed. Here are a few examples of populations:

- emperor penguins (*Aptenodytes forsteri*) in Amundsen Bay, Antarctica
- bush-crickets (*Phaneroptera falcata*) living in a grassland in southern Europe
- rainbow trout (*Oncorhynchus mykiss*) along the northwest coast of the United States
- dandelions (*Taraxacum officinale*) growing in the same grassland as the bush-crickets.

Consider the emperor penguins. There is more than one population of them in Antarctica. How do we know where to draw the line between one population and the next? We can ask how probable it is that the individuals in population A might interbreed with individuals in population B. If penguin population A is separated by hundreds of kilometres of ice and open sea from population B, it is much more likely that penguins will stay and interbreed in their own population. We would be less sure of this if the populations were geographically closer to each other or if penguins could fly.

A population of emperor penguins (*Aptenodytes forsteri*).

When studying ecology, we can measure many things about a population. The most obvious is population size: the number of individuals. But populations are not fixed, they are dynamic, so we can also measure the change in population size over time, which could go up or down as a result of factors such as immigration and death. We will see that estimating the size of a population is not an easy task.

Other characteristics that can be used to describe a population are population density, its geographical distribution, and the maximum number of individuals that can be supported by the resources available. Once we understand a population, we can start to look at how it interacts with its environment and other species.

C4.1.2 – Estimating population size

C4.1.2 – Estimation of population size by random sampling

Students should understand reasons for estimating population size, rather than counting every individual, and the need for randomness in sampling procedures.

NOS: Students should be aware that random sampling, instead of measuring an entire population, inevitably results in sampling error. In this case the difference between the estimate of population size and the true size of the whole population is the sampling error.

Ideally, to know the size of the population we are studying, we would count every last individual. For emperor penguins, this would involve going to Amundsen Bay in Antarctica and counting each one. What would that look like? Once you started counting, how could you be sure you had not already counted a particular penguin? Some members of the population will be chicks keeping warm near their parents, and it will be hard to see all of them. Some individuals may be out in the water getting food. These are just a few examples of how, most of the time, it is simply impossible to count every last individual in a population. Instead, we need to rely on estimates. We count a sample and use that sample to estimate the overall population size.

There are two types of sampling: systematic and random. **Systematic sampling** is when a line or grid is set up and measurements or counting are carried out only at specified, regular intervals. For example, a 50 m measuring tape can be laid across a rocky shore and seaweed or snails can be counted in a 1 m^2 area every 5 m.

Random sampling is when arbitrarily chosen zones of the population's geographic distribution are sampled. Random directions and random distances between samples are used to try to overcome any bias that the investigators might have that would favour a particular area. Random sampling using quadrats and a method of mark and recapture are discussed in Sections C4.1.3 and C4.1.4.

A biology student using a quadrat on a rocky shore.

Nature of Science

Random sampling inevitably results in sampling error. The difference between an estimated population size and the true size of the whole population is the sampling error. If ecologists randomly take 200 samples from a prairie that are 1 m^2 each, and determine that the average number of ferns in each square metre is 1.3, they can deduce that, if the prairie is 10,000 m^2, the total population should be 13,000 ferns. But this assumes that the ferns are evenly distributed across the prairie. If there were actually only 8,000 ferns in the area studied, the sampling error would be quite large: 5,000 extra ferns were incorrectly estimated. This is one of the many reasons why, when given a number, scientists often ask how it was determined and what the degree of precision is. They want to know how much error might be involved and how close the estimate is to the true value.

Here is an allegory to illustrate sampling error. A man gets in a boat and rows out into the ocean. He fills a bucket with sea water and looks in the bucket to count the number of fish he sees. He sees none, and declares, "There are no fish in this ocean!"

C4.1.3 – Sampling sessile organisms

Standard deviation is a quantity that measures the average difference between the measured values and the mean of those values. It describes how spread out the data are in relation to the mean.

C4.1.3 – Random quadrat sampling to estimate population size for sessile organisms

Both sessile animals and plants, where the numbers of individuals can be counted, are suitable.

Application of skills: Students should understand what is indicated by the standard deviation of a mean. Students do not need to memorize the formula used to calculate this. In this example, the standard deviation of the mean number of individuals per quadrat could be determined using a calculator to give a measure of the variation and how evenly the population is spread.

Follow the downloadable activity in the eBook to use random quadrat sampling to estimate the population size of a local species of plant. Determine the **standard deviation** of the mean number of individuals per quadrat to give a measure of the variation and how evenly the population is spread.

Random sampling can be used to estimate population size for organisms that stay in one place, for example plants, lichens and corals. Such organisms are referred to as **sessile**; for much of their lives they do not change location. The example in the eBook shows how to estimate the population size of a sessile organism using a **quadrat** (see the photo on the previous page). A quadrat is a square of a particular dimension that can be made of a rigid material such as metal, plastic or wood. Using a quadrat means the surface area of the sample size is the same for each count you take.

C4.1.4 – Sampling motile organisms

C4.1.4 – Capture–mark–release–recapture and the Lincoln index to estimate population size for motile organisms
Application of skills: Students should use the Lincoln index to estimate population size. Population size estimate = $M \times N/R$, where M is the number of individuals caught and marked initially, N is the total number of individuals recaptured and R is the number of marked individuals recaptured. Students should understand the assumptions made when using this method.

The **capture–mark–release–recapture method** is a sampling technique that enables you to estimate the number of animals in an ecosystem. It is used instead of quadrats for **motile** organisms, those that move around. The technique involves catching some of a population and marking them. The marked animals are released back into the ecosystem and given a suitable period of time to remix with others in their population. A second sample of the population is then captured. Some in the second sample will be marked and some will be unmarked. The proportion of marked to unmarked individuals in the second sample is assumed to be the same as the proportion of the originally marked individuals to the whole population. The **Lincoln index** is then used to estimate the number of individuals in the population. Here is the formula:

$$\text{total population} = \frac{\text{number of individuals}}{\text{caught and marked initially}} \times \frac{\text{number of all individuals recaptured}}{\text{number of marked individuals recaptured}}$$

The Lincoln index can be written using variables instead of words:

$$\text{total population} = M \times \frac{N}{R}$$

where M is the number of individuals originally caught and marked, N is the total number of individuals recaptured, and R is the number of marked individuals recaptured.

You can try this method at home, in your classroom or on a field trip using the activity in your eBook.

The capture–mark–release–recapture method has some limitations:

- capturing and marking the animals may injure them
- the mark may make an animal more visible to predators and, if the marked animals are eaten, the second sample will not be reliable
- it assumes that the population is closed, with no immigration or emigration.

The last point is a rather big assumption, because very few populations are closed. Other assumptions are that all the individuals in a given area have an equal chance of being captured, that marked individuals will be randomly distributed after release, and that marking individuals will not make them more or less likely to be recaptured.

The mark and recapture technique can be used by experts to estimate the size of a bird population. Small metal bands with identification codes can be attached to a bird's leg to tag it, and when that same bird is recaptured, population ecologists know where and when it was originally captured.

C4.1.5 – Carrying capacity

C4.1.5 – Carrying capacity and competition for limited resources
A simple definition of carrying capacity is sufficient, with some examples of resources that may limit carrying capacity.

No habitat can accommodate an unlimited number of organisms: populations often grow, but they cannot continue to grow forever. There comes a time in the growth of a population when its numbers stabilize. This number, the maximum number of individuals that a particular habitat can support, is called the **carrying capacity**, and it is represented by the letter **K**.

Consider, for example, a given zone in a forest. There is a maximum number of trees that can grow there. This number is attained when enough trees are present to catch all the sunlight, leaving every square metre of the forest floor in shade. New tree seedlings trying to grow under the adult trees will have difficulty getting any sunlight.

In tropical rainforests such as this one in Costa Rica, very little sunlight reaches the forest floor because a maximum number of trees and plants have spread their leaves to catch all the sunlight they can.

Young trees can store energy for years with very little vertical growth, waiting for a larger tree to die, leaving a hole in the canopy. The young trees then compete to use the opening to take the old tree's place. Those that lose this competition usually die. For a young tree to join the mature population, an old tree must die. Hence, there is no net increase in the tree population, at least at the canopy level.

A **limiting factor** is something that can prevent a population from getting bigger or reduce a population's size. Limiting factors that define the carrying capacity of a habitat include:

What factors can limit capacity in biological systems?

- the availability of resources, such as water, food, sunlight, shelter, space and oxygen (the latter notably in aquatic habitats)
- the build-up of waste, such as excrement and excess carbon dioxide
- predation
- disease.

C4.1.6 – Negative feedback

C4.1.6 – **Negative feedback control of population size by density-dependent factors**
Numbers of individuals in a population may fluctuate due to density-independent factors, but density-dependent factors tend to push the population back towards the carrying capacity. In addition to competition for limited resources, include the increased risk of predation and the transfer of pathogens or pests in dense populations.

As a population grows, its population density increases; there will be more individuals per unit of habitat, such as more plants per square metre of prairie or more plankton per litre of seawater. Some factors that change the size of a population are dependent on the density of the population, such as the spread of disease. These are called **density-dependent** factors. They tend to keep a densely populated area at or below its carrying capacity. Other factors are **density-independent**, meaning that it does not matter if there is a sparse population or an overcrowded population. Examples of density-independent factors include climate change, or a nearby forest fire or volcanic eruption. Populations both big and small will be affected, and the change in numbers can be an increase or a decrease, depending on whether the factor helps or hinders a population.

Complex systems require feedback mechanisms to keep parameters from reaching dangerously high or low levels. **Positive feedback** in a population is when something increases the population and the system encourages more of the same. For example, in a breeding population that is initially small, the more individuals that are produced by successful breeding, the more individuals there are to produce the next generation. The higher the population density, the higher the chances of a male encountering a female to produce even more individuals in the population. Positive feedback will help a population grow, but it can get out of control.

Negative feedback prevents a system from going too far in one direction. If a population continued to grow uncontrollably, for example, individuals would start to run out of resources such as space, food or water. The chances of diseases spreading also increase when population sizes are large and density is high. The chances of attracting predators to an area increase as the number of prey individuals increases. Negative feedback in the form of competition for resources and the spread of disease works to control the size of the population so that it cannot go above its carrying capacity. When it comes to population dynamics, positive feedback can be a dangerous thing, leading to food shortages and the spread of disease, whereas negative feedback helps prevent such density-dependent issues from getting worse and threatening a population.

In a positive feedback loop in populations, more offspring grow up and produce additional offspring which, in turn, produce even more offspring.

In a negative feedback loop, something such as a paucity of food or the introduction of a disease will regulate the population and slow down or reverse population growth.

Rabbits have a reputation for being able to quickly increase their population. Thanks to positive feedback, the population gets bigger but will eventually reach the carrying capacity for its environment.

C4.1.7 – Population growth

C4.1.7 – **Population growth curves**
Students should study at least one case study in an ecosystem. Students should understand reasons for exponential growth in the initial phases. A lag phase is not expected as a part of sigmoid population growth.   **NOS:** The curve represents an idealized graphical model. Students should recognize that models are often simplifications of complex systems.   **Application of skills:** Students should test the growth of a population against the model of exponential growth using a graph with a logarithmic scale for size of population on the vertical axis and a non-logarithmic scale for time on the horizontal axis.

In 1980, the ecosystems around Mount Saint Helens were destroyed when the volcano erupted in a sudden, violent blast. The devastation is shown in the picture on the left. Contrast this with the picture on the right, where a few of the dead tree trunks can still be seen, as well as smaller trees that have grown since the eruption.

In 1980 there was a major volcanic eruption of Mount Saint Helens on the west coast of the United States. After the massive event, little was left of the forest and rivers that had existed on and around the mountain. Forest fires and hot gases burned everything in sight. Volcanic ash rained down, smothering the destroyed forest and covering the carcasses of the animals that had died there. Many species that could escape fled the area.

Yet, within months of the eradication of an ecosystem, life was back. Seeds, dropped by birds or blown in by the wind, germinated in the fertile volcanic ash. Little by little, insects, then birds, then small mammals, moved in. Within a couple of decades, a grassland and shrub ecosystem had reappeared. Today, thousands of species flourish in what had been a desolate landscape.

If we look at the tree species now present on Mount Saint Helens, such as conifers and the red alder (*Alnus rubra*), their populations have increased over the decades, but some experienced a decrease in growth when the North American elk (*Cervus elaphus*) population started to thrive there too. Attracted by new food sources, elk browsed (ate) the young tree saplings of deciduous trees and conifers, slowing the growth of the tree populations. However, the elk droppings also helped improve the soil quality, so other vegetation could grow better. These complex interactions are part of the reason why an ideal population growth curve like the one shown in Figure 1 almost never exists in nature.

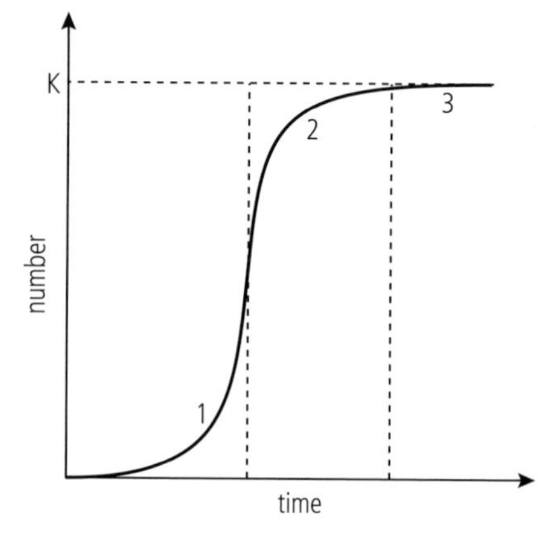

C4.1 Figure 1 An idealized population curve showing growth over time following a sigmoid shape. Phase 1 represents the exponential growth phase, phase 2 the transition phase and phase 3 the plateau phase or stationary phase, in which the population reaches its carrying capacity (K).

The sigmoid (S-shaped) curve of the graph in Figure 1 shows the three stages of population growth.

1. The **exponential phase**, also called the logarithmic phase, during which the number of individuals increases at a faster and faster rate.
2. The **transitional phase**, during which the growth rate slows down considerably; the population is still increasing but at a slower and slower rate.
3. The **plateau phase** or stationary phase, during which the number of individuals stabilizes, and there is no more growth.

So what causes the three different phases of a population growth curve?

The exponential phase

In ideal conditions, a population can double in size on a regular basis. Not counting mortality, for example, a population of bacteria can theoretically double its population every few hours: 1, 2, 4, 8, 16, 32, 64, 128, 256, 512, 1024, and so on. Without predators, introduced species, such as cane toads (*Bufo marinus*) in Australia, can take over habitats with uncontrolled population growth. The reasons for this first phase of exponential growth are:

- plentiful resources, such as food, space and (for photosynthetic organisms) light
- little or no competition from other inhabitants
- favourable abiotic factors, such as temperature and (for aquatic organisms) dissolved oxygen levels
- little or no predation or disease.

The transitional phase

Eventually, after the exponential increase in numbers of individuals, some of the factors listed above no longer hold true. This leads to the transitional phase. The causes of the transitional phase are:

- with so many individuals in the population, there is increasing competition for resources
- predators, attracted by a growing food supply, start to move into the area
- because of the large numbers of individuals living together in a limited space, opportunities for diseases to spread within the population increase.

The plateau phase

Consider the land around a volcano such as Mount Saint Helens, slowly being taken over by vegetation within months of a deadly eruption that wiped out all nearby life. Once all the fertile volcanic ground is covered with plants, the space available will be occupied to its maximum. Thus, there is gradually less and less available space for any seeds produced by the plants to germinate, and the number of plants stabilizes.

With increasing numbers of herbivores, there is a limited supply of food. In response to limited food supplies, animals tend to have smaller numbers of offspring. Some may leave the area and emigrate to a place where there is more abundant food.

Predators and disease increase mortality, and the growth curve tends to level off. In this phase, the number of births plus the number of immigrations is balanced by the number of deaths plus the number of emigrations.

Nature of Science

The population growth curve represents an idealized graphical model. Models are simplifications of complex systems. They help us conceptualize phenomena that otherwise are difficult to understand. They are useful in helping our brains grasp difficult concepts. But because they are simplifications, they have their limits.

What are the benefits of models in studying biology?

You can test the growth of a population against the model of exponential growth using a graph with a logarithmic scale for the size of the population on the vertical axis and a non-logarithmic scale for time on the horizontal axis. Online simulators exist that show how populations grow over time. Do an online search for "Howard Hughes Medical Institute population dynamics logistic growth model". It allows you to determine parameters such as the carrying capacity then launch the simulator. Other simulations show how predators and herbivores can modify a population. Do an online search for "Annenberg learner online ecology lab" to find a simulator of a food web that allows you to introduce plants, herbivores and predators into a habitat and watch what happens to a graph of their populations.

Many biologists, environmental groups, economists and governments wonder what the carrying capacity of Earth is for the human population. Will the number of people continue to increase or will disease, climate change or competition for resources lead to a transitional phase or a plateau? How reliable are mathematical models and what role could they play in shaping the way we make decisions about the future? Do different experts use mathematical models in different ways?

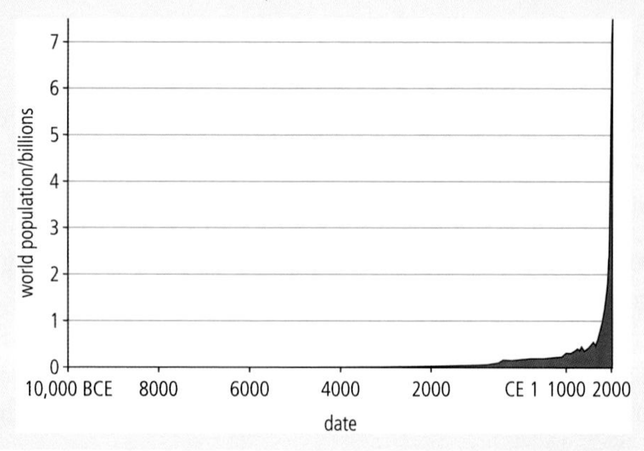

The world human population over the last twelve millennia. In November 2022, the United Nations estimated that the human population reached 8 billion. Compare this observed data with the expected sigmoid population curve.

C4.1.8 – Modelling population growth

C4.1.8 – Modelling of the sigmoid population growth curve
Application of skills: Students should collect data regarding population growth. Yeast and duckweed are recommended but other organisms that proliferate under experimental conditions could be used.

Follow the downloadable activity in the eBook to model the sigmoid population growth curve using baker's yeast or by growing duckweed in pond water. Collect data to determine the population growth and discover whether the population will reach the carrying capacity of the environment that you design.

C4.1.9 – Communities

C4.1.9 – A community as all of the interacting organisms in an ecosystem
Communities comprise all the populations in an area including plants, animals, fungi and bacteria.

A **community** is a group of populations living and interacting with each other in an area. Examples include the soil community in a forest and the fish community in a river.

In ecology, the term "interacting" can mean one population feeding on another, or being eaten by another. It can mean that one species provides vital substances for another, as in the case of certain bacteria that can help plants obtain nitrogen from the air. It can also mean that one species is protected by another, as in the case of aphids protected by ants from predator attacks. Interacting can also mean that one species relies on another for its habitat, as is the case for parasites living on or inside the bodies of other animals.

Challenge yourself

1. Pick three organisms in the figure and determine how many other organisms each one depends on. Which organisms depend on them? What about environmental factors? Which ones does each organism contribute to, and which ones does each depend on?

C4.1.10 – Intraspecific relationships

C4.1.10 – Competition versus cooperation in intraspecific relationships
Include reasons for intraspecific competition within a population. Also include a range of real examples of competition and cooperation.

▲ Social animals such as these Barbary macaques (*Macaca sylvanus*) display intraspecific cooperation in many ways. Nature is not only about competition: often individuals look out for each other.

Intraspecific relationships refer to those that occur between individuals of the same species. Intra means "within". Intraspecific relationships can involve cooperation (helping each other) or competition (competing with other members of the same species for the same resources). Cooperation can be thought of as a "win/win" situation, and competition as a "win/lose" or a "zero sum game", whereby if one wins, the other must lose.

In **intraspecific cooperation**, an individual of a species will help another individual from the same species so that survival is assured not only for the individual but for the group as well. An example is hunting as a pack, as wolves do, rather than hunting solo. Hunting cooperatively has a higher chance of success than hunting alone. Cooperation might mean caring for a neighbour's young while the mother is out looking for food, as vampire bats do. Taking turns helps the survival of both families. Cooperation can also mean that multiple individuals take turns discouraging intruders from approaching their territory, as lions do. If only a single female lion had responsibility for this, she would become exhausted more easily. Another way of thinking about intraspecific cooperation is teamwork; an individual will invest some of their time to help others, which will benefit the whole group.

In **intraspecific competition**, members of the same species compete for the same resources. The resources that are in demand could be space and sunlight for two oak trees growing near each other in a forest, or zones of grazing pastures for herds of bison. Other examples of competition within a species include male gorillas competing to become the alpha male and be able to mate with the females, or lizards competing with other members of their species for the best spot on a sunny rock to bask on. Another way of thinking about intraspecific competition is as a battle; one member of the species will succeed and get the resource, while the other will lose and not get it. But this analogy is not perfect because often there is no aggression or combat, as in the case of an individual finding food and eating it before others of its species can.

C4.1.11 – Interspecific relationships

C4.1.11 – Herbivory, predation, interspecific competition, mutualism, parasitism and pathogenicity as categories of interspecific relationship within communities
Include each type of ecological interaction using at least one example.

Interactions between different species in a community are called **interspecific relationships**. Inter means "between". Different types of interspecific interactions have different effects on the species involved, and can be positive, negative or neutral. Table 1 lists some of the possible ways in which populations can interact.

C4.1 Table 1 Types of interspecific interactions

Interaction	Examples
Herbivory: eating plant material	Snails eating lettuce leaves, sheep grazing on grass, giraffes eating acacia leaves. The photosynthetic organisms and their fruits or tubers are damaged or potentially killed in the process.
Predation: killing and eating prey or eating something that has recently died (scavenging)	Lions hunting and eating zebras, seahawks eating fish, vultures eating the abandoned carcass of a gazelle. In this relationship, only the predator survives: the prey is killed and its body parts benefit others.
Interspecific competition: two species struggle to get the same food resource	An oak tree and a balsam fir tree attempting to get the same soil minerals and sunlight. Spotted hyenas and lions hunting the same population of zebra for food. Humans overfishing a zone where sharks feed. Although one species may get a larger share of the resources, this kind of competition does not necessarily involve death or injury in the species that is less successful.
Mutualism: two species providing food or other resources where both benefit	Lichens are made up of an alga providing food using photosynthesis and a fungus providing minerals. Each one helps the other, and neither is hurt by the relationship; it is an example of mutual benefit. Mutualism is covered in more detail in Section C4.1.12. Lichens growing on a tree branch.
Parasitism: one species living on or in a host and depending on the host for food for at least part of its life cycle. The host can be harmed by the parasite	Parasites belonging to the genus *Plasmodium* cause malaria in humans. They reproduce in the human liver and red blood cells. Part of the life cycle of *Plasmodium* takes place in the body of the *Anopheles* mosquito. The mosquito transmits the malaria parasite from one human to another. An illustration of the parasite *Plasmodium*, a protist that can live in red blood cells and that causes malaria in humans.
Pathogenicity: the ability of microbes such as bacteria and viruses to cause disease in other species	Pneumonia is a transmissible disease caused by a pathogen, either a bacterium or a virus. The host of the pathogen will suffer and can potentially be killed by the microbe infecting it.

Mosquitoes are arguably the deadliest animal to humans, killing between 750,000 and one million people per year with the diseases they spread. When asked to name dangerous animals, we often come up with examples such as sharks. But there are only a handful of documented cases of people being killed by sharks worldwide each year. Mosquitoes are thousands of times more deadly to us than sharks are.

C4.1.12 – Mutualism

C4.1.12 – Mutualism as an interspecific relationship that benefits both species
Include these examples: root nodules in Fabaceae (legume family), mycorrhizae in Orchidaceae (orchid family) and zooxanthellae in hard corals. In each case include the benefits to both organisms.
Note: When students are referring to organisms in an examination, either the common name or the scientific name is acceptable.

Sometimes two species help each other survive and thrive. Neither organism is injured or destroyed. **Mutualism** is a type of interspecific cooperation that benefits both species. We will explore three examples of mutualism.

Plant root nodules and bacteria

Living organisms need the element nitrogen for the formation of amino acids and nucleic acids. Earth's atmosphere is nearly 80% nitrogen gas (N_2). Unfortunately, plants and animals cannot metabolize N_2 directly. The ability to turn gaseous nitrogen into usable nitrogen-rich molecules is called **nitrogen fixation**. One genus of bacteria that can fix nitrogen is *Rhizobium*. *Rhizobium* bacteria live in the **root nodules** of plants in the legume family, Fabaceae, which includes beans, lentils, peanuts and clover. The bacteria convert nitrogen gas from the air into ammonia (NH_3), an organic molecule that acts as a fertilizer for plants, helping them grow better. The bacteria have a mutualistic relationship with the plants: they are **symbiotic**. The *Rhizobium* bacteria receive carbohydrates and a favourable environment in the nodules of their host plant, and the plants receive usable nitrogenous compounds. Farmers and gardeners who want to fertilize their soil without using artificial chemical fertilizers can plant species that have root nodules to enrich the soil for part or all of a planting season.

The spheres protruding from this pea plant root are nodules hosting beneficial bacteria that provide usable nitrogen for the plant. Two species are cooperating for mutual benefit.

The network of fungal hyphae in the forest floor has been called the "wood wide web", in reference to how the internet's world wide web allows the transfer of resources such as information from locations that have it to those that need it. In a mycorrhizal network, it is food and minerals that are exchanged, although information in the form of chemical signals can be sent from one part of the forest to another, for example as a warning of insect attacks.

Mycorrhizae in Orchidaceae

Another example of mutualism is **mycorrhiza**, which occurs when a plant and a fungus help each other. Species of orchids (family Orchidaceae), for example, rely on fungi for one or more stages of their life cycle. Orchid seeds, which do not have enough energy and nutrients to germinate on their own, obtain the nourishment they need to produce a new plant from a fungus. The fungus in the mycorrhizal relationship takes nutrients from its environment and passes them on to the seed.

We usually think of all plants as being autotrophs (organisms that can produce their own food) because the vast majority of them make their own food via photosynthesis. There are a small number of orchid plant species that are non-photosynthetic and therefore qualify as heterotrophs (organisms that cannot produce their own food). They rely on fungi to decompose dead material and pass the nutrients to them via a root-like system of **hyphae** throughout their adult lives. Hyphae are thin filaments produced by fungi to create a network in the soil that allows the transfer of nutrients from one place to another. Sometimes this network is used by trees and plants to transfer the sugars made by photosynthesis from one tree or plant to another.

At first glance, this might not seem to be of mutual benefit. Only the orchid is getting something from the relationship: taking food but not giving it back, whether only at the seed phase or in the adult phase too, depending on the species. Eventually, however, the orchid will die, and the fungi will benefit from the nutrients released as they decompose the orchids.

Fungal mycelia, branches of hyphae filaments, can spread out in soil and are used to transport nutrients not only to the fungi that produce them but between other species, plants and trees, as well.

Zooxanthellae in hard corals

Coral reefs are built by small animals called coral polyps, which are cnidarians, in the same phylum as sea jellies and sea anemones; they all possess stinging cells called nematocysts. Embedded in the tissue of their tentacles are single-celled photosynthetic algae of the genus *Symbiodinium*, a type of dinoflagellate that can photosynthesize and is referred to as zooxanthellae. These dinoflagellates coexist in a symbiotic relationship with the coral polyps, giving them food in the form of the carbon-based energy molecules they make from sunlight. In exchange, the coral polyps give the zooxanthellae carbon dioxide and minerals, which they need to photosynthesize and grow. The coral polyps also provide the zooxanthellae a home on the coral reef. Both species live together for mutual benefit. Neither has to die or be injured for the other to flourish.

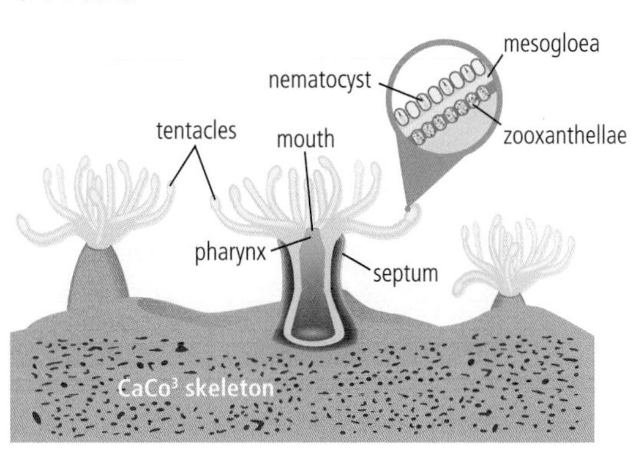

Zooxanthellae living inside the body tissue of the polyp that forms the hard coral. The two species are cooperating for mutual benefit.

Dinoflagellates are colourfully pigmented creatures and are the source of the reds and greens we see in a coral reef. **Coral bleaching** is a phenomenon that happens when the zooxanthellae leave or die. The reef looks white because the calcium carbonate skeleton of the coral does not have any pigments. Bleaching is a sign of a very unhealthy or dead coral, and with changes in the temperature and pH of oceans it is being seen more frequently.

C4.1.13 – Endemic and invasive species

C4.1.13 – Resource competition between endemic and invasive species
Choose one local example to illustrate competitive advantage over endemic species in resource acquisition as the basis for an introduced species becoming invasive.

A species is considered to be **endemic** in an area if it is only found there and nowhere else in the world. **Invasive species** are those that have been introduced into a new area from a distant origin and their populations grow so well that they start to cause problems for the species that are already living there. Without any natural predators to keep their numbers in check, an introduced population can grow exponentially.

Galápagos tortoises are endemic to the Galápagos islands, and in nature are found nowhere else on Earth. The only other populations have been introduced by humans, such as those found in zoos.

Charles Darwin came across the Galápagos tortoises in 1835. These giant tortoises are endemic to the Galápagos islands, and are not found in any other natural ecosystem in the world. Over many decades, the populations of giant tortoises have been decimated. There are many reasons for the decline of these species on the islands (including humans eating them), but one is competition with introduced invasive species such as goats. Goats compete with tortoises for grazing space and tend to destroy their habitats, such as forests, which provide shade and moisture for the tortoises. Goats breed more rapidly than tortoises, giving them an advantage. Local authorities noticed that the tortoises were losing the competition for space and food and were being quickly outnumbered.

Where did all the goats come from? For centuries, humans brought them to the islands to breed as a supply of meat. But some escaped and formed feral populations, which eventually grew to number tens of thousands. In an effort to correct the mistake of introducing goats to the islands and to save the endemic species, culling (reducing population numbers by killing off a certain percentage of the population) and eradication programmes have been implemented over several decades.

Carry out some research in the part of the world where you live or where you are from. What invasive plant or animal species are threatening local populations?

C4.1.14 – Interspecific competition

C4.1.14 – Tests for interspecific competition

Interspecific competition is indicated but not proven if one species is more successful in the absence of another. Students should appreciate the range of possible approaches to research: laboratory experiments, field observations by random sampling and field manipulation by removal of one species.

NOS: Students should recognize that hypotheses can be tested by both experiments and observations and should understand the difference between them.

One technique used to find out whether two species are competing with one another is to observe them in the field and see if they are present in the same zone at the same time. The presence of a species in a zone is recorded as a 1, and the absence of a species is recorded as a 0. A table of results called a **presence–absence matrix** is produced at the end of the study and can be mathematically analysed to see if any patterns exist. If the two species being studied are rarely or never found in the same zone, it might be because one species has out-competed the other.

Ecologists carried out a study of interspecific competition between birds in the Bismarck Archipelago, northeast of Papua New Guinea. Choosing a chain of islands for a study like this is a good way to ensure that the data being collected is from distinct zones that do not overlap. Choosing geologically young islands that were formed by relatively recent volcanic activity ensured that there were no species originally on the island for any arriving birds to compete with. The ecologists analysed which birds were found on which islands and recorded presence–absence matrices. Then they used mathematical models to predict the chances of finding birds in a particular zone by chance. By comparing these predicted values to the ones observed in nature, they could see whether the bird species were distributed by chance or not.

The traditional explanation for why one species is found in an area where similar species are not, is that it has out-competed its rivals. Although the results of this study concluded that interspecific competition could explain some of the results, there were many other factors that had an influence, and chance was one of them. For example, the history of each island was different: by chance, some bird species arrived on certain islands first, giving them an advantage.

Just because the population numbers of some introduced species increase, while the endemic species' population decreases, does not mean that the changes are due to interspecific competition for resources. In science, to find an answer related to ecological phenomena such as competition, we use multiple techniques, including field observations, field manipulation (e.g. removing a species from an area) and laboratory experiments.

A flower called the Antioch Dunes evening primrose (*Oenothera deltoides* subspecies *howellii*) is endemic to central California. Like many endemic species, its numbers are falling, and invasive species of plants have been introduced into its habitat. Originally it was thought that the decline in numbers of this subspecies was the result of competition for resources with the invasive species, but this is a good example of a hypothesis or an assumption rather than an idea established by scientific testing.

▲ The Antioch Dunes evening primrose (*Oenothera deltoides*) is endemic to central California.

In 2008, researchers led by Marc T. J. Johnson published their results after using various scientific approaches to test hypotheses about interspecific competition between the endemic primrose subspecies and an introduced species, smooth brome grass (*Bromus inermis*). Both plants are adapted to have seeds that germinate in disturbed ground, such as sand dunes that are displaced by wind or fields that have been ploughed or trampled by animals. They used two methods to see what would happen when they grew primrose alone, brome grass alone, and primrose and brome grass together.

TOK

Correlation and causation: just because two things are correlated, does not mean that one causes the other. Correlated means two phenomena are connected; when one happens, the other happens. For example, night follows day. But that does not mean that daytime causes night time, or that night causes day. There is a third factor to consider: the spinning of planet Earth on its axis, which causes both. When two things show causation, it is because it is possible to find a mechanism that bridges the correlated phenomena. "Every time I drop my pen, it falls to the floor" is a correlation that actually does show causation, because the laws of gravity explain why this correlation exists. If you drop a pen in outer space, it will float. There is no longer a correlation because there is no gravity. In the natural sciences, we test assumptions whenever possible rather than simply accepting them. But even if we come up with a valid explanation that makes sense, can we be confident that it is the best explanation? To what extent is certainty attainable?

Method 1 involved laboratory experiments. Taking seeds out of their natural habitat and planting them in a laboratory environment, in this case a greenhouse, allows researchers to control variables that would otherwise be uncontrolled in the wild. These variables include light (the researchers used natural sunlight complemented with electric lighting), rainfall (they measured how much water each plant received) and nutrients (the same type of soil was used for all the experiments and any fertilizer used was measured).

Method 2 involved field experiments. The researchers wanted to see what would happen if the primrose seeds were planted in disturbed soil. They wanted to see how well they would colonize freshly dug soil compared to competitors, some of which were invasive or introduced species. Once the plants started to grow, they measured whether the growth was impeded by the presence of other species nearby.

What they found was that primrose's presence reduced the number of plant species near it, and reduced the growth of plant species surrounding it in the wild.

By comparing the plants germinated and grown separately with those grown in the same container, and comparing germination and growth rates in disturbed soil, the researchers came to the conclusion that, although interspecific competition could explain some of the changes in the population, other factors, such as primrose's genetic diversity and the degree to which the soil was disturbed, played a large role in the plant's success. Interestingly, the statistics indicated that primrose could out-compete the brome grass because the grass grew less well in the presence of the primrose, whereas for most measurements the brome grass did not negatively affect primrose's growth.

There is a third possible method: field manipulation. By removing *Bromus* grass from around primrose in the wild, researchers could see whether this will increase or decrease the growth and population size of the primrose. If it increases, it suggests that the brome grass is competing with it for resources. If the population or growth rates continue to decline, something else, such as changes in temperature or presence of pollutants, might be the cause, rather than competition for resources.

Nature of Science

Notice from this example that hypotheses can be tested by both experiments and observations. Experiments in laboratory conditions can help isolate one factor and control all other variables, but the results might not translate into an explanation about the real world. Organisms do not always behave in the same way in laboratories or greenhouses as they do in natural environments.

C4.1.15 – The chi-squared test

C4.1.15 – Use of the chi-squared test for association between two species
Application of skills: Students should be able to apply chi-squared tests on the presence/absence of two species in several sampling sites, exploring the differences or similarities in distribution. This may provide evidence for interspecific competition.

In Section C4.1.3, we explored the use of quadrats for random sampling of sessile species (those that stay in one place). We can take this a step further. If we pick two plant species to count in each randomly sampled area of the quadrat, we can use a statistical test called the **chi-squared test of association** (also called a test of independence) to see whether the two species tend to occur together more often than they would by chance or if, on the contrary, they are never found coexisting or at least less frequently than would be expected by chance. This can help indicate whether the species are in competition with each other or not. The following worked example illustrates how the chi-squared test works.

Worked example

New England aster (*Symphyotrichum novae-angliae*) (top) and a bee collecting pollen from Canada goldenrod (*Solidago canadensis*) (bottom).

A group of students has been working in a prairie in late summer and they have noticed that there are two species of plant that seem to occur together: the New England aster (*Symphyotrichum novae-angliae*) and Canada goldenrod (*Solidago canadensis*). After learning to correctly identify each species, they used random sampling with 1 m^2 quadrats to gather data from 20 quadrat samples. If they found any goldenrod growing in their quadrat, they recorded a 1 in the goldenrod column, if not, they put a 0. They did the same for the aster column, 1 for present and 0 for absent at that sample site (see the table).

Quadrat	Goldenrod	Aster
1	0	1
2	1	1
3	1	0
4	1	0
5	1	0
6	0	1
7	1	0
8	1	1
9	0	0
10	1	0

Quadrat	Goldenrod	Aster
11	1	0
12	1	0
13	0	0
14	1	1
15	0	0
16	1	1
17	1	1
18	1	0
19	0	0
20	1	0

Quadrat data for two plant species. 1=presence of the species in a quadrat; 0=absence of the species in a quadrat.

Apply the chi-squared test to these data to decide whether the two species are associated with each other or occur independently. If you have never done a chi-squared test before or need a refresher, refer to the Skills in the study of biology chapter for an explanation of what the test is, how it works, and what the values for the degrees of freedom should be.

1. State the null hypothesis for this calculation.
2. Determine the number of degrees of freedom for this statistical test.
3. Determine the critical value in order to obtain a 95% certainty that there is a statistically significant difference between these two sets of numbers.
4. Calculate the chi-squared value for these data.
5. Interpret this value. Does it mean we can reject or not the null hypothesis?
6. Is there a statistically significant difference between these two sets of data?
7. Are there enough data points to be confident of the results?

Solution

1. The null hypothesis is that "the two categories (species 1, the goldenrod, and species 2, the aster) are independent of each other". In other words, the distribution of these two species is random, there is no association between them. When one is found, we do not find the other any more than would be expected if it was randomly distributed.
2. Because there are two possible outcomes (plant present or plant not present), the number of degrees of freedom is 2 – 1 = 1.
3. According to the chi-squared table (see the chi-squared eBook activity on page 903), the critical value in order to obtain a 95% certainty is 3.84. This value is found under the column for 0.05, which corresponds to a 95% certainty, and in the row that has a degree of freedom of 1.
4. The chi-squared value is calculated to be 4.91. This is obtained using the following values in a contingency table of observed and expected values.

	Observed goldenrod	**Observed aster**	**Grand total**
Absent	6	13	19
Present	14	7	21
Grand total	20	20	40

The expected value of 9.5 comes from the calculation (20 × 19) ÷ 40, and the expected value of 10.5 comes from the calculation (20 × 21) ÷ 40.

	Expected goldenrod	**Expected aster**	**Grand total**
Absent	9.5	9.5	19
Present	10.5	10.5	21
Grand total	20	20	40

See the Skills in the study of biology chapter for help with this calculation using observed and expected values.

5. Because 4.91 is greater than the critical value of 3.84, this means we can reject the null hypothesis.
6. Yes, the two categories are associated with each other. We can be 95% sure that there is a relationship between the presence of the goldenrod and the aster. In other words, it would be very unlikely (i.e. a 5% chance) that they are independent of each other.
7. Twenty quadrats sounds a bit small. In a random sample, there is always the chance that the sampling is not representative of the zone studied. If the zone in the prairie being studied was the size of a sports field, for example, it would have a surface area of approximately 5000 m^2. Twenty 1 m^2 quadrats represents 20 m^2 of that surface, meaning that only 0.4% of the field was actually sampled. That is the equivalent of finding two pieces of a 500-piece puzzle you have never seen before and declaring that you know what the image will be once the puzzle is complete.

You can be asked about the chi-squared test in IB biology exams. Be sure you know when the chi-squared test can be used, the steps for doing it, and how to interpret the results.

C4.1.16 – Predator–prey relationships

C4.1.16 – Predator–prey relationships as an example of density-dependent control of animal populations
Include a real case study.

Because organisms rely on each other, it would be a disadvantage for one species in a community to completely take over. If an alga floating on the surface of a lake, for example, increased its population until it completely covered the surface, it would cut off light for the photosynthetic plants and phytoplankton below. If a rabbit population on a prairie increased to such a high population density that they were hopping over each other, food supplies would become more and more scarce, their large numbers would attract predators, and their population density would increase the chances of a disease spreading. Density-dependent factors like these affect large populations and small populations differently. A larger, denser population is more likely to experience food shortages, predation, disease or even emigration, because individuals in the overpopulated area may move to other areas in search of more food and space.

The Canada lynx (*Lynx canadensis*), an elusive predator.

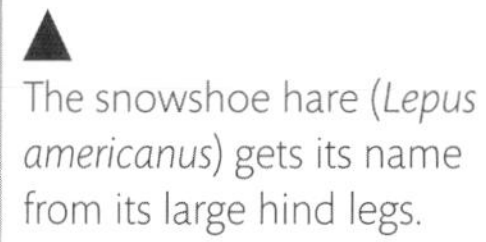

The snowshoe hare (*Lepus americanus*) gets its name from its large hind legs.

Predators provide a form of population control; their activity will increase as a population of prey increases. The Canada lynx (*Lynx canadensis*) and the snowshoe hare (*Lepus americanus*) are often cited as a classic example of a predator–prey interaction. The lynx preys on the hare. Changes in the numbers of the lynx population are followed by changes in the numbers of the hare population.

The graph in Figure 2 overleaf shows a simplified model of the relationship between predator (in this case the lynx) and prey (the hare). Before reading further, try to see if you can describe what each peak on the graph represents and when it occurs compared to the other peaks. Then try to explain the pattern, which repeats one cycle about every 10 years. Now read the paragraph below for an explanation.

C4.1 Figure 2 The cyclical relationship between predator and prey.

population size
time
prey
predator

SKILLS

The online PhET Interactive Simulation called "Natural Selection" was originally designed to show how mutations could help a rabbit population survive and adapt in the presence of changes in the environment, such as the presence of food or of wolves. It is also a good simulation of predator–prey interactions when wolves are introduced. The simulation tracks the size of the population as you modify different parameters. Do a search for it online and give it a try. See what happens to the population with and without wolves.

At the start of the graph in Figure 2, notice that the two lines go up and then down, but the maximum for the predator's line (the red line) is out of sync with the line for the prey (the dashed blue line), because the predator population always reaches its maximum after the prey population. Now focus on the first cycle from the lowest point of the predator line to the next lowest point of that line. At the beginning, the prey line increases to a maximum population for the hare because there are few lynx around to hunt them. This is shown by the predator line, indicating the lynx population, which is low at this point. Over time, as the population of hares increases, the lynxes have more and more food available, which allows them to have more offspring. Recall that the availability of resources such as food acts as a limiting factor for populations, so when more food is available, the population is not as limited.

As the lynx offspring grow and produce young of their own, the population increases to its maximum. However, we notice that by the time the two lines cross, the hare population has declined rapidly. Why? This is because there are so many lynxes hunting that many of the hares are being eaten. Shortly after the lynx population reaches its maximum, the hare population reaches its minimum. During this time we see the predator line going down as the lynx population decreases because of a lack of food. And then we are back at the start of the cycle. This is a density-dependent relationship.

More recent analysis has suggested that this model is oversimplified, however. For one thing, both species live in a food web that involves other species. Snowshoe hares are preyed upon by many other species, such as wolves, foxes, owls and hawks.

C4.1.17 – Control of populations

C4.1.17 – Top-down and bottom-up control of populations in communities
Students should understand that both of these types of control are possible, but one or the other is likely to be dominant in a community.

Communities rely on limiting factors to make sure that no one species takes over completely. In Section C4.1.5 you learned about some limiting factors that define the carrying capacity of a habitat, such as availability of food, predation and disease.

Limiting factors can be top-down or bottom-up. **Top-down controls** are seen when a species' population can be reduced by other species feeding on it. Predation, as seen in the example with the lynx, is a top-down control. Herbivory is a top-down control for plant populations because the more they are eaten, the more their population goes down. **Bottom-up controls** are seen when a species' population can be reduced by a lack of resources such as food, sunlight (for photosynthetic organisms) or minerals. Recall that when there were not enough hares to sustain the lynxes, the lynx population went down.

Challenge yourself

A study of a tropical coral reef revealed the effects of top-down and bottom-up limiting factors. The bottom-up limiting factor was the nutrients that increased algal blooms, which negatively affected the coral. The top-down limiting factor was the fish that ate the algae, so keeping the coral reef healthy. Two study sites on the coral reef, 1 and 2, were isolated for 24 months and their conditions were manipulated. Controlled experiments were performed by pairing high and low herbivory (the amount of algae eaten by fish) with high and low nutrient levels. See the table.

	Study site 1 (low herbivory)		**Study site 2 (high herbivory)**		**Significant differences ($p < 0.05$)**
	Reduced nutrients A	Elevated nutrients B	Reduced nutrients C	Elevated nutrients D	
Crustose corallines	41.2 ± 4.6	1.8 ± 1.8	<0.1	71.7 ± 3.0	D > A > B, C
Frondose macroalgae	20.8 ± 4.3	63.7 ± 8.2	0.6 ± 0.3	16.9 ± 4.1	B > A, D > C
Algal turfs	37.1 ± 3.9	14.5 ± 4.7	<0.1	22.1 ± 2.9	A >D >B >C
Predicted dominants	Turfs	Macroalgae	Corals	Corallines	

Mean percentage cover (with standard error) of benthic functional groups colonizing clay diffusers following 24 months of reduced and elevated nutrients in low- and high-herbivory study sites (n = 4)

Three types of algae were included in the study, as shown in the table:

- crustose corallines, which are beneficial algae that help the coral build the reef
- frondose macroalgae, which are fleshy and filamentous, and can overgrow the coral and prevent healthy reef building because of their algal blooms
- algal turfs, which are microalgae and their blooms are also detrimental to reef building.

The herbivorous fish were parrotfish and surgeonfish.

The question posed by the study was how the effect of top-down herbivores and bottom-up nutrients affected the competition of harmful and beneficial algae. The percentage of reef cover by each type of algae was a measure of its success.

2. For study site 1, compare the mean percentage cover of all three alga types with reduced and elevated nutrients. What were the effects on the coral?
3. For site 1, the prediction was that macroalgae would be dominant in the competition for percentage cover with elevated nutrients. Was that prediction confirmed? Give evidence to support your answer.
4. Describe a benefit to the coral reef that occurred over the 24 months in part D of the experiment.
5. Explain the conditions under which elevated nutrient-induced microalgae blooms decreased the growth of the reef-building corals.

Parrotfish feed on algae.

Although both top-down and bottom-up factors influence ecosystems, one is usually dominant. In marine ecosystems, the limiting factors are usually bottom-up. If phytoplankton are plentiful and able to photosynthesize efficiently, the rest of the ecosystem can usually flourish. Occasionally, because of overfishing, there can be periods of top-down limits to population growth.

C4.1.18 – Allelopathy and antibiotic secretion

C4.1.18 – Allelopathy and secretion of antibiotics
These two processes are similar in that a chemical substance is released into the environment to deter potential competitors. Include one specific example of each—where possible, choose a local example.

Sometimes competition between species generates survival adaptations such as camouflage to hide from predators, aggressive behaviour to defend a territory, or features that allow a plant or animal to obtain a resource before its competitors can. Instead of enhancing their own survival techniques, another way organisms can gain a competitive edge is to release molecules into the surrounding environment to make life difficult or impossible for their competitors. Whereas **primary metabolites** are molecules that are needed for the basic functions of life, such as energy and growth, molecules produced to impede or kill competitors are called **secondary metabolites**. Such molecules are used in a process called **allelopathy**, which is the production of secondary metabolites that influence the growth and success of other organisms.

Examples of allelopathy include:

- inhibiting seed germination in nearby competitors
- interfering with nutrient uptake in roots so that plants cannot grow nearby
- killing bacteria or inhibiting the growth of nearby bacteria.

Garlic mustard (*Alliaria petiolata*) has a chemical weapon against its competitors: sinigrin.

Allelopathy in plants

Garlic mustard (*Alliaria petiolata*) is an introduced and invasive species, and allelopathy may have a role in its success in colonizing new territory. Garlic mustard produces a secondary metabolite called sinigrin. This is the same molecule responsible for the zing in the taste of mustard, horseradish and wasabi. Sinigrin can inhibit the germination of seeds from other plants, and can reduce the growth of roots of plants already growing in the area. Both properties of sinigrin give the garlic mustard a competitive advantage over other plants. Its seeds can germinate unhindered while its competitors' seeds are prevented from germinating, and it can grow its roots down into the soil to obtain water and nutrients, while its competitors' root growth is slowed down by the presence of sinigrin.

Does garlic mustard or any other such plant grow near you? Carry out some research to determine whether your area contains plants or trees that use allelopathy to fight off competition.

Allelopathy in microbes

Competition exists not only in large, visible organisms such as mustard plants and wildcats, but also in the invisible world of microbes. The single-celled fungus *Penicillium rubens* produces a molecule whose identity you have probably already guessed from its name: penicillin. Penicillins are produced by certain species of moulds and have the remarkable ability to stop the growth of bacteria. Molecules that can inhibit the growth of bacteria or that can kill bacteria are called **antibiotics**. In nature, this means that *P. rubens* can compete for space and food sources by releasing the allelopathic molecule penicillin into its surroundings. Antibiotics are used by many species of microscopic organisms to establish a bacteria-free zone around them, allowing their colonies to spread while inhibiting the spread of competing colonies.

Different antibiotics work in different ways. Penicillin blocks bacterial enzymes that are trying to link certain molecules together when a bacterium is building its cell wall. Because the bacterium's enzymes cannot do their job, the cell wall cannot form correctly. This allows water to leak into the cell, causing it to burst and killing the cell. Other antibiotics interfere with protein synthesis in bacteria.

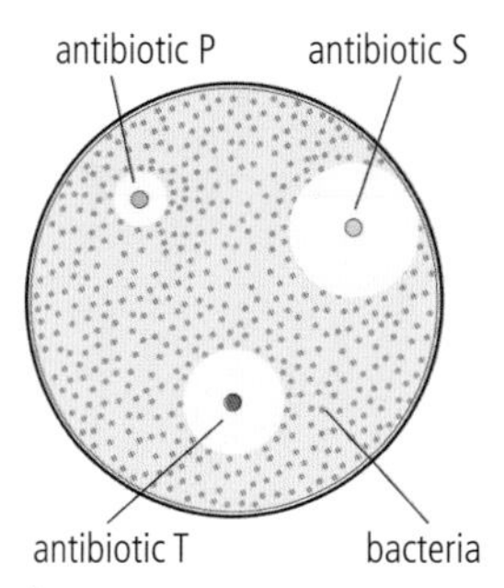

One way to test antibiotics is by putting discs of paper soaked in different antibiotic solutions in a Petri dish, to see whether a bacterial colony is inhibited by any. The bacteria-free zones around the three different types of antibiotics shown in this diagram indicate that the allelopathic molecules being tested can successfully suppress bacterial growth.

Guiding Question revisited

How do interactions between organisms regulate sizes of component populations in a community?

Within this chapter you have learned:

- to estimate a population of sessile organisms, random sampling using quadrats can be used
- for motile organisms, mark and recapture techniques can generate data to use with the Lincoln index
- populations have a maximum size, the carrying capacity, and this is regulated by positive and negative feedback including density-independent factors such as temperature and density-dependent factors such as the spread of disease
- population growth can be modelled using a sigmoid population curve.

Ever since the Nobel Prize winning work of Alexander Fleming, Howard Florey and Ernst Chain in the first half of the 20th century, penicillin has been used as an antibiotic to treat bacterial infections in humans. This is one of the most significant breakthroughs in the history of medicine. The original molecules used in antibiotic medications were extracted from moulds, but today they can be synthesized in the laboratory. In addition to penicillin, doctors today can prescribe dozens of molecules as antibiotics, and new antibiotics are being tested as candidates for future use. Unfortunately, new strains of bacteria that are resistant to the antibiotics we already have are emerging faster than we can find or design new ones.

Guiding Question revisited

What interactions within a community make its populations interdependent?

Within this chapter you have learned:

- when multiple populations interact, a community is formed (interactions include herbivory, predation, interspecific competition, mutualism, parasitism and pathogenicity)
- some populations compete for resources and others cooperate to help each other survive
- to find out if a species is competing with another, various techniques are used, including the removal of the competing species to see if the endemic species' population increases without its competitor
- statistical tests such as the chi-squared test of association can help determine whether two species tend to occur together or whether one tends to dominate and out-compete the other
- population numbers can change over time depending on the relationship between species, as seen in predator–prey relationships or top-down and bottom-up control of population numbers
- some organisms, such as the fungus that makes penicillin, keep away competitors by the use of allelopathy, secreting secondary metabolites into their environment.

Exercises

Q1. A group of individuals belonging to the same species and living in the same area at the same time is called:

A A community.
B Intraspecific competition.
C An ecosystem.
D A population.

Q2. Which population phase is most affected by positive feedback?

A The exponential phase.
B The transition phase.
C The plateau phase.
D The carrying capacity.

Q3. A species found in one geographic area and nowhere else in the world is said to be:

A Native.
B Endemic.
C Invasive.
D Mutualistic.

Q4. Which factors that control populations are density-independent?

I. Forest fires
II. Predation
III. Volcanic eruptions.

A I and II only.
B I and III only.
C II and III only.
D I, II and III.

Q5. Which are examples of mutualism?

I. Zooxanthellae in hard corals
II. Root nodules in Fabaceae
III. *Plasmodium* in *Anopheles* mosquitoes.

A I and II only.
B I and III only.
C II and III only.
D I, II and III.

Q6. Distinguish between the top-down and bottom-up control of populations in communities, giving an example of each.

C4.2 Transfers of energy and matter

Guiding Questions

What is the reason matter can be recycled in ecosystems but energy cannot?

How is the energy that is lost by each group of organisms in an ecosystem replaced?

There is a limited number of carbon and oxygen atoms available on Earth for organisms to use in their bodies. Although cosmic dust may rain down in small quantities from space, it is not enough to meet the needs of life on Earth. As a result, organisms need to reuse and recycle the matter that is available. In contrast, the Sun sends energy to Earth every day and therefore energy does not need to be recycled. In addition, organisms do not have mechanisms for storing heat energy efficiently, so it is lost to their environment. This lost energy is replaced when plants convert the energy from sunshine into food, and that food fuels cells that have lost heat energy through cellular respiration.

C4.2.1 – Ecosystems are open systems

C4.2.1 – Ecosystems as open systems in which both energy and matter can enter and exit
Students should know that in closed systems only energy is able to pass in and out.

In a tropical rainforest, energy enters the ecosystem in the form of sunlight and leaves in the form of heat, essentially from cellular respiration. When the seeds of a tree land in a river that flows through the forest, the river can transport the seeds far from the ecosystem and sometimes all the way to the ocean. Matter that was once in the soil (the minerals the tree used to grow the seeds) or in the air (the carbon dioxide the tree used in photosynthesis) can be removed from the forest ecosystem. Any birds or mammals that migrate into the forest bring with them matter that can be integrated into the ecosystem when they produce waste or die, leaving their carcass. A system that allows matter in and out is considered an **open system**. When humans cut down and remove trees to make paper or wood furniture, we are demonstrating the openness of the forest ecosystem.

In a **closed system**, matter does not enter or leave and so it must be recycled. Energy, however, can still enter and leave the system. An example of this is Earth when viewed on a planetary scale: energy enters in the form of sunlight and leaves in the form of heat, but matter must be recycled on Earth. Only a small quantity of matter enters this closed system, in the form of meteorites or cosmic dust from space, which lands on the surface. Although humans sometimes send equipment into outer space, made from metals, for example, mined from the rocks on Earth, this only represents a tiny fraction of the minerals and other matter recycled and reused on Earth by living organisms. Generally speaking, Earth is a closed system where only energy enters and leaves, while matter is recycled.

Systems theory provides an explanation of how systems interact with each other and with their environment. Scientists use systems theory to explain open and closed systems and to predict what would happen if something in a system or in the environment around a system changed. An ecosystem follows certain laws, such as the law of conservation of mass, which states that matter cannot be destroyed or created. There is also the law of thermodynamics, which describes the flow of energy through a system and how one form of energy can be transformed into another, such as light into chemical energy or chemical energy into heat, but energy cannot be created or destroyed.

Planet Earth can be considered a closed system because only relatively minute quantities of minerals enter or exit it from space. Energy, however, is lost as heat radiating into space, but is added to the system every day in the form of sunlight.

Nature of Science

In science, a **law** is a generalized principle that describes a natural phenomenon. We observe something and try to come up with a general rule that summarizes what is happening. There is no attempt to explain or give a reason why, only an effort to model what we observe. A **theory** does provide an explanation. A theory is used to explain a phenomenon by describing the underlying mechanisms that are responsible for producing it. Laws and theories help us understand the natural world and predict what will happen next. For example, systems theory is being used to predict climate change on Earth.

C4.2.2 – Sunlight sustains most ecosystems

C4.2.2 – Sunlight as the principal source of energy that sustains most ecosystems
Include exceptions such as ecosystems in caves and below the levels of light penetration in oceans. **NOS:** Laws in science are generalized principles, or rules of thumb, formulated to describe patterns observed in living organisms. Unlike theories, they do not offer explanations, but describe phenomena. Like theories, they can be used to make predictions. Students should be able to outline the features of useful generalizations.

The best studied ecosystems are those found on Earth's surface, whether they are on land or in surface water. Such systems rely on sunlight, and they will be the main focus of this section. Be aware, however, that there are other, less well-studied, ecosystems that exist in total darkness, such as those in deep ocean water and those found in dark caves or deep underground; these ecosystems are not well understood because they are so difficult to access and study. One example is the ecosystem created by hydrothermal vents, which exist on the ocean floor below levels where sunlight can penetrate. Minerals dissolved in seawater that has infiltrated cracks in the ocean floor and been warmed by magma, rise to the surface of the ocean floor and provide a rich source of minerals for chemical reactions in microorganisms that use chemosynthesis instead of photosynthesis to make food. You can find out more about this in Section C4.2.7.

All life that you see around you on Earth's surface relies either directly or indirectly on sunlight. Photosynthetic organisms such as plants and phytoplankton take inorganic carbon dioxide (CO_2) and convert it into energy-rich sugar ($C_6H_{12}O_6$). The addition of minerals allows them to synthesize complex molecules such as cellulose, proteins and lipids, which allow them to build stems, leaves, fruit and seeds. Notice what is happening in this process: light energy from the Sun is being converted into chemical energy (food). Chemical energy refers to the fact that carbon compounds, such as carbohydrates, proteins and lipids, are rich in energy, thanks to the chemical bonds that exist between the carbon atoms and other atoms. This is what makes fruits, grains and vegetables good food sources. Consumers cannot "eat" sunlight and air, but they can eat carbohydrates, proteins and lipids. The chemical energy in these carbon compounds can be measured in calories, kilocalories, joules or kilocalories, which we are familiar with on food labels.

Sunlight is the initial source of energy for all ecosystems on the surface of Earth's land and oceans.

C4.2.3 – The flow of energy

C4.2.3 – Flow of chemical energy through food chains
Students should appreciate that chemical energy passes to a consumer as it feeds on an organism that is the previous stage in a food chain.

Photosynthetic organisms provide a remarkable link between the abiotic world (e.g. rocks, water and air) and the biotic world (e.g. plants, bacteria, fungi and animals). They transform air and water into food. This is why they are referred to as **producers**. By feeding on producers, consumers can utilize chemical energy to grow and stay healthy. For example, a cow (the consumer) grazing in a field of grass (the producer) is taking chemical energy from the grass and digesting the carbon compounds to help build meat or milk inside its own body. Humans can consume the meat or milk from the cow, to benefit from the chemical energy the cow has obtained from the grass. Such a pattern of feeding is called a **food chain**. The process of passing energy from one organism to another through feeding is referred to as the flow of energy through a food chain.

Can you identify the consumers and producers in this Peruvian scene?

C4.2.4 – Food chains and food webs

C4.2.4 – Construction of food chains and food webs to represent feeding relationships in a community
Represent relationships in a local community if possible. Arrows indicate the direction of transfer of energy and biomass.

When studying feeding habits, it is convenient to write down which organisms eat which by using an arrow. Thus "herring → seal" indicates that the herring is eaten by the seal. When more of the life cycles of the seal and herring are investigated, new organisms can be added to the chain: copepods (a common form of zooplankton) are eaten by the herring, and great white sharks eat seals. Lining up organisms with arrows between them is how food chains are represented. Here are three examples of food chains from three different ecosystems.

Grassland ecosystem:

clover → grasshoppers → toads → snakes → hawk

River ecosystem:

green algae → mayfly larvae → juvenile trout → kingfisher

Marine ecosystem:

diatoms → copepods → herring → seals → great white shark

The definition of a food chain is a sequence showing the **feeding relationships** and energy flow between species. In other words, it answers the question "What eats what?" The direction of the arrow shows the direction of the flow of energy.

Look at the food chains in Figure 1 showing part of a river ecosystem. Notice how they link together into a **food web**. A food web shows multiple food chains and how they are connected. Often organisms such as juvenile trout eat not only caddis fly larvae but also the larvae of other species.

SKILLS

Can you observe any food chains in your local area? Think about organisms you see everyday and try to construct a food chain with three or four organisms, starting with a producer. If you live in a city, this might be challenging but think about sparrows eating caterpillars, which have fed on the leaves of trees lining the streets. Maybe a neighbour's cat succeeds in hunting and catching a sparrow for a snack.

A food chain is a single linear set of connections from producer to consumers with only one species at each trophic level, whereas a food web shows a more complete picture because often one producer is eaten by more than one consumer, and consumers eat more than one type of organism.

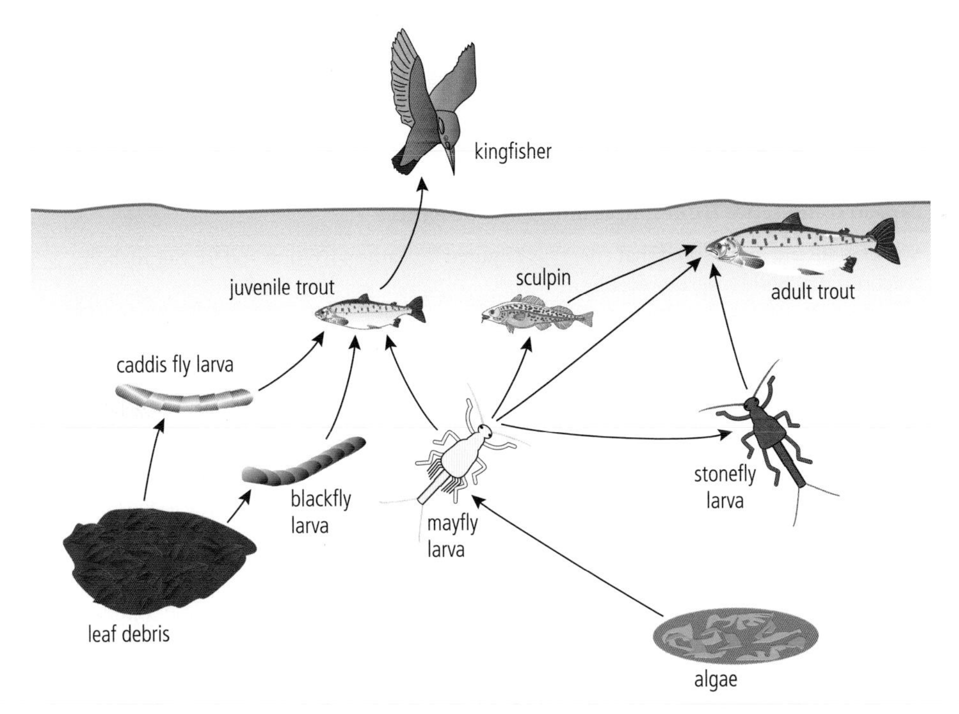

C4.2 Figure 1 A food web from a river ecosystem.

C4.2.5 – Decomposers

C4.2.5 – Supply of energy to decomposers as carbon compounds in dead organic matter
Include faeces, dead parts of organisms and dead whole organisms.

▲ The minotaur beetle (*Typhaeus typhoeus*) is a type of decomposer called a detritivore.

An effective way of unlocking the precious nutrients stored in the cells of plants and animals is through decay. **Decomposers** (saprotrophs and detritivores) break down non-living food sources such as the faeces of organisms, entire dead bodies, or fallen leaves or the skin shed from a snake. Saprotrophs such as fungi secrete enzymes onto dead matter such as a fallen tree and absorb the nutrients. Detritivores such as the minotaur beetle have mouthparts to ingest dead matter and digest it inside their bodies. The digestive enzymes of decomposers convert the organic matter into a more usable form for themselves, and therefore for other organisms. For example, proteins from a dead organism are broken down into ammonia (NH_3) and, in turn, the nitrogen in ammonia can be converted into useful nitrates (NO_3^-) by bacteria.

In this way, decomposers recycle nutrients so that they are available to other organisms and are not locked inside the bodies or waste products of organisms in the ecosystem. Decomposers play a major role in the formation of soil, without which plant growth would be greatly impaired, if not impossible. The rich black layer of soil called **humus** is made up of organic debris and nutrients released by decomposers.

C4.2.6 – Autotrophs

C4.2.6 – Autotrophs as organisms that use external energy sources to synthesize carbon compounds from simple inorganic substances
Students should understand that energy is required for carbon fixation and for the anabolic reactions that build macromolecules.

Some organisms are capable of making their own organic molecules as a source of food. These organisms are called **autotrophs**, and they synthesize their organic molecules from simple inorganic substances. This process involves either photosynthesis or chemosynthesis.

Why do organisms need energy? Metabolic processes such as carbon fixation and other anabolic reactions that build complex molecules from building blocks require energy.

Photoautotrophs can take light energy from the Sun, combine it with inorganic substances (water and carbon dioxide), and obtain a source of chemical energy in the form of a carbon compound (glucose). This ability to convert inorganic carbon dioxide, which is unusable to consumers, into organic molecules that are useful for energy and growth is called **carbon fixation**. Once this is done, some of the molecules can be combined using anabolic reactions to make larger molecules. Plants use these reactions to make macromolecules such as cellulose.

Chemoautotrophs are able to take carbon dioxide and, using inorganic compounds such as hydrogen sulfide (H_2S) as an energy source, build more complex molecules that can be useful to them as food.

Because autotrophs make food that is useful for themselves but also consumed by other organisms, they are also called **producers**.

Examples of photoautotrophs include:

- cyanobacteria
- clover
- algae such as giant kelp
- pine trees.

Examples of chemoautotrophs include:

- sulfur-oxidizing bacteria
- nitrogen-oxidizing bacteria
- iron-oxidizing bacteria.

Electronic monitoring equipment at a hydrothermal vent on the Juan de Fuca Ridge in the Pacific Ocean. Red and white tube worms (*Riftia pachyptila*) host sulfur-oxidizing chemosynthetic bacteria that produce food for them. What looks like black smoke coming out of the vent is, in fact, hot water charged with dark minerals.

TOK

The first hydrothermal vent, and the community of surprising organisms such as giant tube worms surrounding it, was discovered in 1977 just west of the Galapagos Islands. Why had no-one noticed them before? Because they were over 2,000 m below the ocean surface. Think about the kinds of tools and technology that were necessary to make this discovery. How does this show the importance of material tools in the production and acquisition of knowledge for marine biologists?

C4.2.7 – Energy sources

C4.2.7 – Use of light as the external energy source in photoautotrophs and oxidation reactions as the energy source in chemoautotrophs
Students should understand that oxidation reactions release energy, so they are useful in living organisms. Include iron-oxidizing bacteria as an example of a chemoautotroph.

Electrons are needed to produce adenosine triphosphate (ATP) for a cell. During an oxidation reaction, electrons are removed (lost) from atoms. The donated electrons are free to participate in reactions in the cell, such as helping in the production of ATP. Light is used by photosynthetic organisms to oxidize water molecules by photolysis, in order to donate electrons and hydrogen (H^+) ions, and, in the process, the reaction releases oxygen gas. The ATP produced using the donated electrons and the hydrogen ions can then be used to generate the organic molecule glucose with carbon dioxide as the source of carbon.

HL For more on oxidation and reduction (redox) reactions, see Section C1.2.7.

Photoautotrophs get the energy they need for carbon fixation from sunlight, whereas chemoautotrophs get their energy from oxidation reactions that do not require a source of light energy.

Iron-oxidizing bacteria such as *Mariprofundus ferrooxydans*, a bacterium that thrives near hydrothermal vents like those near Hawaii, hundreds of metres below the ocean surface and far from where sunlight can penetrate, are capable of obtaining electrons from iron. They do this by taking one form of iron, iron(II), otherwise written as Fe^{2+}, and removing an electron to transform it into iron(III), or Fe^{3+}. For every negative electron lost, iron will gain a positive charge. The lost (or donated) electron from this oxidation reaction can be used to generate ATP for the cell. These microbes are the producers for (and therefore the starting point of) food chains near the hydrothermal vents where they live.

How does the transformation of energy from one form to another make biological processes possible?

C4.2.8 – Heterotrophs

C4.2.8 – Heterotrophs as organisms that use carbon compounds obtained from other organisms to synthesize the carbon compounds that they require

Students should appreciate that complex carbon compounds such as proteins and nucleic acids are digested either externally or internally and are then assimilated by constructing the carbon compounds that are required.

Heterotrophs cannot make their own food from inorganic matter, and must obtain organic molecules from other organisms. They get their chemical energy from autotrophs or other heterotrophs. Because heterotrophs rely on other organisms for food, they are called **consumers**. Heterotrophs ingest organic matter that is living or has been recently killed.

Examples of heterotrophs include:

- zooplankton
- sheep
- fish
- insects.

The only component in our diet that we can synthesize using sunlight is vitamin D. There are precursors in our skin that absorb ultraviolet (UV) light waves and produce vitamin D. But in order to get all the other types of molecules needed to keep us healthy, we need to consume molecules made by other living things.

Organisms that are not capable of synthesizing their own food from inorganic components of their environment need to get their nourishment by ingesting (eating) parts of other organisms. Consumers take in the energy-rich carbon compounds, such as sugars, proteins and lipids, synthesized by other organisms in order to survive.

When heterotrophs consume food, they digest the proteins into amino acids, the lipids into fatty acids, and the DNA and RNA into nucleic acids. They can then synthesize their own proteins using the amino acids, make their own lipids using the fatty acids, and synthesize their own DNA and RNA using the nucleic acids. This process of integrating nutrients into usable substances in the tissues of the body is called assimilation. When a fox eats a chicken egg, it will digest the chicken protein and then assimilate the amino acids to build its own fox proteins.

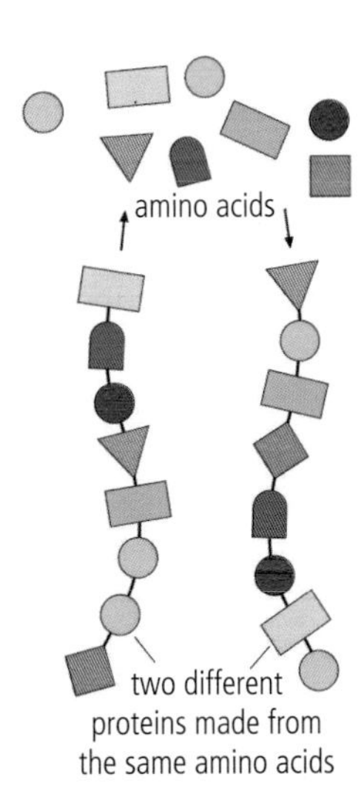

Proteins obtained from food sources (e.g. chicken protein) can be broken down into amino acids and reconnected in the consumer's cells to make new proteins (e.g. fox proteins).

C4.2.9 – The release of energy by cell respiration

C4.2.9 – Release of energy in both autotrophs and heterotrophs by oxidation of carbon compounds in cell respiration
Students are not required to be familiar with photoheterotrophs.

To release energy, carbon compounds such as glucose are oxidized. Cells need a constant supply of energy, which is the role of cellular respiration. Although we often only think about the autotrophs' ability to produce food for consumers, they still need to use the organic molecules they have synthesized as an energy source. So photosynthetic autotrophs and chemoautotrophs both carry out cellular respiration. For example, plants oxidize the glucose they have made in order to release energy for chemical reactions and growth.

In addition to making food for themselves, producers are a food source for heterotrophs, which also use cellular respiration to release energy from carbon compounds such as sugar, proteins and lipids, which the producers have synthesized in their cells. When a bird eats a seed, the carbohydrates and lipids in the seed will be oxidized to release the energy.

C4.2.10 – Trophic levels

C4.2.10 – Classification of organisms into trophic levels
Use the terms "producer", "primary consumer", "secondary consumer" and "tertiary consumer". Students should appreciate that many organisms have a varied diet and occupy different trophic levels in different food chains.

Biologists use the term **trophic level** to indicate how many organisms the energy in the system has flowed through. The first trophic level is occupied by the producers. The next trophic level is occupied by the **primary consumers** (organisms that eat the producers), and the trophic level after that is occupied by **secondary consumers** (organisms that eat primary consumers). If the secondary consumers are eaten by another organism, the next trophic level has been reached: **tertiary consumers**.

Although a food web provides a representative but complicated picture of what is being eaten in an ecosystem (see Figure 2), problems sometimes arise when determining trophic levels. Can you see the following difficulties when you look at the food web in Figure 2?

- An eagle is a tertiary consumer when eating rattlesnakes, but a secondary consumer when eating rabbits.
- A coyote is a primary consumer when it eats the fruit of a cactus, but a tertiary consumer when it eats a rattlesnake.
- A lizard is a tertiary consumer when it eats rattlesnake eggs, but a secondary consumer when it eats insects.

C4.2 Figure 2 A desert food web.

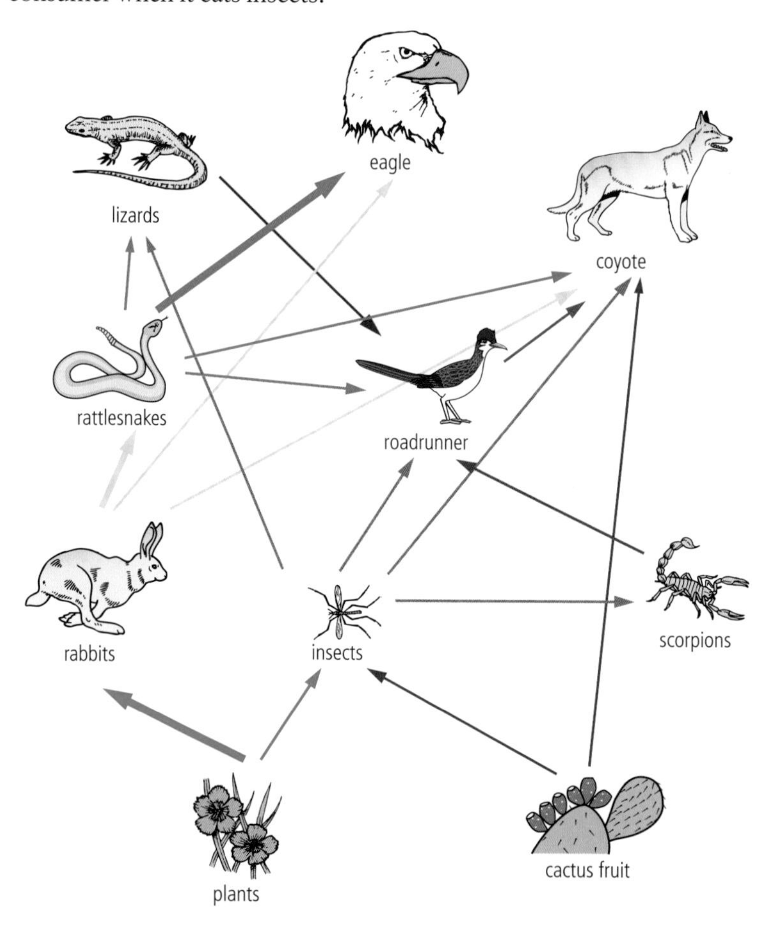

C4.2.11 – Energy pyramids

C4.2.11 – Construction of energy pyramids
Application of skills: Students should use research data from specific ecosystems to represent energy transfer and energy losses between trophic levels in food chains.

Be careful not to confuse pyramids of energy with pyramids of numbers: pyramids of numbers show the population sizes of each trophic level, not the energy.

A **pyramid of energy** is used to show how much and how fast energy flows from one trophic level to the next in a community (see Figure 3). The units used are energy per unit area per unit time: kilojoules per square metre per year (kJ m^{-2} yr^{-1}). Because time is part of the unit, energy pyramids take into account the rate of energy production, not just the quantity. Because energy is lost, each level is always smaller than the one before. It would be impossible to have a higher trophic level wider than a lower trophic level because organisms cannot create energy, they can only transfer it inefficiently.

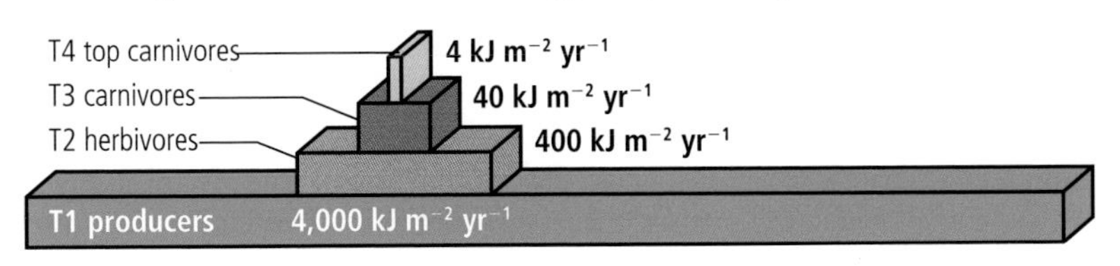

C4.2 Figure 3 A pyramid of energy. The green base is trophic level 1 (T1), the producers. The pink level is trophic level 2 (T2), the herbivorous primary consumers. The purple layer is trophic level 3 (T3), the carnivorous secondary consumers, and the top layer is trophic level 4 (T4), the top carnivores as tertiary consumers.

A pyramid of energy is used to show how much and how fast energy flows from one trophic level to the next in a community.

SKILLS

Use the data below from Silver Springs, Florida (USA), to draw a pyramid of energy. Typically, data for decomposers are not included in pyramids of energy, but you can draw them to one side of the pyramid. Draw the pyramid stepped rather than as a triangle and try to draw the levels to scale. As an extension to this activity, you could research similar data in your area of the world.

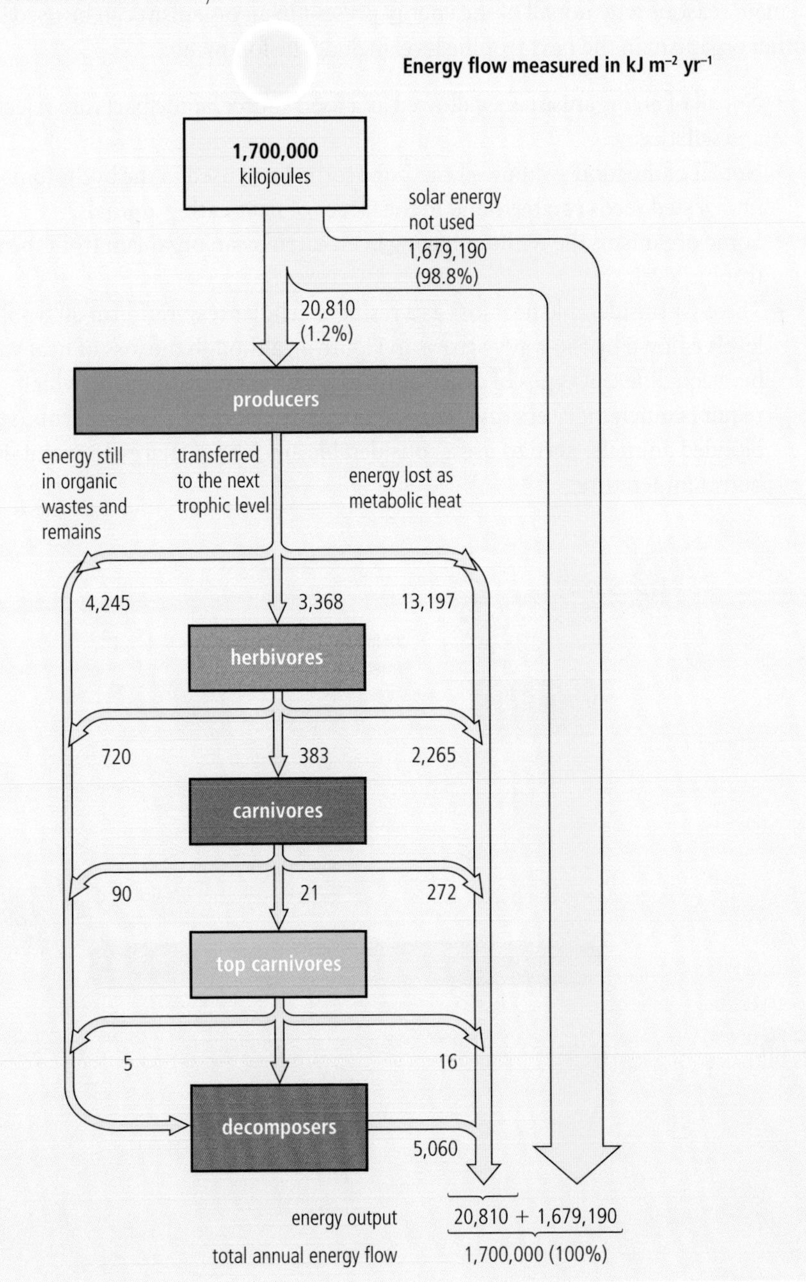

C4.2.12 – Energy loss between trophic levels

C4.2.12 – Reductions in energy availability at each successive stage in food chains due to large energy losses between trophic levels
Decomposers and detritus feeders are not usually considered to be part of food chains. However, students should understand the role of these organisms in energy transformations in food chains. Consider the causes of energy loss.

No organism can use 100% of the energy present in the organic molecules of the food it eats. Typically, only 10–20% of the energy available is used from the previous step in a food chain. This means that as much as 90% is lost at each level.

The main reasons why not all of the energy present in an organism can be used by another organism in the next trophic level include the following.

When an owl swallows a mouse, many of the body parts are retained and processed in the digestive system, but the bird cannot digest bones and hair and will spit them out in an oblong mass called a pellet. When the pellets of undigested material land on the ground, they are decomposed by detritivores and other decomposers.

- Not all of an organism is swallowed as a food source, some parts are rejected and will decay.
- Not all of the food swallowed can be absorbed and used in the body, for example undigested seeds can be found in the faeces of fruit-eating animals.
- Some organisms die without having been eaten by an organism from the next trophic level.
- There is considerable heat loss as a result of cellular respiration at all trophic levels (shown by the wavy arrows in Figure 4), although the loss of heat varies between different types of organism. Most animals have to move, which requires much more energy than a stationary plant needs. Endotherms (warm-blooded animals) need to use a considerable amount of energy to maintain their body temperature.

Decomposers play a key role in an ecosystem. If the nutrients in a dead leaf, for example, stayed locked in the leaf instead of being released back to the soil, the living organisms that need those nutrients would not have access to them.

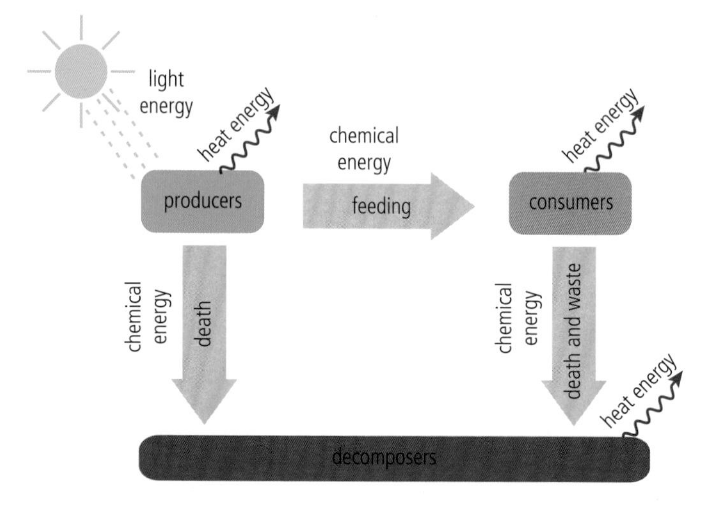

C4.2 Figure 4 How energy moves through an ecosystem in multiple forms, such as light energy, chemical energy and heat energy.

C4.2.13 – Heat loss from cell respiration

C4.2.13 – Heat loss to the environment in both autotrophs and heterotrophs due to conversion of chemical energy to heat in cell respiration
Include the idea that energy transfers are not 100% efficient so heat is produced both when ATP is produced in cell respiration and when it is used in cells.

When studying energy in organisms in a meadow or grassland, inside an animal such as a grasshopper, for example, chemical energy is used for cellular respiration. Glucose originally produced by the grass is converted by the grass into usable energy or, if it is eaten by the grasshopper, it will be converted into carbon dioxide and water. The transfer of energy from one form to another is not 100% efficient. Chemical reactions necessary for the production of ATP generate heat but heat is also generated whenever ATP is used. Any heat generated by cellular respiration is radiated away from the organism and lost to the air, soil or water in which the organism is living. Although this might be more obvious in mammals, which can give off considerable amounts of heat, even grass and grasshoppers will lose heat to the environment. If the grasshopper is eaten, some of the chemical energy in its body (in the form of protein, for example) is passed on to the next organism (a toad, for example). If the grasshopper dies and is not eaten, detritivores and decomposers will use its available energy.

The cells of decomposers also carry out cellular respiration and, as a result, the heat they produce will also be lost to the environment.

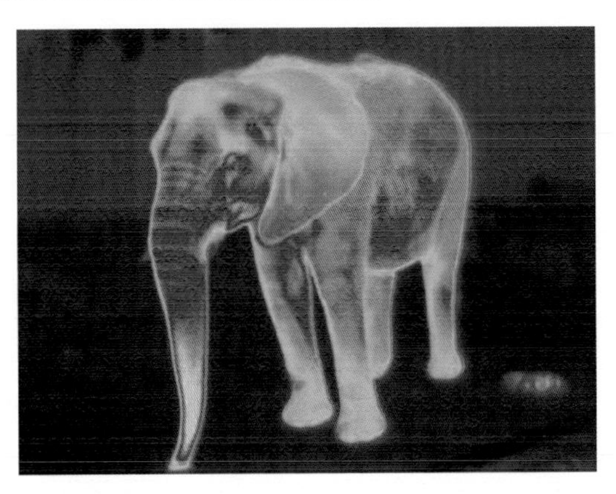

A thermogram is a photo taken with a special camera that can sense heat. Mammals such as this elephant produce heat that is radiated from the body. The red colour indicates hotter parts of the body, while blue indicates cooler parts of the body. Elephants pump blood to their ears to radiate heat and keep cool.

What does it mean when heat is "lost"? As you may already know from other science courses, there is a law stating that energy cannot be created or destroyed, only converted from one form to another. We have seen that light energy can be converted into chemical energy by the process of photosynthesis. We have also seen that, during the process of cell respiration, not all the energy is converted into useful energy (ATP) by the cell: some of it is converted to heat energy. Although this keeps mammals warm, once that heat leaves an organism's body it cannot be used again as a biological energy resource. So, for the organism and its ecosystem, this energy is "lost". It has not disappeared, however; it has simply been converted into a form that the organism can no longer use as a source of energy.

Is this a problem? Usually no, because the Sun is constantly providing new energy to producers. The energy is converted to chemical energy and passed on from one trophic level to the next. However, if, for some reason, sunlight could no longer reach

Earth's surface, because it is blocked by clouds or particles in the sky (as happens after large volcanic eruptions or large asteroid impacts), then the food chain is affected. This can be catastrophic for some organisms, as it was for dinosaurs, for example.

Catastrophic events, such as the asteroid that smashed into Earth marking the end of the Cretaceous period 65 million years ago, can produce enough debris in the atmosphere to greatly reduce the intensity of sunlight energy arriving at the Earth's surface. As a consequence, 65 million years ago the lack of autotroph production led to mass extinctions, including the disappearance of all non-avian dinosaurs. Such major events illustrate just how vital sunlight is to most ecosystems.

C4.2.14 – The number of trophic levels

C4.2.14 – **Restrictions on the number of trophic levels in ecosystems due to energy losses**
At each successive stage in food chains there are fewer organisms or smaller organisms. There is therefore less biomass, but the energy content per unit mass is not reduced.

Although some food chains can have up to six trophic levels, most have four. The number of levels is limited by how much energy enters the ecosystem. Because so much is lost at each level, low energy at the start will be quickly transferred, whereas abundant energy at the start can sustain several trophic levels. The number of organisms in the chain, as well as the quantity of light available at the beginning of the chain, will determine how long a food chain is.

The **biomass** of a trophic level is an estimate of the mass of all the organisms within that level. Biomass is defined as the dry weight of an organism, because the actual mass of an organism includes a large amount of water. Water needs to be removed for the dry weight to be measured. It is expressed in units of mass, but also takes into account area or volume, for example gram per metre squared per year, $g\ m^{-2}\ yr^{-1}$. Although other factors are involved, the amount of sunlight reaching the photosynthetic producers strongly influences the biomass, so sunnier parts of the world can produce more biomass. Phytoplankton nearer the equator generate more biomass than phytoplankton further from the equator, for example.

Examine the following freshwater food chain:

green algae → caddisfly larvae → stickleback fish → pike fish

As we move backwards along a food chain or down an energy pyramid from the top, the number of organisms that occupies the lower trophic level increases. For every predator such as a pike, many dozens of sticklebacks will be eaten by it per year. The sticklebacks will eat hundreds of caddisfly larvae. This is another reason why food chains have a limited length, there is a limited amount of biomass production to start with at the first level, and for each subsequent level there is a continuing appetite for energy. If we examine the biomass, we see that, because there are fewer organisms each time we go up one trophic level, and because not all biomass is consumed or digested, there will be less biomass, and therefore less energy, at that next level. A stickleback weighs less than all the caddisfly larvae and other foods it eats.

It is important to note that, although biomass goes down and energy goes down from a lower trophic level to a higher trophic level, the energy values per unit mass (e.g. J g^{-1}) of the organisms do not go down. This can be seen in Table 1.

	Trophic level	Energy per unit mass / J g^{-1} of wet mass
Algae	1	3,439
Caddisfly larvae	2	3,760
Fish	3	5,341

C4.2 Table 1 Unlike the overall energy in the next level of a food chain or energy pyramid, energy per unit mass does not reduce as the trophic level increases

C4.2.15 – Primary production

C4.2.15 – Primary production as accumulation of carbon compounds in biomass by autotrophs

The units should be mass (of carbon) per unit area per unit time and are usually g m^{-2} yr^{-1}. Students should understand that biomes vary in their capacity to accumulate biomass. Biomass accumulates when autotrophs and heterotrophs grow or reproduce.

In contrast to warmer places in the world, cooler **biomes**, or biomes with fewer hours of sunlight per year, have a lower biomass and therefore cannot support as many organisms. A biome is a large community of plants and animals, such as a desert, a tropical forest or a grassland. Biomes tend to occupy zones that cover wide expanses, often on a continental scale. The grasses in the temperate steppe biome of Mongolia, for example, could be the start of a food chain that includes lemmings, snakes and eagles.

But temperate steppes will accumulate less biomass in a year than a tropical rainforest, where trees, because of their massive size compared to the plants that occupy grasslands, will photosynthesize more and therefore fix more carbon. Bigger trees also produce more food and more habitat space for consumers and the rest of the food web. These heterotrophs will grow and reproduce to contribute to the overall biomass. Each biome varies in its capacity to produce biomass.

Primary production refers to the biomass generated by the activity of producers such as photosynthetic organisms when they fix carbon and make carbon compounds that can be used as a food source. This production is measured as the mass of carbon per unit area per unit of time, or grams per metre squared per year: g m^{-2} yr^{-1}.

Do not confuse units of energy with units of mass in an ecosystem. In an energy pyramid, we use kJ m^{-2} yr^{-1}, whereas when referring to biomass production we use g m^{-2} yr^{-1}.

C4.2.16 – Secondary production

C4.2.16 – Secondary production as accumulation of carbon compounds in biomass by heterotrophs
Students should understand that, due to loss of biomass when carbon compounds are converted to carbon dioxide and water in cell respiration, secondary production is lower than primary production in an ecosystem.

Some molecules along the food chain cannot contribute to the accumulating biomass because they are lost in various forms, for example carbon dioxide is lost from organisms during cellular respiration, and waste products including urea are excreted. So, just as not all energy is passed on from one trophic level to the next, not all biomass is passed on either. When carbon compounds are assimilated by the next trophic level, it is often in the form of proteins and lipids. This conversion of one form of carbon molecule (e.g. glucose from producers) to another (e.g. lipids) inside consumers is called **secondary production**. Primary production is the generation of biomass in the first trophic level; secondary production is the addition of biomass in subsequent heterotrophic levels. As much of the energy is lost from one level to the next, biomass production is always lower in secondary than in primary production.

C4.2.17 – The carbon cycle

C4.2.17 – Constructing carbon cycle diagrams
Students should illustrate with a diagram how carbon is recycled in ecosystems by photosynthesis, feeding and respiration.

The element carbon is the cornerstone of life as we know it. Carbon is such a crucial element for living organisms that it is part of the definition of the term "organic". Hence, life on Earth is referred to as carbon-based life.

Not only is carbon found in the **biosphere** in organic molecules such as carbohydrates, proteins, lipids and vitamins, it is also found in the **atmosphere** as carbon dioxide and in the lithosphere as carbonates and fossil fuels in rocks. The biosphere refers to all the places where life is found; the atmosphere is where the gases that make up air are found; while the **lithosphere** refers to all the places where rocks are found. Petroleum, from which products such as gasoline, kerosene and plastics are made, is rich in carbon because it originates from partially decomposed organisms that died millions of years ago.

As shown in Figure 5, carbon is constantly cycled between living organisms and the inorganic processes that make the carbon available. The carbon atoms that make up the cells of the flesh and blood of a giraffe, for example, came from the vegetation the giraffe ate. Eating organic material provides newly dividing cells in the giraffe's body with a fresh supply of carbon-based energy-rich molecules with which the cells can carry out work. When cellular respiration is complete, carbon dioxide is released into the atmosphere and, when the giraffe dies, its body will be eaten by scavengers and the remains broken down by decomposers. Some of the carbon from the giraffe's body will go back into the atmosphere as carbon dioxide as the decomposers carry out cellular respiration. This section will look at some of the many different forms carbon can take as it is cycled by nature.

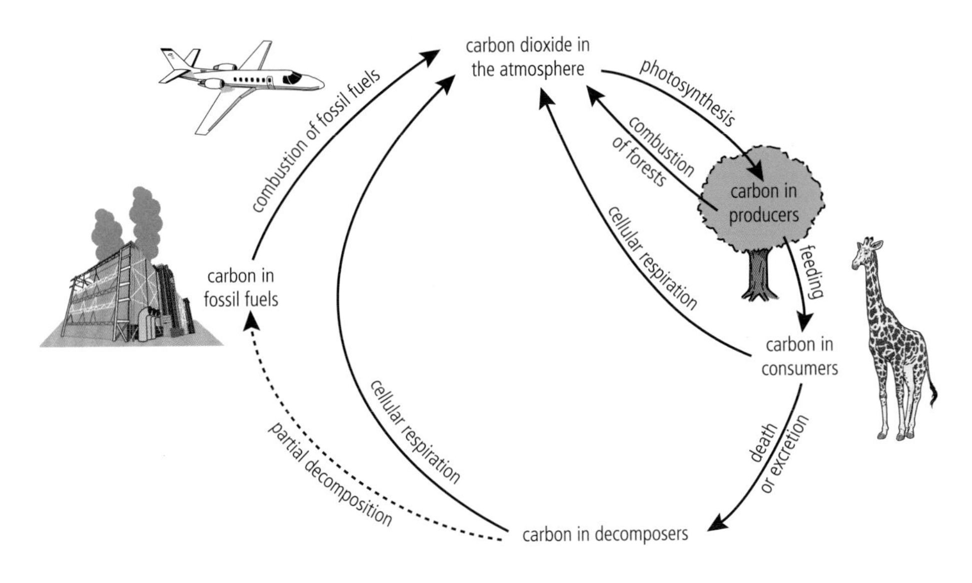

C4.2 Figure 5 Earth's carbon cycle.

We will start with food. Photosynthetic autotrophs take carbon dioxide from the atmosphere and convert it into carbohydrates. The unbalanced chemical equation for photosynthesis is:

$$CO_2 + H_2O \rightarrow C_6H_{12}O_6 + O_2$$

The sugar on the right-hand side of the equation (in green) is a source of food, not only to the autotroph synthesizing it but also to the organisms that feed on the autotrophs. In its inorganic form on the left, as atmospheric carbon dioxide (in black), the carbon is not usable as a food source by the autotrophs or by any consumers. Few people realize how dependent the biosphere is on energy from the Sun for food production. And the biosphere includes us.

Carbon dioxide

Carbon dioxide is absorbed by photoautotrophs such as photosynthetic bacteria, phytoplankton, plants and trees. As you will recall, these producers are eaten by consumers, which then use the carbon in their bodies. Cellular respiration from all trophic levels, including decomposers, produces carbon dioxide, which is released back into the environment. This carbon dioxide diffuses into the atmosphere or into the water, depending on whether the organism is terrestrial or aquatic.

Methane

Other carbon compounds are produced by microbes such as archaea. Some archaea are anaerobic methanogens, meaning they do not require oxygen gas. When these methanogenic archaea metabolize food, they produce methane (CH_4) as a waste gas. You should be familiar with methane because it is the gas used in laboratories (for example the flame of Bunsen burners) and it makes the blue flame used in homes for cooking and heating.

These microbes are common in wetlands, where they produce marsh gas, which can sometimes glow at night, but they are also responsible for producing methane gas in the digestive tracts of mammals, including humans. With large herds of cattle being raised worldwide, there is a concern that the quantities of methane they produce are contributing to the runaway greenhouse effect, which will be discussed later.

SKILLS

Using the ideas listed below, draw a diagram showing the carbon cycle.

- Photosynthesis providing carbon to producers.
- Feeding providing carbon to herbivores.
- Cellular respiration by producers, consumers and decomposers releasing carbon into the atmosphere.
- Excretion and death providing carbon to decomposers.
- Partial decomposition by decomposers providing carbon to fossil fuels.
- Combustion of organic material releasing carbon into the atmosphere.

Compare your diagram with the one in Figure 5. How did you do?

▲ Plant material can be considered a carbon sink or a carbon source depending on what is happening. Combustion releases carbon dioxide into the atmosphere, so a forest on fire behaves as a carbon source (as seen on the left), but when the trees and other plants are growing, they are a carbon sink, absorbing more carbon dioxide than they are releasing (as seen on the right).

C4.2.18 – Carbon sinks and sources

C4.2.18 – Ecosystems as carbon sinks and carbon sources
If photosynthesis exceeds respiration there is a net uptake of carbon dioxide and if respiration exceeds photosynthesis there is a net release of carbon dioxide.

A **carbon source** in an ecosystem is an organism that is a net producer of carbon dioxide. A **carbon sink** in an ecosystem is an organism that absorbs and holds more carbon than it releases. Plants that photosynthesize more than they respire and that hold their carbon in the form of roots, buds, stems, seeds and fruits are carbon sinks. Consumers in the form of herbivores or carnivores produce carbon dioxide through cellular respiration and release it into the atmosphere; they are carbon sources.

It is possible for some organisms to be either a source or a sink depending on the situation. For example, if a tree burns after a lightning strike or if a human cuts down a tree to burn it as firewood, the tree goes from being a carbon sink to being a carbon source.

Often people who choose to study ecology do so because of a profound love of nature, a respect for our planet and a desire to pass on a healthy ecosystem to future generations. They want to understand the things that they admire and love. But would this push them towards only wanting to study parts of the world that are considered remote, pristine or exotic, rather than studying their own local areas? And are they attracted to aesthetically pleasing organisms such as butterflies, wild cats or rare orchids, rather than less admired species such as spiders, parasitic organisms, cockroaches or slime moulds? In what ways do our values affect our pursuit and acquisition of knowledge?

C4.2.19 – The release of carbon dioxide during combustion

C4.2.19 – Release of carbon dioxide into the atmosphere during combustion of biomass, peat, coal, oil and natural gas
Students should appreciate that these carbon sinks vary in date of formation and that combustion following lightning strikes sometimes happens naturally but that human activities have greatly increased combustion rates.

One way to produce carbon dioxide from organic material and release it into the atmosphere is to burn organic matter. This can happen naturally or can be caused by human activity. Forest fires occur naturally when lightning strikes a forest, and it is part of the natural cycle for dead fallen branches to be burned away and forest growth to start afresh. In fact, some seeds cannot germinate until they have been exposed to fire, such as those of the lodgepole pine (*Pinus contorta*). Fires can also be ignited naturally by volcanic eruptions and lava flows cutting through woodlands. However, forest fires can release a considerable quantity of carbon dioxide into the atmosphere.

Humans have changed the natural fire cycle; we ignite forest fires sometimes by accident but often on purpose to clear land for agriculture. Humans and our hominid ancestors have known how to use and control fire for cooking and tool making for at least a million years, but it is only in the most recent decades that there have been enough humans and enough forests burned to have a considerable impact on the quantity of carbon dioxide in the atmosphere. Traditionally, wood has also been used to heat our homes, whether they be 100,000-year-old caves or modern houses with a fireplace.

Since the start of the industrial era, we have used more and more sources of energy other than wood. Organic sources of carbon in the form of **biomass**, coal, peat, oil and natural gas are burned for many purposes, including generating energy for factories, electricity for homes, and fuels for vehicles such as automobiles, trains and aircraft.

Biomass in this context is organic waste and can come from agricultural practices such as growing crops, raising livestock and harvesting trees in forestry. The non-usable parts of plants or the excrement of farm animals can be burned directly for energy, or they can be fermented in large tanks called **biodigesters** and the methane gas produced used as fuel for combustion.

Wood and **peat** form over many decades and, if well managed, forests and peat bogs can be considered renewable resources if the extraction rate does not exceed the production rate. Peat is a form of waterlogged soil found in certain types of wetlands, such as mires and bogs. Although peat is a heterogeneous mixture of many things, at least 30% of its dry mass must be composed of dead organic material for it to be called peat. Because of the high acidity of the soil, the environment is difficult for decomposers, which is why so much energy-rich organic matter remains. In order for it to be usable as a fuel, cut peat is dried out to reduce its high levels of humidity. It is cut into slabs, granules or blocks, and moved to where it is needed for combustion.

Slabs of peat left to dry in Scotland, UK. Think about where peat fits into the carbon cycle.

Some of the sources of energy, such as coal, crude oil and natural gas, used for human activities are considered to be **fossil fuels** because they are mined from the ground and, once removed, they are not renewed. It takes millions of years for these petroleum products to form: partially decomposed organic matter is pushed underground by geological forces. Coal is a hard black rock that has to be mined by breaking apart the layers of sediments it is found in. Crude oil is a viscous black liquid that needs to be pumped out of the ground, where it is trapped between layers of rock. This naturally occurring substance is then refined and transformed into many manufactured products such as fuels and plastics. Natural gas, which has a very low density, bubbles towards the surface and, when it is trapped by a dome of impenetrable rock, can be collected by digging a well.

Type of resource	Time it takes to form	Examples of energy sources used by humans
Renewable	Replenished daily (or almost every day)	Sunshine, wind, river currents, wave motion, geothermal, biomass
Long-term renewable	Decades or hundreds of years	Well-managed hardwood forests and peat bogs
Non-renewable	Millions of years	Coal, crude oil, natural gas, uranium

C4.2.20 – The Keeling Curve

C4.2.20 – Analysis of the Keeling Curve in terms of photosynthesis, respiration and combustion

Include analysis of both the annual fluctuations and the long-term trend.

The longest continuous monitoring of the carbon dioxide concentration in the Earth's atmosphere has been carried out by the National Oceanic and Atmospheric Administration (NOAA) in the US on the Hawaiian island Mauna Loa. This site was chosen because it is in the middle of the ocean far from highly industrialized zones, and because of its high altitude. Mauna Loa is a volcano, so any carbon dioxide released from the volcano is subtracted from the data. The measurements were started under the direction of Charles Keeling, and the graph of the results is called the Keeling Curve.

Figure 6 shows the data from 1958, when Keeling started the measurements, to 2022. You can find the most recent values online. The black line shows the average trend. The wavy red line shows seasonal changes, reflecting the fact that photosynthetic organisms absorb more carbon dioxide in the summer and autumn (the downward pointing spikes under the solid black line) than in the winter and spring (the spikes pointing above the solid black line).

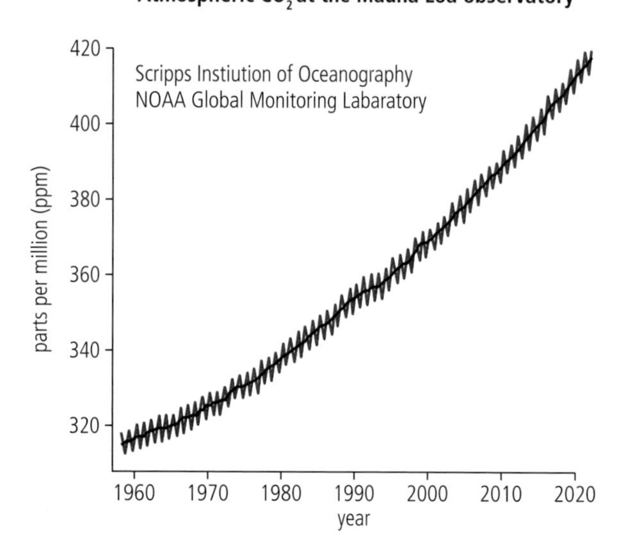

C4.2 Figure 6 The National Oceanic and Atmospheric Administration (NOAA) data on carbon dioxide levels in the atmosphere 1958–2022. The oscillating pattern is caused by seasonal fluctuations in activities such as photosynthesis, but the overall trend largely reflects the results of human activities such as combustion of fossil fuels.

Challenge yourself

1. Using the solid black trend line in Figure 6, determine the atmospheric carbon dioxide concentration for 1960 and for 2020.
2. Calculate the percentage change in atmospheric carbon dioxide concentration between 1960 to 2020.
3. Explain why the measurements have a high point and a low point for each year.

So why has there been such a sudden increase in atmospheric carbon dioxide in less than the duration of one human lifetime? According to the International Panel on Climate Change (IPCC), it is mainly because of human activity such as the combustion of fossil fuels, notably for transporting goods and people within our global economy. This is discussed further in Chapter D4.3, which explores climate change.

Worked example

The figure below, taken from the NOAA website, shows the atmospheric carbon dioxide levels of recent years.

The pattern shown in red (the larger fluctuations) is caused by seasonal fluctuations in carbon dioxide levels. The smoother black line shows the trend corrected for those seasonal fluctuations.

1. **(a)** Determine how the years are divided up on the horizontal *x*-axis of the graph.
   (b) Estimate the level of atmospheric carbon dioxide in April 2017 and in April 2022 using the corrected values on the black trend line.
   (c) Look at the five lowest values on the red line. There is one per year. Determine the month of the year during which this low point most often occurs. Do the same for the high points.
2. **(a)** In terms of cellular respiration and photosynthesis rates in the northern hemisphere, explain the annual downward fluctuations from May to October.
   (b) Do the same for the upward fluctuations from October to May of the following year.

3. Describe the overall trend shown by the graph for the years shown, giving quantitative data in your description.

Solution

1. (a) The years are divided into quarters: January to March, April to June, July to September, and October to December.
   (b) 406 p.p.m. and 417 p.p.m., respectively. It is important to include the units.
   (c) Lows are in October and highs in May.
2. (a) Because plants, phytoplankton and photosynthetic bacteria are generally more active in the summer and autumn months, more carbon dioxide is extracted from the atmosphere and levels drop. During this time, cellular respiration is contributing large quantities of carbon dioxide to the atmosphere, but not as fast as photosynthesis is taking it out.
   (b) Conversely, when photosynthesis is less intense during the winter and spring months, carbon dioxide levels rise and, although organisms are generally less active at colder times of the year, their cellular respiration rates put more carbon dioxide into the air than the photosynthetic organisms can remove.

What are the direct and indirect consequences of rising carbon dioxide levels in the atmosphere?

3. The trend shows an increase from 406 p.p.m. at the beginning of 2017 to 417 p.p.m. in April 2022. This 11 p.p.m. increase represents a percentage change of +2.7% for the 5-year period shown.

C4.2.21 – The dependence on atmospheric oxygen and carbon dioxide

C4.2.21 – Dependence of aerobic respiration on atmospheric oxygen produced by photosynthesis, and of photosynthesis on atmospheric carbon dioxide produced by respiration
The fluxes involved per year are huge, so this is a major interaction between autotrophs and heterotrophs.

If you look at the two summary equations for photosynthesis and aerobic respiration, you will notice that one is the reverse of the other:

Photosynthesis by autotrophs:

carbon dioxide + water → glucose + oxygen

Aerobic cellular respiration by heterotrophs:

glucose + oxygen → carbon dioxide + water

In order for aerobically respiring organisms to have oxygen, photosynthetic organisms need to produce oxygen gas. Life evolved on Earth for about a billion years without oxygen and without photosynthesis. Early bacteria and archaea relied on anaerobic respiration. Photosynthesis came after many hundreds of millions of years of respiration without oxygen. Aerobic cellular respiration was only able to evolve once oxygen gas was produced by photosynthesis.

How much carbon dioxide and oxygen are necessary to keep this cycle between producers and consumers going? The quantities are enormous. The quantity of carbon dioxide released into the atmosphere by all living organisms on Earth every year, for example, is estimated at around 200 GtC yr^{-1} (gigatonnes of carbon per year). A gigatonne is 1,000 million tonnes.

C4.2.22 – The recycling of chemical elements

C4.2.22 – Recycling of all chemical elements required by living organisms in ecosystems
Students should appreciate that all elements used by living organisms, not just carbon, are recycled and that decomposers play a key role. Students are not required to know details of the nitrogen cycle and other nutrient cycles.

We have focused on the recycling of carbon in the form of organic matter in organisms and carbon dioxide in the atmosphere, but all other elements that organisms need are cycled, too. One prominent example is nitrogen. Although it is present in huge quantities in Earth's atmosphere (roughly 79% of air is N_2 gas), like carbon dioxide it needs to be fixed in order to become usable by living organisms. Certain bacteria can fix nitrogen, and the usable molecules that they produce are passed on from one organism in a food chain to the next.

Decomposers are a key part of any nutrient cycle because they make nutrients that are no longer needed by organisms available to others that do need them.

Other elements from the periodic table that are necessary for life and that are cycled and passed on through the food chain include hydrogen, oxygen, calcium, potassium, sodium, iron, phosphorus and many others. Because they cannot be synthesized by living organisms, and because they do not enter the system every day in the way solar energy does, these atoms need to be recycled. Your body is made up of atoms from the foods you have eaten throughout your lifetime. We are what we eat in the sense that we construct our living tissue using atoms borrowed from other sources. And when living organisms die, they pass on their atoms to the next trophic level, or to detritivores and decomposers.

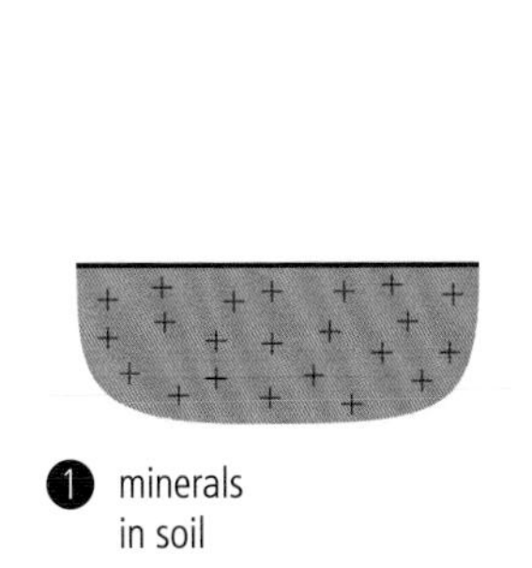

1 minerals in soil

2 minerals incorporated into living tissue

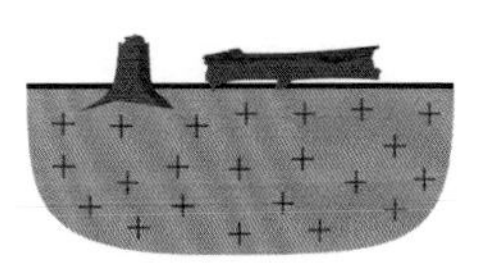

3 minerals returned to soil when organism decays

Chemical elements in minerals are absorbed into living trees and then returned to the soil when the trees die and decay.

Guiding Question revisited

What is the reason matter can be recycled in ecosystems but energy cannot?

In this chapter we have discussed how:

- without new supplies of minerals arriving on Earth in sufficient quantities, organisms need to reuse and recycle the matter that is already available
- decomposers play a key role in returning nutrients to an ecosystem
- carbon cycle diagrams show how carbon is absorbed by producers and released by consumers, or by human activity such as the burning of fossil fuels or deforestation
- food chains and food webs show how energy flows from one trophic level to the next
- energy pyramids show us that energy is lost at each trophic level, and part of this loss is in the form of heat as once the heat leaves an organism, it cannot be reused or recycled.

Guiding Question revisited

How is the energy that is lost by each group of organisms in an ecosystem replaced?

In this chapter you have learned:

- sunlight reaches Earth every day, replenishing the energy requirements of autotrophs
- therefore energy can leave Earth in the form of heat and does not need to be recycled.

Exercises

Q1. Which organism in a food chain is found on the third trophic level?

A Secondary consumer. **B** Tertiary consumer.
C Primary consumer. **D** Producer.

Q2. What is shown in the Keeling Curve?

I. Annual fluctuations in atmospheric carbon dioxide concentration.
II. Primary production of photosynthesis in kJ $m^{-2} yr^{-1}$.
III. An increase in atmospheric carbon dioxide in recent decades.

A I and II only. **B** I and III only.
C II and III only. **D** I, II and III.

Q3. Biosphere 2 is a large research facility, owned by the University of Arizona, USA, used to see whether humans can live for many months inside a sealed set of buildings where all the air, water and matter is recycled. Participants grow all their own food inside greenhouses and recycle all their waste. Explain whether this is an open system or a closed system.

Q4. State the names of the hydrocarbon-rich substances that are described below and used for fuel by humans.

(a) A kind of waterlogged soil found in wetlands and made of partially decomposed plant material.

(b) A hard black rock that can be burned to generate electricity or direct heat.

(c) A black viscous liquid trapped between layers of rock.

(d) Of all the commonly used petroleum products, this one has the lowest density.

Q5. The table shows data from two ecosystems, Cedar Bog and Lake Mendota in Wisconsin, USA.

	Cedar Bog		**Lake Mendota**	
Tropic level	**Productivity/ cal cm^{-2} yr^{-1}**	**Efficiency / %**	**Productivity/ cal cm^{-2} yr^{-1}**	**Efficiency/ %**
Solar radiation	119,000		119,000	
Plants	111	0.1	480	0.4
Herbivores	14.8	13.3	41.6	8.7
Carnivores	3.1	22.3	2.3	5.5
Higher carnivores			0.3	13.0

(a) Construct pyramids of energy for Cedar Bog and Lake Mendota.

(b) Compare and contrast the two energy pyramids.

(c) Describe the level of efficiency from one trophic level to the next.

(d) Suggest, giving a reason, which of the two could support the introduction of a new top predator to the ecosystem.

Q6. (a) From the following information, construct a food web.

- Grass is eaten by rabbits, grasshoppers and mice.
- Rabbits are eaten by hawks.
- Grasshoppers are eaten by toads, mice and garter snakes.
- Mice are eaten by hawks.
- Toads are eaten by hognose snakes.
- Hognose snakes are eaten by hawks.
- Garter snakes are eaten by hawks.

(b) From the food web you have drawn, what is the trophic level of the toad?

C4 Practice questions

1. Explain how a population of grasshoppers could be estimated using the capture–mark–release–recapture technique and outline the assumptions that must be considered.

(Total 6 marks)

2. Distinguish between parasitism and mutualism, giving an example of each.

(Total 2 marks)

3. This energy flow diagram for a temperate ecosystem has been divided into two parts. One part shows autotrophic use of energy and the other shows the heterotrophic use of energy. All values are kJ m^{-2} yr^{-1}.

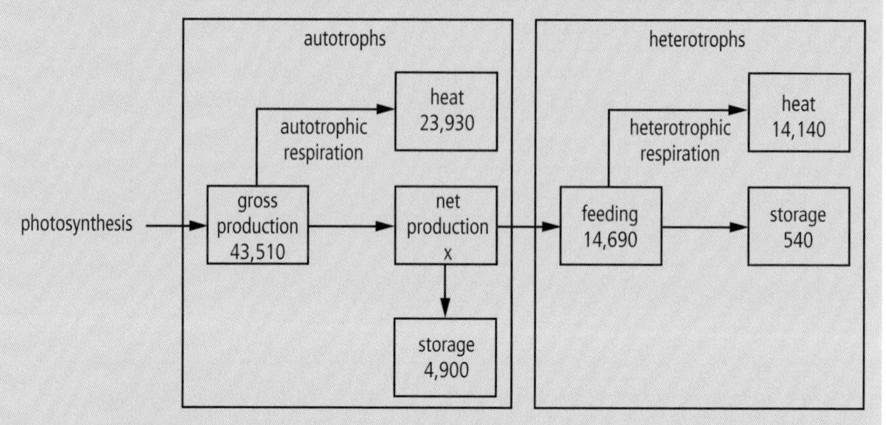

(a) Calculate the net production of the autotrophs (i.e. what is left of the gross production once all the heat loss is subtracted). (1)

(b) **(i)** Compare the percentage of heat lost through respiration by the autotrophs with the heterotrophs. (1)

(ii) Most of the heterotrophs are animals. Suggest one reason for the difference in heat losses between the autotrophs and animal heterotrophs. (1)

The heterotrophic community can be divided into food webs based on decomposers and food webs based on herbivores. It has been shown that of the energy consumed by the heterotrophs, 99% is consumed by the decomposer food webs.

(c) State the importance of decomposers in an ecosystem. (1)

(d) Deduce the long-term effects of sustained pollution that kills decomposers on autotrophic productivity. (2)

(Total 6 marks)

4. Seawater temperature has an effect on the spawning (release of eggs) of echinoderms living in Antarctic waters. Echinoderm larvae feed on phytoplankton. In this investigation, the spawning of echinoderms and its effect on phytoplankton was studied.

In the figure, the top line indicates the number of larvae caught (per 5,000 l of seawater). The shaded bars below show when spawning occurred in echinoderms.

□ = 0% to 25%
▨ = 25% to 75%
■ = 75% to 100%

The concentration of chlorophyll gives an indication of the concentration of phytoplankton. Note: the seasons in the Antarctic are the reverse of those in the northern hemisphere.

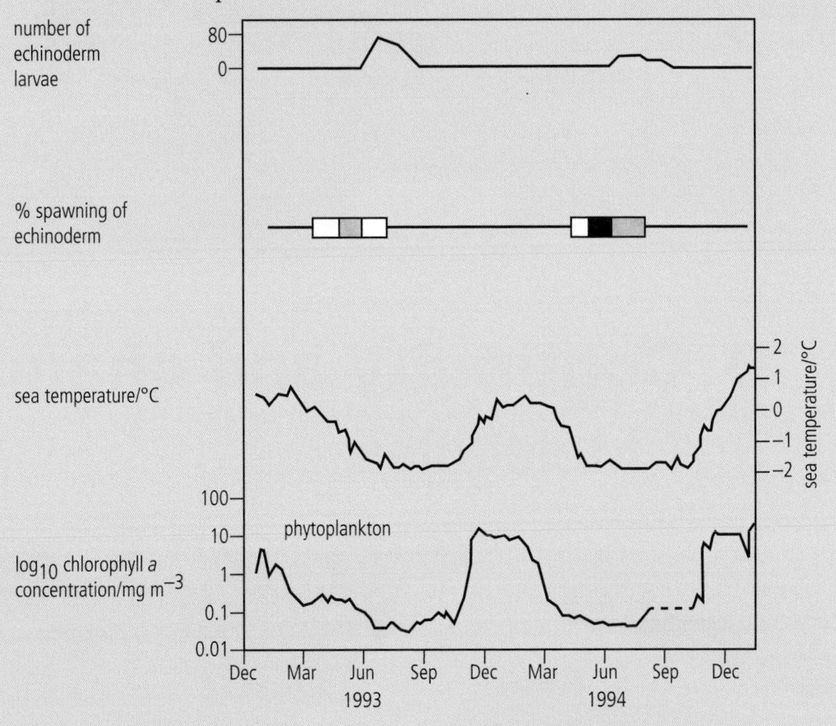

(a) State the trophic level of echinoderm larvae. (1)

(b) Identify the period during which the spawning of echinoderm lies between 25% and 75%. (1)

(c) Explain the relationship between the seasons and the concentration of phytoplankton. (2)

(d) **(i)** Outline the effect of sea water temperature on echinoderm larvae numbers. (2)

(ii) Using the data in the figure, predict the effect of global warming on echinoderm larvae numbers. (2)

(Total 8 marks)

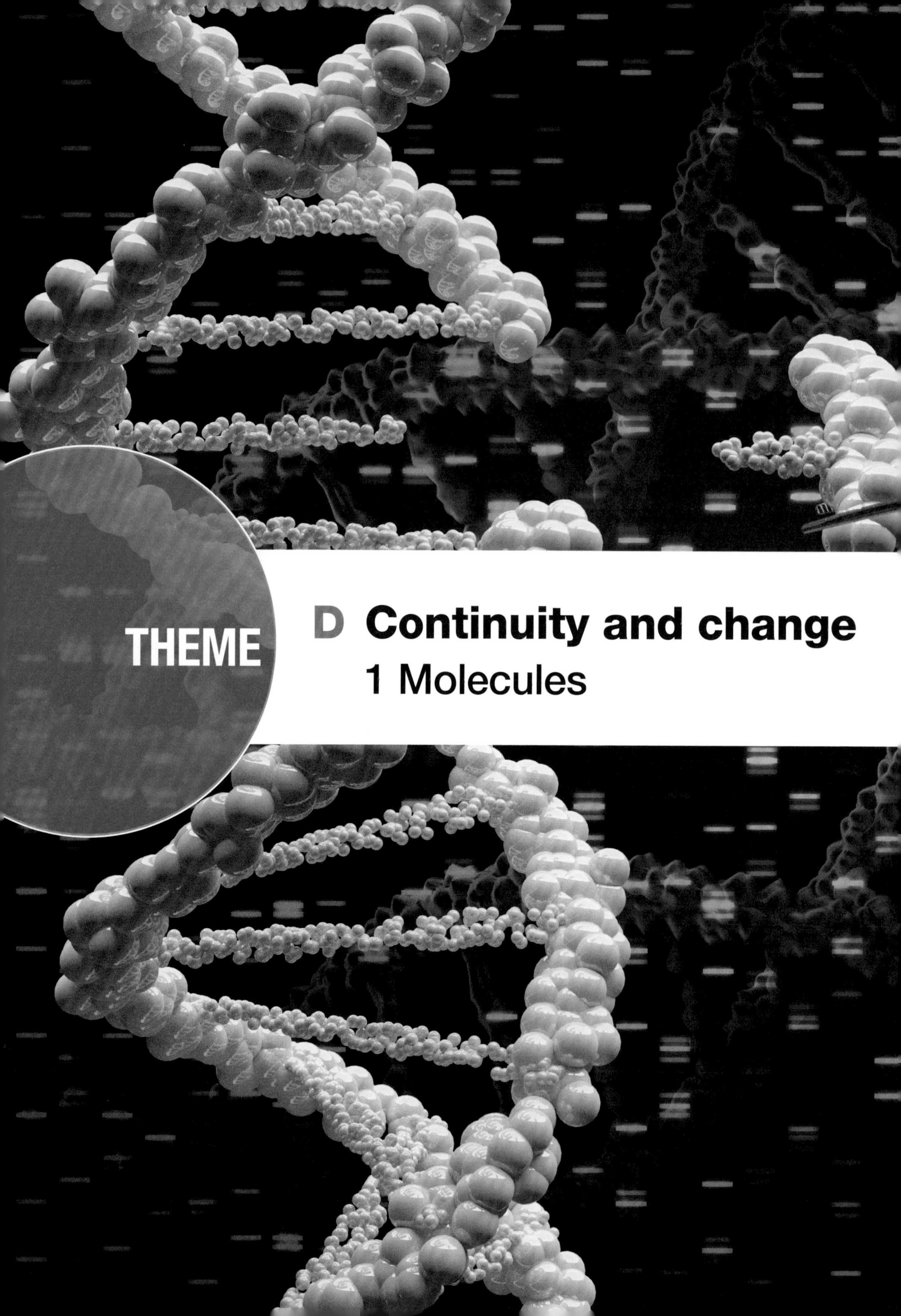

THEME

D Continuity and change

1 Molecules

We are at a critical junction today regarding life on our planet. A recently developed technology called CRISPR provides a revolutionary, inexpensive and effective technique for manipulating the code of life. Immediately, we think of the possibility of not treating but curing genetic diseases such as Huntington's disease, cystic fibrosis and Duchene muscular dystrophy. We see hope for the control of cancers we have had little success treating in the past. There are so many medical possibilities in our future based on this technology.

However, there is a deep concern that this technology will be used inappropriately. Should we allow embryos to be altered to produce the ultimate athlete, an individual with an IQ off the chart, someone with an immune system that can drastically decrease or even eliminate disease, or even someone with "perfect" physical attributes? Who benefits from this technology, the very wealthy? What will happen to the natural evolution of *Homo sapiens*? By changing one attribute, will we also affect others later in the life of an individual?

So many possibilities, and so many questions. If society is to decide the ethics and limits of the use of this technology, we all must have a basic understanding of the genetic code and how it controls life.

D1.1 DNA replication

Guiding Questions

How is new DNA produced?

How has knowledge of DNA replication enabled applications in biotechnology?

When the molecular structure of DNA was first presented by Francis Crick and James Watson in 1953 in the scientific journal *Nature*, it immediately stimulated research into the process of DNA replication. The production of DNA has to be accurate so that new cells and/or organisms could continue the functions of life. Just as important, this production of genetic material has to allow change, mistakes if you prefer, to account for the variation obvious in organisms over time.

Knowledge of this replication process has resulted in tremendous advances in biotechnology. We are seeing applications of DNA replication in many areas of medical science as well as in police investigations involving forensics. DNA analysis has led to revelations concerning family ancestry as well as predictions concerning the inheritance of medical conditions. By understanding the DNA production process, scientists have been successful in replicating extremely small amounts and, in some cases, very old samples of recovered DNA. We have now sequenced the DNA of a million-year-old woolly mammoth, and extracted cartilage from a 125-million-year-old *Caudipteryx* dinosaur. If DNA can be recovered from this cartilage, it might actually be possible to sequence a dinosaur's DNA.

D1.1.1 – The role of DNA replication

D1.1.1 – DNA replication as production of exact copies of DNA with identical base sequences
Students should appreciate that DNA replication is required for reproduction and for growth and tissue replacement in multicellular organisms.

DNA structure is discussed in Chapter A1.2. It is a **polymer** composed of **monomers** called **nucleotides** bonded together. Each nucleotide includes a 5-carbon sugar called deoxyribose, a phosphate group, and a nitrogenous base. There are four nitrogenous bases in DNA: adenine, thymine, cytosine and guanine. The basic structure of DNA is shown in Figure 1.

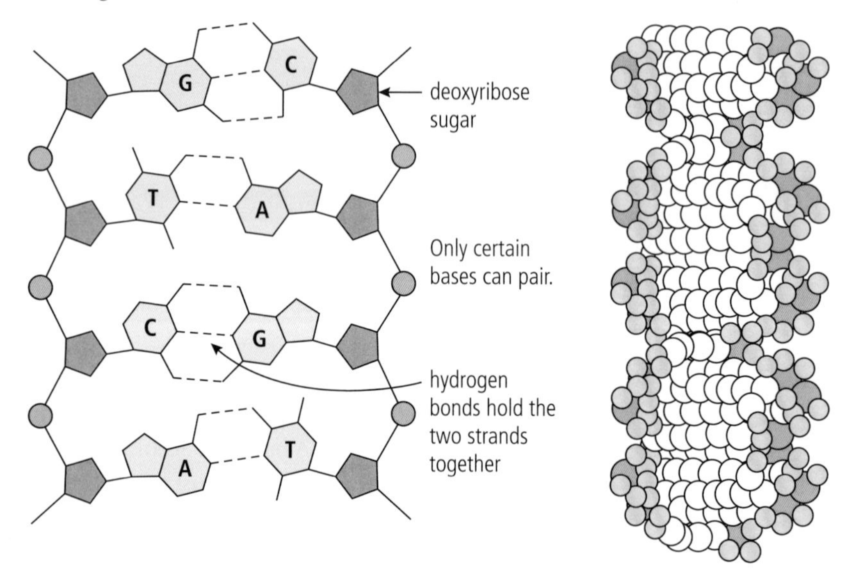

D1.1 Figure 1 The basic structure of a DNA molecule. Note that the bonding of adenine and thymine involves two hydrogen bonds, while the bonding between cytosine and guanine involves three hydrogen bonds. The outside strands are of alternating phosphate groups, shown as circles, and deoxyribose molecules, represented by shaded pentagons.

DNA provides the code that creates the uniqueness of all forms of life on our planet. This code is carried in the sequence of the four possible nitrogenous bases.

Because DNA is the code of life, it must be duplicated so that the reproduction of organisms is possible. DNA duplication is also essential for growth and tissue replacement in multicellular organisms. This duplication is known as **DNA replication**. Replication doubles the quantity of DNA and ensures that exact copies of each DNA molecule are made.

In 2014 a group of Scripps Research scientists introduced two new, unnatural nitrogenous bases into the genetic code of living bacteria. These new bases were incorporated into the DNA code for the bacteria. The research is ongoing: many feel that this approach could provide a means by which new medicines and vaccines can be developed. With six bases instead of four, many more products could be produced.

D1.1.2 and D1.1.3 – Semi-conservative replication and complementary base pairing

D1.1.2 – Semi-conservative nature of DNA replication and role of complementary base pairing
Students should understand how these processes allow a high degree of accuracy in copying base sequences.

D1.1.3 – Role of helicase and DNA polymerase in DNA replication
Limit to the role of helicase in unwinding and breaking hydrogen bonds between DNA strands and the general role of DNA polymerase.

The replication process of DNA is said to be **semi-conservative.** Each strand in the DNA double helix acts as a template for the synthesis of a new, complementary strand. Evidence for a semi-conservative replication process was provided by research using bacteria (see Figure 2).

Figure 2 shows how each daughter DNA double helix contains an old strand from the parental DNA double helix and a new strand. Thus the name semi-conservative replication.

In the nucleus of cells are two types of molecules that are particularly important for the process of DNA replication. They are:

- enzymes needed for replication, which include **helicase** and a group of enzymes collectively called **DNA polymerase**
- free nucleotides, which are nucleotides that are not yet bonded and are found floating freely in the nucleoplasm, some contain adenine, some thymine, some cytosine, and some guanine.

One of the first stages in DNA replication is the separation of the double helix into two single strands. You should remember that the double helix is held together by the hydrogen bonds between the complementary base pairs adenine (A) and thymine (T), and cytosine (C) and guanine (G). The enzyme that initiates this separation into two single strands is called helicase (Figure 3).

D1.1 Figure 2 A diagram demonstrating the general process of semi-conservative replication of DNA.

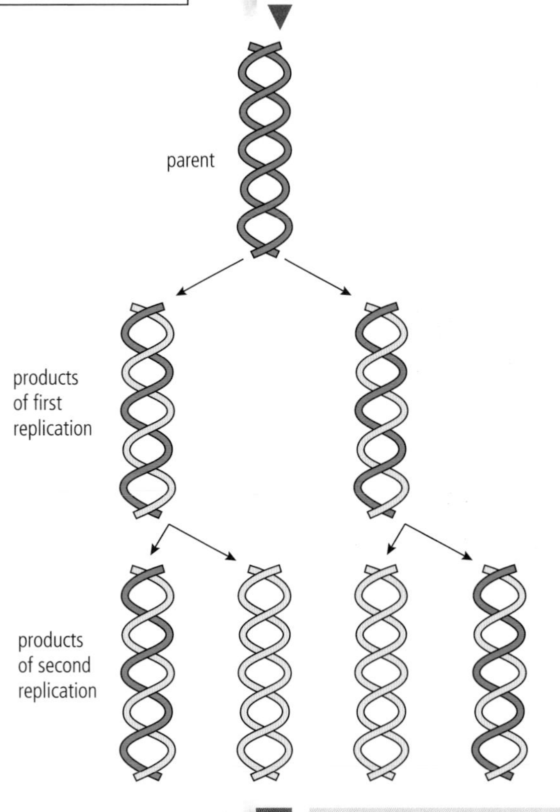

In the 1950s Matthew Meselson and Frank Stahl carried out research using radioactive nitrogen that provided evidence for the semi-conservative replication of DNA. See the eBook for an activity.

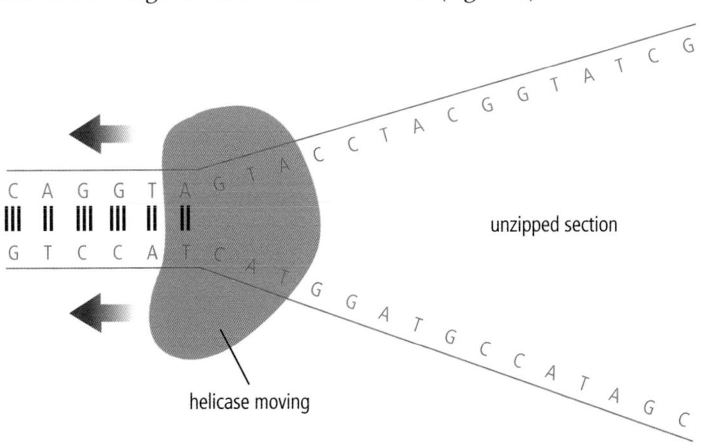

D1.1 Figure 3 The first step of DNA replication is helicase unzipping the double-stranded DNA molecule, forming a section with two single strands.

The process of semi-conservative replication as described allows a high degree of accuracy in copying base sequences and provides stability when passing on the genetic code.

The unpaired nucleotides on each of these single strands can now be used as a template to create two double-stranded DNA molecules identical to the original. Some people use the analogy of a zipper for this process. When you pull on a zipper, the slide mechanism is like helicase. The separation of the two sides of the DNA molecule is like the two opened sides of a zipper.

Formation of two complementary strands

Once DNA has become unzipped, the nitrogenous bases on each of the single strands are unpaired. In the environment of the nucleoplasm, there are free-floating nucleotides available to form complementary pairs with the single-stranded nucleotides of the unzipped molecule. This does not happen in a random fashion. A free nucleotide locates one end of an opened strand, and then a second nucleotide can arrive to join the first. To join, these first two nucleotides must become covalently bonded together, creating the beginning of a new strand. The formation of a covalent bond between two adjoining nucleotides is catalysed by one of the DNA polymerase enzymes, which is an important part of the process.

A third nucleotide then joins the first two, and the process continues in a repetitive way for many nucleotides. The other unzipped strand also acts as a template for the formation of a new strand. This strand forms in a similar fashion, but in the opposite direction to the first strand. In Figure 4, notice that one strand is replicating in the same direction as helicase is moving, while the other strand is replicating in the opposite direction.

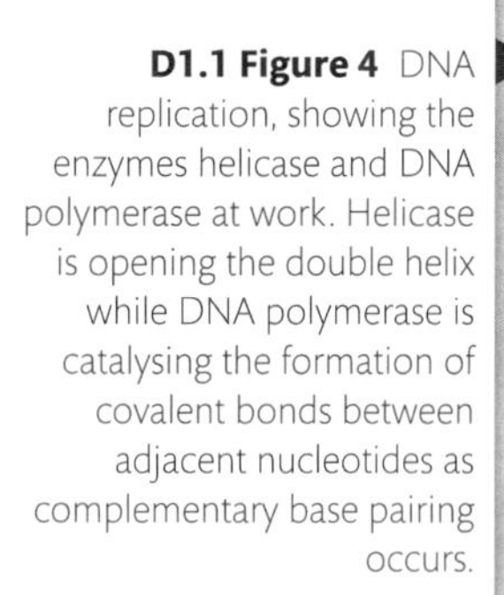

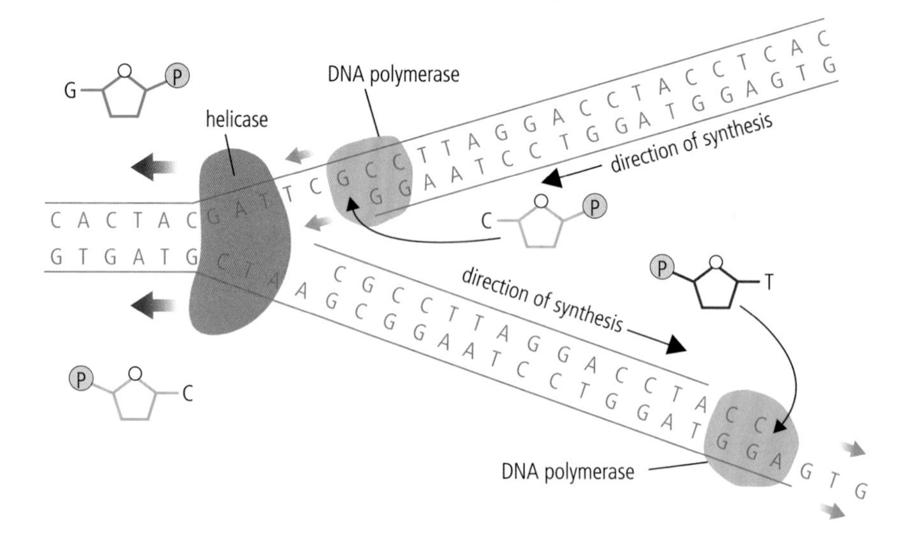

D1.1 Figure 4 DNA replication, showing the enzymes helicase and DNA polymerase at work. Helicase is opening the double helix while DNA polymerase is catalysing the formation of covalent bonds between adjacent nucleotides as complementary base pairing occurs.

The significance of DNA replication is that it ensures two identical copies of DNA are produced from an original strand. Notice that in the area where replication has already taken place, the two strands are identical to each other. This is because the original double-stranded molecule had complementary pairs of nucleotides and it was complementary nucleotides that used the unzipped single-stranded areas as templates.

How is genetic continuity ensured between generations?

The role of helicase and DNA polymerase in DNA replication

Helicase has been found to catalyse the unzipping of DNA at a rate measured in hundreds of base pairs per second.

The enzyme helicase is essential during DNA replication. It separates double-stranded DNA into single strands allowing each strand to be copied. Helicase begins at a point in or at the end of a DNA molecule and moves one complementary base pair at a time, breaking the hydrogen bonds so that the double-stranded DNA molecule becomes two separate strands.

DNA polymerase is another key enzyme in DNA replication. DNA polymerase adds nucleotides one by one to the growing DNA chain.

DNA polymerase has two very important roles in DNA replication. It catalyses the formation of covalent bonds between adjoining nucleotides in the growing DNA strand. It also serves in a proofreading role to ensure each new DNA strand is a near perfect copy of the original.

D1.1.4 – Amplifying and separating DNA

D1.1.4 – Polymerase chain reaction and gel electrophoresis as tools for amplifying and separating DNA
Students should understand the use of primers, temperature changes and *Taq* polymerase in the polymerase chain reaction (PCR) and the basis of separation of DNA fragments in gel electrophoresis.

The exploration and manipulation of DNA have been enhanced by the development of many different techniques during recent decades. In this section we will examine two of these techniques: **polymerase chain reaction (PCR)** and **gel electrophoresis**.

Polymerase chain reaction

PCR is a laboratory technique using a machine called a thermocycler, which takes a small quantity of DNA and copies all the nucleotides to make millions of copies of the DNA (see Figure 5). PCR is specific in that it only amplifies a targeted section of DNA.

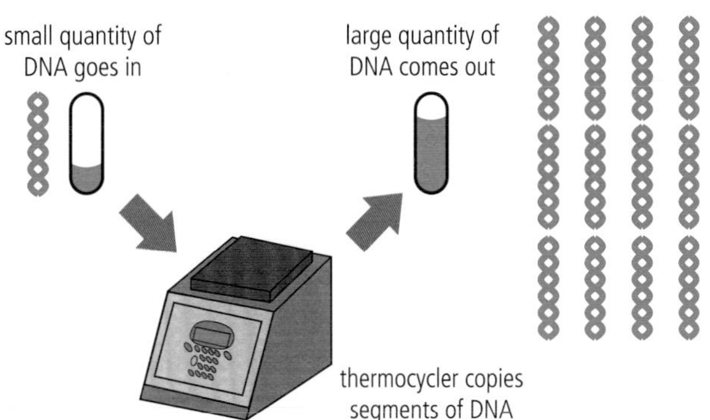

D1.1 Figure 5 A polymerase chain reaction (PCR) is a method of generating enough DNA for analysis.

Various components are needed to run a PCR, including the following.

- **Primers**: single-stranded, short polymers of 15–20 nucleotides that are complementary to the nucleotides at one end of the target DNA to be copied. These primers provide a starting point for DNA synthesis because DNA polymerases can only attach new DNA nucleotides to an existing strand of nucleotides.
- ***Taq* polymerase**: *Taq* is a polymerase from a bacterium that lives in hot springs. This enzyme can withstand high temperatures.
- Free nucleotides: to synthesize the new strands of DNA.

Once all these components have been mixed in a tube, the tube is placed in the thermocycler, which controls the temperature of the reaction.

There are three steps to the PCR process.

1. **Denaturation**: the mixture is heated to a temperature between 92°C and 98°C, which breaks the hydrogen bonds holding the two strands of DNA together.
2. **Annealing**: the mixture is then cooled to between 50°C and 65°C, to allow the primers to bind with nucleotides on both strands at the ends of the target sequence.
3. **Elongation**: The *Taq* polymerase catalyses the building of new DNA strands by extending the primers. A temperature of 70–80°C is needed for this step.

This three-step cycle is repeated over and over until enough target DNA has been produced to accomplish the desired task.

Gel electrophoresis

This laboratory technique is used to separate fragments of DNA in order to identify its origin. Enzymes are used to chop up the long filaments of DNA into varying sizes of fragments. If two different sources of DNA are to be compared, the same enzymes are used to produce fragments. The DNA fragments are then placed into small wells (holes) in a gel placed in an electrophoresis chamber. Figure 6 illustrates an electrophoresis chamber.

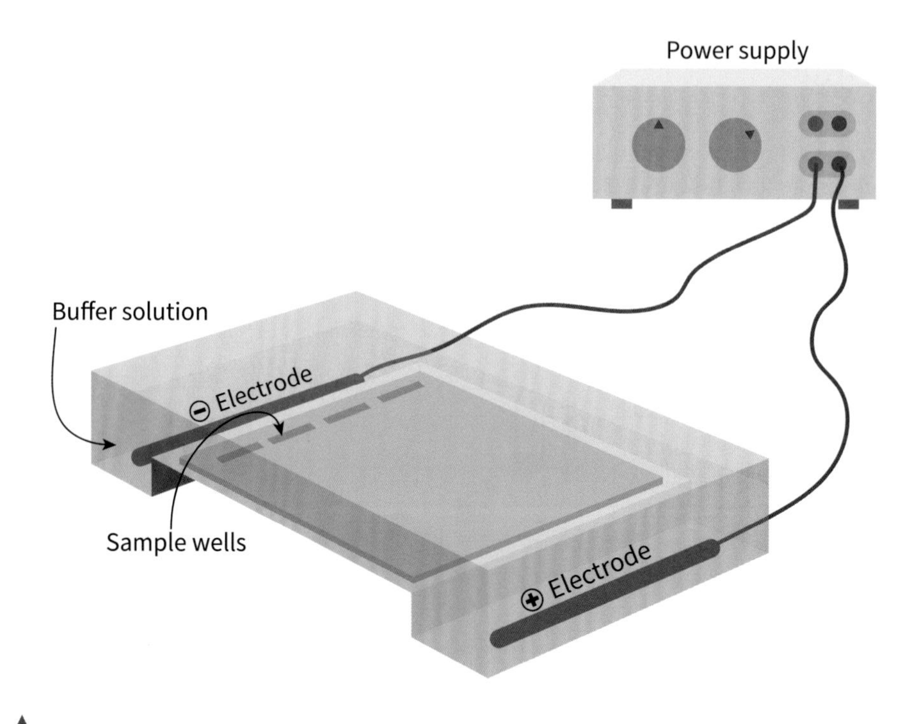

D1.1 Figure 6 An electrophoresis chamber with power supply and electrodes. Notice the buffer solution in the holding tanks at the ends of the chamber. An agarose gel with sample wells is placed on the chamber bed, the middle region of chamber between the end tanks.

D1.1 Figure 7 This autoradiogram or autoradiograph shows banded lines that were formed from nine different DNA samples during electrophoresis. The black traces are left by the radioactivity of the materials used in marking the DNA samples. Smaller chains of DNA travel farther on the gel than larger chains, allowing characteristic banding patterns for a particular sample of DNA.

The gel, with samples of fragments in the wells, is exposed to an electric current, positive on one side and negative on the other. The biggest, heaviest and least charged particles do not move easily through the gel, so they get stuck very close to the wells they were in at the beginning of the experiment. The smallest, least massive and most charged particles pass through the gel to the other side with little difficulty. Intermediate particles are distributed in between. At the end of the experiment, the fragments leave a banded pattern of DNA, like the example shown in Figure 7.

D1.1.5 – Applications of amplifying and separating DNA

D1.1.5 – Applications of polymerase chain reaction and electrophoresis
Students should appreciate the broad range of applications, including DNA profiling for paternity and forensic investigations.   **NOS:** Reliability is enhanced by increasing the number of measurements in an experiment or test. In DNA profiling, increasing the number of markers used reduces the probability of a false match.

The process of matching an unknown sample of DNA with a known sample to see if they correspond is called **DNA profiling**. This is also sometimes referred to as **DNA fingerprinting** because there are some similarities with identifying fingerprints, although the techniques are very different.

The processes of PCR and DNA profiling have been used to sequence the bases of human DNA. This sequencing was the goal of the Human Genome Project.

If, after separation by gel electrophoresis, the pattern of bands formed by two samples of DNA fragments is identical, it means that both must have come from the same individual. If the patterns are similar, it means that the two individuals are probably related.

DNA profiling can be used in paternity suits when the identity of someone's biological father needs to be known for legal reasons. At a crime scene, forensic specialists can collect samples such as blood or semen, which contain DNA. Often such samples only contain very small amounts of DNA. In this case, PCR is carried out to amplify the DNA available for profiling.

Gel electrophoresis can be used to compare DNA collected from a crime scene with that of suspects. If they match, the suspect can be questioned further. If there is no match, the suspect is probably not the person the police are looking for. Criminal cases are sometimes reopened many years after a judgement was originally made, to consider new DNA profiling results. In the USA, this has led to the liberation of many individuals who had been sent to jail for crimes they did not commit.

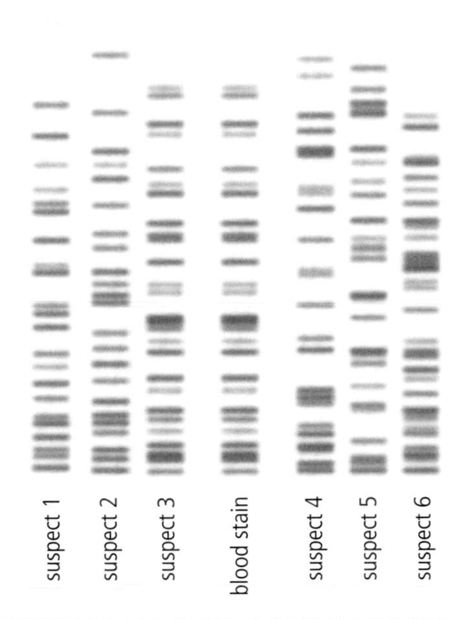

D1.1 Figure 8 These seven tracks were produced by gel electrophoresis to allow investigators to analyse and match DNA samples.

Challenge yourself

1. Using Figure 8, showing the DNA profiles from six suspects, can you identify which one matches the DNA profile of the blood stain found at the crime scene?

DNA profiling is used in other circumstances too, for example in studies of ecosystems, when scientists use DNA samples taken from birds, whales and other organisms to clarify relationships. This has helped establish a better understanding of social relationships, migration patterns and nesting habits, for example. In addition, the study of DNA in the biosphere has given new credibility to theories of evolution: DNA evidence can often reinforce evidence of common ancestry based on anatomical similarities between species.

Nature of Science

How do we decide when evidence is reliable or not? Often when DNA evidence is used in a courtroom trial, it has a certain credibility as scientific fact, yet we know from our own experience in laboratory work that there is a degree of error in any procedure. Whether it be in the laboratory or in a courtroom, it is difficult to imagine evidence that can be considered 100% certain. When a scientist comes up with new evidence, old theories can be challenged or even overturned. Reliability of data can be enhanced by increasing the number of measurements in an experiment or test. In DNA profiling, increasing the number of markers used reduces the probability of a false match. A marker refers to a particular sequence of DNA.

HL

D1.1.6 – DNA polymerases

D1.1.6 – Directionality of DNA polymerases
Students should understand the difference between the 5′ and 3′ terminals of strands of nucleotides and that DNA polymerases add the 5′ of a DNA nucleotide to the 3′ end of a strand of nucleotides.

The bonds of DNA

Each strand of DNA is composed of a backbone of alternating phosphate and deoxyribose molecules. These two molecules are held together by a covalent bond called a **phosphodiester bond** or linkage. A phosphodiester bond in DNA forms between a hydroxyl (–OH) group on the 3′ (three-prime) carbon of deoxyribose and the phosphate ($–PO_4$) group attached to the 5′ (five-prime) carbon of deoxyribose. Study Figure 9 showing a nucleotide.

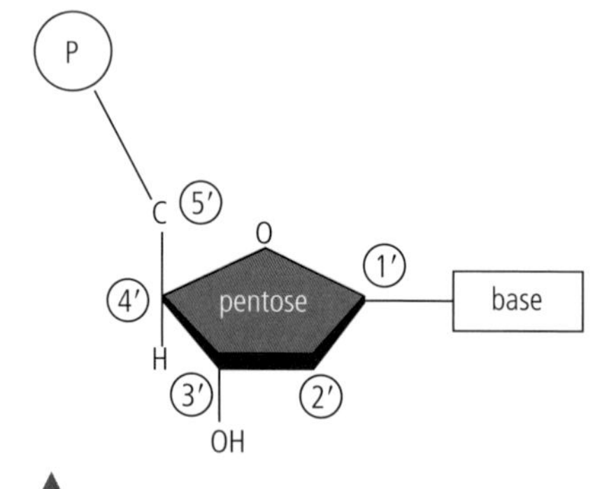

D1.1 Figure 9 A DNA nucleotide is composed of a molecule of deoxyribose (pentose) with a phosphate group attached to the 5′ carbon and a nitrogenous base attached to the 1′ carbon. Notice the hydroxyl (OH) attached to the 3′ carbon.

Each nucleotide is attached to the previous one by a phosphodiester bond. This produces a chain of DNA nucleotides.

D1.1 Figure 10 Five nucleotides bonded to form a very small section of a strand of DNA. Note the carbon numbers of the pentose and what is attached to the 1′, 3′, and 5′ carbons.

Each time a nucleotide is added, it is attached to the 3′ carbon end. Even when thousands of nucleotides are involved, there is still a free 5′ carbon end with a phosphate attached and a 3′ carbon end with a hydroxyl group attached. The reason the DNA chain is built in this 5′ to 3′ direction is that DNA polymerase can only add nucleotides to an existing chain in this direction. This creates the alternating sugar–phosphate backbone of each chain shown in Figure 10.

As nucleotides are linked together, a definite sequence of nitrogenous bases develops. This sequence carries the genetic code that is essential for the life of an organism.

Antiparallel strands of DNA

The two sugar–phosphate backbones are attached to one another by their nitrogenous bases. The two backbones or chains run in opposite directions and are described as **antiparallel**. One strand has the 5′ carbon on the top and the 3′ carbon on the bottom; the other strand is the opposite way round. Study Figure 11.

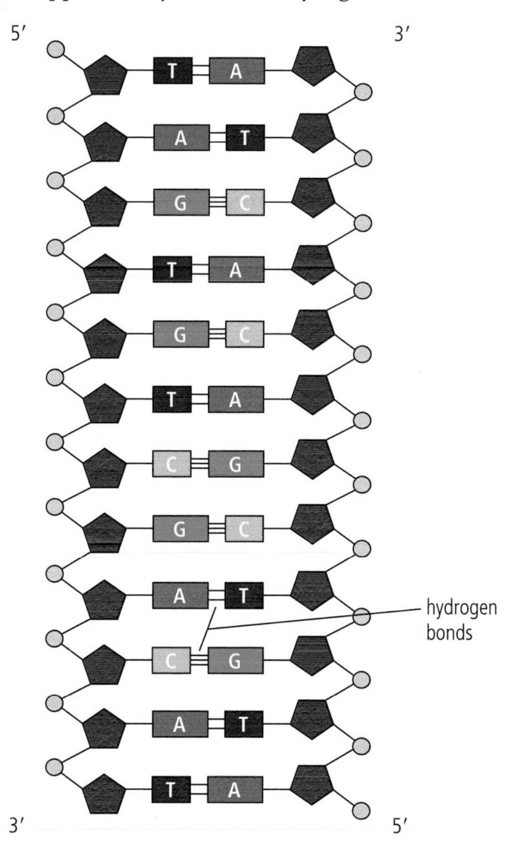

D1.1 Figure 11 The antiparallel strands of DNA run in opposite directions.

The hydrogen bonding between bases is discussed in Chapter A1.2.

D1.1.7 – Leading and lagging strands

D1.1.7 – Differences between replication on the leading strand and the lagging strand
Include the terms "continuous", "discontinuous" and "Okazaki fragments". Students should know that replication has to be initiated with RNA primer only once on the leading strand but repeatedly on the lagging strand.

The first step in the DNA replication process is the separation or "unzipping" of the double helix by the enzyme helicase. Helicase accomplishes this task by breaking the hydrogen bonds connecting the two strands. The point at which helicase is working to unzip the DNA molecule is called a **replication fork**.

DNA strands can only be assembled in the 5′ to 3′ direction because of the specific action of **polymerase III.** Because the two strands of DNA are antiparallel, there is a difference in the process of assembling the two new strands of DNA from the templates. For the 3′ to 5′ template strand, the new DNA strand is formed in a **continuous** pattern and is relatively fast. The strand produced in this continuous way is called the **leading strand.** The other new strand forms more slowly and is called the **lagging strand.**

Replication of a strand of DNA has to be initiated with RNA primer. RNA primer is only needed once on the leading strand, because it is continuously formed. However, the lagging strand does not form continuously and RNA primer has to be added repeatedly as the double helix of DNA is opened by the action of helicase.

The formation of the lagging strand involves fragments and is **discontinuous.** This discontinuous replication requires another enzyme, **DNA ligase.** The replication process is as follows (see Figure 12).

1. The leading strand is assembled continuously towards the progressing replication fork in the 5′ to 3′ direction.
2. The lagging strand is assembled by the production of fragments moving away from the progressing replication fork in the 5′ to 3′ direction.
3. The fragments of the lagging strand are called **Okazaki fragments**, after the Japanese scientist who discovered them.
4. Primer, primase and DNA polymerase III are required to begin the formation of each Okazaki fragment of the lagging strand, and to begin the formation of the continuously produced leading strand.
5. The primer and primase are only needed once for the leading strand because it is produced continuously.
6. Once the Okazaki fragments are assembled, an enzyme called **DNA ligase** attaches the sugar–phosphate backbones of the lagging strand fragments to form a single DNA strand.

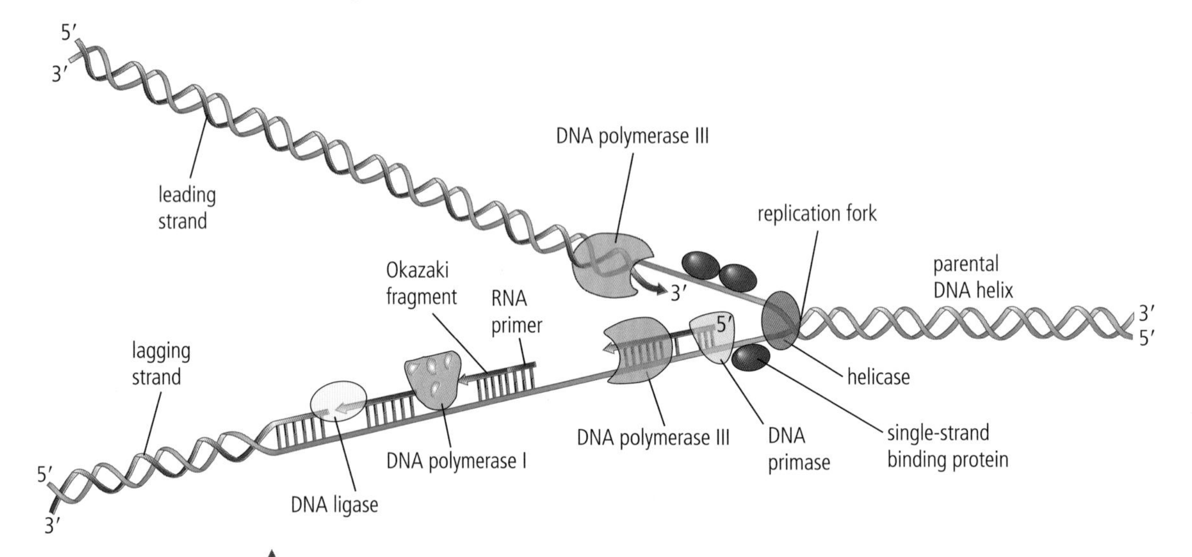

D1.1 Figure 12 At a replication fork, helicase separates the strands of the double helix and binding proteins stabilize the single strands. There are two mechanisms for replication: continuous synthesis and discontinuous synthesis. Continuous synthesis occurs on the leading strand: DNA primase adds an RNA primer and DNA polymerase III adds nucleotides to the 3′ end of the leading strand. DNA polymerase I then replaces the primer with nucleotides. Discontinuous synthesis occurs on the lagging strand: DNA primase adds an RNA primer in front of the 5′ end of the lagging strand and DNA polymerase III adds nucleotides. DNA polymerase I replaces the primer, and finally DNA ligase attaches the Okazaki fragment to the lagging strand.

What biological mechanisms rely on directionality?

D1.1.8 – Functions of specific enzymes and molecules

D1.1.8 – Functions of DNA primase, DNA polymerase I, DNA polymerase III and DNA ligase in replication
Limit to the prokaryotic system.

Table 1 provides a summary of the enzymes and other molecules involved in prokaryotic cell DNA replication. It is important to note the functions of DNA polymerase I, DNA polymerase III and DNA ligase, but they work in conjunction with the other enzymes to result in the successful replication of DNA.

Enzyme/protein	Function
Helicase	Unwinds the double helix at replication forks
DNA primase	Synthesizes RNA primer
RNA primer	Marks the beginning of the replication process
DNA polymerase I	Removes RNA primer at end of replication and replaces it with DNA nucleotides
DNA polymerase III	Synthesizes the new strand by catalysing the addition of nucleotides in a 5′ to 3′ direction
DNA ligase	Joins the ends of DNA segments and Okazaki fragments
Single-stranded binding protein	Keeps bases from reforming hydrogen bonds between them after helicase action

D1.1 Table 1 The function of various enzymes and proteins in DNA replication

D1.1.9 – Removing mismatched nucleotides

D1.1.9 – DNA proofreading
Limit to the action of DNA polymerase III in removing any nucleotide from the 3′ terminal with a mismatched base, followed by replacement with a correctly matched nucleotide.

DNA polymerase III has one other very important function in the replication process. It checks for accuracy in the daughter strand it is synthesizing. It can recognize mismatched nucleotides and remove it from the 3′ end of the daughter strand. After the removal of a nucleotide mismatch, it reverses direction and resumes the synthesis of the new strand. The result of this very efficient proofreading is an error rate in bacteria of only 1 in 100 million base pairs.

DNA replication in eukaryotic cells has many points of origin along each linear chromosome. At each origin a bubble forms from which replication occurs in both directions. Eukaryotic chromosomes have ends composed of short sequences of DNA repeated over and over. These ends are called **telomeres** and they require a special enzyme called telomerase for their replication.

HL end

Guiding Question revisited

How is new DNA produced?

In this chapter we have discussed how:

- DNA replication is essential to produce new cells, both in unicellular and multicellular organisms
- semi-conservative replication results in every new copy of DNA possessing one strand from the parent DNA and one new strand
- complementary base pairing in the DNA replication process ensures a high degree of accuracy
- helicase is required to unwind and break hydrogen bonds so that the process of DNA replication can occur
- DNA polymerase controls the addition of free nucleotides to the developing DNA strand
- HL because of the directionality of DNA replication, one strand of DNA forms faster than the other.

Guiding Question revisited

How has knowledge of DNA replication enabled applications in biotechnology?

In this chapter we have discussed how:

- polymerase chain reactions (PCR) are used to amplify DNA segments so that valid analyses can be carried out
- a thermocycler, *Taq* polymerase, primers and free nucleotides are necessary for the process of PCR
- gel electrophoresis uses an electrical field and agarose gel to separate segments of DNA
- PCR and gel electrophoresis are often used together in attempts to solve crimes where small amounts of DNA have been recovered
- gel electrophoresis is also used in the study of ecosystems
- paternity questions can often be solved using gel electrophoresis.

Exercises

Q1. Using complementary base pairing and the first letter of the nitrogenous base in nucleotides, write the letter of the nucleotide that would base pair with the exposed base in the following sequence.

ATG ACC GCT

Q2. What type of bonds are broken by helicase?

Q3. Rearrange the following steps in DNA replication in their proper order.

I Two new molecules of DNA are created.

II DNA polymerases attach the free nucleotides to the exposed nitrogenous bases.

III Helicase begins to break the hydrogen bonds between nitrogen bases.

IV Free floating nucleotides pair up with exposed nitrogen bases.

HL

Q4. Draw the two strands of a DNA molecule representing their antiparallel relationship.

Q5. Compare the number of primers needed on the leading and lagging strands of DNA during replication.

Q6. Why is one strand slower to form than the other in DNA replication?

HL end

D1.2 Protein synthesis

Guiding Questions

How does a cell produce a sequence of amino acids from a sequence of DNA bases?

How is the reliability of protein synthesis ensured?

DNA controls the production of proteins in a cell. These proteins are specific to each type of cell. It is the sequence of DNA bases in a cell that determines the amino acids and their order in each protein produced. The production of proteins actually occurs in the cytoplasm at organelles called ribosomes. Even though the DNA does not leave the nucleus, it is able to send its message to the ribosomes, directing the synthesis of the exact proteins needed in that cell.

The DNA code produces the proteins by two processes: transcription and translation. Reliability is essential to the successful production of functional proteins, but so is the potential for a degree of variability, to allow evolution to proceed within a population. DNA and the process of protein synthesis fulfil these required characteristics of the genetic code.

D1.2.1 – The synthesis of RNA

D1.2.1 – Transcription as the synthesis of RNA using a DNA template
Students should understand the roles of RNA polymerase in this process.

DNA is sequestered (locked away) in the nucleus of a cell. The organelles essential to protein synthesis are the **ribosomes**, which are located in the cell cytoplasm. The DNA must communicate with the ribosomes to control the production of proteins. It does so by producing a code that is carried from the nucleus to the cytoplasm by **ribonucleic acid (RNA)**. **Transcription** is the synthesis of RNA using the base sequence in DNA as a template.

The set of ideas first proposed by Francis Crick in 1956, called the central dogma of molecular biology, states that information passes from genes (specific base sequences on the DNA) to the RNA copy. The RNA copy then directs the production of proteins at the ribosomes in the cytoplasm by controlling the sequence of amino acids. This mechanism is one-way and fundamental to all forms of life.

The sections of DNA that code for polypeptides (proteins) are called **genes**. Any one gene is a specific sequence of nitrogenous bases found at a particular location in a DNA molecule.

The process of transcription begins when an area of DNA for one gene becomes unzipped (see Figure 1). This is very similar to the unzipping process involved in DNA replication, but in this case only the area of the DNA where the gene is found is unzipped. The two complementary strands of DNA are now single stranded at the gene location. Recall that RNA is a single-stranded molecule. This means that only one of the two strands of DNA will be used as a template to create the **messenger RNA (mRNA)** molecule. This strand is called the **template strand**. An enzyme called **RNA polymerase** is essential to the process of transcription. Transcription can only begin when RNA polymerase binds to a **promoter sequence** near the beginning of a gene on the template strand. RNA polymerase is also involved in linking the RNA nucleotides to form an RNA strand.

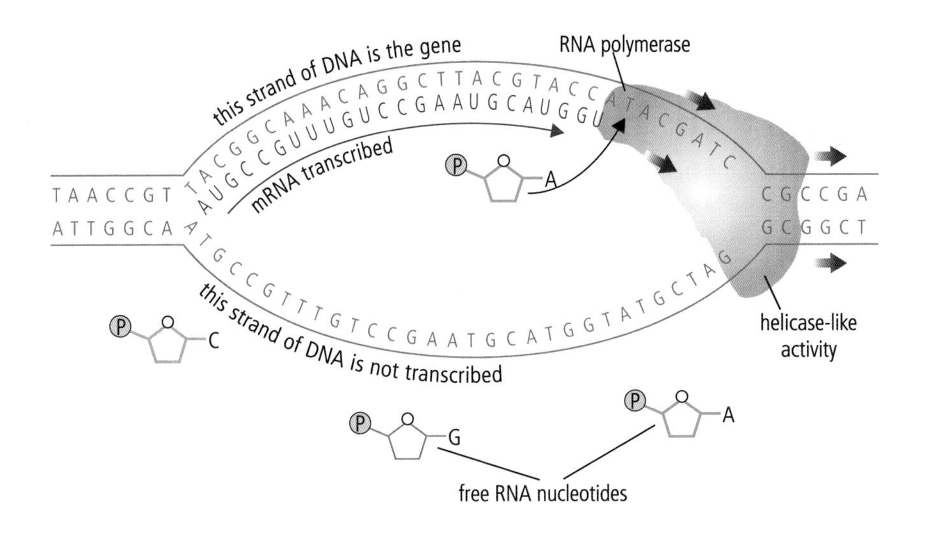

D1.2 Figure 1 The process of transcription (the synthesis of an RNA molecule). RNA polymerase has a helicase-like role, as it plays a part in opening the DNA double helix. It also catalyses the addition and bonding of free RNA nucleotides to the growing messenger RNA (mRNA) strand.

D1.2.2 – Hydrogen bonding and complementary base pairing

D1.2.2 – Role of hydrogen bonding and complementary base pairing in transcription
Include the pairing of adenine (A) on the DNA template strand with uracil (U) on the RNA strand.

RNA polymerase acts as helicase does in DNA replication, to break the hydrogen bonds between the two strands of DNA, unzipping the double helix. As RNA polymerase moves along the DNA template strand, RNA nucleotides that exist in the nucleus float into place by complementary base pairing. The complementary base pairs are the same as in double-stranded DNA, with the exception that adenine (A) on the DNA is now paired with uracil (U) on the newly forming mRNA molecule. Hydrogen bonds form between the bases of the template strand of DNA and the RNA nucleotides undergoing complementary base pairing. The chemical structures of the bases and the number and location of the hydrogen bonds they form ensure that these bonds can only form between specific bases.

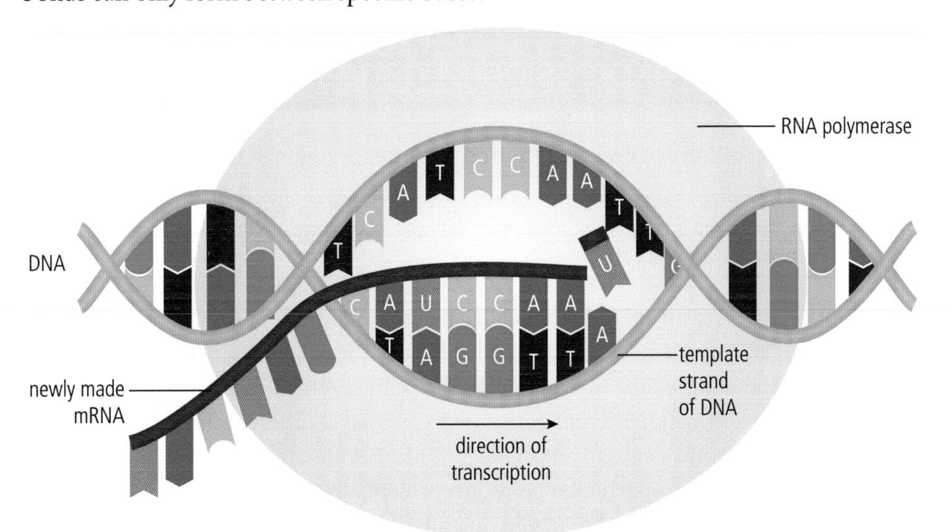

D1.2 Figure 2 Transcription involves RNA polymerase, complementary base pairing, and the formation of hydrogen bonds to produce an mRNA molecule. Notice that the RNA includes the base uracil, in place of the thymine in DNA. mRNA is one of several types of RNA produced by transcription.

Study Figure 2 on the previous page and consider the following facts concerning transcription:

- the process of transcription produces all the RNA in a cell
- only one of the two strands of DNA is "copied", the other strand is not used
- RNA polymerase catalyses bonds between the RNA nucleotides as complementary base pairing occurs
- RNA is always single-stranded and shorter than the DNA that it is copied from, because it is a complimentary copy of only one gene
- the presence of thymine in a molecule identifies it as DNA (the presence of deoxyribose is another clue)
- the presence of uracil in a molecule identifies it as RNA (the presence of ribose is another clue).

What biological processes depend on hydrogen bonding?

D1.2.3 – DNA templates

D1.2.3 – Stability of DNA templates

Single DNA strands can be used as a template for transcribing a base sequence, without the DNA base sequence changing. In somatic cells that do not divide, such sequences must be conserved throughout the life of a cell.

Mutations can lead to the development of cancer. An example of this occurs when genes coding for proteins involved in cellular growth mutate, resulting in cells growing and dividing out of control. Some of these mutations may even be heritable and passed on to future generations.

Chapter D1.1 discusses the fact that DNA is a relatively stable molecule. It can be used as a template for the formation of RNA without undergoing any changes to its base sequence. This is especially important in **somatic** (body) cells such as nerve cells that do not participate in cell division but do produce RNA and need proteins throughout their life.

This stability of DNA may be compromised by, for example, the presence of free radicals, chemicals, cigarette smoke or exposure to ultraviolet (UV) radiation. Cells have specialized proteins that can detect and repair many instances of damage. Permanent changes to DNA are referred to as **mutations** and often negatively impact the cell's ability to produce essential proteins. However, not all mutations are harmful, and some may even contribute to the overall efficiency and/or survival of a species.

D1.2.4 – The expression of genes

D1.2.4 – Transcription as a process required for the expression of genes

Limit to understanding that not all genes in a cell are expressed at any given time and that transcription, being the first stage of gene expression, is a key stage at which expression of a gene can be switched on and off.

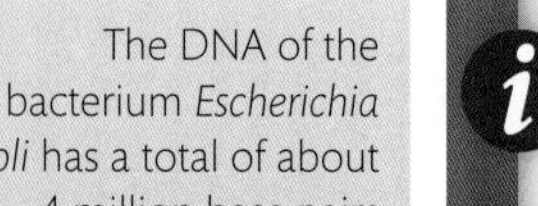

The DNA of the bacterium *Escherichia coli* has a total of about 4 million base pairs and nearly 3,000 genes, together known as its genome. Most bacteria have similar amounts of DNA and genes. Bacteria genomes are only 0.1% the size of ours, and they only have about 10% of the number of genes we do.

Transcription involves an area of DNA called a **gene**. A chromosome and, indeed, the genome of an organism, have thousands of genes. Not all genes go through transcription at the same time. Because of this we can say that not all genes are expressed simultaneously. Different genes are expressed at different times and at different developmental stages in an organism's life. Because transcription is the first step in the expression of a particular gene, it represents the mechanism by which gene expression in an organism can be turned on and off.

D1.2.5 – The synthesis of polypeptides

D1.2.5 – Translation as the synthesis of polypeptides from mRNA
The base sequence of mRNA is translated into the amino acid sequence of a polypeptide.

The mRNA molecule produced by transcription represents a complimentary copy of one gene of DNA. The sequence of mRNA nucleotides is the transcribed version of the original DNA sequence. This sequence of nucleotides making up the length of the mRNA typically provides enough information to make one polypeptide. As you will recall, polypeptides are composed of amino acids covalently bonded together in a specific sequence.

A polypeptide is a polymer of amino acids linked by peptide bonds. A protein is usually more complex than a polypeptide, with folding of one or more polypeptide chains held together by non-covalent bonds.

Once the appropriate mRNA is produced from the DNA template by transcription, the process of producing the protein at the ribosomes can begin. **Translation** is the process by which the information encoded in mRNA directs the synthesis of specific amino acid sequences into polypeptides/proteins. We will be using the terms polypeptide and protein interchangeably as we describe translation.

The overall process that allows the production of a polypeptide/protein is represented in Figure 3.

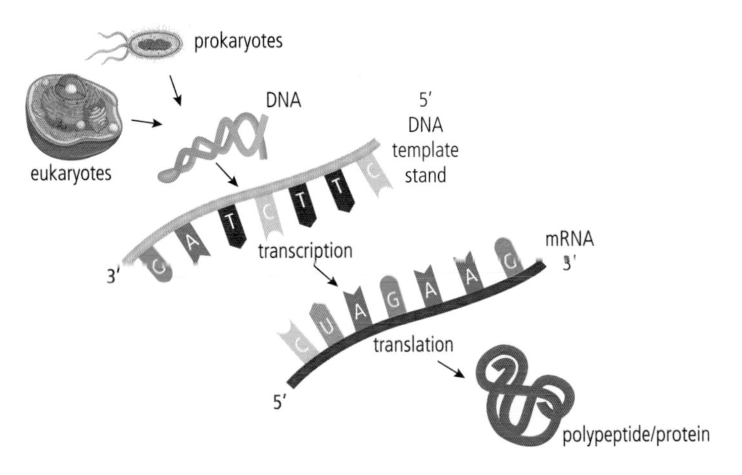

D1.2 Figure 3 The template strand of DNA is transcribed to form mRNA, which is then translated to produce a polypeptide/protein. Transcription occurs in the nucleus, while translation occurs in the cytoplasm at the ribosome.

D1.2.6 – RNA and ribosomes

D1.2.6 – Roles of mRNA, ribosomes and tRNA in translation
Students should know that mRNA binds to the small subunit of the ribosome and that two tRNAs can bind simultaneously to the large subunit.

Three major types of RNA are involved in translation: **messenger RNA (mRNA)**, **transfer RNA (tRNA)** and **ribosomal RNA (rRNA)** (see Table 1).

D1.2 Table 1 The three major types of RNA

Type of RNA	Role in protein synthesis
Messenger (mRNA)	Carries the message from the DNA in the nucleus to the ribosomes in the cytoplasm
Transfer (tRNA)	Functions in the cytoplasm to carry amino acids to the ribosomes
Ribosomal (rRNA)	Combines with ribosomal proteins to construct the cytoplasmic ribosomes

All three types of RNA (Figure 4) are single stranded, and each is transcribed from a gene (a section of DNA). mRNA is a straight chain. rRNA along with proteins makes up the structure of the ribosome. tRNA has a unique shape.

ribosomal RNA (rRNA)

D1.2 Figure 4 Types of RNA. Note the general shapes, and that rRNA is part of the ribosome structure.

Other types of RNA have been found. Most are involved in protein synthesis, but there are a few that function in other ways. Chapter A2.1 mentions how RNA can take part in the control of chemical reactions, thus acting in an enzymatic role. RNA could also have been the first genetic molecule on our planet.

At the centre of the change from the language of DNA to the language of protein, i.e. translation, is the ribosome. Each ribosome consists of a large subunit and a small subunit. The subunits are composed of rRNA molecules and many distinct proteins. Chapter A2.2 discusses how prokaryotic ribosomes are smaller than eukaryotic ribosomes.

Once an mRNA molecule has been transcribed, the mRNA detaches from the single-strand DNA template and floats free in the nucleoplasm. The mRNA will then pass through one of the many pores in the nuclear membrane and enter the cytoplasm. The translation process can then begin.

1. The mRNA binds with the small subunit of a ribosome.
2. A tRNA molecule with a specific amino acid attached now moves in, attaches to the large subunit of the ribosome and, through complementary base pairing, combines with mRNA.
3. A second tRNA with its amino acid follows the first tRNA, complementary base pairs with mRNA, and attaches to the large subunit of the ribosome. Two tRNAs can bind simultaneously to the large subunit.
4. An enzyme then catalyses a condensation reaction between the two amino acids, forming a peptide bond (Figure 5).

D1.2 Figure 5 A peptide bond, bracketed in the middle of this figure, forms when water is released. This process is called condensation. The amino acid on the right of the figure was the first amino acid brought to the ribosome.

5. The first tRNA then breaks free of its amino acid, detaches from the mRNA and floats away into the cytoplasm, where it can usually attach to another amino acid of the same type.
6. The ribosome then moves down the mRNA molecule.
7. The second tRNA molecule is now in the position that the first tRNA originally occupied.
8. A third tRNA floats in and pairs with the next sequence of bases on the mRNA.
9. Another peptide bond forms, and the process continues until the complete polypeptide is assembled.

Figure 6 summarizes the process of translation.

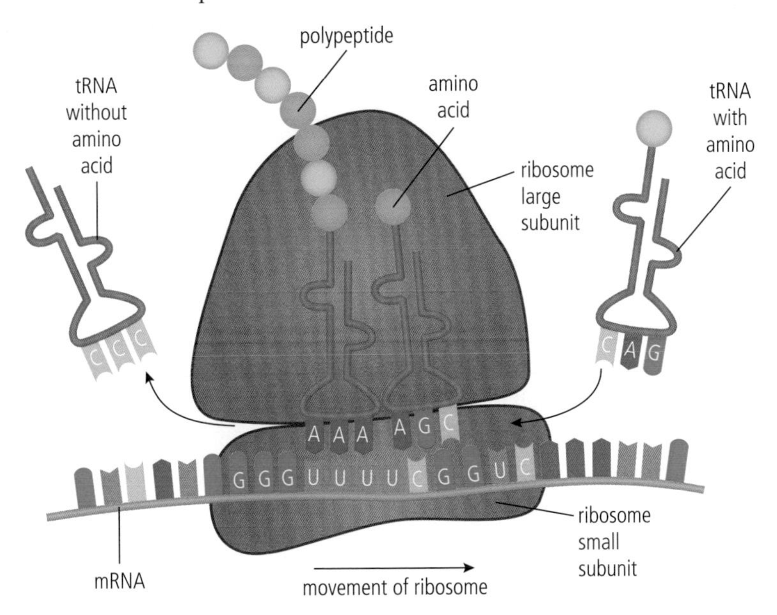

D1.2 Figure 6 Translation at the ribosome. Note the polypeptide chain being assembled, the mRNA, and the tRNAs involved in the process. The mRNA combines with the small subunit of the ribosome. Two tRNAs are capable of attaching simultaneously to the large ribosomal subunit.

D1.2.7 – RNA complementary base pairing

D1.2.7 – Complementary base pairing between tRNA and mRNA
Include the terms "codon" and "anticodon".

The message written into the mRNA molecule determines the order of the amino acids when a protein is assembled.

A set of three bases provides the code for each of the 20 amino acids that make up proteins. Any set of three bases found in DNA that determines the identity of one amino acid is called a **triplet**. When a triplet is found in an mRNA molecule, it is called a **codon**.

DNA triplet → (transcription) → mRNA codon

Translation depends on complementary base pairing between codons on mRNA and anticodons on tRNA. Figure 7 shows a typical tRNA molecule. Notice that the three bases in the middle loop are called the **anticodon** bases and they determine which of the 20 amino acids is attached to the tRNA.

There are 20 different tRNA molecules, one for each of the different amino acids. Each tRNA can be differentiated by its anticodon. The anticodon of a tRNA is complementary to a codon of the mRNA. These match up during the translation process to produce the specific protein coded for by the mRNA.

D1.2 Figure 7 The two-dimensional clover-leaf structure of tRNA, with three loops. The anticodon triplet is unique to each tRNA.

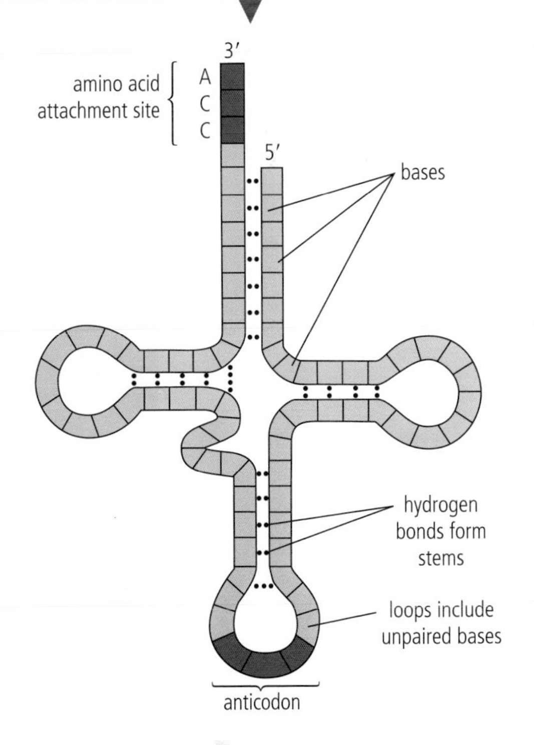

D1.2.8 – The genetic code

D1.2.8 – Features of the genetic code
Students should understand the reasons for a triplet code. Students should use and understand the terms "degeneracy" and "universality".

Researchers have found experimentally that the genetic code is written in a language of three bases. In other words, a set of three bases contains enough information to code for each of the 20 amino acids. There are four possible bases in mRNA (A, U, C and G). In Chapter B1.2, which discusses proteins, it is explained that there are 20 possible amino acids in the human body. So, one base is inadequate to code for 20 different amino acids. If two bases coded for an amino acid, there would be only 16 different combinations possible. However, with three bases coding for an amino acid there are 64 possible combinations. This is an adequate number of combinations to provide a code for all the amino acids.

The genetic code is **degenerate**, which means that, for each amino acid, there may be more than one codon. Also, the genetic code is **universal**, which means that, with only a few minor exceptions, all organisms share the same code. It is this universal aspect of the code that allows us to insert genes from one species to another using genetic engineering techniques. Genetic engineering has enabled us to place the human insulin-coding gene into bacteria so that the bacteria can produce this protein for human use in the treatment of certain cases of diabetes. It is important to note that, even though the code is degenerate, it is not ambiguous or uncertain. A particular codon will always code for the same amino acid or a start or stop message.

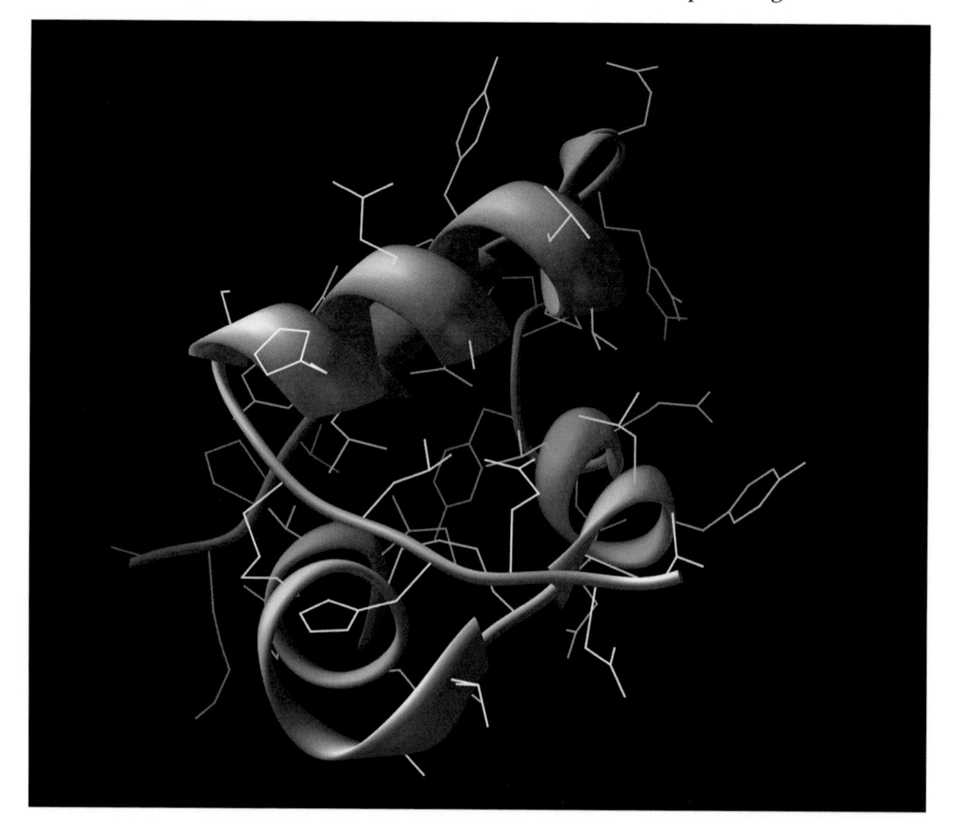

A computer graphic of an insulin molecule. Insulin is a protein hormone produced by protein synthesis and is essential for the control of blood glucose levels. The complex structure of this molecule is dictated by DNA. The genetic code for the production of human insulin has been introduced into bacteria. As a result, the bacteria can produce human insulin because of the universality of the genetic code.

D1.2.9 – mRNA codons

D1.2.9 – Using the genetic code expressed as a table of mRNA codons
Students should be able to deduce the sequence of amino acids coded by an mRNA strand.

The genetic code can be expressed as a table of mRNA codons. Table 2 shows the meaning of the 64 different possible codons.

<table>
<tr><th rowspan="2">First position</th><th colspan="4">Second position</th><th rowspan="2">Third position</th></tr>
<tr><th>U</th><th>C</th><th>A</th><th>G</th></tr>
<tr><td rowspan="4">U</td><td rowspan="2">Phenylalanine</td><td rowspan="4">Serine</td><td rowspan="2">Tyrosine</td><td rowspan="2">Cysteine</td><td>U</td></tr>
<tr><td>C</td></tr>
<tr><td rowspan="2">Leucine</td><td>Stop</td><td>Stop</td><td>A</td></tr>
<tr><td>Stop</td><td>Tryptophan</td><td>G</td></tr>
<tr><td rowspan="4">C</td><td rowspan="4">Leucine</td><td rowspan="4">Proline</td><td rowspan="2">Histidine</td><td rowspan="4">Arginine</td><td>U</td></tr>
<tr><td>C</td></tr>
<tr><td rowspan="2">Glutamine</td><td>A</td></tr>
<tr><td>G</td></tr>
<tr><td rowspan="4">A</td><td rowspan="3">Isoleucine</td><td rowspan="4">Threonine</td><td rowspan="2">Asparagine</td><td rowspan="2">Serine</td><td>U</td></tr>
<tr><td>C</td></tr>
<tr><td rowspan="2">Lysine</td><td rowspan="2">Arginine</td><td>A</td></tr>
<tr><td>*Methionine</td><td>G</td></tr>
<tr><td rowspan="4">G</td><td rowspan="4">Valine</td><td rowspan="4">Alanine</td><td rowspan="2">Aspartic acid</td><td rowspan="4">Glycine</td><td>U</td></tr>
<tr><td>C</td></tr>
<tr><td rowspan="2">Glutamic acid</td><td>A</td></tr>
<tr><td>G</td></tr>
</table>

*And start.

D1.2 Table 2 The genetic code. The first, second and third positions represent the base location in the codon. Twenty amino acids are coded for.

Several important points are illustrated in Table 2. There is a start codon (AUG) that signals the beginning of a polypeptide chain. This codon also encodes the amino acid methionine. Three codons are stop codons (UAA, UAG and UGA). These three codons have no complementary tRNA anticodon and signal the end of a polypeptide chain.

Challenge yourself

Deductions can be made using Table 2. Each amino acid has between 1 and 6 possible codons. Table 2 shows the position of the bases. The left-hand side of the table represents the first base in the codon, the second base is represented on the top of the table, and the third base is represented on the right-hand side of the table.

1. Using this information, deduce the sequence of amino acids coded for in the following example.

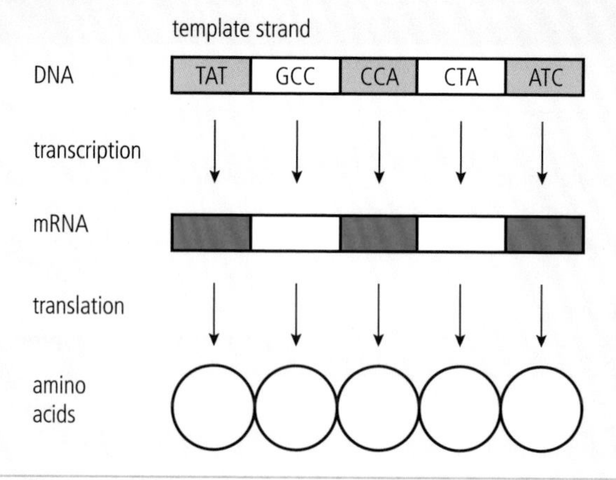

D1.2.10 – Producing a polypeptide chain

D1.2.10 – Stepwise movement of the ribosome along mRNA and linkage of amino acids by peptide bonding to the growing polypeptide chain
Focus on elongation of the polypeptide, rather than on initiation and termination.

The decoding of a strand of mRNA to produce a polypeptide occurs in the space between the two subunits of the ribosome. mRNA binds to the small subunit of the ribosome while tRNA with its attached amino acid binds to the large subunit of the ribosome (Figure 8). Essentially, the ribosome coordinates the functioning of the mRNA and the tRNA at the mRNA–ribosomal complex. The two ribosomal subunits hold the mRNA and tRNA close together so that amino acids can be connected by peptide bonds to produce the specific polypeptide required. This elongation process produces the polypeptide chain.

tRNAs carrying specific amino acids move sequentially through the binding sites of the ribosome as base pairing occurs between

D1.2 Figure 8 The events of translation (synthesis of a polypeptide). The mRNA is held between the large and small subunits of the ribosome.

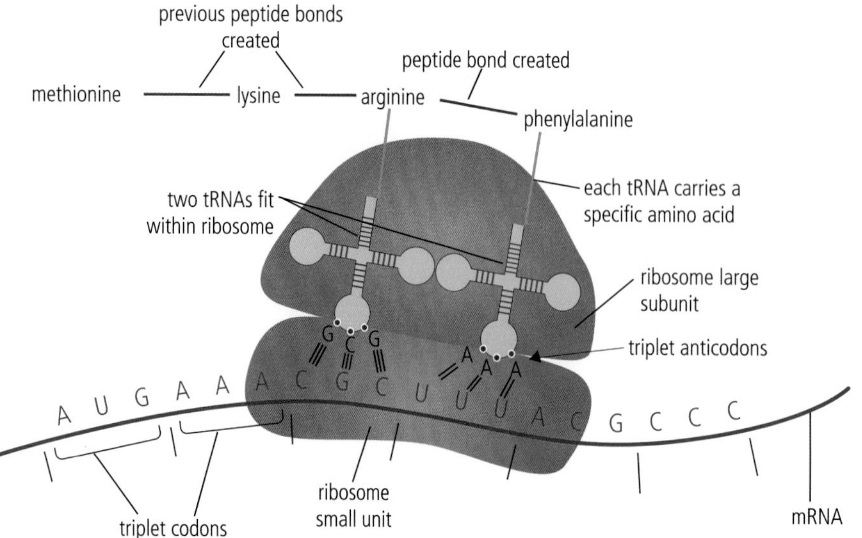

the tRNA anticodons and the mRNA codons, to create the exact sequence of amino acids needed. A continuous cycle of events occurs until the full polypeptide chain has been produced. To accomplish this, the mRNA is moved through the ribosome, one codon at a time.

D1.2.11 – Changing the protein structure

D1.2.11 – Mutations that change protein structure
Include an example of a point mutation affecting protein structure.

Mutations occur when permanent changes to DNA take place. There are many possible causes of mutations and they may be heritable. Even with high levels of accuracy, because of the sheer number of potential mistakes that could occur during DNA replication and protein synthesis, some errors or changes do happen. One type of mistake involving the genetic code is called a **point mutation**. A point mutation involves a change in only one base of a gene. This single nucleotide change alters the transcription and can change the specific amino acid produced and therefore affect protein structure.

A single point mutation is the cause of the genetic disorder known as sickle-cell disease. In this case a single base substitution causes a dramatic change in the shape of haemoglobin, the protein that carries oxygen in the blood. Normally, the shape of haemoglobin is such that red blood cells can move easily through our vascular system. However, with a point mutation of just one amino acid in the haemoglobin, the protein shape is changed. The change in shape of this protein leads to a change in the shape of the red blood cells. Figure 9 shows both normal red blood cells and sickle cell disease red blood cells.

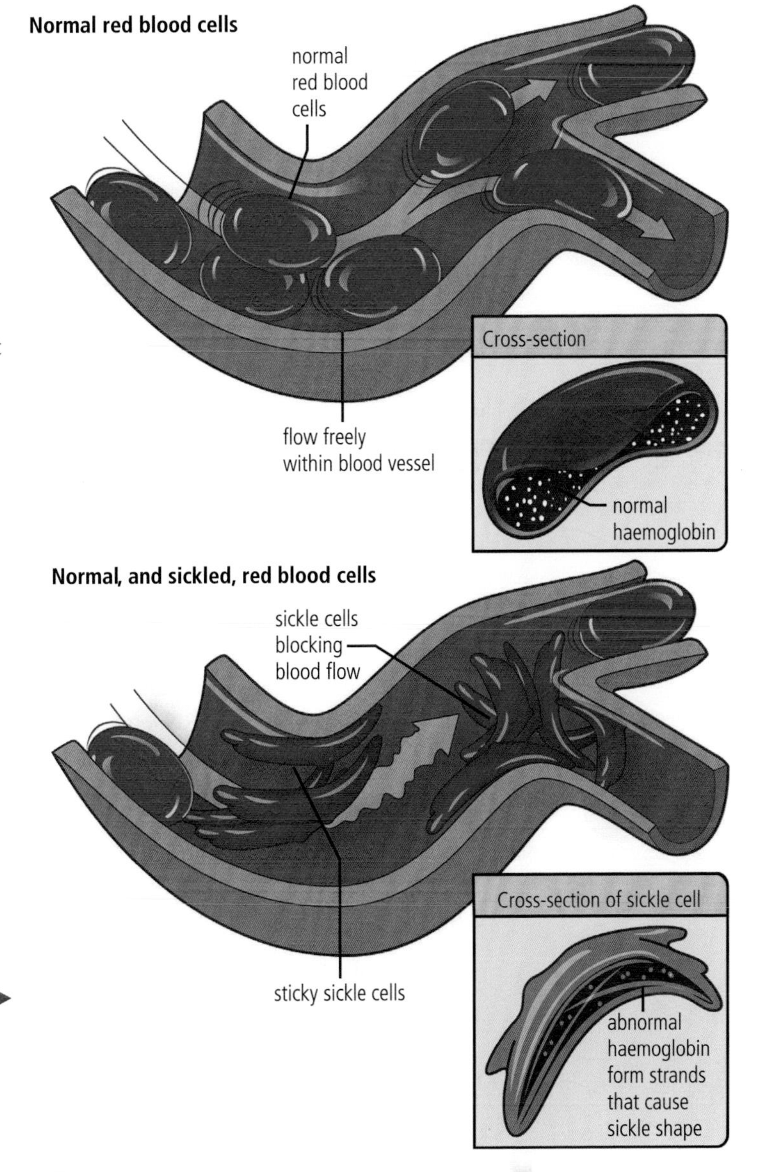

D1.2 Figure 9 Sickle cell disease red blood cells and normal red blood cells. The abnormal shape of the red cells with the mutation means it is difficult for the cells to move through the body's blood vessels. This is a serious medical condition.

HL

D1.2.12 – Directionality

D1.2.12 – Directionality of transcription and translation
Students should understand what is meant by 5′ to 3′ transcription and 5′ to 3′ translation.

You will recall that in DNA replication, DNA polymerase allows assembly of the new strand only in a 5′ to 3′ direction. The same is true with RNA polymerase in transcription. The 5′ ends of free RNA nucleotides are added to the 3′ end of the RNA molecule being synthesized. Also, in translation the mRNA is read in the direction of 5′ to 3′ to produce the polypeptide chain at the ribosome (Figure 10).

D1.2 Figure 10 Directionality of transcription and translation. The direction of transcription to produce mRNA is in the 5′ to 3′ direction. Translation of the mRNA strand to produce the polypeptide also occurs in the 5′ to 3′ direction.

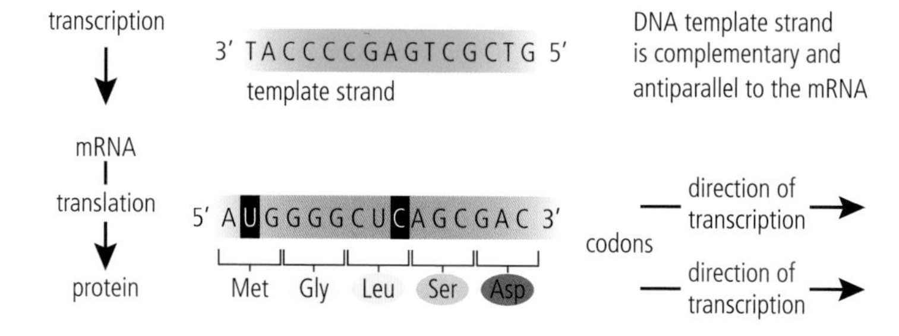

D1.2.13 – Initiating transcription

D1.2.13 – Initiation of transcription at the promoter
Consider transcription factors that bind to the promoter as an example. However, students are not required to name the transcription factors.

The promoter

Transcription has some similarities with replication. For both processes, the double helix must be opened to expose the base sequence of the nucleotides. In replication, helicase unzips the DNA and both strands become templates for the formation of two daughter strands of DNA. However, in transcription, helicase is not involved. Instead, the enzyme RNA polymerase separates the two DNA strands. This same enzyme allows polymerization of RNA nucleotides as base pairing occurs along the DNA template strand. To provide these functions, the RNA polymerase must first combine with a region of the DNA called a **promoter**.

The promoter region for a gene determines which DNA strand will be the template strand. For any gene, the promoter is always on the same DNA strand. The promoter region is a short sequence of bases that is not transcribed.

In bacterial cells, once RNA polymerase has attached to the promoter region for a particular gene, the process of transcription begins. In eukaryotic cells, other transcription factors bind to the promoter region of a particular gene first and then attract RNA polymerase to initiate transcription. The DNA opens and a transcription bubble forms. This transcription bubble contains the DNA template strand, the RNA polymerase, and the growing RNA transcript. Figure 11 illustrates the transcription bubble.

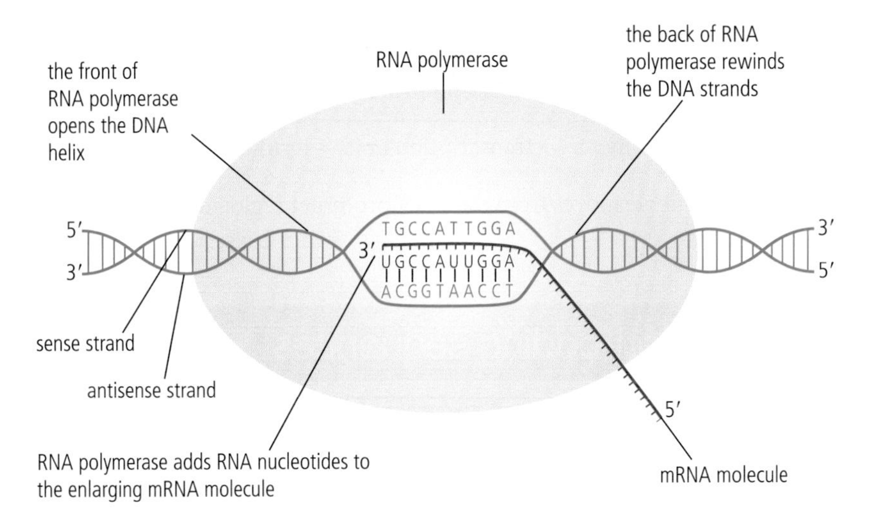

D1.2 Figure 11 An RNA transcription bubble. DNA is opened into two strands by RNA polymerase. The template strand, also known as the antisense strand, is the strand of DNA copied in transcription, so the new mRNA has a base sequence that is complementary to it. The other strand of DNA, labelled as the sense strand in the figure, is not copied in transcription. The mRNA is produced in the 5′ to 3′ direction.

Accurate transcription involves two regions in DNA, the promoter and terminator regions. The promoter region provides a recognition and binding site for RNA polymerase, and it is the site at which transcription begins. The terminator region signals the end of transcription. Transcription is often said to move downstream from the promoter to the terminator region and always moves in a 5′ to 3′ direction.

Transcription factors

Transcription factors are proteins that play a role in the regulation of the transcription process. Some transcription factors attach directly to the promoter region for a gene, while others attach away from the promoter region. They are important in determining which genes are active in each cell of an organism's body. These transcription factors work in eukaryotic cells by first allowing the attachment of RNA polymerase to the promoter region of a gene so that transcription can begin. After this initial binding, other transcription factors move into action. In bacterial cells, RNA polymerase can attach to a promoter region without a transcription factor. However, a transcription factor may then prevent the transcription process in bacteria even with the RNA polymerase attached to the promoter.

It is important to remember that transcription factors are proteins. Therefore, they are produced by a segment of DNA (gene) on a chromosome with transcription factors of its own. This is another method of controlling the expression of genes in cells.

Some transcription factors activate transcription while others prevent or repress transcription. The DNA binding sites for transcription factors are often within or close to a gene's promoter, but not always. The expression of many genes is controlled by multiple transcription factors, in which case often a specific combination of these factors is needed for transcription to take place.

The DNA binding sites for transcription factors represent DNA base sequences that do not code for polypeptides.

In eukaryotic cells, transcription factors known as general or basal factors assist in the binding of RNA polymerase to the promoter. In bacteria, these general or basal factors are not needed for this attachment. Other types of transcription factors are known as activators or repressors.

The terminator

The terminator is a sequence of nucleotides that, when transcribed, causes the RNA polymerase to detach from the DNA. When this happens, transcription stops and the RNA transcript is detached from the DNA. The transcript is released from the DNA strand and carries the code of the DNA and is referred to as messenger RNA (mRNA).

The sections of DNA involved in transcription are:

promoter → transcription unit → terminator

The transcription bubble moves from the DNA promoter region downstream towards the terminator.

D1.2.14 – Non-coding sequences

D1.2.14 – Non-coding sequences in DNA do not code for polypeptides

Limit examples to regulators of gene expression, introns, telomeres and genes for rRNAs and tRNAs in eukaryotes.

Most mutations in non-coding regions of DNA have little or no impact. However, some play an important role in health and disease. Recent research has found DNA variants (mutations) related to heart disease and several types of cancer. Whole genome sequencing (WGS) is a medical test that looks at the entire genome to discover variants that could yield a definitive diagnosis or provide actionable results.

Surprisingly to many people, most of our DNA is not composed of protein-coding genes. Research suggests that as much as 99% of our DNA does not provide instructions for making proteins. Many scientists used to refer to this DNA as "junk" DNA. We now think that much of this DNA is essential to the functions of cells. Examples of DNA sequences that do not code for proteins include the following.

- Regulators of gene expression. Promoters, enhancers, silencers and insulators all play a role in the control of transcription, the first step in the synthesis of a protein.
- Genes for rRNA and tRNA formation. All types of RNA are very important to the essential process of protein synthesis. It is the non-protein coding sections of DNA that direct the synthesis of these types of RNA.
- **Telomeres**. Telomeres are structural features at the ends of chromosomes made up of repetitive non-coding DNA sequences. Telomeres help protect the chromosome.
- **Introns**. These sections of DNA are removed from **primary mRNA** before it leaves the nucleus as **mature mRNA**. Primary mRNA and mature mRNA will be further described in the following section. Many introns appear to have regulatory functions controlling transcription.

D1.2.15 – Post-transcriptional modification

D1.2.15 – Post-transcriptional modification in eukaryotic cells

Include removal of introns and splicing together of exons to form mature mRNA and also the addition of 5′ caps and 3′ polyA tails to stabilize mRNA transcripts.

Eukaryotic cell DNA is different from prokaryote DNA in that within the protein-coding regions there are stretches of non-coding DNA. The stretches of non-coding DNA are called **introns**. As a complete region of a DNA molecule is transcribed to form mRNA, the first RNA formed is called the **primary RNA transcript (pre-mRNA)**. It contains both **exons** and introns. To make a functional mRNA strand in eukaryotes, the introns are removed. The process by which the introns are removed is referred to as **splicing** (Figure 12). The sequences of mRNA that remain after splicing are called exons.

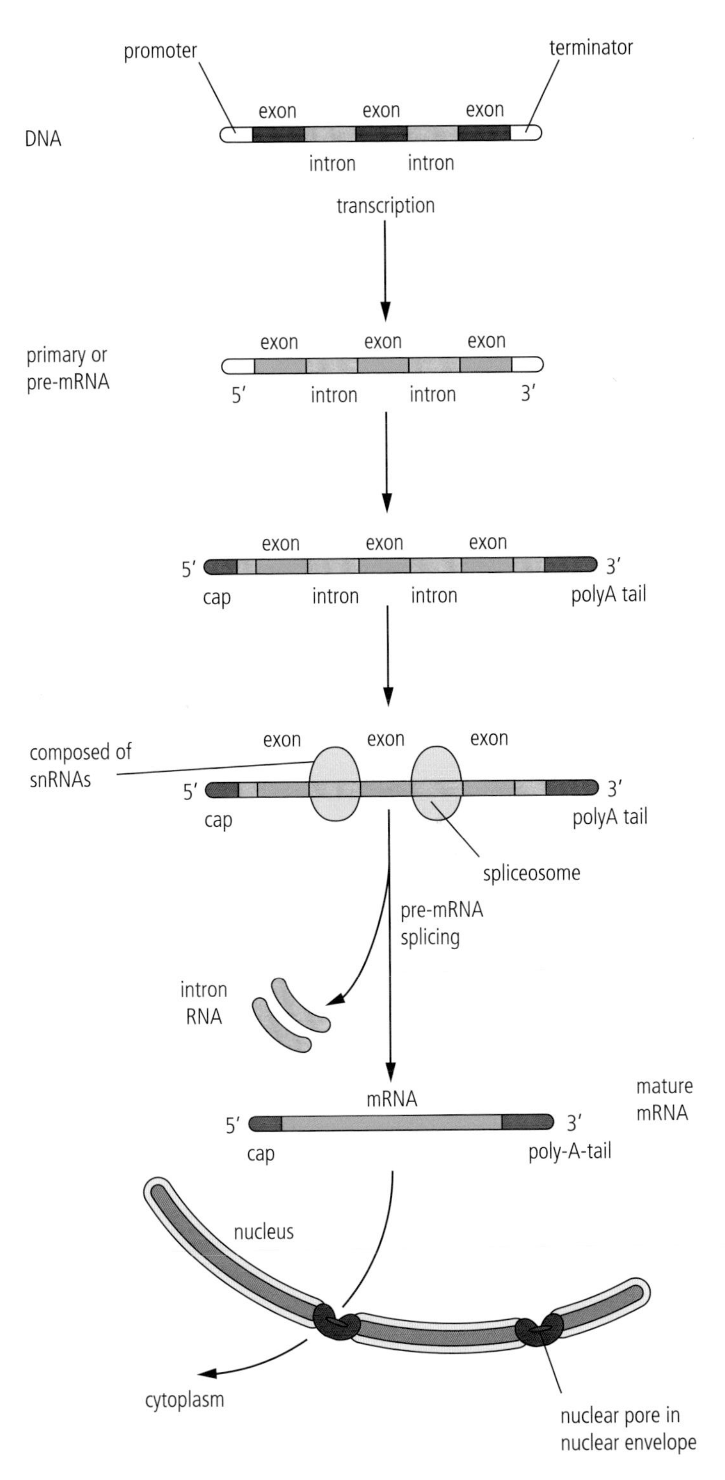

D1.2 Figure 12 Splicing of mRNA in eukaryotes. DNA contains both introns and exons. Both are transcribed in the production of the primary mRNA transcript (pre-mRNA). During splicing the introns are removed, exons may be "rearranged", and a cap and a polyA tail are added to the ends of what is now called mature mRNA. It is this mature mRNA that moves out of the nucleus to the cytoplasm where a protein will be produced.

Spliceosomes (see Figure 12) are composed of multiple **small nuclear RNAs** as well as proteins. They function in the removal of introns from the primary mRNA. When the introns are removed by the spliceosomes, the exons may be rearranged, resulting in the production of different proteins. In some higher eukaryotes different sections of a gene act as introns at different times. Again, this increases the number of possible proteins that can be produced by one gene. The spliceosomes then help join the remaining exons.

An exon is a segment of mRNA that contains the protein-coding portion of a gene. Introns are segments of mRNA found between exons of mRNA that are removed by mRNA processing before translation.

Prokaryotic mRNA does not require processing because no introns are present. Prokaryote mature mRNA does have a cap and polyA tail, but they are not as complex as those in eukaryote mature mRNA.

On the 5′ end of the final mRNA transcript is a cap made of a modified guanine nucleotide with three phosphates. At other end (the 3′ end) is a **polyA tail** composed of 50–250 adenine nucleotides. The cap and polyA tail appear to stabilize the mature mRNA, protecting it from degradation in the cytoplasm, and enhance the translation process that occurs at the ribosome.

D1.2.16 – Alternative splicing of exons

D1.2.16 – **Alternative splicing of exons to produce variants of a protein from a single gene**
Use alternative splicing of transcripts of the troponin T gene in foetal and adult heart muscle as an example.

Biologists at one time firmly believed in the one gene–one protein concept. We now know a gene can produce several, if not many, different proteins**. Alternative splicing** is one of the mechanisms by which this can occur.

As shown in Figure 12, primary mRNA (pre-mRNA) is modified in eukaryotic cells by removing sections called introns and splicing together the remaining exons. However, a primary mRNA transcript can be spliced into different mature mRNA transcripts by the inclusion of different exons. This process is called alternative splicing. Alternative splicing enables cells to increase the number of mature mRNA transcripts and their resulting proteins that can be produced from a single primary mRNA transcript. Alternative splicing has a role in the control of complex biological processes during cardiac development and disease.

Recent studies have revealed other variants of the cardiac troponin T gene (*cTnT*). These variants appear to be correlated with several diseases involving cardiac muscle failure. Further investigations are focusing on sequencing genes such as *cTnT* to diagnose and possibly predict specific cardiac diseases.

Tremendous functional changes occur before and after birth in the human heart. Cardiac troponin T (*cTnT*) is a gene active in heart (cardiac) muscle. In the foetus, a particular exon is included in the mature mRNA produced by this gene that is not included in the mature mRNA produced by the gene in the adult. This exon included in the foetal cardiac muscle makes the heart more sensitive to calcium than adult cardiac muscle.

The Human Genome Project was completed in April 2003. It provided the complete genetic blueprint of *Homo sapiens*. The number of genes found was between 20,000 and 25,000, a surprisingly low number. Most biologists had predicted closer to 100,000 genes would be found. The higher predicted number was because of the vast number of different proteins produced by the human body. We now know there are mechanisms that allow multiple proteins to be made from the same gene, thus explaining the lower number of genes than expected.

How does the diversity of proteins produced contribute to the functioning of a cell?

D1.2.17 – Initiating translation

D1.2.17 – **Initiation of translation**
Include attachment of the small ribosome subunit to the 5' terminal of mRNA, movement to the start codon, the initiator tRNA and another tRNA, and attachment of the large subunit. Students should understand the roles of the three binding sites for tRNA on the ribosome (A, P and E) during elongation.

Initiation is the step that brings all the components of translation together. There are initiation factors that control the assembly of the small subunits of the ribosome, mRNA, initiator tRNA, and the large ribosomal subunit. The ribosomal subunits are composed of rRNA molecules and many distinct proteins. Roughly two-thirds of the ribosome's mass is rRNA.

The decoding of a strand of mRNA to produce a polypeptide occurs in the space between the two subunits of the ribosome. In this area, there are binding sites for mRNA and three sites for the binding of tRNA (Figure 13).

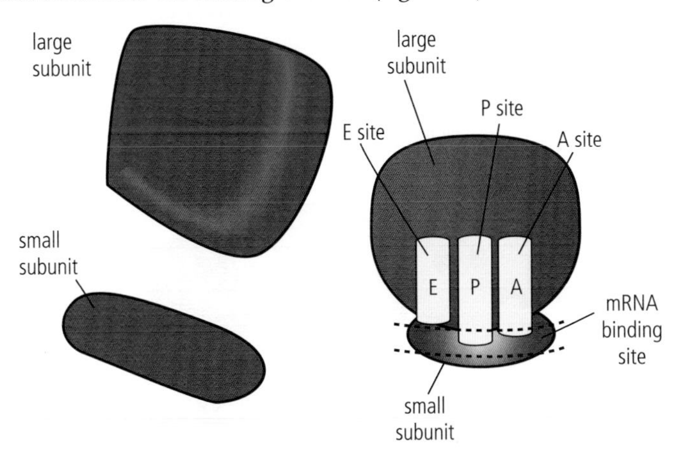

D1.2 Figure 13 A model showing the arrangement of subunits and binding sites in a ribosome.

Site	Function
A	Holds the tRNA carrying the next amino acid to be added to the polypeptide chain
P	Holds the tRNA carrying the growing polypeptide chain
E	Discharges the tRNA that has lost its amino acid

D1.2 Table 3 Ribosomal binding sites for tRNA and their functions.

The steps of initiation are as follows.

1. The initiation complex forms. This complex includes an initiator tRNA that pairs with the mRNA start codon (AUG). The small ribosome subunit attaches to the 5′ terminal of mRNA.
2. Once this initiation complex forms, the large subunit of the ribosome is added.
3. When the complete ribosome is formed, the initiator tRNA is bound to the P site and the A site is empty (see Figure 13 for the structure of a ribosome, and Table 3 for the function of the different ribosomal binding sites).
4. The A site is where the tRNA carrying the next amino acid enters the ribosome. The E site is for any tRNA leaving a ribosome (minus an attached amino acid).
5. Following initiation, translation proceeds with elongation and termination.

A start codon (AUG) is at the 5′ end of all mRNAs. Each codon, other than the three stop codons, attaches to a particular tRNA anticodon. The tRNA has a 5′ and a 3′ end, like all other nucleic acid strands. The 3′ end of tRNA is free and has the base sequence CCA (see Figure 7 showing the structure of tRNA). This is the site of amino acid attachment. Because there are complementary bases in the single-stranded tRNA, hydrogen bonds form at four areas. This causes the tRNA to fold and take on a two-dimensional shape resembling a three-leaf clover. One of the loops of the clover leaf contains an exposed anticodon.

Initiation of translation involves assembling the components that carry out the process. Upon completion of the translation process, disassembling of the components occurs.

Notice the start codon (AUG) is on the 5′ end of all mRNAs. It is this start codon that binds to the initiation complex and begins translation (Figure 14).

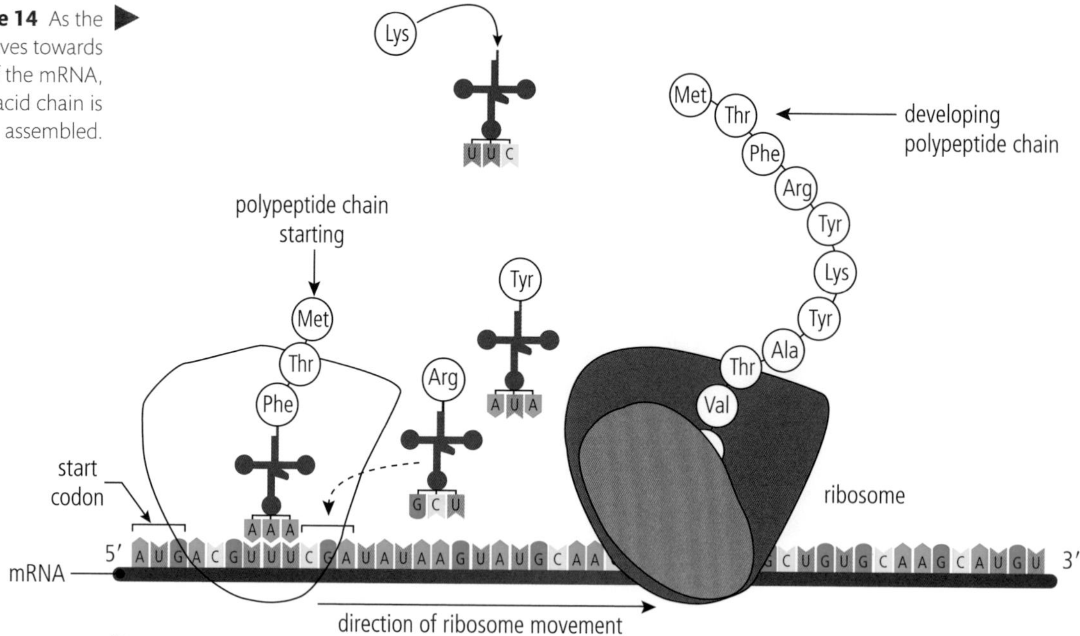

D1.2 Figure 14 As the ribosome moves towards the 3′ end of the mRNA, the amino acid chain is assembled.

Protein synthesis. Full details of how to carry out this experiment with a worksheet are available in the eBook.

The assembling of the polypeptide (protein) chain occurs in the 5′ to 3′ direction. Each chain produced at the ribosome begins with the amino acid methionine, because the codon AUG is the start codon and it attaches to methionine (Met in Figure 14).

D1.2.18 – Modifying polypeptides

D1.2.18 – Modification of polypeptides into their functional state
Students should appreciate that many polypeptides must be modified before they can function. The examples chosen should include the two-stage modification of pre-proinsulin to insulin.

University of Michigan scientists have recently discovered that individuals with prediabetes have large amounts of proinsulin in the endoplasmic reticulum of their pancreatic beta cells. This proinsulin is misfolded, and it is possible this is a trigger for further progression of the disease. Further research has revealed that this build-up was present before any other symptoms of diabetes were detected.

Many polypeptides must be modified before they are able to carry out their function. One example of this modification occurs with the formation of **insulin**. Insulin is a rather small protein, and its precursor, known as **pre-proinsulin**, is produced in the beta cells of the pancreas. Pre-proinsulin goes through two steps of protein modification to become insulin.

1. A signal peptide is removed as the pre-proinsulin enters the endoplasmic reticulum to produce proinsulin. A signal peptide is a short peptide chain that occurs on the end of a polypeptide destined for secretion in some way. In this case, the signal peptide directs the pre-proinsulin to the cell endoplasmic reticulum.
2. Proinsulin is then exposed to enzymes that break peptide bonds. These enzymes result in the removal of a section of peptides known as **C peptide**. This is the final step in the production of the mature form of insulin.

The mature form of insulin, which is composed of two chains held together by disulfide bonds, is then packaged in the Golgi apparatus as secretory granules. The insulin is secreted from the pancreatic cells by exocytosis into the blood when needed.

Other examples of polypeptide modification include the following.

- Molecular chaperones play a role in protein folding. These specialized molecules do not actually fold proteins, but they provide protection from interfering conditions that hinder folding.
- Disulfide bond formation stabilizes the tertiary and often quaternary structure of proteins. Proteins are discussed in Chapter B1.2.
- Glycosylation is the addition of a carbohydrate side chain to a polypeptide. This prevents protein chains from sticking or clumping together, so that they can carry out their functions.

D1.2.19 – Recycling amino acids

D1.2.19 – Recycling of amino acids by proteasomes
Limit to the understanding that sustaining a functional proteome requires constant protein breakdown and synthesis.

The **proteome** is the entire set of proteins that is or can be expressed by a cell, tissue or organism. Proteins are continually being synthesized and degraded. Some proteins become damaged and need to be replaced, while others only provide a cellular function for a short period of time.

Many degenerative diseases, such as Alzheimer's disease, Parkinson's disease and bovine spongiform encephalopathy (mad cow disease), are related to proteins that aggregate to form characteristic plaques (hardened layers) in brain cells.

Enzymes called **proteases** can degrade proteins by breaking peptide bonds. This converts a protein into its component amino acids. The degradative activities of lysosomes are discussed in Chapter B2.2. Proteases must be contained in lysosomes so that cellular proteins that are needed are not broken down.

Eukaryotic cells have a unique way of ridding themselves of damaged or un-needed proteins. They mark these proteins with a chemical called **ubiquitin.** Proteins that have ubiquitin attached are a signal to the cell to destroy those proteins. The cellular organelle that degrades a marked protein is the **proteasome**:

marked protein → proteasome → amino acids

The marked protein enters one end of the proteasome and amino acids exit the other end. The free amino acids can then be reused by the cell in the process of protein synthesis.

HL end

Guiding Question revisited

How does a cell produce a sequence of amino acids from a sequence of DNA bases?

In this chapter you have learned how:

- transcription and translation are involved in the process of producing proteins using a code carried by DNA
- three types of RNA are involved in the process of protein synthesis, mRNA, tRNA and rRNA
- there is a control mechanism for genes in a cell so that only those that need to be expressed are active at any given time
- the genetic code is both degenerate and universal
- transcription produces mRNA
- translation produces chains of amino acids known as polypeptides or protein
- transcription occurs in the nucleus while translation occurs in the cytoplasm at the ribosomes.

Guiding Question revisited

How is the reliability of protein synthesis ensured?

In this chapter we have discussed how:

- complementary base pairing is essential to the high reliability of the processes of transcription and translation, including the base pairing that occurs between codons and anticodons in the translation process
- a triplet code is necessary so that all possible amino acids can be utilized when producing a specific protein
- the sequence of amino acids in a protein can be deduced by examining the base code in an mRNA strand producing that protein
- amino acids are linked by peptide bonds in the elongating polypeptide chain being assembled at the ribosome during translation
- a mutation occurs when there is a change in the amino acid sequence of a protein being assembled at the ribosome
- a point mutation occurs when there is a change in only one base in a gene, sickle cell disease is due to a point mutation that results in a dramatic change in the shape of haemoglobin present in red blood cells

HL

- many factors have a role in the control of transcription, such as transcription factors binding to a promoter region of DNA and post-transcriptional modification in eukaryotic cells
- modification of post-translational proteins is common to ensure a protein is properly structured for its function
- sections of DNA that do not code for proteins often play a role in the control of transcription and translation.

HL end

Exercises

Q1. Describe the functions of the three types of RNA involved in the translation process.

Q2. From the following DNA base sequence, determine the sequence of amino acids that would be assembled.

TACCGTCATAGAAAAATC

Q3. **(a)** On what structures are you most likely to find codons?

(b) On what structures are you most likely to find anticodons?

Q4. Describe two characteristics of the genetic code.

HL

Q5. What type of mRNA requires processing? Explain why.

Q6. In eukaryotes, the regulation of gene expression occurs:

A only by post-transcription modifications

B only at translation

C only at transcription

D only at post-transcription.

HL end

D1.3 Mutation and gene editing

Guiding Questions

How do gene mutations occur?

What are the consequences of gene mutation?

Genes can be modified if one or more letters in a sequence is deleted, inserted or substituted for another letter. Sometimes mutations can be catastrophic for the organism and lead to severe health problems or death, but most have little effect. Some mutations make no difference at all to the organism while a few can be advantageous. Beneficial mutations create a new version of a gene that will result in a trait that gives the organism an advantage for survival.

A mutation is when one or more nucleotides of a DNA code are modified by mistake. In the case of a point mutation, a single letter can be switched during a substitution, removed during a deletion, or added during an insertion.

D1.3.1 – Gene mutations

D1.3.1 – Gene mutations as structural changes to genes at the molecular level
Distinguish between substitutions, insertions and deletions.

A **mutation** is a random, rare change in genetic material. One type involves a change in the sequence of bases in DNA. If DNA replication works correctly, this should not happen (see Chapter D1.1). But nature sometimes makes mistakes. For example, the base thymine (T) might be put in the place of adenine (A) along the DNA sequence. When this happens, the corresponding bases along the messenger RNA (mRNA) are altered during transcription (see Chapter D1.2).

When writing gene names, the convention is to put the name in italics: *GNAT2*. However, when writing the name of the protein that is generated by the gene, we do not italicize it: transducin. Sometimes the two names are the same: later in the chapter we will see that the gene *CCR5* codes for a protein called CCR5.

Three types of mutations are **substitutions**, **insertions** and **deletions**. A substitution happens when one letter is replaced by another. T instead of A, for example. An insertion happens when a letter is added, while a deletion happens when one letter is removed from a sequence. When only one nucleotide (base) is involved, it is called a **point mutation**. But sometimes two or more bases can be involved in a mutation, or even thousands of bases when a segment of a chromosome is involved. We will focus mostly on point mutations.

The **locus** (position) of the gene *GNAT2*, controlling a protein called transducin that enables colour vision in humans, is found on chromosome 1. A mutation of this gene stops an individual from being able to make the protein transducin properly. Transducin is needed to transmit information about colour from the eye to the brain; as a result of the mutation, a person cannot see in colour. This extremely rare genetic condition is called complete achromatopsia. When we say "the ability to see in colour is a genetic trait" we mean that one of two things can happen with a person's DNA: either that person has the DNA code for making colour vision possible, or the person does not have it. Figure 1 illustrates this concept.

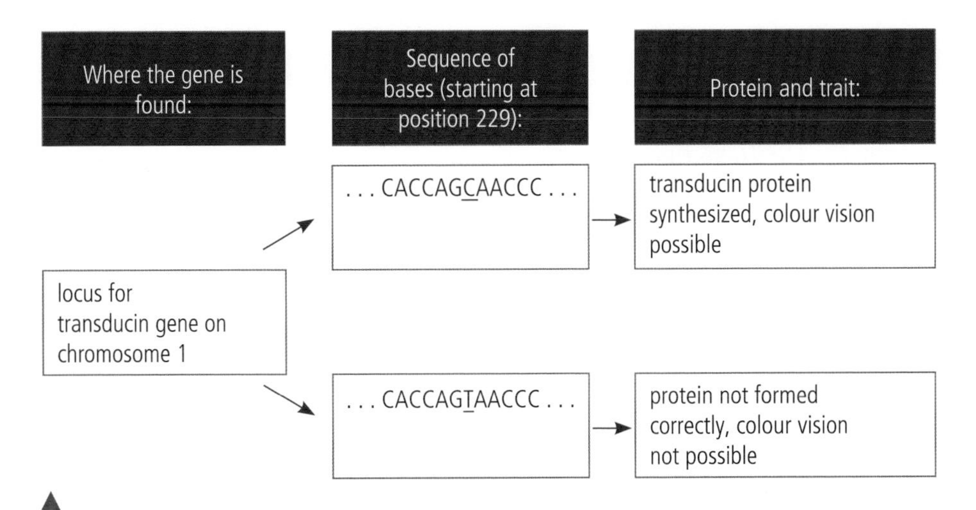

D1.3 Figure 1 The consequences on colour vision of a single base substitution.

Worked example

Look carefully at the two sequences of DNA below. These sequences are from the coding strand of a section of genetic information that helps in the formation of haemoglobin, found in red blood cells. Identify the difference between the two sequences and complete the phrase below.

DNA sequence 1: GTG CAC CTG ACT CCT GAG GAG

DNA sequence 2: GTG CAC CTG ACT CCT GTG GAG

'Codon number __ along the first sequence has the letter __ in position number __, whereas the codon in the same position in sequence 2 has the letter __ instead.'

Solution

Codon number 6 along the first sequence has the letter A in position number 2, whereas the codon in the same position in sequence 2 has the letter T instead.

Worked example

Now look at the effect this has on the mRNA sequences produced from the template strand that is found opposite the coding strand when the DNA is unzipped for transcription:

mRNA sequence 1: GUG CAC CUG ACU CCU GAG GAG

mRNA sequence 2: GUG CAC CUG ACU CCU GUG GAG

Using the figure below, showing which codons are associated with which amino acids, and the mRNA sequences given above, fill in the names of the missing amino acids (a) to (h).

Second base

First base	U	C	A	G	Third base
U	UUU, UUC – phenyl-alanine UUA, UUG – leucine	UCU, UCC, UCA, UCG – serine	UAU, UAC – tyrosine UAA, UAG – stop codon	UGU, UGC – cysteine UGA – stop codon UGG – tryptophan	U C A G
C	CUU, CUC, CUA, CUG – leucine	CCU, CCC, CCA, CCG – proline	CAU, CAC – histidine CAA, CAG – glutamine	CGU, CGC, CGA, CGG – arginine	U C A G
A	AUU, AUC, AUA – isoleucine AUG – methionine start codon	ACU, ACC, ACA, ACG – threonine	AAU, AAC – asparagine AAA, AAG – lysine	AGU, AGC – serine AGA, AGG – arginine	U C A G
G	GUU, GUC, GUA, GUG – valine	GCU, GCC, GCA, GCG – alanine	GAU, GAC – aspartic acid GAA, GAG – glutamic acid	GGU, GGC, GGA, GGG – glycine	U C A G

Sequence 1:	valine	–	histidine	–	(a)______	–	(b)______	–	(c)______	–	(d)______	–	glutamic acid

Sequence 2:	valine	–	histidine	–	(e)______	–	(f)______	–	(g)______	–	(h)______	–	glutamic acid

Solution

Sequence 1:	valine	–	histidine	–	leucine	–	threonine	–	proline	–	glutamic acid	–	glutamic acid

Sequence 2:	valine	–	histidine	–	leucine	–	threonine	–	proline	–	valine	–	glutamic acid

Notice how the error of only one letter in the original DNA code changed the composition of amino acids in sequence 2. This would change the composition and the structure of the resulting protein, in the same way that changing the shape and composition of some of the bricks used to build a house would change the shape (and therefore the structural integrity) of the house. This kind of change in the DNA code is produced by a mutation.

D1.3.2 – Base substitutions

D1.3.2 – Consequences of base substitutions
Students should understand that single-nucleotide polymorphisms (SNPs) are the result of base substitution mutations and that because of the degeneracy of the genetic code they may or may not change a single amino acid in a polypeptide.

One base can make a big difference

The type of mutation that results in a single letter being changed is called a **base substitution mutation**. When a gene sequence is altered by one letter a **single-nucleotide polymorphism (SNP)** results. SNPs are of great interest to geneticists because they define different versions of genes and some can explain genetic diseases and cancers. The consequence of changing one base could mean that a different amino acid is placed in the growing polypeptide chain. This may have little or no effect on the organism, or it may have a major influence on the organism's physical characteristics (see Figure 2).

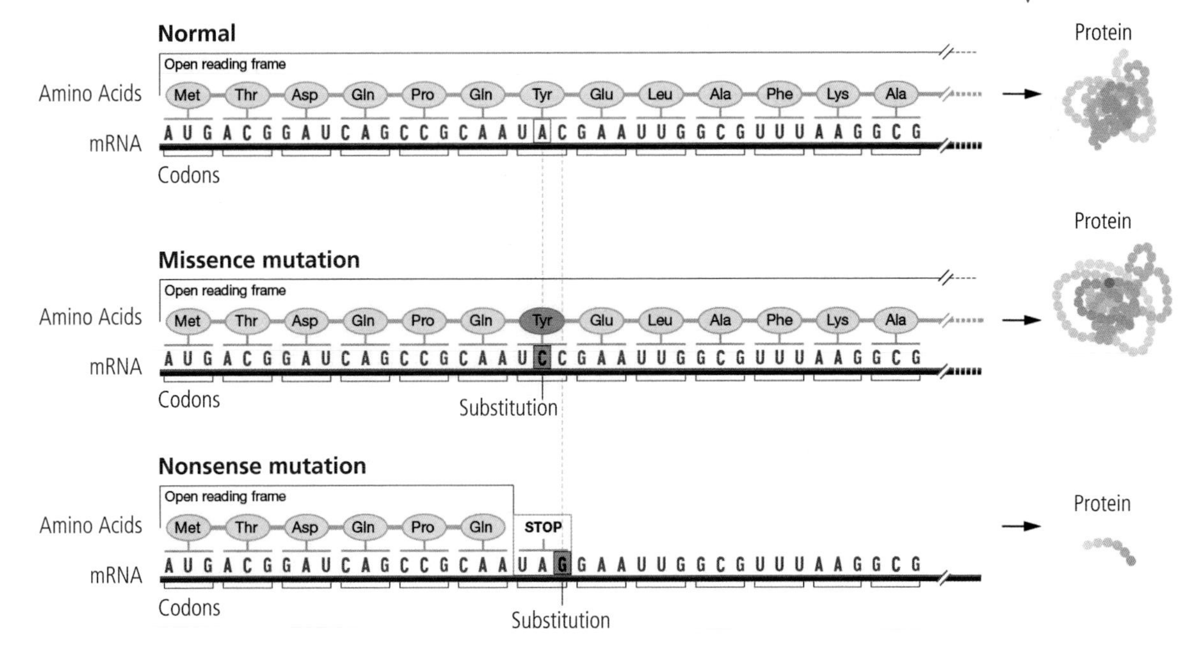

D1.3 Figure 2 Two possible consequences of a base substitution. In the first example, a different amino acid is coded for, which changes the shape and probably function of the protein. This is a **missense** mutation. In the second example, a stop codon is created that cuts off the gene translation, resulting in no functional protein being made. This is a **nonsense** mutation.

There is another possibility not shown in Figure 2. Sometimes changing a letter does not, in fact, change the resulting protein's composition, structure or resulting function. Because of the degenerate nature of the DNA code, the amino acid alanine can be coded for by any codon starting with CG_. CGA works, but if the codon is mutated to CGT, CGC or CGG, the amino acid alanine will still be translated in the final protein.

Sickle cell disease

In humans, a mutation is sometimes found in the gene that codes for haemoglobin in red blood cells. This mutation alters the shape of the haemoglobin molecule. The red blood cells do not look like the usual flattened disc with a hollow in the middle.

The mutated red blood cell has a characteristic curved shape, which made its discoverers think of a sickle (a curved knife used to cut tall plants). The condition that results from this mutation is therefore called **sickle cell disease**, also known as sickle cell anaemia.

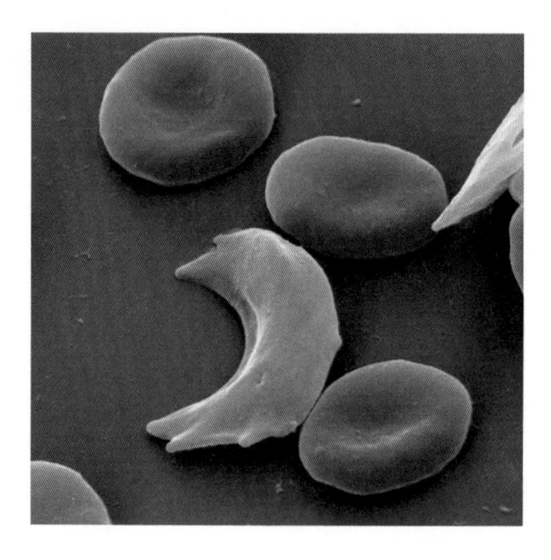

The "sickle" shape of a red blood cell in someone with sickle cell anaemia.

Look at the sequences in Figure 3. The first is for the section of the haemoglobin gene's DNA that codes for standard-shaped red blood cells, and the second shows the mutation that leads to the sickle shape. In this case, one base is substituted for another so that the second codon in this sequence of haemoglobin, GAG, becomes GTG. As a result, during translation, instead of adding glutamic acid, which is the intended amino acid in the sixth position of the sequence, valine is added there instead. You can refer back to the Worked example to see this mutation.

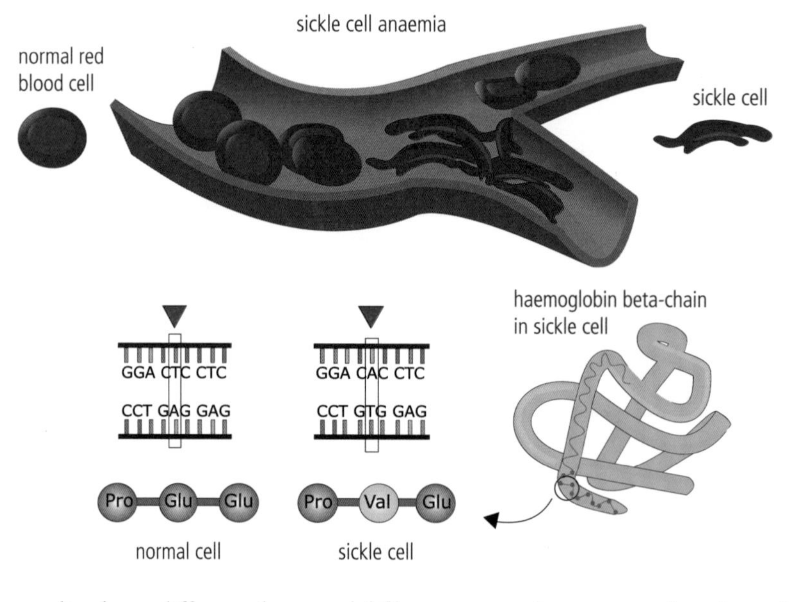

D1.3 Figure 3 The cause of sickle cell disease.

How does variation in subunit composition of polymers contribute to function?

Because valine has a different shape and different properties compared to glutamic acid, the shape of the resulting polypeptide chain is modified. As a result of this, the haemoglobin molecule has different properties that cause the complications associated with sickle cell disease, such as weakness, fatigue and shortness of breath.

Although sickle cell disease is a debilitating condition, those who have it are very resistant to malaria infection. Malaria is an infectious disease that occurs in tropical regions. A parasite of the genus *Plasmodium* is transmitted to human blood by an infected female mosquito of the genus *Anopheles* feeding on a human's blood. The parasite attacks the person's red blood cells and produces symptoms of high fever and chills, which can result in death.

D1.3.3 – Insertions and deletions

D1.3.3 – Consequences of insertions and deletions
Include the likelihood of polypeptides ceasing to function, either through frameshift changes or through major insertions or deletions. Use trinucleotide repeats of the gene *HTT* as an example of insertion and the delta 32 mutation of the *CCR5* gene as an example of deletion.

Consequences of insertions in the *HTT* gene

Huntington's disease is caused by a dominant allele (see Chapter D3.2 for how dominant alleles work). This genetic condition causes severely debilitating nerve damage, but symptoms do not show until a person is about 40 years old. As a result, someone who has the gene for Huntington's disease may not know that they have it when they are younger, before they have perhaps had their own children.

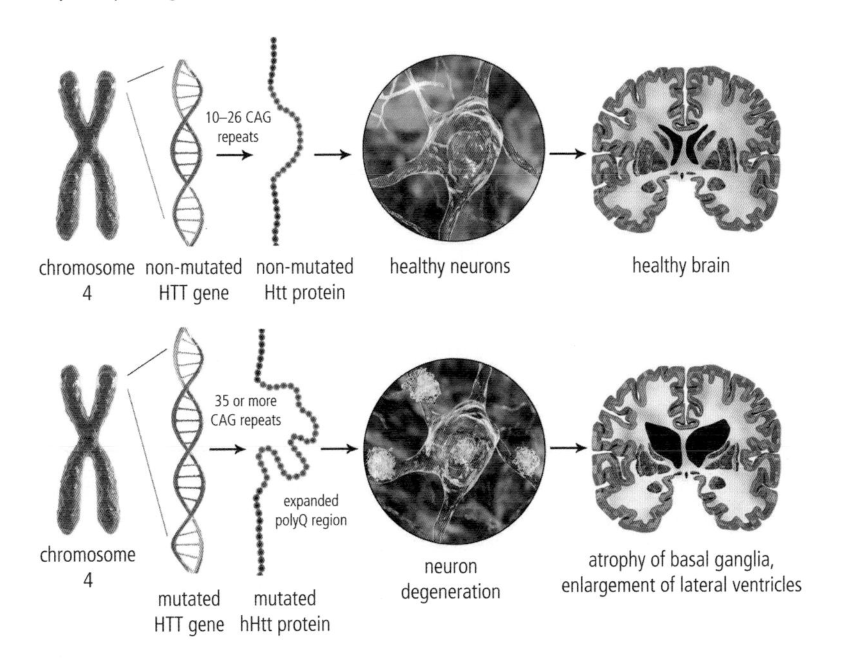

D1.3 Figure 4 Comparison of healthy neurons and a healthy brain to those of a person with Huntington's disease.

Huntington's disease is a life-limiting disease. Symptoms include difficulty walking, speaking and holding objects. Within a few years of the symptoms presenting, a person with Huntington's disease loses complete control of their muscles. This genetic disease is caused when a gene called huntingtin or *HTT*, found on chromosome 4, has an insertion mutation whereby multiple copies of three nucleotides, CAG, are added to the gene. This kind of mutation is known as a **trinucleotide repeat expansion** (see Figure 4).

The resulting mutated protein, mHtt, has an adverse effect on brain cells, causing the symptoms of the disease. Because the CAG trinucleotide repeats the code for the amino acid glutamine, there is much more glutamine in the mutated protein than the normal protein. The more trinucleotide repeats a person has, the more severe their symptoms can be. If a person has any more than 40 repeats, they will be affected. Huntington's disease is not the only polyglutamine disorder that can affect people's nervous system.

To help understand how mutations work, look at these variations of a sentence consisting of only three-letter words, similar to the three-nucleotide codons that code for amino acids:

1. **Sue did ask him why.** (The original, unmutated sentence.)
2. **Ued ida skh imw hy_.** (Deletion mutation: the first letter in the sequence has been removed.)
3. **ASu edi das khi mwh y_ _.** (Insertion mutation: the letter "A" has been added at the start.)
4. **Sue did not ask him why.** (Insertion mutation: the three letters "not" have been added in the middle.)
5. **She did ask him why.** (Substitution of one letter: an "h" has been added instead of "u".)

Grammatically sentences 1, 4 and 5 make sense. But removing or adding one letter and trying to keep the rule that all words must be three letters long results in nonsensical words in sentences 2 and 3 (see Frameshifts below). If there is an addition of letters in multiples of three, the code can still work, but it will be modified. Sentence 4 makes sense, but has a very different meaning compared to the original one. This is how some mutations can cause a genetic disease. A protein can be synthesized from the mutated code, but it might do something very different from the original code. Some substitution mutations have no effect on the result: sentence 5 means the same thing as sentence 1.

Frameshifts

A phenomenon that happens when an insertion or a deletion occurs in non-multiples of three is called a **frameshift**. Normally, the genetic code is read in triplets (codons), but if a letter is added to the sequence as a result of a mutation, the code is shifted. If the code is shifted by three new letters being added or three being deleted, all the other codons will remain unchanged. This is true for any multiple of three.

But if the insertion or deletion is not a multiple of three, the code changes drastically and can often end up not making sense anymore, like sentences 2 and 3 in the Hint for success. Or it could transform a normally coding codon into an unexpected stop codon. This kind of error can happen when the DNA proofreading system attempts to repair a mistake; the DNA polymerase can sometimes reattach in the wrong place along the sequence it is trying to repair.

Consequences of deletions

Leucocytes (white blood cells) need to move towards zones of infection in order to protect the body from disease-causing invaders such as viruses. There are chemical signals called **chemokines** that tell the leucocytes which way to go in order to find the invaders. Special proteins on the surface of the leucocytes act as receptors to pick up these chemicals and follow the message. A molecule that helps form such a receptor is a co-receptor molecule called **C-C chemokine receptor type 5**, or **CCR5**.

There is a virus that can use these receptors in a different way: as a point of entry to infect leucocytes called CD4 cells. This virus is HIV-1, the first of two types of human immunodeficiency virus. As a result, people who have a working set of *CCR5* genes on chromosome 3 and can make a fully functioning version of this protein for their CD4 cells are, unfortunately, at risk of the virus entering their cells if they are exposed to it. People who are HIV-positive slowly experience their leucocytes being destroyed by the virus, and without treatment they will eventually no longer be able to fight off other infections. At that stage, they have AIDS, acquired immunodeficiency syndrome.

There is a mutation of the *CCR5* receptor gene called the **delta 32 mutation**, or **CCR5-Δ32**. It is a deletion mutation whereby 32 nucleotides have been removed.

Because 32 is not a multiple of 3, this deletion causes a frameshift and a stop codon is accidentally formed where it should not be. Because the codon tells the ribosome to prematurely stop making the protein, people with this mutation cannot produce the functioning chemokine receptor protein that HIV needs to infect their leucocytes and, as a result, they are highly protected from HIV infection. Without a working CCR5-enabled doorway to infect the leucocytes, HIV cannot make a person sick.

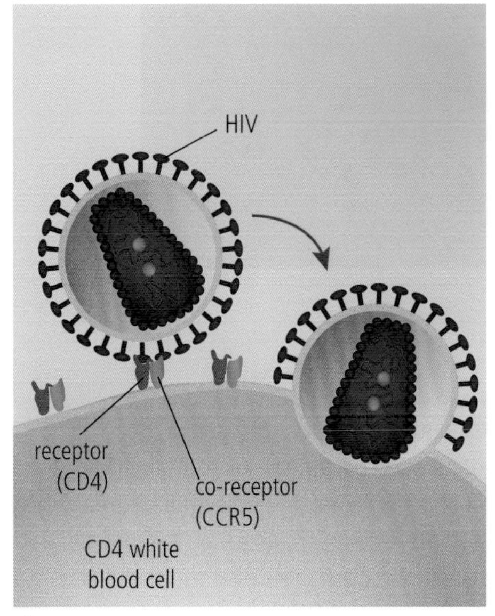

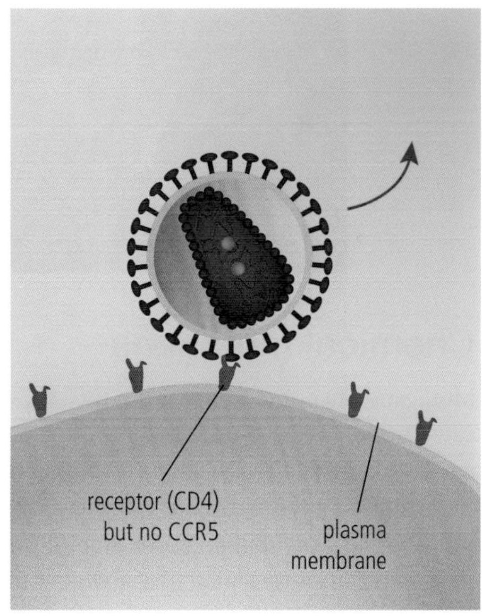

How HIV can enter a cell when the CCR5 protein is present and how it is inhibited from entering the cell when it is absent.

Clearly, this mutation has a beneficial effect. HIV is a major health issue worldwide, and decades of research have not yet found a successful cure or vaccine. To have excellent protection against HIV-1 infections is good news for those who have the mutation. But the mutation can also make people more susceptible to other types of infection, such as the West Nile virus, which is transmitted by mosquitoes and can cause serious illness.

D1.3.4 – Mutagens and replication errors

D1.3.4 – Causes of gene mutation
Students should understand that gene mutation can be caused by mutagens and by errors in DNA replication or repair. Include examples of chemical mutagens and mutagenic forms of radiation.

In principle, DNA is not supposed to be modified during the lifetime of an individual. Normally, the code should be preserved. However, there are exceptions, and exposure to ionizing radiation (from ultraviolet light and radioactive substances) or to **mutagens** (chemicals that can cause a genetic mutation) can sometimes modify the code and cause serious health threats such as cancer. Over exposure to sunlight is linked to melanomas, for example.

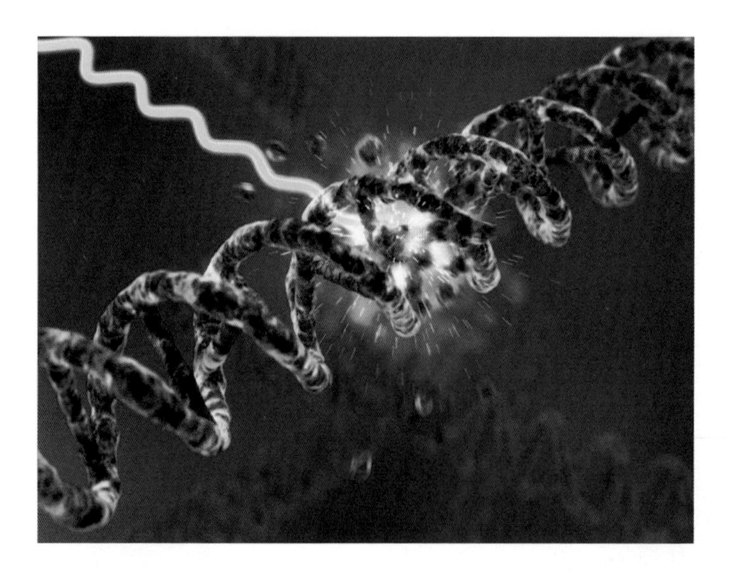

An artist's conception of how DNA can be damaged by radiation.

Chemical mutagens

Mutagens can come from within the cell, such as certain enzymes that attack DNA and can transform the identity of a nitrogenous base. Or they can come from the outside environment, such as benzene (C_6H_6), a chemical used by industries to make other molecules that we use in our everyday lives, such as acetone (for nail polish remover), polystyrene (a rigid foam for packaging) and nylon fibres (for clothing). Benzene is toxic and can cause leukaemia, so industries must handle it carefully and minimize their employees' exposure to it, notably by providing masks that prevent its inhalation.

DNA and ionizing radiation

The world witnessed the terrifying effects of radiation poisoning when the city of Hiroshima in Japan was the target of the first atomic bomb used in warfare, in August 1945. It is estimated that 100,000 people died at its impact or shortly after, but it is difficult to estimate how many in the city died later from the effects of radiation.

When radiation hits a DNA molecule, it can sometimes knock one or more base pairs out of place, modifying the genetic code. This causes a mutation that can sometimes be benign (not harmful) but at other times can be harmful to an organism. When the DNA mutation leads to cancer, as happened to Marie Curie, an organism's health is in jeopardy.

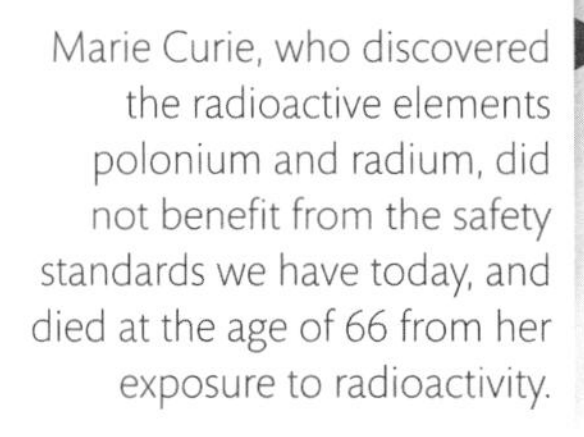

Marie Curie, who discovered the radioactive elements polonium and radium, did not benefit from the safety standards we have today, and died at the age of 66 from her exposure to radioactivity.

As long as nuclear power plants are safe and secure, there should not be any risk of radiation leaking into the environment. But the accidents at Chernobyl in 1986 and Fukushima in 2011 are examples of when people and the environment were exposed to radiation. In both cases, radioactive material leaked out and zones within a radius of tens of kilometres around the power plants were evacuated of all human populations. The disasters would have been much worse if it had not for the heroic efforts of scientists and workers to contain the situation as best they could, many of whom risked their own health and safety.

Scientists assessing the radiation levels in tree trunks near Chernobyl. Notice the dosimeter held by the person standing: it is used to measure radiation levels.

Ecologists are studying the area around Chernobyl to see how ecosystems have responded to the presence of radiation. In some instances, the scientists have been pleasantly surprised to find that nature seems to be doing fine despite the dangerously high radiation levels. In other instances, they have confirmed the presence of mutations in the plants and animals that have colonized the evacuated zone. Cancer studies in the peripheral zones where people are allowed to live, beyond 30 km from the shut-down Chernobyl reactor, suggest that there has been an increase in cancer frequencies. The nuclear power industry has made an effort to isolate the abandoned nuclear power plant at Chernobyl by encasing it in a dome of cement. The hope is that the cement will be thick enough to stop the radiation from continuing to escape into the environment.

Errors in DNA replication or repair

Mutations do not always have to be caused by toxic chemicals in the environment or by radioactive sources. Some mutations are purely random and can occur as easily as a spelling mistake when writing. During DNA replication, letters (nucleotides with nitrogenous bases) are added to the replicated strand and sometimes the wrong letter is added. Most of the time, the errors are corrected during a proofreading process carried out by DNA polymerase. The code is checked and incorrect bases are replaced with the expected ones in a fashion similar to the spellchecker on your phone or computer. Still, even after verification and correction, mistakes can happen. This is rare (in the order of one letter in a million) but does happen.

D1.3.5 – Location of mutations

D1.3.5 – Randomness in mutation
Students should understand that mutations can occur anywhere in the base sequences of a genome, although some bases have a higher probability of mutating than others. They should also understand that no natural mechanism is known for making a deliberate change to a particular base with the purpose of changing a trait.

Just as someone who buys two lottery tickets has twice as much chance of winning as someone who buys only one, some genes or sequences have more than one copy and therefore have a greater chance of being mutated. A gene that has two functioning copies can tolerate mutations in the extra copy without it having an effect on the cell, because the original copy still exists. Also, uncoiled DNA has a higher probability of suffering a mutation than DNA tightly coiled around histones, because the uncoiled DNA is more exposed.

When DNA (in darker blue) is packed closely with histones (in lighter orange) it is better protected from mutations than when it is unravelled and more exposed to mutagenic forces.

Mutations can occur anywhere that sequences of nucleotides can be found. In humans that means they can occur on any of the 22 autosomal (non-sex) chromosomes or on the X or Y sex chromosomes. As only about 1–2% of our genome actually codes for proteins, it is far more likely for mutations to happen in the non-coding zones of the genome. However, such modifications can still sometimes have an effect on the production of proteins because these zones of the genome contain regulatory sequences that can turn coding sequences on and off. **Satellite DNA**, which can be found in the centromere, is an example of non-coding DNA and is used for structural purposes. Surprisingly high mutation rates occur in satellite DNA in humans (on average one mutation per 1,000 base pairs per generation) compared to coding DNA (about one mutation every 500,000,000 base pairs per generation).

Mutations can also be found in mitochondrial DNA and in RNA sequences. In addition to being found in humans, mutations are found in other animals, bacteria, fungi, archaea and plants, and in viruses, where they can be responsible for generating new strains of viruses that are more dangerous or can spread faster.

Another example of **mutation hotspots** (zones where mutations are more frequent) are places where the nucleotide cytosine is followed by guanine. These are called **CpG sites** and, when methylation happens (see Chapter D2.2), the C can mutate into a T. Places where these hotspots repeat are called **CpG islands** and they are more likely to generate mutations. Such mutations are associated with cancers such as colorectal cancer, and therefore researchers are interested in understanding them better.

C → T

cytosine $C_4H_5N_3O$ → thymine $C_5H_6N_2O_2$

Cytosine, when methylated, can sometimes be transformed into thymine when it is deaminated. This is a common mutation.

Can cells invent a mutation on purpose in order to improve?

It would be nice if cells could say, "If I change that C to a T, I could solve this problem I am having", in the same way that a cook changes a recipe or an architect changes a blueprint. No such intentional mechanism for genomic self-improvement has been detected in cells. The DNA repair system can proofread for errors and fix problems to bring the code back to its original sequence, but there is no system to introduce new variations on purpose. Instead, variations only arise as a result of mutations, and then those changes in the genetic sequence are selected for (or against) by natural selection.

One of the most common misconceptions about evolution is that it is somehow aiming for perfection, or that there is a force or desire to improve or to follow a plan. This is a tempting conclusion because we perceive ourselves as being the culmination of evolution. In his book *The Accidental Species*, Henry Gee points out that evolution has no plan: "natural selection cannot be seen as evolution's guiding hand. It has no personality, no memory, no foresight and no end in view". He uses the example of feathers, which did not originally evolve for flight but, because birds use them today for flying, we jump to a conclusion and say how logical and marvellous it was that feathers evolved in order to make flight possible. Gee reminds us of Stephen Jay Gould's thought experiment, imagining undoing millions of years of evolution and then pressing the "play" button to see what would happen if natural selection could start over: we would not see the same creatures evolve on Earth as we have today. How do our expectations and assumptions have an impact on how we perceive things?

D1.3.6 – Mutations in germ cells and somatic cells

D1.3.6 – Consequences of mutation in germ cells and somatic cells
Include inheritance of mutated genes in germ cells and cancer in somatic cells.

Mutations can have very different effects depending on which type of cell they occur in: **germ** cells or **somatic** cells. Somatic cells use mitosis for cell division to grow tissues and organs all over the body, whereas germ cells use meiosis to produce gametes (sperm cells or egg cells).

Germ cells use meiosis to produce egg cells in ovaries and sperm cells in testes. Mutations that happen in germ cells can be passed on to the next generation.

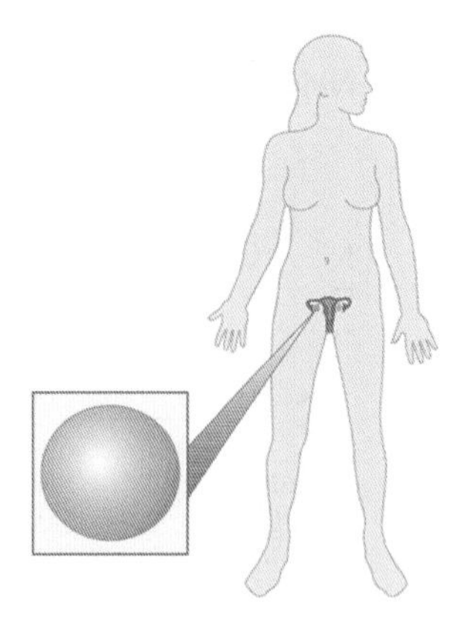

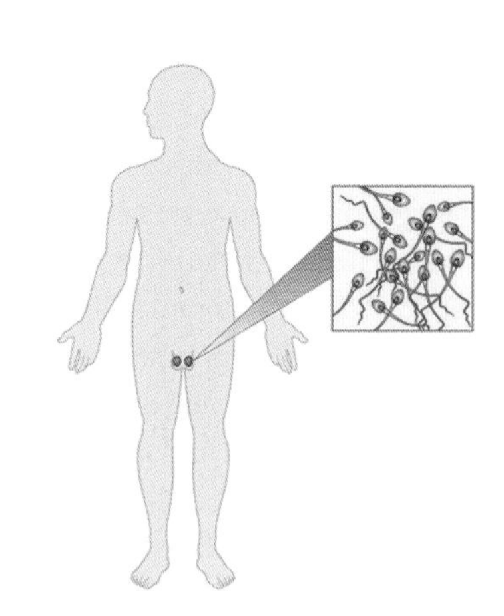

Gametes are sex cells, either sperm cells or egg cells. A zygote is a fertilized egg. Somatic cells are cells that are not gametes: they are the cells that make up the tissues and organs in the rest of the body.

If a mutation happens in germ cells, it can be passed on to a future child. A mutation present in the cells making a male's sperm cells could potentially contribute that mutation to the first cell, the zygote, of a child. As all the child's other cells will be generated from that zygote, they will all contain the mutation. If a female child that has inherited a mutation grows up and has children, the germ cells in her ovaries will contain the mutation and will pass it on to the next generation. Cells involved in passing on genetic information to offspring make up the organism's **germ line**: eggs, sperm cells and zygotes. Each genetic disease that exists in the human population today, such as sickle cell disease, originated as a mutation in the past that was then passed down from generation to generation. Likewise, genetic mutations that were beneficial, such as those that gave us bigger brains or a better-functioning immune system, were also passed down. Remember, mutations are not automatically bad for our health.

If, on the other hand, a mutation happens in a somatic cell, it will not be passed on to the next generation. Mutations in somatic cells are associated with cancer and tumours. These mutations will affect the organ where the mutated cell is found and, as the tumour grows, the mutation will be found in all the cells of the tumour. If the cancer metastasises (spreads out), the cancerous cells can colonize other parts of the body, and each of them will contain the mutation. But the cells that already exist in the parts of the body being invaded by the cancer do not possess the mutation.

If a mutation happens in a somatic cell, for example a skin cell, the mutation will be present in the daughter cells that are produced by mitosis. Such mutations, like the ones that can cause skin cancer, only affect the individual. The red arrow shows ionizing radiation mutating a cell. The daughter cells will also have the mutation.

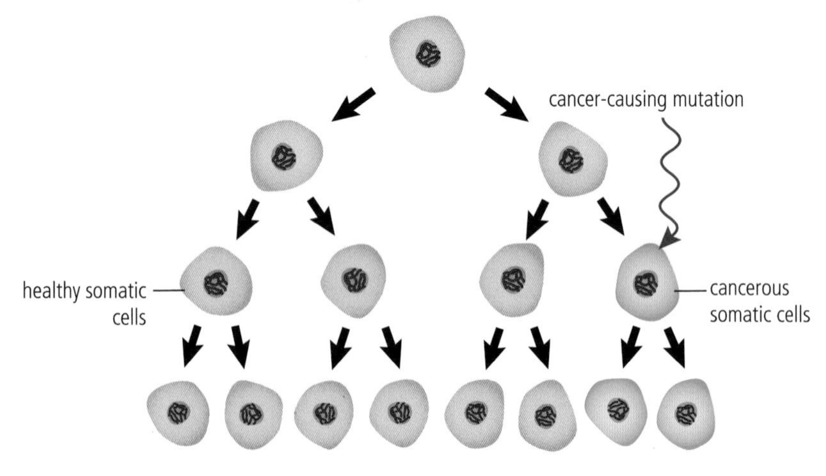

- **Somatic cell mutations only affect the individual; they are not passed on to future generations.**
- **Germ cell mutations affect the individual and their offspring; they can be passed on to future generations.**

D1.3.7 – Genetic variation

D1.3.7 – Mutation as a source of genetic variation
Students should appreciate that gene mutation is the original source of all genetic variation. Although most mutations are either harmful or neutral for an individual organism, in a species they are in the long term essential for evolution by natural selection. **NOS:** Commercial genetic tests can yield information about potential future health and disease risk. One possible impact is that, without expert interpretation, this information could be problematic.

Are mutations good or bad for us?

A mutation that provides an individual or a species with a better chance for survival is considered to be a beneficial mutation, and there is a good chance that it will be passed on to the next generation. In contrast, mutations that cause disease or death are detrimental mutations, and they are less likely to be passed on to future generations, because they decrease the chances of an individual's survival. In addition to beneficial and harmful mutations, there are neutral mutations that do not have an effect on a species' health or chances for survival.

When a mutation is successfully passed on from one generation to the next, it becomes a new **allele**: it is a new version of the original gene. See Chapter D3.2 for more about alleles. This is how new alleles are produced. You and everyone you know possess many mutations. Whether they are harmful, beneficial or neutral depends on what they are and what kind of environment you need to survive in.

It is important for new alleles to be produced in order for evolution to occur. If there were never any mutations, we would only be able to pass on existing genes and nothing new would appear. The problem is that nature and the environment are in constant flux: things are changing all the time and organisms need to adapt. Droughts, temperature changes, alterations in ocean salinity and pH, are all factors that can pose a threat to organisms on land or in the oceans, and having some individuals in a population that possess a slightly different genome can help species survive the changes.

A mutation to help digestion

For most of our existence, humans have been hunter-gatherers and our genes are generally well adapted for this lifestyle. Originally, as for all mammals, the only age at which we drank milk was when we were infants. By the time our ancestors reached adulthood, their bodies had stopped being able to digest milk; more precisely, they could not break down the disaccharide in milk called lactose. This continues to be the case for most people today: more than half of the human population has lactose intolerance, and such people can only digest lactose in their infancy.

▲ In parts of the world where dairy farming had been carried out for thousands of years, genes that enabled adults to digest lactose gave people there a survival advantage.

In the past 10,000 years, however, many human populations have adopted a lifestyle based more on agriculture, raising animals for milk and consuming dairy products on a daily basis. In their genetic makeup, dairy-based agricultural societies show a higher frequency of the genetic code that allows humans to digest lactose throughout adulthood. From an evolutionary point of view, this advantage has increased humans' ability to survive harsh climatic conditions. As these European human populations spread out and established populations elsewhere, notably in North America, they took their lactose tolerance (and their livestock) with them.

Nature of Science

How can you know if you possess a particular gene? In countries where this practice is allowed by law, commercial companies can offer genetic testing kits and provide genome information. Different companies provide different services, such as medical information about the presence of genes that could lead to diseases such as Huntington's or might increase a person's risk of cancer or cardiovascular disease. In association with a doctor who can help interpret what the information means, answer questions and provide advice and medical treatments, such information can be useful to have. But if the information is received via email or text message without a proper context and interpretation, it can be emotionally devastating; the person receiving the information might not understand how to interpret the results.

In addition to medical information, many of these companies provide genealogical data about a person's ancestry. Again, sometimes the news is unexpected and people might learn that their parents or other ancestors were not who they thought they were. This raises many ethical questions, which is one of the reasons why some countries, such as France, have passed laws against using such DNA testing services for paternity and ancestry.

How can natural selection lead to both a reduction in variation and an increase in biological diversity?

HL

D1.3.8 – Gene knockout

D1.3.8 – Gene knockout as a technique for investigating the function of a gene by changing it to make it inoperative

Students are not required to know details of techniques. Students should appreciate that a library of knockout organisms is available for some species used as models in research.

When the Human Genome Project was declared a success in 2003, and we had a map of all the nucleotide sequences in our species, the question became "OK, now what does each sequence mean: what do all these genes do?" We want to know what each sequence codes for, so that we can find those that are responsible for disease or that can provide clues about our past.

One way to find out what role something plays is to remove it and see what happens. To find out what a gene does, we can remove it from an organism or silence it, and observe what effect that has. Or we can find or intentionally breed two organisms, one of which naturally possesses the gene and the other of which does not. By seeing how each organism reacts differently in experiments, we can deduce the role of the gene in question.

Where does the term gene knockout come from? In contact sports such as boxing, a knockout marks the end of the combat, because the boxer who has been knocked out is no longer able to stand and fight. A gene that has been knocked out will no longer be able to make the protein that produced the original effect or trait.

To better understand why we use gene knockout techniques, think about the following method for finding out how a bicycle works. If you remove the chain from a bicycle, the rider can no longer move ahead when pedalling. With this simple experiment, we now know that the role of the chain is to transfer energy from the pedal to the back wheel to move the bike forward. Similarly, if we remove a gene from an organism and see what happens, we can determine the function of the gene.

One such procedure, called **gene knockout**, involves rendering a gene unusable in order to see what effects it has on an organism. The organism whose gene has been knocked out, such as a laboratory mouse, is called a **knockout organism**.

One important use of knockout mice in laboratories is the testing of pharmaceutical drugs. In order to see whether drugs are safe for humans, they are first tested on **model organisms** using species such as mice. A model in biology is an organism that is used in the place of another for practical and/or ethical reasons. Mice have similar metabolic systems to humans, they share many of the same genes, and they are easy to maintain in a laboratory. In addition, they reproduce quickly so that many generations can be tested in a short time. Testing new medicines directly on humans is costly, time consuming, potentially dangerous and unethical, so initial tests are usually carried out using model organisms, before being carried out on human subjects.

In an effort to make the population of laboratory mouse models as similar to the human population as possible, it is important to have mice that show similar genetic variations to humans. Maintaining a library of model organisms with various known knockouts for specific genes allows researchers to carry out tests on animals with a variety of genetic traits. Conditions that have been studied using knockout mice models include:

- obesity
- diabetes
- anxiety (one mouse model for this was called Frantic)
- longevity (one mouse model was named Methuselah for its ability to live longer lives)
- propensity to developing certain types of cancer
- substance abuse/addiction
- cardiovascular disease.

▲ Laboratory mice are used as models for testing medical treatments that could one day be used on humans. Libraries of knockout mice exist for testing specific characteristics such as obesity, as seen in the mouse on the right.

To make a knockout mouse, embryonic stem cells from a mouse are genetically modified *in vitro* to intentionally damage and inactivate the gene under study. The stem cells are introduced into a mouse embryo, and the embryo is placed inside a female's womb for her to carry to term. The resulting mouse may need to be bred with other knockout mice until the desired effects are seen in the offspring.

Today, researchers can choose from these libraries of organisms in order to see whether their drugs can help calm anxious mice, or can prevent heart problems in models that are prone to cardiovascular disease. The system is not perfect, of course, because mice are only models for humans, so eventually clinical trials do need to be carried out using people. But the reasoning is that, if the drug being tested works well in a mouse model and does not cause any concerning side effects, it should be safe to try on humans.

D1.3.9 – CRISPR-Cas9 gene editing

D1.3.9 – Use of the CRISPR sequences and the enzyme Cas9 in gene editing
Students are not required to know the role of the CRISPR–Cas system in prokaryotes. However, students should be familiar with an example of the successful use of this technology. **NOS:** Certain potential uses of CRISPR raise ethical issues that must be addressed before implementation. Students should understand that scientists across the world are subject to different regulatory systems. For this reason, there is an international effort to harmonize regulation of the application of genome editing technologies such as CRISPR.

Every once in a while, a discovery is made in biology that is truly monumental and impactful, a discovery for which there is a "before" and an "after". The invention of the microscope, Mendel's experiments with genetics, the publication of Darwin's *Origin of Species*, the discovery of DNA's double helix and its code, are examples. The development of the gene-editing technology called **CRISPR-Cas9** in 2012 can be added to this list. It not only earned its developers, Jennifer Doudna and Emmanuelle Charpentier, the 2020 Nobel Prize in Chemistry, but regularly makes the news worldwide in the context of breakthrough cures or controversial applications in which not enough attention was paid to ethical concerns.

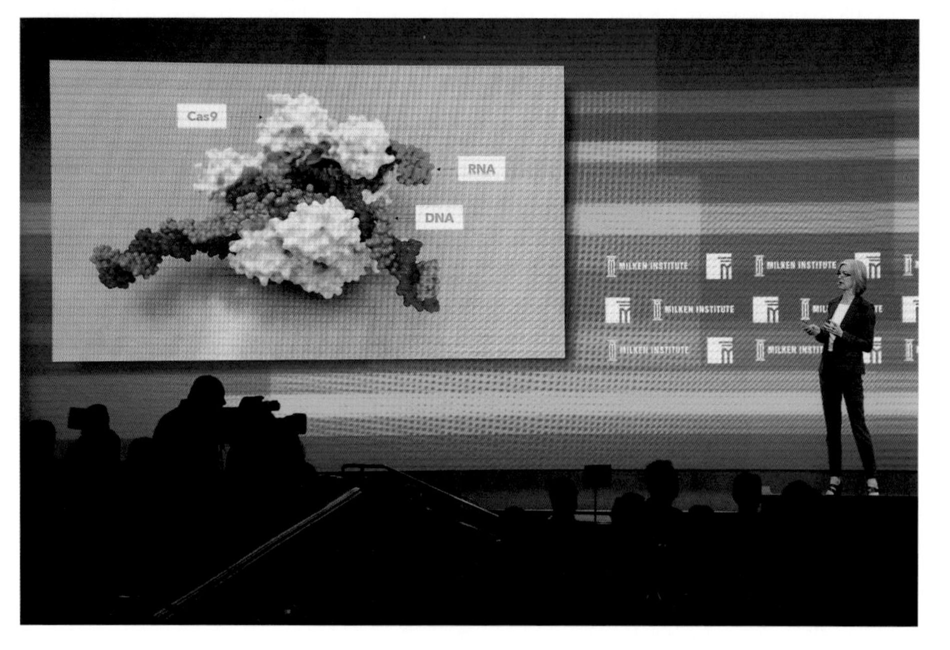

Jennifer Doudna received the Nobel Prize for her work with Emmanuelle Charpentier on CRISPR-Cas9 gene editing.

How CRISPR-Cas9 works

The genius of Doudna's team is that they saw the potential of combining two biochemical concepts to create a toolbox for gene editing. One part of their toolbox uses a specific **Cas protein** (CRISPR-associated protein), the **Cas9 enzyme**, which works like scissors to cleave (cut) DNA. The second part involves building an RNA molecule that acts as a guide to show the enzyme where to cut. This **guide RNA**, or **gRNA**, can be generated in a laboratory to match any desired target sequence. The result is an RNA-guided gene regulation system using a programmable protein (Cas9) to remove or replace genetic sequences.

In short, what your phone or computer can do with text (copy, cut, paste), CRISPR-Cas9 can do with any gene. Because the genetic code is universal, this toolbox for editing genes works in any organism, from bacteria and fungi to plants and animals, including humans. By now, you can hopefully see why this scientific advancement is so monumental.

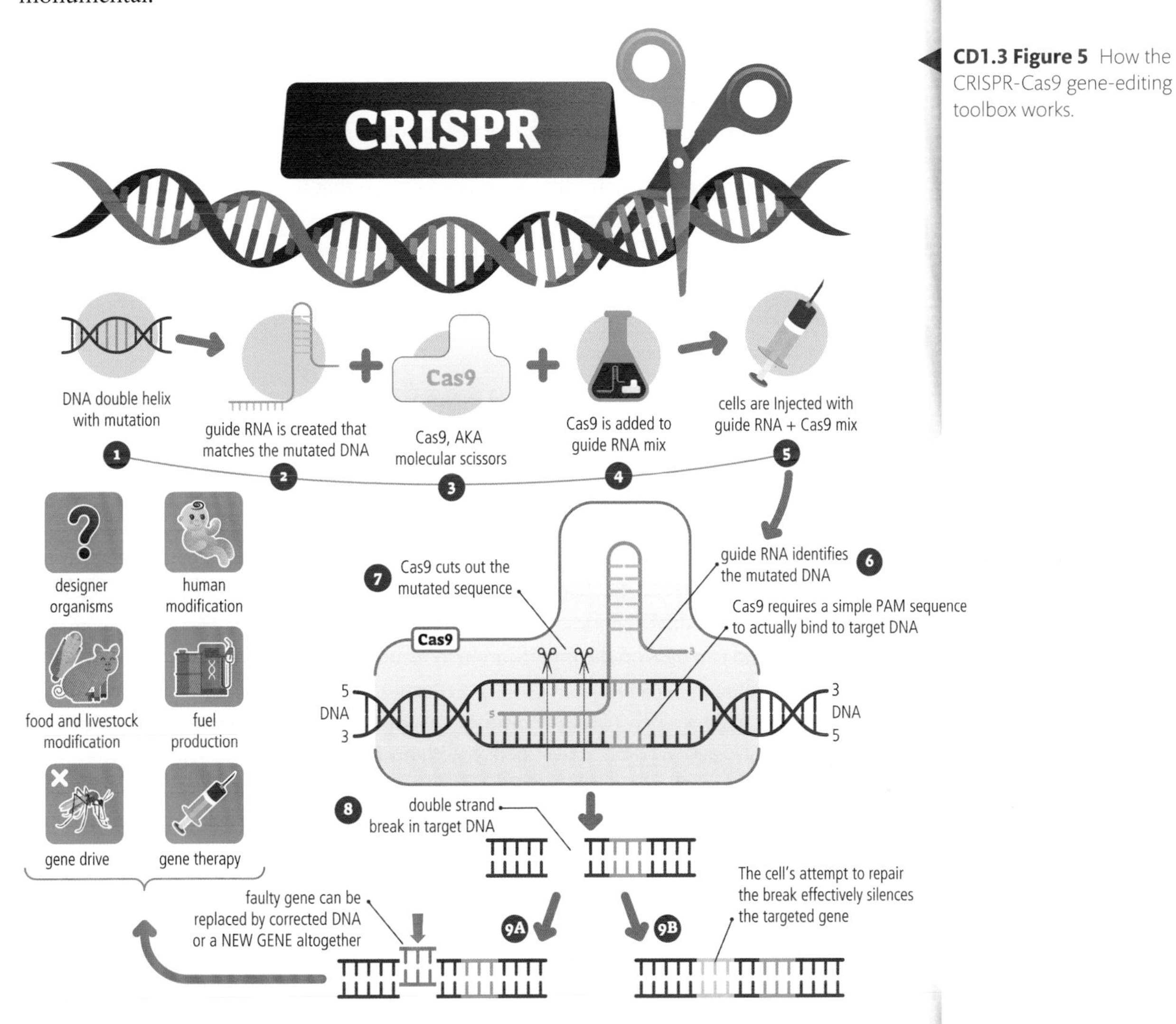

CD1.3 Figure 5 How the CRISPR-Cas9 gene-editing toolbox works.

Figure 5 shows how the CRISPR-Cas9 gene-editing tool functions. The three main components are (1) the DNA sequence to be edited, (2) gRNA and (3) the Cas9 protein. In Figure 5, the DNA strand being modified is one that has an undesired mutation, such as one that can cause a genetic disease. At step (4), the gRNA that has been programmed to recognize the mutated DNA is attached to the Cas9 protein, and at step (5) this is injected into cells, something that can be done with the help of a virus vector. At step (6) the gRNA finds the place on the mutated DNA where it needs to attach. This step requires a **PAM sequence**, a **protospacer adjacent motif**, which is a few nucleotides long (e.g. AGG) and helps the Cas9 protein attach to the DNA. At step (7) , Cas9 cleaves the DNA double strands at the place where the gRNA has been programmed to indicate the cleavage point, seen at step (8). At step (9), either the mutated gene can be removed or replaced, or the attempts by the cell to repair the cut can lead to errors such as frameshifts that render the gene inactive or silence it.

Possible benefits

Because CRISPR-Cas9 can be used to delete a harmful gene, it is logical to apply it to genetic disorders caused by a single gene, such as Huntington's disease or sickle cell disease. The technology is also capable of deleting an entire chromosome, introducing the possibility of treating an embryo with Down syndrome, by eliminating the third 21st chromosome.

Agriculture is an area where we are constantly looking for ways to make food more nourishing or livestock more productive or resistant to disease. One CRISPR-Cas9 project, for example, produced tomato plants that yielded twice as much as non-modified plants.

In terms of human health, modifying mosquitoes so that they can no longer transmit the dengue fever virus, the malaria parasite or West Nile virus is of great interest to public health. This would involve the use of a **gene drive**, a mechanism that increases the chances of a gene being passed on to the next generation. Modifications are made so that a laboratory-modified gene can be spread quickly in a population until all or almost all offspring possess it.

Another concept that has been suggested since the early days of CRISPR-Cas9 is modifying plants or algae to make renewable fuels to replace petroleum-based fossil fuels.

Challenges and risks

As we saw in Section D1.3.6, there is a difference between genetic modifications that happen in somatic cells and those that happen in germ cells. If CRISPR-Cas9 is used in somatic cells, only the individual's genome is modified. But if it is used in germ cells, the individual is modified and can pass on the modification to future generations. Another concern with any gene modification technique is off-target effects. These are when the editing technology accidentally changes a part of the genome that was not intended to be modified. Whatever happens in the years to come, researchers, health professionals and governments will need to engage in dialogue to discuss this exciting, powerful but potentially misused gene-editing tool. Lastly, just because a technique is possible in a laboratory does not mean it is immediately available for anyone who needs it. There are rules about testing and safety that need to be followed and these can take many years to complete.

Advances in genomics and gene editing are raising many questions concerning bioethics.

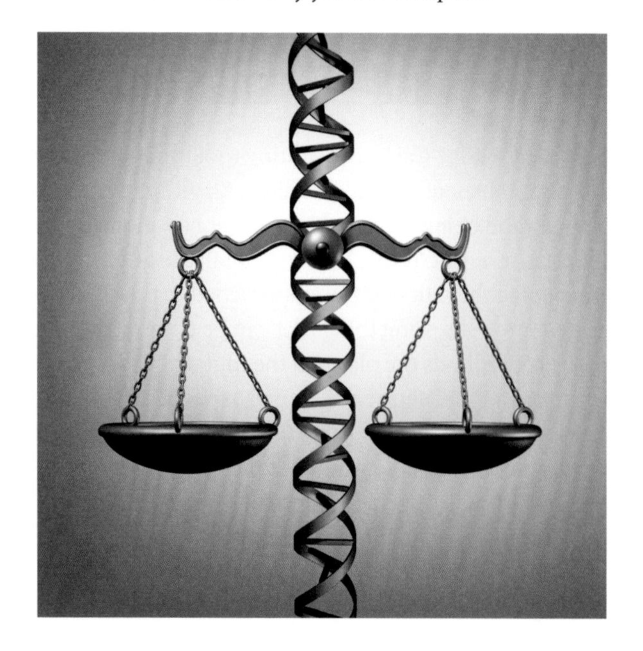

Nature of Science

Experiments involving fundamental aspects of humanity such as human fertility and our genome need to be decided on a humanity-wide scale. Experiments must include informed consent by the participants and must be approved by the ethics committees present at the university or research facility in question. Such committees look to experts worldwide for guidance. Experiments must also follow the laws of the countries where they are taking place. The problem is that the speed at which science is advancing in this domain is much faster than the legislative process that generates new laws.

Fortunately, experts have formed international committees to discuss these issues and make recommendations for clinical researchers and government policymakers worldwide, and to help inform the general public of the benefits and risks. The International Commission on the Clinical Use of Human Germline Genome Editing is one such organization. It is the role of agencies such as the World Health Organization to create guidance for best practice and put guidelines in place. But there is also room for private, non-profit organizations, such as Doudna's Innovative Genomics Institute, to host conferences and debates where experts from diverse countries can share their ideas and work through some of the challenges they face, including concepts related to social responsibility.

Fundamental research versus applied research: one of the fascinating aspects of Doudna's inspiring story of the discovery and development of CRISPR-Cas9 was the fact that she and her team were initially driven by curiosity and not necessarily looking to cure diseases. Many of the applications that CRISPR-Cas9 is used for today and will be in the future are things she and her team never imagined. In order to finance their research, scientists need to write grant proposals asking for funding. Often they are asked to justify why their research is important and what practical (and lucrative) applications will be possible based on their results. Researchers who say "I don't know what will come of this: I am just curious to know how it works" are unlikely to get funding from investors who are interested in making a profit in return for financially supporting the research. What does this say about what types of knowledge we value in society? Is knowledge more valuable if it has a practical application, or is all knowledge inherently equally worth pursuing and valuable?

D1.3.10 – Conserved and highly conserved sequences

D1.3.10 – Hypotheses to account for conserved or highly conserved sequences in genes
Conserved sequences are identical or similar across a species or a group of species; highly conserved sequences are identical or similar over long periods of evolution. One hypothesis for the mechanism is the functional requirements for the gene products and another hypothesis is slower rates of mutation.

Different genes evolve and mutate faster than others and, in some cases, a gene does not change at all or very little over long periods of time. Genetic sequences found in DNA or RNA that show minimal mutations over time are called **conserved sequences**, while those that show no or almost no changes are **highly conserved sequences**. These were first noticed when researchers started sequencing genes and comparing them between species and between individuals.

Examples of sequences that show a remarkably low rate of mutation and that are well conserved include those that are needed for all cells in all species, such as the sequences that make DNA replication, transcription and translation possible. These include the sequences for helicase, for non-coding RNA such as tRNA, and making ribosomes. Other examples include sequences that are necessary for cellular respiration, such as those for cytochrome c and ferredoxin, proteins that participate in key steps along this biochemical pathway.

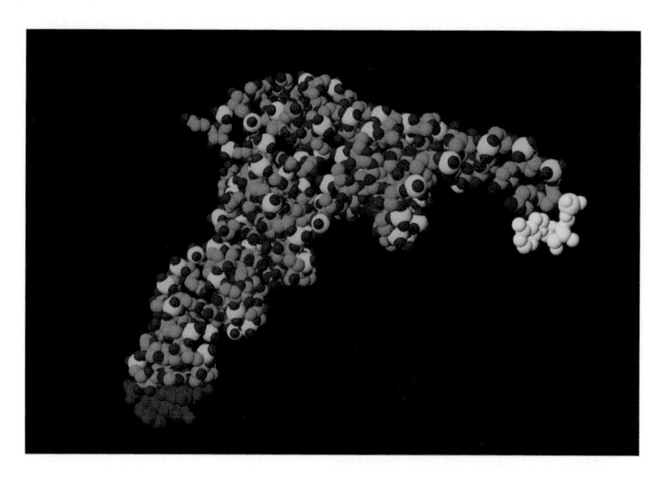

The genetic code for building tRNA is an example of a highly conserved sequence because all cells need tRNA to survive. The lower left (red) part of the model shows the anticodon, and the upper right (yellow) part shows the area involved in delivering the amino acid serine during protein synthesis.

To visualize the sequences, it is necessary to combine the science of **genomics** (gene sequencing) with **bioinformatics** (using computer programs to analyse the enormous data sets). Fortunately, there are online tools for researchers and students, many of which are free to access. One tool that can be used to compare sequences that have been conserved is called the ECR Browser (see Figure 6).

D1.3 Figure 6 Online tools such as the ECR Browser calculate and graph the degree to which sequences have been conserved, by comparing the human genome to several other species of vertebrates.

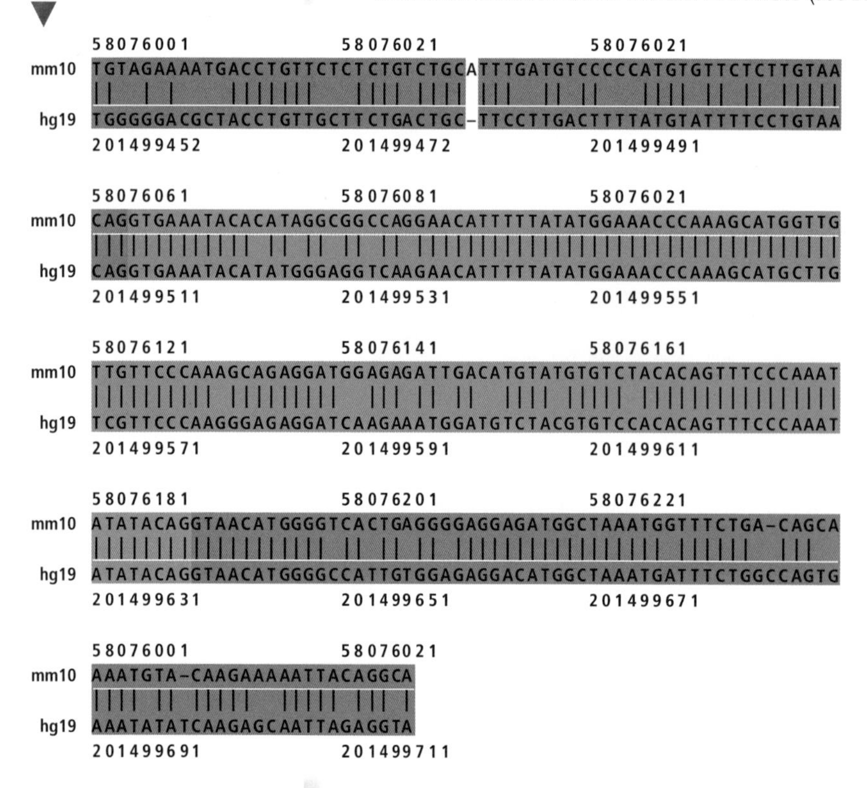

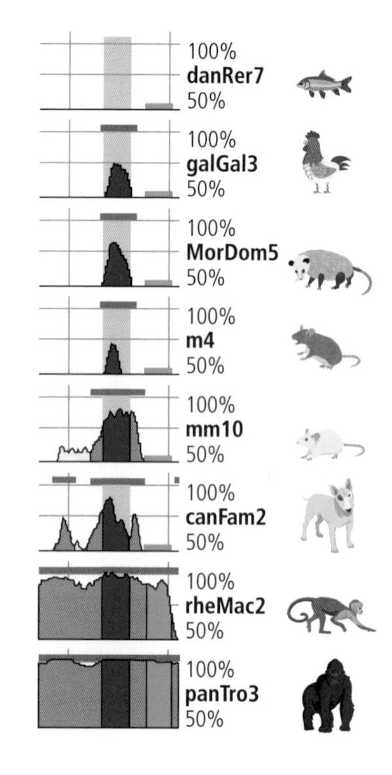

On the left of Figure 6, a mouse's sequence (mm10) is compared to a human's sequence (hg19) for a part of a gene called *AOX1*, which is used to make the protein ferredoxin, necessary for cellular respiration. On the right of Figure 6, the software has calculated and graphed the percentages of nucleotides in the sequence that match humans. The middle blue section represents a sequence that is found in most of the animals selected. The fish at the top, compared to humans, does not demonstrate any significant level of conservation at this part of the sequence (under 50%), whereas the chimpanzee, the species on Earth that is most closely related to us, shows a near 100% conserved sequence compared to humans.

There are two main hypotheses explaining why mutations and evolutionary modifications seem to affect some sequences less than others, despite the fact that some genetic sequences in a species can display multiple mutations. One hypothesis is that the functional requirements of the gene are such that an organism cannot live without it. The other hypothesis is simply that some sequences are subject to slower mutation rates than others. We will examine each hypothesis in turn.

Functional requirements

If we look at examples of highly conserved sequences, such as those for tRNA and for ribosomes, we realize that these substances are vital to a cell's survival. If a cell had a mutation that prevented it from producing the tRNA that could deliver the correct amino acid during translation, or prevented it from adding amino acids to a polypeptide chain, the cell would be incapable of protein synthesis and would not survive. Hence, the functional requirements of the cell maintain the sequence and do not allow mutated sequences to be passed on to the next generation of cells. In short, natural selection conserves these sequences by necessity and does not let any mutations pass down to the next generation. See Chapter D4.1 for more about how natural selection works. The phenomenon of eliminating harmful variations of genes is a type of natural selection called **purifying selection** or **negative selection**.

Slower mutation rates

The second hypothesis is that some zones of a sequence are less prone to mutations and the mutation rate is slower than in other sequences. **Mutation rate** refers to how many changes there are in a genetic sequence over time. It can be expressed as the number of base pairs changing in a single gene at each generation (or each cell division) or as the number of base pairs changing in the whole genome per generation.

Within a species' genome, it is more common to find higher mutation rates in non-coding DNA such as satellite DNA because it does not undergo purifying selection as described above. In addition, the DNA repair system that looks for errors and corrects them is more active in key zones of the genome where errors would have a more severe consequence for the organism. The proofreading function and correction and repair system carried out by DNA polymerase is less active in places where mutations would not lead to life-threatening issues, notably in non-coding DNA and in genes that have been silenced. Zones that have a lower rate of mutation are not necessarily undergoing fewer mutations, it is just that errors are corrected better there than they are elsewhere, and therefore they do not show up when the genes are sequenced.

Although mutation rates are difficult to measure, estimates range from very low, for example one mutation in every 10^{-11} base pairs (one in 100 billion) per generation, to 10^{-3} (one in 1,000) per generation. The rate can become higher if a mutagenic chemical is present or if radiation such as ultraviolet light influences the rate of mutations. This is one of the reasons why researchers are interested in studying mutation rates. If we see the rate of mutation suddenly increase in cells, it could be an indication that something harmful has been introduced to the environment.

HL end

Guiding Question revisited

How do gene mutations occur?

In this chapter you have learned that:

- gene mutations are random and can involve the addition of one or more nucleotides, or can involve deletions or substitutions
- mutations can happen during DNA replication but many are corrected in a proofreading process carried out by DNA polymerase
- mutations can also happen as a result of exposure to chemical mutagens or ionizing radiation.

Guiding Question revisited

What are the consequences of gene mutation?

In this chapter you have learned how:

- some mutations affect the germ line and can be passed down to future generations, whereas others only affect the somatic cells in one individual
- some mutations have a negative effect, some positive, and some have no effect at all on an organism
- natural selection will favour newly formed alleles that help an organism's survival, or reduce the frequency of those that are harmful.

Exercises

Q1. Which terms can be used to describe a mutation that leads to cancer?

I. Somatic.

II. Germ line.

III. Non-heritable.

A I and II only.

B I and III only.

C II and III only.

D I, II and III.

Q2. Which sentences can be used to describe the *CCR5* delta 32 mutation?

I. It is a single base substitution.

II. It affects the immune system.

III. It can protect someone from HIV.

A I and II only.

B I and III only.

C II and III only.

D I, II and III.

Q3. Mutations can happen in two types of cells, somatic cells and germ cells.

(a) Distinguish between the two types of cells in terms of the type of cell division they are capable of and where they are found in the human body.

(b) Distinguish between the consequences of a mutation in each type of cell.

Q4. Give an example of a chemical mutagen and one mutagenic form of radiation.

Q5. HL Describe what a knockout mouse model is and explain its importance to research, giving two examples.

D1 Practice questions

1. Which of the following is an inherited disease as a result of a base substitution mutation in a gene?

   A Trisomy 21.

   B Sickle cell anaemia.

   C AIDS.

   D Type 2 diabetes. *(Total 1 mark)*

2. During the process of replication, which bond(s) in the diagram of DNA below is/are broken?

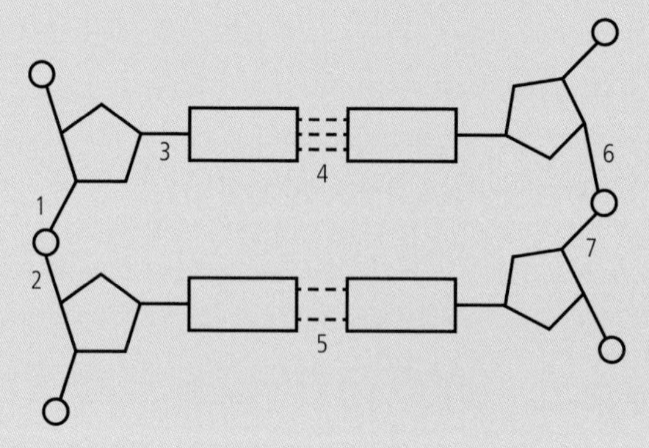

   A 3

   B 4, 5

   C 1, 2, 6, 7

   D 1, 7, 4, 5 *(Total 1 mark)*

3. The diagram below shows a short section of DNA molecule before and after replication. If the nucleotides used to replicate the DNA were radioactive, which strands in the replicated molecules would be radioactive?

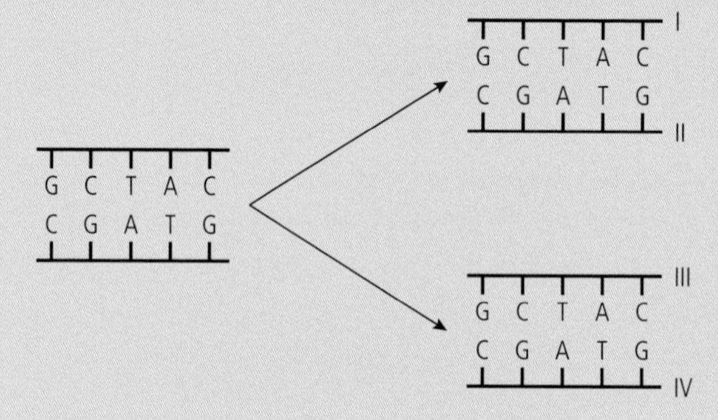

   A II and III only.

   B I and III only.

   C I and II only.

   D I, II, III and IV. *(Total 1 mark)*

4. Outline the process of translation.

(Total 5 marks)

5. A certain gene codes for a polypeptide that is 120 amino acids long. Approximately how many nucleotides long is the mRNA that codes for this polypeptide likely to be?

A 30

B 40

C 360

D 480

(Total 1 mark)

6. Compare DNA transcription with translation.

(Total 4 marks)

7. Describe the genetic code.

(Total 6 marks)

8. Outline how single-nucleotide polymorphisms generate new alleles.

(Total 3 marks)

9. Explain the significance of insertions and deletions happening in multiples of 3.

(Total 4 marks)

HL

10. Which are the correct directions for the following processes?

	Replication	Transcription	Translation
A	5′ to 3′	3′ to 5′	3′ to 5′
B	3′ to 5′	5′ to 3′	5′ to 3′
C	5′ to 3′	5′ to 3′	5′ to 3′
D	3′ to 5′	3′ to 5′	3′ to 5′

(Total 1 mark)

11. Outline how the CRISPR-Cas9 editing tool can be used to silence a gene responsible for a disease in humans.

(Total 5 marks)

HL end

THEME
D Continuity and change
2 Cells

◀ Cells need to be able to divide to ensure the growth and repair of tissue but also to prepare for the next generation. These cells are in various stages of cell division; their chromosomes have been dyed red.

One of the defining characteristics of life is the ability to grow. This requires making copies of cells. Cell division involves a series of steps to make sure that the DNA code is copied into each new cell. In the production of sperm and egg cells, only half of the parent's DNA is passed on, so a special version of cell division is necessary.

HL Just because an organism has DNA coding for a particular protein does not necessarily mean the gene will be expressed. Genes can be activated or silenced, and sometimes an organism's environment can trigger this.

Cell division is just one of the many processes that cells carry out. In order for a cell to maintain all the essential processes of life, it needs to maintain the correct balance of water and solutes. The movement of water from an area of low solute concentration to an area of higher solute concentration through a semipermeable membrane is called osmosis. Experiments can be done in the laboratory that involve placing plant tissues in different solutions to see how water flows in or out of the cells. Understanding how water interacts with cells can be applied to medicine, such as knowing what concentration of solution to bathe transplant organs in.

D2.1 Cell and nuclear division

Guiding Questions

How can large numbers of genetically identical cells be produced?

How do eukaryotes produce genetically varied cells that can develop into gametes?

Cells can divide, and make copies of themselves, in two ways. Mitosis produces two genetically identical cells; this type of cell division is used for growth and repair. Most cells in the body replicate using mitosis. In a developing embryo, after one division there are two cells, but after six divisions there are 64 cells, and after 20 divisions there are over a million cells.

The other type of cell division is called meiosis; this type of cell division is only used for the production of one category of cells, the sex cells. When egg cells or sperm cells are generated, one cell divides twice to make four cells, but each of the four has two exceptional characteristics that make them different from other cells in the body and different from each other. Firstly, each sex cell only contains half the genetic information from the parent cell, and will not have a complete set of DNA until it fuses with another sex cell. Secondly, each of the four cells has a different combination of hereditary information from the parent cell. This increases the genetic variety in the offspring.

D2.1.1 – Generating new cells

D2.1.1 – Generation of new cells in living organisms by cell division
In all living organisms, a parent cell – often referred to as a mother cell – divides to produce two daughter cells.

In order to maintain the population of a single-celled organism, and in order to keep a multicellular organism growing and repairing itself, cells need to make copies of themselves. The process of producing two cells from one is called **cell division**. The role of cell division is to make sure genetic information is passed on to the next generation of cells along with copies of all the organelles necessary to make the cell function.

The cell that produces a copy is called the **parent cell** or **mother cell** and the two new cells that are generated are called the **daughter cells**. This is one of the principles of cell theory: all cells are made from pre-existing cells.

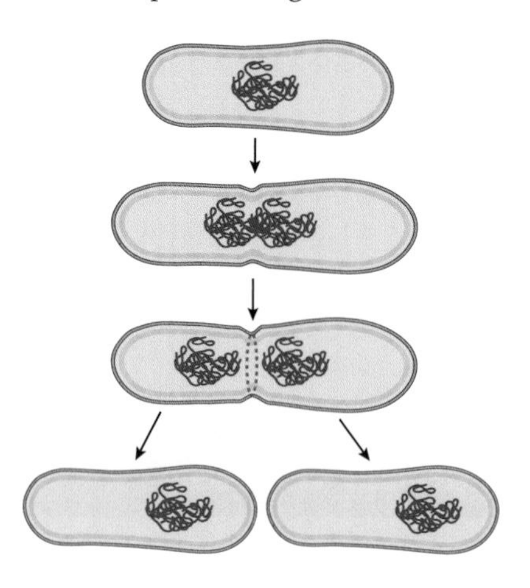

▲ The process of binary fission in prokaryotes. The single DNA molecule is copied before the two cells split apart.

Mitosis produces genetically identical daughter cells, which are necessary for organisms to grow and to repair themselves. Meiosis produces sex cells, which are needed for fertilization and have half the amount of DNA of any other cell in the body.

With only a single chromosome of DNA to replicate, prokaryotic cells divide by **binary fission**. During this process, the DNA is copied, the two daughter chromosomes become attached to different regions on the plasma membrane, and the cell divides into two genetically identical daughter cells. For organisms that have multiple chromosomes, the daughter cells are produced using **mitosis**. For sexually reproducing organisms, egg cells and sperm cells require two divisions that will produce four daughter cells in a process called **meiosis**.

D2.1.2 – Cytokinesis

D2.1.2 – Cytokinesis as splitting of cytoplasm in a parent cell between daughter cells
Students should appreciate that in an animal cell a ring of contractile actin and myosin proteins pinches a cell membrane together to split the cytoplasm, whereas in a plant cell vesicles assemble sections of membrane and cell wall to achieve splitting.

After a cell reaches a certain size, it needs to split in two, a process called **cytokinesis**. The process differs depending on the type of cell. In animal cells, cytokinesis involves an inwards pinching of the fluid plasma membrane to form **cleavage furrows**, a groove along the cell membrane. However, plant cells have a relatively rigid cell wall and during cytokinesis they form a **cell plate** instead. The cell plate is built up by vesicles that collect midway between the two poles of the cell and lay down cell membrane and cell wall cells, which then expand outwards towards the sides of the cell from a central region. Both types of cytokinesis result in two separate daughter cells that have genetically identical nuclei.

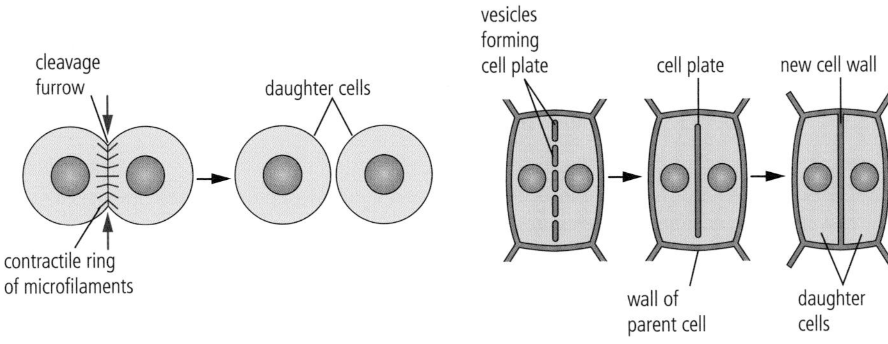

Cleavage of an animal cell

Cell plate formation in a plant cell

Cytokinesis in animal and plant cells.

D2.1.3 – Cytoplasm division

D2.1.3 – Equal and unequal cytokinesis

Include the idea that division of cytoplasm is usually, but not in all cases, equal and that both daughter cells must receive at least one mitochondrion and any other organelle that can only be made by dividing a pre-existing structure. Include oogenesis in humans and budding in yeast as examples of unequal cytokinesis.

In most instances of cell division, the two daughter cells are identical. This is necessary for the process of growth and repair. Each cell receives a full copy of the parent cell's DNA and some of the essential organelles. If all the mitochondria from the parent cell ended up in only one of the two daughter cells, the other would not be able to survive. So at least one mitochondrion from the eukaryotic parent cell needs to end up in each daughter cell to ensure its survival. The same is true of chloroplasts in photosynthetic eukaryotic cells.

But sometimes the daughter cells are not identical, and there is unequal sharing of the parent cell's resources. One example of this is found in the production of eggs, called **oogenesis**. Oogenesis produces four haploid cells. Three of the four cells donate their cytoplasm and organelles to the fourth cell and are not used as eggs because they are much too small to produce a viable **zygote** (fertilized egg). This **unequal cytokinesis** provides the zygote with the resources it needs to survive until it is implanted in the wall of the uterus.

Cytokinesis can also be unequal in yeast cells. Yeast cells divide using a process called budding, which involves generating a small cell from the parent cell. When the daughter cell becomes big enough to survive on its own, cytokinesis closes the cell membranes and each cell is an independent organism.

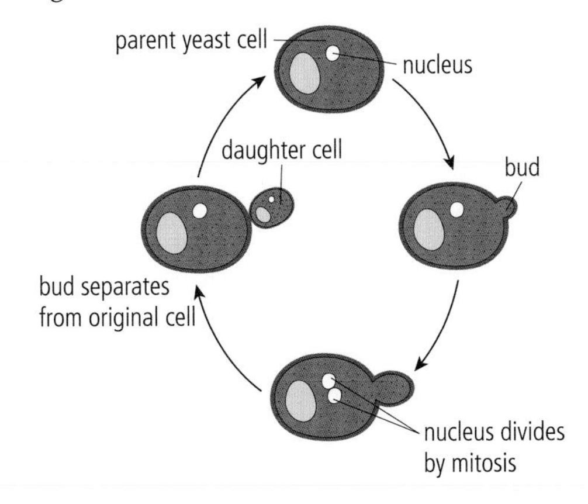

The process of budding in yeast cells is an example of unequal cytokinesis. The daughter cell produced is smaller than the parent cell because of an uneven distribution of resources from the parent cell.

Why do we call them "chromosomes"? The name was coined in 1888 by German biologist Heinrich Wilhelm Gottfried von Waldeyer-Hartz. In Greek "chroma" means colour, and "soma" means body. Because the condensed packages of DNA and proteins require special dyes to make them visible under the microscope, von Waldeyer-Hartz named them chromosomes, meaning "colourful bodies". At the time, there was very little understanding of just how important these little colourful fibres are.

Before cells divide by either mitosis or meiosis, nuclear division must take place: the DNA must be replicated. After replication, each chromosome is made of two strands of DNA held together with a centromere.

D2.1.4 – Nuclear division

D2.1.4 – Roles of mitosis and meiosis in eukaryotes
Emphasize that nuclear division is needed before cell division to avoid production of anucleate cells. Mitosis maintains the chromosome number and genome of cells, whereas meiosis halves the chromosome number and generates genetic diversity.

Mitosis is a type of cell division that results in two daughter cells, each with identical nuclei. Imagine what would happen if a cell divides before it makes a copy of its nucleus. Only one of the daughter cells would have a nucleus. The other cell would be **anucleate**: it would not have the instructions for carrying out its functions or how to divide. The process of mitosis makes sure a full copy of the nucleus is made *before* cytokinesis happens. As a result, each daughter cell not only has the same number of chromosomes as the parent cell, but the same genome. All the genetic information is preserved.

In **gametes**, or sex cells, it is necessary for each sperm and egg cell to have only half the genetic information of the parent. Some organisms, notably certain plants, can thrive with extra copies of chromosomes, but usually animals cannot.

In order to receive 46 chromosomes in total, a human baby needs to receive 23 chromosomes from their mother and 23 from their father. This is why chromosomes come in pairs; one of each pair is from each parent. To ensure that egg and sperm cells only get half the genetic information, a special type of cell division is needed: **meiosis.** Meiosis is a type of cell division that results in four daughter cells that each has a nucleus containing only half the parent cell's DNA. And each of the four receives a different *combination* of genetic information. It is extremely rare, for example, to find two sperm cells from the same male that have the same genetic profile. By putting random combinations of chromosomes into each sperm or egg cell, meiosis helps generate the genetic variety we observe in offspring.

D2.1.5 – DNA replication

D2.1.5 – DNA replication as a prerequisite for both mitosis and meiosis
Students should understand that, after replication, each chromosome consists of two elongated DNA molecules (chromatids) held together by a centromere.

D2.1 Figure 1 Chromosomes make a copy of themselves before cell division happens. In real life, before it is copied the single chromosome does not look as depicted here: it would not be coiled up, but would be unwound so that it can undergo replication.

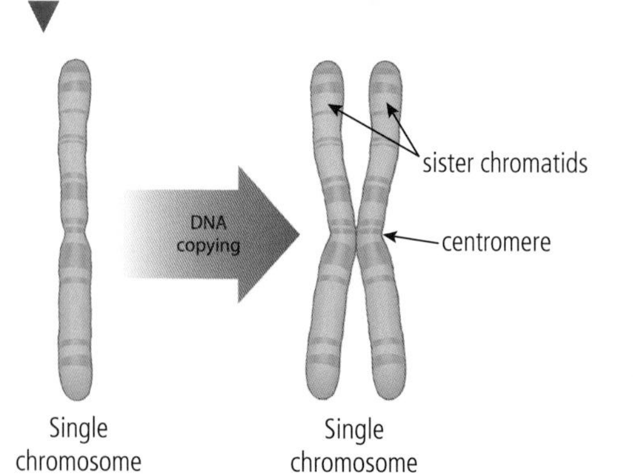

Whether it is in preparation for meiosis or mitosis, cell division cannot happen until a complete copy of the cell's DNA has been made. This process is called **DNA replication**, (see Chapter D1.1). It happens during a phase of the cell's life called the **S phase**, or synthesis phase. The replicated DNA is arranged in a chromosome as two sister **chromatids**, shown as the vertical elongated arms in Figure 1. The chromatid on the left side of the copied chromosome is *one* DNA molecule coiled up. The chromatid on the right side is another, identical, DNA molecule. This can be seen in the banding patterns, which match from top to bottom. The two sister chromatids are attached at the **centromere**. During division, the two sister chromatids are pulled apart. At that moment, they each become an individual chromosome, one for each daughter cell.

D2.1.6 – DNA condensation and chromosome movement

D2.1.6 – Condensation and movement of chromosomes as shared features of mitosis and meiosis
Include the role of histones in the condensation of DNA by supercoiling and the use of microtubules and microtubule motors to move chromosomes.

Preparing DNA for separation

During most of the life of a cell, the DNA is not bunched up as shown in Figure 1. Rather, it is spread out in long unwound chains, which might resemble a plate of spaghetti, inside the nucleus. In order not to misplace any DNA and to prevent the strands tangling up and breaking, condensation of the DNA is necessary before replication. Just as passengers need to gather on the correct platform to get on the right train, DNA molecules group together. The process of condensation (Figure 2) involves the DNA being wrapped around proteins called **histones.** Histones help organize the DNA, which is coiled and then **supercoiled**, so that the coils are stacked on top of each other and form a compact pair of chromatids. When DNA is associated with histone proteins, it is referred to as **chromatin**.

When the supercoiling is complete, the chromosome takes the shape we are familiar with in micrographs. This condensed structure ensures that different DNA molecules can be transported in one package rather than being spread out all through the nucleus.

D2.1 Figure 2 The familiar shape of chromosomes is the result of DNA (the strand) being coiled around histone proteins (the spheres) and then forming a supercoiled shape. Start on the right side of the figure with unwound DNA, and progress to the left to see how it is wound around the histones to become supercoiled.

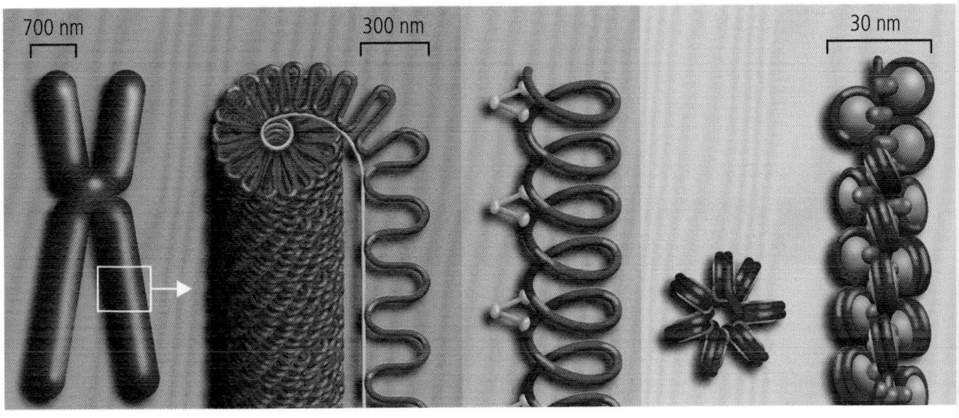

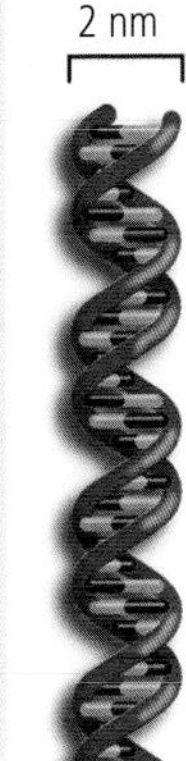

Movement of chromosomes

The **centrosome** is a cell organelle described in Chapter A2.2. This organelle makes the microtubule spindle fibres that are needed to guide the chromosomes to the right place before the cells can divide. Microtubules can be constructed and disassembled as needed, and they have a directionality because one end has a negative charge and the other a positive charge.

Specialized molecules called **motor proteins** push or pull objects around a cell. They use the microtubules as tracks, or they can attach to two microtubules and get one to slide past the other (like muscle fibres do when they are contracting). Motor proteins use adenosine triphosphate (ATP) to produce a **conformational change** (a change in shape) that moves the microtubules.

D2.1 Figure 3 The role of microtubules and motor proteins in the movement of chromosomes during cell division.

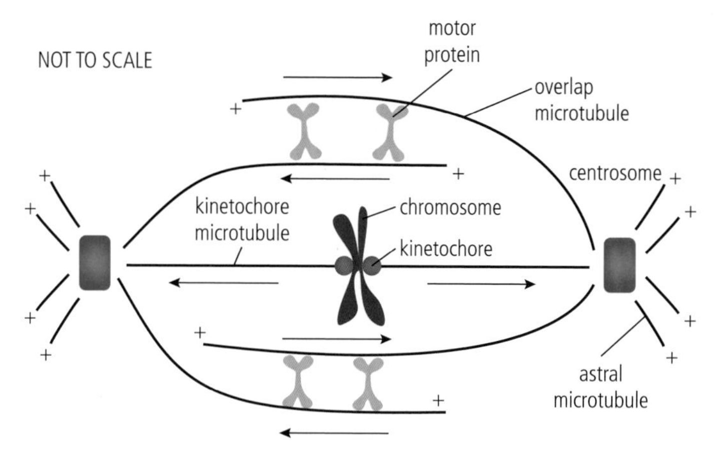

Figure 3 shows a chromosome in the middle of a cell, and three types of microtubule. The **astral microtubules** reach out from the centrosome. The **kinetochore microtubules** attach to the centromere of the chromosome, where the two sister chromatids are attached. The **overlap microtubules** are not attached to the chromosomes but rather pass beside them. Sandwiched between the overlap microtubules are **motor proteins**. These "walk" along the microtubules in such a way that the microtubules are pushed in opposite directions.

In order to ensure that the DNA is replicated successfully, DNA is condensed and coiled. Supercoiled chromosomes are moved around cells by a system of microtubules and microtubule motors.

When a cell is ready to separate its chromosomes, the motor proteins between the overlapping microtubules become active. The action of two microtubules sliding past each other pushes the two poles of the centrosome away from each other. Because the two sister chromatids are attached to the poles via opposite-facing microtubules, they will be pulled away from each other. Each will be transported to one half of the cell, so that when the cell splits in two each newly formed cell has a copy of each chromosome. Once the chromatids have been separated, the microtubules are dismantled and the pieces of track recycled.

D2.1.7 – Mitosis

D2.1.7 – Phases of mitosis
Students should know the names of the phases and how the process as a whole produces two genetically identical daughter cells.

When a cell is not going through cell division, it is in a phase called **interphase**. During interphase, the cell is performing its function in the organism as well as growing and preparing to divide. When the cell is ready, mitosis can begin. Mitosis is used to produce two identical cells, and involves four phases in the sequence:

- prophase
- metaphase
- anaphase
- telophase.

To help memorize the order of things, use mnemonics. You can use IPMAT for interphase and the four stages of mitosis. Make up a sentence with the first letter of each, such as "impatient people mostly arrived today".

Prophase

Figure 4 illustrates prophase. During prophase:

1. the chromatin fibres become more tightly coiled to form chromosomes
2. the nuclear envelope disintegrates and nucleoli disappear
3. a **mitotic spindle** forms as the centrosome builds new microtubules that will be used to pull the chromosomes into position
4. the **kinetochores**, a region in the centromere of each chromosome, attach to the spindle
5. the centrosomes move towards the opposite poles of the cell, as a result of lengthening microtubules.

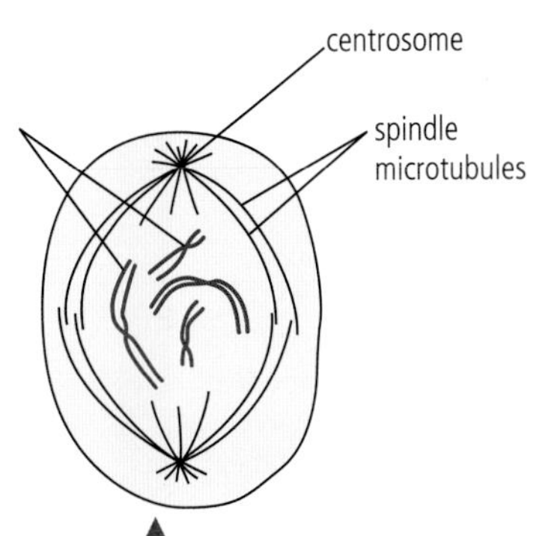

D2.1 Figure 4 This animal cell is in prophase. For clarity, only a small number of chromosomes is shown.

Metaphase

Figure 5 shows a cell in metaphase. During metaphase:

1. the chromosomes move to the middle or equator of the cell, which is called the **metaphase plate**
2. the centromeres of the chromosome align on the plate
3. the chromosomes move as a result of the action of the spindle, which is made of microtubules
4. the centrosomes are at opposite poles.

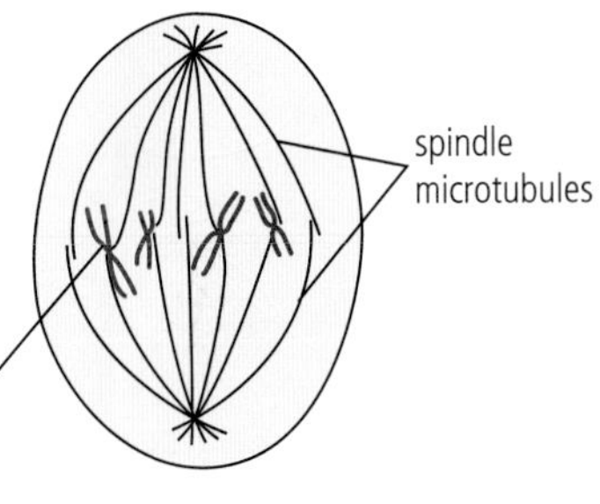

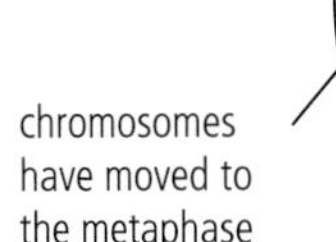

D2.1 Figure 5 The cell is now in metaphase. Again, only a small number of chromosomes is shown.

Anaphase

Figure 6 shows the cell in anaphase. This is usually the shortest phase of mitosis. It begins when the two sister chromatids of each chromosome move apart. During anaphase:

1. the chromatids, now each a chromosome, move towards the opposite poles of the cell
2. they move as a result of motor proteins pushing microtubules in opposing directions
3. because the centromeres are attached to the microtubules, they move towards the poles first
4. at the end of this phase, each pole of the cell has a complete, identical set of chromosomes.

D2.1 Figure 6 The cell is now in anaphase. Again, for clarity, only a small number of chromosomes is shown.

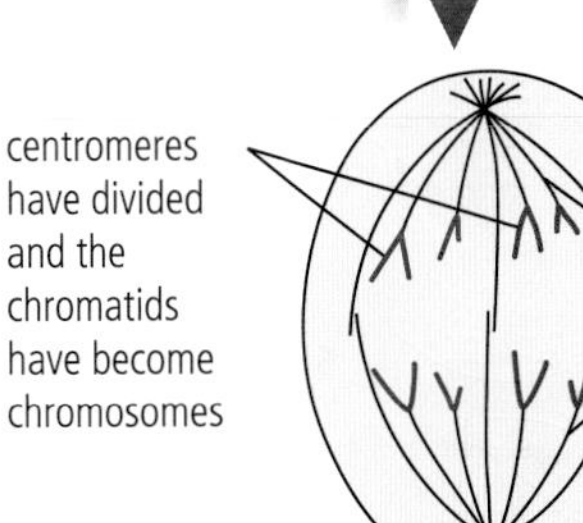

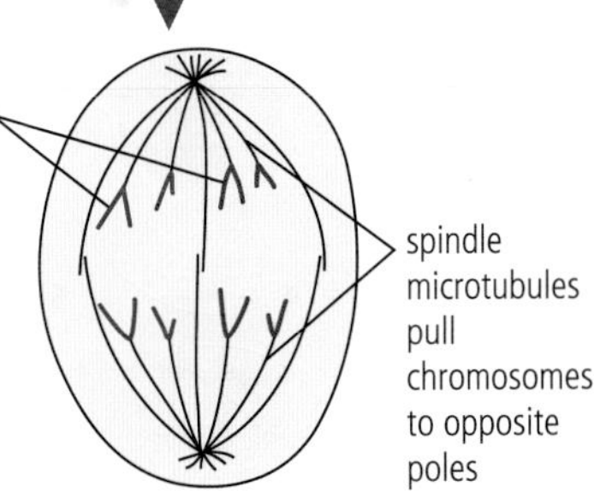

TOK

The IB biology syllabus focuses on the main phases of cell division, but intermediate stages exist such as "prometaphase" between prophase and metaphase. This makes you wonder how we decide where to draw the line when separating the different steps of a repeating process. When is it better to adopt broader terminology and when is it more helpful to devise more specific phases? Are some types of knowledge less open to interpretation than others?

Telophase

Figure 7 shows the cell in telophase. During telophase:

1. a set of chromosomes is located at each pole
2. a nuclear membrane (envelope) begins to re-form around each set of chromosomes
3. the chromosomes start to elongate
4. nucleoli reappear
5. the spindle apparatus disappears
6. the cell is elongated and ready for cytokinesis.

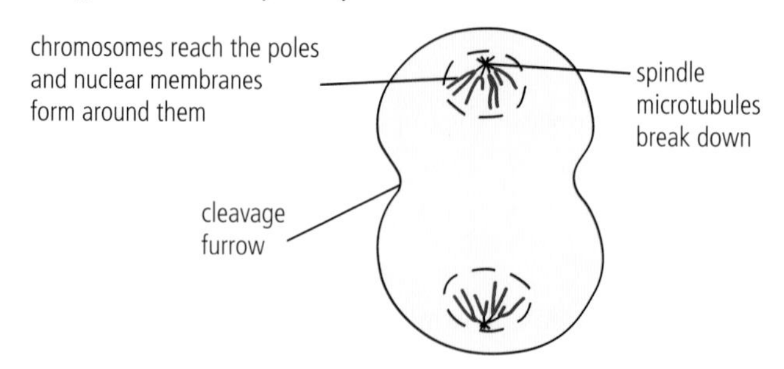

D2.1 Figure 7 Finally, the cell enters telophase.

D2.1.8 – Identifying the phases of mitosis

What processes support the growth of organisms?

D2.1.8 – Identification of phases of mitosis
Application of skills: Students should do this using diagrams as well as with cells viewed with a microscope or in a micrograph.

The IB expects you to be able to identify the different stages of mitosis from both diagrams and photomicrographs. Try the following activities.

SKILLS

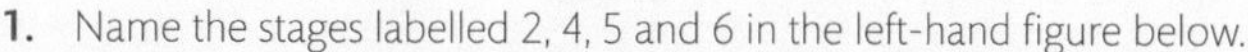

1. Name the stages labelled 2, 4, 5 and 6 in the left-hand figure below.
2. Identify which stage of mitosis the numbered cells in the micrograph in the right-hand figure below are going through.

prophase__ metaphase _ anaphase _ telophase _

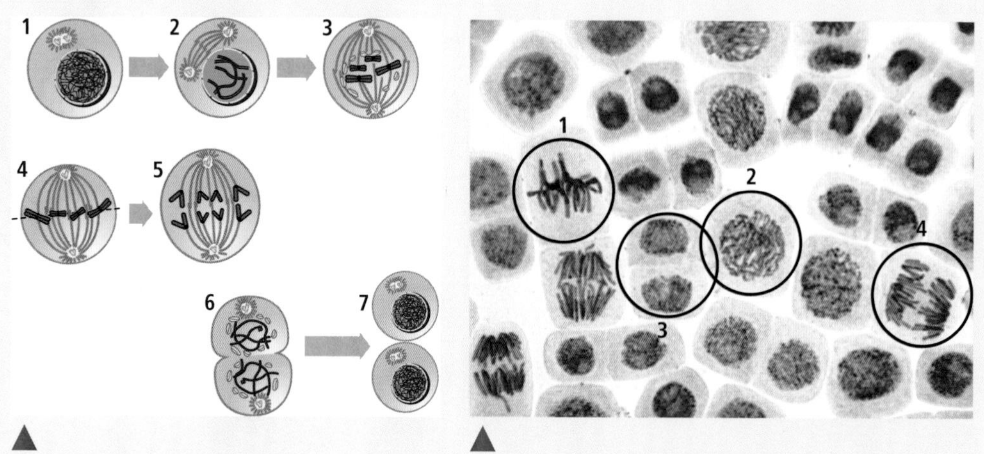

A cell undergoing mitosis.

A photo taken through a light microscope of cells going through various stages of mitosis.

See the eBook for a laboratory activity where you can observe cells undergoing mitosis in a root tip.

D2.1.9 – Meiosis

D2.1.9 – Meiosis as a reduction division

Students should understand the terms "diploid" and "haploid" and how the two divisions of meiosis produce four haploid nuclei from one diploid nucleus. They should also understand the need for meiosis in a sexual life cycle. Students should be able to outline the two rounds of segregation in meiosis.

D2.1 Figure 8 How the chromosome number is halved during meiosis. In this example, the organism has two pairs of chromosomes.

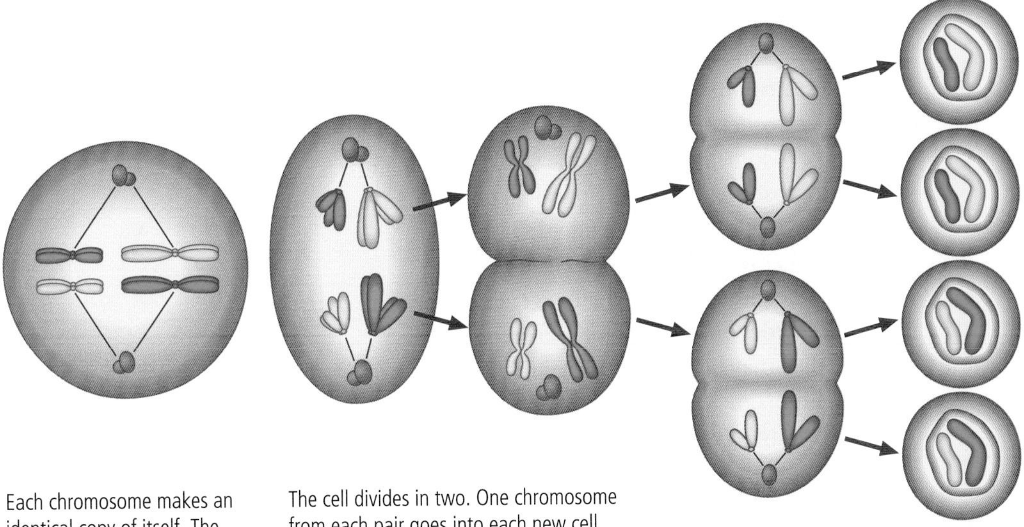

Gametes (sperm cells and egg cells) cannot contain a full set of pairs of chromosomes for the simple reason that, if they did, each new generation would double the chromosome number, creating an impossibly large amount of DNA to deal with. To avoid the problem of accumulating too many chromosomes, humans and other animals produce egg cells and sperm cells in such a way that the number of chromosomes in their nuclei is halved. Hence human sperm and eggs only contain 23 chromosomes, one from each pair, rather than 46 chromosomes arranged in 23 pairs. In order to make such special cells with half the chromosomes, a special type of cell division is needed. This type of cell division is called **meiosis**, and it is a **reduction division**.

Whereas mitosis produces **diploid** ($2n$) nuclei containing 46 chromosomes in humans (organized into 23 pairs), meiosis produces **haploid** (n) nuclei that contain 23 chromosomes, each representing half of one pair. Notice in Figure 8, from a single cell on the left, four cells have been produced on the right. Notice also that the number of chromosomes in this non-human example is four in the parent cell (so $2n = 4$) and two (so $n = 2$) in the daughter cells.

In the human testes and ovaries, respectively, meiosis produces haploid sperm and eggs, so that, when fertilization occurs, the zygote will receive 23 + 23 = 46 chromosomes; half from the mother, and half from the father. This is how the problem of changing chromosome numbers is avoided. As a result, the human chromosome number of 46 is preserved by the sexual life cycle (Figure 9).

D2.1 Figure 9 How chromosome number is maintained in the sexual life cycle of humans.

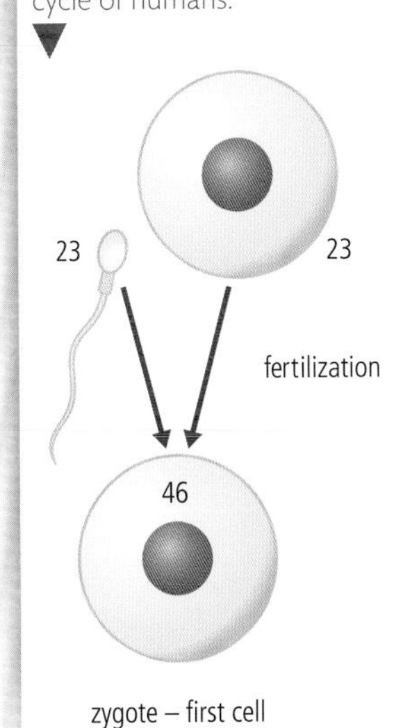

First round of segregation: halving the chromosome number

Meiosis involves two rounds of segregation (shown in Figures 10 and 11), which take place in a series of steps. During the first round of segregation (meiosis I), the chromosome number is halved.

Prophase I

1. Chromosomes become visible as the DNA is arranged around the histone proteins and becomes more compact.
2. **Homologous chromosomes** are attracted to each other and pair up. Homologous chromosomes are pairs of the same chromosome. One of the pair will originally have come from the individual's father, the other from the mother. Homologous chromosomes carry the same genes in the same order but the type of each gene (the **alleles**) may be different. Together, a pair of chromosomes is called a **bivalent**.
3. **Crossing over** occurs. During crossing over parts of homologous pairs can be swapped between chromosomes. This allows the mixing of alleles (see Section D2.1.11 for more on crossing over).
4. Spindle fibres form from microtubules.

Metaphase I

1. The homologous chromosomes line up across the cell's metaphase plate. They are randomly orientated, which means that either of the chromosomes from each pair is equally likely to end up at either pole.
2. The nuclear membrane disintegrates.

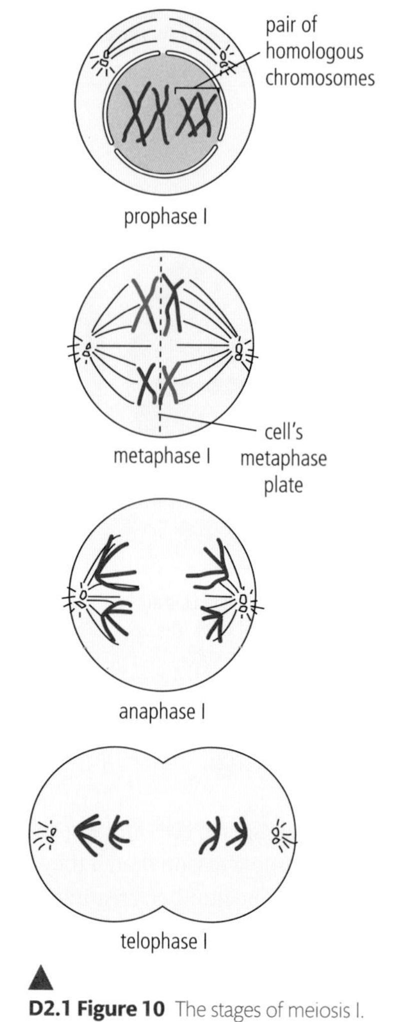

D2.1 Figure 10 The stages of meiosis I.

Anaphase I

Spindle fibres attach to the chromosomes and pull them to the opposite poles of the cell.

Telophase I

1. The spindles and spindle fibres disintegrate.
2. The chromosomes uncoil and new nuclear membranes form.

At the end of meiosis I, cytokinesis happens: the cell splits into two separate cells. The cells at this point are haploid because they contain only one chromosome of each pair. However, each chromatid still has its sister chromatid attached to it, so no S phase is necessary.

Many plants do not have a telophase I stage.

Second round of segregation: separation of sister chromatids

During the second round of segregation (meiosis II), the sister chromatids are separated.

Prophase II

1. The DNA condenses into visible chromosomes again.
2. New meiotic spindle fibres are produced.

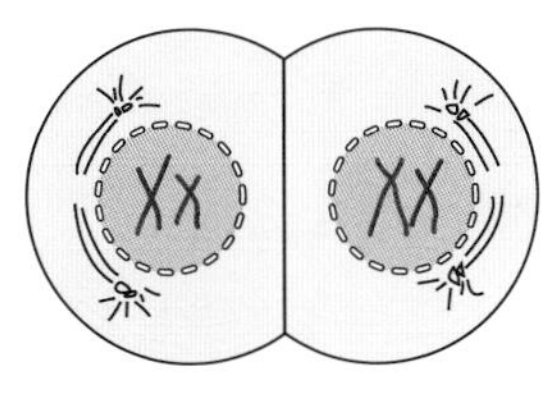
prophase II

Metaphase II

1. Nuclear membranes disintegrate.
2. Individual chromosomes line up along the metaphase plate of each cell in no special order, i.e. random orientation.
3. Spindle fibres from opposite poles attach to each of the sister chromatids at the centromeres.

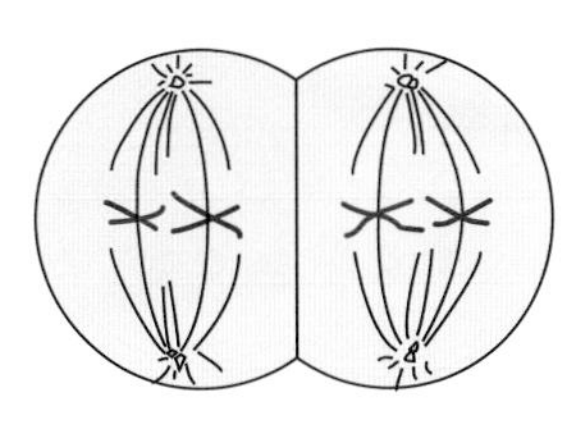
metaphase II

Anaphase II

1. The centromeres of each chromosome split, releasing each sister chromatid as an individual chromosome.
2. The spindle fibres pull individual chromosomes to opposite ends of the cell.
3. As a result of random orientation, the chromosomes can be pulled towards either of the newly forming daughter cells.

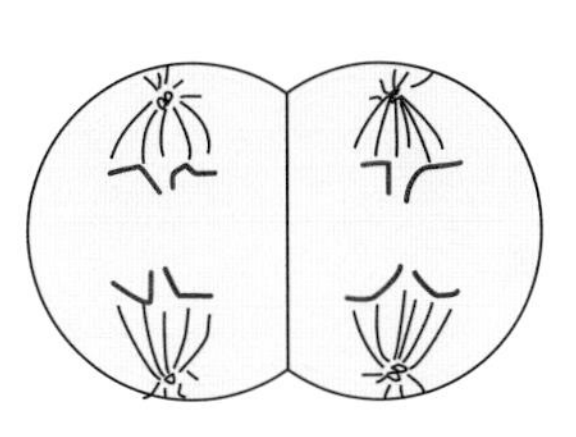
anaphase II

Telophase II

1. In animal cells, cell membranes pinch off in the middle. In plant cells, new cell plates form to make the cell membranes and cell walls of the four new cells.
2. The chromosomes unwind their strands of DNA.
3. Nuclear envelopes form around each of the four new nuclei. Then cytokinesis can take place.

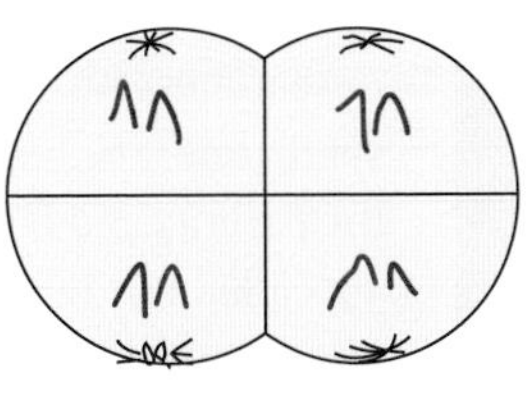
telophase II

D2.1 Figure 11 The stages of meiosis II.

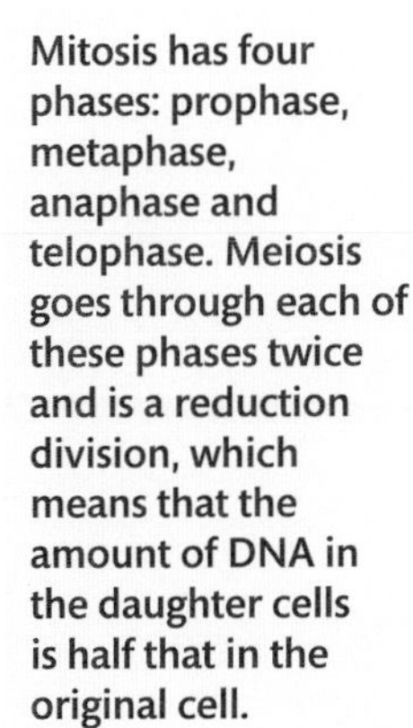

Mitosis has four phases: prophase, metaphase, anaphase and telophase. Meiosis goes through each of these phases twice and is a reduction division, which means that the amount of DNA in the daughter cells is half that in the original cell.

D2.1.10 – Non-disjunction

D2.1.10 – Down syndrome and non-disjunction
Use Down syndrome as an example of an error in meiosis.

Sometimes errors can occur during meiosis, and offspring can receive an atypical number of chromosomes, including extra or missing chromosomes. In humans, **Down syndrome** is caused by an extra copy of chromosome 21; the child has 47 instead of 46 chromosomes. The extra chromosome arises from a phenomenon called **non-disjunction**, which can happen at different times and on different chromosomes but for Down syndrome most often occurs when the 21st pair of homologous chromosomes fails to separate during anaphase I. If this happens in the future mother's ovary, the egg then carries two 21st chromosomes instead of one. When a sperm cell subsequently fertilizes the egg, the total number of 21st chromosomes is three.

The two girls in this photo are twins. They received an extra 21st chromosome and have Down syndrome.

Nature of Science

Researchers wanted to find out what influences the frequency of Down syndrome. Studies collected statistics on the many different characteristics of the parents and families of children born with Down syndrome. Such studies are called **epidemiological studies**, and they look at trends in populations, often examining thousands of cases. The incidence of Down syndrome increases with the age of the mother, particularly over the age of 35 (see the figure). Such data can help doctors and future parents assess the risks.

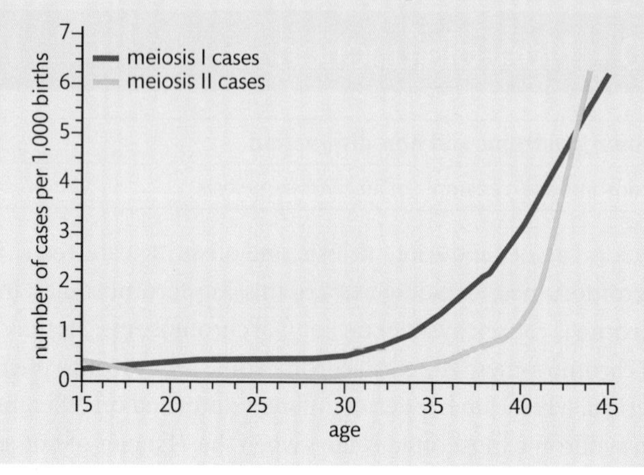

An extra chromosome for the 21st pair can occur during meiosis I or meiosis II, which is why the graph shows both, but the majority of cases happen during meiosis I.

D2.1.11 – Genetic diversity

D2.1.11 – Meiosis as a source of variation
Students should understand how meiosis generates genetic diversity by random orientation of bivalents and by crossing over.

Pairing of homologous chromosomes and crossing over

Meiosis is a step-by-step process during which a diploid parent cell produces four haploid daughter cells. Before the process begins, DNA replication allows the cell to make a complete copy of its genetic information during **interphase** (see Section D2.1.14 for more details of interphase). This results in each chromatid having an identical copy, or sister chromatid, attached to it at the centromere.

In order to produce a total of four cells, the parent cell must divide twice: the **first meiotic division** (meiosis I) makes two cells, and then each of these two cells divides again during the **second meiotic division** (meiosis II) to make a total of four cells.

One of the characteristics that distinguishes meiosis from mitosis is that, during the first step (prophase I) there is an exchange of genetic material between non-sister chromatids in a process called **crossing over** (see Figure 12). This exchange of chromosomal material happens when sections of two homologous but non-sister chromatids break at the same point, twist around each other, and then each connects to the other's initial position.

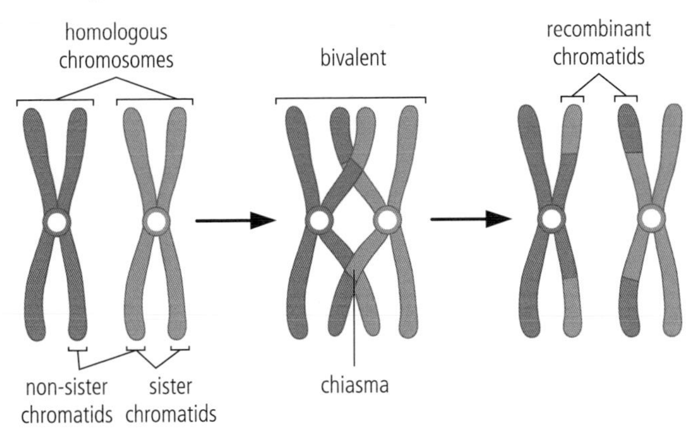

D2.1 Figure 12 Crossing over occurring in a bivalent, a pair of homologous chromosomes.

Crossing over allows DNA from the maternal chromosomes to mix with DNA from the paternal chromosomes. In this way, the **recombinant chromatids** that end up in the sperm or the egg cells are a mix of the parent cells' original chromatids. This helps increase variety among offspring from the same parents, and so increases the chances of survival of at least some offspring if one combination of alleles proves to be more favourable for survival than others.

Random orientation

Figure 13 shows that, during metaphase I, the homologous pairs of chromosomes line up along the centre of the cell. Which of the pairs ends up at which pole is down to chance, which is why it is called **random orientation**. As with crossing over, this is another adaptation that increases variety in the offspring. The result of random orientation is that a male will only very rarely produce two sperm cells that are identical. Likewise, for a female, it is highly unlikely that she will ever produce two eggs with identical chromosomes in her lifetime.

Crossing over and random orientation explain why there is variation between siblings.

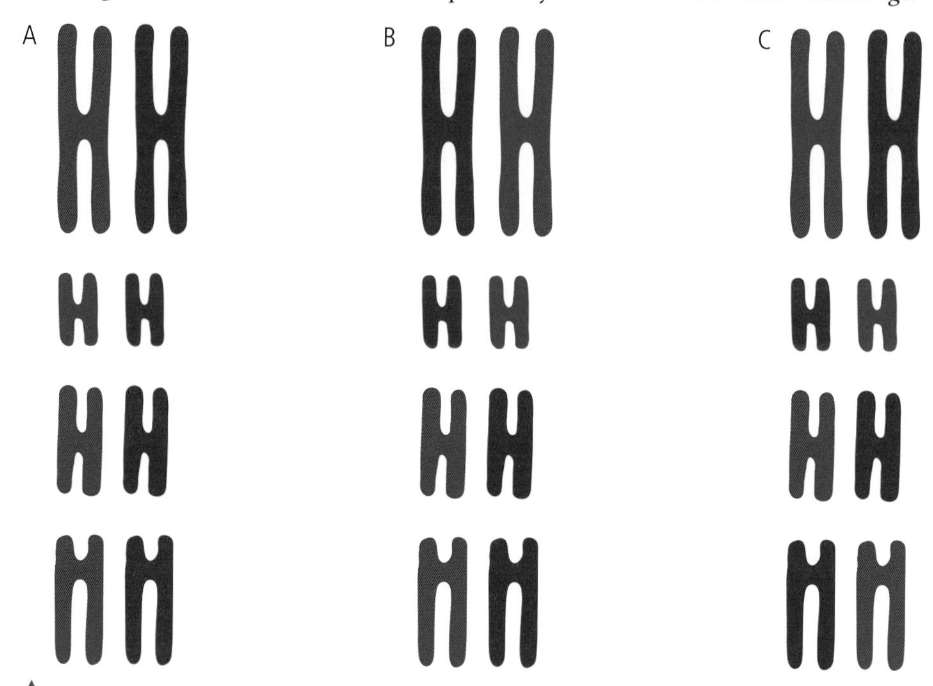

D2.1 Figure 13 Rows A, B, and C show three of the sixteen possible orientations for four pairs of homologous chromosomes. In humans, with 23 pairs of chromosomes, there are more than 8 million possible orientations.

The only way for a couple to have two children with the same combinations of chromosomes is to have identical twins. This is because identical twins form from a single zygote. Non-identical twins or any other siblings will be the product of an egg cell with a unique combination of half the mother's chromosomes and a unique combination of half the father's chromosomes.

How does the variation produced by sexual reproduction contribute to evolution?

How much variation can be generated by random orientation? In each haploid cell (n) the calculation is 2^n because there are two possible chromosomes in each pair (maternal and paternal) and there are n chromosomes in total. For humans, the number is 2^{23} because there are 23 chromosomes in each gamete. So the probability that a woman could produce the same egg twice is 1 in 2^{23} or 1 in 8,388,608. Even this calculation is an oversimplification, however, because it does not take into consideration the additional variety that results from crossing over. In addition, the calculation 2^n only considers one gamete. To produce offspring, two gametes are needed.

HL

D2.1.12 – Cell proliferation

D2.1.12 – Cell proliferation for growth, cell replacement and tissue repair
Include proliferation for growth within plant meristems and early-stage animal embryos as examples. Include skin as an example of cell proliferation during routine cell replacement and during wound healing. Students are not required to know details of the structure of skin.

Mitosis is necessary for the growth of organisms, the development of embryos, and tissue repair. Mitosis is also necessary for asexual reproduction (see Chapter D3.1).

Plant growth

Meristems are areas of special tissue found in plant cells. Meristematic cells are undifferentiated cells that can divide rapidly, allowing growth in plants. Later, these cells will specialize (differentiate) and play a specific role for the plant, such as structural support or transport of liquids as vascular tissue. There are two major types of meristematic tissue, **apical** and **lateral**.

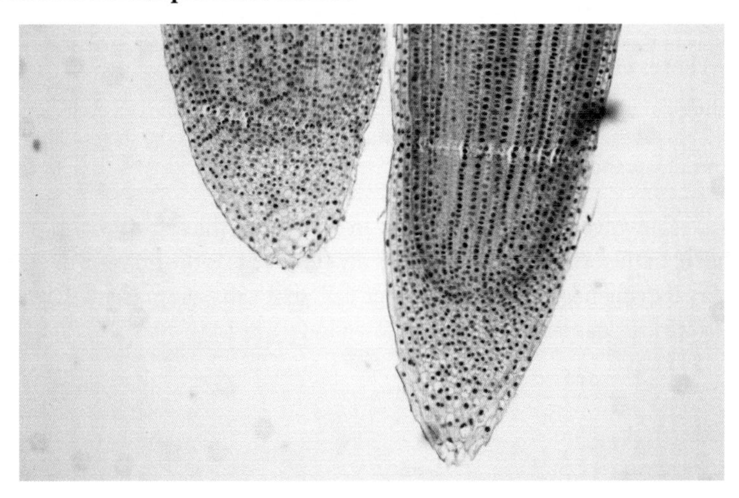

A light micrograph of a root meristem tissue. The dark spots are clumps of chromosomes inside the cells, many of which are going through mitosis.

Apical meristematic tissue occurs in root tips and at the tips of branches. Apical tissue enables a plant to lengthen. You may have been asked to prepare root tip squashes to observe mitosis under the microscope. We expect to find the majority of cells in the tips of roots and shoots to be undergoing mitosis. **Lateral meristematic** tissue occurs in stem tissue and enables the stems to grow in width. In order for growth to be possible, mitosis must be taking place in meristematic tissue. The **zone of cell division** is the area where new, undifferentiated cells are formed.

Animal embryos

A fertilized egg cell, the zygote, is an undifferentiated cell. It will make copies of itself until it forms many thousands of cells, which start to organize themselves into layers and a hollow sphere to make the **embryo**. Making copies of these embryonic stem cells requires mitosis to ensure that full copies of the cell's genome get passed on to all the new cells. The cells then start to differentiate: some become muscle cells, others specialize as skin cells, and still others as gut cells, for example. Once a cell, such as an intestinal cell, has differentiated, it then still needs to undergo mitosis, to make more intestinal cells.

When its outer layer of skin cells is dead, a snake's skin comes off in one long moult, like taking a sock off a foot. In humans, dead skin cells fall off in a constant rain of epidermis. These dead skin cells form dust that can accumulate on your floor and in your bedsheets.

Replace and repair

When cells die, they need to be replaced; when tissue is damaged, new cells need to be produced to repair it. Because skin cells are dying and falling off your body all the time, they constantly need to be replaced. Mitosis is therefore necessary for the growth of new skin cells. Skin has a layered structure; skin grows in the bottom layer, while dead skin cells flake off the top layer.

If your skin is broken or wounded, not only are skin cells damaged but blood can be lost and there is a risk of infection. All damaged and lost cells will be replaced by mitosis in some part of the body.

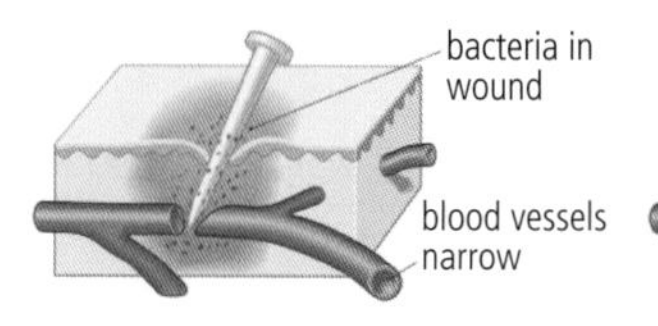

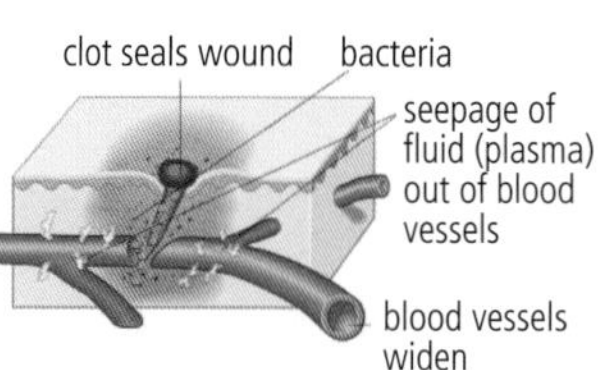

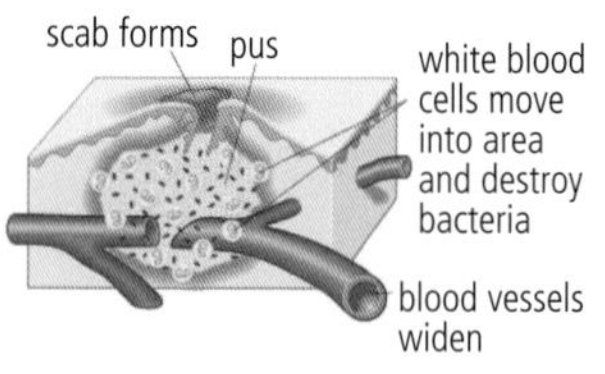

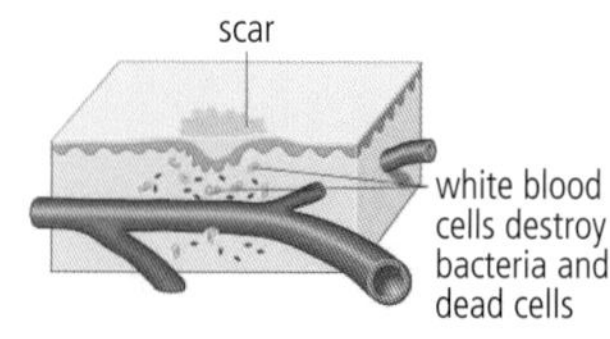

After being damaged, skin reacts quickly to repair itself. White blood cells, blood vessel tissue and skin all rely on mitosis to build new cells.

D2.1.13 – The cell cycle

D2.1.13 – Phases of the cell cycle
Students should understand that cell proliferation is achieved using the cell cycle. Students should understand the sequence of events including G1, S and G2 as the stages of interphase, followed by mitosis and then cytokinesis.

The life of a cell involves two major phases. In one phase, the cell is growing. This is **interphase**. In the other phase, the cell is dividing. This is the mitotic, or meiotic, phase. The **cell cycle** begins and ends as one cell, so it can be represented by a circle divided into various named sections, as shown in Figure 14.

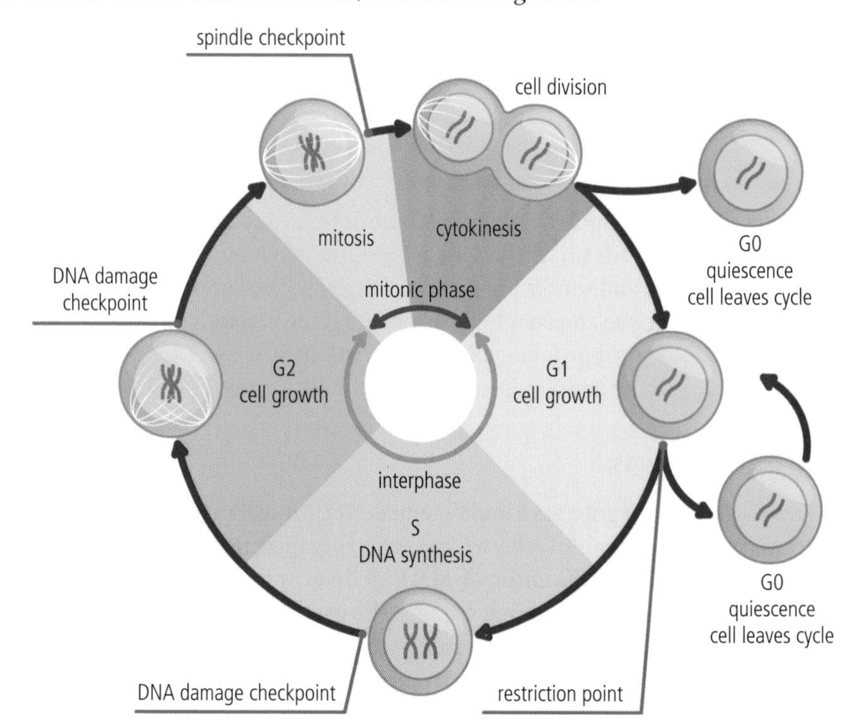

D2.1 Figure 14 The phases of the cell cycle.

D2.1.14 – Interphase

D2.1.14 – Cell growth during interphase
Students should appreciate that interphase is a metabolically active period and that growth involves biosynthesis of cell components including proteins and DNA. Numbers of mitochondria and chloroplasts are increased by growth and division of these organelles.

The longest and most variable phase of the cell cycle in most cells is **interphase**. During interphase the cell is metabolically active and both DNA and proteins are made. Interphase includes three smaller phases: G1, S and G2. During the **G1 phase**, the cell grows. At the beginning of G1, the cell is the smallest it will ever be. As it grows, it accumulates materials such as proteins and nucleotides. G1 is followed by the **S phase**, the synthesis phase, in which the most important activity is replication of DNA. Once the chromosomes have been replicated, the cell enters its second growth phase, called the **G2 phase**. During this phase, the cell grows and makes preparations for mitosis (which is called the **M phase**). During G2, organelles such as mitochondria and chloroplasts can increase in number, DNA begins to condense from chromatin to chromosomes, and microtubules can begin to form. Mitosis is followed by cytokinesis.

D2.1.15 – Cyclins

D2.1.15 – Control of the cell cycle using cyclins
Limit to the concentration of different cyclins increasing and decreasing during the cell cycle and a threshold level of a specific cyclin required to pass each checkpoint in the cycle. Students are not required to know details of the roles of specific cyclins.

Cyclins are a group of proteins that control the cell's progression through the cell cycle. The cyclins bind to **cyclin-dependent protein kinases** (CDKs) and phosphorylate them, enabling the kinases to act as enzymes. Phosphorylation is when a phosphate group is attached to a molecule, and is a common way to activate an enzyme. These activated enzymes then cause the cell to move from G1 to the S phase and from G2 to the M phase. The points where the cyclin-activated CDKs function are called **checkpoints** in the cell cycle (see Figures 14 and 15). Some cells will pause during G1 and enter a separate phase, the **G0 phase**. G0 is a non-growing state and certain cells stay in G0 for varying periods of time. Some cells, such as nerve and muscle cells, never progress beyond the G0 phase.

Cyclin levels change during the cell cycle. When a particular cyclin concentration is low, it binds to so few CDKs that the checkpoint cannot be reached, but when its concentration reaches a certain threshold, the cell can move to the next phase. For example, the **G1 cyclin** tells the cell to grow and to get ready to replicate the DNA (the S phase). It is introduced during the first growth phase. **Mitotic cyclin** tells the cell to start making the microtubules that will form the spindle fibres for mitosis. Figure 15 illustrates how mitotic cyclin reaches its threshold just before mitosis, when it tells the cell it is time to prepare for separating the chromosomes. After anaphase, this cyclin is broken down so that the cell can move on to cytokinesis.

D2.1 Figure 15 Two cyclins are extremely important to the cell cycle: G1 cyclin and mitotic cyclin. Note their location in the cell cycle and that they must combine with a cyclin-dependent protein kinase (CDK) and be phosphorylated (indicated by the letter P) to become active.

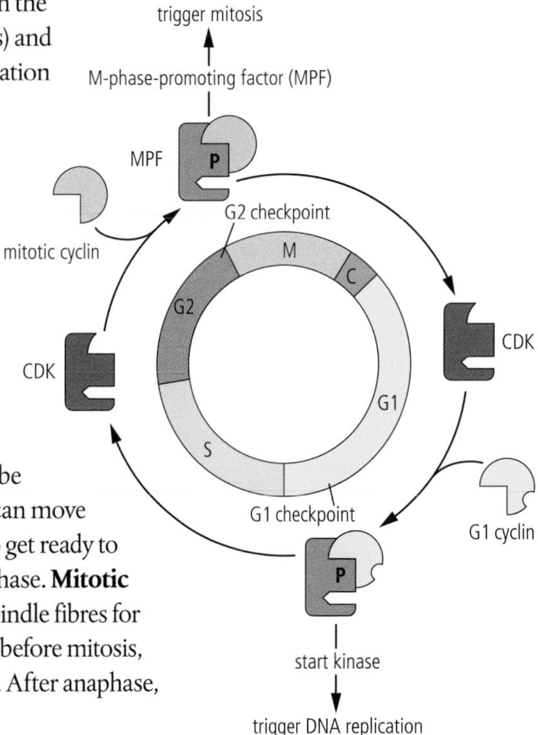

D2.1.16 – The effect of mutations

D2.1.16 – Consequences of mutations in genes that control the cell cycle
Include mutations in proto-oncogenes that convert them to oncogenes and mutations in tumour suppressor genes, resulting in uncontrolled cell division.

Sometimes things go wrong with cell division. **Cancer** cells are cells that undergo extremely rapid and uncontrolled reproduction with very little or improper differentiation. The result is a mass of cells (a tumour) with no useful function to the organism. It appears that any cell can lose its usual orderly pattern of division, because we have found cancer in almost all tissues and organs, and it is observed in many forms of life, not just in humans.

A question to consider is how or why a primary tumour forms. Most organisms have sections of genes that can mutate or can be expressed at abnormally high levels. These sections of genes, called **oncogenes**, contribute to a normal cell becoming a cancer cell. Oncogenes can cause cells to divide more frequently than they should. A gene that can turn into an oncogene is called a **proto-oncogene**. Typically, cells have their own death pre-programmed in their DNA so that, when they no longer function, they can be broken down and their materials recycled in the organism. This programmed death is called **apoptosis**.

Oncogenes can modify apoptosis so that instead of dying and being broken down, the cell keeps dividing using mitosis. The oncogenes may start to change or mutate because they are triggered by an outside agent, referred to as a **mutagen**. One such potential mutagen is cigarette smoke (see the Global context box).

There is a correlation between smoking and the incidence of cancer. This has been shown consistently in many independent studies. Examine the figure below, a graph from a report by the World Health Organization's International Agency for Research on Cancer, and note the positive correlation.

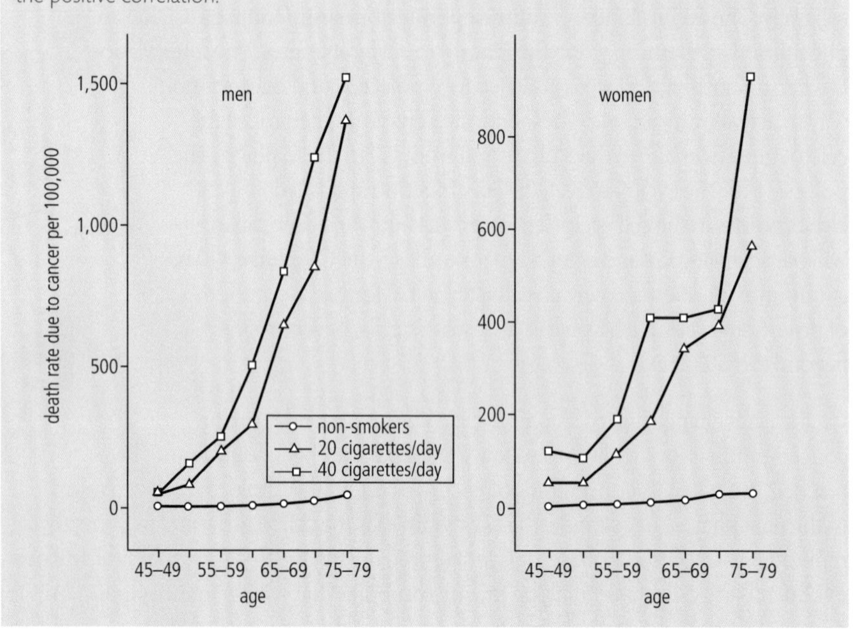

The death rate (for every 100,000 people) per year from lung cancer according to the American Cancer Society cohort (CPS-II) from 1982 to 1988.

Healthy cells have specialized genes called **tumour suppressor genes (TSGs)**, which code for proteins that can regulate the cell cycle. Some of them control cell growth to ensure cells do not grow too quickly, or that apoptosis takes place when necessary, and some are responsible for repairing DNA that otherwise could lead to uncontrolled cell growth. When TSGs are damaged or undergo mutations, the cell cycle can be disrupted: cells can grow unregulated, apoptosis does not take place, and therefore a tumour can form.

Some mutations in TSG can happen during a person's lifetime, while others can be inherited. But sometimes no mutation is present and still a gene is not able to work. Another way TSG can be made less effective is by silencing them. Cells can add chemical tags, in the form of methyl groups, to genetic material that then silences them. If a TSG is silenced, it cannot perform its protective role, and this can lead to cancer. Some researchers have dedicated their careers to studying TSGs and trying to develop techniques of gene repair or gene activation in the hope of getting closer to a cure for cancer.

D2.1.17 – Tumour growth

D2.1.17 – Differences between tumours in rates of cell division and growth and in the capacity for metastasis and invasion of neighbouring tissue

Include the terms "benign", "malignant", "primary tumour" and "secondary tumour", and distinguish between tumours that do and do not cause cancer.

Application of skills: Students should observe populations of cells to determine the mitotic index.

A **primary tumour** is one that occurs at the original site of a cancer. A **secondary tumour** is a **metastasis**, a cancerous tumour that has spread from the original location to another part of the organism. An example of metastasis is a brain tumour that is in fact composed of breast cancer cells. In some cases the metastasis of the primary tumour cells is so extensive that secondary tumours are found in many locations within the organism.

Not all tumours cause cancer. A tumour is said to be **benign** if the cells are forming a mass but are not spreading to the rest of the body. A tumour is said to be **malignant** if the cells rupture the organ they are in and start to spread to the rest of the body. This can cause tumours to grow in tissues far from the primary tumour. Doctors who specialize in cancer prevention and treatment, called **oncologists**, try to detect cancerous growths as early as possible to prevent their spread to other parts of the body. Once a cancer metastasizes, it is much more difficult to treat.

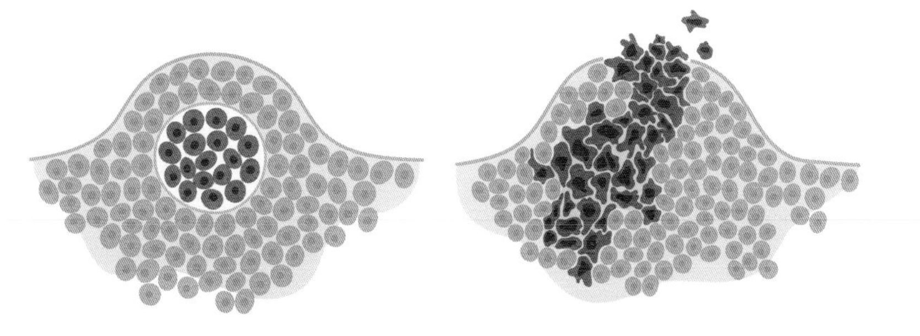

A benign tumour (left) and a malignant tumour (right). Healthy cells are shown in a lighter colour (pink) while the cancerous cells are shown in a darker colour (purple).

Two ways to combat cancer are chemotherapy and radiation therapy: both attempt to stop the cell cycle of cancer cells. The concept behind chemotherapy is to introduce toxins into the bloodstream that will reduce mitosis, damage cells and encourage apoptosis. The concept behind radiation therapy is to shoot a beam of ionizing radiation at a tumour to damage the cells and their DNA enough to encourage apoptosis. Sometimes both treatments are used.

The **mitotic index** is an important tool for predicting the response of cancer cells to chemotherapy. The mitotic index is the ratio of the number of cells undergoing mitosis divided by the total number of cells. A higher mitotic index indicates a more rapid proliferation of cells. It is likely that tumours with higher mitotic indices will be more difficult to control, and a patient with such a tumour may be given a poorer prognosis than a patient with a tumour that has a lower mitotic index.

If a student counts 100 cells and finds the following: 86 are in interphase, 6 are in prophase, 3 are in metaphase, 3 are in anaphase, 2 are in telophase. What is the mitotic index?

Solution:

$$\frac{(6+3+3+2)}{100}$$

A mitotic index of 0.14 or 14% is low. It is unlikely that this tissue is undergoing rapid growth, so this tissue would be diagnosed as cancer-free. If it had a mitotic index higher than 75%, the doctor would be much more likely to consider the tissue as cancerous.

HL end

Guiding Question revisited

How can large numbers of genetically identical cells be produced?

In this chapter you have learned that:

- in prokaryotic cells, binary fission is used to make genetically identical cells
- cytokinesis is the splitting of cells and occurs after nuclear division, and the cytoplasm and cell organelles are usually split evenly between the daughter cells
- in multicellular organisms, mitosis is used to make two genetically identical daughter cells, which are diploid, $2n$

HL

- mitosis allows growth but is also used to repair damaged tissues
- mitosis occurs in meristematic tissue in plants, and occurs during the growth phase of an organism, particularly during the early stages of division in the zygote
- the cell cycle is divided into phases controlled by cyclins, the concentration of cyclin dictating how rapidly or slowly the cell is replicated
- sometimes uncontrolled cell division takes place, leading to a cancerous tumour.

HL end

Guiding Question revisited

How do eukaryotes produce genetically varied cells that can develop into gametes?

In this chapter you have learned:

- in multicellular organisms, meiosis is another type of cell division that is exclusively used for the production of gametes (sex cells)
- sperm cells and egg cells are haploid, n, containing half of an organism's genetic information
- at the end of meiosis, four cells are produced, but each sperm and each egg has a different combination of chromosomes as a result of random orientation and crossing over, which ensures variety in the offspring
- sometimes chromosomes do not separate correctly, and this non-disjunction of chromosomes can lead to genetic conditions such as Down syndrome.

Exercises

Q1. State two reasons why mitosis is necessary.

Q2. A chemical called colchicine disrupts the formation of microtubules. What effect would this drug have on a cell going through mitosis?

Q3. If a parent cell has 24 chromosomes, how many chromatids would be present during metaphase of mitosis?

Q4. Explain when cytokinesis occurs within the cell cycle.

Q5. Compare cytokinesis in plant and animal cells.

Q6. HL Why is it useful to calculate the mitotic index of cancer cells?

HL

D2.2 Gene expression

Guiding Questions

How is gene expression changed in a cell?

How can patterns of gene expression be conserved through inheritance?

Multicellular organisms such as humans start out as a single cell. Over time, the cell divides to make a ball of identical cells. And then something remarkable happens: some cells ignore all their genes except those for making nerve cells. They become the nervous system. Others only turn on the genes for heart tissue, and they start forming the heart. We will explore how this change in gene expression is possible. We will also look at how environmental factors such as air pollution, or experiences such as exposure to famine, can regulate certain genes. In some cases these modifications are erased in future generations, but a few of them can be passed on.

To express a gene, an mRNA copy of it must be made using the process of transcription, then a protein must be generated by ribosomes reading the mRNA and assembling amino acids to make the protein. This protein, such as a pigmentation in the skin or a digestive enzyme in the gut, will contribute to the organism's phenotype, hence the gene is expressed.

D2.2.1 – Phenotype

D2.2.1 – Gene expression as the mechanism by which information in genes has effects on the phenotype
Students should appreciate that the most common stages in this process are transcription, translation and the function of a protein product, such as an enzyme.

A genetic code is useless unless there is a way of transforming it into the molecules needed for living organisms. The term **gene expression** is used to describe the process of reading a gene and building a protein that will then be used by the organism. Gene expression relies on the processes of **transcription** and **translation**, which are studied in Chapter D1.2.

During transcription, a messenger RNA (mRNA) molecule is generated from the DNA in a cell. The next process is translation, during which a copy of the gene is used to build a protein (see Chapter D1.2). The types of proteins our cells build determine our **phenotype**. An organism's phenotype is the set of characteristics that results when the genetic information received from the parents is expressed. Blood type, the ability to develop freckles on the skin, or the ability to digest lactose, are examples of phenotypical characteristics and are all the result of genes being expressed, transcribed and translated, and then the protein produced functioning in a particular way.

The ability to develop freckles is a genetic trait. You may have noticed that freckles are often associated with people who have red hair. Freckles also show up better when the skin has been exposed to sunlight, so genes are not the whole story.

D2.2.2 – Regulation of transcription

D2.2.2 – Regulation of transcription by proteins that bind to specific base sequences in DNA
Include the role of promoters, enhancers and transcription factors.

A gene is a few hundred or a few thousand base pairs long, just a small fraction of the 3,000,000,000 base pairs in the human genome. The **promoter region** is a sequence of DNA found upstream from (in front of) the gene that indicates the gene's starting point and helps to control its transcription. The enzyme in charge of transcription is called **RNA polymerase**, and it attaches to the promoter region along with **transcription factors**. Transcription factors are proteins that bind to the promoter regions and help regulate the transcription of the DNA. **Promoter proteins** are one example of transcription factors, and they encourage the RNA polymerase to attach to the DNA and start transcription. RNA polymerase can then move along the gene to make an mRNA copy, and detach at the end of the gene. The mRNA is now able to leave the nucleus and be translated into a protein.

Analogies can help conceptualize abstract ideas. You can think of the promoter region on a DNA sequence like the street name and number of an address. It helps enzymes find the right place to start transcription. However, only use analogies like this for yourself, not in exams.

But we do not need to express all of our genes all of the time. The enzyme lactase, for example, is only needed when we consume milk or milk-based products such as ice cream. It would be a waste of energy and amino acids to produce lactase when it is not needed. So the system needs a way to turn transcription on or off. This is where enhancers or repressors can help. An **enhancer region** of the DNA is found upstream from the gene (i.e. before the genetic sequence that is being transcribed). The enhancer region allows certain activator proteins to attach to it and encourage transcription.

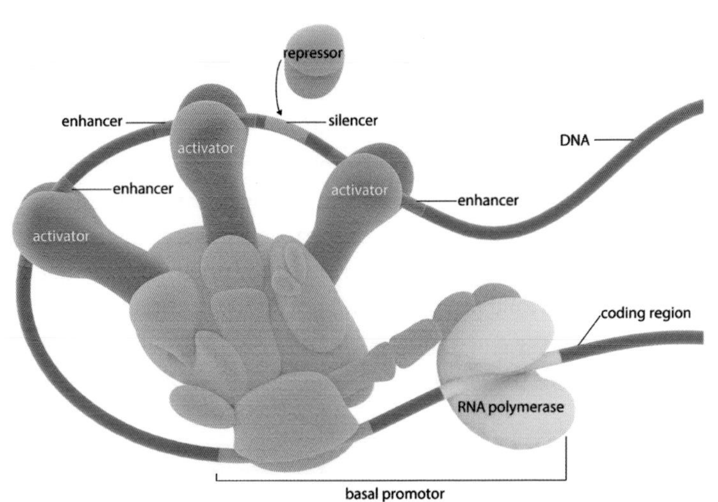

D2.2 Figure 1 A loop of DNA (in blue) along a chromosome is about to be transcribed by the RNA polymerase (in yellow). Transcription of a gene involves regions on the DNA as well as proteins used as transcription factors.

A loop of DNA is shown in Figure 1; enhancer regions are shown along the loop but they are quite a distance upstream from the gene sequence, which is labelled "coding region". The RNA polymerase has attached to the promoter region of DNA at the start of the gene and is ready to start transcription. A collection of proteins makes up the transcription factors that attach to both sides of the loop. The top of the loop contains the enhancer region of the DNA, and the bottom contains the promoter region. Activator proteins attach to the enhancer region and bend the DNA into a loop so that the promoter region ends up on the opposite end of the group of transcription factors.

The length of a promoter region varies between genes. Genes with long promoter regions can bind to more transcription factors and therefore can have complicated control mechanisms for regulating transcription. Simpler promoter regions mean that there is less room for transcription factors to intervene and therefore less fine control.

When the transcription factors are connected to both the enhancer region and the promoter region, they allow the RNA polymerase to start transcription. Notice that the transcription factor labelled "repressor" is not attached to the DNA region called the silencer region. If it did latch on to this part of the DNA, it would block transcription like an "off" switch. Switching on and off RNA polymerase's ability to do its job is an effective way of regulating transcription. This, in turn, regulates translation because, if no mRNA is produced, no translation can occur.

D2.2.3 – Degradation of mRNA to regulate translation

D2.2.3 – Control of the degradation of mRNA as a means of regulating translation
In human cells, mRNA may persist for time periods from minutes up to days, before being broken down by nucleases.

Another way of regulating transcription is to break up the mRNA that is being used to produce a protein. Rendering an mRNA molecule useless in this way is called **mRNA degradation**. Molecules of mRNA that have left the nucleus are available for ribosomes to use for different time periods, anywhere from minutes to days. Some mRNA molecules are quite stable and take a long time to decay. If the cell needs to stop producing the protein in question, it will need to not only stop making new mRNA molecules but also destroy the ones already circulating inside the cell.

To understand how mRNA can be broken down, you need to know about its structure. At one end of an mRNA molecule there is a protective cap, and at the other there is a tail (Figure 2). Because the code it is carrying is an important message, these structures protect it, a bit like front and rear bumpers on a vehicle. The tail end is protected by adding a long sequence of adenine nucleotides, i.e. AAAAAAAAA..., often hundreds of adenines long. This is why it is called the **polyA** tail. It is important that transcription and translation happen in a particular direction, so that the order of amino acids in the resulting protein is correct. If the process accidentally went backwards, the protein would not function. The cap is before the start of the mRNA code and the tail is after the end of the coded message. RNA polymerase translates from the cap end towards the tail end.

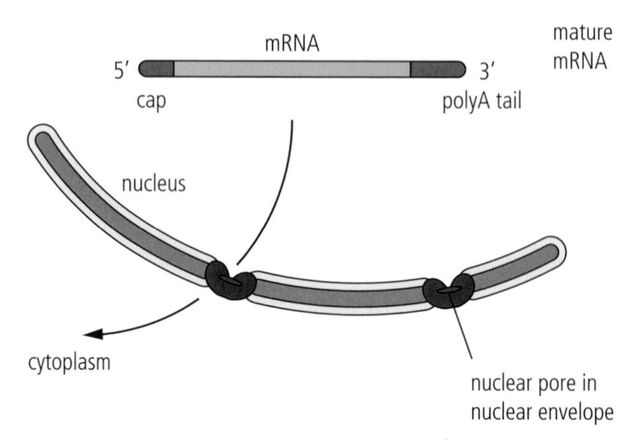

D2.2 Figure 2 The structure of an mRNA molecule. On one end we find the cap, and on the other end the polyA tail. This mRNA strand is ready to leave the nucleus and enter the cytoplasm for translation.

Special enzymes called **exonucleases** have the ability to chop up RNA molecules by removing nucleotides one by one from their extremities. But the exonucleases that digest mRNA molecules seem to have trouble attaching to the mRNA when their caps and tails are intact. To resolve this, other enzymes specialize in removing the caps and tails. The **decapping complex** is a group of enzymes that can remove the cap, while the **deadenylase complex** is a group of enzymes that can remove the adenine nucleotides in the polyA tail.

Exonucleases then break up the mRNA. The nucleotides are removed one by one and reused later. Once the mRNA has been broken up, it can no longer be used for translation, so RNA degradation is another way protein synthesis can be regulated by a cell.

D2.2.4 – Epigenesis

D2.2.4 – Epigenesis as the development of patterns of differentiation in the cells of a multicellular organism
Emphasize that DNA base sequences are not altered by epigenetic changes, so phenotype but not genotype is altered.

During the development of an embryo, cells need to go through the process of differentiation. Certain genes are expressed while others are not, so that each cell can specialize. For example, some cells will become muscle cells specializing in contraction and movement, some will become nerve cells specializing in sending electrical impulses, and others will become white blood cells as part of the immune system. These cells retain all the genetic information of the organism, its **genome**, but only follow the instructions from a small part of the genome.

Epigenesis is the process that results in the formation of organs and specialized tissue from a single undifferentiated cell (stem cells are covered in Chapter B2.3). Humans can make more than a hundred different types of specialized cells. In a human embryo there are three layers of cells:

- the **ectoderm**, which forms tissues such as the skin and brain
- the **mesoderm**, which forms tissues such as the skeleton and circulatory system
- the **endoderm**, which forms tissues such as the lungs and liver.

How does each cell become specialized? **Epigenetics** depends on the influence of non-genetic factors on the expression of genes. This process changes the phenotype of the cell without changing the genotype. Factors that generate epigenetic phenotypes can come from outside the organism (e.g. environmental stressors) or within the organism (e.g. signals from molecules present inside the body).

DNA methylation is the process by which a methyl group ($-CH_3$) is added to a DNA nucleotide such as cytosine. The methyl groups act as **epigenetic tags**: they flag part of the genome. Methylation of DNA in the promoter region of a gene makes the gene inaccessible to RNA polymerase. The gene will not be transcribed and therefore cannot be expressed; this is called **transcriptional silencing**.

Epigenetics modifies which parts of the genome are expressed or not. Epigenetics does *not* modify the genome itself. Be careful not to confuse the two terms epigenesis and epigenetics. Epigenesis is about an undifferentiated cell such as a fertilized egg being able to produce many types of specialized cells, whereas epigenetics is the activation or silencing of genes using methyl tags.

Embryonic stem cells are cells that have not differentiated yet, and the DNA in these cells is not methylated. Such cells are said to be **totipotent**, meaning they can become any category of cell. But as soon as differentiation starts, methylation becomes widespread. In muscle cells, for example, the appropriate genes for muscle will remain free from epigenetic silencing tags so that they can be transcribed and expressed. However, other genes, such as the ones that make digestive enzymes in the small intestine, are methylated to silence them (you would not want your muscles to be digested!).

As an embryo grows, copies of the DNA are made, and these copies contain the patterns of methylation that make each cell specialized for a particular task. This is how the layers, such as the ectoderm and mesoderm, of the embryo can form. With further methylation and specialization, the cells become skin, muscle, blood vessels, nerves, etc.

Epigenetics is a new explanation of how organisms have the traits they have. This is an example of a **paradigm shift**. A paradigm is a way of viewing or understanding a particular phenomenon. The paradigm after the discovery of genes was that they were solely responsible for genetic traits, but the new paradigm is that it is more complex than that and the environment can play a role in which traits are expressed and which are not. Are paradigm shifts more prevalent in the natural sciences compared to other areas of knowledge?

D2.2.5 – Genome, transcriptome and proteome

D2.2.5 – Differences between the genome, transcriptome and proteome of individual cells
No cell expresses all of its genes. The pattern of gene expression in a cell determines how it differentiates.

We do not know for sure how many genes humans have. Decades ago estimates were over a million, but the current estimate is 20,000 protein-encoding genes. Part of the difficulty in determining an exact number is deciding what counts as a gene, but also we have not yet determined whether certain sequences can produce a functional protein or not. The Human Genome Project is an ongoing international collaboration to find out where all the genes are on our chromosomes and to determine what their functions are.

Genome

Cells in the human body generally possess a full copy of the person's **genome**. (Exceptions include red blood cells, which do not have a nucleus, and gametes, which only contain half the genetic material). The genome is all of the genetic information that an organism received from its parents. Remember that no cell expresses all of the genes in its genome. Cells specialize and only express the genes they need to perform their special function.

Transcriptome

The term used to describe all the RNA that a cell makes is its **transcriptome**. Your liver cells have a different transcriptome compared to your muscle cells. They need to follow different sets of instructions from your genome, so they make different copies of RNA such as mRNA. Biologists interested in how an embryo transforms into a foetus measure the number of different **RNA transcripts** (transcribed strands) and the relative quantities produced to learn more about how cells differentiate. It is hoped that by studying the transcriptomes of cells we can better understand different areas of research, such as embryo development and cancer.

The technique used to identify RNA transcripts is called **RNA sequencing** or **RNA-seq**, and researchers use next-generation sequencing techniques similar to those used for DNA (see Chapter A3.1). Libraries of RNA codes exist in online databases, allowing researchers to compare what they have found in the laboratory to known transcripts.

Proteome

The term **proteome** is used to describe all the proteins that a cell, a tissue or an organism can produce. Fireflies and glow worms in the Lampyridae family can produce proteins that generate light. Mice, on the other hand, do not have the requisite genes and therefore do not produce such proteins. If we wanted mice to glow, we would have to introduce the genes for bioluminescence into their genome. Each differentiated cell has a different proteome. Different groups of cells in an organism will have a different proteome. One organism will have a different proteome compared to another.

Cells from different tissues will have different transcriptomes and different proteomes but will have the same genome. You can think about the three terms in order when you think about the central role of DNA: genome > transcriptome > proteome.

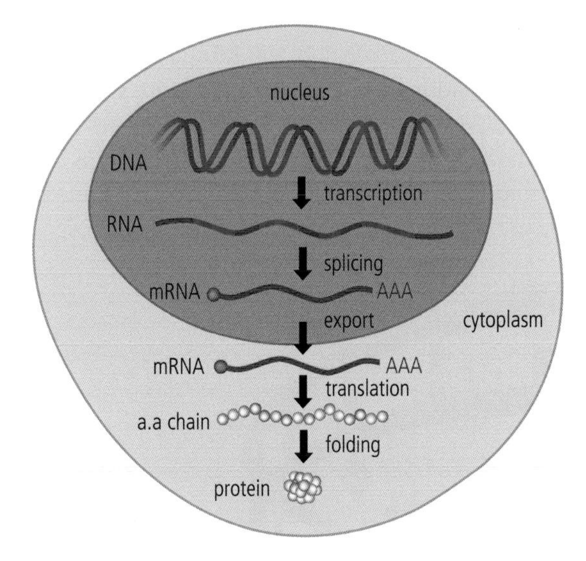

From the top to the bottom of this diagram, we can see the role of a cell's genome (DNA stored in the nucleus), transcriptome (all the RNA strands made) and proteome (all the proteins synthesized).

Knowing a person's proteome could potentially lead to personalized medicine (see Chapter A.3.1). By knowing what proteins a person produces, specific treatments can be tailored to how that person would react.

D2.2.6 – Methylation

D2.2.6 – Methylation of the promoter and histones in nucleosomes as examples of epigenetic tags
Methylation of cytosine in the DNA of a promoter represses transcription and therefore expression of the gene downstream. Methylation of amino acids in histones can cause transcription to be repressed or activated. Students are not required to know details of how this is achieved.

As we saw in Section D.2.2.4, cytosine can be methylated; a methyl group ($-CH_3$) is added to the cytosine, which silences the code at that point. DNA sequences that have undergone methylation will not be transcribed. Often methylation takes place in the promoter region of the DNA (see Section D.2.2.4). The promoter region is normally used by RNA polymerase to start the transcription process. The methyl groups are **epigenetic tags** that work like switches to turn transcription on or turn off.

Another method of regulating transcription is methylation of histones. Recall that histones are proteins around which DNA coils. Eight histone proteins wrapped in two loops of DNA form a **nucleosome**. Figure 3 shows ten nucleosomes. Long chains of nucleosomes form chromatin, which will coil up to make a chromosome.

The proteins in nucleosomes have amino acid sequences that form **histone tails** extending outwards. The amino acid lysine exists in more than one position along the tail of a type of histone called **histone 3** or **H3**. Depending on where the methylated lysine is found, it can either activate or silence gene expression (see the bottom half of Figure 3). If the lysine towards the tip of the histone tail (lysine 4) is methylated, the gene will be transcribed. In contrast, if the lysine along the tail (lysine 9 or 27) is methylated, the gene will be silenced.

When writing out amino acid sequences, single letters are used to represent each amino acid. The letter used to label lysine is K. So the amino acids along the histone tail we are interested in are K4 for activation, and K9 and K27 for silencing. To silence a gene using histone methylation involves making the loops of DNA coil more tightly around the histones so they are more compressed against each other and not accessible, whereas activating a gene involves loosening the loops of DNA to separate the histones from each other and make the DNA available for transcription.

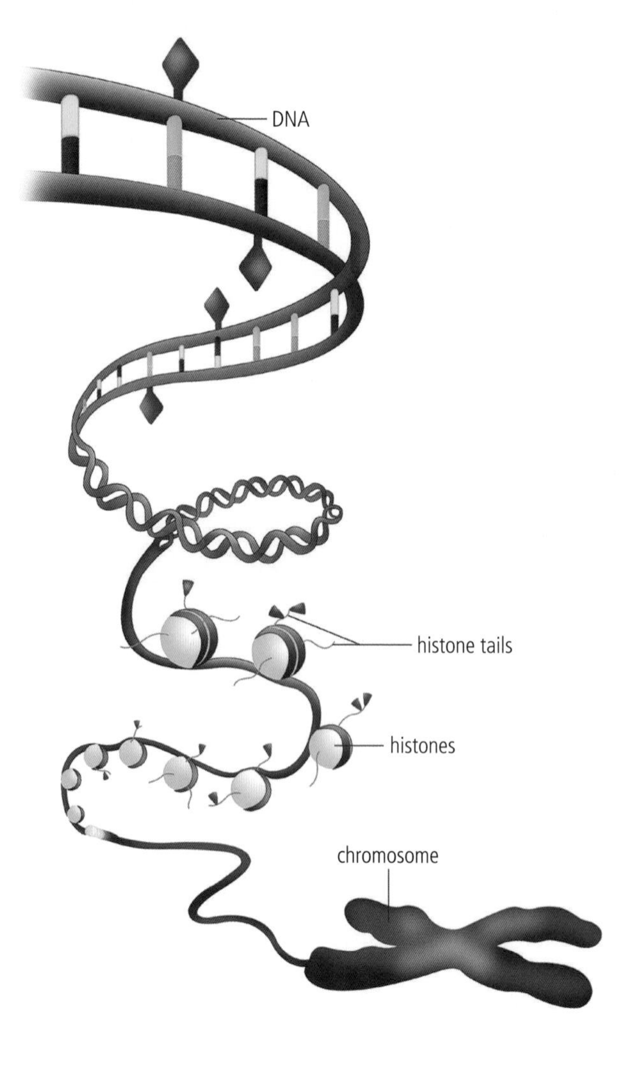

D2.2 Figure 3 Two ways methylation can work as epigenetic tags: at the top of the diagram, it is the DNA itself that has been tagged by a methyl group (purple diamonds) whereas at the bottom of the diagram, histone tails have been tagged by methylation (purple triangles). Notice how some methyl group tags are found towards the tip of the histone tails while others are along the tails.

D2.2.7 – Epigenetic inheritance

D2.2.7 – Epigenetic inheritance through heritable changes to gene expression
Limit to the possibility of phenotypic changes in a cell or organism being passed on to daughter cells or offspring without changes in the nucleotide sequence of DNA. This can happen if epigenetic tags, such as DNA methylation or histone modification, remain in place during mitosis or meiosis.

Three important things to remember about epigenetic modifications are:

- **they are reversible**
- **most are erased, but a few can be passed on to the next generation of cells or to offspring**
- **the DNA code itself is *not* modified by epigenetics.**

Epigenetics is useful for a population because it allows it to adapt its gene expression to different situations. Some populations live in climates where there is food year round, whereas others live where there is plentiful food in the warmer months but very little in the colder months when the ground is frozen and covered in ice. Epigenetics allows a genome, which you will recall is a fixed code, to be applied in different ways in different situations, by turning some genes on and others off. It is a way for the next generation of cells or the next generation of individuals to prepare for adverse conditions in the future. Researchers are extremely interested in epigenetics because of its potential to help us understand health issues related to ageing, obesity, addiction, depression and cancer.

Many epigenetic tags are lost when sperm and egg cells are produced, but some are passed on to the next generation. The implication of this is enormous. It means that some of the experiences an organism's parents or grandparents had could be influencing which genes are being activated or silenced in the current generation.

The care this parent is giving to the young rhesus monkey is methylating its DNA to allow certain genes to be activated and others to be silenced.

One well-studied example of epigenetics observed in humans is the Dutch famine birth cohort from World War II. Between November 1944 and May 1945, food was difficult to obtain in many parts of the Netherlands because of the war. Pregnant women, who should normally increase their caloric intake to provide for their growing baby, saw their daily intake go down to just 500 kcal a day. When they grew up, the children of the women most exposed to hunger in their first trimester of pregnancy had twice as many cases of cardiovascular disease than their siblings born in different years. DNA methylation patterns were different between the siblings who were or were not born during the time of famine.

To better understand whether effects are generated by epigenetics, studies have been carried out on mice, rats and monkeys to measure outcomes of increased disease and changes in phenotype. The way a young monkey is treated by its mother, for example, can generate very different behaviours or disease risks later in life. Such studies demonstrate significant differences in methylation levels depending on the conditions the animals are exposed to.

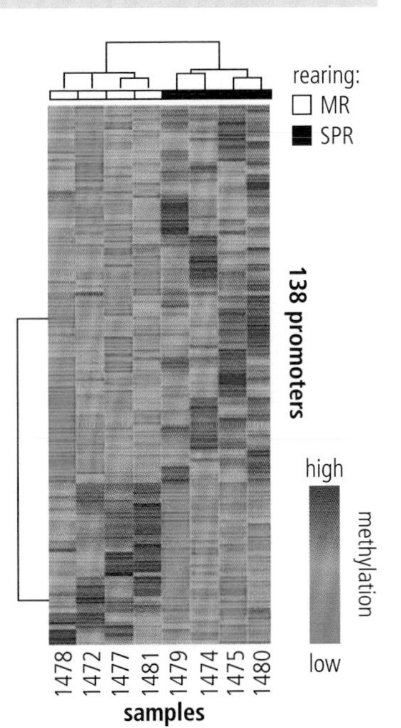

Different methylation levels were observed in rhesus monkeys raised by their own mothers (MR), the four samples on the left, and raised by an inanimate surrogate mother (SPR), the four samples on the right.

TOK

The way an idea is presented can influence how we understand it. Sometimes the use of analogies and models can help, but sometimes they can skew our understanding. Discuss this analogy: the genome versus epigenetics is like a recipe book and a chef. Here are two more to think about: movie script and movie director; video game and gamer. What strengths and weaknesses do such analogies have? Which ones best illustrate the fact that one is fixed and was generated in the past, while the other is open, dynamic and can be changed now and in the future? How can a model be useful even if it is obviously false?

In animals (including us) it appears that certain behaviours and the susceptibility to certain diseases are affected by epigenetics, and can result from the experiences of parents or grandparents, or from experiences shortly after birth (see the Global context box on the previous page). Fortunately, epigenetic modifications are reversible. Researchers are investigating methods to remove the silencing methylation on genes they want to turn back on. The other good news for humans is that, if we understand the methylations that we have inherited from a parent, such as an increased risk of obesity or heart disease, we can learn how to make healthy life choices to counterbalance the negative effects that our methylation patterns might have.

D2.2.8 – Environmental effects

D2.2.8 – Examples of environmental effects on gene expression in cells and organisms
Include alteration of methyl tags on DNA in response to air pollution as an example.

Where we live and what chemicals we are exposed to in our environment can affect our epigenetic tags. Air pollution, for example, can lead to a host of health problems, including asthma, heart disease and lung cancer. The types of air pollutants that are linked to health risks include ground-level ozone (O_3), nitrogen oxides (such as NO_2, or expressed as NO_x), particulate matter (PM) and polycyclic aromatic hydrocarbons (PAHs).

Transportation is one of the leading causes of atmospheric pollution. The smog or haze that you see in the sky above cities is a mixture of multiple chemicals in the air from the combustion engine cars we drive, and the factories that produce the products and energy we use.

DNA methylation has been measured in white blood cells, brain cells and certain genes such as *CCL17*, which plays a role in inflammation and the immune system. Studies have revealed different methylation patterns in the genes of children depending on whether their mother was exposed to high or low levels of PAHs in the air. In addition, females who live closer to major roads, and therefore are exposed to more air pollution, have different methylation patterns in the DNA that is responsible for the formation of the placenta than females who live further away from major roads.

The babies of females exposed to higher air pollution levels have a lower body mass on average compared to those of females exposed to lower air pollution levels. Embryonic cells and the cells of a developing foetus are more susceptible to modifications than cells later in life. This is one of the reasons why pregnant females should be careful about what they eat and why there are so many warnings on products recommending that pregnant females abstain from consuming substances such as alcohol, nicotine and caffeine.

In what ways does the environment stimulate diversification?

D2.2.9 – Removal of epigenetic tags

D2.2.9 – Consequences of removal of most but not all epigenetic tags from the ovum and sperm
Students can show this by outlining the epigenetic origins of phenotypic differences in tigons and ligers (lion–tiger hybrids).

Sperm cells and egg cells come from cells called **primordial germ cells (PGCs)**. These are formed inside an individual when they are a developing foetus. PGCs have their epigenetic tags removed by a process of epigenetic reprogramming. Before fully maturing into egg cells or sperm cells, some of the DNA in developing sex cells is remethylated, so that they can produce a viable zygote. **Imprinted genes** are those that have been silenced in only one of the two copies, either the paternal copy or the maternal copy of the gene. These genes bypass the epigenetic reprogramming process.

The methylation of genes can tag them to be activated or silenced. This epigenetic phenomenon does not modify the person's DNA sequence the way a mutation can, but it can have an influence on an individual's phenotype.

This phenomenon was discovered when researchers tried to combine the nuclei from two mouse eggs to form a zygote, or tried to combine the nuclei from two mouse sperm cells to form a zygote. Even though there was sufficient genetic material to form a new mouse, none of the cells developed into mice embryos. This is because genes in the egg and sperm are imprinted differently. Imprinted genes will be methylated one way in the male copy of the gene, and a different way in the female copy of the gene.

Chapter D3.2 explains how the maternal and paternal copies of a gene typically interact when they are not imprinted: if they match, there is no conflict about which phenotype the offspring will have. If there are different versions of the gene (one for type A blood from the mother and the other for type O blood from the father, for example), one can mask the other and only one version of the gene will be expressed. For example, if a child inherits type A blood from the mother and type O from the father, the child will have type A blood. The expressed gene may be inherited from the person's father (the paternal copy) or mother (the maternal copy). With imprinted genes, however, one parent's copy is silenced using methylation. With one gene silenced, the remaining copy alone will determine the phenotypic outcome.

An example of a pair of genes that can show imprinting is the gene for insulin-like growth factor II (IGF2), which regulates growth, notably in the developing foetus, as well as the gene for the IGF2 receptor, a protein that receives the growth factor's message on the surface of cells in mammals. Each parent can donate a gene and each one can be methylated to be silenced. The table shows what happens when the paternal or maternal genes are turned off in mice.

Maternal copy ***IGF2r* receptor gene**	**Paternal copy** ***IGF2* gene**	**Result**
Off	On	Offspring are too big
On	Off	Offspring are too small

What mechanisms are there for inhibition in biological systems?

Tigers and lions are similar enough species that they can produce hybrids. A female lion and a male tiger can produce a tigon, while a male lion and a female tiger can produce a liger. Imprinted genes for growth seem to function differently in each species, so when the two are mixed there can be some spectacular results. Ligers can grow to an enormous size, over 3 m long and weighing over 500 kg, bigger than each of their parents, making them the largest cats on Earth. It seems that the genes from the father that affect growth are switched on, allowing the offspring to grow.

Interestingly, if the hybrid has the opposite parents: the mother is a lion and the father is a tiger, such large sizes are not observed. This shows that it matters which genes come from the mother and which come from the father, because zygotes tend to activate genes from one parent and silence genes from the other.

A liger will have a mix of genetic characteristics from its father (a lion) and its mother (a tiger), but the way the imprinted genes work in these two species means the size of the animal will be significantly bigger than either parent.

D2.2.10 – Monozygotic twins

D2.2.10 – Monozygotic twin studies
Limit to investigating the effects of the environment on gene expression.

One way to study the effect of the environment on gene expression is to use twin studies. Non-identical or **heterozygotic twins** form when two separate eggs are fertilized at the same time by different sperm cells. Heterozygotic twins do not look any more alike than other siblings might. But identical or **monozygotic twins** are formed from the same zygote: instead of forming a single embryo, two embryos form. The two offspring have the same genetic code in their nuclei because they both originate from the same fertilized egg. However, as we have seen, the genome is only part of the story when it comes to phenotype. The environment can also affect which genes are expressed and which are silenced.

We know that simply because twins have identical genomes does not mean that they are the same and that they will be affected by the same things. It is possible to have one twin who is healthy while the other develops cancer, or one twin who is autistic while the other is not. In order to find out why two individuals with the same genome can have different phenotypes, researchers look for areas along DNA sequences called **differentially methylated regions** (**DMRs**). Looking for DMRs in monozygotic twins allows researchers to find out whether disease is connected to differences in epigenetics. When comparing methylation patterns, there are very few differences in newborn monozygotic twins. However, with adult twins, the differences in methylation patterns increase. This difference is bigger in twins that grew up in different environments compared to those that grew up in the same environment.

TOK

Instead of using human subjects, animal models such as rodents and monkeys are often used to study biology. This generates multiple debatable points. One is about how we decide whether or not it is ethical to manipulate such animals in cages just so that we can pursue knowledge. The other is whether, just because something is true in a mouse or other model organism, that means it will be true in humans. How much validity do claims about humans have if they are based on animal models? How much evidence would we need to conclude that an animal model can predict human traits and behaviour? Is it possible that there is no amount of evidence that can convince people that complex problems can be explained by the presence of molecules in our cells?

D2.2.11 – External factors

D2.2.11 – External factors impacting the pattern of gene expression
Limit to one example of a hormone and one example of a biochemical such as lactose or tryptophan in bacteria.

The hormone insulin has an important role to play in the body: when blood sugar levels are too high, insulin is produced and secreted into the bloodstream to signal to cells that they should allow the sugar to enter into their cytoplasm. For this process to work, the gene for insulin needs to be actively transcribed and translated into insulin when blood sugar levels are high. The presence of glucose in the blood triggers the transcription of the insulin gene *INS*. Transcription factors link to the DNA sequence at the enhancer region, allowing transcription to begin. When blood sugar levels drop, transcription is stopped.

We have mostly focused on gene regulation in eukaryotes, but prokaryotes have techniques for turning on and off transcription too. Lactose is a sugar found in milk. *Escherichia coli* bacteria live in our gut and, when lactose is present, they produce the enzyme lactase. If no lactose is present as a food source, the relevant digestive enzyme does not need to be produced. It would be a waste of valuable protein for the bacteria to produce lactase all the time.

TOK

Although we are only just beginning to understand epigenetics, picture a world where we know everything about it. Would it be worthwhile being able to explain every human disease using a combination of genetics and epigenetics? If there was a test to find out what types of methylation you had, which allowed predictions about your susceptibility to cancer, obesity and depression, would you want to know? Is all knowledge worth pursuing? Think about the types of knowledge that might be detrimental for us to know. Discuss how researchers decide whether something is worth pursuing or whether it should be left alone.

The mechanism that turns on and off the production of lactase in *E. coli* works in the following way. A zone of DNA upstream from the gene but downstream from the promoter acts as a binding site for a repressor molecule called the ***lac* repressor**. This binding site is called the **operator** and, as long as the *lac* repressor is stuck to this sequence of DNA, it acts as a roadblock and stops the RNA polymerase from latching on and transcribing the genetic code.

However, when lactose arrives, it binds to the *lac* repressor and deactivates it so that it detaches from the operator and allows the RNA polymerase to transcribe the genetic code. There is more than one gene for lactase, and the promoter region, operator and gene sequences along the DNA make up what is called the ***lac* operon**.

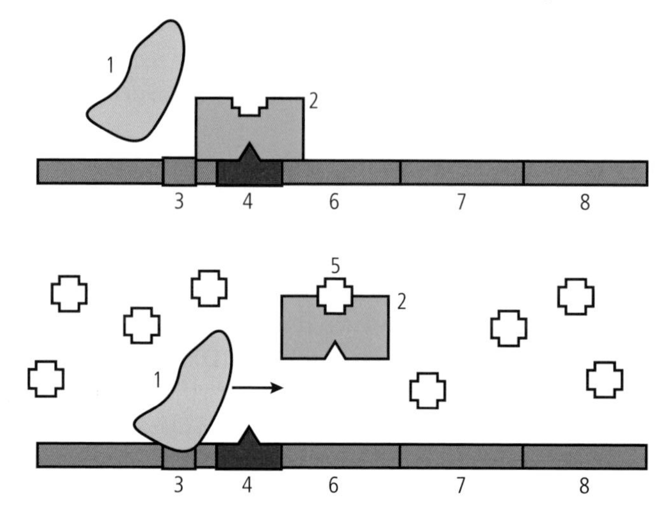

(1) RNA polymerase, (2) repressor, (3) promoter region, (4) operator and (6,7 and 8) genes are all needed to make lactase.

Guiding Question revisited

How is gene expression changed in a cell?

In this chapter you have learned that:

- the genetic code we inherit from our parents is just part of our genetic story, transcription factors and epigenetics can activate some genes and silence others so that, even if we inherit a gene, it does not mean that it will show up in our phenotype
- epigenetic tags can be added via methylation of cystine or histones during a person's lifetime, at any stage between embryo development and adulthood, which explains why identical twins possess the same genes but express different phenotypes
- promoters and repressors can activate or silence gene expression, notably when external molecules are introduced, as seen in the case of the *lac* operon
- experiences and exposure to certain molecules can add or take away certain epigenetic tags.

Guiding Question revisited

How can patterns of gene expression be conserved through inheritance?

In this chapter you have learned that:

- some of the tags acquired through experience or exposure to the environment can be passed on to the next generation so that, potentially, a grandmother's epigenetic tags can be expressed in her granddaughter
- by studying lion–tiger hybrids, patterns of gene expression can be observed from one of the two species but not the other.

Exercises

Q1. If you looked at the tail end of an mRNA molecule, what would you find?

A An abundance of methylated zones.

B A lack of methylated zones.

C A promoter region.

D AAAAAAAAA.

Q2. Which of the following is true of epigenetic tags?

I. They are reversible.

II. They generally cause a modification of the genome.

III. A few can be passed on from one generation to the next.

A I and II only.

B I and III only.

C II and III only.

D I, II and III.

Q3. Which of the following sequences shows how enzymes can be made in a cell?

A Translation of the proteome > transcription of the genome > transcriptome.

B Translation of the genome > transcription of the proteome > proteome.

C Transcription of the genome > translation of the transcriptome > proteome.

D Translation of the transcriptome > transcription of the proteome > genome.

Q4. Explain how a gene can be silenced by methylation.

Q5. Explain why mice and rhesus monkeys are used in studies of epigenetics.

HL end

D2.3 Water potential

Guiding Questions

What factors affect the movement of water into or out of cells?

How do plant and animal cells differ in their regulation of water movement?

Many factors affect the movement of water into and out of cells. Plants must accomplish the seemingly impossible task of getting water from the roots to the top of the plant without a circulating organ. Unlike plants, animals have circulating organs that transport water to all the cells in the organism. It is important to remember, however, that in both plants and animals there are mechanisms that allow water to move through the cell membrane. The environment of organisms is subject to rapid changes, often resulting in varying water needs. If an organism does not adapt to these changes, its life will be at risk.

Plants and animals have different means of regulating water movement. Water delivery to plant cells is influenced by cell walls, which animal cells do not possess. However, the same underlying principles of water movement apply to all life on our planet. All life forms are dependent on water and the substances dissolved in it. Plant cells also require water for support and to maintain shape. The transportation of water and its dissolved content throughout an organism is essential for the maintenance of life.

D2.3.1 – Water as a solvent

D2.3.1 – **Solvation with water as the solvent**
Include hydrogen bond formation between solute and water molecules, and attractions between both positively and negatively charged ions and polar water molecules.

Water is the solvent of life and is known as the **universal solvent**. Living cells typically exist in an environment where there is water both within the cell (as cytoplasm) and outside the cell (as intercellular fluid, freshwater or saltwater, etc.). We refer to solutions as **aqueous** solutions if water is the solvent, no matter what mixture of substances makes up the solutes. Thus cytoplasm and the ocean are both aqueous solutions.

The structure of water molecules

The basic chemical structure of water is presented in Chapter A1.1. In summary:

- water molecules are polar covalent molecules
- the oxygen region of a water molecule has a slight negative charge, while the two hydrogen regions each have a slight positive charge
- water molecules form large numbers of weak hydrogen bonds with other water molecules
- the hydrogen bonds are individually weak, but collectively are strong enough to account for most of the unique properties of water that make it so important to life

- when water is in a liquid state, hydrogen bonds are continually breaking, reforming and moving around
- polar substances, such as salts, alcohols and acids, dissolve in water quite readily
- non-polar substances, such as fats and oils, do not dissolve in water.

Solvation with water as the solvent

Solvation is the interaction of a solvent with a dissolved solute. Many refer to solvation as hydration. There are three steps to solvation:

1. the particles of a solute separate from each other
2. the water particles separate from each other
3. the separated solute and water particles combine to make a solution.

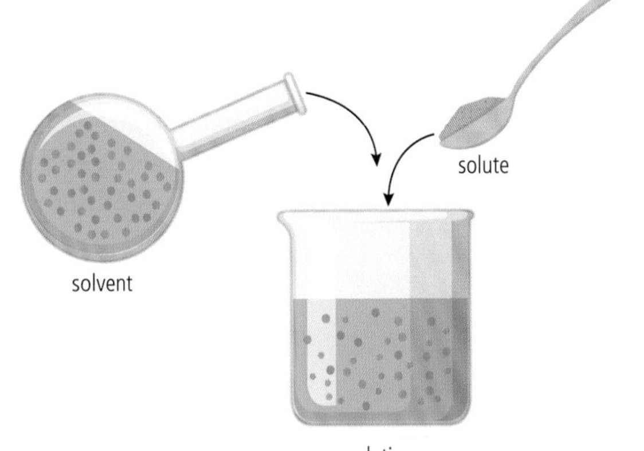

▲ The process of solvation.

Water molecules have hydrogen bonds between them. In order to dissolve a solute the hydrogen bonds between the water molecules must break, as must the bonds between the solute molecules. The water molecules then surround the solute molecules and new hydrogen bonds form. Figure 1 shows the electrical attraction between solute and polar water molecules.

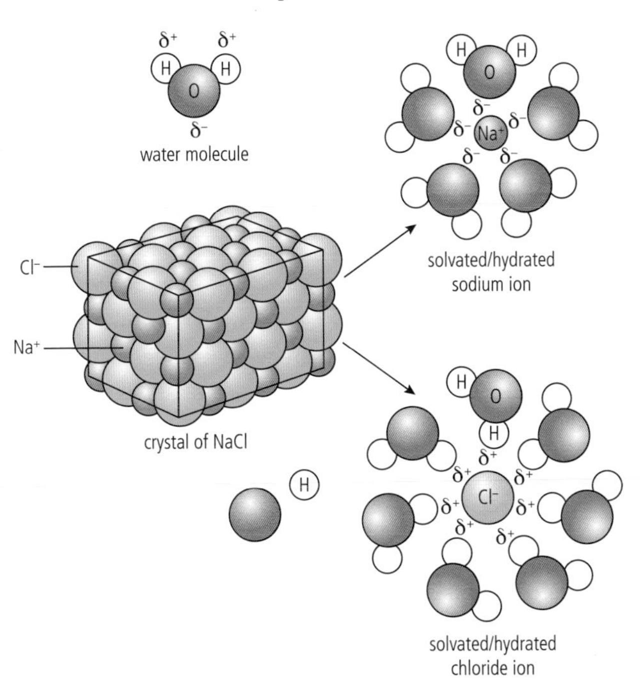

D2.3 Figure 1 Attractions between positively and negatively charged ions and polar water molecules.

Sodium chloride is solvated by polar solvents such as water. The small positive charges on the hydrogen regions of the water molecules surround the negatively charged chloride ions. The small negative charges on the oxygen region of the water molecule are attracted to the positively charged sodium ions. Hydrogen bonds form between the water molecules and the ions. Sodium chloride is relatively soluble in water but other polar molecules may be less soluble. Non-polar molecules such as lipids will not dissolve in water. Each substance has unique chemical properties that affect its solubility in water.

What are the implications of solubility differences between chemical substances for living organisms?

D2.3.2 – Water movement in relation to solute concentration

D2.3.2 – Water movement from less concentrated to more concentrated solutions
Students should express the direction of movement in terms of solute concentration, not water concentration. Students should use the terms "hypertonic", "hypotonic" and "isotonic" to compare concentration of solutions.

As biologists, we generally look at water movement in relation to either cells or the total organism. The cell membrane is very important when considering the movement of substances into and out of cells. It is imperative that this movement of materials is constantly monitored and controlled so that the best possible conditions for life are maintained for the cell and/or organism.

The terms described in Table 1 are introduced in Chapter B2.1, and now is a good time to review them.

D2.3 Table 1 Solution concentration terms

Term	Solute concentration in environment	Water concentration in environment
Hypertonic	Higher	Lower
Hypotonic	Lower	Higher
Isotonic	Equal	Equal

In order to work out which way water is moving between cells and their surroundings, we need to be able to talk about the concentrations of the solutions inside and outside cell membranes. We always talk about the concentration of the solution (not the concentration of the water).

If we say a cell is in a **hypertonic** environment, we mean it is surrounded by a solution that is higher in solutes and lower in water relative to the cytoplasm inside the cell membrane. A cell in a **hypotonic** environment is surrounded by a solution that is lower in solute particles. A cell in an **isotonic** environment has equal concentrations of solute inside and outside the cell.

D2.3 Figure 2 Describing solution concentrations. Notice the direction that the water is flowing in for the three different solution concentrations. The cell membrane is selectively permeable, allowing water to flow through it but not the solute particles.

Solution is hypotonic

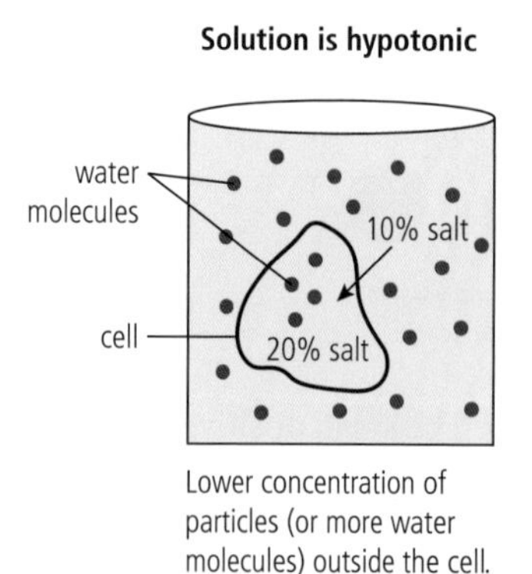

Lower concentration of particles (or more water molecules) outside the cell.

Solution is isotonic

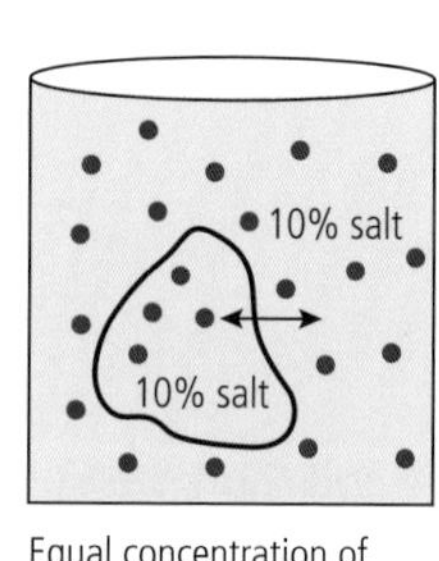

Equal concentration of particles (and water molecules) inside and outside the cell.

Solution is hypertonic

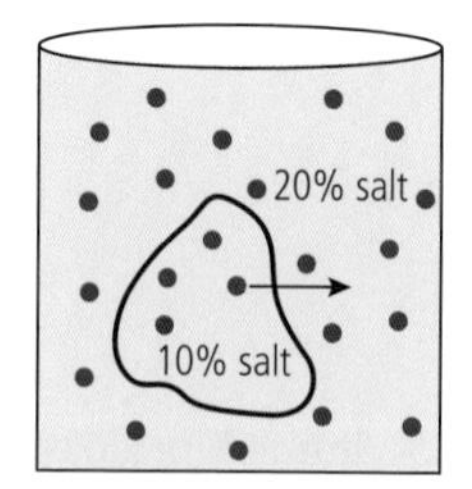

Higher concentration of particles (or less water molecules) outside the cell.

Figure 2 shows water movement in relation to solute concentration. Water moves from an area with a lower solute concentration to an area with higher solute concentration. In the case of the isotonic solutions, water movement is equal in both directions because there are equal concentrations of solute particles on both sides of the cell membrane. In this isotonic condition, water molecules continue to move in both directions as a result of the kinetic energy they possess. There is no net movement of water molecules.

D2.3.3 and D2.3.4 – Hypotonic and hypertonic solutions and osmosis

D2.3.3 – Water movement by osmosis into or out of cells
Students should be able to predict the direction of net movement of water if the environment of a cell is hypotonic or hypertonic. They should understand that in an isotonic environment there is dynamic equilibrium rather than no movement of water.

D2.3.4 – Changes due to water movement in plant tissue bathed in hypotonic and those bathed in hypertonic solutions
Application of skills: Students should be able to measure changes in tissue length and mass, and analyse data to deduce isotonic solute concentration. Students should also be able to use standard deviation and standard error to help in the analysis of data. Students are not required to memorize formulae for calculating these statistics. Standard deviation and standard error could be determined for the results of this experiment if there are repeats for each concentration. This would allow the reliability of length and mass measurements to be compared. Standard error could be shown graphically as error bars.

The passive transport processes of diffusion and osmosis are discussed in Chapter B2.1. The key points about these processes are:

- both diffusion and osmosis are examples of passive transport because they do not require energy in the form of cellular adenosine triphosphate (ATP) to occur
- in both processes, particles of a substance move along a concentration gradient
- osmosis requires a cell membrane, whereas diffusion can occur with or without a cell membrane
- osmosis involves the movement of water across a selectively permeable membrane
- a concentration gradient of water during osmosis is the result of the solute particle concentrations on either side of the cell membrane.

It is important to remember that water molecules are polar, making it difficult for them to pass through the hydrophobic region of a cell membrane. Chapter B2.1 discusses **aquaporins**. Aquaporins are essentially hydrophilic tunnels through the cell membrane that polar water molecules can pass through. Without aquaporins, there is limited water movement across a cell membrane. In osmosis, water moves along a concentration gradient through the aquaporins of the selectively permeable cell membrane. Osmosis will proceed if there is a hypotonic or hypertonic exterior cell environment. When the cell is in an isotonic environment, the water molecules still move but there is no net movement of the water molecules. The cell is said to be in **dynamic equilibrium**.

Challenge yourself

We are going to look at some examples of water movement in and out of cells by osmosis. Consider the three scenarios, and answer the following questions.

1. Raisins in pure water.
2. Human red blood cells in a solution with a high solute concentration.
3. Gargling with saltwater to relieve a sore throat.

For each of these examples, answer the following questions.

(a) What term best describes the external cell environment?
(b) Which way does water move and why?
(c) What is the result of osmosis?

The following activity will allow you to study water movement in plant tissue when it is placed in solutions of varying tonicity. Tonicity refers to the ability of a solution to affect the fluid volume and pressure in a cell. After reading the overall procedure in the downloadable laboratory file accessed from this page of the eBook, write a hypothesis including an explanation.

Follow the directions to determine the osmolarity of potato tissues. Osmolarity refers to the concentration of solute particles. The samples will be bathed in hypotonic, isotonic and hypertonic solutions. Note that tissues from plants other than potatoes may be used.

D2.3.5 – Water movement without cell walls

D2.3.5 – Effects of water movement on cells that lack a cell wall
Include swelling and bursting in a hypotonic medium, and shrinkage and crenation in a hypertonic medium. Also include the need for removal of water by contractile vacuoles in freshwater unicellular organisms and the need to maintain isotonic tissue fluid in multicellular organisms to prevent harmful changes.

Not all cells have a cell wall. Animal cells lack cell walls, which means that osmosis has some particular effects on these cells. To look at this in detail we can use the example of human red blood cells placed in a solution with a high solute concentration (a hypertonic medium). Because of the relatively high number of solute particles outside the cell, water molecules move out of the cell. With no cell wall present, the results are almost immediate, and the cell visibly shrinks, ultimately resulting in **crenation** (extreme shrinkage where the cell becomes crinkled). Figure 3 shows crenation of a red blood cell.

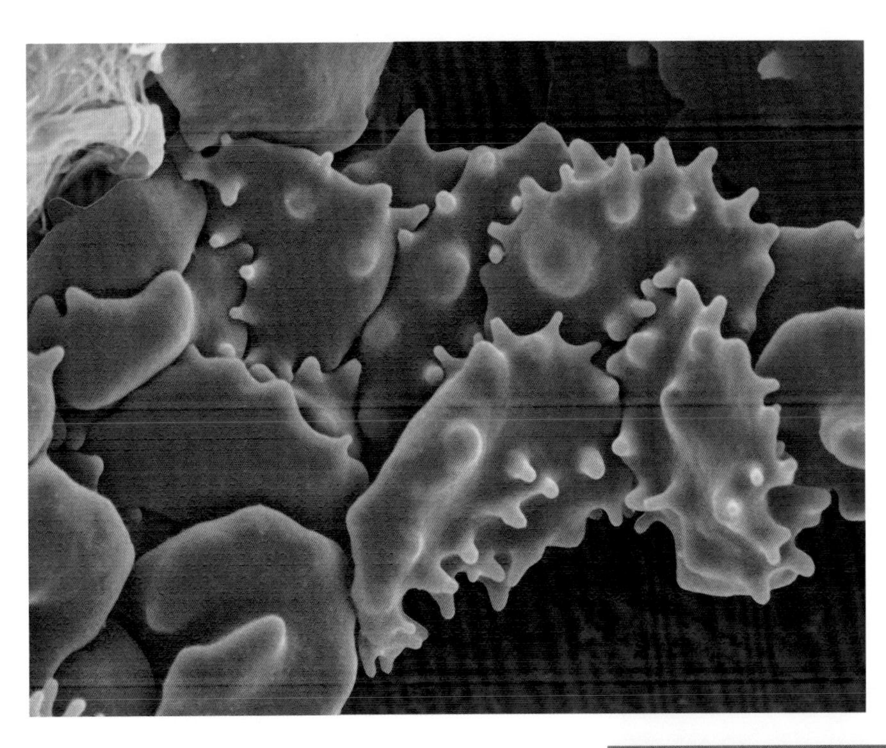

D2.3 Figure 3 Crenation in red blood cells placed in a hypertonic, high solute concentration, environment.

Conversely, if red blood cells are placed in a hypotonic cell environment (for example pure water), water will move into the cell and the solute particle concentrations will become closer to equal. This results in swelling of the red blood cells in the solution, possibly to the point of bursting the cell membrane.

Adaptations have evolved in both single-celled and multicellular organisms to prevent harmful swelling and shrinkage of cells. Aquatic single-celled animals often possess a specialized organelle known as a **contractile vacuole**. This is a regulatory organelle that collects excess water from the interior of the cell and periodically empties it into the surrounding environment. This prevents cells swelling when there is a hypotonic environment.

A contractile vacuole in *Paramecium caudatum*, visible as a circle with radiating canals, in the middle left region of the organism. The contractile vacuole functions in controlling water concentrations within this single-celled organism.

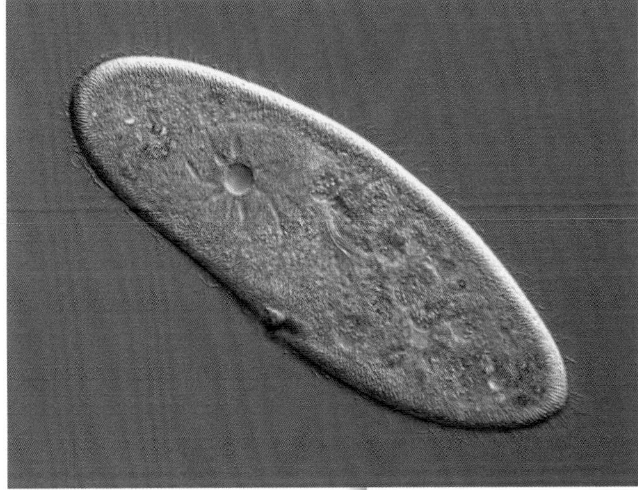

Humans experience the effects of cell shrinkage if they become dehydrated. If dehydration is severe, the cells of the body will begin to shrink and malfunction. In some cases, light-headedness and fainting can occur. Dehydration can cause a build-up of waste products within the body, including in the tubules of the kidneys. Blood flow throughout the body may be diminished, possibly causing muscle cramping and organ malfunctions. To overcome these potential results, it is essential that we drink adequate amounts of water.

Although extremely rare, very rapid water consumption can cause a condition called **hyponatremia**. This condition occurs when water dilutes the blood sodium levels below a certain level. The result is an increased uptake of water into cells, potentially harming those cells.

Challenge yourself

4. The human body is about 60% water, but that water is not quite as salty as seawater. Knowing that our body's water salinity levels are less than seawater, what would be the effect of drinking seawater?

D2.3.6 – Water movement with cell walls

D2.3.6 – Effects of water movement on cells with a cell wall
Include the development of turgor pressure in a hypotonic medium and plasmolysis in a hypertonic medium.

The presence of a cell wall leads to some interesting effects when water moves into and out of plant cells. Most plant cells are hypertonic relative to their environment, and this includes their large central vacuole. Because of this, water tends to move into plant cells. The result is a high **hydrostatic pressure**. Hydrostatic pressure is the pressure that a fluid exerts in a confined space, against the boundary of the space. In plant cells, the pressure is exerted against the cell walls and is called **turgor pressure.** Incoming water swells the cell and presses the cell membrane against the rigid cell wall. Most plants depend on this turgor pressure to maintain their shape and remain upright. Figure 4 represents the different conditions a plant cell may experience as a result of water movement.

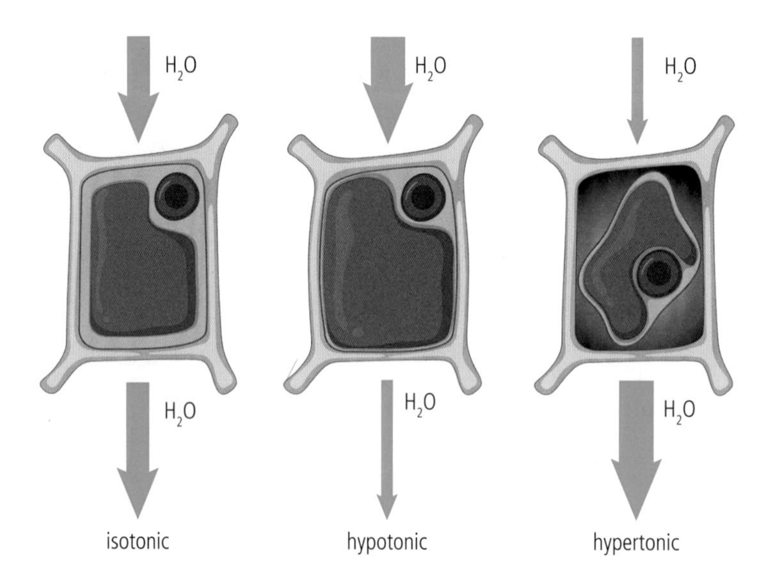

D2.3 Figure 4 The cell wall of a plant makes it difficult to see the many changes that occur inside as a result of water movement. The rigid cell wall resists changes in shape. However, the cell membrane and cell contents are affected by water moving into and out of the cell.

When water is lost from a plant cell, the cell membrane and the cytoplasm shrink away from the cell wall (look at the right-hand cell in Figure 4). This process of cytoplasmic shrinking in plant cells is called **plasmolysis**. When turgor pressure is lost in a plant, plasmolysis occurs and the plant visibly wilts. Large scale and long-lasting plasmolysis can result in cell death and, of course, plant death.

Many people use natural weed killers in their gardens rather than commercial weed killers. These natural weed killers are a mixture of common table salt and vinegar, and a small amount of detergent, in water. They work by causing water to move out of a plant's cells, resulting in massive plasmolysis and death of the weed.

D2.3.7 – Isotonic solutions

D2.3.7 – Medical application of isotonic solutions
Include intravenous fluids given as part of medical treatment and bathing of organs ready for transplantation as examples.

Food, blood products and medication can be administered through a line (catheter) placed in a peripheral vein. These intravenous (IV) fluids are used in many medical treatments for conditions such as haemorrhaging, surgery, cancer and dehydration. IV fluids must maintain a balance of the solutes that exist in the solutions in our cells, called intracellular fluids (ICF), and outside our cells, called extracellular fluids (ECF).

If ECF have a greater solute concentration than ICF, water will leave the body cells, resulting in shrinking and possible crenation. If ECF have a lower concentration of solutes than ICF, water will enter the cells, possibly resulting in swelling and even bursting. When ICF and ECF are isotonic, a dynamic equilibrium is maintained and, while water does move in and out of the cells, there is no net movement. IV solutions of different solute concentrations are used to treat particular medical conditions. Most IV solutions are isotonic so that they do not to cause excess water movement in or out of the body cells.

For organ transplants, there are many steps involved in the preparation and storage of a suitable organ. The organ is immersed in isotonic solutions during the period before it is transplanted, to avoid any cell damage.

Soft contact lens are kept in a physiological saline solution that is isotonic with eye tissue. This solution can also be used to irrigate the eye without damage because it contains the same solute concentrations as the eye tissue.

Examples of IV solutions administered are normal saline, lactated Ringer's and D_5W. All these solutions are isotonic but differ in the types and relative amounts of solutes they contain. Physiological saline solutions are sterile and isotonic to normal body fluids, containing a 0.9% solution of sodium chloride in water.

HL

D2.3.8, D2.3.9 and D2.3.10 – Water movement through plants

D2.3.8 – Water potential as the potential energy of water per unit volume
Students should understand that it is impossible to measure the absolute quantity of the potential energy of water, so values relative to pure water at atmospheric pressure and 20°C are used. The units are usually kilopascals (kPa).

D2.3.9 – Movement of water from higher to lower water energy
Students should appreciate the reasons for this movement in terms of potential energy.

D2.3.10 – Contribution of solute potential and pressure potential to the water potential of cells with walls
Use the equation $\Psi_w = \Psi_s = \Psi_p$. Students should appreciate that solute potentials can range from zero downwards and that pressure potentials are generally positive inside cells, although negative pressure potentials occur in xylem vessels where sap is being transported under tension.

Sequoiadendron giganteum, the redwood sequoia, is the tallest tree in the world. These trees can reach heights of more than 118 m (390 feet) and have a trunk diameter of more than 10.5 m (35 feet). The transport of water with essential solutes is an amazing feat in these plants.

Although lacking a circulatory system with a centralized pumping organ, plants are still able to transport nutrients, waste products and water.

Water potential

Water potential is a measure of potential energy per unit volume in water. Water potential drives the movement of water in living organisms. Measuring the absolute potential of

Potentials are a way of representing free energy. Free energy is the potential to do work. Specifically, we talk of Gibbs free energy when describing the potential to do work by biological systems. Potential energy is therefore the potential to do work.

water is extremely difficult, so the measurements used are relative to pure water energy at atmospheric pressure at a temperature of 20°C. The unit for water potential is usually kilopascals (kPa). However, you may also see MPa or megapascal given as the unit.

$$1 \text{ kPa} = 0.001 \text{ MPa}$$

Water potential is never positive. The water potential of pure water at atmospheric pressure and 20°C is zero. Solutes attract water molecules and make it harder for them to move (see Figure 1), so the water potential of a solution will always be negative (i.e. less than pure water, which has a water potential of zero).

Water potential is represented by the Greek letter psi (Ψ) with a subscript w (Ψ_w). Because of the basic laws of physics and potential energy differences, plants can move water to the tops of the tallest trees. This same water potential is essential for moving water to the leaves so that photosynthesis can occur.

The water potential in plant solutions is influenced by solute potential (Ψ_s) and pressure potential (Ψ_p). The total water potential of a plant cell is the sum of its pressure potential and solute potential, and is represented by the following equation:

$$\Psi_w = \Psi_s + \Psi_p$$

In Section D2.3.6, we mentioned turgor pressure. We said it is the pressure on the cell wall produced by the cell contents pressing outwards. Pressure potential (Ψ_p) and turgor pressure represent the same pressure within the cell. As turgor pressure increases, so does pressure potential. Pressure potential is generally positive within plant cells. However, in plant xylem cells water and nutrients are transported upwards by transpiration, and these cells have negative pressure potentials.

Solute potential is the result of the presence of solutes in a solution. Pure water has a solute potential of zero because there is no solute present. As the solute concentration increases, fewer free water molecules are available and the solute potential will decrease. Solute potential (Ψ_s), also called osmotic potential, is negative in a normal plant cell.

Figure 5 illustrates the water potential at different locations within a plant.

D2.3 Figure 5 Water potential in a tree. Notice the water potential values on the left. The right side of the diagram represents what is happening during the transport of water at a particular location in the tree, starting with water being taken up from the soil. After the water enters the roots, it moves up the tree until it is potentially lost to the atmosphere by transpiration from the leaves.

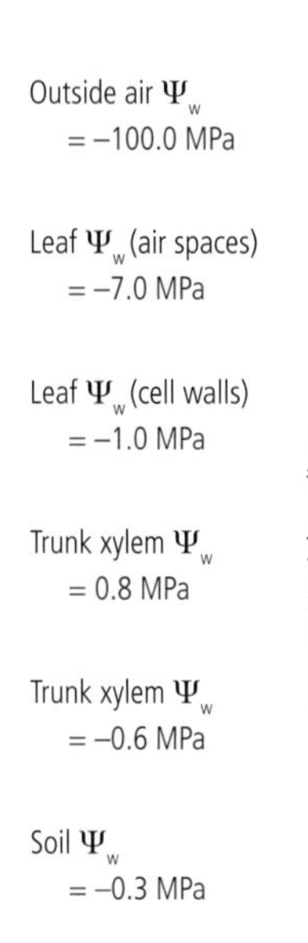

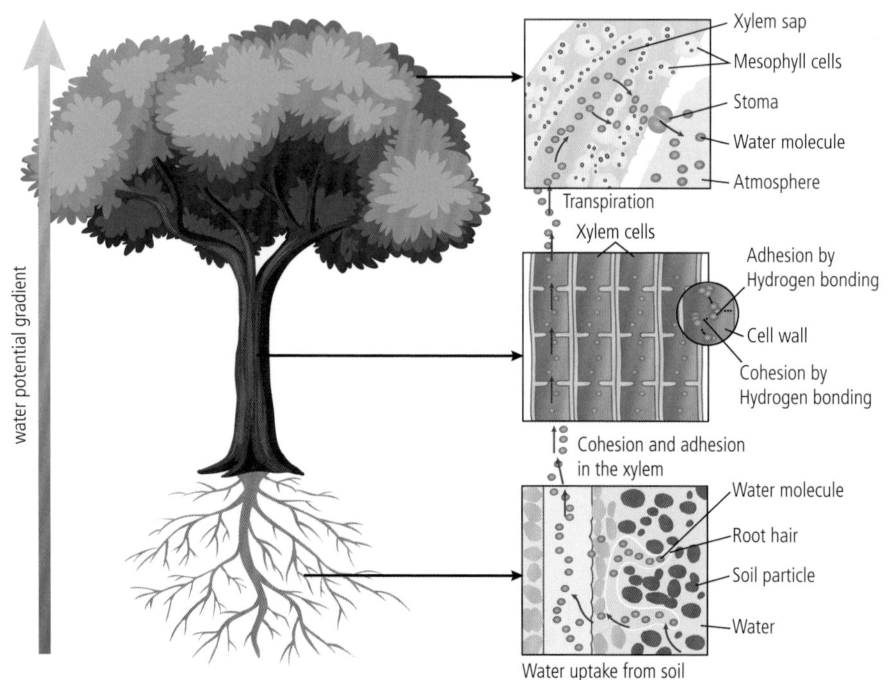

Key points to note about water potential (Ψ_w) while looking at Figure 5 are:

- water potential decreases from the roots to the leaves in a plant
- water moves from a region of higher water potential to a region of lower water potential
- water moves from the soil into the roots, because there is a higher water potential in the soil than in the root cells
- water usually moves from an area of lower solute concentration to an area of higher solute concentration, and solutes are concentrated in plant cells
- transpiration is a process that involves loss of water vapour through specialized plant structures called stomata, and most stomata occur in the leaves of a plant, although many plants have stomata on their stems as well
- water evaporating from the stomata causes the pressure potential to drop in the cells of the leaves, and as the water potential then drops, additional water moves upwards.

In a typical plant more than 90% of the water taken in by the roots is lost by transpiration. A mountain range in the USA called the Great Smoky Mountains takes its name from the continual mist above it created by transpiration from the abundant trees present.

D2.3.11 – Water potential in plant tissue

D2.3.11 – **Water potential and water movements in plant tissue**
Students should be able to explain in terms of solute and pressure potentials the changes that occur when plant tissue is bathed in either a hypotonic or hypertonic solution.

Table 2 compares extracellular and intracellular fluids in a wilted plant and a normal plant.

Wilted plant	Unwilted plant
Extracellular fluid	
Higher water potential Higher pressure potential Higher osmotic potential	Equal water potential inside and outside the plant cells
Intracellular fluid	
Lower water potential Lower pressure potential Lower osmotic potential	Pressure potential increases until the plant cells are turgid. This turgid state is then maintained

D2.3 Table 2 The water potential of wilted and unwilted plants. The osmotic potential is the potential of water to move via osmosis

A wilted plant is certainly able to receive water because of the low water potential within its cells. However, if there is no water available to the plant root system, water cannot move from the extracellular fluid into the cell. The result is continued low turgor pressure and wilting becomes more pronounced. In the unwilted plant, there is water available to the roots and it is free to move in and out of the plant cells, increasing at first and then maintaining pressure potential, turgor pressure, within the cell.

What variables influence the direction of movement of materials in tissues?

In the comparison of a wilted and non-wilted plant it is important to understand what will happen to a plant when it is placed in a hypotonic or hypertonic situation. If a plant's roots are bathed in a hypertonic environment (which has a high solute concentration), there will be a higher water potential inside the roots than outside. The solute potential outside the cell will be lower (more negative) than the solute potential inside the cell. Overall, the water potential will be lower outside the cell. We know that

Whenever confronted with a problem involving water movement in plant tissues, think back to the equation:

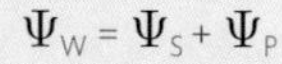

$\Psi_W = \Psi_S + \Psi_P$

Remember that water flows from areas of higher water potential to areas of lower water potential.

Hypertonic solutions have a higher number of solute particles, resulting in a lower water potential. Hypotonic solutions have a lower number of solute particles, resulting in a higher water potential.

water moves from higher water potential to lower water potential. Therefore, in this case, water moves out of the roots into the hypertonic surroundings.

If the plant's roots are placed in a hypotonic environment, the roots will have a lower water potential inside the cells than outside. In this case, water moves into the root cells from the hypotonic surroundings. The solute potential outside the cell will be higher (less negative) than the solute potential inside the cell.

Pressure potential is also involved when plants are placed in hypertonic and hypotonic environments. The effect of a high pressure potential inside a cell is to decrease or stop water flow into that cell. When a plant is placed in a hypotonic environment, water moves in until the pressure potential is high enough to stop this inward movement of water. At this point the cell is isotonic with the surrounding fluid: the water potentials are the same. If the plant is placed in a hypertonic solution, water moves out of the cell and into the solution. The pressure potential within the cell will drop and the plant will start to wilt.

Halophytes are plants that thrive in high salinity conditions. They can concentrate large amounts of solutes in specialized compartments so that they can take up water from their salty surroundings. Recent findings indicate an ability to sequester toxic ions and salts within their cells, especially in leaves, which eventually fall off the plant. Some halophytes even have salt glands that can excrete excess salt from plant tissues. Current research is looking into the gating action of aquaporins in halophytes. Many parts of the world would benefit from the genetic development of plants that can thrive in hypertonic conditions or that flourish while being irrigated with seawater. These plants could contribute to our ever-increasing global need for food.

HL end

These mangroves on the Florida coast of the Gulf of Mexico represent a potential for increased food production in the future because of their ability to thrive in a high saline environment.

Guiding Question revisited

What factors affect the movement of water into or out of cells?

In this chapter you have learned that:

- water is a polar molecule that must pass through the cell membrane via specialized channels called aquaporins
- the hydrogen bonds that water forms with ions and other charged particles contribute to the transport of essential materials into a cell and waste products out of a cell
- solvation is the interaction of a solvent with a dissolved solute
- water will move out of cell when it is placed in a hypertonic environment
- water will move into a cell when it is placed in a hypotonic environment
- the cell membrane contributes to the control of materials moving into and out of the cell
- osmosis requires a selectively permeable membrane to occur

HL

- water potential, which includes solute potential and pressure potential, affects water movement in plants
- water moves from regions of higher water potential to regions of lower water potential.

HL end

Guiding Question revisited

How do plant and animal cells differ in their regulation of water movement?

In this chapter you have learned that:

- turgor pressure (pressure potential) plays a large role in controlling water movement in and out of plant cells
- because animal cells do not have a cell wall, turgor pressure is not a large factor in water movement in and out of animal cells
- human red blood cells and all cells without a cell wall, when placed in a hypotonic solution, will swell and possibly burst
- human red blood cells and all cells without a cell wall, when placed in a hypertonic solution, will shrink and undergo crenation
- plant cells, when placed in a hypotonic solution, will develop high turgor pressure, as the cell contents push against the cell wall, which maintains plant shape
- cells possessing a cell wall will undergo plasmolysis when placed in a hypertonic environment
- when cells are placed in an isotonic environment, there is dynamic equilibrium rather than no movement of water

HL

- water potential is key to the transport of water in plants
- water potential is the sum of the solute potential and the pressure potential.

HL end

Exercises

Q1. Describe the water environment surrounding roots in which plants are most likely to wilt. Explain why.

Q2. Explain the role of aquaporins in the transport of water in and out of cells.

Q3. Explain the consequences of fertilizing a plant too often.

Q4. Compare what would happen to plant and animal cells when placed in a hypotonic environment.

Q5. Which of the following is most likely to increase water uptake by a plant cell?

A Placing a plant cell in hypertonic environment.

B Placing a plant cell in an isotonic environment.

C Placing a plant cell in a hypotonic environment.

D Increasing the turgor pressure within the plant cell.

Q6. HL What two factors have the greatest effect on water potential in plants?

D2 Practice questions

1. Below is a micrograph of an *Escherichia coli* bacterium undergoing reproduction. In the diagram what does label X identify?

   A Nucleoid region.

   B Chromatin.

   C Histones.

   D Endoplasmic reticulum.

(Total 1 mark)

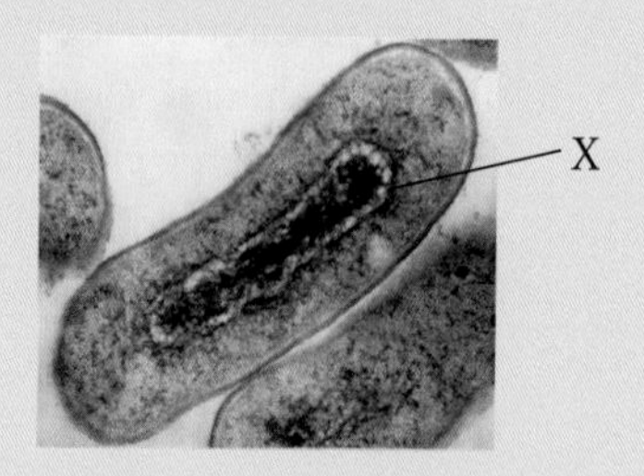

2. Which of the following take(s) place during either interphase or mitosis in animal cells?

   I. Re-formation of nuclear membranes.

   II. Pairing of homologous chromosomes.

   III. DNA replication.

   A I only.

   B I and II only.

   C II and III only.

   D I and III only.

(Total 1 mark)

3. HL The diagram shows the concentration of four cyclins during the cell cycle. Which curve represents the cyclin that promotes the assembly of the mitotic spindle?

(Total 1 mark)

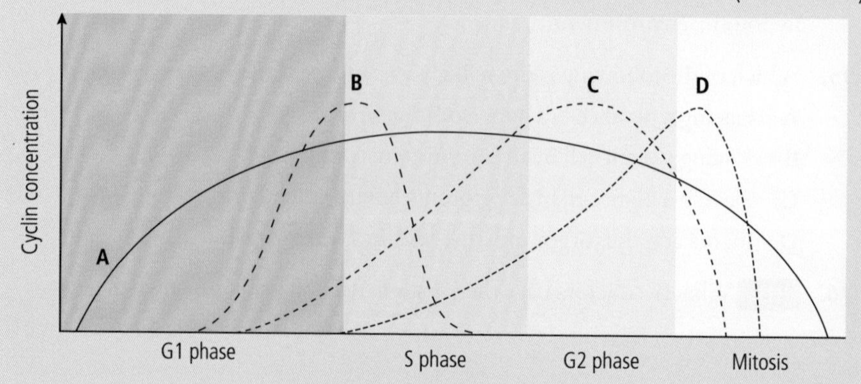

4. Outline how meiosis generates variation in offspring.

(Total 4 marks)

5. HL How can environmental factors affect the expression of genes?

A By promoting the replication of nucleosomes.

B By inactivating epinephrine.

C By making specific changes to the base sequence of genes.

D By causing the pattern of DNA methylation to be changed.

(Total 1 mark)

6. HL Outline the advantages of using monozygotic twins in studies of epigenetics.

(Total 2 marks)

7. HL Explain the effects of air pollution on epigenetics.

(Total 4 marks)

8. Explain the effect of placing the following types of cells in an environment of pure distilled water.

(a) Red blood cell (2)

(b) Plant cell (2)

(c) Unicellular freshwater dwelling organism (2)

(Total 6 marks)

9. Fluid replacements are essential for individuals who have undergone a period of strenuous exercise and heavy perspiring.

(a) Describe the solution of the preferable replacement. (1)

(b) Explain the reason for your answer. (2)

(Total 3 marks)

10. HL Draw an arrow on the diagram to indicate the net flow of water.

(a) Calculate Ψ_w for side A. (1)

(b) Calculate Ψ_w for side B. (1)

(Total 2 marks)

A	B
$\Psi_s = -450$ kPa	$\Psi_s = -550$ kPa
$\Psi_P = 250$ kPa	$\Psi_P = 250$ kPa

THEME D Continuity and change

3 Organisms

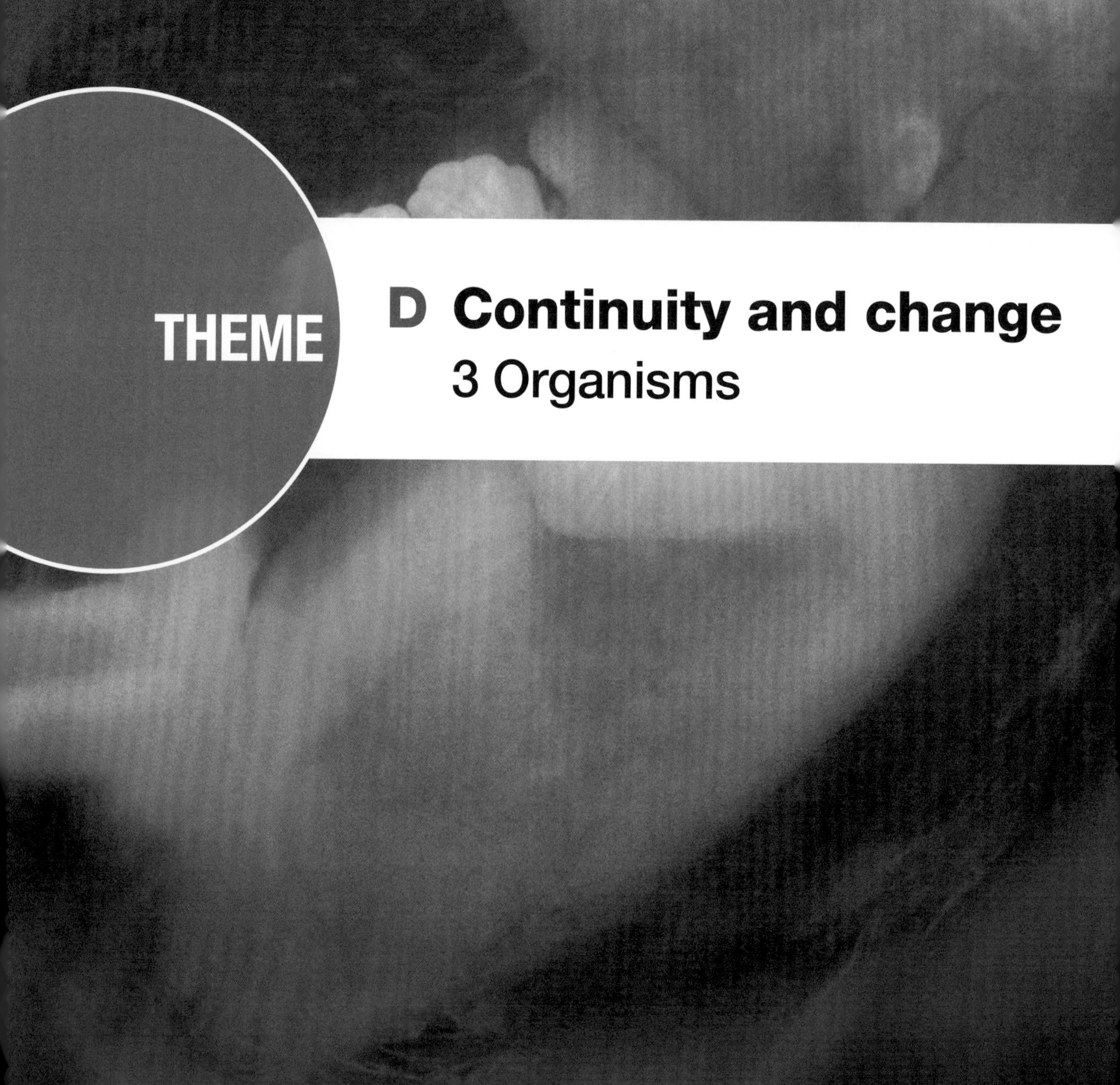

An illustration of a human foetus within the uterus. In this chapter we will explore how organisms reproduce, with a brief overview of asexual reproduction but a focus on the mechanisms of sexual reproduction. An integral part of sexual reproduction is combining the genetic material of two parents into an offspring. Thus, a study of genetics will be a portion of this unit, leading to an understanding of how pairs of chromosomes (one from each parent) interact with each other to give molecular instructions for an entire, complex organism. Finally, in this unit you will study how organisms maintain a homeostatic balance of processes in their bodies by feedback systems involving the nervous system and hormones produced by the endocrine system.

D3.1 Reproduction

Guiding Questions

How does asexual or sexual reproduction exemplify themes of change or continuity?

What changes within organisms are required for reproduction?

Reproduction of living organisms enables both continuity and change. New generations of species allow the continuation of favourable characteristics that have existed in some form for potentially eons, while encouraging new traits through the process of evolution. Asexual reproduction is possible by several mechanisms but only requires one parent. New individuals are formed by cell division. Very little genetic variation is produced, but if an environment is relatively unchanging this strategy is very efficient. Sexual reproduction requires two parents and the union of gametes that each have only half the chromosomes characteristic of the species. A great deal of variation is possible both in the chromosome composition of the gametes and the events associated with choice of mates and biological selection of which gametes are joined together. Offspring show a great deal of genetic variation, and thus sexual reproduction is favoured in environments that change over time.

All organisms must replicate their genetic material (DNA) before reproduction. In sexually reproducing organisms, a cell division called meiosis is used to reduce the chromosome number to half that of the adult. The resulting cells are called gametes and must be joined together to restore the full chromosome number. Often, organisms must reach a certain stage of maturity before they can produce gametes. In humans, this stage is called puberty.

HL Puberty involves a complex set of hormonal changes that results in morphological body changes as well as changes in reproductive tissues to facilitate the formation and use of gametes.

D3.1.1 – Sexual and asexual reproduction

D3.1.1 – Differences between sexual and asexual reproduction
Include these relative advantages: asexual reproduction to produce genetically identical offspring by individuals that are adapted to an existing environment, sexual reproduction to produce offspring with new gene combinations and thus variation needed for adaptation to a changed environment.

Sexual reproduction requires both a male and a female parent. Each parent contributes some, but not all, of their genes to an offspring. This creates a unique genetic makeup that did not exist before. **Asexual reproduction** requires only one parent and results in multiple organisms from that single parent that all have the same genetic makeup.

Some of the major differences between asexual and sexual reproduction

Sexual reproduction	Asexual reproduction
Gametes (usually a sperm and an egg) fertilized	Organism makes a copy of itself
Two parents required	Only one parent required
Offspring are genetically unique compared to both parents	Offspring are genetically identical to parent
Provides new gene combinations and thus promotes genetic variation	Provides no new gene combinations and relatively little genetic variation
Allows adaptations for a changing environment	Promotes little change in adaptations but may be beneficial in an existing non-changing environment

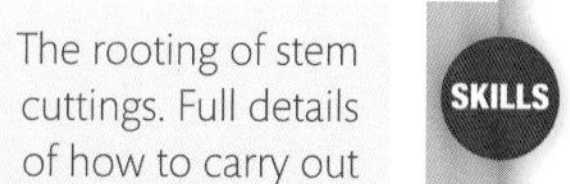

SKILLS

The rooting of stem cuttings. Full details of how to carry out this experiment with a worksheet are available in the eBook.

Asexual reproduction can be accomplished by a variety of mechanisms, for example those listed below.

- **Binary fission**: bacteria replicate their DNA and divide into two cells. In an environment well suited for growth this can occur more than once an hour in some species of bacteria.
- **Mitosis**: unicellular eukaryotic organisms replicate their DNA and many organelles (such as mitochondria). Also leads to growth in these remaining asexual reproduction methods.
- **Budding**: new genetically identical organisms grow directly from an existing organism.
- **Fragmentation**: the body of an existing organism breaks up into several fragments, each growing into a complete organism.
- **Vegetative reproduction**: common in many plants, with new plants emerging from roots, bulbs, tubers or shoots.
- **Parthenogenesis**: in an animal species, growth and development of an egg cell without the involvement of a male gamete.

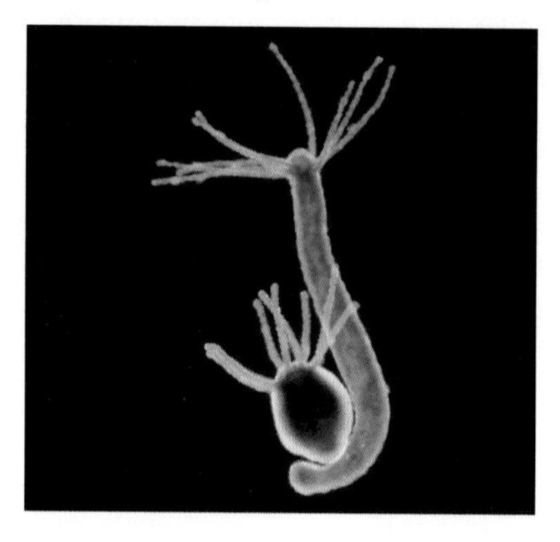

Hydra is a genus of small freshwater animals called cnidarians. This *Hydra* is reproducing asexually by growing a bud. The bud will eventually separate from the original organism and become an entirely separate, but genetically identical, organism.

It is incorrect to consider asexual reproduction as a less important or inferior form of reproduction. It is true that sexual reproduction provides much greater genetic variation, but asexual reproduction has its own place in nature. Asexual reproduction is almost always faster, requires less expenditure of energy, and a single organism can colonize a new area relatively easily. The process of evolution has favoured both types of reproduction in different circumstances. If either sexual or asexual reproduction was always advantageous, evolution would have eliminated the less fit mechanism.

D3.1.2 – The role of meiosis and gametes

D3.1.2 – Role of meiosis and fusion of gametes in the sexual life cycle
Students should appreciate that meiosis breaks up parental combinations of alleles, and fusion of gametes produces new combinations. Fusion of gametes is also known as fertilization.

Meiosis produces haploid cells by a process often described as **reduction division** because the number of chromosomes in each gamete is reduced to one-half of the original number. In a diploid cell, chromosomes exist in homologous pairs. Each chromosome of a homologous pair can be traced back to one of the two parents. In other words, each homologous pair has a **maternal** chromosome and a **paternal** chromosome. Each new individual produced by sexual reproduction will represent a brand new combination of chromosomes.

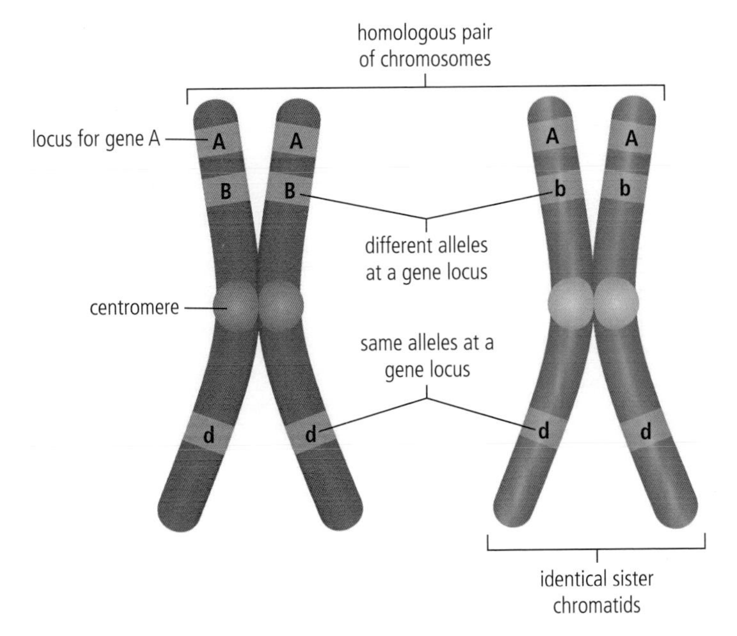

▲ An illustration showing one homologous pair of chromosomes after DNA replication. One of the chromosomes in the homologous pair is from the mother of this individual and one is from the father. Both chromosomes of the homologous pair have the same gene locations (loci) but may or may not have different alleles for any one gene. This is exemplified by genes A, B and D. DNA replication has produced the identical sister chromatids. In a human gonad cell preparing to create a gamete, there would be 22 other homologous pairs all with their own gene loci. As gametes only contain one chromosome from each homologous pair, there is enough genetic material in a gonad cell to make four gamete cells.

Egg(s) are often referred to as ova (plural) or ovum (singular). In humans, an egg does not complete the final stage of meiosis until it is being fertilized. Even so, it is common to refer to the maturing future egg cells within the ovaries as eggs or ova.

Before meiosis begins, DNA replication takes place, producing a pair of chromatids from each chromosome. The cells produced at the end of meiosis do not contain a random split of the chromosomes. Each cell has only one chromosome from each **homologous pair**, so this is referred to as a **haploid** (n) number of chromosomes. Each chromosome from a homologous pair will have the same genes in the same order, but may contain different alleles. The resulting cells are gametes: **eggs** in females and **sperm** in males. In humans the haploid number of chromosomes is 23. Each sperm and egg has 23 chromosomes, one of each type of chromosome.

Gamete genetic variation is provided by:

- crossing over, leading to recombination of alleles during metaphase I
- random orientation and division of chromosomes during metaphase I (often called independent assortment of chromosomes)
- random fertilization, as there are millions of sperm and egg possibilities even for just one set of parents.

Random orientation during metaphase I and crossing over (shown by banding on sister chromatids) promote variety in gametes. Each pair of sister chromatids will be separated into a different haploid cell at the end of meiosis.

Meiosis breaks up parental combinations of alleles and provides new chromosome combinations in offspring.

Homologous pairs of chromosomes assort independently of each other. During human meiosis this means the 23 homologous pairs would have to separate in the same pattern to result in two gametes that are identical. The odds of that happening is one chance in 2^{23}.

Fertilization occurs when gametes fuse into a single cell. The purpose of fertilization is to restore the **diploid** ($2n$) number of chromosomes. This will also restore the chromosomes as homologous pairs. This is necessary as some genes require the interaction of both chromosomes of a homologous pair for gene expression. The diploid number of chromosomes in humans is 46, existing as 23 homologous pairs.

Genetic variation in sexually reproducing organisms is the result of the many chromosome combinations possible during the formation of gametes, and the many possible gamete combinations during fertilization.

D3.1.3 – Male and female gametes

D3.1.3 – Differences between male and female sexes in sexual reproduction
Include the prime difference that the male gamete travels to the female gamete, so it is smaller, with less food reserves than the egg. From this follow differences in the numbers of gametes and the reproductive strategies of males and females.

Even though the process of meiosis is identical in males and females, the resulting morphology of the gamete cells is quite different. Sperm are often motile and have a **flagellum** to help them move towards the egg. They are also very small, to increase swimming efficiency. When a sperm fertilizes an egg, it contributes nothing towards the food reserves for the early embryo. Sperm are adapted to provide an efficient delivery system for a haploid nucleus.

Sperm are designed to travel to the egg. They are very small in size, and have multiple mitochondria and a flagellum, making them efficient "swimmers".

In comparison to sperm cells, an egg is huge. Human eggs are thousands of times larger in volume than human sperm An egg contains all the nutrients needed for early embryonic growth. In some species the egg provides all the nutritional requirements until the young animal hatches from an egg encased in a shell. In humans and other placental mammals, the egg provides the initial source of nutrition for the developing embryo, followed by nutrition from the uterus and then the placenta. In addition, the eggs contain the initial organelles, such as mitochondria, that are needed as the cells grow and divide during development.

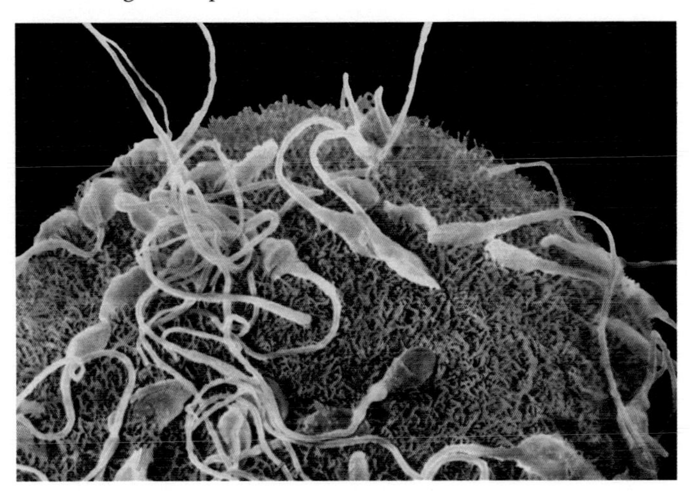

A colorized scanning electron micrograph (SEM) of a human egg with multiple sperm attempting fertilization. Eventually one sperm cell will penetrate the plasma membrane of the egg. A chemical reaction will then occur to prevent any other sperm from fertilizing the egg.

Some animals produce large numbers of eggs, especially if the fertilization process is external. Human females typically release only one egg during the menstrual cycle, although two or more are possible. Human males produce millions of sperm each day, with many of the cells dying. The cellular components of these cells are biologically recycled if they are not used. A single **ejaculation** contains millions of sperm cells in a fluid called **semen**. Males produce such large numbers of sperm cells because the vast majority will never find the egg. Of the millions of sperm that begin swimming towards the egg after an ejaculation, only 100–200 will actually reach the egg and only one will fertilize it.

In many species females have a greater impact on mate selection than the male. This is especially noticeable in bird species where males are more brightly coloured and carry out elaborate mating rituals to attract females. Often it is the female that chooses a mate. Females produce fewer gametes and are often more heavily invested in caring for the young.

The female reproductive strategy is based on producing one or very few, very large stationary eggs, whereas the male strategy is to produce huge numbers of very small and motile sperm.

D3.1.4 – Male and female reproductive systems

D3.1.4 – Anatomy of the human male and female reproductive systems
Students should be able to draw diagrams of the male-typical and female-typical systems and annotate them with names of structures and functions.

The IB requires you to draw diagrams of the male and female reproductive systems and annotate those diagrams with the names of the structures and their functions. Study Figures 1 and 2 and Tables 1 and 2 and practise making your own diagrams.

The structures of the male and female reproductive systems are adapted for the production and release of gametes. In addition, the female reproductive system ensures a suitable location for fertilization and provides an environment for the growth of the embryo/foetus until birth.

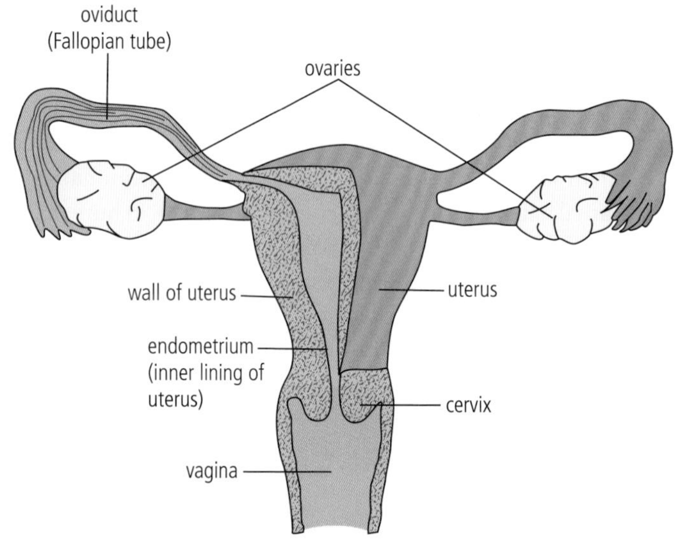

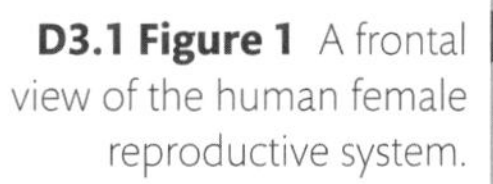
D3.1 Figure 1 A frontal view of the human female reproductive system.

D3.1 Table 1 Anatomy and function of the female reproductive system

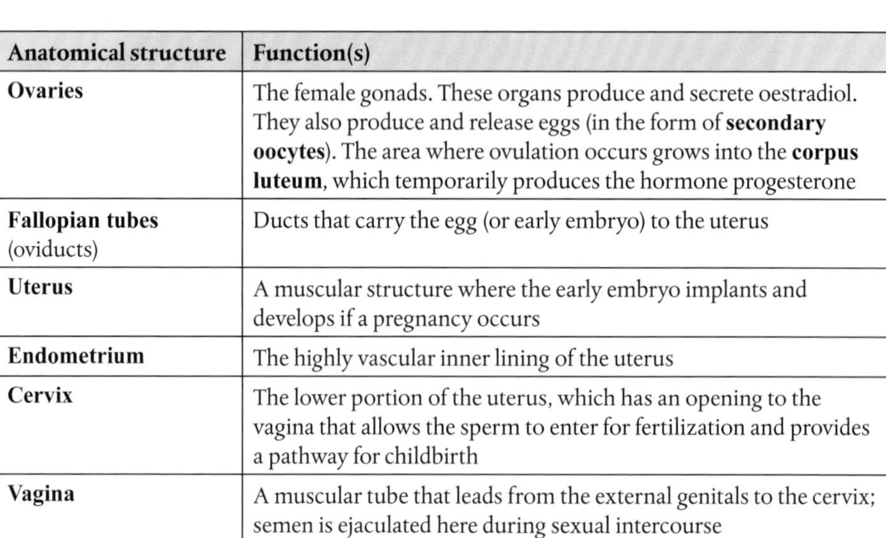

Anatomical structure	Function(s)
Ovaries	The female gonads. These organs produce and secrete oestradiol. They also produce and release eggs (in the form of **secondary oocytes**). The area where ovulation occurs grows into the **corpus luteum**, which temporarily produces the hormone progesterone
Fallopian tubes (oviducts)	Ducts that carry the egg (or early embryo) to the uterus
Uterus	A muscular structure where the early embryo implants and develops if a pregnancy occurs
Endometrium	The highly vascular inner lining of the uterus
Cervix	The lower portion of the uterus, which has an opening to the vagina that allows the sperm to enter for fertilization and provides a pathway for childbirth
Vagina	A muscular tube that leads from the external genitals to the cervix; semen is ejaculated here during sexual intercourse

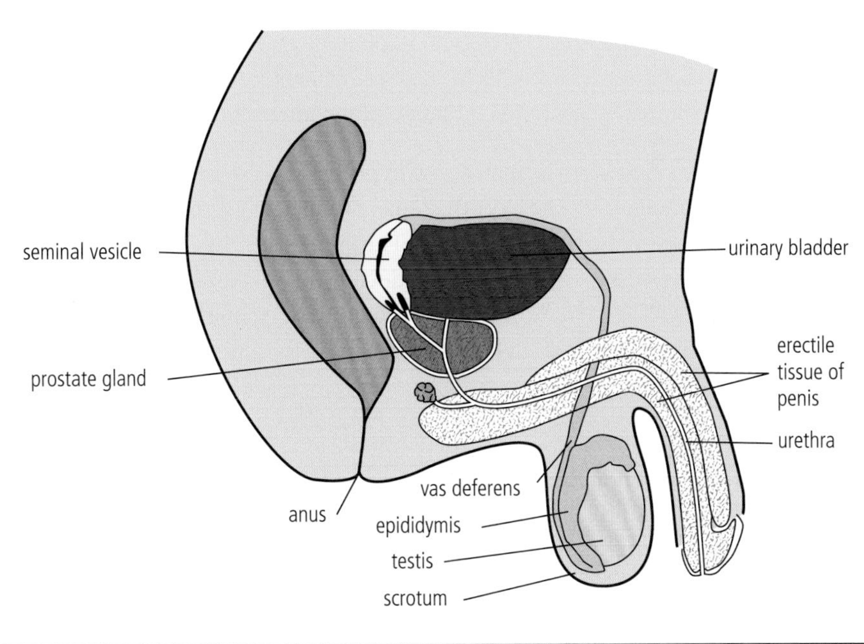

D3.1 Figure 2 A lateral (side) view of the human male reproductive system. The urinary bladder is shown because of its close anatomical proximity to the reproductive structures.

D3.1 Table 2 Anatomy and function of the male reproductive system

Anatomical structure	Function(s)
Testes	The male gonads. Sperm are produced here in small tubes called **seminiferous tubules**
Epididymis	The area where sperm are received, become mature, and become capable of a swimming motion via movement of their flagella
Scrotum	A sac that holds the testes outside the body cavity so that sperm production and maturation can occur at a temperature cooler than body temperature
Vas deferens	A muscular tube that carries mature sperm from the epididymis to the urethra during an ejaculation
Seminal vesicles	Small glands that produce and add seminal fluid to the semen
Prostate gland	A gland that produces much of the semen, including carbohydrates for the sperm
Penis	An organ that becomes erect as a result of blood engorgement in order to facilitate ejaculation
Urethra	After all the glands have added fluids, this is the tube via which the semen leaves the penis

D3.1.5 – Hormonal control of the menstrual cycle

D3.1.5 – Changes during the ovarian and uterine cycles and their hormonal regulation
Include the roles of oestradiol, progesterone, luteinizing hormone (LH), follicle-stimulating hormone (FSH) and both positive and negative feedback. The ovarian and uterine cycles together constitute the menstrual cycle.

At puberty, human females begin a hormonal cycle known as the **menstrual cycle.** Each cycle lasts, on average, 28 days. The purpose of the menstrual cycle is to time the release of an egg (**ovulation**) for possible fertilization and later **implantation** into the inner lining of the uterus. This implantation must occur when the uterine inner lining (the **endometrium**) is rich with blood vessels (i.e. highly vascular). The highly vascular endometrium is not maintained if there is no implantation. The breakdown of the blood vessels of the endometrium leads to the menstrual bleeding (**menstruation**) of a typical cycle. This menstruation is a sign that no pregnancy has occurred.

The part of a female's brainstem known as the hypothalamus is the regulatory centre for the menstrual cycle. The hypothalamus produces a hormone known as **gonadotropin-releasing hormone** (**GnRH**). The target tissue of GnRH is the nearby pituitary gland, and results in the pituitary producing and secreting two hormones into the bloodstream. These two hormones are the **follicle-stimulating hormone** (**FSH**) and **luteinizing hormone** (**LH**). The target tissue for both these hormones is the ovaries.

The hormones FSH and LH have several effects on the ovaries. One is to increase the production and secretion of another reproductive hormone by the ovaries' follicle cells. This hormone is **oestradiol**. Like all hormones, oestradiol enters the bloodstream. Its target tissue is the endometrium of the uterus. One effect of oestradiol is an increase in the density of blood vessels in the endometrium; in other words the endometrium becomes highly vascular. Because of a positive feedback loop, oestradiol stimulates the pituitary gland to release more FSH and LH.

The increase in FSH and LH results in the production of structures called **Graafian follicles.** Within the ovaries are cells known as **follicle cells**, and the reproductive cells (possible future eggs) are at a stage of development called **oocytes.** Under the chemical stimulation of FSH and LH, the somewhat randomly arranged follicle cells and oocytes now take on a particular cellular arrangement known as a Graafian follicle.

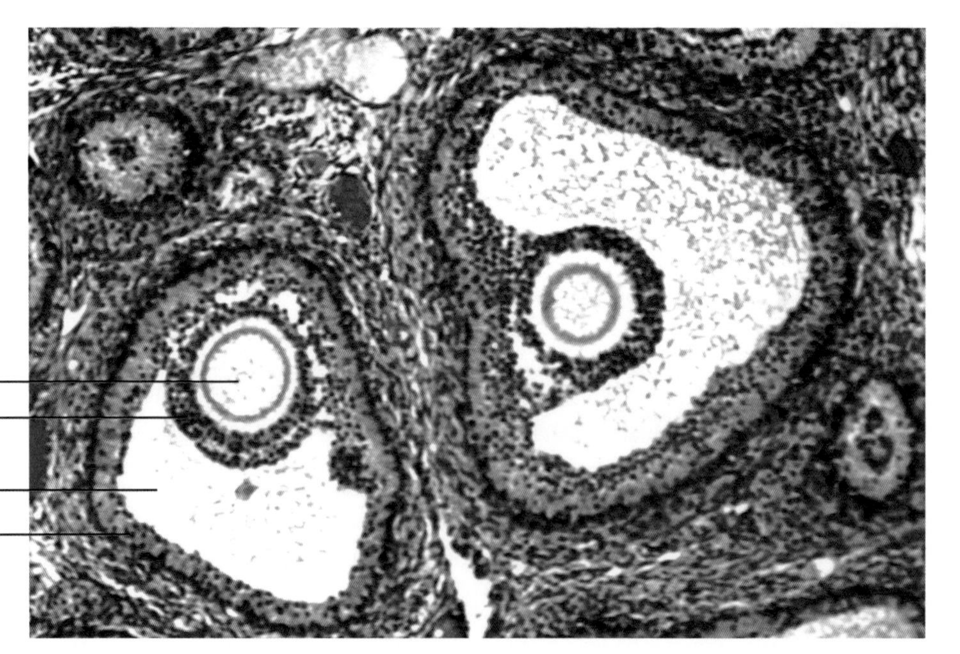

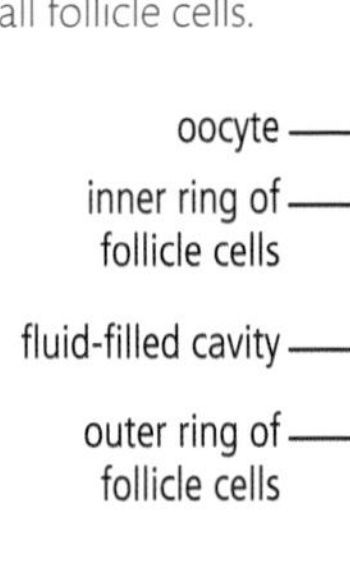

D3.1 Figure 3 A light micrograph of a section of human ovary. Two Graafian follicles are shown with an oocyte at the centre of each. The oocytes are the large, nearly perfect circles, surrounded by a ring of very small follicle cells.

A spike in the level of FSH and LH leads to **ovulation** (as a result of the positive feedback loop between FSH, LH, the pituitary gland and oestradiol). The oocyte is accompanied by the Graafian follicle's inner ring of follicle cells. This structure is referred to as a single follicle, and typically enters a Fallopian tube soon after ovulation. The Graafian follicle's outer ring of follicle cells remains within the ovary. These follicle cells begin to produce and secrete another hormone, **progesterone**. The cells of the outer ring begin to divide and fill the "wound" area left by ovulation. This forms a glandular structure known as the **corpus luteum**. The corpus luteum is only hormonally active (producing progesterone) for 10–12 days after ovulation if fertilization does not occur. Progesterone is a hormone that maintains the thickened, highly vascular endometrium. As long as progesterone continues to be produced, the endometrium will not break down and an embryo will be able to implant. In addition, the high levels of both oestradiol and progesterone at the same time provide a negative feedback signal to the hypothalamus and prevent production of GnRH.

Negative feedback control is exemplified by high levels of oestradiol and progesterone inhibiting the production of GnRH. Without GnRH, FSH and LH are inhibited. As soon as the corpus luteum degenerates, another cycle can begin because the progesterone levels are not being maintained.

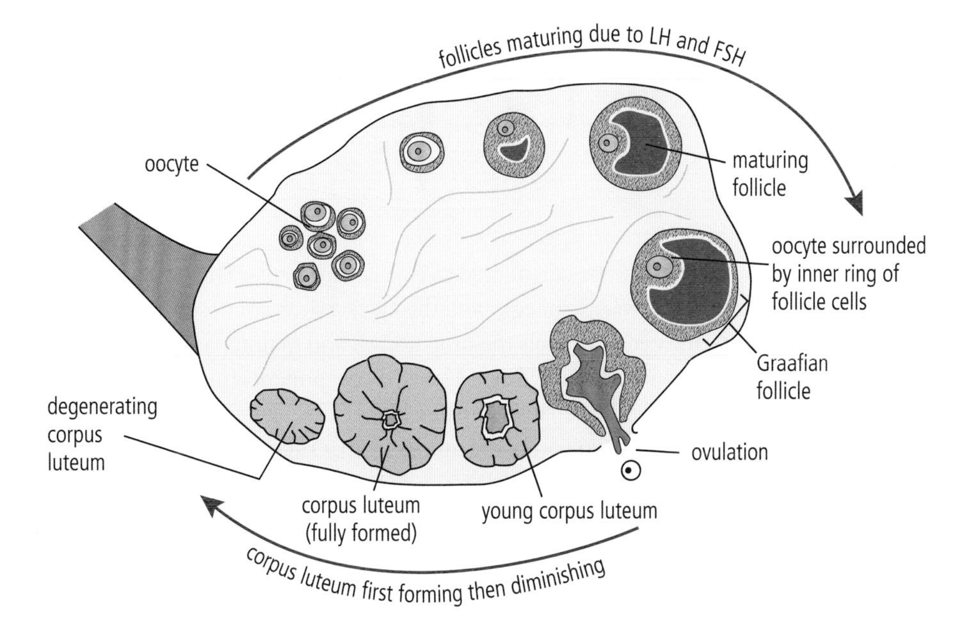

The sequence of events that occur within an ovary during a single menstrual cycle. Twenty-eight days of ovarian events are shown within this single ovary as a time lapse sequence.

Because the hypothalamus does not produce GnRH when oestradiol and progesterone levels are high, FSH and LH remain at low levels, which is not conducive to the production of another Graafian follicle. Assuming there is no pregnancy, the corpus luteum begins to break down after 10–12 days, and this leads to a decline in both progesterone and oestradiol levels. As both of these hormone levels fall, the highly vascular endometrium can no longer be maintained. The capillaries and small blood vessels of the endometrium begin to rupture and **menstruation** begins. The drop in progesterone and oestradiol also signals the hypothalamus to begin secreting GnRH, and thus another menstrual cycle begins. Because the menstrual cycle is a cycle, there is no true beginning or ending point. The first day of menstruation is designated as the first day of the menstrual cycle simply because this is an event that can be recorded fairly easily (see Figure 4).

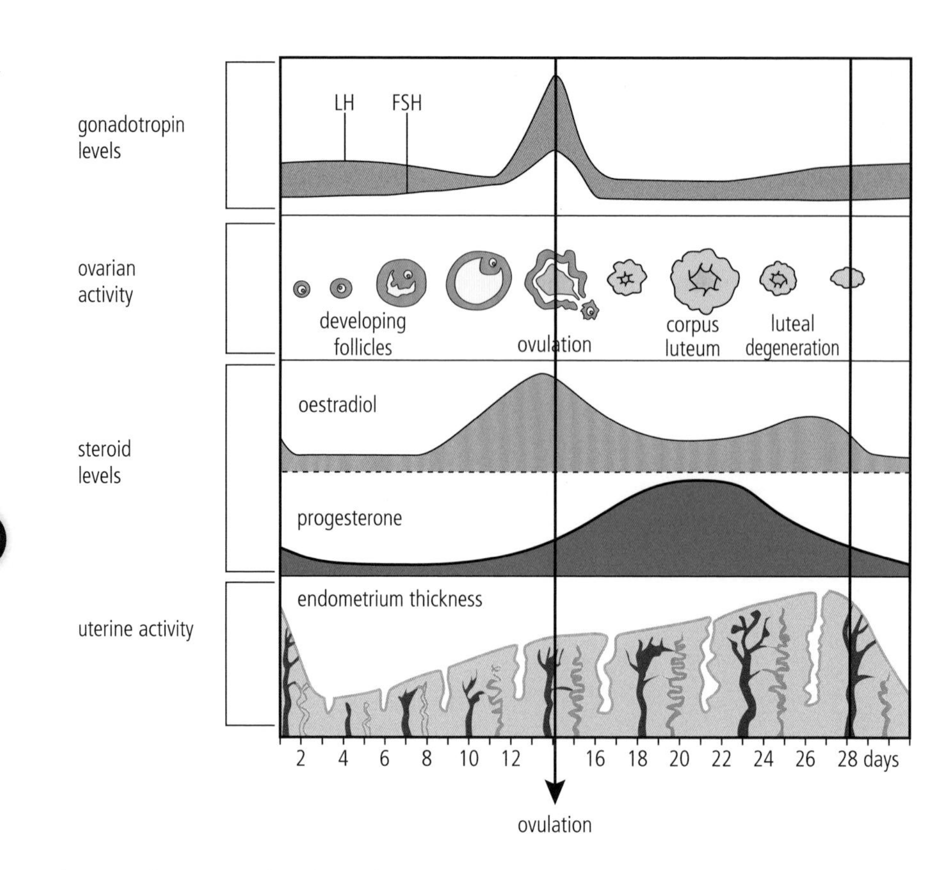

D3.1 Figure 4 The events that occur during a 28-day menstrual cycle. Note that these events are all aligned on the same time scale. Ovulation and possible fertilization occur near the middle of the cycle. The ovarian and uterine cycles together constitute the menstrual cycle. FSH = follicle-stimulating hormone; LH = leutinizing hormone.

When studying Figure 4, use the arrow indicating the time of ovulation as an important marker. Ask yourself: What events led up to and resulted in this ovulation? What will now happen after ovulation?

The overall menstrual cycle can be considered as two components. The ovarian cycle controls the production and release of eggs and the cyclic release of oestradiol and progesterone. The uterine cycle controls the preparation and maintenance of the lining of the uterus to receive a fertilized egg. The two components must be timed synchronously because the endometrium must be highly vascular and ready to receive an early embryo when the embryo enters the uterus from a Fallopian tube.

D3.1.6 – The process of fertilization

D3.1.6 – Fertilization in humans
Include the fusion of a sperm's cell membrane with an egg cell membrane, entry to the egg of the sperm nucleus but destruction of the tail and mitochondria. Also include dissolution of nuclear membranes of sperm and egg nuclei and participation of all the condensed chromosomes in a joint mitosis to produce two diploid nuclei.

As a result of sexual intercourse, millions of sperm are **ejaculated** into a female's vagina. The motile sperm absorb some of the sugar in semen in order to have enough "fuel" for their potentially long journey. Some of the sperm find their way through the cervical opening (the **cervix** separates the vagina and the uterus) and gain access to the uterus. They begin swimming up the endometrial lining, and some enter the openings of the two Fallopian tubes. If the female is near the middle of her menstrual cycle, there may be an egg within one of the two Fallopian tubes. The reason for millions of sperm in each ejaculate becomes clear when you consider that only a very small percentage of the motile sperm will ever reach the location of an egg.

The typical location for fertilization is within one of the Fallopian tubes. No single spermatozoon can accomplish the entire act of fertilization because it takes many sperm to penetrate the follicle cell layer and a coating called the **zona pellucida** (a gel composed of glycoproteins) surrounding the egg. Several sperm release hydrolytic enzymes contained in their acrosomes to help penetrate the egg's plasma membrane. One spermatozoon will reach the plasma membrane of the egg first and gain entry to the egg. This is the sperm cell that fertilizes the egg.

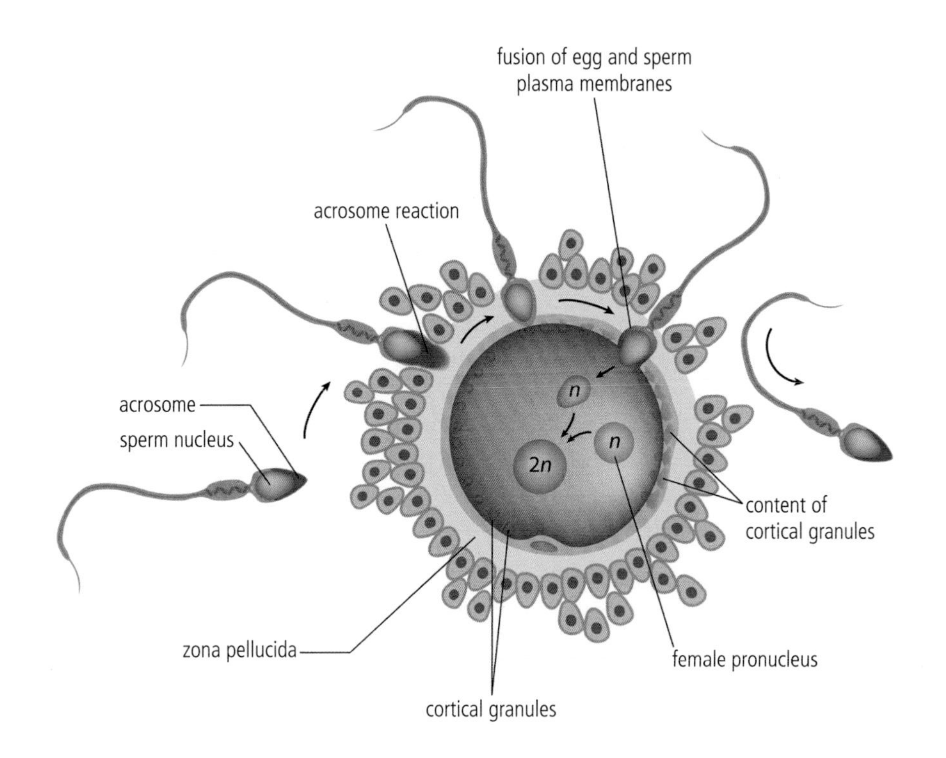

D3.1 Figure 5 The events of fertilization, including the acrosome reaction, fusion of the plasma membranes of the egg and sperm, and fusion of the two haploid nuclei. The cortical granules are described in Section D3.1.15.

Vesicles are released from the egg that destroy the sperm flagellum and mitochondria. The haploid set of chromosomes from the sperm and the egg are now both within the cytoplasm of the egg. The paternal and maternal sets of chromosomes remain separate for a period of time, and membranes form around each. These haploid structures are each called a **pronucleus**. While in this pronuclei stage, the DNA in each undergoes replication in preparation for mitosis. The two pronuclei then come together and the temporary nuclear membranes dissolve. A spindle apparatus typical of mitosis forms as the chromosomes prepare for the first mitotic division of the newly formed diploid cell. It takes about 30 hours after fertilization for this first mitotic cell division to be completed. Subsequent mitotic cell divisions increase in frequency, and by the end of three days 16 cells will have been formed.

D3.1.7 – In vitro fertilization

D3.1.7 – Use of hormones in in vitro fertilization (IVF) treatment
The normal secretion of hormones is suspended, and artificial doses of hormones induce superovulation.

There is a wide variety of possible reasons for infertility, including:

- low sperm counts in males
- impotence (failure to achieve or maintain an erection) in males
- inability to ovulate normally in females
- blocked Fallopian tubes in females.

Reproductive technologies have been developed to help people overcome the problem of infertility. One of the most common of these technologies is **in vitro fertilization** (IVF).

TOK

During IVF, pre-implantation genetic screening is often carried out on embryo cells taken at the **blastocyst** stage (see Section D3.1.16). Chromosome disorders and some genetic diseases can be diagnosed and the prognosis used to decide which embryos to implant. What ethical constraints should there be on the pursuit and application of knowledge?

As part of the IVF procedure, eggs are "harvested" from the female's ovaries. In order to ensure the proper timing for this, and to maximize the number of available eggs, the female undergoes hormone therapy. During the initial stages of the therapy she is treated with a drug that suspends the hormones associated with her natural menstrual cycle. Subsequently she takes hormone injections that include FSH. This ensures that she will produce many Graafian follicles in each ovary and therefore many potential eggs for harvesting. The production of many more eggs than is typical of a normal menstrual cycle is called **superovulation**.

When the time is right, many eggs are harvested surgically. To obtain the sperm cells that are needed for fertilization, the male ejaculates into a container. Harvested eggs are mixed with the sperm cells in separate culture dishes. Microscopic observation reveals which eggs are fertilized, and whether the early development appears normal and healthy. Between one and three healthy **embryos** are later introduced into the female's uterus for implantation. Any healthy embryos from the culturing phase that are not used for the first implantation can be frozen and used later if another implantation procedure is needed.

D3.1.8 – Sexual reproduction in plants

D3.1.8 – Sexual reproduction in flowering plants
Include production of gametes inside ovules and pollen grains, pollination, pollen development and fertilization to produce an embryo. Students should understand that reproduction in flowering plants is sexual, even if a plant species is hermaphroditic.

The gametes of flowering plants are produced within structures called **ovules** (female) and **pollen grains** (male). Meiosis gives rise to the ovules and pollen grains, while the actual gametes are produced by mitosis. Because the reproductive structures are already haploid, reduction division is not required to produce haploid gametes.

Many species of flowering plants produce flowers that are **hermaphroditic**, with both male and female structures. Some hermaphroditic species, including orchids and sunflowers, **self-pollinate**. Self-pollination and fertilization are a form of sexual reproduction because the gametes are produced by meiosis and there is a fusion of gamete nuclei to form an embryo. The disadvantage to this method of reproduction is the loss of genetic variation that is a natural part of combining the chromosomes from two separate individuals.

Cross-pollination is the transfer of pollen produced on one plant with another plant. There are many flower adaptations that facilitate the transfer of pollen from one plant to another. Flowers that have petals often use shapes, markings and colours to attract specific pollinating animals. The magnificent shapes and colours of many species' flowers are not intended to please humans, but to attract animals that can transfer pollen from one plant to the next.

Pollen develops within structures called **anthers**. Anthers are often positioned in a flower so that pollinators can come into contact with them without even realizing it. When that pollinator moves on to a different flower, some of the pollen will be transferred to another structure, called the **stigma**, that is often held upright in the female part of the flower. The stigma is sticky and pollen grains can easily adhere to it.

Pollen that adheres to a stigma will then begin to grow into a structure called a **pollen tube**. In some respects this is equivalent to animal sperm that swim. Rather than swimming, pollen grains grow into tubes that penetrate other parts of the flower, to take male reproductive nuclei to the ovule where female nuclei await fertilization.

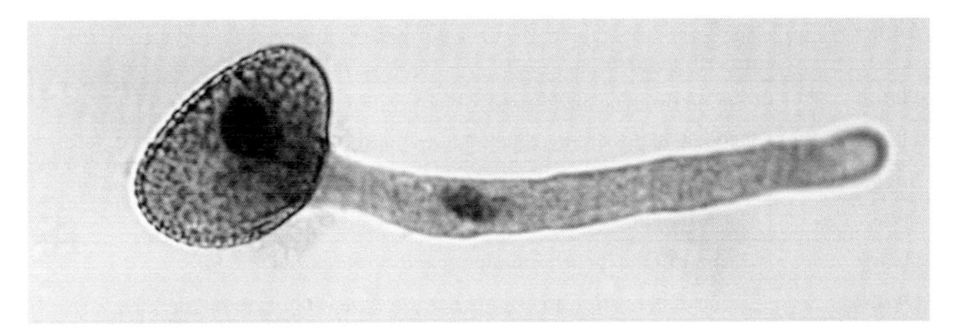

A pollen tube growing from a pollen grain of a lily plant, one of many species within the genus *Lilium*. The growth of the pollen tube is the plant's mechanism for getting the male gametes (nuclei) to the female gametes (nuclei) during a double fertilization process.

One pollen tube carries two male nuclei, and each results in a fertilization, hence plants use a double fertilization process. Within the **ovule** there are three haploid nuclei. One pollen nucleus fertilizes one ovule nucleus to create a zygote. The other pollen nucleus fertilizes the other two nuclei within the ovule to create a tissue called **endosperm**. Because three nuclei were used for this fertilization, the endosperm has the unusual chromosome number of $3n$: it is a **triploid** tissue. Growth of this triploid tissue produces the nutritive endosperm within the seed that will nourish the early plant embryo.

The nutritive value of seeds, such as peas and corn, that we exploit primarily comes from the seed's endosperm tissue.

D3.1.9 – Insect pollination

D3.1.9 – Features of an insect-pollinated flower
Students should draw diagrams annotated with names of structures and their functions.

There are many pollinators of different flower species, including insects, birds, bats and even some mammals. Here we will focus on flowers that are specialized for insect pollination. Common insect pollinators include bees, wasps, flies, butterflies and moths. Insect-pollinated flowers tend to be large and brightly coloured. They give off a strong scent to attract insects, and they offer a reward in the form of nectar at the base of the flower. The stamens of these flowers are often deep inside the flower, so that insects drinking nectar will brush up against the pollen grains. The pollen is often sticky and has numerous spikes in order to adhere easily to the legs or body of the insect. The stigma is also sticky so that the pollen can be transferred from the insect to the stigma as the insect visits other flowers.

How can interspecific relationships assist in the reproductive strategies of living organisms?

A typical insect-pollinated flower. The petal colour(s), flower size and shape are often adapted to attracting a specific pollinator such as a bee. The single ovule of this flower indicates that the resulting fruit that develops from its ovary will only contain one seed.

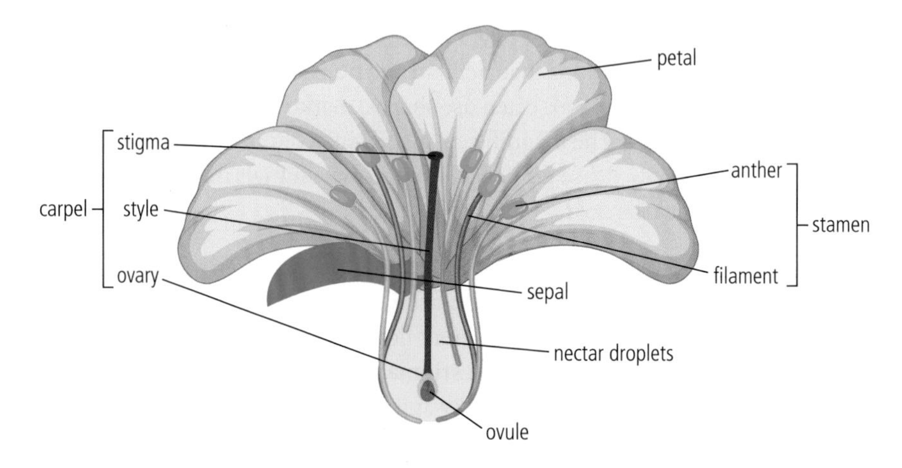

The function of different parts of a flower

Flower part	Function
Sepal	Protecting the developing flower while it is inside the bud
Petal	Often colourful, to attract pollinators
Anther	The part of the stamen (the male portion of a flower) that produces the male sex cells (pollen)
Pollen	Contain the male nuclei used for fertilization
Filament	The stalk of the stamen that holds up the anther
Stigma	The sticky top of a carpel (the female portion of a flower), on which pollen lands
Style	The part of the carpel that supports the stigma
Ovary	The base of the carpel, containing one or more ovules
Ovule	The chamber within an ovary where the female nuclei develop

If you are asked to draw and annotate a diagram, be sure to list the functions of the parts you draw. This can be done directly within your diagram labels rather than providing a separate table, unless you are directed otherwise.

D3.1.10 – Cross-pollination in plants

D3.1.10 – Methods of promoting cross-pollination
Include different maturation times for pollen and stigma, separate male and female flowers or male and female plants. Also include the role of animals or wind in transferring pollen between plants.

The common primrose (*Primula vulgaris*) produces two different kinds of flowers on different plants. The design of the two flower types ensures that the pollen picked up from one type of flower can only be deposited on the other type. This promotes cross-pollination in this species.

Plants benefit from genetic variation within populations just as much as animals do. Even in plant species where flowers are hermaphroditic, a variety of mechanisms have evolved to promote cross-pollination. Such mechanisms include the following.

- Some plant species have different maturation times for the pollen and ovules of the same flower. Maturation at differing times ensures that self-pollination cannot occur.
- In some species, the pollen and stigma of flowers of the same plant use chemical self-incompatibility mechanisms. If pollen lands on the stigma of a flower on the same plant, the pollen tube does not grow because of the chemical incompatibility.

- Some flowering species produce flowers that only have male parts or only have female parts.
- In some species, an entire plant is either male or female and can only produce flowers of their own sex.
- The pollen of some species is transferred by wind, which often takes pollen away from the parent plant.

Pollination and fertilization in plants are different processes and rely on different mechanisms.

D3.1.11 – Self-incompatibility mechanisms

D3.1.11 – Self-incompatibility mechanisms to increase genetic variation within a species
Students should understand that self-pollination leads to inbreeding, which decreases genetic diversity and vigour. They should also understand that genetic mechanisms in many plant species ensure male and female gametes fusing during fertilization are from different plants.

In Section D3.1.8, you learned that when pollen lands on the stigma a pollen tube begins to grow. The pollen tube takes the male nuclei to the ovule, where fertilization can occur. Many plants use the growth of pollen tubes as a mechanism to control self-pollination. Each plant has a set of genes that controls the growth of the pollen tube. When the pollen of a plant lands on the stigma of a flower of the same plant, protein interactions occur that reduce or stop growth of the pollen tube. This is called a **self-incompatibility mechanism.** The specifics of the mechanism differ among plant species, but include the following.

- The pollen grain on the stigma fails to germinate into a pollen tube.
- The pollen grain germinates but does not enter through the stigma into the style.
- The pollen tube enters the ovule but the pollen nuclei degenerate before fertilization can occur.
- Fertilization occurs but the plant embryo degenerates during early growth.

The most successful pollination occurs when the pollen is from one plant and the stigma is in the flower of a completely different plant of the same species. This promotes genetic variation in and healthy growth of the new plant (called **vigour**). Self-pollination leads to inbreeding and a decrease in genetic diversity and vigour.

D3.1.12 – The role of seeds

D3.1.12 – Dispersal and germination of seeds
Distinguish seed dispersal from pollination. Include the growth and development of the embryo and the mobilization of food reserves.

Seed dispersal

Once a successful double fertilization has occurred, a seed will begin to develop. Each fertilization within an ovary will lead to a different seed, and the flowers of some species promote the development of many seeds from the same flower. If a flower has a single ovule within its ovary, then only one seed will develop. If there are many ovules within an ovary, then many seeds will develop. The ovary itself grows (ripens) and becomes a fruit. The number of seeds inside a fruit is an indication of how many ovules the ovary contained.

In biology, the term vegetable has no meaning. Many items commercially marketed as vegetables are really fruits. These include cucumbers, peppers and tomatoes. All were once ovaries of flowers and contain seeds. Other vegetables that do not have seeds include the roots, stems or leaves of a plant.

Plants invest a great deal of chemical energy in the production of fruits. The reward for the plant is that the fruit provides a means of dispersing the seeds away from the parent plant. Many fruits have evolved to be attractive to animals as a source of food. After ingestion and digestion by an animal, the seeds are often still protected within their seed coats, and are deposited in the animal's faeces potentially far away from the parent plant. Successful seed dispersal also depends on available light as well as water and nutrients within the soil. A location will not allow germination or early plant growth unless environmental conditions support that growth. Some seeds, such as a coconut, use water to float to a new location before geminating. Other seeds have structures that allow them to be easily carried by the wind for dispersal. A few species develop pods that dry out as the seeds become ripe. Once the pod is dry enough, it pops open explosively, releasing the seed away from the parent plant.

The seeds of a dandelion plant (*Taraxacum officinale*) develop filamentous structures that allow them to be easily dispersed by the wind.

Seeds can stay dormant for very long periods of time under the right conditions. Seed banks, where seeds are stored by promoting dormancy, have been established in many countries in order to preserve the genetic diversity of plant species.

Seed germination

Once seeds have formed within a flower's ovary, they usually begin a period of dormancy. This requires the seed to lose most of its water content. During dormancy, the seeds display a very low metabolism, with no growth or development. The dormancy period is variable for different types of seeds, but many can remain dormant for years. This represents an adaptation feature to overcome harsh, but potentially temporary, environmental conditions, because it allows seeds to wait and germinate when conditions may be more suitable for growth.

When conditions become more favourable, seeds may **germinate**. Germination is the early growth of a seed as it develops into a plant. Several general conditions must be met for a seed to germinate:

- water is needed, to rehydrate the dried seed tissues
- oxygen is needed, to allow aerobic respiration to produce adenosine triphosphate (ATP)
- an appropriate temperature for the seed is necessary.

The effect of increasing temperature on carbon dioxide output by germinating seeds. Full details of how to carry out this experiment with a worksheet are available in the eBook.

A seed contains a small plant embryo and the food reserves it needs for early growth. The food reserves are called endosperm tissue, and are transferred to the plant embryo through structures called **cotyledons**. As the plant embryo grows larger, the reserves of endosperm tissue become depleted.

Germination inhibition in tomato seeds. Full details of how to carry out this experiment with a worksheet are available in the eBook.

Seeds begin germination by absorbing water in a process called **imbibition**. This activates the biochemistry of the seed. The rate of cell respiration and protein synthesis greatly increases following imbibition, as the embryonic plant prepares to emerge from the seed coat. In most plant species, the portion of the embryo that emerges first is the initial root structure, called the **radicle**. The radicle is positively influenced by gravity and grows down into the soil.

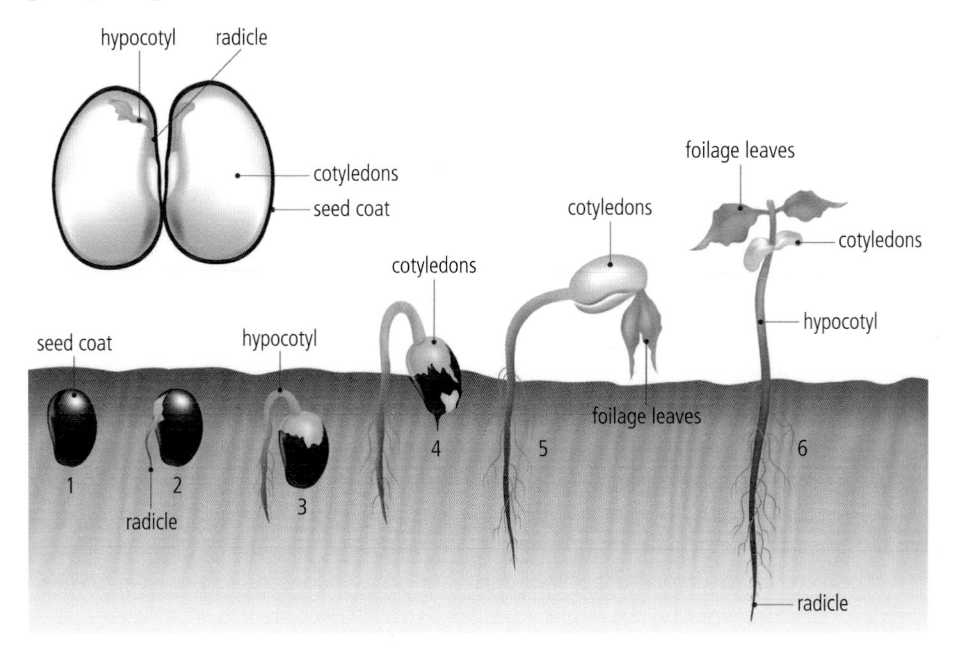

The process of germination, showing the early growth of the radicle and slightly later growth of the shoot.

In many plant species the first structure to appear above ground is called the **hypocotyl**. This is a curved portion of the plant shoot that is found below the cotyledons. The hypocotyl grows in the opposite direction to the force of gravity, and therefore grows up. Once above ground, the early shoot will straighten. This orientates the growing young plant perpendicular to the ground. The first real leaves then develop and begin photosynthesis as the endosperm tissue in the cotyledons is depleted. The root structures during this time continue to develop, forming **secondary roots** as well as **root hairs**. All plant growth from this point on will occur at areas called **meristem tissue** located at the tips of shoots and roots.

HL

D3.1.13 – Developmental changes during puberty

D3.1.13 – Control of the developmental changes of puberty by gonadotropin-releasing hormone and steroid sex hormones
Limit to the increased release of gonadotropin-releasing hormone (GnRH) by the hypothalamus in childhood triggering the onset of increased luteinizing hormone (LH) and follicle-stimulating hormone (FSH) release. Ultimately the increased sex hormone production leads to the changes associated with puberty.

The term gonadotropin means "gonad loving". Any hormone that is a gonadotropin will have either the testes or ovaries as the target tissue.

The hypothalamus of the brainstem increases its production of a hormone called **gonadotropin-releasing hormone** (**GnRH**) during late childhood or early teens in both males and females. The target tissue of GnRH is the nearby pituitary gland. In response, the secretion of two hormones, called **follicle-stimulating hormone** (FSH) and **luteinizing hormone** (LH), by the pituitary gland is enhanced. These two hormones are classified as **gonadotropins** because their target tissue is the gonads, the ovaries in females and the testes in males. The gonadotropins stimulate steroid production by the gonads. Steroids include testosterone, oestradiol and progesterone.

The physiological result of the enhanced production of all of these hormones is the onset of puberty. Puberty produces the changes described in Tables 3 and 4.

Females
Increase in height and body mass
Growth of underarm and pubic hair
Development of breasts
Menstrual cycle begins
Acne may begin
Bone structure of hips widens to prepare for possible childbirth

D3.1 Table 3 Body changes in females as a result of puberty

Males
Increase in height and body mass
Growth of underarm, pubic, facial and chest hair
Acne may begin
Voice becomes deeper
Enlargement of testes and penis
Erections begin

D3.1 Table 4 Body changes in males as a result of puberty

D3.1.14 – The production of gametes

D3.1.14 – Spermatogenesis and oogenesis in humans
Include mitosis, cell growth, two divisions of meiosis and differentiation. Students should understand how gametogenesis, in typical male and female bodies, results in different numbers of sperm and eggs, and different amounts of cytoplasm.

Spermatogenesis

The production of **spermatozoa** (sperm cells; singular **spermatozoon**) occurs within the testes. The testes of human males are located outside the body in order to provide the cooler temperature (lower than the internal body temperature) necessary for production of spermatozoa. Inside each testis, **spermatogenesis** occurs within very small tubes known as **seminiferous tubules**. Near the outer wall of the seminiferous tubules lie cells called **spermatogonia** (singular spermatogonium). Each spermatogonium is capable of undergoing either mitosis or meiosis at any given time.

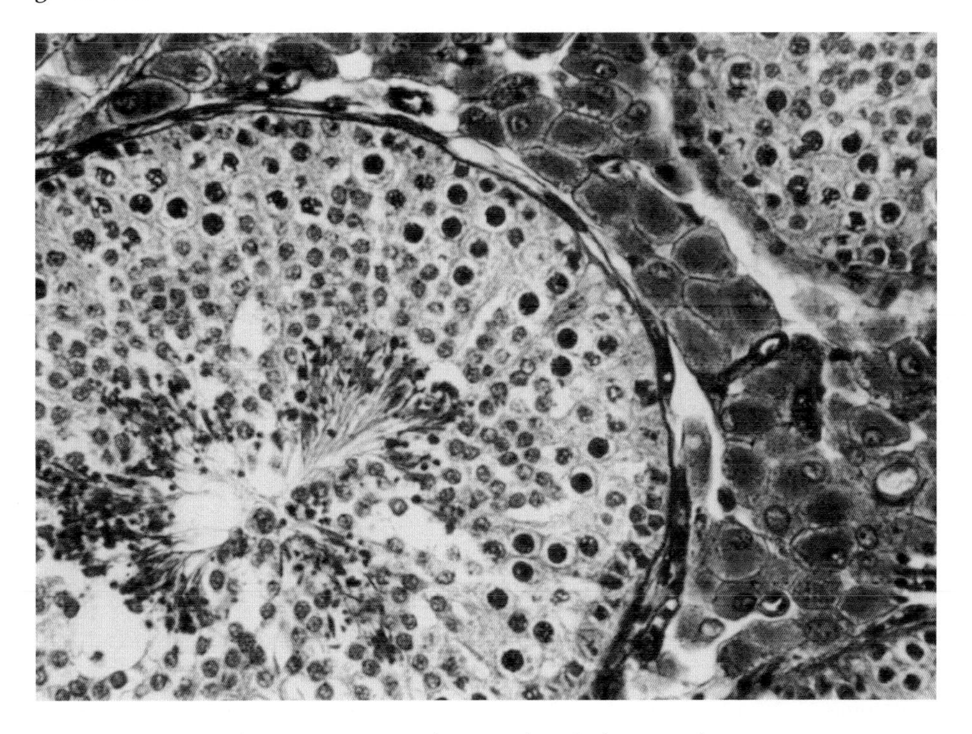

A light micrograph of a nearly complete section of a seminiferous tubule. The cells near the outer edge of the tube (visible as the nearly full circle) are the spermatogonia. As the cells progress through meiosis and differentiation, they move closer to the centre of the tubule. The clear area at the centre is the lumen, where many nearly completed spermatozoa with flagella are located. Soon these spermatozoa will detach and move through the lumen to a storage area called the **epididymis**. Two partial sections of seminiferous tubules are shown. The (darker) cells between the sections are the cells of the testes that produce testosterone.

Spermatogonia undergo mitosis in order to replenish their numbers. Spermatozoa production starts at puberty and continues throughout life. Millions of spermatozoa can be produced in a single day, and mitosis replaces the cells that become spermatozoa.

Spermatogonia undergo meiosis to produce spermatozoa. Meiosis is a reduction division because it reduces the original diploid number of chromosomes to the haploid number in gametes. In humans, 23 homologous pairs of chromosomes (therefore 46 chromosomes in total) become 23 individual chromosomes.

Spermatogonia first replicate the DNA within their still diploid nucleus. At the same time, the spermatogonia are undergoing cell growth in preparation for cell division. If a spermatogonium undergoes mitosis, two half-size cells result, each capable of growing again for a later cell division. If a spermatogonium begins meiosis, four spermatozoa will be produced.

We will follow the steps of meiosis using human cells as an example. Human spermatogonia are diploid and contain 23 homologous pairs of chromosomes. In preparation for meiosis, DNA replication occurs and each of the 46 chromosomes now exists as a pair of chromatids.

The stages of meiosis during spermatogenesis.

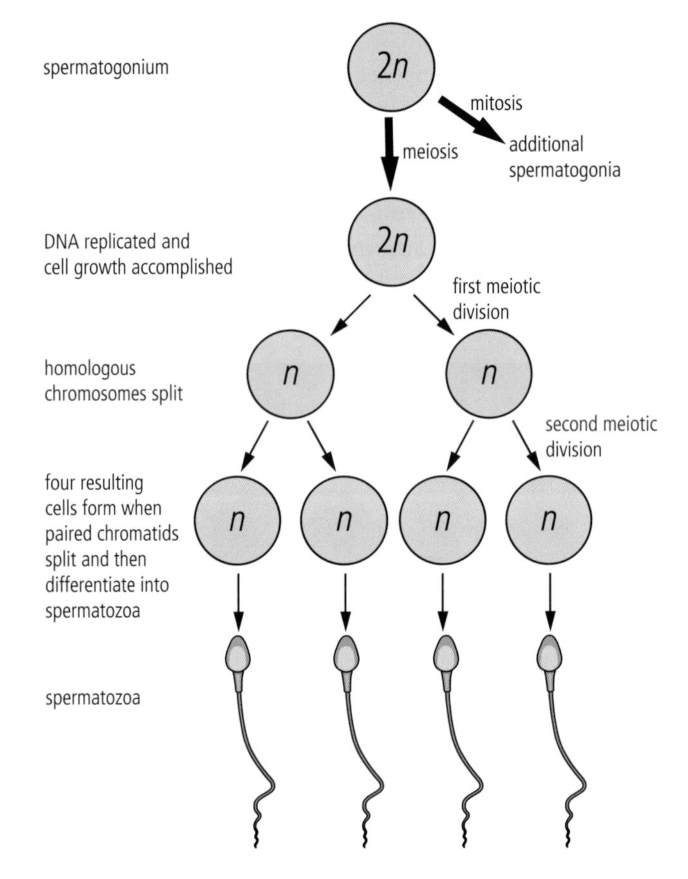

The designation of diploid and haploid is dependent on whether chromosomes are found in a cell in homologous pairs ($2n$) or not (n). Because meiosis I splits the homologous pairs, gametes become haploid before their second cell division occurs.

Meiosis I takes place and two cells result, each with the haploid number of chromosomes (23) because the homologous pairs have been separated. Each chromosome still exists as a pair of chromatids, so there is another cell division, meiosis II. During meiosis II, the chromatids are separated. Four haploid cells, each containing 23 chromosomes, are created from the parent cell that originally contained 23 homologous pairs.

Meiosis is completed for these cells, but each must now **differentiate** into a fully functioning, motile spermatozoon. The cells remain within the interior of the seminiferous tubule while they form the cellular structures characteristic of a mature spermatozoon. These structures include a **flagellum** for motility and an **acrosome** that contains the enzymes necessary for fertilization. The developing spermatozoa need nutrients during this period of differentiation, and thus each remains attached to cells in the seminiferous tubules known as **Sertoli cells**.

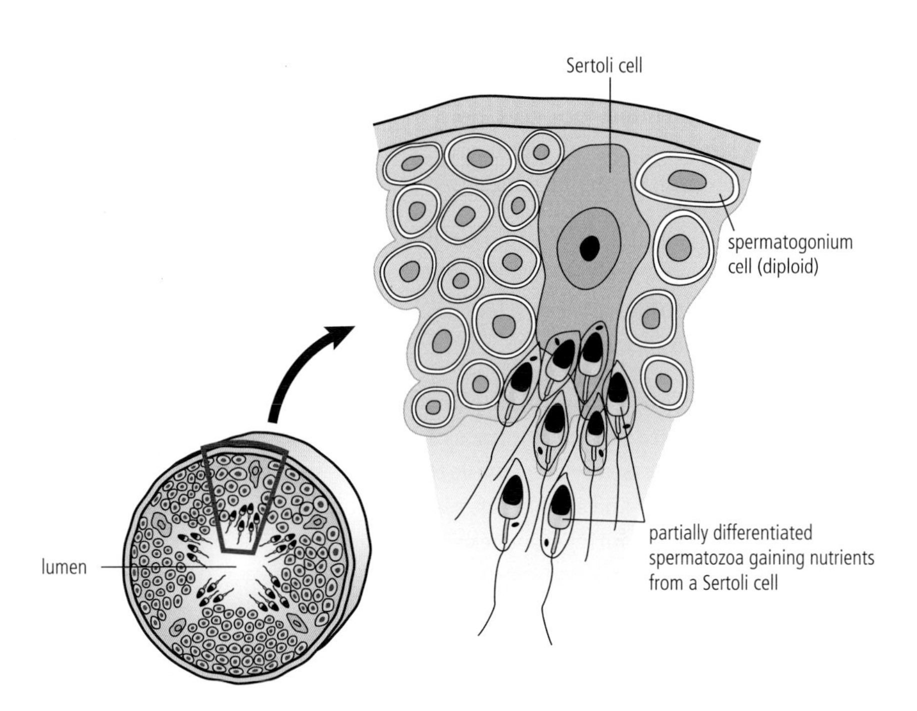

A section view of a seminiferous tubule.

Each of the stages of meiosis moves the resulting cells closer to the interior of the seminiferous tubule. Because the tubule is a small tube, there is a cavity or lumen at the centre. Once spermatozoa have completed differentiation, they detach from their Sertoli cell and move through the lumen to a storage area of the testis called the epididymis.

Oogenesis

Oogenesis and spermatogenesis are the female and male processes of meiosis, respectively. Thus there are many similarities between the two processes, especially regarding the behaviour of the chromosomes. Oogenesis produces four cells as the "end products" of meiosis, as does spermatogenesis. However, three of the four end-product cells of oogenesis are not used as gametes because they are much too small to produce a viable zygote if fertilized. These three cells are called **polar bodies**, and their function is to be a cellular "container" for the divided chromosomes during both meiosis I and meiosis II. The fourth haploid cell produced is very large and is the **ovum** (plural **ova**). As we go through the various developmental stages of an ovum, note the many similarities with spermatogenesis.

Although many of the events described on the next page are relatively similar for other mammals, the details given here are specific to human oogenesis.

Reproductive biology and birth control are important issues from an international perspective. Some countries have achieved zero or negative population growth, but many others have growing populations. At the time of writing, the Earth has just reached 8 billion inhabitants.

Within the ovaries of a female foetus, cells called **oogonia** undergo mitosis repeatedly in order to build up the numbers of oogonia within the ovaries. These oogonia grow into larger cells called **primary oocytes.** Both oogonia and primary oocytes are diploid cells. The large primary oocytes begin the early stages of meiosis, but the process stops (is arrested) during prophase I.

Also within the ovaries, cells called **follicle cells** repeatedly undergo mitosis. A single layer of these follicle cells surrounds each primary oocyte and the entire structure is then called a **primary follicle.** When a female child is born, her ovaries contain nearly half a million primary follicles.

During each menstrual cycle, a few of the primary follicles finish meiosis I. The two haploid cells arising from a primary follicle as a result of meiosis I are not even close to being equal in size. One is very large while the other is very small. The small cell is called the **first polar body** and simply acts as a reservoir for half of the chromosomes. The polar bodies produced during oogenesis later degenerate. The other, very large, cell is a **secondary oocyte**.

Summary of oogenesis. Notice that some of these events occur within the ovary and some occur after ovulation. Oogenesis in humans is a process that takes many years to complete.

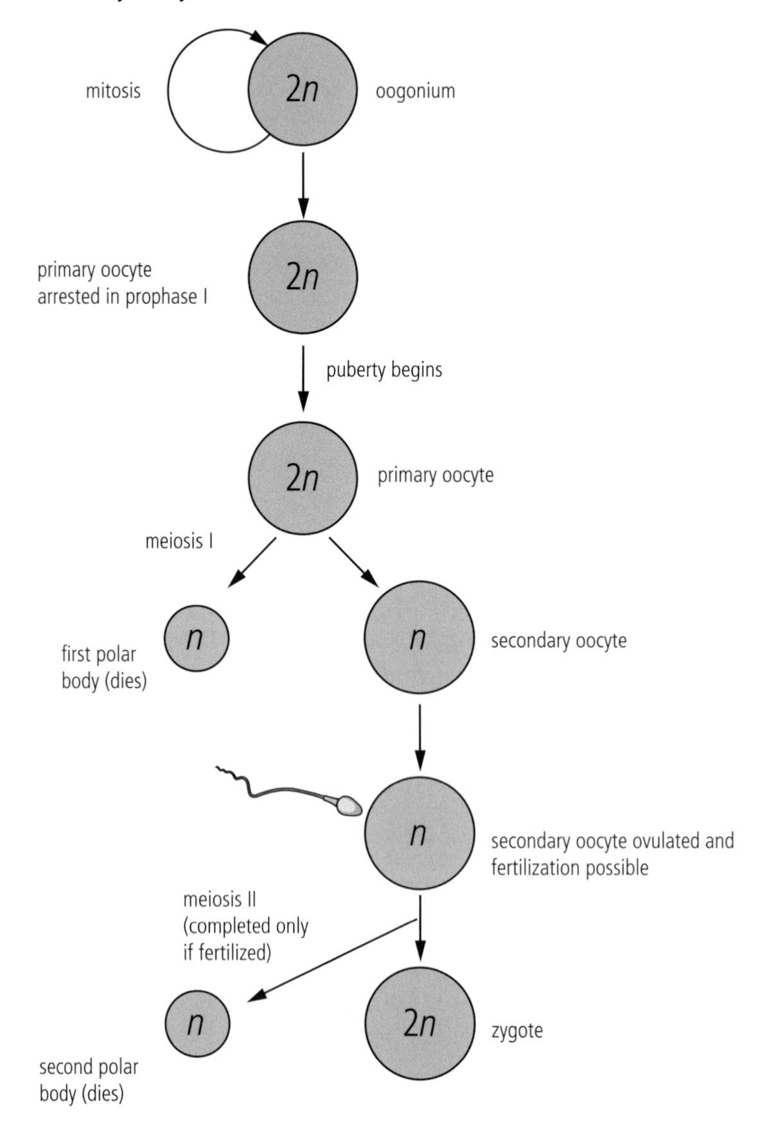

Meiosis in females is a very lengthy process. Meiosis I occurs in foetal ovaries and the oocyte development is then arrested. Meiosis II is completed when the oocyte is being fertilized. These two stages of meiosis can be two or more decades apart.

You will recall that meiosis I produces haploid cells, but each cell has chromosomes existing as paired chromatids. The single ring of follicle cells begins to divide and form a fluid. Two rings of follicle cells are formed, with a fluid-filled cavity separating them. The first (inner) ring of follicle cells surrounds the oocyte, then there is the fluid-filled space, and then the outer ring of follicle cells. This entire structure is now called a Graafian follicle (see Figure 3 on page 692). The increase in fluid between the two follicle cell layers creates a bulge on the surface of the ovary and eventually leads to **ovulation**.

It is a secondary oocyte with the inner ring of follicle cells that is released from the ovary at ovulation, although people often refer to this event as the release of the ovum or egg. The second meiotic division (meiosis II) is not completed until fertilization. If fertilization does not occur, the released gamete remains a secondary oocyte until the cell dies. If fertilization does occur, the subsequent events stimulate the secondary oocyte to complete meiosis II. There is only a true egg for the very brief period between a spermatozoon starting to fertilize the female gamete (the secondary oocyte) and the haploid nuclei fusing to form the zygote nucleus.

Spermatogenesis	Oogenesis
Produces millions of cells daily	Thousands of cells produced in total
Four gametes are produced from each cell	One gamete is produced from each cell (plus two polar bodies)
The gamete cells produced are the smallest cells in body	The gamete cells produced are the largest in body
The gamete cells contain minimal cytoplasm and organelles	The gamete cells contain huge quantities of cytoplasm, organelles and nutrients
The gamete cells are motile	The gamete cells are not motile

A comparison of the cells produced by spermatogenesis and oogenesis

Challenge yourself

1. Identify each of these gamete stages as haploid (n) or diploid ($2n$).

(a) A spermatogonium located in the outer perimeter of a seminiferous tubule.
(b) A spermatozoon stored in the epididymis of a testis.
(c) A recently ovulated oocyte.
(a) A zygote two days after fertilization.

D3.1.15 – Preventing polyspermy

D3.1.15 – Mechanisms to prevent polyspermy
The acrosome reaction allows a sperm to penetrate the zona pellucida and the cortical reaction prevents other sperm from passing through.

Polyspermy is the term used to describe the rare event of more than one spermatozoon fertilizing the ovum. Polyspermy results in the ovum not developing. To prevent polyspermy, the first spermatozoon that uses the enzymes of its acrosome to penetrate the plasma membrane of the ovum initiates a series of events called the **cortical reaction**. Within the cytoplasm of the ovum are many small vesicles called **cortical granules** (see Figure 5 on page 695); they are located around the interior of the plasma membrane. The initial fertilization causes the cortical granules to fuse with

There is an approximately 1 in 250 million chance that any single spermatozoon ejaculated into the vagina will be the sperm cell that fertilizes the ovum. You would not be "you" if you had not won that lottery long ago.

There are two key reaction sequences involved in fertilization. The first is called the acrosome reaction, and involves the release of hydrolytic enzymes from the acrosome of each sperm cell. The second is called the cortical reaction, and is triggered by the first sperm cell to penetrate the ovum. The cortical reaction prevents polyspermy.

the ovum's internal plasma membrane and release their enzymes to the outside. These enzymes result in a chemical change in the glycoprotein layer surrounding the ovum called the **zona pellucida**, making it impermeable to any more spermatozoa. The cortical reaction takes place very quickly after the first spermatozoon gains access. The resulting fertilized ovum is now referred to as a **zygote**. The diploid condition (2*n*) has been restored and a new life has been started.

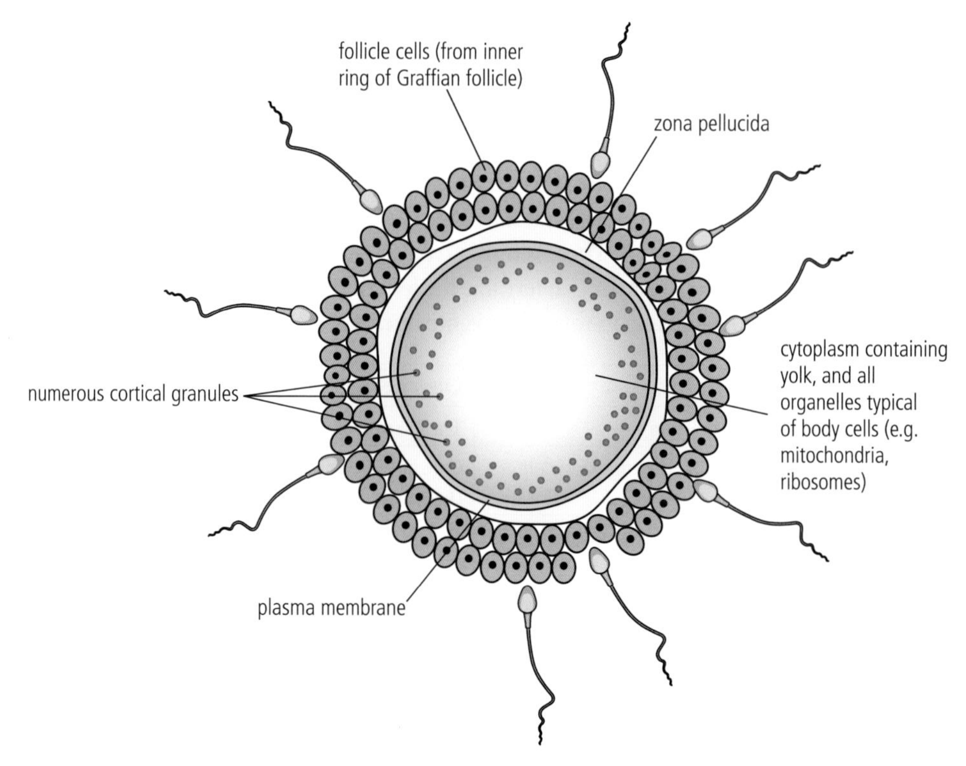

Multiple spermatozoa attempting to fertilize an ovum. Cortical granules release their enzymes to the outside when one sperm cell penetrates the plasma membrane. This triggers a reaction called the cortical reaction, to prevent polyspermy.

D3.1.16 – Embryo development

D3.1.16 – Development of a blastocyst and implantation in the endometrium
Students are not required to know the names of other stages in embryo development.

Fertilization stimulates the embryo to begin mitosis, and the first cell division typically occurs within 30 hours of fertilization. The rate of mitotic division will increase with subsequent divisions. The early embryo continues to move within the Fallopian tube towards the cavity of the uterus as it divides. By the time the embryo reaches the uterine cavity, it is approximately 100 cells in size and is ready to implant itself into the endometrium of the uterus. The embryo at this stage is a hollow ball of cells and is called a **blastocyst** (see Figure 6).

A blastocyst is characterized by:

- a surrounding layer of cells called the **trophoblast**, which will help form the foetal portion of the placenta, embryonic membranes and umbilical cord
- a group of cells on the interior known as the **inner cell mass**, located towards one end of the blastocyst, which will become the body of the embryo
- a fluid-filled cavity.

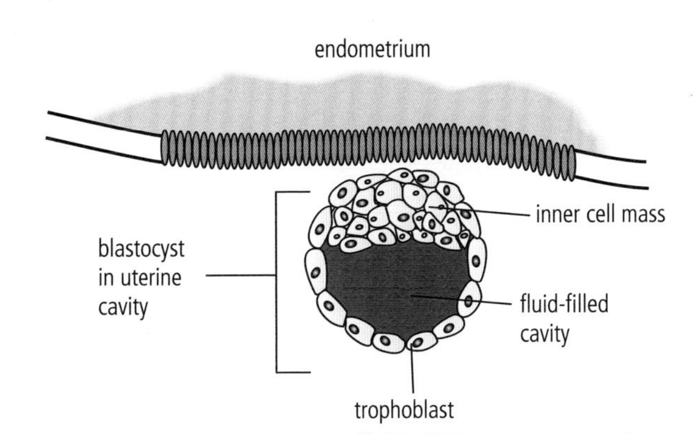

D3.1 Figure 6 A human embryo, now called a blastocyst, after about one week of development. The embryo has passed through the Fallopian tube and entered the uterine cavity. The endometrium is highly vascular and ready to receive the blastocyst.

When the blastocyst enters the inner cavity of the uterus, it is in direct contact with the inner lining called the endometrium. The timing of the menstrual cycle, including ovulation, ensures that the endometrium is highly vascular (has many small blood vessels, including capillary beds) at this point in time. The embryo will eventually stop moving along the endometrium and begin to sink down into the endometrial tissue. This is why this stage of pregnancy is called **implantation**. The primary reason that a human ovum is so large is because it contains the nutrients needed for early embryonic development. During the first week after fertilization, there is no true growth of the embryo. The cell divisions that occur create an embryo of 100 or more cells, but the overall size of the embryo is no larger than that of the original ovum. The nutrients stored within the ovum have been used for metabolism, not for growth. Nutrients from the endometrium will be needed for further development.

D3.1.17 – Pregnancy testing

D3.1.17 – Pregnancy testing by detection of human chorionic gonadotropin secretion
Include the production of human chorionic gonadotropin (hCG) in the embryo or developing placenta and the use of monoclonal antibodies that bind to hCG.

Human embryos, even at the stage of the blastocyst, produce a hormone called **human chorionic gonadotropin (hCG)**. The hormone is produced by the trophoblast layer of embryonic cells (see Figure 6) and then later by embryonic placenta cells. hCG enters the bloodstream of the mother and allows continuation of the **corpus luteum** within the mother's ovaries. The corpus luteum produces progesterone and would begin to recede if no pregnancy occurred. The continuation of the corpus luteum and its secretion of steroids permits the vascular tissue of the uterine endometrium to continue. This is only necessary for the first stage of pregnancy, as later the placenta will produce and secrete its own steroids.

Because hCG is secreted from the embryo, or the placenta formed from the embryo, it is a reliable indicator of a pregnancy and is the basis of early pregnancy tests. Researchers can develop pure cultures of B-lymphocytes that produce only one type of antibody, such as antibodies that recognize hCG as an antigen. The antibodies produced are called **monoclonal antibodies** because the liquid cultures of B-lymphocytes only secrete antibodies capable of detecting a single antigen, in this case hCG. The antibodies are then chemically bonded to an enzyme that changes colour when exposed to a selected substrate. Because the enzyme is bonded to the antibody, the two molecules work together as a unit. The bonded anti-hCG antibodies

The B-lymphocytes that are grown in pure cultures and secrete monoclonal antibodies have been fused to a cancer cell to create a cell called a **hybridoma**. The cancer cell component allows the cell to become long-lived and can be cloned an unlimited number of times.

are placed in small test kits known as early pregnancy tests or EPTs. The urine of a pregnant female will contain hCG. In an early pregnancy test, when exposed to a pregnant female's urine, the anti-hCG antibodies bind to the hCG and a colour change occurs. If there are no hCG molecules, then the anti-hCG antibodies will have no antigen to bind to and a colour change will not occur.

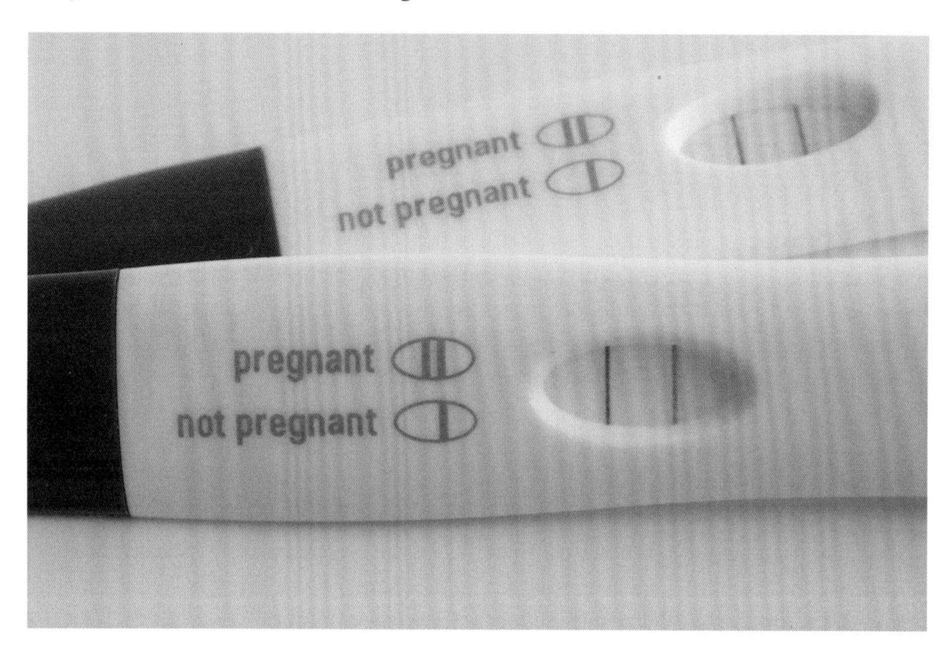

One of the two vertical lines of this early pregnancy test indicates that anti-hCG antibody has attached to hCG antigens. The other vertical line acts as a control to confirm the test is working.

D3.1.18 – The role of the placenta

D3.1.18 – Role of the placenta in foetal development inside the uterus

Students are not required to know details of placental structure apart from the large surface area of the placental villi. Students should understand which exchange processes occur in the placenta and that it allows the foetus to be retained in the uterus to a later stage of development than in mammals that do not develop a placenta.

The **placenta** forms from the **trophoblast** layer of the blastocyst (see Figure 6) and from tissue from the mother. Think of the placenta as being a large pancake-shaped structure. The side of the "pancake" that is located further into the uterine wall is made of connective tissue and blood vessels formed by the mother's body. The side closer to the embryo is formed by the embryo and also contains connective tissue and blood vessels.

On the foetal side of the placenta a protective sheath called the **umbilical cord** will develop, covering three foetal blood vessels. When fully formed, two foetal blood vessels within the umbilical cord carry foetal blood to the placenta. The blood within these two vessels is deoxygenated and carries waste products. This foetal blood exchanges materials with the maternal bloodstream, and the third foetal blood vessel returns the blood to the foetus. The blood that returns to the foetus has been oxygenated, and nutrients have been added while it passes through the placenta.

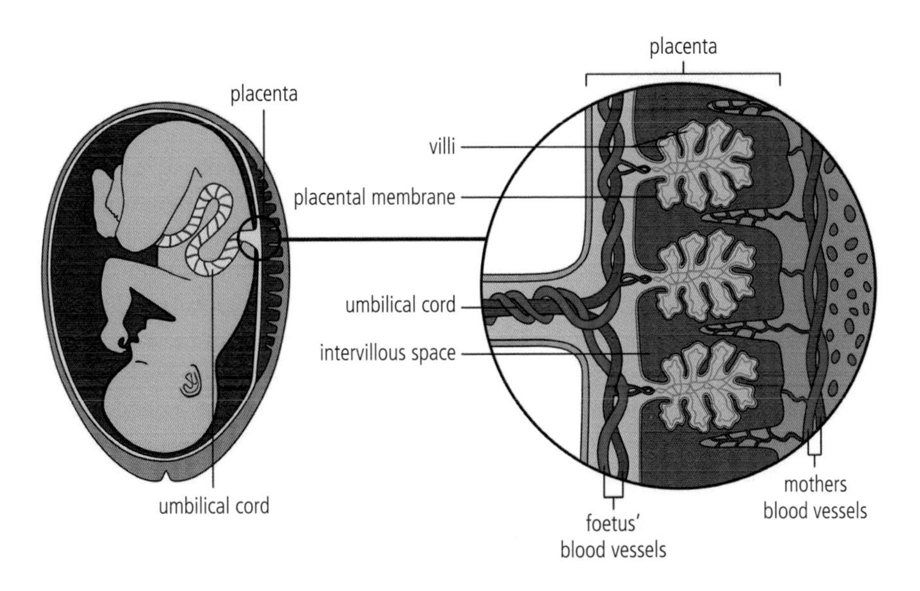

The blood flow pattern of the placenta. The three blood vessels to and from the foetus are encased in a protective sheath called the umbilical cord. The mother's blood vessels directly empty blood into the **intervillous space**. Foetal capillaries in the villi exchange molecules with this pool of maternal blood. The maternal pool of blood is constantly renewed by nearby blood vessels.

No blood is exchanged between the mother and foetus. Only molecular exchanges occur as the foetal capillaries are exposed to the maternal blood within the intervillous space of the placenta. The foetal blood vessels from the umbilical cord divide into numerous projections called **placental villi** that provide a large surface area for molecular exchanges.

Molecules passed from foetus to mother within the placenta	Materials passed from mother to foetus within the placenta
Carbon dioxide	Oxygen
Urea	Nutrients (glucose, amino acids, vitamins, minerals)
Water	Water
Hormones	Hormones
	Alcohol, nicotine and other drugs, if used by the mother during pregnancy

The net diffusion of various substances within the placenta. The direction of exchange is dependent on the concentration gradient of each type of molecule

A placenta is only found in **placental mammals**. Mammals that lack a placenta include **marsupials** and **monotremes**. Female marsupials and monotremes cannot retain their young within their uterus for as long as placental mammals can. Placental foetuses remain in the uterus until a later developmental stage, and as a consequence have a potentially higher survival rate.

What are the roles of barriers in living systems?

D3.1.19 – Pregnancy and childbirth

D3.1.19 – Hormonal control of pregnancy and childbirth
Emphasize that the continuity of pregnancy is maintained by progesterone secretion initially from the corpus luteum and then from the placenta, whereas the changes during childbirth are triggered by a decrease in progesterone levels, allowing increases in oxytocin secretion due to positive feedback.

The steroid hormone progesterone is a key hormone for both maintaining pregnancy and signalling when it is time for birth. Progesterone is initially produced by the corpus luteum of the ovaries, and then later by the placenta. High levels of this steroid hormone maintain the vascular tissue of the placenta. As the foetus reaches full term, less progesterone is secreted by the placenta, which acts as a signal for birthing to begin.

Most physiological feedback mechanisms are designed to work by negative feedback. This works well when there is a physiological factor such as body temperature or blood glucose that needs to be maintained within a fairly narrow homeostatic range. Birth, or **parturition**, is a process that is not part of normal mammalian homeostasis. Parturition is characterized by uterine contractions that begin at a relatively low intensity and occur infrequently. As birth continues, the uterine contractions become more and more intense, and become more and more frequent. The feedback control involved in this process is an example of **positive feedback**. In effect, a previous event results in a more forceful and frequent future event. There is no homeostatic factor being controlled; the series of events will only stop when birth occurs.

A giraffe, *Giraffa reticulata*, a placental mammal, giving birth. The young giraffe will be able to walk and nurse within minutes of birth.

Pregnant women that are past term in their pregnancy are sometimes given a chemical called Pitocin to induce labour. Pitocin is recognized by the receptors of the uterus as oxytocin. This begins the positive feedback loop leading to birth.

The hormone involved in this positive feedback mechanism is **oxytocin**. Oxytocin is a hormone produced by the hypothalamus and secreted by the pituitary gland. When it is time for parturition, a decreased secretion of progesterone stimulates the pituitary to secrete a small amount of oxytocin into the bloodstream. Oxytocin receptors in the muscle of the uterus respond with the first contraction. That first contraction signals the pituitary to release more oxytocin. This happens repeatedly until the uterine contractions are very intense and very frequent. The only event that will terminate the positive feedback loop is birth. Termination occurs because the uterine muscle no longer has anything to contract upon.

D3.1.20 – Hormone replacement therapy

D3.1.20 – Hormone replacement therapy and the risk of coronary heart disease
NOS: In early epidemiological studies, it was argued that women undergoing hormone replacement therapy (HRT) had reduced incidence of coronary heart disease (CHD) and this was deemed to be a cause- and-effect relationship. Later randomized controlled trials showed that use of HRT led to a small increase in the risk of CHD. The correlation between HRT and decreased incidence of CHD is not actually a cause- and-effect relationship. HRT patients have a higher socioeconomic status, and this status has a causal relationship with lower risk of CHD.

Menopause is the period in a female's life when her menstrual cycle first becomes less predictable and then eventually stops altogether. For most females this occurs after the age of 45–55, but the age at which it happens can vary significantly. The reason for menopause is that the ovaries stop producing oestradiol and progesterone. The effects of menopause can include trouble sleeping, hot flushes, some loss of musculature and other symptoms. Some of the symptoms dissipate over time. In order to alleviate symptoms, some females use **hormone replacement therapy** (HRT) in the form of oestradiol (oestrogen).

Different researchers and physicians have taken different positions on the health risks of HRT. Earlier studies reported that HRT reduced the incidence of coronary heart disease (CHD), whereas later studies indicated that HRT led to a small increase in the risk of CHD. Many researchers now believe that there is no connection between reproductive HRT and heart disease.

Nature of Science

Epidemiological studies involving humans are extremely difficult to control and the relationships being explored are correlations rather than cause and effect. Even though some major factors can be used to disqualify test subjects (for example previous heart disease, use of tobacco products, age), there are other factors that are, initially at least, too subtle to be considered. In the study that showed a positive correlation between reproductive HRT and lowered CHD, it was later shown that the test subjects had a generally high socioeconomic status and that this status had health benefits leading to a lower risk of CHD.

HL end

Guiding Question revisited

How does asexual or sexual reproduction exemplify themes of change or continuity?

In this chapter we have described the following:

- asexual reproduction provides nearly genetically identical offspring well adapted to a stable environment
- sexual reproduction provides genetic diversity in populations necessary for adaptations in a changing environment

- genetic diversity is provided by independent assortment of chromosomes during gamete production
- genetic diversity is also provided by the many possibilities inherent in the fertilization of gametes
- genetic variation in flowering plants is enhanced by cross-pollination.

Guiding Question revisited

What changes within organisms are required for reproduction?

In this chapter we have described the following:

- specific reproductive cells in organisms undergo meiosis in order to produce gametes
- gametes have half the number of chromosomes (n) compared to other body cells ($2n$)
- human females, after puberty, undergo menstrual cycles of approximately 28 days to coordinate ovulation with preparation of the uterus for embryo implantation
- flowering plants undergo regular cycles of flower production, often including asynchronous timing of pollen production and stigma maturation to avoid self-pollination
- seed germination is often delayed until environmental conditions can support a young plant

HL

- in humans, males and females undergo a hormonal change early in life known as puberty
- puberty prepares a person to have the physiological and anatomical capability of producing gametes and offspring.

HL end

Exercises

Q1. Arrange the following structures in the sequence that sperm would follow during the process of fertilization:

cervix, urethra, Fallopian tube, testis, vas deferens, uterus, epididymis, vagina.

Q2. What changes does the uterus undergo during a typical menstrual cycle?

Q3. Where in a female's body does fertilization take place?

Q4. Which of these events is not associated with in vitro fertilization (IVF)?

(a) Hormone therapy that suspends the action of the female's own hormone cycle.

(b) Use of FSH to induce superovulation.

(c) Use of hormones to induce males to produce large numbers of sperm.

(d) Testing of embryos following Petri dish fertilization.

HL

Q5. State three environmental conditions necessary for the germination of seeds.

Q6. Many plants are capable of both asexual and sexual forms of reproduction. State one advantage of each of these types of reproduction.

Q7. Identify the term that is incorrectly paired with a chromosome number.

A Spermatogonium (n).

B Sperm (n).

C Egg (n).

D Any one cell of a blastocyst ($2n$).

Q8. Parturition or birth is controlled by hormonal positive feedback. Explain how this feedback works.

Q9. Identify a hormone produced and secreted from an embryo that passes across the placenta into the mother's bloodstream.

Q10. What is inside the umbilical cord of a foetus?

HL end

D3.2 Inheritance

Guiding Questions

What patterns of inheritance exist in plants and animals?

What is the molecular basis of inheritance patterns?

Sometimes a heritable trait is controlled by a single gene, but more often it is controlled by more than one. Certain traits show up more often in males than females depending on which chromosome the gene is located. Some versions of genes mask others, so sometimes a characteristic can skip a generation. With the ABO blood type in humans, there are multiple versions of genes rather than just two, and they can mix in such a way that someone can have type AB blood. In some flowering plant species, plants with red flowers crossed (bred) with plants with white flowers produce plants with either only red flowers or only white flowers, while some species can produce pink flowers.

Genetic variety in humans can be explained by whether or not the gene is found on the first 22 chromosomes or on the X or Y chromosome, what nucleotides are present in the different alleles, and which ones code for which proteins.

D3.2.1 – Haploid gametes and diploid zygotes

D3.2.1 – Production of haploid gametes in parents and their fusion to form a diploid zygote as the means of inheritance
Students should understand that this pattern of inheritance is common to all eukaryotes with a sexual life cycle. They should also understand that a diploid cell has two copies of each autosomal gene.

D3.2 Figure 1 Two haploid gametes meet during fertilization to produce a diploid zygote.

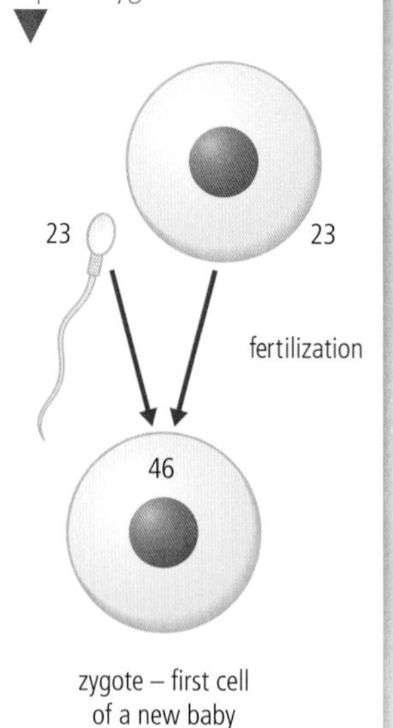

The term **diploid** is used to describe a nucleus that has chromosomes organized into pairs of homologous chromosomes. Most cells in the human body are diploid cells, and in such cells the nucleus contains a set of 23 chromosomes from the mother and 23 from the father (see Figure 1). There is a category of cell that only contains 23 chromosomes in total: the sex cells, also called **gametes.** Because the chromosomes in sperm and egg cells do not come in pairs, but rather only have a single chromosome from each pair, they are said to be **haploid**. The adult form of animal cells is rarely haploid, but there are exceptions, for example male bee, wasp and ant cells are haploid.

The variable n represents the haploid number, and it refers to the number of sets of chromosomes that a nucleus can have. For a human egg cell, $n = 23$. When an egg cell is fertilized by a sperm cell (a sperm is also haploid and therefore contains 23 chromosomes), a **zygote** is formed and the two haploid nuclei fuse together, matching up their chromosomes into pairs. Hence humans generally have a total of $23 + 23 = 46$ chromosomes. This means that, in humans, $2n = 46$, so diploid cells in humans have 23 pairs of chromosomes making a total of 46 chromosomes.

What biological processes involve doubling and halving?

D3.2.2 – Genetic crosses in flowering plants

D3.2.2 – Methods for conducting genetic crosses in flowering plants
Use the terms "P generation", "F1 generation", "F2 generation" and "Punnett grid". Students should understand that pollen contains male gametes and that female gametes are located in the ovary, so pollination is needed to carry out a cross. They should also understand that plants such as peas produce both male and female gametes on the same plant, allowing self-pollination and therefore self-fertilization. Mention that genetic crosses are widely used to breed new varieties of crop or ornamental plants.

Mendel's experiments with pea plants

In 1865, an Austrian monk named Gregor Mendel published the results of his experiments on how garden pea plants passed on their characteristics. At the time, the term "gene" did not exist (he used the term "factors" instead) and the role that DNA played would not be discovered for nearly another century. For thousands of years people have used artificial pollination techniques, deliberately placing pollen from the male parts of one flower on the female part of another flower in order to produce seeds for the next generation. Mendel used artificial pollination to get the sperm cells in the pollen of pea plants into the ovum cells inside the ovaries of other pea plants. He covered the flowers so that bees or other pollinating insects did not interfere with his work.

Gregor Mendel studied the garden pea (*Pisum sativum*).

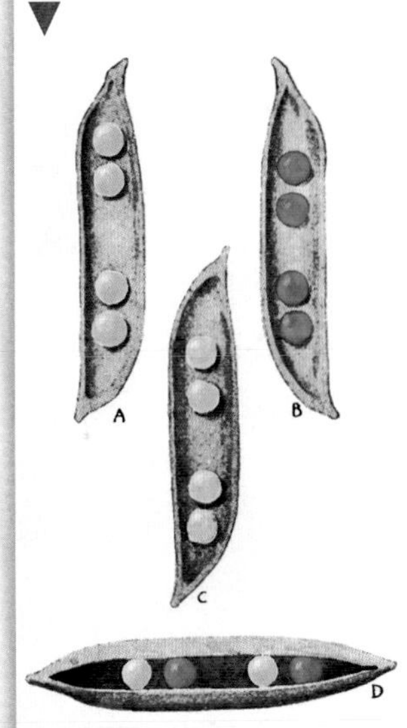

Table 1 shows some of the characteristics Mendel tried to cross. The × in the first column indicates a cross between one variety of pea plant and another, i.e. two varieties were bred together. As we will see later in this chapter, the expected ratio after two generations of crosses is 3:1 (for every 3 of the first type of plant, we expect 1 of the other type): look how close Mendel got.

D3.2 Table 1 Mendel's results

Characteristics in parents	First generation produced	Second generation produced	Ratio of results seen in second generation
Round × wrinkled seeds	100% round	5,474 round 1,850 wrinkled	2.91:1
Yellow × green seeds	100% yellow	6,022 yellow 2,001 green	3.01:1
Green × yellow pods	100% green	428 green 152 yellow	2.82:1
Tall × short plants	100% tall	1,787 tall 277 short	2.84:1

We will consider in more detail one of the crosses that Mendel carried out with his garden pea plants. He took **purebred** tall plants and crossed them with purebred short plants. Purebred means that the tall plants' parents were known to be all tall, and the short plants' parents were known to be all short. In other words, he knew that none of the plants had a mix of short and tall traits. He wanted to find out whether he would get all tall plants, some tall and some short plants, or all short plants, when he crossed tall and short plants.

The answer took months for Mendel to confirm, but theoretical techniques can now be used to get the answer in seconds: the result was 100% tall plants. Why? Because in garden pea plants the **allele** (version of the gene) for tall is dominant over the allele for short plants, thus masking the short trait in **heterozygotes**. A heterozygote is an organism that possesses one dominant allele and one recessive allele for a particular trait, e.g. **Tt**. To show this a **Punnett grid** can be used.

Constructing a Punnett grid

Figures 2 and 3 on the next page show a Punnett grid. A Punnett grid can be used to show how the alleles of parents are split between their gametes and how new combinations of alleles can show up in their offspring. The purpose of a Punnett grid is to show all the possible combinations of genetic information for a particular trait in a **monohybrid cross**. A monohybrid cross is one in which the parents have different alleles, and it shows the results for only one trait. We will see how this works with Mendel's tall and short pea plants. In order to set up a Punnett grid, you need to follow a series of steps.

Get used to saying "big T" and "little t" when reading alleles and genotypes. Also, do not mix letters: for example, you cannot use **T** for tall and **s** for short. Once you have chosen a letter, write down what it means so that it is clear which allele is which.

1. Choose a letter to represent the alleles.

Use the capital (uppercase) and lowercase of the letter to represent the different alleles. Usually, a capital letter represents the **dominant allele** while the lowercase letter represents the **recessive allele**. Recessive alleles can be masked by dominant ones. For example:

- **T** = dominant allele, for example tall pea plants is a trait that is dominant over short pea plants
- **t** = recessive allele, the trait that results in short pea plants.

2. Determine the parents' genotypes.

To be sure that no possibilities are overlooked, write out all three possibilities (based on the possible combinations of two traits) and decide by a process of elimination which genotype or genotypes fit each parent.

The three possibilities here are:

- **TT** –the genotype is **homozygous dominant**, and in this example the phenotype presents as tall plants
- **Tt** – the genotype is heterozygous (with one of each allele), and in this example the phenotype presents as tall plants (heterozygotes carry the recessive allele but display the dominant trait, and they can potentially pass the recessive allele on to their offspring)
- **tt** – the genotype is **homozygous recessive**, and in this example the phenotype presents as short plants.

The easiest genotype to determine by simply looking at a pea plant is **tt**. The other two are more of a challenge. To determine whether a plant is **TT** or **Tt**, we have to look for evidence that the recessive gene was received from a short parent or was passed on to the plant's offspring. The only way to produce a short plant is for each parent to donate one **t**.

If a cross is to be made between **TT** and **tt** purebred parents, they represent the **parental generation**, or **P generation**. The name given to the first generation produced by such a cross is the **first filial generation**, usually referred to as the

F1 generation. Crossing two members of the F1 generation produces the **second filial (F2) generation**.

3. Determine the gametes that the parents could produce.

   An individual with a genotype **TT** can only make gametes with the allele **T** in them.

   Heterozygous carriers can make **T**-containing gametes or **t**-containing gametes. Obviously, individuals whose genotype is **tt** can only make gametes that contain the **t** allele. So you can record and label with **T** or **t** all the possible gametes.

4. Draw a Punnett grid.

   Once all the previous steps have been completed, drawing the actual grid is simple. The parents' gametes are placed on the top and side of the grid. As an example, consider a cross involving a female plant **Tt** crossed with a male short plant **tt**.

   Figure 2 shows a Punnet grid for the parents' gametes.

   Now you can fill in the empty squares with each parent's possible alleles by copying the letters from the top down and from left to right. When letters of different sizes end up in the same box, the big one goes first, as shown in Figure 3.

5. Work out the chances of each genotype and phenotype occurring.

   In a grid with four squares, each square represents one of two possible statistics:

   - the chance that these parents will have offspring with that genotype, each square representing a 25% chance
   - the probable proportion of offspring that will have the resulting genotypes, although this only works for large numbers of offspring.

   The grid in Figure 3 for Mendel's tall and short plants can be interpreted in two ways:

   - there is a 50% chance of producing tall offspring and a 50% chance of producing short offspring
   - 50% of the offspring will be tall and 50% of the offspring will be short.

   All the tall plants are heterozygous.

   In a real experiment, it is unlikely that exactly 50% of the offspring will be short plants. The reason is essentially due to chance. For example, if 89 F2 peas were produced and all of them were planted and grew into new plants, there is no mathematical way that exactly 50% of them would be short. At the very most, mathematically 45 out of the 89 plants would be short, which is 50.6%; that is as close to 50% as is possible in this case.

   Even if a convenient number of plants was produced, such as 100 plants, farmers and breeders would not be surprised if they got, for example, 46, 57 or even 63 short plants instead of the theoretical 50. If the results of hundreds of similar crosses were considered, however, the number would probably be very close to 50%.

	t	t
T		
t		

▲ **D3.2 Figure 2** A Punnett grid showing the parent plants' gametes.

	t	t
T	Tt	Tt
t	tt	tt

▲ **D3.2 Figure 3** A Punnett grid with all the possible genotypes filled in.

Be careful when choosing letters when writing them by hand. Nearly half the letters of the alphabet should in fact be avoided because the capital and lowercase versions are too similar, which can lead to confusion. Try not to use Cc, Ff, Kk, Oo, Pp, Ss, Uu, Vv, Ww, Xx, Yy or Zz. If in an exam one of these letters is used in a question, be sure to clearly differentiate between lowercase and capitals when writing your answer.

Summarizing the five steps of the Punnett grid method:

1. **Choose a letter.**
   T = allele for a tall plant.
   t = allele for a short plant.
2. **Define the parents' genotypes.**
   TT for the purebred tall parent. tt for the purebred short parent.
3. **Determine the gametes.**
   The purebred tall parent can only provide T. The purebred short parent can only provide t.
4. **Draw a Punnett grid.**

	t	t
T	Tt	Tt
T	Tt	Tt

5. **Interpret the grid. 100% will be Tt and therefore will be tall, so 0% will be short.**

Self-pollination

Some plants, such as peas, have flowers that can produce both male pollen and female ova. If they prepare gametes at the same time, it is possible for them to **self-pollinate**. Self-pollination is when a plant's pollen lands on flowers it has produced itself. As a result, it is possible for **self-fertilization** to happen. This will result in less genetic diversity than cross-pollination, which allows genes from different individuals of a species to breed.

When farmers want plants with the same characteristics as previous generations, they can use self-pollination techniques, but when they want to create new varieties with combinations of traits not seen before, they can use cross-pollination techniques. Modern wheat, for example, was produced by crossing breeds that have stalks that are a convenient height for higher yields, that do not drop their seeds before harvest time, and are resistant to disease. It took many years of scientific research and thousands of cross-breeding trials over many years to achieve the desired combination of traits.

A gene is a DNA sequence that codes for a protein that will give an organism a specific trait, whereas alleles are versions of a gene. In peas, for example, all plants have a gene that determines their height, but one allele of the gene is for tall and the other allele is for short.

D3.2.3 – Combinations of alleles

D3.2.3 – Genotype as the combination of alleles inherited by an organism
Students should use and understand the terms "homozygous" and "heterozygous", and appreciate the distinction between genes and alleles.

Alleles: versions of genes

An **allele** is one specific form of a gene, differing from other alleles by one or a few bases. For example, some people are born without the ability to see any colours because of an inability to produce the protein transducin in their retinas. This is caused by a single base pair difference between the most common allele (with a C at position

235) and the rare mutated allele (with a T at position 235). These different forms allow a single trait, such as the trait for the ability to see in colour, to have variants, in this example either colour or grey-scale vision.

Chapter D1.2 discusses transcription and translation of DNA, and shows how important it is for each letter in the genetic code to be in a specific place. If, for whatever reason, one or more of the bases (A, C, G or T) is misplaced or substituted for a different base, the results can be dramatic. The difference between one version of a gene and another (e. g. the mutated and non-mutated alleles of the transducin gene) can mean the difference between fully functional organs and impaired organs.

Genotype

The **genotype** is the symbolic representation of the pair of alleles possessed by an organism, typically represented by two letters. All eukaryotes that reproduce sexually will inherit one allele from the female parent and one from the male parent.

Examples: **Bb**, **GG**, **tt**.

Homozygous refers to having two identical alleles of a gene (see Figure 4).

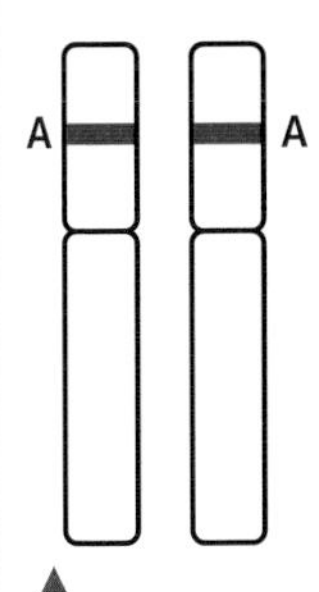

D3.2 Figure 4 The **locus** of a gene on a pair of chromosomes. The locus is the particular position of a gene on homologous chromosomes. Sometimes, as in this example, the allele we inherit from each parent is the same.

Examples: **AA** is the genotype of someone who is homozygous dominant for that trait, whereas **aa** is the genotype of someone who is homozygous recessive.

Heterozygous refers to having two different alleles of a gene (see Figure 5). This results when the paternal and maternal alleles are different.

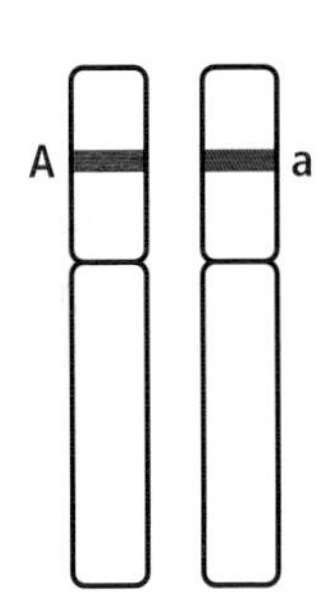

D3.2 Figure 5 In this instance, the individual has inherited **A** on the maternal chromosome and **a** on the paternal chromosome.

Example: **Aa** is a heterozygous genotype.

A **carrier** is an individual who has a recessive allele of a gene that does not have an effect on the phenotype.

Example: **Aa** carries the allele for albinism but has pigmentation, which means an ancestor must have been albino and some offspring might be albino. If both parents are unaffected by a recessive condition yet both are carriers, some of their offspring can be albinos (because they would be **aa,** like the penguin in Figure 6).

D3.2 Figure 6 An albino does not have the genes necessary to produce pigmentation. That is why this Cape penguin (*Spheniscus demersus*) has no black coloration.

D3.2.4 – Phenotype

D3.2.4 – Phenotype as the observable traits of an organism resulting from genotype and environmental factors

Students should be able to suggest examples of traits in humans due to genotype only and due to environment only, and also traits due to interaction between genotype and environment.

The **phenotype** is the observable characteristics or traits of an organism. Examples include being colour blind and having type O blood. But phenotype is not always 100% the result of genes. A person is born with a certain skin colour but that colour can change with exposure to sunlight. A suntan or burn is an acquired trait and is caused by the ultraviolet (UV) rays present in the environment. Can you think of examples of characteristics produced only by genetics? Only by the environment? What about traits that are a mix of genetics and environment? Below are some suggestions.

Phenotypes produced exclusively by genetics include:

- ABO blood type
- genetic conditions such as Huntington's disease, cystic fibrosis and colour blindness.

Phenotypes produced exclusively by the environment include:

- learned behaviour (e.g. birds learning a new song, humans learning mathematics)
- acquired physical traits (such as a scar or large muscles from weightlifting).

Phenotypes produced by an interaction between genes and the environment include:

- height in humans (while the maximum height is genetic, a poorly nourished person might not reach their maximum potential)
- cancer (which can have a genetic component but is often triggered by cancer-causing substances in the environment).

D3.2.5 – Dominant and recessive alleles

D3.2.5 – Effects of dominant and recessive alleles on phenotype

Students should understand the reasons that both a homozygous-dominant genotype and a heterozygous genotype for a particular trait will produce the same phenotype.

A **dominant allele** is an allele that has the same effect on the phenotype whether it is paired with the same allele or a different one. Dominant alleles are always expressed in the phenotype.

Example: the genotype **Aa** gives rise to the dominant **A** trait because the **a** allele is masked; the **a** allele is not transcribed or translated during protein synthesis.

A **recessive allele** is an allele that has an effect on the phenotype only when no dominant allele is present to mask it.

Example: **aa** gives rise to the recessive trait because there is no **A**.

An individual with **Aa** will have the same phenotype as an individual with **AA**.

Codominant alleles are pairs of alleles that both affect the phenotype when present in a heterozygote.

Example: a parent with curly hair and a parent with straight hair can have children with different degrees of hair curliness, because both alleles influence hair condition when both are present in the genotype.

D3.2.6 – Phenotypic plasticity

D3.2.6 – Phenotypic plasticity as the capacity to develop traits suited to the environment experienced by an organism, by varying patterns of gene expression
Phenotypic plasticity is not due to changes in genotype, and the changes in traits may be reversible during the lifetime of an individual.

Phenotypic plasticity is an organism's ability to express its phenotype differently depending on the environment. Not surprisingly, this is an effective way of adapting, and natural selection is all about either adapting or dying out. Birds can activate genes that produce more of the digestive enzyme maltase when fewer insects are available to eat and there are more grains in their diet. A plant can activate genes that produce growth hormones to make thicker leaves when it senses that there is more light available. Animals can modify their foraging behaviour depending on the types of foods available in their environment from one season to the next. Cyclical events such as seasonal food availability are referred to as **phenological**. Phenotypic plasticity can generate changes in physiology (as in the enzyme example), in morphology (the leaf example, or the snail shells we shall look at next), in behaviour (e.g. foraging) or in phenology (e.g. adapting to seasons). Notice that some are permanent changes, such as how an organism grows, but others can change within the organism's lifetime, such as physiological and behavioural changes.

The freshwater snail *Physa virgata* normally has an elongated shell. However, this snail can sense cues from its predator (the bluegill fish, *Lepomis macrochirus*) and express its genes slightly differently in order to grow a less conical shape and a more rounded, shorter shell that is more difficult for its predator to crush. The genotype itself does not change, so this is not a mutation. The snail still has the same alleles. But by expressing and silencing alleles in different combinations, the shell shape can grow differently and produce a modified phenotype.

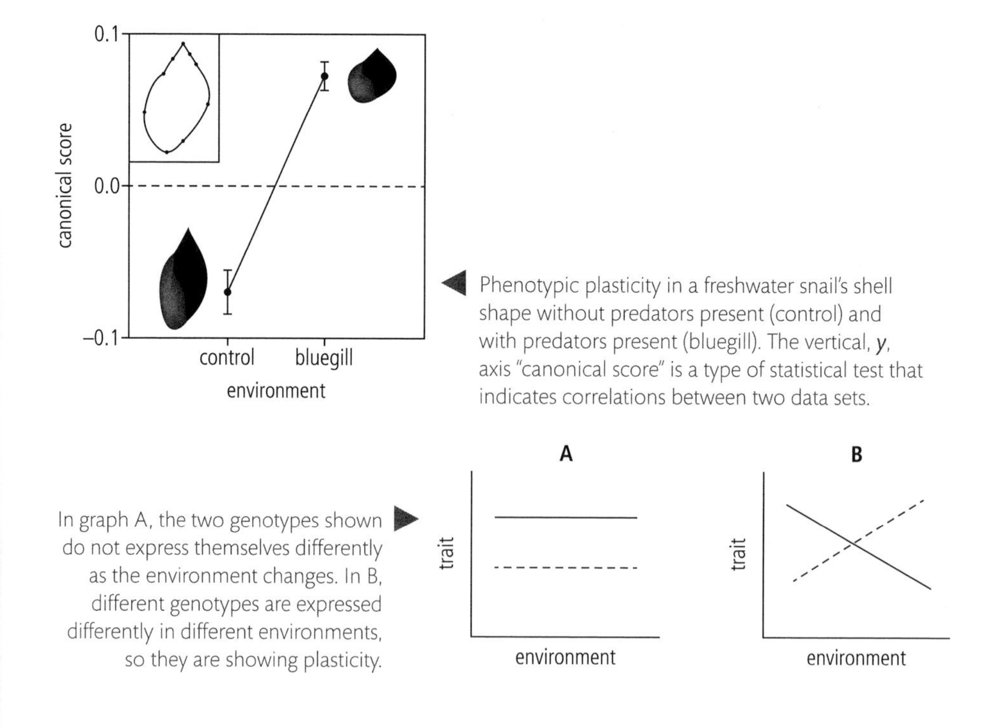

Phenotypic plasticity in a freshwater snail's shell shape without predators present (control) and with predators present (bluegill). The vertical, *y*, axis "canonical score" is a type of statistical test that indicates correlations between two data sets.

In graph A, the two genotypes shown do not express themselves differently as the environment changes. In B, different genotypes are expressed differently in different environments, so they are showing plasticity.

D3.2.7 – Recessive genetic conditions

D3.2.7 – Phenylketonuria as an example of a human disease due to a recessive allele
Phenylketonuria (PKU) is a recessive genetic condition caused by mutation in an autosomal gene that codes for the enzyme needed to convert phenylalanine to tyrosine.

How is it possible for two healthy parents to have a child who is affected by a genetic disease? The disease is caused by a recessive allele and both healthy parents must be carriers of the version of the gene that causes the disease. For example, the genetic disease called **phenylketonuria**, abbreviated to **PKU**, is caused by mutations in the autosomal *PAH* gene, which results in low levels of the enzyme phenylalanine hydroxylase. This enzyme converts the amino acid phenylalanine into tyrosine, which is not toxic. If there is a large quantity of protein in a child's diet, phenylalanine levels can become toxic and can impair brain development. Children diagnosed with PKU are recommended a diet that omits foods rich in phenylalanine, such as eggs, chicken and nuts. The problem is such foods are rich in protein, which is needed for healthy growth and development. As a result, dietary supplements are sometimes needed.

Let us call the allele that produces a functioning enzyme **F**, and the allele that can cause PKU **f**. In Figure 7, showing a family that has the PKU allele, the parents James and Iyana are carriers (**Ff**). The only way to have the disease is to have the genotype **ff**, so James and Iyana do not suffer from PKU but they can pass it on to their children. If you set up a Punnett grid for these parents, you will see that there is a 1 in 4 chance (25%) that they will have a child with PKU, and there are three possibilities for the genotypes in their children: Malik is **Ff**, Jada is **ff**, and Kasim is **FF**.

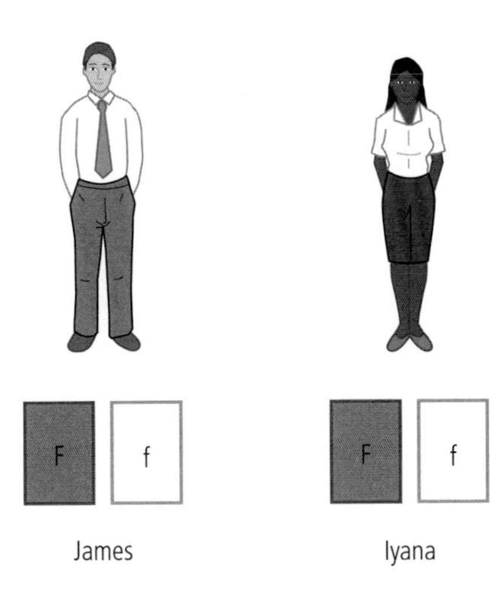

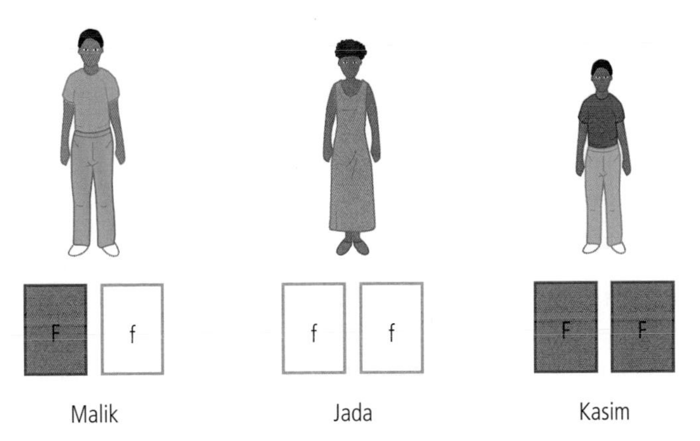

D3.2 Figure 7 How an autosomal gene can be passed on to offspring: Malik is a carrier like his parents, Jada has the autosomal disease caused by the recessive allele, and Kasim not only does not have PKU but cannot pass the recessive allele on to his future children.

Such diseases are called **autosomal recessive diseases** because they are caused by recessive alleles, and the locus of their gene is found on one of the first 22 pairs of chromosomes but not on the sex chromosomes X or Y. Parents can find out if their child has PKU shortly after birth thanks to blood tests that can screen for it. The following are other examples of autosomal recessive diseases, which are all rare in the human population:

- albinism
- cystic fibrosis
- sickle cell disease
- Tay Sachs disease
- thalassemia.

Nature of Science

Students sometimes get the impression that genetics is only about diseases. This is not true. It is just that more is known about disease-causing genes than about genes for characteristics such as eye colour, because researchers spend time and funds studying topics that can help society. Studying diseases and discovering their genetic causes is more useful to medicine than studying eye colour. Governments and university laboratories investing money in research want their work and their discoveries to lead to healthier lives for people. Getting a return on their investment also motivates them. Fundamental research ("I would like to study this just to find out how it works") does not attract as much funding as applied research ("I would like to find out how this disease is caused so that we can develop better medical treatments for it").

D3.2.8 – Single-nucleotide polymorphisms and multiple alleles

D3.2.8 – Single-nucleotide polymorphisms and multiple alleles in gene pools
Students should understand that any number of alleles of a gene can exist in the gene pool but an individual only inherits two.

So far, only two possibilities have been considered for a gene: dominant, for example **A**, and recessive, **a**. With two alleles, three different genotypes are possible, which can produce two different phenotypes. However, genetics is not always this simple; sometimes there are three or more alleles for the same gene. This is the case for the alleles that determine the ABO blood type in humans, which we will look at in detail in Section D3.2.9. In the immune system, there are genes called HLA genes, some of which have dozens of alleles and some thousands. This makes sense when you think of all the different pathogens our bodies have to fight off.

Single-nucleotide polymorphisms, also known as **SNPs**, occur when a nucleotide of the genetic code, such as T, is not found where it is expected, and another, such as C, is found at that position instead. If such a difference is found in a coding part of the genome, it can potentially cause another amino acid to be coded for, and this could modify the structure and properties of the protein that is translated. Such a variation in a nucleotide indicates a different allele.

The ability to taste bitter substances is important to a mammal's survival because bitterness in nature is often a warning that something is toxic. The molecule phenylthiocarbamide (PTC) can be perceived as tasting bitter if a person possesses a particular taste receptor. People with one version of the gene *TAS2R38* can taste PTC while people with a different version cannot taste it. The protein that is encoded by this gene, taste receptor 2 member 38, allows the transduction of a chemical stimulus (bitterness) into a nerve signal for the brain to interpret. If the protein cannot be produced correctly, no signal or a much weaker signal will be passed. Several SNPs have been found on this gene and some account for an allele that is different enough to not allow the person with two copies of the allele to be able to taste PTC. This trait is an example of one that has multiple alleles.

For a person to be able to taste PTC, only one functioning allele needs to be present. It is thought to be a dominant allele, although there is debate about this in the scientific community because in some cases it does not show complete dominance. For our purposes, we will assume it is dominant. Most people can taste PTC and other thiourea compounds similar to it. It is customary to use **T** for "taster" and **t** for "non-taster", to show that three genotypes are possible, **TT**, **Tt** or **tt**. Only the last one would be unable to taste thiourea compounds. In reality, the recessive alleles have more than one version depending on which SNPs they have. But no matter how many alleles are available in the gene pool, each person receives two: one from their mother and one from their father.

D3.2.9 – ABO blood groups

D3.2.9 – ABO blood groups as an example of multiple alleles
Use I^A, I^B and i to denote the alleles.

The ABO blood type system in humans has four possible phenotypes: A, B, AB and O. To create these four blood types there are three alleles of the gene. These three alleles can produce six different genotypes.

The gene for the ABO blood type is represented by the letter **I**. To represent more than just two alleles (**I** and **i**) superscripts are introduced. As a result, the three alleles for blood type are written as follows: $\mathbf{I^A}$, $\mathbf{I^B}$ and **i**. The two capital letters with superscripts represent alleles that are codominant:

- $\mathbf{I^A}$ = the allele for producing proteins called type A antigens, giving type A blood
- $\mathbf{I^B}$ = the allele for producing proteins called type B antigens, giving type B blood
- **i** = the recessive allele that produces neither A nor B antigens, giving type O blood.

Crossing these together in all possible combinations creates six genotypes that give rise to the four phenotypes already listed:

- $\mathbf{I^AI^A}$ or $\mathbf{I^Ai}$ gives a phenotype of type A blood
- $\mathbf{I^BI^B}$ or $\mathbf{I^Bi}$ gives a phenotype of type B blood
- $\mathbf{I^AI^B}$ gives a phenotype of type AB blood (because of codominance, both types of antigens are produced)
- **ii** gives a phenotype of type O blood.

Notice how the genotype $\mathbf{I^AI^B}$ clearly shows codominance. Neither allele is masked: both are expressed in the phenotype of type AB blood. The person makes proteins that include type A antigens and type B antigens.

Worked example

Is it possible for a couple to have four children, each child with a different blood type?

Solution

There is only one way this can happen: one parent must have type A blood but be a carrier of the allele for type O blood, and the other parent must have type B blood and also be a carrier of the allele for type O blood, as shown in the figure below (if necessary, remind yourself of the blood group alleles listed above).

The cross would be $\mathbf{I^Ai} \times \mathbf{I^Bi}$ and the corresponding Punnett grid is shown below. See if you can determine the phenotype of each child before reading on.

Would it be possible for the same couple to have four children and all of them have type AB blood? In theory, yes, but it would be unlikely. This question is similar to asking "Could a couple have four children, all of them girls?" It is possible but statistically less likely than having boys and girls.

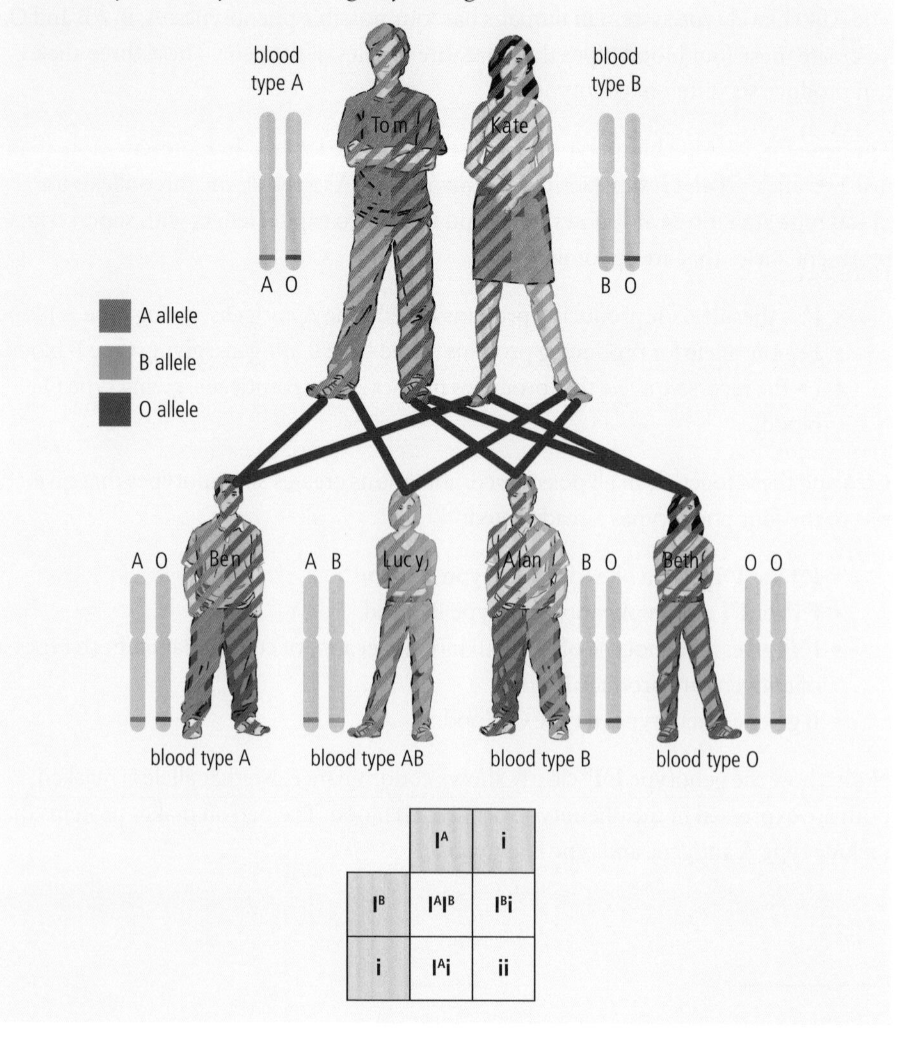

	I^A	i
I^B	I^AI^B	I^Bi
i	I^Ai	ii

D3.2.10 – Intermediate and dual phenotypes

D3.2.10 – Incomplete dominance and codominance

Students should understand the differences between these patterns of inheritance at the phenotypic level. In codominance, heterozygotes have a dual phenotype. Include the AB blood type (I^AI^B) as an example. In incomplete dominance, heterozygotes have an intermediate phenotype. Include four o'clock flower or marvel of Peru (*Mirabilis jalapa*) as an example.

Note: When students are referring to organisms in an examination, either the common name or the scientific name is acceptable.

The four o'clock flower (*Mirabilis jalapa*), also known as marvel of Peru, gets its name because it tends to open its flowers in the later part of the afternoon. Unlike the ABO blood type in humans, which can show codominance with $\mathbf{I^AI^B}$ expressing both phenotypes, this plant shows **incomplete dominance** in its genetics regarding flower pigmentation, resulting in neither one phenotype nor the other but something in between. Here is how it works.

The system of letters for showing colour in four o'clock flowers uses a prefix **C**, which refers to the gene that codes for flower colour, plus a superscript, which refers to the specific colour, **R** (red) or **W** (white).

So the alleles for flower colour are:

- $\mathbf{C^R}$ for red flowers
- $\mathbf{C^W}$ for white flowers.

The genotypes and their phenotypes are:

- $\mathbf{C^RC^R}$ makes red flowers
- $\mathbf{C^WC^W}$ makes white flowers
- $\mathbf{C^RC^W}$ makes pink flowers.

In a cross of purebred four o'clock flowers, white × red = pink (shown in Figure 8). Notice how pink is not a colour present in either of the parents. It is a mix of the two parents' alleles and each one contributes to the phenotype but neither is masked. This is why we call it incomplete dominance.

D3.2 Figure 8 With incomplete dominance, white and red parent flowers can generate pink offspring.

D3.2.11 – Sex determination

D3.2.11 – Sex determination in humans and inheritance of genes on sex chromosomes
Students should understand that the sex chromosome in sperm determines whether a zygote develops certain male-typical or female-typical physical characteristics and that far more genes are carried by the X chromosome than the Y chromosome.

Sex determination

The chromosomes in the 23rd pair in humans are called the sex chromosomes because they determine whether a person is a male or a female. The X chromosome is longer than the Y chromosome, and contains many more genes. Unlike the other 22 pairs of chromosomes, this is the only pair in which it is possible to find two chromosomes that are very different in size and shape (Figure 9).

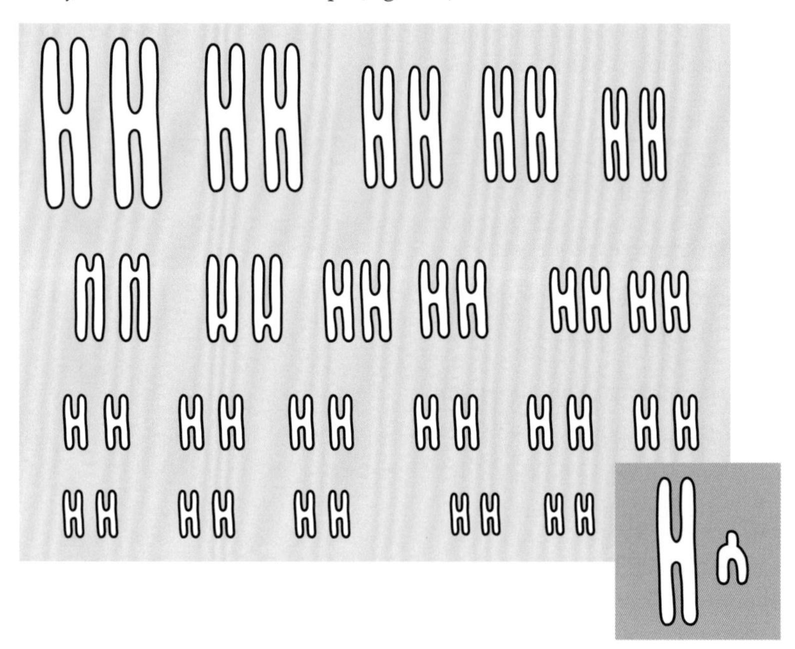

D3.2 Figure 9 Human chromosomes: the majority (grey box) = autosomes, while one pair (shown in a separate purple box) = sex chromosomes.

In human females there are two X chromosomes. When females produce gametes, each egg will contain one X chromosome. Human males have one X chromosome and one Y chromosome. When males produce sperm cells, half of them contain one X chromosome and half contain one Y chromosome. As a result, when an egg cell meets a sperm cell during fertilization, there is always a 50% chance that the child will be a male and a 50% chance that the child will be a female (see Figure 10):

- XX = female
- XY = male.

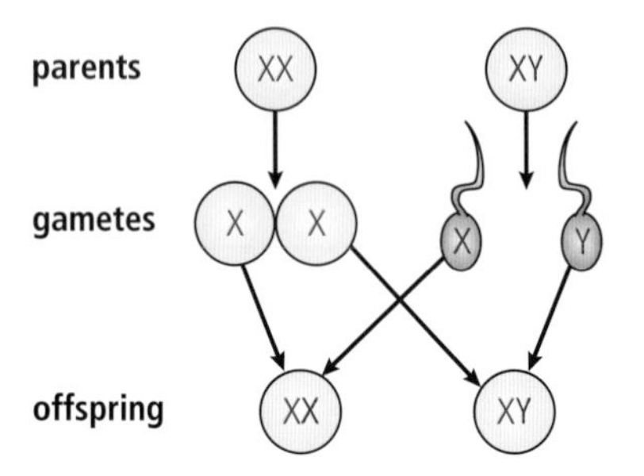

D3.2 Figure 10 How sex is determined: will the baby be male or female?

The chances remain the same no matter how many males and females a family already has. The genetics of being male or female depends on whether you inherit an X or a Y chromosome from your father. If a sperm cell containing an X chromosome fertilizes an egg, a female is produced. Conversely, if a sperm cell containing a Y chromosome fertilizes an egg, a male is produced. Embryos of both sexes are virtually identical until about the eighth week following fertilization. Alleles that interact on both of the X chromosomes of female embryos then result in relatively high oestradiol and progesterone production, resulting in the prenatal development of female reproductive structures. Genes located on the single Y chromosome are responsible for early testes development and relatively high testosterone production, resulting in male reproductive structures during subsequent foetal development.

Inheritance of genes on sex chromosomes

In humans, because the Y chromosome is significantly smaller than the X chromosome, it has fewer loci and therefore carries fewer genes (about 70 in total) than the X chromosome (which carries about 800). This means that most of the alleles present on the X chromosome have nothing to pair up with. For example, a gene whose locus is at an extremity of the X chromosome would have no counterpart on the Y chromosome because the Y chromosome does not extend that far from its centromere.

D3.2.12 – Haemophilia

D3.2.12 – Haemophilia as an example of a sex-linked genetic disorder
Show alleles carried on X chromosomes as superscript letters on an uppercase X.

Sex linkage

Any genetic trait that has a gene locus on the X or the Y chromosome is said to be a **sex-linked trait**. Often genetic traits that show sex linkage affect one sex more than the other. One example of such a genetic trait is **haemophilia**.

Haemophilia is a disorder in which blood does not clot properly. For most people, a small cut or scrape on their skin stops bleeding after a few minutes and eventually a scab forms. This process is called **clotting**. People with haemophilia have trouble with blood clotting and are at risk of bleeding to death from what most people would consider to be a minor injury such as a bruise, which is a rupture of many tiny blood vessels. Bleeding can also occur in internal organs. Medical treatments give people affected by haemophilia a better quality of life.

The letters X and Y refer to chromosomes and not to alleles, so terms such as dominant and recessive do not apply. X and Y should be considered as entire chromosomes rather than alleles of a gene. For sex-linked alleles, the letter that indicates the allele is the superscript after the X or Y. An absence of a superscript means that no allele for that trait exists on that chromosome.

Alleles and genotypes of sex-linked traits

The alleles for haemophilia, represented by **H** and **h**, are found only on the X chromosome:

- $\mathbf{X^h}$ = allele for haemophilia
- $\mathbf{X^H}$ = allele for the ability to clot blood
- **Y** = no allele present on the Y chromosome.

As there is no allele on the Y chromosome, Y is written alone without any superscript. Here are all the possible genotypes for haemophilia:

- $\mathbf{X^HX^H}$ gives the phenotype of a non-affected female
- $\mathbf{X^HX^h}$ gives the phenotype of a non-affected female who is a carrier
- $\mathbf{X^hX^h}$ gives the phenotype of an affected female
- $\mathbf{X^HY}$ gives the phenotype of a non-affected male
- $\mathbf{X^hY}$ gives the phenotype of an affected male.

Notice how only one sex can be a carrier.

Carriers of sex-linked traits

Sex-linked recessive alleles such as $\mathbf{X^h}$ are rare in most populations of humans worldwide. For this reason, it is unlikely for offspring to inherit one and much less likely to inherit two such alleles. This is why so few females have haemophilia: their second copy of the gene is likely to be the dominant allele and will mask the recessive allele.

As you have seen, there are three possible genotypes for females but only two possible genotypes for males. Only females can be heterozygous, $\mathbf{X^HX^h}$, and, as a result, they are the only ones who can be carriers.

Because males do not have a second X chromosome, there are only two possible genotypes, $\mathbf{X^HY}$ or $\mathbf{X^hY}$, in relation to haemophilia. With just the one recessive allele, **h**, a male will have haemophilia. This is contrary to what you have seen up to now concerning recessive alleles: usually people need two to have the trait, while with one they are carriers. In this case, the single recessive allele in males determines the phenotype. Males cannot be carriers for X-linked alleles.

Other examples of sex-linked traits include:

- colour blindness in humans
- Duchenne muscular dystrophy in humans
- white eye colour in fruit flies
- calico–tortoiseshell fur colour in cats.

D3.2.13 – Pedigree charts

D3.2.13 – Pedigree charts to deduce patterns of inheritance of genetic disorders

Students should understand the genetic basis for the prohibition of marriage between close relatives in many societies.

NOS: Scientists draw general conclusions by inductive reasoning when they base a theory on observations of some but not all cases. A pattern of inheritance may be deduced from parts of a pedigree chart and this theory may then allow genotypes of specific individuals in the pedigree to be deduced. Students should be able to distinguish between inductive and deductive reasoning.

The term "pedigree" refers to the record of an organism's ancestry. Pedigree charts are diagrams that are constructed to show biological relationships. In genetics, they are used to show how a trait can pass from one generation to the next. Used in this way for humans, a pedigree chart is similar to a family tree, complete with parents, grandparents, aunts, uncles and cousins.

To build such a chart, symbols are used to represent people. Preparing a pedigree chart helps with the use of Punnett grids for predicting the probable outcome for the next generation. As an example, we will look at the inheritance of Huntington's disease.

Huntington's disease (Huntington's chorea) is caused by a dominant allele that we will refer to by the letter **H**. This genetic condition causes severely debilitating nerve damage but the symptoms do not show until a person is about 40 years old. As a result, someone who has the gene for Huntington's disease may not know they have it until after they have had their own children.

Huntington's disease is life-limiting. Symptoms include difficulty walking, speaking and holding objects. Within a few years of starting to display symptoms, the person loses complete control of their muscles. Because it is dominant, all it takes is one **H** allele in a person's genetic makeup to cause the condition.

The symbols used in pedigree charts are:

- ○ empty circle = female
- □ empty square = male
- ● filled-in circle = a female who possesses the trait being studied
- ■ filled-in square = a male who possesses the trait being studied
- | vertical line = the relationship between parents and offspring
- – horizontal line between a male and a female = the parents of the offspring.

Worked example

The figure below is a pedigree chart showing members of a family affected by Huntington's disease.

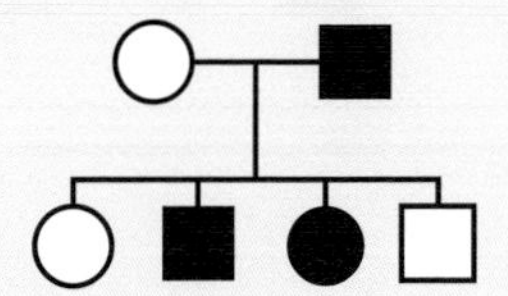

1. Describe the six individuals in the figure, stating who is affected and who is not. See the Key fact box for what the use of shape and colour means.
2. State the genotype for each individual.

Solution

1. The symbols indicate that the unaffected members of the family are the mother, the first child (a girl) and the fourth child (a boy). Those who are affected are the father, the second child (a boy) and the third child (a girl).
2. To work out whether the father is **HH** or **Hh**, consider the fact that some of his children do not have the trait. This indicates that he must have given one **h** to each of them. Hence, he can only be **Hh** and not **HH**. The mother is not affected so she must be **hh**. This is also true for the first daughter and the youngest son. Since the mother always gives an **h**, the two middle children must have at least one **h**, but, because they are affected, they are **Hh**.

From the pedigree shown in the Worked example above, it should be clear why it is taboo, discouraged or illegal in most societies for siblings or other close relatives to have children together. Rare diseases are rare in the general population but among closely related relatives they are much more frequent. In the pedigree example shown, half the individuals (50%) have the genes for Huntington's disease, whereas in the general population only one in about 10,000 (or about 0.01%) have the disease. Long before scientists understood genetics, groups of humans forbade intermarriage between close relatives, undoubtedly because they saw that unfavourable traits or diseases were produced in disturbingly higher frequencies when this was allowed.

Worked example

The pedigree chart below shows how red, white and pink pigmentation is passed on in four-o'clock flowers. For codominant traits, grey is used in pedigree charts rather than black or white. For allele symbols and genotypes, refer to Section D3.2.10.

1. Using the pedigree chart below, state the genotypes for all the plants A to K.
2. What evidence is there that genetic characteristics can sometimes skip a generation?

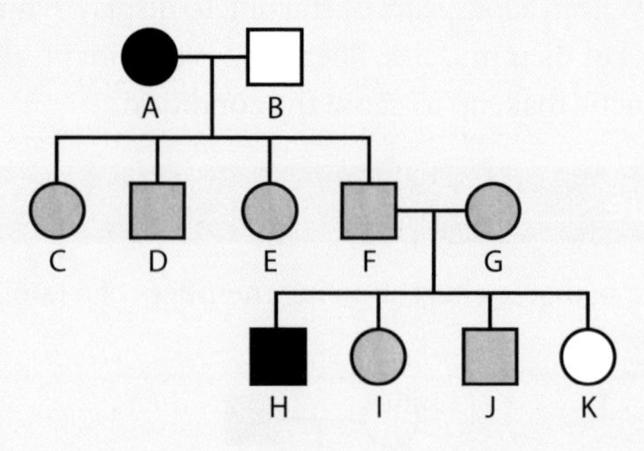

Solution

1. A and H produce red flowers and must be homozygous for red, $\mathbf{C^RC^R}$, because any other combination would give pink or white. B and K produce white flowers and must be homozygous for white, $\mathbf{C^WC^W}$, because any other combination would give pink or red. C to G as well as I and J are pink and must be heterozygous, $\mathbf{C^RC^W}$, because they have one of each allele from each parent plant.
2. It would be impossible for either the colour red or the colour white to be in the middle (F1) generation in this figure. These colours skip a generation and show up again in the last row, the F2 generation.

Nature of Science

When we use a pedigree chart to determine how a gene is passed down, we need to use both inductive and deductive reasoning. **Inductive reasoning** is used when a conclusion or theory is worked out by looking at samples of evidence of a phenomenon. **Deductive reasoning** is when we apply well-established knowledge about a phenomenon to reach a conclusion or theory to explain what is happening. In the pedigree chart above, we can observe which plants have which colour in the phenotype of their flowers, and use inductive reasoning when we notice that all the offspring show a phenotype that is different from the parents (pink flowers instead of red or white). We use deductive reasoning when we apply our knowledge of the phenomenon of inheritance of codominant genes in the genotype, $\mathbf{C^RC^W}$. In a similar fashion we use both inductive and deductive reasoning to explain how the red and white traits can skip a generation.

D3.2.14 – Continuous variation

D3.2.14 – Continuous variation due to polygenic inheritance and/or environmental factors
Use skin colour in humans as an example. **Application of skills**: Students should understand the distinction between continuous variables such as skin colour and discrete variables such as ABO blood group. They should also be able to apply measures of central tendency such as mean, median and mode.

Polygenic inheritance

Polygenic inheritance involves two or more genes influencing the expression of one trait. With two or more allelic pairs found at different loci, the number of possible genotypes is greatly increased. It is believed that most human traits are too complex and show too many combinations to be determined by one gene.

This could partly explain the difficulty in finding out which genes are responsible for traits whose genetic components are poorly understood, for example mathematical aptitude, musical talent or susceptibility to certain illnesses.

Continuous and discrete variation

With dominant and recessive alleles of a single gene, the number of possible phenotypes is limited. For example, a person either has PKU or not. When multiple alleles are introduced, the number of possibilities for a single trait increases accordingly. For example, the ABO blood type has three alleles and four possible phenotypes.

When a second gene is introduced, the number of possible genotypes increases dramatically. If three, four or five genes determine the phenotype, the number of possibilities is so big that it is impossible to see the difference between certain genotypes in the phenotype. When an array of possible phenotypes can be produced, it is called **continuous variation**. The colour of skin in humans is an example of continuous variation and it is thought that the intensity of pigment in our skin is the result of the interaction of multiple genes.

In humans, continuous variation can also be seen in the genetic components of traits such as height, body shape and intellectual aptitude. Each of these is also influenced by environmental components. A person's height, for example, is determined by whether they inherit genes for tallness, but it also depends on their nutrition as they grow.

Polygenic characteristics

When there are many intermediate possibilities in a phenotype, then the trait shows continuous variation. If the results are plotted as a graph, it will produce a bell-shaped distribution curve. There is a smooth transition between the groups of frequencies (see the figure in the Skills box on the next page).

Discontinuous variation:
- in distinct categories which have no transition between them
- the order does not matter
- best plotted as a bar chart
- can determine a mode but not a mean as there is no central tendancy.

Continuous variation:
- not in distinct categories
- the order matters
- smooth transition from one value to the next with no abrupt jumps
- best plotted as a curve or histogram
- can determine a mean to show central tendency.

To help you decide whether or not a trait shows continuous variation, imagine a questionnaire for recording phenotypes. In general, if it is possible to tick "yes" or "no" for a trait, that trait does not show continuous variation, for example grey-scale or colour vision. The same is true for a trait where the possibilities can be represented by just a few choices, such as blood type (A, B, AB or O).

When variation is not continuous, it is referred to as **discrete variation** or **discontinuous variation**. The data for discontinuous variation can be displayed as bar charts (see the figure below). An unbroken transitional pattern from one group to another is not present.

Blood type is an example of discontinuous variation, while height is an example of a trait with continuous variation, with an even distribution around a mean.

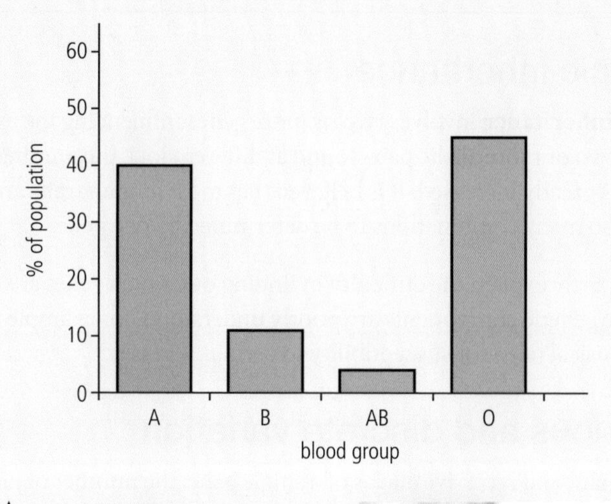

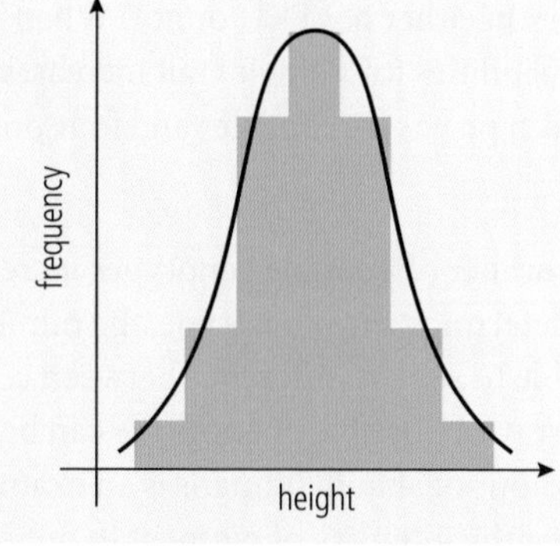

Is it fair to compare heights of humans from different parts of the world? On the one hand, we can argue that we are all the same species and therefore we are comparable. On the other hand, there has been a certain amount of isolation of gene pools over thousands of years, and we also know that different people grow up in very different environments. We are all the same. We are all different.

D3.2.15 – Box-and-whisker plots

D3.2.15 – Box-and-whisker plots to represent data for a continuous variable such as student height

Application of skills: Students should use a box-and-whisker plot to display six aspects of data: outliers, minimum, first quartile, median, third quartile and maximum. A data point is categorized as an outlier if it is more than 1.5 × IQR (interquartile range) above the third quartile or below the first quartile.

One way to show data for a trait that has continuous variation is by using a **box-and-whisker plot**. To generate the graph, we need to calculate some values that help describe the data set (see Figure 11).

Quartiles can be determined by entering the data points in a graphing calculator or a spreadsheet program. They each have a number and a relationship to the **median** value in the data set or to the maximum or minimum value. **Quartile 1** or the first quartile is the middle value between the median and the lowest value in the data set, otherwise expressed as the **25th percentile**. A percentile is a way of expressing the number of data points found below this level, so data in the 25th percentile means a quarter of the data points can be found below this level.

Quartile 2 is the median or the data point at the 50th percentile. **Quartile 3** or the third quartile is the middle number between the median and the highest value in the data set, the **75th percentile**. The **interquartile range, IQR**, measures the spread of the data and is defined as the difference between the 75th and 25th percentiles of the data. The IQR will be the box part of the plotted data. The whiskers show the minimum and maximum values in the data set. Multiplying the IQR by 1.5 and adding this value to quartile 3 or subtracting this value from quartile 1 can be used to determine whether any values should be considered outliers. If it turns out that the minimum value used for the whisker is an outlier, it should not be used for that whisker. Instead, when drawing the whisker, use the lowest number in the data set that is not an outlier. Although outliers can sometimes be the result of faulty data collection, they can also sometimes reveal unexpected results that lead to a whole new investigation.

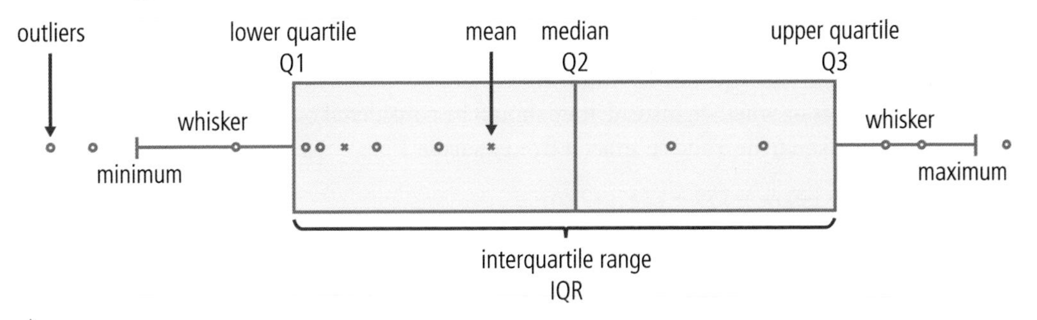

D3.2 Figure 11 This diagram shows which numbers are used to make the box and which are used to make the whiskers in a box-and-whisker plot. This plot is presented horizontally, but box-and-whisker plots can be displayed vertically.

Worked example

Using the data from the Centers for Disease Control and Prevention (CDC) shown in the graph on the right, construct a box-and-whisker plot showing the height of 17-year-old females in the US. Would a height of 183 cm be considered an outlier?

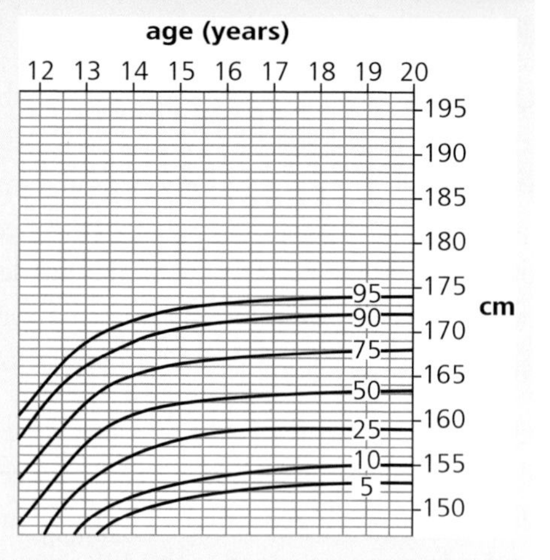

A summary of the data, rounded to the nearest cm, is:

minimum = 150 cm

25th percentile = quartile 1 = 159 cm

50th percentile = quartile 2 (the median) = 163 cm

75th percentile = quartile 3 = 167 cm

maximum = 175 cm

Solution

Calculate the interquartile range (IQR) as well as the IQR × 1.5:

IQR = 167 − 159 = 8

IQR × 1.5 = 12

To decide whether the maximum and minimum values should be used as the whiskers or whether, instead, they should be considered outliers, add the IQR × 1.5 value to quartile 3 and subtract it from quartile 1:

outliers below = 159 − 12 = 147 cm

outliers above = 167 + 12 = 179 cm

Using the numbers you have generated, construct the box-and-whisker plot. The bottom and top whiskers are at 150 and 175 cm because they are the minimum and maximum values and not outliers, the extremities of the box are at 159 and 167 cm and the median is at 163 cm. The figure below presents the plot vertically. A height of 183 cm would be considered an outlier because it is beyond the end of the top whisker.

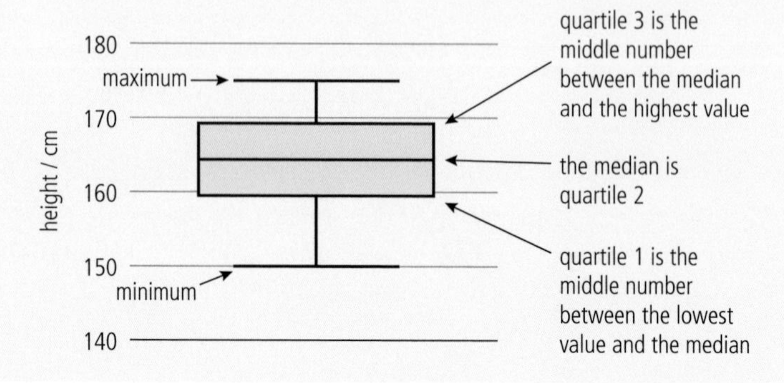

The box-and-whisker plot, also known as a **boxplot**, shows the IQR as well as the highest and lowest values in the data set. It shows the spread of the data and whether or not the data is evenly distributed around the median or **skewed** (biased) towards one side or other of the median. Data points that are outside the IQR zone can be considered outliers.

HL

D3.2.16 – Segregation and independent assortment

D3.2.16 – Segregation and independent assortment of unlinked genes in meiosis
Students should understand the link between the movements of chromosomes in meiosis and the outcome of dihybrid crosses involving pairs of unlinked genes.

Dihybrid crosses

When we looked at monohybrid crosses, we were only considering one genetic trait. Sometimes it is interesting to study two traits at once. We will look at some more of Mendel's experiments with pea plants. In one cross, he examined the following two traits.

- Seed shape: some seeds are round, while others are wrinkled. The allele for round is dominant (see Figure 12).
- Seed colour: some seeds are green inside, while others are yellow. The allele for yellow is dominant (see Figure 12).

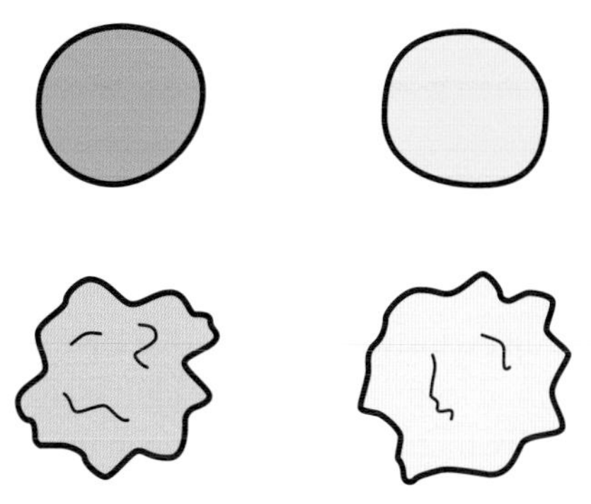

D3.2 Figure 12 Seed colour and seed shape in peas.

If a Punnett grid is set up for this cross and all the alleles can be shuffled in any random order, we find that there are 16 possible random combinations (see Figure 13). As we have seen before with **AA** and **Aa** genotypes, some combinations of alleles can result in the same phenotype. If **R** represents round and **Y** represents yellow, both the genotype **RRYY** and genotype **RrYy** will give seeds that are round and yellow. Normally, **R** and **Y** should be passed on independently. They show no bias in occurrence one way or the other. During the shuffling of alleles in meiosis, they are equally distributed between gametes. The result is that, in the offspring, there should be certain predictable ratios. This can be illustrated by Mendel's cross of two **RrYy** pea plants.

D3.2 Figure 13 A Punnett grid for inheritance of roundness and yellowness in pea seeds. This dihybrid cross shows the following parents: **RrYy** × **RrYy**.

	RY	Ry	rY	ry
RY	RRYY	RRYy	RrYY	RrYy
Ry	RRYy	RRyy	RrYy	Rryy
rY	RrYY	RrYy	rrYY	rrYy
ry	RrYy	Rryy	rrYy	rryy

Phenotypes

= round yellow peas × 9

= round green peas × 3

= wrinkled yellow peas × 3

= wrinkled green peas × 1

Mendel crossed two pure-breeding plants with each other. Pure-breeding means the plants are homozygous for the traits being studied; using them means no surprises will be produced by masked recessive alleles. One parent plant was homozygous dominant for both traits (round and yellow seeds), whereas the other parent was homozygous recessive for both traits (wrinkled and green).

To represent the alleles, Mendel used a system of letters that is incompatible with the system we use today so, in this example, Mendel's original letters have been replaced with the modern convention:

- **R** = allele for round peas
- **r** = allele for wrinkled peas
- **Y** = allele for yellow peas
- **y** = allele for green peas.

Phenotypes and genotypes of round/wrinkled and yellow/green peas

Parent phenotype	Round yellow	Green wrinkled
Parent genotypes	RRYY	rryy
Parent gametes	RY	ry
F1 genotypes	RrYy	–
F1 phenotypes	round yellow	–

F1 refers to the first filial generation and represents the offspring resulting from the cross between the parents. F2 refers to the second filial generation; they are the offspring of the F1 generation.

The F1 generation is made up exclusively of plants that have round yellow peas. When Mendel planted these peas, and let them grow into adult plants and self-pollinate, he expected some of the recessive traits to show up again in the F2 generation. That did happen, and what is interesting is the ratio in which they appeared. From 15 plants, Mendel obtained 556 pea seeds in the following proportions:

- 315 round and yellow (56.6%)
- 101 wrinkled and yellow (18.2%)
- 108 round and green (19.4%)
- 32 wrinkled and green (5.8%).

Segregation and independent assortment

According to the **law of segregation**, Mendel's first law, each gamete will receive only one of the two copies that the parents can give. This law is one of Mendel's key contributions to understanding genetics. The reason we get a distribution ratio of

9:3:3:1 is because of the movement of chromosomes during meiosis. See Section D2.1.11 for an explanation of random orientation and crossing over. Crossing over during prophase I of meiosis and random orientation during metaphase I help shuffle the genetic material of the parent when making sperm cells or egg cells, so that when one gene for one trait is passed on to one of the daughter cells, it separates from the other genes. This is why each sperm cell and egg cell is so unlikely to be identical to another sperm or egg cell from the same person. This phenomenon is referred to as **Mendel's law of independent assortment**, or Mendel's second law. Each gene is passed on independently of other genes. However, we will see that there are exceptions to this.

D3.2.17 – Predicting genotypic and phenotypic ratios

D3.2.17 – Punnett grids for predicting genotypic and phenotypic ratios in dihybrid crosses involving pairs of unlinked autosomal genes

Students should understand how the 9:3:3:1 and 1:1:1:1 ratios are derived.

NOS: 9:3:3:1 and 1:1:1:1 ratios for dihybrid crosses are based on what has been called Mendel's second law. This law only applies if genes are on different chromosomes or are far apart enough on one chromosome for recombination rates to reach 50%. Students should recognize that there are exceptions to all biological "laws" under certain conditions.

If the percentages from Mendel's experiment shown above are converted to ratios, the numbers are close to the expected ratio for alleles that are passed on independently and are not found on the sex chromosomes. This ratio is 9:3:3:1 and is calculated using a 4 × 4 Punnett grid, as shown in Figure 13.

To set up the gametes for an F2 Punnett grid, the **FOIL method** is employed: this is a mnemonic for "first", "outside", "inside" and "last" of the parents' genotypes, in this case **RrYy**. The first gametes take the first letter of each trait in the genotype: **RY**. The second gametes are formed using the outside letters of each trait **Ry**, the third the inside **rY**, and finally **ry**.

The resulting ratio 9:3:3:1 indicates that, for every wrinkled green pea in the F2 generation, there should be three round green peas. This ratio is seen in Figure 13 on the previous page.

Mendel found 3.34 times more round green peas than wrinkled green peas in his experiment. There is often a difference between the theoretical values and actual values obtained in experiments. If thousands of seeds were examined instead of a few hundred, the number would probably be closer to three. Later we will use the chi-squared test to check whether Mendel's dihybrid results show any statistically significant difference between the expected values and the observed values.

Worked example

Using a Punnett grid, show the outcome of a cross between two parents, one heterozygous for both traits and one homozygous recessive for both traits: **AaBb** × **aabb**.

Solution

First we will show the long way, then a shortcut, to working this out. Always start by determining the alleles of the parents. The homozygous recessive parent will only be able to give **ab** no matter what. For the other parent, we use the FOIL method outlined above.

- First = **AB**
- Outside = **Ab**
- Inside = **aB**
- Last = **ab**

	AB	**Ab**	**aB**	**ab**
ab	**AaBb**	**Aabb**	**aaBb**	**aabb**
ab	**AaBb**	**Aabb**	**aaBb**	**aabb**
ab	**AaBb**	**Aabb**	**aaBb**	**aabb**
ab	**AaBb**	**Aabb**	**aaBb**	**aabb**

This cross does not show the 9:3:3:1 ratio we saw before. Instead, it is **1:1:1:1**. This type of cross can be used if a breeder needs to know the genotype of an individual with traits that could be the result of either an **AABB** or **AaBb** genotype. It is called a **test cross** because by testing how it breeds with a homozygous recessive individual and looking at the next generation, the genotype of the parent can be determined. If any **aabb** shows up (individuals with recessive traits in the phenotype for both genes), we know that the parent's genotype must be **AaBb**.

As you can see from the Punnett grid above, the last three rows of the grid are repeats of the first row and so are not necessary. If one parent is homozygous for both traits, only a
4 × 1 grid is necessary instead of a 4 × 4 grid, as shown below.

	AB	**Ab**	**aB**	**ab**
ab	**AaBb**	**Aabb**	**aaBb**	**aabb**

Nature of Science

The ratios we see here, 9:3:3:1 and 1:1:1:1, for dihybrid crosses show independent assortment, also known as Mendel's second law. But some conditions are necessary for the law to function: either the genes need to be on separate chromosomes or, if they are on the same one, they need to be far enough apart on the arms of the chromosome to be shuffled frequently enough by crossing over, so that the recombination rate is at least 50%. In addition, they need to be autosomal, meaning they cannot be present on the X or Y chromosome. It is common in science for laws to not be universal. Just because we find exceptions does not always mean we have to rewrite the law. We simply have to state how this law works under certain conditions and not under others.

Genetic disorders such as Huntington's disease and PKU are autosomal disorders. Both disorders occur in similar proportions in males and females. Colour blindness and haemophilia, in contrast, are sex-linked disorders; their gene loci are both found on the X chromosome and both diseases affect males significantly more than females.

D3.2.18 – Gene loci and polypeptide products

D3.2.18 – Loci of human genes and the polypeptide products
Application of skills: Students should explore genes and their polypeptide products in databases. They should find pairs of genes with loci on different chromosomes and also in close proximity on the same chromosome.

Find an open access gene database online, such as the National Center for Biotechnology Information (NCBI). In the search bar, type ABO to find information about the gene that codes for the ABO blood type in humans. Click on the name of the gene, which links you to the information page about the gene. You should be able to find out the gene's chromosome location, 9q34.2. The first number means it is on chromosome 9, the letter q means it is on the long arm of the chromosome (alternatively, p would mean short arm), and 34.2 is a measure of the distance from the centromere. The larger the number, the farther it is from the centromere. Click on RefSeq proteins and you should get the name of the protein (polypeptide product) that the ABO gene codes for: histo-blood group ABO system transferase. Now you know how to do it, complete the exercise below.

Using the database you found, fill in the blank cells in the table on the following page, showing some genes coding for proteins related to human blood.

SKILLS

	Gene	Description	Location	Polypeptide product
1.	ABO	blood type	9q34.2	histo-blood group ABO system transferase
2.		human haemoglobin subunit	16p13.3	alpha subunit of haemoglobin
3.	HBB	human haemoglobin subunit	11p15.4	
4.		human haemoglobin subunit		delta subunit of haemoglobin
5.		blood clotting (coagulation), a faulty allele causes haemophilia		coagulation factor IX

How likely is it that the genes ABO and HBA1 would be passed on together to the next generation more often than by pure chance? Explain your answer.

D3.2.19 – Gene linkage

D3.2.19 – Autosomal gene linkage

In crosses involving linkage, the symbols used to denote alleles should be shown alongside vertical lines representing homologous chromosomes. Students should understand the reason that alleles of linked genes can fail to assort independently.

Linkage groups

Any two genes that are found on the same chromosome are said to be linked to each other (see Figure 14). **Linked genes** are usually passed on to the next generation together.

A group of genes inherited together because they are found on the same chromosome are considered to be members of a **linkage group**. This applies to genes found on autosomes as well as those on the sex chromosomes. In Figure 14, the genes shown in green and yellow are linked. It is unlikely that crossing over will split them apart because there is so little space between them; they can therefore be considered a linkage group. Neither is linked to the gene shown in red, which is on a separate pair of chromosomes.

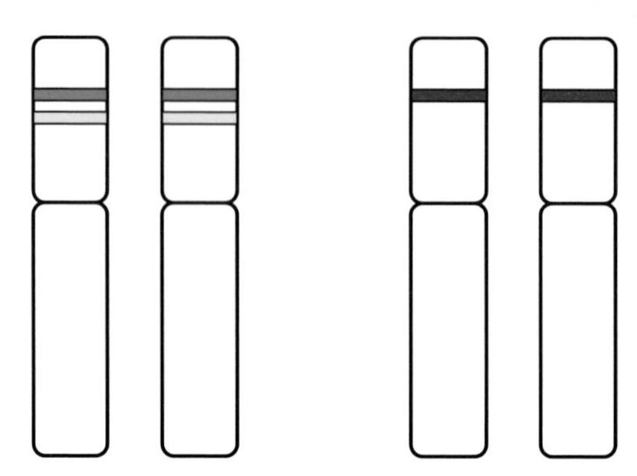

D3.2 Figure 14 In these two pairs of chromosomes, can you see which genes are linked, and which are not?

Linked genes

In the common fruit fly *Drosophila melanogaster*, the gene for body colour (grey or black) is in the same linkage group as the gene for wing length (long or short). This means that the loci for these two genes are located on the same chromosome and are therefore considered to be linked genes. The alleles are:

- **G** = allele for grey body
- **g** = allele for black body
- **L** = allele for long wings
- **l** = allele for short wings.

The genotypes of the pure-breeding (homozygous) parents are:

- **GGLL** = grey-bodied, long-winged
- **ggll** = black-bodied, short-winged.

Fruit flies are often bred in laboratories and used to study genetics. At the top of this tube it is possible to see adult flies, while lower down (near the label) we can see flies at earlier stages in their life cycle.

There is nothing in the genotype's notation **GGLL** that shows that **G** must be inherited with **L**. In order to show linkage, the following notation uses vertical lines to represent the chromosomes:

G ‖ G
L ‖ L

Each of the vertical bars symbolizes homologous chromosomes, or homologues, and shows that the locus of **G** is on the same chromosome as **L**. One **G** is on the maternal homologue, and the other **G** is on the paternal homologue. Likewise, **ggll** is shown as:

g ‖ g
l ‖ l

To read the genotype of the individual for these two linked traits, the pairs of alleles are read across the top first then along the bottom: the above example's genotype is **ggll**.

A cross between a homozygous dominant pure-breeding fruit fly (**GGLL**) and a homozygous recessive pure-breeding fruit fly (**ggll**) will result in flies that are all heterozygous for both of the traits (**GgLl**). The flies will all be grey with long wings, but they will all be carriers for the recessive alleles.

We will examine a cross between a grey long-winged fly (Figure 15), which is heterozygous for both traits, with a black short-winged fly, which is homozygous recessive for the traits.

D3.2 Figure 15 A test cross between a grey long-winged fly of unknown genotype with a black short-winged fly, a homozygous recessive.

Here are the linked genes in the heterozygote:

g ‖ G
l ‖ L

Here is the parent that is homozygous recessive for both traits:

g ‖ g
l ‖ l

One way of showing this cross is by drawing a Punnett grid (see Figure 16). As we saw in Section D3.2.17, a full 4 × 4 grid is not necessary because there are only four possible combinations.

	GL	**Gl**	**gL**	**gl**
gl	**GgLl** G ‖ g L ‖ l	**Ggll** G ‖ g l ‖ l **R**	**ggLl** g ‖ g L ‖ l **R**	**ggll** g ‖ g l ‖ l

D3.2 Figure 16 A Punnett grid showing a test cross. The two offspring labelled R are **recombinants** (see Section D3.2.20). The grid shows both ways of representing the genotypes.

D3.2.20 – Recombinants

D3.2.20 – Recombinants in crosses involving two linked or unlinked genes

Students should understand how to determine the outcomes of crosses between an individual heterozygous for both genes and an individual homozygous recessive for both genes. Identify recombinants in gametes, in genotypes of offspring and in phenotypes of offspring.

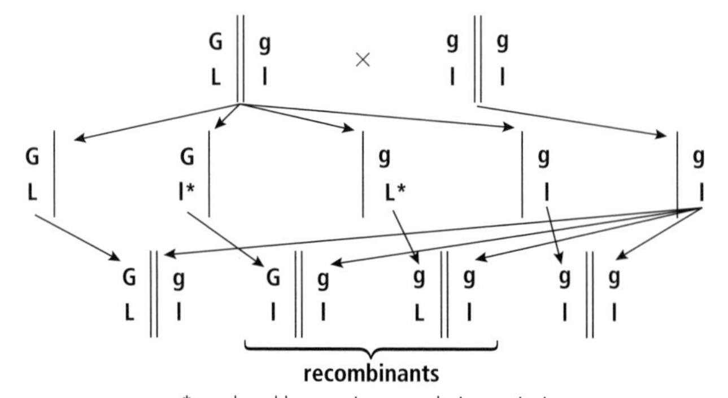

D3.2 Figure 17 A cross between **GgLl** × **ggll** showing recombinants.

Look at the two possibilities for offspring in the middle of Figure 17 (the second and third possibilities marked with an asterisk). By examining the alleles closely, it is possible to see that these offspring are different from either parent. Shuffling of the alleles has created a new combination that does not match either of the parent's genotypes. The term **recombinant** is used to describe both the new chromosome and the resulting organism.

The way these recombinants form is through the process of crossing over. Without crossing over, the allele **G** would always be inherited with **L** for the simple reason that they are linked. Thanks to crossing over, **G** sometimes gets inherited with **l**. In addition, **g** sometimes gets inherited with **L**, as seen in the recombinants in Figure 17.

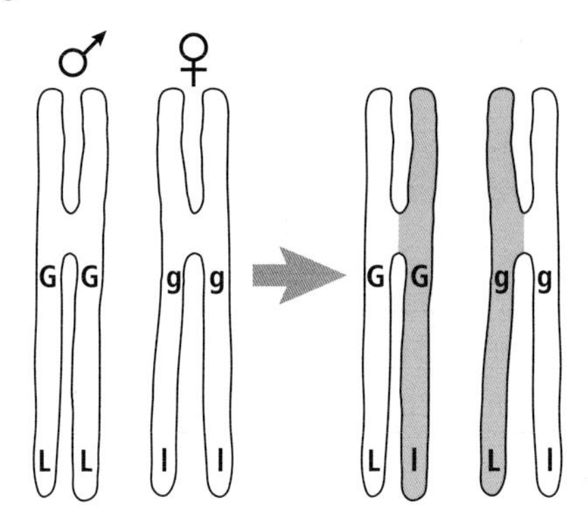

D3.2 Figure 18 The highlighted chromatids show new combinations of alleles that were not observed in the original parents' chromosomes.

When gametes are made from the resulting bivalent shown on the right in Figure 18, two will contain combinations found in the parents (either **GL** or **gl**) whereas two will contain recombinants (**Gl** and **gL**). When gametes form using the recombinants, they could potentially produce a grey-bodied offspring with short wings, or a black-bodied offspring with long wings. Again, neither of these combinations was present in the parents (see Figure 15). Thus, even in linked genes, nature has found a way to increase variety through crossing over.

Outcomes of genetic crosses should typically follow Mendelian ratios of **3:1** for an F2 monohybrid cross or **9:3:3:1** for an F2 dihybrid cross. There is nothing alarming about a slight variation from these expected values, but if there is a significant deviation it suggests that independent assortment is not happening and, instead, the genes of the traits being observed are probably linked.

D3.2.21 – Chi-squared tests

D3.2.21 – Use of a chi-squared test on data from dihybrid crosses

Students should understand the concept of statistical significance, the $p = 0.05$ level, null/alternative hypothesis and the idea of observed versus expected results.

NOS: Students should recognize that statistical testing often involves using a sample to represent a population. In this case the sample is the F2 generation. In many experiments the sample is the replicated or repeated measurements.

When carrying out experiments to study genetic characteristics being passed on from one generation to the next, it is expected, theoretically anyway, that certain ratios will be seen in the results. In practice, however, the sample size is usually far smaller than the whole study population and, hence, the ratios found in experiments are rarely equal to the expected ratios. This can be seen in Mendel's experiments with pea plants. When he counted the characteristics of the 7,324 seeds from his experiment, he got a ratio of **2.96:1**, instead of the theoretical expected value **3:1**. The question is: does that pose a problem? Could this mean that something is wrong, or that another factor besides chance is affecting the distribution of alleles? Could there be some other mechanism that would favour one allele over another and generate results that deviate from the expected ratio?

Note that there are two types of chi-squared test in the IB Biology syllabus. One is a **test of association** to find out if two things occur together or not, and the other one is a **test for goodness of fit** to see how closely the predicted outcome matches the observed outcome.

One statistical test that biologists use to see whether the difference between an observed result and an expected result is statistically significant is the chi-squared (χ^2) test (the Greek letter chi, χ, is pronounced like "sky" without the s). The chi-squared test helps us determine statistically whether or not there is a good fit between a theoretical model (in this case the expected ratios in a Punnett grid) and what really happens in nature. The Worked example below is an application of the χ^2 test on a dihybrid cross.

Worked example

William Bateson and Reginald Punnett (the man whose grids we use to calculate expected offspring) continued Mendel's work on pea plants, and experimented with other traits besides those that Mendel looked at. The table below shows the results of an experiment using purple flowers/red flowers as one trait and long pollen grains/round pollen grains as a second trait in a dihybrid cross (in this plant, purple and long are the dominant traits). The parents that produced this F2 generation were **PpLl**.

	Purple long	**Purple round**	**Red long**	**Red round**	**Total**
Observed (O)	284	21	21	55	381

Use the χ^2 test on Bateson and Punnett's dihybrid cross to determine whether there is a statistically significant difference between the expected ratios and the observed ratios.

Solution

To use this statistical test, it is important to record carefully all the observed results (O) and the expected results (E). The expected results are those can be calculated theoretically and, in genetics exercises, this means using a Punnett grid to determine ratios of offspring. 381 seeds in total were collected, and we expect 9 out of every 16 offspring (or 214 out of 381) to have purple flowers and long pollen grains, three out of every 16 to be purple and round (71), three out of every 16 to be red and long (also 71) and only one out of every 16 to be red and round (24).

	Purple long	**Purple round**	**Red long**	**Red round**	**Total**
Observed (O)	284	21	21	55	381
Expected proportions	9 out of 16	3 out of 16	3 out of 16	1 out of 16	
Expected (E)	214	71	71	24	381
Difference (O – E)	70	−50	−50	31	
Difference squared $(O - E)^2$	4,900	2,500	2,500	961	
$\frac{(O - E)^2}{E}$	22.9	35.2	35.2	40.0	**133.3**

The fourth and fifth data rows of the table above are the intermediate steps taken to find the difference between the observed and the expected values, and their squared values. The values are squared so that a negative sign does not pose a problem.

The bottom right cell of the table is what is needed now: it shows the sum of the last row's values, and this is the χ^2 value we are interested in. The contents of the table can be summarized in the generalized formula for calculating χ^2, which is:

$$\chi^2 = \sum\left(\frac{(O - E)^2}{E}\right)$$

χ = Greek letter chi

O = observed values (the results of the experiment)

E = expected values (calculated theoretically)

Σ = sum of all the calculations for each type of outcome

Now that we know the χ^2 value for this experiment, we need to know what it means. For this there are some concepts that need to be clarified. First of all, there is the concept of the **null hypothesis** (H_0). The H_0 in an experiment of this type is what we would expect: in Bateson and Punnett's example, we expect a **9:3:3:1** ratio. The χ^2 value will help us determine whether the null hypothesis can be rejected or not rejected. Not rejecting the null hypothesis is a way of saying "Yes, the alleles are passed on in a random fashion and there is a high probability that any deviation from the expected values can be attributed to chance". In other words, we have a high level of goodness of fit between the theoretical model and the values obtained, and we do not need to worry about any small deviation from the expected values. On the other hand, rejecting the null hypothesis in a goodness of fit test is equivalent to saying "The observed results do not follow the pattern we expected and it is highly likely that the differences we are seeing are due to more than just chance". This is referred to as the **alternative hypothesis**. This would suggest that a confounding variable (another factor) is altering the results.

Another two important concepts to understand are the **degrees of freedom** (d.f.) and **probability** (p) value. The degrees of freedom are determined by taking the number of classes into which the data fall and subtracting 1 from that number. In the case of round versus long pollen grains and red versus purple flowers, there are four classes into which the data fall, so there are 4 − 1 = 3 degrees of freedom. This number lets us know where to look in a **table of critical values** for χ^2 (see the table below). Notice in this table that, in addition to the degrees of freedom, there are different probability values, p. It is a convention in biology to look for probabilities of 5% (p values of 0.05). This means that there is only a 5% probability that the results are due to chance alone. Another way of looking at this is to think that there is a 95% chance that the results are being caused by something in particular and not just by chance.

Degrees of freedom (d.f.)	Probability values (p): 0.1	0.05	0.025	0.01	0.005
1	2.706	3.841	5.024	6.635	7.879
2	4.605	5.991	7.378	9.21	10.597
3	6.251	7.815	9.348	11.345	12.838
4	7.779	9.488	11.143	13.277	14.86
5	9.236	11.07	12.833	15.086	16.75
6	10.645	12.592	14.449	16.812	18.548
7	12.017	14.067	16.013	18.475	20.278
8	13.362	15.507	17.535	20.09	21.955
9	14.684	16.919	19.023	21.666	23.589
10	15.987	18.307	20.483	23.209	25.188

Look at the critical value table and find the critical value that is of interest to us: it is the one at the cross-section of a probability value of 0.05 and 3 degrees of freedom: 7.815. This means that any value we calculate for χ^2 that is greater than 7.815 tells us to reject the null hypothesis. Conversely, if we end up with a calculated value that is less than 7.815, it means we cannot reject the null hypothesis. In the present case, our calculated value is 133.3, which is much greater than 7.815, so we reject the null hypothesis. This information can be interpreted as:

- there is at least a 95% probability that the deviation from the expected values can be attributed to another factor (a confounding variable)
- there is at most a 5% probability that the results are due to chance alone.

The Worked example illustrates how you need to follow a number of steps in order to use the χ^2 test to help determine whether your results are statistically significant, or whether something unexpected is altering the results more than would be expected by chance. Here is a summary of the steps.

- Determine the expected values (E) (although we sometimes like to use percentages or proportions in science, the chi-squared test requires actual numbers, so do not use percentages or ratios).
- Note down the observed values (O) and decide what the null hypothesis will be (in genetics problems, the null hypothesis usually states that the ratios should be the ones calculated in Punnett grids, such as 3:1 in monohybrid crosses or 9:3:3:1 and 1:1:1:1 in dihybrid crosses).
- Calculate the value for χ^2 by determining the differences between the values (O − E), then squaring them $(O - E)^2$, and finally dividing that answer by E and adding up all the values you have for $(O - E)^2 / E$. This sum is the χ^2 value.
- Determine the degrees of freedom (d.f.) by subtracting 1 from the total number of classes into which the data fall.
- Look up a table of critical values for χ^2 and use the d.f. and desired p value (conventionally we use 0.05 for p) to determine which critical value ($\chi^2_{critical}$) to compare the calculated value of χ^2 ($\chi^2_{calculated}$) to.
- Compare $\chi^2_{critical}$ with $\chi^2_{calculated}$ and decide if the null hypothesis can be rejected or not using the rules shown below.

$$\chi^2_{calculated} < \chi^2_{critical} < \chi^2_{calculated}$$

$\chi^2_{calculated} < \chi^2_{critical}$	$\chi^2_{critical} < \chi^2_{calculated}$
do not reject null hypothesis	reject null hypothesis
any deviations from the expected values are probably the result of chance alone	deviations from the expected values are not the result of chance alone

The chi-squared test should only be used if the size of the observed sample is greater than 30. Ideally, 50 or more observed values should be used. Also, all of the expected values should be equal to or greater than 5.

Challenge yourself

1. Explain why the results of Bateman and Punnett's dihybrid cross do not show the expected 9:3:3:1 ratio. Identify the offspring that represent recombinants.

Nature of Science

It is impossible to observe all the possible outcomes of an experiment. We try to get as much data as we can, but it will only ever be a sample of the whole population. The sample is often the part of an experiment that requires repeated trials, to generate sufficient data. Bateson and Punnett used results from crossing pea plants, as Mendel did. In this case, the F2 generation, the offspring arising from breeding F1 hybrids together to see how the alleles are distributed, was used as the sample. Sufficient mixing should produce the expected ratios.

What are the principles of effective sampling in biological research?

HL end

Guiding Question revisited

What patterns of inheritance exist in plants and animals?

In this chapter you have learned:

- genetic crosses show how haploid gametes can join to form diploid zygotes with various combinations
- dominant alleles mask recessive ones, so usually two copies of a recessive allele are needed in order to show the trait
- with the ABO blood type, there are multiple alleles rather than just two and they show codominance, whereby both alleles are expressed in the phenotype of someone with AB type blood
- in flowering plant species, red flowers crossed with white ones often give either only red flowers or only white, but some species can produce pink flowers, showing incomplete dominance whereby neither allele is completely dominant
- sometimes a heritable trait is controlled by a single gene, such as the colour of the peas Mendel experimented on, whereas others are caused by multiple genes
- haemophilia in humans is caused by sex-linked genes, whereas other conditions, such as Huntington's disease and PKU, are autosomal
- pedigree charts can be used to track and predict patterns of inheritance in a family.

Guiding Question revisited

What is the molecular basis of inheritance patterns?

In this chapter you have learned:

- the X chromosome is much larger than the Y chromosome so it carries more genes
- the X and Y chromosomes from the sperm and egg will determine the sex of the future baby
- polygenic inheritance leads to continuous variation in a trait

- single-nucleotide polymorphisms explain why different alleles can have different outcomes, for example with a C in the place of an A, and the amino acid coded for can therefore be different and give rise to a protein that has different properties

HL

- crosses that consider two traits can lead to ratios of 9:3:3:1 and 1:1:1:1
- crossing over during meiosis allows for a shuffling of certain zones on the chromosomes involved which allows for greater variation in offspring.

HL end

Exercises

Q1. Explain why more human males are affected by colour blindness than females.

Q2. Using the $\mathbf{C^R}$ and $\mathbf{C^W}$ alleles for codominance in four-o'clock flower colour, show how two plants could have some white-flowered offspring, some pink-flowered offspring and some red-flowered offspring within one generation.

Q3. Draw a pedigree chart for the two generations described in question 2.

Q4. Look at the grid below showing the chances that a couple's children might have haemophilia.

(a) State the genotypes of the mother and father.

(b) State the possible genotypes of the children.

(c) State the phenotypes of the children.

(d) Who are carriers in this family?

(e) What are the chances that the parents' next child will have haemophilia?

	X^H	Y
X^H	X^HX^H	X^HY
X^h	X^HX^h	X^hY

Q5. Use the pedigree chart on the next page to answer the following:

(a) Using the numbers, identify the carriers who are male in the pedigree chart.

(b) Using **F** for the functioning version of the gene and **f** for the allele with the PKU mutation, deduce the genotypes of 4, 8 and 11.

(c) Determine the percentage chance that the offspring of 7 and 8 could have a child with PKU.

(d) What are the chances that 11's offspring will receive at least one **f** allele?

(e) Explain why in most societies it is taboo or illegal for close relatives to marry.

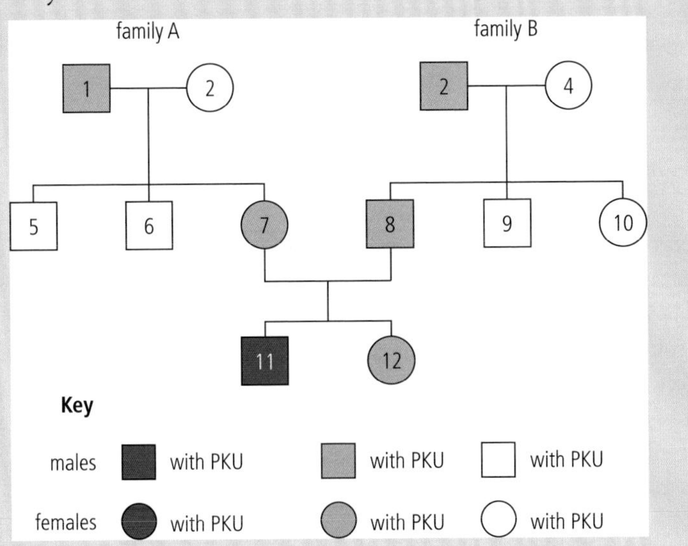

D3.3 Homeostasis

Guiding Questions

How are constant internal conditions maintained in humans?

What are the benefits to organisms of maintaining constant internal conditions?

How does your body stay warm, but not too warm? How do you maintain a healthy amount of water to support your body's needs? Your behaviour influences the answers to these questions, but your body does an amazing job even if your behaviour is not helping. The internal environment of an organism includes regulatory mechanisms to maintain variables within preset limits. We call this maintenance homeostasis. Homeostasis refers to an organism's ability to regulate various physiological processes to keep internal states at or near limits that are optimal. The regulation of these processes takes place mostly without our conscious awareness.

Temperature, blood glucose and water all have set points that are maintained. When a level begins to deviate from a set point, homeostasis works to correct it. Many body systems interact to keep physiological parameters within these set points, including the endocrine, nervous, excretory and circulatory systems. This chapter will look at how these homeostatic mechanisms work.

D3.3.1 – Maintaining the body's internal environment

D3.3.1 – Homeostasis as maintenance of the internal environment of an organism
Variables are kept within preset limits, despite fluctuations in external environment. Include body temperature, blood pH, blood glucose concentration and blood osmotic concentration as homeostatic variables in humans.

Humans and other species have evolved regulatory mechanisms to keep certain physiological factors within preset limits. This is called **homeostasis**. These preset limits are maintained even though our external environment can change considerably. The following are examples of parameters that are maintained by homeostasis:

- internal body temperature
- pH of the blood
- blood glucose concentration
- blood osmotic concentration.

Each of these physiological parameters is influenced by our environment and by what we ingest. For example, the air temperature that we experience can fluctuate from below freezing to very hot. Despite those possible extremes our body uses regulatory mechanisms to keep our internal temperature at or very close to 37°C.

D3.3.2 – Negative feedback mechanisms

D3.3.2 – Negative feedback loops in homeostasis
Students should understand the reason for use of negative rather than positive feedback control in homeostasis and also that negative feedback returns homeostatic variables to the set point from values above and below the set point.

The physiological processes that bring a value back towards a set point are called **negative feedback** mechanisms. Think of negative feedback control as working like a thermostat. The thermostat triggers one set of required actions when a value rises above its set point, and another set of actions when a value falls below its set point. In other words, negative feedback functions to keep a value within the narrow range that is considered normal. Positive feedback mechanisms are not appropriate for homeostasis because that type of feedback amplifies a response.

HL Chapter D3.1 discusses the positive feedback loop where oxytocin amplifies uterine contractions, with birth being the endpoint of the process.

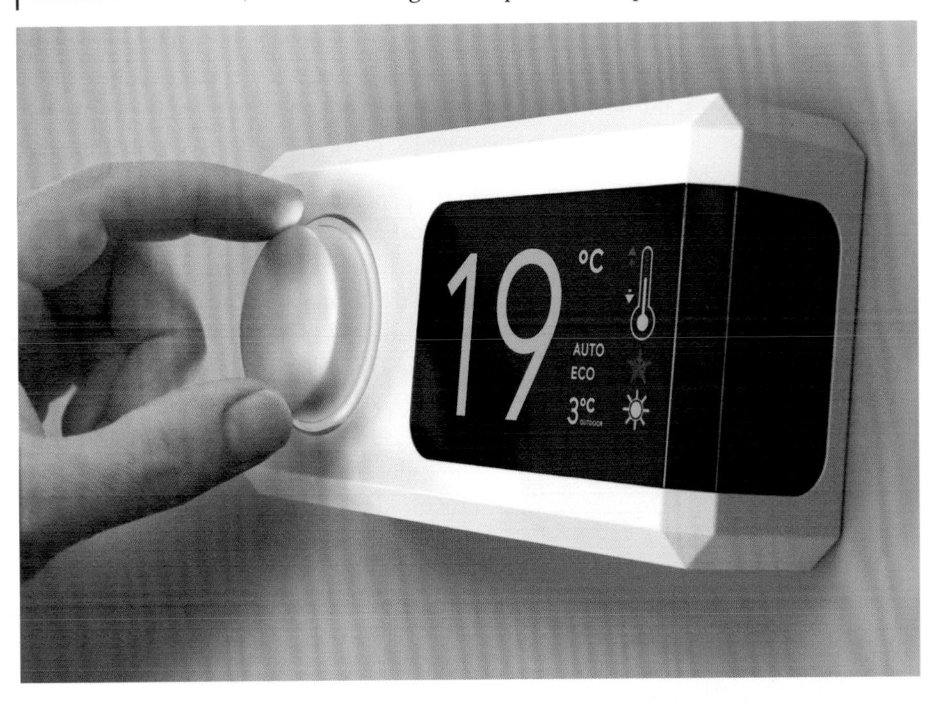

Negative feedback control in the body works like a thermostat. If the controlled variable (temperature) goes above a set point, it will trigger a response (in this case a cooling response). If the variable goes below a set point, the opposite response will be triggered (heating). All actions work towards the set point.

The nervous and endocrine systems often work cooperatively in order to ensure homeostasis. Many of the homeostatic mechanisms initiated by your nervous system are under the control of your autonomic nervous system. The **endocrine system** consists of numerous glands that produce a wide variety of hormones. Each hormone is transported by the bloodstream from the gland where it is produced to the specific cell types, called target cells, in the body that are influenced by that particular hormone.

Endocrine glands produce a hormone(s). The hormone always enters the blood for dispersal to all body cells. The body cells that have protein receptors for that hormone are called the target cells of that hormone. The response of the target cell will depend on the specifics of the hormone and the cell type.

D3.3.3 – The role of hormones

D3.3.3 – Regulation of blood glucose as an example of the role of hormones in homeostasis
Include control of secretion of insulin and glucagon by pancreatic endocrine cells, transport in blood and the effects on target cells.

The pancreas is the only gland in the body that is an endocrine gland and an exocrine gland. Its endocrine secretions (the hormones insulin and glucagon) are secreted into the blood. Its exocrine secretions (digestive fluids) go through ducts into the small intestine.

Insulin and **glucagon** are hormones that are both produced and secreted by the pancreas. In addition, they are both involved in the regulation of blood glucose levels. Cells rely on glucose for the process of cell respiration. Cells never stop cell respiration and thus are constantly lowering the concentration of glucose in the blood. Many people eat three or more times a day, including foods containing glucose, or carbohydrates that are chemically digested to glucose. This glucose is absorbed into the bloodstream in the capillary beds of the villi of the small intestine, and thus increases blood glucose levels. One factor that causes our blood glucose levels to fluctuate is simply that our blood does not receive constant levels of glucose because our ingestion of foods varies by time and content of carbohydrates. The increase and decrease in blood glucose levels caused by eating and digestion goes on 24 hours a day, every day of your life. However, even though blood glucose is expected to fluctuate slightly above and below the homeostatic normal level, it must be maintained reasonably close to the body's set point for blood glucose, and negative feedback mechanisms ensure this.

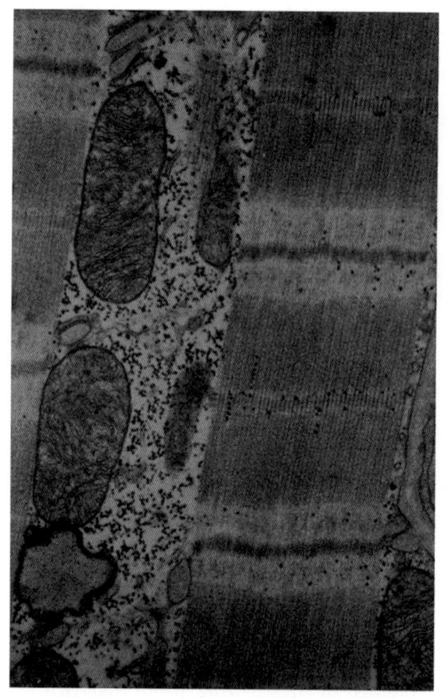

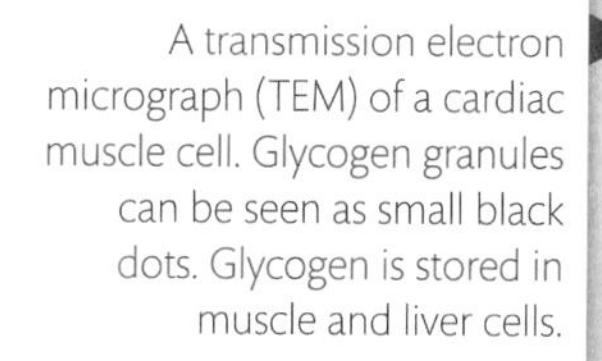

A transmission electron micrograph (TEM) of a cardiac muscle cell. Glycogen granules can be seen as small black dots. Glycogen is stored in muscle and liver cells.

In the pancreas there are cells known as β **(beta) cells** that produce the hormone insulin. Insulin is then secreted into the bloodstream and, because all body cells communicate chemically with blood, all cells are exposed to insulin. Insulin's effect on body cells is to open protein channels in their plasma membranes. These channels allow glucose to diffuse into the cell by the process known as **facilitated diffusion**.

There is another important effect attributed to insulin. Insulin stimulates muscle cells and liver cells to take in glucose (a monosaccharide) and convert it to glycogen (a polysaccharide). The glycogen is then stored as granules in the cytoplasm of these cells. The ultimate effect of insulin is to lower blood glucose levels.

Blood glucose levels begin to drop below the set point if someone does not eat for many hours or exercises vigorously. In either situation, the body needs to use the glycogen made and stored by the liver and muscle cells, as shown in Figure 1. Under these circumstances, α **(alpha) cells** of the pancreas begin to produce and secrete the hormone glucagon. The glucagon circulates in the bloodstream and stimulates hydrolysis of the granules of glycogen stored in hepatocytes (liver cells) and muscle cells. The hydrolysis of glycogen produces the monosaccharide glucose. This glucose then enters the bloodstream. The effect is to increase the glucose concentration in the blood and make that glucose available to body cells.

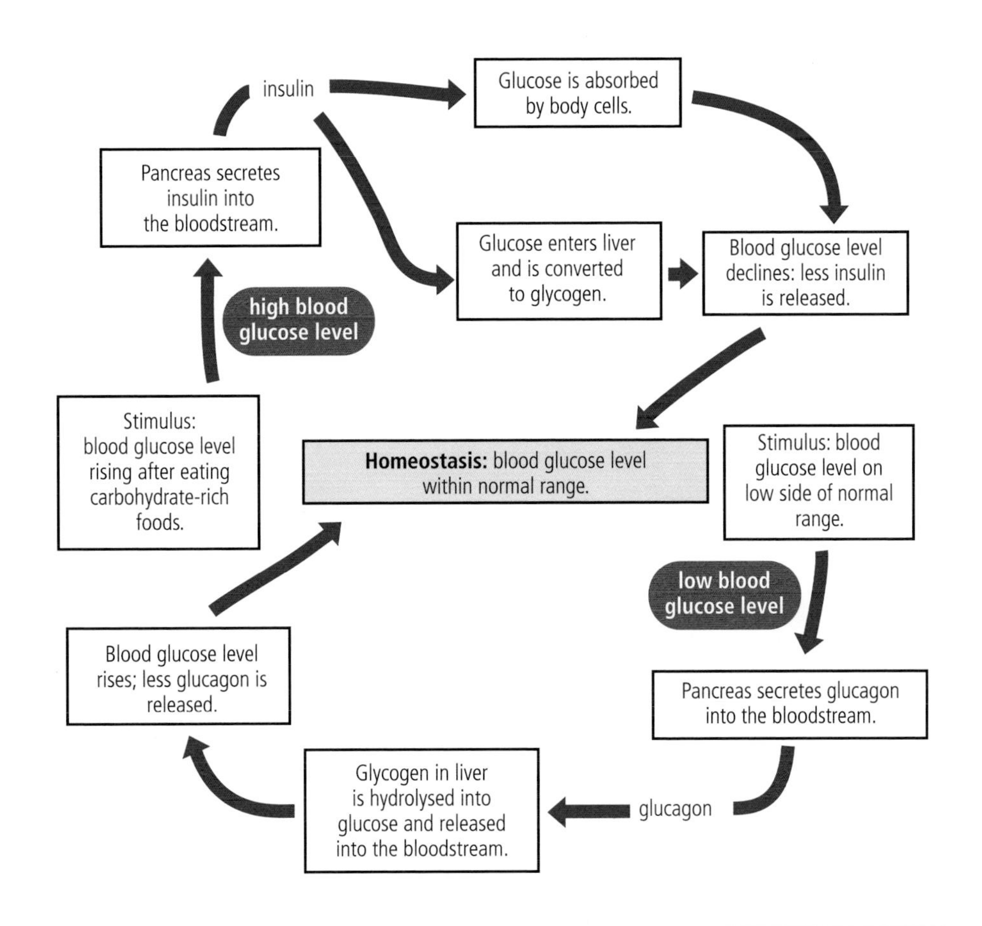

D3.3 Figure 1 A summary of the negative feedback control of blood glucose.

D3.3.4 – Type 1 and type 2 diabetes

D3.3.4 – Physiological changes that form the basis of type 1 and type 2 diabetes
Students should understand the physiological changes, together with risk factors and methods of prevention and treatment.

Diabetes is a disease characterized by **hyperglycaemia** (high blood glucose). The disease exists in two forms.

- Type 1 diabetes is an autoimmune disease where the immune system mistakenly destroys the β (beta) cells of the pancreas. These are the cells that produce insulin.
- Type 2 diabetes is a result of body cell receptors that do not respond properly to insulin and the cells do not take in sufficient glucose.

For what reasons do organisms need to distribute materials and energy?

Type 1 diabetes

Symptoms of type 1 diabetes usually present in children or young adults, but can present in people of any age. Once type 1 diabetes is diagnosed, treatment is based on a controlled diet and injections of insulin as needed. People with type 1 diabetes regularly test their blood to monitor their glucose levels and plan their diet control and timing for insulin injections. Risk factors for type 1 diabetes include family history and age.

Type 2 diabetes

The incidence of type 2 diabetes is almost ten times that of type 1 diabetes.

Low- and middle-income countries account for more than 80% of all deaths related to diabetes.

Type 2 diabetes is the result of body cells no longer responding to insulin as they once did. This is known as **insulin resistance.** Initially, the pancreas continues to produce a normal amount of insulin, but this level may decrease after a period of time. Type 2 diabetes is the most common form of diabetes. Risk factors for developing type 2 diabetes include family history, obesity and lack of exercise. You cannot change your family history but developing good eating habits and regular exercise can prevent or delay the onset of type 2 diabetes.

Uncontrolled diabetes of either type can lead to many serious effects, including:

- damage to the retina of the eye, leading to blindness
- kidney failure
- nerve damage
- increased risk of cardiovascular disease
- poor wound healing.

D3.3.5 and D3.3.6 – Body temperature control

D3.3.5 – Thermoregulation as an example of negative feedback control
Include the roles of peripheral thermoreceptors, the hypothalamus and pituitary gland, thyroxin and also examples of muscle and adipose tissue that act as effectors of temperature change.

D3.3.6 – Thermoregulation mechanisms in humans
Students should appreciate that birds and mammals regulate their body temperature by physiological and behavioural means. Students are only required to understand the details of thermoregulation for humans.
Include vasodilation, vasoconstriction, shivering, sweating, uncoupled respiration in brown adipose tissue and hair erection.

Many animals are **ectothermic**, meaning that their internal temperature equalizes with their environment. The air and water temperatures of their habitat greatly affect the geographical boundaries for many of these ectothermic animal populations, and greatly affect their behaviour. On cold days you may see ectothermic animals sunning themselves to gain body heat. Some fish and marine invertebrates have migratory patterns that help them remain in suitable water temperatures. The advantage of being ectothermic is that these animals do not have to metabolize foods to generate body heat, and as a result do not have to eat as much food.

Caimans (*Caiman yacare*) are ectothermic and often lie in the sun to increase internal body temperature.

Birds and mammals are **endothermic** and maintain a steady internal temperature that is almost always warmer than the environmental temperature. This requires extra nutrition specifically to generate internal body heat.

There are two ways that humans and other endotherms can experience an increase of body temperature above their set point. One is by being in an environment that is warmer than their set point. The other is the result of muscular activity that generates internal body heat. Muscular activity during exercise or work activities often raises the internal body temperature.

The only way to decrease the body temperature below the set point is to be in an environment that is cooler than the body's set point temperature. Once the internal temperature of an endotherm begins to increase or decrease away from its set point, temperature-regulating negative feedback mechanisms are activated.

The two primary temperature sensing tissues in many animals are thermoreceptors located in the skin and a portion of the brain called the hypothalamus. The hypothalamus uses thermoreceptors to sense the temperature of the blood as it passes through that area of the brain. In addition, skin thermoreceptors send impulses to the hypothalamus. The hypothalamus responds by initiating cooling mechanisms or heating mechanisms, as shown in Figure 2.

The location of the hypothalamus in the brainstem. The hypothalamus has control over many autonomic nervous system functions, including many thermoregulatory mechanisms.

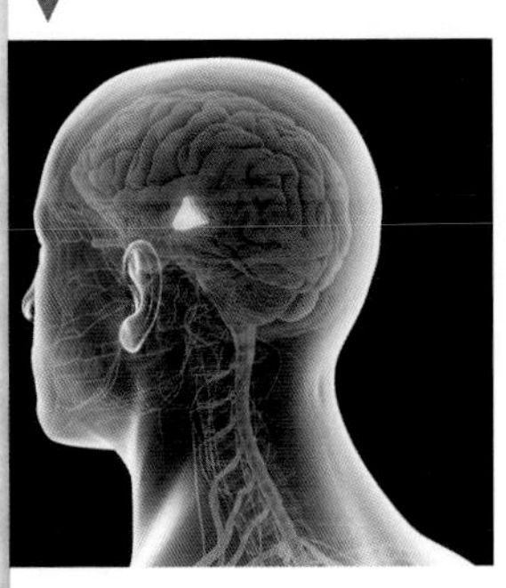

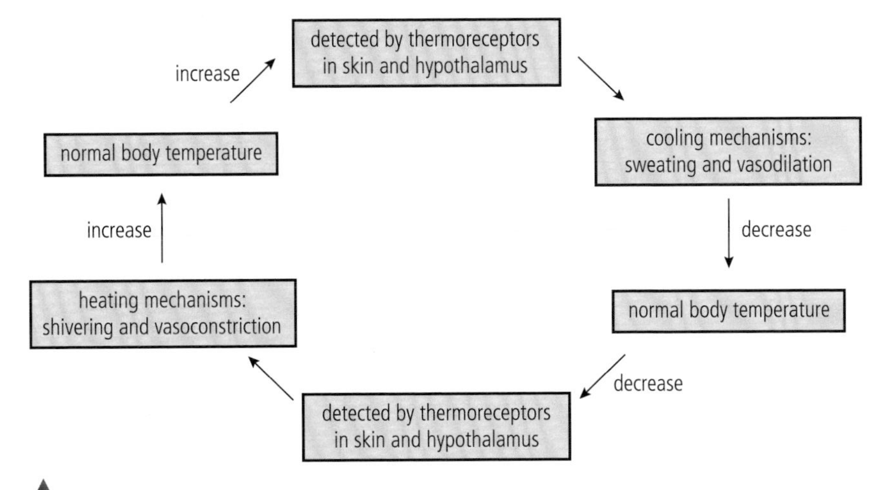

D3.3 Figure 2 A summary of the negative feedback mechanisms involved in thermoregulation. Increased internal body temperature leads to sweating and vasodilation of small blood vessels near the skin. A decrease in internal body temperature leads to shivering and vasoconstriction of small blood vessels near the skin.

The regulation of body temperature is called **thermoregulation**. We will consider two possible situations that require thermoregulation in humans.

If human body and blood temperatures rise above 37°C

When the internal temperature of the body begins to increase above 37°C the hypothalamus sends impulses to the arterioles near the skin that result in vasodilation. This results in more blood travelling through the capillaries in the skin and more heat being released into the surrounding air. The hypothalamus also initiates perspiration. The act of producing perspiration does not cool the body, but evaporation of the sweat from the skin provides evaporative cooling. The heat of the body is transferred to the water in sweat, initiating the phase change of evaporation.

The set point temperature for humans is 37°C. Any internal temperature that is higher or lower than 37°C will initiate negative feedback mechanisms for temperature regulation.

Skin structure adaptations to environmental temperatures that are above (shown on the left of the diagram) or below (shown on the right) the body's set point of 37°C.

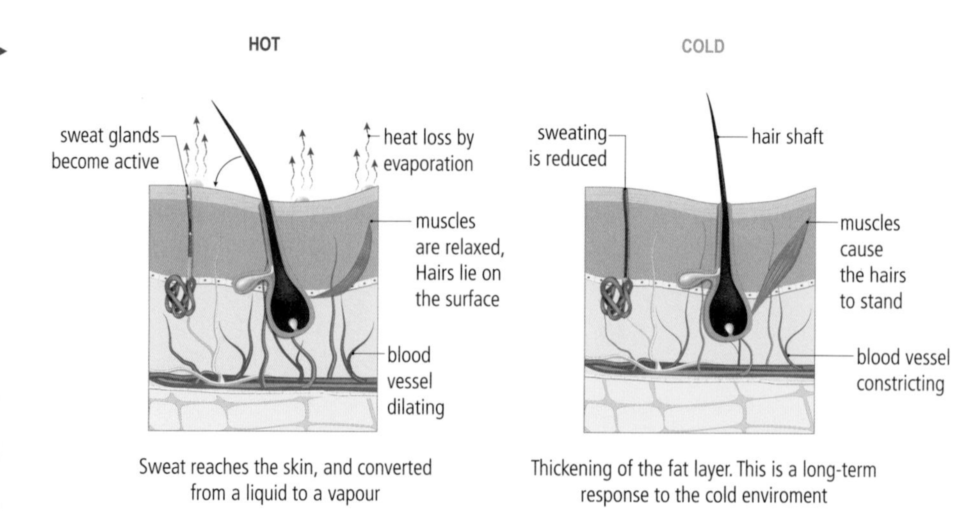

If human body and blood temperatures fall below 37°C

In colder environments, the hypothalamus sends impulses that result in the vasoconstriction of arterioles near the skin. More blood can then be shunted to the internal organs, with less to the skin. This is a protective measure for survival. The capillaries of the skin will receive less blood, and less heat will be released to the environment. In addition, vital internal organs will receive more warm blood. The release of epinephrine (adrenaline) in cold environments results in goosebumps or raised hairs on the skin. In furred mammals, this response creates an insulating layer of air between the fur and the skin to protect the body from the cold air. Humans no longer have enough skin hair for this to be very useful but we still have the **autonomic** response.

When peripheral thermoreceptors in the skin sense cold, the pituitary gland is also stimulated to release hormones that activate the thyroid gland to release a hormone called **thyroxine**, which increases the metabolic rate of all body cells in order to generate heat. Shivering is another autonomic response to cold that is initiated by the hypothalamus. The act of shivering uses muscles and therefore generates body heat.

Some animals, especially marine mammals living in cold waters, have evolved to use blubber as an insulating layer between the water (or air) and their internal organs. The blubber is a form of adipose tissue and helps to retain the warmth generated by the internal metabolic activities of the animal.

An elephant seal (*Mirounga angustirostris*) is an example of an animal that uses blubber as an insulator against the cold air and water typical of its environment in Arctic waters.

Newborns have brown adipose tissue to generate heat when needed

When cold, adults use the rapid muscle contractions of shivering to create heat. Newborns are unable to shiver but instead have a higher proportion of fat that is called **brown adipose** tissue. Brown adipose is visibly darker than other adipose tissue because it contains many more mitochondria compared to other fat cells. When needed, the brown adipose cells use their mitochondria to begin cell respiration that is uncoupled from adenosine triphosphate (ATP) production. Glucose is oxidized for the sole purpose of generating body heat. Even though brown adipose tissue is mainly found in infants, adults retain a small amount of brown adipose tissue.

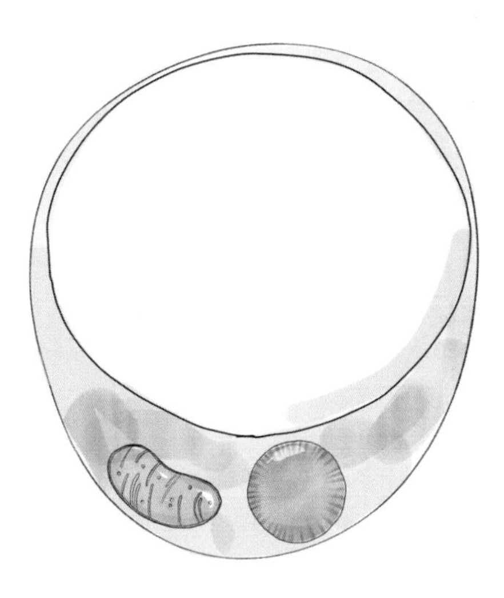

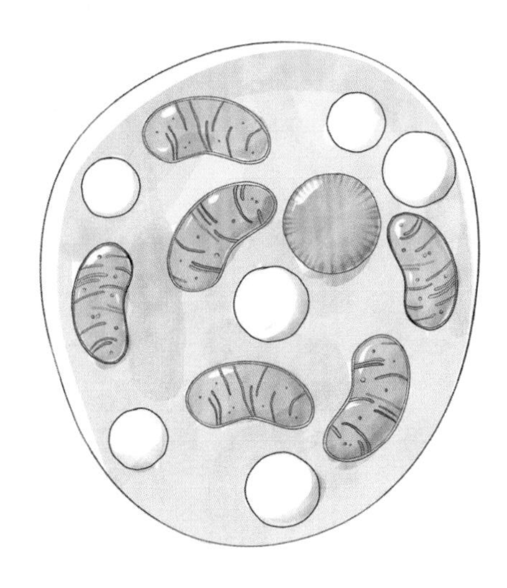

White adipose cells are designed for storing triglycerides (fat). Brown adipose cells have multiple mitochondria and are capable of cell respiration that generates heat uncoupled from ATP production. Newborns have a much higher proportion of brown adipose tissue compared to adults.

When you are cold, not all of your responses may be at the subconscious level: your cerebrum will also help by initiating conscious behaviours that will help you to stay warm. You may decide to move around more to generate muscle heat, dress warmer, or stay inside. Conscious behaviours also help you to stay cool. You may seek shade or air conditioning, wear light clothing and stop exercising. These are all behaviours that help prevent overheating.

What biological systems are sensitive to temperature changes?

HL

D3.3.7 – The role of the kidneys

D3.3.7 – **Role of the kidney in osmoregulation and excretion**
Students should understand the distinction between excretion and osmoregulation. Osmoregulation is regulation of osmotic concentration. The units for osmotic concentration are osmoles per litre (osmol L^{-1}).

The contents of your blood are always changing. Every capillary bed in the body permits some molecules to leave the bloodstream and other molecules to enter. The organs in the human body that regulate the contents of the blood are the kidneys (see Figure 3 on the next page). The kidneys perform two primary functions. One is regulation of the water content of blood using a process called **osmoregulation**. The water content of urine is altered based on water intake, perspiration levels and other factors. The second function of the kidneys is **excretion**. The kidneys filter many solutes out of the blood and then use cellular transport mechanisms to regulate what stays in the urine and what should return to the blood.

The osmolarity of body fluids including urine and blood is defined as the number of osmoles of solute per litre of fluid (osmol L^{-1}). One osmole is one mole of solute particles. Osmolarity depends on the number of particles in a solution but is independent of the identity of those particles.

Anatomical and physiological terms associated with the kidney often contain either "renal" or "nephron" within the terms. Examples include the renal artery and nephron.

The composition of blood plasma is different in the renal artery compared to the renal vein, as a result of the blood filtering action of the kidney. The primary differences between the blood in these two vessels are in the levels of water, mineral ions and urea.

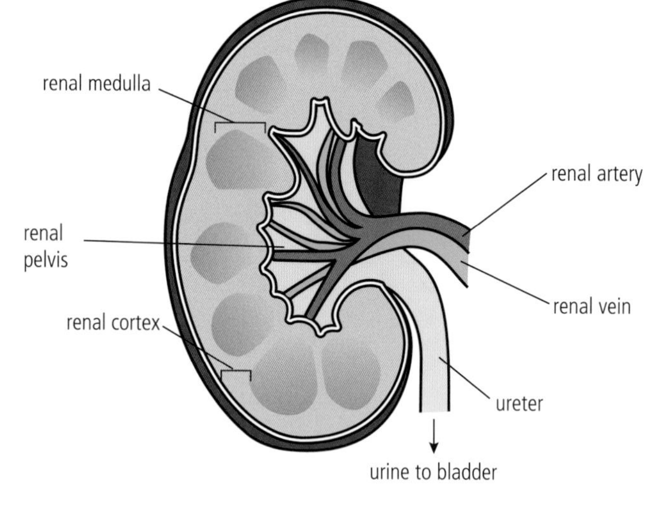

D3.3 Figure 3 A sectioned view of a human kidney. Layers of different tissues can be seen on the left (the cortex, medulla and pelvis). On the right side are the major blood vessels taking blood into the kidney (the renal artery) and taking blood away from the kidney (the renal vein). Also shown is the ureter, a tube that takes urine from the kidney to the urinary bladder.

Each kidney is made up of about 1.25 million filtering units known as **nephrons** (Figure 4). Each nephron consists of:

- a capillary bed, called a **glomerulus**, that filters various substances from the blood
- a capsule surrounding the glomerulus, called the **Bowman's capsule**
- a small tube (tubule) that extends from the Bowman's capsule, consisting of the **proximal convoluted tubule**, **loop of Henle** and **distal convoluted tubule**
- a second capillary bed, called the **peritubular capillary bed**, that surrounds the three-part tubule mentioned above
- **collecting ducts**, which are shared by several nephrons.

D3.3 Figure 4 A single nephron, a filtering unit of the kidney. Refer to this diagram as we follow the production of urine and osmoregulation in the next three sections.

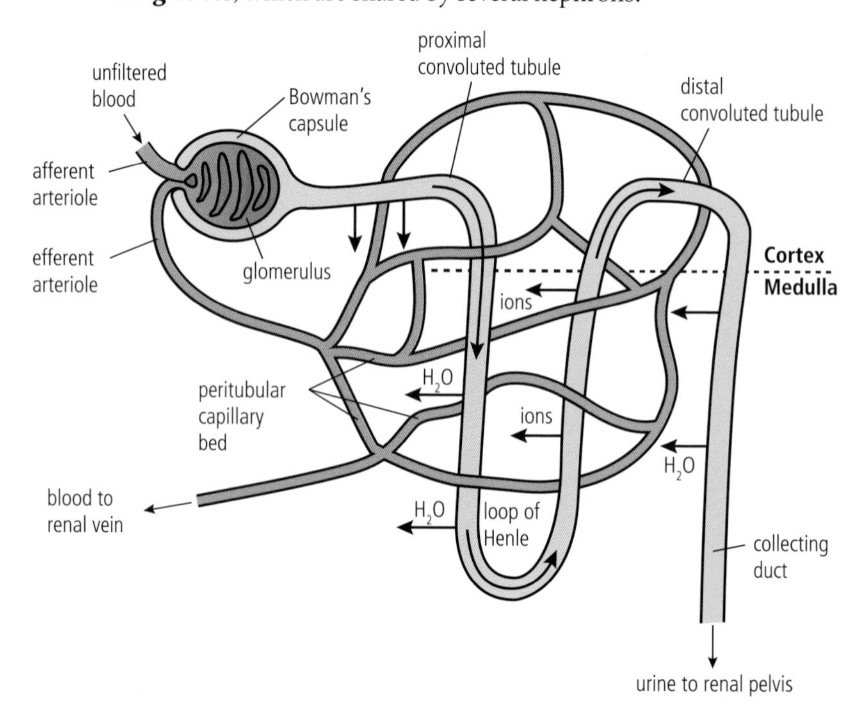

Find the dotted line in Figure 4 that shows the location of the cortex and medulla in relation to an individual nephron. Then go back to Figure 3 and find the (renal) cortex and (renal) medulla in relation to the kidney section.

D3.3.8 – The glomerulus, Bowman's capsule and proximal convoluted tubule

D3.3.8 – Role of the glomerulus, Bowman's capsule and proximal convoluted tubule in excretion
Students should appreciate how ultrafiltration remove solutes from blood plasma and how useful substances are then reabsorbed, to leave toxins and other unwanted solutes in the filtrate, which are excreted in urine.

As shown in Figure 5, each nephron contains a very small branch of the renal artery known as an **afferent arteriole**. This very small artery brings unfiltered blood to the nephron. Inside the Bowman's capsule, the afferent arteriole branches into a specialized capillary bed called the glomerulus.

The glomerulus is similar to most other capillary beds except that the walls of the capillaries have **fenestrations** (small slits) that open under high blood pressure. The increase in blood pressure is caused by the **efferent arteriole**, which drains blood from the glomerulus and has a smaller diameter than the afferent arteriole. Connecting a larger diameter blood vessel to a smaller diameter blood vessel creates a higher pressure where they join, in this case at the glomerulus.

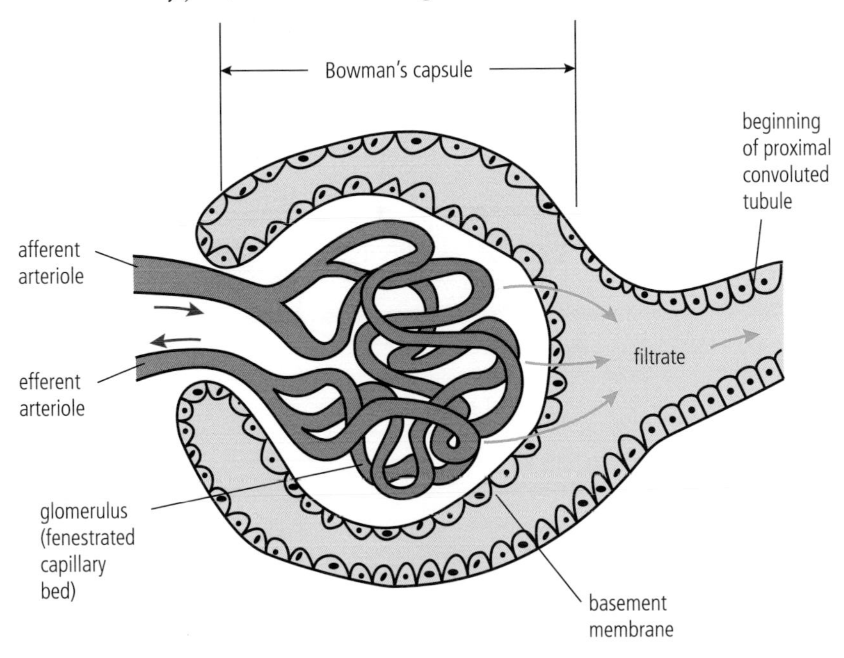

D3.3 Figure 5 The glomerulus, Bowman's capsule and the first portion of the proximal convoluted tubule of a nephron. Notice the larger diameter of the afferent arteriole compared to the efferent arteriole. The filtrate is the initial urine that will be modified in later portions of the nephron. Compare this diagram to Figure 4, which provides a wider perspective.

Ultrafiltration is the term used to describe the process by which various substances are filtered through the glomerulus (and its fenestrations) under high blood pressure in the capillary bed. The fluid that is ultrafiltered from the glomerulus passes through the **basement membrane**, which helps prevent large molecules such as proteins becoming part of the filtrate. This **filtrate** will later become urine once the solute and water content have been adjusted as it travels through subsequent portions of the nephron.

The filtrate from the glomerulus then enters the proximal convoluted tubule. The blood that did not get filtered, including all the cells and proteins as well as many other molecules, exits the Bowman's capsule in the efferent arteriole.

The blood vessel pattern within each nephron is unusual. Blood moves from afferent arteriole to glomerulus (capillary bed) to efferent arteriole to peritubular capillary bed to a venule. In other words, there are two separate capillary beds in the same circuit. This is known as a **portal system of circulation**.

Always compare the individual diagrams of the portions of nephrons to Figure 4, which shows an entire nephron. This will help you gain a perspective of the overall process taking place in nephrons.

The filtrate that leaves the Bowman's capsule contains many substances that the body cannot afford to lose as part of the urine, as well as some molecules that must be excreted in the urine. The mechanism of filtering molecules under pressure results in too many substances in the original filtrate. While urea, some mineral ions and excess water must stay in the filtrate to be excreted in the urine, the body needs to keep a great deal of the water, many of the mineral ions, and all of the glucose that is in the filtrate. Much of the necessary **reabsorption** process occurs along the proximal convoluted tubule. Substances leave the filtrate and are taken back into the bloodstream via the **peritubular** capillary bed. This capillary bed is so named because it surrounds the tubule.

The human kidneys filter about 180 L of fluid a day out of the bloodstream. The average volume of urine excreted each day is 1.5 L. This means that 178.5 L of filtrate are reabsorbed into the bloodstream every day.

The wall of the proximal convoluted tubule is a single cell thick. As you can see in Figure 6, the tubule is composed of a single ring of cells. The interior of the resulting tube is called the lumen, and the filtrate flows within this lumen. The inner portion of each of the tubule cells has **microvilli** that increase the surface area for reabsorption. Microvilli are microscopic projections found only in the plasma membrane portion facing the lumen of the tubule. The surface area for reabsorption is greatly increased by these tiny projections.

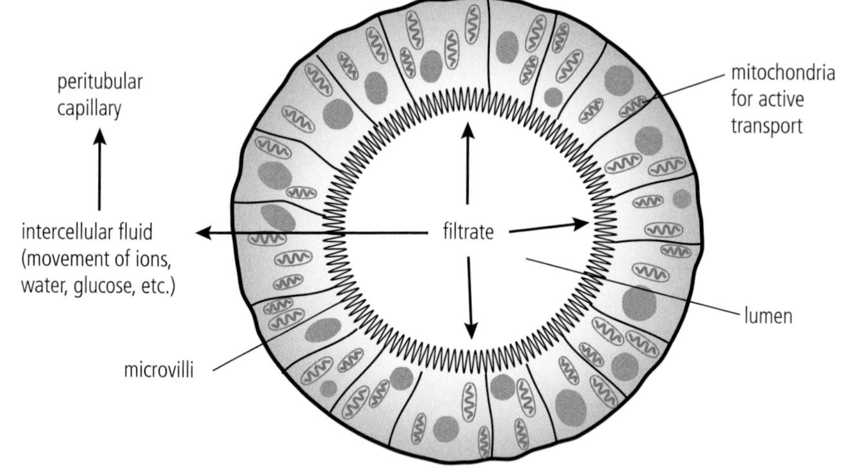

D3.3 Figure 6 A sectioned view of the small tube known as the proximal convoluted tubule. The filtrate from the Bowman's capsule flows through this tube. Cellular transport mechanisms are used to remove excess ions, glucose and some water from the filtrate. Those substances are returned to the bloodstream by the peritubular capillary bed.

The main waste product of protein metabolism in the body is urea; 25–30 g of urea are excreted in urine each day. Eliminating urea each day allows the body to keep the concentration of this molecule at a level that is not toxic.

Three waste products are produced by animals as a result of protein metabolism: ammonia, urea and uric acid. The production of these molecules is a way of disposing of the amine group found in all amino acids. There are advantages and disadvantages to producing each of these three as a waste product. Ammonia requires the least energy to produce but is the most toxic and therefore must be expelled continuously. Uric acid is a complex and energy-requiring molecule to produce but it is not toxic to living tissues because it is insoluble in body fluids. Urea is the molecule produced by organisms that have to make a compromise between toxicity and the need for temporary storage of a waste product (such as urine in mammals).

D3.3.9 – The loop of Henle

D3.3.9 – Role of the loop of Henle
Limit to active transport of sodium ions in the ascending limb to maintain high osmotic concentrations in the medulla, facilitating water reabsorption in the collecting ducts.

Water is the solvent of life. It is the solvent present in almost all body fluids, including cytoplasm, blood plasma, lymph and intercellular fluid. Some water needs to be eliminated in the urine each day, but the total volume of water eliminated depends on many physiological factors. These physiological factors include:

- the total volume of water ingested recently, as liquid and in food
- perspiration rate, which is influenced by both exercise level and environmental temperature
- ventilation (breathing) rate, which is largely dependent on the activity/exercise level (a significant amount of water is exhaled when we breathe out)
- the volume of urine eliminated.

Much of the water in the original filtrate remains after the filtrate has left the proximal convoluted tubule. This water, and the remaining solutes, enters the descending portion of the loop of Henle (see Figure 7). This segment of the loop of Henle is permeable to water but relatively impermeable to mineral ions. The filtrate then enters the ascending portion of the loop of Henle, where the tubule is relatively impermeable to water but permeable to mineral ions. As the filtrate moves up the ascending portion of the loop, sodium ions are actively transported out and enter the fluid surrounding the loop of Henle.

D3.3 Figure 7 The action of the loop of Henle, creating a high osmotic concentration in the medulla of the kidney. The dotted line indicates the portions of the nephrons that are located in the renal cortex and the portions that are located in the renal medulla. The ions are primarily sodium ions actively transported out of the ascending limb of the loop. The high osmotic concentration of the fluid in the medulla encourages water reabsorption from the collecting duct by osmosis.

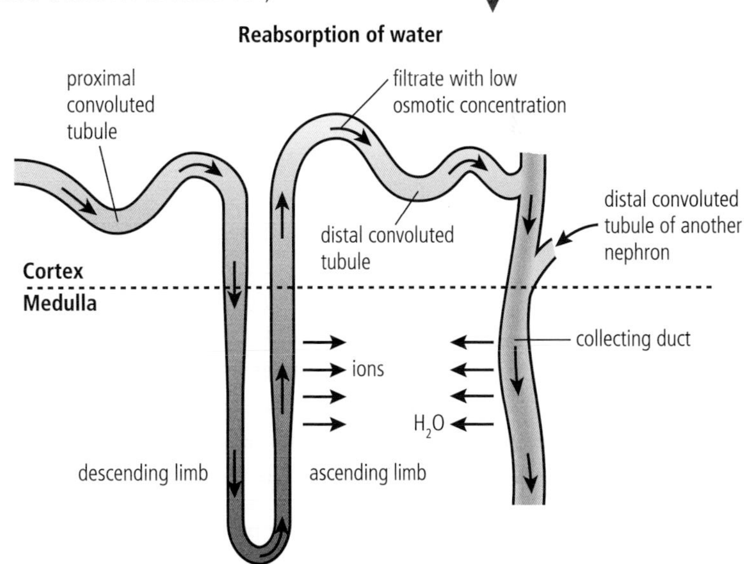

D3.3.10 – Osmoregulation

D3.3.10 – Osmoregulation by water reabsorption in the collecting ducts
Include the roles of osmoreceptors in the hypothalamus, changes to the rate of antidiuretic hormone secretion by the pituitary gland and the resultant switches in location of aquaporins between cell membranes and intracellular vesicles in cells of the collecting ducts.

The filtrate that enters a collecting duct can be thought of as urine, but it is in a dilute form. The water content is quite high, especially in comparison to the fluid of the medulla. If this volume of water was to leave an individual's body consistently as urine, that individual would need an extremely high water intake to make up for the water loss (not to mention many, many trips to a toilet). Thus, under most circumstances, much of this water is reabsorbed through the wall of the collecting duct.

Any regulatory mechanism that affects the water content in an animal's body is part of osmoregulation.

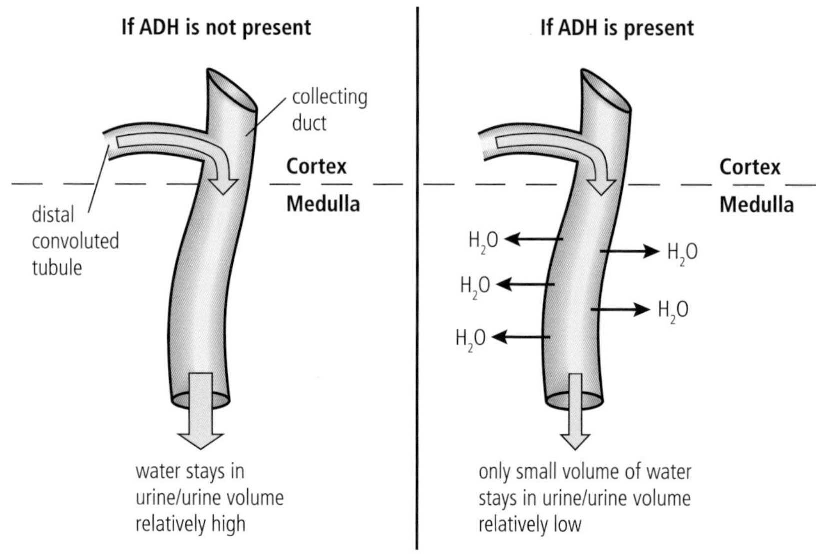

D3.3 Figure 8 A small section of a collecting duct showing what happens when the hormone ADH is or is not present. Because of the high ion concentration outside the collecting duct, the fluid found there has a high osmotic concentration. The fluid inside the collecting duct has a low osmotic concentration. Osmosis will occur only when ADH permits water flow through the collecting duct wall.

The collecting duct is differentially permeable to water. Its permeability depends on the amount of **antidiuretic hormone (ADH)** present. ADH is secreted from the pituitary gland and, like all hormones, circulates in the bloodstream. The target tissue of ADH is the kidney collecting ducts. If ADH is present, the collecting duct becomes permeable to water and water moves by osmosis out of the collecting duct and into the medullary interstitial fluid. From there, water enters the peritubular capillary bed and is returned to the bloodstream to be made available to all body tissues. If ADH is not present, the collecting duct becomes impermeable to water. Water then stays in the collecting duct, along with the various waste solutes, and the urine is more dilute.

The mechanism that allows water to exit the collecting duct begins with **osmoreceptors** in the hypothalamus that monitor the water content of the blood. The hypothalamus then controls the rate of release of ADH from the pituitary gland. As you can see from Figure 8, if little to no ADH is secreted into the bloodstream the collecting ducts remain relatively impermeable to water. Most water stays in the urine and is taken to the urinary bladder for release. If the osmoreceptors of the hypothalamus sense that water needs to be retained in the body, more ADH will be secreted from the pituitary, travel in the bloodstream and affect its target tissue, the cells of the collecting ducts. Water will then pass through the collecting duct and enter the peritubular capillary bed to become part of the bloodstream once again.

The colour of urine, from relatively colourless to dark yellow, provides a clue to how hydrated you are. When your urine is nearly colourless, this means your body has abundant water and is eliminating excess water as dilute urine. After exercising and perspiring a great deal of water, your urine does not contain nearly as much water and the more concentrated solutes will give your urine a yellow colour.

The cells that make up the collecting duct have proteins called **aquaporins** stored in vesicles inside the cells. Aquaporins can be inserted into the plasma membrane to allow polar water molecules to move through the membrane. When ADH is present, the aquaporin vesicles merge with the plasma membrane, leaving the aquaporins inserted into the membrane. Water can then move through the plasma membrane by osmosis and out through the other side of the cell, where aquaporins are permanently fixed. When the rate of ADH production is decreased, the plasma membrane will **invaginate** and store the aquaporins in intracellular vesicles once again. Water will then remain in the collecting duct and be removed in the urine.

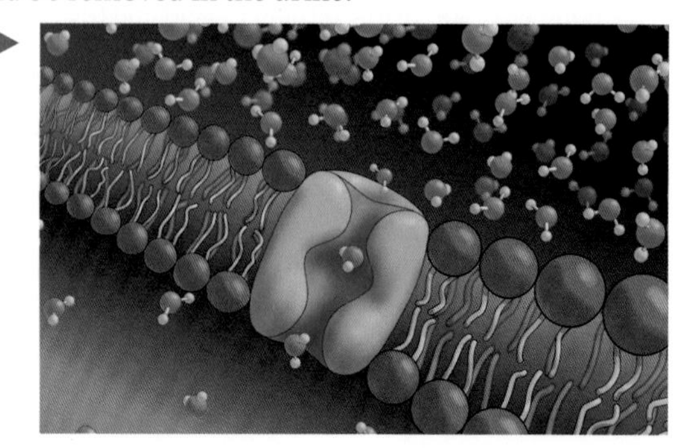

An aquaporin controlling osmosis of water through a plasma membrane. Even though water is a relatively small molecule, its polarity does not allow it to diffuse through the plasma membrane without passing through an aquaporin. Collecting duct cells incorporate aquaporins in the membrane adjacent to the dilute urine in the collecting duct when the hypothalamus senses a need based on the water content of the blood.

D3.3.11 – Variable blood supply dependent on activity

D3.3.11 – Changes in blood supply to organs in response to changes in activity
As examples, use the pattern of blood supply to the skeletal muscles, gut, brain and kidneys during sleep, vigorous physical activity and wakeful rest.

Blood supply to some organs in the body can be adjusted based on body activity. This is done by vasodilation and vasoconstriction of the smallest arteries known as arterioles (see Figure 9). Arterioles are the blood vessels that feed blood directly into capillary beds. If a particular body organ temporarily requires a greater blood supply, the arterioles feeding blood into the capillary beds of that tissue will vasodilate. This requires the arterioles of another area of the body to vasoconstrict so that the overall blood pressure can be maintained.

The two main tissue types that enable the largest blood volume changes are skeletal muscles and the digestive system or gut. The brain and kidneys can tolerate only small decreases in their blood supply.

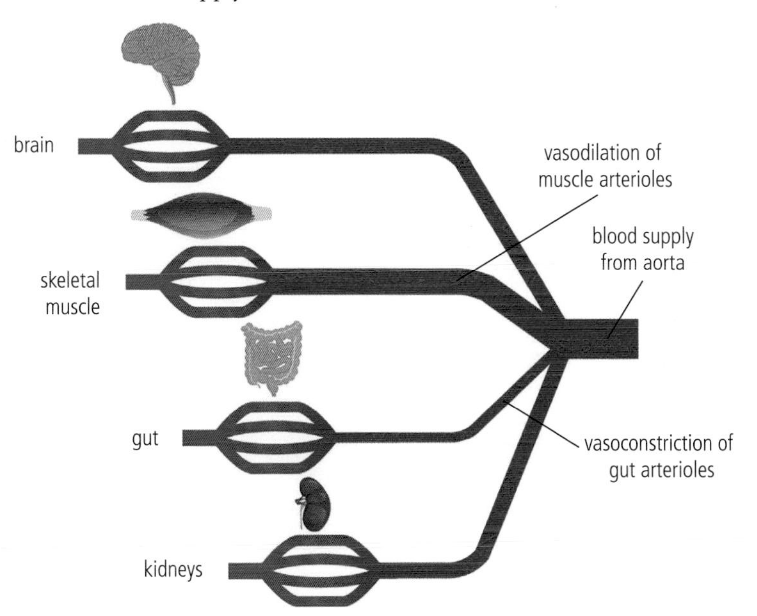

D3.3 Figure 9 During exercise the small arteries feeding oxygenated blood into skeletal muscle become vasodilated to increase blood supply. At the same time vasoconstriction occurs in the small arteries feeding into the gut. Smaller changes can occur in the blood supply to the brain and kidneys.

The following blood volume changes are also known to occur:

- the kidneys receive their maximum blood flow during rest and sleeping
- blood flow to the brain is at its maximum during a stage of deep sleep called rapid eye movement (REM)
- during exercise minimal changes occur to the blood flow to the brain.

HL end

Blood volume increases in one part of the body must be balanced with blood volume decreases in another part of the body.

Guiding Question revisited

How are constant internal conditions maintained in humans?

Within this chapter you have learned:

- homeostasis maintains many body conditions near preset limits despite changes in the external environment
- homeostatic variables include body temperature, blood pH, blood glucose concentration and blood osmotic concentration
- negative feedback control is used to maintain homeostasis
- blood glucose homeostasis is maintained by two pancreatic hormones, insulin and glucagon
- failure to control blood glucose levels is characteristic of both type 1 and type 2 diabetes
- birds and mammals are endothermic and control a set internal temperature by various mechanisms
- receptors inform the nervous system of current values of body conditions that are maintained by homeostasis
- the nervous system and endocrine systems often work cooperatively to ensure body conditions are maintained within set limits
- maintenance of homeostasis is primarily controlled by autonomic functions integrating the nervous and endocrine systems

HL

- kidneys play a major role in maintaining water balance (osmoregulation) and excretion of several body wastes
- the body can regulate blood volume to various organs by vasodilation and vasoconstriction of small arteries.

HL end

Guiding Question revisited

What are the benefits to organisms of maintaining constant internal conditions?

Within this chapter you have learned:

- homeostasis allows an organism to live in a wide range of environments and still maintain vital physiological parameters within set limits
- negative feedback control provides mechanisms for a near steady blood glucose level even when food is not readily available
- maintenance of blood glucose levels within a healthy range avoids the damage to body tissues characteristic of type 1 and type 2 diabetes
- endotherms can remain active in cold and hot temperature extremes because of internal mechanisms that return the internal temperature to a set point
- HL urea and other potential toxins are removed from the blood by the filtering activity of kidney nephrons.

Exercises

Q1. Which of these statements best describes negative feedback control?

A Mechanisms that lower a body variable until it is within physiologically normal levels.

B Mechanisms that allow an organism to change its internal environment to match its outside surroundings.

C Mechanisms that regulate a body variable to keep it within the limits of a set point.

D Mechanisms that are used to change behaviours to adapt to differing environmental challenges.

Q2. Identify the pancreatic hormone that would be secreted under each of these conditions.

(a) Shortly after eating a sugar-filled dessert.

(b) One hour into a very active sports workout.

Q3. Type 1 diabetes is an autoimmune disease.

(a) What cells in the body are destroyed by the immune system in someone with type 1 diabetes?

(b) Identify two specific tissues that are damaged when diabetes is not controlled.

Q4. Identify whether each of these actions has a "warming" or "cooling" effect during thermoregulation.

(a) Perspiration.

(b) Shivering.

(c) Secretion of epinephrine (adrenaline).

(d) Vasodilation of skin arterioles.

Q5. Mitochondria are known for oxidizing glucose to create ATP molecules. What is the purpose of oxidizing glucose in mitochondria found in brown adipose tissue?

HL

Q6. The ultrafiltration of substances through the fenestrations of each glomerulus in a nephron requires high blood pressure. What causes the enhanced blood pressure to allow ultrafiltration?

Q7. Which of these would not be found in the filtrate within a Bowman's capsule under normal physiological conditions?

A Proteins.
B Ions.
C Glucose.
D Urea.

Q8. What happens to the majority of molecules that are reabsorbed from the filtrate initially found within a Bowman's capsule?

Q9. Antidiuretic hormone (ADH) is known to control urine volume output. Under what physiological circumstances will a high concentration of ADH be secreted from the pituitary into the bloodstream?

Q10. Which of these body tissues would be most likely to experience arterial vasoconstriction when someone is heavily exercising?

A Brain.
B Gut.
C Kidneys.
D Skeletal muscle.

HL end

D3 Practice questions

1. Hormones are distributed throughout the body by the blood. Outline the roles of **two** reproductive hormones during the menstrual cycle in women.

   (Total 2 marks)

2. Draw a labelled diagram of the human adult male reproductive system.

   (Total 5 marks)

3. The figure shows a pedigree chart for the blood groups of three generations.

   (a) Deduce the possible phenotypes of individual X. (1)

   (b) Describe ABO blood groups as an example of codominance. (1)

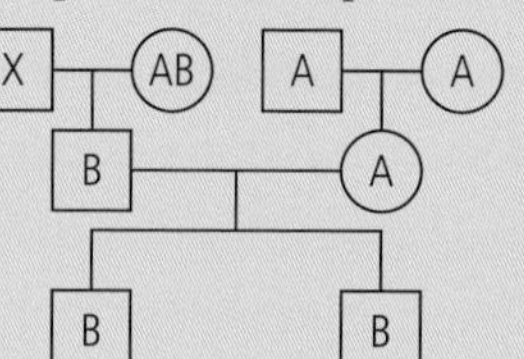

   (Total 2 marks)

4. In the pea plant (*Pisum sativum*), the allele for tall plants is A and the allele for short plants is a. The allele for green plants is B and the allele for yellow plants is b.

   (a) Determine the phenotype of Aabb. (1)

   (b) HL Compare the information that could be deduced when the genotypes are presented as AaBb or (2)

   A || a
   B || b

   (Total 3 marks)

5. Outline how the human body responds to high blood glucose levels.

   (Total 5 marks)

HL

6. Outline the process used to test for human pregnancy.

   (Total 2 marks)

7. Thomas Hunt Morgan established that genes for body colour and wing size in *Drosophila* are autosomally linked. The allele for grey body (b^+) is dominant over that for black body (b) and the allele for normal wing size (vg^+) is dominant over that for vestigial wing (vg).

(a) A fly that is homozygous dominant for both body colour and wing size mates with a fly that is recessive for both characteristics. In the table, draw the arrangement of alleles for the offspring of this mating and for the homozygous recessive parent. (2)

Heterozygous offspring (grey body, normal wings)	**Homozygous recessive parent (black body, vestigial wings)**

(b) The offspring, which were all heterozygous for grey body and normal wings, were crossed with flies that were homozygous recessive for both genes. The table shows the percentage of offspring produced.

grey body, normal wings	48%
grey body, vestigial wings	3%
black body, normal wings	2%
black body, vestigial wings	47%

Explain these results, based on the knowledge that the genes for body colour and wing size are autosomally linked. (2)

(Total 4 marks)

8. Label region X and structure Y on the diagram of the kidney.

(Total 2 marks)

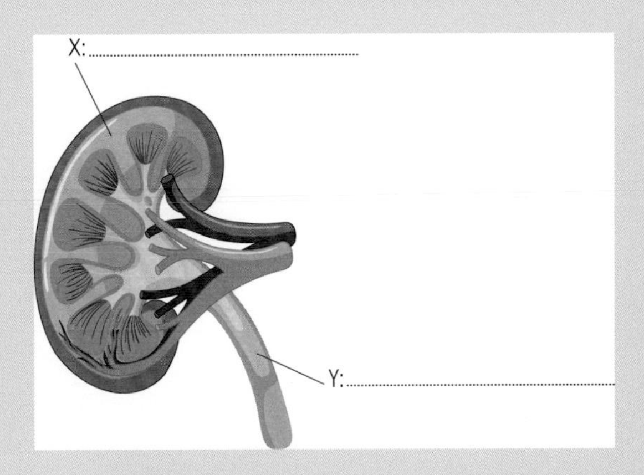

9. All parts of the body change the composition of the blood. Explain how the nephron changes the composition of blood.

(Total 7 marks)

HL end

THEME
D Continuity and change
4 Ecosystems

With the choices we make every day, we can either contribute to climate change and the reduction of biodiversity or we can contribute to the sustainability of ecosystems.

Healthy ecosystems contain diversity; this gives them the ability to adapt to gradual change and can lead to changes in species. If the environment changes, populations will need to adapt. Through natural selection, organisms with higher fitness have a better chance of survival and therefore a higher chance of passing on their genes. Ecosystems are naturally self-sustaining, even when humans harvest some of the resources. However, the impact of humans on ecosystems has escalated in the last century to such an extent that natural processes can no longer maintain a balance. The consequences of human activity are climate change, extinction and pollution. By changing our daily behaviours such as reducing how much non-renewable energy we use as well as rewilding and setting up ecological reserves, we can help reduce climate change.

D4.1 Natural selection

Guiding Questions

What processes can cause changes in allele frequencies within a population?

What is the role of reproduction in the process of natural selection?

Some types of genes help organisms survive better, while others can be detrimental. Organisms that survive are more likely to reproduce. Their offspring are more likely to survive if they have inherited favourable traits from their parents. Natural selection can modify the frequency of alleles over time if the environment changes or if predators or diseases arrive in a population.

HL Artificial selection can also alter the frequency of alleles within a population.

Natural selection depends on the production of viable offspring, so organisms must reproduce successfully. Certain physical characteristics can make a mate more attractive and this will help in the production of offspring. Sometimes physical or behavioural traits are used as an indicator of overall fitness. There is a balance to be struck, however, because traits that include bright showy colours, for example, may also attract the attention of predators.

D4.1.1 – Evolutionary change

D4.1.1 – Natural selection as the mechanism driving evolutionary change

Students should appreciate that natural selection operates continuously and over billions of years, resulting in the biodiversity of life on Earth.

NOS: In Darwin's time it was widely understood that species evolved, but the mechanism was not clear. Darwin's theory provided a convincing mechanism and replaced Lamarckism. This is an example of a paradigm shift. Students should understand the meaning of the term "paradigm shift".

Arguably, once evolution by natural selection is understood, many of the mysteries of nature are revealed. It was Charles Darwin and Alfred Russel Wallace who suggested natural selection as a mechanism for evolution.

Here is a quick overview of the process of natural selection:

- overproduction of offspring
- variation within the population, as a result of meiosis, sexual reproduction and mutations
- struggle for survival, because there are not enough resources for all members of the population to survive
- differential survival, those individuals which are the best fit for their environment tend to survive better
- reproduction, those who survive can pass on their genes to the next generation.

Evolution is the change in heritable characteristics of a population over time. Natural selection explains how the changes occur through a struggle for resources and differential survival, allowing some individuals to pass on their genes but not others.

It is through these steps that populations evolve. Remember that, even though the changes can be observed in individuals from generation to generation, what is of importance is what happens at the level of populations rather than at the individual level.

The Museum of Comparative Anatomy in Paris, France. One indication that life evolves is the fact that the only descendants of dinosaurs still alive on Earth today are birds.

TOK

Evolution is a law of nature because it describes what we observe. Natural selection is a theory because it explains the phenomenon. Laws cannot be modified over time because they simply describe what nature is doing, but theories can be modified over time as we understand mechanisms better.

Nature of Science

When a new idea allows us to see a phenomenon in a different way, it is considered a **paradigm shift**. Before Darwin, the accepted theory explaining evolution was that of Lamarck (see Section D4.1.6).

D4.1.2 – Sources of variation

D4.1.2 – Roles of mutation and sexual reproduction in generating the variation on which natural selection acts
Mutation generates new alleles and sexual reproduction generates new combinations of alleles.

There are three main sources of variation in a species:

- mutations in DNA
- meiosis
- sexual reproduction.

Mutation

Changes in DNA often have no effect on the phenotype of the organism. This might be because the mutation takes place in a piece of DNA that is not used to produce a protein. At other times mutations can produce genes that lead to genetic diseases, and can have devastating effects on the survival of some individuals in a species. However, sometimes a mutation can produce a characteristic that is advantageous. Mutation rates are generally low in populations, so sexual reproduction is a much more powerful source of variation in a population because thousands of genes are mixed and combined. But sexual reproduction is only possible thanks to meiosis.

Meiosis

Meiosis enables the production of haploid cells to make gametes (sperm cells and egg cells). At the end of meiosis, four cells are produced that are genetically different from each other and only contain 50% of the parent cell's genome.

The variety in gametes arises mainly from the process of random orientation during metaphase I (see Figure 1). The lining up of chromosomes in a random order is like shuffling a deck of cards, and it greatly promotes variety in the egg cells and sperm cells produced. In addition to this, the process of crossing over helps shuffle the genetic material and increase the genetic variety further (meiosis is discussed in Chapter D2.1).

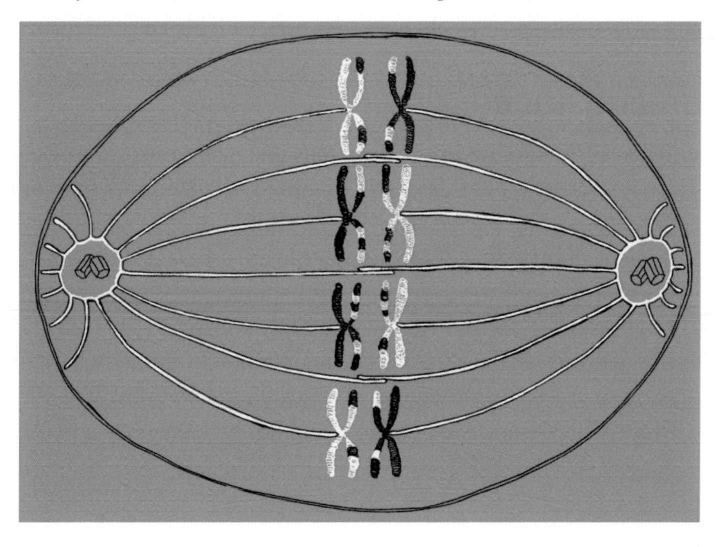

D4.1 Figure 1 Random orientation during metaphase I and crossing over (shown by dissimilar banding on sister chromatids) promote variety in the gametes. Each sister chromatid will separate into a different haploid cell at the end of meiosis.

One of the causes of the Great Famine in Ireland in the mid-1800s was that the potatoes had been produced asexually and were all clones, making them all susceptible to the same infection by a microorganism that causes potato blight.

Sexual reproduction

In an asexually reproducing population, all the members of the population are identical. There may be rare example of mutations or gene transfer, but overall such populations remain identical generation after generation. If there is no variety in a population, there is only a very limited number of outcomes in the event of a change in the environment: the whole population either survives or it dies. Variety in a population allows some individuals to be better adapted to whatever change in the environment is harmful to others. Meiosis means that offspring are varied because of the potential mix of alleles, but sexual reproduction itself also produces variety. Of the many sperm cells that may be present during sexual intercourse, only one will penetrate the egg. Each egg is also different. In flowering plants, which bees will land on which flowers, for example, is also a matter of chance.

Which of these yellow pollen grains on the bee's body will pollinate the next flower it visits?

Although mutation, meiosis and sexual reproduction generate variation, there is another aspect to natural selection that has little to do with chance and allows systematic accumulations of small changes to produce highly adapted forms of life.

Variation and success

There are three main sources of variation in a population:
- **mutations in DNA**
- **meiosis**
- **sexual reproduction.**

How frequently an allele is found in a population can change over time because of changes in the environment. But this is only possible if there is more than one form of a gene, i.e. different alleles. In bacteria, for example, there are essentially no differences within a population: all members of the population are genetically identical copies of each other. This means that if an adverse change happens in the environment, such as a change in pH, if one bacterium is susceptible to the change in pH and dies, they in fact all die because they all have the same vulnerability. In species where there is variation, a change in the environment will eliminate some but not all members of the population. This is why variation is a strength and not a weakness in a population.

D4.1.3 – Overproduction and competition

D4.1.3 – Overproduction of offspring and competition for resources as factors that promote natural selection
Include examples of food and other resources that may limit carrying capacity.

Darwin noticed that plants and animals produce far more offspring than could ever survive. Plants often produce hundreds or thousands more seeds than necessary to propagate the species. Mushrooms produce millions more spores than ever grow into new mushrooms. A female fish lays hundreds or thousands of eggs but only a handful survive to adulthood. This overproduction of offspring allows natural selection to occur, the selection of individuals that are fittest to survive.

Too many offspring and not enough resources cause a problem of supply and demand. If there is high demand for resources such as water, space, nutrients or sunlight, but a limited supply, then there will be competition between individuals to obtain those resources. If there are not enough resources for a growing population, the population will be limited as those who are outcompeted will not survive. The maximum number of individuals that an environment can provide for is called the **carrying capacity**. For example, less food in an area will reduce the carrying capacity, whereas an increase in food supply would raise the carrying capacity. The consequence of supply and demand is competition for resources in order to stay alive. This is called the **struggle for survival** and is a key component of natural selection.

What mechanisms minimize competition?

D4.1.4 – Selection pressure

D4.1.4 – Abiotic factors as selection pressures
Include examples of density-independent factors such as high or low temperatures that may affect survival of individuals in a population.

The main driving force of evolution is change in the environment. If nothing in a population's environment changes, there is no need to adapt. A **selection pressure** is a factor that can influence the success of parts of a population, and thereby influence changes in allele frequencies.

The mass extinction event that occurred 66 million years ago and wiped out so many dinosaurs and other species on Earth is an extreme example of a selection pressure. The environment changed rapidly when the collision of an asteroid with Earth released dust particles that then blocked out much of the Sun's radiation. Producers could no longer be as productive and, as food chains and food webs were disrupted, ecosystems collapsed all over the world. This sudden change in light levels, and subsequently temperature, was a selective pressure on populations. Many species did not make it. It is estimated that three-quarters of all life forms became extinct as a result.

An **abiotic factor** in an environment is something that is not living. Abiotic factors can be part of the physical environment, such as temperature, humidity or the availability of light. They can also be parts of the chemical environment that can affect organisms, such as the availability of certain minerals, the pH of water or soil, and the concentrations of gases such as oxygen and carbon dioxide in the atmosphere or in water. These factors are considered **density-independent factors** because they affect the population no matter how big or small the population is. An increase in the acidity of ocean water (a decrease in pH) will be as harmful to a large coral reef as it will be to a small one.

Magellanic penguins (*Spheniscus magellanicus*) live along the coasts of the southern tip of South America and are well adapted to surviving in the snow and cold. The penguin chicks are covered with down feathers that allow them to maintain a healthy body temperature by trapping warm air near their bodies. If snow lands on them, they can shake it off and keep dry and warm. They are not as well adapted for rain, however. Rain causes the downy feathers to stick to the chicks' bodies and the thermal insulation effect is lost because no more warm air surrounds their bodies. Chicks then die of hypothermia because of their inability to keep warm. With global climate change, higher temperatures in winter means that it now rains more frequently in parts of the range of Magellanic penguins, causing a decrease in some populations.

A Magellanic penguin (*Spheniscus magellanicus*) with young.

Do not confuse density-independent factors with density-dependent factors. A density-dependent factor is one that affects a population more when the population numbers are higher. Disease, for example, will spread faster in a highly populated area where many organisms are close together, but more slowly in a smaller population that is spread over a wider area. With density-independent factors, it makes no difference if the population density is dense or sparse.

Snow crabs, of the genus *Chionoecetes*, thrive in cold water at northern latitudes of the Atlantic and Pacific Oceans. In the Labrador Sea, fishermen have observed an increase in populations of snow crabs in recent years that could be being caused by cooling water temperatures. Although it might seem counterintuitive that cold is better, colder water can dissolve more oxygen, and therefore can potentially provide resources for more individuals of species that can tolerate colder water. Increased dissolved oxygen availability and colder temperature are abiotic factors that can greatly increase the chances of survival of species that thrive under both conditions.

D4.1.5 – Intraspecific competition

D4.1.5 – Differences between individuals in adaptation, survival and reproduction as the basis for natural selection
Students are expected to study natural selection due to intraspecific competition, including the concept of fitness when discussing the survival value and reproductive potential of a genotype.

The noun "adaptation" and the verb "to adapt" are used freely when talking about evolution. However, the terms have very precise meanings within the framework of natural selection and should not be confused with other uses of the term, notably for human behaviour. For example, humans can consciously decide to adapt to a situation: think of a student learning the language of a country they have just moved to. This is a conscious adaptation made by an individual. In nature, the vast majority of adaptations referred to in evolution are unconscious, non-intentional adaptations that, although they take place on an individual level, are only meaningful if they affect the population.

Adaptation and survival

Exactly which individuals survive and which ones do not is not based on chance alone but determined by their surroundings and the compatibility of their characteristics with those surroundings. Competition between individuals of the same species is called **intraspecific competition.**

An organism that has characteristics that means that it is well adapted for its environment is said to be **fit** for its environment. Organisms with **high fitness** (those which possess characteristics that work well in their environment) have a higher chance of survival than those with **low fitness** (those which possess characteristics poorly suited for their environment). Natural selection tends to eliminate individuals from the population that show low fitness, whereas the fittest individuals in a population have a higher likelihood of surviving. Although there are rare exceptions, individuals are usually incapable of changing themselves to adapt to a particular factor. For example, a hummingbird with a short bill that cannot reach the nectar at the base of a flower cannot force itself to intentionally grow a longer bill. In offspring:

- useful variations allow some individuals to have a better chance of survival (e.g. hiding from predators, fleeing from danger or finding food)
- harmful variations make it difficult to survive (e.g. inappropriate colour for camouflage, heavy bones for birds, having such a big body size that there is not enough food to survive).

Because they survive to adulthood, successful organisms have a better chance of competing successfully with other members of the population, reproducing and passing on their successful genetic characteristics, their genotype, to the next generation. Over many generations, the accumulation of changes in the heritable characteristics of a population results in evolution.

How do intraspecific interactions differ from interspecific interactions?

An organism that is well adapted to its environment is not guaranteed success, it simply has a higher probability of survival than another that is less well adapted. Dinosaurs such as the sauropods were among the biggest, strongest animals ever to walk the planet. But they did not survive the environmental changes that drove them to extinction. In fact, the fossil record indicates that more than 99.99% of all life that has ever existed on Earth is now extinct.

▲ **D4.1 Figure 2** Plover eggs show adaptations that have been acquired by natural selection. Their colour and speckles help camouflage them from predators.

In Figure 2, the colours and speckles of the plover eggs act as effective camouflage, making these eggs difficult to spot by predators. Plover chicks are also speckled for camouflage. Such adaptations are good examples of traits that have **survival value**. If a mutation caused a shell to be bright white and/or the chicks to be bright yellow, the mutation would be unlikely to confer an advantage to this species and would have low survival value because these colours would attract the attention of a predator.

Only producing two eggs in a generation is risky. To increase their **reproductive potential**, organisms such as fish produce hundreds of eggs per generation. Reproductive potential is the maximum number of offspring an organism can produce in the absence of offspring mortality. Hypothetically, if 300 fish eggs are produced and they all develop into adults, the reproductive potential will have been reached. In reality, predation of eggs and young will reduce the survival rate and only a handful will reach adulthood.

It is crucial that you remember Darwin's steps of how natural selection leads to evolution:

- **overproduction**
- **variation within the population**
- **struggle for survival**
- **survival of those best suited to the environment**
- **reproduction.**

It is through these steps that populations evolve. Remember that, even though the changes can be observed in individuals from generation to generation, what is of importance is what happens at the level of populations rather than at the individual level.

D4.1.6 – Heritable traits

D4.1.6 – Requirement that traits are heritable for evolutionary change to occur
Students should understand that characteristics acquired during an individual's life due to environmental factors are not encoded in the base sequence of genes and so are not heritable.

In all the examples of traits given so far, you may have noticed that they are all heritable. **Heritable** means that the trait is encoded in the organism's DNA and can, therefore, be passed on to the next generation. Something that is acquired during the lifetime of an organism is considered to be an **acquired characteristic** and is not coded in the DNA. Acquired characteristics of organisms do not result in evolutionary changes. They only affect the individual and not their offspring.

▲ The pink pigmentation in flamingos (*Phoenicopterus ruber*) is not due to genetics and DNA, so it is not heritable. The colour comes from eating plankton rich in beta carotene.

Flamingos, also known as pink flamingos, are not pink because of pigments generated by their DNA. The pink colour comes from ingesting plankton that are rich in a molecule called beta carotene, the same one that contributes to the coloration of sweet potatoes and carrots. A flamingo that has a diet poor in beta carotenes will not turn pink. The offspring of a deeply coloured flamingo like the one in the image will not inherit the colour. It will have to find sufficient beta carotene in its diet if it wants to be as brightly pigmented as its parent.

Before Darwin, the accepted theory explaining evolution was that of Lamarck. **Lamarckism** states that organisms can inherit acquired characteristics from their parents and that there is a trajectory towards complexity and improvement. Lamarck explained his idea with the use and disuse of body parts. For example, ruminants who use their horns in combat will develop bigger stronger horns and pass that trait on to their offspring, whereas the mole, living underground, has eyes weakened by disuse and weaker eyes are passed on to the next generation. These are compelling and plausible explanations but they have not been supported by experiments that can be replicated. In science we keep theories that stand up to testing and reject those that do not. Lamarck's theory has been refuted and rejected but Darwin's has been repeatedly confirmed for over a century and a half.

How would the two theories, Lamarckism and Darwinism, explain the atrophied (weakened) eyes of the European mole (*Talpa europaea*)?

D4.1.7 – Sexual selection

D4.1.7 – Sexual selection as a selection pressure in animal species
Differences in physical and behavioural traits, which can be used as signs of overall fitness, can affect success in attracting a mate and so drive the evolution of an animal population. Illustrate this using suitable examples such as the evolution of the plumage of birds of paradise.

One of the best criteria for measuring long-term success in populations is reproductive success: more offspring means more success, less offspring means less success. Higher numbers in the present generation means more individuals that can reproduce and produce the next generation. Since natural selection is constantly removing members of the population, the ability to maintain healthy numbers is a sign of success.

There are different selection pressures that act upon organisms within a population. For example, lack of water may be a selection pressure in arid areas. **Sexual selection** occurs when the reproductive success of an individual results in more offspring compared to others in the population who do not have as much success in finding a mate. Sexual selection drives the evolution of a population along with other selection pressures.

In birds, exceptionally brightly coloured or shiny feathers can be a way of showing potential mates that they are in such good health that they can afford to use some of their nutritional resources on pageantry. Birds of paradise are remarkable in this way. There are often striking differences between the male and female of the same species. Females have **cryptic** coloration, meaning they have dull or dark colours that blend in with the shadows of the forest and are not ornate or eye-catching. This helps them avoid being noticed by predators.

The lesser bird of paradise (*Paradisaea minor*) lives in the forests of Papua New Guinea. This is a male with ornate colours. The female of this species is brown with a white underside and has no brightly coloured feathers.

When a bird has exaggerated, colourful and long tail feathers, it makes it difficult for the bird to fly, and proteins and minerals have to be invested in pretty colours and extravagance instead of being used for the bird's immune system or for energy. Any individuals who produce ornate plumage are clearly healthy because they can use valuable resources to produce good looks and they are agile enough to overcome any difficulties in flight that the bigger feathers create. This indicates that the bird has strong and healthy genes. The females of the species find this irresistible.

When there is a morphological difference between males and females like this, it is called **sexual dimorphism**. The differences are not visible in the young but appear when secondary sex characteristics develop. Competition between males for access to females is referred to as **intrasexual competition**. Male lesser birds of paradise with duller colours or shorter tails will not be as successful in attracting a mate.

In addition to colourful feathers, males can use behavioural characteristics to attract females. These could be building elaborate nests or performing courtship rituals that might include well-rehearsed choreographies. You may have seen humorous videos online of the male western parotia (*Parotia sefilata*) dancing with head bobs, fancy footwork and twirls to impress the females. Most of its feathers are jet black but some are colourful and reflective and can be orientated to suddenly flash into view as part of the choreography. Such a show sends a clear signal to the female that this individual is very healthy.

It is very tempting to choose human examples for natural selection but try not to when answering exam questions. For example, it is easy to imagine people with money, fame, dance moves and fancy cars attracting members of the opposite sex who are looking to pass on healthy and successful genes, but such clichés should not be used as examples in arguments for your coursework or for exams. Such ideas are useful analogies but should not replace examples from the natural world, such as birds with impressive feathers or courtship rituals.

D4.1.8 – Modelling selection pressures

D4.1.8 – Modelling of sexual and natural selection based on experimental control of selection pressures
Application of skills: Students should interpret data from John Endler's experiments with guppies.

In the 1970s, John Endler carried out a series of experiments to see if the presence of predators modified the bright colours seen in guppies (*Poecilia reticulata*). Guppies live in streams on the island of Trinidad off the coast of Venezuela in South America. The genes for the bright colours you can see in Figure 3 are only expressed in the males. Female guppies, like birds of paradise, show cryptic coloration that provides camouflage. The male coloration is highly diverse and made up of a mosaic of spots. Endler studied the fish for the following pigments: yellow, red, blue, iridescent and black. (Iridescent means that light glitters and is reflected in different ways depending on the angle at which it is viewed, and can look like different colours depending on the way light strikes the surface.) He carried out experiments in two ways: in the field (in the streams of Trinidad) and in 10 artificial ponds in a greenhouse at Princeton University, USA, where more variables could be controlled than in nature. In the field and in the greenhouse, some populations of guppies were kept in water that had predatory fish present that could eat them, while others were in water with fish that were not their predators and therefore harmless.

D4.1 Figure 3 Guppies (*Poecilia reticulata*) are popular aquarium fish and the males can show very flashy colours.

One of the reasons guppies thrive in the streams of Trinidad is that the sloped landscape is broken up by ledges of rock that form ponds and waterfalls. The waterfalls are like steps up the slope of the island; some fish species can swim upstream from the ocean, up the waterfalls, while other fish are blocked by them. Endler's hypothesis was that guppies in pools protected from predatory fish would show more ornate colours, whereas those in pools where predators were present would be less colourful because their ability to hide would lead to better chances of survival.

There are two separate and opposing selective pressures on male guppies.

- Predation: In this case high fitness includes more cryptic colours that blend in with the background. Brightly coloured males tend to be noticed more easily by predators and are selected against. The allele frequency for bright colours will decrease over time.
- Sexual selection: In this case the males need to stand out in order to be attractive to females, so high fitness includes bright conspicuous colours that make them stand out against the background. The allele frequency for bright colours will increase over time.

How did he test this hypothesis? Endler's independent variable was the presence of predators. In the greenhouse ponds, he introduced predatory fish species that eat guppies to some of the bodies of water and in others he introduced fish that were not predators. Then he waited to see what would happen over 15 generations. The predator was *Crenicichla alta*, a type of cichlid; the non-predatory fish was *Rivulus hartii*. In the field, he found ponds along streams where either *C. alta* or *R. hartii* was present, and introduced guppy populations that had the same level of genetic variety as used in the greenhouse experiments. To see the results of the greenhouse and field experiments, see the data interpretation exercise on the next page.

Look at the graph of Endler's results below and answer the questions. Use the following legend to understand his graph:

K = ponds with no other fish

R = ponds with *R. hartii*, the harmless species

C = ponds with *C. alta*, the dangerous predator

Shaded bars = artificial ponds in the greenhouse at the university

Unshaded bars = natural ponds in the field (on Trinidad)

Error bars = two standard errors

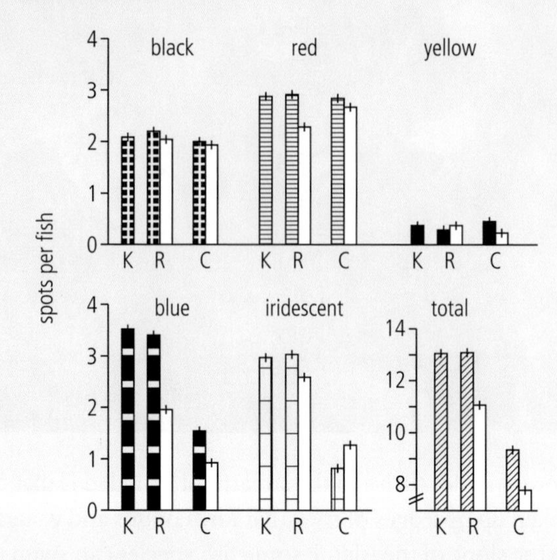

1. Identify the group with the most spots per fish.
2. Explain why there is only one shaded bar for the results in group K.
3. Compare the number of spots on fish in ponds with *R. hartii* and ponds with *C. alta*.
4. Think about the controlled variables of the experiments carried out in the greenhouse and explain why it would be a poor idea to set up the ponds outdoors instead of inside the greenhouse.

HL

D4.1.9 – Gene pools

D4.1.9 – Concept of the gene pool
A gene pool consists of all the genes and their different alleles, present in a population.

Interbreeding populations

The **gene pool** is all the genetic information present in the reproducing members of a population at a given time. The gene pool can be thought of as a reservoir of genes. All the various characteristics present in the population are present in the gene pool. A population that shows substantial variety in its traits has a large gene pool, whereas a population whose members show little variation has a small gene pool. **Inbreeding** occurs when closely related organisms mate with each other. Inbreeding narrows the gene pool.

Allele frequency is a measure of the proportion of a specific version of a gene in a population. The allele frequency is expressed as a proportion or a percentage. For example, it is possible that a certain allele is present in 25% of the chromosomes studied in a population. This would mean that one-quarter of the loci for that gene are occupied by that allele, and the other three-quarters are occupied by a different allele for the same gene.

This could also be interpreted as a 25% chance that a chromosome in that population has the allele at that particular locus. Note that this does not mean that 25% of the members of the population have the allele. Later in this section we will see how these numbers play out in a diploid situation where two chromosomes in each organism can carry a version of the gene.

Look at Figure 4. The gene pool in this population of 16 people is made up of 32 genes. Count the number of **T'**s. You should get 16. Do you get the same for the number of **t'**s? You should. Because half the alleles are **T'**s and half are **t'**s, the allele frequency for each is 50%, or 0.50. Does this mean that half the people have the phenotype caused by the recessive allele? No. Only four people are homozygous recessive and have the phenotype, which is 25% of the population. Be careful not to confuse allele frequency with the number of people who show a particular trait.

D4.1 Figure 4 In this gene pool, the frequencies of each allele, **T** and **t**, is 50%.

Evolution and allele frequencies

Gene pools are generally relatively stable over time. But not always. New alleles can be introduced as a result of mutation, and old alleles can disappear when the last organism carrying the allele dies. After many generations of natural selection, some alleles prove to be advantageous and tend to be more frequent.

Conversely, some alleles are disadvantageous to the survival of the organisms in the population and are not passed on to as many offspring. From this it should be clear that, at any time an allele frequency is estimated, it is only a snapshot of the alleles at that time. Several generations later, the proportions of alleles may not be the same.

In addition, if populations mix as a result of immigrations, there will probably be a change in allele frequencies. The same is true for emigrations, when one group with a particular allele leaves the population. Whatever the reason, if a gene pool is modified and the allele frequencies change, we know that some degree of evolution has happened. No change in allele frequencies, however, means no evolution.

D4.1.10 – Geographically isolated populations

D4.1.10 – Allele frequencies of geographically isolated populations
Application of skills: Students should use databases to search allele frequencies. Use at least one human example.

When a population is split into two separate populations that can no longer interbreed, the isolated groups can start to diverge. This happens if they are living in environments that put different selective pressures on each one. Think back to the guppy experiments: the presence of predators in isolated pools significantly modified the frequency of alleles for ornate colours in the males of those populations. We will now look at an example of when human alleles have different frequencies as a result of geographical separation.

In humans, organ transplants are notoriously difficult because the recipient's immune system tends to reject the donor's organ. The reason for this is that the job of the immune system is to find and respond to any cells with proteins on their surface that reveal they do not belong in the body. Thanks to this system, proteins (antigens) on the surface of viruses or pathogenic bacteria are recognized as "non-self" and are targeted for destruction by white blood cells.

HLA, the **human leucocyte antigen** complex, is a set of genes found on human chromosome 6 that has several roles, one of which is producing proteins that will then either produce molecules that sit on the cell surface to identify it as "self" (these are Class I HLA genes) or molecules that bring to the surface of a white blood cell "non-self" antigens that signal an attack by a pathogen (Class II). These genes, when compatible between donor and recipient, will not reject a transplanted organ. If, on the other hand, the two genes are different enough between the two people, the donor's organ will be rejected by the recipient.

The example in Figure 5 shows some of the versions of the HLA gene of a kidney donor. The letters, A, B and DR refer to zones along the chromosome where HLA genes can be found. The numbers indicate which allele of the genes the person has inherited. To help decide which of the patients that are in need of a kidney transplant should be given the organ, doctors look at the versions of the genes present in the patients' HLA. Knowing that the more matches there are, the less chance of rejection there is, the best choice is patient C, with 6 out of 6 matches.

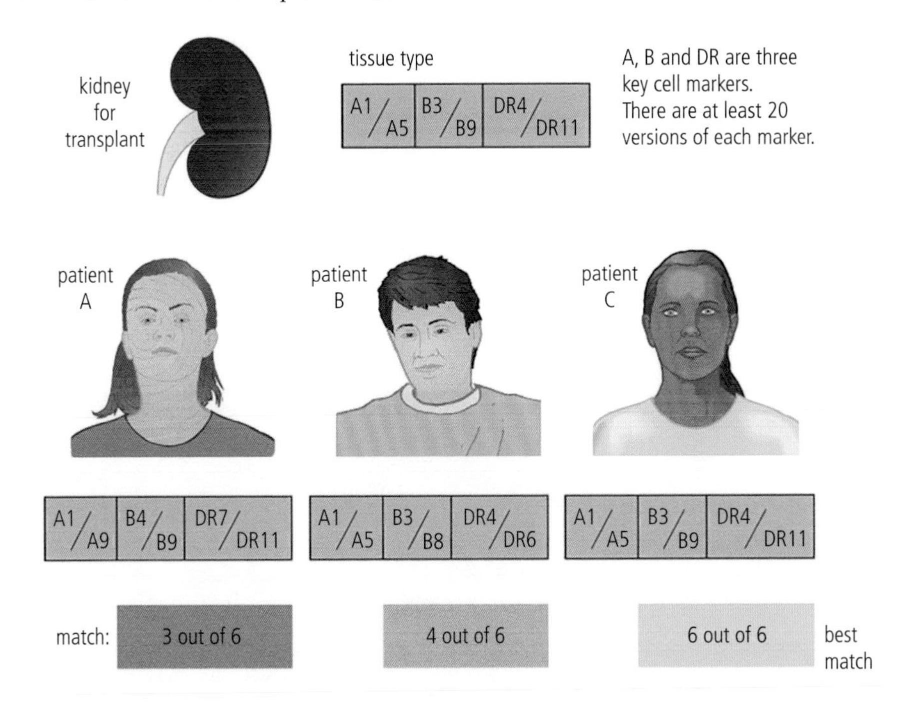

D4.1 Figure 5 Alleles from the HLA complex of genes can be used to find the best match between a recipient and a donated organ.

When a gene or group of genes shows multiple variations, we say it is **polymorphic**. There is a large variety of polymorphisms in the HLA system in humans. In fact, it is the most polymorphic set of genes we possess. The differences can partly be explained by the fact that human populations have spread out and lived in very different parts of the world and therefore evolved differently over many tens of thousands of years. When comparing allele frequencies in the HLA system, two geographically isolated populations will tend to have fewer polymorphisms in common than those that are geographically closer. To see some examples for yourself, try the exercise in the Skills box using data accessible in an online database.

SKILLS

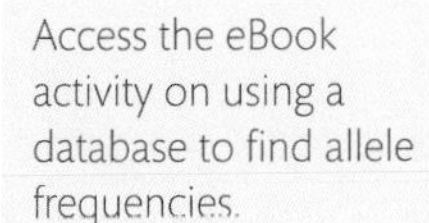

Access the eBook activity on using a database to find allele frequencies.

D4.1.11 – Changes in allele frequency

D4.1.11 – Changes in allele frequency in the gene pool as a consequence of natural selection between individuals according to differences in their heritable traits

Darwin developed the theory of evolution by natural selection. Biologists subsequently integrated genetics with natural selection in what is now known as neo-Darwinism.

When Darwin was defending his theory of natural selection, he had a major disadvantage compared to modern evolutionary biologists: DNA and how its genetic code works had not been discovered yet. Although Gregor Mendel had proposed his ideas about how traits get passed down from one generation to the next, specialists in the mid-19th century had not yet put natural selection together with genetics. Today, DNA evidence provides support for natural selection and is referred to as the **modern synthesis** or **neo-Darwinism**, a combination of Darwin's ideas with the idea of genetics that Mendel started, although this was only confirmed long after both men had died.

An example showing a change in allele frequency in a gene pool over time is the peppered moth (*Biston betularia*) population in the UK before, during and after the industrial revolution. The species has two polymorphisms, one a light-coloured speckled grey phenotype and the other a black phenotype (see Figure 6). On light-coloured surfaces the grey moths are better adapted at hiding from predators. On dark-coloured surfaces (for example trees covered with industrial soot during the industrial revolution), the rarer black moths are better adapted. When the environment was more polluted, the darker allele was selected for and the darker alleles became more frequent in the population. After the Clean Air Act of 1956 was passed, the pollution levels fell and the allele frequency for light-coloured moths increased. Explaining this observed phenomenon using the modern synthesis is more consistent with the data than explaining it with ideas that predated Darwinism, notably Lamarckism.

D4.1 Figure 6 The light-coloured peppered moths (*Biston betularia*) on the left are better camouflaged on a light-coloured background, whereas the rarer dark version (the melanic version) is better camouflaged on a dark-coloured background.

D4.1.12 – Reproductive isolation of populations

D4.1.12 – Differences between directional, disruptive and stabilizing selection
Students should be aware that all three types result in a change in allele frequency.

In some situations, populations of members of the same species (and thus of the same gene pool) can be stopped from reproducing together because there is an insurmountable barrier between them. Such a barrier can be geographical, temporal, behavioural or related to the infertility caused by hybridization.

Directional selection takes place when one phenotype is favoured over another by natural selection. In such a case, the frequency of one phenotype is seen to increase over time, whereas the other phenotype decreases. This can occur when an organism's environment changes, as in the example of the peppered moth. Another way of thinking about directional selection is to think of it as selection away from one extreme.

When one phenotype is favoured over two extreme phenotypes, it is called **stabilizing selection**. For example, a flowering plant might produce some flowers with more nectar and some flowers with less nectar. Producing excessive nectar would be a drain on the plant's sugar resources, and producing too small a quantity would discourage insects from visiting it. The best solution is an intermediate quantity, providing a balance between too much and too little nectar. Another way of thinking about stabilizing selection is to think of it as selection away from two extremes or selection towards the mean.

When two extreme phenotypes are favoured by natural selection, rather than one intermediate phenotype, it is called **disruptive selection**. Sometimes it is an advantage to have two opposing varieties of a phenotype rather than only one. For example, tadpoles of spadefoot toads (*Spea multiplicata*) in the Americas have two very different morphologies, one for an omnivorous diet and one for a strictly carnivorous diet that includes cannibalism if food sources are scarce. Having separate morphologies gives this species a better chance of survival in places where food sources are variable because if each morphologically different tadpole eats different things, there is less overlap in diet so less competition. Another way of thinking about disruptive selection is to think of it as selection against the mean. The idea is to maintain extremely different phenotypes. If the differences caused by disruptive selection are extreme and the two populations occupy different niches, it is possible for speciation to occur.

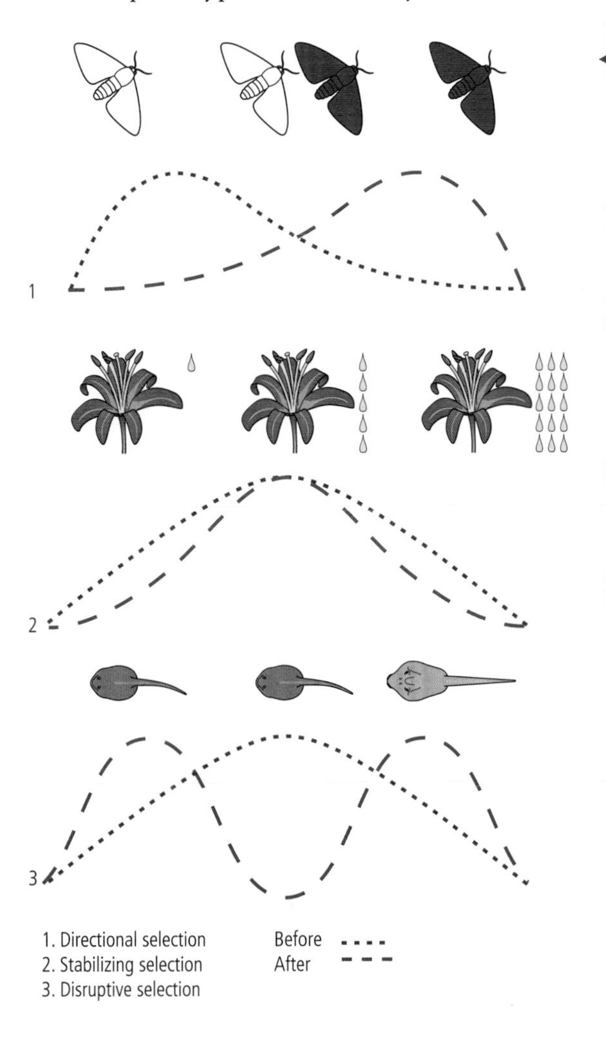

There are three types of selection: (1) directional selection, as in the frequency of the two forms of the peppered moth; (2) stabilizing selection, as in the amount of nectar in lilies; and (3) disruptive selection, as in two types of spadefoot tadpoles (omnivore or carnivore).

An organism's phenotype is the expression of its genotype. Its genotype is the set of alleles it received from its parents, such as Aa, RR or X^bY. When we say that natural selection selects one phenotype over another, it is thanks to the genotype being selected that we see the phenotype.

D4.1.13 – The Hardy–Weinberg equation

D4.1.13 – Hardy–Weinberg equation and calculations of allele or phenotypic frequencies
Use p and q to denote the two allele frequencies. Students should understand that $p + q = 1$ so genotype frequencies are predicted by the Hardy–Weinberg equation: $p^2 + 2pq + q^2 = 1$. If one of the genotype frequencies is known, the allele frequencies can be calculated using the same equations.

In order to calculate the frequencies of alleles within a population, the **Hardy–Weinberg equation** can be used. This equation is designed to mathematically show stability in the distribution of alleles. If the equation gives unexpected ratios, this can indicate that a population is changing. To understand how it is used, it is best to start with understanding how it was derived.

A Punnett grid (discussed in Section D3.2.2) is used to show the genotypes of the parents and offspring in a cross. For the Hardy–Weinberg equation, we need to look at the cross in a new way: as a model for the allele frequencies. To do this, we need the variables $\boldsymbol{p}$ and $\boldsymbol{q}$:

- p = frequency of the dominant allele (allele **T** in the example below, remember to read this as "big T")
- q = frequency of the recessive allele (allele **t** in the example below, remember to read this as "little t").

When looked at individually, the frequencies of the alleles on chromosomes must add up to 1. So $p + q = 1$. For example, if $p = 0.25$ (or 25%) frequency, then $q = 0.75$ (or 75%) because chromosomes that do not carry the dominant allele must carry the recessive one.

What complicates things is the fact that we usually want to consider diploid organisms that carry two copies of any particular gene. As a result, the equation becomes $(p + q)^2 = 1$.

$(p + q)^2$ can be expanded to $p^2 + 2pq + q^2$

D4.1 Figure 7 An annotated Punnett grid showing allele frequencies.

	T	t
T	TT	Tt
t	Tt	tt

p^2 (TT); $2pq$ (Tt, Tt); q^2 (tt)

We can now deduce that $p^2 + 2pq + q^2 = 1$. This mathematical representation for the allele frequencies is known as the **Hardy–Weinberg equilibrium** and it is reached after only one generation of random interbreeding (see Section D4.1.14).

If this is still a bit confusing, try looking at it this way. You can summarize the Punnett grid in Figure 7 as shown in Table 1.

D4.1 Table 1 The Hardy–Weinberg equilibrium

	One square	Two squares	One square
Genotypes	TT	2Tt	tt
Proportions	$\frac{1}{4}$	$\frac{1}{2}$	$\frac{1}{4}$

Looking again at the Punnett grid in Figure 7, in terms of the allele frequencies rather than genotypes, the following can be deduced:

- the frequency of **TT** = p^2
- the frequency of **Tt** = $2pq$
- the frequency of **tt** = q^2.

By adding up all the possible proportions, we can see that $\frac{1}{4} + \frac{1}{2} + \frac{1}{4}$ comes to a total of 1. We can now describe the proportions, as shown in Figure 8.

Remember that allele frequency and genotype frequency are not the same thing. The double letters **TT**, **Tt** or **tt** are the genotypes but the letters p and q represent allele frequencies.

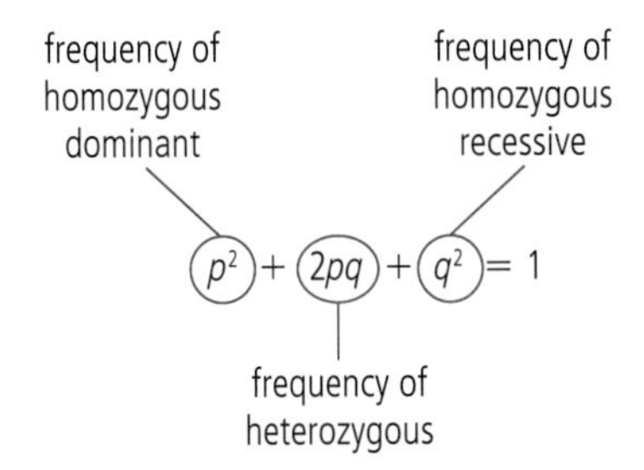

D4.1 Figure 8 An annotated Hardy–Weinberg equation.

Before you move on to Section D4.1.14, try the equation. If you substitute 0.25 for p and 0.75 for q in the equation, you should get 1. Be careful with the order of operations: anything that needs squaring should be done first, then perform any multiplications, then the additions.

Below is a series of worked examples to help you get some hands-on experience: you will not be able to fully understand how to use the Hardy–Weinberg equation without practice.

Worked example 1

Consider a disease caused by a recessive allele, which we shall assign the letter **t**. The predicted frequency of this allele in the population being studied is 10%. Calculate the frequency of the healthy allele in the population.

Solution

From the information given, we know that q is 0.10, and as the proportions of p and q must always add up to 1, we can say that $p = 1 - q$. So p must be 0.90, which means that, in the gene pool, 90% of the alleles are **T**.

Remember that this does not mean that 90% of the population is healthy, because we are calculating an allele frequency, not a genotype frequency.

Worked example 2

In a study of 989 members of the population from example 1, it was found that 11 people had the disease. Calculate the frequency of the recessive allele **t**.

Solution

First, calculate the proportion of people who had the disease (which in this case is the proportion of **tt** genotypes). To do this, divide 11 by 989 to obtain 0.011. This means that 1.1% of the population has the genotype **tt**. Hence, $q^2 = 0.011$.

So to calculate q, we take the square root of 0.011, which gives us 0.105. This means that the frequency of the recessive allele **t** is 10.5% in this population. Note that this is very close to the predicted value of 10% in example 1.

Worked example 3

Using the information from example 1, consider the following in order to calculate the genotype frequency.

1. Fill out a copy of the table.

Allele frequencies	Recessive **t**	q	
	Dominant **T**	p	
Genotype frequencies	Homozygous recessive **tt**	q^2	
	Heterozygous **Tt**	$2pq$	
	Homozygous dominant **TT**	p^2	

2. Calculate the frequency of carriers in 500 members of the population.

Solution

1. We know from the data given in Worked example 1 that $q = 0.10$ and that $p = 0.90$. These values will complete the first two rows of the last column of the table.

   To fill in the other cells, simply perform the mathematical operations in the third column.

   $q^2 = 0.01$, so 1% of the population is **tt**.

   $2pq = 0.18$, so 18% of the population is **Tt**.

   Lastly $p^2 = 0.81$, so 81% of the population is **TT**.

2. To find the number of carriers (heterozygotes) in 500 members of this population, multiply 500 by 18% to get 90. So 90 people would be carriers.

Worked example 4

Using information from your completed table, calculate the number of people among 500 members of the population who do not have the disease.

Solution

Using the numbers calculated in Worked example 3, we can complete the table as shown below.

Allele frequencies	Recessive **t**	q	0.10
	Dominant **T**	p	0.90
Genotype frequencies	Homozygous recessive **tt**	q^2	0.01
	Heterozygous **Tt**	$2pq$	0.18
	Homozygous dominant **TT**	p^2	0.81

We know that, in order to not have this disease, a person must be either **TT** or **Tt**. The combined percentages of these genotypes is 81% + 18%, which gives 99%. 99% × 500 = 495. So 495 people out of the 500 should have the healthy phenotype.

D4.1.14 – Genetic equilibrium

D4.1.14 – Hardy–Weinberg conditions that must be maintained for a population to be in genetic equilibrium

Students should understand that if genotype frequencies in a population do not fit the Hardy–Weinberg equation, this indicates that one or more of the conditions is not being met, for example mating is non-random or survival rates vary between genotypes.

When the Hardy–Weinberg equation was derived, a number of assumptions were made about the populations it can be used for: a population needs to have reached a stable equilibrium in which the allele frequency does not change over time. The conditions that must be met for this to be the case include the following:

- the organisms studied must be diploid
- the population being studied has to be large
- the organisms must reproduce sexually
- the gene being studied cannot be a sex-linked gene, because if an allele is more frequent in one sex than another, the equation will not work
- mating must be random
- there can be no migration into or out of the population
- none of the alleles being studied can reduce the chances of survival (e.g. cause a deadly genetic disease or cause a lack of pigmentation making it impossible to hide from predators).

If the genotype frequencies do not fit the equation, that means that one of the above conditions is not true for the population studied. For example, in a population of farm animals such as dairy cows, because mating is controlled by artificial selection rather than being random, it is unlikely the equation will work with the genotype frequencies present in the cows.

D4.1.15 – Artificial selection

D4.1.15 – Artificial selection by deliberate choice of traits
Artificial selection is carried out in crop plants and domesticated animals by choosing individuals for breeding that have desirable traits. Unintended consequences of human actions, such as the evolution of resistance in bacteria when an antibiotic is used, are due to natural rather than artificial selection.

Humans can breed organisms to increase what we see as desirable characteristics, for example higher yields in crops such as wheat, or greater milk production in cattle. This process, known as **artificial selection,** is performed by selective breeding: humans decide which organisms have the most desirable traits and breed them together, hoping for offspring with enhanced features.

Natural selection and antibiotic resistance

Antibiotic resistance in bacteria is a modern example of natural selection. What is striking is its rapidity. Although evolution is generally considered to be a long-term process, the mechanism of natural selection can sometimes be quick, taking place over months, years or decades, rather than millennia. As you read the description below, see if you can identify the main features of how natural selection works.

Antibiotics are medications such as penicillin that kill or inhibit the growth of bacteria. They are given to patients suffering from bacterial infections. However, overuse of antibiotics has led to resistant strains of bacteria.

Antibiotic resistance in bacteria develops over several steps. Consider the following scenario.

1. A woman gets tuberculosis, which is a bacterial infection.
2. Her doctor gives her an antibiotic to kill the bacteria.
3. She gets better because the vast majority of bacteria are destroyed.
4. Thanks to a pre-existing variation in its genetic makeup, however, one bacterium is resistant to the antibiotic.
5. That bacterium is not killed by the antibiotic and it later multiplies in the patient's body, making her sick again. With all the other bacteria dead, there is little competition for space and food so the mutant strain is able to flourish.
6. She feels unwell again and goes back to the doctor and gets the same antibiotic.
7. This time, the antibiotics do not make any difference: she is still sick and asks her doctor what is wrong.
8. The doctor prescribes a different antibiotic that (hopefully) works. But if the population of bacteria continues to contain mutations or develop new mutations, new strains could display resistance to all the antibiotics available.

Notice how, unlike a soybean plant that has been intentionally artificially bred to have beneficial characteristics such as high protein yields, the production of antibiotic resistant bacteria has happened by natural selection because of decisions humans have made: the intention had not been to generate superbugs.

The development of antibiotic-resistant bacteria has happened more than once.

New strains of syphilis, for example, have adapted to antibiotics and show multiple resistance. Some strains of tuberculosis are resistant to as many as nine different antibiotics. There may be no cure for people who get sick from such super-resistant germs: they may have to rely on their own immune system to recover.

Finding new antibiotics is only a temporary solution, and pharmaceutical companies cannot find new medications fast enough to treat these super-resistant germs. As a result, the best way to stop their expansion is to make sure that doctors minimize the use of antibiotics and that patients realize that antibiotics are not always the best solution to a health problem.

Nature of Science

A *Staphylococcus* bacterium discovered in a hospital is suspected of being resistant to a certain number of antibiotics. To test this hypothesis, the bacterium is introduced into a Petri dish along with small discs of paper that are soaked in different types of antibiotics. In an experiment like this, when the colonies of bacteria grow close to the discs, they show resistance to the antibiotic, whereas when wide, clear circles of inhibited bacterial growth are present, they show that the antibiotic is stopping the bacteria the way it should. Can you interpret the results of the experiment shown in the photo?

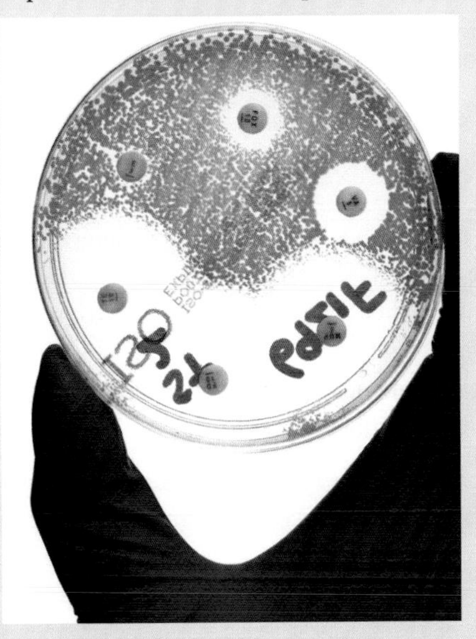

Doctors use such tests to help decide which medications to prescribe. In this case, they should prescribe the antibiotics that the bacteria do not show a resistance to, preferably the three at the bottom of the image. This resistant bacterium is part of a growing number of super bugs, among which we find MRSA, which stands for methicillin-resistant *Staphylococcus aureus*. Resistant bacteria have evolved because of the way humans use antibiotics.

In some countries, there is an intense debate about whether the concept of evolution should be taught in schools. To support the critics of evolution, there are thousands of websites and publications that carefully try to dismantle and disprove the arguments of evolutionary biologists. What criteria are used to determine whether these criticisms are valid or not? What kind of evidence would be necessary to refute Darwin's theory?

An experiment simulating natural selection can be found on this page of your eBook.

HL end

Guiding Question revisited

What processes can cause changes in allele frequencies within a population?

In this chapter you have learned that:

- which versions of genes (alleles) are present in a population and the proportions in which they are found can change over time
- if the environment changes or there is another selective pressure on a population, the frequencies of alleles can be modified by natural selection
- abiotic changes, such as in temperature, humidity or pH, can contribute, as can biological factors, such as the presence of predators.
- organisms within a species compete with each other for resources, which is intraspecific competition
- the individual that is best suited to the environment is more likely to survive to reproduce and pass on its genes.

Guiding Question revisited

What is the role of reproduction in the process of natural selection?

In this chapter you have learned that:

- if something such as ornate colours in male birds or fish makes them more attractive to females, they have a higher chance of reproducing and passing on their genes
- which males mate with which females will determine which genes are passed on and therefore which alleles are present in a population
- organisms with higher fitness tend to be able to pass on their genes more frequently than those with lower fitness
- HL as allele frequencies change, evolution happens, which can be quantified using the Hardy–Weinberg equation.

Exercises

Q1. Mutation is responsible for increasing variation in a population. List two other processes that also increase variation.

Q2. Ground-nesting birds such as grouse lay their eggs in a nest made on the ground. The eggs of this species are generally speckled dark brown. If a mutation occurred causing the eggs to be brightly coloured, how would the change in colour affect their chances of survival?

Q3. Explain how a population of insects could develop resistance to the insecticides sprayed on them.

Q4. In his study about guppies, Endler said this about the males, "The color patterns in a particular place represent a balance between selection for crypsis by predators and selection for conspicuousness by sexual selection." Using the results in the Skills box in Section D4.1.8, explain what he meant by this.

HL

Q5. In a population of 278 mice, 250 are black and 28 are brown. The alleles are **B** = black and **b** = brown.

Fill in the table.

Allele frequencies	Recessive **b**	q	
	Dominant **B**	p	
Genotype frequencies	Homozygous recessive **bb**	q^2	
	Heterozygous **Bb**	$2pq$	
	Homozygous dominant **BB**	p^2	

Q6. When a person's blood type is O^+, the positive sign refers to a genetic characteristic called the Rhesus factor, Rh. **D** denotes the allele for Rh^+ and **d** denotes Rh^-. Anyone with the genotype **dd** has a Rhesus negative (Rh^-) phenotype and anyone with the genotypes **DD** or **Dd** has the Rhesus positive (Rh^+) phenotype. Below are some results from two studies about the allele frequency of this gene in two different parts of the world.

- In Lagos, Nigeria, a study of 23,832 people revealed that 3% of the population was Rh^-.
- In Abha, Saudi Arabia, a study of 944 males revealed that 7.2% of the population was Rh^-.

(a) Using this information and applying the Hardy–Weinberg equation, calculate the frequencies of the alleles **D** and **d** in each of the two countries.

(b) In the south-west of France, a study of 127 French Basques found an allele frequency for **d** to be 0.51. How does this compare with the frequency of **d** in the two other populations?

Q7. Distinguish between artificial selection and natural selection.

HL end

D4.2 Stability and change

Guiding Questions

What features of ecosystems allow stability over unlimited time periods?

What changes caused by humans threaten the stability of ecosystems?

Ecosystems on Earth are remarkably stable when left free from the influence of humans. Evolutionary processes have taken place over eons of time and the result is sustainable and stable ecosystems. Stable ecosystems cycle nutrients efficiently, contain genetically diverse organisms, and have the means to collect sufficient energy through photosynthesis. A well-established ecosystem can also help keep abiotic factors within a tolerable range, for example transpiration from forests can influence rainfall in their area. Some ecosystems have existed for millions of years. The Daintree rainforest on the northeast coast of Queensland, Australia, is considered to be the oldest surviving ecosystem, at 180 million years old. The Amazon rainforest by comparison is approximately 55 million years old.

Unfortunately, human intervention in ecosystems can be devastating. Fuelled by the exponential growth of the human population, disruption of ecosystem processes is often getting worse rather than better. Each day many thousands of hectares of Amazon rainforest are burned in order to plant crops. The result is short-term profit but a devasting loss of biodiversity. Our reliance on fossil fuels continues to add carbon dioxide into the atmosphere. The resulting climate change as a result of global warming is measurable, and the consequences include severe weather, forest fires and crop failures. Plastic pollution is having a negative impact on both terrestrial and marine environments. Science is sounding the alarm but humans need to learn to listen and, most importantly, we need to change our wasteful habits and actions that are rapidly damaging our environment.

D4.2.1 – Sustainability of natural ecosystems

D4.2.1 – **Sustainability as a property of natural ecosystems**
Illustrate ecosystem sustainability with evidence of forest, desert or other ecosystems that have shown continuity over long periods. There is evidence for some ecosystems persisting for millions of years.

Evolution does not lead to unchanging ecosystems. There is no location on Earth where the living organisms have remained static for millions of years or even any time period that is close to that. However, there are many places on Earth where ecosystems have been healthy for very long periods of time. This does not mean that these ecosystems have not changed; it means that there have been slow adaptive changes and living organisms have thrived. A sustainable ecosystem is one that supports itself without any outside influences. Everything that the organisms within the ecosystem need is provided.

The Amazon rainforest

The Amazon rainforest in South America is the largest rainforest in the world. There is evidence that most of the area that is now rainforest was a marine lake as recently as 14 million years ago. Since then, the area has gained sediments and elevation and transitioned into tropical rainforest. What remains of the original lake is now the Amazon River.

At some point the rainforest largely became its own source of moisture, as transpiration from such a vast area of dense plant life became sufficient to provide water for the rainforest. The process of forming this amazing natural wonder has been happening for millions of years. Evolution has led to the amazing diversity of life forms that exists there today. Estimates of its biodiversity include 2.5 million insect species, 40,000 plant species and 1,300 bird species. These species are not static, they have evolved and will continue to evolve over time. This illustrates the importance of protecting sustainable ecosystems. Individual species are the genetic source for life in the future, and loss of biodiversity, and especially the loss of an ecosystem, reduces the possibilities of what that life can become.

▲ A jaguar (*Panthera onca*), an apex predator found in the Amazon rainforest. This one was photographed in a rainforest area of Peru.

Apex predators are also called top predators because they are at the top of a food chain. This trophic level does not have a high population as each level of a food chain has a declining biomass and energy availability.

D4.2.2 – Requirements for sustainability

D4.2.2 – Requirements for stability in ecosystems
Include supply of energy, recycling of nutrients, genetic diversity and climatic variables remaining within tolerance levels.

In order for an ecosystem to be sustainable in the long-term, specific requirements need to be met.

- A sufficient supply of energy
  The energy supply for ecosystems originates with light, which is harnessed by photosynthesis. Thus plants or algae must be abundant and productive. Energy can then pass through food chains and food webs in the form of organic molecules, until it reaches all trophic levels.

- Nutrient recycling
  Every ecosystem has a finite supply of nutrients. Since existing life forms are temporary, the nutrients they contain must be recycled. Nature has provided for this through specific nutrient cycles, such as the carbon, phosphorus and nitrogen cycles. If an organism is removed from an ecosystem, the nutrient stores within that ecosystem are reduced. For example, if a tree is logged in the Amazon and then removed from the forest, the carbon, phosphorus and nitrogen within the tree are no longer available to the forest ecosystem.

- Genetic diversity
  Healthy populations of a species need genetic diversity. Genetic diversity within a species provides protection against cataclysmic events decimating an entire population. A varied gene pool provides alleles that may allow a species to survive disease, harsh climatic events or a sudden increase in predators.

Healthy ecosystems undergo changes over time; however, they are self-supporting.

Nutrient recycling is highly reliant on species of bacteria within soil.

- Response to climatic change
  Ecosystems have responded to changes in climate for millions of years. A high genetic diversity within a species can help the survival of that species as long as the variables that contribute to the climate remain within the species' **tolerance levels**. Currently, human activities are causing global climate changes that are pushing climatic variables outside many tolerance levels, e.g. severe droughts and temperature increases.

D4.2.3 – Tipping points

D4.2.3 – Deforestation of Amazon rainforest as an example of a possible tipping point in ecosystem sustainability
Include the need for a large area of rainforest for the generation of atmospheric water vapour by transpiration, with consequent cooling, air flows and rainfall. Include uncertainty over the minimum area of rainforest that is sufficient to maintain these processes. **Application of skills:** Students should be able to calculate percentage change. In this case the extent of deforestation can be assessed by calculating the percentage change from the original area of forest.

Severe deforestation of the Amazon rainforest has been occurring for the last 60 years. Some of that has been sanctioned by countries and some has been illegal. The motivation to cut and clear the rainforest is the short-term realization of money from logging, agriculture and ranching. At least 17% of the original Amazon rainforest has already been cleared.

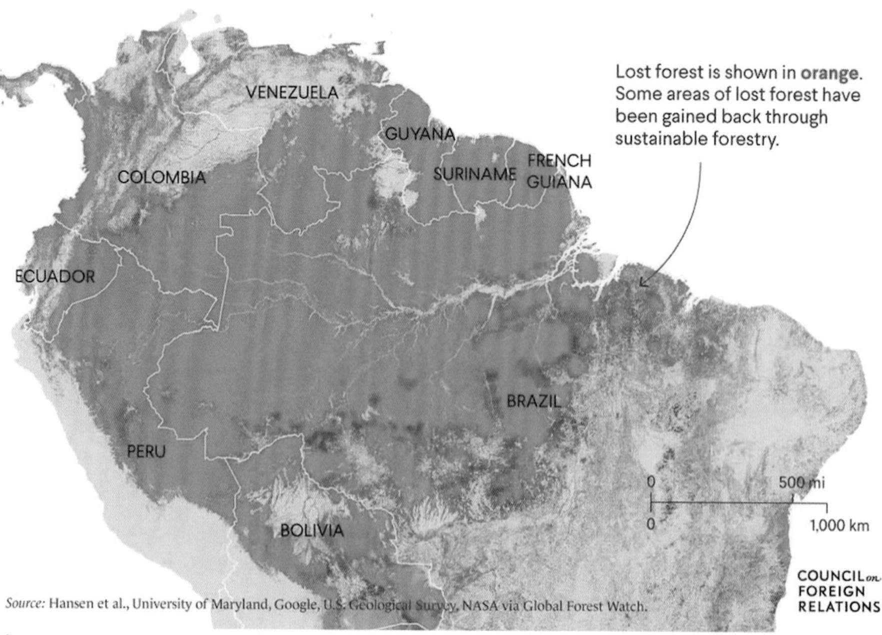

A graphic showing the areas of the Amazon rainforest that have suffered deforestation since 2001. The spread of deforested areas occurs as new edges are exposed to create new areas to cut or burn.

An incredibly large ecosystem like the Amazon rainforest influences its own weather and climate. For example, the rainforest vegetation produces huge amounts of water vapour by transpiration. This evaporation of water produces a cooling effect in the same way that evaporation of perspiration cools your body. This natural cooling effect influences the air flow and rainfall of the entire region. Researchers have already correlated increased temperatures and lower volumes of rainfall with deforestation. The effect is more pronounced near the edges of deforested areas.

No one yet knows how much of the Amazon rainforest can be destroyed before a tipping point is reached where the ecosystem will no longer be able to sustain itself. Past that point the ecosystem may also not be able to recover.

What is not known is how much rainforest can be cleared before it reaches a tipping point, after which it will not be able to sustain itself as an intact ecosystem. No one knows how much of the rainforest must remain in order to maintain its self-propagating climate.

Reaching the tipping point may also remove the ability of the forest to re-establish itself. The Amazon rainforest appears to be capable of restoring itself at the edges of destroyed areas, but is not capable of regrowing from anywhere near the centre of deforestation. It must use the remaining edge vegetation to slowly grow back over areas that have been cut or burned. At some point that healing process may no longer be possible, especially if human activities and settlements are fighting for land in the opposing direction.

Fires are now quite common in the Amazon rainforest. Many fires are purposely set to clear the forest for farming and cattle ranching. Brazil is the largest beef exporter in the world, and much of the cleared land there is used to grow soy, a large percentage of which is used to feed the cattle.

Challenge yourself

In 1970, the estimated remaining forest cover of the Brazilian Amazon rainforest was 4,100,000 km^2.

1. Using the data in the table below, determine the extent of deforestation for each of the given decades by calculating the percentage change compared to the 1970 area of rainforest cover.

Decade	Remaining forest cover (km^2)
1980	3,845,000
1990	3,692,000
2000	3,524,000
2010	3,359,000
2020	3,290,000

A Winogradsky column can be used to demonstrate how bacterial ecosystems can become sustainable based on environmental resources. Think of the many variables that could be altered in order to make this laboratory exercise experimental. For example, you could vary the items you mix into the mud, vary the lighting conditions, vary environmental temperature; there are many other factors that could act as the independent variable in your investigation.

D4.2.4 – Mesocosms

D4.2.4 – Use of a model to investigate the effect of variables on ecosystem stability
Mesocosms can be set up in open tanks but sealed glass vessels are preferable because entry and exit of matter can be prevented but energy transfer is still possible. Aquatic or microbial ecosystems are likely to be more successful than terrestrial ones.
NOS: Care and maintenance of the mesocosms should follow IB experimental guidelines.

A **mesocosm** is a self-contained system that provides a living environment for organisms. If taken to an extreme, air, water and food would all be self-generated by the organisms living in the mesocosm. There are several types of mesocosms; they can be set up in open tanks, but loosely sealed glass or plastic vessels are preferable. The aim is to produce something that allows energy transfer but prevents the entry and exit of matter. Aquatic or microbial ecosystems will probably be more successful than terrestrial ones. Directions for making a microbial mesocosm known as a **Winogradsky column** (see Figure 1) are given in the eBook. This type of mesocosm creates an environment that encourages the growth of different types of bacteria in various layers. Over time (approximately 2 months), different environments will form as a result of the combination of resource availability and the waste products released by different species of bacteria.

You will find a laboratory for creating your own mesocosm in the eBook.

For example, sulfur compounds produced by sulfate-reducing bacteria will accumulate in the lower portions of the column. Oxygen will be used up quickly in the lower portions of the container, as air will not be able to penetrate to replace the used oxygen.

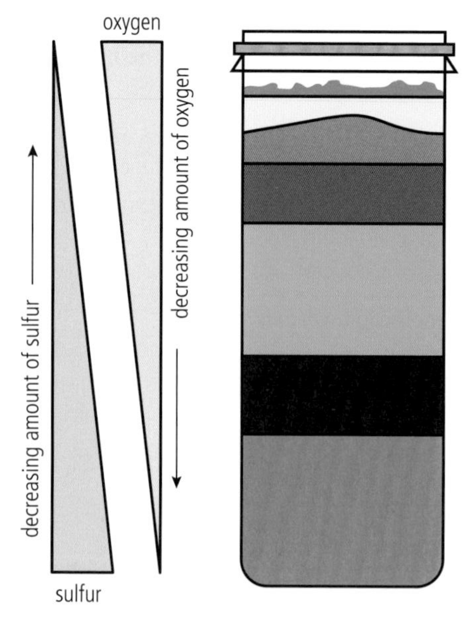

D4.2 Figure 1 A schematic showing the environmental gradients that you can expect over time within a Winogradsky column. The bacterial species more suited to the bottom of the column will be those that use sulfur to produce hydrogen sulfide. These are **anaerobic bacteria**, as opposed to the **aerobic bacteria** that will grow closer to the top. Some algae will probably grow in the water at the top of the column. The colours shown in this artwork are not meant to be the exact colours that you will see, as every column is a little different. Be patient as you wait for the coloured layers to emerge, it may take 6–8 weeks.

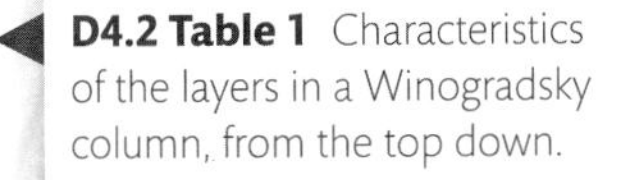

Depending on the bacteria in your mud sample, and the temperature and lighting conditions, your results after a few weeks of growth may resemble those shown in Table 1. All sulfur-oxidizing layers will produce a smell like rotten eggs.

Bacteria type	Visual indication
Aerobic and photosynthetic	Green or red-brown layer
Non-photosynthetic sulfur oxidizers	White layer
Non-sulfur oxidizers	Red, purple, orange or brown layer
Purple sulfur oxidizers	Purple or purple-red layer
Green sulfur oxidizers	Green layer
Sulfate reducers	Black layer
Methanogens	Various dark colours, and small bubbles of methane

D4.2 Table 1 Characteristics of the layers in a Winogradsky column, from the top down.

Nature of Science

The IB experimental guidelines do not permit the use of animals in an environment that is possibly harmful to their health. When you set up a Winogradsky mesocosm, you must take care to remove any animals from the mud first and return them to their habitat. The same applies if you decide to try a terrarium type of mesocosm.

D4.2.5 – Keystone species

D4.2.5 – Role of keystone species in the stability of ecosystems
Students should appreciate the disproportionate impact on community structure of keystone species and the risk of ecosystem collapse if they are removed.

Keystone species are organisms of any type that play an important role in the biodiversity of their ecosystem. The effect they have is not caused by their numbers, but by their impact on the prevalence and population levels of other species within their community. One way to determine whether an organism is a keystone species is to perform a removal experiment (see Figure 2). Ecologist Robert Paine was the first to apply this method. He was studying an intertidal area of western North America. When Paine removed the sea star *Pisaster ochraceous* manually from the intertidal area, a mussel, *Mytilus californianus*, was able to take over the rocky area and exclude algae and other invertebrates from that zone. The mussel simply used all the space available when there was no sea star to keep it in check. It was evident that it was the sea star that limited the number of mussels that could reproduce and attach to the rocks. Paine collected data that showed that, when sea stars were present, 15–20 different species of invertebrates and algae were present. Without the sea star, the diversity rapidly declined to less than five species. This supported the hypothesis that the sea star was a keystone species. When it was present, it controlled the diversity of the community. When it was absent, diversity was lost.

A keystone species has a major impact on other species of an ecosystem regardless of their own population size.

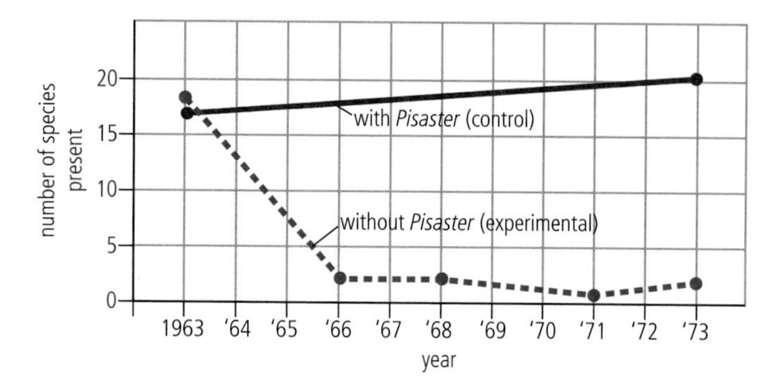

D4.2 Figure 2 Testing a keystone species hypothesis: the effect of removing the sea star *Pisaster ochraceous* from an intertidal area over a 10-year period. Notice that the **species richness** was greatly enhanced when the sea star was not excluded from the tidal area.

Loss of a keystone species can lead to collapse of an ecosystem. Wolves (*Canis lupus*) were largely eradicated from many areas of the western United States by bounty hunters. One area that suffered ecological effects as a result of this was Yellowstone National Park in north-western Wyoming. The park's last natural wolf was killed in 1926. The effect of losing the wolf as a keystone predator species resulted in what is called a **trophic cascade**. The wolves were predators of herbivorous species such as elk. When the wolves were eradicated, the herbivore populations increased. The increase in herbivores greatly decreased the availability of grasses and other vegetation, including young willow and aspen trees. Without the young trees, songbird populations began to decline and beavers had no wood to build dams. Riverbanks started to erode without the roots of trees to stabilize them.

The grey wolf (*Canis lupus*).

In mid-1990s, the US National Park Service decided to reintroduce 31 wolves to Yellowstone. Recently, the wolf count in Yellowstone was over 400. The herbivore population has been reduced, trees are regrowing and the entire ecosystem appears to be on its way to becoming sustainable once again.

D4.2.6 – Sustainable harvesting of natural resources

D4.2.6 – Assessing sustainability of resource harvesting from natural ecosystems
Sustainability depends on the rate of harvesting being lower than the rate of replacement. Include one terrestrial plant species and one species of marine fish as examples of renewable resources and how sustainability of harvesting can be assessed.

Natural resources include non-living and living materials that can be used by humans. This includes everything from trees to marine organisms. The problem is that we harvest many natural resources in such a way that the resource virtually disappears. Sustainability depends on the rate of harvesting being lower than the rate of replacement. Too many times humans have learned this the hard way. Two examples are given below of natural resources that have suffered high rates of harvesting but are now being closely monitored and harvested sustainably to provide a renewable resource.

Chilean sea bass

In the 1970s a seafood merchant discovered a delicious fish that was being sold in a market in Chile. The fish was called the Patagonian toothfish (*Dissostichus eleginoides*). Knowing that this name was unappealing to consumers, he renamed it the Chilean sea bass. In only a few years the fish gained a huge market in many areas of the world.

In longline fishing, all of these hooks are baited and put into the water connected to a single fishing line. The hooks and bait are left in the water for a period of time, and then brought up one at a time to remove the caught fish.

By the 1990s the wild population of Chilean sea bass was collapsing. Consumer demand for the fish resulted in overfishing and much of it was not regulated. In addition, the fishing technique used was **longlining**. This technique sets hundreds of hooks and bait on a single line. Marine birds, like the albatross, often try to eat the bait and die after becoming ensnared by the hooks. The bad publicity surrounding the catch of these fish became so negative that many chefs refused to serve it and some seafood markets refused to sell it. However, illegal catches of the Chilean sea bass continued to be a serious problem until the year 2000.

Seafood packaging with the Marine Stewardship Council label indicating that sustainable fishing practices were used to obtain the seafood. There are other non-profit and governmental organizations that also provide certifications.

Since then, the fishing industry that specializes in catching Chilean sea bass has agreed to make changes to its fishing techniques and to be monitored by an organization that certifies sustainable fishing practices that have minimal effects on the bycatch (species that are not wanted but are caught at the same time as the target species). As part of the certification process, a number of regulations detailing how many fish can be caught, the permitted age distribution of the fish in the catch, and when fishing can take place, have been applied to make the Chilean sea bass catch sustainable. Strict limits on damage to seabirds have also been applied, and changing the fishing season has drastically lowered the number of seabirds affected. The fish population is monitored to make sure sufficient repopulation is taking place. An observer from the certification organization is required to be on board the fishing vessels to document the fishing practices and make sure they are sustainable. The fleet of uncertified, illegal fishing vessels has been reduced to zero.

Sustainable harvesting is promoted by informed consumer demand. Many people will avoid purchasing a product if they are convinced the product is not being harvested or grown using sustainable practices.

Overfishing is a global problem and cannot be tackled by just a few countries. There are countries that enforce fishing restrictions in their own territorial waters, but it is a difficult to apply them outside those boundaries. There are a variety of international non-profit organizations that attempt to regulate and improve fishing practices. One of the most effective ways appears to be educating the public about purchasing seafood from responsible sources that use sustainable fishing practices. The Marine Stewardship Council (MSC) is an example of an international non-profit organization that oversees fishing practices and provides certification to more than 25,000 seafood products that meet their standards. Responsible buyers of marine products look for their label or other certifications of sustainable fishing practices.

Black cherry trees

The black cherry (*Prunus serotina*) is a hardwood tree species that grows in many areas of North America including parts of New York, Pennsylvania, Ohio and West Virginia, collectively known as the Allegheny. Black cherry is usually found in hardwood forests mixed with other hardwood species. The light red of its wood makes it highly sought after for furniture and cabinet making.

Mixed hardwood forests are ecologically diverse communities. They are frequently so dense that availability of light for new growth is only provided when a mature tree falls or is removed as a result of age, disease, windstorm or selective logging.

Large areas of hardwood forest in the Allegheny region were once clear-cut to make farming fields. This type of logging completely removes almost all the trees in an area. The hardwood forests currently found in the Allegheny region are mostly second growth forests that are regrowing in some of the once clear-cut areas. Today, the logging of black cherry trees is usually done by selection of large specimens from the mixed forest. Current logging practices do not remove the surrounding trees. In most cases, an **arborist** will select the trees that can be removed sustainably. Removing a single tree leaves light gaps in the forest; black cherry and other seedlings can grow and establish in the light gaps left by logged specimens and naturally fallen trees. Black cherry trees produce a fruit that is consumed by a variety of bird species; germination and seed dispersal are enhanced by passage of the fruit through a bird's digestive tract and subsequent deposition in the bird faeces.

Hardwoods of all species are slow growing trees. It is imperative that selective logging occurs at a pace that does not exceed the growth of the remaining trees. The Forest Stewardship Council (FSC) was formed in 1993 by environmental, business and community leaders to regulate and certify sustainable logging practices. This organization provides regulations that promote sustainability and is relied on to certify wood supplies as being harvested by renewable logging practices.

The logging of black cherry trees is only sustainable if:

- selective logging is used, not clear-cut logging
- trees are selected rather than cut down randomly
- enough trees are left to produce fruit and seeds for the next generation
- data is collected to compare the quantity of wood being removed with current growth
- logging, processing companies and public buyers all appreciate and abide by the certification awarded by the Forest Stewardship Council or other organizations dedicated to sustainable harvesting.

D4.2.7 – Sustainability of agriculture

D4.2.7 – Factors affecting the sustainability of agriculture
Include the need to consider soil erosion, leaching of nutrients, supply of fertilizers and other inputs, pollution due to agrochemicals, and carbon footprint.

The goal of sustainable agriculture is to meet the food and textile needs of the world today without endangering the ability of future generations to do the same. In many areas of the world, agriculture has evolved from small-scale operations to huge corporate enterprises. However, whether a farm is family operated or a large business, agriculture must be sustainable.

To make agriculture sustainable, the most pressing issues that need to be addressed are soil erosion, leaching, fertilizer use, pollution and the carbon footprint.

Soil erosion

Loss of the upper soil layer of a field greatly reduces its productivity. The upper layer is called **topsoil** and should be rich in organic nutrients. Erosion of topsoil can be caused by excess rainwater and sometimes wind. Soil erosion is most often a problem when a field is left bare without a crop, because there will be no plant roots to hold on to the soil. Farmers sometimes plant rye or clover crops simply to cover the soil when the weather is not suitable for other crops. These cover crops reduce the penetrating force of heavy rainfall and their roots help hold the soil in place. Later, the cover crop can be ploughed back into the soil to increase the organic matter in the topsoil.

At the start of the 1930s, a combination of drought conditions and poor farming practices led to the loss of tons of topsoil from land in the central United States. Huge windstorms blew away soil that had previously been held in place by the natural grasses of the prairie. The 1930s became known as the "dust bowl" years.

Leaching of nutrients

The nutrients needed by plants must be water soluble for the plants to make use of the dissolved nutrients. **Leaching** occurs when rain or irrigation water dissolves nutrients (usually nitrogen and phosphorus compounds) in the soil and then carries them away from the root zone of a crop. These dissolved chemicals often end up in the water supply of the area. Leaching cannot be prevented but it can be minimized by applying appropriate amounts of fertilizer and irrigation water at optimum times, taking the seasons and crop requirements into account.

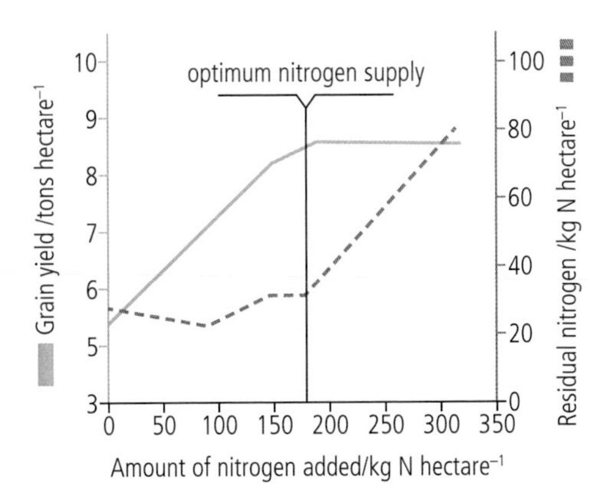

This graphic shows that the crop yield of grain is improved by adding nitrogen fertilizers, but only up to a certain point. The application of any excess nitrogen is not only expensive but is likely to be leached from the soil because it has not been taken up by the crop.

Many foods are shipped around the globe in refrigerated containers on huge cargo ships. As many as 740 million bananas can be shipped in 15,000 containers on some of the largest cargo ships.

Fertilizer supply

Many farms use chemical fertilizers to enrich their soils with nitrogen, potassium and phosphorus compounds. Chemical fertilizers are often a farm's largest single expense. The sources of fertilizers are limited and demand often exceeds supply. The manufacture of chemical fertilizers is also energy intensive. This drives up the cost of the fertilizers and that cost is then passed on to consumers. Some farmers minimize the use of chemical fertilizers by planting crops of beans or clover (legumes) that have nitrogen-fixing bacteria within their roots.

Pollution from agrochemicals

Water runoff and the leaching of both fertilizers and chemical pesticides results in pollution of water bodies. This is especially true when crops are overfertilized and excessive pesticide chemicals are applied.

Carbon footprint

The carbon footprint is the total amount of greenhouse gases (including carbon dioxide and methane) that is generated by an activity. Currently the estimated percentage of greenhouse gases attributable to agriculture is about 12% of the total, although that number is thought by many to be too low. The carbon footprint of agriculture includes:

- the use of petroleum products to run farm machinery
- the addition of fertilizers, which often are made from petroleum products
- clearing natural forest land and other ecosystems to create farmland
- the transportation of crops grown in one area of the world to another.

It is estimated that at any given moment there are over 50,000 cargo ships of various sizes carrying goods across the world's oceans.

D4.2.8 – Eutrophication

D4.2.8 – **Eutrophication of aquatic and marine ecosystems due to leaching**
Students should understand the effects of eutrophication resulting from leaching of nitrogen and phosphate fertilizers, including increased biochemical oxygen demand (BOD).

Fertilizers containing phosphorus (phosphates) and nitrogen (nitrates) must be water soluble in order to be absorbed by plants. However, plants do not always absorb all the nutrients added to the soil. Irrigation and rainwater can leach these compounds through the soil and eventually into nearby streams, rivers and lakes. Sometimes excess minerals are washed into a saltwater (marine) environment. In both freshwater and saltwater, an influx of fertilizers can stimulate algae to grow excessively. This is the start of the process of **eutrophication.** The greatest growth occurs at or near the top of the water, because that is where there is the most sunlight. It does not take long for the algae to form a continuous thick layer across the water's surface. At that point very little light is able to penetrate through the water, and algae growing lower down in the water column begin to die. Aerobic bacteria begin decomposing the excess algae growth.

Eutrophication of a body of water is often measured indirectly by digital meters that measure dissolved oxygen in the water.

There is always some dead organic matter within all bodies of water. Decomposition by aerobic bacteria is normally a process that helps keeps an aquatic environment healthy. This decomposition uses oxygen dissolved in the water. The oxygen needed by bacteria in a body of water is called the **biochemical oxygen demand** or **BOD**. The need for extra decomposition caused by excess algal growth increases the BOD, and the amount of dissolved oxygen may become depleted. During the process of eutrophication, all the aerobic organisms in the water may die because the oxygen content of the water is no longer sufficient to support them. The only way to stop the eutrophication process is to stop nitrate and phosphate compounds from entering the system.

The thick algae growth on this pond is evidence of eutrophication.

Eutrophication is the entire process that begins with water being over-enriched with nutrients, leading to the overgrowth of algae. It also includes the phases of algae die-off and bacterial decomposition, which deplete oxygen in the body of water leading to the death of aerobic organisms.

D4.2.9 – Biomagnification

D4.2.9 – Biomagnification of pollutants in natural ecosystems
Students should understand how increased levels of toxins accumulate in the tissues of consumers in higher trophic levels. Include DDT and mercury as examples.

Biomagnification is the phenomenon that occurs when harmful substances in the environment build up in the organisms towards the top of a food chain. Each trophic level typically consumes many organisms from the previous trophic level. If the organisms that are consumed contain a toxin that does not break down, then the substance becomes more concentrated in the living tissues of the organisms that eat them. Two examples of biomagnification are discussed below.

Mercury

Mercury is a naturally occurring element found throughout the world. Food chains displaying mercury accumulation are rare and almost always are located near an industrial source that emits mercury compounds. However, it now appears that mercury levels are increasing in organisms towards the top of aquatic food chains, and the source of the mercury seems to be the burning of coal and the production of cement. Mercury compounds released by these processes enter the atmosphere and are then washed into the oceans.

Microorganisms at the bottom of an aquatic food chain convert elemental mercury and inorganic mercury compounds into **methyl mercury**. As methyl mercury moves up through an ocean food chain, it becomes more and more concentrated in the tissues of animals.

In 1956, an epidemic of a disease of the central nervous system was reported in an area of Japan called Minamata Bay. A local company there produced acetaldehyde using a mercury compound as a catalyst. The mercury and many other heavy metal compounds were being dumped as wastewater into Minamata Bay. Bacteria in the water transformed the inorganic mercury into an organic form, methyl mercury. Methyl mercury was incorporated into the local food chains, and became concentrated by biomagnification in the shellfish and fish. Local people depended on shellfish and fish as a major component in their diet. It may never be known exactly how many people died or became severely disabled because of the mercury levels in this body of water, but it was in the many thousands.

Humans are exposed to high levels of mercury by eating fish that are long-lived and at or near the top of a food chain. Mercury accumulation in the body can have many severe health effects, including neurological damage. Figure 3 shows the relationship between size of fish and mercury content in their tissues.

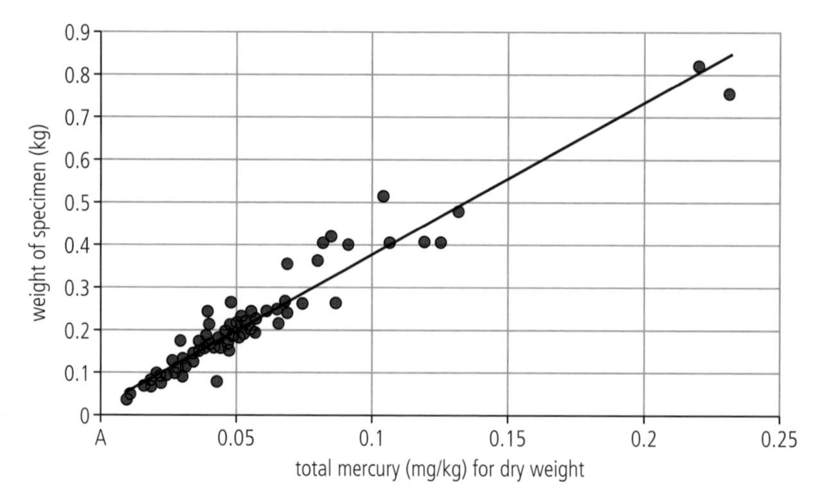

D4.2 Figure 3 In the Madre de Dios river basin of Southeastern Peru small-scale gold mining uses mercury as part of its mining operations. A study was done to measure the mercury content of various freshwater fish in the immediate area. As seen in this graphic, larger fish correlated with higher mercury content within their tissues. Larger and older fish have more time to accumulate higher levels of mercury from their diet.

DDT

DDT (dichloro-diphenyl-trichloroethane) is a synthetic insecticide developed in the 1940s. It was widely used for many years because it was effective, inexpensive to produce and long lasting (persistent). It was usually used against mosquitos and other insects that acted as vectors for diseases such as malaria and typhus. However, DDT was indiscriminate in which insects it killed.

After more than a decade of use, problems began to emerge from the widespread application of DDT. Beneficial insects were being killed and mosquitos were becoming resistant to it. Because DDT was often mass-sprayed from aircraft, some entered bodies of water in water runoff. In marine environments, DDT was absorbed by phytoplankton. As in the mercury example , DDT became more concentrated at each level of the food chain, and collected in the fatty tissues of consumers at higher trophic levels.

DDT being sprayed for mosquito control on a beach in Long Island, New York, USA, in 1945.

The greatest documented effect was on birds such as brown pelicans, osprey and bald eagles. Each of these species feeds on fish as a primary food source. The DDT did not immediately kill the birds, but it did alter their metabolism of calcium. The result was production of thin-shelled eggs that could not withstand the mass of a parent bird during incubation of the eggs. Researchers began to notice that the age structure of the birds was changing, with very few young birds in the populations.

Regulatory actions regarding DDT use began in the 1950s and by 1972 the Environmental Protection Agency in the United States banned the use of DDT for almost all applications, and the populations of predatory birds affected by DDT largely recovered. The United Nations Environment Programme negotiated a treaty to ban the use of DDT world-wide, with the exemption of areas where malaria is known to be a greater health concern than the effects of DDT.

TOK

What criteria should be used when weighing the risk of using DDT against the benefit of controlling a deadly disease such as malaria?

D4.2.10 – Microplastic and macroplastic pollution

D4.2.10 – Effects of microplastic and macroplastic pollution of the oceans
Students should understand that plastics are persistent in the natural environment due to non-biodegradability. Include examples of the effects of plastic pollution on marine life. **NOS:** Scientists can influence the actions of citizens if they provide clear information about their research findings. Popular media coverage of the effects of plastic pollution on marine life changed public perception globally, which has driven measures to address this problem.

Plastics are not **biodegradable**. A plastic object disposed of in an environment will remain there virtually forever. The object may be broken down mechanically, but that just creates smaller pieces of plastic. Plastic pollution is a global problem, especially the plastic objects of various sizes that end up in our oceans. Some plastic is washed up on shorelines, but some floats on the water surface and some sinks to the ocean floor. Plastic debris can be categorized into two main types based on size.

Macroplastics are any plastic debris that is larger than 5 mm. This includes water bottles, grocery bags, food containers and numerous other relatively large plastic objects. Many of these items are single-use items that are carelessly discarded.

Microplastics are plastic debris smaller than 5 mm. Some of this debris was originally a macroplastic object that has since been mechanically broken down into smaller pieces. Other microplastic pollution arises from the very small plastic granules used in face scrubs, cosmetics and hand cleansers. Many countries have now prohibited the use of microplastics in these products because they have been linked to cancer and shown to stimulate some harmful genetic conditions in humans, both as the original products and as pollutants.

D4.2 Figure 4 The five major ocean garbage gyres formed by ocean currents. Each is composed of both macroplastics and microplastics.

▲ The skeletal remains of a young Laysan albatross (*Phoebastria immutabilis*), interspersed with plastics. It is likely that the plastic items were fed to the young bird by its parents. It is also likely that the plastics led to the death of the animal, as many dead young birds have been found with plastics in their bodies.

Plastic items in the oceans tend to get caught up in large ocean vortices known as **gyres**. Each gyre is formed by prevailing winds and ocean currents, and creates a swirling mass of debris that is funnelled towards a centre. The largest plastic gyre is the Great Pacific Garbage Patch, which has two vortex centres. One is off the coast of California, USA, and the other is off the coast of Japan. Five major garbage plastic gyres are shown in Figure 4.

Plastic pollution in the oceans is not just ugly, it is killing wildlife.

- Sea turtles sometimes eat plastic bags thinking they are jellyfish.
- Plastic rings from six-packs of canned drinks entrap sea birds and other wildlife.
- Albatrosses pick up plastics from the ocean surface and feed it to their chicks.
- Plastic fishing nets are often lost by fishing boats and are a death trap for a variety of fish, sea turtles and marine mammals.
- Microplastics are filling the stomachs and intestines of marine organisms after accidental ingestion.

Nature of Science

Scientists can influence the actions of others if they provide clear information about their research findings. Media coverage of the effects of plastic pollution on marine life has changed public perception globally, which has led to measures being taken to address the problem.

D4.2.11 – Rewilding

D4.2.11 – Restoration of natural processes in ecosystems by rewilding
Methods should include reintroduction of apex predators and other keystone species, re-establishment of connectivity of habitats over large areas, and minimization of human influences including by ecological management. Include the example of Hinewai Reserve in New Zealand.

Rewilding activities are conservation efforts aimed at restoring and protecting natural processes and wilderness areas. Rewilding is a form of ecological restoration that leaves an area to nature, as opposed to active natural resource management. Successful long-term rewilding projects require little ongoing human attention, because successful reintroduction of keystone species can create a self-regulated and self-sustaining stable ecosystem.

Methods of rewilding include:

- the reintroduction of **apex predators** and other keystone species (see Section D4.2.5)
- establishing wildlife corridors, to connect habitats over larger areas
- stopping agriculture and resource harvesting such as logging and hunting
- minimizing human influences on an ecosystem, including using ecological management techniques.

Hinewai Reserve in New Zealand

Hinewai Reserve is a 30-year-old rewilded area on the South Island of New Zealand. The reserve started as 109 hectares of land that had previously been farmland. Subsequent land acquisitions have expanded the reserve to 1,250 hectares. The reserve is privately owned by a trust but is freely open to the public via numerous walking paths.

The goal of the reserve is to foster regeneration of native vegetation and wildlife. Minimal intervention was used to remove a few invasive tree and animal species. The strategy is to allow nature to take its own course. The goal is for native vegetation to repopulate the area, along with many native species of animals. The Hinewai Reserve continues to be a work in progress but is an example of rewilding success.

What is the distinction between artificial and natural processes?

One of many waterfalls within Hinewai Reserve, New Zealand.

HL

D4.2.12 – Ecological succession

D4.2.12 – Ecological succession and its causes
Succession can be triggered by changes in both an abiotic environment and in biotic factors.

Ecological succession is the change over time in the species that live in an area. Succession is the reason why some species gradually replace others in one particular area. It is triggered by changes in the abiotic (non-living) and biotic (living) factors in an ecosystem.

All succession events involve changes in an ecosystem. Sometimes one or more abiotic factors trigger the changes, for example when new land is formed from a volcanic lava flow or the receding of a glacier. Sometimes an existing ecosystem is drastically altered by an event such as a forest fire. Even biotic events such as a disease killing a dominant plant species in an area can trigger succession. Succession events can be classified as:

- **primary succession**, when new land is created and a series of communities emerge, any one community prepares the land for the next type of community
- **secondary succession**, when an existing ecosystem is drastically altered by, for example, fire, flood or human intervention, and the remains of the previous ecosystem are used as a starting point for further changes.

A comparison of primary and secondary succession

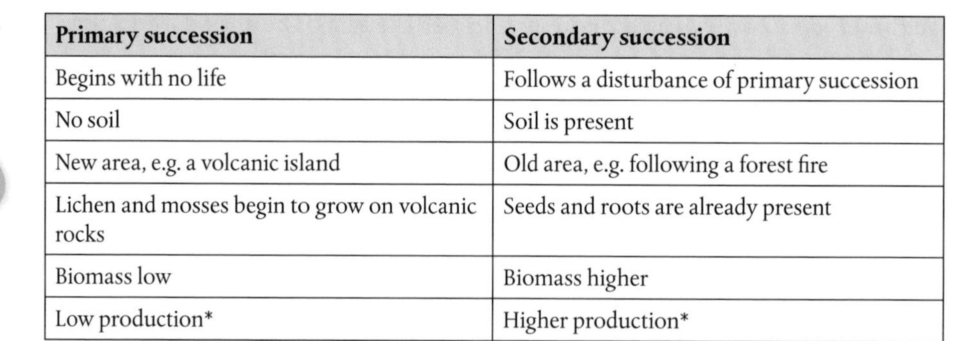

Primary succession	Secondary succession
Begins with no life	Follows a disturbance of primary succession
No soil	Soil is present
New area, e.g. a volcanic island	Old area, e.g. following a forest fire
Lichen and mosses begin to grow on volcanic rocks	Seeds and roots are already present
Biomass low	Biomass higher
Low production*	Higher production*

Primary succession takes much longer to occur than secondary succession. The most time-consuming stage of primary succession is early in the process, when soil is being created.

*Production is the increase in biomass or energy $m^{-2} yr^{-1}$. When production is low, it is because there are only a few plants; higher production occurs when many plants are present.

D4.2.13 – Primary succession

D4.2.13 – Changes occurring during primary succession
Use any suitable terrestrial example to illustrate these general principles: increases in size of plants, amount of primary production, species diversity, complexity of food webs and amount of nutrient cycling.

In some areas of the world, such as the Hawaiian islands, new land is being formed from volcanic lava. Initially, while the lava is cooling, no life is possible. Eventually, the first living organisms, known as **pioneer species**, appear. Often these are lichens and mosses, both of which are photosynthetic and do not grow root systems. Over a long period of time the growth and death of multiple generations of lichens and moss help build up a soil that can be used by other producers.

Moss and lichens growing on volcanic rocks. Over a long period of time, pieces of volcanic rock break off, and the addition of dead moss and lichens enrich the rocky substrate to form soil. This may take several hundred years.

During the early stages of succession, levels of biodiversity and primary productivity are very low. The emergence of a thin soil permits other plants to grow, typically small annual shrubs. These small shrubs do have a root system. The roots find little crevices to extend into and help break up the volcanic rock further, forming more soil. Bacteria will have begun metabolizing organic particles and will further enrich the soil and begin nutrient cycling. Small invertebrates will start to colonize the area and form the basis of more complex food chains. The deepening and enriched soil will allow grasses, larger shrubs and eventually some trees to grow. More animals will join the growing community. After many hundreds of years, a mature ecosystem known as a **climax community** may have developed.

The following characteristics are typical of primary succession progression over time:

- increasing species diversity
- increasing size of plants in the community
- increasing primary production
- more complex food webs
- increasing nutrient cycling.

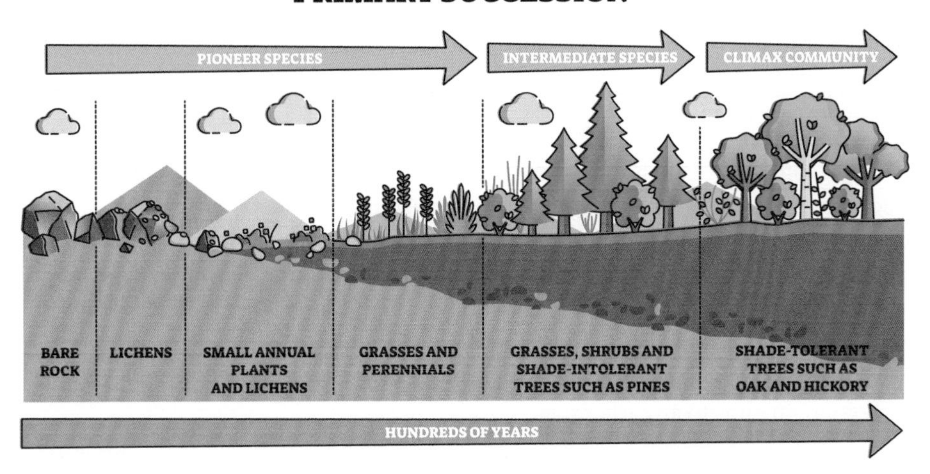

A timeline of primary succession measured in hundreds of years. In this example, the climax community is a shade-tolerant forest.

D4.2.14 – Cyclical succession

D4.2.14 – Cyclical succession in ecosystems
Students should appreciate that in some ecosystems there is a cycle of communities rather than a single unchanging climax community. Students should refer to an example.

Although succession in most ecosystems leads to a unique set of organisms at each subsequent stage, sometimes there is a cyclical pattern that emerges over a very long period of time. **Wood pasture** is an ecosystem that is characterized by open pasture interspersed with various species of large shade trees. This type of ecosystem has existed for a very long time and is thought to have been propagated by large grazing animals. Similar to the way that big herds of herbivores such as wildebeests, elephants and zebra influence vegetation patterns in the current African savannah, herds of similar-sized animals like bison and mammoths could have maintained the wood pastures of long ago.

It is thought that succession in wood pastures begins with the growth of thorned, unpalatable thickets of brush in what was open pasture. One or more tree begins to grow using the thicket as protection from herbivores. At a later stage, the shade of the tree(s) restricts the growth of the surrounding thicket. The tree(s) eventually dies of natural causes and the area is once again open pasture. The cycle of thicket growth can begin again. This cycle is dependent on the action of large herbivores maintaining most of the area as grassland pasture at any given time. Today, that function falls to sheep, cows and horses in the few areas of Europe, Asia and North America where wood pasture still exists.

D4.2.15 – Climax communities

D4.2.15 – Climax communities and arrested succession
Given any specific environmental conditions, ecological succession tends to lead to a particular type of climax community, but human influences can prevent this from developing. Use grazing by farm livestock and drainage of wetlands as examples.

Over what timescales do things change in different biological systems?

Ecological succession usually proceeds through a series of species replacements until a particular stable community is formed. This final community is called a **climax community**. The environmental conditions of an area determine the characteristics of the climax community.

A climax community may take many centuries to develop and will be characteristic of the biome typical of the area.

A climax community tends to be ecologically stable and will not usually change unless environmental conditions change or there is human interference. Most climax communities are a reflection of the biome they are found in, characterized by particular levels of temperature and rainfall (see Section B4.1.6). However, there are many ways in which humans have altered climax communities, including removing forests for livestock grazing and draining wetlands for development. Human activities in an area can **arrest** the natural development of an ecosystem and reset the succession timeline to an earlier stage.

HL end

Guiding Question revisited

What features of ecosystems allow stability over unlimited time periods?

Within this chapter you have learned that healthy, sustainable ecosystems must have:

- the capability to change gradually over time
- a constant supply of energy
- the ability to recycle limited nutrients
- genetic diversity
- a reasonably stable climate
- keystone species that give structure to the ecosystem
- freedom from human interference so that a tipping point is not reached, where the ecosystem becomes incapable of healing itself
- resource harvesting by humans that does not exceed the rate at which the resources are replaced.

Guiding Question revisited

What changes caused by humans threaten the sustainability of ecosystems?

Within this chapter you have learned that humans are threatening the sustainability of ecosystems through activities such as:

- the depletion of natural resources
- lowering the genetic diversity within ecosystems

- pollution by chemical wastes that can build up within food chains and result in the loss of apex predators
- pollution of aquatic ecosystems by macroplastics and microplastics
- the addition of extra nutrients to watercourses resulting in eutrophication and the subsequent depletion of dissolved oxygen (which, in turn, leads to loss of aquatic organisms).
- deforestation of areas for logging and agriculture
- the removal of keystone species
- rapid climate change caused by the release of greenhouse gases
- HL interference of natural succession stages for example by over-harvesting and pollution.

Exercises

Q1. A keystone species is one best described as:

A the most abundant species in an ecosystem

B a species that has the highest primary productivity

C a species that has a dramatic impact on its ecosystem no matter what its population size is

D a species that should be removed from an ecosystem.

Q2. What is BOD and how is it correlated with eutrophication of a body of water?

Q3. Why do some species of oceanic fish accumulate more mercury in their body compared to other species?

Q4. Explain why a farmer might plant a cover crop such as clover.

Q5. Which of these is *not* characteristic of ecological restoration by rewilding?

A Reintroduction of keystone species.

B Establishment of wildlife corridors between sections of preserves.

C Minimal active management.

D Minimal logging and hunting.

Q6. Ecologists say that the loss of Amazon rainforest could reach a “tipping point”. What does the term tipping point mean in this context?

HL

Q7. Name a pioneer organism typical of terrestrial primary succession.

Q8. Explain why primary succession can take a much longer period of time to reach a climax community compared to secondary succession.

HL end

D4.3 Climate change

Guiding Questions

What are the drivers of climate change?

What are the impacts of climate change on ecosystems?

It is crucial you understand that climate change is an enormous issue: our survival and existence as a species depend on what we decide now. Rarely in the human story has there been such a pivotal moment. Earth's climate is complex because it involves multiple systems interacting with each other. A change in one part of the system can have consequences across the globe. Energy transfers involving sunlight and heat determine the climate on Earth. Adding greenhouse gases to the atmosphere as a consequence of human activities such as burning fossil fuels modifies how much heat is retained. Melting snow and ice decreases the surface area of Earth that is reflecting sunlight. The ground and ocean surface that used to be covered with snow and ice is warming up more than usual. This is warming the atmosphere and modifying ocean currents. Changes in atmospheric temperature are leading to more drought in some areas, and increasing the number of forest fires. When climate changes, ecosystems are impacted. Changes to coral reef systems, breeding grounds, migration patterns and the availability of food can greatly disrupt food chains and ecosystems.

D4.3.1 – Human activity and climate change

D4.3.1 – **Anthropogenic causes of climate change**
Limit to anthropogenic increases in atmospheric concentrations of carbon dioxide and methane. **NOS:** Students should be able to distinguish between positive and negative correlation and should also distinguish between correlation and causation. For example, data from Antarctic ice cores shows a positive correlation between global temperatures and atmospheric carbon dioxide concentrations over hundreds of thousands of years. This correlation does not prove that carbon dioxide in the atmosphere increases global temperatures, although other evidence confirms the causal link.

The role of carbon dioxide and methane

We live at the bottom of an ocean of air we call the **atmosphere**. The atmosphere plays a vital role in regulating the temperature of Earth's surface. The **greenhouse effect** refers to the way that the atmosphere retains heat and keeps the planet warm even when no sunlight is reaching the surface. When sunlight hits an object, some of its energy is absorbed, and the object warms up. The energy is re-radiated in the form of **infrared radiation**, which has longer wavelengths than energy in the form of visible light. The glass in a greenhouse traps infrared radiation.

▲ A greenhouse protects plants from the cold in a similar way to the atmosphere stabilizing Earth's temperature.

Greenhouse gases (GHGs), such as carbon dioxide and methane in Earth's atmosphere, can be thought of as the glass of a greenhouse, although, like many models, this is not a very accurate representation of the natural phenomenon. GHGs have the ability to absorb and radiate infrared radiation. When such gases are present, they keep the atmosphere near Earth's surface warm by converting some of the short-wave radiation from the Sun into long-wave radiation, and radiating it in all directions, including back down towards the surface.

Levels of some of the main greenhouse gases have increased as a result of human activities. Carbon dioxide is produced when fossil fuels are burnt to generate electricity or power vehicles. Natural carbon sinks such as forests are being cleared, which means that carbon dioxide is not being absorbed at the same rate. Methane is produced by livestock such as cattle and in some types of agriculture such as rice cultivation. The increased levels of greenhouse gases are causing the atmosphere to retain more and more heat. Something caused by human activity is said to be **anthropogenic**. According to the International Panel on Climate Change (IPCC) the climate change we are seeing in recent decades is not a natural phenomenon but instead is anthropogenic in origin.

▲ A summary of the greenhouse effect: short-wave radiation (the yellow arrow on the left) hits the surface and much of it bounces off the atmosphere and surface of the planet, but some is converted into long-wave radiation (heat, shown as the red arrow on the right) as continents and oceans warm. Some of this infrared heat escapes into space, but some is radiated to the surface by greenhouse gases that can trap and emit the heat.

Global climate change

Climate refers to the patterns of temperature and precipitation, such as rainfall, that occur over long periods of time. Whereas weather can change from hour to hour, climate changes generally occur over thousands or millions of years. Climatologists and palaeoclimatologists collect data about atmospheric conditions in recent decades and the distant past, respectively. As thermometers have only been around for a few hundred years, temperatures on Earth from thousands or millions of years ago have to be inferred from **proxies** (see the Nature of Science box).

Nature of Science

A proxy is a measurement that is used in place of another one. Because it is impossible to go back in time and measure the temperature of the atmosphere 15,000 years ago, climatologists use proxies such as tree rings, coral reef growth and the presence of certain fossils to estimate what the climate was like many thousands of years ago.

Layers found in thick sheets of ice that have been formed by annual snowfall can also provide information on past climates. By drilling into the ice and taking cylinder-shaped samples, called **ice cores**, scientists can study the substances trapped in the layers, such as air bubbles from the year when the layer was deposited. Researchers at Vostok Station in Antarctica have collected layers of ice from more than 3,000 m down, yielding climate information going back more than 400,000 years.

But we must always be careful about making the distinction between correlation and causation. Just because we see a positive correlation between carbon dioxide (CO_2) levels and global temperatures we cannot assume that a rise in one *causes* a rise in the other. Correlation is not the same as causation. In the case of climate change, there is further evidence to link carbon dioxide levels to an increase in global temperatures, and the greenhouse effect also provides a possible mechanism.

A positive correlation means as one factor goes up, the other also goes up. In contrast, a negative correlation would mean that as one variable increases, the other decreases.

Do not confuse the greenhouse effect with the depletion of ozone. Although both are the result of human activity, and both influence the atmosphere, they are not interchangeable phenomena. They have different causes and different effects on the environment.

Do you want to see the data for yourself? One of the principles of science, especially research funded by taxpayers, is to make data available to the public. This is to allow verification, critique and sharing of data, so that scientists with many different approaches can combine their findings and advance our understanding of the topics being studied.

One organization that does this is the National Oceanic and Atmospheric Administration (NOAA) in the US. If you go to the NOAA Earth System Research Laboratory Global Monitoring Division's website, you will find maps, graphs and databases of measurements of carbon dioxide and other atmospheric gases over many decades. Check out the section called Products.

As shown in Figure 1, there is a correlation between temperature increase and carbon dioxide increase. As discussed earlier, it is clear that an increase in carbon dioxide levels will lead to warming of the atmosphere, because there will be an increase in the greenhouse effect. Having said this, closer inspection of the data shows that the increase in temperature happens first and then the carbon dioxide concentration rises. This lag time is partly explained by the fact that, as oceans warm up, they release carbon dioxide, because gases dissolve less well in warm water than in cold water. This leads to further increases in temperatures over time: warmer temperatures → more carbon dioxide → even warmer temperatures → even more carbon dioxide, and so on.

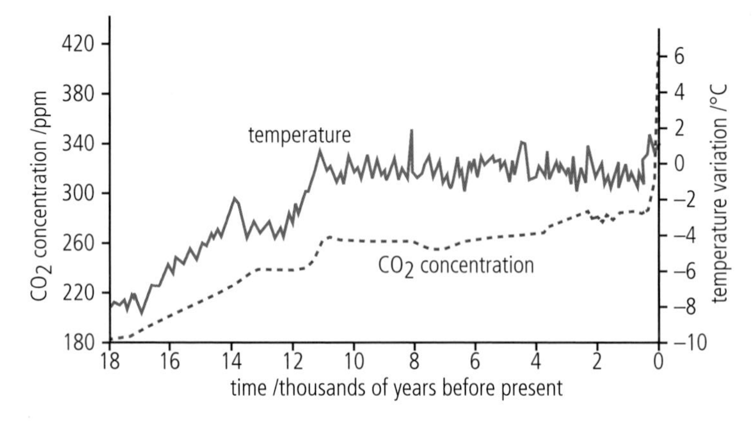

D4.3 Figure 1 18,000 years of atmospheric changes in temperature and carbon dioxide (CO_2) concentrations measured mainly from air bubbles trapped in ice.

Challenge yourself

Figure 1 shows the results of collected data representing thousands of years trapped in ice core samples.

The lower (dashed red) line shows carbon dioxide concentrations that were measured from air bubbles trapped in the ice.

The upper (continuous blue) line shows fluctuations in temperature, with zero representing pre-industrial climatic conditions.

1. Is there a strong or a weak correlation between carbon dioxide levels and atmospheric temperatures over the last 18,000 years?
2. Can scientists conclude that there is causality from the data in Figure 1, i.e. that rising carbon dioxide levels cause global temperatures to go up?
3. What further evidence would be necessary to confirm or refute causality?

Earth has shown many fluctuations in global temperatures over millions of years, long before humans started producing excessive greenhouse gases. The changes being observed now are alarming scientists because they are happening so quickly.

Climate change, especially as portrayed in the press or on social media, raises many issues about how science works, notably concerning the scope and limitations of science.

The fact that there are sceptics and critics of the IPCC reports on global climate change is a good thing. Science encourages constructive criticism and verification, and is open to modification if the criticisms are valid. But some critics will look at a few inaccurate predictions that have been made and say, "See? Your models are wrong. Therefore, no one should listen to you". This is an example of **cherry picking**. Cherry picking is a form of **confirmation bias** caused by only looking at the evidence supporting your side of the argument, and ignoring or downplaying the evidence that hurts your argument. To what extent is objectivity possible in the production or acquisition of knowledge?

D4.3.2 – Global warming

D4.3.2 – Positive feedback cycles in global warming
Include release of carbon dioxide from deep ocean, increases in absorption of solar radiation due to loss of reflective snow and ice, accelerating rates of decomposition of peat and previously undecomposed organic matter in permafrost, release of methane from melting permafrost and increases in droughts and forest fires.

Release of carbon dioxide

When a marine organism dies, it sinks, and either its corpse will be eaten on the way down or it will reach the bottom, called the **benthic zone**. Decomposers, archaea and bacteria digest and break down the organic material, and in so doing can produce carbon dioxide. If surface waters are warmer, phytoplankton can produce more biomass, which can then be passed on to the rest of the marine food chain. And when those organisms die, more organic matter will reach the ocean depths and the organisms living in the benthic zone will produce even more carbon dioxide.

A system that amplifies an effect like this is said to be a **positive feedback loop**. The more something happens, the more it allows the next step to happen, which, in turn, encourages the first step to keep going. Notice that the word "positive" does not necessarily mean it is a good thing, it just means the effect is amplified. Negative feedback loops also exist. A combination of positive and negative feedback loops helps a human maintain steady internal conditions such as the correct body temperature and blood sugar levels. Earth also maintains a balance using feedback loops. Unfortunately, an increase in carbon dioxide levels is increasing the influence of positive feedback loops, throwing the system off balance.

Increase in absorption of solar radiation

The ability of a surface to reflect light is called its **albedo**. Light-coloured objects, such as ice and white sand, have a higher albedo, so very little light is absorbed, meaning such objects do not heat up as much as dark objects. Think about walking barefoot on light-coloured cement on a hot and sunny day, compared to walking barefoot on black asphalt on the same day. Dark-coloured substances such as asphalt have a lower albedo, and can absorb lots of light and convert it into heat.

Snow and ice have a high albedo and tend to reflect so much light that they do not heat up very much. Soil, rocks and open ocean water have a lower albedo and tend to warm

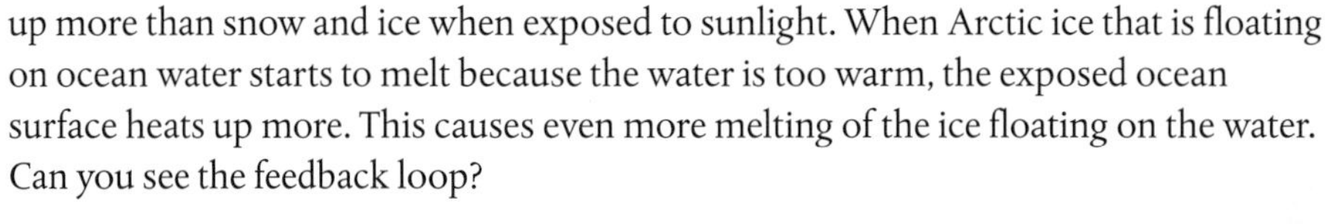

up more than snow and ice when exposed to sunlight. When Arctic ice that is floating on ocean water starts to melt because the water is too warm, the exposed ocean surface heats up more. This causes even more melting of the ice floating on the water. Can you see the feedback loop?

The darker zones between the ice sheets that are melting and breaking up have a lower albedo and so the water warms up much faster than it did before, when high-albedo ice was covering it.

Positive feedback loops intensify or exacerbate a situation. The message is "Yes, keep going. Amplify the effect!" Negative feedback loops prevent a process from going too far. The message is "Do not overdo it! Slow down. No need to keep amplifying the effect".

Decomposition of organic matter

Healthy soil hosts huge numbers of small invertebrates and microbes such as worms, fungi and bacteria that help decompose organic matter. If the temperature is cold, they are less active because their enzyme activity and other metabolic processes slow down at lower temperatures. Because of climate change, some parts of the world are becoming warmer. When peat bogs are in such zones, their decomposition rates increase and the faster decomposition releases more carbon dioxide than before.

If undisturbed, peat bogs are considered to be **carbon sinks**. They trap organic matter and the conditions prevent it from being fully decomposed. But when climate change results in decomposition happening faster, peat bogs can become carbon sources, releasing carbon dioxide into the atmosphere.

A peat bog can switch from being a carbon sink to a carbon source if temperatures rise.

The same phenomenon is happening in the coldest zones of the world, where we find the **cryosphere**, places on Earth where water is in solid form, such as in **permafrost**. Permafrost is a type of soil that exists in very cold climates and is frozen solid. Sometimes, layers above the permafrost can thaw out for a few weeks or months in a year. Generally only low-lying plants with shallow roots can grow in such soil. Tonnes of organic matter are locked up in the permafrost because the microbes that carry out the decomposition cannot survive at such low temperatures.

Permafrost under the surface soil creates these bumps on the surface, called hillocks, when ice, which is less dense than water, rises towards the surface, pushing the soil up. The growing season is so short that plants do not have time to grow very tall.

In zones where temperatures are rising, the permafrost is melting. As it warms up, microbes can become active and decompose the trapped organic matter, and as a result generate more carbon dioxide. The carbon dioxide that is released into the atmosphere is now able to act as a greenhouse gas and contribute to even more warming. Again, you should notice the positive feedback loop here, because the further increase in temperature will cause the permafrost to melt to an even greater depth, allowing even more carbon dioxide to be released (see Figure 2).

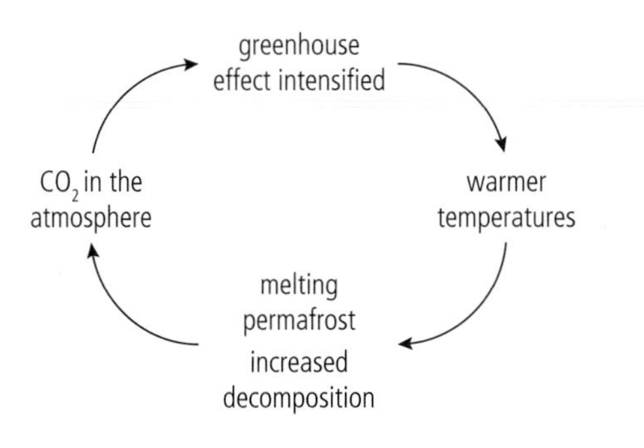

D4.3 Figure 2 Climate change creates a positive feedback loop when permafrost melts and releases greenhouse gases.

In addition to carbon dioxide, permafrost can release methane. This is because some of the microbes in the soil are **methanogenic archaea**, meaning that they generate methane. As seen previously, methane is also a greenhouse gas and so releasing more of it into the atmosphere will create a positive feedback loop similar to the one for carbon dioxide. Warmer temperatures will allow the methane-producing microbes to be even more active.

Methane is an invisible gas but one way to observe it is in frozen lakes, where it bubbles up from the sediments and gets trapped in the ice.

Global climate change can also lead to extreme weather events that cause droughts and more frequent forest fires. Forest fires release carbon dioxide into the atmosphere.

Drought conditions occur when there is not enough rainfall to sustain the water demands of an ecosystem or the needs of humans, notably for agriculture.

When fire destroys a forest, the carbon in the trees is released as carbon dioxide into the atmosphere. This is yet another positive feedback loop, exacerbating the problem of excess greenhouse gases being added to the atmosphere.

D4.3.3 – Tipping points

D4.3.3 – Change from net carbon accumulation to net loss in boreal forests as an example of a tipping point

Include warmer temperatures and decreased winter snowfall leading to increased incidence of drought and reductions in primary production in taiga, with forest browning and increases in the frequency and intensity of forest fires, which result in legacy carbon combustion.

We usually think of forests as getting water from rainfall, but in many parts of the world melted snow provides most of the water that the trees absorb. One consequence of global climate change is that, where temperatures are getting warmer, winters are shorter and there is less snow. Less snow in the winter means less snowmelt in the spring and summer, and therefore less water for **boreal forests**, otherwise known as **taiga**. Such forests are blanketed in snow for much of the year and are the coldest land biome, with average temperatures ranging from −5° C to 5° C.

A frozen river cutting through a boreal forest.

The lack of water in the soil from reduced snowmelt water means that the trees cannot photosynthesize as much as they could before. Primary production in the forest is reduced, which means less carbon dioxide is being removed from the air. This is another example of a positive feedback loop. If coniferous trees cannot get enough water for photosynthesis, their needles lose their green pigment, turn brown and fall off. When this happens on a massive scale, it is called **forest browning**, and if it continues without any drought relief, the trees will die. This is a form of deforestation.

When coniferous trees in the taiga do not get enough water, they start to lose their needles and die in a process called forest browning.

Drought conditions and dying wood are perfect conditions for forest fires. When a wildfire takes hold in a boreal forest, the carbon that has accumulated and been locked up for centuries in the trees' organic matter (as wood and needles) is suddenly released as carbon dioxide into the atmosphere. There is carbon locked up in the soil, too. This carbon is from past ecosystems and can be hundreds or thousands of years old. It is called **legacy carbon**.

Normally, the organic matter holding legacy carbon cannot decompose because of the prevailing cold temperatures, but if a fire burns the organic material trapped in the soil, it can release the carbon into the air in a process called **legacy carbon combustion**.

Usually, forests are considered to be carbon sinks because they remove carbon from the atmosphere during photosynthesis, which converts the carbon dioxide into organic molecules that are locked up in the trees for centuries. But when thousands of decaying trees or forest fires release the carbon back into the atmosphere, the forest becomes a carbon source. The switch from sink to source because of an imbalance in the system is considered a **tipping point**. A tipping point is a certain threshold that if reached or surpassed causes massive changes, some of which could be irreversible.

D4.3.4 – Polar habitat change

D4.3.4 – Melting of landfast ice and sea ice as examples of polar habitat change

Include potential loss of breeding grounds of the emperor penguin (*Aptenodytes forsteri*) due to early breakout of landfast ice in the Antarctic and loss of sea ice habitat for walruses in the Arctic.

Note: When students are referring to organisms in an examination, either the common name or the scientific name is acceptable.

The emperor penguin (*Aptenodytes forsteri*) lives along the coasts of Antarctica and migrates dozens of kilometres over the ice to reach its breeding grounds each winter. Emperor penguins prefer to breed on **sea ice** (ice formed when ocean water freezes) that is connected to the mainland. These **landfast** sheets of ice start to break up and detach from the shore when the Antarctic summer arrives in December and warms up the waters. By then, the eggs, which have been incubated for two months by the males, have hatched and the chicks will have grown big enough to find their own food.

▲ The habitat of emperor penguins (*Aptenodytes forsteri*) is being modified by climate change.

With global climate change, some zones of sea ice are being exposed to warmer temperatures and starting to break up and separate from the shore earlier in the breeding season, making it impossible for the penguins to raise their young. A few penguin colonies have moved to ice found on land, but the reduction of landfast ice means there are concerns about the future of the emperor penguin population.

At the other pole, Arctic ice is also melting and having an impact on ecosystems. Walruses (*Odobenus rosmarus*) also prefer sea ice shelves as breeding grounds and a place to rear their young. The mother walruses suckle their babies then dive into the water to get more food. The fact that the sea ice is on the water makes it convenient, because they never have to leave their young for very long. With melting ice, however, reduced breeding grounds and less space to rear young has meant that either populations have to move poleward (in this case nearer the North Pole) or find places other than sea ice to rear their young. In the latter case, the mothers are not as close to the water and need to leave their babies for longer periods of time This, in turn, makes the young more vulnerable to attack by predators such as polar bears, who are also having trouble finding food as a result of climate change and melting ice.

◀ Walruses (*Odobenus rosmarus*) in the Canadian Arctic.

Low pressure systems tend to form cyclones that spin air masses. These turn in a clockwise fashion in the southern hemisphere but anticlockwise in the northern hemisphere (when observed from a weather satellite). The direction is determined by Coriolis forces generated by the rotation of planet Earth.

D4.3.5 – Ocean current change

D4.3.5 – Changes in ocean currents altering the timing and extent of nutrient upwelling
Warmer surface water can prevent nutrient upwelling to the surface, decreasing ocean primary production and energy flow through marine food chains.

The world's oceans are in perpetual motion. Currents move water that has been warmed near the equator to the poles where the water is cooled and sinks. The water is moved around by prevailing winds and also spins as it moves due to the **Coriolis effect**, which is caused by the rotation of Earth on its axis.

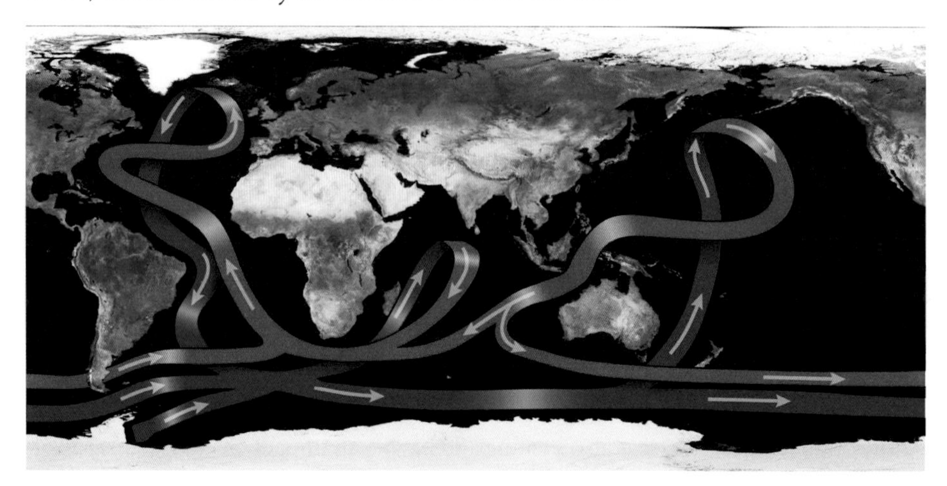

The world's oceans have currents that transport heat, oxygen and nutrients between ecosystems. The ribbon-like arrows in the figure can be thought of as conveyor belts.

Some currents exist year round and others can change or intensify with the seasons. Some do not necessarily happen every year, such as the **El Niño Southern Oscillation**, which occurs in the Pacific Ocean between Indonesia and South America. During years without an El Niño event, prevailing trade winds blow air towards the west of the Pacific Ocean. As coastal surface water is pushed seaward by the wind, the water below it is pulled upwards to take its place. This cold water is full of nutrients. Movement of nutrient-rich water from deeper parts of the ocean towards the surface is called **nutrient upwelling** and it plays an important role in bringing nutrients to the food webs that exist along the coasts. In years when there is no El Niño event, large schools of sardines can be found off the coast of South America. The sardines are feeding off organisms that are thriving because of the nutrients that have been pulled up from the deeper water.

When an El Niño event occurs, however (roughly every 2 to 7 years), the currents are modified. Winds arriving from Southeast Asia push against the normal prevailing winds and create an upwards air movement in the middle of the Pacific Ocean. This causes the warm water along the Central and South American coast to stagnate instead of being pushed seaward.

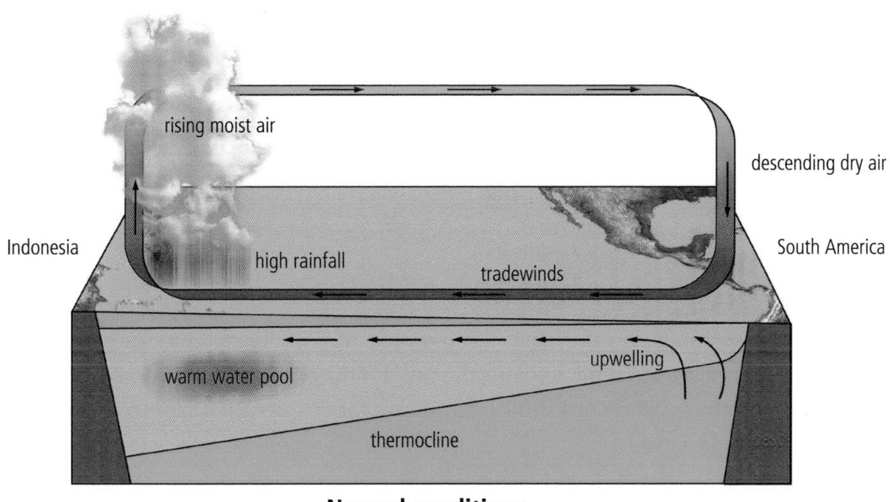

Normal conditions

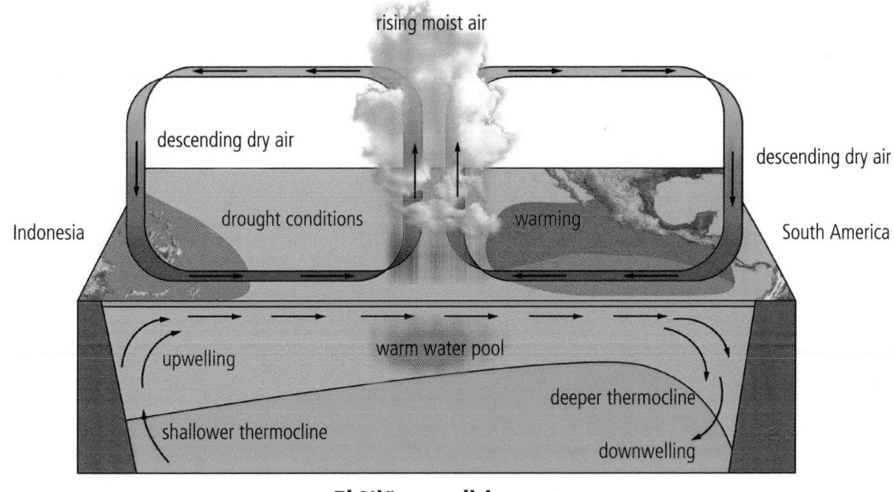

El Niño conditions

Conditions in the Pacific Ocean in years without (top) and with an El Niño event (bottom). Upwelling is cancelled and vital nutrients do not reach the surface during an El Niño event.

When there is an El Niño event, instead of a nutrient upwelling along the Central and South American coast, water is pushed down (downwelling). As a result, the marine ecosystems that depend on cool water upwelling do not receive the nutrients they otherwise would. Primary production is the production of organic molecules by living organisms. In the oceans, algae are the producers. Without the nutrients that are provided by cold water, primary production falls, and less energy flows through the ecosystem. The whole food chain is affected, and fewer fish are seen during El Niño years.

Studies indicate that climate change is increasing the severity of El Niño events, as well as producing warmer surface temperatures in many other places in the oceans, which could reduce nutrient upwelling and throw marine ecosystems off balance.

Nutrient upwelling does not only happen in the Pacific Ocean. Plankton use the nutrients and and proliferate in blooms so big they are visible from space, as here in the Bay of Biscay. This primary production will nourish the rest of the marine ecosystem in the area. Without the upwelling, not only will phytoplankton numbers be lower, the rest of the ecosystem that depends on it will also suffer.

D4.3.6 – Range shifts

D4.3.6 – Poleward and upslope range shifts of temperate species
As evidence-based examples, include upslope range shifts for tropical-zone montane bird species in New Guinea and range contraction and northward spread in North American tree species.

Poleward range migration of trees

Tree species in northern latitudes have been monitored over decades to see if their ranges are moving, for example towards the poles. This process is called **range migration**, and the fossil record shows that such a range migration happened just after the most recent glaciation.

Generally what stops a tree population in northern latitudes from expanding its northern upper range limit is extreme cold. Many seeds or saplings cannot tolerate the cold that intensifies nearer the North Pole, but as global temperatures increase and winters are becoming less harsh at many northern latitudes, it is hypothesized that some tree populations are spreading poleward, like they did when the Earth warmed after the most recent ice age.

A limiting factor for the northern boundary of a tree range is cold temperatures. As global temperatures increase, certain tree species can change their range and move into zones that were once too cold for them.

Figure 3 shows the results of a study published in 2014 by the Ecological Society of America, which compared the ranges of 11 species in North America between the 1970s and the early 2000s. To understand the graph, you need to know what a **percentile** is. When collecting data, if 50% of the data points can be found below a certain value, we say that value is the 50th percentile. So if half the trees are found below a certain latitude, that latitude is the 50th percentile latitude. The graph shows that black spruce trees (*Picea mariana*) in the 50th percentile latitude have shifted 11 km to the north (shown in green). The 90th percentile latitude for black spruce trees has a similar value, with a 10 km northward shift. However, the 90th percentile for the same species' saplings show an overall southward range migration of similar magnitude, which makes it difficult to draw decisive conclusions about what is happening to the black spruce's range in recent decades.

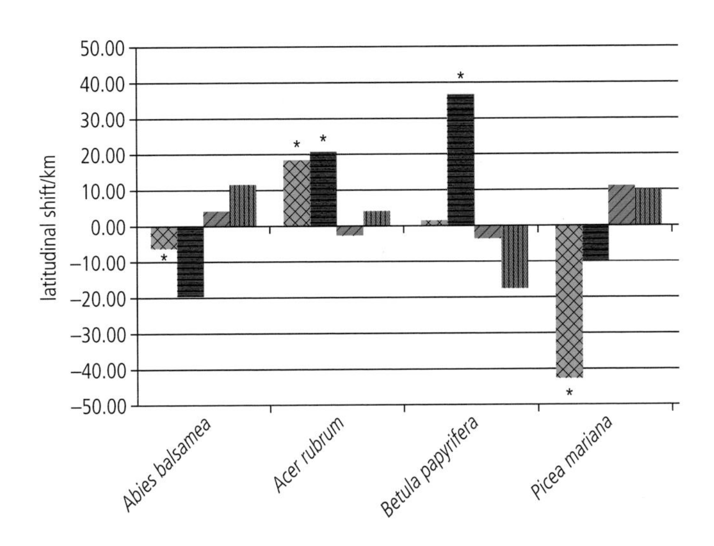

D4.3 Figure 3 Latitudinal range shifts (in kilometres) for four species. Saplings shown in blue (with criss-cross hatching) are at the 50th percentile of latitudinal distribution; saplings shown in red (with horizontal hatching) are at the 90th percentile; mature trees shown in green (with diagonal hatching) are at the 50th percentile; and trees shown in purple (with vertical hatching) are at the 90th. An asterisk indicates where the shifts are statistically significant.

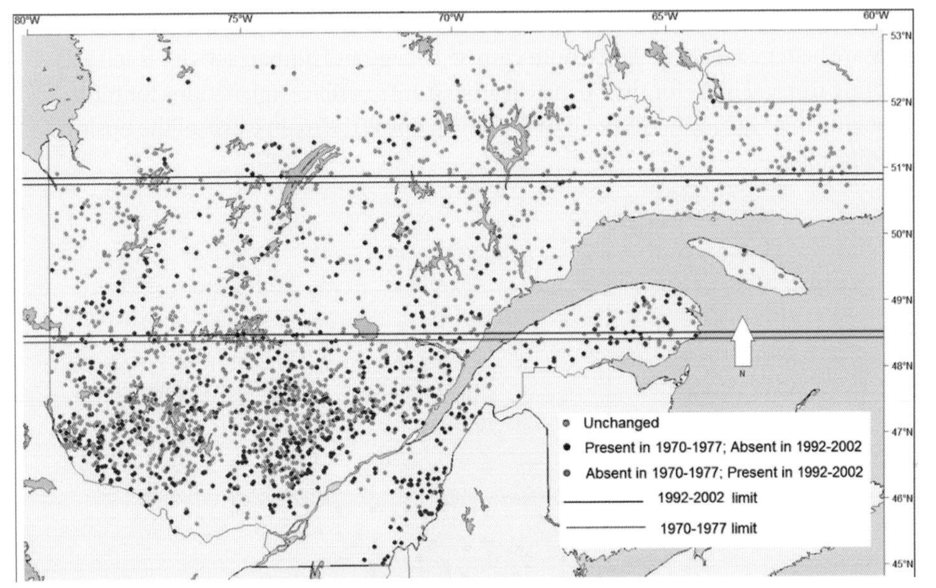

A map showing the distribution of black spruce (*Picea mariana*) trees. The pairs of lines towards the top represent the 90th percentile of latitude and the lines towards the bottom show the 50th percentile latitude for this species, meaning 90% and 50% of the trees sampled were below these latitudes. Both pairs of lines show a northward migration of about 10 km over the decades of the study.

Several challenges and limitations make it difficult to answer the question of whether or not tree ranges are moving poleward as a result of human-induced climate change. One is that, for the same species, the pattern can change as we move from west to east: some zones might show a poleward range change and some might show no significant change or even a southward change. Patterns can be contradictory for a species when looking at the statistics for both full-grown trees and saplings. Another is that the range changes could be caused by factors other than temperature, such as ice storms, forest fires or logging activity.

TOK

Scientists use data and measurable evidence to reach conclusions. But are there some types of evidence that are less reliable than others? What features of evidence in the natural sciences make them convincing and valid, and what features make them unreliable? Often when evaluating the reliability of data and assessing the level of confidence, scientists look at the overall characteristics of the study, including sample size, range of results or the methodologies used, such as whether or not the data was measured directly or by proxy. What features of evidence in the natural sciences make them convincing and valid and what features make them unreliable?

▲ The white-winged robin (*Penoethello sigillatus*) in Papua New Guinea has moved its habitat range upslope in recent decades.

Upslope range migration of birds

Papua New Guinea is a mountainous country in which some habitats are temperate and others tropical. The country is home to some visually spectacular and unique birds, including colourful birds of paradise, parrots and cockatoos. In the 1960s, detailed surveys of **montane** bird species (birds that live on mountains) were carried out. All birds have a certain altitude range they live in; some montane species prefer lower altitudes, some prefer living high up near the peaks, and others are found somewhere else along the slope of the mountain. In 2014, nearly 50 years after the original survey, a second survey looked at the zones where the birds were found. A few bird species had stayed in the same ranges, and some were observed at lower altitudes, but a surprising number had increased their range upslope (see Figure 4).

One such bird, the white-winged robin (*Penoethello sigillatus*), lives on Mt Karimui in the Eastern Highlands, and was found to have moved upslope more than 100 m since the previous survey. Tropical birds are very sensitive to temperature, so if their habitat warms up, they could go to higher altitudes to find the relatively cooler temperatures they are better adapted to. Since both climate change and human activity (such as hunting tropical birds for their ornate feathers) are possible explanations for birds changing their range, it is difficult to say which one is the main cause of the upslope modification.

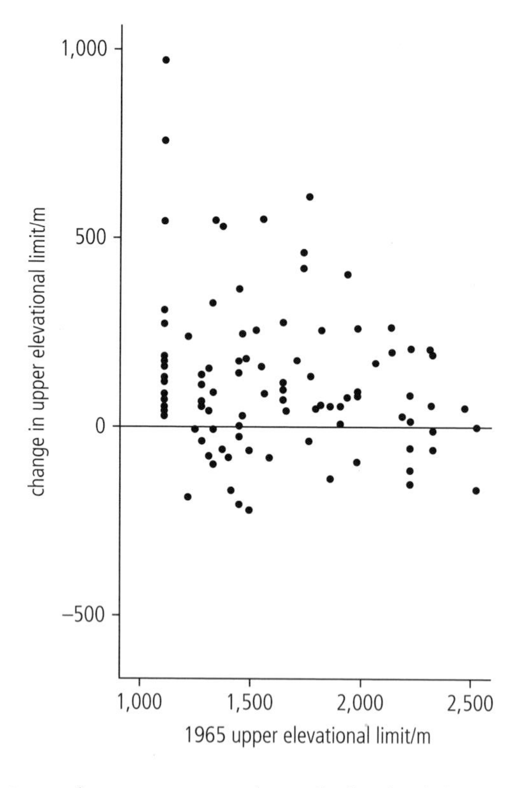

D4.3 Figure 4 The results of the part of the study looking at upper elevational limits. Each montane bird species is represented by a dot. The *x*-axis shows the highest elevation at which the bird was observed back in 1965, and the *y*-axis shows the change since then. Although a few have stayed at the same altitude or moved down a few dozen metres, more species have moved upslope in recent decades, some by over 50 m. The average was more than 100 m upslope.

This phenomenon is much more extensively studied in birds living in temperate zones of the world and we know with more certainty that climate change is a major contributor to alterations in range both poleward and upslope.

D4.3.7 – Ecosystem collapse

D4.3.7 – Threats to coral reefs as an example of potential ecosystem collapse

Increased carbon dioxide concentrations are the cause of ocean acidification and suppression of calcification in corals. Increases in water temperature are a cause of coral bleaching. Loss of corals causes the collapse of reef ecosystems.

The organisms that build coral reefs are very sensitive to water temperature, water acidity and the depth of the water. Unfortunately, all three factors are changing in the world's oceans as a result of human activities. Increased carbon dioxide concentrations in the air lead to increased dissolved carbon dioxide in the oceans (see Figure 5). This lowers the pH of seawater because, as the carbon dioxide dissolves in the water, it forms **carbonic acid**:

$$CO_2 + H_2O \rightarrow H_2CO_3 \text{ (carbonic acid)}$$

$$H_2CO_3 \rightarrow H^+ + HCO_3^- \text{ (hydrogen carbonate ion)}$$

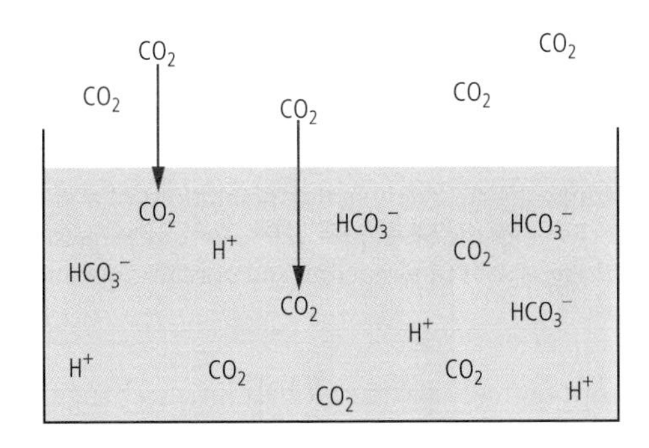

D4.3 Figure 5 Increased dissolved carbon dioxide will lower the pH of ocean water and increase the concentration of hydrogen carbonate ions.

When dissolved in water, the carbonic acid forms **hydrogen ions** (H^+ ions), which reduces the pH. **Calcium carbonate ($CaCO_3$)** is a key component in the skeletons of coral reefs and the hard shells of marine organisms. It is formed when **calcium ions** (Ca^{2+}) dissolved in ocean water are combined with **carbonate ions** (CO_3^{2-}):

$$Ca^{2+} + CO_3^{2-} \rightarrow CaCO_3$$

This process is called **calcification** and it is as important to the growth of coral reefs as the growth of bones is to us. Acid can dissolve calcium carbonate, so if calcification is disrupted by low pH levels, either the reef's growth rate or its ability to form strong, dense skeletons will be reduced. Beyond certain thresholds, ocean acidification and changes in temperature can lead to the death of coral polyps and algae, and when they die the reefs no longer thrive. As a result, the colour of the reef changes from being richly multi-coloured to as white as bone (see Figure 6).

D4.3 Figure 6 A healthy coral reef on the left and a dead reef on the right.

The United Nations' Sustainable Development Goal 14, entitled "Life below Water", lists reducing ocean acidification as one of its targets because ongoing acidification of ocean water will continue to disrupt and destroy marine ecosystems. Coral reefs are also discussed in Chapter B4.1.

This coral reef death is called **coral bleaching** and it interrupts the reef's food chain, causing many of the organisms that live there to seek food and shelter elsewhere or die out. A bleached coral reef can no longer support the rich ecosystem that once lived there. This is an example of **ecosystem collapse**.

TOK

The scope and methodologies of scientists are different from those of politicians, and they do not always share the same perspectives. This is not a new phenomenon. The astrophysicist Carl Sagan, for example, testified before a US Senate hearing in 1985, warning the government that action needed to be taken to avoid irreversible changes to our planet's climate. The 2022 IPCC report on climate change was about 4,000 pages long. It is unlikely many politicians read the whole document. In addition to scientists, other experts have tried to warn governments of the dangers. Indigenous knowledge could hold possible solutions to bring nature back into balance, starting with rebuilding a relationship with the planet and showing more respect and humility. How can we decide between the judgements of experts if they disagree with each other?

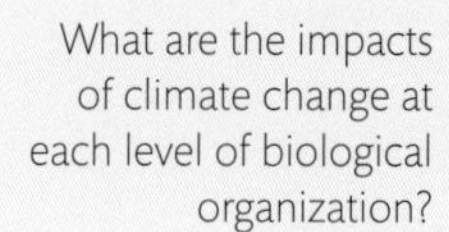

What are the impacts of climate change at each level of biological organization?

D4.3.8 – Carbon sequestration

D4.3.8 – Afforestation, forest regeneration and restoration of peat-forming wetlands as approaches to carbon sequestration
NOS: There is active scientific debate over whether plantations of non-native tree species or rewilding with native species offer the best approach to carbon sequestration. Peat formation naturally occurs in waterlogged soils in temperate and boreal zones and also very rapidly in some tropical ecosystems.

One way to mitigate and potentially reverse global climate change is to find ways of removing the carbon we have added to the atmosphere. This process, called **carbon sequestration**, involves taking atmospheric carbon dioxide and locking it up somewhere. Although engineers have proposed many high-tech solutions using costly industrial technologies for carbon sequestration, some low-cost solutions are natural and use systems that have been proven to work for millions of years: forests and wetlands.

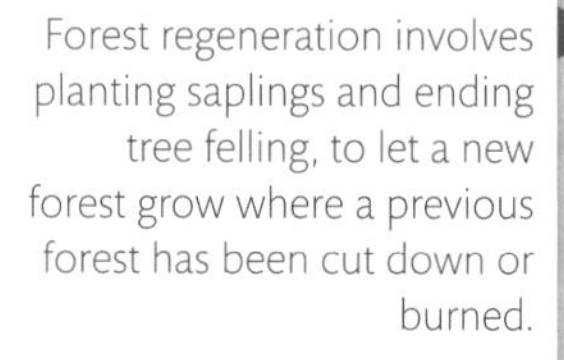

Forest regeneration involves planting saplings and ending tree felling, to let a new forest grow where a previous forest has been cut down or burned.

For many thousands of years humans have cut down trees to use the wood, or burnt forests to clear land for agriculture and other land uses. Humans have also become good at draining wetlands for agriculture or building. **Reforestation** is the process of replanting deforested areas, which would allow trees to remove carbon from the atmosphere. Likewise, **afforestation**, the process of planting trees where no forest has previously existed, can increase carbon capture. And if we stopped filling in or draining wetlands and instead allowed new wetlands to form, wetlands could fulfil their role as carbon sinks. These are long-term processes, but some governments and non-governmental organizations have instigated such activities.

Arguably, the government that can claim to plant the most trees in the world is China. In 1981, China passed a law mandating that one tree should be planted every year by every student over the age of 11. They continue to fund and implement forest improvement programmes but, as is the case with many countries, they also get outside financial help for such projects, notably from the World Bank. Just as with any project, if it is to be successful, an afforestation programme needs follow-up and maintenance. The programmes that work the best include provision for irrigation and for educating the local population about the need to preserve forests.

Nature of Science

As with many scientific endeavours, there is active debate when it comes to the best way to achieve carbon sequestration. Some experts advise countries to plant non-native species of trees that will grow faster than local trees and therefore capture more carbon more quickly. Others say that rewilding is the best approach, bringing back the native species that once flourished in an area. For wetland restoration or the introduction of carbon sinks such as peat bogs, once stagnant water has been introduced or reintroduced, plant species tend to arrive by natural means and start carbon sequestration rapidly, notably in tropical zones, for example the Eastern Congolian swamp forests, which contain peatlands in the Congo River basin that cover an area of 145,000 square kilometres and contain an estimated 30 billion tonnes of carbon.

Wetlands can be used as carbon sinks. Instead of draining it to use the land for agriculture or urban development, this wetland in Poland has been set aside as a natural reserve where human activity is minimized.

The US Fish and Wildlife Service reported in 1990 that over 40 million hectares of wetlands had been destroyed since 1770, which was half of all the wetlands in the country. There are now protection policies in place to make sure that any wetlands that are destroyed are replaced. Restoring or introducing wetlands involves excavation (digging out) of the land to provide low-lying zones where rainwater can accumulate. Sometimes the types of plants that are best adapted to wet soil need to be deliberately introduced, such as marsh grasses or, in the case of peat bogs, *Sphagnum* mosses. Active management might also need to include the removal of invasive species of plants, but sometimes just leaving natural colonization and succession to do their work is enough once the water starts to accumulate.

Although many human activities tend to damage ecosystems and exacerbate climate change, efforts have been made to restore land to its natural state, improve biodiversity and bring the ecosystem back into balance. One of the best ways of doing this is to minimize the impact of human activities in an area. Nature reserves are places set aside by governments or private landowners to let nature do what it does best: optimize biodiversity and reach a sustainable equilibrium. By reducing access to motorized vehicles and prohibiting land use, natural restoration can take place, and sometimes it occurs over a surprisingly short period of time.

Gorse (*Ulex europaeus*).

Gorse (*Ulex europaeus*) is a small thorny plant that produces yellow flowers. It was introduced to New Zealand by Europeans during British colonization of the islands. The plant quickly covers any open land such as abandoned farmland or pastures. Although it is considered an invasive species in New Zealand today, it has one advantage that favours the succession process: as it grows, its dense vegetation thins out, leaving spaces between branches for young saplings of native plants and trees to take hold and mature. The saplings benefit from the nurse canopies maintained by the gorse as it holds the soil in place, blocks the wind and shades the ground to reduce evaporation of water. As they grow in this nursery environment, native plants can eventually replace the gorse by growing above it, shading it out and taking its place.

This phenomenon has been observed over a very short period of time in the Hinewai Reserve in New Zealand. The reserve was started in 1987 and currently hosts many native plants that have not grown in the area for many decades. Although trees covered the region before humans lived there, farming in recent centuries has greatly reduced the original vegetation. Since the establishment of the reserve a few decades ago, a rapid regeneration of native plants is taking place. Today, if you visit the reserve or look at its website, you can see that native *kānuka* trees and podocarp trees such as the *tōtara* are repopulating the area.

What processes determine the distribution of organisms on Earth?

HL

D4.3.9 – Phenology

D4.3.9 – Phenology as research into the timing of biological events
Students should be aware that photoperiod and temperature patterns are examples of variables that influence the timing of biological events such as flowering, budburst and bud set in deciduous trees, bird migration and nesting.

The word "nature" has the same root word as "natality", and both refer to birth. The natural world is in a cycle of constant rebirth. Flowers burst open and desolate landscapes become green and lush again. The study of the timing of such periodical events is called **phenology**. How do plants synchronize with the season to end their dormant period and become active again at just the right time? How do seeds time their germination so that they are not too early or too late? Other periodical events that are synchronized with the seasons include migration, mating periods and bird nesting.

This insectivorous bird depends on the availability of larvae to feed its hungry chicks.

Birds that migrate to warmer zones in the winter need to make sure that when they come back to build a nest and start a family, the timing is just right. If they arrive back too early, temperatures might be dangerously low and the choice of possible mates might be reduced, as not all individuals from the population will have returned yet. If, on the other hand, returning birds wait too long to make their homeward journey, the best nesting sites and mates might already be taken.

Timing must also be right so that when baby chicks hatch there is enough food for them. If the chicks hatch too early in the season, and insect eggs have not hatched, there will be no larvae for the adult birds to feed to their offspring. Thankfully, natural selection helps. Those birds whose instinctive behaviour led them to leave too early or too late have a higher chance of being selected out of the population, while those that fly back at optimal times of the year will have a higher likelihood of survival.

In order for the bee and the flower to be ready for each other, their dormancy and active cycles need to be synchronized.

Insect-pollinated flowers and pollinating insects have an intimate reproductive relationship. Insects provide the transportation for the pollen containing the male sex cells of the flowering plants they visit and, in return, they get a sweet drink of nectar. Timing is crucial here, too. If bees that are hibernating come out of dormancy too early in the spring, they might not find any open flowers: all the buds will still be closed and no nectar will be ready for them. If they wait too long, their winter caloric energy stores might run out. Plants in temperate zones use the length of day and the temperature to work out the best time to bloom.

▲ Plants produce buds during the growing season that will then open up and produce either flowers or leaves or both, depending on the species, in the subsequent season.

The **photoperiod** is the number of hours in a day when sunlight is shining. In temperate zones the photoperiod changes through the year. Different plants have different thresholds, but each will start to produce flowers when a certain length of day is reached. As the days get longer in the springtime, temperatures rise. The greater warmth enables enzymes to be more active and promotes the flowering process.

Bud setting is the process by which a plant uses current stores of energy to prepare leaf and flower buds for the next season. Inside a **leaf bud** is the **leaf primordia** (structures that will become leaves) and shoot apical meristems. Inside a **flower bud** is an **embryonic flower**. Depending on the species, some buds can contain both leaves and flowers. Deciduous trees prepare resting buds at the end of one growing season in preparation for the beginning of the next.

Once the photoperiod is long enough and temperatures are warm enough in the spring, the process of **bud burst** can happen and the young leaves can unfold or the flower petals open. The young tender leaves are often fed on by organisms such as caterpillars, which find them easier to chew than older, tougher leaves.

D4.3.10 – Disruption of phenological events

D4.3.10 – Disruption to the synchrony of phenological events by climate change

Students should recognize that within an ecosystem temperature may act as the cue in one population and photoperiod may be the cue in another. Include spring growth of the Arctic mouse-ear chickweed (*Cerastium arcticum*) and arrival of migrating reindeer (*Rangifer tarandus*) as one example. Also include a suitable local example or use the breeding of the great tit (*Parus major*) and peak biomass of caterpillars in north European forests as another.

Note: When students are referring to organisms in an examination, either the common name or the scientific name is acceptable.

For migrating animals, the return from migration is synchronized with peak resource availability (e.g. maximum plant production for herbivores, maximum availability of insect larvae for insectivorous birds). If the two fall out of sync, there is a **trophic mismatch**. Trophic levels are steps along the food chain, and a mismatch happens when one organism is missing, has not arrived yet or is in low numbers. A trophic mismatch can endanger the reproductive success of the new generation in the migrating population.

Often migrating animals know when to return to their warm season habitats by using day length as a cue. Reindeer (*Rangifer tarandus*), also called caribou, are herbivores that live in northern latitudes in places such as Greenland and Norway. Populations of reindeer that migrate seasonally use day length to know when to return to their breeding grounds, to graze and start the next generation.

A reindeer (*Rangifer tarandus*) in Svalbard (Spitsbergen, Norway).

Often the plants that reindeer eat, however, use temperature as a cue to come out of dormancy, rather than day length. One such plant is the Arctic mouse-ear (*Cerastium arcticum*). If there is good synchronization between the peak production of plants like this and the arrival of reindeer, the reindeer will have enough nourishment to produce a healthy number of calves to maintain their population. If there is a trophic mismatch because the reindeer have not changed the timing of their migratory return (day lengths have not changed in recent decades) but the Arctic mouse-ear has changed its peak production time (temperatures have warmed in many places in the Arctic as result of climate change), this will have a negative impact on reindeer calving.

The Arctic mouse-ear (*Cerastium arcticum*) is one of the plants that reindeer graze on, and uses temperature as a phenological cue to stop its dormant period.

Researchers working in Greenland saw such a modification in a study carried out between 2002 and 2006. Arctic temperatures had risen by more than 4° C in the areas they studied, and peak plant production advanced by about 2 weeks. The graph in Figure 7 shows that, over the time period studied, the date of emergence of plants that reindeer eat were affected by the temperature increases and emerged earlier, whereas there was much less difference in the timing of the earliest births of reindeer calves.

D4.3 Figure 7 The results of a study carried out from 2002 to 2006, showing early emergence of plants (when they reached 5% of their maximum presence in the areas sampled) as empty circles and earliest births of reindeer calves (when the number of births reached 5% of the total) as filled circles. The dates on the *y*-axis are the day of the year (data points at the bottom are in early May and at the top early June).

Trophic mismatches can be calculated using a mismatch index. The researchers wanted to know whether or not the calf production would be influenced if the mismatch was more severe. Figure 8 confirms that the greater the magnitude of the mismatch, the lower the calf production. Calf production is expressed as the final proportion of calves in the new generation.

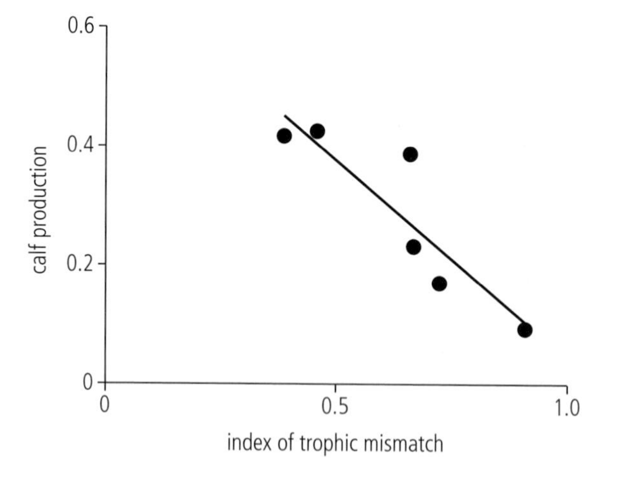

D4.3 Figure 8 Magnitude of trophic mismatch plotted against calf production expressed as the final proportion of calves in the new generation.

If you live in Europe or temperate parts of the Middle East or Asia, you probably have noticed the beautiful yellow colour of the great tit (*Parus major*). It can eat many different types of food but prefers eating insects. Caterpillars are particularly favoured for feeding to their young, because the caterpillars are rich in protein and easy to digest thanks to their soft bodies. Being able to match the timing of hatching with the maximum availability of caterpillars has a significant impact on the survival of the hatchlings (see Figure 9, graph A).

The great tit (*Parus major*).

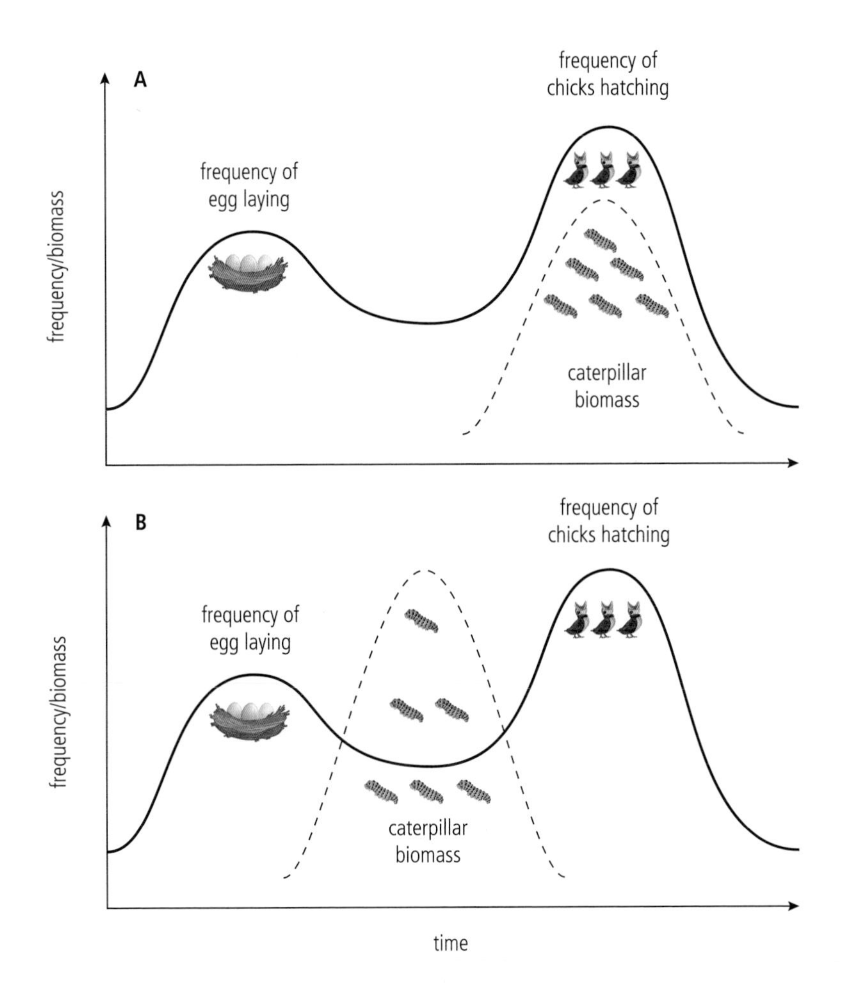

D4.3 Figure 9 Natural selection allows the maximum caterpillar biomass to be synchronized with the hatching of great tit chicks (graph A). However, climate change is producing more and more mismatches between the timing of egg-laying and the availability of caterpillars, as measured by their maximum biomass (graph B).

With milder winters, however, caterpillars studied in Germany had advanced the date they reached their maximum biomass by more than a week between the years of 1973 and 1995. Great tits whose eggs hatched at dates similar to previous years had a mismatch between when the food source was at its peak and when the chicks were ready to eat (see Figure 9, graph B). Once the chicks hatched, either the caterpillars had been eaten by other predators or they had progressed to the next stage of their life cycle and were more challenging to catch and digest. This is an example of how species have trouble adapting to temperature changes that occur in just a few decades.

D4.3.11 – Insect life cycles

D4.3.11 – Increases to the number of insect life cycles within a year due to climate change
Use the spruce bark beetle (*Ips typographus* or *Dendroctonus micans*) as an example.
Note: When students are referring to organisms in an examination, either the common name or the scientific name is acceptable.

The spruce bark beetle (*Ips typographus* or *Dendroctonus micans*) has an ingenious adaptation for escaping predators: it burrows under the bark of spruce and other coniferous trees and makes tunnels to lay its eggs. When the eggs hatch, the larvae feed on the wood just under the bark. As a single female can produce up to 300 eggs, a spruce bark infestation can lead to serious structural damage and eventually the death of a tree. Spruce bark beetles can survive winter either in the pupal stage of their life cycle or in their adult stage. To survive over the winter, they enter a phase of dormancy, which pauses their growth.

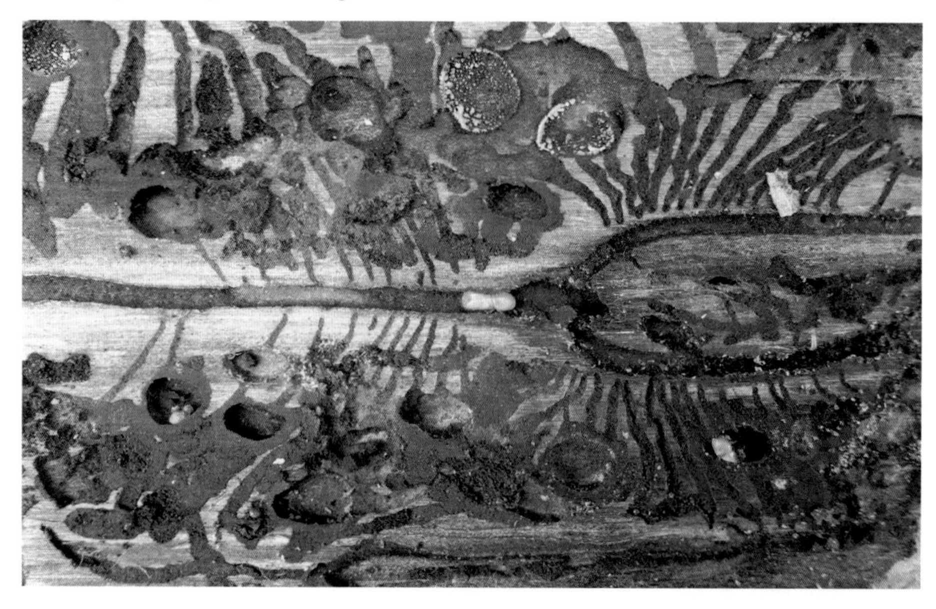

Galleries in a tree trunk produced by spruce bark beetles, in which they lay their eggs and where the pupae can spend the winter.

The life cycle of the beetle can vary between a **monovoltine cycle**, when one generation is produced in one year, and a **bivoltine cycle**, when two generations are produced in the same year. In a bivoltine cycle, the first generation starts as eggs in the spring and matures into reproducing adults by the summer, so that they can produce eggs for the next generation. The eggs from the generation that began in the summer will develop into the pupal stage in the autumn but become dormant over the winter to survive. When those individuals come out of dormancy in the following year and mature, another generation can start. So within the same year, two generations lay eggs. Because of this overlap in generations, it is possible to find beetles under the bark of a tree at multiple stages of their cycle.

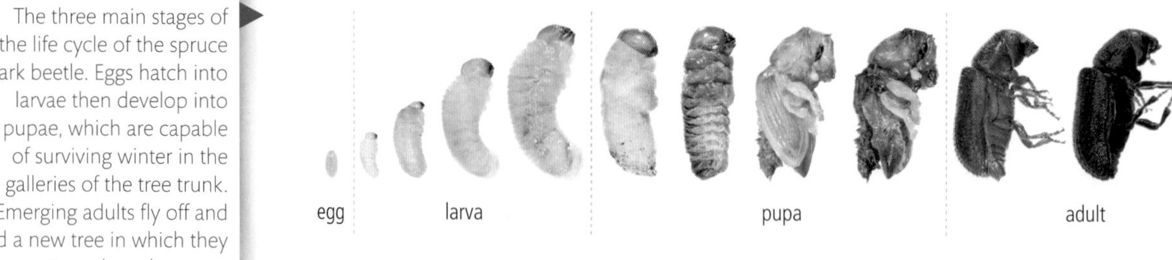

The three main stages of the life cycle of the spruce bark beetle. Eggs hatch into larvae then develop into pupae, which are capable of surviving winter in the galleries of the tree trunk. Emerging adults fly off and find a new tree in which they can mate and produce new eggs. Adults can also go into a dormant state to survive the winter.

What determines how fast a life cycle can be? Although genetics sets the cycle, temperature can alter the number of life cycles in a year. Colder temperatures lead to just one generation of spruce bark beetles per year, whereas warmer temperatures with milder and shorter winters enable more life cycles per year (see Figure 10). Within the same boreal forest, for example, north-facing slopes will tend to support only one generation per year, whereas south-facing slopes can support two. With global warming affecting many forests, the number of life cycles is increasing, which is also increasing the total population of beetles and therefore the number of trees being attacked. This can have a negative economic impact on timber industries if temperatures continue to rise, because for them the beetle is a pest that is destroying their livelihood.

From an ecological viewpoint, the beetles tend to attack older trees, ones that have been damaged or ones that are already in a poor state of health. Hardier and healthier trees are better at defending themselves and resisting beetle attacks. As a result, the spruce bark beetle plays an important role in killing off less healthy trees and clearing some of the canopy for a new generation of trees to flourish.

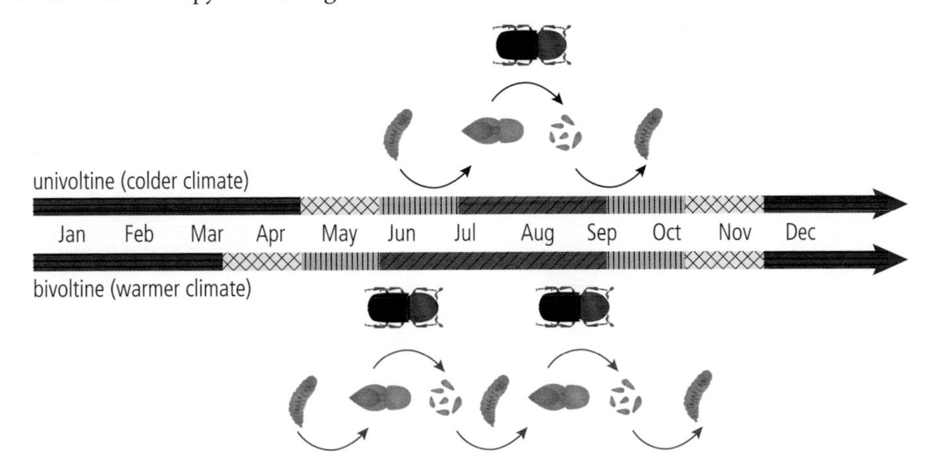

D4.3 Figure 10 When winters are colder (the top left longer blue line), spruce bark beetles have a monovoltine cycle, but when winters are milder (the bottom left shorter blue line), spruce bark beetles can have a bivoltine cycle.

D4.3.12 – Evolution and climate change

D4.3.12 – Evolution as a consequence of climate change
Include changes in the fitness of colour variants of the tawny owl (*Strix aluco*) as a consequence of changes in snow cover. *Note: When students are referring to organisms in an examination, either the common name or the scientific name is acceptable.*

Microevolution in a bird species

The tawny owl (*Strix aluco*), also known as the brown owl, has two colorations, brown and grey. When genetics can produce two or more phenotypes in a population, the phenomenon is called **polymorphism**. Each variation is referred to as a **morph**; in this case there is a grey and a brown morph. Natural selection plays a role in determining which colour dominates in a population and, if a shift occurs over time whereby one morph is found in increasing frequencies, it is considered to be an example of evolution, or at least a microevolution. A study published in *Nature Communications* in 2011 examined data collected over 48 years and found that climate change is the most likely driving force in the increased frequency of the brown morph of the tawny owl in Finland (see Figure 11). The researchers investigated snow depth, which has decreased in recent years as a result of milder winters and changing precipitation rates in Finland.

The tawny owl (*Strix aluco*), otherwise known as the brown owl. The owl pictured is a brown morph.

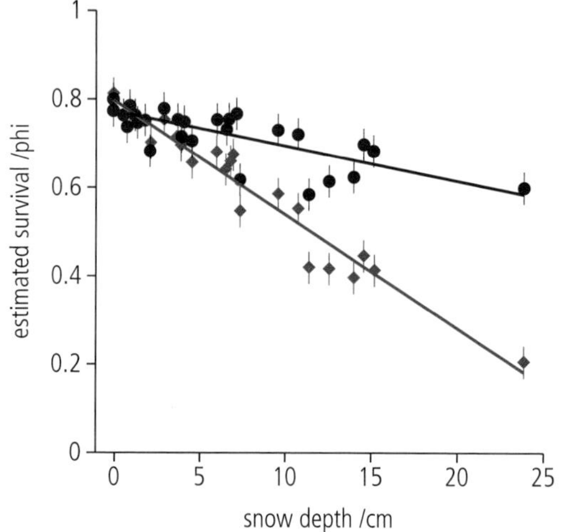

D4.3 Figure 11 Survival rates of grey tawny owls (black circles) and brown tawny owls (red diamonds) compared to snow depth.

Why does the depth of snow make a difference, and how does the survival rate interact with the colour of the owl? As with anything associated with ecosystems, the answer is not a single factor because, in an ecosystem, everything is connected. Tawny owls eat small rodents such as voles. The warmer temperatures associated with milder winters and less snow have altered vole population numbers: more voles are available to eat in the autumn now compared to previous decades. Harsh winters are associated with deep snow so perhaps the brown owls are more visible to predators on a snowy background. Whatever the cause, warmer temperatures make it more likely that the brown morph will survive, especially the ones that are not as well-adapted to colder climates. The top left of the graph suggests that, with very mild winters and little snow, the red trend line shows brown owls are better adapted than grey owls (the black trend line). The researchers concluded that survival selection was responsible for the change in polymorphism frequencies over time.

HL end

Guiding Question revisited

What are the drivers of climate change?

In this chapter you have learned that:

- human activity is the main driver of climate change
- activities such as burning fossil fuels or deforestation add carbon dioxide to the atmosphere
- methane is also a greenhouse gas and is added by farming methods such as raising cattle and rice farming
- there are positive feedback loops in global warming
- melting snow and ice decreases albedo so that the water underneath warms up more than usual, in a positive feedback loop
- in addition, warming oceans causes an accelerated release of carbon dioxide
- warmer temperatures mean that permafrost and peat also release carbon dioxide
- some forests may reach a tipping point, and rather than being carbon sinks they may become a carbon source.

Guiding Question revisited

What are the impacts of climate change on ecosystems?

In this chapter you have learned that:

- changes in ocean currents and atmospheric temperatures can lead to extreme weather events, drought and forest fires
- when the climate changes there are impacts on habitats
- warmer polar habits means that there is less ice to form breeding grounds for polar species
- changes to ocean currents can alter the timing of nutrient upwelling, which disrupts marine food chains
- an increase in carbon dioxide in oceans causes acidification and warming, which leads to bleaching of coral reefs and subsequent collapse of ecosystems
- species may change where they live, for example, montane birds may move further upslope
- HL ecologically timed events, such as when insects start key stages in their life cycles or when birds migrate, can be disrupted by the changes we see in the climate.

Exercises

Q1. Which gases are contributing to the anthropogenic runaway greenhouse effect?

I. Carbon dioxide.
II. Ozone.
III. Methane.

A I and II only. **B** I and III only.
C II and III only. **D** I, II and III.

Q2. The El Niño Southern Oscillation modifies which of the following?

I. Nutrient upwelling.
II. Precipitation near the coasts.
III. Primary production in marine ecosystems.

A I and II only. **B** I and III only.
C II and III only. **D** I, II and III.

Q3. Which Arctic animals are seeing their habitat modified by melting sea ice?

I. Polar bears.
II. Walruses.
III. Emperor penguins.

A I and II only. **B** I and III only.
C II and III only. **D** I, II and III.

Q4. Explain the positive feedback loop related to climate change and methane production in regions with permafrost.

Q5. Distinguish between how a garden greenhouse works and how the greenhouse effect on Earth works.

Q6. A scuba diver returns to her favourite coral reef only to find it empty of life and all the corals turned white. She asks you if you know what this phenomenon is: what do you tell her?

Q7. HL Maximum caterpillar biomass is available 10 days before an insectivorous bird's chicks hatch. This is an example of ___.

A Trophic mismatch. **B** Polymorphism.
C Artificial selection. **D** Assisted migration.

D4 Practice questions

1. Which statements are characteristics of alleles?

I. Alleles differ significantly in number of base pairs.

II. Alleles are specific forms of a gene.

III. New alleles are formed by mutation.

A I and II only.

B I and III only.

C II and III only.

D I, II and III.

(Total 1 mark)

2. What is a direct consequence of the overproduction of offspring?

A Individuals become more adapted to the environment.

B They will be subject to intraspecific competition.

C They will diverge to produce different species.

D They will suffer mutations.

(Total 1 mark)

3. What would restrict evolution by natural selection if a species only reproduced by cloning?

A Too few offspring would be produced.

B Mutations could not occur.

C The offspring would show a lack of variation.

D The offspring would be the same sex as the parent.

(Total 1 mark)

4. Some lice live in human hair and feed on blood. Shampoos that kill lice have been available for many years but some lice are now resistant to those shampoos. Two possible hypotheses are:

Hypothesis A	Hypothesis B
Resistant strains of lice were present in the population. Non-resistant lice died with increased use of anti-lice shampoo and resistant lice survived to reproduce.	Exposure to anti-lice shampoo caused mutations for resistance to the shampoo and this resistance is passed on to offspring.

Discuss which hypothesis is a better explanation of the theory of evolution by natural selection.

(Total 3 marks)

5. Explain how evolution by natural selection depends on mutations.

(Total 4 marks)

6. Extensive areas of the rainforest in Cambodia are being cleared for large-scale rubber plantations. Distinguish between the sustainability of natural ecosystems such as rainforests and the sustainability of areas used for agriculture.

(Total 3 marks)

7. Adélie penguins *(Pygoscelis adeliae)* are only found in Antarctica and need sea ice for feeding and nesting. Biologists are able to deduce how these penguins have responded to changes in their environment for the last 35,000 years, as the Antarctic conditions have preserved their bones and their nests. The image is a map of Antarctica and the surrounding Southern Ocean. It shows the trends in the length of the sea ice season (days of the year when sea ice is increasing) and the sites of nine Adélie penguin colonies.

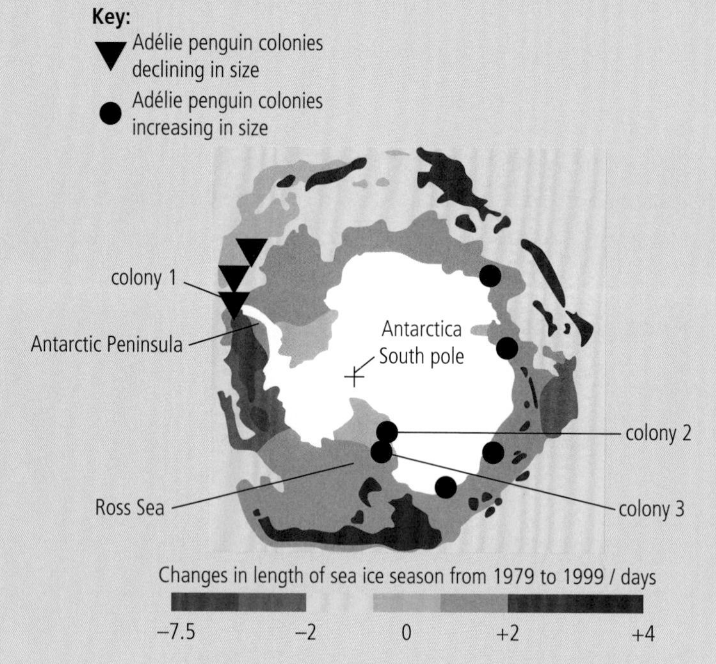

(a) Describe the trends in the length of the sea ice season around the Antarctic Peninsula and in the Ross Sea. (2)

The graphs shows the changes in penguin population in three of the colonies shown on the map.

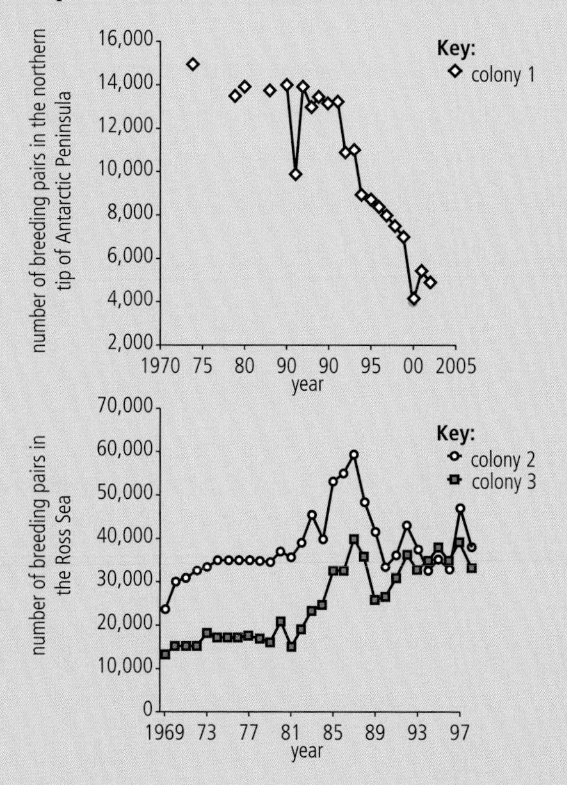

(b) Analyse the trends in colony size of the Adélie penguins in relation to the changes in the sea ice. (3)

(Total 11 marks)

8. Several greenhouse gases occur in the atmosphere. Carbon dioxide (CO_2) is one of them but so are methane (CH_4) and oxides of nitrogen (NO_x). Why are oxides of nitrogen classed as greenhouse gases?

A They trap some of the long-wave radiation emitted by Earth's surface.

B They prevent short-wave radiation from reaching Earth's surface.

C They dissolve in rainwater to produce acid rain.

D They are only produced by human activity, whereas CO_2 and CH_4 are also produced naturally.

(Total 1 mark)

9. The oceans absorb much of the carbon dioxide in the atmosphere. The combustion of fossil fuels has increased carbon dioxide ocean concentrations. What adverse effect does this have on marine life?

A Heterotrophs consume more phytoplankton.

B Phytoplankton have increased rates of photosynthesis.

C Corals deposit less calcium carbonate to form skeletons.

D Increased pH reduces enzyme activity in marine organisms.

(Total 1 mark)

10. Describe the relationship between the rise in the concentration of atmospheric carbon dioxide and the enhanced greenhouse effect.

(Total 5 marks)

11. The incidence of white syndrome, an infectious disease of coral, was investigated in a six-year study on Australia's Great Barrier Reef. The map shows disease conditions on coral reefs at six study sites.

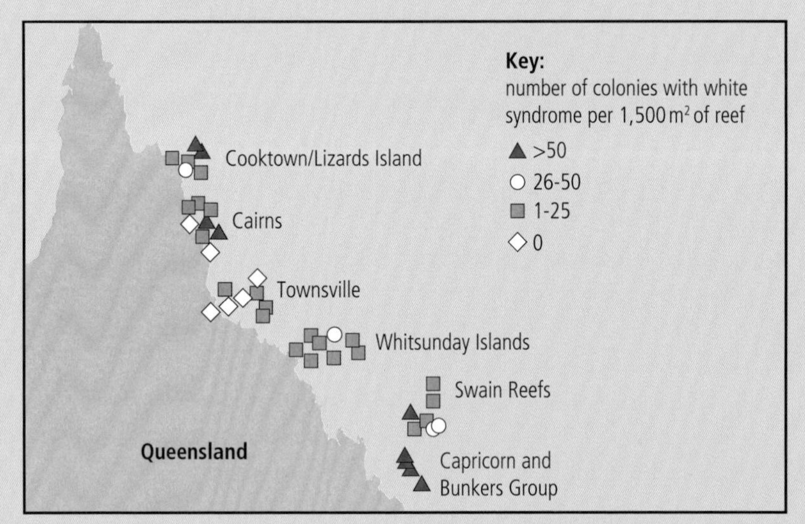

(a) Identify the site with the highest incidence of diseased colonies. (1)

(b) Deduce whether there is a trend in the incidence of white syndrome over the north–south range of latitude. (1)

(c) Satellites were used to record sea surface temperatures. The temperatures each week above a reef were compared with mean temperatures for that week between 1985 and 2004. If the sea surface temperature was 1°C or more above the mean, this was recorded as a weekly sea surface temperature anomaly (WSSTA). The number of WSSTAs was calculated for the twelve months preceding the date on which a reef was surveyed for white syndrome.

On each reef, the number of cases of white syndrome in a 1,500 m² sample area was surveyed once per year. The table shows these cases in relation to numbers of WSSTAs and coral cover on the reef. Low coral cover was 0–24% and high coral cover was 50–75%.

	Mean number of corals with white syndrome per 1,500 m²		
	Coral cover / %		
WSSTAs per year	**0–24**	**25–49**	**50–75**
0	0.9	0.9	10.4
1 to 5	3.6	9.8	23.3
>5	4.7	4.5	80.1

Describe the evidence that is provided by the data in the table for the harmful effects of rising sea temperatures on corals. (2)

(d) The researchers concluded that there was a threshold coral cover percentage, below which infection rates tended to remain fairly low. Using the data in the table, identify this threshold level. (1)

(e) Suggest a reason for a larger percentage of corals being infected with white syndrome on reefs with a higher cover of corals. (1)

(f) The graphs show the relationship between the WSSTAs and coral cover during two twelve-month periods (1998–99 and 2002–03), which were the warmest in the six-year study. Each dot represents one studied reef. (1)

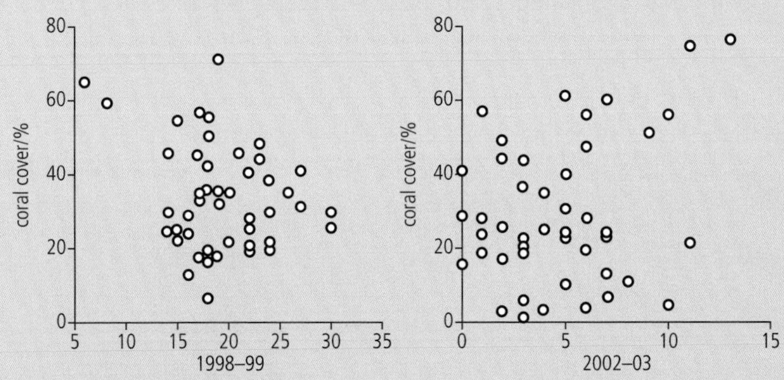

Compare and contrast the data for 1998–99 and 2002–03. (2)

(g) Suggest a reason for the correlation between coral cover and WSSTA in 1998–99. (1)

(h) Some scientists predict that, if humans continue to produce carbon dioxide at the current rate, the pH of the oceans will become more acidic. Suggest possible effects on the coral reefs. (3)

(Total 12 marks)

Theory of Knowledge in biology

Theory of Knowledge (TOK) is a way of thinking about our thinking. We ask ourselves questions about knowledge, about how we acquire, refute or confirm what we know. This chapter will guide you through how to think about TOK from a biologist's perspective in particular, and from the natural sciences' perspective in general. We can ask ourselves questions including "What makes biology different from the other natural sciences such as chemistry or physics in the way we acquire knowledge?" and "In what ways are the natural sciences different from other areas of knowledge: the human sciences, history, mathematics or the arts?" TOK is meant to challenge your brain and push you outside your comfort zone. Enjoy the ride.

An astronomer, a physicist and a mathematician are on a train going to a conference in Edinburgh, Scotland. Out of the window, they see a solitary black sheep.

- Astronomer: That's interesting, sheep in Scotland are black.
- Physicist: It would be more prudent to say that *some* of the sheep in Scotland are black.
- Mathematician: To be more precise, we can say that in Scotland there exists at least one field in which there is at least one sheep, which is black on at least one side.

What does this story reveal about scientific observations, hypotheses and conclusions? What does it reveal about the nature of each of the disciplines represented?

Is biology less exact than physics or mathematics? If there had been a biologist on board the train, what would they have said about the sheep?

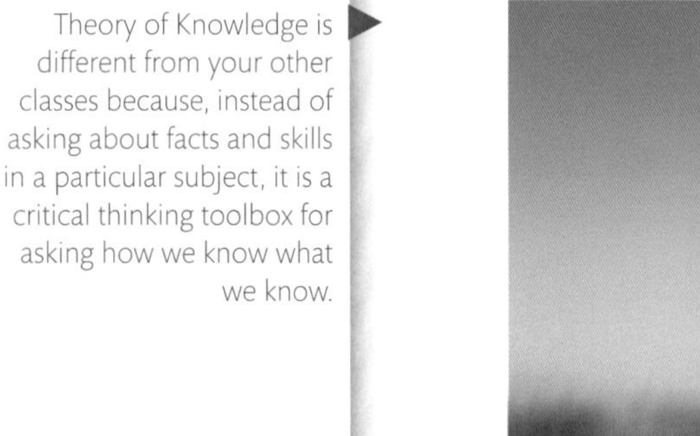
Theory of Knowledge is different from your other classes because, instead of asking about facts and skills in a particular subject, it is a critical thinking toolbox for asking how we know what we know.

What is this chapter all about?

A paradigm is a way of seeing the world, and a paradigm shift is when new information transforms that view.

This chapter contains ideas, quotes, anecdotes, case studies and many unanswered questions. With TOK, it is important to develop your own ideas and arguments. Start with a **knowledge claim** such as "all sheep in Scotland are black" or a **knowledge question** such as "What role does imagination play in the pursuit of knowledge in the natural sciences?"

The study of biology is filled with objects that should help you see connections to TOK. For your TOK exhibition, you are asked to pick three such objects. Many examples are included in this chapter, and the microscope is one of them. The microscope represents a **paradigm shift** in biological thinking because it opened up a whole new category of life: organisms that occupy the microscopic world. A paradigm is a way of thinking about something, and a paradigm shift is when something new completely changes the old way of conceptualizing things. The microscope allowed biologists to see for the first time that there is a parallel universe of life living right under our noses.

An antique microscope.

Knowledge claim: a statement that someone declares as being true from their perspective. Upon verification, it may be confirmed, but a knowledge claim can also be refuted.

Knowledge question: a question that asks how we know something, how we acquire, confirm or refute knowledge. Knowledge questions are open-ended (they do not have a single correct answer), contain TOK concepts and vocabulary and, ideally, can be applied to a wide variety of situations in the real world.

Debates

Consider the following two knowledge claims about the nature of all human beings on Earth.

A: We are all the same.

B: We are all different.

Use your biological knowledge to support or refute these two claims. Choose one, and try to imagine someone saying to you, "That's not true! How can you say that?" How would you respond to that person?

Now try these two statements.

X: Biology is a collection of facts about nature.

Y: Biology is a system of exploring the natural world.

Use your critical thinking. Critical thinking is characterized by reflective inquiry, analysis and judgement. Ask yourself: "Should I believe this?" "Am I on the right track?" "How reliable is this information?" In short, you are deciding whether or not you should accept something as valid. If you think it is an easy, quick decision, then you are not treating the question in the way that you should.

Critical thinking

Is the statement valid? What is its source? Is the person who said it reliable? Do they have a **bias** that I should know about? Bias refers to a type of prejudice whereby a person gives an unfair preference to one perspective over another, rather than giving a balanced argument.

Coming back to the pairs of statements on the previous page, what would lead someone to believe one or the other? For each pair, could it be possible that both statements are valid? Or are they necessarily mutually exclusive? What about the following two statements?

- There is only one scientific method that is universal throughout the world: only by following the same method can scientists reach the same results and conclusions.
- Different scientists and different cultures in different regions of the world use different versions of the scientific method to obtain valid results and conclusions.

Is there just one scientific method? Do all scientists always follow the same steps?

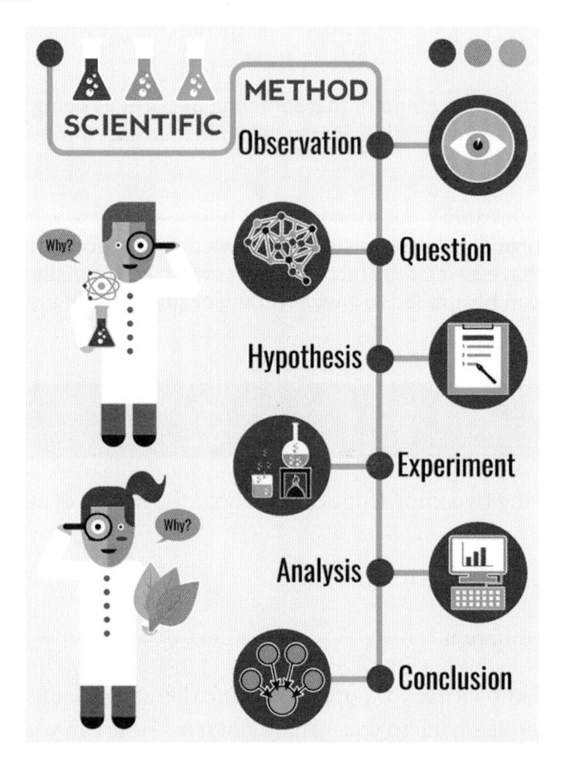

TOK knowledge framework

One of the most important skills students are asked to develop in the IB programme is **analysis**. Analysis is the process of breaking apart something complex in order to see how it works. In biology, "lysis" refers to splitting, as in *cytolysis* or *hydrolysis*, words that are based on the ancient Greek word for "loosening" or "releasing". When you analyse something, hopefully you can pull out or release some meaning.

When you want to "analyse" something, where should you start? How should you break a concept up into its constituent parts? There is no single answer, but the TOK framework is a useful tool for analysing knowledge questions. When analysing a particular scenario, you are not expected to address each of the four points below exhaustively, but this framework can be a good place to start.

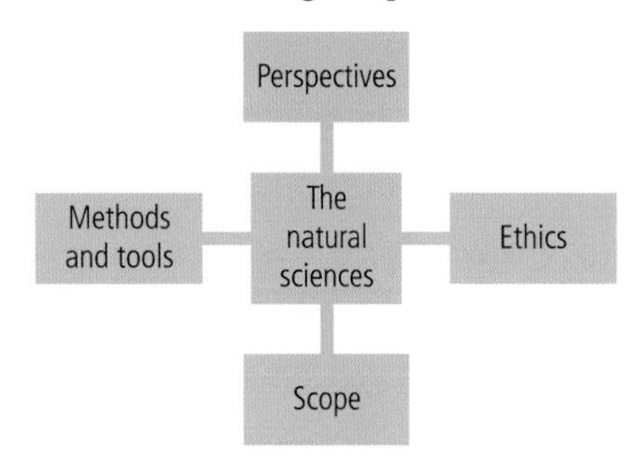

This framework is a useful tool for analysing knowledge questions.

In the TOK knowledge framework:

Perspectives refer to someone's point of view. It might be the perspective of an individual or it might be that of a group, such as molecular biologists or neo-Lamarckists.

Ethics refers to questions of responsibility. In TOK we ask how we know if it is morally right or wrong to genetically modify crops or to experiment on animals.

Scope concerns what questions the natural sciences can and cannot answer. The scientific method is only good at answering scientific questions, for example.

Methods and tools gets us to think about how scientists produce, acquire or refute knowledge. What is the scientific method and how is it different from the approach that historians or mathematicians use in their area of expertise?

Try the framework out on the sheep example at the beginning of the chapter. There may be some aspects of the framework that apply nicely, and others that do not fit well. Throughout this chapter there are a number of case studies; as practice, it is worth analysing them based on the knowledge framework. This is good practice for your essay writing because you are encouraged to incorporate ideas from the framework in your TOK essay.

What is knowledge?

Consider the knowledge questions below.

- What counts as knowledge in biology?
- How does biological knowledge grow?
- What are the limits of knowledge in biology?
- Who owns biological knowledge?
- What is the value of knowledge in biology?
- What are the implications of having or not having biological knowledge?
- Is there one way that is best for acquiring knowledge in biology?
- Where is biological knowledge? Is it a "thing" that resides somewhere: is it in books, in your head, in a computer database?

Exercise

Look at the following images. Use the list of questions on the previous page and your critical thinking to evaluate whether some or all of these are valid as scientific knowledge. For example, can mythology count as scientific knowledge? To help guide your discussion, consider the following key concepts in TOK:

Evidence Certainty Truth Interpretation Power Justification
Explanation Objectivity Perspective Culture Values Responsibility

How do we know?

An example of scientific knowledge in biology is: "The organelle in a plant cell that is responsible for photosynthesis is the chloroplast". How could you verify this? How can you be sure that there is not another part of the cell that performs photosynthesis? Is it a falsifiable idea? Such questions are **second-order questions**. They are not about the thing we want to know, they are about how we know it. In your IB Diploma subject exams and internal assessments, you are asked lots of first-order questions, but in TOK you need to focus on second-order questions.

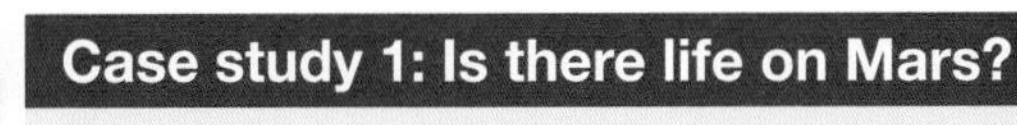

Case study 1: Is there life on Mars?

This is not a simple question. Despite several visits by space probes, it has been extremely difficult to find conclusive evidence on Mars that can lead scientists to declare that there is life on its surface. And yet the search continues. One piece of compelling evidence that there was once life on Mars comes from a meteorite found in Antarctica that The National Aeronautics and Space Administration (NASA) claims came from Mars and contains fossils of bacteria.

If you apply your critical thinking, some knowledge questions should pop into your mind. How do we know that this chunk of rock is really from Mars? How does NASA know that the "fossils" are from bacteria? How certain are we that they were not formed from non-living chemical reactions?

Knowledge framework: Scope The question about life on other planets falls within the realm of knowledge that can be acquired using the natural sciences. Decades of data collection from probes sent to the surface did not reveal the presence of life, but is this an indication that there is no life on Mars, or does it show the limitations of our technology?

Knowledge framework: Methods and tools Although no humans have gone to Mars, researchers collect photos and soil samples, as well as measure the temperature and composition of the atmosphere, using various technologies on robotic probes. NASA analyses the data and proposes hypotheses. Subsequent missions are sent to collect more samples and test those hypotheses.

Knowledge framework: Ethics How do we decide that it is acceptable to spend millions of dollars on space missions when there are still so many problems to solve here on Earth? At what point do we accept that the amount of pollution here on Earth, in space and on Mars produced by the rockets and probes is worth the knowledge we gain from such missions?

Knowledge framework: Perspectives Such debates will have different points of view depending on which stakeholder you ask. NASA is likely to view the missions differently compared to environmentalists or groups fighting for social justice. Biologists will argue that it would be difficult to imagine any greater discovery in human existence than finding life on another planet.

From this specific example of life on Mars, two more general questions arise. Is it possible to really "know" the truth? Is information absolute or relative?

Are you an empiricist or a rationalist?

Critical thinking does not mean you criticize everything and refuse to ever accept anything as valid. It means you are aware of questions of validity. You are not being negative, you are just being inquisitive and prudent.

Empiricism = the belief that our senses allow us to acquire knowledge.
Rationalism = the belief that reason allows us to acquire knowledge.

Case study 2: Babies born on a full moon

Knowledge framework: Perspectives

Ask an experienced midwife "Are more babies born on a night when there is a full moon?" and the chances are she will say yes. You would have no reason to challenge her: she is the expert. She is an eyewitness to this phenomenon.

But knowledge questions arise. Is this verified knowledge or is it opinion? How does she know? Is it just a feeling, an intuition, a belief? Or is this knowledge claim based on carefully analysed statistics comparing birth numbers with a lunar calendar?

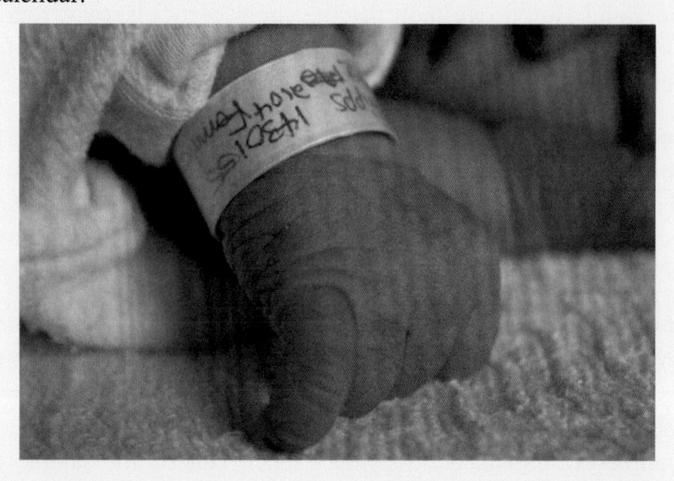

As it turns out, statistics do not support this knowledge claim. The evidence from maternity ward numbers does not show a correlation between births and the full moon. So what is going on? Is the midwife lying? In fact she is probably the victim of something we all are susceptible to: **confirmation bias**. Confirmation bias happens when we only remember the times when something confirmed our beliefs, and ignore the times when something refuted them. In the case of the midwife, on a busy night she might look out the window, see a full moon, and cry out to her colleagues, "See? I was right! More babies on nights when there is a full moon". Two weeks later, on another busy night, she looks out the window and what does she see? No moon at all. It is unlikely that she will now tell her colleagues, "Sorry, I was wrong: it's a busy night and yet there is no full moon". It is more likely that she will forget this negative result and only remember the positive result, thereby showing a bias for confirmation.

Confirmation bias happens when we only remember the times when something confirmed our beliefs and ignore the times when something refuted them.

Should we tell her she is wrong? It could be argued that she is not hurting anyone and that it is lots of fun to have these sayings in our culture. Having shared beliefs unites people and strengthens a sense of community and belonging. Is it better to be right or to belong?

TOK journal

Do you keep a TOK journal? You should. Write down any time one of your teachers mentions TOK, or any time you see a situation that could lend itself well to a knowledge question. For example, on a field trip you might be asked to determine how polluted a body of water is. You do some chemical tests and you catch some larvae that you identify as being sensitive to pollution. But then, as a good biologist and good TOK student, you start asking yourself how certain you can be that your measurements will allow you to conclude that the body of water is "clean" or "safe" (**Scope**, **Methods and tools**). What if you found that the water was contaminated by a local industry that makes products you like to buy? Is it your responsibility to inform the local authorities (**Ethics**)? What if you get different results and draw different conclusions compared to your classmates, or if you think that it does not matter what the data shows, it is more important that we keep industry productive in the region for financial success and access to local products (**Perspectives**)? How do you decide whose results are more valid?

Biology fieldwork should get you thinking about knowledge questions.

Catching a cold

Despite the biological evidence that colds are caused by viral infections, many people believe that you can catch a cold from being exposed to low temperatures or changes in humidity.

Who is right? Where does the truth lie? For something to be considered "true", does it have to be formally proven using a scientific method? Is a profound conviction that something is true good enough to make it valid? If one person believes that something is true, does that make it true or does there have to be a certain number of believers before the idea can be considered true? What if my grandmother says it and she learned it from her grandmother? Does that make it a valid claim?

Theories versus laws

A **law** is a generalization used to describe a phenomenon. The laws of thermodynamics describe the flow of energy in a food chain, for example. We make measurements and observations and realize that energy cannot be created or destroyed, only transformed from one type (e.g. light energy from sunlight) to another (e.g. chemical energy in sugar) and another (e.g. heat energy released during cellular respiration reactions). This is the concept of conservation of energy. It is a law of nature. But there is nothing in the law that attempts to explain the phenomenon. Laws only describe. They are not falsifiable because they make no contestable claims.

A law describes a phenomenon and makes no attempt to explain it. Laws are not contestable because they make no claims. A scientific theory is a proposed explanation. A theory is contestable, and by challenging and testing it, the theory can be confirmed or refuted. A theory can be modified over time, or can be replaced with a better theory as new evidence appears.

A **theory** is an explanation of a phenomenon. It can start out as a hypothesis, but to qualify as a theory it must be thoroughly tested and have ample evidence to support it, with little or no evidence to refute it. A theory is testable and can be refuted if enough evidence builds up against it, or can be replaced if a new theory does a better job of explaining the phenomenon. Robust theories are those that have withstood many attempts to falsify them. Darwin's theory of natural selection to explain evolution, for example, is one of the most tested and robust theories ever published in biology. For over a century and a half, it has withstood countless attempts to refute it. If someone finds one exception that goes against a theory but there is still overwhelming evidence for it, we do not abandon the theory.

The word "theory" can be problematic because different groups of people use it in different ways. Sometimes when someone says, "My theory is that this is caused by ...", they are really referring to a hypothesis. They are suggesting a possible cause for a phenomenon. When it is used this way, it is sometimes qualified as "just a theory" meaning that it is, in fact, just a hypothesis. How can the language of science contribute to or detract from clarity when communicating knowledge?

Phrenology

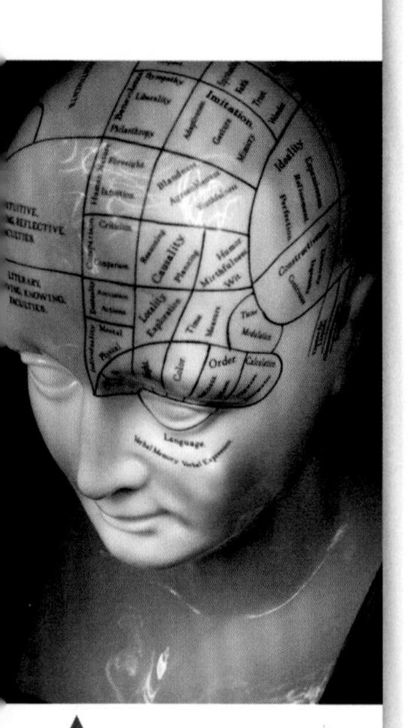

The study of phrenology has been discredited.

The **pseudo-science** of **phrenology** claimed that the shape of a person's skull, and the bumps and indentations on it, determined a person's intelligence, personality and talents. A pseudo-science is a discipline that is presented as if it is a science but does not follow the rules of scientific inquiry and scrutiny. For example, in science, if a large body of evidence refutes an idea, the idea should be abandoned. Phrenology persisted for decades despite mounting evidence that it was not confirmed by observations and experiments. More controversially, it was used by some to justify the superiority or inferiority of "races" of humans.

How do you think it was demonstrated that the "laws" of phrenology were not, after all, scientifically valid? What needed to be done to disprove them? How can the social context affect what is studied and how? If scientists today wanted to use genome analysis to compare "races" the way phrenology did, how would the scientific community or the rest of society react? The word "race" has been used freely to refer to different groups of humans for hundreds of years. Linnaeus used the term in the 1700s. Although we still use the concept in words such as "racism" or "racist", scientists very rarely if ever use the term "race" when referring to subgroups of humans. This is an example of a **historical development**. It also shows how language changes over time as ideas about what the words represent change.

The term historical development in TOK refers to a change in methods and tools in an area of knowledge. For example, the use of the microscope to discover bacteria was an astounding historical development. More recently, genome sequencing has helped us see evolutionary relationships with much more detail. Do not confuse a historical development with a historical event, such as the signing of a treaty or the first time humans walked on the Moon.

Tongue map

As students and teachers, what do we claim to know about biology? Are we justified in making such claims? How?

What experiences have you had that give you insight concerning these issues? Consider the following example.

For generations of students, the idea of a "tongue map" (i.e. certain zones of the tongue relate to certain tastes) was propagated by biology textbooks, and taste-test investigations were suggested as laboratory work in schools. It has since been shown that all parts of the tongue can taste sweet, sour, bitter and salty.

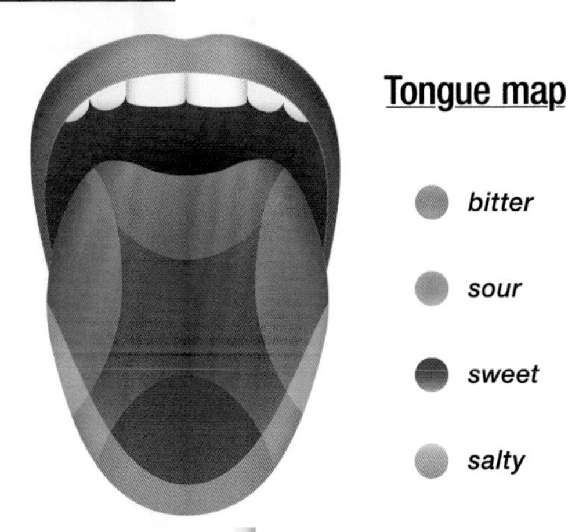

▲ The idea of a tongue map, with zones where certain flavours are sensed, has been discredited.

> *There must be no barriers for freedom of inquiry. There is no place for dogma in science. The scientist is free, and must be free to ask any question, to doubt any assertion, to seek for any evidence, to correct any errors.*
>
> Robert Oppenheimer
> 1949

Art and imagination

Is there a place for creativity and emotional expression in science? Are there any parallels between biology and the visual arts? Could it be argued that just as an artist sees the natural world in their own way, so a scientist sees nature in their own way? Or, on the contrary, are science and the visual arts diametrically opposed ways of interpreting nature? Was someone like Ernst Haeckel, who drew visually striking drawings of what he observed, just as much a visual artist as a biologist?

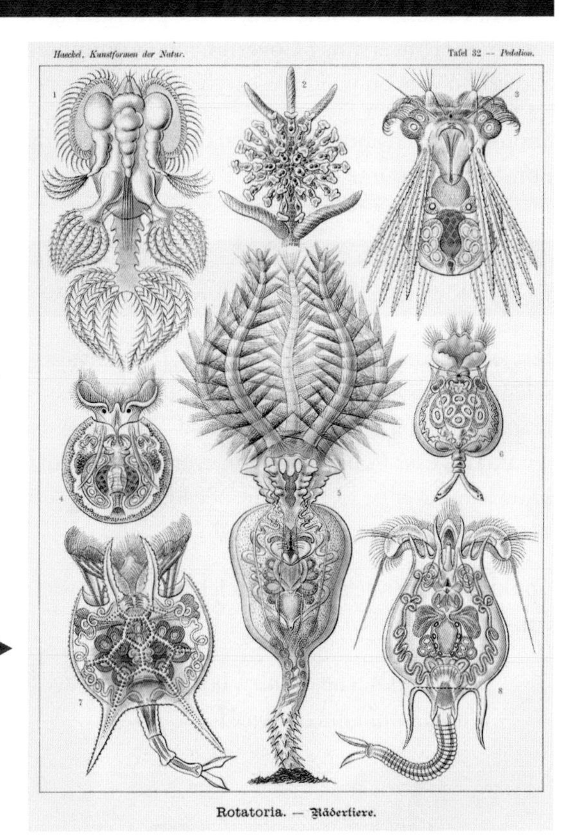

Plate 32. *Rotatoria*, from Haeckel's *Kunstformen der Natur* (*Artforms of Nature*), published in 1904, showing members of the genus *Rotatoria* as observed under a light microscope. ▶

Case study 3: Spirit/soul

Knowledge framework: Scope

In 1907, Dr Duncan MacDougall conducted experiments to determine whether people lost mass after death. His results seemed to suggest that they did, and led him to the conclusion that the human soul weighs 21 g. As his experiments (some of which did not give conclusive results) were carried out with scales of questionable accuracy, and he had only six subjects, his conclusions are widely criticized and are not taken seriously by the scientific community today.

Will questions about souls always remain beyond the capabilities of science to investigate or verify? Why has no one repeated this experiment in over a century? What do you think the reaction of the religious community would be if scientists repeated MacDougall's experiment?

Decisions, decisions ...

Should experiments be performed to answer fundamental questions, or should they only be done if they have a useful application in our everyday lives?

Who should decide which research pursuits are of the most value? Who should decide on how funding is distributed, or the prioritizing of the use of laboratory space and resources? Universities? Governments? Committees of scientists? Should taxpayers be allowed to vote on which research projects receive public funding?

Should research about a tropical disease such as malaria be paid for by tax money from non-tropical countries?

Is there an end?

Is scientific knowledge progressive? Has it always grown? Imagine a graph with scientific knowledge on the *y*-axis and time on the *x*-axis. How would you draw the graph? Would it be a curve or would it be linear? Is it always increasing? What units would you use? Could the graph ever go down? In other words, could scientific knowledge ever be lost (maybe because of the outbreak of war, a laboratory burning down, a brilliant scientist dying)?

Could there ever be an end to science? If there was an end, what would be the consequences?

> *Science knows no country, because knowledge belongs to humanity, and is the torch which illuminates the world.*
>
> Louis Pasteur

Doctor, which drug treatment is best for me?

How do doctors know which medication is best for their patients? One way is for them to keep up with the latest breakthroughs and developments published in scientific and medical journals. Doctors put their faith in these prestigious peer-reviewed journals and, because they do not have the time or the budget to do all the clinical trials themselves, they trust that the researchers doing the work are following sound practice. One problem is that a large percentage of these studies are being funded by the companies that make the drugs and, according to epidemiologist Dr Ben Goldacre's 2012 book *Bad Pharma*, it is common practice in the pharmaceutical industry to use a wide variety of tricks and manipulations to make a new drug look good in clinical trials. One trick follows this type of pattern: a company sets up a 2-year trial to test a new drug and then, after only 6 months, it decides to stop the trial and publish the data because the numbers show that its drug is performing well. This is advantageous to the company because it saves money (trials are very costly), and it reduces the chances that participants develop side effects or show negative results. Doctors reading about the clinical trials will never be informed, however, that the trials were stopped early. Another trick is to not report in the published study any participants who dropped out of the trial because they felt ill from side effects. By only mentioning the people who stayed in the study, they can report that, at the end of the trial, none of the participants complained of any major side effects. In short, Goldacre claims that the studies being published are not showing all the data and that, in order for doctors to decide whether a drug is safe to prescribe, they need to see both the positive and the negative results. What knowledge questions does Goldacre's book raise about the highly competitive pharmaceutical industry (that earns hundreds of billions of dollars annually)? Do the practices he denounces sound like the kinds of things your biology teacher encourages you to do in your laboratory investigations? If you were a scientist working at one of these companies, and you decided to complain and point out that some of the trials seemed unfair, what do you think your boss's reaction would be? If a company decided to publish the positive as well as the negative results of its drug trials, what do you think might happen to the sales of its drugs? Lastly, as doctors find out more about the practices followed, what will happen to their faith in the data presented by the medical journals? What kind of critical thinking or TOK questions should doctors apply when they pick up a medical journal and read about the latest breakthroughs in drug research?

The placebo effect

One of the ways that scientists test a drug is to compare it with a **placebo**. A placebo contains no active ingredients: it is often just a sugar pill. To find out if a new drug is effective, one group of volunteers in a study is given the drug and another group is given the placebo. Neither group knows whether it is taking an active pill or a placebo. Surprisingly, even in the group taking the placebo, there are patients who report that they feel better. This is called the **placebo effect**.

A placebo is a helpful tool in understanding the efficacy of drug treatments.

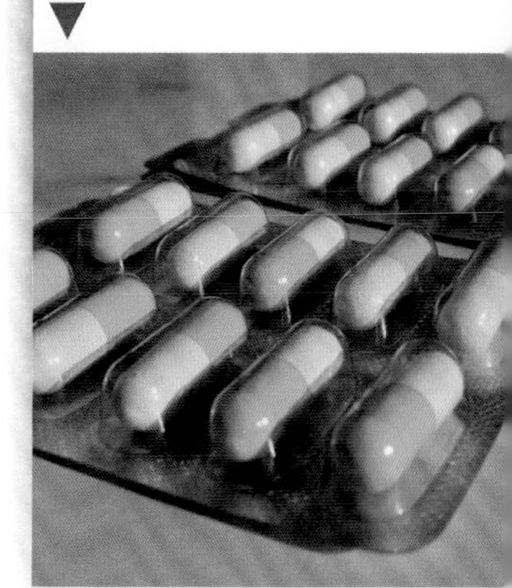

According to Goldacre, researchers studying the placebo effect have observed that the following things have a positive influence on how effective the patients thought the pill was:

- the doctor was wearing a white laboratory coat
- there were diplomas on the wall in the doctor's office
- the doctor sat down and listened attentively to the patient.

The medical community is essentially unanimous on the validity and power of the placebo effect, and yet the mechanism of how it works is poorly understood. For example, astounding as it may seem, placebos seem to have an effect even when people are told that they are receiving a placebo. Some participants still feel better, even when they are aware that they have not been given any active drugs.

Inhabitants of industrialized countries often discredit herbal medicine and healers in indigenous peoples. And yet, those same critics may very well accept the effectiveness of the placebo effect: an effect that appears to be produced essentially by ritual (laboratory coat, diplomas, attentive listening). What knowledge questions are raised by this puzzling effect? What does it say about the limits of modern medicine?

Models

The double-helix shape for DNA and the fluid mosaic membrane are examples of models that were created in order to explain observed phenomena. Are such models just inventions of our imagination? If so, how is it that they can be used to make predictions or explain natural phenomena?

> *All models are wrong, but some are useful.*
>
> George E. P. Box
> (innovator in statistical analysis)

Who's right?

Among all the points of view that are available to you in the classroom, at home, in the media, on websites, how do you know which to trust?

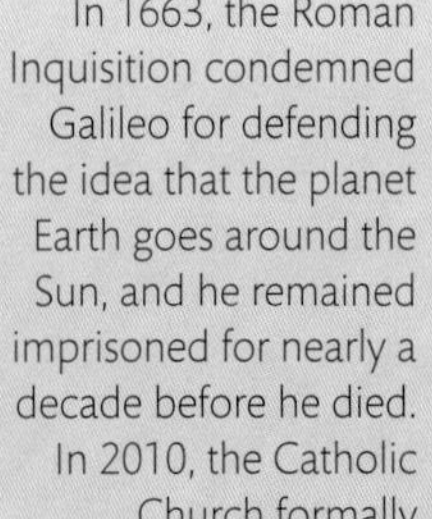

In 1663, the Roman Inquisition condemned Galileo for defending the idea that the planet Earth goes around the Sun, and he remained imprisoned for nearly a decade before he died. In 2010, the Catholic Church formally apologized for Galileo's condemnation.

Religion in an age of science

In what ways could someone's cultural or religious background influence their acceptance of certain scientific theories?

There was a time when scientists hesitated to publish their works out of fear of the church's reaction. Have the tables turned? Are there religious writers who fear scientific criticism if they publish their ideas?

If a student refuses to answer questions about evolution by natural selection on an IB exam because of their religious beliefs, should they get any marks?

Ockham's razor

Simply put, the **principle of Ockham's razor** states that, all other things being equal, the simplest explanation should be preferred. This is reflected in the idea of **parsimony**: seeking out the least convoluted solution. Scientists take this principle very seriously and yet some aspects of science seem to be extremely complex. Is there a conflict here?

Limits of perception

Can we, here on Earth, possibly know of worlds beyond our own? Can we know what the distant past was like, or what the distant future will hold? Or are we like a frog at the bottom of a well trying to understand what the ocean might be like?

> *You cannot speak of the ocean to a well frog …*
>
> Chuang Tzu Taoist text
> (written more than 2,000 years ago)

The eye is not a camera

A fun classroom activity is to have someone unknown to the students barge in during a lesson, say something, take something off the teacher's desk, and leave, after which the teacher asks each student to write down a description of the person, and what they said and did. The students' observations are often remarkable in their diversity, and the activity demonstrates how human perception is notoriously bad at picking up crucial details, and notoriously good at filling in missing information. "The eye is not a camera" is a good example of a knowledge claim that can be explored and discussed after such an observation activity.

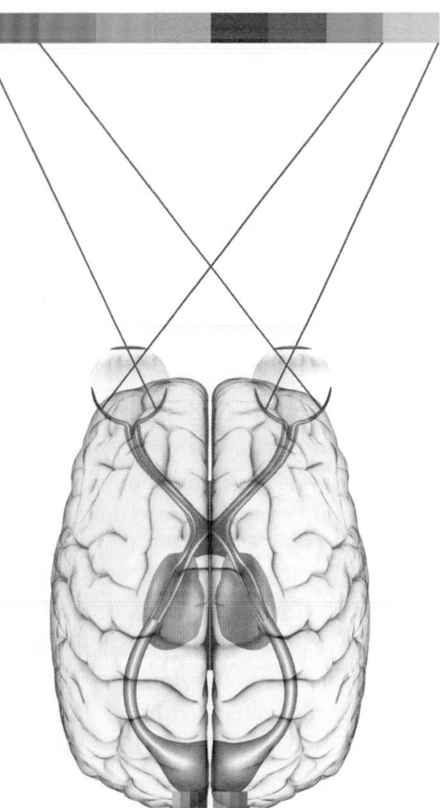

Do we see with our eyes or with our brain?

Although we associate eyes with vision, truly seeing something means interpreting the signals that arrive at the photoreceptors in the eye. This interpretation is done in the brain, so in fact, do we see with our eyes or our brain?

Another knowledge claim on this theme also relates to eyewitnesses: "all memories are reconstructed memories". Have you ever had a story in your family that was told time and time again for years, and then one day you found out that the event in question never actually happened? And yet, you could swear that you remember the event clearly. Or have you ever watched a video recording of something you experienced and thought to yourself, "That's funny, I don't remember it being like that: my memory of that event is very different". Such examples put into question the validity of eyewitness testimonies in a court of law. Given that we know that our memories and observations can trick us, should we trust an eyewitness's account as irrefutable evidence during a trial? Are such testimonies reliable enough to put defendants in jail or to sentence them to death? This theme is explored well in Sydney Lumet's 1957 film *Twelve Angry Men*.

We were wrong, here's the real story ...

In palaeontology, it seems that every time a new hominid fossil is dug up, we must redraw the human family tree. If you search the internet for human phylogeny, you will probably find that few sources agree with each other. Likewise, every few years nutrition experts change their minds about dietary advice.

Does this frequent revision give credibility to science, or does this make science less credible?

Perspective 1: It is important for scientists to be able to modify ideas as new evidence is revealed. This is how science grows and progresses and, without such a system, we would be intellectually stuck.

Perspective 2: Why can so-called experts not make up their minds? One year they say one thing, and then a year or two later they say "Oh, we were wrong, here's the real story".

Archaeopteryx

Here is an object that should help you see connections between TOK and the real world: *Archaeopteryx*, one of the most famous fossils in the world. It has some features of a dinosaur, such as reptile teeth and a bony tail, but it also has some bone structures similar to a bird and the most bird-like feature of all: feathers. It did not take long for observers to jump to the conclusion that *Archaeopteryx* is the "missing link" between dinosaurs and birds. Can we be so sure that this fossil is the transition between the two? Are physical features enough to base such a decision on? What kind of evidence would give more credibility to this claim?

Many palaeontologists shy away from phrases like "missing link": what features of this use of language make the term unscientific? One way to approach this question is to examine the **assumptions** and **implications** of the phrase "missing link". It assumes species like dinosaurs and birds have enough similar features to say the latter descended directly from the former. In other words, you could link the various transformations together over millions of years into an unbroken chain. This implies that if we find an organism that is half-and-half, it must be one of the links in the chain. Can you see the limitations of jumping to conclusions like this? One perspective to consider is given by palaeontologists who say, "What if *Archaeopteryx* is on a branch of the tree of life that led nowhere and all its descendants are now extinct?" It is as plausible as the hypothesis that it is the "first bird".

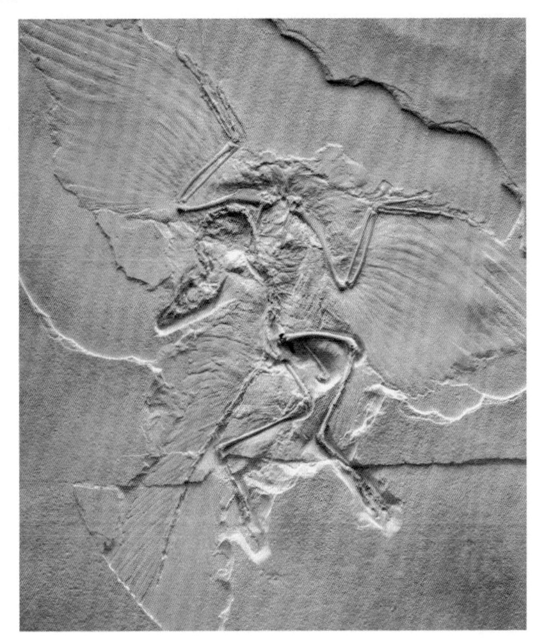

Archaeopteryx, a fossil with features of both dinosaurs and birds. Can we call it a missing link in evolution? It depends on your perspective.

Assumption: an idea that is likely to be true and presumed to be true despite the fact that we are not certain that it is true. Assumptions are based on previous observations and experience and should be reasonable. Assumptions are often tacit, meaning that they are not stated explicitly. You have to read between the lines to find them.

Implication: a logical consequence. If the assumption turns out to be true, then this logical consequence will be true, too.

Here is an example.

- Observation: I see only white swans.
- Assumption: All swans are white.
- Implication: The next swan I observe will be white and I will never observe a swan of a different colour.

Assumptions are often a necessary starting point in an argument but we need to be careful to recognize that we are only presuming that they are true, and we should not be surprised when they sometimes turn out to be false. You are encouraged to consider assumptions and implications in TOK.

Nature of science?

For centuries, it was firmly believed that rats, maggots and mould sprang from rotting meat and vegetable matter. This was called spontaneous generation. It took tireless experiments by Louis Pasteur and others to refute this idea and demonstrate that the rats, maggots and mould came from the surrounding environment

The end of spontaneous generation

The idea of spontaneous generation has been shelved as unscientific. It has no value as biological knowledge, but it does have historical value and it helps to illustrate how science works.

This is a good example of an original hypothesis that was disproved and falsified by experimentation. It can be argued that, in order for something to be considered valid as scientific knowledge, it has to be verifiable. If experiments show that the results do not support the hypothesis, or even refute it, the idea is falsified. This assumes that the experiment is repeatable. Other scientists should be able to do the same experiment and get similar results.

Case study 4: Science and government

Knowledge framework: Perspectives

Trofim Denisovich Lysenko was a Soviet biologist who opposed the ideas of Mendel and Morgan concerning genetics. Instead, he promoted the Lamarckian idea that acquired characteristics could be passed on from one generation to the next. Under Stalin, he was promoted to a high-ranking post in agronomy and given his own scientific journal to publish his ideas. The agricultural techniques he developed were used to feed the Soviet population and the Red Army. Once Stalin and Khrushchev were no longer in power, however, his methods were widely criticized and his theories attacked for lack of scientific validity. An inquiry revealed that, in order to retain his powerful position and promote his ideas, he had intimidated and removed scientists who questioned his theories. He was finally fired from his post at the Institute of Genetics in 1965 and his reputation was crushed. What does this story reveal about the influence of politics on scientific theories? In what ways does it reveal scientific bias? How do we know that Lysenko's critics were not simply trying to push their own opposing political agenda?

Unprovable assumptions?

Does biology make any assumptions that are impossible to prove? Consider this knowledge claim: all events in nature are caused by physical phenomena.

In other words, every natural event can be explained by the interactions between atoms and molecules. Is such a statement provable? If we find enough examples of instances where this is true, can we proceed by **induction** that it is true for all phenomena? Induction, or inductive reasoning, is when we look at many examples of a phenomenon and try to come up with a general pattern. This seems reasonable, and yet the philosopher David Hume criticized induction, saying that there is no logical reason to assume that it is the case. Consider Karl Popper's quote about swans, which illustrates clearly the problem of induction. Interestingly, because the claim can be tested and refuted, it makes it a scientific claim.

> *No matter how many instances of white swans we may have observed, this does not justify the conclusion that all swans are white.*
>
> Karl Popper
> 1992

For a long time, Europeans thought there were only white swans. Black swans do exist. They are native to Western Australia.

Scientific science

To what extent is there an overlap between biology and the social sciences? Are the latter "less scientific"? Consider psychology, sociology, anthropology and economics.

Science vocabulary

Does scientific language and vocabulary have a primarily descriptive or interpretive function? Consider the following expressions.

- Natural selection
- Concentration gradient
- Artificial intelligence

Wikis

Online wikis are filled with user-generated content on a wide range of subjects, including scientific ones. Wikis have been created for scientists to upload their latest laboratory findings. In what ways is this useful to scientists wanting to publish their results? In what ways is this useful to the general public? In what ways does this go against the very nature of peer-reviewed scientific publications, which is the norm today for sharing experimental results? For example, are such wikis just as valid as traditional scientific journals? How does the scientific community look upon these? Speaking of which, is there any such thing as a "scientific community"? Who belongs to it and how do we decide? When a specialist says something controversial and the media says that "the scientific community" does not agree with them, does that imply that they are not part of the scientific community? What about a wiki or a scientific journal for failed experiments? By seeing a list of published failures, would that not save researchers time by not repeating the same mistakes? Or, could it be that, if another scientist reads what one team thought was a failure but sees it for what it really is, a breakthrough, would that not help science advance?

Prediction is very difficult, especially about the future.

Niels Bohr
(a Danish physicist who helped us understand how atoms work) 1970

Case study 5: Herbal medicine

Knowledge framework: Methods and tools

An Indigenous woman collecting an uncultivated fruit in a forest in Jharkhand, India.

If a traditional treatment using a medicinal plant was tested in the laboratory and found to be effective, would that make the knowledge of the plant's properties more true? What if a single study disproved the plant's effectiveness as a medical treatment? Would that erase thousands of years of traditional knowledge? Turmeric, for example, has been used in herbal medicine for thousands of years to relieve digestive problems and inflammation. It is also widely used in savoury culinary dishes. Multiple scientific studies have been carried out to test its medicinal properties but none has conclusively seen any medical benefits. Have some types of knowledge been devalued by the perceived primacy of modern science? Conversely, today, some conventional Western medicine practitioners prescribe treatments that are categorized as complementary and alternative medicine, such as herbal remedies. What evidence would doctors need to be convinced that a herbal remedy is effective?

Seeing is believing: but what if you cannot see?

There is a story about a small group of blind men who encounter a tame work elephant, a creature none of them has ever had contact with before.

- One blind man touches the elephant's side and says "It's like a wall".
- Another grabs the end of its tail and says "It's covered in long hairs".
- Another feels a leg and says "Elephants are round and vertical like a pillar".
- A fourth holds his ear and says "It's like a sail".
- A fifth holds the animal's trunk and exclaims "Elephants are like snakes".

None of the men is wrong but no one is completely correct. This story illustrates how easy it is to jump to conclusions before having all the evidence. In science, is it possible to have all the evidence of any particular phenomenon?

Case study 6: Science deniers

Knowledge framework: Perspectives

In an editorial to the *Wall Street Journal* in March 2021, astrophysicist Neil deGrasse Tyson wrote about how surprised he was at the number of people who were sceptical of science when it came to dealing with the COVID-19 coronavirus pandemic: "If the enterprise of science were some new fangled, untested way of knowing, one might empathize with these sentiments. But the people who battle against science are the same ones who, for instance, wield and embrace their pocket-sized smartphones, which merge state-of-the-art engineering, mathematics, information technology and space physics". He noted that, for certain aspects of their lives, people trusted science, but for others they used non-scientific ways of thinking, even when considering science-related topics such as the immune system. He mentioned the influences of religion, culture and politics, and postulated that scientists were being out-competed by more savvy communicators from other domains. Is it the responsibility of knowers to share their knowledge with the public in a clear and convincing way? Should scientists be held responsible for not being better communicators?

Case study 7: CRISPR Cas9 technology, somatic cells versus germ-line cells

Knowledge framework: Ethics

In general, researchers working on gene-editing techniques in humans focus on somatic cells. They want to modify some tissue or an organ in a single individual. Modifying germ-line cells is much more controversial. Few researchers think it is a good idea to pursue knowledge about modifying the human species by intervening at the germ-cell level. The fear is that this would lead to "designer babies" with certain "desirable" traits, such as being taller, having a certain eye colour or being more intelligent. This raises many controversial issues and may lead to a return of eugenics, the practice of trying to improve the genetic makeup of humans. We genetically edit plant genomes to give them the characteristics we want, so why not do the same with our babies? How is such a distinction made: what causes a group of scientists to agree that this is where they will draw the line and say "no"?

Can you think of other situations in which a researcher might say, "Sorry, I refuse to do that on ethical grounds. Even though it is an interesting question to pursue, it goes beyond what I am willing to do."? How do scientists choose to draw such an ethical line between what knowledge is acceptable to pursue and what is not?

Perception

Which red circle is bigger? Judge using your eye first and then use a ruler to check your answer. What does this say about our perceptions and reality?

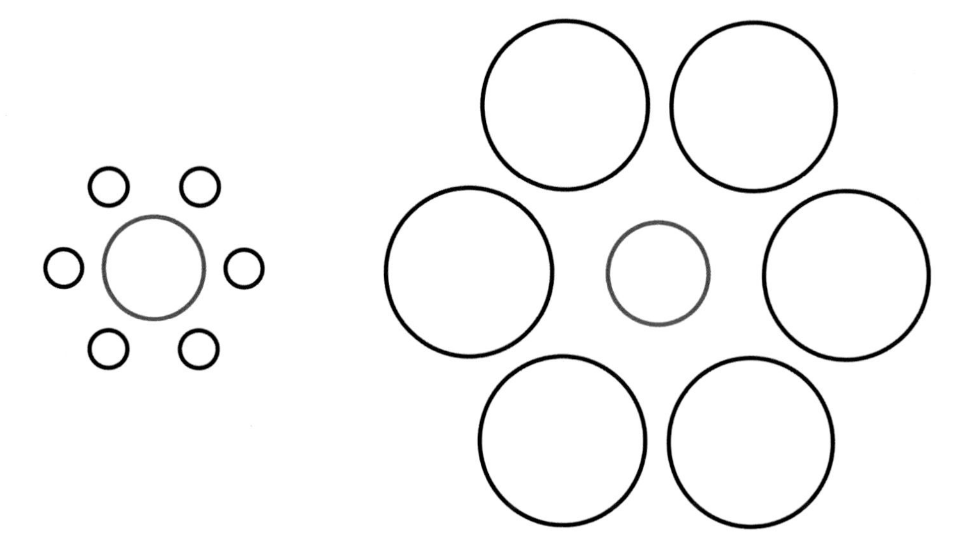

Science may be described as the art of systematic oversimplification.

Karl Popper
1992

What qualifies as an experiment?

Biology is an experimental science, but what constitutes an "experiment"? Do you have to have a hypothesis, controlled variables, a laboratory? What if you just have people filling out questionnaires? Is that an experiment? What about digging up fossils?

Theory versus myth

In what ways are theories and myths similar and different? Consider the similarities and differences when comparing and contrasting the two.

Is it based on well-substantiated facts? Is it passed on from generation to generation? Can it be modified over time? Can it be used to predict future events? Has it been tested repeatedly? Is it widely accepted as being true? Is it considered to be a supposition? Is it considered by many to be false?

Irrationally held truths may be more harmful than reasoned errors.

Thomas Henry Huxley

Biology and values

Do the ends justify the means? Consider the following domains of research in biology. What are the ethical issues associated with each?

- Gene therapy
- Vaccine tests
- Experimentation on human volunteers, notably prisoners
- Research involving human embryos

Nothing in this world is to be feared ... only understood.

Marie Curie

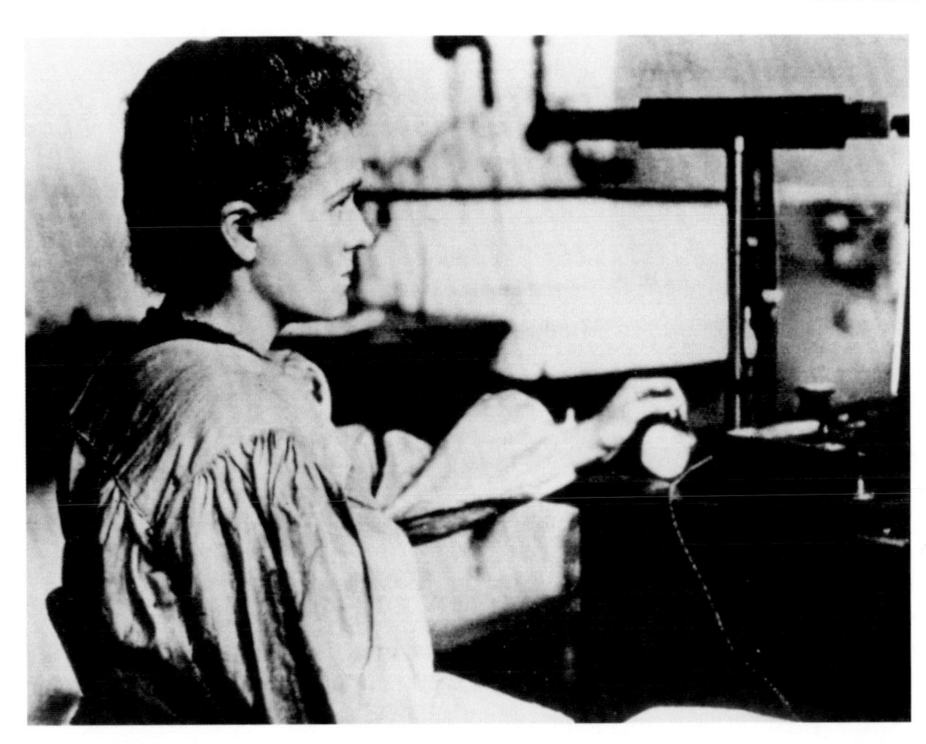

Marie Curie was the first woman to be awarded a Nobel Prize, and the first person to be awarded two.

Science and technology

Is scientific knowledge valued more for its own sake or for the technology that it makes possible?

Reading your mind

With modern technology tracking everything we do on our computers and smartphones, it can be argued that the kind of privacy our grandparents had no longer exists. Can we at least say that our private and personal thoughts are still safe within our minds and cannot be tracked and monitored?

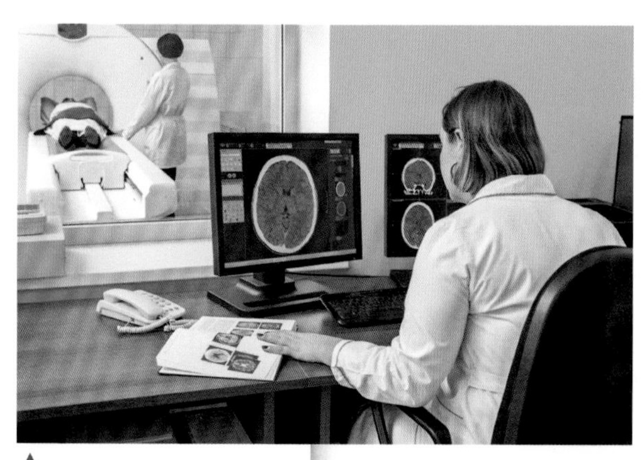

▲ Some companies claim they can use fMRI technology to identify peoples' thoughts or intentions.

Functional magnetic resonance imaging (fMRI) technology allows researchers to see which parts of the brain are active when a person is thinking a specific thing or performing a specific task. This has led to the possibility of identifying thoughts or, as some call it, "mind reading". For example, researchers have shown a series of images to participants and recorded the patterns that show up on the fMRI scanner for each image. Later, they pick an image at random and show it to the participant while they are still in the scanner. A computer can match the current brain scan pattern with one of the patterns observed before and determine which image the person's brain is perceiving. Experts claim that they can use this technology to see whether someone is lying or to see whether someone recognizes a crime scene that they claim they have never been to. Marketing agencies are interested in seeing how the brain reacts to different advertising campaigns.

Some major knowledge issues and knowledge questions arise from this. How can we know if such claims are valid? How do we test them and decide if the scanner and computer are accurate? How can we decide if the evidence collected in this way is credible enough to be used legally in court as evidence? Could complex thought patterns be identified, such as musical creativity or cruel intentions? Who should decide whether such experimentation and exploration into our private thoughts should be pursued or banned? Would you want a scan done of your thoughts?

Inaccessible worlds

Some scientific fields of exploration have only been possible since suitable technology has been invented, for example genetic engineering has only existed since the technological developments of the 1970s and 1980s. Could there be problems with knowledge that are unknown now because the technology needed to reveal them does not yet exist? Remember that, despite the fact that bacteria are all around us, we were not able to see them until the microscope was invented in the 1600s. Perhaps there are other phenomena that we simply cannot observe because no one has invented an apparatus to detect them yet.

Is there any science that can be pursued without the use of technology?

> *The most important discoveries will provide answers to questions that we do not yet know how to ask …*
>
> John Bahcall (commenting on the Hubble Space Telescope's capabilities)

> *My business is to teach my aspirations to conform themselves to fact, not to try and make facts harmonize with my aspirations.*
>
> Thomas Henry Huxley

Internal Assessment

The scientific investigation

This chapter is not meant to replace the section in the official IB subject guide for biology about the Internal Assessment (IA) component of the course, the **scientific investigation**. Please be sure to read that first and consult it regularly, so that you keep in line with the criteria and the requirements. This chapter aims to help you understand the criteria and give you some guidance about how to fulfil them.

The IA component is worth 20% of your final grade and it is the only one in biology that assesses all four of the assessment objectives (AO):

AO1	Demonstrate knowledge
AO2	Understand and apply knowledge
AO3	Analyse, evaluate, and synthesize
AO4	Demonstrate the application of skills necessary to carry out insightful and ethical investigations

In other words, this is your opportunity to shine. Be sure to apply what you have been learning about the topic you choose to investigate and the laboratory skills and analytical techniques you have been acquiring during the course. And always keep in mind safety, ethics, environmental issues and academic integrity. These are all **transferable skills**: skills you can use later in a variety of contexts in your future studies and/or career.

Ways of obtaining your data

There are multiple ways of gathering the data you will need.

1. Laboratory work: carrying out an experiment you have designed. This allows you to manipulate a variable in controlled conditions.
2. Fieldwork: collecting data in streams, meadows woodlands or an urban park, for example. This allows you to select a variable such as sunlight or temperature to see if it affects the organisms you are studying.
3. Group data: analysing and modelling data collected by a group, either in the laboratory or during a field trip, and recorded in a shared spreadsheet. This allows you to choose one of the variables and see how it affects another in the spreadsheet. Each student needs to present different sets of data extracted from the spreadsheet.
4. Database: extracting and analysing data from a database such as a medical or environmental data set. This allows you to use data you would not otherwise be able to obtain on your own, such as numbers of patients with a particular tropical disease or statistics on deforestation in a distant country.
5. Simulation: using a computer model that allows you to change the parameters to see what happens. This could allow you to do things you would not be able to do in the laboratory or in the field, such as performing a genetics experiment on fruit flies or introducing new predators to an ecosystem.

The process from start to finish

Below is an overview of the steps to consider in the IA process in order to complete your scientific investigation.

1. Advance preparation: acquiring skills, learning how to design an investigation, learning the use of various laboratory techniques, databases, simulations or fieldwork techniques, practice data processing and writing a conclusion and evaluation.
2. Launch of the scientific investigation: criteria understood, calendar outlined.
3. Explore and brainstorm: initial research and narrowing down a topic followed by the formulation of a research question (RQ) and research about the methodology.
4. Plan and test: designing and trying out the method.
5. Data collection: this can be in the form of laboratory work, fieldwork, work with a database or with a computer simulation.
6. Data processing and analysis: calculations, graphs, statistical tests.
7. Writing, part 1: draft writing, including guidance and feedback from your teacher.
8. Writing, part 2: perfecting the draft of the report to produce the final submission, maximum word count 3,000 words. (See the official IB Biology guide for what is included in the word count.)

The criteria your work will be assessed on are listed here and total 24 marks (each is equally weighted).

- Research design: maximum 6 marks
- Data analysis: maximum 6 marks
- Conclusion: maximum 6 marks
- Evaluation: maximum 6 marks

Some parts of the work, notably the design of the method and the data collection (steps 4 and 5 listed in the overview above), can be done in small groups but most of the work is individual, notably the initial research, developing an RQ, the data processing, and all the writing involved in producing the report. Collaborative work can be done with up to three students as long as each has a different RQ and uses different data. Two examples of how collaborative work might be done are discussed below.

Collaborative work: example 1

Two students want to use human volunteers for their investigations and decide it would be a good idea to work as a team. The first student has chosen a pupil reflex investigation and the second is looking at pulse rates in response to different types of exercise. They design their respective investigations collaboratively, each giving the other advice and suggestions, including how they are going to recruit their volunteers. They carry out the investigations collaboratively, asking each volunteer to take part in the first investigation and then the second. The students help each other with the setup and the recording of data. Each student uses the results collected to write their own report individually.

Collaborative work: example 2

Three students use nets to collect invertebrates from a stream during a class field trip. They assist each other with using the nets, identifying the invertebrates using dichotomous keys, and measuring the dissolved oxygen and flow rate in lotic (faster flowing) and lentic (slower moving) zones of the stream. Other measurements, such as water temperature, pH and light levels, are recorded using probes. They share all their raw data and qualitative observations but each student has a different RQ. For example, one might look at the relationship between dissolved oxygen levels and the presence of three species of invertebrates, another might look at the relationship between flow rate and the presence of all the species observed, and another could see how the invertebrates' anatomy is adapted to water speed. Each student has a separate RQ, each presents different raw data tables using the data they collected collaboratively, and each writes up a separate report individually.

In a similar fashion to the second example, large data sets collected by an entire class during fieldwork or laboratory work can be used as a database but the resulting investigations should be considered a database investigation. Each student presents only the data they have used. See the official IB Biology guide for more information on collaborative work.

Getting started

Once your teacher has launched the IA process, you need to come up with some possible ideas. Your early ideas may be rejected by your teacher because of safety or environmental reasons, or maybe just because the logistics will not work in the time given.

Some students start with a big topic, such as digestion, and narrow it down, for example to one enzyme being affected by one factor, after which they need to decide on what method to use. The danger with that approach is that not all themes or big topics will lead to an investigation that can be carried out. A student might be fascinated by parasites but the study of that topic in the laboratory raises serious logistical and ethical questions. It could be a database analysis instead of laboratory work, but it is not always easy to find a freely accessible database on a specialized topic.

Other students start with a particular method that they enjoyed, such as chromatography, and then work from there. If you are not sure where to start, one way is to make a list of the investigations you have already done and think about the techniques you are familiar with. Applying a familiar technique to a new situation is a great way to get started.

Example 1: applying a familiar technique to an unfamiliar situation

Lysa enjoyed an experiment using a delivery tube to measure the volume of gas produced when hydrogen peroxide was added to liver samples. She decided to apply the same measuring technique to yeast cells when feeding them different types of sugar.

Example 2: combining two familiar techniques

Matteo had already used the eyepiece graticule on a microscope to measure cell sizes in a previous laboratory, and in another microscope session while looking at living pondweed cells he observed cytoplasmic streaming of the chloroplasts. He thought, why not put the two together and measure the speed of cytoplasmic streaming? All he had to do then was come up with an independent variable. A bit more research on what factors activate or hinder cytoplasmic streaming, and he was on his way.

Example 3: database

Elif preferred research over laboratory work. She used the UniProt protein database and some online genomic databases, such as The National Center for Biotechnology Information (NCBI), to test different hypotheses about common ancestry in primates by comparing mutations in humans and non-human primates.

Example 4: trying to invent a technique unfamiliar to the student (unwise)

Fred wanted to measure stress levels in students who were given difficult mathematics problems to solve. His teacher asked, "Have we ever measured stress in the laboratory? Do we have a stress detector that you can point at people to measure their stress levels?" There is no such thing, and stress is subjective, so it was going to be very challenging to find a reliable measurement technique. Fred's teacher asked him to go back and look at things they had measured in previous experiments and come up with a more realistic idea, so he started to look into factors that influence heart rate.

Some questions to ask yourself when brainstorming topics and ideas for methods can include the following.

- Is the main focus on biology? Does it connect with a topic in the syllabus? *It is okay for the investigation to have some overlap with another subject, such as chemistry, psychology or geography, but the main focus should be on biology.* Where in the investigation is the living organism or the organic molecule(s) made by living organisms? *If there is not a connection to an organism, it is not biology.*
- Is it safe, ethical and environmentally friendly? Does it break any rules from the IB, my school or laws in my country?
- Does it involve a technique I am already familiar with? *If so, it will probably go more smoothly than if you try to invent a new technique no one has ever tried.*
- Can I actually measure (or count) the dependent variable? Do I have access to the necessary equipment to obtain number values that I will need later for data processing?
- Is it doable in the time frame on my school's calendar?
- Is it a relatively simple, straightforward idea that has lots of little details I can change or control? *Those work the best.*
- If I need human volunteers to participate, will the investigation follow the IB and my school guidelines for using human subjects?
- If I am using animals such as snails or mealworms, am I following the guidelines for the ethical use of animals in the laboratory?

- If it is fieldwork, are the logistics going to work? For example, can I carry all the equipment I need to the site?
- Will I be able to collect enough data and do a sufficient number of trials to get reliable data?
- If I am getting my numbers from a database or computer simulation instead of producing the data myself, is the source I am using going to give me usable data?
- Is the topic of interest to me and is it worthy of investigation?

Once you have narrowed down a topic and have some ideas of what you might measure and how, you will need to continue your background research. Reread the section in your textbook and in your notes that deal with your chosen topic. This will help you select key biological vocabulary to search for, such as "osmolarity", "membrane integrity", "patellar reflex" or "phototropism". With guidance from your teacher and your school or local librarians, try to find a variety of resources.

A general online search using a search engine is often not as productive as a search within a curated collection of resources that a librarian can direct you to. Ask your teacher if the science department has any books on your topic, or if there are resources at school or online that are specifically for high school-level biology. Use the research papers found in the list of sources at the bottom of Wikipedia pages to help guide you too. Many are flagged as being open access to the public without cost. If some are locked behind a paid subscription, ask your school librarian if the school has a subscription. Keep a list of all the books, articles and websites you access, including the date. That way, not only can you find them later, you can include the ones you use in your list of cited sources at the end of your report.

It is mandatory to submit your IA work to complete your course. Failing to produce and upload a laboratory report for biology will result in no grade being awarded for the subject, even if you get fantastic results on all your exams. A missing grade could jeopardize your diploma, so be sure to give this work all the time and effort its needs.

We will now examine each of the four criteria one at a time. Questions will help guide you.

Criterion 1: Research design

Writing an RQ

- If you are really stuck and do not know where to start, one formula for an RQ is to ask "What is the influence of X on Y?", where X and Y are factors or variables that can be measured, controlled, modified or counted.
- Be as precise as possible, even if it means that the RQ is quite long.
- The RQ should contain precise and focused words to describe both your independent and dependent variables (see the next page). For example, "patellar reflex times in adolescents from 16 to 18 years of age" is more focused than "knee reflex".
- If you are using any living organisms, or products from living organisms, such as seeds from a certain plant, give the most precise name you can and give the scientific name as well (e.g. *Pisum sativum* for garden peas).
- Even if your RQ is the title of your investigation, be sure to restate it clearly early on in your report. It is helpful if you state, "The research question is . . .".

Types of variables to consider

- The **dependent variable** is what you will be measuring as the results of your investigation. It is what changes in the experiment because of the manipulations of the experimenter. One way to think of the dependent variable is as "nature's answer": it is how the natural world's laws respond to your RQ.
- The **independent variable** is what is changed on purpose by the investigator to see effects it will produce. It is what you are testing to find out what happens. It should be the only thing that is different from one part of the experiment to another. For example, in an experiment testing the effect of different amounts of fertilizer on the growth of bean plants, a range of five different concentrations of fertilizer would represent the independent variable. Everything else must be the same: the type of plant, type of soil, age of plants, light conditions, etc. The one thing that you can vary on purpose is the concentration of fertilizer.
- The **controlled variables** are the things that are kept the same in all parts of the experiment in order to be sure that the experiment is fair. Controlled variables ensure that the independent variable really is solely responsible for any changes recorded. There is no need to make an exhaustive list, just be sure to identify the controlled variables that would most dramatically affect the results in an undesirable way.
- Remember: do not confuse **controlled variables** with **the control**. The control of an experiment is a variant of the experiment that is set up in order to have something to compare the other results with.

Writing about the context surrounding your RQ

- In addition to describing your RQ, have you given your reader some background information to put it into context? What are the main biological ideas surrounding your topic?
- What inspired you to pursue this investigation? Why is it of interest to you?
- What properties or laws of nature will you be investigating? How does it connect to what you have been studying?
- Have you explained why you chose what you did for your independent variable and your dependent variable?
- Why have you chosen this method for answering the RQ compared to other possible methods?
- For database or computer simulation investigations, how did you select your sources? What decisions did you make to select the data or the conditions for the simulation?

Writing a step-by-step method

When writing your method, take inspiration from methods your teacher has already introduced you to, or think about the recipe you use to prepare your favourite dessert. You should be as precise and concise as possible. Here are some things to consider.

- Could your method be read by someone else and be fully understood by that person?
- Have you clearly described how the independent variable is integrated into the steps? How about the dependent variable?

- For the controlled variables, have you explained how they will be controlled? If it is impossible to control one or more of them, have you described a method for monitoring them?
- In selecting the materials, apparatus and glassware, have you chosen the equipment with a degree of precision that is appropriate for your investigation?
- For glassware such as beakers and flasks, be sure to indicate the volume in millilitres (ml). If you just ask for test tubes, the standard size will be given but be aware that there are some with smaller or wider diameters.
- If the glassware is going to be heated, think of what you might need when moving it once it is hot, such as wooden pinchers or metal tongs.
- If the experiment involves cutting something, do not forget to ask for a knife (or scalpel if necessary).
- For chemical solutions, you must be precise about the concentration (in % or in moles per litre) as well as the volume (in ml) that you will need.
- Think about the materials used to transport things: the manipulation of liquids will probably require the use of pipettes or syringes, the manipulation of powdered chemicals will require a spatula, and, if you need to weigh the powder, how will you put it on the balance? Did you remember to ask for a balance?
- If you ask for any electronic probes (for temperature, light, humidity, etc.), be sure to ask for an interface for connecting them to your computer, or a data-logging device.
- Thermometers come in multiple forms, including glass, electronic and temperature probes. Be sure to state clearly what kind you need.
- If an experiment needs to be saved overnight from one lesson to the next, did you ask for a tray or a box to keep the samples in? Is it labelled?
- Will the methodology you are planning result in sufficient numerical data so that techniques of analysis such as standard deviation can be used? Clearly state how many different experimental variations (scope, range) you will have and state how many trials are needed for sufficient data to be collected.
- Have you explained how you have modified a standard method and made it your own design? For details such as the concentrations you chose for the independent variable, did you explain why you chose those particular increments? If you got your inspiration from outside sources, have you cited them?
- Have you mentioned safety, environmental and ethical concerns? Not all investigations will require comments about all three, so adapt your report accordingly.
- Is your step-by-step method concise? In other words, check that you have not repeated yourself and that you have given all necessary information, but not more.

Level descriptor for top marks

Here is what is expected if you want to get full marks for Criterion 1: Research design.

- *The research question is described within a specific and appropriate context.*
- *Methodological considerations associated with collecting relevant and sufficient data to answer the research question are explained.*
- *The description of the methodology for collecting or selecting data allows for the investigation to be reproduced.*

Criterion 2: Data analysis

This criterion asks to what extent your report provides evidence that you have presented, processed, analysed and interpreted the data in such a way that a conclusion can be reached that is in line with the proposed RQ.

- Have you selected and recorded raw data, including the uncertainties and appropriate qualitative data? Is everything you have chosen to put in the results directly relevant to the RQ?
- Have you selected an appropriate method for analysing the data? The Skills for biology chapter explores mathematical tools for analysing and graphing data, as well as statistical tests you can use.

Setting up effective raw data tables

- Give the table a number and a title (e.g. Table 2: Pea seed characteristics).
- Set up the rows and columns in an orderly way to facilitate interpretation, for example values that have been measured using the same tool, such as a thermometer, should be aligned in the same column.
- In the headings of each column, put three things: the name of what was measured, the appropriate units, and the degree of precision.
- Put only numbers in each box (cell) of the table, no units, and be sure to have only one value in each box of the table. Do not include symbols such as ± or ≈ in the cells with the raw data. An exception is the negative sign: that can be used.
- The number of decimal places after the decimal point should be in accordance with the degree of precision, for example if a thermometer is precise to ±0.5°C, then all the numbers in the column should end in .0 or .5 and not have any more or any fewer decimal places after the decimal point (even for 0.0).
- How reliable are the accuracy and precision of your results? (See Figure 1). Are there other uncertainties in the measurements that you can point out?
- Align the decimals, even when there are negative signs in front of some of the numbers.
- Is there a clearer or more concise way of showing the results?
- Have you followed all the conventions for presenting graphs and tables?

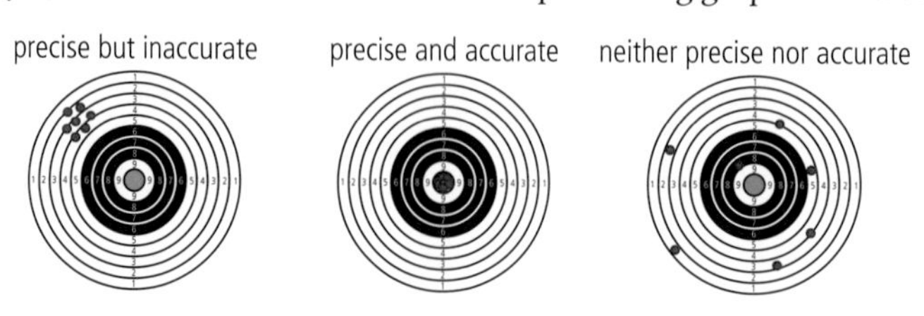

Figure 1 Accuracy is how close an investigator gets to measuring the accepted value that is reliable and verifiable. Precision is how close the data points are to each other. Precise measuring instruments do not give different values each time the same thing is measured. It is possible to measure something very precisely (getting the same results each time) but be very inaccurate, as seen in the first illustration of the target. This might happen if a balance was not set to zero after placing a recipient on it, giving the mass of the substance being weighed but also including in that value the mass of the container, thus falsifying the measurement.

Types of data to consider: quantitative and qualitative, raw and processed

Figure 2 shows qualitative data in purple. Such data cannot be expressed in numbers. Raw quantitative data are shown in green, and processed quantitative data are shown in orange.

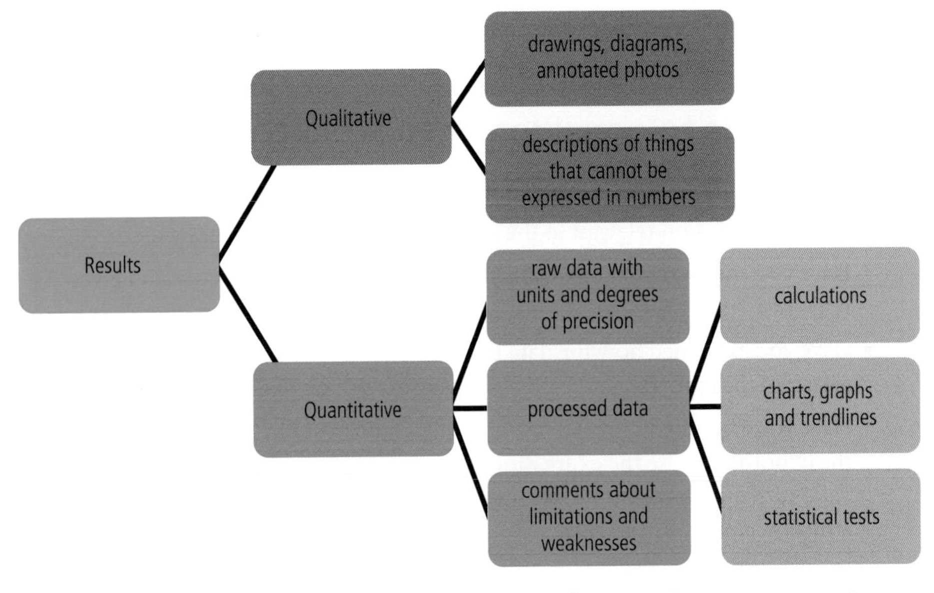

Figure 2 Your results should show all three types of data: qualitative, quantitative and processed data.

- Have you successfully analysed the data in a relevant and appropriate way?
- Is the analysis accompanied by consideration of the uncertainties?
- Do you have enough data to carry out sufficient data processing? Data processing ideally includes mathematical and statistical work as well as graphical work to support a valid conclusion fully. (Note: only graphing raw data is not considered to be data processing.)
- Do you need to use a null hypothesis test to demonstrate whether or not the differences you see in the data are statistically significant?
- For any processing you have done, have you made it clear what steps were taken?
- For any statistical tests, have you justified why this test was chosen over others?

Level descriptor for top marks

Here is what is expected if you want to get full marks for Criterion 2: Data analysis.

- *The communication of the recording and processing of the data is both clear and precise.*
- *The recording and processing of data shows evidence of an appropriate consideration of uncertainties.*
- *The processing of data relevant to addressing the research question is carried out appropriately and accurately.*

Criterion 3: Conclusion

This criterion measures how well you have answered the RQ. You will need to show that the conclusion is entirely in line with your RQ and has been described fully. In addition, the conclusion must be justified by the data you have collected, as well as justified through relevant comparison with the accepted scientific context.

Academic integrity is of the utmost importance in the IB. At the end of the process, your teacher needs to authenticate your work. This means that your teacher will declare that the final submission is your own work and that any ideas, data, images, etc., that are not your own have been cited correctly. Throughout the process, be sure to keep a good record of where you got your ideas, and keep updating your list of cited sources.

Students are encouraged to do research to find similar investigations and see how their results and conclusions compare with those of other scientists who have published their findings. Such comparisons do not necessarily have to be quantitative but full citations of any sources used are required. If a comparison with a similar investigation is not possible, there could at least be a comparison with the current scientific understanding of the theories or laws governing the phenomenon being investigated. Always cite your sources.

Some things to consider when writing a conclusion

- Have you interpreted the analysis to form a conclusion?
- Is your conclusion relevant to the purpose of the investigation?
- Have you compared your conclusion to accepted scientific theory and given references?
- Have you explained how the data that was collected has answered the RQ stated earlier in the report?
- Have you explained how the results either confirmed the hypothesis or refuted the hypothesis? Use the expressions "confirmed by the data" or "refuted by the data" rather than "right" or "wrong". The latter two terms should be reserved for ethical arguments in science.
- Have you described any unexpected results: were there any outliers in the data, or any surprises?
- Have you commented on how much the measurement uncertainties may have influenced the results? In other words, how confident are you that the measurements gave results you can consider to be reliable?
- Have you explained what can be learned from the results? This is where you can usually connect the theory from class with your laboratory work. When possible, compare your first-hand data with literature values (secondary sources).

Level descriptor for top marks

Here is what is expected if you want to get full marks for Criterion 3: Conclusion.

- *A conclusion is justified that is relevant to the research question and fully consistent with the analysis presented.*
- *A conclusion is justified through relevant comparison to the accepted scientific context.*

Criterion 4: Evaluation

This criterion requires you to be reflective about the method of your investigation and to suggest possible improvements. Here are some things to ask yourself in your evaluation.

- Have you mentioned some of the strengths? Have you looked back at the list of controlled variables to get inspiration: were they, in fact, controlled?
- Do you feel the approach that you chose was an effective one to answer the RQ? Or should you have chosen a different method?

- Have you discussed the limitations and/or likely sources of error in the method?
- Have you discussed the reliability of the data? What weaknesses and limitations did you see in the data collection?
- Have you evaluated the level of impact of the sources of error? Are they minor, major or moderate?
- Have you explained how the source of error might generate unexpectedly high or low results?
- Have you commented on the range or spread of data? Do the data points follow a clear trend, for example, or are they very irregular and difficult to interpret?
- Have you assessed your sample size? It is big enough to have confidence in the conclusion?
- Have you commented on any assumptions that are being made? For example, if you used seeds for a germination experiment and all the seeds were from the same source, is it safe to assume they had similar genetics or ages or were kept in the same conditions? If those assumptions turn out to be incorrect, they could have an unexpected effect on the results.
- Have you suggested relevant and feasible modifications to the method? Have you explained how these would help improve the method? These must be things you could realistically do in a high school laboratory, so do not suggest that multi-million dollar laboratory equipment be used next time.
- Have you demonstrated that you understand the implications of the conclusion? For example, just because you saw a correlation in your results between the independent and dependent variables, does that mean that one causes the other to change?
- Have you suggested relevant and feasible extensions to the investigation? During investigations, often new questions arise. An extension is an idea for a future investigation to answer such new questions.

Level descriptor for top marks

Here is what is expected if you want to get full marks for Criterion 4: Evaluation.

- *The report explains the relative impact of specific methodological weaknesses or limitations.*
- *Realistic improvements to the investigation, that are relevant to the identified weaknesses or limitations, are explained.*

General questions to ask about your report overall

Have you communicated the focus, process and outcomes of your investigation clearly?

You need to write in such a way that you can communicate effectively to your reader.

- Have you written your report on the investigation in a concise, clear and logical format?
- Can your written explanation of data analysis be easily followed?

- Are your graphs, tables and images unambiguous? Photos should be annotated, have a legend describing what should be observed, and the source should be cited. If it is your own photo, put "author's photo" or "investigator's photo" as the citation.
- Is subject-specific notation used throughout?
- Have you used subject-specific terminology throughout?

Formatting checklist before submitting the final version

A downloadable version of this checklist is available on this page of the eBook.

Think about your IA work in the IB like an exam that you get to take home, work on for as long as you like, submit once to the teacher for feedback, and then get back to correct it before the final submission. IA work should take priority over other homework assignments because it counts directly towards your diploma. The points you earn will already be in your pocket as you walk into the final exams.

- ☐ Check your document format: The IB only accepts Word, RTF or PDF. PDF is preferable as there is a lower risk of losing formatting.
- ☐ Check the overall word count is under 3,000 words. Make sure to include the word count in the report. The following are not included in the word count: charts and diagrams, data tables, equations, formulae and calculations, citations/references, bibliography, headers.
- ☐ Do you have a title? There is no need for a separate title page. You may use your RQ as the title, but this is not always practical as some are rather long.
- ☐ Include your IB candidate code (e.g. xyz123). Other than this code, the document should be anonymous. None of the following should be on any pages of the document: your name, your school's name, your teacher's name or anything else that could take away the anonymity of your work.
- ☐ Group work: if you collaborated with other students to collect data, their IB candidate codes should be listed too.
- ☐ Font: use a standard font and font size.
- ☐ Make sure the pages are numbered.
- ☐ Tables, graphs, photos and illustrations should all have titles or legends stating what they are, and any sources should be appropriately cited.
- ☐ Citations: cite all referenced works. Check with your school to see if there is a preferred citation format. Otherwise, you are allowed to use any format you like as long as it is a recognized standard that is used in the academic world, and as long as you stick to the same format throughout the document.
- ☐ Footnotes: using footnotes to explain ideas or define terms is seen as an attempt to get around the word limit. Put the ideas directly in your main text instead.
- ☐ Make sure you have a References or Works cited section: this is mandatory, but only works that are actually cited within the report can be in the list. Even if you got your inspiration from other works and think they are interesting, they do not belong in this list. You could add a separate section called Further reading or Additional sources, but only if you have space.

The process is long and challenging but, if you choose a topic you are really interested in, it will also be extremely rewarding. The ability to collect information, connect it to a larger context, analyse it, draw a conclusion and evaluate the process, are skills you will be able to apply for many years to come, whether at home, school or work.

Skills in the study of biology

This chapter presents three sets of **tools** (experimental techniques, technology and mathematics) followed by three **inquiry skills** (exploring and designing, collecting and processing data, concluding and evaluating). You will need these to complete the biology course successfully but also to navigate today's world, which, more than ever, is filled with data and ideas that are presented as scientific but sometimes are not.

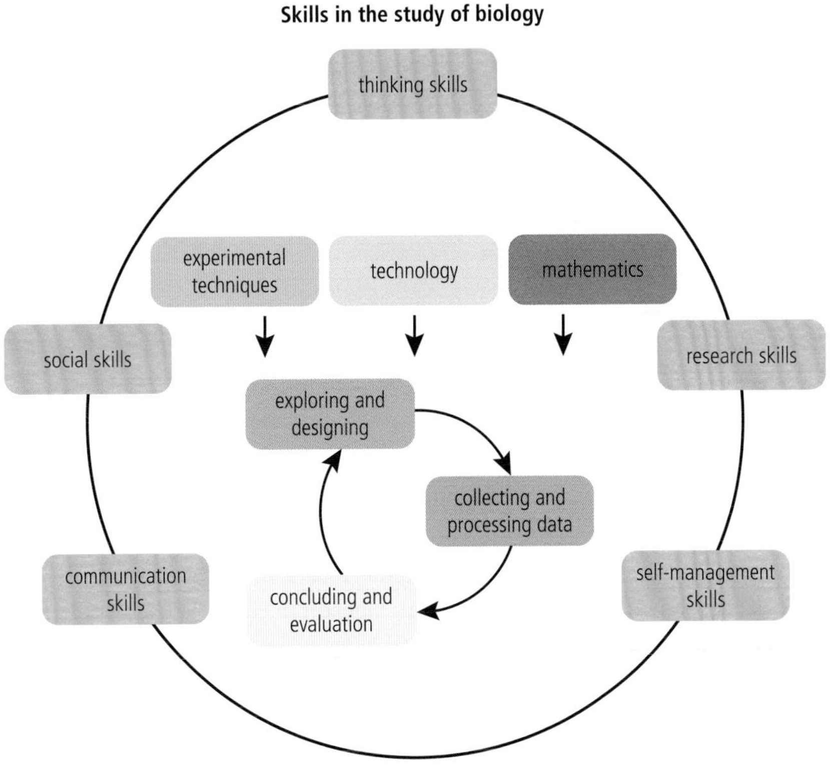

The approaches to learning (ATLs), shown in the grey boxes, should be integrated into the skills needed in biology, shown in the coloured boxes.

Tool 1: Experimental techniques

Addressing safety of self, others and the environment

When performing experiments in the laboratory, top priority should be given to safety. You should be mindful of preventing accidents such as burns, cuts, splashes and poisoning. Whenever possible, materials should be chosen that will not harm yourself, others or the environment. Students should never grow microbes without knowing what strains are present and must ensure that the colonies could not pose any health threats. Chemicals should be chosen in accordance with local safety and environmental rules. For example, if you need acidic buffers for an investigation, consider how much is necessary so that there is minimal waste, and think about how the chemicals will be disposed of at the end of the experiment. If you are using aquatic plants, choose a species that is non-invasive and consider what will happen to the plant at the end of the investigation. Evaluate whether using potentially toxic or dangerous materials is truly necessary. Be sure to consider ethical issues as well, such as the treatment of any animals you might use, respecting the privacy of human volunteers, and the impact of collecting samples from the field.

The first three rules of laboratory work: (1) safety, (2) safety and (3) safety.

Measuring variables

You should be able to measure basic variables such as mass, volume, time, temperature and length. All measurements are approximations, so you should also write down the degree of precision of your measurements. For example, when determining the mass of an object using a balance, look at the specifications of the balance, which will be in accompanying documentation, on a label on the balance itself or on the manufacturer's website. It might say "d = ±0.01 g" for example, which means it is precise to one one-hundredth of a gram. This means you can include up to 2 decimal places after the decimal point for each measurement.

When measuring volume, use the lowest part of a meniscus on a graduated cylinder and check whether the cylinder has a degree of precision printed on it. If not, you can estimate the degree of precision by looking at the smallest measurement interval and dividing that by 2. So if the graduated cylinder has graduations 2 ml apart, the degree of precision can be declared as ±1 ml, which means you cannot include anything after the decimal point when expressing your measurements in ml. However, a graduated cylinder with lines 1 ml apart can have a degree of precision that is ±0.5 ml, which would mean that all your measurements should end in either .0 or .5. If you need a more precise volume measurement, a syringe, burette or volumetric pipette would be better. When measuring time, ask yourself if you should take into account the reaction time to push the stopwatch button on and off. For temperature, glass thermometers are less precise but more practical to use than data-logging temperature probes. Always read a glass thermometer straight on, never at an angle. For measuring length, ask yourself how precise the ruler is and whether you need to consider the fact that you are actually making two approximations, one at one end of the object being measured and another at the other end. If the ruler is precise to ±1 mm, maybe your measurement is only precise to ±2 mm. For high precision measurements of length, consider using callipers.

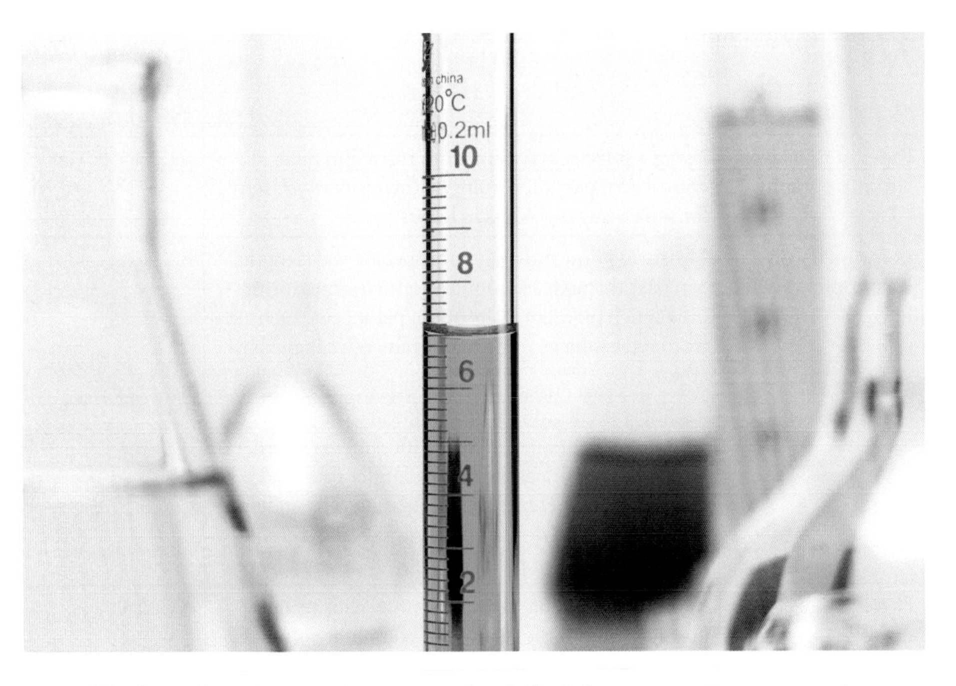

Be as precise as you can. Read the bottom of the meniscus by looking at the graduated cylinder straight on. In this example, you should get a measurement of 7.0 ml.

One of the basic foundations of science is the skill of observation. Be sure to write down what you observe. Table 1 indicates which parts of the course will help you practise the types of observation skills you need.

Table 1 Skills covered by the syllabus

Skill	Example	Relevant section of the syllabus
Counts	Counting plants found in a quadrat, or bubbles per minute to determine the rate of photosynthesis.	C4.1.3 and C1.3.7
Drawing annotated diagrams from observation	Drawing and annotation of cell organelles, distribution of tissues in stems and roots.	A2.2.11, B3.2.9 and B3.2.10
Making appropriate qualitative observations	Observations of tropic responses in seedlings: although measuring the angle is feasible, sometimes only qualitative descriptions are possible.	C3.1.17
Classifying	Working with dichotomous keys.	A3.1.14 HL

Applying techniques

Table 2 on the following page lists some techniques that are useful in biology and gives a possible application of each. You might consider using one or more of these in your internal assessment work. The last column indicates which part of the syllabus covers the application of a technique: look at the relevant chapters for further details.

Table 2 Techniques covered by the syllabus

Technique	Description and purpose	Relevant section of the syllabus
Paper or thin layer chromatography	Using a solvent to separate mixtures into their constituent parts according to their polarity. Useful for separating pigments in leaves.	C1.3.4
Colorimetry or spectrophotometry	Measuring how much light of specific wavelengths can pass through a solution. Useful for measuring how much pigment leaks out of plant cells, such as from red cabbage, if the membrane is weakened by ethanol.	B2.1
Serial dilutions	Adding a small volume of a solution to larger volumes of water in calculated increments in order to reduce the concentration. Example: 10% > 1% > 0.1%. Useful for counting free floating cells under a microscope that would be too numerous to count directly from a sample.	A2.2.2
Physical and digital molecular modelling	Building models of sugars or amino acids using plastic, paper or metal, or with computer software. Useful for visualizing molecules that are otherwise invisible.	A1.2.13
Using a light microscope and eyepiece graticule	Viewing a slide under a microscope or using an eyepiece graticule as a ruler to measure the size of objects observed at a microscopic scale. Useful for both qualitative observations of cells and tissue and for comparing sizes.	A2.2.2
Preparation of temporary mounts	Preparing a microscope slide by sampling tissue, placing it on a slide and staining it. Useful for observing cells that are still alive in order to observe movement such as cytoplasmic streaming.	A2.2.2 HL
Identifying and classifying organisms	Dichotomous keys can be used to identify organisms. Useful for determining the family, genus or species of an unknown organism while doing fieldwork.	A3.1.14
Using random and systematic sampling techniques	During fieldwork, tools such as quadrats and transects can be used for taking random or systematic samples of an area. Useful for determining what plant species are present and where, as well as testing a hypothesis about the association of two species.	C4.1.3 and B4.1.4
Karyotyping and karyograms	Reading karyograms can help determine the karyotype of a person. Useful in detecting the sex of a future baby or the presence of any genetic anomalies such as an extra or missing chromosomes.	A3.1.7
Cladogram analysis	By looking at the branches of a cladogram, it is possible to see where and when speciations most likely occurred. Useful in proposing or testing hypotheses about which species are the most closely or distantly related.	A3.2.7 HL

Tool 2: Technology

Applying technology to collect data

Data logging using probes and sensors

Studying a forest, stream, grassland or marine environment poses some unique challenges. Abiotic factors, such as temperature, air humidity and light, can vary considerably, and could have an influence on what is being studied. Because they cannot be controlled, abiotic factors should be monitored and data should be collected.

A student using a hand-held data-logging device to measure the pH and temperature of a sample of water in a river.

Temperature probes connected to **data-logging devices** can automatically record temperatures at particular intervals. Such devices can be equipped with probes and sensors for:

- temperature
- light intensity
- relative humidity
- flow rate (to see how fast water is flowing)
- dissolved oxygen.

Data loggers can also include a global positioning system (GPS) to record the exact location of each measurement.

There are various modes that can be used to collect data, including real-time or intervals, for example taking a measurement every 5 m or recording measurements over a 10-minute interval. Many of these devices allow you to graph and analyse the data directly on the screen. This is especially useful when doing fieldwork without a readily available computer.

Databases

Sometimes it is impossible to collect your own data, or you may want to compare what you have found with other sources. This is where **databases** can be helpful. Examples include comparisons of genome sizes (see Section A3.1.10), loci of human genes and their polypeptide products (see Section D3.2.18), and allele frequencies in populations (see Section D4.1.10). When using a database, it is useful to consider the following.

- Is it possible to extract the data you need from this database? If the data is sortable and searchable, it will make your work a lot easier.
- The reliability of the source: is it an authoritative academic or scientific source?
- How up to date it is: is the information outdated and can you find more recent data elsewhere?
- Is it possible to compare more than one database to see if they agree or disagree?

If you decide to use a database for your internal assessment or for an extended essay in biology, be sure to use some of the ideas above to justify why you chose a particular database.

Models and simulations

Some experiments are too dangerous, too time-consuming or too costly to carry out in the laboratory; **computer simulations** of those experiments can be performed safely and in a time-saving fashion. Because variables can be manipulated within them, simulations are often used to predict an outcome or to find out what the optimum parameters are for a system. Experiments based on mating fruit flies, for example, would take many weeks and do not comply with the IB rules about the treatment of animals in experiments. However, computer simulations allow us to collect data and perform experiments virtually. If you choose a simulation for your internal assessment or for an extended essay in biology, make sure it is possible to modify variables to get different outcomes. The simulation should also be accessible to your teacher and IB examiners for verification.

Applying technology to processed data

Spreadsheets

Make sure you know how to do the following with a **spreadsheet** program such as Excel, Numbers or LibreOffice Calc.

- Understand the system of identifying cells as A1, B2, C3, etc.
- Format: changing the format of the cell to match the type of data, such as number, date, percentage, text, time, scientific notation, etc.
- Format: changing the number of decimal places to correspond to the desired degree of precision.
- Insert: using maths operations by inserting an equals sign "=" followed by a formula, using "A1 + A2" to add, or "B3/B2" to divide, or "(A1 + A2 + A3)*B1" to combine more than one operation in the same formula.
- Insert: inserting predefined formulas such as sum, average, maximum or minimum, standard deviation, chi-squared, etc. As an example for Excel: typing "=max(A1:A100)" in cell A101 finds the maximum value between A1 and A100. Replacing the term "max" with "min" in the formula finds the minimum value.

- Converting a relative reference into an absolute reference by adding $, for example B2 does not behave the same way as $B2 or B$2 or B2 when the formula it is in is copied and pasted to another place on the sheet.

The worked example in the eBook shows how to use a spreadsheet program to calculate mean, mode and median from a data set.

Graphing

Make sure you know how to use a spreadsheet program to select a type of graph that will lead to useful analysis and insert a trend line and error bars. The worked example in the eBook will walk you through the steps needed and show you how to add labels so your graph is clear.

Just because you spend time making a graph look great does not mean it is worth including in a laboratory report. Sometimes it will inspire you to look for other patterns. In the example given in the eBook, further processing could be done by plotting soil temperature against light levels to see whether they are correlated.

Image analysis

Programs such as ImageJ, available through the National Institutes of Health, or LoggerPro from Vernier Software, have a function that allows you to measure things on a computer image or video. For ImageJ, once you have taken a photo and saved it to your computer, you can open it in ImageJ for analysis. Figure 1 shows an image of an iris and pupil taken during an investigation to measure pupil reflex after exposure to light. Tracing points around the iris allows the software to generate a measurement of area in arbitrary units. It is possible to convert this to centimetres if a scale bar is placed in the image, such as on the subject's face below the eye (e.g. a sticky note with a 1 cm scale bar drawn on it) and used to calibrate the tool.

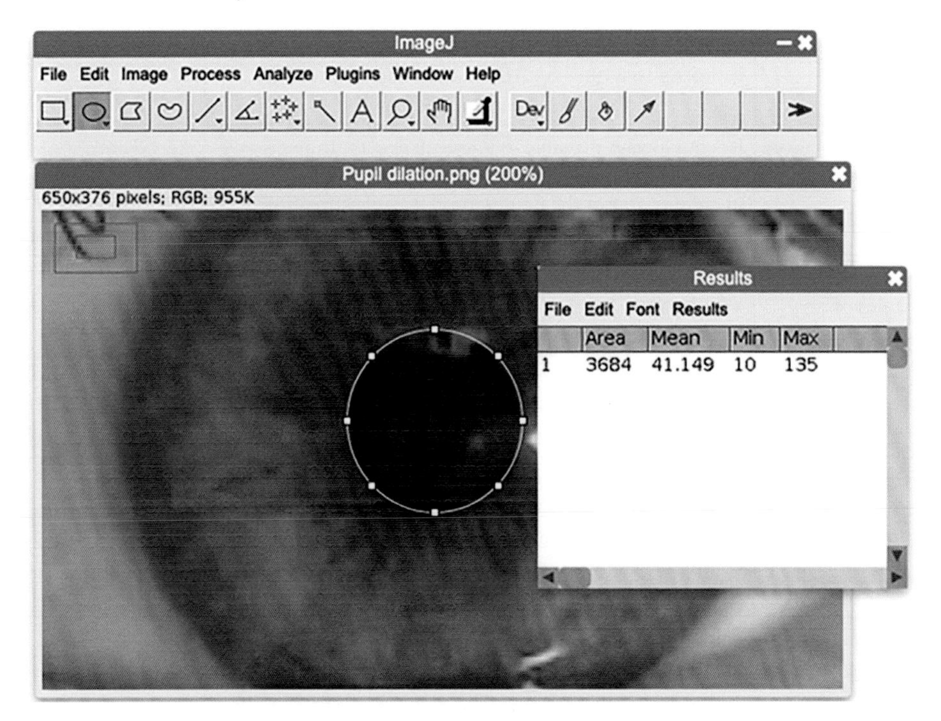

Figure 1 Image analysis software can help you generate quantitative data from images.

Tool 3: Mathematics

Applying general mathematics

You do not need to be a maths genius to be good at biology, but you do need to master several basic tools and techniques to understand other scientists' data and make sense of your own. Table 3 provides some examples of the kinds of mathematical tools used in the syllabus.

Table 3 Mathematical tools covered by the syllabus

Tool	Example	Relevant part of syllabus
Percentages	Punnett grids use percentages to express the chances that a genetic cross will produce certain traits in the offspring.	D3.2.2
Ratios	The surface area-to-volume ratio of an organism or cell decreases with increasing size, so bigger organisms will need to adapt their gas exchange system accordingly.	B2.3.6
Frequencies	How often something is found is its frequency and can be expressed as a percentage or decimal. Useful for allele frequencies in a population where, for example, 25% or 0.25 of the population has a particular allele.	D4.1.13 HL
Densities	Divide the number of things you are observing by the surface area to get the density. Useful under the microscope when determining stomatal density on a leaf.	B3.1.10
Proportions	Calculating how much of the whole a subcategory occupies. The mitotic index shows the proportion of observed cells going through mitosis and can be used to diagnose cancer.	D2.1.17 HL
Scientific notation	Instead of writing out 3,500,000,000, for example, it is shorter to write 3.5×10^9. Useful when expressing tonnes of carbon in the carbon cycle.	C4.2.15
Approximation and estimation	It is not always possible to know an exact number, so we try to give a number that is as close as we can get. Useful in giving dates of the first living cells and estimations of population sizes.	A2.1.8 and C4.1.2
Calculate scales of magnification	The image size divided by the true size of an object in a microscope is its magnification. Useful in using scale bars to calculate magnification.	A2.2.2
Rates of change	By dividing quantities such as distance, mass or volume by time, it is possible to get the rate. Useful for transpiration rates in plants or enzyme reaction rates.	B3.1.9 and C1.1.8
Direct and inverse proportionality	If two variables are proportional to each other, they show a correlation. This can be positive (as X goes up, Y goes up) or negative (as X goes up, Y goes down). Useful for seeing connections in coronary heart disease statistics or anthropogenic causes of climate change.	B3.2.6 and D4.3.1
Percentage change or difference	Subtract the "before" value, V_1, from the "after" value, V_2, and divide by the "before" value, V_2, then multiply by 100 to get the percentage change. Useful for calculating the deforestation of the Amazon rainforest.	D4.2.3

Tool	Example	Relevant part of syllabus
Continuous and discrete variables	Continuous means that values do exist between data points (even if we did not measure them) whereas discrete means the data comes in separate categories and there is nothing in between. Height in humans is continuous whereas the ABO blood type has only four categories, A, B, AB and O, so it is discontinuous or discrete.	D3.2.14
The Lincoln index	Calculating the number of captured, marked and recaptured organisms. Useful in estimating the population size of a non-sessile (freely moving) organism.	C4.1.4

Measures of central tendency

You will be required to calculate measures of **central tendency** (mean, median and mode). The worksheet in the eBook explains how these measures are calculated.

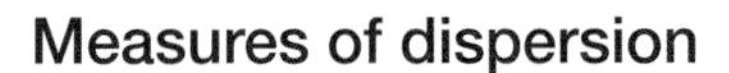

Measures of dispersion

You will also need to be able to apply **measures of dispersion** (range, standard deviation, standard error and interquartile range). The worksheet in the eBook explains how these measures are applied.

Simpson reciprocal index

The **Simpson reciprocal index** can be used to estimate the biodiversity of a habitat. It takes into account the number of species found in an area and the number of individuals in each species. Table 4 on the following page shows the numbers of benthic invertebrates identified in the Triouzoune river.

Table 4 Data set 1: the numbers of benthic invertebrates identified in the Triouzoune, France, June 2015

Benthic invertebrates	Total
Plecoptera	2
Ephemeroptera	41
Trichoptera with cases	9
Trichoptera without cases	8
Megaloptera	1
Crustacea	8
Gastropoda	12
Diptera	7
Hirudinea	21
Oligochaeta	4
Total species found	**10**
Total number of organisms	**123**

The formula for calculating the Simpson reciprocal index is:

$$D = \frac{N(N-1)}{\sum n(n-1)}$$

where n is the number of organisms that belong to one species and N is the total number of organisms. Applying this formula to data set 1 gives a value for D of 6.17. The higher this number, the more biodiversity is present. Lower numbers would suggest an unhealthy ecosystem.

Comparing the means and spread of data between two or more samples

Remember that in statistics we make inferences about a whole population based on just a sample of the population. Table 5 shows the results from an experiment growing bean plants in the sunlight and shade, and the data can be used to show how standard deviation is useful for comparing the means and the spread of data between two samples.

Table 5 Data set 2: the results of a bean plant experiment

Height of 10 bean plants grown in sunlight, in centimetres ±1 cm	Height of 10 bean plants grown in shade, in centimetres ±1 cm
125	131
121	60
154	160
99	212
124	117
143	65
157	155
129	160
140	145
118	95
Mean 131 cm	Mean 130 cm

The standard deviation of the bean plants grown in sunlight is 17.68 cm, while the standard deviation of the bean plants grown in shade is 47.02 cm. Looking at the means alone, 131 cm and 130 cm, it appears that there is little difference between the two sets of bean plants. However, the high standard deviation of the bean plants grown in the shade indicates a very wide spread of data around the mean.

Significant difference between two data sets using a *t*-test

In order to determine whether the difference between two sets of data is a **statistically significant** difference, *t*-tests are commonly used. The **Student's *t*-test** compares two sets of data, for example the heights of bean plants grown in sunlight and the heights of bean plants grown in shade. Look at the top of Table 6, and you can see values for the **probability** (p) that chance alone could make a difference. If $p = 0.50$, it means the difference could be the result of chance alone 50% of the time.

Statistical significance refers to how probable it is that a relationship is caused by pure chance. If a relationship is statistically significant, it means that it is very unlikely that the relationship is caused by chance. We can also use this idea to see whether the differences between two populations are random or not.

Bean plants being grown for an experiment.

If you reach $p = 0.05$, the probability that the difference is caused by chance alone is only 5%. This means that there is a 95% likelihood that the difference has been caused by something other than chance. A 95% probability is statistically significant in statistics. Statisticians are rarely completely certain about their findings, but they like to be at least 95% certain before drawing conclusions.

The formula used to compare two populations that are assumed to have equal variance is:

$$t = \frac{\overline{X}_1 - \overline{X}_2}{\sqrt{\left(\frac{(N_1 - 1)s_1^2 + (N_2 - 1)s_2^2}{N_1 + N_2 - 2}\right)\left(\frac{1}{N_1} + \frac{1}{N_2}\right)}}$$

$\overline{X}_1$ = the mean of population 1
$\overline{X}_2$ = the mean of population 2
N = sample size of the population
s = standard deviation

Variance is a measure of variability. It gives a value to how much the numbers in a set of data vary from the mean.

Note that you will not be asked this formula in exams: your graphing calculator can do it automatically.

If you put in the values from data set 2, you should get $t = 0.06$. You can use a table of critical t-values (Table 6) to find out what this number means. To do this, look at the left-hand column headed "Degrees of freedom", then look across to the given t-values. For a two-sample t-test like the one we are doing, the **degrees of freedom** (**d.f.**) are the sum of the sample sizes of the two groups minus two: 10 + 10 – 2 = 18.

If d.f. = 18, we need to look at the row in the table of t-values that corresponds to 18. We see that our calculated value of t (0.06) is less than 0.69 in Table 6, indicating that the probability that the differences between the two populations of plants are due to chance alone is greater than 50%. In other words, we can safely declare that there is no statistically significant difference in the data collected from the bean plants in the sunlight and those from the shade. The differences are most likely due to chance. In order to be able to declare that our two populations show a level of 95% significance in their differences, we would need a t-value of 2.10 or more (look up d.f. = 18 and p = 0.05 (5%) in Table 6). Interpretations of such data processing can be a crucial addition to an effective conclusion in a laboratory report because just looking at numbers or graphs cannot tell us if the differences we see are statistically significant or not.

When something is considered to be statistically significant, it means that there is a strong probability that it is not caused by chance alone. When something could be caused by chance, we say in statistics that the difference observed between two results is not statistically significant.

Table 6 *t*-values

Degrees of freedom	Probability (*p*) that chance alone could produce the difference					
	0.50 (50%)	0.20 (20%)	0.10 (10%)	0.05 (5%)	0.01 (1%)	0.001 (0.1%)
1	1.00	3.08	6.31	12.71	63.66	636.62
2	0.82	1.89	2.92	4.30	9.93	31.60
3	0.77	1.64	2.35	3.18	5.84	12.92
4	0.74	1.53	2.13	2.78	4.60	8.61
5	0.73	1.48	2.02	2.57	4.03	6.87
6	0.72	1.44	1.94	2.45	3.71	5.96
7	0.71	1.42	1.90	2.37	3.50	5.41
8	0.71	1.40	1.86	2.31	3.37	5.04
9	0.70	1.38	1.83	2.26	3.25	4.78
10	0.70	1.37	1.81	2.23	3.17	4.59
11	0.70	1.36	1.80	2.20	3.11	4.44
12	0.70	1.36	1.78	2.18	3.06	4.32
13	0.69	1.35	1.77	2.16	3.01	4.22
14	0.69	1.35	1.76	2.15	2.98	4.14
15	0.69	1.34	1.75	2.13	2.95	4.07
16	0.69	1.34	1.75	2.12	2.92	4.02
17	0.69	1.33	1.74	2.11	2.90	3.97
18	0.69	1.33	1.73	2.10	2.88	3.92
19	0.69	1.33	1.73	2.09	2.86	3.88
20	0.69	1.33	1.73	2.09	2.85	3.85
21	0.69	1.32	1.72	2.08	2.83	3.82
22	0.69	1.32	1.72	2.07	2.82	3.79
24	0.69	1.32	1.71	2.06	2.80	3.75
26	0.68	1.32	1.71	2.06	2.78	3.71
28	0.68	1.31	1.70	2.05	2.76	3.67
30	0.68	1.31	1.70	2.04	2.75	3.65
35	0.68	1.31	1.69	2.03	2.72	3.59
40	0.68	1.30	1.68	2.02	2.70	3.55
45	0.68	1.30	1.68	2.01	2.70	3.52
50	0.68	1.30	1.68	2.01	2.68	3.50
60	0.68	1.30	1.67	2.00	2.66	3.46
70	0.68	1.29	1.67	1.99	2.65	3.44
80	0.68	1.29	1.66	1.99	2.64	3.42
90	0.68	1.29	1.66	1.99	2.63	3.40
100	0.68	1.29	1.66	1.99	2.63	3.39

You will find a second worked example in your eBook.

Using units, symbols and numerical values

You should be familiar with the units shown in Table 7; some are SI (Système International) base units, commonly referred to as the metric system, while others are not. See Table 7 for examples.

Table 7 Basic units

Measurement	SI base units	Other examples
Length	metre (m)	
Time	second (s)	days, months, years
Amount of substance	mole (mole)	
Temperature	kelvin (K)	°C
Mass	kilogram (kg)	tonne
Luminous intensity	candela (cd)	
Illuminance	–	lux
Volume	–	litre (L)

Units can have prefixes added to them to make bigger or smaller units, such as millilitres, gigatonnes, nanoseconds and micrometres. Table 8 summarizes some of the prefixes used.

Table 8 Common prefixes for units

Prefix	Base 10	Decimal	Name
giga	10^9	1,000,000,000	billion
mega	10^6	1,000,000	million
kilo	10^3	1,000	thousand
centi	10^2	100	hundred
milli	10^{-3}	0.001	thousandth
micro	10^{-6}	0.000 001	millionth
nano	10^{-9}	0.000 000 001	billionth

All measurements are approximations. Be sure to express the degree of precision by choosing how many decimal places are used. For example, if the balance you are using is precise to ±0.01 g, all your recorded measurements should have two decimal places after the decimal point. Sometimes temperature probes will give results such as "20.83510226°C". This does not mean that we should keep all those decimal places after the decimal point. Find out what the true precision is from the documentation provided by the manufacturer of the device.

Processing uncertainties

Why are decimal places important? If you want to show that there is a temperature increase during an experiment and the readings you get on your electronic temperature probe are 20.8351022°C for the before value and 20.8416491°C for the after value, it might look like the temperature has increased. But because the difference is in the second decimal place and is beyond the degree of precision of the temperature probe (±0.1°C) both values are, in fact, 20.8°C. So no increase can be declared.

This is why it is crucial to declare what the degree of precision is for your measurements. This can be included in the headings of data tables, as shown in Table 5 for data set 2, measured with a degree of precision of ±1 cm. It is recommended that in your reports you comment on the limitations of the measurement uncertainties and what impact they can have on the results of your investigations.

Error bars

Error bars are a graphical representation of the variability of data. Error bars can be based on several different values, such as the range of data, the standard deviation or the standard error. Notice the error bars representing standard deviation on the bar chart in Figure 2 and the graph in Figure 3.

The value of the standard deviation above and below the mean is shown extending above and below the top of each bar of the chart. As each bar represents the mean of the data for a particular tree species, the standard deviation for each type of tree can be different, but the value extending above and below a particular bar will be the same for normally distributed results. The same is true for the line graph in Figure 3. As each point on the graph represents the mean data for each day, the bars extending above and below the data point are the standard deviations above and below the mean.

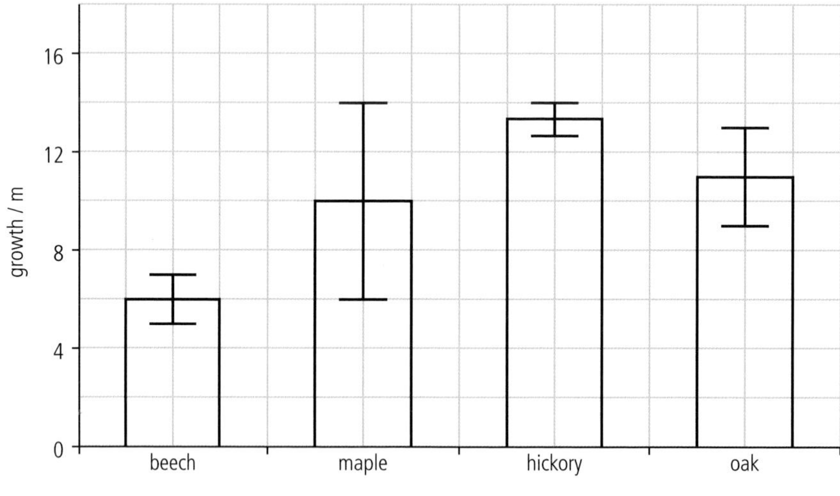

Figure 2 Rate of tree growth on an oak–hickory dune in 2004–05. Values are represented as the mean ±1 SD of 25 trees per species.

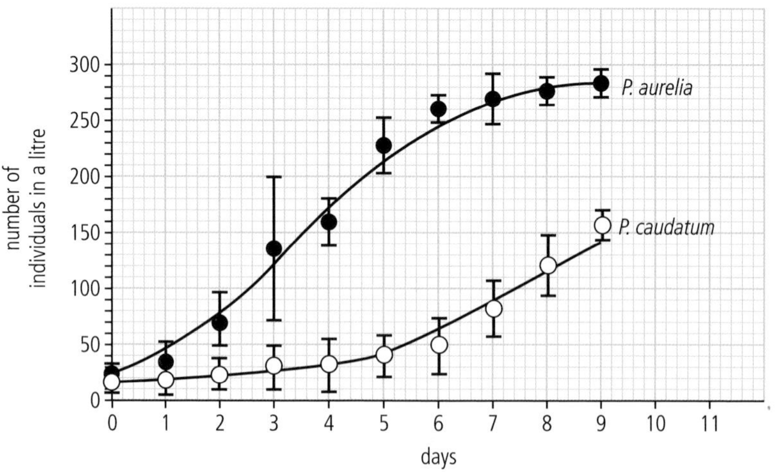

Figure 3 Mean population density ±1 SD of two species of *Paramecium* grown in solution.

Regression models and coefficient of correlation

When scientists measure something, often they are looking to see whether they can demonstrate that the phenomenon is following a law of nature. Sometimes laws of nature follow patterns that can be expressed in mathematical equations. For example, when measuring the light that a leaf might use for photosynthesis, a scientist knows that the intensity of the light varies according to the distance to the light source. In Figure 4, the graph on the left illustrates the "pure" mathematical law for light intensity and distance from the light source. On the right is the same graph superimposed with measurements taken in a laboratory. Because of any number of things, including limitations in the equipment and human error, the laboratory measurements do not fit the mathematical model perfectly.

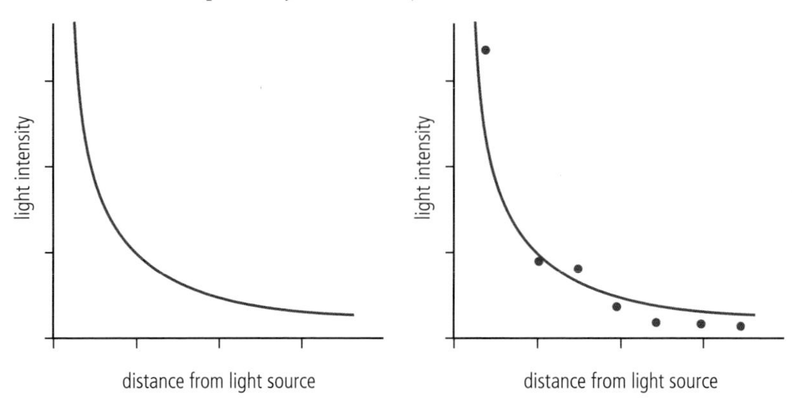

Figure 4 A model of what the data should show (on the left) and the actual collected data (dots on the right), of which only one point is actually where it was expected to be.

Now imagine the opposite. You take some measurements and wonder if there is some kind of mathematical equation that could act as a model of your data. You make a scatter plot and then see if there is a trend line that fits the data reasonably well. You might start with a straight line because that is the simplest relationship between two variables. This is called a simple **linear regression model**. But if that does not fit the data well, you could try other regression models that are not straight lines. Fortunately, statistical functions in a calculator, or a spreadsheet program on a computer, can do this for you in an automated fashion.

How can we know if the trend line's regression model is the best one for the data we have collected? The **squared correlation coefficient**, R^2, also called the **coefficient of determination**, is used to see how well a regression model matches the data collected. A value of $R^2 = 0$ means the regression model does not fit the data at all, whereas a value of $R^2 = 1$ means a perfect fit. Note that R^2 cannot be a negative number. Figure 5 shows some examples of R^2 values calculated using Microsoft Excel for three data sets and their trend lines.

Figure 5 Three examples of data that have been modelled with a linear regression. The R^2-value is calculated to see how closely the linear regression model matches the data.

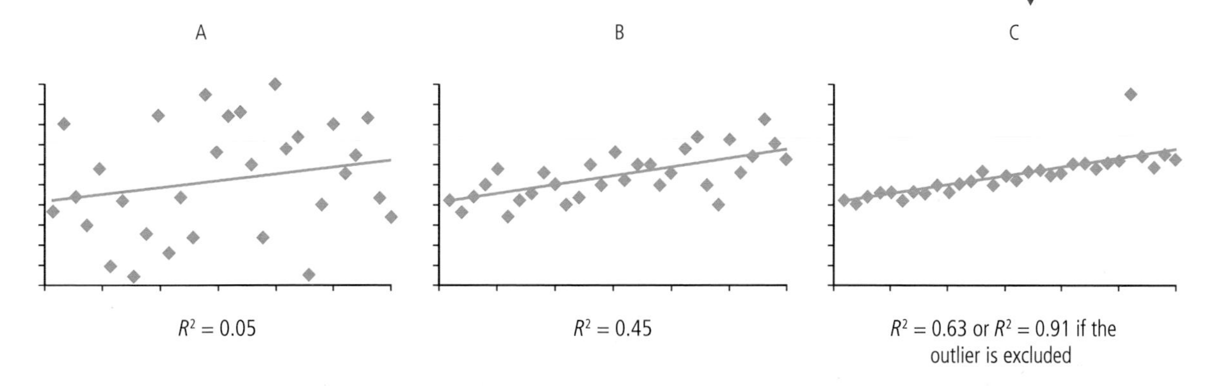

- **Trend lines are useful for seeing whether there is an overall pattern or tendency in the data points.**
- **The R^2-value, the coefficient of determination, is useful for seeing if the trend line matches the data points closely or not. It indicates how good the model is. The closer it is to 1, the better the model. Values close to 1 reveal that there is a strong correlation between the *x*- and *y*-values.**
- **If the regression model fits the data well, it can be used to predict values that were not measured.**

A cormorant (*Phalacrocorax carbo*).

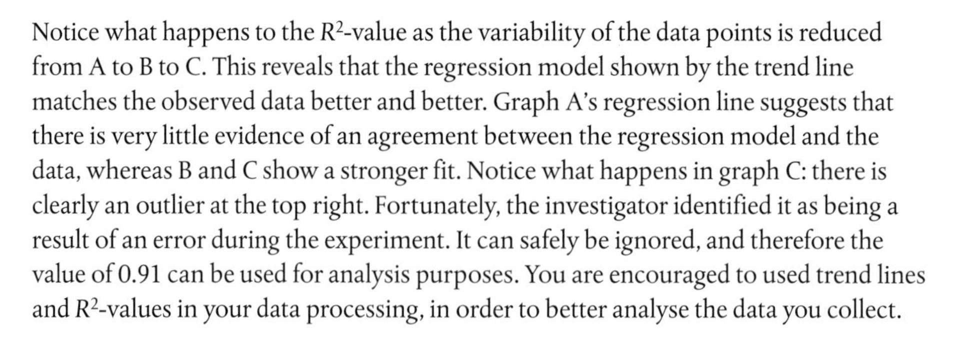

Notice what happens to the R^2-value as the variability of the data points is reduced from A to B to C. This reveals that the regression model shown by the trend line matches the observed data better and better. Graph A's regression line suggests that there is very little evidence of an agreement between the regression model and the data, whereas B and C show a stronger fit. Notice what happens in graph C: there is clearly an outlier at the top right. Fortunately, the investigator identified it as being a result of an error during the experiment. It can safely be ignored, and therefore the value of 0.91 can be used for analysis purposes. You are encouraged to used trend lines and R^2-values in your data processing, in order to better analyse the data you collect.

In addition to simply seeing whether the data points follow a predictable pattern, a regression model can be used to predict values that were not measured. Knowing the equation of the line or the curve allows us to add hypothetical values and get a prediction from the model. For example, changes in the human population in the coming decades can be predicted based on a regression model of current trends in the population. When using a regression model for prediction purposes, the R^2-value can help give a sense of how reliable the prediction will be. For example, predicting an outcome using graph A above would be extremely unreliable. However, using C's regression model would be more likely to give reliable results.

Correlation does not mean causation

Observing that wilting occurs in our house plants every time the soil is dry is a simple **correlation**, but a controlled experiment measuring wilting at different frequencies of watering provides us with evidence that the lack of water is the **cause** of the wilting. Experiments provide a test that shows cause. Observations without an experiment can only show a correlation. Also, in order for there to be evidence of causality, there must be a mechanism to explain how one phenomenon can cause the other. Knowing the properties of osmosis and turgidity in plant cells would explain the causality associated with the correlation, thus giving it greater scientific plausibility.

When using a mathematical correlation test, the value of the **correlation coefficient**, *r*, is a measure of the degree of linear relationship or **linear dependence** between two variables. This can also be called the Pearson correlation coefficient. The value of *r* can vary from +1 (completely positive correlation) to 0 (no correlation) to −1 (completely negative correlation). For example, we can measure the size of breeding cormorant birds to see whether there is a correlation between the size of males and females that breed together (Table 9).

Table 9 Cormorant size data

Pair number	Size of female cormorants/cm	Size of male cormorants/cm
1	43.4	41.9
2	47.0	44.2
3	50.0	43.9
4	41 .1	42.7
5	54.1	49.5
6	49.8	46.5
$r = 0.88$		

The *r*-value of 0.88 shows a positive correlation between the sizes of the two sexes: large females mate with large males. However, correlation is not causation. To find the cause of this observed correlation requires experimental evidence. There may be a high correlation, but only carefully designed experiments can separate causation from correlation. Causality requires that the mechanism of exactly how X causes Y needs to be demonstrated. For example, the mathematics here does not explain whether it is the males choosing the females or the females choosing the males. Correlation says nothing about the direction of the influence.

Correlation does not necessarily mean causation. Just because two things show a relationship and have a strong *r*-value, does not mean one causes the other. To jump from one to the other requires a scientific explanation outlining the mechanism that causes the relationship.

Expected versus observed values: first application of the chi-squared test for goodness of fit

The **chi-squared (χ^2) test for goodness of fit** calculates how close our observed results are to the expected values. Chi is the Greek letter χ and is pronounced like the word "sky" without the s at the beginning.

In the activity accessed from this page of your eBook, we will use the χ^2 test to compare our observed results with what we can theoretically calculate the results should be (the "expected" results). To use this statistical test it is important to note down carefully all the observed results (O) and the expected results (E). In the case of genetics exercises, for example, the expected results would be the proportions of phenotypes as determined by a Punnett grid, such as 25%/50%/25% or 25%/75%, although it is important to use the actual numbers of offspring rather than percentages or ratios. Setting up a table to keep track of the numbers is helpful.

For more about using the chi-squared test for association between two species, see Section C4.1.15.

Graphing

Scientists use graphs extensively because they are useful tools for presenting data and identifying relationships that might otherwise remain hidden. Graphs are instrumental in analysing data, and if you know how to make accurate and appropriate graphs, your conclusion and evaluation will be greatly enhanced.

The most common forms of graphs you are expected to be able to use are:

- bar charts
- histograms
- line graphs
- scatter plots.

Occasionally, you may also need to use pie charts or box-and-whisker plots. Not all graphs are plotted using data. Sometimes you might be asked to sketch a graph rather than plot one. This means that you do not need to have numbers on your axes, only labels. Examples include sketching graphs of enzyme activation energy or the effect of temperature on photosynthesis. Sketches are assessed on the shape of the graph and the labels.

In your eBook, you will find descriptions of the graphs listed above.

Data analysis exercises

Both for your internal assessment work and in data-based sections of exams, you will be required to interpret sets of data presented either as tables or as graphs. Being able to extract scientific information from data is a key skill in biology.

The first thing to look for on a table or a graph is a title. When titles are not available, often the text before or after the tables and graphs will reveal some key information about what they are showing. The next clues to look for in order to interpret the data correctly are labels and units in the headings of tables, or labels and units on the axes of graphs. In both cases, the labels are often the dependent and independent variables of the investigation that generated the data. Knowing these will help you reach conclusions about the investigation. The units might be familiar to you, such as grams, millilitres or °C, but sometimes they are units you have never heard of. In such cases, do not panic, just be sure to include those unfamiliar units in your answers and in your analysis. The same goes for arbitrary units, which are sometimes used to avoid using confusing units.

Next, look at the scales on the axes of graphs. Do they show regular intervals (10, 20, 30, 40) or is there an atypical scale, such as a logarithmic scale (1, 10, 100, 1,000)? If two graphs are being compared, do they use the same scales and the same maximum and minimum values? If not, be careful how you compare the two because they may look the same but in fact be very different.

In the eBook, you will find a worked example of the analysis of a graph.

Throughout this book, there are examples of past paper questions, and some of them have graphs or tables of numbers that need to be interpreted and analysed. Be sure to practise analysing them because that is what you will be asked to do in exams.

Applying skills

Skill	Example	Relevant part of syllabus
Interpolate values from a graph	To estimate or predict values between measured data points. Useful in estimating an enzyme rate at pH 5 even though you only have data points on your graph for pH 2, 4, 6, 8 and 10.	C1.1.9
Extrapolate values from a graph	To estimate or predict values beyond measured data points. Useful in predicting future atmospheric CO_2 concentrations based on current trends in the Keeling Curve.	C4.2.20
Quantifying energy flow	Represent energy flow in the form of food chains, food webs and pyramids of energy. Useful for showing the dependencies of species on others for energy.	C4.2.11
Pedigree charts	Construct a diagram to show which parents produced which offspring over many generations. Useful in tracing how genetic traits get passed down and the likelihood of a trait being passed on to the next generation.	D3.2.13
Dichotomous keys	Build your own identification key	A3.1.14 HL

Inquiry 1: Exploring and designing

Here we will briefly introduce the three inquiry skills you need in biology; they are covered in more depth in the Internal Assessment chapter.

Exploring

It is hoped that when you are learning about topics in biology, you will have a desire to know more about certain concepts that spark your curiosity. You are encouraged to demonstrate independent thinking in your exploration as well as a good amount of initiative and insight. Insight is when you have an "A-ha!" moment and you see something in a new light. If you do things right, you should see a new connection or something will gain new meaning.

When researching a topic for investigation, a considerable amount of background research is necessary. It is unlikely that you will find what you are looking for just using a search engine. Talk to your teacher, and ask your school or local librarian if there are resources available on your topic.

Have you consulted your school librarian or local librarian about what resources are available on your topic?

Once you have done sufficient background research, you are ready to formulate a research question. This might take the form of "What is the effect of X on Y?" where X and Y are measurable quantities, at least one of which is connected in some way to a living organism. A hypothesis should be formulated based on your research, past experiences and perhaps some initial trials. A hypothesis must be stated as a claim rather than as a question. It is one possible answer to your research question.

Designing

Once you have a research question, you can design your method. You are encouraged to show some creativity in the design, implementation or presentation of the investigation. You are discouraged from trying to invent a brand new investigation using techniques no one has ever tried before. It is best to apply techniques you have used in the past but in a different way. There is more than one way of obtaining data for an investigation in biology, for example:

- a hands-on experiment in the laboratory
- fieldwork
- a database
- a computer simulation
- surveys.

You will need to identify and justify your variables.

- The independent variable is the thing you change on purpose in order to see what will happen.
- The dependent variable is the results of the investigation: what is measured or counted at the end.
- The controlled variables are a list of conditions that should stay the same in order to make the experiment a fair test, but also to show that it is only the independent variable that is causing the changes in the results and nothing else.

A method for observing the rate of photosynthesis by capturing oxygen gas bubbles produced by an aquatic plant. When designing an investigation, careful attention must be paid to every detail and your choices should be explained.

To justify the variables, say why it is a good idea to have chosen them. You will need to decide how much of a range you want to test with your independent variable. For example, if you are investigating the effects of different coloured light on photosynthesis, how many different colours are you going to test? What range would provide a graph that could answer the research question? For the dependent variable, you will need to decide how many trials are required in order to get a satisfactory result. With only two trials for each colour, for example, it would be difficult to know if one of them has been affected by an unexpected source of error. On the other hand, you are limited in time so you need to find a balance between the size of the range and the number of trials. When you choose a variable, explain why it was chosen. Say why it is important to keep controlled variables the same and explain what would happen if you did not control them.

Once you have made these decisions, you are ready to write your method. This should be written step-by-step with enough clarity and detail that any other student could pick it up, follow the instructions and get similar results as you.

For more information on how to design a laboratory investigation, see the chapter on internal assessment.

Controlling variables

In the laboratory and during fieldwork, some probes and sensors require calibration. Think of calibration in the following way: the difference between a glass of water and a graduated cylinder is that one has markings on the side that were set by comparing volumes to a known volume. Without the calibrated markings, you can only say if one volume is relatively bigger than another but not by how much.

If you have declared that temperature is one of your controlled variables, you are going to have to try your best to keep the temperature constant. In ideal conditions, all the environmental conditions for your investigation should be maintained at a steady state.

Sometimes an investigation requires measuring a sample of a population because measuring the whole population would be unrealistic, for example counting the number of dandelions in a field that is 20,000 m^2. We usually take **random samples** or **systematic samples**. If you do that, you should justify why one is a better choice than the other and this will depend on the nature of your research question. If, during your fieldwork, in a field that is 20,000 m^2 you counted dandelions in 200 sampled metres, is that enough? It represents only 1 in 100 square metres of the field's surface area, or 1%. But if that is all you can do in the time given, that will have to suffice. One temptation when performing a random sample is to favour zones that are easy to access, such as those on flatter or drier ground and with fewer obstacles. Unfortunately, this does not make the sampling truly random. If you do run into situations that could qualify as **sampling errors**, be sure to mention them in your report.

If you are testing the effects or influence of something, it is a good idea to set up a **control**. This is a version of your experiment without the characteristic you are using as an independent variable. For example, when testing the effect of different concentrations of salt solutions on seed germination, the version of the experiment that has no salt added at all is the control. Some investigations, notably in fieldwork, do not lend themselves to having a control but if it is possible, it is often useful.

For more information on how to keep controlled variables constant, see the chapter on internal assessment.

Inquiry 2: Collecting and processing data

Collecting data

The best laboratory investigations and fieldwork will have both qualitative and quantitative observations. Things like colour, texture and smell can be described with words. An annotated drawing of cells observed under a microscope, or a sketch of a zone studied along a river bank, are also qualitative observations.

Quantitative results are those that can be expressed with numbers (e.g. temperature, number of plants, volume of solution) whereas **qualitative observations** are those that cannot.

When collecting data, it is important to be careful about being precise and getting sufficient results in order to be able to process the data. If you notice that there are some difficulties during the process and something is not going as planned, be sure to note it down so you can talk about it in your report. The data points collected at that stage will be less reliable but perhaps still exploitable. If you can modify the method to avoid repeating the problem, you should. This is why it is best to trial an experiment before undertaking data collection to answer the research question.

Investigations can be carried out in the field as long as the particular aspects of safety, ethics and environmental issues associated with working in nature are addressed.

Processing data

There are several types of **data processing**, such as the mathematical operations, graphing or statistical analyses explained in the tools section earlier in this chapter. Be sure to pay close attention to how precise you can be and how confident you are of the reliability of the data points you use.

Here are some possible questions to ask yourself to make sure that your data processing and comments about your data are complete and helpful in reaching a conclusion.

- Is my data categorical, meaning it is in categories? (type A blood/type O ...)
- Is my data normally distributed? (Meaning if graphed, it follows a Gaussian curve in the shape of a bell. See the central tendency worksheet linked to from page 895 of the eBook.)
- How representative is my sample of the population?

- Is my data ordinal: does the order matter (e.g. a questionnaire asking people to answer "always", "sometimes", "rarely", "never")?
- Is my data continuous or discrete?
- Do I have a way of showing the range of my data?
- Is the data well-suited for calculating the standard deviation?
- Do I have a way of determining outliers in the data? (Be sure to state how you determined outliers and what you decided to do with them.)
- If I see a correlation, am I careful about explaining whether there is a causation? (If so, explain how.)
- In my report, have I made comments about how confident I am of my results?
- Have I considered any confounding variables? (These are things that were not measured but which, in the end, had an unexpected influence on the results.)
- Should I do any hypothesis testing using my null hypothesis (H_0)?
- Should I calculate the regression on a regression line, R^2?

Interpreting results

Think about what comments you could make about your tables of quantitative results or your qualitative results. Say if you see any patterns that emerge or connections whereby qualitative and quantitative results agree or disagree with each other. Look at your graphs and comment on any trends they show. If one of your variables is supposed to show a relationship with another, comment on whether or not this is the case. If there are some data points that are very unreliable because something went wrong, what did you decide to do with them? How confident are you concerning your results? Comment on the accuracy, precision, reliability and validity of your results. For example, did the data points gathered from multiple trials of the same part of the experiment give similar results? Is the trend or relationship you are seeing supposed to be there?

Inquiry 3: Concluding and evaluating

Concluding

Now you are ready for the conclusion. The conclusion is an opportunity to give the answer to the research question. It says what can be learned from the results. You should say whether your hypothesis was supported or refuted by the results you obtained. You should include an answer to your research question and you should support your conclusion with examples from your results. How do you know if the conclusion you have reached is a valid one? You need to compare your results to the scientific context of your topic. For example, you could compare your results to what others have found previously in a similar investigation. It does not have to be identical to yours. You can do some research to find out if the outcome of your investigation can be explained by the current scientific understanding of your topic. In other words, say if your results fit with theories scientists have published. Cite your sources.

Evaluating

To evaluate something means to determine something's value or worth. An evaluation is an opportunity to comment on what might not have gone as you expected and suggest what to do differently next time. If your hypothesis was confirmed by the data, assess how fully it was confirmed or not. For example, were you way off or were you close to what you expected?

All investigations have their limitations. Obviously, if we had more time to collect results we could get more data, but that is not necessarily possible. Were there things on your list of controlled variables that turned out to be impossible to control? Did they have an influence on the reliability of the data? Were there any assumptions that need to be addressed? For example, if you assumed that an experiment left overnight was not subjected to any temperature changes in the laboratory, how sure can you be of that assumption? Be sure to consider the impact of uncertainties on the conclusions. Think about how confident you are of the precision of your measurements and decide if the precision is enough to give you confidence in your conclusion. If a result shows a difference that is ten times larger than the degree of precision, you can be more confident that the results are not simply due to the lack of precision of the measuring device than if the difference was smaller.

Comment on any sources of error that gave unreliable results. Some might be **random errors** and some might be **systematic errors**. Brainstorm all the possible sources of error you can think of and then put them in order based on which ones would have the biggest impact. Then focus on the ones that are specific to your investigation rather than ones that could be true for any investigation. For example, "maybe the steps of the method were not followed correctly" is too generic.

Random errors are ones that happen by chance. For example, a temperature probe might be accurate to ±0.1°C and fluctuate between 21.7°C and 21.6°C but it is not because the temperature is changing, it is just the device. Systematic error would be if the temperature probe was defective or improperly calibrated and gave results that were always 2°C too low no matter what.

For each source of error, describe what the source is and then comment on whether it would tend to give unreliably high results or unexpectedly low ones. In a good evaluation, sources of error should be assessed for how severely they affect the data. Is the source of error minor, moderate or major? How would you fix the problem? Suggested improvements to the method or the types of equipment used should be realistic in terms of time available, resources you can access and safety. Some suggestions might solve one problem but create another one, such as being more time consuming.

For more specific details on what to put in a laboratory report, see the chapter on internal assessment.

Advice on the Extended Essay

Introduction

One of the requirements of the IB Diploma is to write an Extended Essay. An Extended Essay is an in-depth study of a limited topic within a particular subject area. It provides the opportunity to carry out independent research within a subject of your choice. Biology is a subject often selected by students for their Extended Essay. It is a popular subject because many of the topics studied in IB biology stimulate further research ideas. Laboratory work carried out in the IB biology course also provides a basis for student ideas involving possible research questions. A good research question is essential for Extended Essay success.

General guidelines for all Extended Essays

The following guidelines apply to Extended Essays in all subjects. You should:

- not exceed the upper word limit of 4,000 words, not including the bibliography, contents page, appendices, or any labelling or captioning of graphs, diagrams, illustrations or tables
- spend around 40 hours working on your essay
- be assigned a teacher with Extended Essay training who will act as your supervisor and provide general guidance for the project
- pay attention to your school's set deadlines (your essay will be assessed externally using published criteria but your school will set its own deadlines)
- ensure your essay represents a unique approach to addressing a specific research question
- work independently
- select a topic and research question that is of interest to you, and be certain to show that interest in your writing
- complete the Reflections on Planning and Progress Form (RPPF) at the end of the process, which focuses on your final insights and methods evaluation.

General guidelines for biology Extended Essays

The following guidelines apply specifically to biology Extended Essays.

- A biology Extended Essay should involve biological theory and concepts. It will be assessed on the nature of biology.
- Most successful biology Extended Essays involve some sort of independent, hands-on experimental work along with literature-based research.

- A detailed procedure representing the exact steps carried out in any experimental work must be given. This detailed procedure is known as the **protocol**.
- Biology Extended Essays may involve data collected through experimentation, survey, microscopic observations, fieldwork or some other appropriate biological approach. It is essential that a proper analysis of this primary data is presented.
- Students taking an experimental approach to the biology Extended Essay must also consult and reference secondary data resources.
- Some biology Extended Essays do well when they are mostly literature based. These literature or secondary data-based Extended Essays in biology should include a unique analysis of raw data generated by reputable protocols and procedures.
- A successful Extended Essay will clearly illustrate access to sufficient data or information to effectively research the topic selected.
- The main body of the biology Extended Essay should centre on an argument or evaluation based on the primary and/or secondary data presented. Any research involving organisms must be ethical. Any animal research must follow the IB guidelines concerning the use of animals in IB World Schools.
- It is essential that no research be done that may inflict pain or cause stress to any living organisms. Also, research must not be conducted that may have a harmful effect on health, such as culturing micro-organisms at or near body temperature (37°C). Confidential medical information may not be accessed or utilized in an Extended Essay. If humans are to be used as subjects in any way, evidence of informed consent must be provided. Your instructor will provide added clarity regarding the ethical limits of potential research involving organisms.
- IB states clearly that biology topics dealing with symptoms and treatment of particular human diseases are very rarely successful Extended Essays.
- Topics dealing with ethical issues, different general approaches to medical treatments, and surveys involving attitudes or opinions concerning science research, are rarely successful biology Extended Essays.
- Extended Essays based on experimental or practical work at a laboratory outside school must have a cover letter submitted with the Essay detailing your role in the protocol design. This letter must also specifically describe any guidance received. This is especially important when the research is done at a university or research institution. For safety and/or academic honesty reasons, some schools do not allow students to work outside the school, so check with your teacher.

Suggested steps towards a successful essay

The following steps are a suggested way in which to approach an Extended Essay in biology.

I. Initial research and planning

(a) Decide on your subject of interest.

(b) Think of potential research questions.

(c) Make sure your research question has a "biological" focus. It must directly relate to an organism in some specific way.

(d) Meet with your supervisor to discuss your proposed topics and research questions. One of the most important functions of the supervisor is to help develop a proper research question for the Extended Essay.

II. Continued research involving your chosen research question

(a) Research should involve a survey of the topic literature, keeping a detailed account of the sources from which ideas and/or data are used.

(b) Plan your procedure for any experimental work.

(c) Discuss your proposed research and procedure with your supervisor.

(d) It may be necessary to refine your topic and research question as more information is gathered. Proper focus on the research question is essential.

III. Experimental work and data collection

(a) Make certain your experimental procedure is safe and ethical in the opinion of your supervisor before beginning the procedure.

(b) Arrange for all necessary equipment, chemicals and specific needs before beginning the experimental work. This may involve sources outside your school. Be certain all sources of materials outside your school are acceptable to your supervisor.

(c) It is extremely important to consider the independent variables, the dependent variables, and the controlled variables in your procedure or procedures.

(d) An essential part of all experimental work is to ensure an adequate sample size. It is important to discuss sample size with your supervisor.

(e) Control groups and experimental groups must be carefully considered.

(f) A plan should be in place for recording the raw data before the procedure begins. Qualitative data and quantitative data should both be considered in the data collection and recording stage.

(g) Processing and presentation of data is an essential part of the experimental work. Careful consideration should be given to tables, graphs and statistical tests so that data will allow meaningful and proper conclusions to be reached.

(h) If taking a non-experimental approach to the Extended Essay, it is essential there is sufficient secondary data to research the topic effectively. It is also important with this approach to use the secondary data and manipulate or analyse it in an original way.

(i) Your supervisor may give general suggestions throughout this experimental work.

IV. Writing the essay

(a) Your essay should have a structure that allows for an acceptable and appropriate presentation. An acceptable Extended Essay organization is as follows.

- Title page
- Table of contents
- Introduction with research question stated early and clearly
- Hypothesis and explanation of hypothesis
- Background information
- Presentation of variables
- Materials used

- Protocol of experimental procedures
- Data collection and presentation
- Data analysis
- Evaluation
- Conclusion
- Bibliography
- Appendix (This is optional, and may include details of protocols, raw data, or any calculations using the raw data. It is important to note that the essay should be sufficient without the presence of an appendix.)

(b) A first draft should be submitted to your supervisor so that general directions may be given for writing the final draft.

(c) The first draft should be checked against the IB marking criteria by you and your supervisor.

(d) The bibliography style should be one used at your school. There is not a specific form of bibliography to use. It is important that some reference in the essay is made to each bibliography source provided. Information about any online sources used must be appropriate and complete.

V. Final draft

(a) Make changes generally suggested in the first draft by your supervisor.

(b) Proofreading is essential.

(c) Double check your final essay against the "presentation" criterion in the Extended Essay marking criteria.

(d) Arrange a meeting with your supervisor to submit your final essay. Your supervisor should go over the final essay with you, making certain the major sections have been included.

The Extended Essay criteria and advice for each criterion

Criterion	Advice
Criterion A: Focus and method – 6 marks possible	• Explain the topic in a clear and focused manner. • The research question must be effectively stated early in the paper. • The research question should lend itself to discussion and even debate within the Extended Essay. • A sound research question can only be presented when appropriate sources are utilized and properly cited. • The research question should be utilized to formulate a hypothesis, or hypotheses, that can be tested. • Methods used in the Extended Essay should be obviously well planned and allow gathering of data that is relevant to the research question. • Methods used should involve controls and allow adequate data collection.

Criterion	Advice
	• It is essential the research question and the method are biological in nature. • If an investigation is conducted in an external laboratory, you must clearly demonstrate your understanding of the methods and materials utilized. In this situation, you should also clearly present *your role* in choosing and applying any methods and materials utilized.
Criterion B: Knowledge and understanding – 6 marks possible	• The essay must demonstrate a thorough understanding of the topic. • Your essay should flow in a logical way towards the development of a proper conclusion concerning the research question. • Clearly demonstrate in the Extended Essay that you understand all aspects of the essay. • Your analyses should represent an obvious understanding of the topic and research question. • Sources utilized in the Extended Essay must be appropriate and contribute to analyses and the conclusion of the essay. • It is essential the terminology used in the essay is accurate, focused and relevant. • Any technical terms used in the essay should be explained and used appropriately within the text. • Symbols, equations, significant digits and International System of Units (SI) units should be utilized throughout the essay.
Criterion C: Critical thinking – 12 marks possible	• Be certain to present a convincing argument in your Extended Essay. • All arguments or data presented must relate logically to the research question. • It is wise to present alternative views as you develop your research question. • Carefully analyse all of the sources used in your essay. • Evaluate all aspects of the argument/experiment for appropriateness. • All data must be analysed. This analysis may involve mathematical transformations, statistical analysis, tables and/or graphs. • Tables and graphs should be presented logically and appropriately. Each table and graph used in the essay must relate to the research question and to the conclusion. • You must comment on the quality and quantity of the secondary sources and data used. • Limitations, validity and reliability of data should be critically commented on within the text of the essay.
Criterion D: Presentation – 4 marks possible	• Proper structure and layout are essential for high marks with this criterion. • The structure and layout of the essay should add to the presentation and development of the argument. • Scientific and annotated representations of equipment and experimental setups should be present and clear. • Procedural steps should be appropriately summarized. • Irrelevant details should be minimized. • Clarity of diagrams, graphs and tables is essential. They should add to the effective communication essential in the development of the analyses towards a logical conclusion.

Criterion	Advice
	• Raw and processed data tables must be clearly displayed in the most appropriate form. • Any mathematical processes used in analysing raw data should be illustrated/explained clearly. • The essay must not exceed 4,000 words. It is essential to include the following in your Extended Essay: title page, table of contents, page numbers, appropriate illustrations, proper citations and bibliography, and appropriate appendices, if needed. • Graphs, figures, calculations, diagrams, formulas and equations are not included in thc word count. • Examiners will not assess any material presented after the 4,000 word upper limit.
Criterion E: Engagement – 6 marks possible	• The examiner will utilize your RPPF after the assessment of the essay to determine the marks achieved for this criterion. • Your reflections should explain how and why the topic was determined. • Your reflections should clearly demonstrate how methods and approaches were determined in developing the Extended Essay. • Reflections should demonstrate knowledge gained by performing this activity. It is essential that you explain suggested changes in further work on the topic chosen.

Final advice for your Extended Essay

1. Be cautious concerning plagiarism. Presenting someone else's ideas or work as your own is plagiarism. Be certain to give proper credit to all ideas or work that has been used in any way in your Extended Essay.
2. Your title page should include:
   - the title of the essay
   - the research question
   - the subject for which the essay is registered
   - the word count, within the 4,000 limit.
3. Neither your name nor your school's name should appear on the title page or any page headers.
4. The introduction must present a strong reasoning for pursuing a conclusion to the presented research question, which should be stated clearly and early. Why the research question is significant for your Extended Essay should also be stated.
5. Any experimental procedures must be clearly and appropriately presented in a way they can be easily replicated.
6. Ethical and safety factors must be thoroughly addressed. It must be obvious that ethical and safety factors have been seriously considered in pursuing the research question.
7. Pages should be clearly numbered. The title page is not numbered. Sections of the essay should be clearly and appropriately labelled.
8. Footnoting and the bibliography must be proper and consistent. Sources in the bibliography not specifically used in the paper should be minimal.

9. All visual presentations must be clear, labelled appropriately, and must pertain to the research question and conclusion.
10. The conclusion must be clearly related to the research question. Limitations to the conclusion should be discussed. It is suggested that you present a brief plan for possible further development of a conclusion to your research question.
11. Any appendices used must be appropriate. Large tables of raw data collected are best included in an appendix. However, a representative sample of collected raw data should be included in the core of the essay in a data table.
12. Be certain the essay is based on suitable biological topics. Stay away from psychology- or medicine-focused topics.
13. Remember that a strong evaluation includes a comparison of your results with that of the literature. This evaluation may be addressed at several points in the essay besides the conclusion. An evaluation only presented in the conclusion is often rather shallow and ineffective.
14. Be certain your Extended Essay does not duplicate any other work being submitted for the Diploma programme. The Extended Essay should not take the form of an internal assessment. The Extended Essay assesses the ability to analyse and evaluate scientific arguments.

Viva voce and the RPPF

The completion of your Extended Essay is signified by the viva voce (concluding interview). This is a 10–15 minute interview with your supervisor. It provides an opportunity to reflect on successes and what has been learned.

After completing the viva voce, you must complete your RPPF. This is your final reflections on the topic, methods, analyses and conclusion of the Extended Essay.

Enjoy your research.